BIOLOGY
THE DYNAMICS OF LIFE

GLENCOE
McGraw-Hill

New York, New York   Columbus, Ohio   Woodland Hills, California   Peoria, Illinois

A GLENCOE PROGRAM
BIOLOGY: THE DYNAMICS OF LIFE

Student Edition
Teacher Wraparound Edition
Laboratory Manual, SE and TE
Study Guide, SE and TE
Chapter Assessment
Lesson Plans
Videodisc Correlations
Science and Technology Videodisc Series,
    Teacher Guide
Transparency Package

Transparency Masters
Critical Thinking/Problem Solving
Spanish Resources
Concept Mapping
Biolab and Minilab Worksheets
Exploring Applications of Biology
Great Developments in Biology
Biology Projects
Computer Test Bank
    IBM/APPLE/MACINTOSH
English/Spanish Audiocassettes

Glencoe Science Professional Series:
    Exploring Environmental Issues
    Performance Assessment in the Biology Classroom
    Alternate Assessment in the Science Classroom
    Cooperative Learning in the Science Classroom

Send all inquiries to:
Glencoe/McGraw-Hill
8787 Orion Place
Columbus, OH 43240-4027

ISBN 0-02-826647-1
Printed in the United States of America.

10 11 12 13 14 15   071/046   04 03 02 01 00

Authors

Alton Biggs teaches biology at Allen High School, Allen, Texas, where he also served as Science Department Chairperson for 16 years. He has a B.S. in Natural Sciences and an M.S. in Biology from East Texas State University. Mr. Biggs received the Outstanding Biology Teacher Award for Texas in 1982, he was the founding president of the Texas Association of Biology Teachers in 1985, and in 1992 was elected president of the National Association of Biology Teachers. Mr. Biggs has led several excursions to the Galapagos Islands, the Amazon River basin, and the Andes. He is also a contributing author to Glencoe's *Biology: Living Systems*.

Chris Kapicka is a research scientist at the Veteran's Administration Hospital in Boise, Idaho. Previously, she was a biology teacher at Boise High School, Boise, Idaho. She has a B.S. in Biology from Boise State University, an M.S. in Microbiology from Washington State University, and a Ph.D. in Cell Physiology and Pharmacology from the University of Nevada-Reno. Dr. Kapicka received the Presidential Award for Excellence in Science Teaching in 1986, the National Association of Biology Teachers Biology Award in 1987, and the Sigma Xi Distinguished Science Teaching Award in 1987.

Linda Lundgren has taught biology at Bear Creek High School, Lakewood, Colorado, for eight years. She has a B.A. in Journalism and Zoology from the University of Massachussets and an M.S. in Zoology from Ohio State University. In 1988, while on sabbatical leave, Mrs. Lundgren was awarded a research fellowship as a visiting scientist at the National Renewable Energy Laboratory, Golden, Colorado. In 1991, she was named Colorado Science Teacher of the Year by the Colorado Association of Science Teachers. Mrs. Lundgren is also the author of a workbook on Cooperative Learning and a contributing author to Glencoe's *Biology: Living Systems*.

Contributing Authors

Daniel Blaustein
Science Writer
Evanston, IL

Rebecca Johnson
Science Writer
Sioux Falls, SD

Devi Mathieu
Science Writer
Sebastopol, CA

Susan Offner, Ph.D.
Teacher, Milton High School
Sudbury, MA

Contents in Brief

Contents in Brief

Contents

Contents

Contents

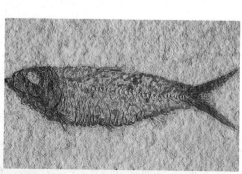

Contents

Contents

Contents

Contents

Epilogue

Contents

BioLabs

Working in the lab is often the most enjoyable part of biology. BIOLABS give you an opportunity to act like a biologist and develop your own plans for studying a question or problem. Whether you're designing experiments or following well-tested procedures, you'll have fun doing these lab activities.

Chapter

MiniLabs

Do you often ask how, what, or why about the living world around you? Sometimes it takes just a little time to find out the answers for yourself. These short activities can be tried on your own at home or with help from a teacher at school. When you're feeling inquisitive, try a MINILAB.

Thinking Labs

Sharpen up your pencil and your wit because you'll need them. THINKING LABS *offer a unique opportunity to evaluate another scientist's experiments and data without lab bench mess.*

Interdisciplinary Connections

It may not have occurred to you that biology is connected to all your courses. Learn in these features how biology is connected to art, literature, and other subjects that interest you.

people in biology

What are biologists like? What excites them about biology, and how did they ever get interested in biology in the first place? Read these interviews with PEOPLE IN BIOLOGY, *and find out what makes a biologist "tick."*

Some topics of biology deserve more attention than others because they're unusual, informative, or just plain interesting. Here are several features that **FOCUS ON** *topics from dinosaurs to your brain.*

A BROADER VIEW

*The seemingly small events, anecdotes, and details found in the following articles will give you **A BROADER VIEW** and better understanding of the big picture.*

How does biology impact you and society? Consider some of the issues that have been covered by the media. Take this opportunity to understand the many different sides of issues, learn how technology may change your life, and develop your own viewpoints with BIOLOGY & SOCIETY.

ISSUES

ETHICS

TECHNOLOGY

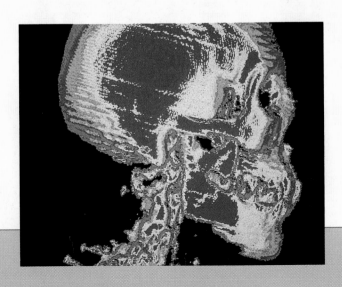

Dear student biologist:

One thing all humans share is a curiosity about the living world. Questions puzzle us continually, like why does a dog turn in circles and scratch the carpet before it lays down, or do horses dream? You may even ponder questions such as why the ozone is so important, why AIDS is such a devastating disease, and even how life on Earth began. Questions like these have been considered by biologists for centuries. As in all other sciences, the answer to one question often leads to another. Biology at work is the probing and searching for answers to question after question.

As you watch television or read the newspaper, you will hear about new discoveries in biology—such as how cells work, new possible cures or treatments for AIDS, and new genetic treatments for diseases. Throughout this year, we challenge you to discover biology in action around you. As you ride in a car, watch the landscape go by and think about how living things are related to each other—again, biology at work. Perhaps you will watch a hummingbird drinking the nectar from a petunia flower, or a squirrel picking up an acorn and scurrying up the trunk of a large oak tree. As you watch a basketball game, notice biology at work as the basketball players rush down the gym floor while excited fans cheer them on.

Our goal is that you experience the joy and awe of learning about the living world. Take time to marvel at the complexity of living things, as well as the beauty in biological forms. As you consider the tiniest of bacteria to the largest dinosaur, remember that all life is connected.

We hope that you do not get caught up in just learning the terms and facts of biology, but can appreciate the beauty of the processes and orderliness of living things. Continue to ask "Why...?", "How...?", and "What if...?". Seeking answers to questions about the living world will help us understand how we are connected to the world around us. We wish you a successful year in your course of biology. As you continue the study of life, we challenge you to keep asking questions and searching for answers. That is what the science of biology is all about.

Sincerely,

Alton L. Biggs   *Chris L. Kapicka*   *Linda Lundgren*

Authors of *Biology: The Dynamics of Life*

1 What Is Biology?

The eagle's sharp eyes gaze into the lake, searching for signs of food. Without warning, the great bird of prey hurtles toward a movement in the water. Seconds later, the eagle flies back to its nest, a small fish gripped tightly in its powerful talons. What factors are responsible for this single dramatic moment in the life of an eagle?

Answering questions about the natural world involves methods of investigation and discovery that are at the core of biology. As biologists go about their work, they use scientific processes, special tools and techniques, and their knowledge of the basic principles of biology and other sciences.

An eagle's great strength comes from the energy in its food—a steady supply of fish and other small prey. In what other ways are eagles and other living things dependent on their environment?

The ability to single out a movement in the water from a great height is just one adaptation that has evolved among all birds of prey. How do biologists find out which adaptations are important to a species's survival?

The idea for the invention of flying machines almost certainly came from observations of feathers and birds in flight. What are some other applications of biology in our daily lives?

Unit Contents

1 Biology: The Science of Life

The world's oceans teem with many unusual animals, but perhaps none are as unusual as the sea slug. Often brilliantly colored, sea slugs appear as floating ribbons to the observer, but these slow-moving, delicate-looking relatives of snails and clams are anything but helpless. All known species of sea slugs are carnivores and prey on animals such as sea anemones, corals, and sponges. A remarkable characteristic of some species of sea slugs is the ability to incorporate the stinging cells of their prey into their own bodies. They then use these cells for defense.

You won't always find sea slugs living alone. Several species interact in a curious relationship with another organism. Living within the skin of some species of sea slugs are microscopic algae—green, single-celled organisms that produce some of the sugars and other nutrients sea slugs need for their activities.

In some ways, the world of the sea slug mirrors what you can expect as you study biology—a world of fantastic living things, unusual behaviors, and unexpected relationships. Welcome to the world of biology!

Sea slugs, sponges, sea anemones, and algae may look vastly different, but they are all living things and share many traits in common. Living things are the subjects of biology. What are the characteristics of life, and what are the benefits of studying biology?

Section Preview

Objectives

Identify the topics studied in biology.

Recognize some possible benefits from studying biology.

Key Term

biology

Why don't people get goose bumps on their faces? How do plants that make seedless fruits reproduce? Why are there so many different kinds of insects? For humans who share this planet with an amazing diversity of living things, the natural world often poses questions that arouse our curiosity. More often than not, such questions have simple explanations, but sometimes nature defies common sense. Whether nature's puzzles are simple or complex, many may be explained with the concepts and principles of biological science.

The Science of Biology

People have always been curious about living things—how many different species there are, where they live, what they are like, how they relate to each other, and how they behave. These and many other questions about life are answerable, and the concepts, principles, and theories that allow people to understand the natural environment form the core of **biology,** the study of life. What will you, as an amateur biologist, learn about in your study of biology?

A key aspect of biology is simply learning about the different types of living things around us. With all the

Figure 1.1

None of the creatures that you read about in works of fantasy and science fiction are as unusual as some of the kinds of living things that actually live on Earth.

▶ **Mudskippers are a type of fish that crawl over land on strong, muscular fins.**

Figure 1.2

Questions about the features and behaviors of living things can sometimes be answered only by finding out about their interactions with their surroundings.

▶ The banded pipefish can hide horizontally in its environment of seaweeds. Its stripes blend in with the stems of the weeds.

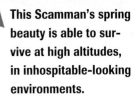

▲ This Australian frog burrows underground and encases itself in a waterproof envelope to prevent water loss during dry weather.

▲ This Scamman's spring beauty is able to survive at high altitudes, in inhospitable-looking environments.

facts in biology textbooks, you might think that biologists have answered almost all the questions about life. Of course, this is not true. Millions of life forms haven't even been named yet, let alone studied. The ones that are studied are often distinctive looking organisms that have unusual behaviors, *Figure 1.1.*

When studying the different types of living things, you'll ask what, why, and how questions about life. That is, you may ask, "What are some of the features of this living thing? Why

does this living thing possess such features? How do these structures work?" By asking such questions, you will develop general principles and rules, which indicate that, as strange as some forms of life appear to be, there is order in the natural world.

Biologists study the interactions of life

One of the most general concepts in biology is that living things do not exist in isolation; they are all functioning parts in the delicate balance of nature. As you can see in *Figure 1.2,* living things depend upon other living things in a variety of ways and for a variety of reasons.

◀ The meat-eating sundew plant lures insects with a powerful fragrance and then traps them on hundreds of sticky hairs.

◀ When threatened, the Texas horned lizard raises its blood pressure and squirts blood out of the corners of its eyes.

Nesting Geese

by Robert Bateman (1930–)

The best things in life—clean air, clean water, and the song of a bird—are not free anymore. They will be beyond price in the next century if we don't start rethinking our priorities now.

Internationally known painter Robert Bateman is an artist with an environmental message. Through his dramatic and powerful portrayals of wildlife in their natural settings, he raises our awareness of a vanishing world in conflict with the 20th century. Toronto-born Bateman travels the world to exhibit and lecture on the plight of our endangered planet, its species, and their habitats.

A vanishing world   Critics note that many of Bateman's wildlife paintings have a disturbing sense of immediacy; the viewer often feels as if she or he has stumbled into the presence of some wild animal or bird and been given a rare glimpse of a fleeting moment in its world.

The painting "Nesting Geese" clearly shows the fierce territoriality of nesting Canada geese. Bateman's attention to the minute details of the feathers, as well as to the accuracy of the bird's behavior, give the painting realism.

CONNECTION TO   Biology

Bateman has donated many works to the conservationist cause, helping to raise millions of dollars for the preservation of endangered animals and their habitats. What other ways do the paintings of naturalists such as Bateman contribute to the preservation of our wildlife and their surroundings?

Why Study Biology?

Many people study biology simply for the pleasure of learning about the world of living things. As you've seen, the natural world is filled with examples of living things that can be amusing, or amazing, or with other examples that challenge one's thinking. Through your study of biology, you will come to appreciate the great diversity of species on Earth and the way each species fits into the dynamic pattern of life on our planet.

Biology and the future

Human existence depends on the existence of all living things on Earth. Living things are our supply of food and raw materials, such as wood, cotton, and oil. Plants replenish the essential oxygen in the air, *Figure 1.3.* Only with a thorough understanding of living things and the intricate web of nature can humans expect to understand the future health of our planet.

Figure 1.3

By understanding the interactions of living things, we are better able to impact our planet in a positive way.

The future of the human species holds many promises, but problems will also arise, *Figure 1.4*. For instance, scientists may one day be able to produce complex living things in their labs, but many species that already exist will go extinct. New agricultural techniques may help farmers see insects, droughts, and floods become problems of the past, but the number of people to be fed is expected to rise sharply in the near future. With a basic knowledge of biology, you'll be able to make critical choices relevant to our future on Earth.

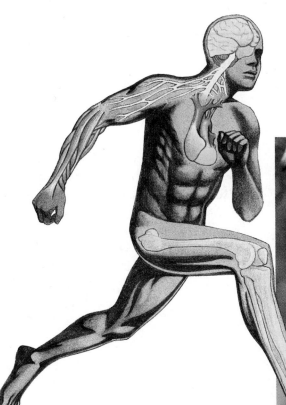

Figure 1.4

Biology will teach you about how humans function and fit in with the rest of the natural world. It will also equip you with the knowledge needed to handle any future biological problems of Earth.

Section Review

Understanding Concepts

1. What kinds of questions are answerable using the science of biology?
2. Identify an important reason for studying biology.
3. Identify at least two ways in which other living things affect you every day.

Thinking Critically

4. Describe a situation you have been in where knowledge of human biology was important.

Skill Review

5. **Observing and Inferring** Choose a type of living thing that you either live with or have seen on television or in a magazine, and identify three questions that you would like answered about this living thing. For more help, refer to Thinking Critically in the *Skill Handbook*.

Biologists in Action

*B*iologists who do research follow the general methods followed by all scientists. But the ways in which they gather, record, and analyze data range from simple to high-tech. The environments in which biologists carry out their investigations are as varied and diverse as Earth itself.

Biologists work in a variety of environments Some biologists work in clean, air-conditioned laboratories. But others carry out their research in the field, which includes places ranging from mountaintops to the ocean floor—and everything in between. Field work often involves doing experiments and recording observations in remote and inhospitable places. In some cases, special equipment is essential in order to venture into environments where human beings would not normally be able to survive.

Arctic environments In order to collect the data they need, biologists working in Antarctica must face extreme cold, fierce winds, and blizzards that can last many days. These harsh conditions affect instruments as well as people. Why do you think experienced antarctic researchers always bring along several sets of spare parts for all of their instruments?

Technology in the field
With portable laptop computers, biologists can begin analyzing data while still in the field. What advantage can you see in being able to analyze data at the site where you carried out an experiment or made your observations, before returning to your office or laboratory?

Radio tracking How far does a wolf travel in a day, a month, or a year? By using instruments like this radio-transmitter collar, biologists can track animals from a great distance over long periods of time. The data collected in this type of experiment can be used to produce maps that show daily movements and seasonal migration patterns. Why do you think it would be important to use long-life batteries in such a radio-transmitter collar?

Research in the laboratory A microbiologist counts bacterial colonies growing in a culture dish in which antibiotics are present. Different antibiotics inhibit the growth of different kinds of bacteria. By counting the number of colonies in the dish, this researcher can tell how effective this antibiotic is against the bacteria. What would it mean if no colonies of bacteria were growing on this petri dish?

Deep-sea studies To study the deep sea, biologists rely on complex equipment like this deep-diving research submersible. These small subs can descend hundreds of meters below the surface. They are equipped with mechanical arms, claws, and suction tubes that are used to collect organisms that live in the darkness at these great depths. How might collecting organisms with a submersible be better than using nets lowered from ships at the surface?

EXPANDING YOUR VIEW

1. **Writing in Biology** In a short paragraph, summarize methods of research used by biologists.
2. **Research** Conduct library research on current studies of marine environments. What risks might be associated with some of these studies?

1.2 What Is Life?

Section Preview

Objectives

Summarize the characteristics of living things.

Relate the characteristics of life to specific examples in organisms.

Recognize the major themes of biology.

Key Terms

organism
organization
reproduction
species
growth
development
environment
stimulus
response
adaptation
homeostasis
energy
evolution

*I*t was just another hot summer day along Mexico's Sonoron coast when marine biologists Katrina Mangin and Pete Raimondi took time out from their research to view something curious. Here and there, amid the dense growth of acorn barnacles that normally covers the coastal rocks, were bare areas totally devoid of barnacles. These bare areas seemed to occur in a regular pattern. Upon closer examination, the scientists discovered a tiny, hard, coal-black spot in the center of each of the bare areas. Could these black spots be living things? Could they be causing the formation of the bare areas?

Characteristics of Living Things

Most people feel confident that they could identify a living thing from a nonliving thing, but sometimes it's not so easy. In identifying life, you might ask, "Does it move? Does it grow? Does it reproduce?" These are all excellent questions, but consider a flame. A flame can move, it can grow, and it can produce more flames. So are flames alive?

Biologists have formulated a list of characteristics by which we recognize living things. Sometimes, nonliving things have one or more of life's characteristics. But only when something has all of them can it be considered alive. Anything that possesses all of the characteristics of life is known as an **organism**, *Figure 1.5.* What are the characteristics of living things?

Figure 1.5

These plants are called *Lithops* from the Greek *lithos,* meaning "stone." Although they don't appear to be so, *Lithops* are just as alive as elephants. They both possess all of the characteristics of life.

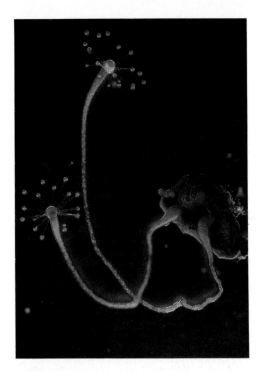

Figure 1.6

Like all organisms, hydroids show organization—they possess structures for every function. Each function that it performs, such as feeding or digestion, is vital to its existence, but these functions don't occur independently. In all organisms, body functions interact with one another to create a single, orderly, living system.

Living things are organized

When biologists Mangin and Raimondi were searching for signs of life in the little black spots they collected, one of the first things they did was put them under a microscope to observe their structure. That's because they knew that all living things show an orderly structure, or **organization.**

After careful scrutiny under the microscope, the black spots were identified as tiny relatives of jellyfish, known as hydroids, *Figure 1.6.* As Mangin and Raimondi suspected, they found organization within these little animals. Like other organisms, hydroids have specialized parts that perform particular functions.

Although the living world is filled with many examples of organisms, life may be defined on the basis of several characteristics shared by all organisms. One of these characteristics is organization. Whether it is unicellular or multicellular, all structures and functions of an organism come together to form an orderly living system.

Living things make more living things

Perhaps the most obvious of all the characteristics of life is **reproduction,** the production of offspring, *Figure 1.7.* Organisms don't live forever. For life to continue, organisms must replace themselves. Biologists Mangin and Raimondi realized that their little black spots reproduced because they found many bare areas within the patches of barnacles. The tiny hydroids had produced more hydroids.

Figure 1.7

Living things reproduce to make more of their own kind. Organisms have evolved a variety of mechanisms for reproducing and ensuring the continuation of their own species. Some organisms, such as rabbits, tend to produce many offspring in one lifetime.

time, it grew and took on the characteristics of its species. **Growth** results in an increase in the amount of living material and the formation of new structures.

All organisms grow, as shown in *Figure 1.8,* and different parts of organisms may grow at different rates. Organisms made up of only one cell may change little during their lives, but they do grow.

On the other hand, organisms made up of numerous cells go through many changes during their lifetimes, *Figure 1.9.* Think about some of the structural changes your body has already undergone in your short life. All of the changes that take place during the life of an organism are known as its **development.**

Figure 1.8

All life begins as a single cell. As cells multiply, organisms grow and develop and begin to take on the characteristics that identify them as members of a particular species of organisms.

desiccate:

desiccatus (L) to dry up
Desiccate means to dry up.

Reproduction is not essential for the survival of an individual organism. However, it is essential for the continuation of an organism's **species,** a group of similar-looking organisms that can interbreed and produce fertile offspring. If individuals in a species never reproduced, it would mean an end to that species's existence on Earth.

Living things change during their lives

The hydroids studied by Mangin and Raimondi were adults, but they didn't always look like this. A hydroid's life begins as a single cell, as do the lives of all organisms, and over

Living things adjust to their surroundings

One of the things that really perplexed scientists Mangin and Raimondi about the new hydroids was the fact that these animals were able to survive full exposure to the sun for up to six hours at temperatures reaching up to 100°C. Until the discovery of the black spots, no known hydroid was able to withstand these dry, hot conditions without desiccating. How did they do that?

Figure 1.9

At times during an organism's life span, development occurs rapidly. At other times, little change occurs for many years.

Figure 1.10

Living things are able to respond to stimuli and make adjustments to environmental conditions. Adaptations such as long hind legs enable rabbits to quickly avoid predators in their environment. Fur is an adaptation that allows rabbits and other mammals to regulate their body temperatures.

Living things live in a constant interface with their surroundings, or **environment,** which includes the air, water, weather, temperature, any organisms in the area, and many other factors. But anyone who has ever lived through the freezing temperatures of winter or has had to cross a busy street knows that it's not enough to just exist in an environment; sometimes you must adjust to the environment, *Figure 1.11*. Any condition in the environment that requires an organism to adjust is known as a **stimulus.** A reaction to a stimulus is called a **response.**

The ability to respond to stimuli in the environment is an important characteristic of living things, as shown in *Figure 1.10,* and it's one of the more obvious ones, as well. That's because many of the structures and behaviors that you see in organisms enable them to make adjustments to stimuli in the environment.

Consider the hydroid animals, for example. The delicate, flowerlike forms that they display occur only when the tide is up and they are underwater. When the tide is out and conditions are hotter and drier, hydroids withdraw their bodies into shell-like structures that effectively seal them off from the external environment—the black spots!

Figure 1.11

Humans adjust to weather changes for survival and for comfort.

Microscopic organisms known as rotifers create a water current with their wheels of cilia. They feed on microscopic food particles from the water.

Many nocturnal animals, such as this tarsier, possess large eyes for efficient vision at night.

The sharp spines of the cactus are reduced leaves. This adaptation enables cacti to conserve water in their desert environment.

Figure 1.12

Living things adapt to their environments in diverse ways.

Any structure, behavior, or internal process that enables an organism to respond to stimuli and better survive in an environment is known as an **adaptation.** *Figure 1.12* shows some other adaptations in organisms.

An organism must respond to stimuli from its internal environment, as well. Factors such as external temperature or infection from bacteria can cause changes in body temperature; the quantities of water, nutrients, and minerals inside the body; or other internal changes. Such changes can disrupt proper functioning. Adjustments to internal stimuli help organisms maintain a steady internal environment.

An example of this kind of adjustment is human sweating. When it's hot outside or after strenuous physical activity, your body temperature increases slightly. In response to this internal change, you begin to sweat

and your face gets flushed as many tiny blood vessels fill with blood. Both of these responses have the effect of cooling the body, and these adaptations help your body maintain its proper internal temperature necessary for metabolism. The regulation of an organism's internal environment to maintain conditions suitable for life is called **homeostasis.**

As you learn more about Earth's organisms in the chapters of this book, always reflect on these general characteristics of life, and ask yourself questions about how each organism meets the requirements for life. By asking questions such as, "How does this organism reproduce?" "What is the organization of this organism?" and "What are the adaptations of this organism that enable it to survive in its environment?" you'll learn that there are more similarities in the natural world than differences.

The Themes of Biology

In this course, you will be presented with many facts about organisms. Such facts are useful for gaining a good, working vocabulary of biology. However, biology, like other sciences, isn't merely a collection of isolated facts. Several major themes in biology serve to unify it as a science by linking isolated facts and ideas.

Energy

Energy is a central concept of the physical sciences, but it also pervades other sciences, including biology. Defined in physical terms, **energy** is the ability to do work or the ability to make things move.

Energy is important because it powers life processes. It provides organisms with the ability to maintain homeostasis, grow, reproduce, move, and carry out other life functions. Organisms obtain energy from the foods they eat or, in the case of plants and several other types of organisms, the foods that they produce.

As you'll learn, energy doesn't just flow through individual organisms; it also flows through communities of organisms, or ecosystems, and determines how organisms interact with each other and the environment. *Figure 1.13* shows one example.

Systems and interactions

You're probably familiar with a variety of systems in your life: the telephone system, a stereo system, the public-transportation system. As you know, each of these systems is made up of separate parts interacting to form a functioning whole.

In biology, you'll come across the idea of systems frequently. Organisms themselves may be thought of as systems. You're probably aware that your body contains several systems, including a nervous system, digestive system, and circulatory system. Each of these systems, as you'll learn, does not function independently; they interact in some rather complex ways to help perform the functions of life.

Unity within diversity

As you study the various types of ecosystems in the natural world, you'll be introduced to the different kinds of organisms that live there and how they interact to form a stable system. The theme of unity within diversity reflects the idea that, although ecosystems contain countless numbers of species—each with their own structural and behavioral specializations—all life is unified by the general characteristics you learned about earlier.

Figure 1.13

In coastal communities, such as the one where hydroid animals live, plants and algae convert the sun's energy into energy that can be used by other organisms in the community.

Have you ever wondered what happens to animals in a pond when it dries up, when it gets cold, or when some other unfavor-

How does temperature affect a living thing?

able condition occurs? Some animals are capable of burrowing into the mud; others can form a capsule around themselves when conditions are unfavorable. Assume that the capsule you have been given is an animal that has formed a protective cover under cold conditions in the pond in which it lives. It will come out of its capsule when conditions are favorable again.

PREPARATION

Problem
Under what conditions will an "animal" in a protective capsule emerge?

Objectives
- **Determine** the temperature of the water that causes the animal to come out of its protective capsule.
- **Compare** the time it takes for the animals to come out of capsules under different conditions.

Possible Materials
plastic cups or beakers of warm water and ice water
stirring rods or coffee stirrers
thermometers
Instant Sealife toy capsules

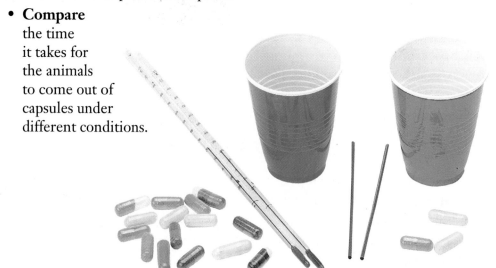

PLAN THE EXPERIMENT

1. In your group, make a prediction about how temperature might affect your capsule animal.

2. Design a way, based on your prediction, to test your animal with the materials provided by your teacher.

3. Make a numbered list of directions.

4. Make a list of the materials you will use.

5. Design and construct a table in which to record what happens during your experiment. In your table, you may want to record how long it takes for your animals to come out of their capsules, the temperature of the water, and details of the appearance of the animal.

Check the Plan

1. ***Make sure your teacher has approved your experimental plan before you proceed further.***

2. Carry out your experiment.

ANALYZE AND CONCLUDE

1. **Calculating Results** How long did it take for your animals to come out of their capsules?

2. **Analyzing Data** What was the relationship between water temperature and the length of time it took for your animals to come out of their capsules?

3. **Relating Concepts** How might your results compare with those of real animals in a pond that gets cold during the winter?

4. **Making Inferences** Why would it be important for an animal in a protective capsule to stay in the capsule until water temperatures become warm?

Going **Further**

Project Plan a study in which you could test the effects of warm and cold temperatures on a variety of insects.

Figure 1.14

Organisms have a variety of adaptations, both behavioral and structural, that help them maintain homeostasis.

◀ Many animals, such as this box turtle, dig shallow burrows to escape heat and maintain optimum body temperatures.

▼ The normally jet-black darkling beetle produces a yellowish, waxy covering over its body that acts as a natural sunblock when temperatures rise.

▶ The camel's fatty hump acts as a storage compartment for water. Water is released from the fat through metabolism.

Homeostasis

You've learned that homeostasis is the regulation and maintenance of the internal environment of organisms. The concept of energy is involved here, as well. Without energy, hydroid animals wouldn't be able to withdraw into their tiny, black shells to maintain homeostasis. Nor could you perspire or shiver to maintain homeostasis in your body without energy. *Figure 1.14* gives several examples of these mechanisms.

Evolution

Have you ever noticed that the paws of cats, the flippers of frogs, and the hands of people, although different, are overall similar in structure, *Figure 1.15?* Clues to the diversity of life on Earth may be understood through the study of **evolution,** the gradual change in the characteristics of species over time.

Evolution is another important theme in biology and is perhaps the major unifying one. This is because all of the structures, behaviors, interactions, and internal processes observed in the millions of species of organisms on Earth are the result of the process of evolution.

In the science of biology, organisms are the principle objects of study, just as elements are in the science of chemistry and numbers are in the science of mathematics. But biologists have seen only the tip of the iceberg in terms of what the forces of evolution can produce. Tens of millions of organisms await discovery

Figure 1.15

The paws of cats, the feet of frogs, and the hands of people, although appearing different on the outside, contain similar sets of bones. This suggests that these animals all share a common ancestry.

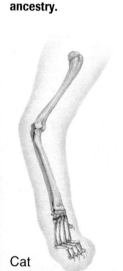
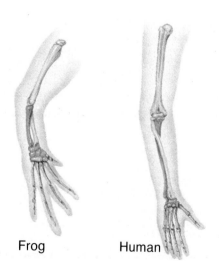

Cat   Frog   Human

and will only add to the great diversity that we already know is present on Earth, *Figure 1.16.*

The nature of science

Biology, like all sciences, is a continuous process that seeks to discover facts about the natural world. The discovery of a new species of hydroid with a new set of behaviors and characteristics may have invalidated earlier ideas about hydroids, but the scientists who developed such ideas were not wrong. Scientific facts can be determined only by making careful observations of present phenomena and by building on previous knowledge. The modification of ideas, rather than their outright rejection, is the norm in science. As new species are discovered and studied, more change is inevitable.

Figure 1.16

By most estimates, biologists have identified and studied about 1.5 million different types of organisms; however, the number of undiscovered species is estimated by some scientists to be in the tens of millions.

Connecting Ideas

Our world abounds with a great variety of life. Giant tube worms thrive at the bottom of our oceans near bubbling underwater volcanoes, colonies of green algae populate the tips of polar bear fur, and microscopic bacteria make their home in the pores of your skin. These and the millions of other organisms that inhabit the natural environment form the core of the science of biology.

However, biology is not just a body of knowledge. Biology is a process. It is a way of knowing. In the next chapter, you'll see that biologists, like other scientists, have an organized way of finding out about the natural world.

Section Review

Understanding Concepts

1. What is homeostasis, and how is it maintained in hydroid animals?
2. In what ways does the theme *systems and interactions* apply to both the living and the nonliving world?
3. What is meant by *unity within diversity?*

Thinking Critically

4. Why is evolution considered to be a unifying theme in biology? Explain your answer.

Skill Review

5. **Observing and Inferring** Suppose you discover an unidentified object on your way home from school one day. What steps would you take to determine whether the object is a living or nonliving thing? For more help, refer to Thinking Critically in the *Skill Handbook.*

Reviewing Main Ideas

1.1 What Is Biology?
- Biology is the organized study of living things and their interactions with the natural and physical environments.
- Biology can teach you about yourself and is important for preserving the environment and for developing useful medical and agricultural techniques.

1.2 What Is Life?
- All living things are united on the basis of four general characteristics: organization, reproduction, growth and development, and the ability to adjust to the environment.

- Six general themes are woven throughout the science of biology: Energy, Homeostasis, Unity Within Diversity, Systems and Interactions, Evolution, and The Nature of Science.

Key Terms
Write a sentence that shows your understanding of each of the following terms.

adaptation	homeostasis
biology	organism
development	organization
energy	reproduction
environment	response
evolution	species
growth	stimulus

Understanding Concepts

1. Why is reproduction an important life characteristic?
2. Why is it important for organisms to maintain homeostasis?
3. To what kinds of environmental stimuli must organisms respond?
4. Your heart beats more quickly and you breathe more rapidly after exercising. What characteristic of life does this illustrate?
5. Flames can grow, reproduce, move, and respond to stimuli such as wind. Why aren't flames considered to be alive?
6. Explain how an organism is an example of a system.
7. Why is energy important for organisms?
8. The series of changes a caterpillar goes through when it turns into a butterfly is an example of what life characteristic?

Using a Graph

9. The new species of hydroid discovered along the Sonoran coast was tested to see how long a dry period it could survive. Study the graph of the data below. What is the maximum length of time these organisms survived in dry conditions? Why wouldn't these organisms have evolved mechanisms to survive longer?

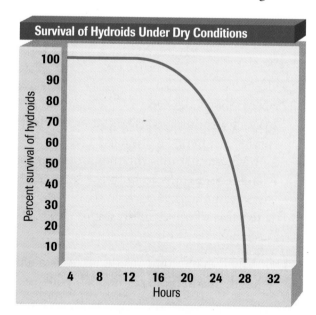

Survival of Hydroids Under Dry Conditions

Percent survival of hydroids: 100, 90, 80, 70, 60, 50, 40, 30, 20, 10

Hours: 4 8 12 16 20 24 28 32

Relating Concepts

10. Make a concept map that relates the following terms and phrases. Supply the appropriate linking words for your map.

    biology, homeostasis, organism, growth, development, adaptation, evolution

Applying Concepts

11. Hydroid animals may be a new addition to the beach, washed in from another area during a storm. If they have no natural enemies in the new area, what might happen to the environment on the beach rocks?
12. Describe a situation in which knowledge of biology was helpful to you.
13. Describe how humans show the life characteristic of organization.
14. Identify several different types of natural communities. What do they have in common? How do they differ?
15. What effects, if any, might the removal of weeds from a garden have on other organisms?

Art Connection

16. How does the realistic style of the paintings of Robert Bateman help increase our knowledge and understanding of our natural world?

Thinking Critically

Making Inferences

17. **Biolab** Assume that you tested tiny animals in protective capsules to see how quickly they would come out of the capsules under dry and wet conditions. You found that they emerged quickly when totally submerged in water. They did not emerge at all in conditions of low to moderate moisture or in dry conditions. What might you conclude about the requirements of these organisms?

Making Comparisons

18. Examine the following items: a flame, bubbles being blown from a bubble wand, and a balloon released into the air. List characteristics that each show that might indicate life. List the characteristics that indicate they are not alive.

Applying Concepts

19. You may have seen squirrels such as this one around your home or school. Examine the photograph and identify two adaptations that enable squirrels to survive in their environment.

Connecting to Themes

20. **Evolution** Hydroid animals have been shown to be the cause of bare areas within patches of acorn barnacles. If this trend continues, how might it affect the evolution of acorn barnacles?
21. **The Nature of Science** Why would knowledge about other organisms be important in human medicine?
22. **Systems and Interactions** Sea slugs receive food from the algae that live within their skin. What benefits, if any, do you think algae receive from sea slugs?

CHAPTER 2 Scientific Methods in Biology

Lions in the wild are free to wander the landscape, chase prey, eat, breed, take care of cubs, and sleep, which is one of their major occupations. The behavior of a caged lion is much different from the behavior of lions in their natural environment. Most modern zoos provide open spaces for large animals, but sometimes they must be confined to cages while cleaning and maintenance are performed. Have you ever seen a lion or tiger pacing back and forth in a cage? Did you wonder why they pace in their cages when they don't behave this way in nature? Biologists have found that pacing is a reaction many animals have to restricted space.

How do biologists know these things? They've acquired this knowledge using well-established methods of study. In this chapter, you'll learn more about experimenting as well as the other methods that scientists use to study the natural world.

Concept Check

You may wish to review the following concepts before studying this chapter.
- Chapter 1: biology, themes of biology

Chapter Preview

2.1 **Problem-Solving Methods in Biology**
Observing and Hypothesizing
Experimenting

2.2 **The Nature of Biology**
Kinds of Research
Science and Society

Laboratory Activities

Biolab: Design Your Own Experiment
- How does fertilizer affect early plant development?

Minilabs
- How can you use inductive reasoning?
- How do you decide which paper towel to buy?

Scientists do experiments to discover the reasons for differences in animal behavior. If you've ever done something as simple as test two different types of dog food to see which one your dog prefers, then you've performed an experiment.

25

SECTION 2.1 Problem-Solving Methods in Biology

Have you ever watched elephants in a zoo? Maybe you've heard them trumpet and snort or watched them swing their heads, examine things with their sensitive trunks, or flare their large ears. Have you ever seen them do things that make you think they are communicating with each other? In this section, you will see how one scientist applied scientific methods to learn how elephants communicate.

Observing and Hypothesizing

Why are biologists interested in answering a question such as, "How do elephants communicate?" For a scientist, the simple reason is curiosity about how and why things happen in nature. In addition, the answer to the question will lead to a better understanding of elephant behavior. This knowledge, in turn, may enable zookeepers and wildlife conservationists to better care for elephants.

The methods that biologists use

To solve problems, different biologists may take different approaches, yet there are some steps that are common to these different approaches. The common steps that biologists and other scientists use to gather information to solve problems are **scientific methods.**

Scientists often discover problems to solve—that is, questions to ask and answer—simply by observing the world around them. For example, a scientist may be working on the reproduction of mosses in the laboratory and come up with another question. Another scientist may ask a question about the feeding habits of prairie dogs after making a series of observations of prairie dog behavior in the field.

The question of elephant communication

For many years, observers had noticed that elephant herds appear to move, turn, or stop suddenly and all together without any apparent audible or visual signal. For example, if one elephant noticed a lion, all the elephants in the herd seemed to become alarmed. Some might even charge, as shown in *Figure 2.1*.

Figure 2.1

Your first thought about elephant communication might be of the trumpeting sound they make when alarmed. A biologist might ask, "Is this the only sound elephants make to communicate?"

One biologist, Katharine Payne of Cornell University, made an important observation about elephant communication at the Metro Washington Park Zoo in Portland, Oregon. While visiting the zoo, Payne felt the air throbbing around her. It reminded her of the rumbling of thunder, a sound more felt than heard. Payne had spent 17 years studying the calls of whales and that some whales make sounds that are too low-pitched for people to hear. When she felt the vibrations, she also noticed that the skin on an elephant's forehead seemed to be fluttering. Look at *Figure 2.2*. Putting this observation together with her past experiences, she suspected that the vibrations were generated by the elephants and that they might be using the sounds to communicate. In other words, Payne hypothesized that elephants communicate by means of low-frequency sounds. A **hypothesis** is a testable explanation for a question or problem.

As you can see from Katharine Payne's example, a scientist's hypothesis is usually not just a random guess or a "shot in the dark." More likely, before a scientist makes a hypothesis, he or she has some idea what the answer to a question might be because of experience, extensive reading, and previous experiments. Applied to all this knowledge is the scientist's reasoning powers.

Using inductive reasoning

Stop to think for a moment about how you solve problems in your everyday life. For example, suppose you can't find your house key. The last time you had it, you were wearing your gray jacket; and on two earlier occasions, coins and a pen had slipped through a hole in the pocket of that jacket into its lining. So you hypothesize that that's where the key is. You've used inductive reasoning. **Inductive reasoning**, the most often used method of reasoning in science to

Figure 2.2

Payne hypothesized that elephants communicate by means of low-pitched sounds when she saw the skin on the elephant's forehead fluttering. These sounds are too low-pitched to be heard by humans.

The Sounds of Life

Chuff–chuff–chuff–chuff, chuff, chuff! Resounding through the thick northern woods, these sounds signal the presence of a male ruffed grouse claiming his territory. To other males, it is a warning and a challenge; to females, it is an invitation.

Waves of energy The ruffed grouse makes sounds by spreading its wings and rapidly moving them down and in. This compresses the air trapped between its wings and body. Each beat of the wings produces a pulse of high pressure followed by a period when the air springs back to normal pressure. Each compression is a sound wave. But there is more to sound than just compressions. Sounds can be high or low, loud or soft. What causes these differences?

High or low? The pitch of a sound depends on its frequency—how many compressions are produced each second. The ruffed grouse's sound is low pitched because its wings beat only a few times each second. The rapidly beating wings of a mosquito produce many more air compressions per second; thus, a high-pitched sound results.

Loud or soft? The more energy a sound wave has, the more it compresses the air and the louder it sounds. In fact, some sounds such as over-amplified music or sharp explosions such as those from a gun have enough energy to severely damage your hearing.

The amount of energy in a sound wave also determines its range. Sound waves spread out in a spherical pattern as they travel. Therefore, the energy of the original sound becomes diluted in a larger space. A sound that has more energy will be heard at a greater distance. Thus, the powerful bellow produced by a 400-kg moose will be heard farther away than a chirp produced by a 250-g chipmunk.

CONNECTION TO Biology

Bats and dolphins make sounds and locate objects by the reflection of those sounds. In what ways do humans use this or similar techniques for practical purposes?

develop a hypothesis, is reasoning from a particular set of facts to a general rule. Payne used inductive reasoning to form a hypothesis about the problem of how elephants communicate.

Using deductive reasoning

Sometimes a general rule is known before a particular case is apparent. For example, you know that dogs pant when they are hot and thirsty. One day, you see your dog panting heavily and think, "If the dog is panting, she must be hot and need water." So you check her water bowl and, sure enough, it's dry. You've used deductive reasoning. **Deductive reasoning** involves suggesting that something may be true about a specific case from known general rules. This kind of reasoning is often expressed as an "If . . . then" statement. Suppose you live in an area

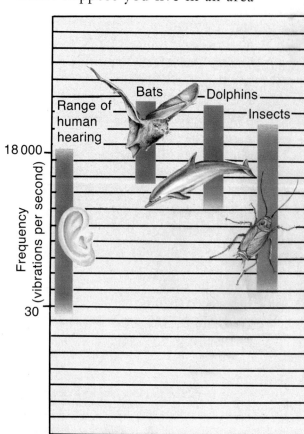

Range of human hearing

Bats Dolphins Insects

18 000

Frequency (vibrations per second)

30

that has a history of flash floods. Then you would use deductive reasoning to say, "If it rains another two inches in the next hour, we'll have a flood."

After stating a hypothesis, scientists use deductive reasoning. In Payne's case, she applied some general rules from physics to the hypothesis she had made about the elephant sounds. She already knew that humans can hear sounds beginning around 20 to 30 vibrations per second, but that lower-pitched sounds can be produced by animals. Look at *Figure 2.3* for a comparison. She also knew that lower-pitched sounds aren't easily absorbed by objects they strike, and therefore can travel farther than higher sounds before they become too faint to hear. Thus, Payne suggested that *if* elephants produce low-pitched sounds to communicate with other elephants, *then* there should be some evidence of elephants' reacting to the low-pitched sounds made by elephants that are far away. Payne's next step was to design a controlled experiment to test her hypothesis.

Experimenting

People do not always use the word *experiment* in their daily lives in the same way as scientists do in their work. As an example, you may have heard someone say that he or she was going to experiment with a cookie recipe. The person is planning to substitute raisins for chocolate chips, use margarine instead of butter, add cocoa powder, reduce the amount of sugar, and bake the cookies for a longer time. This is not an experiment in the scientific sense because there is no way to know what effect any one of the changes alone has on the resulting cookies. To a scientist, an **experiment** is a procedure that tests a hypothesis by the process of collecting information under controlled conditions.

What is a controlled experiment?

Some experiments involve two groups: the control group and the experimental group. The **control** is the standard, in which all conditions are kept the same. The experimental group is the test group, in which all conditions are kept the same except for the single condition being tested.

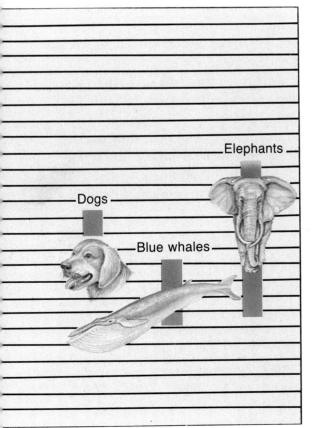

Dogs

Blue whales

Elephants

Figure 2.3

The normal range of human hearing is from about 20 to 30 vibrations per second up to about 18 000 to 20 000 vibrations per second. Some animals, such as bats and dolphins, can make sounds that are higher-pitched than humans can hear. Elephants and whales, on the other hand, can make audible sounds as well as sounds that are too low-pitched for humans to hear.

MiniLab

How can you use inductive reasoning?

Scientists must be excellent observers. They need both their observation skills and their reasoning power to form hypotheses. This procedure will give you a chance to use your own.

Procedure

1. Obtain a single specimen of a flower, leaf, or small insect.

2. With the specimen in front of you, observe as many characteristics as you can about its shape, size, color, texture, or any other features that you can observe.

3. Write as many observations as you can in five minutes.

4. Now write two or three questions you have about your specimen.

5. After you have made and recorded your observations and questions, use inductive reasoning to form one hypothesis that you could test.

Analysis

1. How do you think your observations were similar to those a scientist might make?

2. Explain how your observations led you to form a question.

3. Why do scientists use inductive reasoning?

Suppose you wanted to learn how salt water affects a new strain of bluegrass. The control group would consist of several grass plants watered with plain water. The experimental group might consist of plants watered with several different strengths of salt water. The condition being tested is the concentration of salt in the water. All other conditions would be held constant for both groups. For example, in *Figure 2.4* you can see a control and an experimental group for an experiment on growing soybeans.

Designing an experiment

Katharine Payne used insight and imagination in developing her hypothesis. Most scientists would agree that both of those qualities are also needed to design an experiment to test a hypothesis.

In a controlled experiment, only one condition is changed at a time. The condition in an experiment that is changed is the **independent variable.** While changing the independent variable, the scientist observes or measures a second condition that results from the change. This condition is the **dependent variable.** In the experiment to test the effect of salt water on bluegrass, the concentration of salt is the independent variable. The resulting growth rate of the grass is the dependent variable.

Figure 2.4

Shown here are the results of an experiment to test the effect of various soil bacteria on the growth of soybeans. The center rows are the experimental plants. The left and right pairs of rows are controls. All other conditions in the field—soil, light, water, and fertilizer—are the same.

In the experiment designed by Payne, the production of low-frequency sound was the independent variable, and observable change in the behavior of the elephants was the dependent variable. To perform her experiment, Payne went back to the zoo with a tape recorder and microphone. She recorded hours of what seemed to her to be silence among the elephants. At the same time, Payne was making careful notes about the behavior of the elephants in their compound.

Just as problems may be arrived at differently, the approaches taken to solve a particular problem can vary widely. The experimental design that a scientist selects depends on what other experimenters have done and what information the scientist hopes to gain. Sometimes, a scientist will design a second experiment even while a first one is being conducted if the scientist thinks the new experiment will help answer the question.

Using tools

In order to carry out her experiment, Payne required tools that would enable her to record the sounds of the elephant. For instance, her tape recorder and microphone had to be able to respond to low-pitched sounds.

Biologists use a wide variety of tools to obtain information in an experiment. Some of the common tools are beakers, test tubes, hot plates, petri dishes, balances, thermometers, metric rulers, and graduated cylinders. More complex tools include specialized microscopes, centrifuges, radiation detectors, spectrophotometers, DNA analyzers, and gas chromatographs. *Figure 2.5* shows some of these more complex tools.

Figure 2.5

The microscope (below left), the gas chromatograph (right), and gel electrophoresis (below right) are three of the many tools that biologists can use in their studies.

▲ The gas chromatograph can be used to detect and measure pesticide residues in plants or fish.

▼ The microscope magnifies organisms or parts of organisms, making small details visible.

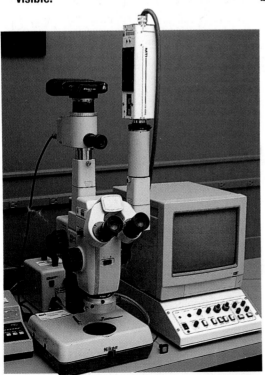

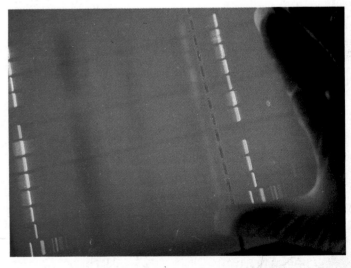

▲ Gel electrophoresis can be used to analyze DNA, producing a DNA print as shown. Comparing DNA can tell a biologist how closely related two organisms are.

Meet Katharine Payne, Naturalist

Even with all of the high-tech equipment used by modern biologists, sometimes the simplest tools are the most valuable to Katharine Payne, who studies animal behavior in the wild. She may carry a tiny bag of ashes from her morning campfire as she approaches a herd of African elephants. It's important to stay downwind of these sensitive animals, so she shakes out the ashes and watches the way they drift in the barely detectable breeze.

In the following interview, Ms. Payne describes her life as an observer of animals.

On the Job

Q Ms. Payne, could you describe your work?

A I'm a freelance biologist, and I study elephants, whales, and beavers. I make my living by getting grants to study things that interest me. The National Geographic Society is one of the agencies that provides funds for me to set up a study. Each of my studies has two objectives. The first is to learn something new and fascinating about an animal whose behavior is important to people. The second is to use what I learn to get people interested in my discoveries so that they will become advocates for wildlife.

Q Why is your work important?

A My discovery that elephants communicate with one another has called attention to the species at a time when people are furiously arguing over how to assure its survival. People disagree on whether a ban on trading ivory, the tusks of elephants, will ultimately save or doom the African elephant. I hope that my work reminds people on both sides of the debate that elephants are more than an economic issue; they are sensitive, social living things that lead extraordinary lives.

Q Could you tell us what you are working on right now?

A I'm writing up the results of an expedition to Africa in which I studied how elephants drill wells through the sand in dry river basins. They get water for themselves, but other animals also depend on these elephant wells. This elephant behavior is quite fascinating because we can't yet understand how elephants seem to know just where to dig.

Early Influences

Q How did you get interested in your field?

A I spent my whole childhood exploring the outdoors. I lived on a farm outside of Ithaca, New York, where glaciation left high waterfalls. My family lived in a city for a year when I was about ten. I felt like an alien there and was glad to get back to the country.

Q Were there books that influenced your career choice?

A As a child, I probably read *The Jungle Books* by Rudyard Kipling 100 times. The stories are fanciful ones in which a young boy, Mowgli, can communicate with animals. He understands the separate languages of the bat, bear, tiger, and other animals. I've wanted to do that all my life.

Q What people were special influences on you?

A Since my college degree was in music, you could say my training in biology was an old-style apprenticeship, working alongside the animal behaviorist, Roger Payne, who was my husband. In the 1960s, we studied the humpback whales in the Bermuda area and discovered that these whales sing long, complex, very beautiful songs that are continually changing. I started as a volunteer to study these songs, and after a few years I was able to get grants to study communication among whales. After 20 years of this work, I guess I had become a professional in this field of biology.

Personal Insights

Q Has your background in music been useful in your career in biology?

A Yes, both in studying whale songs and in my work with elephant communication. In the case of the elephants, I first noticed on a visit to the zoo that there was a kind of throbbing in the air around the elephant enclosure, even though I heard nothing. I remembered a similar feeling from my childhood, when I sang in choir and stood right next to the largest, deepest organ pipe. That connection led me to the idea that elephants might communicate in sounds too low for humans to hear.

Q Do you have any advice for a student who might like a career as a naturalist?

A It's harder today to come into the field through a "back door," as I did without a degree in biology. So, I encourage young people to get the training they need, but to start by going out into the natural world and experiencing it. Then they need to think hard about what they've seen and heard and decide what interests them the most. And I especially urge students to choose areas of study that have value to our environment.

Does a weed killer perform as advertised?

Scientific methods are useful for solving problems and answering questions that arise in everyday life. One method all scientists use is observation. A gardener made an observation that dandelions grow as unwanted weeds in many lawns. On television, the gardener heard about a new chemical, Kilimal, that is supposed to be useful for ridding lawns of weeds. The gardener decided to do an experiment to find out if the chemical was really effective on dandelions.

Analysis

The diagram below represents a controlled experiment to examine the effects of Kilimal on dandelions.

1. Observation: A manufacturer claims that its herbicide will get rid of dandelions.

Original Lawn

2. Hypothesis: Kilimal will kill dandelions while leaving grass healthy.

3. Experiment: Two identical sections of a lawn overgrown with dandelions are selected. Kilimal is applied to experimental section; other section is left untreated as a control.

4. Results (data) are obtained.

Without Kilimal

With Kilimal

Thinking Critically

According to your interpretation of the illustration, what is your conclusion about the effect of the weed killer on dandelions? What can you say about its effects on other weeds?

 SHARP OBJECT SAFETY: This symbol appears when a danger of cuts or punctures caused by the use of sharp objects exists.

 RADIOACTIVE SAFETY: This symbol appears when radioactive materials are used.

 CLOTHING PROTECTION SAFETY: This symbol appears when substances used could stain or burn clothing.

 EYE SAFETY: This symbol appears when a danger to the eyes exists. Safety goggles should be worn when this symbol appears.

 CHEMICAL SAFETY: This symbol appears when chemicals used can cause burns or are poisonous if absorbed through the skin.

Maintaining safety

Safety is another important factor that scientists consider when carrying out experiments. Biologists try to minimize hazards to themselves, the people working around them, and the organisms they are studying.

In the experiments in this textbook, you will be alerted to possible safety hazards by safety symbols like those shown in *Figure 2.6*. A **safety symbol** is a symbol that warns you about a danger that may exist from chemicals, electricity, heat, or procedures you will use. Refer to the safety symbols in the *Skill Handbook* at the back of this book before beginning any lab activity in this text. It is your responsibility to maintain the highest safety standards to protect yourself as well as your classmates.

Data gathering

To answer their questions about scientific problems, scientists seek information from their experiments. This information is called **data.** Sometimes, data from experiments are referred to as experimental results.

Often, data are in numerical form, such as the number of elephant calls per minute or the length that corn plants grow per day. Numerical data may be measurements of time, temperature, length, mass, area, volume, or other factors. Numerical data may also be counts, such as the number of bees that visit a flower per day or the number of wheat seeds that germinate.

Figure 2.6

Which symbol warns you about dangers from radiation? Which tells you to wear eye protection? Which alerts you to hazards from acids? You can review all the safety symbols used in this textbook in Appendix D.

Figure 2.7

In the field, Payne recorded elephant sounds and then played them into a machine that produced a sonogram like the one shown here. A sonogram is a picture of a sound. You may have seen sonograms of bird songs in some field guides. With the sonogram, Payne was able to see when the elephants made sounds and how their pitch varied.

Sometimes data are expressed in verbal form, using words to describe observations during an experiment. The data that Payne gathered were the tape recordings and observations about elephant behavior. When she first played the tape recordings she had made of the elephants, Payne heard only background noise. This result was not unexpected, for she hadn't heard any other sound when she made the recordings. However, when the tapes were played faster to increase the frequency of the sound waves, Payne discovered that she had recorded hundreds of elephant calls. Some of these calls are represented by the sonogram in *Figure 2.7*

Thinking about what happened

Having the data from the experiment did not end the scientific process for Katharine Payne. Often, the thought process that goes into analyzing experimental data takes the greatest amount of a scientist's time. After careful review of the results, the scientist must come to a conclusion: Was the hypothesis supported by the data? Or was it not supported? Or are more data needed? Data from an experiment may be considered confirmed only if repeating that experiment several times yields similar results.

Payne wanted to learn more than she could from her experiment at the zoo. After analyzing her data, she, like most scientists, had more questions than she had before her experiment. In order to compare her results and conclusions with studies by other scientists in the same field, she researched the published literature for more information on elephant communication. She also began to think of experiments she might carry out.

Why do people fertilize plants? You may have observed someone fertilize a houseplant and wondered why he or she went to the trouble. Does fertilizer really make any difference in the growth of a plant? Does it make flowers bloom more rapidly or vegetable plants grow larger and produce more vegetables? These questions are all within the realm of science because they are testable by controlled experiments, observation, and data gathering.

How does fertilizer affect early plant development?

PREPARATION

Problem

Do seeds germinate faster if fertilizer is applied? Do more seeds germinate when fertilizer is applied? Do different strengths of fertilizer cause different rates of growth in plants? Do roots, stems, or leaves grow faster or bigger if fertilizer has been applied to the plant?

Hypotheses

Make a group decision about which of these questions you will test, or make up your own questions. Finally, form testable hypotheses about the questions.

Objectives

In this Biolab, you will:
- **Carry out** a controlled experiment.
- **Observe** the effect of fertilizer on plant growth or seed germination.

Possible Materials

seeds
plant seedlings
water
plastic trays
sand or potting soil
foam cups
fertilizer
balance
metric ruler
graduated cylinder

Safety Precautions 🧪 🥽 🧤

Do not eat seeds or taste chemicals.

PLAN THE EXPERIMENT

1. Write your experimental plan in the form of a numbered list similar to a recipe. First, list the materials you will need. Then, give details of steps you will take to collect your data.

2. Work in a group. Each group member should have one or more clearly defined tasks such as collecting materials, making observations, recording observations, or cleaning up.

3. Identify the conditions you will hold constant and a single independent variable. The independent variable could be how much fertilizer you apply or the strength of the fertilizer. Decide which dependent variable you will measure and how you will know if your data support your hypothesis.

Check the Plan

1. Review the summary of scientific methods in *Figure 2.11* to see if you have included most of the scientific methods.

2. Does your plan test only one variable, such as the amount of fertilizer added?

3. Have you determined how many seeds or plants you will use in each group and what dependent variable you will measure? Also, have you decided how often you will make measurements?

4. Did you make a data table that compares the observations you will make on the control and experimental groups?

5. ***Make sure your teacher has approved your experimental plan before you proceed further.***

6. Carry out the experiment.

ANALYZE AND CONCLUDE

1. **Identifying Variables** What factors did you hold constant during this experiment?

2. **Checking Your Hypothesis** What conclusions can you draw by using the data you collected?

Does your conclusion support your hypothesis?

3. **Interpreting Data** Make a statement explaining how your data did or did not support your hypothesis.

Going Further

Changing Variables Design another experiment to test the effects of fertilizer on seeds or plants of different species. With your teacher's permission, carry out the experiment. Can you conclude that fertilizer is always beneficial to plants?

Figure 2.8

This elephant herd in Africa is making its daily trip to the water hole. If these elephants in nature make sounds similar to the ones produced by elephants in zoos, do you think those sounds might have similar meanings? How could you find out?

It is clear that the elephants in the zoo make sounds that people cannot hear, but do elephants in nature, like those shown in *Figure 2.8*, make similar sounds? If the sounds are communication, how does the behavior of the elephants change as a result of the sound? To conduct field experiments to answer these questions, Payne and several colleagues traveled to Namibia in southwest Africa, where elephant herds roam.

Reporting results

Results and conclusions of experiments are reported in scientific journals, where they become open to examination by other scientists. Hundreds of scientific journals, such as those shown in *Figure 2.9*, are published weekly or monthly. In fact, scientists usually spend a large part of their time just reading journal articles trying to keep up with new information being reported.

Figure 2.9

The amount of information published every day in scientific journals is more than any single scientist could read. Fortunately, scientists also have access to computer databases that contain summaries of scientific articles, both old and new. It is important for a scientist to report data in the most easily understood way. Therefore, data are usually presented in tables, charts, and graphs such as those shown here.

Verifying results

Data and conclusions are shared with other scientists for an important reason: after results of an experiment have been published, other scientists may try to verify the results by repeating the experiment. If the results from the original experiment occur again, the later experiments provide support for the hypothesis. When a hypothesis is supported by additional data from the same or other scientists, the hypothesis is considered valid and is accepted by the scientific community.

To verify her original data from the Portland Zoo, Payne and her research team in Namibia set up equipment similar to that used at the zoo. As shown in *Figure 2.10*, the researchers placed their equipment at distances of more than a mile apart. In that way, they could observe whether widely separated elephants heard and responded to the sounds made by other elephants. The team made many trials before reporting their results.

When scientists publish the results of their experiments, other scientists, such as Katharine Payne, can relate their work to the published data. For example, other biologists studying elephants in Africa had published observations of their behaviors. Thus, Payne knew what behaviors might be the result of sound communication. Her team found that female elephants emit certain sounds in order to attract mates. Other sounds are produced by bull elephants to warn other males away from receptive females.

Theories and laws

People use the word *theory* in everyday life much differently from the way scientists use this word. You may have heard someone say that he or she has a theory that a particular football team will win the Super Bowl this year. What the person really means is that he or she believes one team will play better for some reason.

Figure 2.10

Payne's research team observed groups of elephants from a tower. Microphones were set up at various locations in the field. Elephant sounds were recorded and activity was videotaped at the same time to see if there was a relationship between elephant sounds in one location and elephant behavior in another location.

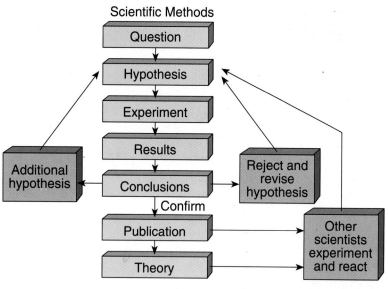

Scientific Methods

Figure 2.11

This chart presents an idealized version of the scientific method. Not all scientists carry out every step in every experiment. Not all experiments have hypotheses, and certainly not all experiments lead to published theories.

theory is an explanation of a natural phenomenon that is supported by a large body of scientific evidence obtained from many different experiments and observations. A theory results from continual verification and refinement of a hypothesis.

For example, the theory of plate tectonics explains how continents have moved together and apart several times during Earth's history. It also explains why similar fossils and related organisms can be found on continents that are now widely separated. A valid theory enables scientists to predict new facts and relationships. The theory of plate tectonics, for example, predicts that fossils of organisms found on the east coast of the United States should resemble those found in northeast Africa.

Scientists also recognize certain facts of nature called laws or principles. The fact that a dropped apple falls to Earth is an illustration of the law of gravity. *Figure 2.11* presents a schematic diagram of scientific processes.

Of course, much more evidence is needed to support a theory in science.

In science, a hypothesis that is supported by many separate observations and experiments, usually over a long period of time, becomes a theory. A

2.2 The Nature of Biology

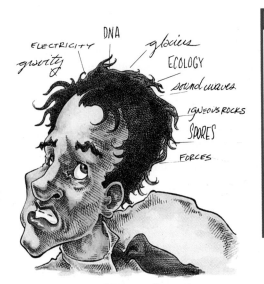

*P*art of science is learning many of the known facts about the world around us. More important, though, is that scientists use these known facts to discover new problems, make hypotheses, design experiments, interpret data, and draw conclusions. In short, science is not just a body of facts and ideas but also a process of study by which we come to understand the natural world.

Kinds of Research

You learned in the first part of this chapter that scientists use a variety of methods to test their hypotheses about the natural world. Scientific research can usually be classified into one of two main types, quantitative or descriptive.

Quantitative research

Most biologists conduct controlled experiments that result in counts or measurements—that is, numerical data. These kinds of experiments occur in quantitative research.

Data obtained in quantitative research may be used to make a graph or table. Suppose, for example, that a biologist is conducting research to count the number of microscopic organisms called *Paramecium* that survive at a given temperature. The study is an example of quantitative research. *Figure 2.12* shows the data as a graph.

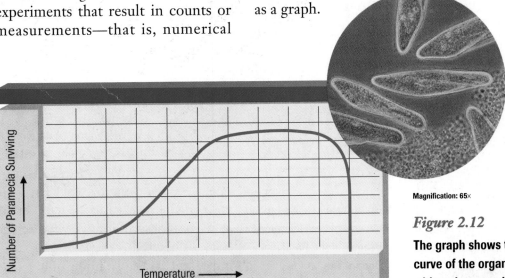

Magnification: 65×

Figure 2.12

The graph shows the survival curve of the organism *Paramecium* with an increase in temperature.

How do you decide which paper towel to buy?

Suppose that you are the new manager of a restaurant. The kitchen staff has complained that the paper towels they're given to clean up spills are not absorbent. The procedures manual for your restaurant chain says that you are allowed to choose from among the four lowest-priced brands available in your area. You decide to conduct an experiment to decide which brand of paper towel you should purchase.

Procedure

1. Cut a 5cm × 5cm piece from each of four brands of paper towel. Lay each piece on a smooth, level, waterproof surface.
2. Add one drop of water to each square.
3. Continue to add drops of water until the square can absorb no more.
4. Record your observations and make a graph of your results.

Analysis

1. Did all the squares absorb equal amounts of water?
2. If one brand of paper towel absorbs more water than the others, can you conclude that it is the towel you should buy? Explain.
3. Which scientific methods did you use to answer the question of which paper towel is most absorbent?

Figure 2.13

Each of these animals can be seen in zoos. Do you think these animals behave in the same way in zoos as in nature?

▼ **Toucans live in the rain forests of South America.**

◀ **Giant tortoises live on several of the Galapagos Islands in the Pacific Ocean. The bird is a Galapagos hawk.**

Measuring in the International System

It is important that scientific research be understandable to scientists around the world. For example, what if scientists in the United States reported quantitative data in inches, feet, yards, ounces, pounds, pints, quarts, and gallons? People in many other countries would have trouble understanding these data because they are unfamiliar with the English system of measurement. Instead, scientists always report measurements in a modern form of the metric system called the International System of Measurement, abbreviated SI.

One advantage of SI is that there are only a few basic units, and nearly all measurements can be expressed in these units or combinations of them. The greatest advantage, though, is that SI is a decimal system. Measurements can be expressed in multiples of tens or tenths of a basic unit by applying a standard set of prefixes to the unit. In biology, the metric units you will encounter most often are meter (length), gram (mass), liter (volume), second (time), and Celsius degree (temperature). For a thorough review of measurement in SI, study Appendix E.

Descriptive research

Do you think the behavior of the animals shown in *Figure 2.13* would be easier to explain with numbers or with written descriptions of what the animals did? Observational data—that is, written descriptions of what scientists observed—are often just as important in the solution of a scientific problem as numerical data.

Figure 2.14
Gorillas in nature need large areas in which to roam. They also form family groups. This grouping makes foraging for food more efficient and helps protect them from enemies. Describe another biological study that could better be conducted in the field than in the laboratory.

When scientists use purely observational data, they are carrying out descriptive research. Descriptive research is useful because some phenomena aren't appropriate for quantitative research. For example, how a particular wild animal reacts to events in its environment cannot easily be illustrated with numbers.

An example of descriptive research in the field would be a study of how gorilla families behave. Look at *Figure 2.14* and you will quickly realize that it would be very difficult to duplicate the natural environment of the gorillas in a laboratory.

Science and Society

The road to scientific discovery includes making observations, formulating hypotheses, doing experiments, collecting and analyzing data, drawing conclusions, and reporting those conclusions in scientific journals. No matter what methods scientists choose, their research provides society with important information that can be put to practical use.

Maybe you have heard people blame scientists for the existence of nuclear bombs or controversial drugs. To truly comprehend the nature of science in general and biology in particular, people must understand that knowledge gained through scientific research is never inherently good or bad. Notions of good and bad arise in the context of human social, ethical, and moral concerns. If scientists had to worry what might be done with their discoveries when deciding what questions to work on, probably no research would be done at all.

Can science answer all questions?

Some questions are simply not in the realm of science. Many of these involve questions of good versus evil, ugly versus beautiful, or similar judgments. If a question is not testable using scientific methods, the question is not science. However, this does not mean that the question is unimportant.

Life as a Field Biologist

Because many species of plants and animals, such as the giant panda, are endangered, biologists sometimes attempt to save endangered animals by breeding them in captivity.

Giant clams of the Pacific With a splash, Dr. Suzanne Williams rolls backward over the side of the boat and begins her descent to the coral reef below. Dr. Williams is a marine biologist who studies giant clams, the largest living mollusks. Some species grow to five feet across and weigh 1000 pounds. Several species of giant clams have nearly disappeared because of over-fishing for their meat and shells. Dr. Williams is working to restore giant clam populations by breeding them in captivity.

Working underwater Because a giant clam closes its shell rapidly when disturbed, Dr. Williams and the divers working with her approach each clam slowly. They cautiously move to within inches of it before quickly inserting a metal wedge between the two halves of the shell. Now, try as it might, the clam cannot close its shell all the way. Dr. Williams uses scissors to snip away a small piece of tissue. She places the sample in a self-sealing bag while another diver measures the clam and records data about its size, species, and location on an underwater slate.

In the ship's laboratory In the ship's laboratory, Dr. Williams carefully lowers the samples into a container of liquid nitrogen, where they freeze instantly. By the time she has finished preserving the samples, the scuba tanks have been refilled and everyone is ready for another dive.

Back on land After ten rigorous days at sea, Dr. Williams returns to her laboratory on land. There she runs a variety of genetics tests on all of the frozen samples. This information will help biologists select the best clams to use in captive-breeding programs.

Thinking Critically
Thousands of coral reefs are scattered throughout the Pacific Ocean. Why is it important for marine researchers like Dr. Suzanne Williams to accurately record information about where and how samples were collected?

Consider a particular question that is not testable. Some people assert that if a black cat crosses your path, you will have bad luck as shown in *Figure 2.15.* On the surface, that hypothesis appears to be one that you could test. But what is bad luck, and how long would you have to wait for the bad luck to occur? How would you distinguish between bad luck caused by the black cat and bad luck that occurs at random? Once you examine the question, you can see there is no way to test it scientifically because you cannot devise a controlled experiment that would yield valid data.

Can technology solve all problems?

Science attempts to explain how and why things happen. Scientific study that is carried out mainly for the sake of knowledge—with no immediate interest in applying the results to daily living—is called pure science.

However, much of pure science eventually does have an impact on people's lives. Have you ever thought what it was like to live in the world before refrigerators, electric lights, and modern lifesaving medical equipment existed? These and other inventions are indirect results of research done by scientists in many different fields.

Figure 2.15

In cartoons, black cats always cause bad luck when they cross someone's path. If bad luck due to black cats occurred as reliably and as swiftly in real life as it does in cartoons, it really would be scientifically testable.

Other scientists work in research that has obvious and immediate applications. **Technology** is the application of scientific research to society's needs and problems. It is concerned with making improvements in human life and the world around us. Technology has helped increase the production of food, reduced the amount of manual labor needed to make products and raise crops, and aided in the reduction of wastes and environmental pollution.

The advance of technology has benefited humans in numerous ways, but it has also resulted in some serious problems. For example, suppose irrigation technology is used to boost the production of food crops in one area. If the irrigation is used for too long a time, the soil may become depleted of minerals or the evaporation of the irrigation water may leave deposits of mineral salts in the soil,

making it useless for growing crops, as illustrated in *Figure 2.16.*

Considering ethics in science

Most scientists would state that scientific research and its results are neither right nor wrong, good nor bad. **Ethics** is a study of the standards of what is right and wrong.

Figure 2.16

One example of a harmful side effect of technology is the deterioration of soil due to a buildup of salts to such high levels that crops cannot grow. In this case, technology appeared to solve one problem—that of low crop yield—but actually caused a different problem.

research. In the 1980s, scientists were forbidden to use such tissue in research on the treatment of diseases such as Alzheimer's disease because many people consider abortion to be immoral. To many victims of these diseases, this ban seemed unjust. The ban was later lifted because different political leaders believed that the possible benefits of the research outweighed the negative effects. *Figure 2.17* illustrates a different ethical problem.

Today, more scientists feel that they should become involved in the decisions about the consequences of their work and how that work is applied. Even so, once a scientific discovery is made, it is the people of a society who make that decision.

Figure 2.17

Steroids and tranquilizers are drugs that have been developed by scientists to treat certain disorders. Sometimes these drugs are used illegally and unethically to artificially enhance the performance of racehorses. This treatment may produce short-term benefits at the expense of the long-term health of the animal.

Frequently, ethics issues move from the scientific arena to the political, moral, and social arenas and back, as is the case with the issue of using tissue from aborted fetuses in

Connecting Ideas

Science is not only a body of facts, but also a process. The methods that scientists employ to help them explore and understand the physical world include observing, forming hypotheses, experimenting, and publishing results and conclusions so that they may (or may not) be verified. What kinds of hypotheses do biologists form about the living world? How do they use scientific methods to learn about the interactions among organisms and between organisms and their environment? Answers to these questions will be revealed as you study and understand basic ecology.

Section Review

Understanding Concepts
1. Why is it important that scientists repeat their experiments whenever possible?
2. Compare and contrast quantitative and descriptive research.
3. Why is science considered to be a combination of fact and process?

Thinking Critically
4. Biomedical research has led to the development of technology that can keep very old, very ill patients alive. How does the statement "The results of research aren't good or bad; they just are" apply to such research?

Skill Review
5. **Making and Using Graphs** Look again at *Figure 2.12*. Why do you think the high-temperature side of the graph drops off more sharply than the low-temperature side? For more help, refer to Organizing Information in the *Skill Handbook*.

Computers in Biology

Today, you can buy a personal computer that will fit on your lap. This small device contains a silicon chip that has more computing power than a machine that occupied a large room just a few years ago. Still, it is instructions (a program) written by humans that make the computer perform a useful task.

Computer biology Biologists use computers to eliminate much of the drudgery in analyzing the reams of data they collect. For example, an ecologist may need to know how many organisms are in a population. Before computers, the ecologist counted a small sample and then calculated a likely population by using complex mathematics. Now, an ecologist can enter data into a computer program that quickly calculates a population.

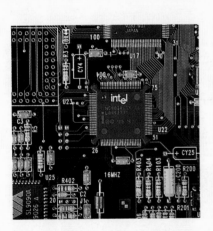

Sometimes biologists aren't even required to enter data. By using sensors, they can let the subjects of their research enter their own data. For example, a computer can measure and record the amount of light that passes through a bacterial culture growing in a liquid medium. Using this data, the computer can monitor bacterial growth and determine the population.

The past is prologue When designing research, a scientist must carry out a literature search to determine what has already been done in the field. Before computers, a literature search often required months of library study. Today, summaries of most research papers are stored in giant computer data banks. Using a computer, a search program, and a telephone network, a scientist can do literature searches in a few days without ever leaving the laboratory.

Applications for the Future

As more becomes known about biological systems, it is sometimes possible to describe them mathematically, enabling programmers to write software that makes the computer simulate the system.

For example, a computer program might simulate a temperate forest ecosystem. How will a long drought affect the owl population? How will a forest fire affect the growth of mosses? If all the larger trees are removed, how will the kinds of smaller plants change? All these questions would take years to answer by direct observation. Using the computer simulation, biologists can predict answers to these and many other questions.

INVESTIGATING the Technology

1. **Research** One of the most exciting research projects in history is the human genome project. What effect does the use of computers have on this project? Find out by reading "The Genome Finds its Henry Ford" by Tony Dajer, *Discover*, January 1993, pp. 86-87.
2. **Apply** Pick a specific area of biology that interests you. Go to a library that has a computer database to find the most recent research on your topic.

Reviewing Main Ideas

2.1 Problem-Solving Methods in Biology

- Based on observations of the natural world, scientists use inductive reasoning to form hypotheses. They then use deductive reasoning to help them develop the hypothesis and design an experiment to test it.
- Biologists use controlled experiments to obtain data that either do or do not support a hypothesis. By publishing the results and conclusions of an experiment, a scientist allows others to try to verify the results. Repeated verification over time leads to the development of a theory.

2.2 The Nature of Biology

- Biologists do their work in laboratories and in the field. They collect both quantitative and descriptive data from their experiments.

- Scientists conduct experiments in order to increase knowledge about the natural world. Questions that are not testable are not in the realm of science. Scientific results may help solve some problems, but not all, and ethical issues must be decided by all of society, not just scientists.

Key Terms

Write a sentence that shows your understanding of each of the following terms.

control	hypothesis
data	independent variable
deductive reasoning	inductive reasoning
	safety symbol
dependent variable	scientific methods
	technology
ethics	theory
experiment	

Understanding Concepts

1. In what ways are science and society related?
2. What is a scientist's usual next step if data support a hypothesis?
3. If a scientist conducts an experiment, and the data do not support the hypothesis, what is the scientist likely to do next?
4. A student experiments to study the growth of radish roots at ten different temperatures from 5°C to 45°C. What are the dependent and independent variables in the experiment?
5. What is the goal of scientists?
6. Give an example of an experiment that would best be performed in the field.
7. How do ethics affect science?
8. What kind of questions are not legitimate for science to consider?

Using a Graph

9. Do the following data support the hypothesis that leaves fall from trees only in the autumn? Explain your answer.

Relating Concepts

10. Make a concept map that relates the following terms and phrases. Supply the appropriate linking words for your map.

    inductive reasoning, observation, deductive reasoning, scientific methods, control, experiment, hypothesis, experimental group, data

Applying Concepts

11. Give an example of a problem that technology may not solve.

12. Give an example of how you have used inductive reasoning in your everyday life.

13. If your friend said that she had a theory about who would win the Academy Award for best actor, how would you know that she wasn't speaking scientifically?

14. List two ways that technology has benefited society.

15. Every day, Carmina observed the sparrows that lived in her neighborhood and wrote in her journal what they were doing. What kind of research was Carmina doing? Explain your answer.

Physics Connection

16. How does a high-pitched sound differ from a low-pitched sound? How does a loud sound differ from a soft sound?

Thinking Critically

Interpreting Data

17. **Biolab** A team of students measured the number of seeds that germinated over ten days in a control group and in an experimental group that contained added fertilizer. The following graph of the data was plotted by the students.
    (a) When was the earliest appropriate day to end their experiment?
    (b) Would you conclude that the fertilizer used influences the number of seeds that germinate when planted? Explain your answer.

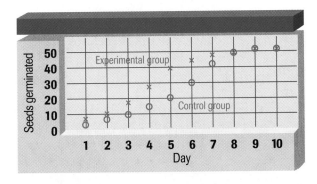

Interpreting Data

18. **Minilab** Suppose your data from the Minilab, "How do you decide which paper towel to buy?" looked like that in the bar graph below.
    (a) Which brand of paper towel were the employees probably complaining about in the beginning?
    (b) Which brand would you switch to? Explain your answer.

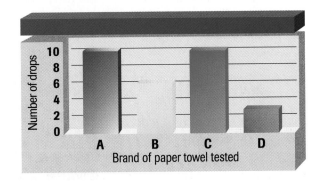

Interpreting Data

19. **Thinking Lab** Suppose the final two boxes both contained grass and a few shriveled dandelions. What would you conclude about your hypothesis that Kilimal is an effective weed killer for dandelions?

Connecting to Themes

20. **Nature of Science** How are scientific methods useful in everyday life?

21. **Energy** Why do you think biologists consider the ability to use energy the most important characteristic of life?

22. **Unity Within Diversity** Scientists do research in many different fields from astronomy to zoology. What characteristics do you think scientists have in common?

A tropical rain forest—wet, warm, and, above all, green. At first glance, you may equate this vast expanse of green-ness with sameness. Nothing could be further from the truth. In fact, rain forests are home to more species per square meter than any other environment on Earth, although most of them are hidden from view. One scientist counted more than a hundred species of beetles living among the leaves of just one species of tree.

Even more hidden than the diversity of life in the rain forest is the complex web of interactions that exists among all these different living things and their environment. It is this web of interactions that defines and sustains a rain forest ecosystem and all other ecosystems. If enough of these interactions are disrupted, the whole ecosystem could change drastically.

The unique shape and color of this orchid attracts just one kind of wasp, which visits the flower for its nectar. What are other ways that plants and animals depend on each other?

This frog's bright coloration warns predators of poisonous chemicals in its body. What other adaptations do animals and plants have to avoid being eaten?

Well over half of the world's species of plants and animals live in tropical rain forests. What are the effects of deforestation on these organisms? What can be done to save them?

Unit Contents

3 Principles of Ecology

Are mosquitoes of any use to anyone or anything? If all mosquitoes were killed, would any negative effects result? You might say they're just pests, but if you could ask the opinion of a sunfish, a tadpole, a dragonfly, or a swallow, you'd get a different answer. For these animals and others, mosquitoes and their larvae are a major food source.

Every organism is connected in some way to many other organisms. In nature, living organisms interact in a variety of ways. Some relationships are complex such as the pollination of a flower by a hummingbird, while others are quite simple such as the dispersal of seeds from a plant by a passing animal. In this chapter, you'll learn how groups of organisms interact and depend on each other for survival. You'll also consider how organisms are affected by physical factors of the environment.

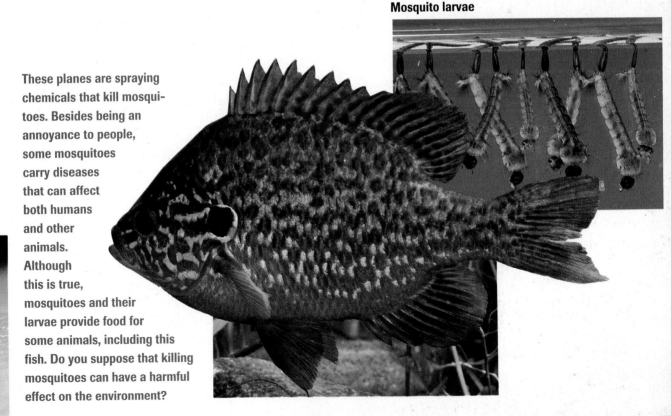

Mosquito larvae

These planes are spraying chemicals that kill mosquitoes. Besides being an annoyance to people, some mosquitoes carry diseases that can affect both humans and other animals. Although this is true, mosquitoes and their larvae provide food for some animals, including this fish. Do you suppose that killing mosquitoes can have a harmful effect on the environment?

Organisms and Their Environments

Section Preview

Objectives

Distinguish between biotic and abiotic factors in the environment.

Compare the different levels of biological organization used in ecology.

Explain the difference between a niche and a habitat.

Key Terms

ecology
biosphere
biotic factor
abiotic factor
population
community
ecosystem
niche
habitat

When was the last time you heard or read something about the environment? Almost every evening newscast, daily newspaper, and news magazine presents you with stories about recycling efforts, campaigns to save tropical forests, concerns about water or air pollution, and many other environmental issues. Learning more about basic ecological principles that explain how organisms interact with each other and the environment can help you understand these issues and form your own opinions about them. Ecology is both an old and a new science. In this chapter, you will see how the observations of early and present ecologists, and the use of technology and new research tools, help us understand the interactions in nature.

The Beginnings of Ecology

Many people make a hobby out of observing and studying plants and animals in their natural habitats. Plant enthusiasts explore fields, forests, city parks, and even vacant lots to learn about the different trees and wildflowers that grow there, such as the goldenrods in *Figure 3.1.* Plant identification guides help these people identify different species of plants and learn about a plant's habitat requirements and flowering times.

Casual backyard bird observers may notice a nest of robins and the behavior patterns of the male and female. While avid bird-watchers would certainly be interested in a nest of robins, their hobby is to identify and observe as many bird species as possible. In order to find new birds, bird-watchers listen for bird songs and investigate all types of environments from woodlands to fields to wetlands. The pursuits of bird watching and plant identification are known as nature study or natural history.

Figure 3.1

Goldenrod plants flower at different times. Depending on rainfall and temperature, goldenrods may flower as early as August or as late as November in different parts of the United States.

Natural history led to ecology

The goal of natural history is to find out as much as possible about living things. Many early natural historians spent their lives studying how various organisms live out their life cycles and how they depend on one another and on their environment. Natural historians then began to report their findings in a systematic way.

The branch of biology that developed from natural history is called ecology. **Ecology** is the scientific study of interactions between organisms and their environments. Ecological study reveals inter-relationships between living and nonliving parts of the world. Ecology combines information from many scientific fields including chemistry, physics, geology, and other branches of biology.

You may remember from Chapter 2 that scientific methods include both descriptive and quantitative research. Most ecologists use both types of research. They obtain descriptive information by observing organisms in the field and laboratory. They obtain quantitative data by making measurements and carrying out carefully controlled experiments in both field and lab. Using these methods, ecologists can learn a great deal about relationships, such as how pesticides seep through soil from a farm to a nearby stream, how day length influences the behavior of migrating birds, how tiny shrimp help rid ocean fish of parasites, or how acid rain threatens some of Earth's forests.

Literature

A Sand County Almanac
by Aldo Leopold

The idea of unity of organisms and environment is one of long standing. However, the concept that humans are not separate from nature was not seriously considered by biologists until the American naturalist, Aldo Leopold, wrote *A Sand County Almanac* over a period of years. It was published in 1949, a year after his death. Leopold described himself as one who could not live without wild things.

On a farm in Wisconsin The book describes the annual events of nature Leopold observed on an abandoned farm in Wisconsin that he purchased in 1935 and visited during weekends over a ten-year period. Like many naturalists, Leopold describes nature as a continuum of events from the past to the future, with every organism relying on its environment. He not only describes the appearance of plants and animals, but also gives them life with needs as relevant as our own. When Leopold describes a rough-legged hawk sailing over the meadow in January in search of a mouse for lunch, he points out that the hawk has no opinion about why grass grows. The hawk, like the one shown here, only knows that snow melts to reveal his prey. It migrated south from the Arctic in hope of such a thaw. For the hawk, a thaw means a freedom from want and fear.

A plea for the community Leopold wrote in his journal about the effects that improved mechanization would have on the environment. Leopold was concerned about human abuse of the land because he realized land is too often considered a commodity. He pleaded for an understanding that land is not just something to be bought and sold. He wanted people to understand that the land is essential to our psychological and spiritual well-being, as well as to our physical survival, and that we must take good care of it.

CONNECTION TO Biology

How might observations of naturalists such as Aldo Leopold lead to the conservation of lands destined for development?

ecology:
oikos, eco (GK) house-hold, habitat
The study of organisms and their environment.

Figure 3.2

All organisms are adapted for life in a particular environment. They have adaptations for obtaining food, for protecting themselves, and for reproduction.

The living environment: Biotic factors

As far as we know, life exists only on Earth. Living things can be found in the air, on land, and in both fresh water and salt water. The **biosphere** is the portion of Earth that supports life. It extends from high in the atmosphere to the bottom of the oceans. This life-supporting layer may seem extensive to us, but if you could shrink Earth down to the size of an apple, the biosphere would be thinner than the apple's peel.

Many different environments, both aquatic and terrestrial, exist in the various regions of the biosphere. Each environment includes both living and nonliving factors that affect the organisms living there. All the living organisms that inhabit an environment are called **biotic factors.**

Different organisms are adapted for life in different parts of the biosphere. For example, the mountain goats shown in *Figure 3.2* are adapted to climb on steep mountainsides and withstand freezing temperatures and strong winds. Animals like the mole live in, rather than on, the soil. Fishes that live in the darkest depths of the ocean have adaptations for finding food and mates in darkness. They also have adaptations to withstand cold temperatures and tremendous water pressure.

▼ Moles are small mammals that dig tunnels in the soil. Their digging provides the soil with aeration, which is beneficial for soil organisms and plant roots. However, moles also eat roots as they dig, and can damage large populations of plants such as grasses.

▼ Food is scarce in the dark waters of the deep sea. Some animals feed on scraps, waste, and dead organisms that drift downward. Others such as this swallower fish have elastic, expandable stomachs and huge mouths to be able to consume prey organisms that may be larger than themselves.

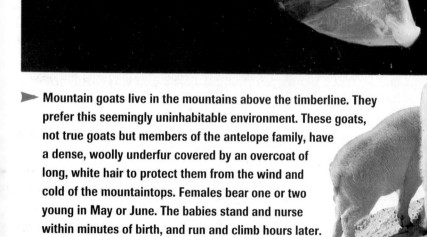

▶ Mountain goats live in the mountains above the timberline. They prefer this seemingly uninhabitable environment. These goats, not true goats but members of the antelope family, have a dense, woolly underfur covered by an overcoat of long, white hair to protect them from the wind and cold of the mountaintops. Females bear one or two young in May or June. The babies stand and nurse within minutes of birth, and run and climb hours later.

The nonliving environment: Abiotic factors

An ecological study of moles wouldn't be complete without an examination of the types of soil in which these animals dig their tunnels. Similarly, research on the life cycle of trout would need to include where these fish lay their eggs—on rocky or sandy stream bottoms. Ecology includes the study of features of the environment that are not living. Ecologists study how the nonliving factors in the environment affect living things. **Abiotic factors** are the nonliving parts of the environment. Examples of abiotic factors include air currents, temperature, moisture, light, and soil.

Abiotic factors can have obvious effects on living things and often determine which species can survive in a particular environment. For example, lack of rainfall can cause drought in a grassland, as shown in *Figure 3.3.* Can you think of changes in a grassland that might result from a drought? Grasses would grow less quickly, wildflowers would not produce as many seeds, and the animals that depend on plants for food would find it harder to survive.

MiniLab

How can you measure water loss by plants?

All organisms lose water by respiration. Plants also lose water by a process called transpiration. Transpiration occurs when water travels to a plant's leaves and escapes through pores as water vapor.

Procedure

1. Allow the soil of a potted plant to dry out so that the plant just begins to wilt.
2. Add water in measured amounts to the pot until it just begins to drip from the bottom.
3. Allow the water to drip back into the measuring cup.
4. When the water stops dripping, record the amount of water that remained in the pot by subtracting the amount of water that dripped back into the measuring cup from the original amount.
5. Place the potted plant in a plastic bag and tie the top with string.
6. Put the plant in sunlight.
7. When water has collected on the bag, estimate the amount of water that has transpired from the plant.

Analysis

1. If the potted plant had not been placed in the bag, where would the water have gone?
2. Why do you not have to water plants in a terrarium as often as plants in a pot?
3. How might an ecologist alter the setup you used to measure transpiration of plants in an ecosystem?

Figure 3.3

Droughts are not uncommon in grasslands. As the grasses dry out, they turn yellow and appear to be dead, but new blades grow from the low-lying growing points when it rains again. Some animal species are adapted to living in grasslands by their ability to burrow underground and sleep through the dry period.

Figure 3.4

These marsh marigolds represent a population of organisms. What characteristics are shared by this group of flowers that make them a population?

Levels of Organization: The Hierarchy of Life

The study of an individual organism, such as a male deer, or buck, might reveal what food items he prefers, how often he eats, and how far he roams to search for food or shelter. However, even though he spends a large part of his time alone, he does interact with other individuals of his species. For example, he periodically goes in search of a mate, which may require battling with other bucks.

All organisms depend on others for food, shelter, reproduction, or protection. So you can see that the study of an individual would provide only part of the story of its life cycle. To get a more complete picture requires studying its relationships with other organisms.

Ecologists study interactions among organisms at several different levels as shown in The Inside Story. They study individual organisms, interactions between organisms of the same species, and interactions between organisms of different species. Ecologists also study how abiotic factors affect groups of interacting species.

These levels of organization provide ecologists with a tool to use in planning their research. For example, one ecologist might focus on the interactions among the deer within a single herd, while another might want to study how the herd interacts with other animals living in the same area. Yet another ecologist might investigate how the deer are affected by severe winter storms or by extremely hot, dry summer weather.

Interactions within populations

The marsh marigolds in *Figure 3.4* form a population. A **population** is a group of organisms of one species living in the same place at the same time that interbreed.

Members of the same population compete with one another for food, water, mates, and other resources. The ways in which organisms in a population share the resources of their environment determine how far apart populations are and how large each population can become.

The Organization of Life

White-tailed deer

All living matter can be organized into levels starting with the smallest subatomic particles and going all the way to Earth as a whole. Ecology deals with the levels of organisms, populations, communities, ecosystems, biomes, and the biosphere.

Organisms

Ecologists may study the behavior of an individual organism. They may study its daily movements, feeding, or breeding behavior.

Populations

Ecologists may study the effects of populations of organisms on the environment. They may also study growth rates of populations and the future of certain populations.

Communities

All organisms in a community depend in some way on the other organisms living there. Ecologists are concerned with studying the effects on the community when a new species is added or one is removed.

Ecosystems

All the biotic and abiotic factors in an area form an ecosystem. Ecologists are concerned with ecosystem stability and knowing what keeps ecosystems stable.

Biosphere

The biosphere is the highest level of organization. It is made up of the entire planet and all its living and nonliving parts. Ecologists are concerned with all interactions within the biosphere.

Figure 3.5

Adult frogs and their young have different food requirements. This limits competition for food resources for the species.

Eggs that adult frogs lay in the water hatch into tadpoles. Tadpoles have gills, live in water, and eat algae and small aquatic creatures.

Adult frogs live both on land and in the water. They breathe air and eat insects such as dragonflies, grasshoppers, and beetles.

Some species have adaptations that reduce competition within a population. One example can be seen in the life cycle of a frog, shown in *Figure 3.5*. The juvenile stage of the frog is the tadpole, which not only looks very different from the adult but also has completely different food requirements. Many species of insects, including dragonflies and moths, also have juveniles that differ from the adult in body form and food requirements.

Individuals interact within communities

No population of organisms of one species lives independently of other species. Just as a population is made up of individuals, a community is made up of several populations. A **community** is a collection of interacting populations. An example of a community is shown in *Figure 3.6*.

A change in one population in a community will cause changes in the other populations. Some of these changes can be subtle, as seen when a small increase in one population causes a small decrease in another.

Sugar maple leaf

Figure 3.6

Beech and maple trees dominate this forest community; therefore, it is called a beech-maple forest. Beech-maple forests are found in the eastern United States, Europe, and northeast China. Many other populations are also found in this community. These populations interact with one another and with the abiotic factors of this environment.

For example, if the population of mouse-eating hawks increases slightly, the population of mice will, as a consequence, decrease slightly. Other changes might be more dramatic, as when the size of one population grows so large it begins affecting the food supply for another species in the community.

Interactions between living things and abiotic factors form ecosystems

In a healthy forest community, interactions among populations include birds eating insects, squirrels eating nuts from trees, mushrooms growing from decaying leaves or bark, and raccoons fishing in a stream. In addition to population interactions, ecologists also study interactions between populations and their physical surroundings. An **ecosystem** is made up of the interactions among the populations in a community and the community's physical surroundings, or abiotic factors.

Terrestrial ecosystems are those located on land. Examples include forests, meadows, and desert scrub.

Not all ecosystems occur on land. Aquatic ecosystems may occur either in fresh water or salt water. Freshwater ecosystems include ponds, lakes, and streams.

Saltwater ecosystems, also called marine ecosystems, occupy approximately 75 percent of Earth's surface. *Figure 3.7* gives an example of both a marine and a freshwater ecosystem.

Figure 3.7

There may be hundreds of populations interacting in a pond or tide pool. How do you think the abiotic factors in these environments affect the biotic factors?

▶ Dragonflies live near moist meadows and ponds. They feed on small insects they catch while flying. Dragonflies lay their eggs in the pond or on pond plants.

▲ Tide pools occur along the seashore, often in rocky areas. Organisms living in tide pools must survive dramatic changes in abiotic factors. When the tide is high, ocean waves replenish the water in the pool. When the tide is low, water in the pool evaporates. Oxygen levels decrease, temperatures increase, and some organisms may even be exposed to the air.

Figure 3.8

This series of photographs shows how a habitat can be seen as a collection of several niches. As you can see, each species uses the available resources in a different way. Can you think of other ways the resources could be shared? Further division of resources would provide additional niches.

A worm obtains nourishment from the organic material it eats as it burrows through the soil.

A centipede is a predator that captures and eats beetles and other animals.

Where and how organisms live

Every species has a particular function in its community. In a grassland community, for example, the role of fungi is the breakdown of organic matter contained in the bodies of dead organisms. This process recycles nutrients and materials, releasing them back into the environment where they can be used by other organisms. What do you suppose is the role that coyotes play in a grassland community? They keep down populations of rodents by eating them on a regular basis. Both fungi and coyotes help maintain homeostasis in their communities. A **niche** is the role a species plays in a community. The space, food, and other conditions an organism needs to survive and reproduce are part of its niche. A species niche includes more than just feeding relationships. It also includes how a species uses and affects its environment.

A prairie dog living in a grassland makes its home in burrows it digs underground. Many birds make their homes in the trees of a beech-maple forest. These are examples of habitats. A **habitat** is the place where an organism lives out its life.

"You call this a niche?"

A millipede eats decaying leaves near the log.

Although several species may share a habitat, the food, shelter, and other resources of that habitat are divided into separate niches. For example, if you turn over a log like the one shown in *Figure 3.8,* you will find millipedes, centipedes, insects, and worms living there. At first, it looks as though all these animals are competing for food because they live in the same habitat. But close inspection reveals that each feeds in different ways, on different materials, and at different times. These differences lead to distinct behaviors and reduced competition.

Section Review

Understanding Concepts

1. List several different biotic and abiotic factors in an ecosystem.
2. Compare and contrast populations and communities.
3. Give examples that would demonstrate the differences between the terms *niche* and *habitat.*

Thinking Critically

4. On a visit to a desert, you observe a coyote, three snakes, several small types of cactus, and hundreds of creosote bushes. What name would an ecologist give to such an area?

Skill Review

5. **Forming a Hypothesis** Cape May warblers, bay-breasted warblers, and yellow-rumped warblers all feed in the same kind of spruce tree. Hypothesize how these birds reduce competition in the spruce tree habitat. For more help, refer to Practicing Scientific Methods in the *Skill Handbook.*

3.1   Organisms and Their Environments   **63**

3.2 How Organisms Interact

Wile E. Coyote and Roadrunner weren't just characters in a Saturday morning cartoon. In the deserts of the American southwest, roadrunners are one source of food for coyotes. Coyotes will usually eat any animal they can catch, which is good for Wile E. Coyote because he never seems to have much luck catching the roadrunner! The connection between a coyote and its prey is only one example of a feeding relationship in a desert.

herbivore:
herba (L) grass
vorare (L) to devour
Herbivores feed on plants.

Species Relationships

Coyotes, roadrunners, and other living things interact in a variety of ways. These interactions allow organisms to obtain energy and materials necessary for life processes. These interactions also maintain homeostasis within populations, communities, and ecosystems.

Feeding relationships: How organisms obtain energy

You have learned that the role an organism plays in its community is its niche. Feeding relationships between organisms reflect these niches, or roles in the community. *Figure 3.9* shows a grassland community in Africa. The plants in the photograph are autotrophs. **Autotrophs** are organisms that use energy from the sun or energy stored

in chemical compounds to manufacture their food. Autotrophs are also called producers. Plants are the most common terrestrial autotrophs, but some chlorophyll-containing, single-cell organisms also produce their own food. All other organisms depend on autotrophs for nutrients and energy.

Organisms that depend on autotrophs as their source of nutrients and energy are **heterotrophs.** Some heterotrophs, such as grazing, seed-eating, and algae-eating animals, feed directly on autotrophs. The wildebeests depend on plants for their food. Heterotrophs are also known as consumers because they can't make their food; they must consume it. A consumer that feeds only on plants is called an herbivore. Herbivores include grazing animals such as rabbits, cows, and grasshoppers, as well as rodents such as rabbits, mice, and squirrels.

Carnivores and scavengers

Some heterotrophs eat other heterotrophs. Animals that kill and eat only other animals are carnivores. Some animals do not kill for food; instead, they eat animals that have already died. **Scavengers** such as black vultures are animals that feed on carrion, refuse, and similar dead organisms. Scavengers play a beneficial role in the ecosystem. Imagine for a moment what the environment would be like if there were no vultures to devour animals killed on the African plains, buzzards to clean up dead animals along roads, or ants and beetles to remove dead insects and small animals from sidewalks and basements.

Omnivores and decomposers

Humans are an example of yet another type of consumer. The teenagers in *Figure 3.9* are eating a variety of foods that include both animal and plant materials. They are omnivores. Raccoons, coyotes, and bears are other examples of omnivores.

A fungus is another type of consumer. Organisms that break down and absorb nutrients from dead organisms are called **decomposers.** Decomposers break down the complex compounds of dead and decaying plants and animals into simpler molecules that can be absorbed. Many bacteria, some protozoans, and most fungi carry out this essential process of decomposition.

carnivore:
caro (L) flesh
vorare (L) to devour
Carnivores feed on other animals.

omnivore:
omnis (L) all
vorare (L) to devour
Omnivores eat both plants and animals.

Figure 3.9

Feeding relationships determine how organisms interact and govern how energy flows through the ecosystem.

◀ People are omnivores because we eat both producers and consumers.

▼ Without fungi and other decomposers, dead organic material would pile up in all ecosystems.

◀ The grass plants in this African savanna are producers. They use energy from sunlight to convert inorganic molecules into the nutrients necessary for their own survival. They also provide energy and nutrients for all heterotrophs. Wildebeests are herbivores that eat the grass and make its energy available for carnivores, such as this lion. The vultures are scavengers, consuming organisms that have already died. They will eat anything left by the carnivores.

Why don't prey populations disappear when predators are present? Prey organisms have evolved a variety of defenses to avoid

How can one population affect another?

being eaten. For example, some caterpillars are distasteful to birds, and some fish confuse predators by appearing to have eyes at both ends of their bodies. Just as prey have evolved defenses to avoid predators, predators have evolved mechanisms to overcome those defenses.

Even single-celled protists such as *Paramecium* have predators. *Didinium* is another unicellular protist that attacks and devours a *Paramecium* larger than itself. Do populations of *Paramecium* change when a population of *Didinium* is present? In this investigation, you will use various methods to determine how both of these species interact.

Didinium

PREPARATION

Problem
How does a population of *Paramecium* react to a population of *Didinium?*

Hypotheses
Have your group agree on a hypothesis to be tested. Record your hypothesis.

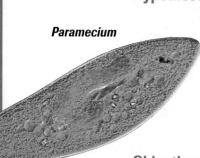

Paramecium

Objectives
In this Biolab, you will:
- **Design** an experiment to establish the relationships between *Paramecium* and *Didinium*.

- **Use** appropriate variables, constants, and controls in experimental design.

Possible Materials
microscope
microscope slides
coverslips
culture of *Didinium*
culture of *Paramecium*
beakers or jars
eyedroppers
sterile pond water

Safety Precautions 🖐️
Take care when using electrical equipment. Review the Care and Use of a Microscope section in the *Skill Handbook*.

1. Review the discussion of feeding relationships in this chapter.

2. Decide which materials you will use in your investigation. Record your list.

3. Be sure that your experimental plan contains a control, tests a single variable such as population size, and allows for the collection of quantitative data.

4. Prepare a list of numbered directions. Explain how you will use each of your materials.

Check the Plan

Discuss the following points with other group members to decide final procedures. Make any needed changes to your plan.

1. What will you measure to determine the effect of the *Didinium* on *Paramecium*?

2. What single factor will you vary? For example, will you put no *Didinium* in one culture of *Paramecium* and 5 mL of *Didinium* culture in one culture of *Paramecium*?

3. How long will you observe the populations?

4. How will you estimate the changes in the populations of *Paramecium* and *Didinium* during the experiment?

5. **Make sure your teacher has approved your experimental plan before you proceed further.**

6. Carry out your experiment.

7. Make a data table that has Date, Number of *Paramecium*, and Number of *Didinium* across the top. Place the data obtained for each culture in rows. Consider making a different table for each of your cultures. Complete your data table. Design and complete a graph of your data.

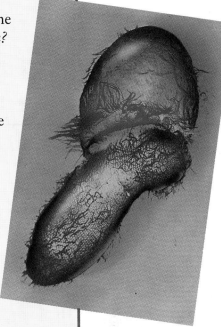

A *Didinium* captures a *Paramecium*.

ANALYZE AND CONCLUDE

1. **Analyzing Data** What differences did you observe among the experimental groups? Were these differences due to the presence of *Didinium?* Explain.

2. **Drawing Conclusions** Did the *Paramecium* die out in any culture? If they did, why do you think this happened?

3. **Checking Your Hypothesis** Was your hypothesis supported by your data? If not, suggest a new hypothesis that is supported by your data.

4. **Thinking Critically** List several ways that your methods may have affected the outcome of the experiment. Suggest ways that you might improve your methods. You may suggest other equipment or materials.

5. **Drawing Conclusions** Write a brief conclusion to your experiment.

Going Further

Application Based on this lab experience, design another experiment that would help you answer any questions that arose from your work. What factors might you allow to vary if you kept the number of *Didinium* constant?

Figure 3.10

Red-breasted geese (left) and peregrine falcons (right) both nest in the Siberian arctic in the spring. They share a symbiotic relationship. The falcon is a predatory bird that includes geese among its prey. The falcon also fiercely defends the area around its nest from predators. Since the falcon hunts away from its nest, geese nesting nearby are not attacked by the falcon. The falcon's behavior also protects the geese from other predators.

Close relationships for survival

Biologists once assumed that all organisms living in the same environment are in a continuous battle for survival. Some interactions are harmful to one species while beneficial to another. Predators such as lions and insect-eating birds kill and eat other animals—prey. Predator-prey relationships such as the one between wildebeests and lions do involve a fight for survival. But there are also other relationships among organisms that help maintain survival in many species. The relationship in which there is a close and permanent association between organisms of different species is **symbiosis.** *Symbiosis* means "living together."

There are several kinds of symbiosis. A symbiotic relationship between the peregrine falcon and red-breasted goose, shown in *Figure 3.10,* has evolved in the cold arctic region of Siberia in Russia. During the nesting season, the falcon's behavior protects nearby geese from predators. The geese benefit from the relationship, while the falcon is neither benefited nor harmed. This relationship is called a commensal relationship. **Commensalism** is a symbiotic relationship in which one species benefits and the other species is neither harmed nor benefited.

Commensal relationships also occur among plant species. The Spanish moss in *Figure 3.11* is a kind of flowering plant growing on the branch of a tree. Orchids, ferns, moss, and other plants sometimes grow on the branches of larger plants. The larger plants are not harmed, but the smaller plants benefit from the additional habitat.

Figure 3.11

Spanish moss grows on the limbs of trees but does not obtain any nutrients or cause any harm to the trees unless it is in large enough numbers. This commensal relationship, like all symbiotic interactions, occurs between two organisms of different species.

Sometimes, two species of organisms benefit from living in close association. A symbiotic relationship in which both species benefit is called **mutualism.** *Figure 3.12* illustrates mutualism between a type of ant and a species of acacia tree living in the subtropical regions of the world. The ant protects the tree by attacking any herbivore that tries to feed on it. The ants also clear vegetation away from the trunk of the plant and kill any plant that comes too close to the acacia. The tree provides nectar and a home for the ants.

Sometimes, one organism harms another. Have you ever owned a dog or cat that had been attacked by ticks or fleas? Ticks and fleas are examples of parasites. A symbiotic relationship in which one organism derives benefit at the expense of the other is called **parasitism.** Parasites have evolved in such a way that they harm, but usually do not kill, the host. Why would it be a disadvantage for a parasite to kill its host? If the host dies, the parasite also will die unless it can quickly find another host. Some parasites live inside other organisms. Examples of parasites that live inside the bodies of their hosts are tapeworms and roundworms, which live in the intestines of dogs, cats, and other vertebrates.

What type of symbiosis is found in a lichen?

You may have seen lichens growing on bare rock, open soil, or tree bark. A lichen looks like a gray-green, orange, or yellow patchwork of spots on rocks. However, a lichen is a type of organism. In fact, it is more than one! It is a combination of a fungus and either an alga or cyanobacterium. Algae and cyanobacteria are producers that usually exist alone. Fungi, such as mushrooms, also usually exist alone.

Procedure

1. Take a bit of lichen that you have found on tree bark or a rock, or one provided by your teacher, and tease it apart with straight pins.
2. Prepare a wet mount slide of the lichen pieces.
3. Observe the lichen under the microscope.

Analysis

1. Describe the appearances of the two kinds of organisms that make up the lichen.
2. How could the alga or cyanobacterium benefit the fungus, and how could the fungus benefit the alga or cyanobacterium?
3. What kind of symbiosis exists between the fungus and the alga or cyanobacterium?

Figure 3.12

These ants and acacia trees both benefit from living in close association. In an experiment, ecologists removed the ants from some acacia trees. Results showed that the trees with ants grew faster and survived longer than trees without ants. This mutualistic relationship is so strong that in nature, the trees and ants are never found apart.

Matter and Energy in Ecosystems

When you pick an apple from a tree and eat it, you are consuming carbon, nitrogen, and other elements the tree has used to produce the fruit. That apple also contains energy from the sunlight trapped by the tree's leaves while the apple was growing and ripening.

Matter and energy are constantly cycling through stable ecosystems. You have already learned that feeding relationships and symbiotic relationships describe ways in which organisms interact. Ecologists study these interactions to make models that trace the flow of matter and energy through ecosystems.

Food chains: Pathways for matter and energy

The wetlands community pictured in *Figure 3.13* illustrates an example of a food chain. A **food chain** is a simple model that scientists use to show how matter and energy move through an ecosystem. Nutrients and energy proceed, from autotroph to heterotroph and, eventually, to decomposers.

A food chain is typically drawn using arrows to indicate the direction in which energy is transferred from one organism to the next. One food chain in *Figure 3.13* could be shown as

$$\text{algae} \rightarrow \text{fish} \rightarrow \text{heron}$$

Food chains can consist of three links, but most have no more than

Figure 3.13

A community is able to function because each organism within the ecosystem depends on other organisms. In order for a wetland ecosystem to function, its organisms must depend on each other for a supply of energy. Energy must first pass from autotrophs to heterotrophs in a food chain. Follow the steps in the wetland food chain shown here.

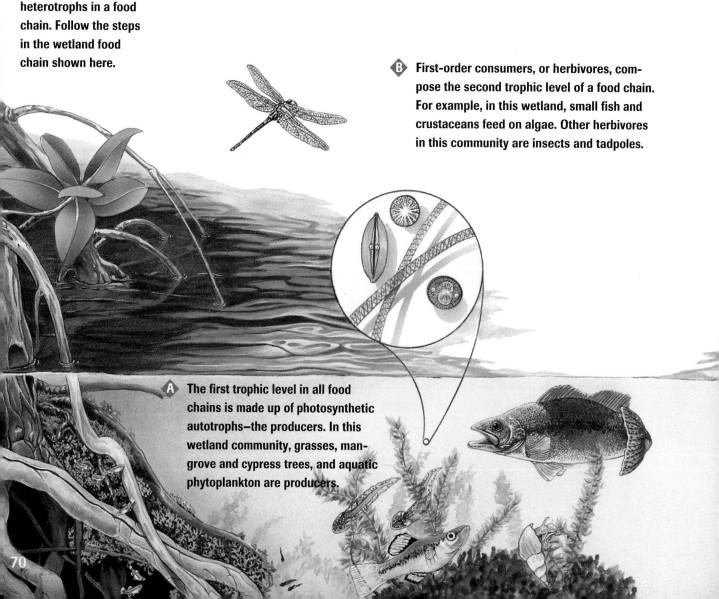

B First-order consumers, or herbivores, compose the second trophic level of a food chain. For example, in this wetland, small fish and crustaceans feed on algae. Other herbivores in this community are insects and tadpoles.

A The first trophic level in all food chains is made up of photosynthetic autotrophs—the producers. In this wetland community, grasses, mangrove and cypress trees, and aquatic phytoplankton are producers.

five links. This is because the amount of energy left by the fifth link is only a small portion of what was available at the first link. A portion of the energy is lost as heat at each link. It makes sense, then, that typical food chains are three or four links long.

Trophic levels represent links in the chain

Each organism in a food chain represents a feeding step, or **trophic level,** in the passage of energy and materials. A food chain represents only one possible route for the transfer of matter and energy in an ecosystem. Many other routes exist. As *Figure 3.13* indicates, many different species occupy each trophic level in a wetlands ecosystem. In addition, many different kinds of organisms eat a variety of foods, so a single species may feed at several trophic levels. For example, the great blue heron eats largemouth black bass, but it also eats minnows, bluegills, and frogs. The alligator may feed on the heron, fish, or even a deer that comes too close. Can you think of other possible food chains?

C Second-order consumers, or carnivores, make up the third trophic level. Carnivores feed on first-order consumers; they are meat eaters. The heron is a carnivorous bird that feeds on fish, frogs, and other small animals of the wetland habitat. Can you think of other animals that might occupy the third trophic level?

D Third-order consumers, carnivores that feed on second-order consumers, make up the fourth trophic level. An alligator eating a shorebird is one example of a third-order consumer. Bacteria and fungi decompose all the links of the food chain when organisms die and can function when necessary at any point in a food chain.

Food webs

Simple food chains are easy to study, but they cannot indicate the complex relationships that exist among organisms that feed on more than one species. Notice how the food web of the forest ecosystem in *Figure 3.14* represents a network of interconnected food chains. In an actual ecosystem, many more plants and animals would be involved in the food web. Ecologists who are particularly interested in energy flow in an ecosystem set up experiments with as many organisms in the community as they can. The model they create, a **food web,** expresses all the possible feeding relationships at each trophic level in a community. Food webs are a more natural model than a food chain since most organisms depend on more than one other species for food.

Figure 3.14

A forest community food web includes many organisms at each trophic level. Arrows indicate the flow of materials and energy. How many food chains can you find in this food web?

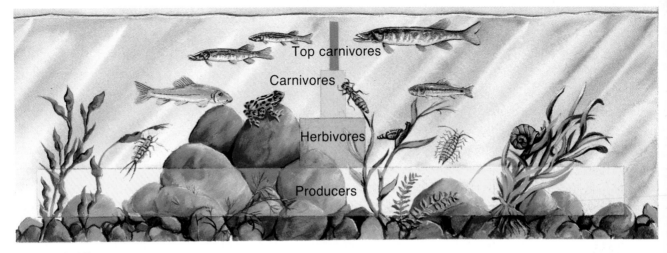

Figure 3.15

Pyramid of energy A pyramid of energy illustrates that energy decreases at each succeeding trophic level. The total energy transfer from one trophic level to the next is only about ten percent. What happens to the other 90 percent of the energy? Organisms fail to capture and eat all the food available at the trophic level below them. Not all the food that is captured and eaten gets digested. And some of the digested food is used by the organism as a source of energy. When the organism is eaten, that energy isn't available—it has already been used. The energy lost at each successive trophic level enters the environment as heat.

Energy and trophic levels: Ecological pyramids

How can you show how energy is used in an ecosystem? Ecologists use food chains and food webs to model the distribution of matter and energy within an ecosystem. They also use another kind of model, called an ecological pyramid, to depict energy conversions in an ecosystem. The base of the ecological pyramid represents the producers, or first trophic level. Higher trophic levels are layered on top of one another. Examine each type of ecological pyramid in *Figures 3.15, 3.16 and 3.17*. Each pyramid gives different information about an ecosystem. Observe that each summarizes interactions of matter and energy at each trophic level. Notice that the source of energy for all ecological pyramids is energy from the sun.

Figure 3.16

Pyramid of numbers Ecologists construct a pyramid of numbers based on the population sizes of organisms in each trophic level. This pyramid of numbers shows that population sizes decrease at each higher trophic level. This is not always true. For example, one tree can be the home for 50 000 insects. In this case, the pyramid would be inverted. Can you think of another situation where the pyramid would be inverted?

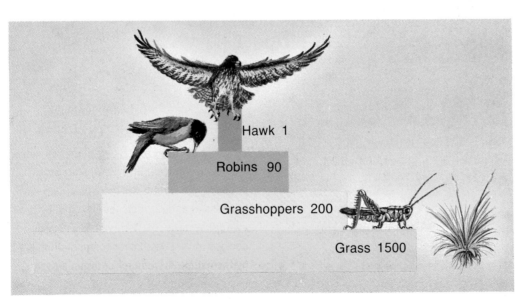

Ecological Efficiency in Catfish Farming

Fish raised on fish farms account for about ten percent of the world's fish harvest. In the United States, shellfish such as shrimp are raised on fish farms, as are catfish and trout. Fish farms provide a large yield because fish are kept free from predators and can be kept under optimal conditions. Some of the difficulties that must be overcome include limited space and lack of a large enough natural food source. To provide enough food for the fish, farmers supply the food the fish will eat.

Catfish farmers feed their fish grain. Catfish are omnivores that readily feed on smaller fish, crayfish, frogs, or other animals, as well as on plants. Why don't catfish farmers provide their fish with frogs for food instead of grain? The reason is that grain provides more energy than they would get from frogs.

Using ecological efficiency By studying an ecological pyramid of energy, it is obvious that lower trophic levels provide greater amounts of energy. This concept is important to any managed system such as a catfish farm. Farmers can raise more catfish by feeding them from

lower trophic levels because more energy is available in these levels. This concept is true for humans also. For example, it takes 10 kg of grain to build 1 kg of human tissue if the grain is directly consumed by the human. It takes 100 kg of grain to build 1 kg of human tissue if the grain is first eaten by a cow, and then the beef is eaten by the human.

Thinking Critically

How can this knowledge of the efficiency of eating on the lower trophic levels of a food chain be applied to agriculture and feeding the world's population?

Figure 3.17

Pyramid of biomass A pyramid of biomass expresses the weight of living material at each trophic level. Ecologists calculate the biomass at each trophic level by finding the average weight of an organism of each species at that trophic level and multiplying by the estimated number of organisms in each population.

Cycling maintains homeostasis

Food chains, food webs, and ecological pyramids all show how energy moves in only one direction through the trophic levels of an ecosystem, and how energy is lost at each transition from one trophic level to the next. This energy is lost to the environment as heat generated by the body processes of organisms. Keep in mind that sunlight is the source of all this energy, so energy is always being replenished.

Matter, in the form of nutrients, also moves through the organisms at each trophic level of an ecosystem. But matter cannot be replenished like the energy from sunlight. The atoms of carbon, nitrogen, and other elements that make up the bodies of organisms alive today are the same atoms that have been on Earth since life began. Matter is constantly recycled. Because of this recycling, some of the atoms in your body right now could once have been part of a dinosaur! *Figures 3.18, 3.19, and 3.20* show how three important materials—water, carbon, and nitrogen—cycle through ecosystems.

The water cycle

Water (H_2O) occurs on Earth as a liquid or a solid and in the atmosphere as a gas. Water begins its cycle through the ecosystem community when plants absorb it through their roots. Animals drink water or get it indirectly with the food they consume. As water moves through the ecosystem, plants and animals lose it back to the atmosphere through respiration. Organisms also lose water through excretion. After an organism dies, decomposition releases water back into the environment.

Figure 3.18

The Water Cycle

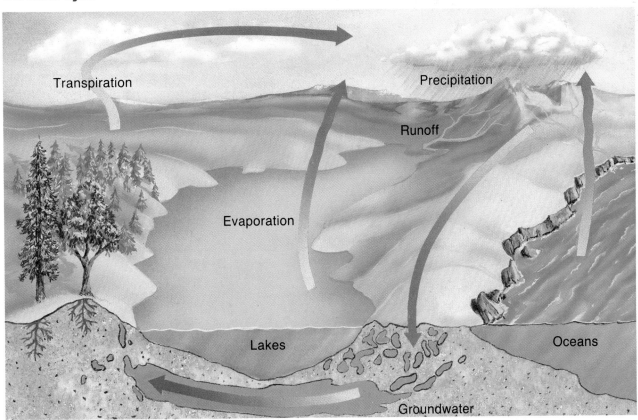

The carbon cycle

Carbon is found in the environment as carbon dioxide gas (CO_2) in the atmosphere and ocean. From the atmosphere, carbon dioxide moves to aquatic and terrestrial producers. Producers use carbon dioxide in photosynthesis, a process that chemically combines carbon dioxide with water to make sugar. Photosynthesis changes these molecules from low- to high-energy forms. Energy from the sun joins carbon dioxide, oxygen, and hydrogen from water into energy-rich sugars. Organisms obtain carbon when they consume producers or other consumers. Respiration and decay are two processes that usually return carbon to the atmosphere in the gas CO_2. Decay sometimes occurs in swamps, bogs, or other areas that have low amounts of oxygen. If decay occurs without oxygen, the carbon can be bound up in a fossil fuel formed over time by geological processes. Carbon also returns to the atmosphere in large amounts as carbon dioxide when fossil fuels are burned.

Figure 3.19

The Carbon Cycle

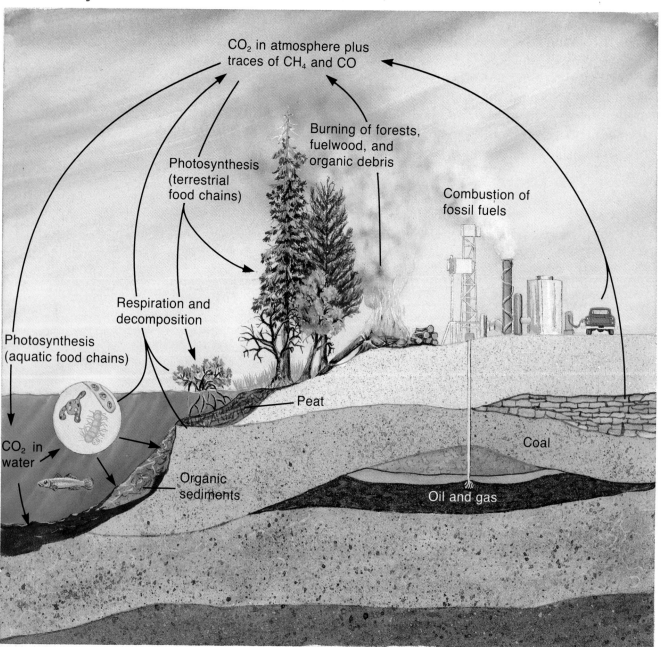

CO$_2$ in atmosphere plus traces of CH$_4$ and CO

Burning of forests, fuelwood, and organic debris

Photosynthesis (terrestrial food chains)

Combustion of fossil fuels

Respiration and decomposition

Photosynthesis (aquatic food chains)

Peat

CO$_2$ in water

Coal

Organic sediments

Oil and gas

The nitrogen cycle

Atmospheric nitrogen makes up nearly 78 percent of air. However, living things cannot use nitrogen in the atmospheric form. Notice that lightning and some bacteria convert atmospheric nitrogen into usable nitrogen-containing compounds that can be used by living things. Plants take up nitrates made by bacteria and lightning. Plants convert the nitrates into molecules that contain nitrogen. Herbivores eat plants and convert nitrogen-containing plant proteins into nitrogen-containing animal proteins. During digestion, you convert plant proteins and animal proteins to forms that combine to make human proteins. Organisms return nitrogen to the atmosphere when they die and decay. Can you trace the roles of producers, consumers, and decomposers in the cycling of nitrogen?

ThinkingLab Make a Hypothesis

Rates of Nutrient Cycling

Chemical cycling rates differ among ecosystems. Once decomposition releases chemicals into the environment, they can be taken up again by living organisms. The amount of nitrates in soils, for example, is generally reduced if the amount of decomposition in an area decreases. The amount of biomass, any form of living material, in an ecosystem also determines the amount of soil nitrates. If there is a lot of biomass, then there are fewer nitrates in the soil.

Analysis

The table below presents nitrate data from some hypothetical communities. Compare the amounts of nitrates in the soil and biomass of each community.

Table 3.1

Ecosystem	Soil	Biomass	Atmosphere
	Kilograms/Hectare		Percent
#1	25	85	0%
#2	10	163	0%
#3	1	275	0%

Thinking Critically

Hypothesize which community has the fastest cycling rate for nitrogen. Which of the numbered ecosystems is a tropical rain forest, a desert, and a grassland?

Figure 3.20

The Nitrogen Cycle

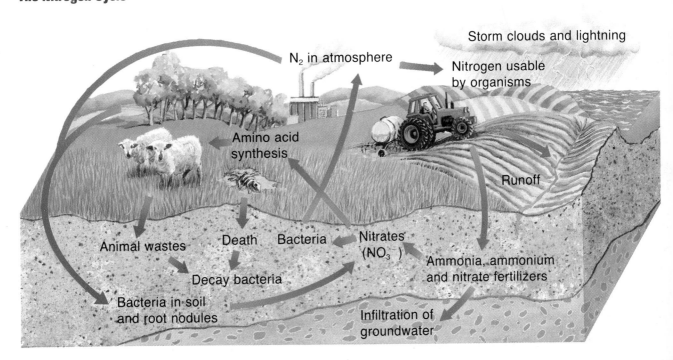

calcium, and phosphorus, as well as others, must also cycle through an ecosystem. One essential element, phosphorus, cycles in two ways. Plants use phosphorus in the soil in their body tissues. Animals get phosphorus by eating plants, *Figure 3.21.* When these animals die, they decompose and the phosphorus is returned to the soil to be used again. This is the short-term phosphorus cycle. Phosphorus can also have a long-term cycle. Phosphates washed into the sea become incorporated into rock as insoluble compounds. Millions of years later, as the environment changes, the rock containing phosphorus is exposed and the phosphorus can again be made a part of the local ecological system.

Figure 3.21

Phosphorus is an element essential to all organisms, such as this moose and the plants it feeds on, because it is part of all their cell membranes. Phosphorus is important in energy transport in a cell.

The cycling of phosphorus

Materials other than water, carbon, and nitrogen cycle through ecosystems also. Substances such as sulfur,

Connecting Ideas

Organisms in different ecosystems are affected by changes in the abiotic factors of the environment as well as the biotic. The interactions among organisms and the physical factors in the environment result in a balance in each ecosystem.

In this chapter, you found out how ecologists model the balanced movement of matter and energy in ecosystems.

Homeostasis in ecosystems is essential to the health of all species. What might happen when ecosystems are disrupted? Communities of organisms react in different ways to changes in their environments. Populations of different species in communities have evolved a variety of adaptations that enable them to survive these changing conditions.

Section Review

Understanding Concepts
1. What is the difference between a parasitic relationship and a commensal relationship?
2. Why do producers always occupy the lowest layer of ecological pyramids?
3. Give two examples of how nitrogen cycles from the abiotic portion of the environment into living things and back.

Thinking Critically
4. Clownfish are small, tropical marine fishes usually found swimming among the stinging tentacles of sea anemones. What type of symbiotic relationship do these animals have if the clownfish are protected by the sea anemone, but the anemone does not benefit from the clownfish?

Skill Review
5. **Designing an Experiment** Design an experiment to test whether or not clownfish are immune from the stinging cells of the sea anemone. For more help, refer to Practicing Scientific Methods in the *Skill Handbook.*

Saving the Everglades

Can the Everglades Still Be Saved?

The Nature Watch   April 2, 1990

Close to a million people visit Everglades Park each year to marvel at the wildlife and gaze at the steamy landscape that is part African veld, part tropical swamp. Life plays out here as it has over the millennia: primal, unchanging, indestructible. At least, that is how it seems.

The year is 1930. The place is the incredibly expansive wetlands region in southern Florida. Everywhere you look, wildlife abounds: vast colonies of wading birds, alligators, panthers, manatees, fish, and other species. Together with mammals and birds, the number and variety of insects and plants make this massive ecosystem—which covers 4 million acres—America's greatest.

Endangered ecosystem Today, the entire ecology of the region has been altered by the damming, draining, and poisoning of its precious water supply for agriculture and urban development. Acres of natural sea grass have been overrun by exotic vegetation, which thrives on the pollutants fed into the water system by surrounding farms. The number of wading birds has dropped from a million to fewer than 4000, as their nesting areas have been drained for flood control and urban development. Thirteen native species including the snail kite, wood stork, manatee, and American crocodile are classified as either threatened or endangered. The only Florida panthers left in the world—fewer than 50—are near extinction a few miles south of downtown Miami.

Different Viewpoints

Everglades restoration In 1978, a coalition of state and national organizations began a multi-million-dollar project to revive the Everglades Park by restoring natural water flow to the park south of Lake Okeechobee. This project will require more than just money. It will require the support of farmers and urban citizens in South Florida, and their willingness to conserve water and limit environmental pollution.

Distribution of funds: Park restoration or tourism? Hurricane Andrew in August 1992 did not help the cause of Everglades restoration. Homestead Florida, the Everglades park headquarters, suffered extensive damage from the hurricane. The cost of repairing the infrastructure of the park will divert millions

of dollars away from necessary park restoration research. Money for repairing structures for park visitors will come before research money.

INVESTIGATING the Issue

1. **Writing About Biology** Write a survey that will determine people's opinions on urban development at the expense of the environment. Collect and analyze the data.
2. **Debating the Issue** Tourism is an important source of income for most coastal cities and states. Debate with classmates the pros and cons of tourism.

3.1 Organisms and Their Environments

- Natural history, the observation of how organisms live out their lives in nature, led to the development of the science of ecology—the study of the interactions of organisms and the environment.
- Ecologists classify and study the biological levels of organization from the individual up to the ecosystem. The role of an organism in its habitat is its niche.

3.2 How Organisms Interact

- Species that live in the same community have evolved feeding and symbiotic relationships. Energy and matter are transferred through ecosystems via these relationships.
- Ecologists use models to illustrate the flow of matter and energy from one trophic level to the next. Food chains are simple models that show one way that materials move from producers to consumers and eventually to decomposers.

- Food webs illustrate all of the possible ways materials are transferred within an ecosystem. Ecological pyramids model the transfer of energy from trophic level to trophic level. Homeostasis requires a constant energy input and is maintained in ecosystems by cycling matter through biotic and abiotic portions of the ecosystem.

Key Terms

Write a sentence that shows your understanding of each of the following terms.

abiotic factor	food web
autotroph	habitat
biosphere	heterotroph
biotic factor	mutualism
commensalism	niche
community	parasitism
decomposer	population
ecology	scavenger
ecosystem	symbiosis
food chain	trophic level

Understanding Concepts

1. How are naturalists and ecologists similar and different in what and how they study?
2. Describe the difference between a population and a community.
3. Identify and list three abiotic factors and three biotic factors in a wetland community.
4. Why do two species in a niche compete?
5. Why are food webs more realistic models than food chains?
6. What does an ecologist mean when he or she says that competition leads to only one species per niche?
7. Use a pyramid model to explain what happens to energy as it flows through a community.
8. Describe the relationship between abiotic factors and biotic factors in the nitrogen cycle.

Using a Graph

9. The following graphs show the results of an experiment with two species of *Paramecia*. What type of relationship is shown between these two closely related species?

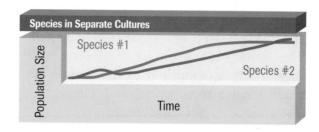

Species in Separate Cultures
Species #1
Species #2
Population Size
Time

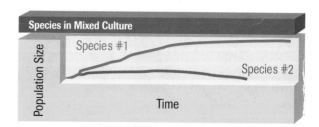

Species in Mixed Culture
Species #1
Species #2
Population Size
Time

Relating Concepts

10. Make a concept map that relates the following terms and phrases. Supply the appropriate linking words for your map.

    mutualism, predator, vulture, biotic relationships, symbiosis, parasitism, lichen, scavenger, tapeworm, Spanish moss, commensalism, feeding relationships, tiger

Applying Concepts

11. Why don't food chains have 20 trophic levels?
12. Explain how pesticides sprayed on the water in a wetland ecosystem could affect a different ecosystem.
13. Sloths are herbivores that move so slowly that algae grow in their fur. Caterpillars of some moths eat the algae. Birds eat the moths. Using this example, draw two food chains and explain one symbiotic relationship.
14. Deserts contain high levels of chemicals in soil, but low levels of biomass. Would you expect deserts to have large numbers of decomposers? Why or why not?
15. Why do fertilizers contain nitrates?

A Broader View

16. What foods might you consume in an average day if you were "eating low on the food chain"?

Thinking Critically

Applying Concepts

17. **Biolab** How can predators be beneficial to a prey population?

Making a Prediction

18. **Minilab** Predict what would happen if you tried to grow the lichen's fungus separately from the unicellular producer.

Relating Concepts

19. If parasites typically do not kill their hosts, then why is it important to keep your pets free from parasites?

Interpreting Data

20. Does this diagram represent a stable community? Why or why not? What do you predict will happen to this community?

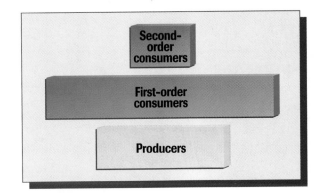

21. **Identifying Patterns** Place the following organisms in correct order in a food chain: mouse, hawk, wheat, snake.

Connecting to Themes

22. **Homeostasis** What three factors are necessary to maintain homeostasis in an ecosystem?
23. **Systems and Interactions** Describe some examples of feeding relationships that cycle matter through an ecosystem.
24. **Energy** Assume that 1000 kilocalories of solar energy are converted to biomass by algae in a pond. Assume that a minnow consumes the algae. How many kilocalories would be available to a bluegill that consumes the minnow?

4 Community Distribution

Have you ever wondered why plants, animals, and other organisms live where they do? Why do lichens grow on bare rock or wooden fences but not in rich soil? Why do polar bears and walruses live only in cold, snowy polar regions? How do catfish manage to survive in waters that are too warm for trout to survive? Some of the most fundamental questions ecologists try to answer have to do with why populations of organisms live where they do, and what kinds of biotic and abiotic factors are important to their survival. Answers to questions like these form the basis for decisions humans must make about our future and the future of our planet.

In this chapter, you will learn how ecological communities are formed, and how and why they change. You will find out why some communities remain stable for long periods of time, while others change from year to year. You will learn how to recognize some of Earth's major aquatic and terrestrial ecosystems.

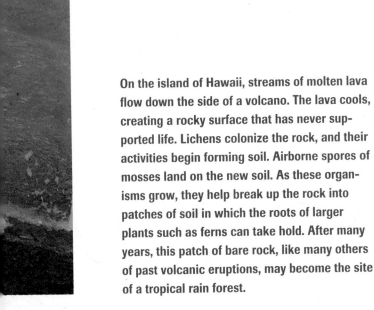

On the island of Hawaii, streams of molten lava flow down the side of a volcano. The lava cools, creating a rocky surface that has never supported life. Lichens colonize the rock, and their activities begin forming soil. Airborne spores of mosses land on the new soil. As these organisms grow, they help break up the rock into patches of soil in which the roots of larger plants such as ferns can take hold. After many years, this patch of bare rock, like many others of past volcanic eruptions, may become the site of a tropical rain forest.

4.1 Homeostasis in Communities

Section Preview

Objectives

Explain how limiting factors and ranges of tolerance affect distribution of organisms.

Sequence the stages of succession in different communities.

Key Terms

limiting factor
succession
primary succession
climax community

A cactus can live in the driest desert, but it still needs water to survive. Its cells and tissues can absorb and store large amounts of water. Chipmunks can survive cold winters in the forest by going into hibernation. Most organisms are adapted to maintain homeostasis within the environments in which they live.

But what if the ecosystem changes? What happens when a flash flood sends torrents of water through the desert, uprooting the cacti that have lived there for a hundred years? What happens if a new group of chipmunks moves into the forest and begins competing with the other chipmunks for food? Even more dramatically, what happens when a forest fire destroys hundreds of acres of trees?

Changing with the Environment

Ecosystems are always changing. Sometimes they change quickly and dramatically, as with fire or flood. They also can change slowly. As young saplings grow into mature trees that shade the ground below them, grasses are slowly replaced by shade-loving plants. Changing conditions within an ecosystem affect the communities of organisms that live there. As ecologists study these changes, they discover patterns that help explain how the ecosystem has developed. These patterns can be used not only to deepen our understanding of the interactions that take place within an ecosystem, but also to predict what might happen if an ecosystem is disturbed.

Limiting factors

Why do more people live in middle latitudes than near the north pole? You would be correct if you pointed out that there isn't much to eat and it's too cold there! Obtaining food and warmth in the frozen reaches of the Arctic is difficult for humans, but the polar bears in *Figure 4.1* thrive in this icy environment. Environmental factors, such as food availability and temperature, that affect an organism's ability to survive in its environment are limiting factors. A **limiting factor** is any biotic or abiotic factor that restricts the existence, numbers, reproduction, or distribution of organisms. The timberline in *Figure 4.1* illustrates how limiting factors affect the plant life of an ecosystem.

Figure 4.1

Polar bears, among the largest carnivores on Earth, survive in the frozen lands and waters near the north pole (top). Their white fur offers camouflage against the ice and snow, enabling them to stalk the seals and walruses that serve as their primary food. The forest stops at the timberline on this mountainside (bottom). At high elevations, temperatures are too low, winds too strong, and the soil too thin to support the growth of large trees. Vegetation is limited to small, shallow-rooted plants, mosses, ferns, and lichens.

Factors that limit one population in a community may also have an indirect effect on another population. For example, a lack of water could limit the growth of grass in a grassland, reducing the number of seeds produced. The population of mice dependent on those seeds for food will also be reduced. What about hawks that feed on mice? Their numbers may be reduced too as a result of a decrease in their food supply.

Ranges of tolerance

Corn plants need two to three months of sunny weather and a steady supply of water to produce a good yield. Corn grown in the shade or during a long dry period may survive, but probably won't produce much of a crop. The ability to withstand fluctuations in biotic and abiotic environmental factors is known as tolerance. *Figure 4.2* illustrates

how the size of a population varies according to its tolerance for environmental change.

Some species can tolerate conditions that another species cannot. For example, catfish can live in warm water with low amounts of dissolved oxygen, which other fish species, such as bass or trout, could not tolerate. The bass or trout would have to swim to cooler water with more dissolved oxygen to avoid exceeding their range of tolerance.

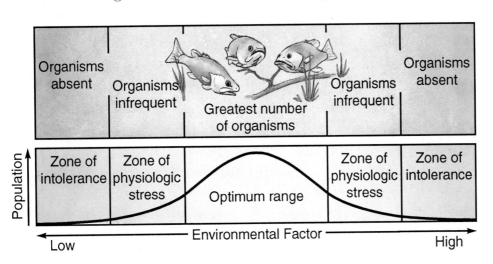

Figure 4.2

Limits of tolerance are reached when an organism receives too much or too little of some environmental factor. Notice that organisms become fewer and fewer as conditions move from optimal toward either extreme of the range of tolerance.

How can you create a closed ecosystem?

Earth can be thought of as a giant, closed system in which all living things interact with the environment within their ranges of tolerance. It seems like an easy matter to create a similar closed system in a bottle, but is it?

Procedure

1. Remove the black plastic base from an empty, 2-L plastic soda pop bottle and add some potting soil to it.

2. Choose some plants such as grass and plant them in the soil.

3. Determine the number of insects, snails, or similar small animals for your closed ecosystem. Place them in.

4. Cut the bottle about four inches from the spout. Place it over the plastic bottom.

5. Tape the ecosystem closed, and observe to see how long it appears to remain in balance.

Analysis

1. What process releases the nutrients of organisms that die?

2. Did your ecosystem remain in balance? For how long? What happened to the plants and animals?

3. How is your closed ecosystem different from Earth?

Succession: A Change in Communities over Time

Changes in an ecosystem also take place as populations grow and diminish in size or move in and out of the area. Ecologists refer to orderly, natural changes that take place in the communities of an ecosystem as **succession.** Succession is often difficult to observe. It can take decades, or even centuries, for one type of community to completely succeed another.

Primary succession

Lava flowing from the mouth of a volcano is so hot it destroys everything in its path, but when it cools it forms new land. Streams gradually deposit silt

Figure 4.3

The first organisms to colonize a new, rocky site are hearty pioneer species such as lichens. Larger plants such as mosses, ferns, and shrubs replace the pioneer species.

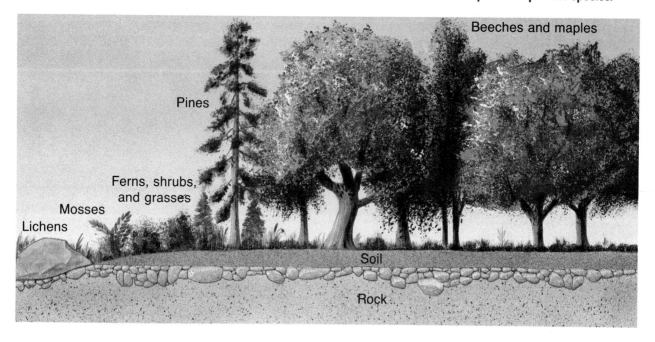

Beeches and maples

Pines

Ferns, shrubs, and grasses

Mosses

Lichens

Soil

Rock

along their banks, creating new soil in which plants can take root. The colonization of new sites by communities of organisms is called **primary succession.**

After some time, primary succession slows down, and the community becomes fairly stable. A stable, mature community that undergoes little or no succession is called a **climax community.** *Figure 4.3* illustrates primary succession of bare rock into a climax community.

As pioneer organisms die, their decaying bodies add to the bits of rock accumulating in cracks and crevices, initiating the first patches of soil. The presence of soil makes it possible for weedy plants, small ferns, and insects to move in. The soil builds up, and seeds borne by the wind blow into these larger patches of soil and begin to grow.

Over time, as the community of organisms changes and develops, additional habitats emerge and new species move in. Eventually, the area becomes a forest of vines, trees, and shrubs inhabited by birds and other forest-dwelling animals.

Secondary succession

What happens when a natural disaster such as a forest fire or hurricane destroys a community? What happens when farmers abandon a field or when a building is demolished in a city and nothing is built on the site? Secondary succession is the sequence of community changes that takes place when a community is disrupted by natural disasters or human actions.

Surtsey, the Newest Place on Earth

On a cold November morning in 1963, a sailor stood on the deck of a fishing boat bobbing in the sea off the southern coast of Iceland. A strange smell—a little like rotten eggs—was in the air. Rubbing his eyes in disbelief, the sailor witnessed land emerging from the sea. He saw the birth of a new volcanic island, named Surtsey after a giant in an Icelandic myth.

Lava flows; An island grows For nearly two years, lava flowed and volcanic rocks the size of automobiles shot into the sky. When the lava stopped, the barren island measured almost one square mile.

Because the island presented a rare chance for scientists to study how life colonizes a new area, Iceland designated it a nature preserve. Access was strictly controlled, and scientists were forbidden to take away even a pebble.

New life on Surtsey Even before the island cooled, seabirds were visitors. Borne by the wind, the ocean currents, or these birds, seeds reached Surtsey, and in 1965 the first plant, a sea rocket, bloomed. Spiders soon glided to the island on silken threads. Lichens and moss grew on the lava, and their remains formed soil. Later, seals sought Surtsey's beaches to bear their young.

A natural laboratory
Today, eroding winds and waves are nibbling away at Surtsey. One-fourth of the island's area has disappeared. One day, Surtsey may vanish. Until then, biologist and ecologist Sturla Fridriksson says that scientists "will be able to tell how a society is built up—and how it is destroyed."

CONNECTION TO **Biology**

In what other places might conditions similar to those on Surtsey provide a chance to study succession in a natural environment?

Succession is the replacement of one community by another as environmental conditions change. You have read about the processes of primary succession and secondary succession in terrestrial communities. For most communities, succession takes a long time, but succession occurs much more rapidly in microscopic populations of small aquatic communities. In this Biolab, you'll investigate succession in a puddle community.

Do abiotic factors affect succession in a puddle community?

PREPARATION

Problem

How can you determine the effects of abiotic factors on succession? Among your group, determine one abiotic factor that you can measure in a controlled experiment.

Hypotheses

Decide on one hypothesis that you will test. Your hypothesis might be that added fertilizer increases the speed of succession, or that a higher or lower temperature affects the speed of succession.

Objectives

In this Biolab, you will:
- **Analyze** how abiotic factors affect the rate of succession in a small aquatic community.
- **Observe** how populations change during succession.

Possible Materials

plastic cups
pond or puddle water
fertilizer
hay
rice
thermometer
graduated cylinder
microscope
coverslips
microscope slides
eyedropper

Safety Precautions

Always wash your hands after handling materials that contain live bacteria. Wear eye protection to keep pond water or fertilizer from splashing into your eyes.

1. Decide on a way to test your group's hypothesis. Keep the available materials in mind as you plan your procedure. Be sure to include a control. For example, you might use two cups—one with untreated water and one with added rice, hay, or fertilizer.

2. Decide how long you will observe your ecosystem. Prepare a table that includes a place to record data from your observations, such as population size, and an area to draw each of the kinds of organisms you observe. You will want to record the date, number of drops of water observed from each cup, and numbers of each kind of organism you observe. It is not necessary to know the name of the organism if your drawings are clearly different.

Check the Plan

Discuss the following points with other group members to decide the final procedure for your experiment.

1. What is your one independent variable?

2. What control will be used?

3. How much water will be put in each cup? How many drops of water will you observe each day?

4. What data will you collect, and how will it be recorded?

5. *Make sure your teacher has approved your experimental plan before you proceed further.*

6. Carry out your experiment.

1. **Checking Your Hypothesis** Was your hypothesis supported by your data? Explain using specific experimental data to confirm your hypothesis about how the abiotic factor you chose affected succession in the puddle.

2. **Interpreting Observations** How many different species did you observe?

3. **Analyzing Data** Were the numbers and kinds of species the same in both cups? If not, how did they differ?

4. **Thinking Critically** Do you think other abiotic factors would have provided the same results? Explain.

Going Further

Project Design an experiment that you could perform to learn whether succession could be accelerated in a similar community of soil microorganisms. If you have all of the materials you will need, you may want to carry out this experiment.

During secondary succession, as in primary succession, the community of organisms inhabiting an area gradually changes. But the species involved in secondary succession are different. Pioneer species, for example, are not the same because the area has not reverted back to the original conditions. Secondary succession also takes less time to reach a climax community than primary succession.

In 1988, a forest fire burned out of control for two months in Yellowstone National Park. Thousands of acres of trees, shrubs, and grasses were burned. As you can see in *Figure 4.4,* the fire has given biologists an excellent opportunity to study secondary succession in a community. Biologists have been observing succession from the time the ground was still smoking. Biologists have been able to observe and compare secondary succession in areas that suffered damage of different levels of severity.

Figure 4.4

The first plants to grow in the newly bare soil following the great fire of 1988 in Yellowstone National Park were annual wildflowers that need plenty of sunshine and were not able to grow in the shade of the forest. Within three years, perennial wildflowers, grasses, ferns, and pine seedlings began coming up through the annuals. Once the pine seedlings grow above the shade cast by the grasses and these herbaceous perennials, the trees will grow more quickly, and eventually a mature forest will once again develop.

Section Review

Understanding Concepts

1. Compare the limiting factors and ranges of tolerance of a lichen and a pine tree.
2. How can a pond community be altered by a dry season?
3. Compare and contrast primary and secondary succession.

Thinking Critically

4. Explain how the success of one species can bring about the disappearance of other species during succession.

Skill Review

5. **Making and Using Graphs** Using the following data, graph the limits of tolerance for temperature for carp. The first number in each pair is temperature in degrees C; the second number is the number of carp surviving at that temperature: 0,0; 10,5; 20,25; 30,34; 40,27; 50,2; 60,0. For more help, refer to Organizing Information in the *Skill Handbook.*

4.2 Biomes

C limate—a combination of temperature, sunlight, prevailing winds, and precipitation—is an important factor in determining what climax community will develop in any one ecosystem. Soil type is also important. Many regions of the world share similar soil and climate characteristics and, as a result, also share similar types of climax communities. Although the species of organisms living in each desert ecosystem may vary, all are adapted for life in an environment with dry weather, poor soil, and hot daytime temperatures.

Section Preview

Objectives

Compare and contrast the euphotic and aphotic zones of ocean biomes.

Identify the major limiting factors affecting distribution of terrestrial biomes.

Distinguish among the terrestrial biomes.

Key Terms

biome
photic zone
aphotic zone
estuary
intertidal zone
plankton
tundra
permafrost
taiga
desert
grassland
temperate forest
tropical rain forest

Aquatic Biomes: Life in the Water

Ecosystems that have similar kinds of climax communities can be grouped into a broader category of organization called a biome. A **biome** is a large group of ecosystems that share the same type of climax community. Biomes located on land are called terrestrial biomes; those located in oceans, lakes, streams, ponds, or other bodies of water are called aquatic biomes.

As a human who lives on land, you may tend to think of Earth as primarily a terrestrial planet. But one look at a globe, a world map, or a photograph of Earth taken from space tells you there is an aquatic world too; approximately 75 percent of Earth's surface is covered with water. Most of that water is salty. Oceans, seas, and even some inland lakes contain salt water. Fresh water is confined to rivers, streams, ponds, and most lakes. Saltwater and freshwater environments have similarities, but they

also have important differences. As a result, aquatic biomes are separated into marine biomes and freshwater biomes.

Marine biomes

If you've watched TV programs about ocean life, you may have gotten the impression that the oceans are mostly full of great white sharks, whales, and other large animals. The oceans contain the largest amount of biomass, or living material, of any biome on Earth, but most of this biomass is made up of extremely small, often microscopic organisms that humans usually don't see.

One of the ways ecologists study marine biomes is to separate them into shallow, sunlit zones and deeper, unlighted zones. The portion of the marine biome that is shallow enough for sunlight to penetrate is called the **photic zone.** Deeper water that never receives sunlight makes up the **aphotic zone.** Different parts of the ocean differ in physical factors and in the organisms found there.

aphotic:
a (GK) without
phos (GK) light
photic:
phos (GK) light
Aphotic means without light, photic means with light.

Figure 4.5

Shallow marine environments exist along the coastlines of all land masses on Earth. These coastal ecosystems include rocky shores, sandy beaches, and mud flats, and all are part of the photic zone.

A mixing of waters

If you were to follow the course of any river, you eventually would reach a sea or ocean. Wherever rivers join oceans, fresh water mixes with salt water. In many such places, an estuary is formed. An **estuary** is a coastal body of water, partially surrounded by land, in which fresh water and salt water mix. Salinity ranges between that of seawater and of fresh water, and depends on how much fresh water the river brings into the estuary. Salinity in the estuary also changes with the tide. Estuaries contain salt marsh ecosystems, which are dominated by grasses, as illustrated in *Figure 4.5.* These plants grow so thick that their stems and roots form a tangled mat that traps food material and provides additional habitat for small organisms. Decay of dead organisms proceeds quickly, recycling nutrients through the food web.

Twice a day, the gravitational pull of the sun and moon causes the rise and fall of ocean tides. Tides vary in height, depending on the season and the slope of the land. The portion of the shoreline that lies between high and low tide lines is called the **intertidal zone.** *Figure 4.6* illustrates organisms living in the intertidal zone of a sandy beach and a rocky shoreline.

Figure 4.6

Intertidal ecosystems have high levels of sunlight, nutrients, and oxygen, but productivity is limited by waves crashing against the shore.

If the shore is rocky, wave action constantly ▷ threatens to wash organisms into deeper water. Snails, starfish, and other intertidal animals of the rocky shore have body parts that act as suction cups for holding onto the wave-beaten rocks. Other animals, such as mussels and barnacles (inset), secrete a strong glue that helps them stay put.

In the light

As you move away from the intertidal zone and into deeper water, the ocean bottom is no longer affected by waves or tides. Many organisms live in this shallow-water region that surrounds most continents and islands. Nutrients washed from the land by rainfall contribute to the abundant life and high productivity of this region of the photic zone.

The photic zone of the marine biome includes the vast expanse of open ocean that covers most of Earth's surface. Most of the organisms that live in the marine biome are plankton. **Plankton,** shown in *Figure 4.*7, are microscopic organisms that float in the waters of the photic zone. Unicellular algae, including diatoms, are the producers of the plankton. Consumers include tiny shrimplike creatures, jellyfishes, worms, and the juvenile stages of animals such as crabs, snails, jellyfish, and marine worms.

In the dark

What is the ocean like in the dark depths of the aphotic zone? Imagine a darkness blacker than night and pressure so intense it exerts hundreds of pounds of weight on every square centimeter of your body's surface. Does this sound like a hospitable place to live? Almost 90 percent of the ocean is more than a mile deep. In some places, the ocean may extend miles below the sunlit surface. Even though the animals living here are very far below the photic zone where plankton abound, they still depend on plankton for food, either directly or indirectly by eating organisms that feed directly on plankton. Fish living in the deep areas of the ocean are adapted to a life of darkness and a scarcity of food. What adaptations might help these organisms survive in this environment?

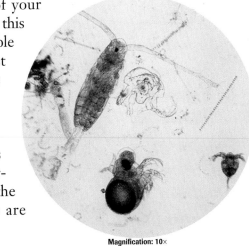

Magnification: 10×

*Figure 4.*7

Plankton form the base of all marine food chains, but not all organisms that eat plankton are small. Baleen whales and whale sharks, some of the largest organisms that have ever lived, consume vast amounts of plankton.

If the shore is sandy, wave action keeps the bottom in constant motion. Most of the clams, worms (inset), snails, crabs, and other organisms that live along sandy shores survive by burrowing into the sand.

Meet Dr. Sharyn Richardson, College Professor

A typically busy afternoon finds Dr. Sharyn Richardson conducting a tour through the Florida Everglades for two Chinese engineers who are planning mass-transit systems in their country. She is pleased that they want to see the difference between old-style, dredge-and-fill operations and the newer, environmentally sensitive methods that allow a more natural flow of water. Environmental interest is growing all over the world, thanks in part to teachers like Dr. Richardson.

In the following interview, Dr. Richardson talks about her work and interest in the environmental sciences.

On the Job

Q Dr. Richardson, would you tell us about your career?

A I teach environmental science at Barry University in Miami. A favorite course is "Ocean World," which focuses on the interrelationships among the four disciplines of biology, chemistry, physics, and geology.

Q You also have a related outside interest, don't you?

A Yes. I'm the national secretary of Friends of the Everglades, founded by environmental activist Marjory Stoneman Douglas. The group promotes the protection and restoration of the Florida Everglades. One current issue is preservation of the panther. Schoolchildren lobbied the legislature to make the panther the state animal, but it may be the first state animal to become extinct! However, the real issue isn't any one individual species; it's the entire habitat. The Florida panther represents what can happen to all kinds of final feeders at the top of their food pyramids, from predator animals in the wild, and particulary to human beings if we fail to understand the intricate workings of an ecosystem. One theory suggests that if we select stock out of the wild and breed it in captivity, we would have animals to put back. But because the cause of the dwindling numbers of many animals is habitat loss, restocking the wild doesn't address this fundamental problem.

Early Influences

Q Could you tell us how you became interested in the field of biology?

A When I was a high-school sophomore, I did a project for the school's science fair. My kitchen chemistry project dealt with the differences between the ways proteins survive cooking in hard water and in soft water. Although I won the competition, the runner-up was sent to the county fair because she was a senior. I didn't give up, even though I was hurt. The next year I worked hard to improve my project, and this time I was invited to the science fair at the state level. I finally won fourth place in the international division. Life throws curves like that at all of us, and it's how we deal with the problem that counts.

Later, when I moved to Florida and began teaching in a middle school, I saw changes happening in the Everglades and wanted to know why. I went back to school to find the answers through environmental science courses.

Q What people have influenced your life?

A Marjory Stoneman Douglas certainly has. Even before I met her, I stood in awe of the beautiful prose in her book *The Everglades: River of Grass*. She has devoted so much of her life to saving the Everglades. One lesson I've learned from Marjory, who is now 103, is to be an optimist—always.

Marjory was involved in one of the singular pleasures of my life. She mentioned that she was disappointed at never having heard the alligators roar. I arranged to transport her in her wheelchair to the Everglades on a beautiful, moonlit night. I got a huge, electronic device to amplify the sounds for Marjory, who is nearly deaf. Being blind, she couldn't see the alligators, but she could hear their roar and smell the aroma of the night-blooming moon vine. It was a wonderful moment for both of us.

Personal Insights

Q What makes you tick, both as a person and as a teacher?

A I've never stopped asking questions. The more changes I see in the world, the more I want to know what caused them. To me, the world is a great puzzle, and science is basically the asking of questions. There are very few careers in which one can be paid to continue asking questions and

guiding other people in finding answers to their own questions.

Q Do you have any advice for students who are beginning to think about choosing a career?

A I always advise young people to find work they can play at, because we spend more of our lives doing that than almost anything. For those who choose a career in sciences, I encourage them to find a discipline in science that encourages them to think and grow in an interdisciplinary fashion. The glasses I put on as an environmentalist require me to understand Earth's interrelations and interactions. That's where our environment will succeed or fail: on the interrelations, not on any single part.

Figure 4.8

Cattails and sedges are the plants most commonly found growing around the edges of lakes. As you move from the margins of a lake or pond toward the center, you find concentric bands of different species of plants. The shallow water in which these plants grow serves as home for tadpoles, aquatic insects, turtles basking on a partially submerged rock, and worms and crayfish burrowing into the muddy bottom. Insect larvae, whirligig beetles, dragonflies, and fish such as minnows, bluegill, or carp also live here.

Freshwater biomes

Have you ever gone swimming or boating in a lake or pond? If so, you may have noticed concentric rings of different kinds of plants growing around the shoreline and even into the water, as shown in *Figure 4.8*. The shallow waters on which these plants grow are highly productive and include fish, algae, protists, mosquito larvae, tadpoles, and crayfish.

If you ever jumped into a deep lake on a warm summer day, you probably got a cold surprise the instant you entered the water. Though the summer sun heats the lake's surface water, a foot or so below the surface remains cold. If you were to dive all the way to the bottom of the lake, you would discover more layers of increasingly cold water as you descended. These temperature variations within a lake are an abiotic factor that limits the kinds of organisms that can survive in deep lakes.

Another abiotic factor that limits life in deep lakes is light. Not enough sunlight penetrates to the bottom to support photosynthesis, so few aquatic plants or algae grow. As a result, the density of populations is lower in deeper waters. Decay takes place at the bottom of a lake. As dead organisms drift to the bottom, bacteria break them down and recycle the nutrients they contain.

Terrestrial Biomes

If you were to set off on an expedition beginning at the north pole and travel south to the equator, what kinds of environmental changes would you notice? The weather would get warmer, of course. You also would see a gradual change in the kinds of plants that cover the ground. At the snow- and ice-covered polar cap, temperatures are always freezing and no plants exist. A little farther south, where temperatures sometimes rise above freezing but the soil never thaws, you might see soggy ground with just a few small cushions of low-growing lichens and plants.

As you continue on your journey, temperatures rise a little higher and you enter forests of coniferous trees. Farther south are grasslands and deserts with scorching summertime temperatures and little rain. Finally, as you approach the equator, you find yourself surrounded by the lush growth of a tropical forest, where it rains almost every day.

As you move south from the north pole, you find yourself traveling through one climax community after another. The graph in *Figure 4.9* shows how two abiotic factors—temperature and precipitation—influence the kind of climax community that will develop in a particular part of the world. These climax communities can be used to group terrestrial ecosystems into six major biomes, as you can see from the Focus On World Biomes.

Figure 4.9

If you know the average annual temperature and rate of precipitation of a particular area, you should be able to determine the climax community that will develop. Predict which climax community would result from an area that has an annual precipitation of 150 cm and an annual average temperature of 15°C.

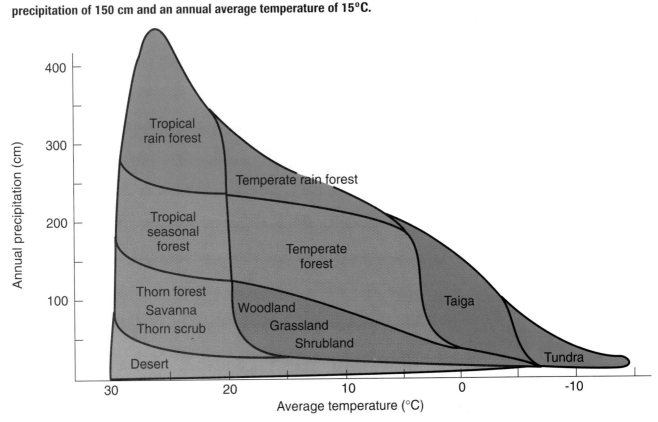

World Biomes

W hen you think of a biome, your first images may be of a land with one or two prominent animal species—elephants or lions on an African plain, musk oxen or reindeer on the tundra, or monkeys in a rain forest. Actually, it is the climax community of plants that ecologists use to characterize a biome. Plants don't migrate and are a better indicator of the long-term characteristics of the biome.

Earth is marvelously diverse. Millions of species are spread over its surface. But this distribution is not random. As shown on the map of the world, there is a pattern to the biomes of Earth.

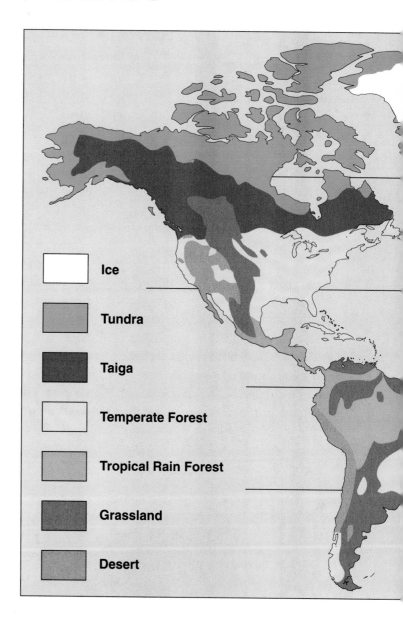

Ice

Tundra

Taiga

Temperate Forest

Tropical Rain Forest

Grassland

Desert

Terrestrial biomes

What defines a biome? In general, three factors determine which biome will be dominant in a terrestrial location—latitude, altitude, and precipitation. A low-lying area near the equator that gets lots of rain will have a tropical rain forest as climax vegetation. Just a few kilometers away on the side of a mountain, you may find plants typical of a biome thousands of kilometers to the north or south. Similarly, latitudes that produce rain forests in the western hemisphere can produce deserts or tropical savannas and grasslands in Africa, where rainfall patterns are different.

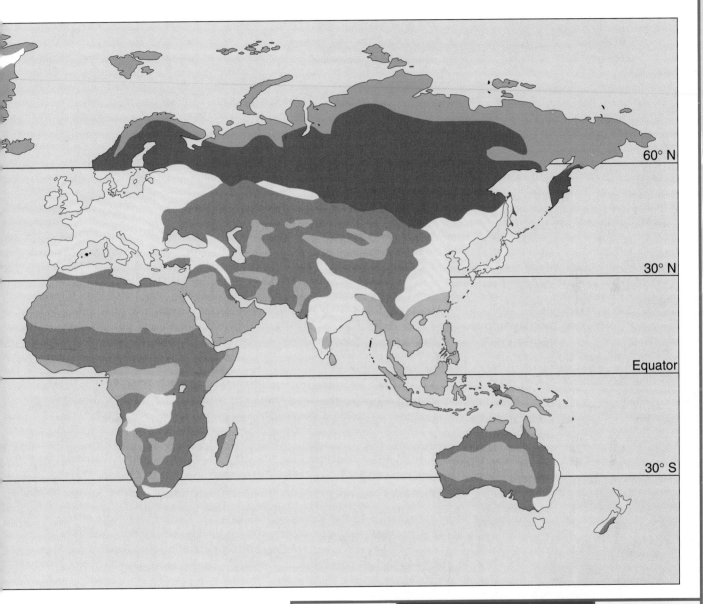

60° N

30° N

Equator

30° S

EXPANDING YOUR VIEW

1. **Making Inferences** Which biome would recover the most slowly from destruction due to natural or human-caused events? Explain.

2. **Thinking Critically** Think about the general pattern of biome types that exists from the equator to the poles. Do you think you would find a similar pattern if you were to climb from the base to the top of a very high mountain? Explain your answer.

Figure 4.10

Grasses, grasslike sedges, small annuals, and reindeer moss, a type of lichen on which reindeer feed, are the most numerous producers of the tundra. The short growing season may last less than 60 days.

Life on the tundra

As you begin traveling south from the north pole, you reach the first of two biomes that circle the pole. This community is the **tundra,** a treeless land with long summer days and short periods of winter sunlight.

Because temperatures in the tundra never rise above freezing for long, only the topmost layer of soil thaws during the summer. Underneath this topsoil is a layer of permanently frozen ground called **permafrost.** Some areas of permafrost have remained frozen for so long that the frozen bodies of animals extinct for thousands of years, such as the elephant-like mammoth, have been found there.

In most areas of the tundra, the topsoil is so thin it can support only shallow-rooted grasses and other small plants. The soil is also lacking in nutrients. The process of decay is slow due to the cold temperatures, so nutrients are not recycled quickly.

Summer days on the tundra may be long, but the growing season is short. Since all food chains depend on the producers of the community, the short growing season is a limiting factor for life in this biome. For example, a typical flowering tundra plant, *Figure 4.10,* is a grass, a dwarf shrub, or a cushion plant. These organisms live a long time and are resistant to drought and cold.

Mosquitoes and other biting insects are some of the most common tundra animals, at least during the short summer. The tundra is also home to a variety of small animals including ratlike lemmings, weasels, arctic foxes, snowshoe hares, snowy owls, and hawks. Musk-oxen, caribou, and reindeer are among the few large animals that inhabit this biome.

Figure 4.11 shows some common tundra animals.

Figure 4.11

Lemmings are the most numerous mammals living in tundra communities (left). Populations of these small, furry, ratlike animals sometimes rise to exceedingly high numbers. At those times, populations of lemming predators, such as snowy owls and arctic foxes, also increase (right).

Life on the taiga

Just south of the tundra lies another community that circles the north pole. The **taiga,** also called the northern coniferous forest, is a land of mixed pine, fir, hemlock, and spruce trees, as shown in *Figure 4.12.* How can you tell when you leave the tundra and enter the taiga? The line between any two biomes is indistinct, and patches of one blend almost imperceptibly into the other. For example, if the soil in the taiga is waterlogged, a peat swamp habitat develops that looks much like tundra. Taiga communities are usually somewhat warmer and wetter than tundra, but the prevailing climatic conditions are still harsh, with long, severe winters and short, mild summers.

In the taiga, which stretches across much of Canada, Northern Europe, and Asia, permafrost is usually absent. The topsoil, which develops from

decaying coniferous needles, is acidic and poor in minerals. When the taiga community is disrupted by fire or logging, the first trees to recolonize the land may be birch, aspen, or other deciduous species. The abundance of trees in the taiga provides more food and shelter for animals than the tundra. More large species of animals are found in the taiga as compared to the tundra. *Figure 4.13* shows some animals in the taiga.

Figure 4.12

The dominant climax plants of the taiga are primarily fir and spruce trees. The needlelike leaves of these conifers are drought resistant.

Figure 4.13

The taiga stretches across most of Canada, Northern Europe, and Asia.

◄ **The lynx is a predator that depends on the snowshoe hare as its primary source of food.**

The snowshoe hare, one of the taiga's smaller herbivores, also lives in this biome all year long. During the winter, it grows a thick, white winter coat that includes extra hair on its feet for warmth.

Large, herbivorous mammals of ▶ the taiga include caribou, which migrate down from the tundra during the winter, and moose (right), which may be found in the taiga during most of the year.

Figure 4.14

Creosote bushes cover many square miles of desert in the south-western United States. These plants, shown above in bloom with yellow flowers, have small leaves coated with a waxy resin that helps reduce water loss. The roots give off a growth-inhibiting chemical that prevents other plants from growing too close, reducing competition for scarce water resources.

Life in the desert

The driest of the biomes south of the taiga is the desert biome. A **desert** is an arid region with sparse to almost nonexistent plant life. Deserts usually get less than 25 cm of precipitation annually. One desert, the Atacama Desert in Chile, is the world's driest place. This desert receives an annual rainfall of zero.

Vegetation in deserts varies greatly, depending on precipitation levels. Areas that receive more rainfall produce a shrub community that may include drought-resistant trees such as mesquite. Less rainfall supports scattered plant life and produces an environment with large areas of bare ground. The driest deserts are drifting sand dunes with virtually no life at all.

Plants have evolved various adaptations for living in arid areas, as shown in *Figure 4.14*. Many desert plants are annuals that germinate from seed and grow to maturity quickly after sporadic rainfall. Cacti have leaves reduced to spines, photosynthetic stems, and thick waxy coatings that reduce water loss. The leaves of some desert plants curl up, or even drop off altogether, to reduce water loss during extremely dry spells. Desert plants sometimes have spines, thorns, or poisons that act to discourage herbivores.

Most desert mammals are small herbivores that remain under cover during the heat of the day, emerging at night to forage on plants. The kangaroo rat is a desert herbivore that does not have to drink water. These rodents obtain all the water they need to live from the water content in their food. A few larger herbivores such as pronghorn antelopes may be found in American deserts. Foxes, coyotes, hawks, owls, and roadrunners are carnivores that feed on the snakes, lizards, and small mammals of the desert. Scorpions are an example of a desert carnivore that uses venom to capture prey. *Figure 4.15* illustrates some of the many reptiles that make the desert their home.

Figure 4.15

Lizards, tortoises, and snakes are numerous in desert communities. Desert tortoises (left) feed on insects and plants. Venomous snakes such as the diamondback rattlesnake are major predators of small rodents (right).

Life in the grassland

If an area receives between 25 and 75 cm of precipitation annually, a grassland usually forms. **Grasslands** are large communities covered with grasses and similar small plants. The grasslands like the ones shown in *Figure 4.16* principally occur in climates that experience a dry season, where insufficient water exists to support forests. This biome occupies more area than any other terrestrial biome, and it has a higher biodiversity than deserts, often with more than 100 species per acre.

The soils of grasslands have considerable humus content because many grasses die off each winter, leaving decay products to build up in the soil. Grass roots survive through the winter, enlarging every year to form a continuous underground mat called sod.

Because they are ideal for growing cereal grains like oats, rye, and wheat, which are different species of grasses, grasslands have become known as the breadbaskets of the world. Some species of grasses grow as high as eight feet or more. Many other plant species live in this environment, including drought-resistant and late-summer-flowering species of wildflowers such as blazing stars and sunflowers.

Figure 4.16

Summers are hot, winters are cold, and rainfall is often uncertain in a grassland. Called *prairies* in Australia (left), Canada, and the United States (right), these communities are called *steppes* in the former Soviet Union, *savanna* in the Serengeti of Africa, and *pampas* in Argentina. Grasslands contain fewer than ten to 15 trees per hectare, though larger numbers of trees are found near streams and other water sources.

Figure 4.17

The prairies of America support bison as well as many species of birds and insects.

Most grasslands are populated by large herds of grazing animals. *Figure 4.17* shows bison, a species of mammal that roams the American prairies. Millions of bison, commonly known as buffalo, once ranged over the American prairie where they were preyed upon by wolves, coyotes, and humans. Other important prairie animals include prairie dogs, seed-eating rodents whose underground "towns" have been known to stretch across mile after mile of grassland, and the foxes and ferrets that prey on them. Many species of insects, birds, and reptiles including tortoises, lizards, and snakes also make their homes in grasslands.

Life in the temperate forest

When precipitation ranges from about 70 to 150 cm annually in the temperate zone, temperate deciduous forests develop. **Temperate forests** are dominated by broad-leaved hardwood trees that lose their foliage annually, as shown in *Figure 4.18.*

When European settlers first arrived on the east coast of North America, they cleared away large tracts of temperate forest for farmland. The thin soil of the mountainous regions was soon depleted by crops, and farmers abandoned their land. Since then, secondary succession has restored much of the original forest.

The soil of temperate forests usually consists of a top layer that is rich in humus and a deeper layer of clay. If mineral nutrients released by the decay of the humus are not immediately absorbed by the roots of the living trees, they may be washed into the clay and lost from the food web for many years.

Figure 4.18

Temperate forests occur where an even amount of precipitation falls during each of the four seasons. There are many types of temperate forests, each described by the two or three dominant species of trees. Typical trees of the temperate forest include birch, hickory, oak, beech, and maple.

Figure 4.19

Black bears and deer have always been residents of temperate forests in the United States. The numbers of these large animals are growing today because their habitat is protected in many places. Other abundant animals in temperate forests are squirrels and salamanders.

The animals that live in the temperate deciduous forest, as shown in *Figure 4.19,* include familiar squirrels, mice, rabbits, deer, and bears. Many birds live in the forest all year long, while others migrate south to live in tropical regions during the winter.

Life in tropical rain forests

The most biologically diverse of the terrestrial biomes, the **tropical rain forest** is a region of uniformly warm, wet weather dominated by lush plant growth. These forests occupy the equatorial regions around the world. The rain forest receives at least 200 cm of rain annually, while some rain forests can receive as much as 400 cm. Temperatures remain warm throughout the year, about 25°C.

Rain forests form jungles of dense, tangled vegetation only near stream banks or in areas that have been disturbed by fire, logging, or agriculture. In mature, undisturbed rain forest communities like the one in *Figure 4.20,* the leafy tops of tall trees shade the forest floor so completely that few plants grow there.

Much of the precipitation the rain forest receives is generated by water given off by the heavy canopy of leaves. When trees are removed over a large area, this type of precipitation

Figure 4.20

Warm temperatures, high humidity, and abundant rainfall allow the lush growth and great species diversity found in tropical rain forests.

What factors affect the rate of decomposition?

Decomposition occurs when any organism dies; it is the result of both abiotic and biotic factors. Most decomposition is the result of the feeding of bacteria and fungi. Nutrients released by decomposition either build up in soil, are washed away, or are absorbed by producers for recycling through the food chain. The rate of decomposition varies with several factors, including temperature and amount of water.

Analysis

Study the table below. The table shows the approximate rates of decomposition in each of the terrestrial ecosystems.

Relative Rates of Decomposition	
Tundra	− −
Taiga	−
Desert	−
Temperate Forest	+
Tropical Rain Forest	+ + +
Grassland	+ +

Thinking Critically

Design an experiment to determine what factor or factors have the most effect on the rate of decomposition for a specific terrestrial ecosystem.

ceases, and the area may receive too little rain for regrowth.

Tree roots are often shallow, forming a thick mat on the surface of the soil. Roots that support tall trees are sometimes greatly enlarged or may form buttresses, which resemble scaffolding used to support buildings.

Most of the nutrients of the rain forest are tied up in living material. The hot, humid climate enables ants, termites, fungi, and other decomposers to break down leaves and dead animals quickly, before they can become humus. As a result, in some areas of the rain forest, the ground is completely bare.

Many animals live in dense vegetation of tropical rain forests. *Figure 4.21* shows some common rain forest animals.

Figure 4.21

Sloths (left), monkeys, and other mammals, as well as a multitude of bird species such as this black-headed cacique (top), live in the rain forest canopy. Insects such as this Hercules beetle are also numerous in the rain forest (bottom).

Biodiversity in Rain Forests

Tropical rain forests are among Earth's richest ecosystems in terms of their biodiversity. Biodiversity refers to the variety of animals, plants, fungi, and microorganisms that make up a community.

Take a walk through a South American rain forest, and the first thing you notice is the rich plant life. Ferns and mosses grow densely at your feet. Flowering vines and colorful orchids cling to the tall, straight trunks of the rain forest trees. In just one hectare (about 2.5 acres) of this forest, you might find nearly 500 different species of trees! High overhead, the bushy tops of these trees grow close together, forming the uppermost part of the forest called the canopy.

Life in the canopy Many different kinds of animals inhabit the canopy high above the forest floor. Dozens of species of colorful butterflies flit through the treetops. Monkeys move with ease among the tangled branches as they search for flowers, fruits, and seeds. Three-toed sloths hang upside down from branches with long, hooklike claws. Their thick fur sheds rain, and in it green algae grows, giving these slow-moving animals a greenish tinge—excellent camouflage in a world of leaves. It's needed, too, because harpy eagles patrol the canopy, looking for suitable prey like monkeys and sloths.

The dark understory Beneath the canopy but above the forest floor is the dimly lit understory. Here, plants have large leaves to help capture as much sunlight as possible. Their flowers are usually large, showy, and strongly scented to attract the attention of jewel-like hummingbirds, bees, fruit bats, and other pollinators. Brilliantly colored parrots clamber through the vegetation in search of fruit and seeds. Tree snakes lie in wait among the branches for small lizards, birds, and frogs to pass close enough to become a meal. Most insects in the understory fool predators with cunning disguises— some look like sticks, others like tree bark, and still others like flowers. In just one hectare of rain forest, there can be 40 000 different insect species.

On the forest floor Leaves, flowers, and fruits fall frequently to the forest floor where they are quickly eaten or decomposed. Along rotting logs, masses of fungi grow. Giant millipedes and slugs slowly maneuver through the decaying litter, while six-inch-long centipedes and tarantulas the size of your hand hunt for insects, lizards, and birds. Pacas and agoutis, small mammals related to guinea pigs, scurry from shadow to shadow. Along with piglike peccaries, they are prey for ocelots, jaguars, and venomous snakes.

Thinking Critically

Tropical rain forests are being destroyed at a rate of ten to 20 hectares *per minute* for the valuable lumber they contain and to clear land for farms, ranches, and mining operations. Can we afford to lose the tremendous biodiversity found in tropical rain forests?

Figure 4.22

This blue morpho butterfly lives in the canopy of a tropical rain forest.

More species of reptiles, amphibians, and birds are found in the tropical rain forest than in any other terrestrial biome. Although large animals like gorillas and cougars are ground dwellers, most rain forest organisms live in the trees rather than on the forest floor. Butterflies such as the one in *Figure 4.22* and other insects are by far the most numerous animals in the rain forest. Biologists estimate that there may be as many as 3 million species of insects in the tropical rain forest.

Connecting Ideas

A population's range of tolerance determines whether or not it will exist in a particular community. As limiting factors change, and populations of different species of organisms succeed one another, a climax community eventually results. Climax communities are in homeostatic balance. Large climax communities, called biomes, remain stable over long periods of time. However, changes occur within populations of species. How do ecologists determine how interactions among populations can change the structure of a community?

Section Review

Understanding Concepts

1. Explain why the photic and aphotic zones of marine biomes are interdependent.
2. What is the most important abiotic factor that limits distribution of the tundra biome?
3. Describe some common plants and animals from a tropical rain forest and a grassland terrestrial biome.

Thinking Critically

4. Shaneka and her family were planning a trip to a foreign country. In her reading, Shaneka found that the winter was cold, the summer was hot, and most of the land was planted in fields of wheat. Infer which biome Shaneka's family would visit. Explain your answer.

Skill Review

5. **Making and Using Tables** Make a table to show the climate, plant types, plant adaptations, animal types, and animal adaptations for the terrestrial biomes. For more help, refer to Organizing Information in the *Skill Handbook*.

The View from on High

In the early 1970s, NASA launched a satellite that would scan our planet from a distance of 800 km. Many people wondered what this distant spacecraft, called Landsat, could reveal that wouldn't be just as clear from the ground.

The big picture Landsat sent back some of the most amazing pictures that scientists had ever seen. Huge areas could be viewed in detail, and the relationships of distant landforms were seen for the first time.

Landsat used a television camera called a Return Beam Vidicon (RBV) and a multispectral scanner (MSS) that recorded the reflectivity of objects on Earth. The multispectral scanner, far more sensitive to color than the human eye, proved invaluable in locating minerals, tracing pollutants, studying crops, and locating ancient topographical features. The RBV could pinpoint objects as small as a football field.

Mapping the colors of Earth More than 20 years later, successors to the first Landsat are relaying even more detailed information to Earth. The MSS views Earth in the green and red regions of the visible spectrum and two regions of the infrared spectrum. These regions of the spectrum are able to distinguish features on Earth such as soil, vegetation, and water.

Newer Landsats also carry an instrument called the Thematic Mapper (TM). The TM detects colors in the blue-green range. Geographers and mapmakers can fine tune their studies of Earth to include soils and vegetation.

Tracking pollution In their efforts to clean up polluted waterways, scientists have used Landsat data to evaluate the effectiveness of their work. One particular lake studied was Lake Chicot in Arkansas. Lake Chicot was once a fishing paradise, but it had become polluted with sediment. Changes in drainage patterns and runoff from farm fields were responsible. In 1985, the U.S. Army Corps of Engineers began a cleanup of the lake.

Applications for the Future

Scientists at the Canada Centre for Remote Sensing used TM data to check for features that indicate precious metals. Through their studies, they found possible locations for metals such as copper, gold, and molybdenum.

Often, maps of developing countries contain errors. Resources for costly land surveys are not available. Landsat can survey these areas easily and record information about soils, minerals, and vegetation.

INVESTIGATING the Technology

1. **Making Inferences** A river in the northeastern United States is polluted. Officials are unable to locate the source of the pollution. Explain how information from Landsat could help.
2. **Thinking Critically** Builders want to develop an area near a river. Environmentalists are fearful of harm to nearby wetlands. How could Landsat give each side a realistic view?

4.1 Homeostasis in Communities

- Communities, populations, and individual organisms occur in areas where biotic or abiotic factors fall within their range of tolerance. Abiotic or biotic factors that define whether or not an organism can survive are limiting factors.
- Primary succession is the development of living communities from bare rock. Secondary succession occurs when communities are disrupted. Left undisturbed, both primary succession and secondary succession will eventually end in a climax community.

4.2 Biomes

- Biomes are large areas that have characteristic climax communities. Aquatic biomes may be marine or fresh water. Estuaries occur at the boundaries of marine and freshwater biomes. Approximately three-quarters of Earth's surface is covered by aquatic biomes, and the vast majority of these are marine communities.
- Terrestrial biomes include tundra, taiga, desert, grassland, deciduous forest, and tropical rain forest. Climate is the major limiting factor for the formation of terrestrial biomes.

Key Terms

Write a sentence that shows your understanding of each of the following terms.

aphotic zone
biome
climax community
desert
estuary
grassland
intertidal zone
limiting factor
permafrost
photic zone
plankton
primary succession
succession
taiga
temperate forest
tropical rain forest
tundra

Understanding Concepts

1. Explain the difference between a photic zone and an aphotic zone.
2. What are the main abiotic factors that limit the extent of the tundra?
3. Describe the sequence of events that might occur if a large slab of rock fell off the side of a mountain in a temperate deciduous forest.
4. What are pioneer species?
5. Why would you expect more bass in a deep lake than a shallow one?
6. How does the climate of the grassland differ from the climate of a temperate forest?
7. What is a climax community?
8. What are some adaptations that would permit a successful existence of bottom-dwelling organisms in the aphotic zone of the ocean?

Using a Graph

9. From the graph below, determine the upper and lower temperature tolerance limits for a species of minnow.

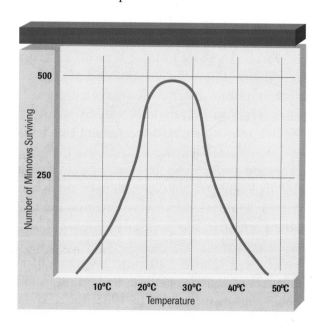

Relating Concepts

10. Make a concept map that relates the following terms and phrases. Supply the appropriate linking words for your map.

    marine biome, lake, river, pond, freshwater biome, sandy shore, estuary, rocky shore, stream, euphotic zone, plankton, aphotic zone

Applying Concepts

11. Why would a cleared area in a tropical rain forest make poor farmland?
12. If you were lost in a desert, list at least two reasons why you should travel only at night.
13. How may agriculture lead to soil erosion in a grassland biome?
14. How might annual fires affect the succession of a temperate deciduous forest ecosystem?
15. Compare the biodiversity of the temperate forest biome with that of a tropical rain forest biome.
16. Many plants of the tropical rain forest are successful houseplants. What characteristics make these plants successful indoors?

History Connection

17. The land area of Surtsey is very small. How might its small size limit the biodiversity of the island?

Thinking Critically

Drawing Conclusions

18. **Biolab** Two students performing an experiment dealing with succession in a puddle added rice only to the water in their experimental cup. The water remained clear. After several days, they saw no signs of living organisms. What might the organisms in their sample have needed as food? Explain.

Making Inferences

19. **Minilab** Suppose you doubled the number of animals and halved the number of plants you enclosed in your soda pop-bottle ecosystem. Would you expect the ecosystem to remain in balance for a longer or shorter period of time? Explain.

Making Predictions

20. **Investigation** What do you think the results would be if you added hay and fertilizer to your bottle of water?

Relating Cause and Effect

21. **Thinking Lab** Look at the photograph of the floor of the tropical rain forest. Notice that there is practically no buildup of dead leaves. What does this tell you about the combined effect of warm, wet air and a tremendous variety of bacteria and fungi?

Connecting to Themes

22. **Systems and Interactions** In what way are limits of tolerance and succession similar?
23. **Energy** Give three reasons why tropical rain forests are considered to be energy-efficient biomes.

CHAPTER 5 Population Biology

Consider the large cities of the world like New York City, Los Angeles, and Tokyo—cities populated by millions of people. Teeming streams of people bustle along the crowded streets, and it seems as if every last inch of space is occupied by stores, offices, and apartments. Now contrast one of these cities with a rural area populated by only a few people. Living in a rural area, you wouldn't experience traffic jams, crowded streets, or polluted air from automobile fumes. On the other hand, you probably wouldn't have your choice of concerts, movies, or sporting events to attend either. Cities can be exciting places, and according to the most recent census, the population of these and other cities is increasing.

Why do populations grow, and how can we predict future population trends? Can we learn how to deal with the problems of larger populations? Many characteristics of population growth are the same for other organisms as they are for humans. By studying the principles of population growth in other species, you may be able to suggest possible answers to these questions.

Concept Check

You may wish to review the following concepts before studying this chapter.
- Chapter 3: population
- Chapter 4: limiting factor, range of tolerance

Chapter Preview

5.1 **Population Dynamics**
Principles of Population Growth
Interactions Among Organisms that Limit Population Size

5.2 **Human Population Growth**
Demographic Trends

Laboratory Activities

Biolab
- Population Growth in *Paramecium*

Minilabs
- How fast do populations use resources?
- How does the population density of weeds in an area affect other plants?

What additional problems might be created if the human population of New York City doubled? How would increased population impact trash removal or electrical supply? Would clean water and food be as easy to obtain?

113

Weeds! *Those pesky little plants are a real problem to gardeners. You've probably observed a scene like this before. What was recently a clear, grass-filled field or lawn is now crowded with hundreds, perhaps thousands, of bright yellow dandelions. Why do these plants appear so quickly and in such large numbers?*

Dandelions are a problem to gardeners, but to population biologists, they are a useful species because they're easy to see and study. By studying organisms such as dandelions, scientists can infer much about human population growth. Do similarities exist between human population growth and population growth in other species? In many ways, the changes that occur in human populations are identical to those of other populations. All populations—whether human, plant, or animal—are in a state of constant change.

Principles of Population Growth

How and why do populations grow? Population growth is defined as the change in the size of a population with time. Scientists use a variety of methods to investigate population growth in organisms, *Figure 5.1.* One method involves placing a microorganism, such as a bacterium or yeast cell, into a tube or bottle of nutrient solution, and observing how rapidly the population grows. Another interesting method involves introducing a plant or animal species into a new environment that contains abundant resources, and observing the population growth of that species. Through studies such as these, scientists have identified clear patterns of how and why populations grow.

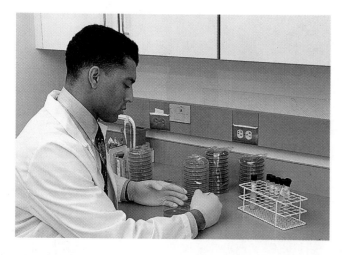

Figure 5.1

Ecologists can study population growth by inoculating a Petri dish containing nutrient medium with a few organisms and watching their growth.

How fast do populations grow?

What's interesting about the growth of a population of living organisms is that it is unlike the growth of some other familiar things. Consider, for example, the growth of a weekly paycheck for an after-school job. Suppose that you are working for a company that pays you $5 per hour. You know that if you work for two hours, you will be paid $10; if you work for four hours, you will be paid $20; if you work for eight hours, you will be paid $40; and so on. If you were to plot this rate of increase on a graph, as shown in *Figure 5.2*, you can see that the result is a steady, linear increase; that is, growth occurs in a straight line.

Populations of organisms do not experience this linear growth. Rather, the resulting graph of a growing population first resembles a J-shaped curve. Whether the population is one of weeds in a field, of frogs in a pond, or of humans in a

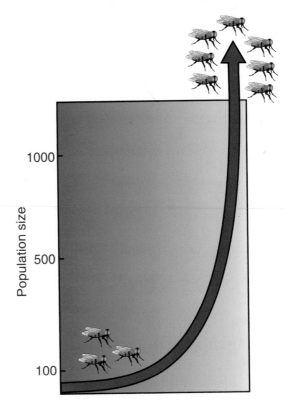

city, the initial increase in the number of organisms is slow because the number of reproducing organisms is small. Soon, however, the rate of population growth increases rapidly because the total number of potentially reproducing organisms increases. This pattern illustrates the exponential nature of population growth. **Exponential growth** of a population of organisms occurs when the number of organisms increases by an ever-increasing rate. Exponential growth, *Figure 5.3*, results in a population explosion.

Figure 5.3

Because they grow exponentially, populations such as houseflies have the potential for explosive growth. One estimate is that a single pair of houseflies could have more than 6 trillion descendants in one summer if all of their eggs hatched and survived to reproduce. This type of population growth results in a J-shaped curve, showing no limit to a population's size.

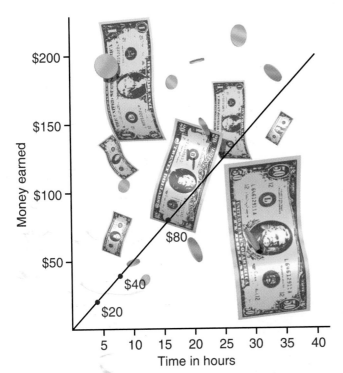

Figure 5.2

This graph shows that the way you earn money at an hourly rate is a straight line. Other examples might include the growth of your weekly allowance or the number of cars produced by an assembly line each month.

BioLab

Paramecium is a unicellular species of protist that lives in freshwater environments. They consume bacteria that live on decaying material in the water. Although a paramecium can reproduce sexually, its usual method of reproduction is by simple cell division from one to two cells. Division occurs when there is a favorable environment with a good supply of food. When conditions such as crowding and decreasing food availability occur, *Paramecium* often begins to reproduce sexually. In this activity, you will determine the population of these protozoans and compare their growth with models of population growth.

Population Growth in *Paramecium*

PREPARATION

Objectives
In this Biolab, you will:
- **Calculate** the population size of *Paramecium*.
- **Compare** growth rate of a *Paramecium* population to models of population growth.
- **Relate** the reproductive strategy of *Paramecium* to environmental factors.

Materials
microscope
microscope slides
coverslips
eyedropper
10-mL graduated cylinder
plastic cup
    rice grains
        Paramecium culture

Safety Precautions
Use care when handling the microscope.

PROCEDURE

1. Measure 10 mL of *Paramecium* culture, and pour into a plastic cup.

2. Add two grains of rice to the cup. Rice grains serve as a food source for the bacteria, which are a food for *Paramecium*.

3. Make a data table like the one shown.

4. Make three wet mount slides using one drop of culture in each. Using the microscope under low power, count the number of *Paramecium* on each slide, and record the number and average of the three slides in your data table.

5. Twenty drops of culture equals 1 mL of liquid, so multiply the average by 20 to find the approximate number of *Paramecium* in 1 mL. Now multiply the number per mL by 10 to estimate the number of *Paramecium* in the culture.

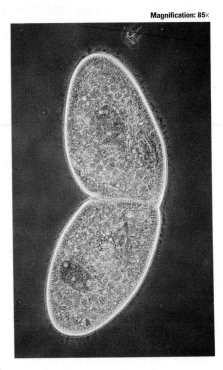

6. Repeat steps 4 and 5 every school day for two weeks.

7. Use the data in your table to graph the population growth of *Paramecium* in your culture.

Number of *Paramecium* in Culture				
	Slide 1	Slide 2	Slide 3	Average	Total in Culture
Mon.					
Tues.					
Wed.					
Thurs.					
Fri.					

Going Further

Application What do you think would happen if you added more rice or hay to the water when the population began to decrease? Do you think dividing the culture and adding more water and rice every few weeks would make a difference? How could you find out?

ANALYZE AND CONCLUDE

1. **Drawing Conclusions** From your data, what reproductive strategy would you say *Paramecium* was using?

2. **Analyzing Data** What was the shape of your graph?

3. **Thinking Critically** When did the population appear to grow most quickly? Can you suggest reasons why the population may have grown most rapidly then?

4. **Interpreting Observations** If you observed *Paramecium* reproducing either sexually or asexually, make a drawing of your observation.

How fast do populations use resources?

Fruit flies and similar insects have rapid life-history patterns. Insects are ideal organisms for population studies because they reproduce quickly and are easy to count. In this activity, you will observe the growth of insect populations as they exploit a food supply.

Materials
jar
half of a banana

Procedure

1. Place half of a banana in a jar and allow it to sit outside in a warm, shaded area.

2. Observe the jar each day for insects that feed off the banana.

3. Record the type of insect and the numbers of each type that appear in the jar. Be sure to record this information each day. If you do not know the name of the insect, draw it so that someone else will be able to recognize it.

Analysis

1. What types of insects were attracted to your jar?

2. How soon did insects appear, and how long were insects present?

3. Why are insects considered to have a rapid life-history pattern?

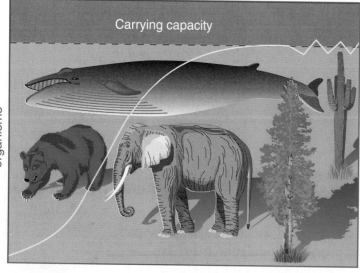

Carrying capacity

Number of individual organisms

Time

Figure 5.4

Populations may follow an S-shaped curve. There is an initial slow-growth phase, a period of exponential growth, and a plateau where the number of organisms the environment can support is reached.

Limits of the environment

Can a population of organisms grow indefinitely? What prevents the world from being overrun with all kinds of living things? Through population experiments, scientists have found that, fortunately, population size does have a limit. Eventually, populations do have limiting factors in their environment, such as availability of food and space. This leveling off of population size results in an S-shaped growth curve. The number of organisms of a population a particular environment can support over an indefinite period of time is known as its **carrying capacity.** Carrying capacity is often represented by the letter K. When populations are under the carrying capacity of a particular environment, births will exceed deaths until the carrying capacity is reached. If the population temporarily overshoots the carrying capacity, deaths will exceed births until population levels are once again at carrying capacity. *Figure 5.4* shows the S-shaped curve followed by populations.

Types of Population Growth

In nature, many animal and plant populations remain in equilibrium. However, this is not true for other organisms. Why, for example, does it seem like mosquitoes are more numerous at some times of the year than others? Why don't all populations of organisms reach the carrying capacity and become stable? To answer these questions, population biologists study the most important factor that determines population growth—an organism's reproductive pattern.

Populations of organisms usually follow one of two growth patterns. Some species, such as mosquitoes, reproduce very rapidly and produce many offspring in a short period of time. Other species such as elephants have a slow rate of reproduction and produce relatively few young over their lifetimes. What causes species to have either one of these life-history strategies?

The kind of strategy a species has depends mainly on environmental conditions. For example, species such as mosquitoes are successful in environments that are unpredictable and change rapidly. This life-history pattern is found in organisms with similar adaptations. Typically, these organisms have a small body size, mature rapidly, reproduce early, and have a short life span. Populations of these organisms increase rapidly and then decline rapidly as environmental conditions suddenly change and become unsuitable. The small surviving population will begin reproducing exponentially when conditions are again favorable.

Species of organisms such as elephants that live in more stable environmental conditions have a different life-history pattern. Elephants, bears, whales, and long-lived plants such as cacti and California redwoods are large in size, and reproduce and mature slowly. These organisms maintain population sizes near the carrying capacity of the environment.

Theoretically, any population could follow either a rapid or slow life-history pattern depending on the environmental conditions. Under uncrowded conditions, such as the pioneer stage in succession, rapid population growth seems to be effective. The organisms

ThinkingLab Design an Experiment

How can you control an organism's life-history pattern?

Populations with a rapid life-history pattern, such as minnows, tend to grow exponentially until they reach the limits of the environment. This results in most or all of the population dying. But if organisms are given unlimited resources and environmental factors are controlled, populations can reach a maximum reproduction rate. Minnows give no parental care to their offspring, and female minnows may bear as many as 100 offspring in their lifetimes.

Analysis

Minnows will grow and reproduce in an aquarium or other large container of water. If you feed them commercial fish food, bits of crusty bread, or small insects, most minnows will live for several months and begin reproducing when they are only a month old.

Thinking Critically

Hypothesize how you can get minnows to achieve a maximum reproductive rate. Consider the environmental factors you will need to control in your aquarium.

shown in *Figure 5.5* represent both types of life-history patterns. Which of these organisms would be most successful in a rapidly changing environment?

Figure 5.5

Wild mustard plants taking over an abandoned field represent a species with a rapid life-history pattern. These plants also take advantage of any waste area, such as a vacant lot or a roadside. Organisms that have a slow life-history pattern, such as these Canada geese, provide much parental care for their young. They stay with them from the time they hatch through their migration to wintering grounds.

Mussel Beach

In 1989, the residents of Monroe, Michigan, made a startling discovery—zebras were living in the city's water pipes! Not just a few zebras, either, but tens of thousands—enough zebras, in fact, to shut down the city's water supply! Needless to say, normal activity in Monroe came to a screeching halt.

Bothersome bivalves   The zebras that shut down schools, businesses, and industries in Monroe were not the striped mammals that roam the African plains. They were mussels, little bivalve mollusks 5 cm and smaller that take their name from the distinctive dark-brown stripes that band their light-tan shells. Marine biologists speculate that the first zebra mussels arrived in the Great Lakes as stowaways on ships from Europe navigating the St. Lawrence Seaway.

Unlike their edible cousins, zebra mussels are not considered a viable food source for humans. Their chief occupation seems to be clogging water inlet pipes in the rivers and lakes.

The mussels attach themselves to pipes and other surfaces by means of sticky, elastic strands called byssal fibers. Once attached, the mussels take advantage of strong currents to filter passing phytoplankton through their ciliated gills. The Great Lakes are now home to billions of zebras, and the mussels' voracious appetites leave very little food for other plankton-reliant species.

A population explosion   An adult female can produce up to 50 000 eggs each year. In the summer of 1989, officials inspecting the pipes and water-treatment machinery in Monroe counted about 1350 mussels per square yard. Before the year was over, the count per square yard had reached 600 000!

Controlling the zebra mussel population is proving no easy task. Toxic chemicals such as chlorine will kill both adults and larvae, but are equally fatal to other harmless, lake-dwelling species. Removing the mussels manually is a slow and largely ineffective task.

Perhaps the most promising solution is biological deception. Zebra mussels spawn when they detect odors coming from phytoplankton. The larvae need this rich source of food to survive to adulthood.

Thinking Critically

If biologists can successfully simulate that same odor, and release it when phytoplankton are actually scarce, what is likely to happen to the zebra mussel population? Might this kind of biological interference be dangerous?

Environmental limits to population growth

Limiting factors, you remember from Chapter 4, are biotic or abiotic factors that determine whether or not an organism can live in a particular environment. Limiting factors also regulate the size of a population. Limited food supply, extreme temperatures, and even storms can affect population size. Ecologists have recognized two kinds of limiting factors: density-dependent and density-independent factors.

Density-dependent factors include disease, competition, and parasites, which have an increasing effect as the population increases in size, as seen in *Figure 5.6*. Disease, for example, spreads more quickly in a population whose members live close together than in smaller populations whose members live farther from each other. In very dense populations, disease may wipe out an

Figure 5.6

Corn smut is a fungus that produces large, deformed growths on the ears of corn. To prevent it from spreading through a cornfield, affected plants must be burned or buried before the fungi reproduces.

entire population. In crops such as corn or soybeans, in which large numbers of the same plant are grown together, a disease will spread rapidly throughout the whole crop. In less-dense populations, fewer individuals may be affected.

Density-independent factors affect all populations, regardless of their density. Most density-independent factors are abiotic factors such as temperature, storms, floods, drought, and habitat disruption, *Figure 5.7.* No matter how many earthworms live in a field, they will all drown in a flood.

Figure 5.7

Populations are affected by both density-dependent and density-independent factors.

▲ **Hurricane Andrew, which hit south Florida in 1992, did extensive damage to both heavily populated areas and less populated ones.**

➤ As a population of foxes increases, all foxes in the population begin to compete for the available food, such as mice and rabbits. When food, a density-dependent factor, is in short supply, young foxes begin to die, and population growth slows. Problems for one population of organisms also can affect populations of other organisms. What do you think will happen to populations of mice and rabbits in the foxes' area?

MiniLab

How does the population density of weeds in an area affect other plants?

An ecological study of your front yard, the school grounds, or a similar area can reveal the density of organisms in the area. Population density is the measure of the number of individuals in a given area. When population densities of weeds are high, it tends to affect populations of other plants. In this lab, you will calculate the average density of weeds in an area, and observe how this affects other plants.

Materials
string
small sticks
meterstick or sewing tape measure

Procedure
1. Identify a weed plant in your study area.
2. Randomly choose three plots. For each plot, place four sticks in the ground so that they form one square meter each.
3. Mark off your plots with string.
4. Count and record the number of weed plants in each square.
5. Calculate the average number of weed plants in your three plots.

Analysis
1. What was the average density of weeds in the area you measured?
2. Did the density of weed plants in your plots affect the populations of other plants in the same plot? Explain your answer.
3. Why do you suppose a high density of weeds can affect the way other plants grow?

Interactions Among Organisms that Limit Population Size

Population sizes are not limited only by environmental factors, but are also controlled by various interactions among organisms that share a community.

Predation affects population size

Predation of one organism by another is important for the proper functioning of a community. Recall that energy is moved through a community by way of food chains and food webs. Predation ensures the continuation of the flow of energy, but it may also be a limiting factor of population size.

Populations of predators and prey experience changes in their numbers over a period of years. Many predator-prey relationships show a cycle of population increases and decreases over time. The rise and fall of populations of snowshoe hare and lynx over a 90-year period is shown in *Figure 5.8.* The populations of both animals rise and fall almost together. Does this mean that the lynx overhunted their prey, losing their food supply? Or, did the food supply of the snowshoe hare become less abundant, causing their numbers to decline? While the graph can't indicate what factors caused the population fluctuations, it is clear that most populations are controlled in some way by predators.

Prey-predator relationships are important for the health of natural populations. Usually, in prey populations, the young, old, or unfit members are caught. This leaves the strongest and most adapted to reproduce and continue the population.

Figure 5.8

The data in this graph came from the number of hare and lynx pelts sold to the Hudson Bay Company in northern Canada. Notice that as the number of hares increased, so did the number of lynx. The number of hares is regulated by their food supply. The number of lynx is controlled by their food supply also—the number of hares.

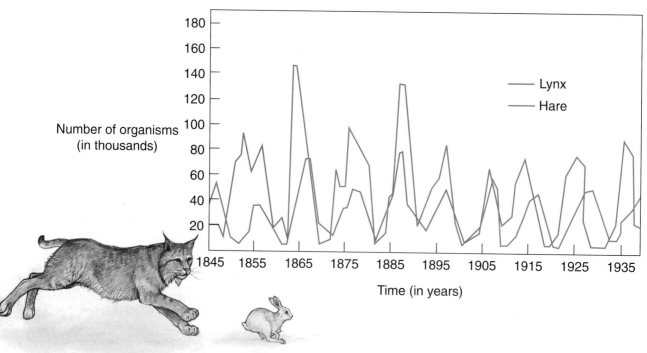

The effects of competition

Organisms within a population constantly compete for resources. When population numbers are low, resources are plentiful. However, as population size increases, competition for resources such as food, water, and territory can be fierce. Competition is a density-dependent factor. When only a few individuals compete for resources, no problem arises. When population size increases to the point where demand for these resources exceeds the supply, the population size decreases.

The effects of crowding and stress

When populations of organisms become crowded, individuals may exhibit stress. The factors that create stress are not well understood, but the effects have been documented experimentally in populations of rats and mice, *Figure 5.9*. As populations increase in size, individual animals begin to exhibit a variety of symptoms, including aggression, decrease in parental care, decreased fertility, and decreased resistance to disease. All of these symptoms can lead to a decrease in population size.

Section Review

Understanding Concepts

1. How are the graphs of exponential growth and linear growth different?
2. Explain how rapid and slow life-history patterns differ.
3. Describe how density-dependent and density-independent factors regulate population growth.

Thinking Critically

4. Discuss how an environmental factor such as a hurricane can affect a population of organisms.

Skill Review

5. **Making and Using Graphs** Graph the following population growth for the unknown organism, and state whether the organism has a rapid or slow life-history pattern. For more help, refer to Organizing Information in the *Skill Handbook*.

Population Size of Unknown Organism				
Year	Spring	Summer	Autumn	Winter
1992	564	14 598	25 762	127
1993	750	16 422	42 511	102
1994	365	14 106	36 562	136

5.2 Human Population Growth

*D*oes Earth have a carrying capacity for the human population? How many people can live on Earth? No one knows how many people Earth can support, and it is presently impossible to tell when the human population will stop growing. However, an increasing number of scientists suggest that food production will not always keep pace with the population increase.

Demographic Trends

A good way to predict the future of the human population is to look at past population trends. For example, are there observable patterns in the growth of populations? That is, are there any similarities among the population growths of different countries—similarities that might help scientists predict, and therefore control, future population catastrophes? As you have seen, populations tend to increase until the environment cannot support any additional growth. The study of population growth characteristics is the subject of **demography.** Demographers study such population characteristics as growth rate, age structure, and geographic distribution.

What is the history of population growth for humans? Although local human populations often show fluctuations, the worldwide human population has increased exponentially over the past several hundred years, as shown in *Figure 5.10.* Unlike other organisms, humans are able to reduce environmental effects by eliminating competing organisms, increasing food production, and controlling disease organisms.

Effects of birthrates and death rates

How can you tell if a population is growing? A population's growth rate is the difference between the birthrate and the death rate. In many industrialized countries, such as the United States, declining death rates have a greater effect on total population

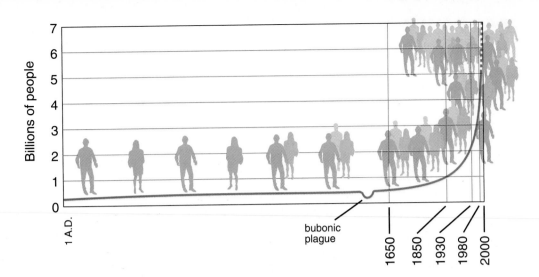

Figure 5.10

Ten thousand years ago, approximately 10 million people inhabited Earth. Today, there are more than 5 billion, and scientists estimate that by the year 2000, there will be more than 6 billion people on Earth.

growth than increasing birthrates. For example, in the United States, life expectancy increases almost every year. This means that you probably will live slightly longer than students who are presently in college. Interestingly, although people in the United States are living longer, the fertility rate is decreasing. This is because people are waiting until their thirties and forties to have children.

Childbirth in women in their twenties was more common just a few decades ago. Today's families also have fewer children than they did in previous decades.

What are the birthrates and death rates of some countries around the world? Look at *Table 5.1* to see the birthrate, death rate, and fertility rate of some rapidly growing and slower-growing countries. Fertility is the

Table 5.1				
	Birthrate (per 1000)	Death Rate (per 1000)	Fertility (per woman)	Population Increase (percent)
Rapidly Growing Countries				
Gaza	50	7	7.0	4.3
Iraq	46	7	7.3	3.9
Kenya	46	7	6.7	3.8
Rwanda	51	16	8.5	3.7
Uganda	52	17	7.4	3.6
Slowly Growing Countries				
Hungary	12	12	1.8	-0.2
Germany	12	12	1.5	0.0
Italy	10	9	1.3	0.1
Sweden	14	11	2.0	0.2

Source: 1990 World Population Data Sheet (Washington D.C.: Population Reference Bureau, Inc., April, 1990).

How many are there?

Every ten years, the U.S. Government undertakes one of the most complicated counting jobs in the world. Every decade, as called for in the Constitution, a census is carried out. The goal is to count every individual person residing in the country and gather other information that measures progress or helps anticipate future needs.

The people who study populations and their characteristics and needs are mathematicians called demographers. Demographers can study any type of population—plant, animal, or human.

Using demographics   One group of scientists that uses demographics as a foundation for their work is wildlife managers. They keep track of populations on a regular basis and use the information to make decisions in managing the size of those populations. One of the most important functions of wildlife managers is to determine the carrying capacity of the lands under their care. Another responsibility may be to protect populations from predators, people, or other factors that are stressing them. Wildlife managers gather facts about weather conditions, water availability, habitat, and many other conditions in areas in their care. Then they use these facts to balance one with another to produce the healthiest conditions possible for all populations.

They also use these facts to see that the needs of all are being met. For example, orchard owners want to control populations of deer feeding on their trees. Hunters want more rabbits. It is the goal of the wildlife manager to work out a population size to benefit all. Wildlife managers try to balance and control wild populations of organisms by various means.

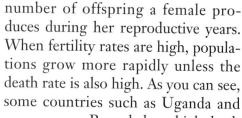

CONNECTION TO   Biology

In many areas of the United States, deer populations have increased to the point where many starve each winter due to a lack of food. It has been suggested that surplus farm products such as corn or other grains could be used to save these animals. Discuss whether this is or is not a good solution to the problem.

number of offspring a female produces during her reproductive years. When fertility rates are high, populations grow more rapidly unless the death rate is also high. As you can see, some countries such as Uganda and Rwanda have high death rates among children because of disease and malnutrition. However, both countries have extremely high birthrates, and they are both growing rapidly. Some other countries such as Sweden and Italy have low death rates, but their birthrates are also low, so these countries grow slowly, if at all.

The birthrate, death rate, and fertility rate of a country provide clues to that country's rate of population growth. As you can see, different combinations of birthrates and death rates have different effects on populations.

Does age affect population growth?

Imagine a country filled mostly with teenagers! Will it make a difference to population growth if the largest proportion of the population is in one age group? In order to make predictions about populations of the future, demographers must know the age structure of a population. **Age structure** refers to the proportions of a population that are either in their pre-reproductive years, their reproductive years, or their post-reproductive years. By knowing the age structure, you can tell if a population is growing rapidly, growing slowly, or not growing at all, as shown in *Figure 5.11.* What would happen to the population growth rate if you and other students your age from around the country lived to be 100? You can see from studying the age structure distribution that if most of the population is living longer, then the overall population will be growing slowly, if at all, depending on the proportion of individuals in their pre-reproductive and reproductive years.

Mobility has an effect on population size

Unlike most organisms, humans can move in and out of different communities. The effects of human migrations can make it difficult for a demographer to make predictions, but patterns do exist. Movement of individuals into a population is **immigration.**

Figure 5.11

Notice that in a rapidly developing country such as Mexico, the large number of individuals in their pre-reproductive years will add significantly to the population when they reach reproductive age. Populations that are not growing, such as in Germany, have an almost even distribution of ages among the population.

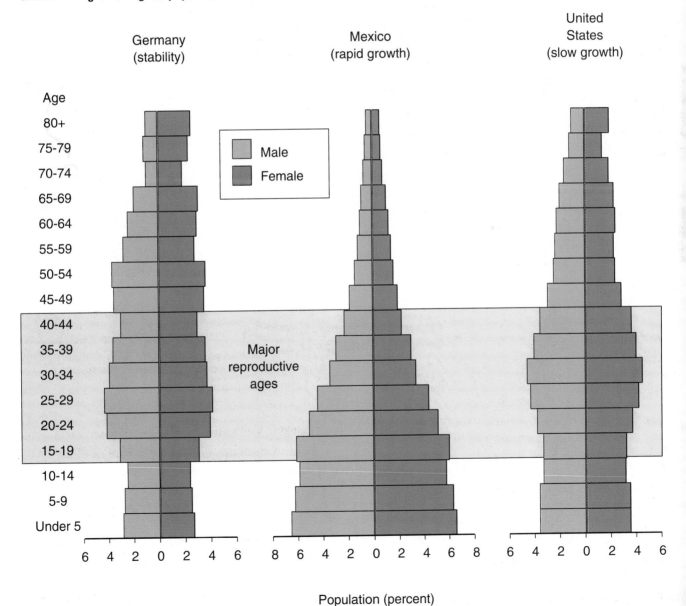

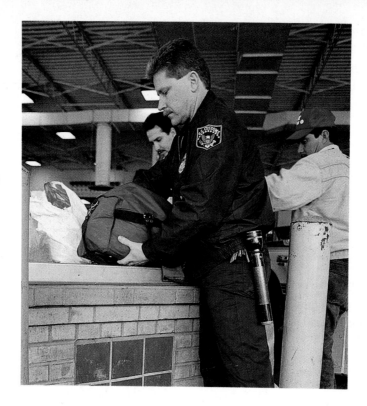

Figure 5.12

More people immigrated to the United States during the past 165 years than to any other country in the world, mostly from European countries. How do you think these countries were affected by this emigration of people?

Movement from a population is **emigration.** Obviously, movement of people between countries has no effect on total world population, but it does affect national population growth rates. Local populations can also feel effects of a moving population. Many suburbs of large cities are expanding rapidly. This places stress on schools, roads, and police and fire services. Can you think of any other problems caused by suburban growth?

Connecting Ideas

In this chapter, you investigated some characteristics of populations and learned how environmental factors place limits on population growth. Unlike other organisms, humans can manipulate and regulate some of these biotic and abiotic factors, and this has resulted in a large increase in the world's human population.

Many ecologists suggest that the pollution observed today is directly related to the world's increasing human population. If human population growth continues at the present pace, additional environmental problems will result from such things as a diminishing energy supply and increasing pollution.

Section Review

Understanding Concepts
1. What characteristics of populations do demographers study?
2. What clues can an age structure graph provide about a country's population growth?
3. Discuss some possible problems for local populations caused by immigration and emigration of people.

Thinking Critically
4. Hungary has the world's slowest-growing human population. Which age groups do you think make up the largest portions of Hungary's population?

Skill Review
5. **Making and Using Graphs** Construct a bar graph showing the age structure of Kenya using the following data: pre-reproductive years (0-14)—42 percent; reproductive years (15-44)—39 percent; post-reproductive years (45-85+)—19 percent. For more help, refer to Organizing Information in the *Skill Handbook.*

The High Cost of Living Longer

Staying Young Longer

World Dispatch   April 10, 1993

In an internist's office in the year 2020, a young physician writes out several prescriptions for this patient, a slender and energetic man of 75. He enjoys an active life, and is anxious to get the results of his physical. "Well, your hormone profile is holding steady at about the 25-year-old level," the doctor reassures him, "so we'll continue the same therapy. And keep your calories down. Have a nice day."

Healthy lifestyles and improved health care have already helped increase America's elderly population so that centenarians have become the nation's fastest-growing age group.

The war on aging Researchers in genetics, endocrinology, and molecular biology lead the fight in the war against aging. The most significant of their findings has been to discover ways to manipulate genetic mutation and use hormones. Geneticists have discovered a genetic clock that measures the rate of aging. They have also identified genes that limit our life span.

Endocrinologists are working to create antiaging products that both extend and improve life by attacking the diseases common to old age. One of the most effective antiaging treatments is an adrenal hormone that is prevalent during youth but diminishes as we age.

Different Viewpoints

The health-care crisis As the elderly population continues to grow, so will the number of frail or disabled. Our bodies, like those of all other animals, cannot last forever. One health-care study dismally predicts a six-fold rise in Medicare costs by the year 2040; dementia will become the leading affliction in more than 28 percent of the population; and more than 800 000 hip fractures will occur annually by the year 2040. The advances made to prolong life become almost cruel if medical researchers don't also address the disabling diseases that extended longevity brings.

An optimistic prognosis Some health-care experts point to two optimistic trends: a slowing of the increases in longevity and a delay in the age of disability. Many nonfatal diseases can be prevented by weight control and exercise. Fractures can also be prevented by exercise, hormone replacement, and proper diet.

Current trends that help the elderly to manage old age better include healthy lifestyles, which can preserve independence as well as fitness.

INVESTIGATING the Issue

1. **Applying Concepts** Research the state of elderly health care in the United States today. Include the care and treatment of diseases such as Alzheimer's disease, osteoarthritis, and osteoporosis.
2. **Writing About Biology** Design a survey describing your opinions on research efforts to extend longevity. If given the opportunity to live for 150 to 200 years, would you do it?

Reviewing Main Ideas

5.1 Population Dynamics

- Populations grow exponentially until they reach the carrying capacity of the environment. Organisms have one of two life-history patterns: slow growth that tends to approach the carrying capacity with minor fluctuations, and rapid growth that tends to grow exponentially and then experience massive diebacks.
- Density-dependent factors such as disease and food supply, and density-independent factors such as weather, have effects on population size. Interactions among organisms such as predation, competition, stress, and crowding also limit population size.

5.2 Human Population Growth

- Demography is the study of population characteristics such as growth rate, age structure, and movement of individuals. Birthrate, death rate, and fertility differ considerably among different countries, resulting in uneven population growth patterns across the world.

Key Terms

Write a sentence that shows your understanding of each of the following terms.

age structure
carrying capacity
demography
density-dependent
  factor

density-independent
  factor
emigration
exponential growth
immigration

Understanding Concepts

1. Explain why two different environments can have different carrying capacities for the same type of organism.
2. An environment has food resources that are abundant for only a short time during the year. Which life-history pattern would be most efficient in this environment? Why?
3. Another environment has a steady supply of food and other resources available throughout the year. Which life-history pattern would be most efficient in this environment? Why?
4. How does the availability of resources relate to the carrying capacity of a particular environment?
5. Describe some possible effects of disease in a population of organisms.
6. What is the relationship between fertility and age structure in a population?
7. Can hunting be beneficial for a population of oganisms?

Using a Graph

8. For several years, a scientist studied the population of a fruit-eating insect in a tropical rain forest. Based on the graph, what conclusion can you make about the insect's life-history pattern? What environmental factors may have caused the population sizes seen in the graph? Predict what may happen next.

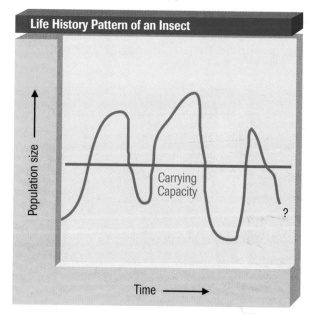

Life History Pattern of an Insect

Population size / Time / Carrying Capacity

Relating Concepts

9. Make a concept map that relates the following terms and phrases. Supply the appropriate linking words for your map.

   environmental factors, demography, age structure, carrying capacity, emigration, population growth, exponential growth, predation, crowding, immigration

Applying Concepts

10. Sweden and Norway have populations that are in equilibrium. Discuss some advantages of a stable population size.
11. Which environmental factors would most affect the populations of developing countries?
12. As human populations grow, what might happen to the populations of other species? Discuss some possible effects.

Math Connection

13. The population of a small town is beginning to grow rapidly due to a new highway that provides access to a larger city. Who in this community would be most interested in the changing demographics of the area? How would they use information supplied to them by demographers to provide services for the growing population?

Thinking Critically

Making Predictions

14. **Biolab** What might happen to the *Paramecium* population if another bacteria-eating protist were added to the culture?

Making Inferences

15. **Minilab** Why are rapid life-history species, such as mosquitoes and some weeds, very successful organisms even though they often experience massive population declines?

Interpreting Data

16. **Minilab** The following bar graph was presented by groups of students to depict the data they obtained while measuring the number of dandelions on the school grounds. Each group measured a different area. Assume that the school grounds have 21 950 square meters planted in grass. What is the approximate population size of dandelions on the school grounds?

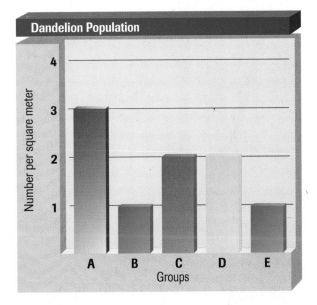

Dandelion Population

Connecting to Themes

17. **Energy** How are the energy resources of fossil fuels, such as petroleum and coal, affected by the increasing world population?
18. **The Nature of Science** How might technology increase the carrying capacity of the world in the future?
19. **Systems and Interactions** How can predation help keep a population at the carrying capacity of the environment?

6 Wise Use of Our Resources

All living things have a basic set of needs that must be met. If you were to make a list of basic human needs, what would be on your list? You need food, water, clothing, shelter, and oxygen. How do human needs and demands affect the environment? We burn oil, gas, and coal to fuel power plants and cars and to provide electricity. We dam rivers for irrigation water. We dig deep into the ground to mine metals and minerals. We clear forests to build shopping centers and housing developments.

Does our use of Earth's natural resources always have to be destructive? In this chapter, you'll look more closely at the natural resources needed to support modern human life. You will investigate how humans have influenced Earth's ecosystems, and explore some of the methods we already use to maintain a healthy environment.

Clearing forests for new buildings and mining the earth for minerals, coal, and oil are not the only human activities that have an effect on our natural resources. Many commonly used products are unnecessarily overpackaged. Single-serving packages, layers of plastic wrap, and nonrecyclable materials contribute to an overuse of natural resources. Products packaged in recyclable materials or biodegradable materials place less stress on resources and the environment.

6.1 Effects of Human Activities on Our Resources

Section Preview

Objectives

Compare and contrast renewable and nonrenewable resources.

Determine the effects of increasing demand and decreasing supply on natural resources.

Key Terms

natural resource
renewable resource
nonrenewable resource
fossil fuel
extinction
threatened species
endangered species

The sight of a quiet, misty lake, a snow-capped mountain, the geraniums growing in a flower box on your windowsill, or a squirrel chewing on a pine cone in the park usually make people stop to admire the natural resources of Earth. Our survival and the survival of all living things on Earth is dependent on our wise use of these resources.

Earth's Resources

Whenever you ride a bus, turn on a light, read a book, or eat lunch, you are using a **natural resource.** A natural resource is any part of the natural environment used by humans for their benefit. Natural resources include soil, water, crops, wildlife, oil, gas, and minerals, as shown in *Figure 6.1.* The gasoline that powers a bus or automobile, and the plastics and metals the vehicle is made of, are examples of natural resources. So is the wood pulp used to make the paper for this book, and the coal or natural gas that supplied the electricity for the printing press. Wilderness and recreation areas such as beaches, mountains, forests, and lakes are also natural resources.

Renewable resources

Why don't we run out of oxygen to breathe or water to drink? Many organisms produce oxygen during a process called photosynthesis, which constantly replenishes the oxygen consumed by all aerobic organisms. The physical materials on which life depends are limited to what is presently on Earth. You learned in Chapter 3 that water is naturally recycled from the atmosphere to the surface of Earth, through food webs and back to the atmosphere. Nitrogen, carbon, and other essential substances are cycled in a similar manner. A natural resource that is replaced or recycled by natural processes is called a **renewable resource.** Other examples of renewable resources include plants, animals, food crops, sunlight, and soil.

Figure 6.1

Different regions are rich in different types of natural resources.

▲ Corn you eat for dinner may have been grown on a farm in Kansas.

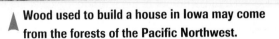

▲ Wood used to build a house in Iowa may come from the forests of the Pacific Northwest.

▲ If you live in Los Angeles or some other part of the arid Southwest, the water that runs from your kitchen faucet may come from a river hundreds of miles away.

▲ Many of the resources you use every day are imported from outside the United States. For example, the aluminum in your soft drink can may have come from a mine on the island of Jamaica.

Nonrenewable resources

What does the aluminum in a soda can have in common with the plastic used to make grocery bags or the gasoline that fuels a car? These are examples of nonrenewable resources. A **nonrenewable resource** is available in only limited amounts and is not replaced or recycled by natural processes. Metals—including aluminum, tin, iron, silver, uranium, gold, and even the copper used to make wires and coins—are nonrenewable. Some minerals such as phosphorus, which is essential for plant growth, are recycled so slowly in the natural environment that they are considered nonrenewable. It is estimated to take between 500 and 1000 years for a 2.5-cm-deep layer of topsoil to develop from decomposed plant material. Because it would take several generations to replace, topsoil is also described as a nonrenewable resource.

Fossil fuels and petroleum products

You are probably using a nonrenewable resource whenever you watch TV or use some other electrical appliance. Electricity is produced in power plants, many of which are

What do we do with millions of tires?

They don't decay, are practically indestructible, are piled up in heaps six stories high, and can be found submerged in ponds more than 100 feet deep. Of the 240 million tires discarded in the U.S. every year, about 11 percent are burned as fuel and 15 percent are retreaded. The remainder are simply dumped. Each tire has the energy potential of 2.5 gallons of oil. This potential became obvious when a tire dump in Virginia burned for 275 days. The 690 000 gallons of oil that oozed out of the burning tires were collected and sold.

Tires as fuel In 1982, a tire-collection company realized the energy potential in that 2.5 gallons of oil and began selling tire chips to paper mills to be burned as fuel. Even though tires produce about the same polluting emissions as coal, they contain more energy than the same weight of bituminous coal. Only a few power plants in the United States sell electricity generated by the combustion of tires due to regulatory concerns and low profitability. Whole or shredded tires are ignited at high temperatures, and the heat from their combustion is used to produce steam that powers turbines. The process is controlled by computers, which also monitor emissions. The ash components are used as gravel and to enrich soil.

Tires and cement Tires are also used as a source of direct heat in the manufacture of cement. Burning tires circulate with the rock while supplying heat for the roasting process. While only a few cement plants currently use tires, the process has high economic potential because the cement-producing industry could use 3 billion tires per year in this way, with no significant change in emissions.

Thinking Critically

What other uses might there be for waste tires besides using them for fuel? What steps might be taken to make using tires for fuel more economically and environmentally acceptable?

Figure 6.2

Fossil fuel deposits are always being formed, but humans are consuming them far faster than they can be replaced.

fueled by coal, oil, or natural gas. These **fossil fuels** are substances made up of the remains of organisms that have been buried underground for millions of years. Fossil fuels are nonrenewable resources because they are produced only over long spans of time. *Figure 6.2* illustrates how coal is mined.

Extinct organisms

A living population of organisms is a renewable resource; new individuals are born as older ones die. But what happens when all members of a species die? No individuals are left to reproduce, and the species is lost forever. Have you ever seen a flock of passenger pigeons? How about a woodland caribou, relic leopard frog, or Louisiana prairie vole? Unless you've seen a photograph or a stuffed museum specimen, the answer is no. These animals are extinct. **Extinction** is the disappearance of a species when the last of its members dies. In the last 20 years, almost 30 species of plants and animals living in the United States have become extinct. Although extinction can occur as a result of natural processes, humans have been responsible for the extinction of many species.

Most extinctions come about because of destruction of the natural habitat of a species. Because the environment is always changing, there will always be species in danger of extinction. However, when you look at the increasing number of species that are becoming extinct every year, an unpleasant picture emerges. It has been estimated that, during the extinction of the dinosaurs, about one species was lost every year. Some scientists hypothesize that today, species are being lost at the rate of one per day. Human activities—including hunting, the building of cities and housing developments, and the destruction of forests to create farmland—are primarily responsible for this high extinction rate.

When the population of a species begins declining rapidly, the species is said to be a **threatened species.** African elephants, for example, are listed as a threatened species. In the early 1970s, the wild elephant population numbered about 3 million. Twenty years later, they numbered only about 700 000. Elephants have traditionally been hunted for their ivory tusks, which are used to make jewelry and ornamental carvings. Many countries in the world have banned the importation and sale of ivory, and this has eased some of the hunting pressure on the elephant population. *Figure 6.3* describes some threatened plant and animal species under protection in the United States.

Figure 6.3

Several plant and animal species in the United States are threatened.

▶ **The bald eagle is threatened in Michigan, Minnesota, Oregon, Washington, and Wisconsin. It is endangered in the remaining states except Alaska.**

▲ **Wildlife experts classify loggerhead turtles as well as more than 40 other turtle species as threatened.**

▶ **Sea otters have been hunted for centuries for their fur.**

◀ **Grizzly bears are threatened in every state in the United States except Alaska.**

Figure 6.4

A species becomes endangered when its numbers are so low that they are in danger of extinction. In the United States, scientists have developed programs designed to save endangered species.

A species is considered to be an **endangered species** when its numbers become so low that extinction is possible. In Africa, the black rhinoceros has become an endangered species. Like the African elephant, the black rhino is protected in national parks and preserves in several African countries. However, poachers continue to hunt and kill these animals for their horns. Rhinoceros horns are composed of matted hair rather than bone or ivory. In the Middle East, the horns are carved into handles for ceremonial daggers. In parts of Asia, they are used to make traditional medicines. *Figure 6.4* shows several endangered species found in the United States. In the United States, endangered species receive protection by law. They cannot be hunted or trapped, and their habitats are protected.

▼ Manatees, sometimes called "sea cows," are endangered due to loss of habitat and injury from barges and motorboats.

▼ In 1850, approximately 20 million bison roamed the western plains. Today, approximately 15 000 of these animals live on fenced game preserves.

◀ The Florida panther, *Felix concolor coryi,* is a subspecies of the mountain lion population that once roamed most of North America.

► Urban growth has destroyed much of the California condor's habitat. To protect the species, all California condors were captured.

When Demand Exceeds Supply

In Chapter 4, you learned that the human population is growing at an increasingly rapid pace. The more people that live on Earth, the more demand there is for food, water, living space, clothing, transportation, and other essentials. How long will Earth's resources last? At some point, will we start running out of the things we need to live?

When demand for a resource exceeds the available supply, competition for that resource increases. As a result, the cost of that resource goes up. For example, the price of housing in prosperous metropolitan areas has increased drastically in the last 30 years. A house in Levittown, New York, a suburb of New York City, that cost $8000 when it was built in 1950 is now valued at $200 000. A similar home in a small, midwestern city might cost about half of that. Increasing demand for resources means not only higher prices; it may also mean that people simply have to do without, as shown in *Figure 6.5.*

Figure 6.5

The demand for resources in over-populated areas can have drastic effects on people's lives.

▲ The need for living space so far exceeds supply in some parts of the world that portions of the population live in makeshift homes of cardboard, plastic, or other discarded materials. Some people have no homes at all.

▲ In parts of Africa, drought and political strife have created food shortages so severe that thousands of people have starved.

Section Review

Understanding Concepts
1. What is the difference between a renewable and a nonrenewable resource?
2. How are renewable resources similar to recyclable, nonrenewable resources?
3. Explain some possible effects of increasing demand and decreasing supply of a nonrenewable resource.

Thinking Critically
4. Many people value elephant tusks as decorative objects in their homes. How would this affect the growth rate of the endangered elephant population in Africa?

Skill Review
5. **Observing and Inferring** Infer what might happen to the price of an electrical component if a new metal that is abundant is substituted for a rare metal that has been used exclusively in the manufacture of the electrical component. For more help, refer to Thinking Critically in the *Skill Handbook*.

Section Preview

Objectives

Identify major sources of air, water, and land pollution.

Differentiate between biodegradable and nonbiodegradable pollutants.

Distinguish between methods of preservation and conservation.

Key Terms

pollution
particulate
smog
acid precipitation
ozone layer
greenhouse effect
groundwater
biodegradable
nonbiodegradable
preservation
conservation

All living things produce waste, but humans produce a huge variety of wastes from many different activities. The wastes produced are solids, liquids, and gases and come from homes, factories, automobiles, and numerous other sources. The average American produces so much solid waste each year that we are running out of places to put it.

Effects of Pollution on the Biosphere

All living organisms produce waste products, which are usually recycled by natural processes. What happens when so much waste is produced that natural recycling processes are overloaded? Wastes build up faster than they can be broken down. **Pollution** is the contamination of any part of the environment—air, water, or land—by an excess of waste materials. A pollutant is a waste product that causes pollution. *Figure 6.6* describes how nitrogen, a nutrient essential to all living organisms, can become a pollutant.

Figure 6.6

Cattle manure contains nitrogen and other nutrients that make it valuable as a plant fertilizer. But too much of a good thing can cause pollution problems.

▶ The cattle on this crowded feed lot produce more waste than the decomposers in the soil can handle. As a result, much of the waste is washed away by rainfall.

▶ The large amounts of nitrogen in runoff from the feed lot stimulate the rapid growth of algae in waterways downstream. This lush growth of algae consumes all or most of the oxygen in the water, making it impossible for insects, fish, and other animals to live there.

Air pollution

Although pollutants can enter the atmosphere in many ways—including volcanic eruptions, forest fires, and the evaporation of volatile chemicals—the burning of fossil fuels is by far the greatest source of air pollution problems.

Figure 6.7

Smog is a term first used in the early 1900s to describe the mixture of smoke and fog that often covered industrial England.

Take a moment to think about some of the reasons why humans burn fuel: to heat homes and businesses; to run plane, car, bus, and train engines; to produce electricity in power plants; and to drive manufacturing and industrial processes of all kinds. The smoke released by burning fuels contains gases and **particulates,** which are solid particles of soot that can harm living organisms directly, or change the environment in ways that are later harmful to life. Some major air pollutants released into the air by fossil fuel combustion are carbon monoxide, carbon diox-

ide, nitrogen oxides, sulfur oxides, and hydrocarbons. Some of these chemicals interact to form the smog that hangs over many of the world's large cities. **Smog** is a form of air pollution found in cities, *Figure 6.7.* It consists mostly of particulates, sulfur dioxide, and other chemicals.

MiniLab

Can you observe particulates from car exhaust?

Particulates reduce visibility, and so are the most apparent form of air pollution. You can measure and compare the amounts of particulates in air and a car's exhaust without much trouble. For instance, you may find dust, ash, soot, lint, smoke, pollen, and other particulates suspended in air. Human activity produces approximately 100 million tons of particulates per year.

Procedure

1. Obtain two microscope slides, and coat one side of each slide lightly with petroleum jelly.

2. Place one of the slides with the petroleum side up on a table or window ledge for several hours.

3. Have an adult start a car engine in an open area, not a garage. While the engine is idling, place one of the microscope slides about six inches from the exhaust stream for 30 s. Set the slide on a bucket or box near the tailpipe. Move away before the adult starts the car. Wait until the engine is turned off before retrieving your slide.

CAUTION: *Car exhaust is hot and poisonous. Do not set the slide so close that you burn yourself.*

4. Cover both slides, and keep them covered until you observe them with a microscope or magnifying glass.

Analysis

1. What was the difference between what you observed on the two microscope slides?

2. Suggest what the particulates on the slide might be.

3. How can air pollutants from cars be reduced?

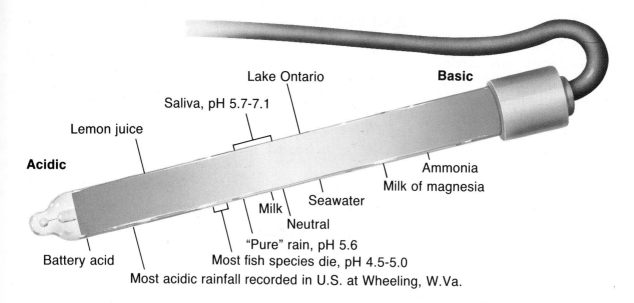

Lake Ontario

Basic

Saliva, pH 5.7-7.1

Lemon juice

Acidic

Ammonia

Milk of magnesia

Seawater

Milk

Neutral

"Pure" rain, pH 5.6

Battery acid

Most fish species die, pH 4.5-5.0

Most acidic rainfall recorded in U.S. at Wheeling, W.Va.

Figure 6.8

The pH scale indicates whether a solution is an acid or a base. Which is more acidic: lemon juice or soft drinks? Which solutions are bases?

Acid precipitation

The atmosphere contains moisture, in the form of water vapor, which condenses and returns to Earth's surface as rain, snow, and other forms of precipitation. While in the atmosphere, water molecules can come into contact with air pollutants.

Carbon dioxide gas dissolves into water droplets as they form into rain, producing weak carbonic acid. As a result, rain is normally slightly acidic. But air pollutants, particularly sulfur dioxide and nitrogen oxides, enhance the acidity of rain. In the presence of sunlight, these pollutants react with water and oxygen in the air to form sulfuric acid and nitric acid. By definition, **acid precipitation** is rain or snow that is more acidic than unpolluted rainwater. Sulfur is released primarily by coal-burning factories and power plants. The major source of nitrogen oxides is automobile exhaust.

The acidity of normal rain, acid rain, and some common substances is compared in *Figure 6.8.* In some cities and heavily industrialized areas, the amount of these pollutants released into the air is so great that the rain or snow may become as

acidic as vinegar. Even fog and dew can become acidic as a result of air pollution. Acidity is measured in units called pH. The pH scale ranges from 0 to 14; the lower the pH number, the more acid the substance. Nonpolluted rainfall has a pH of about 5.6 to 5.7. Precipitation with a pH below 5.7 is considered acidic. Most of the rain that now falls in the northeastern portion of the United States has a pH between 4.0 and 4.5. Rain with a pH as low as 1.9 has been detected.

Effects of acid precipitation

Acid precipitation leaches calcium, potassium, and other valuable nutrients from the soil. As these substances are washed away, the soil becomes less fertile. This loss of nutrients can lead to the death of trees, especially conifers. Acid rain also damages plant tissues and interferes with their growth and nitrogen fixation. Many trees in the forests of the northeastern United States, as well as in many European countries where the concentration of industry is high, are dying as a result of acid rain.

Acid precipitation also has severe effects on lake ecosystems. Acid rain falling into a lake, or entering it as

Figure 6.9

Acid precipitation has damaging effects on trees, organisms in affected lakes, and even on stone buildings and statues. This lake in Adirondack State Park shows the effects of acid precipitation.

runoff from streams or neighboring slopes, causes the pH of the lake water to fall below normal. The excess acidity can severely disrupt the lake ecosystem. *Figure 6.9* shows damage due to acid precipitation.

Ozone depletion

The atmosphere contains a sort of sunscreen—known as the **ozone layer**—that prevents living organisms on Earth's surface from receiving lethal doses of ultraviolet radiation. Ozone is a molecule made up of three oxygen atoms. Too near to Earth's surface, it is an air pollutant that causes lung damage and contributes to smog formation. But a thin layer of ozone located high in the stratosphere layer of the atmosphere, shown in *Figure 6.10*, absorbs most of the sun's harmful radiation.

Figure 6.10

The atmosphere extends several miles above Earth's surface and is made up of four layers. The layer closest to the surface, the troposphere, contains the air we breathe. The next layer is the stratosphere. At the top of the stratosphere lies the ozone layer. CFC molecules gradually rise up into the stratosphere, where they are exposed to ultraviolet light. The absorption of UV light releases chlorine atoms, which react with and destroy ozone molecules.

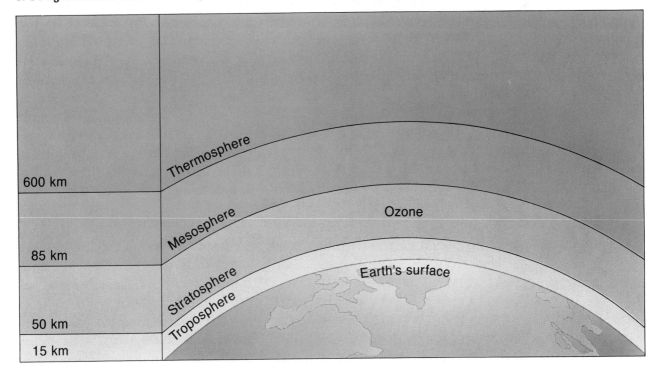

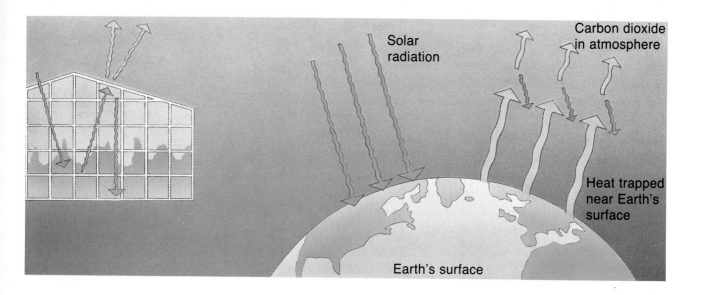

Solar radiation

Carbon dioxide in atmosphere

Heat trapped near Earth's surface

Earth's surface

Figure 6.11

The greenhouse effect is a normal phenomenon and a major reason why life has been able to survive on Earth. Carbon dioxide and other atmospheric gases, including CFCs and nitrous oxide, prevent heat from escaping into space. Otherwise, Earth would probably be a cold, lifeless planet.

Scientists have recently discovered that this protective layer of the atmosphere is becoming thinner. This depletion of the ozone layer is caused by a number of air pollutants that are not necessarily associated with fossil fuel combustion. The primary culprits are chlorofluorocarbons, or CFCs. These chemicals do not occur naturally on Earth, but are manufactured for a variety of purposes. They are used as coolants in refrigerators and air conditioners, and in the production of Styrofoam.

The ozone layer is extremely important. Without it, large amounts of ultraviolet radiation could reach the surface of Earth and cause significant damage to living organisms. As a result, governments around the world have agreed to eliminate the manufacture of CFCs by the year 2000. Efforts are underway to regulate the recycling of existing CFCs rather than releasing them into the atmosphere.

The greenhouse effect

What happens to the air inside a closed car that's been parked in the sun for a few hours? The radiant energy of sunlight heats the air inside the car, making it much warmer than the air outside. The glass of the car windows, like the glass walls of a greenhouse, keeps much of that heat trapped inside.

Similarly, gases in the atmosphere trap much of the radiant energy from the sun that reaches the surface of Earth. The land, water, and all the things on Earth absorb the sun's energy. These warmed objects radiate heat energy back into space. The atmosphere prevents a good part of this heat from escaping back into space. This process of heat retention by atmospheric gases is called the **greenhouse effect.** Without the greenhouse effect, all of the sun's energy would be radiated back into space, and Earth would be too cold for living things to survive and evolve. Gases that contribute to the greenhouse effect are called greenhouse gases and include carbon dioxide, as shown in *Figure 6.11.*

Water pollution

Water covers more than two-thirds of Earth's surface. Only about three percent of it is fresh water, and most of that is frozen in the polar ice caps. The fresh water of our lakes, rivers, and underground wells represents only about 0.1 percent of all the water on Earth. This is the water supply that plants, animals, humans, and other organisms depend on. It's also the water supply humans use to carry away many of our wastes. Without adequate care, those wastes can make the water unsuitable for other uses, such as drinking.

Have you ever heard travelers use the phrase, "Don't drink the water"? In many parts of the world, pure drinking water is not available. The same rivers and streams in which sewage and industrial wastes are discharged also serve as the source for drinking and washing. Waterborne diseases such as cholera, dysentery, and hepatitis are common in these areas. Other major pollutants of water include inorganic substances such as fertilizers that provide excessive amounts of nutrients, sediments including silt and small particles from runoff, and heat from warm or hot water that has been used to cool machinery in power plants and factories. *Figure 6.12* shows the causes of two forms of water pollution.

Figure 6.12

Many power plants draw water from a river or lake to cool their machinery (left). The discharged water is much warmer and can alter or destroy the habitat for many organisms in the body of water. Sediments washed away from farm fields can enter a river or lake. Water thick with sediments can clog fish's gills and kill them (right).

How many pesticides are used on our food?

In July of 1993, reports were filed by the National Research Council and the Environmental Working Group that cite problems that even minute amounts of pesticides pose to young children. The NRC study found that it is difficult to set safety standards without valid information on what infants and children actually eat or what constitutes safe limits for the chemicals. The Food and Drug Administration sets limits of pesticide residues that may be found on fruits and vegetables, but the NRC suggests that the limits should be redefined at lower levels.

Analysis

The chart shows ten types of produce samples tested for pesticides. For each fruit or vegetable, the percent that contained pesticide is shown. The chart also shows the number of types of pesticides detected on the ten kinds of produce. Compare the numbers of pesticides detected.

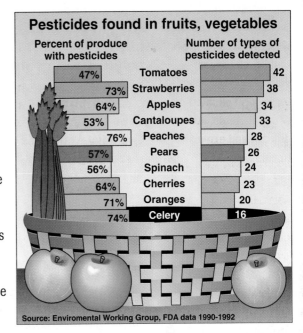

Pesticides found in fruits, vegetables

	Percent of produce with pesticides		Number of types of pesticides detected
Tomatoes	47%		42
Strawberries	73%		38
Apples	64%		34
Cantaloupes	53%		33
Peaches	76%		28
Pears	57%		26
Spinach	56%		24
Cherries	64%		23
Oranges	71%		20
Celery	74%		16

Source: Enviromental Working Group, FDA data 1990-1992

Thinking Critically

Conclude from the chart which kinds of produce are at greatest risk of having pesticide residues, and suggest what might be done to protect young children and infants from exposure.

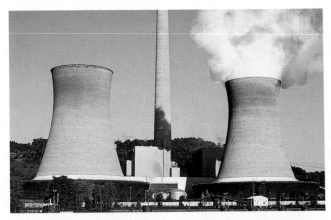

Groundwater pollution

Water in lakes and streams is known as surface water. Fresh water that is found underground is called **groundwater.** Do you know whether your community water supply comes from surface water or groundwater? About half the population of the United States depends on groundwater for drinking water. Until the 1970s, it was assumed that soil would filter out any pollutants before they reached underground reservoirs, so groundwater could not become polluted. *Figure 6.13* illustrates how pollutants can enter underground reservoirs.

Other problems are also associated with the use of groundwater. The Ogallala Aquifer is an underground reservoir that lies beneath much of the farmland in the midwestern United States. So much water is being removed from this aquifer for irrigating farmland and supplying water to cities and towns that the underground water level is going down faster than it can be recharged by rainfall. The total amount of water removed from this aquifer every year is estimated to equal the annual flow of the Colorado River. As the water level goes down, the soil above it may collapse, creating a giant sinkhole. Water from the Ogallala Aquifer and other underground reservoirs contains large amounts of dissolved salts. As this water is used to irrigate farmland, these salts gradually accumulate in the soil, eventually making it unsuitable for plant growth.

Figure 6.13

It takes hundreds, even thousands, of years for water to accumulate in underground reservoirs. Once groundwater is contaminated with pollutants, it is difficult and expensive to clean.

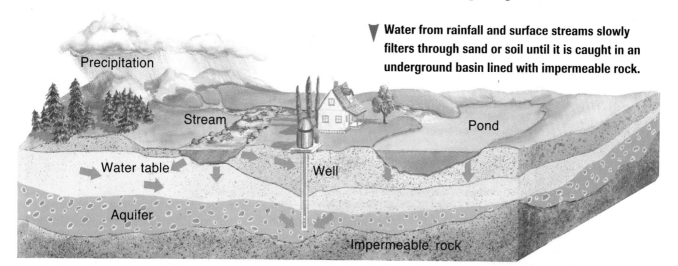

Water from rainfall and surface streams slowly filters through sand or soil until it is caught in an underground basin lined with impermeable rock.

Precipitation

Stream

Pond

Water table

Well

Aquifer

Impermeable rock

Pollutants from leaking storage containers or agricultural runoff also travel through the soil, eventually reaching and contaminating nearby groundwater (left). It is estimated that between 20 and 45 percent of the water wells in some areas of the United States may be polluted with agricultural or industrial chemicals. Efforts are being made to develop safer methods for storing hazardous materials to prevent them from contaminating groundwater supplies (right).

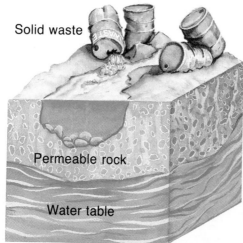

Solid waste

Permeable rock

Water table

Figure 6.14

Many landfills are being filled to capacity and closed.

Land pollution

How much garbage do you and your family produce every day? Remember to include the notebook paper you use for schoolwork, and the plastic bags or other containers you discard after you've eaten your lunch. Do you put your aluminum soda cans in the trash or in the recycling bin?

Trash, or solid waste, is made up of the cans, bottles, paper, plastic, metals, dirt, and spoiled food that people throw away every day. The average American produces about 1.8 kg of solid waste daily. That's a total of about 657 kg of solid waste per person per year. Where do we put all this garbage? Does it ever decompose? Your trash becomes part of the billions of tons of solid waste that are burned or buried in landfills, *Figure 6.14*, all over the world every year.

Chemistry

Separating Plastics

It is predicted that 75 billion tons of plastic will be discarded by the year 2000. Currently, only about three percent of this plastic gets recycled. Unlike aluminum, which is easily separated from trash and reclaimed, the problem in recycling plastic is that the plastics have to be sorted by type, by hand, which is an expensive process. Different types of plastic polymers cannot be mixed together to form a strong, stable plastic.

Chemical sorting Researchers have devised a method by which chemical solvents sort the six plastics found in typical household trash. In a pilot program, researchers placed shredded, mixed plastics in a vat with a common solvent called xylene. Immediately, one type of plastic polymer, polystyrene, the polymer that makes Styrofoam, was separated from the other polymers.

When the temperature in the vat reached 170 C°, xylene dissolved another plastic polymer, polyethylene. Polyethylene is used in plastic grocery bags and plastic wrap. At 220 C°, another form of polyethylene used in milk jugs and detergent bottles was dissolved. Polypropylene, the polymer that is used in labels and some types of bottles, was dissolved at 250 C°. The last two polymers, PET (polyethylene terephthalate) and PVC (polyvinyl chloride), required a different solvent because xylene is not effective on them. With the different solvent, they too were separated into their pure forms and were ready to be used again.

An efficient process Researchers have estimated that this process could yield nearly four tons of plastic polymer per hour. This process could also cost almost 30 percent less than producing the polymers from new raw materials.

CONNECTION TO | Biology

How could chemical research save other kinds of Earth's resources and limit our waste production?

Figure 6.15

Scientists who dug up garbage from old landfills were surprised to find newspapers from the 1950s that were still readable. They also discovered that food thrown away along with those newspapers hadn't decomposed at all. Neither had grass clippings from lawns mowed 40 years ago.

You might expect that the trash that goes into landfills is decomposed by bacteria. But scientists who have dug up and sifted through the garbage in old landfill sites as shown in *Figure 6.15* have discovered that much of this material is not broken down by natural processes. Because oxygen is sealed out of landfills when they are covered with soil, bacteria and other decay-promoting organisms cannot survive there. As a result, materials that could be recycled back into the environment are trapped for decades.

Some types of solid waste—such as wood products, food, animal wastes, dead leaves, and other yard wastes—are called **biodegradable** wastes because they can be broken down by natural processes. Other types of waste—such as pesticides, toxic metals, and radioactive residues from nuclear power plants—are not easily broken down. These materials are considered **nonbiodegradable**

because they can persist in the environment for hundreds or even thousands of years. There is currently much debate over where and how to store toxic, nonbiodegradable wastes. One method that is receiving a great deal of attention involves burial of these wastes in geologically stable areas.

What can we do to help?

How can we continue to meet the human population's increasing demand for resources? Modern technology offers some solutions. For example, selective breeding of plant crops—a topic you'll learn more about in Chapter 14—has resulted in the development of strains of wheat that produce more grain per wheat plant. Such developments not only increase food production, but also reduce the amount of land needed for growing crops. Recycling, as

Figure 6.16

Recycling metals, plastics, paper, and other materials can help conserve important resources. If all Americans recycled just one-tenth of their newspapers, we could avoid harvesting about 25 million trees every year. We also could save energy. Making recycled paper uses only about half the energy it takes to make paper from wood pulp.

described in *Figure 6.16,* also helps meet demand without using up our nonrenewable resources.

Preservation of natural areas

How can human demands for additional living space and other resources be met without loss of habitat? **Preservation** is the act of keeping an area or organism from harm or destruction. One way of preserving wildlife habitats is by setting aside parks, refuges, and other areas that can be regulated through conservation. **Conservation** is the planned management of a natural area to prevent exploitation or destruction. Yellowstone National Park, the world's first national park, was founded in 1872 when Congress set aside 2 million acres to protect natural features and meet public

recreational needs. Today, almost 800 million acres of U.S. land have been set aside, *Figure 6.17.*

About three percent of Earth's land has been set aside by world governments as parks or preserves.

Figure 6.17

Millions of tourists visit the national parks in the United States each year.

In the United States, approximately one ton of municipal waste is dumped into landfills each year for every man, woman, and child. A significant portion of this waste is in the form of packing materials for food, clothing, and other household items. As suitable places for waste disposal become increasingly scarce, the waste stream must be slowed or changed. One way to do this is to be sure that packing materials can be decomposed or recycled.

Degrading Time of Packing Materials

PREPARATION

Problem
How long does it take for packing materials to degrade?

Objective
In this Biolab, you will:
- **Determine** the conditions for rapid degradation of certain packing materials.

Materials
soil
Styrofoam peanuts
cornstarch packing material
popcorn
paper packing material
paper garbage bags
plastic garbage bags
4 large beakers or clay pots

Safety Precautions
Be sure to wash your hands after doing this Biolab.

1. Collect four samples of each type of packing material.

2. Make four piles of packing material, mixing each kind of material in each pile.

3. Place each pile of material in a different beaker or clay pot.

4. Add soil to two of the pots so that all the samples are well covered and mixed with the soil.

5. Place a label that says DRY on one of the pots with soil and one of the pots without soil.

6. Place a label that says MOIST on the remaining two pots.

7. Keep the second two pots moist at all times.

8. Make a data table like the one shown.

9. Observe each pot once a week for four weeks, and record in your data table any evidence of decomposition.

Pot #	Observations
#1 Dry with soil	
#2 Dry without soil	
#3 Moist with soil	
#4 Moist without soil	

ANALYZE AND CONCLUDE

1. **Interpreting Observations** Describe your observations of decomposition of any of the packing materials.

2. **Analyzing Data** Did any of the packing materials appear to be nonbiodegradable? Explain.

3. **Thinking Critically** Suggest ways that commercial producers could help to reduce the waste stream.

4. **Drawing Conclusions** Which type of packing material appears to decompose fastest, and which type decomposes most slowly?

Going Further

Application Find out how your community disposes of its wastes and how much disposal costs. Do the disposal costs reflect stresses placed on the environment's food webs or only the monetary cost of disposal?

Figure 6.18

The American alligator became endangered due to commercial exploitation and habitat destruction.

Preserving threatened and endangered species

Recovery programs have been developed to keep some endangered species from becoming extinct. One of the greatest success stories is that of the American alligator. Once endangered throughout the entire United States, the species, shown in *Figure 6.18,* has made a comeback and is no longer listed as endangered or threatened in some parts of its range.

At present, far more species are becoming extinct than we can possibly save. Which species should be rescued? How do we decide whether one species is more important than another? These are not questions that ecologists can answer because they realize that every organism is important to the fabric that makes up the tapestry of life.

Connecting Ideas

The human population is increasing at a staggering rate. Because of the increased demands this places on the biosphere, natural resources once considered inexhaustible are now in short supply. Human activities have caused pollution of air, water, and land.

How can these trends be reversed? What can humans do to lessen stress on the environment while maintaining our standard of living? Finding the answers will require learning as much as possible about our environment and the organisms it supports. One area of study is the homeostatic balance in an ecosystem, which depends on a fundamental balance of chemicals and chemical reactions within each community of organisms.

Section Review

Understanding Concepts

1. What are two major sources of air pollution?
2. How are biodegradable and nonbiodegradable pollutants different?
3. How can the application of fertilizer and pesticides to a farm field affect drinking water?

Thinking Critically

4. Are all recyclable substances biodegradable? Explain.

Skill Review

5. **Designing an Experiment** Design an experiment, using a terrarium, to test the effects of different concentrations of acid precipitation. For more help, refer to Practicing Scientific Methods in the *Skill Handbook.*

Clean Water

Have you ever thought about the water that flows out of your faucets every day? Most people are aware of the lake, river, or reservoir that supplies them with water, but where does the used water go and how is it cleaned? The used water is called wastewater. It is estimated that each day, every individual in the United States contributes 50 to 100 gallons of wastewater to a community's wastewater flow. Wastewater is formed as we use water to wash dishes, wash cars, flush toilets, and do laundry. Businesses and industry create wastewater when they use water to transport wastes away from offices and factories.

What does wastewater contain? Wastewater is composed of 99.94 percent water. The other 0.06 percent consists of solids that are dissolved or suspended in the water. Wastewater may also contain harmful, disease-causing pathogens such as bacteria, viruses, and other microorganisms. Pesticides, heavy metals, and nutrients are also substances that may be a part of wastewater.

Wastewater treatment The purpose of wastewater treatment is to remove the natural or human-made impurities from the water so that it will be safe to return to the environment. In cities and towns, each house or business has a pipe or sewer that carries wastewater out to a series of larger sewers that eventually lead to a treatment plant. What happens to the water at a treatment plant is almost the same as what naturally occurs in a stream or lake. Wastewater-treatment plants speed up the natural processes. Primary treatment removes most of the solids by screening and settling or floating. Secondary treatment uses hundreds of species of bacteria, protozoa, fungi, and algae to digest the waste. Before being discharged from the plant, a chemical, usually chlorine, is added to disinfect the water to kill disease-carrying organisms. The water is then discharged from the plant into a receiving stream, which eventually returns it to the original water supply.

Applications for the Future

New methods of treatment Constructed wetlands are providing an inexpensive alternative to conventional water-treatment systems. A constructed wetland is defined as a system specifically designed for wastewater treatment at a site where a wetland previously did not exist. Construction costs for a wetland are around $2 per gallon of water in contrast to conventional systems, which can cost between $5 and $10 per gallon. Wastewater is piped into a large, constructed wetland containing aquatic plants and microbes that utilize the waste materials in the water for growth. These artificial wetlands also allow for settling to occur, and the result is thousands of gallons of naturally treated water per day. These artificial systems are provided with an ungrounded barrier that prevents groundwater contamination. Approximately 142 constructed wetland facilities currently operate in the United States.

INVESTIGATING the Technology

1. **Research** Find out what happens to wastewater in your area.
2. **Apply** What effects could undertreated water have if discharged into a lake or river?

Reviewing Main Ideas

6.1 Effects of Human Activities on Our Resources

- Resources can be classified as renewable or nonrenewable. Renewable resources may be replaced if managed properly. Nonrenewable resources cannot be replaced by natural means, but many of them can be recycled.
- Trends in resource use suggest that as demand increases, supply tends to diminish. Substitution, recycling, and technology can reduce demand for some resources.

6.2 Maintaining the Natural Balance

- The use of resources faster than the environment's ability to recycle them has resulted in problems of air, water, and land pollution. Air pollution is caused by particulates and gases such as carbon dioxide and nitrogen oxides being added at a rapid rate. Water pollution is caused by additions of organic and inorganic substances, sediments, heat, and radioactive substances. Land pollution occurs when solid wastes are dumped and not decomposed by natural systems.
- Human activities have increased the rate of habitat destruction and caused the extinction of many species of wildlife. The creation of national parks, forests, and similar areas helps reduce the number of species lost to extinction.

Key Terms

Write a sentence that shows your understanding of each of the following terms.

acid precipitation	nonrenewable
biodegradable	resource
conservation	ozone layer
endangered species	particulate
extinction	pollution
fossil fuel	preservation
greenhouse effect	renewable
groundwater	resource
natural resource	smog
nonbiodegradable	threatened species

Understanding Concepts

Using a Graph

1. The following bar graph indicates the sources of sulfur, nitrogen, and carbon in the air. From the graph, which element is probably most out of balance? Why?

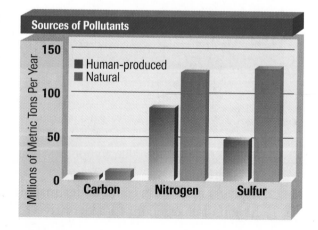

Sources of Pollutants

Millions of Metric Tons Per Year

■ Human-produced
■ Natural

Carbon   Nitrogen   Sulfur

2. How can nonrenewable resources be made to last longer?
3. What is the effect of acid precipitation on sensitive ecosystems?
4. Why is the destruction of tropical rain forests considered disturbing?
5. How could a thinning of the ozone layer in the Antarctic affect humans and animals in South America?
6. What would be one result of increasing the participation in efforts to recycle metals, plastics, and paper?
7. How does agricultural runoff cause water pollution?
8. What are two effects of using water from aquifers for irrigation?

Relating Concepts

9. Make a concept map that relates the following terms and phrases. Supply the appropriate linking words for your map.

   endangered species, smog, threatened species, extinction, fossil fuel, acid precipitation, air pollution, pollution, particulates, ozone layer, greenhouse effect, natural resources

Applying Concepts

10. Why do those who fish for sport not consider temperature pollution a problem in some lakes?
11. How would putting milk into glass bottles instead of plastic bottles be a possible benefit to the environment?
12. Why should you be concerned about the kinds of material placed in the sanitary landfill near where you live?
13. How can increasing the price of a natural resource save it for future generations?
14. Why is it possible to read a newspaper that has been in a landfill for 25 years?

Chemistry Connection

15. By separating plastics to their original form, the same packing products can be made from the same materials indefinitely. What effect would this have on our landfills?

Thinking Critically

Relating Cause and Effect

16. **Biolab** What was the effect of moisture on the cornstarch packing material? Why do you think this material is environmentally friendly?

Making Predictions

17. **Minilab** Suppose that the car had not been tuned properly. Predict how your results would have been different. Explain.

Interpreting Data

18. **Minilab** Assume that Akira had the following results from a survey of his home. Explain how Akira's data were similar and different from yours.

Akira's Home List of Possible Pollutants		
Room	**Organic products**	**Inorganic products**
Garage	Gasoline Motor oil, used Motor oil, new	Paint Turpentine
Bathroom	Toothpaste Mouthwash Soap	
Kitchen	Food scraps Furniture polish	Window cleaner Scouring powder Insecticide
Bedroom	Medicine Nosedrops	

Connecting to Themes

19. **Evolution** In what way is the present extinction rate different from previous mass extinction events?
20. **Homeostasis** Explain how pollution indicates that ecosystems are out of balance.
21. **Systems and Interactions** What kinds of interactions are out of balance when biodegradable materials build up in landfills?
22. **Energy** How does recycling aluminum cans and other nonrenewable minerals save energy?

3 The Life of a Cell

Greenish, rectangular shapes cram into every available space, filling your view. Within each rectangle, microscopic structures of varying shape and color float like tiny islands in a flowing sea of colorless liquid. At first glance, this scene may look like an alien landscape, but what you are actually looking at are the cells of a common aquarium plant.

Every living thing is made up of one or more cells. A cell is the smallest unit that can carry on all the processes of life. Living things, like nonliving things, are composed of chemicals such as carbon, hydrogen, and oxygen, but it is the organization of these elements into cells that distinguishes living things from all other matter.

Magnification: 4500×

All organisms need energy to carry out life processes. How is this chloroplast important in providing the energy needed for the lives of all cells?

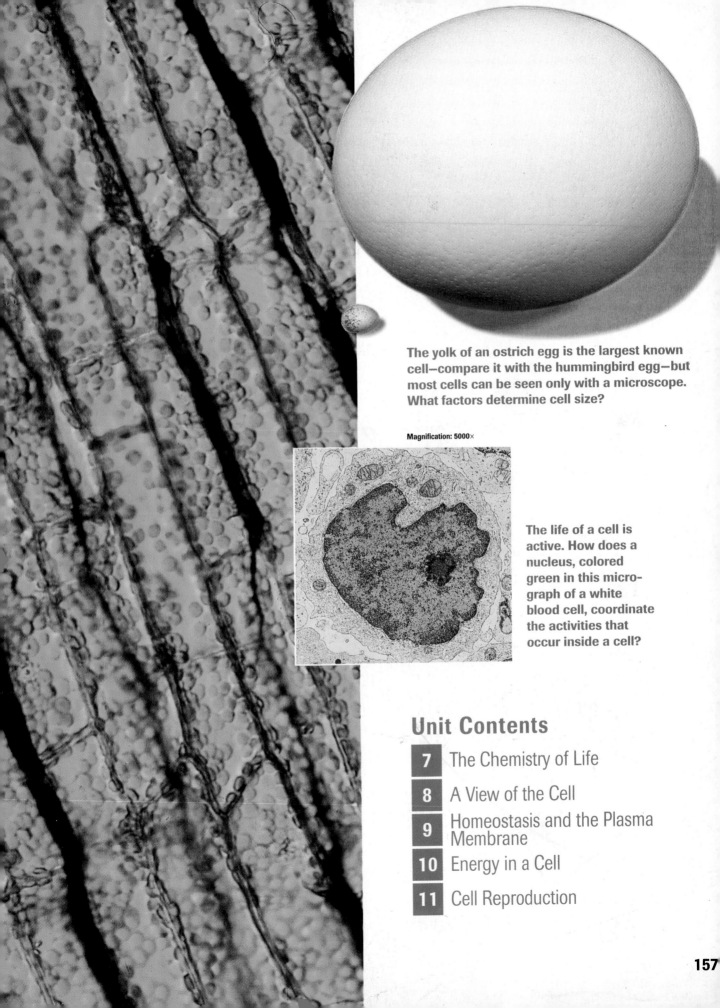

The yolk of an ostrich egg is the largest known cell—compare it with the hummingbird egg—but most cells can be seen only with a microscope. What factors determine cell size?

Magnification: 5000×

The life of a cell is active. How does a nucleus, colored green in this micrograph of a white blood cell, coordinate the activities that occur inside a cell?

Unit Contents

157

CHAPTER
7 The Chemistry of Life

What makes a living thing unique? Suppose you held a rock in one hand and a bouquet of flowers in the other. How would you explain that the flowers are alive, and the rocks are part of the nonliving world? Are living and nonliving things composed of entirely different substances?

Early biologists struggled with the same questions. Two hundred years ago, many biologists believed that some mysterious life force guided all life processes. They also thought that this life force somehow made the matter of living things special so that it could never be prepared in the laboratory. If you've seen a Frankenstein movie, you know that Dr. Frankenstein was searching for this life force so that he could restore life to dead tissue. Scientists now know that there is no life force, and that living things actually have a great deal in common with rocks, cassette tapes, computer chips, and other nonliving objects.

Living and nonliving things are composed of the same basic materials, and the substances found in horses, grass, and trees obey the same laws of chemistry and physics as those found in rocks. Why, then, do you suppose that living and nonliving things seem so different?

7.1 Elements and Atoms

Industrial robots are often used to do work that humans could do if it weren't for danger, hazardous materials, or the need for a strictly clean environment. While the behavior of industrial robots may seem human in many ways, they are still nonliving things. Even so, they are composed of the same basic building blocks as humans and other living organisms—elements.

Elements

Everything—whether it is a rock, robot, or flower—is made of substances called elements. Suppose you found a nugget of pure gold. You could grind it into a billion bits of powder and every particle would still be gold. You could treat the gold with every known chemical, but you could never break it down into two simpler substances. That's because gold is an element, one of the simplest chemical substances. An element is a substance that can't be broken down into simpler substances.

On Earth, 90 elements occur naturally. The periodic table, shown in Appendix E, lists all currently known elements. With the exception of elements 43 and 61, elements numbered 1 to 92 are found in nature. Elements numbered 93 and above are synthetic elements.

Natural elements in living things

Of the 90 naturally occurring elements, only about 25 are essential to living organisms. *Table 7.1* lists several elements found in the human body. Notice that only four elements—carbon, hydrogen, oxygen, and nitrogen—make up more than 96 percent of the mass of a human. Each element is abbreviated by a one- or two-letter symbol. For example, the symbol C represents the element carbon, Ca represents calcium, and Cl represents chlorine.

Trace elements

Notice that some of the elements listed in *Table 7.1,* such as iron and magnesium, are present in living things in very small amounts. Such elements are known as trace elements. They play a vital role in maintaining healthy cells in all organisms, as shown by the examples in *Figure 7.1.*

Table 7.1 Elements that Make Up Living Things

Element	Symbol	Percent by Mass in Human Body	Element	Symbol	Percent by Mass in Human Body
Oxygen	O	65.0	Iron	Fe	trace
Carbon	C	18.5	Iodine	I	trace
Hydrogen	H	9.5	Copper	Cu	trace
Nitrogen	N	3.3	Manganese	Mn	trace
Calcium	Ca	1.5	Molybdenum	Mo	trace
Phosphorus	P	1.0	Cobalt	Co	trace
Potassium	K	0.4	Boron	B	trace
Sulfur	S	0.3	Zinc	Zn	trace
Sodium	Na	0.2	Fluorine	F	trace
Chlorine	Cl	0.2	Selenium	Se	trace
Magnesium	Mg	0.1	Chromium	Cr	trace

Atoms—The Building Blocks of Elements

Elements, whether they are found in living things or not, are made up of atoms. An **atom** is the smallest particle of an element that has the characteristics of that element.

The structure of the atom

Each element has distinct characteristics because of the structure of the atoms of which it is composed. For example, iron differs from aluminum because the structure of iron atoms differs from that of aluminum atoms. Still, all atoms have the same general arrangement. The center of an atom is called the **nucleus** (plural, nuclei). It is made of positively charged particles called protons (p^+) as well as particles called neutrons (n^0), which have no charge. All nuclei are positively charged because of the presence of protons.

Forming a cloud around the nucleus are even smaller particles called electrons (e^-), which have negative charges. If you've ever looked at a spinning fan, you've probably noticed that as the fan blades turn, they appear to form a blurry disk that occupies a space around the center of the fan. An electron cloud is simply the space around the atom's nucleus that is occupied by these fast-moving electrons.

Figure 7.1

Plants obtain trace elements by absorbing them through their roots. Animals get these important elements in the foods they eat.

Mg^{++} ions ⟶ ⟵ Mg^{++} ions

Plants must have magnesium (Mg) in order to form the green pigment chlorophyll that captures light energy for the production of sugars.

Iodine (I) is used in mammals to produce a substance that affects the rates of growth, development, and chemical activities in the body.

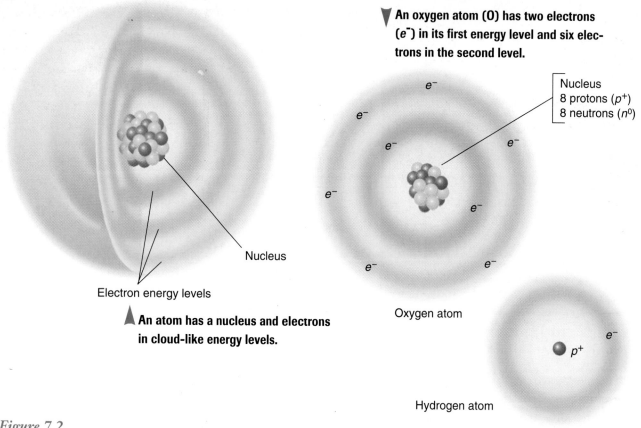

An oxygen atom (O) has two electrons (e^-) in its first energy level and six electrons in the second level.

Nucleus
8 protons (p^+)
8 neutrons (n^0)

Nucleus

Electron energy levels

▲ An atom has a nucleus and electrons in cloud-like energy levels.

Oxygen atom

p^+

Hydrogen atom

Figure 7.2

Electrons move rapidly around atoms, forming electron clouds. Electron clouds are actually composed of several energy levels, which are regions in which electrons travel.

▲ Hydrogen (H), the simplest atom, has just one electron in its first energy level and one proton in its nucleus.

Electron energy levels

Electrons travel around the nucleus in certain regions known as energy levels, **Figure 7.2**. Each energy level has a limited capacity for electrons.

Figure 7.3

The properties of an element are determined by the structure of its atoms. Copper atoms in the coin have 29 protons and 29 electrons; gold atoms in the ring have 79 protons and 79 electrons. The diamond is nearly pure carbon, with atoms that contain six protons and six electrons. As you can see, these three elements have very different properties.

Because the first energy level is the smallest, it can hold a maximum of two electrons. The next level is larger and can hold a maximum of eight electrons. The third level is larger yet and can hold 18. For example, the oxygen atom in **Figure** 7.2 has a total of eight electrons. Two electrons fill the first energy level. The remaining six electrons must, therefore, occupy the second level.

Atoms contain equal numbers of electrons and protons; therefore, they have no net charge. The hydrogen (H) atom in **Figure** 7.2 has just one electron and one proton. Oxygen (O) has eight electrons and eight protons. **Figure 7.3** compares the atomic structure and properties of three other elements.

Figure 7.4

Radioactive isotopes are used in medicine to diagnose and/or treat some diseases.

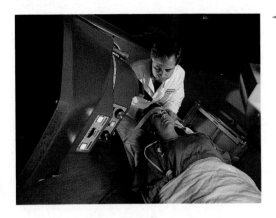

◀ Neutrons given off when radioactive isotopes break apart are deadly to rapidly growing cancer cells. The patient is being treated with radiation from a radioactive isotope of cobalt (Co).

▼ The thyroid gland in mammals uses iodine to make thyroid hormones. Radioactive iodine (I) introduced into the body is absorbed by the thyroid gland. By detecting the radioactive iodine taken up, the function of the thyroid gland can be measured.

Isotopes of an Element

Atoms of an element sometimes contain different numbers of neutrons in the nucleus. Atoms of the same element that have different numbers of neutrons are called **isotopes** of that element. For example, most carbon nuclei contain six neutrons. However, some have seven or eight neutrons. Each of these are isotopes of the element carbon. Scientists refer to isotopes by giving the combined total of protons and neutrons in the nucleus. Thus, the most common carbon atom is referred to as carbon-12 because it has six protons and six neutrons. Other isotopes of carbon include carbon-13 and carbon-14.

Isotopes are often useful to scientists. The nuclei of some isotopes are unstable and tend to break apart. For instance, carbon-14 is unstable. As nuclei break, they give off radiation. These isotopes are said to be radioactive. Because radiation is detectable and can damage or kill cells, scientists have developed some useful applications for radioactive isotopes, as you can see in *Figure 7.4*.

Section Review

Understanding Concepts

1. A nitrogen atom contains seven protons, seven neutrons, and seven electrons. Describe in detail the structure of a nitrogen atom. Use a labeled drawing if you wish.
2. An atom of aluminum has 13 protons in its nucleus. How many electrons does the atom have? Explain how you know.
3. How do two isotopes of the same element differ?

Thinking Critically

4. A fluorine atom has nine electrons. Make an energy level diagram similar to the one in *Figure 7.2*. How many electrons would be needed to fill its outer level?

Skill Review

5. **Making and Using Graphs** Make a pie graph of the elements listed in *Table 7.1*. Group all the elements below phosphorus in one segment. For more help, refer to Organizing Information in the *Skill Handbook*.

Section Preview

Objectives

Relate the formation of covalent and ionic chemical bonds to the stability of atoms.

Interpret the formulas of chemical compounds and the meaning of chemical equations.

Relate water's polarity to its ability to dissolve substances and to the formation of acids and bases.

Key Terms

compound
covalent bond
molecule
ion
ionic bond
mixture
solution
polar molecule
hydrogen bond
metabolism
pH
acid
base

For many people, nothing beats the delight of biting into a fresh, hot pizza cooked to perfection and built according to your instructions. Pizza is fun to eat because it can be made in so many different ways. Each of the individual ingredients in pizza may be boring to eat alone, but when assembled in the right combination and amounts, they become a personalized meal. In many ways, elements are similar to the ingredients of pizza. Standing alone, most elements are unremarkable. But when combined with one or more other elements, they form a large variety of useful substances.

Compounds and Bonding

Water is a substance everyone is familiar with. Is it an element? If you pass an electric current through it, water breaks down into hydrogen and oxygen. Neither hydrogen nor oxygen can be broken down further, so they must be elements. From this description, you can infer that water is not an element. Rather, it is a type of substance called a compound. A **compound** is a substance that is composed of atoms of different ele- ments chemically combined. Water (H_2O) is a compound composed of the elements hydrogen and oxygen. Just as the combined in- gredients in a pizza result in a tasty meal, you can see in *Figure 7.5* that the properties of a compound are dif- ferent from those of its individual elements.

Figure 7.5

Table salt is made from the elements sodium (Na) and chlorine (Cl). The flask contains the poisonous, yellow-green gas chlorine. The lump of silver-white metal is the element sodium. When sodium and chlo- rine are placed together, they combine. One atom of sodium combines with one atom of chlorine to form sodium chloride (NaCl), commonly known as table salt. Table salt consists of white crystals that no longer resemble either sodium or chlorine.

How covalent bonds form

Most matter is in the form of compounds, but how and why do atoms combine? Atoms combine with other atoms only when conditions are right, and they do so because they become more stable by combining.

Remember electron energy levels? For most elements, an atom becomes stable when its outer energy level has eight electrons. An exception is hydrogen, which becomes stable with two electrons in the first energy level. One way that atoms can become stable is by sharing electrons with other atoms.

For example, two hydrogen atoms can combine with each other by sharing their electrons, as shown in *Figure 7.6.* As you know, individual atoms of hydrogen contain only one electron. Each atom becomes stable by sharing its electron with the other atom. The two shared electrons move about the first energy level of both atoms. The attraction of the positively charged nuclei for the shared, negatively charged electrons holds the atoms together. When two atoms share electrons, the force that holds them together is called a **covalent bond.**

Most compounds in organisms have covalent bonds. Examples of these compounds include sugars, fats, proteins, and water. A **molecule** is a group of atoms held together by covalent bonds and having no overall charge. A molecule of water is represented by the chemical formula H_2O. The subscript 2 represents two atoms of hydrogen (H) combined with one atom of oxygen (O). As you will see, many compounds in living things have more complex formulas.

How ionic bonds form

Not all atoms bond with each other by sharing electrons. Sometimes, atoms combine with each other by gaining or losing electrons in their outer energy levels. An atom (or group of atoms) that gains or loses electrons has an electrical charge and is called an **ion.**

Figure 7.6

Sometimes atoms combine by sharing electrons to form covalent bonds.

Hydrogen molecule

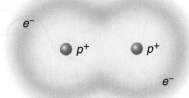

▲ Hydrogen gas exists in nature as two hydrogen atoms sharing electrons with each other. The electrons move around the nuclei of both atoms. Because of this bonding, hydrogen gas is written as H_2 rather than just H.

▼ When two hydrogens share electrons with oxygen, they form covalent bonds to produce a molecule of water. Each hydrogen atom shares one electron with the single oxygen atom.

Water molecule

e^-
p^+
$2e^-$
$8p^+$
$8n^0$
p^+
e^-
$6e^-$

A different type of chemical bond holds ions together. The bond formed between a sodium atom (Na) and a chlorine atom (Cl) provides a good example of this. Sodium becomes stable by losing the one electron in its outer energy level, and chlorine becomes stable by accepting this lost electron. Because sodium has lost one negatively charged electron, it now has one more proton than it has electrons. Thus, it is a positively charged ion. Chlorine has gained one electron, and it now has one more electron than it has protons. Thus, it is a negatively charged ion. Because opposite charges attract, there is an attractive force between two ions of opposite charge known as an **ionic bond**. The compound formed when sodium and chlorine react to form an ionic bond is known as sodium chloride, as shown in *Figure 7.7*. Sodium chloride is represented by the chemical formula NaCl.

Ionic compounds are less abundant in living things than are covalent molecules, but ions are important in biological processes. For example, sodium and potassium ions are required for transmission of nerve impulses. Calcium ions are necessary for muscles to contract. Plant roots absorb needed minerals as ions.

Mixtures and Solutions

When elements combine to form a compound, they no longer have their original properties. What happens if substances are just mixed together and do not combine chemically? A **mixture** is a combination of substances in which individual substances retain their own properties. For example, if you stirred sand and sugar together, they would neither change nor combine chemically.

Figure 7.7

A sodium atom (Na) contains 11 electrons–two in the first energy level, eight in the second, and one in the third. A chlorine atom (Cl) has 17 electrons, with the outer level holding seven electrons (below). When sodium and chlorine combine, the sodium atom loses one electron, and the chlorine atom gains it. Thus, the chlorine ion formed is stable with eight electrons in its outer level and has a negative charge. Sodium has lost the one electron that was in its third energy level. Thus, the sodium ion has a positive charge and is stable. The actual reaction is shown at right.

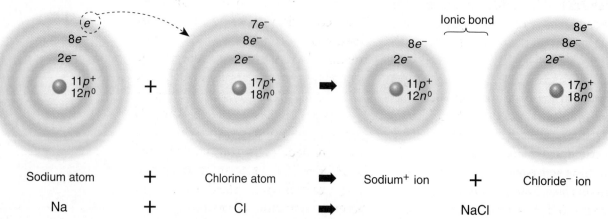

Figure 7.8

The sugar molecules in the Kool-Aid mix (left) spread evenly throughout the water, making a kind of mixture called a solution. The substance being dissolved, sugar, is called the solute; water is called the solvent. Neither the sugar nor the water changes chemically. Water molecules simply attract the sugar molecules and pull them into solution (below). The sugar can be recovered easily by evaporating the water.

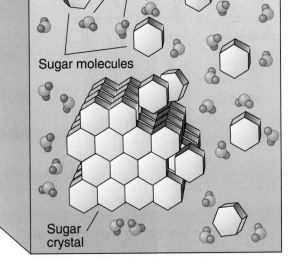

Water molecules

Sugar molecules

Sugar crystal

A solution is a kind of mixture that is very important in living things. A **solution** is a mixture in which one or more substances are distributed evenly in another substance. You may remember making Kool-Aid when you were younger. The sugar molecules in Kool-Aid will dissolve easily in water to form a solution, as shown in *Figure 7.8.*

In organisms, many important substances, such as sugars and mineral ions, are dissolved in water. The more solute that is dissolved in a given amount of solvent, the greater is the solution's concentration (strength). The concentration of a solute is important to organisms. Organisms can't live unless the concentration of dissolved substances stays within a narrow range. Organisms have many mechanisms to keep the concentrations of molecules and ions within this range. For example, the amount of sugar dissolved in your bloodstream must stay within a critical range. Substances produced in the pancreas and other organs cause sugar to be stored or released in order to keep blood sugar levels in this range. This regulation is another example of homeostasis.

Water and Its Importance

Water is perhaps the most important compound in living organisms. Most life processes can occur only when molecules and ions are free to move and collide with one another. This condition exists when they are dissolved in water. Water serves as a means of transport of materials in organisms. For example, both blood and plant sap are mostly water. In fact, water makes up from 70 to 95 percent of most organisms. The water molecule, as you have learned in *Figure 7.6,* is composed of two atoms of hydrogen linked by covalent bonds to one atom of oxygen.

How does liquid soap affect the surface tension of water?

Surface tension of water is a force exerted by the surface of water on the particles below. It results from the attraction of water molecules to other water molecules. The force of surface tension tends to pull drops of water into spherical (ball) shapes. When drops of falling water are photographed with a high-speed camera, their spherical shape is visible.

Procedure

1. Place one drop of water on a piece of waxed paper.
2. Look carefully at the water drop from the side and draw its shape.
3. Dip a toothpick into liquid soap.
4. Touch the toothpick to the water drop while viewing from the side.

Analysis

1. What was the shape of the original water drop?
2. How did the shape change when the soap was added?
3. What do you think caused the drop to change?
4. What can you conclude about the effect of soap on the surface tension of water?

Water is polar

Sometimes, when atoms form covalent bonds, they do not share the electrons equally. In the water molecule, shown in *Figure 7.9,* you can see that the oxygen atom attracts the shared electrons more strongly than the hydrogen atoms do. As a result, the electrons spend more time near the oxygen atom than they do near the hydrogen atoms.

Water is an example of a polar molecule. A **polar molecule** is a molecule with an unequal distribution of charge; that is, each molecule has a positive end and a negative end. Polar water molecules attract each other as well as ions and other polar molecules. Because of this attraction, water has the ability to dissolve many ionic compounds, such as salt. Water also dissolves many polar molecular substances, such as sugar.

Water molecules also attract each other. The positively charged hydrogen atoms of one water molecule attract the negatively charged oxygen atoms of another water molecule. This attraction of opposite charges between hydrogen and oxygen forms a weak bond called a **hydrogen bond.** Hydrogen bonds are important to organisms because they help hold many large molecules, such as proteins, together.

Figure 7.9

In water molecules, the electron-rich oxygen atom has a slight negative charge. Because they lack electrons, the hydrogen atoms have a slight positive charge. Because of water's bent shape, the protruding oxygen end of the molecule has a slight negative charge, while the end with protruding hydrogen atoms has a slight positive charge.

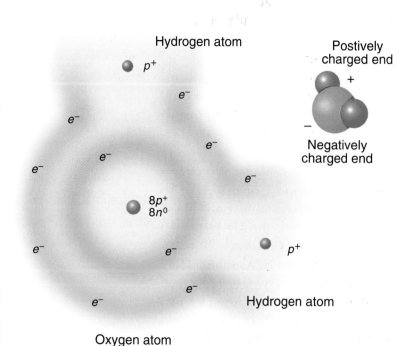

Hydrogen atom
p^+
e^-
e^-
e^-
e^-
e^-
$8p^+$
$8n^0$
e^-
e^-
p^+
e^-
Hydrogen atom
Oxygen atom

Postively charged end
+
−
Negatively charged end

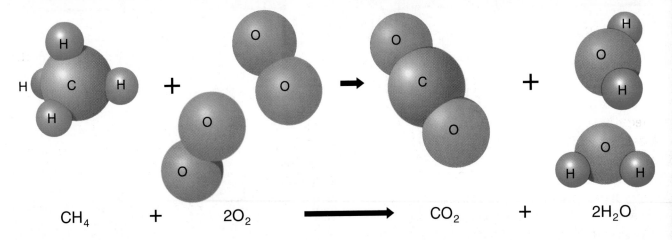

$$CH_4 \quad + \quad 2O_2 \quad \longrightarrow \quad CO_2 \quad + \quad 2H_2O$$

Figure 7.10

When methane burns in air, it reacts with two oxygen molecules. Notice how all the atoms that were present in the reactants (methane and oxygen) have been rearranged to produce the products (carbon dioxide and water). No atoms were lost, gained, or changed to different atoms. The equation tells you that one molecule of methane reacts with two molecules of oxygen to yield (produce) one molecule of carbon dioxide and two molecules of water.

Chemical Reactions

When chemical reactions occur, bonds between atoms are formed or broken, and substances change into different substances. In organisms, chemical reactions occur over and over inside cells. All the chemical reactions that occur within an organism are referred to as that organism's **metabolism.** These reactions break down and build molecules that are important for the functioning of organisms. Scientists represent a chemical reaction by writing a chemical equation. Chemical equations use chemical symbols and formulas to represent each element or substance.

Writing chemical equations

The events that take place when methane burns in air are pictured in *Figure 7.10.* Substances that undergo chemical reactions, such as the methane and oxygen in *Figure 7.10,* are called reactants. Substances formed by chemical reactions, such as the carbon dioxide and water in the figure, are called products.

It's easy to tell how many molecules are involved in the reaction because the number before each chemical formula indicates the number of molecules of each substance. The subscript numbers in a formula indicate the number of atoms of each element in a molecule of the substance. A molecule of table sugar can be represented by the formula $C_{12}H_{22}O_{11}$. The lack of a number before a formula or under a symbol indicates that only one atom or molecule is present.

Looking at the equation in *Figure 7.10,* you can see that methane is composed of one atom of carbon and four atoms of hydrogen. Oxygen is composed of two-atom molecules. Perhaps the easiest way to understand chemical equations is to know that atoms are never created or destroyed in chemical reactions. They are simply rearranged. Therefore, an equation is written so that the same numbers of atoms of each element appear on both sides of the arrow. In other words, equations must always be written so that they are balanced.

metabolism:
metabole (GK) change
Metabolism involves many chemical changes.

The Uniqueness of Water

*W*ater is the most common liquid on Earth. It covers three-fourths of Earth's surface and makes up from 70 to 95 percent of the weight of living things. Water has some extraordinary properties. In particular, a water molecule's ability to attract and bond to other water molecules makes it especially important to living organisms. You already know that water is an excellent solvent. Here are a few more examples of how water is important to living systems.

Water has a high surface tension

Because water is a polar molecule, it easily attracts other water molecules. This attraction causes the surface layer of water molecules to act like a stretched film over the surface of the water. This property of water is called surface tension. Thus, the water strider can stand on top of the water, with surface tension supporting its body weight. The nonpolar wax molecules on the surface of a leaf have little attraction for water, so the water beads up because of its surface tension.

Water creeps up in thin tubes

Plants also take advantage of the great attraction water has for itself and other molecules. In fact, it is this property of water that allows plants to get water from the ground. Water creeps up the thin tubes in plant roots and stems. This property is called capillary action and plays a major role in getting water from the soil to the tops of even the tallest trees.

Water has a high heat of vaporization

Because water strongly attracts other water molecules, it takes a lot of energy to make water evaporate. As water heats, its molecules vibrate faster and faster. As more heat is applied, molecules begin to separate from the rest and evaporate into the air. This high heat of vaporization helps cool the human body. When sweat evaporates from human skin, it obtains heat of vaporization from the body itself, thus helping keep the body cool.

Water resists temperature change

Water requires more heat to increase its temperature than do most other common substances. Likewise, water must lose a lot of heat when it cools. This property of water is extremely important because temperature variations have a strong effect on the kinds of organisms that can live in a given region. Because they are near large bodies of water, coastal areas vary less in temperature than inland areas at the same latitude. Thus, organisms such as wine grapes that cannot tolerate temperature extremes can live in these areas.

Water expands when it freezes

Water is one of the few substances that expands when it freezes. Because of this property, ice is less dense than liquid water and floats as it forms in a pond. Water expands as it freezes inside the cracks of rocks. As it expands, it often breaks apart the rocks. Over long time periods, this process will form soil.

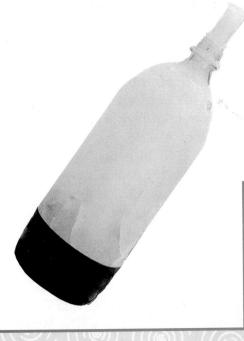

EXPANDING YOUR VIEW

1. **Applying** If you observe plants after a rain or in the early morning dew, you'll see water beaded up on the leaves. What does this tell you about the surface of plant leaves?

2. **Understanding Concepts** What characteristic of water molecules helps explain all the properties discussed here?

Earth Science

Exploding Sodas, Potholes, and Soil

Have you ever put a warm can of soda in the freezer to cool and then forgotten it? If you have, you know that water expands as it freezes. The can may only bulge, but it may explode, spraying soda all over the inside of the freezer. What's happening? How is it related to potholes and the formation of soil?

The pothole effect   Most substances contract, taking up less space as they freeze. But, unlike other substances, water expands as ice crystals form. Potholes in streets and highways are created by the same action that caused the can of soda to explode. Water seeps into cracks in the road surface. When the water freezes, it expands, exerting a great deal of force, which makes the crack wider and deeper. With enough freezing and thawing, an entire chunk of roadway breaks loose. Also because of freezing and thawing—plus traffic—the chunk becomes pulverized, leaving a pothole in the road.

Making soil   In the same way, exposed rocks are cracked and pulverized, eventually weathering into small particles of soil. The soil in an area can be less than a centimeter to more than 5 m thick. Soil holds moisture, provides minerals and nutrients for plant growth, and acts as a filter for water flowing through it to underground reservoirs.

CONNECTION TO Biology

Cells can be thought of as sacks containing mostly water. Why do strawberries and some other fruits become soft and mushy when frozen and then thawed?

Figure 7.11

The pH scale has a range from 0 to 14. Pure water has a pH of 7. A solution with a pH of 7 is neutral because there are equal numbers of hydrogen and hydroxide ions. As the pH of a solution becomes lower, the concentration of hydrogen ions becomes greater. A solution with a pH below 7 is acidic. As the pH of a solution becomes higher, the concentration of hydroxide ions becomes greater. A solution with a pH above 7 is basic.

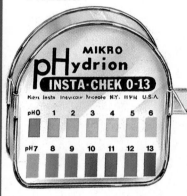

Acids and bases

Chemical reactions can occur only when conditions are right. For example, a reaction might depend on temperature, the availability of energy, or a certain concentration of a substance dissolved in solution. Chemical reactions in organisms also depend on the pH of the environment. The **pH** is a measure of how acidic or basic a solution is. A scale with values ranging from 0 to 14 is used to measure pH. *Figure 7.11* depicts a typical dispenser of pH paper.

In Chapter 6, you learned how acids and bases are important in the environment. For example, some plants grow well only in acidic soil, while others require soil that is basic. An **acid** is any substance that forms hydrogen ions (H^+) in water. When the compound hydrogen chloride (HCl) is added to water, hydrogen ions (H^+) and chloride ions (Cl^-) are formed. Thus, hydrogen chloride in solution is called hydrochloric acid. This solution contains an abundance of H^+ ions and has a pH below 7.

A **base** is any substance that forms hydroxide ions (OH^-) in water. For example, if sodium hydroxide (NaOH) is dissolved in water, it forms sodium ions (Na^+) and hydroxide ions (OH^-). This solution contains an abundance of OH^- ions and has a pH above 7.

MiniLab

How can you determine the pH of common household items?

The pH of a solution is a measurement of how acidic or basic that solution is. An easy way to measure the pH of a solution is to use pH paper. This paper has been treated with chemical indicators whose color varies according to pH.

Procedure

1. Pour a small amount (about 5 mL) of each of the following into separate clean, small beakers or other small glass containers: lemon juice, household ammonia solution, liquid detergent, shampoo, and vinegar.
2. Dip a fresh strip of the pH paper into each solution briefly and remove.
3. Compare the color of the wet paper with the pH color chart, and record the pH of each material.

Analysis

1. Which solutions were acids?
2. Which solutions were bases?
3. What ions in the solution were causing the pH paper to change? Which solution contained the highest concentration of hydroxide ions? How do you know?

Strong acid — Stomach acid — Soft drinks — Tomatoes — Coffee — Urine — Milk — Pure water — Blood — Sea water — Detergent — Household ammonia — Hair remover — Oven cleaner

0 1 2 3 4 5 6 7 8 9 10 11 12 13 14

Acidic — Neutral — Basic →

Section Review

Understanding Concepts

1. Why do atoms combine chemically with other atoms?
2. How does the formation of an ionic bond differ from the formation of a covalent bond?
3. What can you say about the proportion of hydrogen ions and hydroxide ions in a solution that has a pH of 2?

Thinking Critically

4. Explain why water dissolves so many different substances.

Skill Review

5. **Interpreting Scientific Illustrations** Study the diagram in *Figure 7.8*, which shows the process of a polar compound dissolving in water. In your own words, describe the process step-by-step. Tell what the water molecules are doing and why. Describe what is happening to the sugar molecules and why. Describe the nature of the mixture after the compound dissolves. For more help, refer to Thinking Critically in the *Skill Handbook*.

Meet Dr. Baldomero Olivera, Biochemist

Have you ever had a chance to look at the circuit board of a computer? All that complicated circuitry is simple when compared to the working of your nervous system! Dr. Baldomero Olivera is conducting research that may someday help unravel the puzzle of how the nervous system works.

In the following interview, Dr. Olivera describes his research and tells how his early childhood interests pointed the way to his career as a biochemist.

On the Job

Q Dr. Olivera, could you tell us about your current work?

A I'm a biochemist, a chemist who deals with chemical processes that take place in living things. In our lab, we research the biochemistry of the nervous system. We're interested, ultimately, in how the nervous system works. We'd like to understand how the components are put together to enable nerve cells to signal each other. A problem in studying a whole system is that we can't break it apart to take a close look because then, of course, it is no longer a system. So we've had to find a way of studying the system while keeping it intact.

Q What approach is your research taking?

A It's actually somewhat unusual. Years ago, we began studying the effects of venoms of marine snails on nerve cells. Initially, we were just trying to figure out why these venoms could kill people. We soon discovered that the venoms are loaded with components that very specifically interfere with the function of particular molecules in nerve cells. That led to an entirely new study, the one we're concerned with now, which is tracking the detailed functions of all the component molecules of the nervous system. Basically, what we do is use molecules from venoms as a way to enter the molecules of the nerve cells. We mark molecules from the venom with radioactive tags, and then trace the radioactivity to the specific targets in the nervous system. The process is rather like a tracking device to pinpoint areas that are crucial for a particular function in the nervous system.

Q What kinds of practical applications might come from your research?

A It's pure research, so our main goal is, first and foremost, to understand how nervous systems work. Of course, that kind of understanding has a lot of potential applications because knowing how the brain works could lead to figuring out what goes wrong to cause mental illness. Then there would be a chance of coming up with ways to alleviate the symptoms of these illnesses.

Q Do you think your current research will be wrapped up in a few years?

A No, certainly not within my lifetime. To discover how the brain works is a really big job—one we're just starting.

Early Influences

Q Can you remember how you first got interested in science?

A Even though it was a long time ago, I still remember very clearly an experiment that my second-grade teacher demonstrated to our class. She brought in a bunch of substances—ground-up chalk, sugar, flour, and a few more. As she put each powderlike substance into a tube with water, she said that some things dissolve in water and some don't. She suggested that we could figure out for ourselves which ones do and which ones don't. I still remember the excitement of discovery of that day.

Q Did you have any interests when you were growing up that directly relate to your present research?

A Growing up in the Philippines, I used to collect snails. That's how I knew about venomous sea snails. If I hadn't had that hobby, the present research using snail venom might never have begun.

Personal Insights

Q Do you have hobbies today that relate to other fields of science?

A My family and I got interested in fossil hunting when we were in Germany several years ago. And then we discovered that here in Utah, there are a lot of interesting opportunities for amateur paleontologists. Many biochemists are excited about trying to extract DNA from old bones. It's a situation in which different disciplines of science are interfacing.

Q What advice would you give to students who are interested in a career in biochemistry?

A I think research is a lot of fun. I like discovering something new. If kids have a strong streak of curiosity, they might enjoy the field. As preparation for any kind of career, I would advise kids to spend time developing hobbies and doing the things that interest them, because sometimes hobbies can lead to a profession.

7.3 Life Substances

Section Preview

Objectives

Explain why there is a large variety of organic compounds.

Explain how polymers are formed and broken down in organisms.

Compare the chemical structures of carbohydrates, lipids, proteins, and nucleic acids, and explain the importance of these substances in living things.

Key Terms

isomer
polymer
carbohydrate
monosaccharide
disaccharide
polysaccharide
starch
glycogen
cellulose
lipid
protein
amino acid
peptide bond
enzyme
nucleic acid
nucleotide
DNA
RNA

Did you ever hear the saying, "You are what you eat"? It's at least partially true because the compounds that form the cells and tissues of the body are produced from similar compounds in the foods you eat. Common to most of these foods and to most substances in organisms is the element carbon. The first carbon compounds that scientists studied came from organisms. Because of this, they were called organic compounds.

Role of Carbon in Organisms

A carbon atom has four electrons available for bonding in its outer energy level. In order to become stable, carbon atoms form four covalent bonds. Follow the illustration of carbon atoms and bonding in *Figure 7.12.* Carbon can bond with other carbon atoms, as well as with many other elements. When carbon atoms bond to each other, they can form straight chains, branched chains, or rings. In addition, these chains and rings can have almost any number of carbon atoms and can include atoms of other elements as well. This property makes a huge number of carbon structures possible.

A great variety is possible in organic molecules. In fact, compounds with the same simple formula often differ in structure. Compounds that have the same simple formula but different three-dimensional structures are called **isomers.** Both

Single bond	Double bond	Triple bond
e^- e^-	e^- e^- e^- e^-	e^- e^-
		e^- e^-
		e^- e^-
$\geq$C—C$\leq$	$>$C$=$C$<$	$-$C$\equiv$C$-$

Figure 7.12

When two carbon atoms bond, they share one electron each and form a covalent bond. They can also share two or three electrons. When each atom shares two electrons, a double bond is formed. A double bond is represented by two bars between carbon atoms. When each shares three electrons, a triple bond is formed. Triple bonds are represented by three bars drawn between carbon atoms.

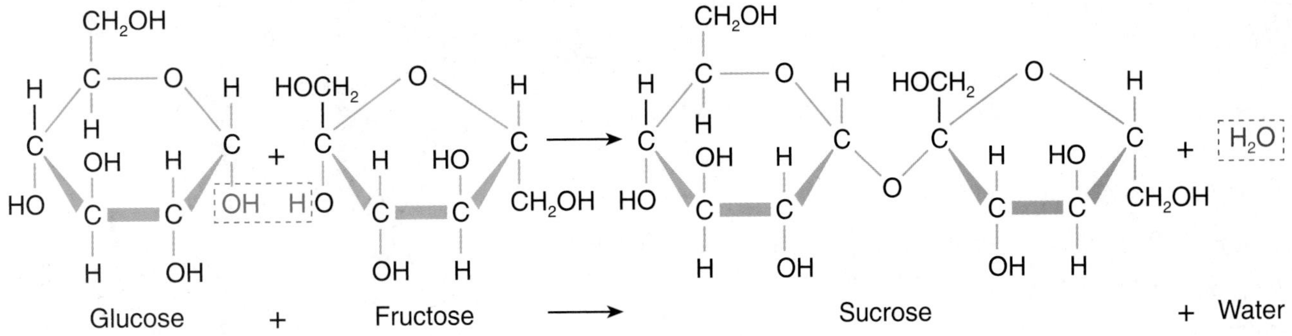

Glucose + Fructose → Sucrose + Water

Figure 7.13

The different arrange-
ment of hydrogen and
oxygen atoms around
each carbon atom
gives glucose and
fructose molecules dif-
ferent chemical prop-
erties. When glucose
and fructose combine,
they form the disac-
charide, sucrose, also
known as table sugar.

glucose and fructose, shown in *Figure 7.13*, have the same simple formula, $C_6H_{12}O_6$, but different structures.

Molecular chains

Carbon compounds also vary greatly in size. Some compounds contain one or two carbon atoms, while others contain tens, hundreds, or even thousands of carbon atoms. These large molecules are called macromolecules. Proteins are examples of macromolecules in organisms. Cells build macromolecules by bonding small molecules together to form chains called polymers, as shown in *Figure 7.14*. A **polymer** is a large molecule formed when many smaller molecules bond together, usually in long chains.

Figure 7.14

Condensation and Hydrolysis

Ⓐ In condensation, the small molecules that are bonded together to make a polymer have an −H and an −OH group that can be removed to form H−O−H, a water molecule. The subunits become bonded by a covalent bond.

Ⓑ Hydrolysis involves the breaking apart of a polymer by the addition of water. Hydrogen ions and hydroxide ions from water attach to the bonds between the subunits that make up the polymer. Hydrolysis takes place during the digestion of most food molecules.

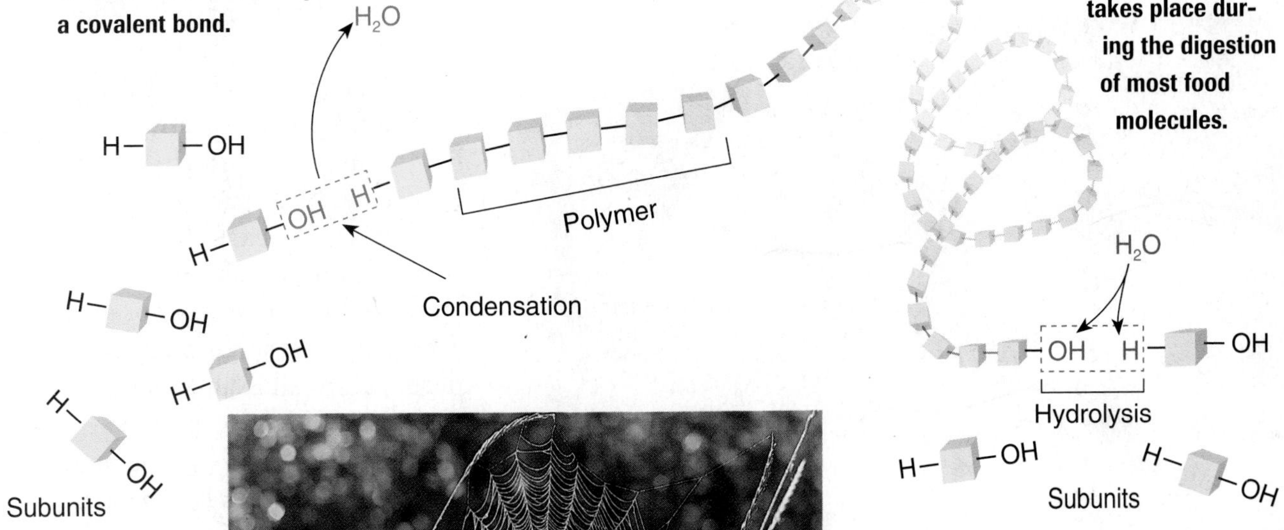

Ⓒ Spider silk is one example of protein, a biological polymer formed by condensation reactions.

A Trip to Mars

In the summer of 1975, two *Viking* spacecraft, each consisting of an orbiter and an attached lander, were launched to Mars from Cape Canaveral. One of their missions was to look for signs of life.

Signs of life Something that has the capacity to replicate or reproduce, repair, evolve, and adapt is considered to be alive. These properties appear to originate in large, organic (carbon-based) molecules. The Martian atmosphere consists of 95% carbon dioxide, 2.5% nitrogen and 1.5% argon, with traces of oxygen, carbon monoxide, and inert gases. The elements necessary for life are available on the planet, but there is virtually no water vapor, a substance essential to life as we know it. The only water on Mars is locked up in the form of ice in the polar caps. However, Mars has channels resembling dry river beds that cut through its desert terrain, suggesting that the atmosphere of Mars was once much different than it is today. The instruments that the landers carried were designed to look for evidence of life: complex, carbon-based molecules and by-products of metabolic activity in the soil.

Looking for life processes The robot arm of the lander scraped up soil. The soil was placed in a container, where it was mixed with nutrients and incubated under conditions that the scientists believed would encourage growth of microorganisms. Seven months later, no evidence of living things had been found.

As interesting as it would be to have found evidence of life on Mars, the findings indicate that there are no clear signs of life. These tests do not completely rule out the possibility of life on Mars, but the current data make it less likely.

Thinking Critically
Today, Mars is a cold, dry, and windy planet. Why would the appearance of dry river beds and features caused by erosion suggest that the climate of Mars was much different in the past?

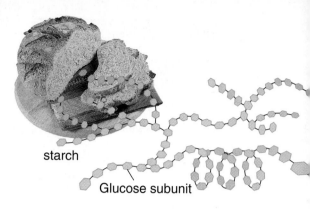

starch

Glucose subunit

The structure of carbohydrates

You may have heard of runners eating large quantities of spaghetti or other starchy food the day before a race. This practice is called carbohydrate loading and works because carbohydrates are used by cells to store and release energy. A **carbohydrate** is an organic compound composed of carbon, hydrogen, and oxygen with a ratio of about two hydrogen atoms and one oxygen atom for every carbon atom.

The simplest type of carbohydrate is a simple sugar called a **monosaccharide.** Common examples are the isomers glucose and fructose, which were shown in *Figure 7.13*. Two monosaccharide molecules can link together to form a **disaccharide**, a two-sugar carbohydrate. When glucose and fructose combine by a condensation reaction, a molecule of sucrose is formed, as shown in *Figure 7.13*. Sucrose is more commonly known as table sugar.

The largest carbohydrate molecules are **polysaccharides**, polymers composed of many monosaccharide subunits. Starch, cellulose, and glycogen are all examples of polysaccharides, as shown in *Figure 7.15*. **Starch** consists of highly branched chains of glucose units and is used as food storage by plants. Animals store food in the form of **glycogen,** another glucose polymer similar to starch but more highly branched. **Cellulose** is another glucose poly-

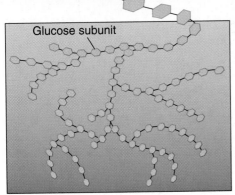

Glucose subunit

Glycogen

Cellulose

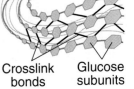

Crosslink bonds   Glucose subunits

Figure 7.15

Notice the structural differences among the polysaccharides starch, glycogen, and cellulose. Notice that all three are polymers of glucose.

mer that forms the cell walls of plants and gives plants structural support. Cellulose is also made of glucose units hooked together somewhat like a chain-link fence.

The structure of lipids

If you've ever tried to lose weight, you may have wished that lipids (fats) never existed. Actually, they are extremely important for the proper functioning of organisms. **Lipids** are organic compounds that have a large proportion of C–H bonds and less oxygen than carbohydrates. For

example, a common lipid found in beef fat has the formula $C_{57}H_{110}O_6$.

Lipids are commonly called fats and oils. They are insoluble in water because their molecules are nonpolar and, therefore, are not attracted by water molecules. Cells use lipids for long-term energy storage, insulation, and protective coatings. In fact, lipids are the major components of the membranes that surround all living cells. The most common type of lipid, shown in **Figure 7.16**, consists of three fatty acids bonded to a molecule of glycerol.

▼ Lipids that contain fatty acid chains of carbon with only single bonds are referred to as saturated fats. Examples include the fats in butter and steak. These fats are generally solid at room temperature.

Figure 7.16

Glycerol is a 3-carbon molecule that serves as a backbone for the lipid molecule. Attached to the glycerol are three fatty acids.

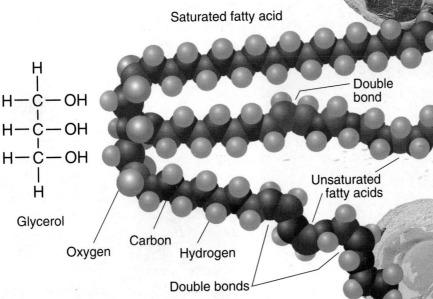

Saturated fatty acid

Double bond

H
|
H—C—OH
|
H—C—OH
|
H—C—OH
|
H

Glycerol

Oxygen   Carbon   Hydrogen

Double bonds

Unsaturated fatty acids

◄ Lipids that contain fatty acid chains of carbon with double bonds are called unsaturated fats. Unsaturated fats are usually liquids at room temperature. Examples are peanut, corn, and olive oils.

Peptide bond

$$H-N-C-C-OH + H-N-C-C-OH \longrightarrow H-N-C-C-N-C-C-OH + H_2O$$

Figure 7.17

Each amino acid contains a central carbon atom, to which is attached a carboxyl group (–COOH), a hydrogen atom, and an amino group (–NH₂). Also attached to the central carbon atom is a group (–R) that makes each amino acid different. To form protein chains, the hydrogen from an amino group and a hydroxyl group from a carboxyl group are removed to bond the amino acids together.

hydrolysis:

hydor (GK) water
lysis (GK) to split, loosen
In hydrolysis, molecules are split by water.

polymer:

poly (GK) many
meros (GK) part
A polymer has many bonded subunits (parts).

The structure of proteins

Proteins are essential to all life. They build structure and carry out cell metabolism. A **protein** is a large, complex polymer composed of carbon, hydrogen, oxygen, nitrogen, and sometimes sulfur. The basic building blocks of proteins are called **amino acids,** as shown in *Figure 7.17.* There are 20 common amino acids. Because there are 20 types of building blocks, proteins come in a large variety of shapes and sizes. In fact, proteins vary more in structure than do the other classes of biological organic molecules.

Amino acids are linked together by condensation, the removal of an –H and –OH group to form a water molecule. The covalent bond formed between amino acids is called a **peptide bond,** as shown in *Figure 7.17.* Protein chains contain hundreds of amino acids, and it is the order of amino acids that determines the kind of protein. Many proteins consist of two or more amino acid chains that are held together by hydrogen bonds.

Proteins are the building blocks of many structural components of organisms, as shown in *Figure 7.18.* Proteins are also important in muscle contraction, transporting oxygen in the bloodstream, providing immunity, and carrying out chemical reactions.

Figure 7.18

Besides lipids, proteins are also a major component of the membrane coverings of all cells. Proteins make up much of the structure of organisms, such as hair, horns, hoofs, and nails.

Enzymes are important types of proteins found in living things. An **enzyme** is a protein that speeds up a chemical reaction. In some cases, enzymes cause reactions that would otherwise occur so slowly that you could say they didn't occur at all. Some of these reactions *would* take place at high temperatures or in strongly acidic or basic solutions. Enzymes enable these reactions to occur under the conditions present in living cells. Enzymes are involved in nearly all metabolic processes. They speed the reactions in digestion of food, synthesis of molecules, and storage and release of energy. *Figure 7.19* shows how enzymes work to speed up chemical reactions.

The structure of nucleic acids

Nucleic acids are another important type of organic compound that is necessary for life. A **nucleic acid** is a complex macromolecule that stores information in cells in the form of a code. Nucleic acids are polymers made of smaller subunits called

ThinkingLab Draw a Conclusion

How does dilution affect enzyme action?

You have read that starch digestion begins in the mouth with enzymes that are secreted in saliva. You decide to see what effect diluting the saliva will have on the digestion of soda crackers, which are mostly baked flour.

Analysis

To verify that crackers contain starch, you grind up a cracker and add 5 mL of plain water. You add a drop of iodine solution, and it produces a dark blue color, indicating the presence of starch. Next, you repeat the test but use 5 mL of saliva instead of water. This time, the iodine produces a grayish-purple color instead of the blue color of a positive starch. Apparently, your saliva enzymes had broken down all the starch in the time it took to make the mixture and test it.

Now you decide to see what effect dilution of saliva will have on the digestion of the starch. You place one drop of saliva into 5 mL of water in a test tube, and add the same amount of ground cracker as before. You test the tube at various time intervals and obtain the data shown in the following table.

Time	Iodine Test for Starch
20 Seconds	Blue
40 Seconds	Blue
60 Seconds	Blue
80 Seconds	Blue
100 Seconds	gray-blue
120 Seconds	gray-purple

Thinking Critically

Draw a conclusion about the effect of dilution of the enzyme found in saliva on the digestion of starch, and explain the results.

Figure 7.19

Enzymes enable molecules, called substrates, to undergo a chemical change to form new substances, called products.

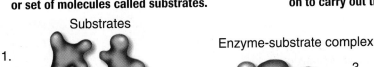

1 **Each enzyme acts on a specific molecule or set of molecules called substrates.**

4 **After the reaction, the enzyme releases the products and can go on to carry out the same reaction again and again.**

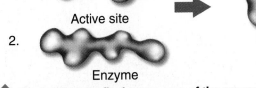

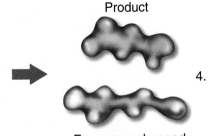

2 **Each substrate fits into an area of the enzyme called the active site. This fitting together is often compared to a lock-and-key mechanism. However, the enzyme changes shape a little to fit with the substrate.**

3 **In the enzyme-substrate complex, the enzyme holds the substrate or substrates in a position where a reaction can occur easily.**

The compound hydrogen peroxide, H_2O_2, is a by-product of metabolic reactions in most living things. However, hydrogen peroxide is damaging to delicate molecules inside cells. As a result, nearly all organisms contain the enzyme peroxidase, which breaks down H_2O_2 as it is formed. Potatoes are one source of peroxidase. Peroxidase speeds up the breakdown of hydrogen peroxide into water and gaseous oxygen. This reaction can be detected by observing the oxygen bubbles generated.

Does temperature affect an enzyme reaction?

PREPARATION

Problem

Does the enzyme peroxidase work in cold temperatures? Does peroxidase work better at higher temperatures? Does peroxidase work after being frozen or boiled?

Hypotheses

Make a hypothesis regarding how you think temperature will affect the rate at which the enzyme peroxidase breaks down hydrogen peroxide. Consider both low and high temperatures.

Objectives

In this Biolab, you will:
- **Observe** the activity of an enzyme.
- **Compare** the activity of the enzyme at various temperatures.

Possible Materials

thermometer
ice
beaker, 400-mL
hot plate
potato slices, 5-mm thick
kitchen knife
waxed paper
3% hydrogen peroxide
clock or timer

Safety Precautions

Be sure to wash your hands before and after handling the lab materials.

1. Decide on a way to test your group's hypothesis. Keep the available materials in mind.

2. When testing the activity of the enzyme at a certain temperature, consider the length of time it will take for the potato to reach that temperature, and how the temperature will be measured.

3. To test for peroxidase activity, add 1 drop of hydrogen peroxide to the potato slice and observe what happens.

4. When heating the potato slice, first place it in a small amount of water in a beaker. Then heat the beaker slowly so that the temperature of the water and the temperature of the slice are always the same. Try to make observations at several temperatures between 10°C and 100°C.

Check the Plan
Discuss the following points with other groups to decide on the final procedure for your experiment.

1. What data will you collect, and how will you record it?

2. What factors should be controlled?

3. What temperatures will you test?

4. How will you achieve those temperatures?

5. ***Make sure your teacher has approved your experimental plan before you proceed further.***

6. Carry out your experiment.

---

1. **Checking Your Hypothesis** Does your data support or reject your hypothesis? Explain your results.

2. **Analyzing Data** At what temperature did peroxidase work best?

3. **Identifying Variables** What factors did you need to control in your tests?

4. **Recognizing Cause and Effect** If you've ever used hydrogen peroxide as an antiseptic to treat a cut or scrape, you know that it foams as soon as it touches an open wound. How can you account for this observation?

Going Further

Changing Variables
To carry this experiment further, you may wish to use hydrogen peroxide to test for the presence of peroxidase in other materials such as cut pieces of different vegetables. Also, test raw beef and diced bits of raw liver.

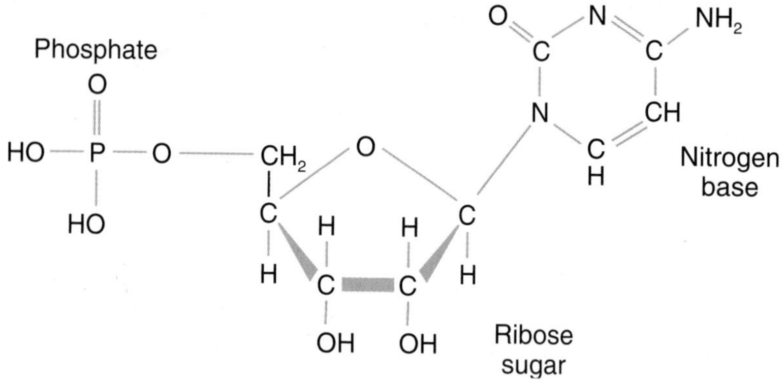

Phosphate

Nitrogen base

Ribose sugar

Figure 7.20

Each nucleic acid is built of subunits called nucleotides, which are formed from a sugar molecule bonded to a phosphate group and a nitrogen base.

nucleotides. Nucleotides consist of carbon, hydrogen, oxygen, nitrogen, and phosphorus atoms arranged in three groups—a base, a simple sugar, and a phosphate group, as shown in *Figure 7.20*. You have probably heard of the nucleic acid DNA, which stands for deoxyribonucleic acid. **DNA** is the master copy of an organism's information code. The information coded in DNA contains instructions used to form nearly all of an organism's enzymes and structural proteins. Thus, DNA determines how an organism looks and acts. DNA's instructions are passed on every time a cell divides and from one generation of an organism to the next. Thus, we say that DNA forms the genetic code.

Another important nucleic acid is RNA, which stands for ribonucleic acid. **RNA** is a nucleic acid that forms a copy of DNA for use in protein synthesis. The chemical differences between RNA and DNA are minor but important. Chapter 13 will discuss how DNA and RNA work together to produce proteins.

Connecting Ideas

Organic molecules found in cells are used to build and repair cell parts and to carry out the life processes in cells. Organic compounds in living things behave according to the same laws of chemistry and physics as compounds found in nonliving objects. How do you think the properties of organic compounds relate to their functions in cells? In the next three chapters, you will learn more about cell structures, their chemical makeup, and the reactions that store and release energy for cell metabolism.

Section Review

Understanding Concepts

1. List three important functions of lipids in living organisms.
2. Describe the process by which polymers in living things are formed from smaller molecules.
3. If there are only 20 different common amino acids, why is it possible to have thousands of different proteins?

Thinking Critically

4. As polymers containing carbon atoms pass through the carbon cycle, what happens to the covalent bonds between carbon atoms?

Skill Review

5. **Making and Using Tables** Make a table comparing polysaccharides, lipids, proteins, and nucleic acids. List these four types of biological substances in the first column. In the next two columns, list the subunits that make up each substance and the functions of each in organisms. In the last column, provide some examples of each from the chapter. For more help, refer to Organizing Information in the *Skill Handbook.*

Scanning the Mind

Advancements in medical technology have made medical tests such as X rays and magnetic resonance imaging (MRI) available for examining the human body in a noninvasive way. However, another technology has been added to the medical toolbox. Positron emission tomography (PET) scanners are a relatively new medical tool. This instrument is unique in that it allows a physician to see activity in the body.

How PET scans are made PET scanners are excellent tools for study of the human brain and its operation. Monitoring either the blood flow to an area or the amount of glucose being metabolized pinpoints the active sections of the brain.

The patient receives a compound containing a radioactive isotope by inhalation or injection. Because these isotopes emit detectable radiation, they can be tracked by the sensitive PET scanner. Computers create a "picture" of the brain activity by converting the energy emitted from the radioisotopes into a map. A colorful palette represents the various levels of activity.

Only a few PET scanners exist in the United States today because the machines can cost more than $6 million. Some insurance companies still question the need for expensive tests using PET technology.

Applications for the Future

PET scans have established which regions of the brain take part in a particular activity, such as listening to music. In addition, these scans reveal changes in the areas of the brain involved as a task is learned. Subjects perform a particular activity while the PET scanner records events in the brain. The computer then prepares a visual record of the blood flow or glucose consumption in areas of the brain during the activity. Current research suggests that learning at an early age, especially of subjects like foreign languages, may be critical to future learning for an individual.

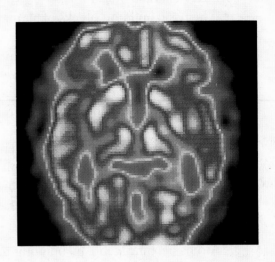

PET scans are also proving useful in the study of drug and alcohol addiction. The subject is given the addictive drug and then answers questions about his or her physical and emotional status while the scanner records metabolic activity in the brain. Researchers hope that information gained from the study of drug addiction will provide evidence for diagnosing and treating manic-depressive psychosis and schizophrenia.

INVESTIGATING the Technology

1. **Research** Read the March 1991 issue of *Scientific American*, pages 94 through 103. How can PET technology determine the mental mechanisms involved in the craving for a drug like cocaine?
2. **Debate** Hold a debate on the question, "Under what circumstances do you think an insurance company should pay for a PET scan?"
3. **Write About Biology** Research to find more information on noninvasive techniques. Include those mentioned here. Prepare a report or an illustrated poster describing these techniques, telling how they work, and discussing the most common applications.

Chapter 7 Review

7.1 Elements and Atoms

- The basic building blocks of all matter are atoms.
- Atoms consist of a nucleus containing protons and neutrons. The positively charged nucleus is surrounded by a cloud of rapidly moving, negatively charged electrons.
- Isotopes, atoms of the same element that differ in the number of neutrons in their nucleus, are useful in diagnosing and treating some diseases.

7.2 Interactions of Matter

- Atoms become stable by bonding to other atoms through covalent or ionic bonds.
- Water is the most abundant compound in living things. The polar property of water molecules makes it an excellent solvent.
- Chemical reactions are expressed by a balanced chemical equation.

7.3 Life Substances

- All organic compounds contain carbon atoms. There are four principal types of organic compounds that make up living things: carbohydrates, lipids, proteins, and nucleic acids.

Key Terms

Write a sentence that shows your understanding of each of the following terms.

acid	lipid
amino acid	metabolism
atom	mixture
base	molecule
carbohydrate	monosaccharide
cellulose	nucleic acid
compound	nucleotide
covalent bond	nucleus
disaccharide	peptide bond
DNA	pH
enzyme	polar molecule
glycogen	polymer
hydrogen bond	polysaccharide
ion	protein
ionic bond	RNA
isomer	solution
isotope	starch

Understanding Concepts

1. What is the charge on an ion formed when an atom loses two electrons? Explain your answer.
2. Why is the oxygen atom in a water molecule slightly more negative than the hydrogen atoms?
3. During a reaction, three amino acids were linked by peptide bonds. How many molecules of water were formed?
4. Why are foods with a large proportion of starch considered to be good energy sources?
5. What is the difference between a covalent bond and an ionic bond?

Using a Graph

6. A biologist measured the rate of enzyme activity at various temperatures and produced the following graph. What can you conclude about the relationship between temperature and enzyme activity?

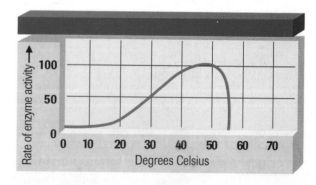

7. How do elements and compounds differ?

8. What is the function of an enzyme?

9. What is the importance of DNA to organisms?

Relating Concepts

10. Make a concept map that relates the following terms and phrases. Supply the appropriate linking words for your map. atom, element, compound, amino acid, peptide bond, protein, enzyme

Applying Concepts

11. A magnesium atom has 12 electrons. When it reacts, it usually loses two electrons. How does this loss make magnesium more stable?

12. Explain why water and a sponge would not be very effective in cleaning up a grease spill.

13. What is the purpose of the large amount of starch found in the seeds of many plants?

14. Heating a white substance produces a vapor and black material. Was the substance an element or a compound? Explain your answer.

15. Digestion of food cannot occur without sufficient water. Explain why this is so.

Earth Science Connection

16. Read about other properties of water in the Focus On feature on pages 170–171. What property is responsible for water entering the smallest cracks in the rocks?

Thinking Critically

Making Inferences

17. People sometimes take milk of magnesia to counteract an "acid stomach." If you checked the pH of milk of magnesia, would you expect it to be more or less than 7? Explain your choice.

Interpreting Data

18. The following graph compares the abundance of four elements in living things to their abundance in Earth's crust, oceans, and atmosphere. Which element is the most abundant in organisms? What can you say about the general composition of living things compared to nonliving matter near Earth's surface?

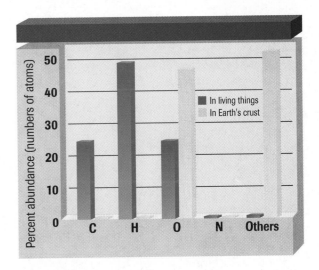

Extension

19. **Biolab** Consider that the action of peroxidase on hydrogen peroxide produces oxygen gas. Devise a way to perform a quantitative experiment that measures the rate of the reaction.

Making Inferences

20. **Minilab** Chemical indicators usually change color sharply at a certain pH. The paper you used to test pH contained a mixture of several indicators. Why do you think several indicators are necessary in pH paper?

Connecting to Themes

21. **Energy** Where in the carbon cycle do you find carbon compounds that are important to living organisms?

22. **Unity Within Diversity** It is said that all life on Earth is based on carbon. What feature of carbon makes this statement true?

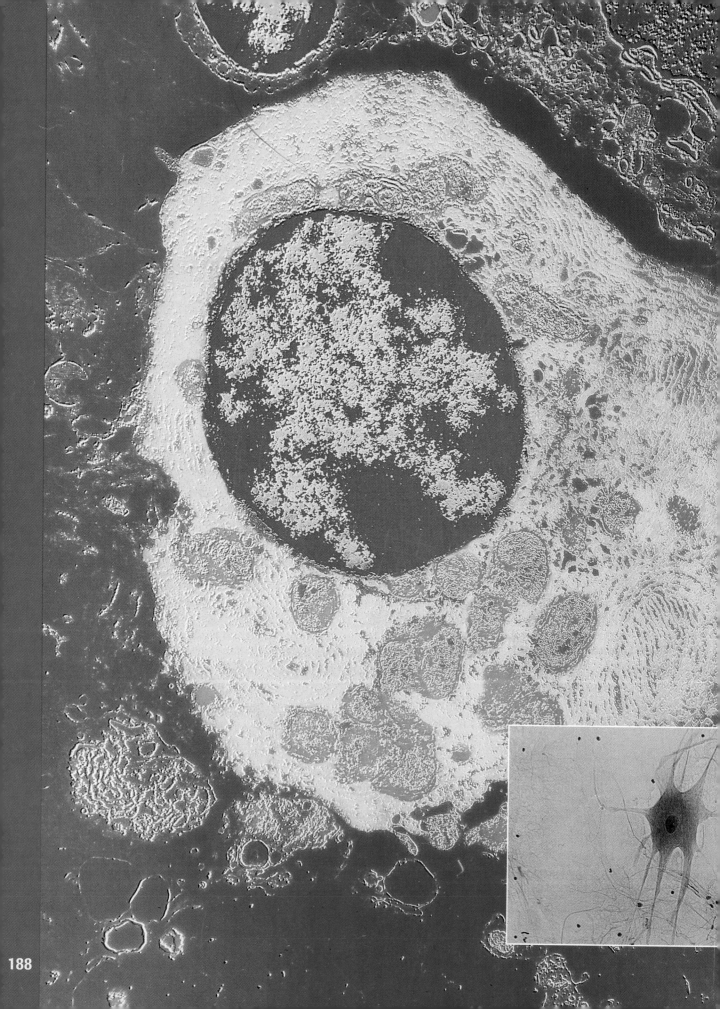

8 A View of the Cell

What comes to mind when you think of visiting a busy pizza parlor during a fall evening? Can you smell the aroma of cheese, onions, and pepperoni or picture the employees busy making pizzas? Can you hear music playing in the background as well as the sounds from people laughing and video games being played? As the front door opens, cold crisp air rushes in. People come in and go out, taking boxes of pizza-to-go. What a flurry of activity is going on!

Cells, the basic units of life, can be likened to a pizza parlor. They, too, are a constant flurry of activity. These cells, so small that at least 50 000 of them will fit into this letter O, are busy building and breaking down macromolecules. They are at work releasing energy from foods, and then using that energy to make needed cell parts. Together, your cells function to make your body operate like a well-run pizza business.

Magnification: 2800×

This large, white blood cell, magnified 3325 times, is one of many millions and trillions of cells that make up a mature human. The nerve cell in the inset photo (enlarged 150 times) has a shape that is adapted for transmitting electrical impulses to the brain. The euglenas shown on this page are unicellular organisms. The single cell performs all the functions necessary for the organism to survive.

8.1 The Discovery of Cells

From the smallest bacterium to the largest redwood tree, *all living things are made of cells. Each cell carries out life functions. However, microscopes had to be invented before scientists could study—or even discover—cells, because most cells are so tiny that they cannot be seen without the aid of a microscope. The unaided human eye can see something as small as 0.01 cm, or about the size of a tiny pencil dot on a piece of paper. An average-sized cell, however, has a diameter of about 0.002 cm. No wonder most cells can't be seen.*

Magnification: 4000×

The History of the Cell Theory

Before microscopes were invented, people believed that diseases were caused by curses and supernatural spirits. They had no idea that organisms such as bacteria existed. The invention and development of the microscope enabled scientists to discover and study **cells,** the basic units of living organisms.

Development of light microscopes

The first microscopes used for viewing cells were simple microscopes—basically a single magnifying lens. In the 1600s, Anton van Leeuwenhoek described living cells as seen through a simple microscope.

During the next 200 years, microscopes were greatly improved. Biologists began to use compound light microscopes. A **compound light microscope** has a series of lenses that magnify an object in steps. Visible light is passed through the object and then through the lenses. The object may be a bacterium or a cell part as small as 0.00002 cm (0.2 μm) in diameter. The compound light microscope allowed biologists to achieve greater magnification and to see more detail inside cells.

The cell theory is developed

In 1665, Robert Hooke, an English scientist, used a crude compound microscope to examine thin slices of cork from the bark of an oak

microscope:

mikros (GK) small
skopein (GK) to
  look
A microscope is used to examine small objects.

tree, as shown in *Figure 8.1.* He observed that the cork was composed of tiny, hollow boxes similar to a honeycomb. Hooke called the boxes *cells* because they reminded him of the small, boxy living quarters of monks. Their rooms were, in fact, called cells. Today, we know that the structures Hooke observed were the walls of dead plant cells. Hooke published his drawings and descriptions, which led other scientists of that time to look for evidence of cells.

In the 1830s, two German scientists, Matthias Schleiden and Thomas Schwann, were able to view many different organisms with microscopes and draw some important conclusions. Schleiden used a microscope to observe plants and concluded that all plants are composed of cells. Schwann made similar observations about animal cells.

The observations and conclusions of scientists from the late 17th century to the time of Schwann and Schleiden are summarized into the cell theory, one of the most basic ideas of modern biology.

Figure 8.1

Robert Hooke observed cork cells such as these using a crude compound microscope that magnified only 30 times.

Magnification: 250×

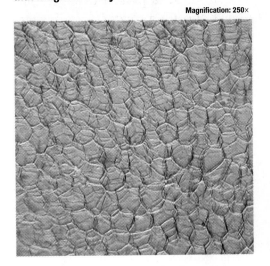

MiniLab

What do cork cells look like?

Robert Hooke discovered and named cells when he looked at cork with a crude microscope. Cork is a dead, protective tissue that is abundant on the trunk of some trees such as cork oak. It is periodically stripped off for making commercial products, such as bottle corks.

Procedure

1. Carefully shave a paper-thin piece of cork off a bottle cork with a razor blade.
2. Place the piece of cork in a drop of water on a microscope slide. Add a coverslip.
3. Observe the cells at the edge of the piece of cork on low and high power. Make a drawing of the cells at each power.

Analysis

1. What shape are the cork cells?
2. What is inside the cork cells?
3. What function might the cork provide for the tree?

The **cell theory** is made up of three main ideas.

1. *All organisms are composed of one or more cells.* An organism may be a single cell, such as an amoeba or a bacterium. Larger organisms, such as humans, are composed of many cells.
2. *The cell is the basic unit of organization of organisms.* In the same way that the basic unit of matter is the atom, the basic unit of life is the cell.
3. *All cells come from preexisting cells.* Cells come from the reproduction of previously existing cells, a process in which cells reproduce to make exact copies of themselves.

Development of electron microscopes

During the 1940s, more powerful microscopes were developed. Rather than using light, they used electrons that passed over or through an

The History of Microscopes

1600   1665   170●

The invention and development of the light microscope about 300 years ago enabled scientists to discover and study cells. Improved versions of the microscope have allowed scientists to increase the range of visibility enormously, and they can now see molecules.

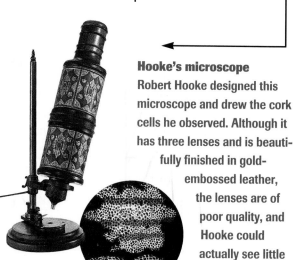

Hooke's microscope
Robert Hooke designed this microscope and drew the cork cells he observed. Although it has three lenses and is beautifully finished in gold-embossed leather, the lenses are of poor quality, and Hooke could actually see little detail.

TEM A transmission electron microscope (TEM) aims a beam of electrons through an object in a vacuum. The more dense portions allow fewer electrons to pass through. These denser areas appear darker on the screen. Transmission electron microscope images are two dimensional. The TEM is used to study the details of cell parts. It can magnify objects hundreds of thousands of times.

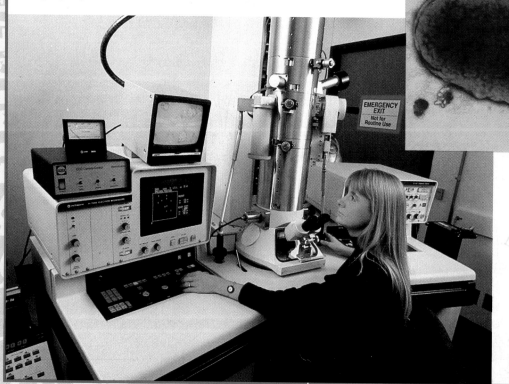

Magnification: 67 500×

TEM of the bacterium, *E. coli*

Transmission electron microscope

1800   1900   1940

Electron Microscope

Modern light microscopes A compound light microscope uses two or more glass lenses to magnify objects. Light microscopes can be used to look at living cells and small organisms, as well as at preserved cells. These microscopes can magnify up to about 1500×.

Magnification: 63×

Light micrograph of paramecium

Better lenses Anton van Leeuwenhoek improved upon the simple microscope. He built more than 240 microscopes, grinding the lenses himself. Although his microscopes had only a single lens, the quality was far better than that of Hooke's compound microscope. Leeuwenhoek was the first to discover and describe red blood cells and bacteria taken from scrapings off his teeth.

STM The scanning tunneling microscope (STM) can show the arrangement of atoms on the surface of a molecule, such as this DNA molecule. A metal probe is brought near the object's surface. Electrons flow from the surface to the probe. In this way, the hills and valleys of the surface can be mapped.

Magnification: 1 000 000×

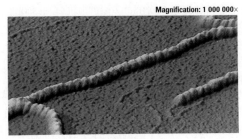

STM of DNA

SEM A scanning electron microscope (SEM) sweeps a beam of electrons over the surface of a specimen. This causes electrons to be emitted from the specimen. Scanning electron microscopes produce a realistic, three-dimensional picture of the object, but only the surfaces of specimens can be observed. The SEM can magnify only about 60 000 times without losing clarity.

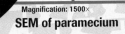

Magnification: 1500×
SEM of paramecium

EXPANDING YOUR VIEW

1. **Comparing and Contrasting** Compare the images seen with an SEM and a TEM.
2. **Thinking Critically** Can live specimens be examined with an electron microscope? Explain. Consider how the specimen must be prepared for viewing.

Focus On the History of Microscopes   **193**

Resolution of Microscopes

The quality of a microscope and its ability to magnify depend on its resolving power. Resolution is the minimal distance that two points can be separated and still be seen as two separate and distinct points. If two details inside an organism are closer than the resolution of the microscope, they will be seen as one.

The cells of the retina in the eye either "fire" if light strikes them, or do not fire if light does not strike them. As a result, the image that is produced in the retina is a pattern of light and dark spots. These spots resemble the appearance of a magnified newspaper photograph, above.

Light from two points separated by a distance of less than 0.01 cm activates the same retinal cell. The two points then are perceived as only a single point. It is only when light from two points activates two retinal cells separated by at least one unactivated cell that the two points can actually be seen as two distinct points.

Resolution and the compound light microscope   One way to increase resolution is to add a lens between your eye and the object. If small details can be made to appear larger, the distance between points will be increased. The theoretical limit of the compound light microscope depends on the wavelength of light. Light must be able to go between two points in order for them to be seen distinctly. One can calculate that the maximum magnification available to a compound light microscope is about 1500x. If objects are magnified further, the details look bigger but not clearer. Finer detail cannot be resolved.

Resolution and the electron microscope   The electron microscope allows for greater resolution because it uses a beam of electrons, which have a much shorter wavelength than that of light. The theoretical resolving power of an electron microscope is about 1000 times greater than that of the light microscope.

CONNECTION TO   Biology

Why was the development of the electron microscope with its greater resolving power so important to the advancement of cell biology?

---

object. These powerful microscopes are called electron microscopes. An **electron microscope** aims a beam of electrons through a magnetic field to focus them, then through or over the surface of a specimen in a vacuum, and finally onto a fluorescent screen where it forms an image. The image can be viewed or photographed. Because the specimen must be in a vacuum, only dead cells or organisms can be viewed. Several types of electron microscopes have been developed. The development of the electron microscope has provided much detail about the structure of cell parts and has allowed scientists to view objects even as small as molecules.

Two Basic Cell Types

As biologists studied cells, they found two basic kinds—prokaryotic and eukaryotic, depending on their internal organization. Organisms that

Figure 8.2

Prokaryotic cells do not have a true nucleus or other internal organelles surrounded by a membrane. Their DNA is concentrated in a region called the nucleoid, but it is not separated from the rest of the cell by a membrane. Most of a prokaryote's metabolic functions take place in the cytoplasm.

Magnification: 30 500×

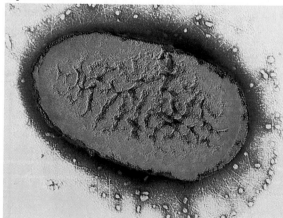

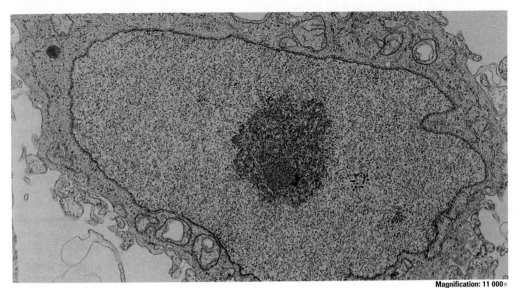

Figure 8.3

Eukaryotic cells have a true nucleus and discrete membrane-bound organelles, which allow different parts of the cell to specialize in different functions. The majority of cells in the living world are eukaryotic.

Magnification: 11 000×

are made of these two kinds of cells are called prokaryotes and eukaryotes, respectively. A **prokaryote,** shown in *Figure 8.2*, is an organism with a cell that lacks internal structures surrounded by membranes. Most prokaryotes are single-celled organisms. A **eukaryote,** shown in *Figure 8.3*, is an organism that has cells containing internal, membrane-bound structures. These structures are called **organelles.** Each organelle has a membrane surrounding it, isolating it from the rest of the cell. The largest organelle is a membrane-bound **nucleus,** which contains the cell's DNA and manages the cell's functions. Eukaryotes are either single celled or made of many cells.

The evolution of organelles in eukaryotes provided a way for the cell to be divided into compartments. Most of the internal organelles are surrounded by membranes, making them separate "rooms" and isolating them from the rest of the cell's contents. Thus, the activity of an individual compartment can become specialized. Chemical reactions that could not occur in the same area can be carried out at the same time in different compartments.

organelle:
organon (GK) tool, implement
ella (GK) small
Organelles are small, membrane-bound structures in cells.

Section Review

Understanding Concepts

1. Why was the development of microscopes necessary for the study of cells?
2. Compare the way images are formed in light microscopes and electron microscopes.
3. How are prokaryotic and eukaryotic cells different?

Thinking Critically

4. Suppose you were to discover a new microorganism in some ocean water. Applying the cell theory, what can you say for certain about this organism?

Skill Review

5. **Care and Use of a Microscope** A compound light microscope has three objective lenses that magnify 6, 40, and 95 times. What magnifications are available if an eyepiece that magnifies 15 times is used? For more help, refer to Practicing Scientific Methods in the *Skill Handbook*.

8.2 Eukaryotic Cell Structure

During the late 1800s and early 1900s, scientists began to observe the internal organization of the eukaryotic cell and learn the functions of each organelle. The amazing organization of a cell enables it to function in an efficient manner—somewhat like the well-run pizza business you read about earlier. Think about what organization is needed to prepare a pizza. Raw materials such as cheese, pepperoni, and mushrooms must be delivered to the pizza business. Someone has to assemble these raw materials, bake them, and slide the pizza into a box. A cell also takes in raw materials and assembles them into a usable form. The cell puts together proteins, carbohydrates, and lipids to build cell structures and to carry out its life processes.

Boundaries and Control

When you enter a pizza parlor, you go through a door in the wall—the boundary of the business. Cells also have boundaries with passageways for materials to enter or exit.

Cells must have boundaries

A cell has a **plasma membrane** that serves as a boundary between the cell and its external environment. Unlike the walls of the pizza business, though, the plasma membrane is quite flexible and allows the cell to vary its shape if needed. The plasma membrane, shown in *Figure 8.4*, controls the movement of materials that enter and exit the cell. It allows useful materials such as oxygen and nutrients to enter, and waste products such as excess water to leave. Some materials enter and leave through protein passageways. Other materials pass directly through the membrane. The plasma membrane helps maintain a chemical balance within the cell.

Some cells have another external boundary outside their plasma membrane. This additional boundary, the **cell wall,** shown in *Figure 8.5,* is a relatively inflexible structure that surrounds the plasma membrane. The cell wall is much thicker than the plasma membrane and is made of different substances in different organisms. The cells of plants, fungi,

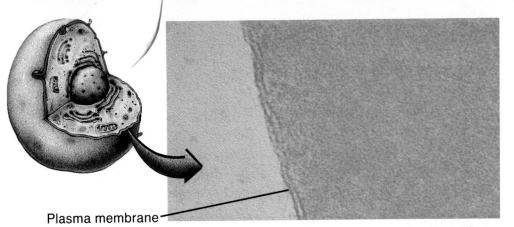

Plasma membrane

Magnification: 415 000×

Figure 8.4

The plasma membrane that surrounds a cell is made of two layers of lipid and protein, which you can distinguish in the photomicrograph. Two-layered membranes similar to the plasma membrane also enclose all of the membrane-bound organelles inside a cell. Is the photo a TEM or an SEM? How do you know?

almost all bacteria, and some protists have cell walls. Animal cells have no cell walls. Plant cell walls contain cellulose molecules, which form fibers. The fibers are interwoven to produce a strong network that protects the cell and gives the plant support. It is this fibrous cellulose of plants that provides the bulk of the fiber in our diets. Chitin, a nitrogen-containing polysaccharide, makes up the cell walls of fungi.

Organelles that control cell functions

Just like the pizza business, the cell needs a manager within its boundaries. The manager controls and directs the affairs of the pizza business. The nucleus of the cell is the organelle that manages cell functions in a eukaryotic cell. The nucleus is surrounded by a nuclear envelope, which is a double membrane; each membrane is made up of two layers. Thus, the nuclear envelope is four

Figure 8.5

The cell wall is a firm structure that protects the cell and gives the cell its shape. Plant cell walls are made mainly of multiple layers of cellulose, top right, but also contain pectin, the substance that causes jams and jellies to thicken. The fibrous nature of cellulose, magnified 20×, can be seen in the photomicrograph, bottom right.

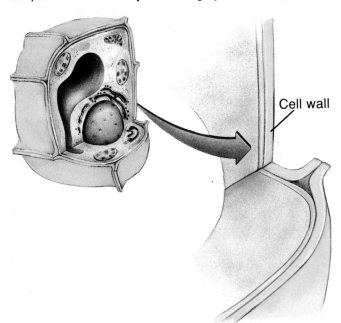

Cell wall

Magnification: 3000×

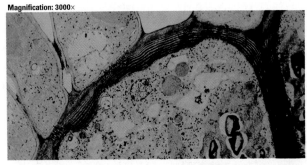

What organelle directs cell activity?

Acetabularia, a type of marine algae, grows as single, large cells 2 to 5 cm in height. The nuclei of these cells are in the "feet." Different species of these algae have different kinds of caps, some petal-like and others that look like umbrellas. If a cap is removed, it quickly grows back. If both cap and foot are removed from the cell of one species of algae and a foot from another species is attached, a new cap will grow. This new cap will have a structure with characteristics of both species. Then, if this new cap is removed, the cap that grows back will be like the cell that donated the nucleus.

The scientist who discovered these properties was Joachim Hämmerling. He wondered why the first cap that grew had characteristics of both species, yet the second cap was clearly like that of the cell that donated the nucleus.

Analysis
Look at the diagram below and interpret the data to explain the results.

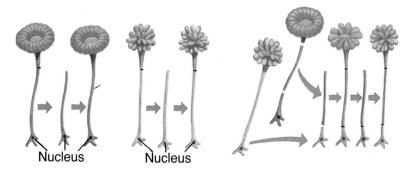

Nucleus   Nucleus

Thinking Critically
Why is the final cap like that of the cell from which the nucleus was taken? (HINT: Recall the function of the nucleus.)

layers thick. The nuclear envelope has large pores so materials can pass back and forth between the nucleus and the rest of the cell. The nucleus contains DNA, the master instructions for building proteins. DNA forms tangles of long strands called **chromatin,** which is packed into identifiable chromosomes when the cells are ready to reproduce. Also within the nucleus, *Figure 8.6,* is the **nucleolus,** a region that produces tiny cell particles that are involved in protein synthesis. These particles, called **ribosomes,** are the sites where the cell assembles enzymes and other proteins according to the directions of the DNA. Although ribosomes are considered cell organelles, they are not bounded by a membrane.

Assembly, Transport, and Storage

A major function of most cells is to make proteins and other materials. Many of the cell organelles are involved in protein synthesis or storage of materials. Much of this assembly and storage takes place in the fluid inside the cell—the cytoplasm.

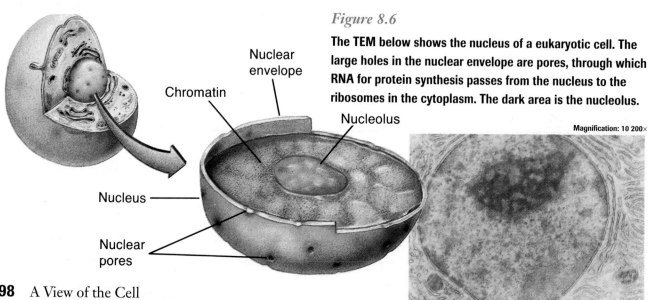

Figure 8.6

The TEM below shows the nucleus of a eukaryotic cell. The large holes in the nuclear envelope are pores, through which RNA for protein synthesis passes from the nucleus to the ribosomes in the cytoplasm. The dark area is the nucleolus.

Nuclear envelope

Chromatin

Nucleolus

Nucleus

Nuclear pores

Magnification: 10 200×

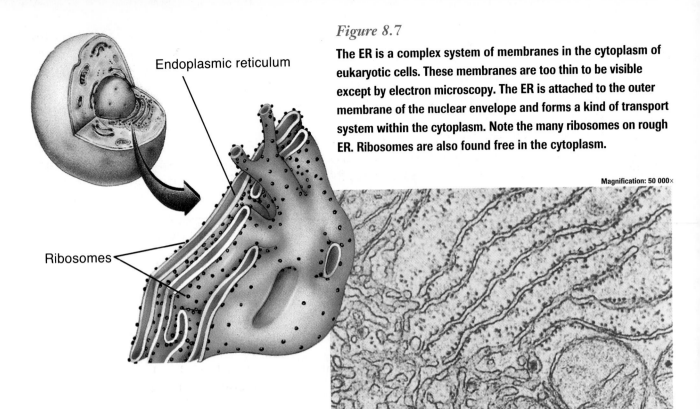

Endoplasmic reticulum

Ribosomes

Figure 8.7

The ER is a complex system of membranes in the cytoplasm of eukaryotic cells. These membranes are too thin to be visible except by electron microscopy. The ER is attached to the outer membrane of the nuclear envelope and forms a kind of transport system within the cytoplasm. Note the many ribosomes on rough ER. Ribosomes are also found free in the cytoplasm.

Structures for assembly and transport of proteins

The material that lies outside the nucleus and surrounds the organelles is the **cytoplasm,** a clear fluid that is a bit thinner than toothpaste gel. It usually constitutes a little more than half the volume of a typical animal cell. Many important chemical reactions, such as protein assembly, take place in the cytoplasm.

Much of the cytoplasm is occupied by a folded system of membranes. The **endoplasmic reticulum** (ER), shown in *Figure 8.7,* is a folded membrane that forms a network of interconnected compartments inside the cell. Suppose you took all of the tissues out of a box of facial tissues and spread them out on a table. They would cover a large surface area, probably several square meters. Yet, all of that surface area was packed into a small box. A large surface area can be packed into a small area by folding the surfaces. Like facial tissues that are folded into a box, the ER is folded into the cell with cytoplasm surrounding it. This system of membranes provides a large surface area on which chemical reactions can take place. The ER membranes also contain the enzymes for almost all of the cell's lipid synthesis, so they serve as the site of lipid synthesis in the cell.

Some of the ER is coated with ribosomes. The parts of the ER that are studded with ribosomes have a bumpy appearance when viewed with an electron microscope, and are referred to as rough ER. In areas without ribosomes, the ER is called smooth ER. The ER functions as the cell's delivery system, much like the trucks that deliver the raw products such as cheese and mushrooms to the pizza parlor. To make the pizza, these raw products must be assembled on a counter or workbench in the pizza parlor. In the cell, the sites of protein assembly are the ribosomes.

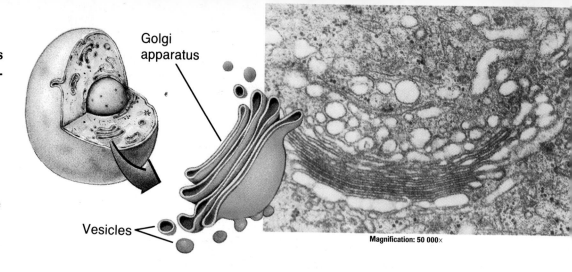

Golgi apparatus

Vesicles

Magnification: 50 000×

Figure 8.8

The Golgi apparatus, as seen by electron microscopy (right), looks like a side view of a stack of pancakes. Associated with the Golgi apparatus are small, membrane-bound vesicles that are involved in protein packaging. The various proteins in the vesicles are sorted and sent to their final destination, like mail moving through a post office.

Structures for protein storage

Businesses usually have a storage room for keeping extra materials. Cells, too, have storage areas. Organelles that store materials include the Golgi apparatus, vacuoles, and lysosomes. The **Golgi apparatus** is a series of closely stacked, flattened membrane sacs that receives newly synthesized proteins and lipids from the ER and distributes them to the plasma membrane and other cell organelles. Proteins are transferred from the ER to the Golgi apparatus in small, membrane-bound transport packages. These packages, called vesicles, have pinched off from the membrane of the ER and contain proteins, *Figure 8.8.* The Golgi apparatus modifies the proteins chemically, then repackages them in new vesicles for their final destination in the cell. They may be incorporated into cell structures, expelled, or remain stored for later usage.

Vacuoles and storage

In the pizza parlor kitchen, the cheese and other ingredients are stored in bins. Cells have spaces, called vacuoles, for temporary storage of materials. A **vacuole,** like those in *Figure 8.9,* is a sac of fluid surrounded by a membrane. Vacuoles often store food, enzymes, and other materials needed by a cell, and some vacuoles store waste products. In some single-celled organisms, a specialized vacuole collects excess water and pumps it out of the cell. A plant cell has a single, large vacuole that stores water and other substances.

Figure 8.9

The vacuole is a membrane-bound, fluid-filled space within the cytoplasm, like a microscopic water balloon. Plant cells (left) usually have one large vacuole; animal cells (right) contain many smaller vacuoles.

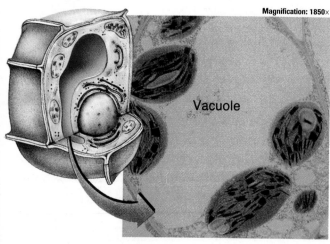

Magnification: 1850×

Vacuole

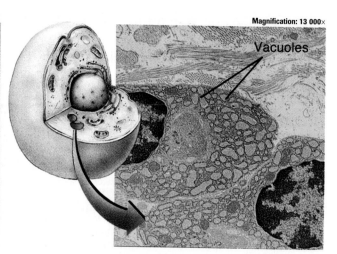

Magnification: 13 000×

Vacuoles

Lysosomes and recycling

In addition to the assembly and storage of macromolecules, cells also can disassemble things. **Lysosomes,** organelles that contain digestive enzymes, digest excess or worn out cell parts, food particles, and invading viruses or bacteria. The membrane surrounding a lysosome prevents the digestive enzymes inside from destroying the cell's proteins. Lysosomes can fuse with vacuoles and dispense their enzymes into the vacuole, digesting its contents. For example, when an amoeba engulfs a food morsel and encloses it in a vacuole, a lysosome fuses with the vacuole, releases its enzymes, and digests the hapless prey. Sometimes, lysosomes digest the cells that contain them. When a tadpole develops into a frog, lysosomes within the cells of the tadpole's tail cause its digestion. The molecules thus released are used to build different cells, perhaps in the newly formed legs of the adult frog.

Energy Transformers

When you entered the pizza parlor, lights were on inside. The ovens in the kitchen were filled with baking pizzas. All of this equipment requires energy with which to run. The cell also requires energy to carry out its many functions.

Mitochondria and energy

Eukaryotic cells have membrane-bound organelles that transform energy for the cell. **Mitochondria,** shown in *Figure 8.10,* are organelles in which food molecules are broken down to release energy. This energy is then stored in other molecules that can power cell reactions easily.

A mitochondrion has an outer membrane and a highly folded inner membrane. As with the ER, the folds of the inner membrane provide a large surface area in a small space. Here on the inner folds, energy-storing molecules are produced.

Figure 8.10

Mitochondria are found in every cell except prokaryotes, in varying numbers (up to 2500 per cell in liver cells). They are granular and rod or thread shaped, with an inner membrane that forms long, narrow folds called cristae. The cristae are covered with enzyme systems involved in producing energy molecules for many cell functions. The TEM (right) is a section through a mitochondrion.

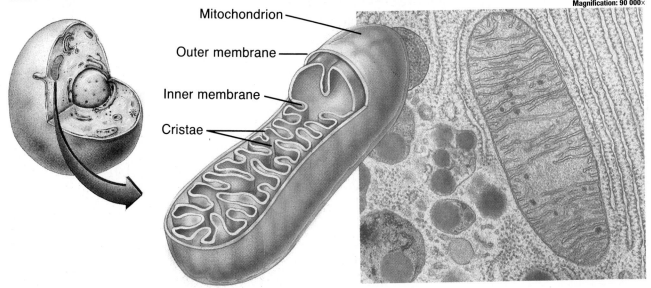

Mitochondrion

Outer membrane

Inner membrane

Cristae

Magnification: 90 000×

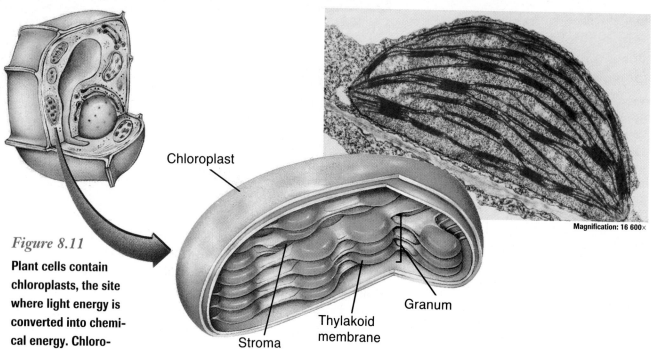

Chloroplast

Granum

Thylakoid membrane

Stroma

Magnification: 16 600×

Figure 8.11

Plant cells contain chloroplasts, the site where light energy is converted into chemical energy. Chloroplasts are usually disc shaped but have the ability to change shape and position in the cell as light intensity changes. The pigment chlorophyll is embedded in the inner series of membranes, called thylakoid membranes, where light energy is trapped.

chloroplast:

chloros (GK) green
platos (GK) formed object

A chloroplast is a green structure found in plant cells.

Chloroplasts and energy

Cells of green plants and some protists have organelles, called **chloroplasts,** that transform light energy directly into usable chemical energy and store that energy in food molecules. These foods include sugars and starches. Chloroplasts contain the molecule **chlorophyll,** a green pigment that traps the energy from sunlight and gives plants their green color.

A diagram and a TEM of a chloroplast, which has a double outer membrane and a folded inner membrane system, are shown in *Figure 8.11.* It is within the inner membranes that the energy from sunlight is trapped. These inner membranes are arranged in stacks of membranous sacs called grana, which resemble stacks of coins. The fluid that surrounds the grana membranes is called stroma.

The chloroplast belongs to a group of plant organelles called **plastids,** which are used for storage. Some plastids store starches or lipids, whereas others contain pigments,

molecules that give color. Plastids are named according to their color or the pigment they contain. The chloroplast contains the green pigment chlorophyll, which gives leaves and stems their green color. Other plastids have different pigments and give flowers and fruits their beautiful colors of red, purple, blue, and yellow.

Structures for Support and Locomotion

Less than 50 years ago, scientists thought plastids and other organelles just floated in a sea of cytoplasm. In the last 20 years, however, scientists have discovered that cells have a support structure called the cytoskeleton within the cytoplasm. The cytoskeleton is composed of tiny rods and filaments that form a framework for the cell, like a bony skeleton that forms the framework for your body. However, unlike your skeleton, the cytoskeleton is a dynamic and constantly changing structure.

Cytoplasm support

The **cytoskeleton,** shown in *Figure 8.12,* is a network of thin, fibrous elements that act as a sort of scaffold to provide support for organelles. It also helps maintain cell shape in a manner similar to the way poles maintain the shape of a tent. The cytoskeleton is usually composed of microtubules and microfilaments. **Microtubules** are thin, hollow cylinders made of protein. **Microfilaments** are thin, solid protein fibers. Microtubules and microfilaments make up most of the cytoskeleton.

Cilia and flagella

Some cells have cilia and flagella, which are structures adapted for locomotion. The two types of structures can be distinguished by the

Figure 8.12

This photograph of a eukaryotic cell has been treated with fluorescent dye to show the cytoskeleton. The microtubules are stained yellow and the microfilaments are red. The cytoskeleton is constantly being formed and taken apart by the cell, depending on its needs.

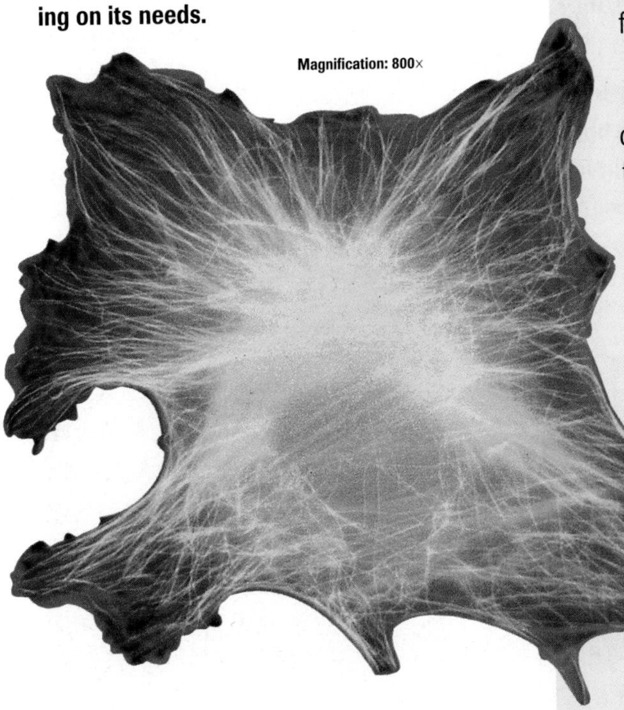

Magnification: 800×

Getting from Here to There

Biologists have known for many years of the existence of microtubules and microfilaments and the role they play in maintaining the shape of the cell because they could be seen in dead cells with electron microscopes. But in 1980, a malfunction in a video camera mounted on a light microscope increased the contrast between cell parts and their background. Suddenly, all the bustling activity of the living cell came into sharp focus, and a new role for microtubules as a network of highways within the cell was discovered.

Highways of the cell In addition to maintaining shape, the microtubules of the cytoskeleton act as tracks on which organelles move from place to place. Think of the cytoskeleton as a highway system on which traffic flows. A group of molecules called protein motors moves organelles and other particles along the protein highways. One protein motor, kinesin, has a long tail that attaches to the particle to be moved, while the front end of the motor molecule grabs onto the microtubule. Kinesin acts in a clawlike motion by repeatedly pulling forward on the microtubules, letting go, and then repositioning its hold farther up the microtubule—much like the motion of a rope climber. With this motion, kinesin is able to move particles such as mitochondria and large protein molecules. Kinesin is a one-way motor—it moves particles only in one direction. Particles that have to move in the opposite direction are hauled by a different kind of motor molecule. Several other protein motors have been discovered, and researchers predict that they will soon find a specialized motor for each kind of particle.

Cell locomotion Microtubules are also important in muscle cell contraction and in cell locomotion—the movement of whole cells from place to place. White blood cells use their microtubules to crawl toward and engulf the bacteria and viruses that invade our bodies. The crawling motion is caused by the expansion and contraction of various parts of the cytoskeleton, which changes the shape of the cell. The cell moves forward by expanding and then reinforcing the cytoskeleton to form a bulge in the plasma membrane—called a pseudopodium. At the same time, the cytoskeleton at the back of the cell contracts, and the tightening action squeezes the cell contents forward. Thus, it appears as though the role of the microtubules in cell locomotion is similar to that in movement of particles within the cell.

Thinking Critically

Why would being able to see living cells at a magnification of 100× be an advantage over using the electron microscope, which can magnify dead cells more than 100 000 times?

Which parts of a sperm cell are visible in a transmission electron micrograph?

Sperm cells are small, motile cells that combine with female reproductive cells. Careful inspection of longitudinal and cross sections of a sperm will allow identification of some of the cell structures.

Procedure

1. Look at the longitudinal section of a sperm flagellum shown here and the cross section shown below in *Figure 8.13.*
2. Identify the cell structures that are visible.

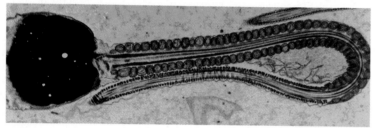

Magnification: 18 000×

Analysis

1. Along the tail of the sperm are many oval-shaped structures. What are they? What is their cellular function?
2. The oval structures are used to power what cellular structure?
3. On the head of the sperm is a large lysosome. What do you think the function of this structure might be?

nature of their action. **Cilia** are short, numerous, hairlike projections out of the plasma membrane. As shown in *Figure 8.13,* cilia tend to occur in large numbers on a cell's surface, and their beating activity is usually coordinated, much like the movement in a stadium "wave." **Flagella,** also shown in *Figure 8.13,* are longer projections that move with a whiplike motion. Cells that have flagella have only one or two per cell.

In single-celled organisms, cilia and flagella are the major means of locomotion. Sperm cells of animals and some plants move by means of flagella. Organisms that contain many cells, including humans, have cilia that move fluids over a cell's surface, rather than moving the cell itself.

Plant and animal cells are both adapted to carry out the functions of the organisms that they make up. Their structures and functions are similar, yet different. Many of the cell structures discussed in this chapter are found in both plant and animal cells, but each cell type also has some unique structures. Compare and contrast the drawings of the animal and plant cells in *Figure 8.14.*

Figure 8.13

Even though cilia and flagella project out of the plasma membrane, they still are covered by the membrane.

▼ In eukaryotic cells, both cilia and flagella are composed of microtubules arranged in a ring of nine pairs surrounding two single microtubules, as seen in this cross section of a sperm flagellum.

Magnification: 75 000×

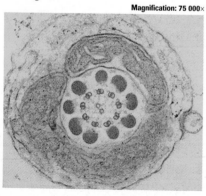

▼ Cilia in the windpipe beat and propel particles of dirt and mucus toward the mouth and nose, where they are expelled.

Magnification: 3840×

▼ The flagella of these sperm cells move the cells forward with their whiplike action.

Magnification: 4500×

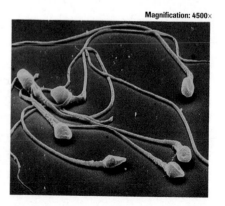

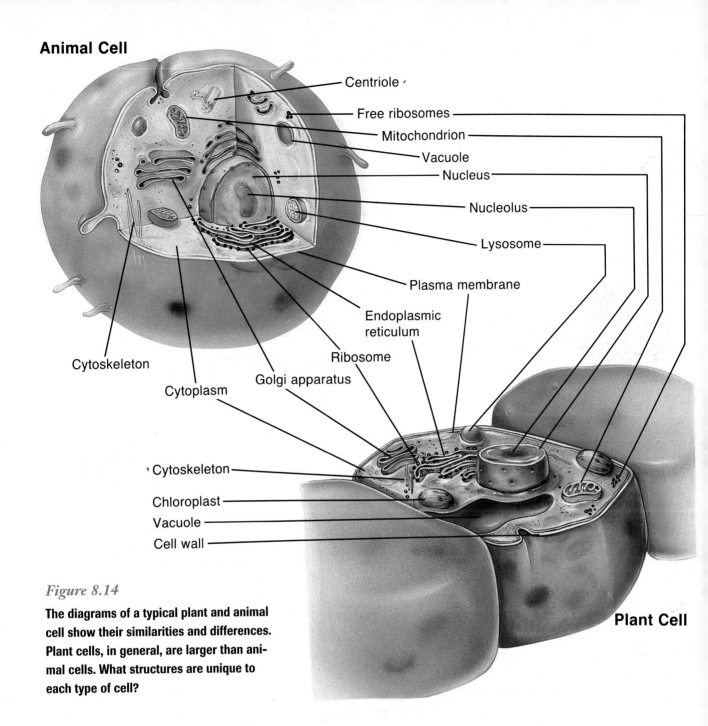

Animal Cell

- Centriole
- Free ribosomes
- Mitochondrion
- Vacuole
- Nucleus
- Nucleolus
- Lysosome
- Plasma membrane
- Endoplasmic reticulum
- Ribosome
- Golgi apparatus
- Cytoplasm
- Cytoskeleton

- Cytoskeleton
- Chloroplast
- Vacuole
- Cell wall

Plant Cell

Figure 8.14

The diagrams of a typical plant and animal cell show their similarities and differences. Plant cells, in general, are larger than animal cells. What structures are unique to each type of cell?

Cellular Organization

Some cells exist as single-celled organisms that perform all of the organism's metabolism within a single cell. Such single-celled organisms are called **unicellular.** Other organisms are made up of many cells, with their cells specialized to perform distinct metabolic functions. One cell within an organism may be adapted for movement, while another cell carries out digestion. The individual cells no longer carry out all life functions, but rather depend on each other. Many-celled organisms are called **multicellular.**

When a group of cells functions together to perform an activity, they form a **tissue.** The cells of your body are organized into tissues such as muscle and nerve tissues. Plant tissues include those of the stem and root. Many cells in tissues are linked

uni-, multi-:
uni (L) one
multi (L) many
Unicellular organisms have one cell; multicellular organisms have many cells.

BioLab

Sizing Cells and Cell Structures

All eukaryotic cells have membrane-bound organelles. Textbooks such as this one often have many electron micrographs of cells and cell structures so that the readers can see the shape and structure. It is often helpful for the reader to know the size of the actual object. How are the sizes of these cells and cell structures determined? By using the information given in a scale bar or the magnification of the photo, the real size of an object can be calculated.

PREPARATION

Problem
How can you determine the size of cell structures?

Objectives
In this Biolab, you will:
- **Measure** cells and organelles.
- **Calculate** the real size of objects.

Materials
small metric ruler
calculator

PROCEDURE

1. Copy the data table.
2. Measure to the nearest 0.1 cm the length and width of the "graphic square" labeled A, and record the measurements in the data table.
3. Magnification can be represented by a scale bar. Measure the length of the scale bar beneath the graphic sausage.
4. Calculate the real size of the graphic square by taking each of the dimensions and dividing them by the length of the scale bar. Multiply the answer by the scale (in this case 1 µm). Keep track of

the units of measurement. Record your answer in the data table.

A. Graphic square

B. Graphic sausage

├───── 1 µm ─────┤

5. Measure to the nearest 0.1 cm the length and width of the "graphic sausage" labeled B, and record the measurements in the data table.

6. A second way that magnification can be represented is by the number of times the object has been magnified. The graphic sausage has been magnified 30 000×. Calculate the real size by dividing the length and width measurements by 30 000. Keep track of the units of measurement. Record your answer in the data table.

7. Calculate the real sizes of the chloroplast (C) and the red blood cell (D). Record your measurements in the data table.

8. Calculate the real length and height of a granum in the chloroplast. Record your measurements in the data table.

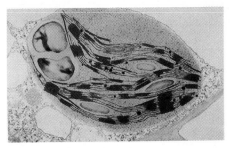

C. Chloroplast

2 μm

Magnification: 7300×

D. Red blood cell

Object	Length	Width
Graphic square measurement	_____	_____
Graphic square scale bar	_____	_____
Graphic square real size	_____	_____
Graphic sausage measurement	_____	_____
Graphic sausage real size	_____	_____
Chloroplast measurement	_____	_____
Chloroplast real size	_____	_____
Red blood cell measurement	_____	_____
Red blood cell real size	_____	_____
Granum measurement	_____	_____
Granum real size	_____	_____

ANALYZE AND CONCLUDE

1. **Observing** How many granum packages can you count in the chloroplast? What is their function?

2. **Calculating Results** Look at page 198 and calculate the real diameter of a cell nucleus. How does its size compare with the size of the chloroplast?

3. **Calculating Results** Sometimes it is difficult to realize how greatly cell parts are magnified in electron micrographs. If your real height were magnified 30 000×, how tall would you appear in kilometers? (1 km = 1000 m)

Going Further

Application Make a table listing all of the cell parts from the chapter. Use the electron micrographs in this chapter to help you calculate the real size of each cell part. Place the sizes in your table.

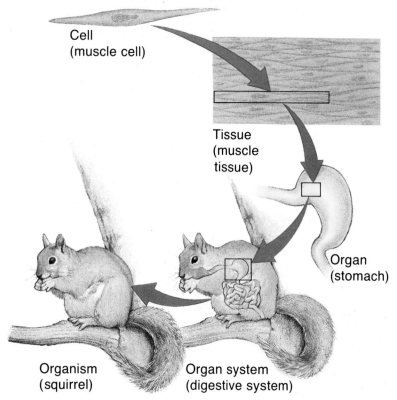

Cell
(muscle cell)

Tissue
(muscle
tissue)

Organ
(stomach)

Organism
(squirrel)

Organ system
(digestive system)

Figure 8.15

Multicellular organisms are highly organized into tissues, organs, and organ systems. Why can't a cell from an organ carry out all life processes?

to each other at contact sites called cell junctions. Cell junctions help maintain differences in the internal environment between adjacent cells, help anchor cells together, and allow cells to communicate with one another by passing small molecules from one cell to another.

Groups of two or more tissues that function together make up **organs.** Your stomach and the leaf of a plant are organs. Cooperation among organs makes life functions within an organism efficient. An **organ system** is a group of organs that work together to carry out major life functions. Your nervous system and the flower of a plant are examples of organ systems. *Figure 8.15* presents the organization of cells to form a multicellular organism.

Connecting Ideas

Cells are the basic building blocks of both unicellular and multicellular organisms. Cells share basic "business" tasks that keep them alive. Each cell must interact with its environment directly through the plasma membrane by transporting needed materials into the cell and getting

rid of harmful waste products. How does the cell monitor what comes in and what goes out? Understanding how the plasma membrane interacts with its external environment will help you understand how cells maintain homeostasis.

Section Review

Understanding Concepts
1. What is the advantage of highly folded membranes in a cell?
2. What organelles would be especially numerous in a cell that produces large amounts of a protein product?
3. Why are digestive enzymes in a cell enclosed in a membrane-bound organelle?

Thinking Critically
4. Compare the functions of mitochondria and chloroplasts. Why are they referred to in

the text as energy transformers rather than as energy producers or energy generators?

Skill Review
5. **Observing and Inferring** Some cells have large numbers of mitochondria with many internal folds. Other cells have few mitochondria with few internal folds. What can you conclude about the functions of these two types of cells? For more help, refer to Thinking Critically in the *Skill Handbook.*

Natural Fiber–Natural Color

Cotton is one of the most widely produced fabrics in the world. Before the Civil War, cotton was widely grown in the southern United States. Cotton production was very labor intensive as the cotton bolls—the white, fluffy seed pods—had to be hand picked from the plant, and the seeds had to be removed by hand.

Production of cotton By the late 18th century, technological developments began to change the way cotton was grown, harvested, and treated. In 1793, Eli Whitney invented the cotton gin, which automated the removal of seeds from the cotton fibers and made the production of cotton far less backbreaking and costly. Today, cotton gins employ much the same principles and technology as Whitney's original machine. Additional machines are used today to cultivate and harvest cotton, eliminating all the tedious hand labor.

Structure of cotton fibers Cotton fibers are unicellular hairs that project from the outer covering of cotton seeds. The walls of the hairs are almost pure cellulose, which is arranged in interwoven layers and gives cotton thread its strength. Seed hairs of cotton range in length from 16 mm to almost 50 mm. The longer the fibers, the higher the quality and strength of the thread that can be manufactured. When the cotton seeds mature, the seed pod bursts open, displaying a mass of soft, white fluff.

Cotton cloth After the cotton fibers are spun into thread, the thread is woven into cloth; the cloth then undergoes bleaching, dyeing, or printing. The dyeing process has been especially important in producing the colors and patterns of fabric that make cotton so desirable.

Applications for the Future

Until recently, the color of cotton was always white. An Arizona cotton breeder, Sally Fox, is using modern knowledge of genetics and technology to breed plants that produce fibers in a variety of colors. The newest colors are green and brown, with other colors still being developed.

Why would anyone want to produce cotton plants that grow colored fibers? Naturally colored cotton eliminates one step in the production of cloth. There's no need for a dyeing process, and naturally colored cotton fibers don't fade as dyed fibers do. But perhaps the most important reason is that dyeing fabric uses water and chemicals that create pollution. Naturally colored fibers decrease the pollution, decrease the use of water, and protect the environment.

A small amount of naturally colored cotton fabric is already on the market, and several companies are using the cloth in their ready-to-wear items. Because they are new and in relatively short supply, clothing with naturally colored cotton fiber is more expensive. As Fox's research progresses, the price of naturally colored cotton clothing undoubtedly will drop.

INVESTIGATING the Technology

1. **Research** Are other naturally colored, natural fibers being produced? Research such fabrics as silk, wool, and linen.
2. **Hypothesize** What properties of cotton fiber other than color might be developed through breeding of cotton plants?

Meet Dr. Marian Diamond, Neuroscientist

Brains and basketballs both have places in Dr. Marian Diamond's office. In a bucket of formaldehyde beside her desk is a human brain, and in the corner is a basketball, ready for her weekly game with students. Dr. Diamond is a professor of anatomy at the University of California in Berkeley, as well as executive director of the Lawrence Hall of Science, a unique, hands-on center for learning.

In the following interview, Dr. Diamond talks about her research on aging brain cells, as well as her enthusiasm for the museum she directs.

On the Job

Q Dr. Diamond, what is your specialty in the field of biology?

A I study the effects of the environment on the anatomy of the brain as it develops from before birth through old age. I've discovered that brain mass can increase with learning, and this can happen at any age. The more the brain is stimulated, the more the neurons increase in size. Within the brain, the dendrites become thicker, synapses expand, and the basic chemistry changes.

Q How can we keep our brains functioning at full capacity?

A The view that there is a loss of brain cells in advanced old age is a myth. From all my research on the development of rats' brains, I would recommend keeping your brain challenged with new experiences for a lifetime—not just in school but in every situation. If you sit back and let life pass you by, it's a strong probability that your brain may even shrink. You can avoid the possibility of a shrinking brain by simply using it and challenging yourself daily. Most people seem unaware of the importance of what they carry in the top of their heads. I like to show people the brain because I want them to see it as the place where education takes place.

Q What about treatment of damaged brains in the future?

A Doctors now can transplant pieces of brains, such as using embryonic brain tissue to treat Parkinson's disease. I think Alzheimer's may be treated when this complicated disease is better understood. Perhaps research into what triggers the expression of particular genes will be the key.

Q How did you become the director of the Lawrence Hall of Science?

A I guess I was asked because of my experience with both science and education. I've always had my own programs, using my university students to teach in grade schools and high schools and help other students enjoy the human body and brain. At this museum, I can put together my theories of education with my teaching of brain functions and brain research.

Q Do you find it difficult to get kids interested in science?

A Not at all. Every child is naturally a scientist. At the Lawrence Hall of Science, children can find whatever fascinates them and play with an exhibit all day if they want. They discover the joy of doing something themselves. That's one of life's greatest pleasures.

Early Influences

Q Could you talk about how you became interested in brains?

A I saw my first human brain when I was in my teens, and I fell in love with brains because this mass represented cells that could think. I wanted to find out how cells could think and what was going on behind people's eyes. I've been studying human brains for 40 years. I even keep one beside my desk.

Q Do you remember any one person who may have introduced you to a love of biology?

A I especially remember a junior-high school teacher who taught me about the intricate structure of a fly's wing. Until then, I just thought a fly was for swatting. I never thought of a fly wing being built for special aerodynamics until she helped me see it differently.

Personal Insights

Q Do you have any advice for students who might be interested in a career in biology?

A You're surrounded by biology for a lifetime. In fact, you *are* biology. Everything around you is biology. It's an exciting thing to find out that there are chemical simi-

CAREER CONNECTION

To follow in Dr. Diamond's footsteps, you would need a Bachelor of Science degree in biology and a PhD in anatomy. Science museums like the one Dr. Diamond directs hire artists and writers to work on their displays and prepare written material, such as brochures and study guides.

Science Writer *College degree in English; courses in general science*

Neuroscientist *A doctoral degree in anatomy and physiology; courses in zoology*

Science museum volunteer *An interest in science and volunteering*

larities in, say, a leaf, an ant, a monkey, and a person. Understanding that makes you look at the world differently. I would advise kids to follow their own hearts. Just as no two brains are alike, no two loves are the same. If you have a passion for something, follow that passion. You'll probably take a lot of courses that might not seem relevant at the time, but later on you'll understand.

Q How do you personally keep a balance in your life?

A I eat right and do the things I love to do. That includes playing a mean game of basketball with my students every week, even though they complain about my sharp elbows!

8.1 The Discovery of Cells

- The development of microscopes enabled biologists to develop the cell theory.
- The cell theory states that all organisms are composed of one or more cells, the cell is the basic unit of organization in organisms, and all cells come from pre-existing cells.
- Using electron microscopes, scientists have studied detailed cell structure.
- Cells can be classified as prokaryotic or eukaryotic, based on whether they have membrane-bound organelles.

8.2 Eukaryotic Cell Structure

- Eukaryotic cells are composed of many specialized cell parts.
- Cells are enclosed by a plasma membrane. Many cells have cell walls that provide support and protection.
- The nucleus is the control center of eukaryotic cells.
- Cells make their proteins on ribosomes, which are often attached to the highly folded endoplasmic reticulum.
- Cells store materials in the Golgi apparatus, vacuoles, and lysosomes.
- Mitochondria break down food molecules to release energy.
- Chloroplasts convert light energy into chemical energy.

- The cytoskeleton helps maintain cell shape by providing a framework for the cell.
- Cilia and flagella, structures made of microtubules, allow cells to move or to move materials past the cell surface.
- The cells of most multicellular organisms are organized into tissues, organs, and organ systems.

Key Terms

Write a sentence that shows your understanding of each of the following terms.

cell	lysosome
cell theory	microfilament
cell wall	microtubule
chlorophyll	mitochondria
chloroplast	multicellular
chromatin	nucleolus
cilia	nucleus
compound light microscope	organ
cytoplasm	organ system
cytoskeleton	organelle
electron microscope	plasma membrane
endoplasmic reticulum	plastid
	prokaryote
eukaryote	ribosome
flagella	tissue
Golgi apparatus	unicellular
	vacuole

Understanding Concepts

1. How does the nucleus control a cell?
2. What is the importance of the plasma membrane to cell function?
3. Suggest a reason why large pores are present in the nuclear envelope but are absent in the plasma membrane.
4. Why can't living cells be examined with an electron microscope?
5. What functions do plant cell walls perform?

6. How are cilia and the cytoskeleton related?
7. Why are cilia and flagella especially important in unicellular organisms?
8. Why are folded membranes in cells more effective than non-folded membranes?
9. How do scanning electron microscopes and transmission electron microscopes differ in operation?

Using a Graph

10. A biologist measured the amounts of several ions in pond water and in the cytoplasm of a pond plant. He found that the pond water and the cytoplasm had different concentrations of some ions, as shown in the graphs below. Consider the cell structures that you have studied in this chapter. How is it possible that the environment inside the cell is different from the outside environment?

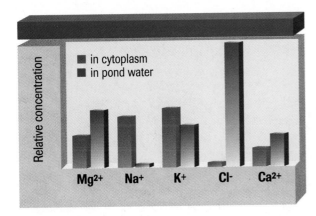

Relating Concepts

11. Make a concept map that relates the following terms and phrases. Supply the appropriate linking words for your map.

nucleus, nucleolus, chromatin, ribosome, endoplasmic reticulum, Golgi apparatus

Applying Concepts

12. Why did it take almost 200 years after Hooke discovered cells for the cell theory to be developed?

13. Sometimes packets of proteins collected by the Golgi apparatus merge with a lysosome. Suggest reasons for this activity.

14. Why did Schleiden and Schwann conclude that cells are the basic units of life?

15. Why must plant cells have both mitochondria and chloroplasts?

16. During human development, the embryo's hands develop from solid, paddle-shaped parts. How do you think fingers might be formed from these parts?

Biology & Society

17. Technology Why do you think thread made from long cotton fibers is stronger than thread made from short fibers?

Thinking Critically

Interpreting Data

18. Biolab Sometimes the size of an object in a photograph is indicated by a known object, such as a dime, being placed beside the object to be photographed. How does this procedure allow you to calculate the real size of the object being photographed?

Making Inferences

19. Minilab Why do you think cork cells appear to be empty?

Interpreting Scientific Diagrams

20. The inner membrane of a mitochondrion is highly folded. Estimate the length of the folded membrane in illustration A, and estimate how much longer it is compared with the membrane in illustration B.

A. 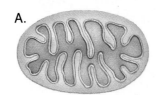   B.

Connecting to Themes

21. Unity Within Diversity Cells such as nerve cells, muscle cells, and skin cells have different functions. Yet, they are all basic units of life. Name three features that all of these cells have in common.

22. Evolution Mitochondria are thought to have once been separate cells. As cells evolved, mitochondria are thought to have taken up residence inside other cells. Why would this combination of cells have been an advantage for both cells?

Homeostasis and the Plasma Membrane

The sun sparkled on the waves as they swept against the boat. The shark cage stood in position above the deep blue water. There were no sharks in sight, but I knew they were down there, and now it was time to step into the cage and find out firsthand. My stomach had butterflies as I stepped off the stern of the boat and swam into the strong, steel cage. Inside, the bars gave me assurance of protection while I filmed the great white sharks as they swam by.

As the day went on, I became less tense as I filmed the details of the shark—its eyes, its sleek body, and those wicked white teeth. I sensed the strength and power of this animal—and its instinct to find food. Several times, a shark struck the cage, causing it to jerk wildly. I might have become the shark's food if I had not been inside a cage that let me look out but did not let the sharks inside.

Magnification: 180×

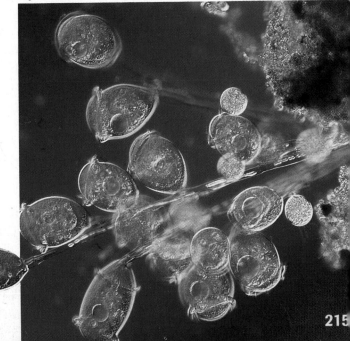

A steel cage allows a diver to observe sharks in their own environment because the cage will not allow the sharks to enter. In a similar manner, the plasma membrane of a cell enables it to exist in a hostile environment. The plasma membranes of these unicellular organisms called *Vorticella* protect the contents of cells and regulate what can come into and out of each cell. How is the structure and function of the plasma membrane similar to that of the shark cage?

9.1 The Plasma Membrane

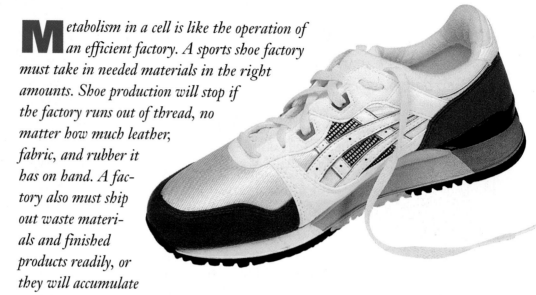

Metabolism in a cell is like the operation of an efficient factory. A sports shoe factory must take in needed materials in the right amounts. Shoe production will stop if the factory runs out of thread, no matter how much leather, fabric, and rubber it has on hand. A factory also must ship out waste materials and finished products readily, or they will accumulate and force the factory to stop production. At the same time, though, the factory must keep its raw materials, plans, tools, and machinery inside. Therefore, a factory, like a cell, must control what leaves, as well as what enters.

Maintaining a Balance

All organisms are subject to constant changes in their environment. Aquatic organisms are subjected to changes in water temperature and the chemicals dissolved in the water. Terrestrial organisms can be subjected to changes in air temperature and amount of sunlight. Organisms must adjust to changes in their environment; failure of living things to adjust means death.

Why cells must control materials

Living cells maintain a balance by controlling materials that enter and leave. Without this ability, the cell cannot maintain homeostasis and will die.

It's important for a cell to keep its internal concentrations of substances, such as water, glucose, and other nutrients, while eliminating wastes as they are produced. Concentrations of these materials in a cell's external environment often change. It is the plasma membrane that maintains the proper concentrations of materials inside a cell by controlling the passage of materials into and out of the cell. The job of the plasma membrane is illustrated in *Figure 9.1.* Thus, homeostasis in a cell is maintained by the plasma membrane, which allows only certain particles to pass through.

The property of a membrane that allows some materials to pass

Figure 9.1

Without the property of selective permeability, all substances could pass in and out of the cell freely, and its contents would always be the same as its surroundings.

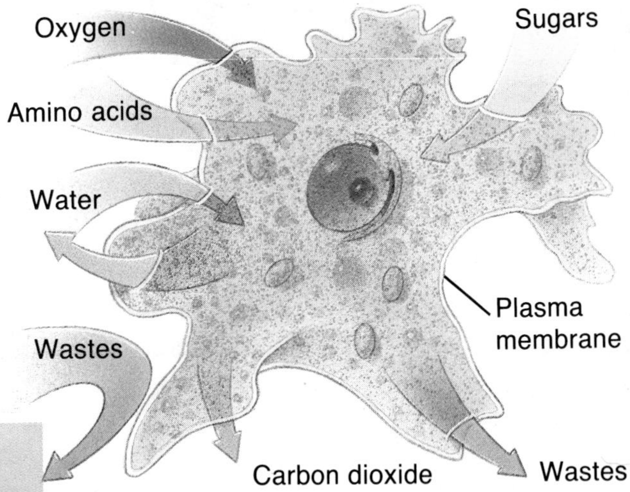

Oxygen

Sugars

Amino acids

Water

Plasma membrane

Wastes

Carbon dioxide

Wastes

▲ A cell must remove wastes and keep harmful substances from entering. Likewise, it must take in and keep needed materials, such as glucose, while maintaining a balance with its surroundings.

▲ Like a plasma membrane, the strainer is selectively permeable because it allows water to run off while holding back the cooked spaghetti.

through while keeping others out is known as **selective permeability.** *Figure 9.1* also shows another familiar example of selective permeability.

Selective permeability also allows different cells to carry on different activities within the same organism. For example, only nerve cells in your body may respond to a certain chemical, even though the chemical is present in the bloodstream and all cells in the body are exposed to it. The membranes of the nerve cells admit the chemical, but the membranes of other cells do not.

Structure of the Plasma Membrane

The structure and properties of a plasma membrane show how it can be selective and maintain cell homeostasis. The plasma membrane is a bilayer, meaning a structure made up of two layers. Powerful electron microscopes have revealed many of the details of this two-layer structure, shown in *Figure 9.2*. Each layer is made up of a sheet of lipid molecules. Protein molecules are embedded in the lipid bilayers like raisins in a slice of raisin bread.

Magnification: 100 000×

permeable:

per (L) through
meare (L) to glide
Materials move easily (glide) through permeable membranes.

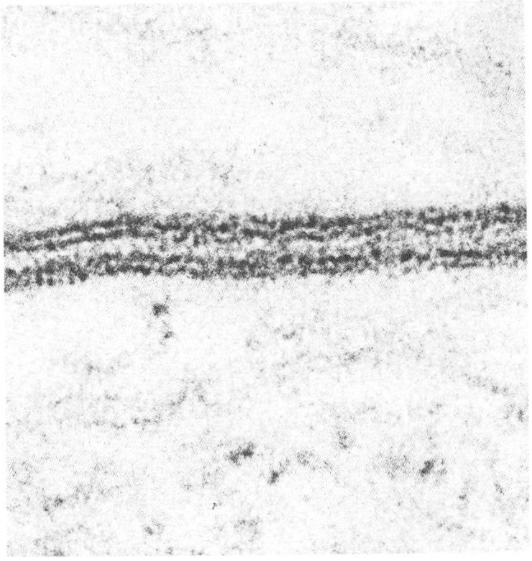

Figure 9.2

The transmission electron micrograph shows two cells side by side. Each cell is clearly bounded by a plasma membrane made of two layers with a space between the two cells.

BioLab | Design Your Own Experiment

Plastic bags, such as sandwich bags, are made of thin, plastic membranes. If the plastic is selectively permeable, it will allow certain ions and molecules to diffuse across it but will hold back others. If particles diffuse through, they will have to pass between the molecules of polyethylene that make up the membrane. Molecules that are small enough to pass between the polyethylene molecules and cross the membrane should diffuse, but molecules that are too large will be held back.

Are plastic bags selectively permeable?

PREPARATION

Problem
Will the polyethylene membrane allow iodine to cross the membrane? Will the polyethylene membrane allow starch, a larger molecule than iodine, to cross the membrane?

Hypotheses
Formulate a hypothesis that predicts the movement of starch and iodine across the plastic membrane. Consider all possible movements, as well as what factors are important in diffusion across a selectively permeable membrane.

Objectives
In this Biolab, you will:
- **Experiment** to determine whether diffusion occurs across a plastic membrane.
- **Interpret** the results to determine whether the plastic membrane was selectively permeable.

Possible Materials
400-mL beakers
small plastic bags
twist ties
graduated cylinder
starch solution
iodine solution
masking tape

Safety Precautions
Be sure to wash your hands thoroughly after you have started your experiment.

1. When you plan your experiment, be sure to consider how you will know whether starch and/or iodine crosses the membrane.

2. After you put solution in the bag, carefully seal the bag so that the solution does not leak out the top of the bag. Wash the bag under running water to clean off any solution that may be on the outside.

Check the Plan

1. Remember that when iodine combines with starch, the color of the starch solution will change to bluish-black.

2. Allow the bag to remain immersed in the solution overnight.

3. ***Make sure your teacher has approved your experimental plan before you proceed further.***

4. Carry out your experiment.

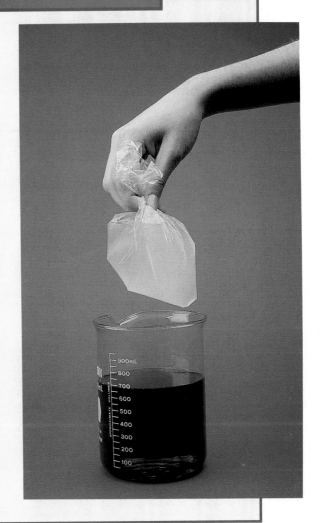

ANALYZE AND CONCLUDE

1. **Analyzing Data** Did iodine molecules move through the membrane? Explain how you know.

2. **Analyzing Data** Did starch molecules pass through the membrane? Explain how you know.

3. **Drawing Conclusions** What can you infer from this experiment about movement of large and small molecules through a thin polyethylene membrane?

4. **Checking Your Hypothesis** Do your data support your hypothesis? Explain why or why not.

Going Further

Changing Variables Will glucose pass through the polyethylene membrane? Set up an experiment that will answer this question. Use glucose test strips to detect the presence of glucose.

Figure 9.3

The structure of a plasma membrane (below) is a bilayer of phospholipids with proteins inserted into either side of or completely penetrating the membrane. Notice also that the phospholipid molecules (right) are not chemically bonded to each other; they are free to move sideways within the layer. The proteins within the membrane poke out like icebergs in a sea of phospholipids.

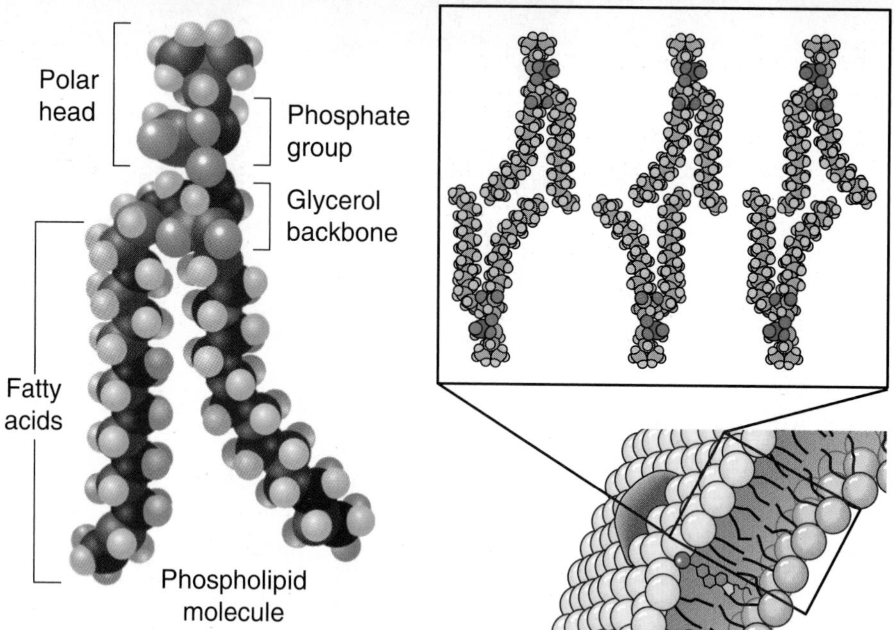

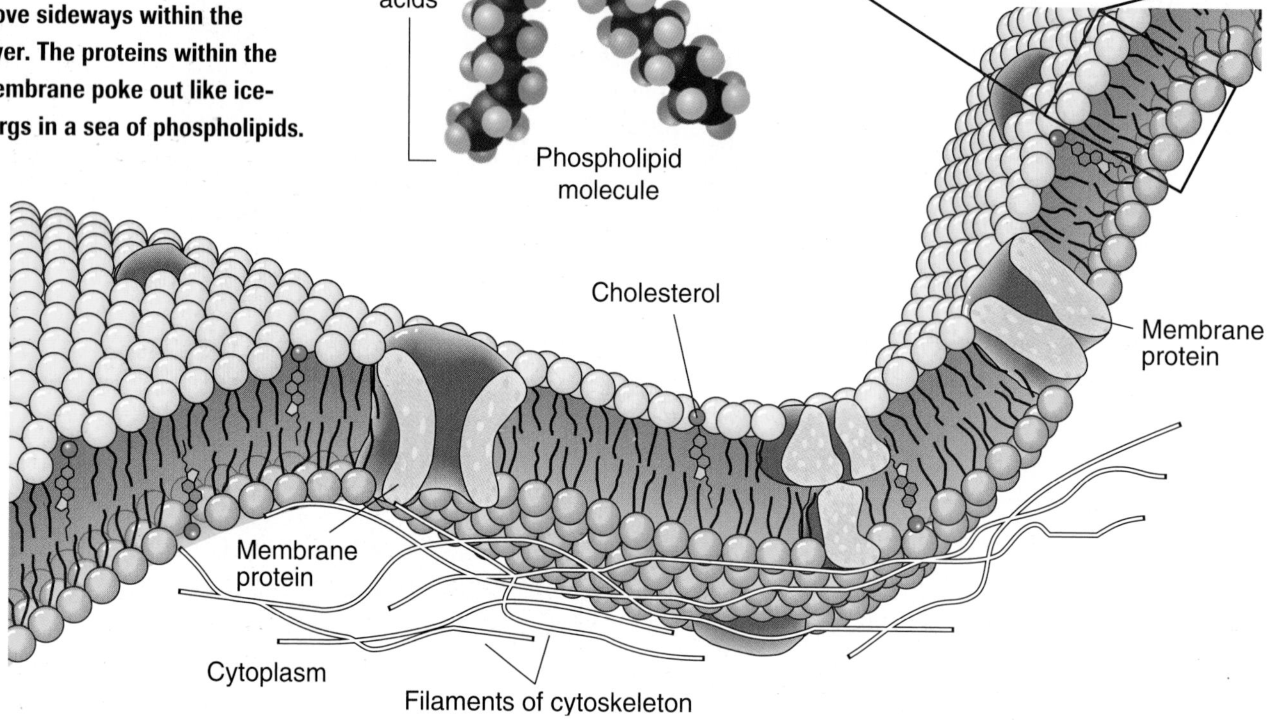

Polar head

Phosphate group

Glycerol backbone

Fatty acids

Phospholipid molecule

Cholesterol

Membrane protein

Membrane protein

Cytoplasm

Filaments of cytoskeleton

Makeup of the lipid bilayer

Unlike typical triglyceride lipids, most of the lipids that make up the two layers in the plasma membrane have two fatty acids attached to glycerol instead of three. In place of a third fatty acid, a membrane lipid has a small organic section attached to a phosphate group and thus is called a **phospholipid.** As you can see in *Figure 9.3,* phospholipids have polar, water-soluble heads attached to long, nonpolar, insoluble tails. The phosphate group is soluble in water because it is polar and therefore is attracted to water molecules. The fatty acid chains are not soluble in water because they are nonpolar.

Most cells have a watery environment both on the inside and outside. Because water attracts the phosphate ends, the phospholipids align to form a double layer with the water-soluble phosphate ends toward the outside of each layer, as in *Figure 9.3.* The nonpolar tails lie inside the bilayer.

Within each layer, phospholipid molecules can move sideways through their layer. These properties make the bilayer behave like a fluid, a material that flows. This description of a plasma membrane as a structure made up of many similar molecules that are free to move sideways within the membrane is called the **fluid mosaic model.** Some cell organelles, such as the nucleus, vacuoles, mitochondria, and chloroplast, also are enclosed by membranes that have a fluid mosaic bilayer structure. Because the kinds and arrangements of proteins and lipids vary from one membrane to another, each type of membrane has its own permeability properties.

Saturated versus unsaturated fatty acids

The fatty acids that make up the phospholipids of the plasma membrane can be saturated or unsaturated. When a fatty acid tail contains unsaturated fatty acids, the fatty acid chain bends at the double bond, like a knee or elbow joint. As discussed in *Figure 9.4,* the more unsaturated fatty acids a membrane has, the more fluid it is.

Figure 9.4

In a caribou, an arctic animal, the membranes of cells near the animal's hooves have phospholipids with many unsaturated fatty acids. Unsaturated fatty acids remain liquid at low temperatures. The unsaturated lipids allow the caribou's feet and legs to drop to almost 0°C in the arctic winters and still maintain plasma membrane function. The cell membranes of the rest of the caribou's body have more saturated lipids.

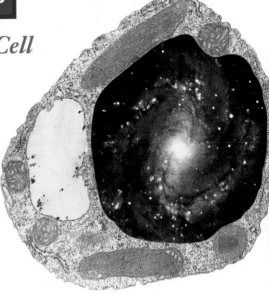

Figure 9.5

Eukaryotic plasma membranes can contain large amounts of cholesterol—up to one molecule for every phospholipid molecule. Although cholesterol tends to make lipid bilayers less fluid, it prevents the fatty acid chains from sticking together and makes the bilayer more stable. Cholesterol also decreases the permeability of lipid bilayers to water-soluble substances.

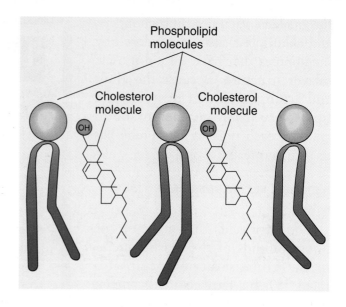

Membranes that are very fluid are not always an advantage. In many eukaryotes, especially animals, cholesterol is an important part of the plasma membrane, as shown in *Figure 9.5*. Fatty acid chains are flexible, but the cholesterol molecule is rigid. Therefore, the presence of cholesterol molecules strengthens the fluid mosaic and makes it more stable. In addition, the cholesterol helps keep the fatty acid tails of the phospholipids separated.

The function of membrane proteins

Although the basic structure of a plasma membrane is a lipid bilayer, most of the functions of the membrane are carried out by proteins. Some of the proteins extend through the bilayer and on both sides of the membrane, as you saw in *Figure 9.3*, while others do not extend into the interior of the lipids.

Many of the proteins determine which particles can pass across the membrane. Some proteins serve as enzymes. Others act as markers that are recognized by chemicals from both inside and outside the cell. Some of these markers are involved in fighting disease. For example, your immune system can distinguish your own cells from cells of a bacterium by certain membrane proteins. Thus, in many respects, the plasma membrane can be thought of as a communication center between a cell and its environment.

Section Review

Understanding Concepts

1. Why is the structure of the plasma membrane referred to as a bilayer and as a fluid mosaic structure?
2. Explain why selective permeability is necessary for homeostasis within the cell.
3. What is the function of cholesterol in the plasma membrane?

Thinking Critically

4. Suggest what might happen if cells grow and reproduce in an environment where no cholesterol is available.

Skill Review

5. **Recognizing Cause and Effect** Consider that plasma membranes allow materials to pass through. Explain how this property contributes to homeostasis. For more help, refer to Thinking Critically in the *Skill Handbook*.

9.2 Cellular Transport

Section Preview

Objectives

Explain how the processes of diffusion, passive transport, and active transport occur and why they are important to cells.

Predict the direction of diffusion of a dissolved substance.

Key Terms

diffusion ✓
dynamic equilibrium
osmosis ✓
isotonic solution ✓
hypotonic solution
turgor pressure
contractile vacuole
hypertonic solution ✓
plasmolysis ✓
passive transport
transport protein
facilitated diffusion
active transport
endocytosis ✓
exocytosis ✓

From your study of simple chemistry in Chapter 7, you learned that the electrons of atoms are constantly moving about the nucleus. In fact, all particles of matter are in constant motion. The atoms, ions, and molecules that make up all materials are moving. It makes no difference whether a material is solid, liquid, or gas; much like bees in a swarm, its particles move constantly in a totally random fashion. This random movement helps explain how materials enter or leave cells.

Diffusion

All objects in motion, like the fish in *Figure 9.6,* have energy of motion called kinetic energy. A moving particle of matter moves in a straight line until it collides with another particle, much like balls on a pool table. After the collision, both particles rebound. Imagine a room full of Ping-Pong balls, all in constant motion, colliding with no loss in energy. Particles of matter move in the same way.

The discovery of Brownian motion

In 1827, Robert Brown, a Scottish scientist, used a microscope to observe pollen grains suspended in water. He noticed that the grains moved constantly in little jerks, as if being struck by invisible objects. This motion, he thought, was the result of the life hidden within the pollen grains. However, when he repeated his experiment using dye particles,

which are nonliving, he saw the same erratic motion. This motion has been called Brownian motion ever since. Brown had no explanation for the motion he saw, but today we know that Brown was observing evidence of the random motion of molecules. These were the invisible objects that were moving the tiny visible particles.

Figure 9.6

These leaping salmon have kinetic energy, the energy of motion. Like atoms and molecules, all moving objects have kinetic energy.

Figure 9.7

A few crystals of copper(II) sulfate were dropped into a beaker.

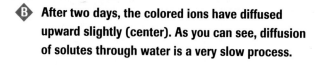

B After two days, the colored ions have diffused upward slightly (center). As you can see, diffusion of solutes through water is a very slow process.

A In a few minutes, the blue color formed by the ions of the dissolving compound has begun to diffuse.

C After a much longer time—sometimes months or even years—the ions will have diffused completely throughout the beaker if it is left undisturbed and covered.

The process of diffusion

Most substances in and around a cell are in water solution, a mixture in which the ions or molecules of a solute are distributed evenly among water molecules. All of these particles, both water and solute, move randomly, colliding with each other. When a soluble substance is placed in water, these random collisions tend to scatter particles of solute and water until they are evenly mixed. One way of observing this effect is to drop a few crystals of a colored substance into a beaker of water, as shown in *Figure 9.7.*

The movement of individual particles is random. The overall movement of the dissolved ions in the beaker in *Figure 9.7* is from a region of high concentration (near the crystals) to a region of lower concentration (throughout the water). **Diffusion** is the net movement of particles from an area of higher concentration to an area of lower concen-

tration. Diffusion results because of the random movement of particles.

The results of diffusion

Eventually, the colored ions will become evenly distributed throughout the molecules of water in the beaker. After this point, the ions continue to move randomly and collide with one another. However, no change will occur in concentration throughout the beaker. This condition in which there is continuous movement but no overall change is called **dynamic equilibrium** as illustrated in *Figure 9.8.* The word *dynamic* refers to movement or change, while *equilibrium* refers to balance. Maintaining a dynamic equilibrium is one of the characteristics of homeostasis.

kinetic:

kinein (GK) to move
Moving objects have kinetic energy.

Figure 9.8

When a cell is in dynamic equilibrium with its environment, materials move into and out of the cell at equal rates. As a result, there is no net change in concentration inside the cell.

Material moving into cell = Material moving out of cell

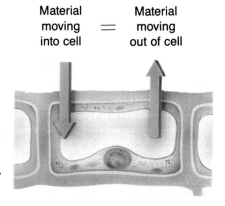

Diffusion depends on concentration gradients

Diffusion cannot occur unless a substance is in higher concentration in one region than it is in another. For example, oxygen diffuses into the capillaries of the lungs because there is a greater concentration of oxygen in the air sacs of the lungs than in the capillaries. The difference in concentration of a substance across space is called a concentration gradient. Because ions and molecules diffuse from an area of higher concentration to an area of lower concentration, they are said to move *with* a gradient. If no other processes interfere, diffusion will continue until there is no concentration gradient. At this point, dynamic equilibrium occurs.

The selectivity of membranes

You can compare this process to the deflation of a helium-filled balloon. The balloon slowly loses helium because the helium atoms are tiny and diffuse out by passing between the molecules that make up the membrane.

A plasma membrane is selectively permeable. The lipid bilayer makes it difficult for charged ions or polar molecules to pass through by diffusion because they are not attracted to the nonpolar structure of the fatty acid tails. Only molecules of water, oxygen, nitrogen, carbon dioxide, and a few other small, nonpolar molecules can diffuse directly across the lipid bilayer. As you will see, cells have specific ways to allow needed molecules, such as amino acids, to enter.

Spartina

The Salt of the Earth

You have seen what happens to cells exposed to hypertonic solutions such as salt water. You may be surprised to learn that life can exist in some of the most saline (salty) environments on Earth such as the Great Salt Lake.

Salt-loving bacteria   A few species of bacteria thrive in the hypertonic solution that makes up the Great Salt Lake in Utah and the Dead Sea on the border between Israel and Jordan. Thus, the Dead Sea is not really dead. However, these species of bacteria are the only prokaryotic organisms that survive in these hostile environments.

Magnification: 10×

Brine shrimp   If you own an aquarium, you've probably fed brine shrimp to your fish. These tiny, multicellular organisms are present by the millions in the Great Salt Lake. Spending their entire lives in brine, they start out as eggs carried in a pouch in the female's body. If part of the lake dries up, these eggs can survive for many years until water reactivates them, causing them to develop and hatch.

Salt marsh plants   Salt marshes are common at the edges of salt lakes and the shores of the oceans. These marshes are productive and important to the survival of birds and many of the fish we use for food. Examples of plants found in these marshes are grasses of the genus *Spartina.* These plants are unique in that they survive in a salty environment by excreting excess salt from specially adapted glands.

In addition to plants, the lake contains at least 29 species of algae. The algae that live near the surface are most noticeable. These species sometimes grow into huge populations that can color the entire lake red, pink, or orange.

Thinking Critically

One problem in some countries is that their farmland has been poisoned by salt left by centuries of irrigation with saline water. Suggest ways these lands might be put to productive use and that future soil problems might be minimized.

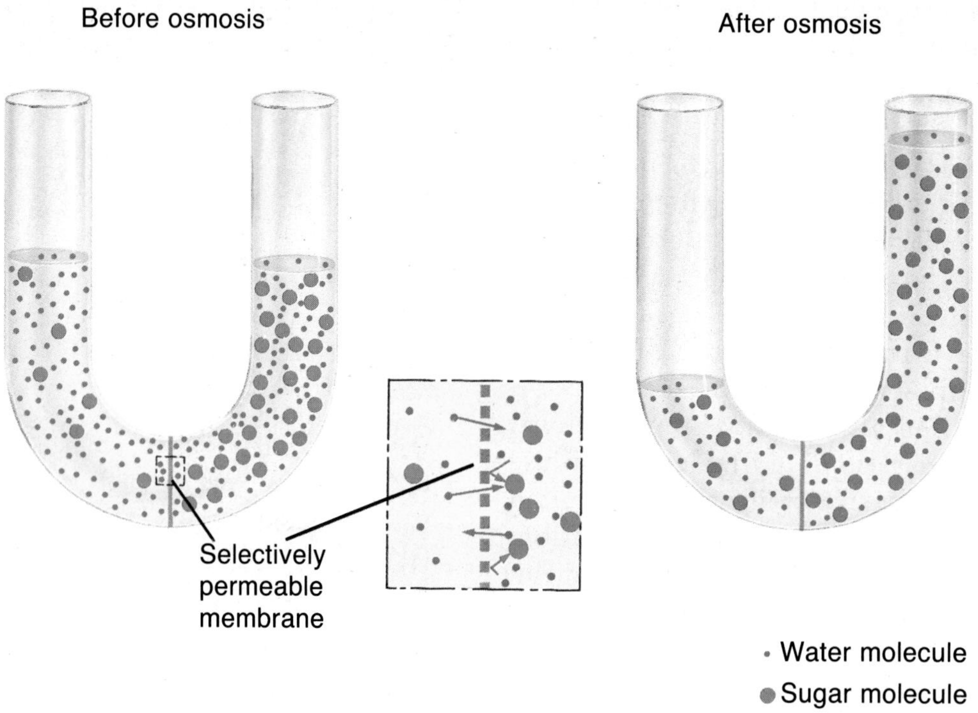

Figure 9.9

During osmosis, water diffuses across a selectively permeable membrane when one side has a higher concentration of a dissolved material that cannot pass through the membrane. Notice that the number of sugar molecules (orange dots) did not change on each side of the membrane, but the number of water molecules (blue dots) did change.

Selectively permeable membrane

• Water molecule

● Sugar molecule

osmosis:

osmos (GK) pushing
Osmosis can push out a cell's plasma membrane.

iso-, hypo-, hyper-:

isos (GK) equal
hypo (GK) under
hyper (GK) over

Osmosis—Diffusion of Water

The diffusion of water molecules into and out of cells is so common that the process is given its own name, osmosis. **Osmosis** is the diffusion of water molecules through a selectively permeable membrane from an area of higher water concentration to an area of lower water concentration.

How osmosis occurs

A strong sugar solution has a lower concentration of water than a weak sugar solution. If these two solutions are placed in direct contact, water molecules diffuse in one direction and sugar molecules diffuse in the other direction. Now, suppose you separate the two solutions with a selectively permeable membrane that will allow water molecules to cross but will not allow sugar molecules to cross. What would happen to the solutions? Study *Figure 9.9.* Osmosis would occur as water molecules diffused across the membrane toward the more concentrated sugar solution on the right. The result would be a buildup of water on the right side of the membrane.

A plasma membrane has properties similar to those of the membrane shown in *Figure 9.9.* As you can see, a cell will lose water by osmosis if it is placed in an environment in which the water concentration is lower than that of the cell contents. Likewise, it will gain water if the water concentration is greater than that of the cell contents.

Osmosis in an isotonic solution

Most cells, whether in multicellular or unicellular organisms, are subject to osmosis because they are surrounded by water solutions. An **isotonic solution** is a solution in which the concentration of dissolved substances is the same as the concentration inside the cell. Likewise, the concentration of water is the same as inside the cell. Do you think osmosis will occur in a cell placed in an isotonic solution?

If a cell is placed in an isotonic solution, water molecules still move into and out of the cell at random, but there is no net movement of water. Therefore, no osmosis occurs. The cell is in dynamic equilibrium with the surrounding liquid because there is movement of molecules, but no overall change is taking place. Cells in isotonic solution have their normal shape as shown in *Figure 9.10*. Most solutions injected into the body are isotonic, so that cells are not damaged by the loss or gain of water.

Cells in hypotonic solution

A **hypotonic solution** is a solution in which the concentration of dissolved substances is lower than the concentration inside the cell. If a cell is placed in a hypotonic solution, osmosis will cause water to move through the plasma membrane into the cell. As water diffuses into the cell, the cell swells and its internal pressure increases. The pressure that exists in a cell is called **turgor pressure.** This pressure increases in a hypotonic solution.

MiniLab

How many contractile vacuoles does a paramecium contain?

Paramecium is an organism that has contractile vacuoles. The vacuole fills with water, contracts, and expels its contents to the outside of the paramecium. When the vacuole contracts, it disappears from view but then refills with water and repeats the process.

Procedure

1. Make a ring of methyl cellulose on a slide. Add a drop of *Paramecium* culture in the center of the ring. Add a coverslip.

2. Find a paramecium, observe it under high power, and locate a contractile vacuole. You may need to adjust your light source or microscope mirror for best results. A contractile vacuole looks like a clear, star-shaped structure that will appear and disappear.

Analysis

1. How many contractile vacuoles does a paramecium have?

2. Do the contractile vacuoles move throughout the cell?

3. How do contractile vacuoles help a paramecium survive in fresh water?

Figure 9.10

Cells in Isotonic Solution

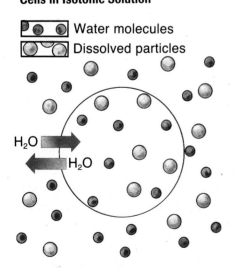

Water molecules

Dissolved particles

▲ **In an isotonic solution, water molecules move into and out of the cell at the same rate.**

Magnification: 2000×

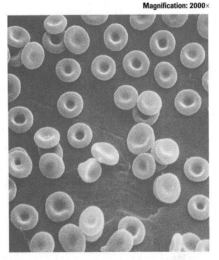

▲ **Cells have their normal shape in isotonic solutions. Notice the concave disc shape of a red blood cell.**

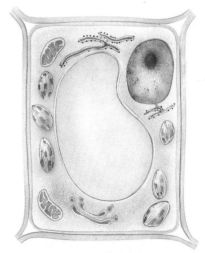

▲ **A plant cell has its normal shape and turgor pressure in isotonic solution.**

Figure 9.11

Cells in Hypotonic Solution

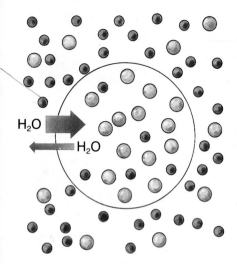

In a hypotonic solution, water enters a cell by osmosis, causing the cell to swell.

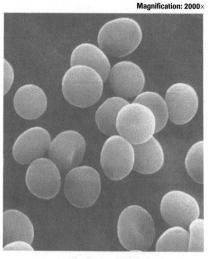

Magnification: 2000×

Animal cells, like these red blood cells, may continue to swell until they burst.

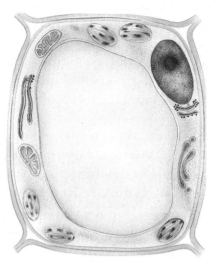

Plant cells swell beyond their normal size as turgor pressure increases.

You will recall from Chapter 8 that plant cells have strong cell walls. The rigid walls can resist the turgor pressure caused by osmosis. Turgor pressure causes the cytoplasm and plasma membrane to press outward against the cell wall, making the cell rigid. This action gives shape and support to plants. The rigidity produced by turgor pressure in cells gives shape and support to plants that are not woody, such as tulips and tomatoes. Plants wilt when they are deprived of water. Wilting occurs because the plant cells lose turgor pressure. When a wilting plant is supplied with water, its cells regain their turgor. Because plain water is hypotonic in comparison to cell contents, turgor pressure can increase to greater than normal, causing the cell to swell as shown in *Figure 9.11.* Grocers take advantage of this process by spraying their produce with water so that the vegetables stay crisp and appealing to customers.

Animal cells do not have cell walls. Therefore, an animal cell in a hypotonic solution will swell and may burst, as shown in *Figure 9.11.* Even so, many organisms that lack cell walls are adapted to live in hypotonic solutions such as fresh water. These organisms have adaptations that keep their cells from bursting. One mechanism is the continual excretion of excess water from cells. Some protists contain organelles, called **contractile vacuoles,** that work like medicine droppers. They collect excess water from the cell and then contract, squeezing the water out of the cell through the outer membrane as shown in *Figure 9.12.*

Figure 9.12

Paramecium contains contractile vacuoles, the star-shaped structures in the photo. Water is collected in ducts (the rays of the star) and passed to the center vacuole. When the vacuole contracts, the water is expelled through a pore in the plasma membrane.

Magnification: 100×

Osmosis in a hypertonic solution

Organisms that live in salty water have the opposite problem. A **hypertonic solution** is a solution in which the concentration of dissolved substances is higher than the concentration inside the cell. If a cell is placed in a hypertonic solution, osmosis will cause water to leave the cell.

Animal cells placed in a hypertonic solution will shrivel because of decreased pressure in the cells, as shown in *Figure 9.13*. Cookbooks suggest that you not salt meat before cooking. The salt forms a hypertonic solution on the meat's surface, and the water inside the meat's cells diffuses out. The result is cooked meat that is dry and tough.

If a plant cell is placed in a hypertonic environment, it will lose water, mainly from its central vacuole. The plasma membrane and cytoplasm will shrink away from the cell wall, as shown in *Figure 9.13*. Loss of water from a cell resulting in a drop in turgor pressure is called **plasmolysis.** This process causes a plant to wilt.

MiniLab

What happens to plant cells in a hypertonic solution?

Plant tissue is normally rigid or crisp because of the turgor pressure within the cells. When plant cells lose their turgor pressure, the plant wilts. What happens on the cellular level when a plant cell is exposed to a hypertonic solution?

Procedure

1. Use forceps to take a leaf from the tip of an *Elodea* sprig, and place it on a slide with a drop of water. Apply a coverslip.

2. Observe it first with low power and then with high power. Draw a cell and its contents.

3. Place a drop of 3% salt solution on the slide near one side of the coverslip. Place a piece of paper towel on the opposite side of the coverslip to draw out the plain water and pull the salt water underneath the coverslip.

4. Observe the cell for a few minutes and then sketch it.

Analysis

1. What happened to the cells when they were exposed to the 3% salt solution?

2. Based on what you have learned about diffusion and osmosis, explain what you observed.

Figure 9.13

Cells in Hypertonic Solution

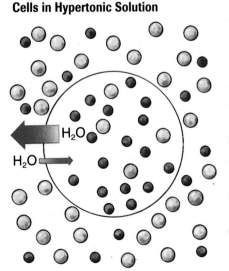

▲ In a hypertonic solution, water leaves a cell by osmosis, causing the cell to shrink.

Magnification: 2000×

▲ Animal cells like these red blood cells shrivel up as they lose water.

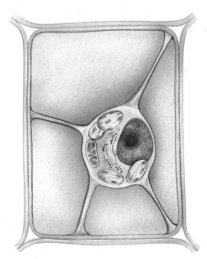

▲ Plant cells lose turgor pressure as the plasma membrane shrinks away from the cell wall.

Figure 9.14

There are two main types of transport proteins: channel proteins and carrier proteins. Channel proteins (left) are tubelike and provide openings through which small, dissolved particles, especially ions, can diffuse. In general, each type of carrier protein (right) has a shape that fits a specific molecule or ion. When the proper molecule binds with the protein, it changes shape and moves the molecule across the membrane.

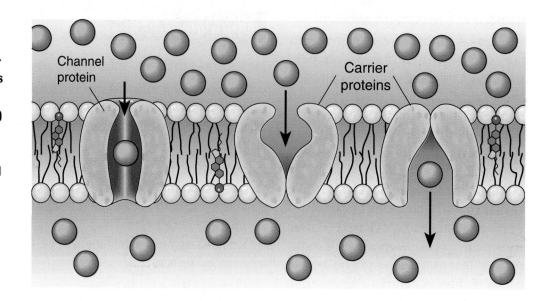

Channel protein

Carrier proteins

Passive Transport

Water, lipids, and lipid-soluble substances can pass through plasma membranes by diffusion. The cell does no work in moving the particles. This movement of particles across membranes by diffusion is called **passive transport** because the cell uses no energy to move the particles.

However, only a few substances are able to pass directly through a phospholipid bilayer. Particles of many other substances are not attracted to the lipid bilayer or are too large to pass through. Passive transport of other substances occurs in different ways.

Passive transport by proteins

Many kinds of proteins are embedded in the lipid bilayer of the plasma membrane like icebergs in the ocean. Some of these are **transport proteins,** which allow needed substances or waste materials to move through the plasma membrane. These proteins function in a variety of ways to transport molecules and ions across the membrane.

The passive transport of materials across the plasma membrane by means of transport proteins is called

facilitated diffusion. As illustrated in *Figure 9.14,* the transport proteins provide convenient openings for particles to pass through. The process of facilitated diffusion is common in the movement of sugars and amino acids across membranes. Movement of materials by facilitated diffusion is the same as in any other diffusion because random motion of particles brings them into the transport proteins.

Active Transport

Can a cell ever move particles from a region of lesser concentration to a region of greater concentration? Yes, but to do so, the cell must expend energy to counteract the random motion of the particles that tends to make them diffuse in the opposite direction. For example, cells require nutrients, such as minerals, that are scarce in the environment.

Cells are adapted to move these nutrients from areas of lower concentration (the environment) to areas of higher concentration (inside the cell). Transport of materials against a concentration gradient requires energy and is called **active transport.**

How active transport occurs

In active transport, a transport protein first binds with a particle of the substance to be transported. Chemical energy from the cell is then used to change the shape of the proteins so that the particle to be moved is released on the other side of the membrane. Once the particle is released, the protein's original shape is restored, as shown in *Figure 9.15*. Thus, a molecule or an ion can be moved into or out of a cell against a gradient.

Transport of large particles

Some cells can take in large molecules, groups of molecules, or even whole cells. **Endocytosis** is a process in which a cell surrounds and takes in material from its environment. This material does not pass directly through the membrane. Instead, it is engulfed and enclosed by a portion of the cell's plasma membrane. That portion of the membrane then breaks away, and the resulting vacuole with its contents moves to the inside of the cell.

ThinkingLab — Draw a Conclusion

How does fertilizer affect earthworms in soil?

Earthworms play an important role in soil ecosystems. As earthworms burrow through the soil, their tunnels allow air to enter the soil. The earthworm, on the other hand, must respond to changes in the soil environment.

Analysis

A group of students placed 50 earthworms on a balance and recorded their mass. They then calculated the average mass of one worm. The students placed 25 of the earthworms into ordinary soil and 25 into soil that had been heavily fertilized. After one day, they removed the earthworms, and again determined the average mass of a worm. The data they obtained is shown in the following table.

Soil Conditions	Average Mass	
	Start	End
Unfertilized	2.4 g	2.5 g
Fertilized	2.4 g	1.8 g

Thinking Critically

Interpret the results shown in the table using what you have learned about osmosis.

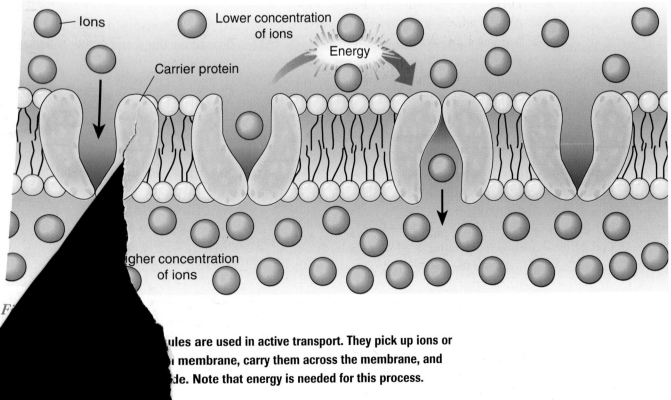

Figure 9.15

Ions — **Lower concentration of ions** — **Energy** — **Carrier protein** — **Higher concentration of ions**

...ules are used in active transport. They pick up ions or ... membrane, carry them across the membrane, and ...le. Note that energy is needed for this process.

Figure 9.16

Phagocytosis is a type of endocytosis in which large particles move into a cell. Some unicellular organisms, such as amoebas, ingest food by phagocytosis. Exocytosis can involve the release of wastes or cell products from a vacuole. The vacuole fuses with the cell membrane, and the contents are released to the outside.

Nucleus

Waste

Digestion

Phagocytosis—a form of endocytosis

Exocytosis

endo-, exo-:
endon (GK) within
exo (GK) out
An endoscope is used to look within the body.

The reverse process of endocytosis is exocytosis, as shown in *Figure 9.16.* Cells use **exocytosis** to expel wastes, such as indigestible bits, from the interior to the exterior environment. They also use this method to secrete substances produced by the cell, such as hormones. Because endocytosis and exocytosis both move masses of material, they both require energy and are therefore classified as active transport.

Connecting Ideas

The plasma membrane plays an important role in maintaining cell homeostasis. Its bilayer structure of phospholipids and embedded proteins allows it to regulate ions and molecules that enter and leave the cell. Passive transport does not require any energy from the cell, but active transport does. How does the cell get energy for this process and all other life processes? What is the source of the energy needed by all living things? Answers to these questions depend on further understanding of how cells function.

Section Review

Understanding Concepts
1. Explain how osmosis and diffusion are alike and how they are different.
2. What factors affect the diffusion of a dissolved substance through a membrane?
3. Compare and contrast active transport and facilitated diffusion.

Thinking Critically
4. Cells in the human body constantly use oxygen during metabolism. Blood contains

oxygen. Why does oxygen dif[ ] into the cells from the bloodstream?

Skill Review
5. **Observing and Inferring** Kineti[ ] molecules increases as temper[ ] increases. What effect do yo[ ] increase in temperature has [ ] For more help, refer to T[ ] Critically in the *Skill H[ ]*

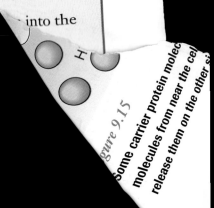

Some carrier protein molec[ ] molecules from near the ce[ ] release them on the other s[ ]

Frozen in Time

Frostbite is a condition in which body tissues, usually in fingers and toes, freeze and die. So how can a process that naturally kills healthy tissue be used to preserve living cells and tissues such as blood, skin, and sperm for later use?

Ordinary freezer burn Your freezer at home cools materials slowly, allowing the water in cells to form large, slowly grown ice crystals that may puncture the cell's plasma membrane. Also, if foods are not properly protected or are frozen for an extended period, the ice crystals evaporate, leaving the cells dried up. This phenomenon is called freezer burn.

Methods of preserving tissue To avoid the problems of large ice crystals and freezer burn, biological specimens are frozen rapidly. An antifreeze such as glycerol is added to lower the temperature at which freezing occurs. The ice crystals formed by this method are much smaller and less damaging. The antifreeze and rapid temperature drop solidify the materials in the cell and the surrounding fluids into a glasslike rather than crystalline state. The fact that the cells are intact helps keep the ice within the cells from evaporating. These procedures, called cryopreservation, produce cells that regain their normal functions when thawed, even after long periods of storage.

Applications for the Future

Cryopreservation allows blood banks to maintain a supply of less common blood types. Freezing organs is not yet possible because freezing does not occur uniformly in a large object, and the antifreeze does not penetrate completely into the tissue. Research is continuing to find ways to overcome these problems.

Another application of this technology is cryosurgery, which kills cells by rapid freezing. Using liquid nitrogen as a cooling agent, surgeons can destroy a specific area by placing it in contact with a freezing apparatus the size of a large knitting needle. This method is precise and results in little or no bleeding, making it a good choice for neurosurgery.

INVESTIGATING the Technology

1. **Debating the Issue** Controversy and lawsuits have arisen over the ownership of frozen human embryos. If a couple involved gets a divorce, to whom do the embryos belong? Debate this issue, presenting arguments for both sides.
2. **Research** Use a chemistry book to find out how glasslike materials differ from crystalline materials.

9.1 The Plasma Membrane
- The plasma membrane controls what enters and leaves a cell.
- The plasma membrane is a phospholipid bilayer with embedded proteins.

9.2 Cellular Transport
- Particles of matter are in constant motion and diffuse from higher concentration to lower concentration.
- Osmosis is the diffusion of water through a membrane and depends on the concentrations of solutes on both sides of the membrane.
- Facilitated diffusion uses transport proteins in the plasma membrane.
- Diffusion, osmosis, and facilitated diffusion are forms of passive transport, which requires no energy from the cell.

- Active transport uses energy and can move materials against a concentration gradient.
- Large particles may enter a cell by endocytosis and leave by exocytosis.

Key Terms
Write a sentence that shows your understanding of each of the following terms.

active transport
contractile vacuole
diffusion
dynamic equilibrium
endocytosis
exocytosis
facilitated diffusion
fluid mosaic model
hypertonic solution
hypotonic solution

isotonic solution
osmosis
passive transport
phospholipid
plasmolysis
selective
   permeability
transport protein
turgor pressure

Understanding Concepts

1. Describe the structure and arrangement of the molecules that make up the plasma membrane.
2. Describe three mechanisms of passive transport.
3. How do cholesterol and unsaturated lipids function in plasma membranes?
4. Describe how dissolved particles move across a plasma membrane from regions of lower to higher concentration.
5. How does an amoeba ingest food particles?
6. What kinds of particles cross the lipid bilayer easily?
7. Under what conditions does diffusion occur? Describe how diffusion takes place.
8. Why must a cell regulate the substances that enter and leave?
9. Explain how active transport operates.

Using a Graph

10. Study the graph and describe how the rate of diffusion increases as the concentration of the diffusing substance increases for both simple diffusion and facilitated diffusion. Explain why the two are different.

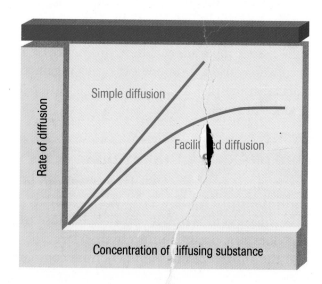

Relating Concepts

11. Make a concept map that relates the following terms and phrases. Supply the appropriate linking words for your map.

    diffusion, active transport, passive transport, facilitated diffusion, water, dissolved ions and polar molecules, osmosis, transport protein

Applying Concepts

12. How would you expect the number of mitochondria in a cell to be related to the amount of active transport it carries out?

13. Explain why drinking ocean water to quench thirst would be dangerous to humans. Hint: The body must excrete salt as a water solution.

14. How does the structure of the plasma membrane allow materials to move across it in both directions?

15. When an amoeba from the sea is placed in fresh water, it forms a contractile vacuole. Explain how this ability is an adaptation for survival.

16. Describe the process by which a stalk of celery becomes limp and rubbery when placed in salt water.

Literature Connection

17. Why do you think Dr. Thomas compares writing to being in a laboratory?

Thinking Critically

Formulating Models

18. When phospholipid molecules are placed in water, they form bilayer membranes. Describe what you think happens to cause the formation of bilayer membranes.

Making Inferences

19. Three funnels containing three different starch solutions were placed for 24 hours into a beaker that contained a starch solution of unknown concentration. The end of each funnel was covered by a selectively permeable membrane. What can you say about the concentration of the solution in the beaker based on the results shown in the diagram?

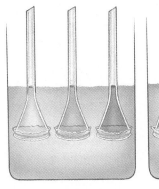
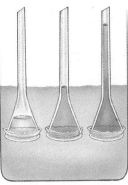

Start   After 24 hours

Making Predictions

20. **Biolab** Suppose that the membrane used in the Biolab was permeable to both iodine and starch. Describe what you would see in the bag and in the beaker after 24 hours.

Making Predictions

21. **Minilab** When a freshwater paramecium is placed in salt water, what do you think will happen to the rate at which the contractile vacuole pumps? Explain.

Connecting to Themes

22. **Homeostasis** Describe briefly the role that a plasma membrane plays in maintaining cellular homeostasis.

23. **Energy** The plasma membrane uses energy for part of its processes. Which processes use energy, and why?

24. **Systems and Interactions** Why must an antibiotic solution to be injected into the body be isotonic?

10 Energy in a Cell

Marathon runners participate in a strenuous race of 26.2 miles, pounding their feet on the pavement continuously for more than two hours. As some runners cross the finish line, they collapse onto the ground, gasping for breath. Where do these runners get energy for their muscles to finish such a race? Perhaps you have heard of carbohydrate loading in preparation for a marathon. Why would a runner wish to "carbo-load" by eating platefuls of carbohydrates such as pasta before a race? Why do carbohydrates represent energy to the marathoner?

Life on Earth—whether it is a marathon runner, a flashing firefly, or a growing tomato plant—depends on a flow of energy. This flow begins some 93 million miles away with the sun. Each day, the sun delivers to Earth an amount of energy that is 1.5 billion times the amount of electrical energy generated in the United States each year. A small fraction of this solar energy—less than one percent—powers the activities performed by the cells of Earth's living things.

This marathon runner required an incredible amount of energy to finish the 26.2-mile race. The muscle cells of the runner obtain much of their energy from the carbohydrates in the food eaten earlier. Considering that the sun is the ultimate source of energy on Earth, how did its energy become stored in carbohydrates?

10.1 ATP: Energy in a Molecule

Section Preview

Objectives

Explain why organisms need a supply of energy.

Explain how energy is stored in ATP and released from ATP.

Key Terms

ATP
ADP

You wouldn't think of trying to make a car run by dropping a lighted match into the gasoline tank. So much energy would be released at once that the car would explode. Instead, gas must be delivered to the engine in small, measured amounts so that the energy is released from the fuel in a useful, controlled way. Likewise, living organisms have mechanisms that release energy from food molecules in a controlled way.

Cell Energy

Work is done whenever anything is moved. In animals, muscle cells contract in order to move the body and pump blood. In all cells, life processes constantly move and rearrange atoms, ions, and molecules. All this biological work demands energy. Energy is the ability to do work, so organisms and their cells need a steady supply of energy in order to function. Just consider the amount of energy the leafcutter ant in *Figure 10.1* needs.

Energy for biological work

Your muscle cells need energy to contract as you walk, bend, and exercise. Energy is used to move ions and molecules across membranes in your nerves and kidneys. Your cells use energy even when you sleep. Your heart muscle continues to pump blood. Your brain is actively processing the events of your day. Where do your cells get the energy they need to do this biological work? They get it from the food you eat.

Figure 10.1

Energy can take different forms, such as chemical, mechanical, or radiant energy. This leafcutter ant has transformed chemical energy from food into mechanical energy by moving the large piece of leaf.

Releasing food energy

Although your diet may vary from spaghetti to spinach, almost all the food you eat is broken down by your body into smaller and smaller bits. By the time food reaches your bloodstream, it has been broken down into nutrient molecules that can enter your cells. Cell reactions then break down the food molecules, releasing energy for the biological work your cells need to perform.

ATP as energy currency

A cell can't use all the energy available in food molecules at once. Consider the following analogy. You go to the bank to withdraw $100, and the teller gives you a $100 bill. You stop on the way home to buy a drink from a vending machine, but you can't stuff the $100 bill in the slot. Smaller bills and coins are much more convenient for such purchases.

In the same way, cell processes need "smaller change" for their energy expenses. A mechanism has evolved that balances energy supply and demand. The cell "makes change" by breaking down food molecules bit by bit and distributing the energy to other molecules, as shown in *Figure 10.2*. These energy-storing molecules are adenosine triphosphate, or **ATP** for short. ATP can be called the cell's energy currency because it is from ATP molecules that the cell gets energy for its work. In fact, without a constant and abundant supply of ATP, a cell will die.

Forming and Breaking Down ATP

Cells store energy by bonding a third phosphate group to adenosine diphosphate, **ADP,** to form ATP. When the phosphate-phosphate bonds form, energy is stored. ADP contains adenine, a sugar called ribose, and two phosphate groups. ATP molecules are made of adenine, ribose, and three phosphate groups bonded in a row. When the third phosphate group breaks off an ATP

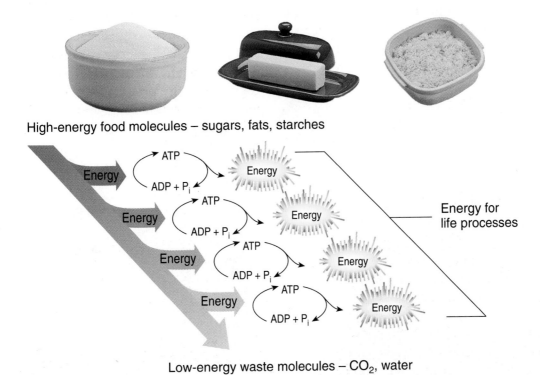

High-energy food molecules – sugars, fats, starches

Energy for life processes

Low-energy waste molecules – CO_2, water

Figure 10.2

Food molecules are broken down in several steps. At each step, the energy released is stored in ATP, the cell's energy currency. This energy can be released from ATP for use in cell work.

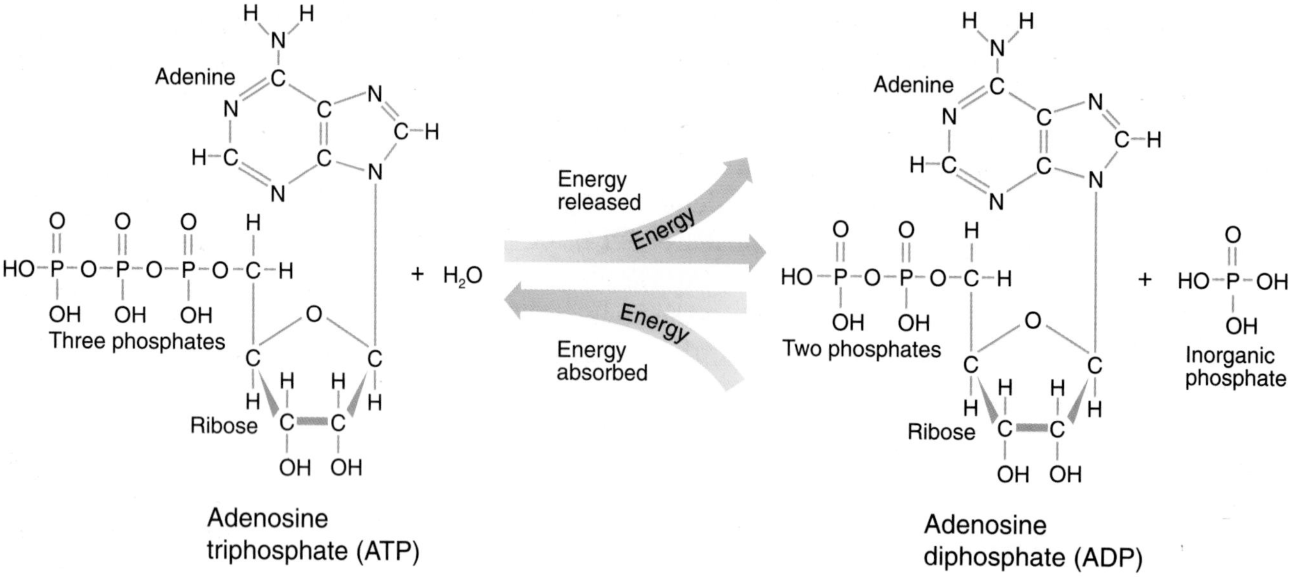

At the top of the page is a chemical diagram showing the ATP ⇌ ADP reaction.

Left structure (ATP):
Adenine

Three phosphates

$HO-P-O-P-O-P-O-C-H$ (with OH and O groups)

Ribose

Adenosine triphosphate (ATP)

$+ H_2O$

Center arrows:
Energy released — Energy
Energy absorbed — Energy

Right structure (ADP):
Adenine

Two phosphates

Ribose

Adenosine diphosphate (ADP)

$+ HO-P-OH$ (OH)

Inorganic phosphate

Figure 10.3

Energy is stored when ATP is made from ADP and phosphate. Energy is released when ATP breaks down to ADP and phosphate. It is estimated that in a resting adult human, this ATP ⇌ ADP cycle processes more than 40 kg of ATP each day.

molecule, the result is ADP, a phosphate group, and the release of energy, *Figure 10.3*. ADP is available again to store energy by forming ATP. Within a cell, formation of ATP from ADP and phosphate occurs over and over, storing energy each time. As the cell uses energy, ATP breaks down repeatedly to release energy, ADP, and phosphate. The conversion reactions of ADP and ATP are shown by the following equation and in *Figure 10.3*. The symbol P_i represents an inorganic phosphate group. As you can see, the formation and breakdown form a cycle.

$$ADP + P_i + energy \leftrightarrows ATP$$

Lower energy   Higher energy

ATP links energy use and energy release

Biological work, as you know, takes energy. Energy that is available to do work is called free energy. Energy that is locked up in molecules is unavailable for work; it is potential energy. Unlocking this potential energy to produce free energy is vital to a cell.

Consider the reaction shown in *Figure 10.4,* in which a plant combines glucose and fructose to make the disaccharide sucrose. Making sucrose requires energy to form the bond between glucose and fructose. As a result, the product, sucrose, has more potential energy than the reactants. This energy comes from the breakdown of some of the plant's supply of ATP.

Figure 10.4

As the cells of a plant such as this holly tree make sucrose from glucose and fructose, energy must be released from the breakdown of ATP to ADP and inorganic phosphate.

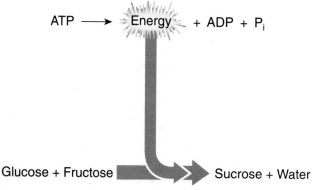

$$ATP \longrightarrow Energy + ADP + P_i$$

Glucose + Fructose $\longrightarrow$ Sucrose + Water

ATP seems to have evolved as the main energy link between energy-using and energy-releasing reactions. That's because the amount of free energy released when it breaks down is suitable for use in most cellular reactions.

The Uses of Cell Energy

Making sucrose is an example of one way in which cells use energy—making new molecules. Some of these molecules are enzymes, which carry out cell reactions. Others build membranes and cell organelles. In what other ways do cells use energy that is available from the breakdown of ATP to ADP?

Cells use energy to maintain homeostasis. Your kidneys—and those of other animals—use energy to move molecules and ions in order to eliminate waste substances while keeping needed substances in the bloodstream. *Figure 10.5* shows other ways that cells use energy.

Figure 10.5

Some Ways that Cells Use Energy

▶ **Warm-blooded animals make use of the heat given off in cell reactions to maintain a warm body temperature. With mechanisms of homeostasis to control body temperature, warm-blooded animals such as this polar bear can be active in many different climates.**

◀ **Nerve cells are able to transmit impulses by using ATP to power the active transport of certain ions.**

Magnification: 7000×

▶ **Some cells such as muscle cells and cells with cilia or flagella (right) use energy from ATP in order to move.**

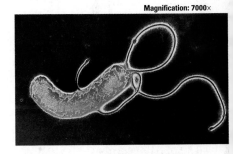

◀ **Fireflies, many deep-sea animals, and some caterpillars such as the one shown here produce light by a process called bioluminescence. The light results from a chemical reaction that is powered by the breakdown of ATP.**

Section Review

Understanding Concepts

1. How does ATP store energy?
2. Why must there be a mechanism for storing energy that is given off during the breakdown of food molecules?
3. Suppose a cell combines molecules A and B to produce molecule C in an energy-requiring reaction. Explain how ATP may be used in this type of reaction.

Thinking Critically

4. Make a list of possible ways that an eagle might use energy from the breakdown of ATP.

Skill Review

5. **Observing and Inferring** When animals shiver in the cold, muscles move almost uncontrollably. Suggest how shivering helps an animal survive in the cold. For more help, refer to Thinking Critically in the *Skill Handbook*.

Photosynthesis: Trapping Energy

Section Preview

Objectives

Relate chlorophyll to the process of photosynthesis.

Explain how the light reactions and Calvin cycle are related.

Key Terms

photosynthesis
light reactions
Calvin cycle
electron transport chain
photolysis
chemosynthesis

*Y*ou say you're hungry? Well, how about a nice hot, steaming plate of—carbohydrates? Mmm. That doesn't sound appetizing? Well think again. You may be turning down a deep-dish pizza, a seafood pasta plate, or a Danish pastry. All those foods contain carbohydrates. Not only can carbohydrates taste delicious, but they are vital.

photosynthesis:

photos (GK) light
syntithenai (GK)
to put together
Photosynthesis
puts together
sugar molecules
using energy
from light.

Autotrophs and Sunlight

Your cells break down carbohydrates to release energy and form ATP. Of course, a chef doesn't make these carbohydrates, so where do they come from? Think about the kinds of organisms called autotrophs. Autotrophs—green plants, algae, and some bacteria—are found in nearly every ecosystem. These organisms, too, must break down carbohydrates to form ATP. These carbohydrates are usually in the form of simple sugars, especially glucose. However, autotrophs don't take in these sugars as food; they make the sugars themselves.

Sunlight and chlorophyll

How do autotrophs make sugars? A major ingredient in their recipe is light. Autotrophs trap energy from sunlight and use this trapped energy to build carbohydrates in a process called **photosynthesis.** Sunlight, then, is the natural energy source for photosynthesis.

You may already know that white light from the sun consists of a mixture of colors ranging from red through orange, yellow, green, blue, and indigo to violet. Objects that appear colored, *Figure 10.6,* do so because they reflect some of the colors of white while absorbing other colors.

The green energy trap

What color are almost all the plants you see? Green, of course. Most plants are green because they contain chlorophyll. Chlorophyll reflects green and some yellow light and absorbs the energy of other colors of sunlight. It is this energy that is stored in the carbohydrates produced by photosynthesis.

Chlorophyll is found in the chloroplasts of green plants and algae and on membranes in the cytoplasm of photosynthetic bacteria. A chloroplast is surrounded by two lipid bilayer membranes. Internally, the chloroplast has a series of membranes called thylakoid membranes. It is within these membranes that the energy from sunlight is trapped by chlorophyll.

Figure 10.6

As white light from the sun passes through a prism, it is separated into the colors of the visible spectrum. As white light strikes a green leaf, most of the colors except green and yellow are absorbed. Green and yellow are reflected by the leaf and are ~~ the viewer.

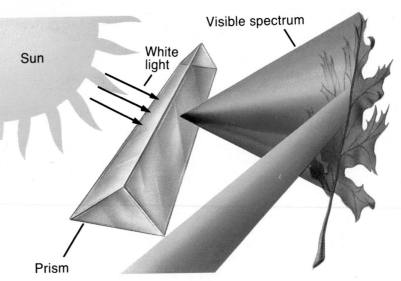

Sun

White light

Visible spectrum

Prism

Other pigments may be found in the same membrane layers that hold chlorophyll. These pigments trap energy from colors of light that chlorophyll does not absorb well. They give cells the colors red and yellow that you see in carrots, tomatoes, some fruits, and fall leaves, *Figure 10.7*.

The general equation for photosynthesis follows.

$$6CO_2 + 6H_2O + \text{light energy} \xrightarrow{\text{chlorophyll}} C_6H_{12}O_6 + 6O_2$$

You can see that the overall reaction of photosynthesis is simple. However, it represents the sum of many separate chemical reactions that take place when simple sugars are formed.

What happens during photosynthesis?

Trapping the energy of the sun in the chloroplasts is the first step of photosynthesis. This light energy is converted to chemical energy and then stored. Water and carbon dioxide are the other two ingredients needed for photosynthesis to occur.

The reactions of photosynthesis

The process of photosynthesis includes two main groups of reactions. These are the light reactions and the Calvin cycle. **Light reactions** are the reactions in which light energy is converted to chemical energy. The light reactions are the *photo* part of photosynthesis. These reactions split water molecules, providing hydrogen and an energy source for the Calvin cycle. Oxygen from water is given off. The **Calvin cycle** is the series of reactions that forms simple sugars using carbon dioxide and hydrogen from water. The Calvin cycle is the *synthesis* part of photosynthesis.

chlorophyll:

chloros (GK) pale (yellowish) green
phyllon (GK) leaf
Chlorophyll is a green substance found in leaves.

Figure 10.7

The red, yellow, and purple pigments in plants are usually masked by the green color of chlorophyll. However, they are visible in the autumn when the leaves of trees lose their chlorophyll.

How does photosynthesis vary with light intensity?

Photosynthesis is the process in which autotrophs synthesize organic compounds from water and carbon dioxide using energy absorbed by chlorophyll from sunlight. This process is basic for the existence of nearly all forms of life.

Analysis

Green plants were exposed to increasing light intensity, as measured in candelas, and the rate of photosynthesis was measured. The temperature of the plants was kept constant during the experiment. The following graph depicts the data obtained.

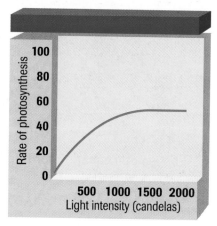

Thinking Critically

Considering the overall equation for photosynthesis, make a statement summarizing what the graph shows. Under normal field conditions, what factors may limit the relative rate of photosynthesis when the light intensity is increased and temperature remains constant?

Light Reactions

The light reactions of photosynthesis are quite complex, and scientists are still studying the process. Many reactions and enzymes are involved. You can follow along with the important steps of the light reactions in *Figure 10.8*.

Electrons in chlorophyll absorb light energy

When light strikes chlorophyll, electrons within the chlorophyll molecule absorb energy. When electrons absorb enough energy, they leave the chlorophyll molecule and are passed along a series of molecules in the thylakoids, releasing energy as they go. This series of molecules is known as the **electron transport chain.** As the electrons pass down the chain, the extra energy they received from light is stored in the bonds of ATP. In other words, energy from light becomes available to do biological work. The ATP produced will be used in the second stage of photosynthesis, the Calvin cycle.

Figure 10.8

In the light reactions of photosynthesis, there are two events involving chlorophyll and light energy. Between these reactions is another reaction involving water.

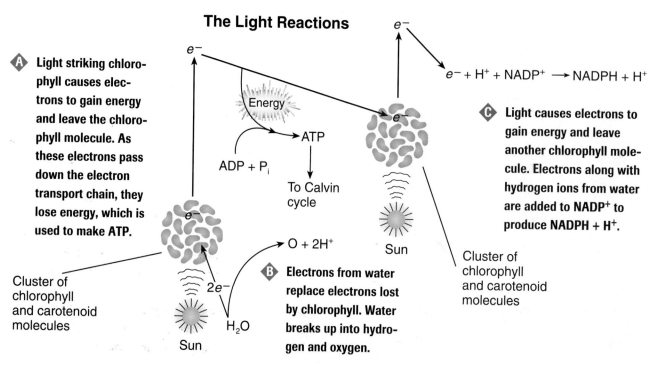

The Light Reactions

Ⓐ Light striking chlorophyll causes electrons to gain energy and leave the chlorophyll molecule. As these electrons pass down the electron transport chain, they lose energy, which is used to make ATP.

Cluster of chlorophyll and carotenoid molecules

ADP + P$_i$

ATP

To Calvin cycle

O + 2H$^+$

2e$^-$

H$_2$O

Sun

Energy

e^-

Ⓑ Electrons from water replace electrons lost by chlorophyll. Water breaks up into hydrogen and oxygen.

$e^- + H^+ + NADP^+ \longrightarrow NADPH + H^+$

Ⓒ Light causes electrons to gain energy and leave another chlorophyll molecule. Electrons along with hydrogen ions from water are added to NADP$^+$ to produce NADPH + H$^+$.

Cluster of chlorophyll and carotenoid molecules

Sun

Water molecules are split

Notice in *Figure 10.8* that water is one of the reactants in photosynthesis. Chlorophyll that has lost electrons now picks up electrons from water, splitting the water into hydrogen ions and oxygen. The splitting of water during photosynthesis is called **photolysis.** A freed hydrogen ion and an electron are picked up by the carrier ion $NADP^+$ (nicotinamide dinucleotide phosphate) to form $NADPH + H^+$, and carry them to the Calvin cycle.

NADP is one of a class of organic molecules called coenzymes. Coenzymes act as carriers in many biological processes. The oxygen freed during photolysis is given off as a waste product. This process is the source of nearly all the oxygen in Earth's atmosphere. Note that the light reactions trap energy but do *not* involve CO_2, and no sugars are produced.

The Calvin Cycle

Take a deep breath of air. That oxygen you're inhaling probably was released during the light reactions of photosynthesis. This oxygen is vital to much of life on Earth, for a reason that you will soon learn. During the light reactions, a plant produces ATP and $NADPH + H^+$. It is now able to use CO_2 molecules to produce food molecules in the second stage of photosynthesis—the Calvin cycle.

Reactions of the Calvin cycle

In the Calvin cycle, *Figure 10.9*, an enzyme adds the carbon atom of carbon dioxide to a 5-carbon molecule. Because the carbon is now fixed in place in an organic molecule, the process is called carbon fixation. Carbon fixation is vital because it takes carbon dioxide from the air and converts it to a form that is usable by living things.

photolysis:
photos (GK) light
lyein (GK) to split
Water molecules are split in photolysis.

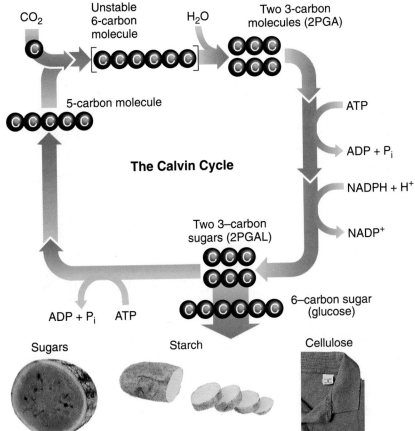

The Calvin Cycle

Figure 10.9

When carbon dioxide combines with the 5-carbon molecule, a 6-carbon molecule forms and splits immediately into two 3-carbon molecules. The two 3-carbon molecules formed are phosphoglyceric acid (PGA) molecules. These molecules are converted into two 3-carbon sugars, phosphoglyceraldehyde (PGAL), using the hydrogens of NADPH + H^+ and energy from ATP. Some of these sugars leave the cycle and are used to form glucose, fructose, starch, cellulose, or other complex carbohydrates.

Water molecules act as a source of electrons in the light reactions. The resulting products are hydrogen ions and oxygen gas.

What factors influence photosynthesis?

Oxygen is one of the end products of photosynthesis. Because oxygen is only slightly soluble in water, visible bubbles of oxygen are formed by aquatic plants, such as *Elodea*, as they carry out photosynthesis. By measuring the rate at which bubbles form, you can measure the photosynthetic rate.

PREPARATION

Problem

What colors of light are most effective in photosynthesis? What environmental factors, such as temperature and the availability of CO_2, influence the rate of photosynthesis in aquatic plants?

Hypotheses

Choose one factor that may influence the rate of photosynthesis. Hypothesize as to how it will influence that rate. Be sure to consider factors that you can easily change using the aquatic aquarium plant *Elodea*.

Objectives

In this Biolab, you will:
- **Identify** a factor that influences photosynthesis.
- **Predict** how the factor will influence the rate of photosynthesis.
- **Measure** the rate of photosynthesis.

Possible Materials

250-mL Erlenmeyer flask
3 sprigs of *Elodea*
string
washers
1000-mL beaker
hot plate
ice
water
oven mitts
thermometer
colored cellophane, assorted colors
lamp with reflector and 150-watt bulb
0.25 percent sodium hydrogen carbonate (baking soda) solution
ring stand with clamp
watch with second hand
metric ruler

Safety Precautions

Exercise caution when using hot plates.

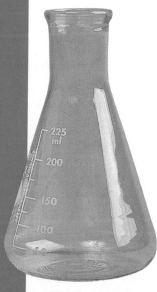

1. Study the diagram for the basic setup to measure photosynthesis from *Elodea*, and decide how your group will modify this setup to measure the factor you have chosen.

2. The *Elodea* needs to be lightly tied together with a weight and placed in sodium hydrogen carbonate solution inside the flask so that the plants are underneath the solution. Sodium hydrogen carbonate solution provides CO_2 for the aquarium plants.

Check the Plan

1. Consider all factors that you believe may influence the rate of photosynthesis. All of these factors, except the one you are testing, need to be kept constant throughout your experiment.

2. Consider how you will set up a control for the factor that you are testing. Be sure to run a control setup.

3. Design and construct a data table for recording your observations.

4. Plan to measure the number of bubbles of oxygen generated under both control conditions and experimental conditions at various time periods for a total of at least five minutes.

5. ***Make sure your teacher has approved your experimental plan before you proceed further.***

6. Carry out your experiment.

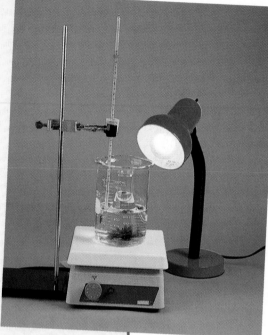

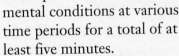

1. **Identifying Variables** What conditions did you keep constant in the experiment? What one condition did you vary?

2. **Interpreting Observations** From where did the bubbles of oxygen emerge?

3. **Making Inferences** Explain how counting bubbles measures the rate of photosynthesis.

4. **Analyzing Data** Make a graph of your data with the photosynthetic rate per minute plotted against the factor tested for both the control and experimental setups. Write a sentence or two explaining the graph.

5. **Checking Your Hypothesis** Did your results support your hypothesis? If not, suggest a new hypothesis that is supported by your data.

Going Further

Application Design a method to measure the rate of photosynthesis of a land plant, such as a potted geranium.

10.2   Photosynthesis: Trapping Energy   **247**

Too Much Sun, Too Little Water

As you have learned, plants require light, carbon dioxide, and water. Carbon dioxide is taken into the plant through pores in leaves and stems. However, when these pores are open, water is lost from the plant. This loss presents little problem for plants that live in mild climates with adequate rainfall. Plants that are successfully adapted to hot, dry climates must have a means of getting carbon dioxide while avoiding excessive water loss.

Special CO₂ delivery A more efficient way of trapping CO_2 has evolved in some tropical grasses, including corn, crabgrass, and sugarcane. These plants have cells that contain 3-carbon molecules with a strong attraction for CO_2. When CO_2 bonds to these 3-carbon molecules, 4-carbon molecules are formed. These molecules move to the cells in which the Calvin cycle is taking place and release CO_2. Thus, the 4-carbon molecule acts as a specialized delivery system to get CO_2 molecules to the Calvin cycle. As a result, the Calvin cycle can operate at high efficiency while conserving water.

Living like a cactus Another specialized method for trapping CO_2 evolved among plants, such as cactus, which continue to thrive throughout long periods of hot, dry conditions, such as in deserts. These species of plants take in CO_2 at night and bond it to an organic substance. In the daytime, the CO_2 is already stored inside the plant, available for use in the Calvin cycle. With this mechanism for getting and storing CO_2, the desert plants do not have to open their pores during the heat of the day in order to take in CO_2. This adaptation allows them to photosynthesize with minimum water loss.

Thinking Critically

In much of the interior United States, the climate is cold and wet in winter with hot, dry periods in the summer. Many people in this region consider crabgrass to be a weed that can take over lawns during the summer. Why does crabgrass have this ability?

Figure 10.10

This diagram of a chloroplast summarizes the process of photosynthesis. In photosynthesis, the light reactions furnish energy and some raw materials to the remainder of the process, the Calvin cycle.

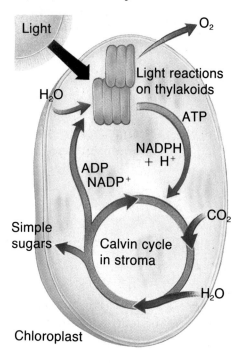

Light

O_2

Light reactions on thylakoids

H_2O

ATP

NADPH + H⁺

ADP NADP⁺

CO_2

Simple sugars

Calvin cycle in stroma

H_2O

Chloroplast

The Calvin cycle reactions take place in the stroma of the chloroplasts. The overall effect of the Calvin cycle is that carbon dioxide combines with hydrogen to form simple sugars that are used to make other carbohydrates such as more complex sugars, starch, and cellulose.

Carbon dioxide enters the leaves and stems of green plants through hundreds of small pores. Besides carbon dioxide, ATP and hydrogen from the light reactions are needed for the Calvin cycle. The Calvin cycle reactions, shown in *Figure 10.9*, continue as long as the materials are available.

You can get a good overall view of the relationship between the light reactions and the Calvin cycle, as well as the reactants and end products of the whole process of photosynthesis, in *Figure 10.10.*

During the Calvin cycle, a plant produces carbohydrate molecules—mainly sugars, starch, and cellulose. Some of these molecules are now available as sources of energy to the plant and to organisms that eat the plant. In the cells of producers and consumers, carbohydrates are broken down to release the energy that originally radiated from the sun and make it available for life processes on Earth.

Figure 10.11

Most of these sea-vent communities rely on energy from reactions of inorganic sulfur compounds that come from openings in the ocean floor. The sulfur compounds are metabolized by bacteria that live symbiotically in the gills of clams, mussels, and tube worms.

Life Without Light

Some bacteria are able to capture energy from sources other than the sun. Instead of being photosynthetic, these bacteria are called chemosynthetic. **Chemosynthesis** is a process by which an autotroph obtains energy from inorganic compounds instead of from light.

One group of chemosynthetic prokaryotes is methane-producing bacteria. These organisms are adapted to live under conditions where there is no oxygen. In fact, they are poisoned by oxygen. They are found in marshes, lake sediments, and the digestive tracts of ruminant mammals, such as cows. These bacteria are the final participants in the decomposition process. They react CO_2 and H_2 to produce methane (CH_4). They are also important in the breakdown of sewage at sewage-disposal plants.

Scientists have discovered thriving animal communities, as shown in *Figure 10.11,* near cracks deep on the floor of the Pacific Ocean. In fact, these communities are a part of a whole ecosystem that doesn't need the sun to survive.

Section Review

Understanding Concepts

1. Why do you see green when you look at a leaf on a tree?
2. How do the light reactions of photosynthesis relate to the Calvin cycle?
3. What is the function of water in photosynthesis?

Thinking Critically

4. In the general equation for photosynthesis, why is chlorophyll written over the arrow rather than on the left side of the equation?

Skill Review

5. **Designing an Experiment** Design an experiment that would compare photosynthesis in red, green, and blue light. What would you use as a control? For more help, refer to Practicing Scientific Methods in the *Skill Handbook.*

10.3 Getting Energy to Make ATP

Section Preview

Objectives

Compare and contrast aerobic and anaerobic processes.

Explain how cells obtain energy from respiration.

Key Terms

respiration
aerobic process
anaerobic process
glycolysis
citric acid cycle
lactic acid fermentation
alcoholic fermentation

anaerobic:

an (GK) without
aeros (GK) air
Anaerobic organisms can live without air, i.e., oxygen.

Food molecules produced by photosynthesis contain energy from the sun converted to chemical energy. But like the person who wants to buy only a can of soda but is stuck with a $100 bill, a cell can't use all the energy in a food molecule at once. As you will see, organisms have processes to "make change"—to break down food molecules to yield usable amounts of energy.

The First Steps of Respiration

When scientists talk about food molecules in cells, they don't mean hamburgers or asparagus. Molecules of glucose and other 6-carbon sugars are the major source of energy for most organisms, although fatty acids and amino acids can also be used as energy sources. Cells break down glucose through a series of chemical reactions. The stored energy of glucose is released bit by bit and used to attach phosphate groups to ADP molecules to form ATP molecules, the cell's energy currency.

Figure 10.12

Notice that the first step of glycolysis uses the energy from two molecules of ATP. However, four ATP molecules are formed using the energy released by the second step of glycolysis. Thus, in glycolysis, the energy payoff for the breakdown of one molecule of glucose to two molecules of pyruvic acid is only two molecules of ATP. Notice that the intermediate product, PGAL, is the same 3-carbon sugar that was formed in photosynthesis.

Glycolysis

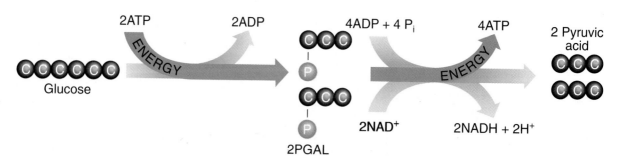

The process by which food molecules are broken down to release energy is called **respiration.** In many cells, both aerobic and anaerobic processes take place during respiration. **Aerobic processes** require oxygen in order to take place. **Anaerobic processes** do not require oxygen. Anaerobic processes are simple and yield energy quickly. However, as you will see, the energy payoff is much greater when molecules are broken down aerobically.

The first step—Glycolysis

When a 6-carbon glucose molecule breaks down, it is changed into pyruvic acid molecules, which are 3-carbon molecules. This breakdown occurs without oxygen. The anaerobic process of splitting glucose and forming two molecules of pyruvic acid is called **glycolysis,** which you can see in *Figure 10.12.*

Glycolysis also produces hydrogen ions and electrons. These combine with organic carrier ions called NAD^+ (nicotinamide dinucleotide) to form NADH $+ H^+$. The compound NAD, like NADP used in photosynthesis, is a coenzyme. Glycolysis occurs in the cytoplasm of cells and does not require oxygen. However, if oxygen is present, energy-yielding aerobic processes may follow.

Releasing Energy with Oxygen

The process of breaking down glucose begins with glycolysis, yielding two molecules of ATP for each molecule of glucose broken down. When

Joseph Priestley's Experiments

English scientist Joseph Priestley was trained as a minister, but was unhappy in his profession because a speech impediment caused him to stammer in front of a congregation. Instead, he turned to what he called his "taste for science" and began experimenting with "different kinds of air." (Nowadays, the term is *gases.*)

Electricity and sodas   With the encouragement of his American friend Benjamin Franklin, Priestley wrote a book on the history of electrical research. Then he returned to his experiments with "airs," as well as the production of CO_2 with sulfuric acid and calcium carbonate, which led to today's soda water.

Of mice and mint   A few years later, in 1773, Priestley performed an experiment to test his idea that animal and plant metabolism "used up" air. To do this, he put a sprig of mint into an air-filled glass jar inverted in a vat of water. After a while, he inserted a candle and found that it would still burn. He knew that a candle would not burn in used-up air, so maybe the air hadn't been used up after all, he thought. Next, he replaced the candle with a mouse. The mouse continued to live, or, as Priestley put it, the air "was not at all inconvenient to a mouse." However, when the mint was removed, the mouse died. What Priestley had discovered was the gas, oxygen, and a process by which it is produced, photosynthesis.

Priestley believed that a burning candle or a living mouse used up air somehow. Today, we know that air is a mixture of 78 percent nitrogen, 21 percent oxygen, 0.03 percent carbon dioxide, and other gases in small amounts. What was happening when air became "used up"?

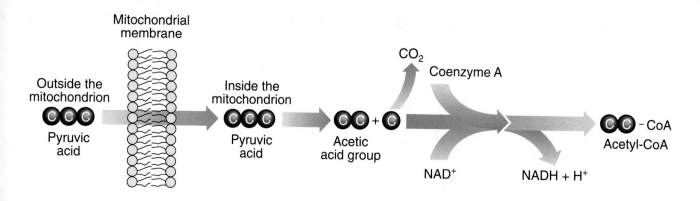

Figure 10.13

Pyruvic acid enters a mitochondrion. Here in the mitochondrion, pyruvic acid changes to acetic acid by losing a CO_2 molecule. It then forms acetyl-CoA before entering the third step of aerobic respiration.

oxygen is present, aerobic respiration, which uses oxygen, takes place in a mitochondrion, the powerhouse of the cell.

Breakdown of pyruvic acid

Aerobic respiration involves two more steps after glycolysis. The first step includes reactions that change pyruvic acid, as shown in *Figure 10.13*. Pyruvic acid, a 3-carbon compound, is changed to acetic acid, a 2-carbon compound. The third carbon atom forms carbon dioxide, CO_2. The acetic acid from the reaction is combined with a substance called coenzyme A to form a compound called acetyl-CoA.

The citric acid cycle

The second step of aerobic respiration is the **citric acid cycle,** shown in *Figure 10.14.* This cycle of chemical reactions produces more ATP and releases additional electrons. The electrons are picked up by NAD^+ and FAD (flavin adenine dinucleotide, an electron carrier similar to NAD^+).

In the citric acid cycle, acetyl-CoA combines with a 4-carbon molecule to form a 6-carbon molecule, citric acid. Locate acetyl-CoA and citric acid in *Figure 10.14* and follow each of the reactions. Notice that citric acid is broken down first to a 5-carbon molecule and then to a 4-carbon molecule, releasing CO_2 at each step.

Figure 10.14

Citric acid is changed to a 5-carbon molecule and then to a 4-carbon molecule. Notice that a carbon dioxide molecule is given off in each of these reactions. This loss of two CO_2 molecules produces a 4-carbon molecule that reacts to form the original 4-carbon substance again. As a result, a new 4-carbon molecule is available to combine with each new acetyl-CoA that enters the cycle. Follow the sequence of reactions in the cycle and locate where NAD^+ and FAD pick up electrons.

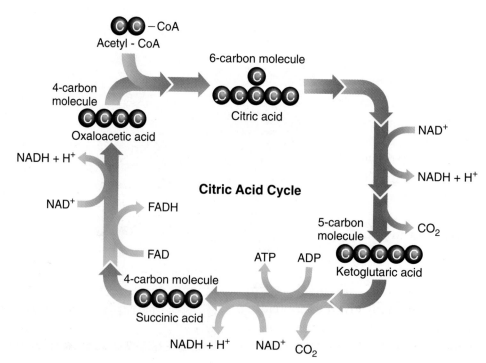

The electron transport chain

In both glycolysis and the citric acid cycle, some energy is trapped in the ATP that is formed. However, important sets of reactions also release electrons to NAD+ and FAD. The fourth part of aerobic respiration is an electron transport chain.

Remember that an electron transport chain is a series of molecules along which electrons are transferred, releasing energy. Carrier molecules bring electrons from the reactions of both glycolysis and the citric acid cycle to the electron transport chain, as shown in *Figure 10.15*. The molecules of the electron transport chain are located on the inner membranes of mitochondria.

This process is aerobic because the final electron acceptor in the transport chain is oxygen. When oxygen

MiniLab

Do aquariums contain measurable amounts of CO_2?

Bromothymol blue solution is an acid-base indicator solution. When CO_2 dissolves in H_2O, they combine to form a weak acid, carbonic acid. Bromothymol blue changes to yellow color in the presence of carbonic acid.

Procedure
1. Place 50 mL of aquarium water from a fish tank into a 100-mL beaker.
2. Add six to ten drops of bromothymol blue.

Analysis
1. Were there detectable amounts of CO_2 in the aquarium water? How do you know?
2. Did the aquarium contain numerous living plants? Would you expect these plants to affect the amount of CO_2 in the water? Explain.
3. Why would the number of fish in the aquarium affect the amount of CO_2 in the water?

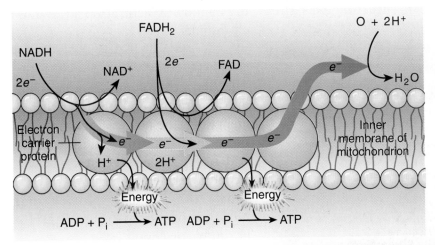

Figure 10.15

At the electron transport chain, the carrier molecules NADH and FADH$_2$ give up electrons that pass through a series of reactions. At least 15 carrier proteins are involved in the electron transport chain. At the top of the chain, the electrons have high energy. As the electrons pass down the chain, the energy given off is captured in molecules of ATP.

accepts electrons, it combines with two hydrogen ions to form a molecule of water, H_2O. Much of the water vapor and carbon dioxide that you exhale is the result of aerobic respiration in your cells. You can see in *Figure 10.16* another example of the aerobic breakdown of sugar.

What happens if no oxygen is present? If the final electron acceptor, oxygen, is used up, the chain becomes jammed. It is somewhat like a

Figure 10.16

When sugar burns in air, the reaction releases CO_2, H_2O, and a lot of energy at once. In aerobic respiration, the same overall process occurs, but step-by-step. The energy available from the breakdown of each glucose molecule can be used to make as many as 38 ATPs.

parking lot with a blocked exit. If the first car can't exit, the cars behind can't get out either. Also, just as additional cars can't enter, neither can electrons. As you can see, the reactions of the electron transport chain cannot take place in the absence of oxygen.

Releasing Energy Without Oxygen

Remember that at the end of glycolysis, each glucose molecule has been broken down to two pyruvic acid molecules. If no oxygen is available to carry on aerobic respiration, the pyruvic acid molecules can be changed further without oxygen in an anaerobic process called fermentation. No additional ATP is formed during the process of fermentation. However, fermentation uses the electrons carried by NADH + H^+ and thus regenerates NAD^+ so that glycolysis can continue.

There are two major types of fermentation. In one, pyruvic acid changes to lactic acid. In the other, it changes to ethyl alcohol and carbon dioxide.

Lactic acid fermentation

When lactic acid is formed as an end product of fermentation, the process is called **lactic acid fermentation**. *Figure 10.17* shows what happens during lactic acid fermentation.

Do your muscles become fatigued after you've run for a while or lifted heavy objects? Anytime your muscle cells require energy at a faster rate than it can be supplied by aerobic respiration, they begin to carry out lactic acid fermentation. The resulting buildup of lactic acid causes muscle fatigue.

Alcoholic fermentation

When ethyl alcohol and carbon dioxide are the end products of fermentation, the process is called **alcoholic fermentation**. Many bacteria and fungi such as yeast carry out alcoholic fermentation. *Figure 10.18* shows the process. Notice that NAD^+ is produced from NADH + H^+ in this reaction.

Figure 10.17

Lactic Acid Fermentation

▶ Strenuous exercise has caused lactic acid to build up inside this racer's muscle cells, resulting in muscle fatigue. When oxygen is available, the body will break down the lactic acid. For this reason, we say that he has built up an oxygen debt.

Lactic Acid Fermentation

▶ Lactic acid fermentation occurs in some bacteria, in plants, and in most animals, including humans.

NAD⁺ → NADH + H⁺   NADH + H⁺ → NAD⁺

Glucose — Glycolysis → 2 Pyruvic acid — Lactic acid fermentation → 2 Lactic acid

2ADP + 2P$_i$ → 2ATP

Comparing Photosynthesis and Aerobic Respiration

Both photosynthesis and aerobic respiration are complex sets of reactions that are alike in several ways. Both involve energy, require enzymes, occur in specific organelles, and involve the movement of electrons in transport chains.

In many ways, photosynthesis is the opposite of aerobic respiration. The end products of photosynthesis are the starting materials of respiration. In photosynthesis, starch or sugars are eventually formed, and oxygen is released as a waste product. In aerobic respiration, oxygen is used to break down sugars. Carbon dioxide and water are given off as wastes. So, the end products of aerobic respiration are the starting materials of photosynthesis. Compare the two reactions in the summary diagram, *Figure 10.19,* on the next page.

MiniLab

Will apple juice ferment?

Organisms such as yeast have the ability to break down food molecules and synthesize ATP when no oxygen is available. When the appropriate food is available, yeast can carry out alcoholic fermentation, producing CO_2. Thus, the production of CO_2 can be used to judge whether alcoholic fermentation is taking place.

Procedure
1. Carefully study the diagram and set up the experiment as shown.
2. Hold the test tube in a beaker of warm (not hot) water and observe.

Analysis
1. What were the gas bubbles that came from the plastic pipette?
2. Predict what would happen to the rate of bubbles given off if more yeast were present in the mixture.
3. Why was the test tube placed in warm water?

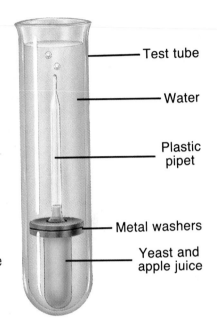

Test tube

Water

Plastic pipet

Metal washers

Yeast and apple juice

Figure 10.18

Alcoholic Fermentation

► Yeast is used in fermentation to produce alcohol. Alcoholic fermentation also produces CO_2, which can be used to raise bread dough.

Alcoholic Fermentation

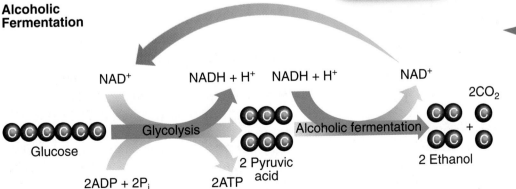

NAD⁺   NADH + H⁺   NADH + H⁺   NAD⁺

Glucose   Glycolysis   2 Pyruvic acid   Alcoholic fermentation   2 Ethanol $2CO_2$

2ADP + 2Pᵢ   2ATP

◄ The ethyl alcohol produced by alcoholic fermentation is the alcohol in beer, wine, and other alcoholic beverages.

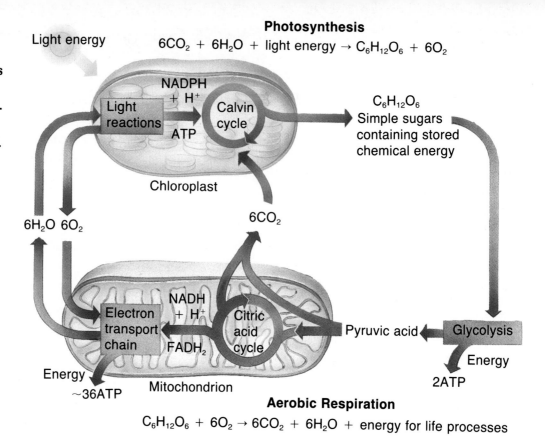

Figure 10.19

Photosynthesis in eukaryotic cells occurs in the chloroplast, whereas aerobic respiration occurs in the mitochondrion. In photosynthesis, energy from sunlight is converted to chemical energy in the bonds of carbohydrate molecules. In respiration, chemical energy from the bonds of carbohydrate molecules is released and used to make ATP.

Photosynthesis

$$6CO_2 + 6H_2O + \text{light energy} \rightarrow C_6H_{12}O_6 + 6O_2$$

Light energy

NADPH + H$^+$

Light reactions

Calvin cycle

ATP

Chloroplast

$C_6H_{12}O_6$ Simple sugars containing stored chemical energy

$6H_2O$ $6O_2$

$6CO_2$

NADH + H$^+$

Electron transport chain

FADH$_2$

Citric acid cycle

Pyruvic acid

Glycolysis

Energy

Energy

2ATP

~36ATP

Mitochondrion

Aerobic Respiration

$$C_6H_{12}O_6 + 6O_2 \rightarrow 6CO_2 + 6H_2O + \text{energy for life processes}$$

Connecting Ideas

Nearly all of the millions of present-day species break down organic nutrients for energy, but only about half a million species are able to build nutrients through photosynthesis. If photosynthesis stopped today, both heterotrophs and autotrophs alike would die from lack of energy from food. Life on Earth depends largely on light energy channeled through photosynthesis.

One of the uses of this flow of energy by living cells is in the progress of cell reproduction. You will see that cell structures, chemistry, and cell energy also play a role in cell reproduction.

Section Review

Understanding Concepts

1. Compare the ATP yields of glycolysis and aerobic respiration.
2. How do alcoholic and lactic acid fermentation differ?
3. How is most of the ATP from aerobic respiration produced?

Thinking Critically

4. Compare the energy-producing processes in a jogger's leg muscles to those of a sprinter's leg muscles. Which is likely to build up more lactic acid? Which runner is more likely to be out of breath after running? Explain.

Skill Review

5. **Making and Using Tables** Use the section called Comparing Photosynthesis and Aerobic Respiration to make a table summarizing the similarities and differences between the two processes. For more help, refer to Organizing Information in the *Skill Handbook*.

Collecting the Sun

Sunlight provides energy for nearly all life on Earth, so why can't we use sunlight for all the other energy needs of human society? After all, sunlight is the ultimate energy resource. The sources of energy that we use most—coal, oil, and natural gas—will eventually be used up. Burning them to release energy causes pollution. If we could obtain energy from the sun instead, these fuels would be conserved and pollution would be reduced.

Flat-plate collectors Gathering solar energy requires collectors with large surface areas exposed to the sun, just as many plants have large leaf surfaces for efficient photosynthesis.

The most popular type of solar collector is the flat-plate collector. You may see this type on the roof of a house. It consists of a flat, black surface that absorbs sunlight and changes it to heat. Air or water flows behind the plate, absorbing the heat. This type of collector is used most often to heat homes and provide hot water.

Concentrating collectors A second type of collector uses an array of mirrors to focus sunlight on pipes containing water. The water boils into steam that drives a turbine connected to a generator which produces electricity.

Photovoltaic cells Do you have a light-powered calculator? This device contains a third kind of collector, a photovoltaic cell, which converts light into electricity. Arrays of these cells are used to provide electric power for satellites, some highway signs, and communications units in remote areas.

Applications for the Future

Scientists are designing a satellite system to collect solar energy in space without atmospheric interference, convert it to microwave radiation, and transmit it to Earth, where a receiver will convert the energy into electricity. Engineers are testing electric cars and buses on our highways. They run on batteries charged by large, stationary photovoltaic arrays. These vehicles typically can go a distance of less than 100 miles before they must be recharged, but may be a good solution for commuters.

INVESTIGATING the Technology

1. **Apply** A problem that must be solved in all solar energy technology is how to store the energy for times when sunlight is not available. How have green plants solved this problem?
2. **Research** In regions that have cold, cloudy winters, solar energy is usually not sufficient to heat homes. Do research to compile a list of ways that people in these regions can use solar energy to minimize fuel bills.

Reviewing Main Ideas

10.1 ATP: Energy in a Molecule
- ATP is the molecule that receives energy released in cell reactions.
- ATP is formed when a phosphate group is added to ADP. When ATP is broken down, ADP and phosphate are formed, and energy is released.
- ATP is the main link between energy-releasing and energy-using reactions.

10.2 Photosynthesis: Trapping Energy
- Photosynthesis is the process by which most producers use energy from light to make their own food. Green chlorophyll in the chloroplasts of a plant's cells traps light energy needed for photosynthesis.
- The light reactions of photosynthesis use light to produce ATP and result in the splitting of water.
- The reactions of the Calvin cycle make carbohydrates using CO_2 along with ATP and hydrogen from the light reactions.
- Certain autotrophs can obtain energy from inorganic compounds by chemosynthesis.

10.3 Getting Energy to Make ATP
- Respiration is the process in which cells break down molecules to release energy.
- Aerobic respiration takes place in mitochondria, uses oxygen, and yields many more ATPs than anaerobic processes.
- Energy can be released anaerobically by glycolysis followed by fermentation.
- The end products of photosynthesis are the starting products for respiration.

Key Terms
Write a sentence that shows your understanding of each of the following terms.

ADP	electron transport
aerobic process	chain
alcoholic	glycolysis
fermentation	lactic acid
anaerobic process	fermentation
ATP	light reactions
Calvin cycle	photolysis
chemosynthesis	photosynthesis
citric acid cycle	respiration

Understanding Concepts

1. What three substances must be supplied to the Calvin cycle?
2. What happens to the products of photosynthesis in plants?
3. Why is it important that plants have a constant supply of water?
4. How do aerobic and anaerobic processes differ?
5. What is the function of coenzymes?
6. What is the source of heat in the human body?
7. What is oxygen debt? Explain how it develops.
8. If four ATPs are produced by glycolysis, why is the net yield only two ATPs?
9. What is the function of oxygen in aerobic respiration?

Using a Graph
10. The absorption spectrum of a pigment can be measured with a spectrophotometer. This machine directs a spectrum of the visible colors at a substance and records what percentage of light of each color is absorbed by the sample. The graph here shows the absorption spectrum of carotenoids, pigments that are found in green plants along with chlorophyll. Explain from the graph what colors are absorbed and what colors are reflected by the carotenoid pigments.

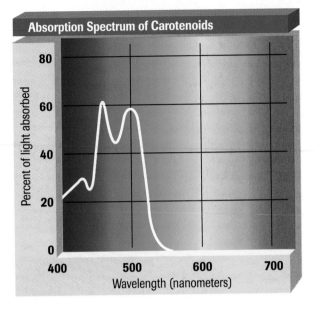

Absorption Spectrum of Carotenoids

Percent of light absorbed (vertical axis): 0, 20, 40, 60, 80

Wavelength (nanometers) (horizontal axis): 400, 500, 600, 700

sis was measured by counting the number of oxygen bubbles per minute for ten minutes as shown in the graph. Predict what would happen to the rate of photosynthesis if a piece of red cellophane were placed over the white light.

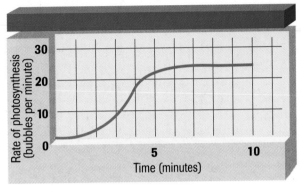

Rate of photosynthesis (bubbles per minute) (vertical axis): 0, 10, 20, 30

Time (minutes) (horizontal axis): 5, 10

Relating Concepts

11. Make a concept map that relates the following terms and phrases. Supply the appropriate linking words for your map.

    ATP, alcoholic fermentation, anaerobic processes, citric acid cycle, electron transport chain, glycolysis, lactic acid fermentation, respiration

Applying Concepts

12. Why would human muscle cells contain many more mitochondria than skin cells?
13. Why are aerobic respiration and photosynthesis opposite processes?
14. A person tells you that plants "live off light energy instead of food." What could you say to help the person's understanding?
15. What happens to sunlight that strikes a leaf but is not trapped by photosynthesis?

History Connection

16. What mistaken idea did Priestley have about plants?

Thinking Critically

Making Predictions

17. **Biolab** *Elodea* sprigs were placed under a white light, and the rate of photosynthe-

Making Inferences

18. **Minilab** The yeast cells used in the Minilab were forced to ferment by placing them in an environment without any oxygen. Why would the yeast cells carry out aerobic respiration rather than fermentation when oxygen is present?

Recognizing Cause and Effect

19. A "window" plant native to the desert of South Africa has leaves that grow almost entirely underground with only the transparent tip of the leaf protruding above the soil surface. Suggest how this adaptation aids the survival of these plants.

Making Predictions

20. What do you think would happen to Earth's atmosphere if photosynthesis suddenly stopped?

Connecting to Themes

21. **Energy** Explain why it is accurate to say that heterotrophs indirectly rely on the sun for their energy.
22. **Homeostasis** Muscle cells are able to carry out both aerobic respiration and lactic acid fermentation. How does this ability serve as a survival adaptation for animals?

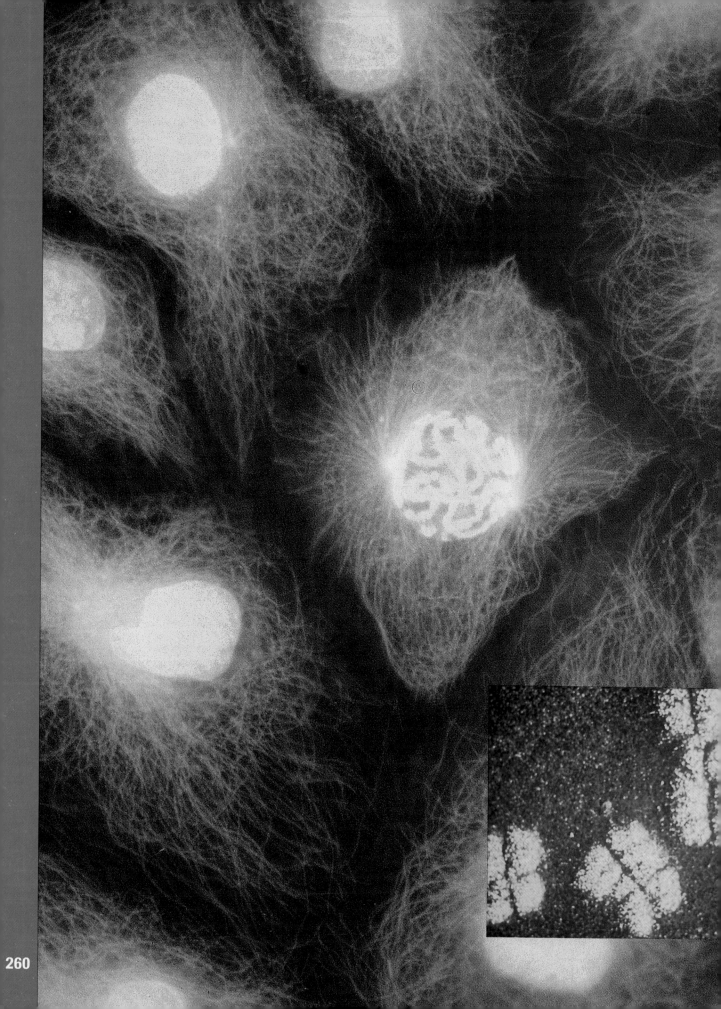

When you see a newborn baby, is it hard for you to imagine that just nine months ago, that baby was a single cell? Or can you imagine that a small acorn will eventually grow into a towering oak tree? When you mow the lawn every week, do you sometimes wonder how it grows so fast? Multicellular organisms such as humans, oaks, and grass contain billions and trillions of cells. Where do they all come from?

Living organisms are always making new cells. A healthy human adult produces 25 million new cells each second. Every time an organism grows in size or repairs worn out or damaged tissues, new cells are made. Regardless of whether new cells are being made in plants or in animals, the method by which they are produced is remarkably similar.

If you were to observe cells of a growing organism, you would see that they are reproducing in the same manner as the cell in the center of the large photo. The inset photo on the left shows several chromosomes—key players in the drama of cell reproduction.

Magnification: (large) 300×, (inset) 6000×

(Large photomicrograph courtesy of J.M. Murray, Department of Cell and Developmental Biology, University of Pennsylvania)

Section Preview

Objectives

Analyze the reasons why cells are small.

Sequence the events of the cell cycle.

Key Terms

chromosome
cell cycle
interphase
mitosis
prophase
sister chromatid
centromere
centriole
spindle
metaphase
anaphase
telophase

gene
cancer
- DNA

*I*f you could examine the cells of this lion and her cub, how do you think they would compare in size? Would you be surprised to learn that the cells of both mother and cub are, on average, exactly the same size? The cells are too small to be seen without the aid of a microscope. Why do organisms have such small cells?

Cell Size Limitations

Cells come in a wide variety of sizes. Some, such as red blood cells, measure only 8 µm in diameter. Others, such as nerve cells in large animals, can reach lengths of up to 1 m. The largest known cell is the yolk of an ostrich egg—measuring a whopping 8 cm in diameter! Most living cells, however, are between 2 and 200 µm in diameter. The body of an adult lion contains trillions of these small cells. Why can't organisms be just one giant cell?

Why diffusion limits cell size

To understand why cell size is limited, think of the materials that cells need to stay alive and do their jobs. You know that cells require a constant supply of glucose and oxygen to carry out cellular respiration and produce large amounts of ATP. These substances, and waste products such as carbon dioxide, move through the cytoplasm by diffusion. It follows, then, that cells can metabolize only as quickly as they receive raw materials by diffusion.

It has been estimated that it takes a molecule of oxygen only a small fraction of a second to diffuse through the cytoplasm from the plasma membrane to the center of a typical cell. Therefore, the innermost mitochondrion in the center of an average-sized cell with a diameter of 20 µm will receive supplies of oxygen and other important molecules a fraction of a second after they pass through the plasma membrane.

What would happen if the cell got bigger? Although diffusion is a fast and efficient process over short distances, it becomes slow and inefficient as the distances become larger. A mitochondrion at the center of a hypothetical cell with a diameter of 20 cm would have to wait months before receiving molecules that enter the cell. Because of this time restriction, organisms can't be just one giant-sized cell. They would die long before the oxygen ever got to the mitochondria.

Why a cell's DNA content limits size

A second reason why cells are small is because most cells contain only one nucleus. Molecules of RNA that are involved in protein production are made by the DNA in the nucleus. The RNA then leaves the nucleus and moves through the cytoplasm to the ribosomes, where it directs the production of enzymes and other proteins. If a cell doesn't have enough DNA to program its metabolism, it cannot live.

What happens in larger cells where an increased amount of cytoplasm requires increased supplies of enzymes? In many large cells such as the giant amoeba *Pelomyxa* shown in *Figure 11.1,* more than one nucleus has evolved. Large amounts of DNA in many nuclei ensure that cell activities will be carried out quickly and efficiently.

Surface area-to-volume ratio

The third reason why cells are limited in size is that as a cell's size increases, its volume increases much faster than its surface area.

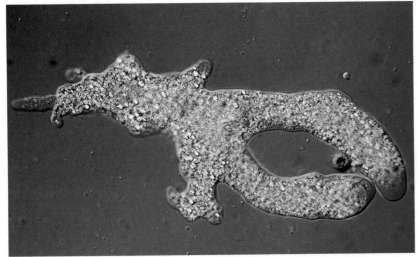

Magnification: 100×

Figure 11.1

This giant amoeba, *Pelomyxa,* is several millimeters in diameter. It can have up to 1000 nuclei.

What happens to the surface area of a cube as the volume increases?

One reason cells are small is that, as they grow, their volume increases faster than does their surface area. In this lab, you will measure how much greater the increase in volume is than the increase in surface area of an object.

Procedure

1. Using graph paper and the diagram at the right, cut out two cubes. The first cube should be 1 cm on each side. The second cube should be 2 cm on each side.

2. Put together each cube using tape.

3. Calculate the surface area of each cube by multiplying the area of one face by six. Express your answers in square centimeters.

4. Working with other groups, assemble your 1-cm cubes to form a cube as large as your 2-cm cube. Observe how many small cubes are needed to obtain a volume equal to the larger cube.

Analysis

1. How many small cubes were required to equal the volume of the large cube?

2. How much greater is the total surface area of the assembled cubes than the surface area of the single 2-cm cube?

3. As the assembled cube became larger, which increased more quickly: the surface area or the volume? What is the significance of this for cell size?

Picture a cube-shaped cell like those shown in *Figure 11.2*. The smallest cell has 1-mm sides, a surface area of 6 mm^2, and a volume of 1 mm^3. If the side of the cell is doubled to 2 mm, the surface area will increase fourfold to $6 \times 2 \times 2 = 24$ mm^2. Observe what happens to the volume; it increases eightfold to 8 mm^3.

Is big better?

What does this mean for cells? How does it affect cell function? If cell size doubled, the cell would require eight times more nutrients and would have eight times more waste to excrete. The surface area, however, would increase by a factor of only four. Thus, the plasma membrane would not have enough surface area through which oxygen, nutrients, and wastes could diffuse. The cell would either starve to death or be poisoned from the buildup of waste products. In fact, though, cells divide before they reach this point.

Figure 11.2

Surface area-to-volume ratio is one of the major factors limiting cell size. Note how the surface area and the volume change as the sides of a cell double in length from 1 mm to 2 mm. Calculate the change in surface area and volume as the cell doubles in size again to 4 mm on a side.

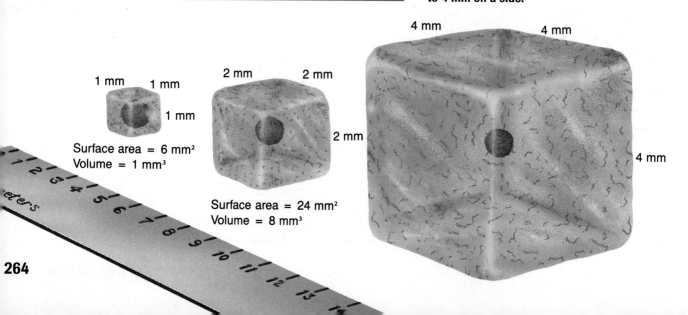

1 mm 1 mm
1 mm
Surface area = 6 mm^2
Volume = 1 mm^3

2 mm 2 mm
2 mm
Surface area = 24 mm^2
Volume = 8 mm^3

4 mm 4 mm
4 mm
4 mm

Figure 11.3

Cells reproduce at different times in the life of an organism, depending on their function in the organism.

Cells reproduce when a leaf bud grows to be a leaf.

Cells reproduce when an insect egg grows first to be a caterpillar and then to be an adult butterfly.

Cells reproduce when one amoeba divides to become two amoebas.

Magnification: 150×

When Do Cells Divide?

Right now, as you are reading this page, many of the cells in your body are growing, dividing, and dying. Old cells on the soles of your feet and on the palms of your hands are being shed and replaced, cuts and bruises are healing, and your intestines are producing millions of new cells each second.

Other organisms go through similar changes. New cells are produced as tadpoles become frogs, and as a tomato plant grows and wraps around a garden stake. All organisms grow and change; worn out tissues are repaired or are replaced by newly produced cells. *Figure 11.3* shows other instances when new cells are made.

Cell Reproduction—The Role of Chromosomes

By the middle of the 19th century, scientists knew that cells divide to form more cells. As microscopes improved, the details of cell structure were discovered, allowing scientists to interpret and better understand the process of cell division.

Figure 11.4

This bacterial cell was ruptured osmotically, and its DNA molecule leaked out of the cell. Although prokaryotic cells such as this one contain only one-thousandth as much DNA as a typical eukaryotic cell, you can still see that uncoiled DNA molecules are very long structures.

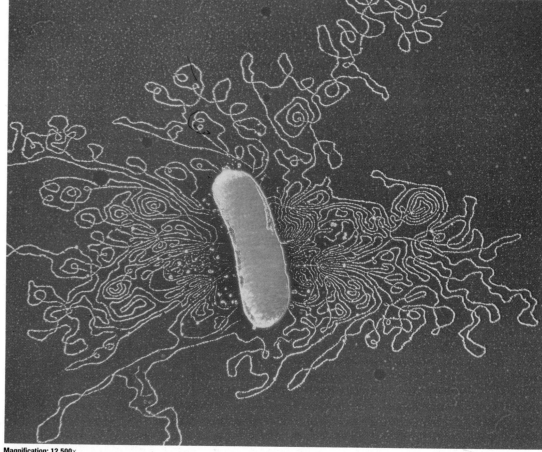

Magnification: 12 500×

chromosome:

chroma (GK) colored *soma* (GK) body Chromosomes are dark-staining structures that contain genetic material.

The discovery of chromosomes

Most interesting to the early biologists was their observation that just before cell division, several short, stringy structures suddenly appeared in the nucleus. What was strange was that these structures seemed to vanish, as mysteriously as they appeared, soon after division of a cell. These structures, which contain DNA and become darkly colored when stained, are now called **chromosomes.**

With the introduction of more advanced microscopes, scientists eventually came to understand that chromosomes are the carriers of genetic material, which is copied and passed from generation to generation of cells. Scientists also learned that chromosomes are always present in cells, but most of the time they exist in a form too thin to be seen with simple microscopes.

The structure of eukaryotic chromosomes

During most of a cell's life cycle, chromosomes exist as chromatin, strands of DNA wrapped around protein molecules. Under an electron microscope, chromatin resembles a plate of tangled-up spaghetti. How can such long structures function without becoming tangled? Before a cell begins to divide, the threads of chromatin begin to coil. The coils shorten and thicken, forming distinct chromosomes that are tightly packed with chromatin.

The electron micrograph in *Figure 11.4* shows a single, uncoiled DNA molecule from a bacterium. Why is it necessary for DNA to coil and uncoil during the life cycle of a cell? When DNA is uncoiled, it can be actively involved in making RNA and in replicating itself.

The Cell Cycle

Fall follows summer, night follows day, and low tide follows high tide. Many events in nature follow a recurring cyclical pattern. Living organisms are no exception. One cycle common to most living things is the cycle of the cell, *Figure 11.5*. The **cell cycle** is the sequence of growth and division of a cell.

As a cell proceeds through its cycle, it goes through two general periods: a period of growth and a period of division. The growth period of the cell cycle is known as **interphase.** Most of a cell's life is spent carrying on the activities of interphase. During interphase, a cell grows in size and carries on metabolism. Also during this period, chromosomes are duplicated in preparation for the period of division.

As you recall, chromosomes are made of DNA. Because DNA contains the master instructions for the cell, it is important that new cells have complete copies of DNA when parent cells divide. Accurate replication of the parent cell's chromosomes, and thus its DNA molecules, accomplishes this.

Following interphase, a cell enters its period of division, in which its nucleus and then its cytoplasm divide to form two daughter cells, each containing a complete set of chromosomes. The process of nuclear division followed by division of the cytoplasm, when chromosomes are distributed equally to daughter cells, is known as **mitosis.** Interphase and mitosis together make up the cell cycle.

mitosis:
mitos (GK) thread
Mitosis is a process that divides threadlike nuclear material equally between two daughter cells.

Interphase—A Busy Time

Interphase is the busiest part of the cell cycle. Cells are making ATP, repairing themselves, and excreting their wastes. They are making proteins, producing new organelles such

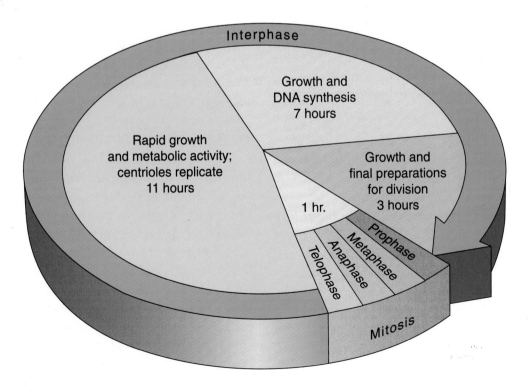

Interphase

Growth and
DNA synthesis
7 hours

Rapid growth
and metabolic activity;
centrioles replicate
11 hours

Growth and
final preparations
for division
3 hours

1 hr.

Prophase

Metaphase

Anaphase

Telophase

Mitosis

A 22-hour cell cycle

Figure 11.5

The cell cycle is divided into phases representing the important events in the life of a cell. As soon as the cell completes one phase of the cycle, it proceeds to the next phase. During which phase does a cell spend most of its time?

How does the length of the cell cycle vary?

The cell cycle varies greatly in length from one kind of cell to another. Some kinds of cells divide rapidly, while others divide more slowly.

Analysis

Examine the cell cycle diagrams, shown below, of two different types of cells. Observe the total length of each cell cycle, as well as the length of time each cell spends in each phase of the cell cycle.

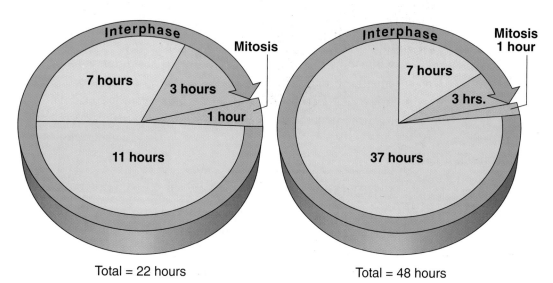

Total = 22 hours

Total = 48 hours

Thinking Critically

Which part of the cell cycle is most variable in length? What inferences can you make concerning the functions of these two types of cells? Why do you think the cycle of some types of cells is faster than in others? Explain your answer.

In the next part of interphase, the cell copies its chromosomes. DNA synthesis does not occur all through interphase but is confined to this discrete time.

After the chromosomes have been duplicated, the cell enters another shorter growth period in which mitochondria and other organelles are manufactured and cell parts needed for cell division are assembled. Following this activity, interphase ends and mitosis begins.

The Phases of Mitosis

Although cell division is a continuous process, biologists recognize four distinct phases—each phase merging into the next. The four phases of mitosis are prophase, metaphase, anaphase, and telophase. Refer to the photos and diagrams in *Figure 11.8* on pages 270–271 as you read about mitosis.

Prophase—The first phase of mitosis

During **prophase,** the first and longest phase of mitosis, the long, stringy chromatin coils up into visible chromosomes. As you can see in *Figure 11.6,* each duplicated chromosome is made up of two halves. The two halves of the doubled structure are called **sister chromatids.** Sister chromatids and the DNA they contain are exact copies of each other and are formed when DNA is copied during interphase. Notice also that sister chromatids are held together by a structure called a **centromere.**

as ribosomes and mitochondria, and copying their chromosomes. A cell from a plant leaf might be photosynthesizing, a liver cell might be storing extra sugar from lunch in the form of glycogen, and an onion root tip cell might be taking in water and minerals from the soil.

If you look back at the cell cycle diagrammed in *Figure 11.5,* you can see that interphase is divided into three parts. During the first part, the cell grows in size and protein production is high.

As prophase continues, the nucleus begins to disappear as the nuclear envelope and the nucleolus disintegrate. By late prophase, these structures are completely absent. Two important pairs of structures, the centrioles, begin to migrate to opposite ends of the cell. **Centrioles** are small, dark, cylindrical structures that are made of microtubules and are located just outside the nucleus, *Figure 11.7.* Only animal cells contain centrioles; plant cells do not.

As the pairs of centrioles move to opposite ends of the cell, another important structure, called the spindle, begins to form between them. The **spindle** is a football-shaped, cagelike structure consisting of thin fibers made of microtubules. The spindle fibers play a vital role in the separation of sister chromatids during mitosis.

Magnification: 18 000×

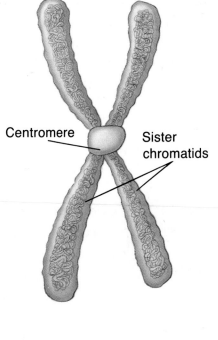

Figure 11.6

The fully coiled chromosome in this electron micrograph appears "hairy" because it consists of a single, long chromatin fiber that is coiled up. The two sister chromatids are held together by a centromere. Each sister chromatid is a complete copy of the original parent chromosome, and the two chromatids are, therefore, identical.

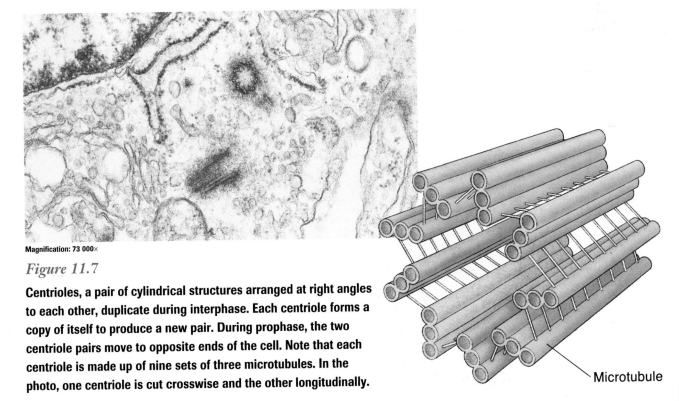

Magnification: 73 000×

Figure 11.7

Centrioles, a pair of cylindrical structures arranged at right angles to each other, duplicate during interphase. Each centriole forms a copy of itself to produce a new pair. During prophase, the two centriole pairs move to opposite ends of the cell. Note that each centriole is made up of nine sets of three microtubules. In the photo, one centriole is cut crosswise and the other longitudinally.

Microtubule

Figure 11.8

The cell cycle is a complex series of continuous events consisting of inter-phase and mitosis. Identify the characteristics of each phase in the cell cycle, and note the differences between mitosis in animal cells (diagrams) and in plant cells (photos). What two differences do you notice?

The Cell Cycle

INTERPHASE

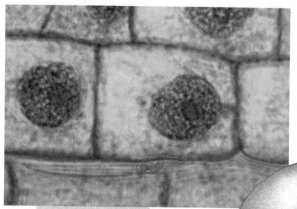

Magnification: 1000×

A **INTERPHASE** During interphase, the nucleus and the darker-staining nucleolus can be clearly seen. The DNA is copied. The chromosomes cannot yet be seen because they are still in the form of uncoiled chromatin. In animal cells, the centrioles duplicate themselves.

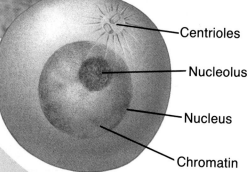

— Centrioles

— Nucleolus

— Nucleus

— Chromatin

PROPHASE

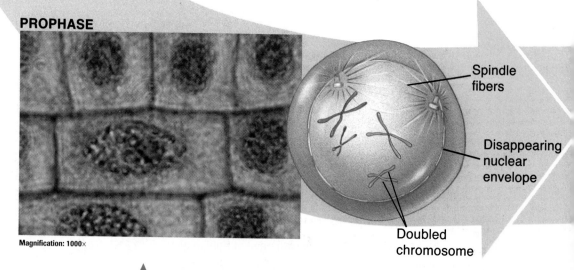

Magnification: 1000×

Spindle fibers

Disappearing nuclear envelope

Doubled chromosome

B **PROPHASE** In prophase, the chromatin coils to form visible chromosomes, the nuclear envelope and nucleolus disappear, and a spindle forms between the pairs of centrioles, which have moved to opposite ends of the cell.

E **TELOPHASE** In this final phase of mitosis, two daughter cells are formed. With a complete set of chromosomes at each end of the cell, the cytoplasm divides, the nucleolus and nuclear envelope reappear, and the chromosomes begin to uncoil. When the new cells are separated, they enter interphase and a new cycle begins.

TELOPHASE

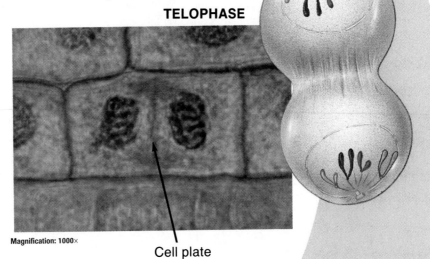

Magnification: 1000×

Cell plate

D **ANAPHASE** During anaphase, the centromeres split and the sister chromatids are pulled apart to opposite poles of the cell. Each chromatid is now a separate chromosome.

ANAPHASE

Magnification: 1000×

Centromere

Sister chromatids

METAPHASE

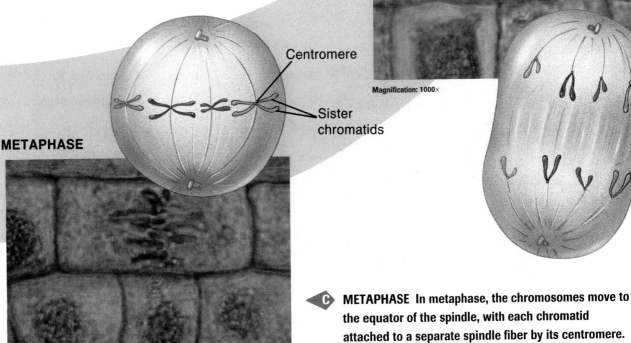

C **METAPHASE** In metaphase, the chromosomes move to the equator of the spindle, with each chromatid attached to a separate spindle fiber by its centromere.

Magnification: 1000×

11.1   Cell Growth and Reproduction   **271**

The Spindle

The spindle is an important cell structure that appears in late prophase of mitosis and plays a major role in the distribution of chromosomes. The events of mitosis are similar among all organisms, but the details of spindle formation differ between plants and animals.

Microtubules and the spindle
Scientists have discovered that, in addition to forming the cell's cytoskeleton, microtubules are involved in spindle formation. At the beginning of prophase, the cytoskeleton breaks apart, releasing tubulin protein molecules. The tubulin molecules then reassemble to form the spindle fibers.

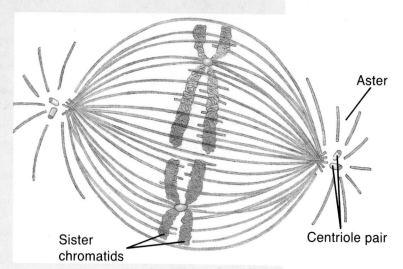

Sister chromatids

Aster

Centriole pair

Centrioles and the spindle   In animal cells, the spindle first appears as a star-shaped group of microtubules called an aster, which radiates from each pair of centrioles. As the two pairs of centrioles move to opposite poles of the cell during prophase, spindle fibers form between them. Some of the fibers extend from pole to pole; others extend from a pole to the equator. It is these shorter fibers that attach to the chromosomes and pull the chromatids apart.

Because of the close relationship between centrioles and the spindle, scientists inferred that the centrioles directed the formation of the spindle, but until recently, it was not clear how this was accomplished. This question was also complicated by the fact that plant cells don't have centrioles and asters, but they form spindles.

New evidence shows that in animal cells, the centrioles are surrounded by a substance that appears to direct the formation of the spindle from microtubules. Scientists have proposed that a similar substance exists in the cytoplasm of plant cells. Before scientists can fully understand spindle formation in animal and plant cells, the nature of this substance and how it induces formation of the spindle must be further researched.

Thinking Critically
The microtubules of the asters appear to serve a mechanical function by bracing the centrioles against the plasma membrane and giving the cell support. Why would plant cells not need asters?

Metaphase—The second stage of mitosis

During **metaphase,** the short second phase of mitosis, the doubled chromosomes become attached to the spindle fibers by their centromeres. The chromosomes are pushed and pulled by the spindle fibers and begin to line up on the midline, or equator, of the spindle. As you saw in *Figure 11.8,* each sister chromatid is attached to its own spindle fiber. One sister chromatid's spindle fiber extends to one pole, and the other extends to the opposite pole. This arrangement is an important one because it ensures that each new cell will get an identical and complete set of chromosomes.

Anaphase—The third phase of mitosis

The separation of sister chromatids marks the beginning of **anaphase,** the third phase of mitosis. During anaphase, the centromeres split apart, and chromatid pairs from each chromosome separate from each other. How the chromatids move is not fully understood, but most scientists believe that spindle fibers shorten when the microtubules forming the spindle fibers begin to break down. The chromatids are pulled apart by the force of shortening.

Telophase—The fourth phase of mitosis

The final phase of mitosis is **telophase.** Telophase begins as the chromatids reach the opposite poles of the cell. During telophase, many of the changes that occurred during

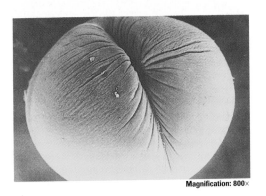

Figure 11.9

At the end of telophase in animal cells such as this frog egg, contractile proteins positioned just under the plasma membrane at the equator of the cell slide past each other to cause a deep furrow. The furrow deepens until the cell is pinched in two.

Magnification: 800×

prophase are reversed as the new cells prepare for their own independent existence. The chromosomes, which had been tightly coiled since the end of prophase, now unwind so they can begin to direct the metabolic activities of the new cells. The spindle begins to break down, the nucleolus reappears, and a new nuclear envelope forms around each set of chromosomes. Finally, the plasma membrane begins to separate the two new nuclei.

Division of the cytoplasm

The final event to occur in mitosis, the division of the cytoplasm, differs between plants and animals. Toward the end of telophase in animal cells, the plasma membrane pinches in along the equator, as shown in *Figure 11.9*. Two new cells are formed, each identical to the original.

The division of the cytoplasm is different in plants. Because plant cells have a rigid cell wall, the plasma membrane does not pinch in. Rather, a structure known as the cell plate forms across the cell's equator, as shown in *Figure 11.10*.

Results of mitosis

Mitosis is a process that guarantees genetic continuity, resulting in the production of two new cells with chromosome sets that are identical to those of the parent cell. These new daughter cells will carry out the same cellular processes and functions as their parent cell and will grow until the limitations of cell size force them to divide.

Figure 11.10

In plant cells, the cytoplasm is divided in half by formation of a cell plate. Vesicles from the Golgi apparatus fuse to form the cell plate, which grows outward and joins the old cell wall. New cell wall material is secreted on each side of the cell plate until separation is complete.

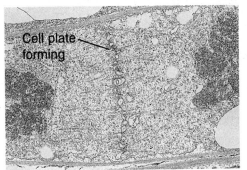

Cell plate forming

Magnification: 20 000×

Section Review

Understanding Concepts
1. Why is cell size limited?
2. Why is it necessary for a cell's chromosomes to be distributed to its daughter cells in such a precise manner?
3. How is the division of the cytoplasm different in plants and in animals?

Thinking Critically
4. At one time, interphase was referred to as the resting phase of the cell cycle. Why do you think this description is no longer used?

Skill Review
5. **Making and Using Tables** Make a table showing the phases of the cell cycle. Mention one important event that occurs at each phase. For more help, refer to Organizing Information in the *Skill Handbook*.

BioLab

The Time for the Cell Cycle

Onion root tips are regions of rapid growth, so cells in all phases of the cell cycle can be observed. Cells in interphase have an intact nucleus. The nucleus is evenly stained, and individual chromosomes cannot be distinguished. In prophase, the nucleus appears enlarged. Darkly stained chromosomes are visible. In metaphase, the chromosomes are lined up at the equator of the spindle. During anaphase, the chromosomes are found at both sides of the equator, the centromeres leading and the arms of the chromosomes trailing behind. In telophase, two dark, compact nuclei are visible, one at each pole of the cell. A cell plate may be visible between the nuclei.

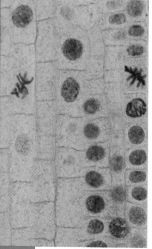

Magnification: 160×

Onion root tip cells

PREPARATION

Problem
How long do onion root tip cells spend in each phase of the cell cycle?

Objectives
In this Biolab, you will:
- **Identify** the phases of the cell cycle.
- **Determine** the time a cell spends in each part of the cell cycle.

Materials
compound microscope
prepared onion root tip slides
compass
protractor

Safety Precautions
Always use care when handling microscopes and glass slides.

Germinating onion seeds

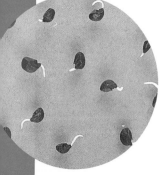

PROCEDURE

1. Examine a prepared onion root tip slide. Using the low power of your microscope, find an onion root tip. Then find the area just behind the tip of the root. This is where the cells are dividing most rapidly.
2. Now, switch your microscope to high power, and start looking at individual cells. You'll learn to recognize each phase of the cell cycle.
3. Copy the data table. Working with a partner, examine each cell in your field of view, and identify the phase of the cell cycle that each cell is in. Have your partner

record in the data table the phase of each cell you examine.

4. Now switch with your partner. Have your partner examine a different field of view while you record the phase of each cell.

5. Continue in this manner until you have counted at least 100 cells.

6. Add the number of cells in each phase, and record these numbers in the row marked *Total*. Total the number of cells you counted.

7. Calculate the fraction of cells in each phase by dividing the number of cells in that phase by the total number of cells you counted. Express your answer as a decimal, then as a percentage.

8. Calculate the number of degrees in a circle graph for each phase by multiplying the fraction of time the cell spends in each phase by 360°.

9. Make a circle graph showing the percentage of time a cell spends in each phase of the cell cycle. Label each portion of the circle with the phase it represents.

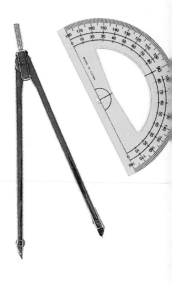

	Interphase	Prophase	Metaphase	Anaphase	Telophase
Number of cells					
First field					
Second field					
Total					
Fraction of cells in this phase					
Percent of time cell spends in this phase					
Number of degrees in a circle graph for this phase					

ANALYZE AND CONCLUDE

1. **Drawing Conclusions** In which phase of the cell cycle does a cell spend most of its time? The least time?

2. **Making Inferences** Explain why a slide of stained cells can be used to estimate how much time a living cell spends in each phase of the cell cycle.

3. **Comparing and Contrasting** Describe the appearance of the cell plate in the telophase cells you observed. How would the division of the cytoplasm appear in an animal cell?

Going Further

Application
Examine a prepared slide of a whitefish blastula to observe mitosis in animal cells. How are dividing animal cells different from dividing plant cells?

11.1   Cell Growth and Reproduction   **275**

11.2 Control of the Cell Cycle

How fast do cells divide? This depends on the particular type of cell involved. Cells that line the intestine will typically complete a cell cycle in 24 to 48 hours, whereas cells of frog embryos reproduce in less than an hour. Cells that divide slowly, such as those in an adult liver, reproduce only once a year. Other cells, such as nerve cells, do not divide at all once they mature. Despite this diversity, the factors that control the cell cycle are generally similar among all organisms.

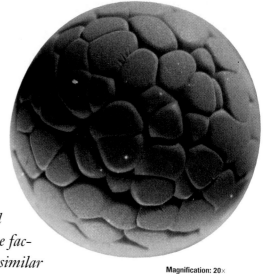

Magnification: 20×

Dividing frog embryo

Normal Control of the Cell Cycle

For the past 25 years, intensive research has been conducted into what causes a cell to divide. Despite the research, the full story still is not known. However, scientists do have some clues.

Enzymes control the cell cycle

Most biologists agree that it is the production of a series of enzymes that monitors a cell's progress from phase to phase during the cell cycle. Some enzymes are necessary to trigger the progression of the cell cycle, whereas other enzymes function to inhibit progression. Cell death or an uncontrolled dividing of cells, known as **cancer,** can result from the failure to produce certain enzymes, or from the overproduction or production at the wrong time of other enzymes. Enzyme production is directed by genes located on the chromosomes.

A **gene** is a segment of DNA that controls the production of a protein.

Many studies point to the portion of interphase just before DNA replication as being a key control period in the cell cycle. Scientists have identified several enzymes that act to trigger DNA replication.

Contact inhibition

Another factor involved in control of the cell cycle is the phenomenon of contact inhibition, as shown in *Figure 11.11.* Many studies have shown that when normal cells are allowed to grow in a glass culture dish, they will stop dividing when they cover the bottom of the dish and come in contact with each other.

Contact inhibition is a type of cell-to-cell communication. Cells communicate with each other by producing and secreting chemical signals. When changes occur in the genetic control of these chemicals, cancer can result.

Figure 11.11

In experiments demonstrating contact inhibition, cells are allowed to grow in a dish containing nutrients. Normal cells divide until they form a continuous layer on the bottom of the dish (a). When several rows of cells are removed, the cells at the edge of the gap begin to divide (b). They stop dividing when the bottom of the dish is covered once again (c). Cancer cells continue to divide, piling on top of each other (d).

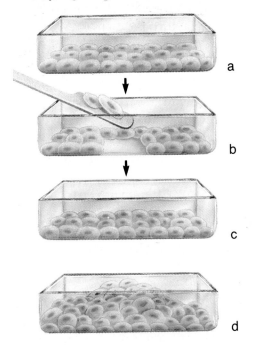

Cancer—A Mistake in the Cell Cycle

The current view of cancer is that it is caused by changes in one or more of the genes controlling production of enzymes involved in the cell cycle. These changes are expressed as cancer when environmental factors trigger the damaged genes into action. Cancerous cells affect normal cells, forming masses of tissue called tumors that deprive normal cells of nutrients. In the final stages, cancer cells enter the circulatory system and spread throughout the body, forming new tumors that disrupt the functioning of organs.

Health

Skin Cancer

Skin cancer makes up one-third of all malignancies diagnosed in the United States, and its incidence is increasing. Most cases are caused by exposure to harmful ultraviolet rays emitted by the sun, so skin cancer most often develops on the exposed face or neck. The people most likely to develop skin cancer are those whose fair skin contains smaller amounts of a protective pigment called melanin.

Types of skin cancers   Basal cell carcinoma, the most common type of skin cancer, is relatively harmless. Basal cells damaged by ultraviolet light form precancerous growths that can be removed easily in a doctor's office. Precancerous growths produced by sun-damaged basal cells can also become squamous cell carcinomas, the second most common type of skin cancer. Squamous cell cancers take the form of red or pink tumors that can grow rapidly and spread. Both types of skin cancer respond to treatment such as surgery, chemotherapy, and radiation therapy. By far the most lethal skin cancer is malignant melanoma. Although melanomas currently make up only three percent of all skin cancers, their incidence is growing faster than that of any other kind of skin cancer. Melanomas develop in the pigment-producing cells found deep within the skin.

Two kinds of UV rays   There are two kinds of ultraviolet rays: UV-A and UV-B. Sunlight contains both kinds of rays, but it contains much more UV-A than UV-B. Both kinds can cause melanoma. Most sunscreens don't protect the skin from UV-A rays, although they do block UV-B. Thus, sunbathers might falsely think they are protected from damaging radiation.

Since suntanning salons have become increasingly popular, there has been concern over the fact that they claim to use only safe wavelengths of ultraviolet light. Many medical authorities believe this radiation is still potentially dangerous and will increase the risk for skin cancer and premature aging of the skin.

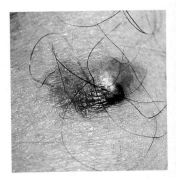

Melanoma

CONNECTION TO **Biology**

Many people still consider a tan both healthy and attractive. What are some ways these perceptions might be changed?

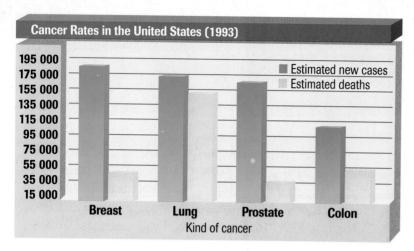

Cancer Rates in the United States (1993)

195 000
175 000
155 000
135 000
115 000
95 000
75 000
55 000
35 000
15 000

■ Estimated new cases
■ Estimated deaths

Breast   Lung   Prostate   Colon
Kind of cancer

Figure 11.12

Some types of cancer are more treatable than others. Compare the number of new cases with the number of deaths for each type of cancer.

Cancer is the second leading cause of death in the United States, exceeded only by heart disease. Cancer can affect any tissues of the body. In the United States, the main cancer killers are lung, colon, breast, and prostate cancers. *Figure 11.12* shows the estimated number of people in the United States who will develop each of these kinds of cancer in 1993 and the estimated number of people who will die from them.

The causes of cancer

The causes of cancer are difficult to pinpoint because both genetic and environmental factors are involved. The environmental influences of cancer are clear when it is noted that people in different countries get different types of cancers at different rates. For example, the rate of breast cancer is relatively high in the United States but relatively low in Japan. Similarly, stomach cancer is common in China but rare in the United States. Yet, when people move from one country to another, their cancer rates follow the pattern of the country in which they are currently living, not their country of origin. Environmental factors such as cigarette smoke, air and water pollution, and exposure to ultraviolet radiation from the sun are all known to damage the genes that control the cell cycle. Cancer may also be caused by infection with particular viruses.

Connecting Ideas

Mitosis is an orderly cell process that assures the genetic continuity of the cells and tissues of an organism. If executed without problem, mitosis results in newly formed daughter cells that are genetically identical to the parent cell. Another orderly cell division process that produces sex cells not only guarantees the continuity of the cells involved, but also promotes the genetic continuity of a species.

Section Review

Understanding Concepts

1. Describe the way genes control the cell cycle.
2. How can disruption of the cell cycle result in cancer?
3. How does cancer affect normal cell functioning?

Thinking Critically

4. How do we know that the environment influences the formation of cancer?

Skill Review

5. **Observing and Inferring** Although there are more cases of breast cancer than lung cancer per year, more deaths are caused by lung cancer than breast cancer. Using your knowledge of how cancer spreads and factors influencing cancer, provide an explanation for this difference. For more help, refer to Thinking Critically in the *Skill Handbook*.

Replacement Skin

Skin is the body's first defense against injury and infection. When both the outer and inner layers of the skin are burned away, as they are in third-degree burns, bacteria and other harmful organisms have access to the body's systems. Burn victims are also in incredible pain from exposed nerve endings, and they must receive volumes of fluids to stay hydrated. Every minute counts; the sooner the burn sites are covered, the greater the chances of survival.

The need for new skin Until recently, skin to cover burns came from cadavers, pigs, or uninjured areas of the body. The immune system rejects "foreign" skin as an invader. In addition, transferring skin from unburned areas simply isn't possible for patients who have lost most of their skin. Hospital stays and therapy can last more than a year. Because the cells that usually produce new skin cells by mitosis have been destroyed in third-degree burn patients, new skin cells cannot be made. Thus, young people face multiple surgeries as skin grafts are outgrown.

Cloned skin Now, burn specialists can make as much skin as is needed without being concerned about rejection. A postage stamp-size piece of unburned skin is cut into tiny pieces, then suspended in a solution of proteins and nutrients similar to those found in the blood vessels of the skin. These skin cells each grow into a colony by repeated mitosis in the protein-rich environment. The colonies continue growing together until a thin sheet is formed that can be used for grafting. The process takes 17 to 24 days to grow several square feet of skin; in the meantime, antibiotics, painkillers, and vast quantities of fluids keep the patient alive.

While cloned skin is a perfect match for the recipient, it is not without problems. Grown only from the living epithelial cells of the dermis, the skin develops without sweat glands or pores. Sweat normally rises through pores to the surface of the skin, where it evaporates and cools the body. Burn patients must take care not to get overheated.

Applications for the Future

Research continues on improving cloned skin and on the formulation and use of artificial skin. Artificial skin, a flexible gelatin film made up of collagen composites, cannot replace human skin but is useful for covering open wounds temporarily. It adheres spontaneously, is permeable to body fluids, and releases medications over a period of four to five days.

INVESTIGATING the Technology

1. **Writing About Biology** Find out why cultured skin is grown in a medium of proteins and nutrients similar to blood. Write a journal entry summarizing your research.
2. **Apply** What applications other than for burn victims exist for commercially produced, artificial skin?

11.1 Cell Growth and Reproduction

- Cell size is limited because of physical factors, such as the rate of diffusion of materials into and out of the cell, the amount of DNA available to program the cell's metabolism, and the cell's surface area-to-volume ratio.
- The life cycle of a cell can be divided into two general periods: a period of active growth and metabolism known as interphase, and a period of cell division known as mitosis.
- Mitosis is divided into four phases: prophase, metaphase, anaphase, and telophase.

11.2 Control of the Cell Cycle

- The cell cycle is controlled by the production and secretion of key enzymes at specific times.

- Cancer is caused by an interaction between environmental factors and changes in the genes that control the cell cycle.

Key Terms

Write a sentence that shows your understanding of each of the following terms.

anaphase	interphase
cancer	metaphase
cell cycle	mitosis
centriole	prophase
centromere	sister chromatid
chromosome	spindle
gene	telophase

Understanding Concepts

1. How does surface area-to-volume ratio influence cell size?
2. Why is it important for nutrients to reach the center of a cell quickly?
3. Why is cell size limited by the rate of diffusion?
4. Why might the amount of DNA in a large cell be a limiting factor to cell size?
5. Why is it important that a cell's DNA be copied before mitosis can occur?
6. Why can't chromosomes be seen during interphase?
7. Why is interphase an important period in the cell cycle?
8. Describe the movement of chromosomes during metaphase.
9. Compare spindle formation in plants and in animals.
10. Explain how a change in a gene controlling the cell cycle can result in cancer.

Using Photomicrographs

11. Identify the phase of mitosis shown by the two cells in the center of this photomicrograph of a whitefish blastula, magnified 320×.

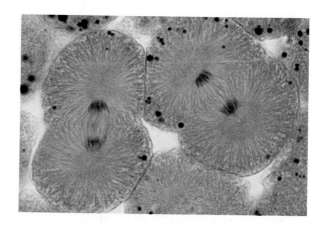

Relating Concepts

12. Make a concept map that relates the following terms and phrases. Supply the appropriate linking words for your map.

cell cycle, interphase, prophase, metaphase, anaphase, telophase, chromosome replication, spindle, chromosome, cancer

Applying Concepts

13. The body cells of a dog contain 78 chromosomes. How many chromosomes do new dog cells have following mitosis?
14. Human muscle cells and nerve cells seldom divide after they are formed. Explain how this fact might affect an individual who has received a spinal cord injury.
15. Explain why animal cells change their shape during mitosis. Refer to the normal function of the cytoskeleton and what happens to it during mitosis.
16. Compare the division of the cytoplasm in plants and in animals. Why do you think different mechanisms might have evolved?
17. Suppose that all of the enzymes that control the normal cell cycle were identified. Suggest some ways that this information might be used to fight cancer.

Health Connection
18. Based on this feaure, what can you hypothesize about the incidence of melanoma among African Americans?

Thinking Critically

Interpreting Data
19. **Biolab** An unidentified group of plant cells was found to have a cell cycle averaging 24 hours. A second set of plant cells had a cell cycle averaging 88 hours. As a scientist, you were asked to identify the tissues from which the cells came. Using your knowledge of mitosis, predict which set of cells came from the plant's root tip. From which part of the plant might the other set have come? Explain.

Making Inferences
20. Fertilized eggs of animals have a large amount of cytoplasm, and the first few mitotic divisions occur rapidly. What do you think happens to the size of the cells during this series of divisions? Explain.

Making Predictions
21. **Minilab** What do you think would have happened if you had left the potato cube in the purple solution overnight?

Making Inferences
22. Some cells of fungi have many nuclei. Suggest a way such a cell could have evolved. Use your knowledge of mitosis to help you.

Magnification: 5×

Multinucleate fungus, *Rhizopus*

Connecting to Themes

23. **Evolution** What are the advantages to a cell of having a mechanism for coiling its chromosomes before cell division?
24. **Homeostasis** Why is cell division in adult animals needed to maintain homeostasis?
25. **Unity Within Diversity** Both plants and animals have basically the same method of cell division. What can you infer from this about the relationship between plants and animals?

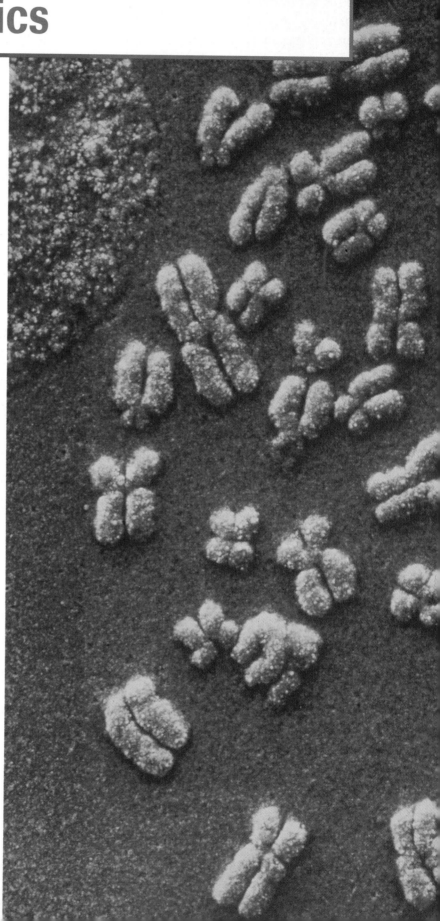

I t's difficult to believe that these simple looking cellular structures will one day be the ultimate source for understanding human biology and inherited genetic disorders. Contained within these human chromosomes is a lengthy chemical message that makes up the spiraling DNA strands inside the nucleus of each of your cells—a message that scientists hope will one day help explain birth, development, growth, and death.

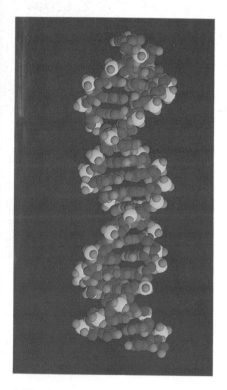

DNA, the basis of all life, resides in every cell of every organism on Earth. How is the chemical code in the double helix of DNA translated into cells, tissues, and organs?

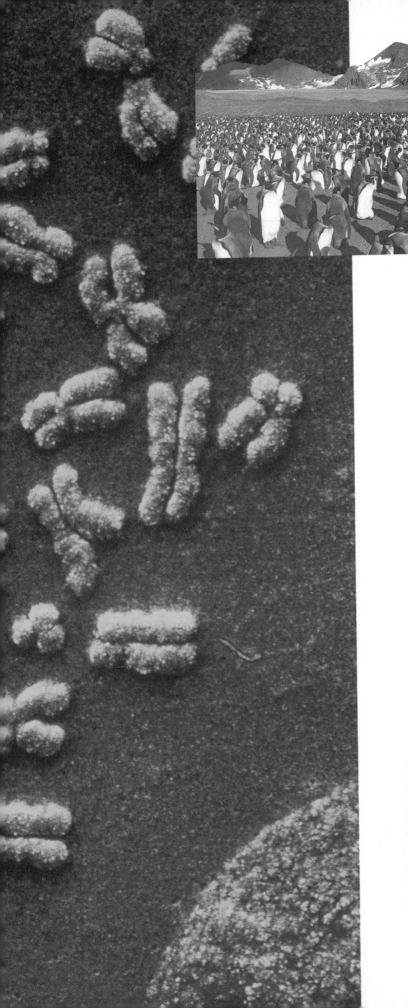

Although individuals in a population, such as a penguin colony, may appear identical, each has almost undetectable differences. How is DNA related to the variation in a species? How is it inherited?

Through the study of DNA, scientists have already learned how to manipulate the fundamental building blocks of life. This tobacco plant has been given the genes of a firefly. What are some of the possible uses for genetic engineering techniques?

Unit Contents

283

12 Mendel and Meiosis

Do you look just like one of your parents? Have you ever heard a teacher absentmindedly call a student by the name of an older brother or sister whom he or she resembles? For thousands of years of human history, people have noticed family resemblances and wondered at them. People have even noted physical characteristics that are transferred from parent to offspring in other species, and have learned to produce unique strains of plants and breeds of animals.

Yet, it was not until the mid-1800s that a scientific study of how physical characteristics are inherited was carried out. Surprisingly, the first answers came, not from the study of family resemblances in people or even in dogs, but from the study of resemblances among garden peas. The scientist who did this pioneering work was an Austrian monk named Gregor Mendel.

Concept Check

You may wish to review the following concepts before studying this chapter.
- Chapter 8: nucleus
- Chapter 11: chromosomes, mitosis

Chapter Preview

This Shar-Pei dog and her puppies are the result of our knowledge of inheritance. What biological mechanism lies behind the transfer of physical characteristics from parent to offspring? The answer lies at the much less obvious, microscopic level of sex cells and the bits of hereditary material within them called chromosomes.

Magnification: 3500×

12.1 Mendel's Laws of Heredity

Section Preview

Objectives

Analyze the results obtained by Gregor Mendel in his experiments with garden peas.

Predict the possible offspring of a cross by using a Punnett square.

Key Terms

heredity
genetics
trait
gamete
fertilization
pollination
allele
dominant
recessive
law of segregation
phenotype
genotype
homozygous
heterozygous
dihybrid cross
law of independent
  assortment

An Austrian monastery in the mid-19th century might seem an unusual place to begin your search for the answer to why offspring resemble their parents. Yet, during the 1800s, the Augustinian monastery in Austria was an important center of scientific study. It was this community of scholars that Gregor Mendel joined as a young man in 1843.

Why Mendel Succeeded

Gregor Mendel carried out the first important studies of **heredity,** the passing on of characteristics from parents to offspring. Although people had noticed that family resemblances were inherited from generation to generation, an explanation required the careful study of **genetics**—the branch of biology that studies heredity. Characteristics that are inherited are called **traits.** Mendel was the first person to succeed in predicting how traits are transferred from one generation to the next. How was he able to solve the problem of heredity?

Mendel chose his subject carefully

Mendel studied many plants before deciding to use the garden pea in his experiments. Garden pea plants reproduce sexually. That means they have two distinct sex cells—male and female. Sex cells are called **gametes.** In peas, both male and female gametes are in the same flower. The male gamete is in the pollen grain, whereas the female gamete is in the ovule. **Fertilization,** the uniting of male and female gametes, occurs when the male gamete in the pollen grain meets and fuses with the female gamete in the ovule. The transfer of the male pollen grains to the female organ of a flower is called **pollination.** After the ovule is fertilized, it matures into a seed.

To study inherited traits, Mendel had to transfer pollen from one plant to another plant with different traits, *Figure 12.1.* This is called making a cross. In these crosses, he had to be sure of the identity of the parents. Mendel chose the pea for the crosses because the structure of the plant's flower made this identity relatively easy.

Mendel was a careful researcher

Mendel carefully controlled his experiments and the peas he used. He studied only one trait at a time to control variables, and he analyzed his data mathematically. The tall pea plants he worked with were from populations of plants that had been tall for many generations and had always produced tall offspring. Such plants are said to be true breeding for tallness. Likewise, the short plants he worked with were true breeding for shortness.

Mendel's Monohybrid Crosses

What did Mendel do with the tall and short pea plants he so carefully selected? He crossed them to produce new plants. Mendel's first experiments are called monohybrid crosses because the two parent plants differed by a single trait—height.

The first generation

Mendel selected a six-foot-tall pea plant that came from a population of pea plants, all of which were over six feet tall. He cross-pollinated this tall pea plant with a short pea plant that was less than two feet tall and which came from a population of pea plants that were all short. He found that all of the offspring in the first generation were as tall as the taller parent. It was as if the shorter parent had never even existed!

Figure 12.1

The reproductive parts of the pea flower are tightly enclosed in petals, preventing the pollen of other flowers from entering. As a result, peas normally reproduce by self-pollination. In many of Mendel's experiments, this is exactly what he wanted. When he needed to cross one plant with another, Mendel removed the male parts from an immature flower. He then dusted the female part of the flower with pollen from the plant he wished to cross it with and covered the flower with a small bag. In this way, Mendel could be sure of the parents in his cross.

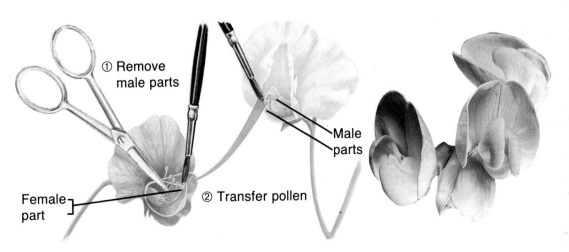

① Remove male parts

Male parts

Female part

② Transfer pollen

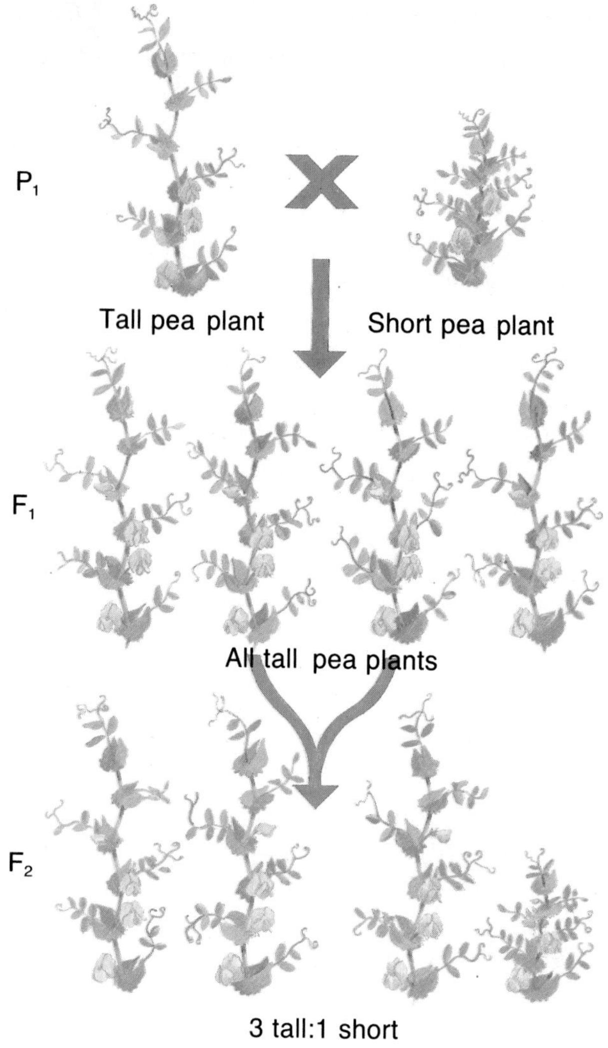

Figure 12.2

When Mendel crossed true-breeding tall pea plants with true-breeding short pea plants, all the offspring were tall. When he allowed first-generation tall plants to self-pollinate, three-fourths of the offspring were tall and one-fourth were short. Mendel counted more than 1000 tall and short plants in this second generation and found that tall and short plants occurred in a ratio of approximately three tall plants to one short plant.

P_1

Tall pea plant   Short pea plant

F_1

All tall pea plants

F_2

3 tall:1 short

The second generation

Next, Mendel crossed two tall offspring plants with each other. This was easy to do because he could simply let them self-pollinate. Mendel found that three-fourths of the offspring in the second generation were as tall as the tall plants in the parent and first generations, whereas one-fourth of the offspring were as short as the short plants in the parent generation, as shown in *Figure 12.2.* It was as if the short trait had reappeared from nowhere!

The original parents, the true-breeding tall and short plants, are known as the P_1 generation. The *P* stands for "parent." The offspring of the parent plants are known as the F_1 generation. The *F* stands for "filial"—son or daughter. When you cross two F_1 plants with each other, their offspring are called the F_2 generation—the second filial generation.

Mendel did similar monohybrid crosses with a total of seven pairs of traits, shown in *Figure 12.3.* In every case, he found that one trait seemed to disappear in the F_1 generation,

only to reappear unchanged in one-fourth of the F_2 plants.

The rule of unit factors

Mendel concluded that each organism has two factors for each of its traits. We now know that these factors are genes and that they are located on chromosomes. Genes exist in alternative forms. We call these different gene forms **alleles.** For example, each of Mendel's pea plants had two alleles that determined its height. A plant could have two alleles for tallness, two alleles for shortness, or one allele for tallness and one for shortness. Alleles are located on different copies of a chromosome—one inherited from the female parent and one from the male parent.

allele:
allelon (GK) of each other
Genes exist in alternative forms called alleles.

The rule of dominance

Remember what happened when Mendel crossed a tall P_1 plant with a short one. The F_1 offspring were all tall. In other words, only one trait was observed. In such crosses, Mendel called the observed trait **dominant** and the trait that disappeared **recessive**. We now know that in Mendel's pea plants, the allele for tall plants is dominant to the allele for short plants. Pea plants that had two alleles for tallness were tall, and those that had two alleles for shortness were short. Plants that had one allele for tallness and one for shortness were tall because the allele for tallness is dominant to the allele for shortness. Expressed another way, the allele for short plants is recessive to the allele for tall plants. You can see in *Figure 12.4* on the following page how the rule of dominance explained the resulting F_1 generation.

ThinkingLab — Analyze the Procedure

How did Mendel analyze his data?

In addition to crossing tall and short pea plants, Mendel crossed plants that formed round seeds with plants that formed wrinkled seeds. He found a 3:1 ratio of round-seeded plants to wrinkled-seeded plants in the F_2 generation.

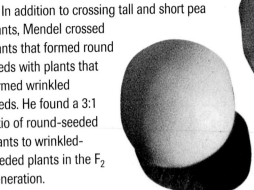

Analysis

Mendel's actual results in the F_2 generation are shown below.

Kind of Plants	Number of Plants
Round-seeded	5474
Wrinkled-seeded	1850

1. Calculate the actual ratio of round-seeded plants to wrinkled-seeded plants. To do this, divide the number of round-seeded plants by the number of wrinkled-seeded plants. (Round to the nearest hundredth.) Your answer tells you how many times more round-seeded plants resulted than wrinkled-seeded plants.

2. To express your answer as a ratio, write the number from step 1 followed by a colon and the numeral *1*.

Thinking Critically

What was the actual ratio Mendel observed for this cross? How does this compare to the expected 3:1 ratio? Why was the actual ratio different from the expected ratio?

Figure 12.3

Mendel chose seven traits of peas for his experiments. Each trait had two clearly different forms; no intermediate forms were observed.

	Seed shape	Seed color	Flower color	Flower position	Pod color	Pod shape	Plant height
Dominant trait	round	yellow	purple	axial (side)	green	inflated	tall
Recessive trait	wrinkled	green	white	terminal (tips)	yellow	constricted	short

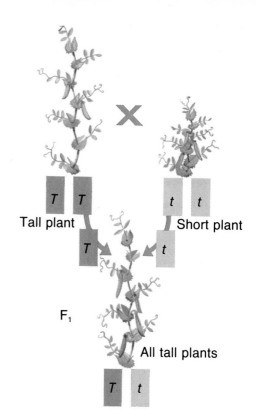

History

Mendel's Life

Johann Gregor Mendel was an Austrian peasant who entered a monastery for financial security. Born July 22, 1822, Mendel was the middle child of a family in which all his siblings were girls. His father was a farmer and his mother was the daughter of a gardener.

Too poor to continue When Mendel's father was seriously injured in a farming accident, he turned the land over to his son-in-law. Left without financial support, Johann had to earn his living by private tutoring. The physical and mental strain of teaching so affected Mendel that he dropped out of his studies for several months to rest and recover. Despite his poor health, he managed to complete his studies and enroll at a university. Two years later, in 1843, at his physics professor's urging, he entered the Augustinian monastery in Brno where he took the name Gregor.

Monastic freedom The monastic life took care of Mendel's basic needs, leaving him free to pursue his scientific studies. The monastery was a center of learning and scientific endeavor, and many of its monks were teachers. A large garden was cultivated for the study of variation in plants. Mendel was put in charge of this garden; he later raised and studied almost 30 000 pea plants over a period of ten years.

Mendel's genius Gregor Mendel's genius lay in his ability to simplify a problem enough to make it testable, and his habits of meticulously collecting and recording data, thinking in quantitative terms, and subjecting his large groups of data to statistical analysis. His careful study of inheritance in pea plants was completed in 1863. Ten years of observing, planting, gathering, labeling, and counting pea seeds led to one of the great discoveries of science—the basic laws of genetics.

CONNECTION TO Biology

How would you maintain records on the characteristics and offspring of 30 000 pea plants?

Figure 12.4

The rule of dominance explains the results of Mendel's cross between P$_1$ tall and short plants. When recording the results of crosses, it is customary to use the same letter for different alleles of the same gene. An uppercase letter is used for the dominant allele, and a lowercase letter for the recessive allele. The dominant allele is always written first. Therefore, the allele for tallness is represented by *T* and the allele for shortness by *t*.

The law of segregation

Now recall the results of Mendel's cross between F$_1$ tall plants when the trait of shortness reappeared. To explain this result, Mendel formulated the first of his two laws of heredity. He concluded that the two alleles for each trait must separate when gametes are formed. A parent, therefore, passes on at random only one allele for each trait to each offspring. This conclusion, illustrated in *Figure 12.5*, is called the **law of segregation.**

Phenotypes and Genotypes

Mendel's experiments showed that tall plants are not all the same. Some tall plants, when crossed with each other, yielded only tall offspring. These were Mendel's original P_1 true-breeding tall plants. Other tall plants, when crossed with each other, yielded both tall and short offspring. These were the F_1 tall plants that came from a cross between a tall plant and a short plant.

Two organisms, therefore, can look alike but have different underlying gene combinations. The way an organism looks is its **phenotype.** The phenotype of a tall plant is tall, regardless of the genes it contains. The gene combination an organism contains is known as its **genotype.** The genotype of a tall plant that has two alleles for tallness is *TT.* The genotype of a tall plant that has one allele for tallness and one allele for shortness is *Tt.*

You can see that you can't always know an organism's genotype simply by looking at its phenotype. An organism is **homozygous** for a trait if its two alleles for the trait are the same. The true-breeding tall plant that had two alleles for tallness (*TT*) would be homozygous for the trait of height. Because tallness is dominant, a *TT* individual is homozygous dominant for that trait. A short plant would always have two alleles for shortness (*tt*). It would, therefore, always be homozygous recessive for the trait of height.

An organism is **heterozygous** for a trait if its two alleles for the trait are different. Therefore, the tall plant that had one allele for tallness and one allele for shortness (*Tt*) is heterozygous for the trait of height.

homozygous:

homo (GK) same
zygotos (GK) joined together
Homozygous individuals have two identical alleles for a trait.

heterozygous:

hetero (GK) the other of two
zygotos (GK) joined together
Heterozygous individuals have two different alleles for a trait.

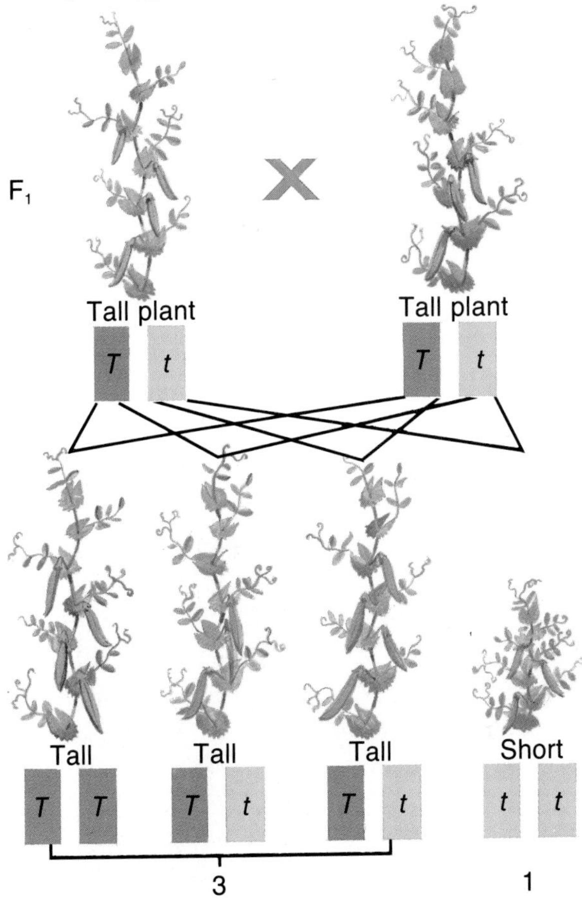

Mendel's Dihybrid Crosses

Next, Mendel performed another set of crosses in which he used peas that differed from each other in two traits rather than only one. A cross involving two different traits is called a **dihybrid cross.**

Figure 12.5

Mendel's law of segregation explains the results of his cross between F_1 tall plants. Each tall plant in the F_1 generation carried one dominant allele for tallness and one unexpressed recessive allele for shortness. It received the allele for tallness from its tall parent and the allele for shortness from its short parent in the P_1 generation. Because each F_1 plant has two different alleles, it can produce two different types of gametes— "tall" gametes and "short" gametes. During fertilization, gametes randomly pair to produce four combinations of alleles.

It's difficult to predict the traits of plants if all that you see is their seeds. But if these seeds are planted and allowed to grow, certain traits will appear. By observing these traits, you might be able to determine the possible phenotypes and genotypes of the parent plants that produced these seeds. In this lab, you will determine the genotypes of plants that grow from two groups of tobacco seeds. Each group of seeds came from different parents. Plants will be either green or albino (white) in color. Use the following genotypes for this cross. CC = green, Cc = green, and cc = albino

How can phenotypes and genotypes of plants be determined?

PREPARATION

Problem
Can the phenotypes and genotypes of the parent plants that produced two groups of seeds be determined from the phenotypes of the plants grown from the seeds?

Hypotheses
Have your group agree on a hypothesis to be tested that will answer the problem question. Record your hypothesis.

Objectives
In this Biolab, you will:
- **Analyze** the results of growing two groups of seeds.
- **Draw conclusions** about phenotypes and genotypes based on those results.

Possible Materials
potting soil
small flowerpots or seedling flats
two groups of tobacco seeds
hand lens
light source
thermometer
plant-watering bottle

Safety Precautions
Always wash your hands after handling plant materials.

PLAN THE EXPERIMENT

1. Examine the materials provided by your teacher. As a group, make a list of the possible ways you might test your hypothesis.

2. Agree on one way that your group could investigate your hypothesis.

3. Design an experiment that will allow you to collect quantitative data. For example, how many plants do you think you will need to examine?

4. Prepare a list of numbered directions. Include a list of materials and the quantities you will need.

5. Design and construct a data table for recording your observations.

Check the Plan

1. Carefully determine the data you are going to collect. How many seeds do you think you will need? How long will you carry out the experiment?

2. What variables, if any, will have to be controlled? (Hint: Think about the growing conditions for the plants.)

3. Does your table allow for all the various kinds of data you need?

4. **Make sure your teacher has approved your experimental plan before you proceed further.**

5. Carry out your experiment. Make any needed observations, such as the numbers of green and albino plants in each group, and complete your data table. Design and then complete a graph or other visual representation of your results.

ANALYZE AND CONCLUDE

1. **Thinking Critically** Why was it necessary to grow plants from the seeds in order to determine the phenotypes of the plants that formed the seeds?

2. **Drawing Conclusions** Using the information provided in the introduction, describe how the gene for green color (C) is inherited.

3. **Making Inferences** For the group of seeds that yielded all green plants, are you able to determine exactly the genotypes of the parents that formed these seeds? Can you determine the genotype of each plant observed? Explain.

4. **Making Inferences** For the group of seeds that yielded some green and some albino plants, are you able to determine exactly the genotypes of the plants that formed these seeds? Can you determine the genotypes of each plant observed? Explain.

Going Further

Journal Report
From your results, design an experiment that would help you determine some of the genotypes you were unable to determine from the experiment you just performed. Write up your experiment in your journal.

How does sample size affect results?

The ratio of males to females born in a population should be 1:1. If progressively larger populations of males and females are counted, the actual numbers approach this theoretical ratio.

Procedure

1. Count the number of males and females in your class. Determine the ratio.
2. Add the brothers and sisters of your classmates to your total, and again count the number of males and females. Determine the male:female ratio.
3. Redo the count, this time adding all aunts, uncles, and cousins of the members of your class. Determine the male:female ratio.

Analysis

1. How many people were included in each of your three samples?
2. How did the male:female ratio vary as the sample size became larger?
3. Based on your data, how large a sample do you think you need in order to be fairly sure of obtaining numbers that are close to the expected ratio?

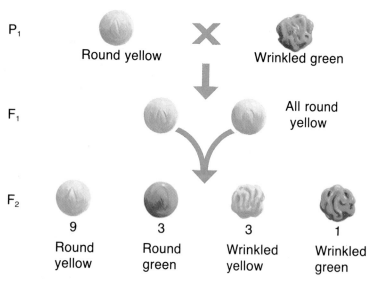

Figure 12.6

When Mendel crossed true-breeding plants with round yellow seeds and true-breeding plants with wrinkled green seeds, the seeds of all the offspring were round and yellow. When the F₁ plants were allowed to self-pollinate, they produced four different kinds of plants in the F₂ generation. If the alleles for seed shape and color were inherited together, only two kinds of pea seeds would have been produced: round yellow and wrinkled green.

The first generation

Mendel took true-breeding pea plants that had round yellow seeds ($RRYY$) and crossed them with true-breeding pea plants that had wrinkled green seeds ($rryy$). He already knew that when he crossed plants that produced round seeds with plants that produced wrinkled seeds, all the plants in the F_1 generation produced seeds that were round. In other words, round trait was dominant. Similarly, when he crossed plants that produced yellow seeds with plants that produced green seeds, all the plants in the F_1 generation produced yellow seeds—yellow was dominant. Therefore, Mendel was not surprised when he found that the F_1 plants of his dihybrid cross all had round yellow seeds, as *Figure 12.6* shows.

The second generation

Mendel then let the F_1 plants pollinate themselves. As you might expect, Mendel found some plants that produced round yellow seeds and others that produced wrinkled green seeds. But that's not all. He also found some plants with round green seeds and others with wrinkled yellow seeds. Mendel sorted and counted the plants of the F_2 generation. He found they appeared in a definite ratio—9 round yellow:3 round green:3 wrinkled yellow:1 wrinkled green. To explain the results of this dihybrid cross, Mendel formulated his second law.

The law of independent assortment

Mendel's second law states that genes for different traits—for example, seed shape and seed color—are inherited independently of each other. This conclusion is known as the **law of independent assortment.** When a pea plant with the genotype $RrYy$ produces gametes, the alleles *R* and *r* will separate from each other (the law of segregation) as well as from the alleles *Y* and *y* (the law of independent assortment), and vice versa. These alleles can then recombine in four different ways. We now know that this principle is true only if genes are located on different chromosomes or are far apart on the same chromosome.

Punnett Squares

In 1905, Reginald Punnett, an English biologist, devised a shorthand way of finding the expected proportions of possible genotypes in the offspring of a cross. This method is called a Punnett square. It takes account of the fact that fertilization occurs at random, as Mendel's law of segregation states. If you know the genotypes of the parents, you can use a Punnett square to predict the possible genotypes of their offspring.

Monohybrid crosses

Consider the cross between two F_1 tall pea plants, each of which has the genotype *Tt*. Half the gametes of each parent would contain the *T* allele, and the other half would contain the *t* allele. A Punnett square for this cross would be two boxes tall and two boxes wide because each parent can produce two kinds of gametes. Refer to the Punnett square on the next page as you determine the possible genotypes of the offspring.

The Laws of Probability

Punnett squares are good for showing all the possible combinations of gametes and the likelihood that each will occur. In reality, however, you don't always get the exact ratio of results shown in the square. That's because, in some ways, genetics is like flipping a coin—it follows the rules of chance.

Rules of chance At the start of every football game, the captain of the visiting team calls either *heads* or *tails* in a coin toss. The probability of his choice coming up is expressed as a fraction—the numerator is the number of ways he could be right; the denominator is the total possible events. In this case, the probability is 1/2, or one in two.

Boys and girls Now, consider the chances of parents having a girl or a boy. Every time two parents have a child, there is one chance in two that the child will be a girl, and one chance in two that it will be a boy. Yet, you must know families that have three or even four girls and no boys, or vice versa. However, just look at a large number of people. You will find that the number of each sex is close to a 1:1 ratio. When dealing with events governed by the laws of probability, the results will approach the expected ratios when large numbers of events are counted.

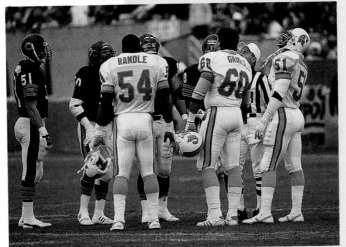

Mendel and probability
When Mendel crossed his F_1 heterozygous tall pea plants, three-fourths of the offspring in the F_2 generation were tall and one-fourth were short. In other words, they were in a ratio of 3 tall:1 short. Yet, if you were to look at Mendel's actual data, you would see that he counted 787 tall plants and 277 short plants—a ratio of 2.84:1. This ratio was observed because events in genetics are governed by laws of probability. A 3:1 ratio really means that every time an F_2 plant is produced, there are three chances in four that it will be tall and one chance in four that it will be short. Each new plant is an independent event and is not affected by what previous plants were. When you count large numbers of plants, you will get close to the expected ratio of 3:1.

Thinking Critically
The probability of independent events happening together is the product of their individual probabilities. If you tossed three coins at the same time, what is the probability of getting all heads?

After the genotypes have been determined, you can determine the phenotypes. Looking at the Punnett square for this cross in *Figure 12.7*, you can see that three-fourths of the offspring are expected to be tall and one-fourth are expected to be short. Of the tall offspring, one-third will be homozygous dominant (*TT*) and two-thirds will be heterozygous (*Tt*). Note that whereas the genotype ratio is 1 *TT*:2 *Tt*:1 *tt*, the phenotype ratio is 3 tall:1 short.

Dihybrid crosses

What happens in a Punnett square when two traits are considered? Think again about Mendel's cross between peas with round yellow seeds and peas with wrinkled green seeds. All the seeds of the F$_1$ plants were round and yellow, and were heterozygous for each trait (*RrYy*). What kind of gametes will they form?

Mendel explained that the genes for seed shape and seed color would be inherited independently of each other. This means that each F$_1$ plant will produce gametes containing the following combinations of genes with equal frequency: round yellow (*RY*), round green (*Ry*), wrinkled yellow (*rY*), and wrinkled green (*ry*). A Punnett square for a dihybrid cross will then need to be four boxes on each side for a total of 16 boxes, as *Figure 12.8* shows.

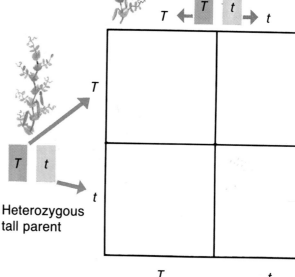

Figure 12.7

This Punnett square predicts the results of a monohybrid cross between two heterozygous tall pea plants.

A The two kinds of gametes from one parent are listed on top of the square, and the two kinds of gametes from the other parent are listed on the left side. It doesn't matter which set of gametes is on top and which is on the side, that is, which parent contributed the *T* and which contributed the *t*.

B To find the possible genotypes of the offspring from a Punnett square, each box is filled in with the gametes above and to the left side of that box. Each box then contains two alleles—one possible genotype. You can see that there are three different possible genotypes—*TT*, *Tt*, and *tt*—and that *Tt* can result from two different combinations.

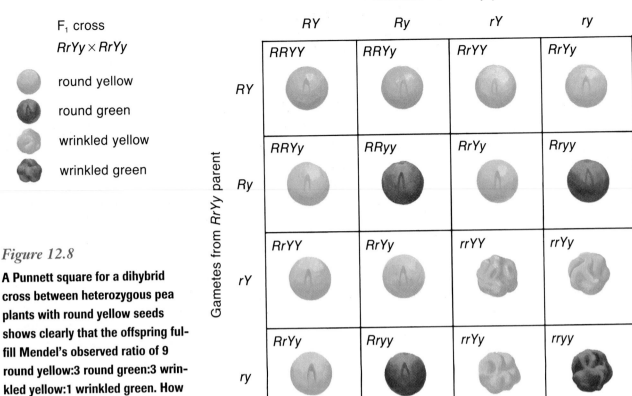

Gametes from *RrYy* parent

	RY	*Ry*	*rY*	*ry*
RY	RRYY	RRYy	RrYY	RrYy
Ry	RRYy	RRyy	RrYy	Rryy
rY	RrYY	RrYy	rrYY	rrYy
ry	RrYy	Rryy	rrYy	rryy

F₁ cross

RrYy × RrYy

- round yellow
- round green
- wrinkled yellow
- wrinkled green

Gametes from *RrYy* parent

Figure 12.8

A Punnett square for a dihybrid cross between heterozygous pea plants with round yellow seeds shows clearly that the offspring fulfill Mendel's observed ratio of 9 round yellow:3 round green:3 wrinkled yellow:1 wrinkled green. How many different genotypes and phenotypes result from this cross?

Section Review

Understanding Concepts

1. Why did the flower structure of pea plants make them suitable for Mendel's genetic studies?
2. What is the genotype of a homozygous and a heterozygous tall pea plant?
3. One parent is homozygous for a certain trait and the other parent is heterozygous. Make a Punnett square to determine what fraction of their offspring is expected to be heterozygous.

Thinking Critically

4. In garden peas, the allele for yellow peas is dominant to the allele for green peas. Suppose you have a plant with yellow peas but you don't know whether it is homozygous dominant or heterozygous. What experiment could you do to find out? Draw a Punnett square to help you.

Skill Review

5. **Observing and Inferring** The offspring of a cross between a purple-flowered plant and a white-flowered plant are 23 plants with purple flowers and 26 plants with white flowers. What are the genotypes of the parent plants, and which allele is dominant? Use the letters *W* and *w* to represent the alleles. For more help, refer to Thinking Critically in the *Skill Handbook*.

SECTION

12.2 Meiosis

Mendel succeeded in describing how inherited traits are transmitted from one generation to the next. Yet he had no way of explaining what his factors might be, or in what physical way a factor could be passed from parent to offspring. Pieces of the heredity puzzle were still missing. As a result, Mendel's work went relatively unnoticed until almost 20 years after his death.

Today, more than 125 years later, we have a much greater understanding of what Mendel's factors are, how they determine inherited traits, and how they are transmitted between generations of organisms.

Genes, Chromosomes, and Numbers

Organisms have tens of thousands of genes that determine individual traits. Genes do not exist free in the nucleus of a cell, but are lined up in linear fashion on chromosomes. Typically, a thousand or more genes are arranged on each chromosome.

Diploid and haploid cells

If you examined the nucleus in a cell of one of Mendel's pea plants, you would find it had 14 chromosomes—seven pairs. In the body cells of animals and most plants, chromosomes occur in pairs. One chromosome in each pair came from the male parent, and the other came from the female parent. A cell with two of each kind of chromosome is called a **diploid** cell and is said to contain a diploid, or $2n$, number of chromosomes. This pairing supports Mendel's conclusion that organisms have two factors—alleles—for each trait. One allele is located on each of the paired chromosomes.

Organisms produce gametes that contain *one* of each kind of chromosome. A cell with one of each kind of chromosome is called a **haploid** cell and is said to contain a haploid, or n, number of chromosomes. This fact supports Mendel's conclusion that parent organisms give one factor, or allele, for each trait to each of their offspring.

Each species of organism contains a characteristic number of chromosomes. *Table 12.1* shows the diploid and haploid numbers of chromosomes of some species. Note the large range of chromosome numbers. Note also that the chromosome number of a species is not related to the complexity of the organism.

Table 12.1 Chromosome Numbers of Some Common Organisms

Organism	Body Cell (2*n*)	Gamete (*n*)
Human	46	23
Garden Pea	14	7
Fruit fly	8	4
Tomato	24	12
Dog	78	39
Chimpanzee	48	24
Leopard frog	26	13
Corn	20	10
Apple	34	17
Indian fern	1260	630

Fruit fly

Leopard frog

Indian fern

Homologous chromosomes

Together, the two chromosomes of each pair in a diploid cell help determine what the individual organism looks like. These paired chromosomes are called **homologous chromosomes.** Homologous chromosomes have genes for the same traits arranged in the same order. However, because there are different possible alleles for the same gene, the two chromosomes in a homologous pair are not identical to each other.

For example, look at one of the seven pairs of homologous chromosomes in Mendel's peas. These chromosome pairs are numbered 1 to 7.

Each pair contains certain genes, located at specific places on the chromosome. Chromosome 4 contains the genes for three of the traits that Mendel studied. You can see the positions of these genes in *Figure 12.9.* Many other genes can be found on this chromosome as well.

Every pea plant has two copies of chromosome 4. It received one from each of its parents and will give one at random to each of its offspring. Remember, however, that the two copies of chromosome 4 in a pea plant may not have identical alleles. Each can have one of the different alleles possible for each gene.

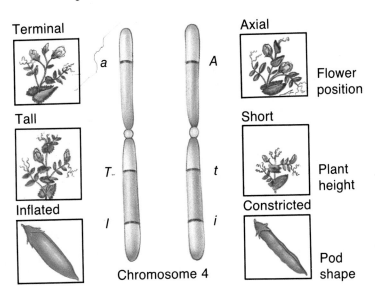

Terminal / Tall / Inflated / Chromosome 4 / Axial / Short / Constricted — a, A, T, t, l, i / Flower position / Plant height / Pod shape

Figure 12.9

Each chromosome 4 in garden peas contains genes for flower position, height, and pod shape. The homologous chromosomes in the diagram show both alleles for each trait. Thus, the plant represented by these chromosomes is heterozygous for each of the three traits. Flower position can be either axial—flowers located along the stems, or terminal—flowers clustered at the top of the plant. As you know, plant height can be either tall or short. Pod shape can be either inflated or constricted.

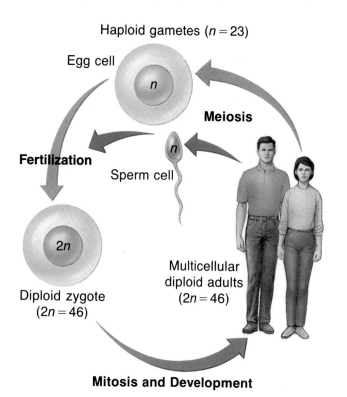

Haploid gametes (*n* = 23)

Egg cell

n

Meiosis

n

Fertilization

Sperm cell

n

Multicellular
diploid adults
(2*n* = 46)

2*n*

Diploid zygote
(2*n* = 46)

Mitosis and Development

Figure 12.10

In sexual reproduction, the doubling of the chromosome number that results from fertilization is balanced by the halving of the chromosome number that results from meiosis.

zygote:

zygotos (GK) yolk
A zygote is formed by the fertilization of a gamete by another gamete.

Why meiosis?

When cells divide by mitosis, the new cells have exactly the same number and kind of chromosomes as the original cells. Imagine if mitosis were the only means of cell division. Each pea plant parent would then give each offspring a complete set of 14 chromosomes. That means each offspring would have twice the number of chromosomes as each of its parents. The F_1 pea plants would have cells with 28 chromosomes, and the F_2 plants would have cells with 56 chromosomes. They wouldn't be pea plants!

Clearly, there must be another form of cell division that allows offspring to have the same number of chromosomes as their parents. This kind of cell division, which produces gametes containing half the number of chromosomes as a parent's body cell, is called **meiosis.** Meiosis occurs in the specialized body cells that produce gametes.

Meiosis consists of two separate divisions, known as meiosis I and

meiosis II. Meiosis I begins with one diploid (2*n*) cell. By the end of meiosis II, there are four haploid (*n*) cells. In animals and most plants, these haploid cells are called sex cells—gametes. Male gametes are called **sperm.** Female gametes are called **eggs.** When a sperm fertilizes an egg, the resulting cell, called a **zygote,** once again has the diploid number of chromosomes. It can then develop by mitosis into a multicellular organism. The pattern of reproduction that involves the production and subsequent fusion of haploid sex cells is called **sexual reproduction.** This pattern is illustrated in *Figure 12.10.*

The Phases of Meiosis

In meiosis, a spindle forms and cytoplasm divides in the same way as occurs in mitosis. However, what happens to the chromosomes in meiosis is very different. *Figure 12.11* illustrates interphase and the phases of meiosis.

Interphase

Recall from Chapter 11 that during interphase, the cell carries out its usual metabolic activities and replicates its chromosomes. During interphase that precedes meiosis I, the cell also replicates its chromosomes. Each chromosome then consists of two identical sister chromatids, held together by a centromere.

Prophase I

The chromosomes coil up and a spindle forms. Then, in a step unique to meiosis, each pair of homologous chromosomes comes together to form a four-part structure called a tetrad. A tetrad consists of two homologous chromosomes, each made up of two sister chromatids.

Figure 12.11

Follow the diagrams showing the process of meiosis as you read about each phase. Compare these diagrams with those of mitosis on pages 270–271 in Chapter 11. How are the processes different?

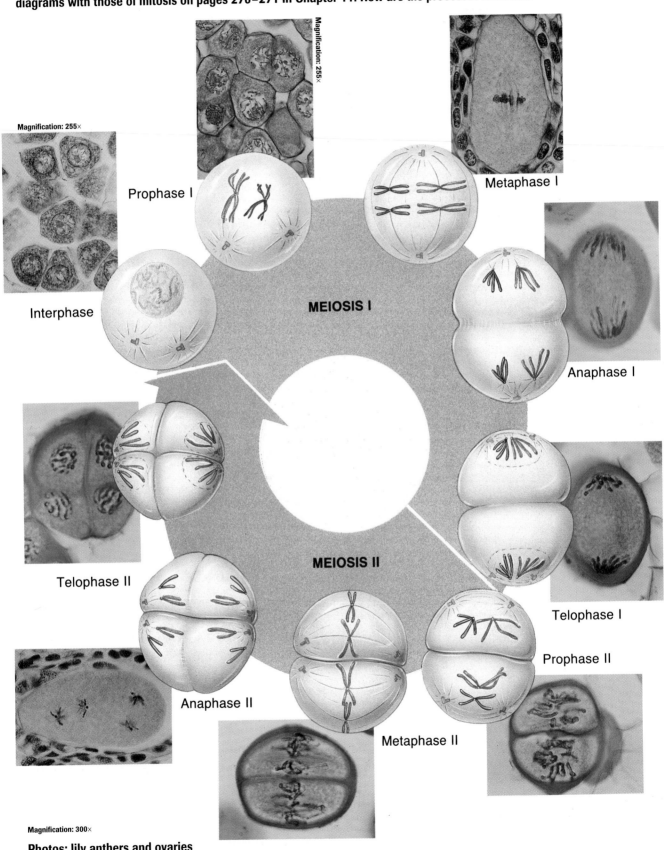

Magnification: 255×

Magnification: 255×

Prophase I

Metaphase I

Interphase

MEIOSIS I

Anaphase I

Telophase II

MEIOSIS II

Telophase I

Prophase II

Anaphase II

Metaphase II

Magnification: 300×

Photos: lily anthers and ovaries
Diagrams: animal cells

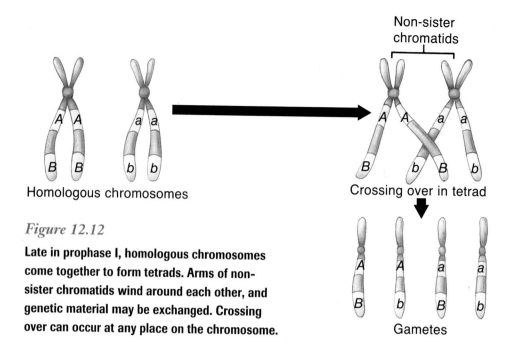

Non-sister chromatids

Homologous chromosomes

Crossing over in tetrad

Gametes

Figure 12.12

Late in prophase I, homologous chromosomes come together to form tetrads. Arms of non-sister chromatids wind around each other, and genetic material may be exchanged. Crossing over can occur at any place on the chromosome.

The chromatids in a tetrad pair tightly. In fact, they pair so tightly that non-sister chromatids from homologous chromosomes sometimes actually exchange genetic material in a process known as **crossing over.** You can see this exchange in *Figure 12.12.* Crossing over results in new combinations of alleles on a chromosome.

Metaphase I

Tetrads line up on the midline, or equator, of the spindle. This is an important step unique to meiosis. Note that homologous chromosomes are lined up together in pairs. In mitosis, on the other hand, homologous chromosomes line up on the equator independently of each other.

Anaphase I

Homologous chromosomes separate and move to opposite ends of the cell. This occurs because the centromeres do not split as they do during anaphase in mitosis. This critical step ensures that each new cell will receive only one chromosome from each homologous pair.

Telophase I

Events occur in the reverse order from the events of prophase I. The spindle is broken down, the chromosomes uncoil, and the cytoplasm divides to yield two new cells. Each cell has only half the genetic information of the original cell because it has only one chromosome from each homologous pair. However, another cell division is needed because each chromosome is still doubled, containing two identical sister chromatids.

The phases of meiosis II

The newly formed cells may undergo a short interphase in which the chromosomes do not replicate, or the cells may go from meiosis I directly to meiosis II. The second division in meiosis consists of prophase II, metaphase II, anaphase II, and telophase II.

During prophase II, a spindle forms in each of the two new cells. The chromosomes, still made up of sister chromatids, line up at the equator during metaphase II, just as they do in mitosis. Anaphase II begins as

inter-, pro-, meta-, ana-, telo-:

inter (L) between
pro (GK) before
meta (GK) after
ana (GK) up, onward
telos (GK) end
Each division of meiosis consists of interphase, prophase, metaphase, anaphase, and telophase.

the centromere of each chromosome splits, allowing the sister chromatids to separate and move to opposite poles. Finally, nuclei reform, the spindles break down, and the cytoplasm divides during telophase II. These events are identical to those of mitosis.

At the end of meiosis II, four haploid cells have been formed from the original diploid cell. Each haploid cell contains one chromosome from each homologous pair. These haploid cells will become gametes, transmitting the genes they contain to offspring.

Meiosis Provides for Genetic Variation

Cells that are formed by mitosis are identical to each other and to the parent cell. Meiosis, however, provides a mechanism for shuffling the chromosomes and the genetic infor-

mation they carry. *Figure 12.13* shows how this shuffling of chromosomes can occur.

Genetic recombination

How many different kinds of sperm can a pea plant produce? Each cell undergoing meiosis has seven pairs of chromosomes. Because each of the seven pairs of chromosomes can line up in two different ways, 128 different kinds of sperm are possible ($2^7 = 128$).

By the same reasoning, any one pea plant can form 128 different eggs. Because any egg can be fertilized by any sperm, the number of different possible offspring is 16 384 (128×128).

These numbers increase greatly as the number of chromosomes in the species increases. In humans, $n = 23$, so the number of different kinds of eggs or sperm a person can produce is more than 8 million (2^{23}). When fertilization occurs, 70 trillion different

Figure 12.13

If a cell has two pairs of chromosomes ($n = 2$), four kinds of gametes (2^2) are possible, depending on how the homologous chromosomes line up at the equator during meiosis I (as shown in A or B on the left). This event is a matter of chance. When zygotes are formed by the union of these gametes, 4×4 or 16 possible combinations may occur (right).

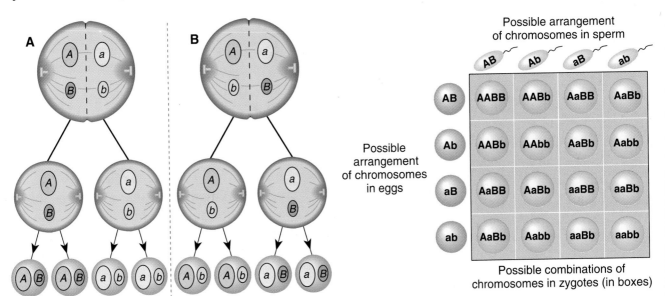

Figure 12.14

Variation is of great importance to a species because some combinations of traits may be more useful to an organism's survival in a changing environment than others.

zygotes are possible, or $2^{23} \times 2^{23}$! It's no wonder that each individual is unique.

In addition, crossing over can occur anywhere at random on a chromosome. Typically, two or three crossovers occur per chromosome during meiosis. This means that an almost endless number of different possible chromosomes can be produced by crossing over, providing additional variation to that already produced by the random assortment of chromosomes. This reassortment of chromosomes and the genetic information they carry, either by crossing over or by independent segregation of

homologous chromosomes, is called **genetic recombination.** It is a major source of variation among organisms, *Figure 12.14.* Why is variation important to a species? Variation is the raw material that forms the basis for evolution.

Meiosis explains Mendel's results

Meiosis explains the physical basis of Mendel's results. The segregation of chromosomes in anaphase I of meiosis explains Mendel's observation that each parent gives one allele for each trait at random to each offspring, regardless of whether the allele is expressed. The random segregation of chromosomes during anaphase I explains Mendel's observation that factors, or genes, for different traits are inherited independently of each other. Today, Mendel's laws of heredity form the foundation of modern genetics.

Connecting Ideas

Genes that are transmitted from parents to offspring are responsible for the physical characteristics that the offspring inherits. You may have wondered just how genes function. How does a gene for tall plants make a plant tall? And how can a gene for wrinkled seeds cause a seed to be wrinkled? The answers to such questions lie in an understanding of the structure and processes of genes and chromosomes. They can be found in learning how to read the genetic code.

Section Review

Understanding Concepts

1. How are the cells at the end of meiosis different from the cell at the beginning of meiosis? Use the terms *chromosome number*, *haploid*, and *diploid* in your answer.
2. What is the role of meiosis in maintaining a constant number of chromosomes in a species?
3. Why are there so many varied phenotypes within a species such as humans?

Thinking Critically

4. How do the events of meiosis explain Mendel's law of independent assortment?

Skill Review

5. **Interpreting Scientific Illustrations** Compare *Figures 12.11* and *11.8*. Explain why crossing over between non-sister chromatids of homologous chromosomes cannot occur during mitosis. For more help, refer to Thinking Critically in the *Skill Handbook.*

A Rose Is a Rose—Or Is It?

The Peace rose was developed in the 1930s, just before the outbreak of the second world war. Fearing that the advent of war might destroy their painstaking work, the French hybridizers sent seedlings of the rose to the United States to be preserved and propagated. The exquisite, cream-colored bloom touched with pink and gold was immediately recognized as something special. Today, the Peace rose is one of the most famous of all roses.

Structure of a rose Just as in many flowering plants, roses bear both male and female sex organs. The rose has many of each type of organ. The delicate, threadlike stigmas that protrude from the center of the flower are the tips of the many female organs called pistils. Circling the pistils are a multitude of stamens, the male organs. Each stamen bears an anther that provides pollen to the stigmas.

Creating the supernatural To create a new rose variety, hybridizers choose the parents of a possible hybrid. The anthers are removed from the female parent before they ripen to avoid any random pollination. The anthers from the chosen male parent are labeled and stored.

Once the pistils of the selected female parent become receptive, pollen from the chosen male parent is carefully hand brushed onto the stigmas. Then, the flowers are covered with bags to prevent further pollination by insects. Each pollinated flower is then labeled with a numbered tag that identifies the parents.

Seeds form Once successfully pollinated, the ovary of the rose swells to form a hip, the fruit of a rose plant. The swollen fruit contains the seeds of the new hybrid generation of roses. The seeds are then harvested and planted. The first flowers of the new rose may form as early as eight weeks after planting. The best seedlings are labeled, propagated, and test grown for several years to confirm their beauty and reliability.

Applications for the Future

It has long been a dream of rosarians to produce a blue rose. Presently, the closest a rose comes to being blue is in the lilac-magenta range. A gene for blue color has not yet been found in roses. If this sounds challenging, perhaps you could be the first to breed this new variety.

INVESTIGATING the Technology

1. **Thinking Critically** People who develop new plants are protected by the United States Plant Patent Act of 1930. This means that they alone control the propagation and distribution of the plant. To obtain a patent, hybridizers must show how the plant differs significantly from others. Make a list of at least three traits that would differentiate a new rose from existing varieties.

2. **Writing About Biology** New varieties of roses are grown from seeds. But most of the roses available to consumers are grafted. A branch of the new variety is attached to the root of an existing rose. Research the reasons for grafting, and write a paragraph that explains why this is done.

Reviewing Main Ideas

12.1 Mendel's Laws of Heredity

- Genes are located on chromosomes and exist in alternative forms called alleles. A dominant allele can mask the expression of a recessive allele.
- When Mendel crossed pea plants differing in one trait, one form of the trait disappeared until the second generation of offspring. To explain his results, Mendel formulated the law of segregation.
- Mendel formulated the law of independent assortment to explain that two traits are inherited independently.
- Events in genetics are governed by the laws of probability.

12.2 Meiosis

- In meiosis, one diploid (2n) cell produces four haploid (n) cells, providing a way for offspring to have the same number of chromosomes as their parents.
- Mendel's results can be explained by the distribution of chromosomes during meiosis.

- Random assortment and crossing over during meiosis provide for genetic variation among the members of a species.

Key Terms

Write a sentence that shows your understanding of each of the following terms.

allele
crossing over
dihybrid cross
diploid
dominant
egg
fertilization
gamete
genetic
 recombination
genetics
genotype
haploid
heredity
heterozygous

homologous
 chromosome
homozygous
law of independent
 assortment
law of segregation
meiosis
phenotype
pollination
recessive
sexual reproduction
sperm
trait
zygote

Understanding Concepts

1. What were some of the factors responsible for Mendel's success? Why do you think these were important?
2. How did Mendel cross one pea plant with another? Why was it easy for him to let the F_1 plants self-pollinate to produce the F_2 generation?
3. Why was it important for Mendel to count large numbers of plants?
4. How are the chromosomes of a homologous pair similar? Different?
5. If a female chimpanzee has 24 pairs of chromosomes in each of her body cells, how many are in one of her eggs?
6. Why is the second meiotic division necessary if reduction of chromosome number is accomplished by the first division?

7. Why do you think it would be advantageous to organisms to be diploid? Can you see any survival value in having two alleles rather than one for each trait?
8. In guinea pigs, the allele for black coat color, *B*, is dominant to the allele for white coat color (albino), *b*. The allele for rough coat, *R*, is dominant to the allele for smooth coat, *r*. If you cross two guinea pigs that are heterozygous black and homozygous rough, what will be the phenotypes and genotypes of the offspring? If you cross a heterozygous black rough guinea pig with a homozygous recessive guinea pig, what will be the phenotypes and genotypes of the offspring? Show all your work, including a Punnett square for each of these crosses.

Interpreting Scientific Illustrations

9. In fruit flies, the allele for long wings is dominant to the allele for short wings. Study the following Punnett square.

	W	w
?	Ww	ww
?	Ww	ww

(a) What was the genotype and phenotype of each parent?
(b) What are the phenotypes and genotypes of the offspring?

Relating Concepts

10. Make a concept map that relates the following terms and phrases. Supply the appropriate linking words for your map.

   P_1 generation, F_1 generation, F_2 generation, dominant, recessive, tall pea plants, short pea plants

Applying Concepts

11. Why do you think the results Mendel obtained are also valid for humans? List and explain at least three similarities between peas and humans.
12. On the average, each human has about six recessive alleles that, if expressed, would be harmful or even lethal. Why do you think that most human cultures have laws against marriage between close relatives?
13. Using Mendel's dihybrid cross between pea plants with round yellow seeds and pea plants with wrinkled green seeds, explain what the law of independent assortment means.

A Broader View

14. What is the probability of tossing a head if you got heads on four previous tosses?

Thinking Critically

Interpreting Data

15. **Biolab** How many generations of tobacco plants would you have to observe to determine the actual genotype of a parent plant in this experiment?

Making Inferences

16. **Minilab** Why is it possible to have a family of six girls and no boys, but extremely unlikely that there will be a public school with 500 girls and no boys?

Interpreting Scientific Illustrations

17. The following drawing shows one pair of homologous chromosomes pairing tightly during prophase I. Observe the places where the chromosomes are actually touching each other. What is the significance of this contact?

Relating Cause and Effect

18. Describe how each of the following events either increases or decreases variation among offspring.
   (a) One allele is dominant to another allele in determining phenotype.
   (b) Separation of homologous chromosomes during anaphase I of meiosis.

Connecting to Themes

19. **The Nature of Science** Why do you think it was important for Mendel to study only one trait at a time?
20. **Homeostasis** Discuss how meiosis, by maintaining a constant number of chromosomes in individuals of a species, contributes to homeostasis in the species.

The forest canopy shakes violently as a pair of blue guenon monkeys reacts to the sound of a predator. Lower in this arboreal environment, one of their close relatives—a crowned guenon with its bushy, yellow hair—rests quietly after devouring a leafy dinner. Nearby on the forest floor, a young, moustached guenon squeaks and devours a fruit it has just picked.

In fact, there are at least 19 species of guenons, all remarkably diverse in color, appearance, and behavior but similar in size and body proportions. Guenons demonstrate how only a few small changes in the genetic blueprint of living organisms, DNA, can produce a variety of species, all with a common body plan. In this chapter, you'll investigate the role and structure of DNA and discover how its remarkable properties can produce the diversity of life you see on Earth.

Chapter Preview

13.1 DNA: The Molecule of Heredity
Structure and Function of DNA
Replication of DNA

13.2 From DNA to Protein
Genes and Proteins
The DNA Code
Transcription—From DNA to RNA
Translation—From RNA to Protein

13.3 Genetic Changes
Mutation—A Change in DNA
Chromosomal Mutations
Errors in Disjunction
Causes of Mutations

Laboratory Activities

Biolab
- RNA Transcription

Minilabs
- What does DNA look like?
- How do gene mutations affect proteins?

The many species of guenons differ in coloration, behavior, and diet, but they are remarkably similar overall. The nucleic acid DNA determines the inherited traits and behaviors of all species. How can one molecule result in the millions of species seen on Earth?

DNA: The Molecule of Heredity

Section Preview

Objectives

Analyze the structure of DNA.

Determine how the structure of DNA enables it to reproduce itself accurately.

Key Terms

nitrogen base
double helix
replication

Can you imagine all of the information that could be contained in 1000 books the size of this textbook? Remarkably, at least this much information is carried by the genes of an organism, even one as simple as a bacterium. Scientists have found that the substance DNA, contained in genes, is responsible for this control.

Structure and Function of DNA

In Chapter 7, you learned that DNA is an example of a complex biological polymer called a nucleic acid. You also learned that nucleic acids are made up of many smaller subunits called nucleotides. The compo-

nents of a DNA nucleotide are deoxyribose (a simple sugar), a phosphate group, and a nitrogen base. A **nitrogen base** is an organic ring structure that contains one or more atoms of nitrogen. In DNA, there are four possible nitrogen bases—adenine (A), guanine (G), cytosine (C), and thymine (T). Thus, in DNA,

Figure 13.1

DNA Nucleotides

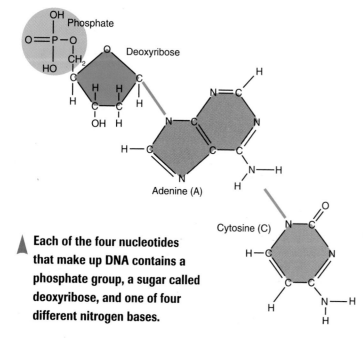

▲ Each of the four nucleotides that make up DNA contains a phosphate group, a sugar called deoxyribose, and one of four different nitrogen bases.

▼ Adenine and guanine (top) are double-ring bases called purines. Cytosine and thymine (bottom) are smaller, single-ring bases called pyrimidines.

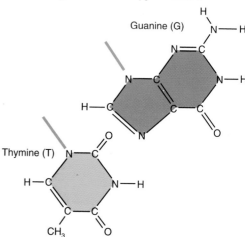

there are four possible nucleotides as shown in *Figure 13.1*—each containing one of the four bases just mentioned.

Chains of nucleotides

In nucleic acids, nucleotides do not exist as individual molecules. In DNA, nucleotides combine by condensation to form two long chains, producing one large molecule, *Figure 13.2.* Each chain of nucleotides contains single nucleotides connected to each other. The two chains

What does DNA look like?

DNA is pictured as a coiled helix in most diagrams. But what does the substance DNA really look like? You can extract DNA from salmon testes using the following procedure.

Procedure

1. Add 1 mL of cold ethanol to the cold test tube of salmon testes preparation given to you by your teacher. Add the ethanol by letting it run slowly down the inside of the test tube. A layer of ethanol should float on the salmon testes preparation.

2. Use a stirring rod to stir at the point where the two liquids come together. The DNA will cling to the stirring rod.

3. Withdraw the stirring rod slowly, pulling the thread of DNA with it. When the stirring rod is just above the mouth of the tube, start winding the thread of DNA onto the rod.

Analysis

1. How does the physical appearance of DNA relate to its chemical structure?

2. Why do you think that testes are used as a dependable source of DNA?

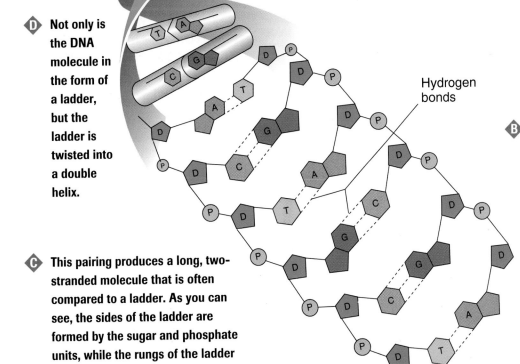

D Not only is the DNA molecule in the form of a ladder, but the ladder is twisted into a double helix.

Hydrogen bonds

B The two chains of nucleotides in a DNA molecule are held together by hydrogen bonds between the bases. The two bases on the same rung of the DNA are referred to as a base pair. In DNA, cytosine always forms hydrogen bonds with guanine, and thymine always bonds with adenine.

C This pairing produces a long, two-stranded molecule that is often compared to a ladder. As you can see, the sides of the ladder are formed by the sugar and phosphate units, while the rungs of the ladder are the pairs of bases.

A In each chain of nucleotides, the sugar of one nucleotide is joined to the phosphate group of the next nucleotide by a covalent bond.

Figure 13.2

The Structure of DNA

The Double Helix

It began as an unlikely scientific partnership. James Watson, a 24-year-old American biochemist, hated to talk to people. In fact, his ambition was to isolate himself on a wildlife refuge. His reasoning: "Birds, no people." Then, in 1951, Watson met English physicist Francis Crick, who loved to talk. Within minutes, the two were deep in conversation about DNA.

From photo to model Watson and Crick's scientific partnership developed for more than two years as they carried out an intensive study of DNA at Cambridge University in England. Words weren't the only means by which the two scientists exchanged ideas. Rosalind Franklin, a crystallographer at King's College in London, had produced a photo made by passing X rays through DNA and recording the interference pattern that resulted. By studying the photo, they deduced that DNA must have a spiral structure. They began trying to build a model that would produce the image they saw.

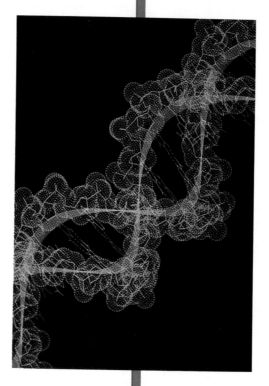

A new theory In February 1953, Watson and Crick's newest model resembled a rope ladder that had been given a twist. The pair of scientists was convinced that DNA had the form of a double helix with chains running in opposite directions. Two months later, the British science journal *Nature* published their article, which contained a tentative-sounding sentence: "It has not escaped our notice that the specific pairing we have postulated immediately suggested a possible copying mechanism for the genetic material." Watson and Crick were awarded a Nobel prize in 1962 for their work. It is thought that Franklin would have shared the prize with them but for her untimely death four years earlier.

CONNECTION TO Biology

Many scientists were working on the DNA puzzle at the same time. Most of them preferred to work alone in the laboratory. How do you think Watson and Crick's working style might have given them an advantage?

are joined by hydrogen bonds between the bases, resulting in a structure that is like a twisted ladder. Whenever something is twisted like a coiled spring, the shape is called a helix. Because DNA is composed of two strands, its shape is called a **double helix.**

The importance of DNA

The genetic material of living things is made of DNA. Thus, if you compare the chromosomes of two different organisms, you will find that both contain DNA made up of nucleotides with adenine, thymine, guanine, and cytosine. How can organisms be different from each other if their genetic material is made of the same molecules? They are different because the order of nucleotides in the two organisms is different. For example, a squirrel differs from a rosebush because the order of nucleotides in its DNA is different.

It is this sequence of nucleotides that forms the genetic information of an organism. The more closely related two organisms are, the more alike the order of nucleotides in their DNA will be. Scientists use this information to determine evolutionary relationships among organisms. The more closely related two organisms are, the more alike their sequence of nucleotides will be. The Human Genome Project is working to determine the sequence for the human species.

Replication of DNA

Of course, there are mechanisms to pass exact copies of all genetic information from one cell to the next. During cell division, the mechanism is mitosis. During production of gametes for sexual reproduction, the mechanism is meiosis. Therefore, there must be a means of making duplicate copies of DNA.

Each organism on Earth today has a nucleotide sequence in its DNA that was obtained from its parents. This sequence goes back some 3.5 billion years to the first organisms on Earth.

Why must DNA replicate?

You learned in Chapter 11 that every time a cell divides, it must first make a copy of its chromosomes. In this way, each new cell can have a complete set of chromosomes. In fact, every time a cell reproduces by mitosis or a gamete is formed by meiosis, DNA is copied in a process called **replication.** Without replication, species could not survive and individuals could not successfully grow and reproduce. *Figure 13.3* shows bacterial DNA replicating.

How DNA replicates

Recall that a DNA molecule is composed of two strands, each containing a sequence of nucleotides. As you know, an adenine on one strand always pairs with a thymine on the other strand. Similarly, guanine always pairs with cytosine. Therefore, if you knew the order of bases on one strand, you could predict the sequence of bases on the complementary strand. In fact, part of the process of DNA replication is done in just the same way. During replication, each strand serves as a pattern to make a new DNA molecule.

What evidence does chemical analysis provide about the structure of DNA?

Much of the early research on the structure and composition of DNA was done by carrying out chemical analyses. The data from these experiments provide evidence of a relationship among the nitrogen bases of DNA.

Analysis

Examine the table below. Compare the amounts of adenine, guanine, cytosine, and thymine found in the DNA of each of the cells studied.

Percent of Each Base in DNA Samples				
Source of Sample	A	G	C	T
Human liver	30.3	19.5	19.9	30.3
Human thymus	30.9	19.9	19.8	29.4
Herring sperm	27.8	22.2	22.6	27.5
Yeast	31.7	18.2	17.4	32.6

Thinking Critically

Compare the amounts of A, T, G, and C in each kind of DNA. Why do you think the relative amounts are so close in human liver and thymus cells? How do the relative amounts of bases in herring sperm compare with the relative amounts of bases in yeast? What fact can you state about the overall composition of DNA, regardless of source?

Figure 13.3

This bacterial DNA is in the process of replication. The two loops are new copies. (The bottom loop is twisted into a figure-8 shape.) Replication is occurring at the intersections of the two loops, as indicated by the arrows.

Magnification: 200 000×

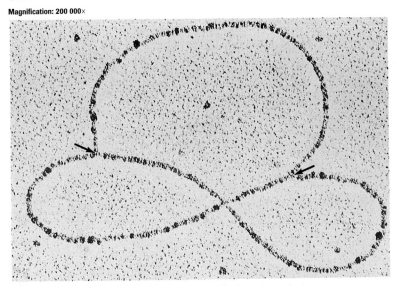

DNA replication begins as an enzyme breaks the hydrogen bonds between nitrogen bases that hold the two strands together, unzipping the DNA molecule. As the DNA continues to unzip, free nucleotides from the surroundings in the nucleus bond to the single strands by base pairing, as shown in *Figure 13.4*. Another enzyme bonds these new nucleotides into a chain.

This process continues until the entire molecule has been unzipped and replicated. As a result, each new strand formed is a complement of one of the original, or parent, strands. The result is the formation of two DNA molecules, each of which is identical to the original DNA molecule.

When all the DNA in all the chromosomes of the cell has been copied by replication, there are two copies of the organism's genetic information. In this way, the genetic makeup of an organism can be passed on to new cells during mitosis or to new generations through meiosis followed by sexual reproduction.

Figure 13.4

Replication of DNA

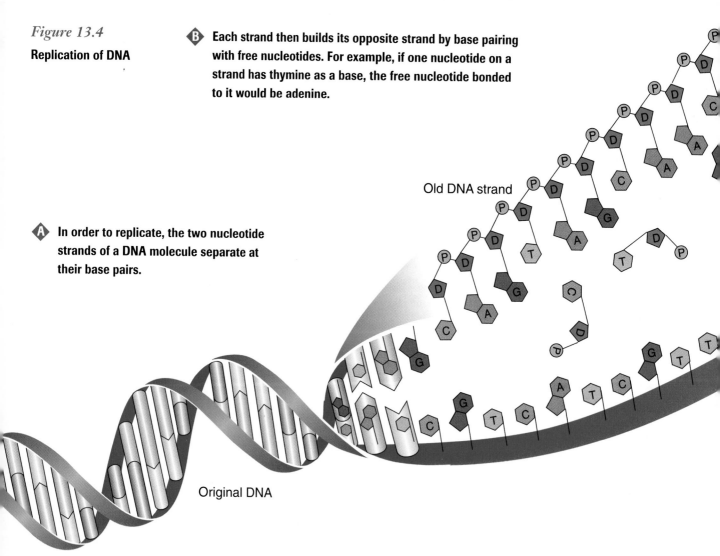

B Each strand then builds its opposite strand by base pairing with free nucleotides. For example, if one nucleotide on a strand has thymine as a base, the free nucleotide bonded to it would be adenine.

Old DNA strand

A In order to replicate, the two nucleotide strands of a DNA molecule separate at their base pairs.

Original DNA

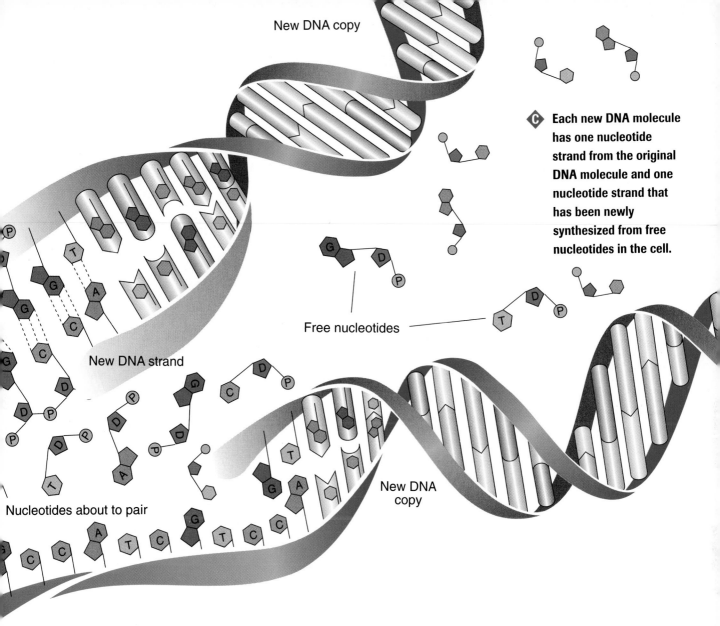

New DNA copy

New DNA strand

Free nucleotides

Nucleotides about to pair

New DNA copy

C Each new DNA molecule has one nucleotide strand from the original DNA molecule and one nucleotide strand that has been newly synthesized from free nucleotides in the cell.

13.2 From DNA to Protein

orse code is a way of communicating that was developed in the 19th century. Even though more sophisticated means have since been developed, Morse code is sometimes used today. Morse code uses a pattern of dots and dashes to represent letters of the alphabet. In this way, long sequences of dots and dashes can produce an infinite number of different messages. Living organisms have their own code, the genetic code, in which the sequence of nucleotides in DNA represents information.

Genes and Proteins

What is the message contained within the DNA code? The answer has to do with proteins. Recall from Chapter 7 that proteins are complex polymers of amino acids. Different proteins have different functions. Some proteins, such as enzymes, control chemical reactions that perform key life functions such as building ATP or digesting food. Other proteins are produced to build and repair cell structures, such as microtubules or transport proteins in membranes. In general, proteins determine the structure and function of organisms. Perhaps you have wondered how the cell was able to make these complex molecules.

Since the time of Mendel, a lot of research has gone into figuring out what genes are and how they function.

A major step toward answering this question occurred in the 1920s when it was hypothesized that each gene is somehow responsible for the production of one protein. We also know that genes are composed of DNA. Today, it's estimated that a human cell contains from 50 000 to 100 000 genes, each of which is made up of DNA. So, the message of the DNA code is information for building proteins.

The DNA Code

How do genes code for proteins? When scientists worked out the structure of DNA, it became clear to them that the sequence of nitrogen bases along one of the two strands was a code for the synthesis of proteins. This code is known as the genetic code.

The genetic code

As you know, proteins are built from chains of smaller molecules called amino acids. There are 20 different amino acids, but DNA contains only four types of bases. How can these bases form a code for proteins?

Scientists working out the code realized that a single base can't represent a single amino acid because that system would code for only four different amino acids. Similarly, a sequence of two bases—such as AT, GC, or AG—yields only 16 possible combinations. Experiments have shown that a sequence of three bases provides more than the 20 combinations needed to code for all amino acids. Each set of three nitrogen bases representing an amino acid is known as a **codon.** Because a sequence of three nitrogen bases forms the code for an amino acid, the DNA code is often called the triplet code.

The order of nitrogen bases in DNA can determine the type and order of amino acids in a protein. Sixty-four combinations are possible when a sequence of three bases is used; thus, 64 different codons are in the genetic code, *Table 13.1.* Of these, 61 code for amino acids, and the remaining three are signals to stop the synthesis of a polypeptide chain. As you can see, more than one codon can code for the same amino acid. However, for any one codon, there can be only one amino acid.

Table 13.1 The DNA Code

First Base in Codon	Second Base in Codon				Third Base in Codon
	A	G	T	C	
A	phenylalanine	serine	tyrosine	cysteine	A
	phenylalanine	serine	tyrosine	cysteine	G
	leucine	serine	stop	stop	T
	leucine	serine	stop	tryptophan	C
G	leucine	proline	histidine	arginine	A
	leucine	proline	histidine	arginine	G
	leucine	proline	glutamine	arginine	T
	leucine	proline	glutamine	arginine	C
T	isoleucine	threonine	asparagine	serine	A
	isoleucine	threonine	asparagine	serine	G
	isoleucine	threonine	lysine	arginine	T
	methionine (start)	threonine	lysine	arginine	C
C	valine	alanine	aspartate	glycine	A
	valine	alanine	aspartate	glycine	G
	valine	alanine	glutamate	glycine	T
	valine	alanine	glutamate	glycine	C

Universality of the genetic code

The genetic code was figured out by studying the DNA of the bacterium *Escherichia coli*. Yet the code is exactly the same in humans and virtually every other known organism. The code is said to be universal because the codons represent the same amino acids in all organisms. The universality of this code is powerful evidence that all organisms alive today shared a common ancestor billions of years ago.

Transcription—From DNA to RNA

As you have read in Chapter 8, proteins are made on ribosomes in the cytoplasm of the cell. Yet DNA, which contains the genetic code, is found only in the nucleus. How is information from the genetic code brought to the ribosomes for protein synthesis? Scientists began to understand how proteins are made when a second kind of nucleic acid, RNA (ribonucleic acid), was discovered.

RNA structure

Like DNA, RNA is a nucleic acid. However, RNA structure differs from DNA structure in three ways, as shown in *Figure 13.5*. First, RNA is usually composed of a single strand of nucleotides, rather than a double strand as in DNA. RNA also contains a different type of sugar molecule, ribose, instead of deoxyribose. Like DNA, RNA also contains four nitrogen bases, but rather than thymine, RNA contains a similar base called uracil (U).

Figure 13.5

The three chemical differences between DNA and RNA are shown here.

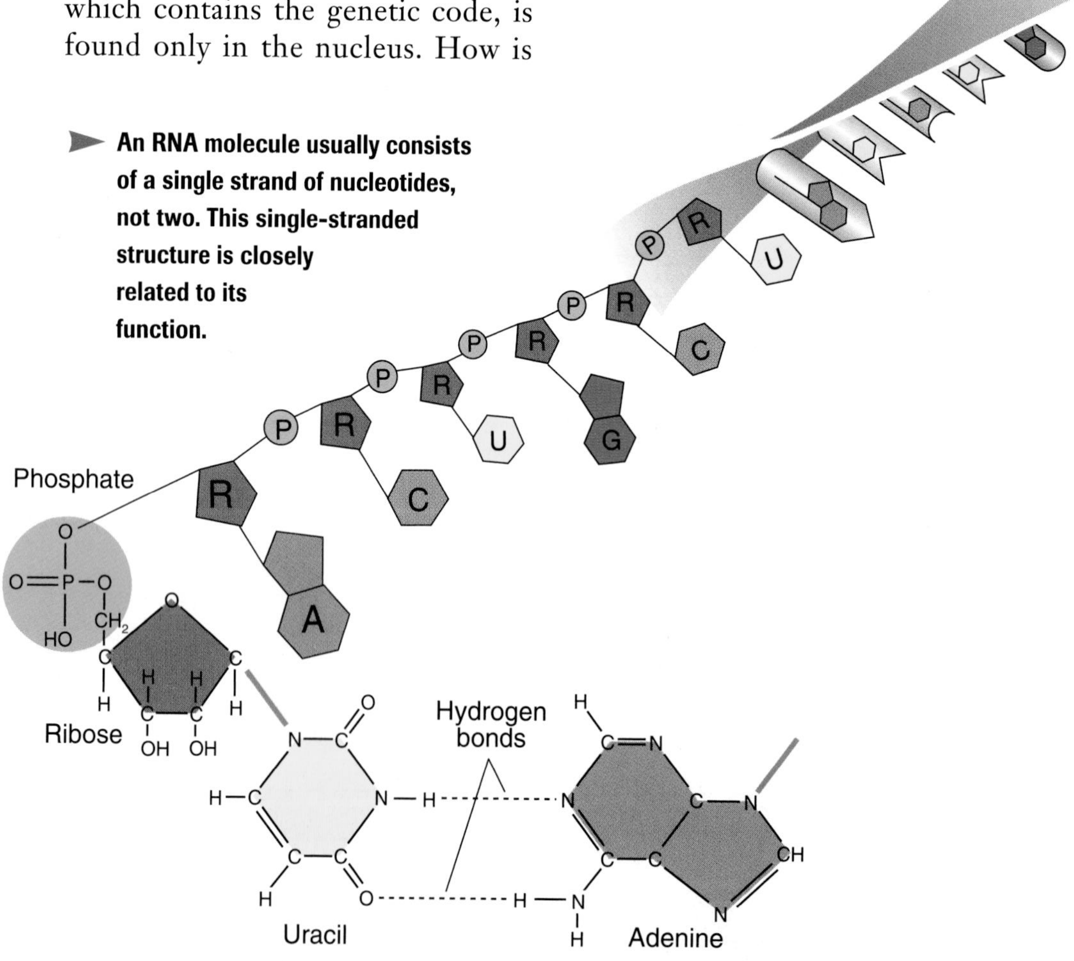

▶ An RNA molecule usually consists of a single strand of nucleotides, not two. This single-stranded structure is closely related to its function.

▶ The sugar in RNA is ribose sugar, rather than the deoxyribose sugar of DNA.

Phosphate

Ribose

▲ RNA contains the nitrogen base uracil (U) instead of thymine (T). Uracil pairs with adenine just as thymine does in DNA.

Making RNA by transcription

Today, we know that RNA is the form in which information moves from DNA in the nucleus to the ribosomes in the cytoplasm. Enzymes make an RNA copy of a DNA strand in a process called **transcription,** *Figure 13.6.* The process of transcription is similar to the process of DNA replication. The main difference is that the process results in the formation of one single-stranded RNA molecule, rather than one new, double-stranded molecule of DNA. This RNA copy that carries information from DNA out into the cytoplasm of the cell is called **messenger RNA (mRNA).**

Messenger RNA carries the information for making a protein chain to the ribosomes, where proteins are synthesized. Some portions of DNA code for the RNA that makes up ribosomes. This type of RNA is called **ribosomal RNA (rRNA).** Recently, scientists have shown that rRNA helps to produce enzymes needed to bond amino acids together during protein synthesis.

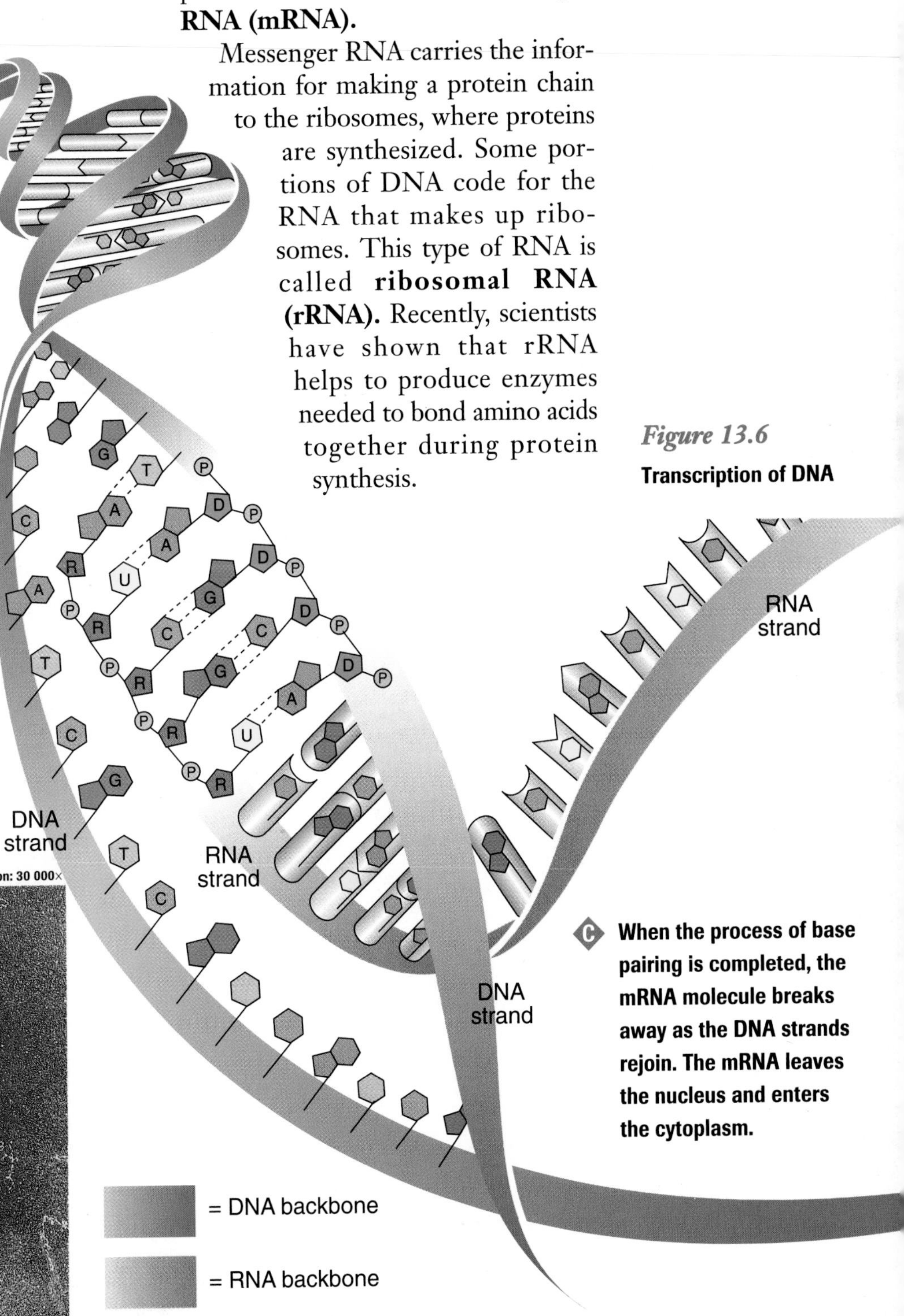

A The process of transcription begins as enzymes unzip the molecule of DNA, just as they do during DNA replication.

B As the DNA molecule unzips, free RNA nucleotides pair with complementary DNA nucleotides on one of the DNA strands. Thus, if a sequence of codons on the DNA strand were AGC TAA CCG, the sequence of codons on the RNA strand would be UCG AUU GGC.

DNA strand

RNA strand

Magnification: 30 000×

RNA strand

Figure 13.6

Transcription of DNA

RNA strand

DNA strand

C When the process of base pairing is completed, the mRNA molecule breaks away as the DNA strands rejoin. The mRNA leaves the nucleus and enters the cytoplasm.

= DNA backbone

= RNA backbone

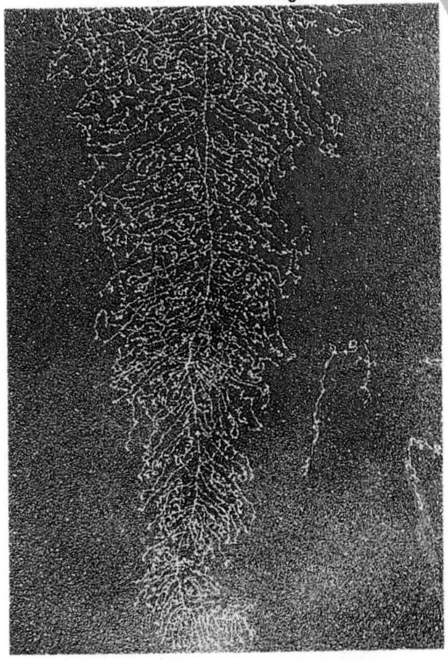

◀ Multiple transcriptions from a single DNA molecule.

13.2 From DNA to Protein **319**

BioLab

RNA Transcription

DNA is the substance that makes up genes. DNA passes its information into the cytoplasm of the cell by coding for another nucleic acid, messenger RNA. This mRNA is made from the DNA pattern by base pairing in the process of transcription. In this activity, you will demonstrate the process of transcription through the use of paper DNA and mRNA models.

PREPARATION

Problem
How does the order of bases in DNA determine the order of bases in mRNA?

Objective
In this Biolab, you will:
- **Determine** how the order of bases in DNA determines the order of bases in mRNA.

Materials
construction paper, 5 colors
scissors
clear tape
pencil

Parts for DNA Nucleotides

Deoxyribose

Phosphoric acid

Thymine

Cytosine

Guanine

Adenine

Extra Parts for RNA Nucleotides

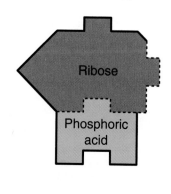

Ribose

Phosphoric acid

Uracil

1. Copy the illustrations of the four different DNA nucleotides onto your construction paper. Make sure that each different nucleotide is on a different color of construction paper. In all, you should make ten copies of each nucleotide.

2. Using scissors, carefully cut out the shapes of each nucleotide.

3. Using any order of nucleotides that you wish, construct a double-stranded DNA molecule. If you need more nucleotides, copy them as before.

4. Fasten your molecule together using clear tape. Do not tape across base pairs.

5. As in step 1, copy the illustrations of A, G, and C nucleotides onto the same colors of construction paper. Use the fifth color of construction paper to make copies of uracil nucleotides.

6. With scissors, carefully cut out the nucleotide shapes.

7. With your DNA molecule in front of you, demonstrate the process of transcription by first pulling the DNA molecule apart between the base pairs.

8. Using only one of the strands of DNA, begin matching mRNA nucleotides with the exposed bases on the DNA model.

9. When you are finished, tape your new mRNA molecule together using cellophane tape.

1. **Observing and Inferring** Does the mRNA model more closely resemble the DNA strand from which it was transcribed or the complementary strand that wasn't used? Explain your answer.

2. **Recognizing Cause and Effect** Explain how the structure of DNA enables the molecule to be easily transcribed. Why is this important for genetic information?

3. **Relating Concepts** Why is RNA important to the cell? How does an mRNA molecule carry information from DNA?

Going Further

Journal Report
Do library research to find out more about how the bases in DNA were identified and how the base pairing pattern was determined.

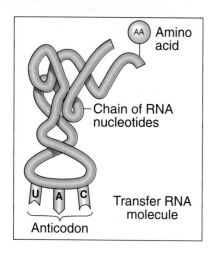

Figure 13.7

A tRNA molecule is composed of about 75 nucleotides. Each tRNA recognizes only one amino acid. The amino acid becomes bonded to the top of the tRNA molecule. Located on the bottom of the tRNA molecule are three nitrogen bases, called an anticodon, that pair up with mRNA codons during translation.

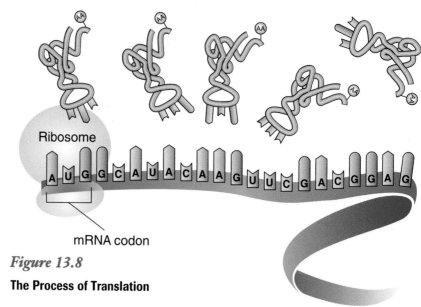

Figure 13.8

The Process of Translation

Ⓐ **As translation begins, the first codon of the mRNA strand attaches to a ribosome. Then, tRNA molecules, each carrying a specific amino acid, approach the ribosome. When the tRNA anticodon pairs with the mRNA codon, the two molecules join and the tRNA molecule remains there.**

Translation—From RNA to Protein

When mRNA is synthesized, it carries a complementary copy of the DNA code for a protein chain. How is mRNA language used to synthesize a sequence of amino acids?

The role of transfer RNA

The process of converting the information in a sequence of nitrogen bases in mRNA into a sequence of amino acids that make up protein is known as **translation.** Translation, which occurs on ribosomes, involves a third kind of RNA. If proteins are to be built, the 20 different amino acids dissolved in the cytoplasm must be brought to the ribosomes. This is the role of transfer RNA, *Figure 13.7.* **Transfer RNA (tRNA)** brings amino acids to the ribosomes so they can be assembled into proteins. How does the cell "know" which tRNA molecules carry the proper amino acids? The answer again involves base pairing.

Translating the mRNA code

Correct translation of the code depends upon joining of each mRNA codon with the anticodons of the proper tRNA molecules. *Figure 13.8* shows how the message in an mRNA molecule is translated into a protein. The end result of translation is the formation of the large variety of proteins that make up the structure of organisms and help them to function.

The importance of proteins

Now that you know proteins are made by bonding together of amino acids during translation of the mRNA copy of the DNA code, can you guess how many types of proteins can be made in this way? The fact is, an almost infinite variety of proteins can be made from the same 20 amino acids. Remember, although there are only 20 amino acids, they can occur in any order and in any number. It is the sequence of amino acids in a protein that determines the characteristics of that protein.

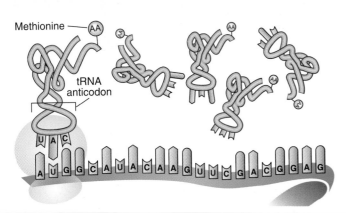

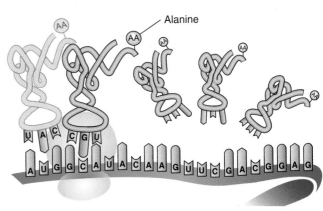

Alanine

B Often, the first codon on mRNA is AUG, which codes for the amino acid methionine. AUG signals the start of protein synthesis. When this signal is given, the mRNA slides along the ribosome to the next codon.

C Again, a new tRNA molecule carrying an amino acid will pair with the mRNA codon.

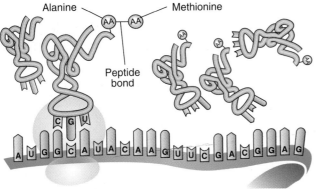

D When the first and second amino acids are in place, an enzyme joins them by forming a peptide bond between them.

Stop codon

E As the process continues, a chain of amino acids is formed until the ribosome reaches a stop codon on the mRNA strand.

Have you ever copied a phone number incorrectly? Perhaps you've written a 2 when it should have been a 3. What are some possible consequences of changing digits in numbers? Sometimes there is little effect, but you might find that you can't call your doctor in an emergency. Mistakes in the DNA code can produce similar results. Sometimes, there is no effect on an organism, but often mistakes in DNA can cause serious consequences for individual organisms.

Mutation—A Change in DNA

The cell processes that copy genetic material and pass it from one generation to the next are usually accurate. Accuracy is important to ensure the genetic continuity of both new cells and offspring. However, sometimes mistakes can occur in the genetic material. Any mistake or change in the DNA sequence is called a **mutation.**

The effects of point mutations

Consider what might happen if an incorrect amino acid were inserted into a growing protein chain during translation of the DNA code. It might affect the synthesis of the entire molecule, right? Such a problem can occur if a point mutation arises. A **point mutation** is a change in a single base pair in DNA.

A simple analogy can illustrate point mutations. Read the sentences below to see what happens when a single letter in a sentence is changed.

THE DOG BIT THE CAT.
THE DOG BIT THE CAR.

As you can see, changing a single letter changes the meaning of this sentence. Similarly, a change in a single nitrogen base can change the entire structure of a protein. The top diagram in *Figure 13.9* shows what can happen with a point mutation.

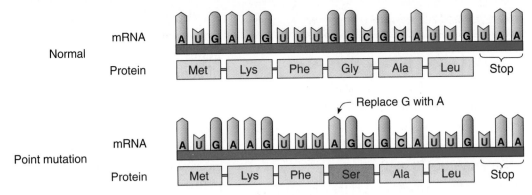

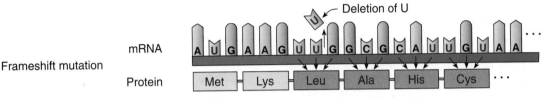

In this point mutation, the base guanine was changed to adenine. This change in the codon caused the insertion of serine, rather than glycine, into the growing amino acid chain. Sometimes, the errors caused by point mutations don't interfere with protein function, but often the effect is disastrous.

Proteins that are produced as a result of frameshift mutations seldom function properly because such a mutation usually changes several amino acids. Adding or deleting one base of a DNA molecule will change every amino acid in the protein after the addition or deletion.

Figure 13.9

Gene Mutations

Frameshift mutations

When the mRNA strand moves across the ribosome, a new amino acid is added to the protein for every codon on the mRNA strand. What would happen if a single base were lost from a DNA strand? This new sequence with the deleted base would be transcribed into mRNA. But then, the mRNA would be out of position by one base. As a result, every codon after that base would be different, as shown at the bottom of *Figure 13.9*. This mutation would cause nearly every amino acid in the protein after the deletion to be changed. The same effect would also result from the addition of a single base. A mutation in which a single base is added or deleted from DNA is called a **frameshift mutation.**

Chromosomal Mutations

Changes may occur at the level of chromosomes as well as in genes. Mutations to chromosomes may occur in a variety of ways. For example, sometimes parts of chromosomes are broken off and lost during mitosis or meiosis. Often, chromosomes break and then rejoin incorrectly. Sometimes, the parts join backwards or even to the wrong chromosome. These changes in chromosomes are called **chromosomal mutations.**

Effects of chromosomal mutations

Chromosomal mutations occur in all living organisms, but they are especially common in plants. Such mutations may be serious because they affect the distribution of genes

How do gene mutations affect proteins?

Gene mutations often have serious effects on proteins. In this activity, you will demonstrate how such mutations affect protein synthesis.

Procedure

1. Copy the following base sequence of one strand of an imaginary DNA molecule: AATGCCAGTGGTTCGCAC

2. Below the first strand, write the base sequence of the complementary DNA strand.

3. Then, write the base sequence that would appear on an mRNA strand after transcription.

4. Use *Table 13.1* on page 317 to determine the order of amino acids in the resulting protein fragment.

5. If the fourth base in the original DNA strand were changed from G to C, how would this affect the resulting protein fragment?

6. If a G were added to the original DNA strand after the third base, what would the resulting mRNA look like? How would this addition affect the protein?

Analysis

1. Which change in DNA was a point mutation? Which was a frameshift mutation?

2. In what way did the point mutation affect the protein?

3. How did the frameshift mutation affect the protein?

to gametes during meiosis. Gametes that should have a complete set of genes may end up with extra copies of some genes or a complete lack of certain genes.

Few chromosome mutations are passed on to the next generation because the zygote usually dies. In cases where the zygote develops, the mature organism is often sterile and thus incapable of producing offspring. The most important of these mutations—deletions, insertions, inversions, and translocations—are illustrated in *Figure 13.10.*

Errors in Disjunction

Many chromosome mutations result from the failure of chromosomes to separate properly during meiosis. Recall that during meiosis I, one chromosome from each homologous pair moves to each pole of the cell. Occasionally, an error occurs in which both chromosomes of a homologous pair move to the same

Figure 13.10

Chromosomal Mutations

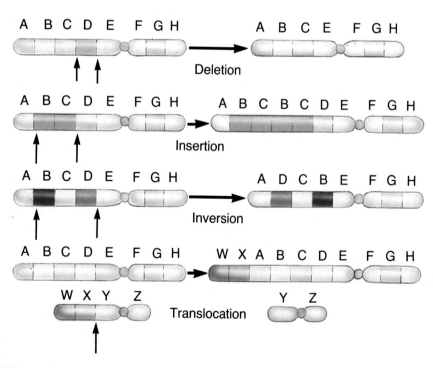

◄ *Deletions* occur when part of a chromosome is left out.

◄ *Insertions* occur when a part of a chromatid breaks off and attaches to its sister chromatid. The result is a duplication of genes on the same chromosome.

◄ *Inversions* occur when part of a chromosome breaks out and is reinserted backwards.

◄ *Translocations* occur when part of one chromosome breaks off and is added to a different chromosome.

pole of the cell. The failure of homologous chromosomes to separate properly during meiosis is called **nondisjunction.**

Trisomy, triploidy, and monosomy

In one case of nondisjunction, two kinds of gametes are formed as a result of nondisjunction. One has an extra chromosome, and the other is missing a chromosome. The effects of nondisjunction are often seen when gametes fuse in fertilization. For example, when a gamete with an extra chromosome is fertilized by a normal gamete, the zygote will have an extra chromosome. This condition is called **trisomy.** In humans, if a gamete with an extra chromosome number 21 is fertilized by a normal gamete, the resulting zygote has 47 chromosomes instead of 46. This zygote will develop into a baby with Down syndrome.

Another case of nondisjunction involves a total lack of separation of homologous chromosomes. When this happens, a gamete inherits a complete diploid set of chromosomes. When a gamete with an extra set of chromosomes is fertilized by a normal haploid gamete, offspring have three sets of chromosomes and are triploid. This condition is rare in animals but frequently occurs in plants. Often, the flowers and fruits of these plants are increased in size. Many triploid plants, such as the banana shown in *Figure 13.11*, are of great commercial value.

While organisms with extra chromosomes often survive, organisms lacking one or more chromosomes usually do not. When a gamete with a

Gene-control Operons

Is the information in genes always expressed, or can genes be turned on and off depending upon the needs of a cell? The most extensively studied system of gene regulation is the mechanism for lactose sugar metabolism in *Escherichia coli.* In order to digest lactose, *E. coli* has to produce two proteins. One of these proteins carries lactose into the cell, and the other is an enzyme that digests lactose. The genes that code for these proteins are always present in *E. coli,* but the proteins are made only when lactose is present. The expression of these genes is controlled by a genetic mechanism called the *lac* operon. How does this system work?

The* lac *operon Genes that code for proteins are called structural genes. In the diagram of the *lac* operon, z is the structural gene that codes for the enzyme that digests lactose; y is the structural gene that codes for the protein that carries lactose across the cell membrane. Next to them on the *E. coli* chromosome are two overlapping sections of DNA, the operator (o) and the promoter (p). This DNA does not code for protein. Instead, these sections determine when the structural genes, z and y, will be transcribed into mRNA. R is a regulatory gene that produces a repressor protein, which normally binds to the operator. This protein prevents the transcription enzyme from attaching to the promoter.

Switch on, switch off When lactose is present, molecules of lactose bind to the repressor protein causing it to leave the operator. Then the transcription enzyme can attach and proceed to transcribe genes z and y into mRNA. The mRNA is then translated into the proteins required to carry lactose into cells and digest it.

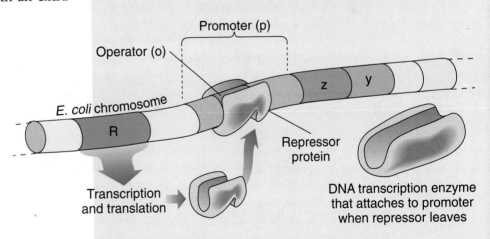

Promoter (p)
Operator (o)
E. coli chromosome
R
z   y
Repressor protein
Transcription and translation
DNA transcription enzyme that attaches to promoter when repressor leaves

Thinking Critically
What do you think is the adaptive value of a regulatory mechanism like this one?

Figure 13.11

The banana plant is an example of a triploid organism.

missing chromosome is fertilized by a normal gamete, the resulting zygote will lack a chromosome. This condition is called **monosomy.** Examples include human females with only a single X chromosome. Zygotes with other types of monosomy usually do not survive.

Causes of Mutations

Mutations are generally random events, and, as you will learn in Chapter 18, errors in DNA provide the variation that enables species to evolve. Mutations that occur at random are called spontaneous mutations.

However, it is known that many environmental agents also cause mutation. Exposure to X rays, ultraviolet light, radioactive substances, or certain chemicals can cause changes in DNA. Mutations often result in sterility or the lack of normal development in an organism. If these mutations occur in human gametes, they can cause birth defects. If they occur in body cells, the mutations can lead to cancer.

Connecting Ideas

The discovery that genes are made of DNA, along with the explanation of how DNA determines inherited traits, is perhaps the greatest achievement of 20th-century biology. Unraveling the mystery of genes and the DNA within them has enabled scientists to better understand how traits are passed from one generation to the next, and has helped them explain patterns of inheritance that do not follow Mendelian principles. What are these patterns? How can they be explained by the properties of genes?

Section Review

Understanding Concepts
1. What is a mutation?
2. Describe how point mutations and frameshift mutations affect the synthesis of proteins.
3. Describe four kinds of chromosomal mutations.

Thinking Critically
4. Why do you think a low level of mutation might be advantageous to a species, while a high level of mutation might be disadvantageous?

Skill Review
5. **Interpreting Scientific Illustrations** Make a chart illustrating and summarizing the different kinds of gene and chromosomal mutations. For each kind of mutation, describe the mutation and discuss one possible effect of the mutation on an organism. For more help, refer to Thinking Critically in the *Skill Handbook*.

Boy or Girl?

Sex Selection Now Available

THE DAILY TIMES   SEPTEMBER 24, 1994

Fertility clinics now offer medical procedures to couples who are interested in choosing the sex of their children. Medical ethicists worry that widespread use of these methods could have serious social consequences.

Traditionally, advances in family-planning techniques create ethical dilemmas when they become available and affordable to the general public. Now there are sex-selection clinics that offer prospective parents the chance to choose the sex of their offspring. In the future, it may be possible to select other traits as well. Is the prospect of "designer" children and families becoming a reality?

Sex-selection techniques   The most popular method of sex selection is a high-tech medical procedure in which sperm cells are separated into X-chromosome-bearing (female) and Y-chromosome-bearing (male) sperm cells. The woman is then inseminated with sperm that will produce the desired gender. Another more controversial method involves fertilizing eggs in a laboratory,

genetically determining the sex of the embryos, and then implanting the selected embryo in the woman's uterus. This method was designed for families at risk for sex-linked genetic diseases such as hemophilia and Duchenne muscular dystrophy.

Techniques for sex selection are now available in fertility clinics across America. What harm could it do to allow parents to choose the sex of their children?

Different Viewpoints

Experts believe that, given the opportunity, many parents would choose what is considered by many to be the ideal household—two offspring, with the boy born first, the girl second. Some people argue that women are now raised to be second to men, and, with this choice, they will be born into such a status as well.

Another concern is that couples unable to afford the expense of selection clinics would use routine prenatal genetic tests, such as ultrasound, to determine gender and then abort the fetus if it were of the "wrong" sex.

Designer children   Supporters of sex selection argue that parents who choose the sex of their children are more likely to treat them better than parents who do not choose. These supporters also claim that couples looking to "round out" their families with either a daughter or a son would not have to endure unwanted pregnancies. Opponents argue that people would use such techniques for frivolous reasons, leading to further erosion of societal values.

INVESTIGATING   the Issue

1. **Research**   What reasons would people have to abuse sex-selection techniques?
2. **Debate**   Organize teams and argue the issue of sex selection in class. Each side should have well-researched, prepared arguments.

13.1 DNA: The Molecule of Heredity

- DNA, the genetic material of all organisms, is a large molecule composed of four different kinds of nucleotides. A DNA molecule consists of two strands of nucleotides with the sugars and phosphates on the outside and the bases paired by hydrogen bonding on the inside. The paired strands form a twisted-ladder shape called a double helix.
- Because adenine can pair only with thymine and guanine can pair only with cytosine, DNA can replicate itself with great accuracy. This process keeps the genetic information constant as it passes from cell to cell during cell division, and from one generation to the next during reproduction.

13.2 From DNA to Protein

- Genes are small sections of DNA code. Nearly every sequence of three bases in DNA codes for one amino acid in a protein, thus forming the genetic code.
- The order of nucleotides in DNA determines the order of nucleotides in messenger RNA in a process called transcription.

- Translation is a process through which the order of bases in messenger RNA codes for the order of amino acids in a protein.

13.3 Genetic Changes

- A mutation is a change in the DNA code. Mutations may affect only one gene, or they may affect whole chromosomes.
- Errors in disjunction result from the failure of homologous chromosomes to separate during meiosis. They result in gametes with too many or too few chromosomes.

Key Terms

Write a sentence that shows your understanding of each of the following terms.

chromosomal mutation	nitrogen base
codon	nondisjunction
double helix	point mutation
frameshift mutation	replication
messenger RNA	ribosomal RNA
monosomy mutation	transcription
	transfer RNA
	translation
	trisomy

Understanding Concepts

1. Name the four nitrogen bases found in DNA and describe how they pair. What holds the base pairs together?
2. Name the nitrogen base found in RNA that is not found in DNA. What base in DNA does it replace?
3. What determines the order of bases in a molecule of mRNA?
4. Which might cause a greater change in a protein—a point mutation or a frameshift mutation? Explain your answer.

5. How is your DNA the same as the DNA of a jellyfish? How is it different from the DNA of a jellyfish? How do these differences in DNA cause you to be different from a jellyfish?
6. How is DNA involved in determining the structure of a particular protein?
7. Explain how point mutations affect the synthesis of proteins.
8. Explain how nondisjunction affects the survival of zygotes.

Using Diagrams

9. The following diagram shows the structures of ribose and deoxyribose. How are these two molecules chemically different? How are they alike?

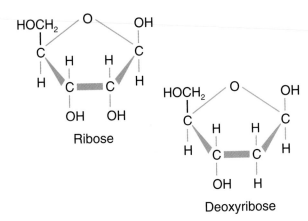

Ribose

Deoxyribose

Relating Concepts

10. Make a concept map that relates the following terms and phrases. Supply the appropriate linking words for your map.

DNA, nucleotide, mRNA, rRNA, tRNA, amino acid, protein, ribosome

Applying Concepts

11. Which would cause a greater change in a protein: the deletion of one base in the DNA that coded for it or the deletion of three sequential bases in the DNA that coded for it? Explain.

12. Describe a gene in terms of DNA.

13. Explain why the genetic code can't have sequences of two nucleotides instead of three nucleotides for each amino acid.

14. Many mutations result in changes in proteins. Can you give an example of a mutation in a gene that would result in no change in the protein for which the gene coded?

History Connection

15. Explain the statement of Watson and Crick that "specific pairing . . . suggested a possible copying mechanism for the genetic material."

Thinking Critically

Analyzing

16. Identify the type of mutation illustrated in each of the following diagrams.

(a)

(b)

(c)

Making Inferences

17. Biolab Suppose your DNA model began with the sequence ATCCTCGGA. Consider what would happen when the resulting mRNA was translated. Why would your sequence be an unlikely one for DNA?

Interpreting Data

18. Minilab The following is the sequence of bases on one strand of a DNA molecule: A A A T G C C A T C C G T C A
(a) Write the sequence of bases that makes up the complementary DNA strand.
(b) Draw this DNA molecule.
(c) What base sequence in mRNA would the first DNA strand code for?
(d) What sequence of amino acids would this mRNA code for?

Connecting to Themes

19. Evolution Explain how the universality of the genetic code is evidence that all organisms alive today evolved from a common ancestor in the past.

20. Systems and Interactions Describe the interactions between the cell organelles involved in transcription and translation of the genetic code.

14 Patterns of Heredity

Do you like to put puzzles together? Geneticists often must feel that they are solving puzzles. They need to explain how unseen factors (genes) cause visible effects (traits). After Mendel laid the groundwork explaining how traits pass from generation to generation, many of the questions that scientists had been asking could be answered. But new mysteries arose. Soon, what appeared to be exceptions to Mendel's laws were found. For instance, moss roses (*Portulaca grandiflora*) have flowers of different colors—pink, white, yellow, red, purple—all in the same species! Simple dominance and recessiveness of two alleles for flower color could not explain this trait.

Geneticists have extended Mendel's laws to explain unusual patterns of inheritance. Because of Mendel's work, it has been easier for others to solve new mysteries such as the inheritance of flower color in moss roses, thus fitting additional pieces into the puzzle of inheritance.

When Heredity Follows Different Rules

Section Preview

Objectives

Distinguish between incompletely dominant and codominant alleles.

Compare multiple allelic inheritance and polygenic inheritance.

Summarize how internal and external environments affect gene expression.

Key Terms

incomplete dominance
codominant alleles
multiple alleles
autosome
sex chromosome
sex-linked trait
polygenic inheritance

Variations in the patterns of inheritance explained by Mendel became known soon after his work was rediscovered. What do geneticists do when observed patterns of inheritance do not appear to follow Mendel's laws? They often employ a strategy of piecing together bits of a puzzle until the basis for the unfamiliar inheritance pattern is understood.

Kernel color in corn is inherited in a complex pattern.

Complex Patterns of Inheritance

Patterns of inheritance that are explained by Mendel's experiments are often referred to as simple Mendelian inheritance—the inheritance controlled by dominant and recessive paired alleles. However, many inheritance patterns are more complicated than those in garden peas studied by Mendel. As you will learn, most alleles are not simply dominant or recessive.

Incomplete dominance—Appearance of a third phenotype

When inheritance follows a pattern of complete dominance, heterozygous individuals and homozygous dominant individuals have the same phenotype. When traits are inherited in an **incomplete dominance** pattern, the phenotype of the heterozygote is intermediate between those of the two homozygotes. For example, if a homozygous red-flowered snapdragon plant is crossed with a homozygous white-flowered snapdragon plant, all of the F_1 offspring will have pink flowers, as shown in *Figure 14.1.* This intermediate form of the trait occurs because neither allele of the pair is completely dominant. Both alleles of the gene produce products, which combine to give a new trait.

Note that the segregation of alleles is the same as in simple Mendelian inheritance observed in garden peas. However, because neither allele is dominant, the plants of the F_1 generation all have pink flowers. When pink-flowered F_1 plants are crossed with each other, the offspring in the F_2 generation appear in a 1:2:1 phenotypic ratio of red to pink to white flowers. This result supports Mendel's law of independent assortment.

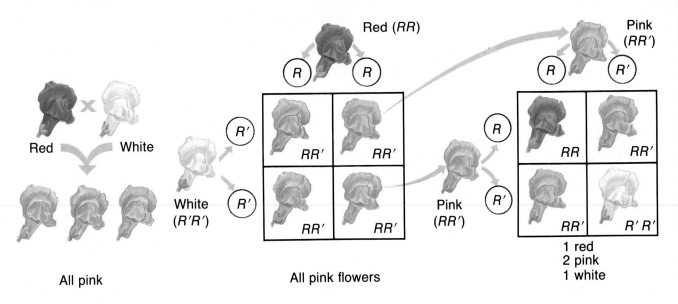

Red (RR)

R R

White (R'R')

R' R'

	R	R
R'	RR'	RR'
R'	RR'	RR'

All pink flowers

Pink (RR')

R R'

Red × White

Red   White

All pink

Pink (RR')

R R'

	R	R'
R	RR	RR'
R'	RR'	R'R'

1 red
2 pink
1 white

Figure 14.1

In snapdragons, the combined expression of both alleles for flower color produces a new phenotype—pink—illustrating incomplete dominance. For this reason, the letters *R* and *R'*, rather than *R* and *r*, are used here to indicate incomplete dominance. A Punnett square shows that the red snapdragon is homozygous for the allele *R*, and the white snapdragon is homozygous for the allele *R'*. All of the pink snapdragons are heterozygous, or *RR'*.

Codominance—Expression of both alleles

In chickens, black-feathered and white-feathered birds are homozygotes for the *B* and *W* alleles, respectively. Two different uppercase letters are used to represent the alleles in codominant inheritance. You might expect that heterozygous chickens, *BW*, would be black if the pattern of inheritance followed Mendel's law of dominance, or gray if the trait were incompletely dominant.

One of the resulting heterozygous offspring in a breeding experiment between a black rooster and a white hen is shown in *Figure 14.2*. Notice that the heterozygote is neither black nor gray. Instead, all of the offspring are checkered; some feathers are black and other feathers are white. In such situations, the inheritance pattern is said to be codominant. **Codominant alleles** cause the phenotypes of both homozygotes to be

produced in heterozygote individuals. In codominance, both alleles are expressed equally.

Multiple phenotypes from multiple alleles

Although each trait has only two alleles in the patterns of heredity you have studied thus far, it is common for more than two alleles to control a trait in a population. This is understandable when you recall that a new

Figure 14.2

When a certain variety of black chicken is crossed with a white chicken, all of the offspring are checkered, black and white, as a result of both feather colors being produced by codominant alleles.

Figure 14.3

In rabbits, a single gene that controls coat color has multiple alleles. An enzyme that activates the production of a pigment is controlled by the *C* allele. This enzyme is lacking in *cc* rabbits.

▲ The *c* allele produces a white coat.

▲ The dominant *C* allele produces the dark-gray coat.

▲ The *c^ch* allele results in a light-gray coat called chinchilla and is dominant to *c^h* and *c*.

▲ The allele *c^h* produces a white coat with black points, a Himalayan, and is dominant to *c*. In *c^h c^h* rabbits, the enzyme works only in cooler regions of the body—the ears, the feet, and the area around the nose.

Figure 14.4

The sex chromosomes are named for the letters they resemble. Why are the X and Y chromosomes not homologous?

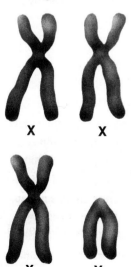

X   X

X   Y

allele is formed every time a mutation occurs in a nitrogen base somewhere within a gene. How can there be several different types of blood among humans? How can fruit flies have so many different eye colors? These questions can be answered because more than two alleles of a gene can exist.

Traits controlled by more than two alleles are said to have **multiple alleles.** Recall that a diploid individual can possess only two alleles of each gene. Therefore, multiple alleles can be studied only in populations. The rabbits in *Figure 14.3* show the effects of multiple alleles for coat color. Four alleles of a single gene govern coat color in rabbits, although each rabbit can have only two of these alleles. The number of alleles for any particular trait is not limited to four, and there are instances in which more than 100 alleles are known to exist for a single trait!

Sex-linked inheritance is determined by genes on sex chromosomes

Each chromosome carries genes for certain traits. Recall that in humans, the diploid number of chromosomes is 46, or 23 pairs. There are 22 pairs of matching homologous chromosomes called **autosomes.** The two chromosomes in a homologous pair of autosomes look exactly alike. The 23rd pair of chromosomes differs in males and females. These two chromosomes, which determine the sex of an individual, are called **sex chromosomes.** In humans, the chromosomes that control the inheritance of sex characteristics are indicated by the letters X and Y. If you are a female, XX, your 23rd pair of chromosomes look alike, as shown in *Figure 14.4.* However, if you are a male, XY, your 23rd pair of chromosomes look different. Traits controlled by genes located on sex chromosomes are called **sex-linked traits.**

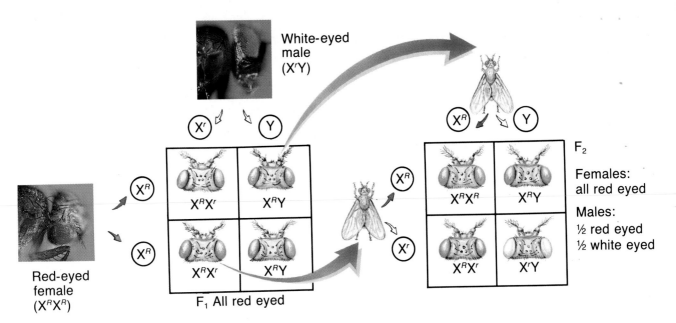

Figure 14.5

Morgan's first cross is shown here. Alleles for sex-linked traits are written as superscripts to the right of an X if the allele is on an X chromosome. If a Y chromosome has no matching allele, no superscript is used. Examples are $X^R X^r$ (female heterozygous for red eyes) and $X^r Y$ (male with white eyes—note only one recessive allele).

The first known example of sex-linked traits was discovered in *Drosophila* (fruit flies). In 1907, genetic researcher Thomas Hunt Morgan noticed that in one of his fruit fly cultures, one male had white eyes instead of the usual red eye color. Thinking that the white-eyed male was homozygous recessive for a new trait, he crossed the white-eyed male with a red-eyed female, as shown in *Figure 14.5*. As Morgan expected, all of the F_1 offspring had red eyes. But when the F_1 flies were allowed to mate among themselves, all of the F_2 female offspring and half of the male offspring had red eyes.

Later, when Morgan crossed white-eyed females and red-eyed males, all the females had red eyes and all the males had white eyes, as shown in *Figure 14.6*. The F_2 generation from this cross resulted in 50 percent of each sex with red eyes and 50 percent

Figure 14.6

If a white-eyed female is mated with a red-eyed male, the female offspring all have red eyes and the males all have white eyes. Note the four types of offspring in the F_2 generation.

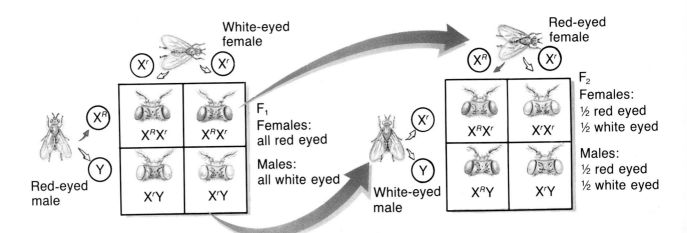

How is leaf width inherited?

Inheritance patterns are often not obvious. Every organism has thousands of genes. Each gene may have several alleles, and the inheritance of a trait may be governed by more than one gene. To determine an inheritance pattern, you must collect quantitative information about a trait and analyze the information carefully.

Procedure

1. Collect five leaves from each of ten trees of the same species.
2. Place the leaves into ten groups.
3. Measure to the nearest millimeter the width of each leaf at its widest point, and record the data.
4. Calculate the mean and range of the widths for each group of five leaves.
5. Make a graph similar to the one below.

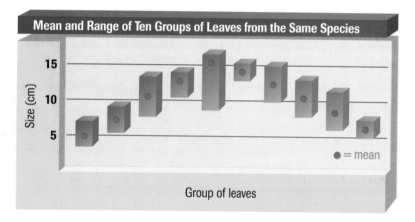

Mean and Range of Ten Groups of Leaves from the Same Species

Size (cm) / Group of leaves / ● = mean

Analysis

1. How did your data compare with those presented in the sample graph?
2. From your data, how do you hypothesize that leaf width is inherited?
3. Design an experiment that would help you determine whether your hypothesis is correct.

of each sex with white eyes. The only way Morgan could explain these results was that the gene for eye color was on the X chromosome. Today, geneticists know that most known sex-linked genes in humans are located on the X chromosome.

Polygenic inheritance explains a continuous range of phenotypes

Some traits, such as skin color and height in humans and cob length in corn, vary over a wide range. Such ranges occur because these traits are governed by many different genes. **Polygenic inheritance** is the inheritance pattern of a trait that is controlled by two or more genes. The genes may be on the same chromosome or on different chromosomes, and each gene may have two or more alleles. For simplicity, uppercase and lowercase letters are used to represent the alleles, as they are in Mendelian inheritance. Keep in mind, however, that the allele represented by an uppercase letter is not dominant. All heterozygotes are intermediate in phenotype.

In polygenic inheritance, each allele represented by an uppercase letter contributes a small, but equal, portion to the trait being expressed. The result is that the phenotypes usually show a continuous range of variability from the minimum value of the trait to the maximum value.

polygenic:
poly (GK) many
genos (GK) gene
Polygenic inheritance involves many genes.

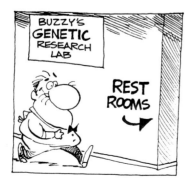

Suppose, for example, that stem length in a plant is controlled by three different genes: *A*, *B*, and *C*. Each gene is on a different chromosome and has two alleles, which are represented by an uppercase letter and a lowercase letter. Thus, each diploid plant has a total of six alleles for stem length. A plant that is homozygous for short alleles (*aabbcc*) at all three gene locations might grow to be only 4 cm tall. A plant that is homozygous for tall alleles (*AABBCC*) at all three gene locations would be 16 cm tall. The difference between the tallest possible plant and the shortest possible plant is 12 cm, or 2 cm per tall allele. You could say that each allele represented by an uppercase letter contributes 2 cm to the total height of the plant.

Suppose a 16-cm-tall plant were crossed with a 4-cm-tall plant. In the F_1 generation, all the offspring would be *AaBbCc*—intermediate (10 cm tall) in height. If they are allowed to interbreed, the F_2 offspring will show a broad range of heights. A Punnett square of this cross would show that 10-cm-tall plants are most often expected, and the tallest and shortest plants are seldom expected. Notice in *Figure 14.7* that, when these results are graphed, the shape of the graph confirms the prediction of the Punnett square.

ThinkingLab Interpret the Data

How many phenotypes result from polygenic inheritance?

Polygenic inheritance can be observed quantitatively, and it is possible to determine the number of gene pairs governing the trait if the number of genes is three or fewer. Once the number of gene pairs increases to four or more, the trait becomes continuous.

Analysis

The graphs below illustrate the number of possible phenotypes in an F_2 generation for one, two, and three pairs of genes that govern grain color in wheat. Colors range from red to white.

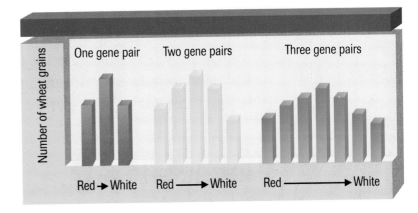

Thinking Critically

Analyze the graphs and calculate the number of phenotypes that would result from four gene pairs. Hint: It may be helpful to use letters to represent the genes and combinations.

Figure 14.7

Polygenic inheritance occurs when many genes interact to produce a single trait. In this theoretical example, three genes each have two alleles that contribute to the trait. Each dominant allele contributes 2 cm to the height of the plant. When the distribution of plant heights is graphed, a bell-shaped curve is formed. Intermediate heights occur most often.

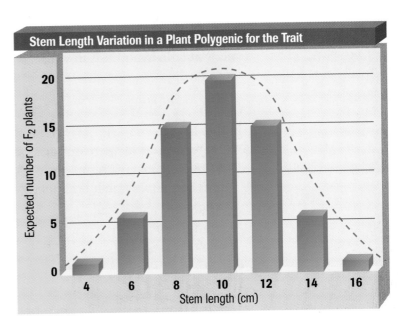

BioLab | Design Your Own Experiment

Mitochondria and chloroplasts contain DNA. This DNA is not coiled into structures called chromosomes, but it still carries genes that control genetic traits. Because mitochondria are the site of aerobic respiration, many of the mitochondrial genes control steps in the respiration process.

What is the pattern of cytoplasmic inheritance?

The DNA in chloroplasts controls traits such as chlorophyll production. Lack of chlorophyll in some cells causes the appearance of white patches in a leaf. This trait is known as variegated leaf. In this Biolab, you will design and carry out an experiment to determine the pattern of inheritance of the variegated leaf trait in *Brassica rapa*.

PREPARATION

Problem
What inheritance pattern does the variegated leaf trait in *Brassica* show?

Hypotheses
Consider the possible evidence you could collect that would answer the problem question. Among the people in your group, form a hypothesis that you can test to answer the question, and write the hypothesis in your journal.

Objectives
In this Biolab, you will:
- **Determine** which crosses of *Brassica* plants will reveal the pattern of cytoplasmic inheritance.
- **Analyze** data from *Brassica* plant crosses.

Possible Materials
Brassica rapa seeds, normal and variegated
potting soil
potting trays
paintbrushes
forceps
single-edge razor blade
light source
labels

Safety Precautions
Handle the razor blade with extreme caution. Always cut away from you. Be sure to wash your hands carefully after handling the plants. Do not eat the seeds of any plants.

1. Decide which crosses will be needed to test your hypothesis.

2. Keep the available materials in mind as you plan your procedure. How many seeds will you need?

3. Record your procedure, and list the materials and quantities you will need.

4. Assign a task to each member of the group. One person should write data in a journal, another can pollinate the flowers, while a third can set up the plant trays. Determine who will set up and clean up materials.

5. Design and construct a data table for recording your observations.

Check the Plan

Discuss the following points with other group members to decide the final procedure for your experiment.

1. What data will you collect, and how will they be recorded?

2. When will you pollinate the flowers? How many flowers will you pollinate?

3. How will you transfer pollen from one flower to another?

4. How and when will you collect the seeds that result from your crosses?

5. What variables will have to be controlled? What controls will be used?

6. When will you end the experiment?

7. *Make sure your teacher has approved your experimental plan before you proceed further.*

8. Carry out your experiment.

1. **Checking Your Hypothesis** Did your data support your hypothesis? Why or why not?

2. **Interpreting Observations** What is the inheritance pattern of variegated leaves in *Brassica?*

3. **Making Inferences** Explain why genes in the chloroplast are inherited in this pattern.

4. **Drawing Conclusions** Which parent is responsible for passing the variegated trait to its offspring?

5. **Making Scientific Illustrations** Draw a diagram tracing the inheritance of this trait through cell division.

Going Further

Application Make crosses between normal *Brassica* plants and genetically dwarfed, mutant *Brassica* plants to determine the inheritance pattern of the dwarf mutation.

Maternal Inheritance

In some cultures, individuals trace their lineage through their mothers rather than through their fathers. This tradition makes a certain amount of sense. The father contributes a tiny sperm that is almost nothing but a nucleus containing DNA. The mother, on the other hand, contributes a much larger egg that has all the attributes of a cell—cytoplasm, nucleus, and other organelles. If both parents contribute the DNA contained in a nucleus, how might inheritance be through the mother?

Does the nucleus always rule? While most traits are indeed determined by nuclear DNA in chromosomes, some traits are controlled by DNA found in the mitochondria that are distributed throughout the cytoplasm. Being circular in shape, this mitochondrial DNA does not look like the DNA from the nucleus, which is packed into rod-shaped chromosomes. In fact, mitochondrial DNA closely resembles the DNA in prokaryotic cells. Mitochondrial DNA codes for the synthesis of specific proteins that are involved in cellular respiration—the process that takes place in the mitochondria. These traits come only from the mother, the parent that contributes the organelles. This pattern of inheritance is called cytoplasmic, or extranuclear, inheritance.

Magnification: 4000×

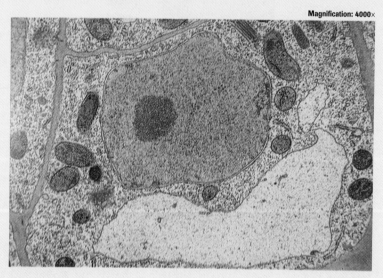

Chloroplasts have genes, too Chloroplasts are the second major organelle that carries extranuclear genes. The genes in chloroplasts control traits involving photosynthesis and the production of chlorophyll. Thus, the genes in mitochondria and chloroplasts, which play a key role in the capture and storage of energy, are as vital as any in existence.

Thinking Critically

Why do you think the genes in chloroplasts are so important to the existence of life on Earth?

Environmental Influences

Even when you have solved the puzzles of dominance and recessiveness and you understand the other patterns of heredity, the inheritance picture is not complete. The genetic makeup of an organism at fertilization determines only the organism's potential to develop and function. As the organism develops, many factors can influence how the gene is expressed, or even whether the gene is expressed at all. Two such influences are the organism's internal and external environments.

Influence of internal environment

The age or gender of an organism can affect gene function. The nature of such patterns is not well understood; however, it is known that the internal environment of an organism

Figure 14.8

In some species of sheep, rams (right) have much heavier and more coiled horns than ewes (left) have. What is the most likely cause of this difference?

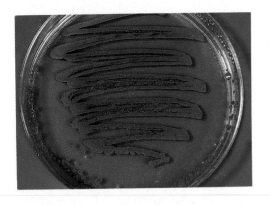

Figure 14.9

Serratia marcescens is a bacterium that forms brick-red colonies at 25°C. However, when the same bacteria are grown at 30°C, the colonies are cream colored.

changes with age. Also, the internal environments of males and females are different because of hormones and structural differences. Thus, some traits, such as horns in the sheep shown in *Figure 14.8,* are expressed differently in the sexes. In many cases, the function of one set of genes that determine a phenotype depends on the function of another set of genes.

Influence of external environment

Sometimes, individuals known to have a particular gene fail to express the phenotype specified by that gene. Temperature, nutrition, light, chemicals, and infectious agents all can influence gene expression. You have already studied the Himalayan trait of coat color in rabbits and learned that temperature has an effect on the expression of the trait. Temperature has a similar effect on the expression of color in certain bacteria, as *Figure 14.9* shows.

You can now see that genes interact with each other and with the environment to form a more complete picture of inheritance. Mendel's idea that heredity is a composite of many individual traits still holds. Later researchers have filled in more details of Mendel's great contributions.

Section Review

Understanding Concepts

1. A cross between a purebred animal with red hairs and a purebred animal with white hairs produces an animal that has both red hairs and white hairs. What type of inheritance pattern is involved?

2. In a cross between individuals of a species of tropical fish, all of the male offspring have long tail fins, and none of the females possess the trait. Mating of the F_1 fish fails to produce females with the trait. Explain the inheritance pattern of the trait.

3. A red-flowered sweet pea plant is crossed with a white-flowered sweet pea plant. All of the offspring are pink. What is the inheritance pattern being expressed?

Thinking Critically

4. Armadillos always have four offspring that have identical genetic makeup. Suppose that within a litter, each young armadillo is found to have a different phenotype for a particular trait. How could you explain this phenomenon?

Skill Review

5. **Forming a Hypothesis** An ecologist observes that a population of plants in a meadow has flowers that may be red, yellow, white, pink, or purple. Hypothesize what the inheritance pattern might be. For more help, refer to Practicing Scientific Methods in the *Skill Handbook.*

14.2 Applied Genetics

Section Preview

Objectives

Interpret testcrosses and pedigrees.

Evaluate the importance of plant and animal breeding to humans.

Key Terms

testcross
carrier
pedigree
inbreeding
hybrid

If you wanted to buy a purebred dog, how could you be certain that the animal does not have a genetic defect commonly found in its breed? Geneticists use methods developed from an understanding of Mendel's work to answer this question. Using these methods, they can find out the genotypes of individuals, anticipate the incidence of phenotypes resulting from crosses, and predict the occurrence of phenotypes and genotypes in populations.

Determining Genotypes

The genotype of an organism that is homozygous recessive for a trait is obvious to an observer because the recessive trait is expressed. However, organisms that are homozygous dominant and heterozygous for a trait controlled by Mendelian inheritance both have the same phenotype. If an organism exhibits a dominant phenotype, how can you determine whether the organism is homozygous or heterozygous for that trait?

Testcrosses can determine genotypes

One way to determine the genotype of an organism is to perform a testcross. A **testcross** is a cross of an individual of unknown genotype with an individual of known genotype in order to determine the unknown genotype. Usually, the known test organism is homozygous recessive for the trait in question.

Many traits, such as disease vulnerability in rose plants and progressive blindness in German shepherd dogs, are inherited as recessive alleles. These undesired traits are maintained in the population by carriers of the trait. A **carrier**, or heterozygous individual, appears the same phenotypically as one that is homozygous dominant. Plant and animal breeders are always cautious about introducing undesired traits into their lines of plant varieties and animal breeds. To check questionable individuals, they rely on testcrosses.

What are the possible results of a testcross? If the known test organism is homozygous recessive and the questionable organism is homozygous dominant, all of the offspring will be heterozygous for the trait and will show the dominant trait (be phenotypically dominant), as shown in *Figure 14.10.* However, if the organism being tested is heterozygous, the predicted 1:1 phenotypic ratio will be observed. If any of the offspring have the undesired trait, the parent in question must be heterozygous.

Figure 14.10

A testcross determines whether an organism is heterozygous or homozygous dominant for a trait.

A In this testcross of Alaskan malamutes, the known test dog is homozygous recessive for a dwarf allele (*dd*), and the other dog's genotype is unknown. It can be either homozygous dominant (*DD*) or heterozygous (*Dd*) for the trait.

Homozygous × Homozygous

	d	d
D	Dd	Dd
D	Dd	Dd

DD × dd

B If the unknown dog's genotype is homozygous for the dominant trait, all of the offspring will be phenotypically dominant.

Offspring: all dominant

Dd   Dd

C If the unknown dog's genotype is heterozygous, half the offspring will be expected to express the recessive trait and appear dwarf. The other half will express the dominant trait and be of normal size.

Heterozygous × Homozygous

Dd × dd

	d	d
D	Dd	Dd
d	dd	dd

Offspring:
1/2 dominant
1/2 recessive

Dd   dd

Pedigrees illustrate inheritance

Pedigree analysis is another way to solve the puzzle of whether an individual possesses a given allele, to predict the chances of an offspring receiving a trait, or to determine the inheritance pattern of a particular trait. A **pedigree** is a graphic representation of an individual's family tree, which permits patterns of inheritance to be recognized.

A pedigree is made up of a set of symbols that identify males and females, affected and unaffected individuals, matings, and other relationships. Some commonly used symbols are shown in *Figure 14.11*. Each horizontal row of circles and squares in a pedigree designates a generation,

☐ Male

○ Female

▨ Affected male

● Affected female

☐—○ Mating

☐—○ Parents
   ○ ☐ Siblings

▧ Known heterozygotes for recessive allele

▨ Death

Figure 14.11

Symbols are used by geneticists to make and analyze a pedigree. A circle represents a female; a square represents a male. Unshaded circles and squares designate individuals that have a normal phenotype for the trait being studied. Shaded or colored circles and squares represent affected individuals. A horizontal line connecting a circle and a square indicates a mating between those individuals. A vertical line connects a set of parents with its offspring.

How can you illustrate a pedigree?

The pedigree method of studying family relationships uses phenotypic records extending over two or more generations. Studies of pedigrees can be used to yield a great deal of genetic information about a related group.

Procedure

1. Choose one trait in a common pet animal that interests you.

2. Ask a pet shop owner or someone who breeds the animal to provide you with information about an animal group. Perhaps you will be able to observe a litter of puppies or gerbils.

3. Collect information about your chosen animal group. Include whether each individual is male or female, does or does not have the trait, and the relationship of the individual to others.

4. Use your information to complete a pedigree for the trait.

Analysis

1. What trait did you study? From your pedigree, what is the apparent inheritance pattern of the trait?

2. How is the study of inheritance patterns limited by pedigree analysis?

with the most recent generation shown at the bottom. The generations are identified in sequence by Roman numerals, and each individual is given an Arabic numeral. Pedigrees are particularly useful if testcrosses cannot be made, if the number of offspring is small, or if the results of a testcross would take too long.

Analyzing a pedigree

Suppose that the pedigree shown in *Figure 14.12* is for a rare, recessive disorder that results in weakened bones in cattle. The owner of bull IV-1 wants to know if the animal possesses the allele for the trait.

Notice that information can be gained about the other cattle in the pedigree. You know that I-1 and I-2 are both carriers of the recessive allele for the rare trait because they have produced II-3, which shows the recessive phenotype. You also know that individuals II-2 and II-5 must be carriers like their parents because they each have passed on the recessive allele to a future generation. Nothing can be said about II-1 or II-6 because their parents are not shown and because their offspring are unaffected. You can't tell the genotype of II-4, but it has a normal phenotype.

The probability that II-4 is a carrier is two out of three because its only two possible genotypes are homozygous normal and heterozygous. (The homozygous recessive genotype is not a possibility because it shows a different phenotype.) The same situation holds for IV-1, IV-3, and IV-5 because they show a normal phenotype but have two carrier parents; therefore, each has a two-in-three chance of being a carrier.

In generation III, individuals III-1, III-2, and III-5 are each normal and have one parent that is heterozygous. These three individuals have either a two-in-three chance or a one-in-two chance of being carriers, depending upon whether II-1 and II-6 were heterozygous or homozygous dominant. Because the trait is rare, it is reasonably safe to assume that II-1 and II-6 are not carriers. Thus, the three individuals in question most likely have a one-in-two chance of being carriers.

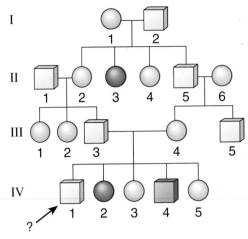

Selective Breeding

Heredity is particularly important in the breeding of domesticated plants and animals. Humans have been using genetics to breed and raise plants and animals for thousands of years. Most of the food you eat and the animals you take in as pets have been changed over time to suit the needs of humans.

Selective breeding produces organisms with desired traits

From ancient times, breeders have chosen the most desired plants and animals to serve as parents of the next generation. Farmers use for seed the largest heads of grain, the juiciest berries, and the most disease-resistant clover. They raise the calves of the best milk producer and save the eggs of the best-laying hen for hatching. Breeders of plants and animals want to be sure that their populations are uniform and breed predictably, that is, they have a desired trait in homozygous condition.

Horses and dogs are two examples of animals that breeders have developed as pure breeds. A breed (called a cultivar in plants) is a selected group of organisms within a species that has been bred for particular characteristics. For example, the German shepherd breed of dog has long hair, is black with a buff base, has a black muzzle, and resembles a wolf.

Inbreeding develops pure lines

Breeders also want to eliminate any undesired traits from their breeding lines. One method employed by breeders in the development of pure lines is inbreeding. **Inbreeding** is mating between closely related individuals. It ensures that offspring are homozygous for most traits. However, inbreeding also brings out harmful, recessive traits because there is a greater chance of matings between carriers of rare recessive alleles than in random matings. Breeders are careful not to continue breeding individuals that produce offspring with undesired traits.

Hybrids are usually bigger and better

Selective breeding of plants can produce plants that have increased value as food for humans. For example, plants that are disease resistant can be crossed with others that produce fruit vigorously. The result is a plant that has greater market value. When two varieties or closely related species are crossed, their offspring are called **hybrids.** Hybrids produced by crossing two purebred plants are often larger and stronger than their parents.

Figure 14.12

This fictional pedigree shows how a rare, recessive gene is transmitted from generation to generation. Note that the trait appears to skip generations, suggesting recessive inheritance. Also note that the trait occurs about equally in both sexes, indicating autosomal rather than sex-linked inheritance.

The Domestication of Cats

Early people had an uneasy relationship with the animals with which they shared Earth's habitats. People hunted animals for food, clothing, and whatever other necessities of life the animals could provide. Understandably, many animals came to fear these early hunters. At the same time, humans lived in fear of the animals. Quite often, the hunter became the hunted; the predator became the prey.

Egyptian mummified cat

History of cats

About 15 000 years ago, people began to domesticate small wildcats for the help they could provide to humans. Cats were valued for their hunting ability. They helped control rodent and reptile populations, keeping both crops and people safe.

Cats in ancient Egypt

Ancient Egyptians were so grateful for the hunting ability of cats that they worshipped cats as sacred beings. When a pet cat died, members of its human family would shave off their eyebrows as a sign of mourning. Anyone who killed a cat in ancient Egypt was usually put to death.

37348

Cats in the Middle Ages

During the Middle Ages in Europe, cats were believed to be evil. Hundreds of thousands of cats were killed. During this time, the rat population exploded because its natural enemy, the cat, was being killed off. Rats transmitted bubonic plague to humans through infected rat fleas. The plague killed about one-fourth of all Europeans living during the 1300s.

Selective breeding

Cats have been bred mostly for their physical features and personalities. Abyssinians, one type of short-haired cat, have been bred in Ethiopia (once called Abyssinia) for thousands of years. Some experts believe that Abyssinians may be the direct descendants of the sacred cats of Egypt. Abyssinians are prized for their melodic voices and their soft, beautifully colored coats.

Big business

Today, cats are big business throughout the developed world. People spend hundreds of millions of dollars each year caring for their feline companions. Cats have their own scientifically developed diets, medical care rivaling that offered to humans, and even their own psychiatrists to help them deal with stress and depression!

EXPANDING YOUR VIEW

1. **Journal Writing** In a short paragraph, summarize the reasons why cats are beneficial to humans.

2. **Understanding Concepts** In what ways do domestic cats reflect the wild nature of their evolutionary ancestors?

3. **Research** Conduct library research on Manx cats. Find out what is unique about their tails, how it is determined, and how this trait appeared.

Figure 14.13

Hybrid wheat is more nutritious and more productive than non-hybrid wheat. It was developed by selective-breeding methods.

Many crop plants such as wheat, corn, and rice and garden flowers such as roses and dahlias have been developed by hybridization, or selective breeding. *Figure 14.13* shows one example.

Animals have been selectively bred for thousands of years. Chickens, turkeys, horses, cattle, sheep, and other livestock have been bred to increase the expression of desired traits in a variety. For example, at one ranch in southern Texas, where the climate is hot and semitropical, English short-horn cattle were being raised. The cattle produced good beef but were not conditioned to the hot climate, so the rancher crossed them with Brahman cattle from India, which could tolerate hot weather. After 35 years of hybridization and selection, a pure breed was produced that combined the traits of good beef production and hot-weather tolerance—Santa Gertrudis cattle. You have benefited from selective breeding by having more eggs, milk, and meat than would have been possible otherwise.

Connecting | Ideas

Geneticists discovered that Mendel had not determined all of the ways traits are inherited. The study of genetics reveals that genes interact in ways that confirm and extend Mendel's work. Animals and plants have traits that are expressed by codominant alleles, incompletely dominant alleles, and sex-linked alleles. Do these same inheritance patterns apply to humans? Can human traits be affected by internal and external environments? Answers to these questions are important as geneticists apply their understanding to traits that affect humans.

Section Review

Understanding Concepts

1. A testcross made on a cat that is suspected of being heterozygous for an undesired, recessive trait produces ten kittens, none of which have the trait. What is the presumed genotype of the cat? Explain.
2. Why is inbreeding rarely a problem among animals in the wild?
3. What effect might selective breeding of plants and animals have on the size of Earth's human population? Why?

Thinking Critically

4. Suppose you wanted to breed a variety of plants with red flowers and speckled leaves. You have two varieties, each having one of the desired traits. How would you proceed?

Skill Review

5. **Interpreting Scientific Illustrations** Examine the following pedigree and explain whether the trait is dominant or recessive. How did you arrive at your conclusion? For more help, refer to Thinking Critically in the *Skill Handbook*.

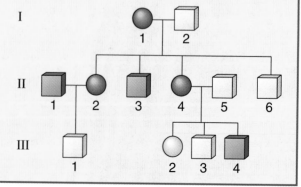

Breeding Pedigreed Dogs

Ancient Dog Found

WORLD DISPATCH   JUNE 14, 1993

Remains of an ancient dog have been found with human artifacts in Iraq. The remains date back 14 000 years, making man's best friend also man's oldest friend.

Dogs are important to the American way of life, making up a $7-billion-a-year industry. There are more than 57 million dogs in over 30 million homes.

Modern dog breeds are relatively new in the history of *Canis familiaris*. They are the result of decades of selective breeding for desired traits in appearance and behavior.

Selective breeding   Dogs have been bred for a variety of purposes. Beagles were bred to track game because of their highly developed sense of smell. Siberian huskies, strong dogs with great stamina and tolerance for bitterly cold weather, were bred to pull sleds across the frozen Arctic. Many breeds have been developed as companions to humans. These dogs, such as poodles and Chihuahuas, are usually small and friendly.

The goal of selective breeding is to eventually fix desired genes or gene combinations in the homozygous state. Taken to its extreme form, selective breeding of closely related individuals— such as breeding to grandam or grandsire, or even son to mother, daughter to father—quickly eliminates the chance of unwanted variables. Such breeding systems are controversial, yet considered the most reliable methods of ensuring the worth of a breeding line. The obvious drawbacks of inbreeding are the harmful, homozygous recessive gene pairings that are likely to manifest genetic defects in the offspring.

Different Viewpoints

Selective breeding may assist the breeder in fixing a winning pedigree line, but the results are not always in the best interest of a breed.

Unnatural selection   One example is the English bulldog, bred for decades to an unnatural standard that has left it dependent on human assistance for its most basic functions. With its massive skull size and low, narrow hips, it must be assisted in mating and in giving birth.

Responsible breeders learn to balance the positive aspects of their trade—developing new, pure breeds that are uniquely fitted to their purpose—with the negative aspects—the problems caused by the appearance of recessive traits.

INVESTIGATING the Issue

1. **Analyzing Consequences** Suppose several newborn dogs have genetic problems at birth. How can a breeder prevent these problems from appearing in future generations?
2. **Research** Choose a dog breed that you particularly admire, and research its history. Include facts concerning its place of origin, original function, introduction to the United States, and any predispositions to genetic disease.

14.1 When Heredity Follows Different Rules

- Alleles can be incompletely dominant or codominant in addition to being completely dominant. There may be many alleles for one trait or many genes that interact to produce a trait. Inheritance patterns of genes located on sex chromosomes are caused by differences in the number and kind of sex chromosomes in males and females.
- Making inheritance more difficult to understand are interactions between genes and the environment. The expression of some traits is affected by internal environments that are governed by age or sex. Expression of other traits is affected by external factors in the environment such as temperature, chemicals, or light.

14.2 Applied Genetics

- Geneticists use testcrosses and pedigrees to determine the genotypes of individuals. Both methods can also be used to predict the probability of offspring having a particular allele.
- Plant and animal breeders use genetics to selectively breed organisms with traits that are desirable to humans.

Key Terms

Write a sentence that shows your understanding of each of the following terms.

autosome
carrier
codominant
  alleles
hybrid
inbreeding
incomplete
  dominance

multiple alleles
pedigree
polygenic inheritance
sex chromosome
sex-linked trait
testcross

Understanding Concepts

1. Why does inbreeding usually result in the appearance of genetic disorders?
2. Explain how incomplete dominance and codominance inheritance patterns are similar.
3. What is the purpose of a testcross?
4. Why can an individual carry only two alleles for a given gene?
5. How did a *Drosophila* cross involving sex-linked genes help geneticists understand that certain genes are found only on certain chromosomes?
6. Explain why traits controlled by polygenic inheritance show more than three distinct phenotypes.
7. When may a testcross be less helpful than a pedigree in explaining an inheritance pattern?

Using a Graph

8. The following graph illustrates the number of flowers produced per plant by a certain plant population. From the graph, suggest a mechanism of inheritance for the trait.

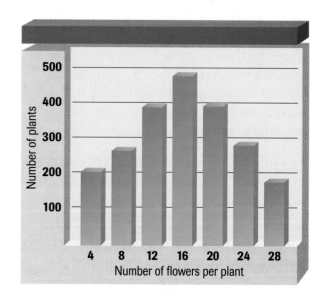

Number of flowers per plant (x-axis)
Number of plants (y-axis)

Relating Concepts

9. Make a concept map that relates the following terms and phrases. Supply the appropriate linking words for your map.

   multiple alleles, polygenic inheritance, sex-linked inheritance, pedigree, test-cross, codominant alleles, incompletely dominant alleles

Applying Concepts

10. In cats, the allele for fur color is sex linked. The allele for black fur is X^C, and the allele for orange fur is X^c. From this information, explain why calico cats, X^CX^c, are always female.

11. An organism has three genes—*e*, *f*, and *g*—for a trait. What are all of the possible genotypes of a random individual?

12. Suppose you mate a black rooster with a white hen. The feathers of all the offspring are "blue," a color that is intermediate between black and white. Explain the inheritance pattern in these chickens.

13. Julia purchased a puppy from a breeder. The breeder explained that the puppy should never be bred with another dog because it was a carrier. What did the breeder mean by this statement?

14. Inheritance of the palomino coat color in horses is a result of incomplete dominance. A white horse is *DD*, a chestnut horse is *dd*, and a palomino horse is *Dd*. What is the expected ratio of coat colors in the offspring of two palomino horses?

A Broader View

15. A well-known mutation of a mitochondrial gene in yeast is *petite*, in which the yeast cells grow very slowly. Suggest a reason for this slow growth.

Thinking Critically

Making Inferences

16. **Biolab** What part, if any, does the male parent play in cytoplasmic inheritance?

Interpreting Data

17. Explain why a male organism with a recessive sex-linked trait usually produces no female offspring with the trait.

Analyzing

18. **Minilab** Suppose that individuals III-1 and IV-1 in the pedigree shown below were mated. What would be the probability that their offspring would be affected with the trait? What did you assume was the pattern of inheritance? Why did you make this assumption?

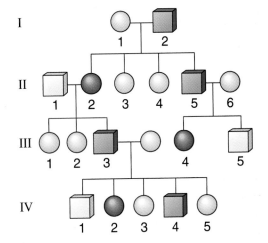

Extension

19. You unexpectedly find a male fruit fly with narrow eyes. Normal flies have round eyes. How could you determine whether the gene for narrow eyes is autosomal or sex linked?

Connecting to Themes

20. **Evolution** Give an example of how selective breeding results in evolution.

21. **The Nature of Science** Explain why Morgan reasoned that the gene for eye color in fruit flies is carried on the X chromosome but not on the Y chromosome.

15 Human Heredity

With one powerful swing of the bat, Ken Griffey, Jr. can send a long, hard drive over the outfield wall for a home run. So, too, could his father, Ken Griffey, Sr., when he starred as a major-league ball-player. That such athletic talent is inherited has long been believed. However, talent is not so clearly inherited as are certain physical traits, such as skin and eye color. Can you see any physical resemblance between Ken Griffey, Sr. and his son? Most people acknowledge that environment, including knowledge and training, plays a key role in developing talent. Evidence even shows that people inherit more potential for developing some traits than others. Human heredity is indeed complex.

Large photo: Ken Griffey, Jr.
Inset: Ken Griffey, Sr.

Folksinger and songwriter Arlo Guthrie's musical talent may have come from his father, folksinger Woody Guthrie, shown on this page. Unfortunately, Arlo also may have inherited a trait from his father that is better understood but highly undesirable—a fatal genetic disorder called Huntington's disease. How exactly are human traits inherited?

Can you curl your tongue the way this girl can? If so, you've inherited the dominant allele for tongue curling from at least one of your parents. If not, you've inherited two recessive alleles for tongue curling—one from each of your parents. Tongue curling is just one of thousands of traits humans inherit. Some traits, like tongue curling, appear to have little significance in our present environment. Others, such as hair color or eye shape, help make up an individual's personal appearance. Still others, such as hereditary disorders like cystic fibrosis, seriously affect a person's life.

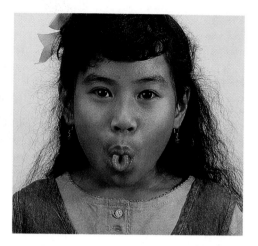

Dominant Autosomal Heredity

Many traits are inherited just as the rule of dominance predicts. Remember that in Mendelian inheritance, a single dominant autosomal allele inherited from one parent is all that is needed for a person to show the dominant trait. Recessive alleles must be inherited from both parents for a person to show the recessive phenotype.

Simple dominant traits

Tongue curling is one of these simple dominant traits. Earlobe type, illustrated in *Figure 15.1,* is also determined by simple autosomal inheritance. Having earlobes that are attached to the head is a recessive trait, whereas heterozygous and homozygous dominant individuals have earlobes that hang freely.

Huntington's disease—A rare genetic disorder

Huntington's disease is a lethal genetic disorder caused by a rare autosomal dominant allele. A lethal disorder is a disease that causes death in most of the individuals who are affected. The nervous system of a person with Huntington's disease undergoes progressive degeneration, resulting in uncontrolled, jerky movements of the head and limbs and mental deterioration. No effective treatment exists.

Ordinarily, a dominant allele with such severe effects would be expected to occur only as a new mutation and not be transmitted to future generations. But because the onset of Huntington's disease usually occurs between the ages of 30 and 50, an individual may have children before knowing whether he or she carries the allele.

Figure 15.1

Many traits are determined by a single dominant allele.

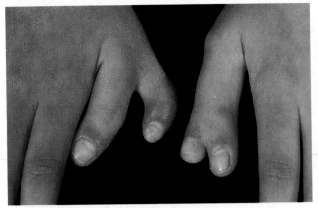

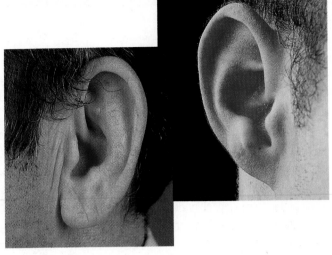

▲ The allele for polydactyly, having more than five fingers or toes, is dominant to the allele that produces only five digits on each hand and foot.

▲ The allele for freely hanging earlobes, *F* (left), is dominant to the allele for attached earlobes, *f* (right).

A biochemical test resulting from DNA technology allows some persons who are at risk to determine whether they are indeed carriers. Persons who test positive and do not have children will eventually cause a decrease in the frequency of the allele in the population. However, not everyone who is at risk wishes to be tested because anyone who tests positive for the allele must then live with the knowledge that he or she will eventually develop the disease. It is a difficult decision. The pedigree in *Figure 15.2* shows a typical pattern of occurrence of Huntington's disease in a family.

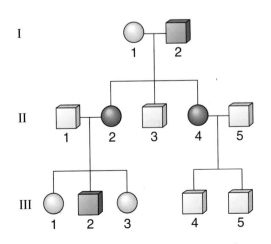

Recessive Autosomal Heredity

Unlike Huntington's disease, most genetic disorders are caused by recessive alleles. Many of these alleles are rare, but a few are common in certain ethnic groups.

Cystic fibrosis

Cystic fibrosis (CF) is the most common lethal genetic disorder among white Americans. Approximately one in 20 white Americans carries the recessive allele, and one in 2000 children born to white Americans inherits the disorder.

In a person with CF, the mucus in the lungs and digestive tract is particularly thick and viscous. Breathing is difficult because the mucus collects in the lungs, and lung infections are frequent. The thickened mucus also

Figure 15.2

A typical pedigree for Huntington's disease shows the trait in each generation and equally distributed among males and females. Every child of an affected individual has a 50 percent chance of being affected and then a 50 percent chance of passing the defective allele to his or her own child.

Cystic Fibrosis

Cystic fibrosis (CF) kills about 500 children and young adults each year. It is caused by a defective gene that disrupts the channel protein regulating the passage of chloride ions out of the body's cells. Chloride ions are a component of table salt and are prevalent in the body. As a result of this disruption, chloride ions build up in the cells. Water then leaves the surrounding fluid and enters the cells by osmosis. This causes the fluid between the cells to become thick and sticky. The abnormal buildup of mucus in the lungs and pancreas clogs airways and disables digestion. CF patients need daily therapy to loosen and remove the mucus; in spite of this therapy, most CF patients develop serious lung infections, caused by growth of bacteria in the thick mucus.

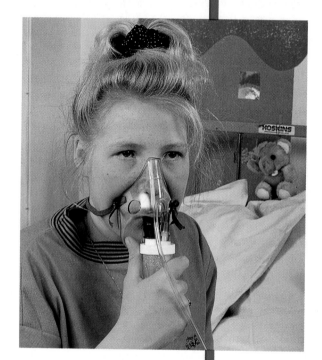

The war against CF A major breakthrough occurred in 1989 when scientists finally located the CF gene on chromosome 7, making genetic testing possible. It is estimated that one in 20 white Americans is a CF carrier. Children of two carrier parents each have a one-in-four chance of inheriting two defective alleles and being born with the disorder. The U.S. Congressional Office of Technology Assessment predicts that routine testing for CF will be offered to all prospective parents (including those not at risk) within the next decade. In 1992, geneticists continued their technological war against the lung disease with the first human gene therapy study of CF. They hope that treated cells will begin excreting salt normally.

CONNECTION TO Biology

How does the defective membrane protein in CF patients affect transport of materials across the plasma membrane? What is the result? ATP is necessary for the function of this protein. By what process does this transport occur?

slows the secretion of some digestive enzymes, so the individual cannot digest food properly. Physical therapy, special diets, and new drug therapies have raised the average life expectancy of CF patients.

Sickle-cell anemia: A blood disorder

Like cystic fibrosis, sickle-cell anemia is inherited as an autosomal recessive trait. Unlike CF, however, the disease is most common in black Americans whose families originated in Africa and in white Americans whose families originated in the countries surrounding the Mediterranean Sea. About one in 12 African Americans, a much larger proportion than in most populations, is heterozygous for the disorder.

In an individual with sickle-cell anemia, the red blood cells are shaped like a sickle, or half-moon, as shown in *Figure 15.3.* Normal red blood cells are disc shaped. Sickled cells contain hemoglobin, the oxygen-carrying protein, that differs from normal hemoglobin molecules in just one amino acid. Because sickled cells have a shorter life span than normal red blood cells, the person suffers from anemia, a low number of red blood cells. Sickled cells also clog small blood vessels, causing tissues to become damaged and deprived of oxygen and nutrients.

People with sickle-cell anemia have other serious health problems in addition to anemia because of their impaired circulation. Treatments include blood transfusions and drug

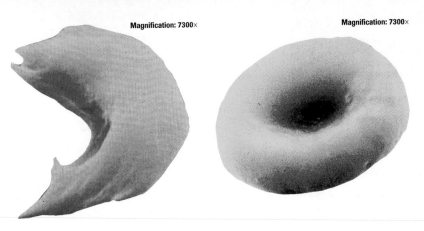

Figure 15.3

In sickle-cell anemia, the hemoglobin forms crystal-like structures that change the shape of the red blood cells (left). The change in shape occurs in the veins after oxygen has been released. Abnormally shaped cells slow blood flow, block small vessels, and result in tissue damage and pain. A normal red blood cell is shown on the right.

therapy. Individuals who are heterozygous for the trait do not have sickle-cell anemia, but because they produce both normal and abnormal hemoglobin, they may show some symptoms if availability of oxygen is low.

Tay-Sachs disease affects the nervous system

Tay-Sachs disease is an autosomal recessive disorder of the central nervous system. In individuals with Tay-Sachs, a recessive allele results in the absence of an enzyme that normally breaks down a lipid produced and stored in tissues of the central nervous system. Therefore, this lipid, which is found in membranes in the brain, fails to break down properly and accumulates in the cells. This accumulation results in blindness, progressive loss of movement, and

mental deterioration—damage for which no treatment is available. The symptoms begin within the first year of life and result in death before the age of five. The allele for Tay-Sachs is especially common in the United States among the Pennsylvania Dutch people and among Ashkenazic Jews, whose ancestors came from eastern Europe. The Pennsylvania Dutch and Ashkenazic Jewish populations may have a Tay-Sachs allele frequency as high as one in 60. By contrast, most other populations have an allele frequency of fewer than one in 100 000. *Figure 15.4* shows a typical pedigree for Tay-Sachs disease.

Phenylketonuria is a treatable genetic disorder

Phenylketonuria (PKU) is a recessive disorder that results from the absence of an enzyme that converts one amino acid, phenylalanine, to a different amino acid, tyrosine. Because phenylalanine cannot be broken down, it and its by-products accumulate in the body and result in

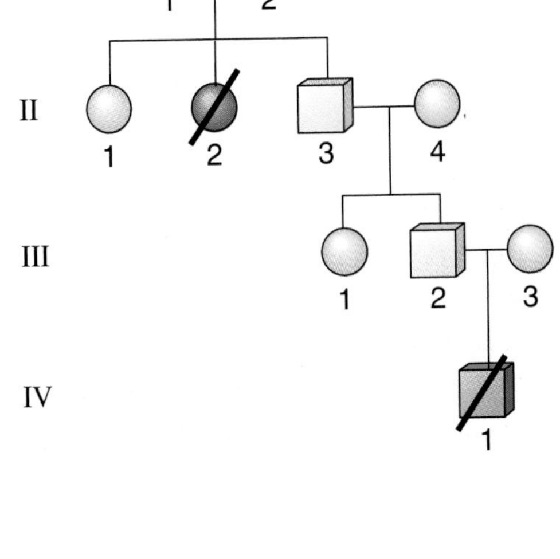

Figure 15.4

A study of families who have children with Tay-Sachs disease shows typical pedigrees for traits inherited as autosomal recessives. Note that the trait appears to skip generations. Why is this a characteristic of a recessive trait?

15.1   Simple Mendelian Inheritance of Human Traits   **359**

Meet Dr. Robert Murray, Geneticist

Suppose that you had a genetic defect that might one day make you very ill. Would you want to know what difficulties your future or your children's futures could hold? Or would you prefer to accept life as it comes, day by day? That's the kind of dilemma that genetic counselors help their patients face.

In the following interview, Dr. Robert Murray tells about his work as Chief of the Division of Medical Genetics in the Department of Pediatrics at Howard University in Washington, D.C.

On the Job

Q Dr. Murray, could you describe your duties as head of your division?

A Although I remain involved directing clinics that diagnose genetic conditions and provide genetic counseling, I also teach graduate courses in human genetics.

Q What are some of the genetic conditions you deal with?

A Sickle-cell anemia is a prominent one because the patient population we serve is primarily African American. We also deal with defects like cystic fibrosis and a whole host of blood disorders that relate to the red blood cells, such as hemophilia.

Q How do you and the other doctors counsel parents who may be faced with having a child with a genetic defect?

A We tell them that we have information and would be happy to share it with them and to help them with the problems that might arise as a result of having that information. Even though we might have strong feelings ourselves, it's our ethical obligation not to express those feelings in a way that would influence people's personal decisions. We're in a sensitive position, where the role of morality and ethics is very strong. Unless people are absolutely opposed, because of their religious or personal beliefs, to having the knowledge we can offer, we feel they should be counseled.

Early Influences

Q Could you tell us how you got into biology in general, and how you narrowed your specialty to genetics?

A I went into biology because of my interest in people. In high school, I completed a test that assessed students' areas of strength. My results showed strong interest in three areas: music, people, and science. For me, "Science" and "People" came together to equal "Medicine." I played piano and trumpet and sang in choruses during high school and college. As a black person, one of the reasons I decided not to go into music, even though I loved it, was because there were many minority people successful in the entertainment world but very few who worked in medicine or science, especially research. There was and still is a great need for minorities in science and medicine.

After medical school, I practiced internal medicine because I wanted to provide clinical care, but I found that many diseases were hopelessly advanced by the time they could be diagnosed. I wanted to be able to predict what might happen before the advanced stage, so that led me to genetics.

Personal Insights

Q What about your third area of strength, music?

A Music remains an important part of my life. At the Unitarian church I attend, I sing spirituals, gospel, and folk music in a group called the Jubilee Singers. The spirituals are closer to me than any other music because I grew up with those.

Q Do you have any advice for students about their future careers?

A It's my personal belief that we Americans need to focus on careers that deal with helping people. In recent years, too much energy, I think, has been focused on attempts to make money and accumulate things—big houses, fancy cars, material goods. It's disappointing to have medical students ask me what specialty pays the most money and requires the least effort, although I know many of them face huge debts in paying for their education.

My advice to high-school students would be to worry less about monetary matters and more about what their lives can contribute. I think that the most satisfying choices in the long run are those that are concerned with a dedication to helping people. For myself, I could have worked in a laboratory without having to worry about dealing with people, but I feel strongly that our society suffers when we don't share what we have with others. That's why I continue to stay involved in patient care.

How is Duchenne's muscular dystrophy inherited?

Muscular dystrophy is often thought of as a single disorder, but it is actually a group of genetic disorders that produce muscular weakness, a progressive deterioration of muscular tissue, and a loss of coordination. Different forms of muscular dystrophy can be inherited as a dominant autosomal, a recessive autosomal, or a sex-linked disorder. Each pattern of inheritance appears differently when a pedigree is made.

One rare form of muscular dystrophy, called Duchenne's muscular dystrophy, affects three in 10 000 American males. People with this disorder rarely live past the age of 20.

Analysis

The pedigree shown here represents the typical inheritance pattern for Duchenne's muscular dystrophy. Refer to *Figure 14.11* if you need help interpreting the symbols. Analyze the pedigree to determine the pattern of inheritance exhibited by this form of the disorder.

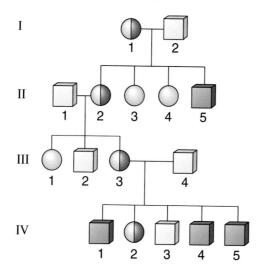

Thinking Critically

What can you infer from the pedigree about the way Duchenne's muscular dystrophy is inherited in families?

severe damage to the central nervous system. A homozygous PKU newborn appears healthy at first because its mother's normal enzyme level prevented phenylalanine accumulation during development. Once the infant begins drinking milk, which is rich in phenylalanine, the amino acid accumulates and damage occurs.

Mental retardation was once the usual result of PKU. Today, however, biochemical tests to detect the problem are available. Therefore, a PKU test is normally performed on all infants a few days after birth. Infants affected by PKU are given a diet that is low in phenylalanine until their brains are fully developed. With this special diet, the toxic effects of the disorder can be avoided. Ironically, the success of treating PKU infants has resulted in a new problem. If a homozygous recessive female becomes pregnant, the high phenylalanine levels in her blood can damage her **fetus**—the developing baby. This problem occurs even though the fetus is heterozygous and would be phenotypically normal. The PKU allele is most common among people whose ancestors came from Norway or Sweden.

Understanding Concepts

1. Describe one genetic disorder that is inherited as a recessive trait.
2. How are the cause and onset of symptoms in Huntington's disease different from those in PKU and Tay-Sachs disease?
3. How does the structure of abnormal hemoglobin cause the symptoms of sickle-cell anemia?

Thinking Critically

4. The organism that causes malaria, a tropical disease, requires oxygen that is carried by the hemoglobin in red blood cells. Why might heterozygotes for sickle-cell anemia be resistant to malaria?

Skill Review

5. **Observing and Inferring** Suppose a child with free-hanging earlobes has a mother with attached earlobes. Can a man with attached earlobes be the father of the child? Explain. For more help, refer to Thinking Critically in the *Skill Handbook*.

Complex Inheritance of Human Traits

*D*id you ever consider whether it would be possible to predict the genetic makeup of your children? Is there some way that you can know beforehand what characteristics your children will have? Most human traits follow complex inheritance patterns, which explains why it is not an easy task to predict the genetic makeup of offspring. Also, environmental factors may influence the expression of genes. For these reasons, it is impossible to predict with any certainty the genetic makeup of a child.

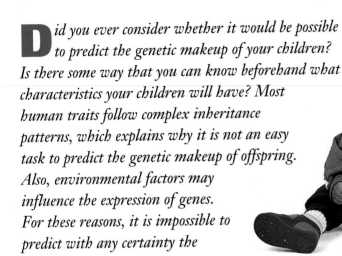

Section Preview

Objectives

Compare multiple allelic, polygenic, and sex-linked patterns of inheritance in humans.

Distinguish between autosomal and sex chromosome aneuploidy.

Key Terms
aneuploidy
karyotype

Multiple Alleles

Traits that are governed by simple Mendelian heredity have only two alleles. However, more than two alleles are possible for certain traits. Blood groups are a classic example of multiple allelic inheritance.

Multiple alleles govern blood type

Human blood types are determined by the presence or absence of certain molecules on the surfaces of red blood cells. As the determinant of blood type, the gene I has three alleles: I^A, I^B, and i, often written A, B, and O. The I^A allele produces surface molecule A, and the I^B allele produces surface molecule B. The i allele produces no surface molecule. The possible combinations of blood type alleles and their phenotypes are shown in *Table 15.1*. The alleles I^A and I^B are both always expressed; that is, they are codominant. In addition, both I^A and I^B are dominant to i. So, if a person inherits an I^A allele from one parent and an I^B allele from the other parent, he or she will produce both molecules A and B and will have blood type AB. A person who is I^Ai will produce molecule A and will

Table 15.1 Human Blood Groups		
Genotypes	**Surface Molecules**	**Phenotypes**
$I^A I^A$ or I^Ai	A	A
$I^B I^B$ or I^Bi	B	B
$I^A I^B$	A and B	AB
ii	none	O

What colors and patterns can you detect in eyes?

Human eye color, like skin color, is determined by polygenic inheritance. You can detect several shades of eye color, especially if you look closely at the iris with a magnifying glass. Often, the pigment is deposited so that light reflects from the eye, causing the iris to appear blue, green, gray, or hazel (brown-green). In actuality, the pigment may be yellowish or brown, but not blue.

Procedure

1. Use a magnifying glass to observe the patterns and colors of pigment you detect in the eyes of five classmates.

2. Make a drawing of the iris using colored pencils.

3. Write a description of your observations in your journal.

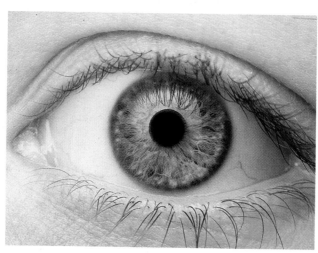

Analysis

1. How many different pigments were you able to detect?

2. From your data, do you suspect that eye color might not be inherited by simple Mendelian rules? Explain your answer.

3. Suppose that two people have brown eyes. They have two children with brown eyes, one with blue eyes, and one with green eyes. What pattern might this suggest?

have blood type A, as will an I^AI^A individual. Only those people who are ii have blood type O because the i allele produces no surface molecules.

The importance of blood typing

Determining blood types is necessary before a person can receive a blood transfusion because incompatible blood types could clump together, causing death. Blood typing can also be helpful in cases of disputed parentage. For example, if a child has type AB blood and its mother has type A, a man with type O blood could not possibly be the father. But blood tests cannot prove that a certain man definitely is the father; they indicate only that he could be.

Polygenic Inheritance

Think of all the traits you inherited from your parents—from obvious ones such as hair and eye color, to more obscure, biochemical traits such as coding for the enzyme that converts the amino acid phenylalanine correctly. While many of your traits were inherited through simple Mendelian patterns or through multiple alleles, many other human traits are determined by polygenic inheritance. These kinds of traits are usually quantitative, representing a measurable range of variation.

Skin color is determined by polygenic inheritance

In the early 1900s, the idea that polygenic inheritance occurs in humans was first tested using data collected on skin color. Scientists found that when light-skinned people marry dark-skinned people, their offspring have intermediate skin colors. When these children marry and produce the F_2 generation, the resulting skin colors range from the light skin color to the dark skin color of the grandparents (the P_1 generation), with most children having an intermediate skin color. As shown in *Figure 15.5*, methods of measuring differences in colors indicate that between three and six genes are involved.

Sex-linked Traits

Several human traits are determined by genes that are carried on the sex chromosomes. The pattern of sex-linked inheritance is explained by the fact that males, who are XY, pass on an X chromosome to each of their daughters and a Y chromosome to each son. Females, who are XX, pass one of their X chromosomes to each child.

Therefore, if a trait is X linked, males pass the X-linked allele to all of their daughters and none of their sons. Heterozygous females have a 50 percent chance of passing on a recessive X-linked allele to each child. If a son receives an X chromosome with a recessive allele from his mother, he will express the recessive phenotype because he has no chance of inheriting from his father a dominant allele that would mask the expression of the recessive allele. Two traits that are known to be governed by X-linked inheritance are certain forms of color blindness and hemophilia.

Red-green color blindness

Imagine what the world would look like if you couldn't differentiate the colors red and green. People who have red-green color blindness don't have to imagine—they know. Many people who are red-green color blind are not even aware of the fact until someone points out that they are not correctly performing a simple task

MiniLab

How is height inherited in humans?

Since people have more than three different heights, height in humans is obviously not inherited as a simple Mendelian trait. Such environmental conditions as diet and general health also affect the expression of the trait. However, by measuring the heights of many people, you can get an indication of the inheritance pattern of the trait.

Procedure

1. Measure the heights of all your classmates to the nearest centimeter.
2. Arrange your data in 10-cm groups beginning with the shortest person and continuing to the tallest person.
3. Make a graph of your results.

Analysis

1. What was the difference in height between the tallest and the shortest person?
2. What was the average height of the group, and what group had the largest number of students?
3. From your graph, what inheritance pattern is suggested?

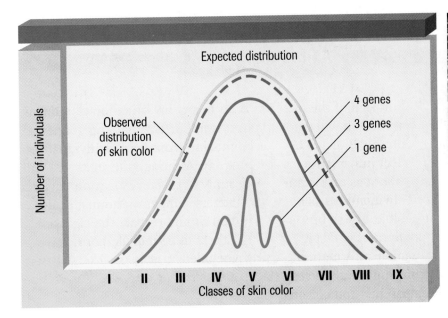

Figure 15.5

This graph shows the expected distribution of human skin color if controlled by one, three, or four genes. Which number of genes has an expected distribution of skin color that most closely matches the observed distribution?

Several human traits are easily observable, and constructing a pedigree is useful to determine how the traits are inherited. For instance, you may have noticed that some members of your family have attached earlobes, whereas others have earlobes that hang freely. By collecting information about family members from two or more generations, you should be able to create and analyze a pedigree that explains the inheritance pattern.

Constructing Pedigrees to Trace Heredity

PREPARATION

Problem
How can the construction of a pedigree illustrate the heredity of human traits?

Objectives
In this Biolab, you will:
- **Construct** pedigrees for five human traits.
- **Interpret** pedigrees to determine inheritance patterns.

Materials
pencil and paper
PTC paper

Safety Precautions
PTC paper should not be ingested. After tasting, dispose of the test paper in a trash can.

PROCEDURE

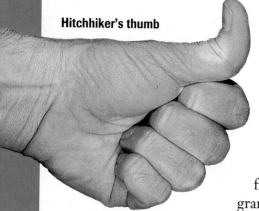

Hitchhiker's thumb

1. For each of the five traits you will study in this Biolab, collect information about as many family members as possible. It will be helpful if you can gather information from aunts and uncles, grandparents, and cousins.

2. As in any investigation of humans, it is important that you get permission from every individual you study. Although the traits you will study in this investigation have no adverse effects on human health, you should respect the rights of others to be included or not included.

Trait Studied:_____			
Generation number	Individual number	Relationship	Trait +/-

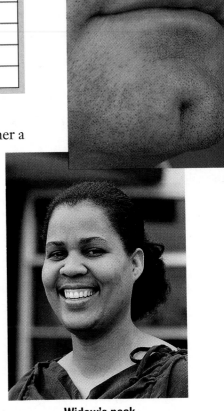

Chin dimple

Widow's peak

3. Set up a data table similar to the one above for each of the traits you study. Record all your data in the tables.

4. For each person, determine where he or she will appear in the pedigree. You need to determine each person's generation position (a Roman numeral) and individual position (an Arabic numeral).

5. Write down the relationship of each person to you.

6. Use the illustrations here and on page 357 to determine whether the person has each of the following traits: widow's peak, hitchhiker's thumb, attached earlobes, and dimple in the chin.

7. To determine whether a person can taste PTC, place the test paper flat on his or her wet tongue. After two or three seconds, the paper should be removed and discarded. PTC tasters will detect a bitter taste, and non-tasters will notice only the taste of the paper.

8. Draw a pedigree based on your data for each trait.

Use the illustrations here and on page 357

ANALYZE AND CONCLUDE

1. **Interpreting Observations** Which phenotype appeared to be dominant in your investigation of the ability to taste PTC?

2. **Interpreting Observations** What appeared to be the inheritance pattern for hitchhiker's thumb?

3. **Making Inferences** If you knew that a particular trait was inherited as a recessive trait, how could a pedigree help you predict the appearance of the trait in the next generation?

4. **Analyzing Data** Were you unable to interpret the heredity pattern for any of the traits from your data? Explain why your results were inconclusive.

Going Further

Project Design a project to determine the inheritance pattern of a human trait that interests you. Include information from several families that express the trait. Carry out your investigation with direction from your teacher and the permission of the individuals included in the study.

Figure 15.6

Red-green color blind-
ness was first
described in a young
boy who could not be
trained to harvest only
the ripe, red apples
from his father's
orchard. Instead, he
chose green apples
just as often as red.
These days, it might be
a minor embarrass-
ment if you didn't
match the color of your
clothes or pick only red
apples, but what might
happen if you were
unable to match col-
ored wires while
repairing a television
set? Serious conse-
quences can result
from being color blind
and not realizing it.

such as matching the colors of their clothes. Other more serious problems can result from this disorder, as shown in *Figure 15.6.* Color blindness is caused by the inheritance of either of two recessive alleles at two gene sites on the X chromosome that affect red and green receptors in the cells of the eyes.

Hemophilia—An X-linked blood disorder

Did you ever wonder at how quickly a cut stops bleeding? This human adaptation can be lifesaving. If your blood didn't have the ability to clot and you bruised yourself or scraped your knee, you would be in danger of bleeding to death, either internally from the bruise or externally from the scrape.

Hemophilia A is an X-linked disorder that causes just such a problem with blood clotting. About one male in every 10 000 has hemophilia, but only about one in 100 million females inherits the same disorder. Why? Because males have only one X chromosome. A single recessive allele for hemophilia will cause the disorder. Females would need two recessive alleles to cause hemophilia. Males inherit the allele for hemophilia on the X chromosome from their carrier mothers. The family of Queen Victoria, shown in *Figure 15.7,* is the best known study of hemophilia A, also called royal hemophilia.

Hemophilia can be treated with blood transfusions and injections of Factor VIII, the blood-clotting enzyme that is absent in people affected by the condition. However, both treatments are expensive. New methods of DNA technology are being used to develop a safer and cheaper source of the clotting factor.

Figure 15.7

Queen Victoria of England was a carrier of hemophilia. Her daughter, Beatrice, transmitted the allele to the Spanish royal family, while her granddaughter, Alix, transmitted the mutant allele to the Russian royal family.

Mistakes in Meiosis

You have been reading about traits that are caused by one or several genes on chromosomes. What would happen if an entire chromosome or part of a chromosome were missing from the complete set? What if there were an extra chromosome? Usually, but not always, abnormal numbers of chromosomes result from accidents of meiosis. Many phenotypic effects result from such mistakes.

Autosomal aneuploidies

You know that a human usually has 23 pairs of chromosomes, or 46 chromosomes altogether. Of these 23 pairs of chromosomes, 22 pairs are autosomes. All humans who have an unusual number of autosomal chromosomes are trisomic—that is, they have three of a particular autosome instead of just two. In other words, they have 47 chromosomes. Trisomy usually results from nondisjunction, which occurs when paired homologous chromosomes fail to separate during meiosis. Failure to separate results in **aneuploidy**—the condition of having an abnormal number of chromosomes. Trisomy is one kind of aneuploidy.

To identify an aneuploidy, a sample of cells is obtained from an individual or from a fetus. Metaphase chromosomes are photographed, and the chromosome pictures are then enlarged, cut apart, and arranged in pairs on a chart according to length and location of the centromere. This chart of chromosome pairs is called a **karyotype,** and it is valuable in pinpointing aneuploidies.

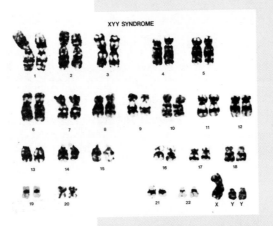

XYY SYNDROME

Sex Chromosome Aneuploidy

Millions of babies are born every year, each having 46 chromosomes in every one of their cells. To produce a baby, each chromosome must duplicate and separate thousands of times without a hitch. But this process doesn't always work perfectly. Most serious genetic mistakes are lethal, and the fetus dies before fertilization is even known to have occurred. Thus, we never see most of the major mistakes. However, we are made very aware of near-perfect successes, in which the fetus completes development and a baby is born. A group of well-studied differences is that produced by changes in the number of sex chromosomes.

Aneuploidy What happens is that the sex chromosomes fail to separate during anaphase I or II, resulting in one daughter cell that lacks a sex chromosome and the other daughter cell that has an extra sex chromosome. The daughter cells mature to form gametes. When a gamete with an unusual number of sex chromosomes fuses with another gamete during fertilization, any one of several sex-chromosome aneuploidies can result. Individuals who have a genotype of XXX, XXY, XYY, or just X are possible.

Turner syndrome The presence of a single X chromosome, rather than the normal pair, results in Turner syndrome. The genotype is written as XO. Turner syndrome occurs approximately once in every 2000 live births. Because these individuals have an X chromosome and no Y chromosome, they are females, although they lack ovaries and their sex characteristics are not fully developed. Turner syndrome individuals are unusually short in stature and are sterile, but they can lead normal lives.

Klinefelter syndrome About one in every 500 males is born with an extra X chromosome, producing an XXY genotype. Because a Y chromosome is present, individuals are male. People with Klinefelter syndrome are usually taller than average, have longer-than-average limbs, and are sterile. Often, men with this syndrome exhibit some degree of mental retardation.

XYY male As many as one in every 1000 males has an extra Y chromosome. This results in individuals who are normal and fertile but usually are taller than average.

Thinking Critically
What genetic events must occur to produce an XYY man?

Figure 15.8

Down syndrome is the only autosomal trisomy in which affected individuals survive to adulthood. Individuals who have Down syndrome characteristically have a large, thick tongue and shortened stature. They have at least some degree of mental retardation. If not profoundly affected, many people with Down syndrome live happy and productive lives.

Down syndrome–Trisomy 21

Down syndrome is a group of symptoms that results from trisomy of chromosome 21. The incidence of Down syndrome births is higher in older mothers, especially those over 40. *Figure 15.8* shows a person with Down syndrome.

Other syndromes due to aneuploidy

In addition to Down syndrome, other autosomal aneuploidies can occur in humans. These syndromes result from trisomy of other chromosomes, and they cause symptoms so severe that the developing fetus dies, many times even before the woman realizes she is pregnant. Fortunately, these syndromes occur only rarely.

aneuploidy:

an (GK) not
eu (GK) true
having an abnormal number of chromosomes

Connecting Ideas

A knowledge of Mendelian inheritance and other, more complex patterns of heredity makes it easier to understand how human traits are inherited. New technology enables us to learn of the inherited traits of individual humans even before they are born. How does the application of technology affect inheritance? Can inherited characteristics be altered in individuals? How can desirable genes be used to improve other organisms? Answers to these questions depend on understanding how genes and chromosomes can be manipulated in organisms.

Section Review

Understanding Concepts

1. Why are sex-linked traits such as red-green color blindness and hemophilia more commonly found in males than in females?
2. In addition to detecting chromosome abnormalities, what information would a karyotype reveal?
3. What would the genotypes of parents have to be for them to have a color-blind daughter? Explain.

Thinking Critically

4. A man is accused of fathering two children, one with type O blood and another with type A blood. The mother of the children has type B blood. The man has type AB blood. Could he be the father of both children? Explain your answer.

Skill Review

5. **Making and Using Tables** Construct a table of the traits discussed in this section. For column heads, use Trait, Pattern of Inheritance, and Characteristics. For more help, refer to Organizing Information in the *Skill Handbook*.

Genetic Screening: Who Has the Right to See Your Genome?

Human Genome Project Begun

MEDICAL NEWS   DECEMBER 24, 1990

Within each nucleus of nearly all of the several trillion cells that make up the human body is a complete copy of the human genome—a master blueprint for building that particular person.

Most of us are not aware of our genetic heritage. Within the next decade, scientists are likely to change all that. The Human Genome Project, an international effort, has already located 3000 of the 100 000 genes carried on our chromosomes. With each discovery, they come closer to identifying the entire human genome.

Genetic tests Prenatal tests are being used to test fetal cells for genetic disorders such as Down syndrome. Carrier screening of prospective parents for defective recessive alleles is becoming common. What is particularly new is the ability to test for a person's susceptibility to diseases, such as cancer, which in many ways is influenced by environmental factors.

Different Viewpoints

Genetic tests for disease susceptibility offer a chance for preventive action—such as exercise, quitting smoking, and a low-fat diet in the case of heart disease—all of which can extend life expectancy. Or it can allow for early treatment and diagnosis, as in the case of a familial tendency for breast cancer, which can save lives. Such tests can also become part of our medical histories and be of significant help in cases of organ or bone marrow transplants.

An instrument of discrimination? Genetic information can be used as an instrument of discrimination. People who carry genes associated with diseases pose a higher risk for insurers and employers. In addition, many inheritable disorders occur more often within ethnic groups, which can make screening programs appear racially biased. Test accuracy can also be a problem in prenatal diagnoses, which can cause extreme emotional trauma for prospective parents.

Concerns for civil rights and the new genetics have prompted new state and national laws to protect individual privacy. It is important that the methods used in genetic testing are accurate because of their serious repercussions.

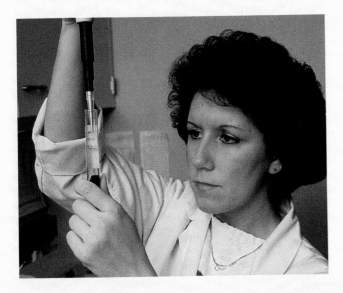

INVESTIGATING the Issue

1. **Analyzing Consequences** Genetic testing for susceptibility to disease is of concern to employers and insurers. Why?
2. **Debating the Issue** Debate in class whether insurance companies should have access to the results of genetic tests.

15.1 Simple Mendelian Inheritance of Human Traits

- Many human traits are inherited according to the simple Mendelian rule of dominance.
- Most human genetic disorders are inherited as rare recessive alleles, but a few are inherited as dominant alleles.

15.2 Complex Inheritance of Human Traits

- The majority of human traits are controlled by multiple alleles or by polygenic inheritance. Because the inheritance patterns of these traits are highly variable, it is almost impossible for individuals to predict the exact phenotype or genotype of their offspring.
- Sex-linked traits are determined by inheritance of sex chromosomes.

X-linked traits are usually passed from carrier females to their male offspring. Y-linked traits are passed only from male to male.

- Mistakes in meiosis result in aneuploidy, an abnormal number of chromosomes. The affected chromosomes can be autosomes or sex chromosomes. Nondisjunction during meiosis is a common cause of aneuploidy.

Key Terms

Write a sentence that shows your understanding of each of the following terms.

aneuploidy

fetus

karyotype

Understanding Concepts

1. Describe the events of meiosis that would result in trisomy 21.
2. How is the inheritance pattern of Huntington's disease different from that of Tay-Sachs disease?
3. What is the relationship between the inheritance patterns of hemophilia and red-green color blindness?
4. Why doesn't hemophilia occur more often in females?
5. What causes sickle-cell anemia, and what are the effects of the disease?
6. What would be the inheritance pattern of a trait that is X linked? Y linked?

Relating Concepts

7. Make a concept map that relates the following terms. Supply the appropriate linking words for your map.

   gamete, meiosis, aneuploidy, homologous chromosomes, nondisjunction, 2n, n

Using a Graph

8. The following graph represents the number of children in a large family who inherit a particular trait. Assume that no one in the history of the family has ever expressed the trait. What is the most probable inheritance pattern for the trait?

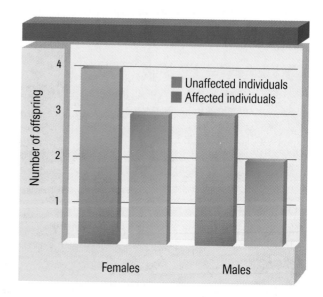

Applying Concepts

9. The brother of a woman's father has hemophilia. Her father is unaffected, but she worries that she might have an affected son. Should she worry? Explain.

10. A pregnant woman has learned that her fetus will have Down syndrome. How might this have been determined?

11. If a child has type O blood and its mother has type A, could a type B man be the father? Why couldn't a blood test be used to prove he is the father?

12. A woman whose father is color blind marries a color-blind man. Can they have an affected son? An affected daughter?

13. If one parent has type O blood and the other has type AB, what is the chance that their first child will have type O blood? Type A? Type B? Type AB?

Health Connection

14. Doctors used to diagnose cystic fibrosis by the presence of unusually salty sweat in an individual. Explain why.

Thinking Critically

Interpreting Tables

15. **Biolab** Analyze the table James completed to explain inheritance of PTC tasting in

Trait Studied: PTC Tasting			
Generation number	Individual number	Relationship	Trait +/-
I	1	Mom's father	-
I	2	Mom's mother	+
II	3	Mom	+
II	4	Dad	+
III	5	Myself-Male	+
III	6	Brother	-
III	7	Sister	-
III	8	Brother	+

his family by completing a pedigree that charts the information.

Drawing Conclusions

16. **Biolab** From James's table and your pedigree, determine the inheritance pattern of PTC tasting. Does this pattern agree with the results you obtained in the Biolab?

Interpreting Data

17. **Minilab** The graph below shows heights of a group of students. How might the group interpret its data in terms of the inheritance pattern for height in humans?

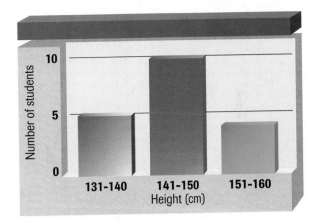

Making Inferences

18. Muscular dystrophy is a genetic disorder in which muscle tissue gradually breaks down. It almost always occurs in males. What does that suggest about its genetic basis? Explain.

Connecting to Themes

19. **Evolution** Why do certain human genetic disorders, such as sickle-cell anemia and Tay-Sachs, occur more frequently among one ethnic group than another?

20. **Systems and Interactions** How can a single mutation in a protein such as hemoglobin affect several body systems?

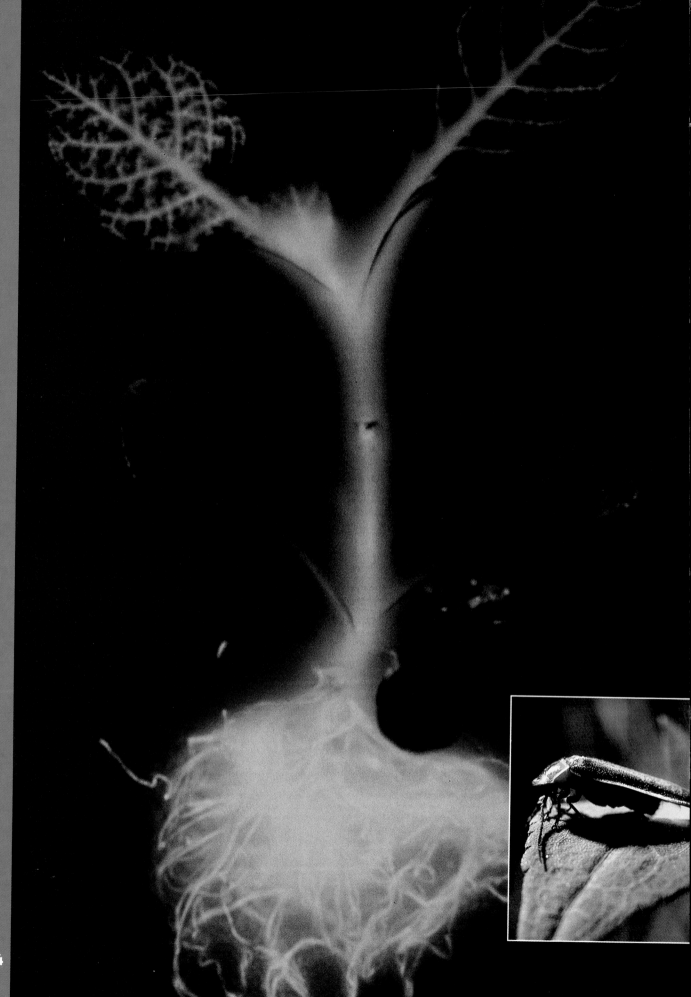

16 DNA Technology

Tobacco plants that glow? How can there be such an unusual organism? You won't find glowing tobacco in nature, but it exists in a lab. Biologists now can alter the heredity of organisms by moving genes from one species to another. In a well-known experiment done in 1986, the gene for light production in fireflies was inserted into the chromosomes of a tobacco plant. When the gene was activated, the plant glowed!

On a more practical note, geneticists can produce crop plants that supply more protein. Through genetic manipulation of certain bacteria, medicines such as insulin for diabetics are being made in large quantities. Other bacteria are being custom produced with the ability to break down pollutants such as pesticides and convert them to harmless substances. Soon it may be possible to alter human heredity and perhaps "cure" such disorders as muscular dystrophy and diabetes by genetic means.

The key that unlocks the door to genetic manipulation is DNA. The strands of synthetic DNA shown at the right were made by copying DNA extracted from animal cells. Technology that would cut, combine, and insert DNA such as this had to be developed before genetic manipulation could become a reality.

Firefly

Magnification: 75×

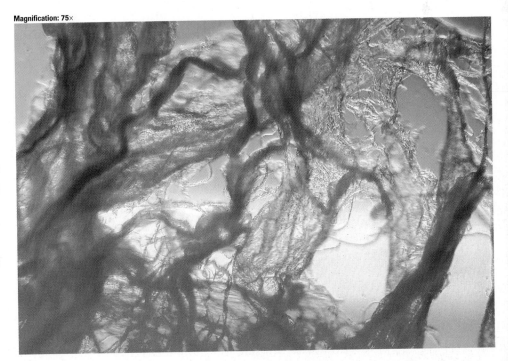

Recombinant DNA Technology

When engineers set out to design a bridge, they consider the function of the bridge and the characteristics of the materials, such as steel and concrete, that will be used to build the bridge. Today, molecular biologists "design" organisms in much the same way, although the process is infinitely more complex than designing a bridge. First, they decide what new trait the organism should have. Then, they find the section of DNA—the gene—that will be used to modify the genetic information of an existing organism. The result of a genetic engineer's work is an organism that has been "custom designed" for a unique purpose.

Gene Manipulation

Research on moving genes in organisms requires the cutting—or cleaving—of DNA into fragments. The procedures for cleaving DNA from an organism into small fragments, and inserting the fragments into another organism of the same or a different species, are called **genetic engineering.** You may also hear genetic engineering referred to as recombinant DNA technology. **Recombinant DNA** is DNA made by connecting fragments of DNA from different sources.

Once any kind of recombinant DNA has been engineered, it can be inserted into an organism; that organism then uses the foreign DNA as if it were its own. In the example of the glowing tobacco plant, a section of DNA that codes for the light-producing enzyme in fireflies is inserted into a piece of bacterial DNA and transferred into a bacterium. When the bacterium carrying the recombinant DNA infects a tobacco plant, the recombinant DNA becomes incorporated into the tobacco plant chromosomes and causes the plant to glow when it is watered with a substrate on which the light-producing enzyme acts. The glow indicates that the cells of the plant are expressing the firefly gene by making the enzyme.

Transgenic organisms contain recombinant DNA

Organisms that contain functional recombinant DNA, such as the glowing tobacco plant, are called **transgenic organisms.** The basic techniques for producing a transgenic organism involve (1) the isolation of the foreign DNA fragment to be inserted, (2) the joining of this DNA with a vehicle to transport it, and (3) the actual transfer of the recombinant DNA into a suitable host. As you read this chapter, you will learn about the many uses of transgenic organisms in agriculture, medicine, and industry.

Restriction enzymes cleave DNA

DNA can be prepared for recombination only after it has been isolated and snipped into smaller fragments. The discovery in the early 1970s of DNA-cleaving enzymes made such cutting possible. These **restriction enzymes** are bacterial proteins that have the ability to cut both strands of the DNA molecule at certain points. There are hundreds of restriction enzymes, each capable of cutting DNA at a specific point in a specific nucleotide sequence. The resulting DNA fragments are of different lengths.

You can see in *Figure 16.1* that the enzyme called *Eco*RI makes staggered cuts in the double strands of DNA. The result is a set of double-stranded DNA fragments with protruding, single-stranded ends. These ends can be joined readily to complementary single strands on DNA molecules from another organism to form recombinant DNA. Because they join so readily, the ends are referred to as being sticky.

Vectors transfer DNA

Once DNA has been cleaved, the fragments can be inserted into a host cell. In order to produce recombinant DNA, the cleaved fragment of DNA must recombine with something else. That something else is a vector. A **vector** is a means by which foreign DNA can be transferred into the host cell.

Vectors may be classified as mechanical or

vector:
vectus (L) carrier
A vector carries the foreign DNA fragment into a host cell.

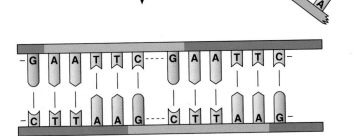

Figure 16.1

In the presence of the restriction enzyme *Eco*RI, a double strand of DNA containing the sequence –GAATTC–is cleaved between the G and the A. Notice that the same sequence of bases is found on both DNA strands, but running in opposite directions–an arrangement called a palindrome. When the DNA is cleaved, double-stranded fragments with single-stranded, sticky ends are formed. If DNA from two different sources is cleaved with the same restriction enzyme, the sticky ends of the fragments will match and pair, forming recombinant DNA.

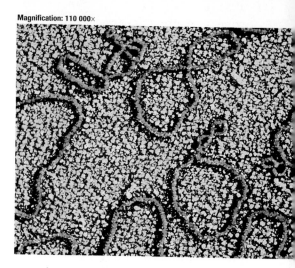

MiniLab

Why does recombinant DNA require palindromes?

A word, group of words or letters, or numerals that read the same forward and backward are called palindromes. Many restriction enzymes cut sequences of DNA that are palindromes. This property is essential for engineering recombinant DNA. As a result of cuts made by restriction enzymes, single-stranded sequences of DNA are left dangling at the ends of a fragment. These ends are available for pairing with their complementary bases in a plasmid or piece of viral DNA.

Procedure

1. Make a drawing similar to the illustration of the DNA sequence shown at the right.

2. Assume that a particular restriction enzyme cuts the chain between -G-G- so that the fragments are -CCTAG and
G-   -G
GATCC-. Use scissors to cut your drawing in the same way that the restriction enzyme would.

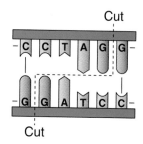

Analysis

1. Restriction enzymes were first isolated from bacteria. How might restriction enzymes be a bacterial defense against viral invaders?

2. Why are the cleaved ends of DNA called sticky ends?

3. Why would a restriction enzyme that makes a cut as shown at right be less useful than the one illustrated above?

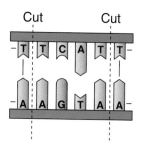

Figure 16.2

Plasmids are small rings of DNA found in bacterial cells. They usually carry non-essential "accessory" genes, such as resistance to antibiotics.

Gene splicing

The plasmid can be cleaved with restriction enzymes. If the plasmid and the foreign DNA have been cleaved with the same restriction enzyme, the sticky ends of each will match and they will join, reconnecting the plasmid ring. Rejoining cut DNA fragments is called **gene splicing.** The foreign DNA is recombined into a plasmid or viral DNA with the help of a second enzyme.

Gene cloning

After the foreign DNA has been spliced into the plasmid or virus vector, the recombined DNA is transferred into a host cell. When the host cell prepares to divide, it copies the recombinant DNA along with its own DNA. The process of making extra copies of recombinant DNA is a form of cloning. **Clones** are genetically identical copies. *Figure 16.3* summarizes cloning of recombinant DNA in a bacterial host cell.

Each identical recombinant DNA molecule is called a gene clone. Gene cloning is an important step in the

palindrome:

palindromos (GK) running back again; a word, number, or phrase that reads the same backward and forward, such as *1881* or *madam*

biological. Two mechanical vectors are used to introduce foreign DNA directly into the nucleus of a cell. A micropipette is inserted into a cell, or tiny metal bullets coated with pieces of DNA are shot into the cell using a device called a gene gun. Biological vectors include viruses and plasmids. A **plasmid,** seen in *Figure 16.2,* is a small ring of DNA found in a bacterial cell. It carries different genes from those of the bacterial chromosome.

process of genetic engineering because multiple copies of the desired DNA are produced.

Why clones are possible

Cloning is possible because a foreign piece of DNA, introduced into a bacterial cell, has been integrated into that cell's DNA so completely that the foreign DNA is replicated as if it were bacterial DNA. Thus, when the bacteria reproduce, millions of bacterial cells are eventually formed, each of which has the desired piece of foreign DNA that can be easily purified, studied, and manipulated.

Genetic-engineering techniques provide pure DNA for use in genetic studies or for producing a gene product in large quantities. Drug manufacturers use genetic-engineering techniques to produce certain drugs, such as human growth hormone.

MiniLab

How is splicing wires similar to splicing genes?

Electricians must often splice wires at terminal boxes. In some ways, splicing wires can be like splicing genes. Instead of using restriction enzymes, electricians use wire cutters and strippers. Instead of using enzymes to splice the wires back together, electricians twist the wires together and tape or cover them.

Procedure

1. Obtain two plastic-wrapped wires of different colors.
2. Using wire cutters or scissors, cut one piece of wire into three sections. Strip the ends of the middle section using wire strippers, or cut around the plastic and twist it off the wire.
3. Using wire cutters or scissors, cut the other piece of wire into two sections. Strip the ends that you just cut in the same way as before.
4. Join the middle section of wire from the first piece to the two sections you cut from the second piece.
5. Twist the wires together.

Analysis

1. How do the wires represent DNA fragments?
2. Explain why the ends must be stripped.
3. How is the process of splicing wires similar to splicing genes?

Figure 16.3

In a recombinant DNA technique using a plasmid vector, foreign DNA is spliced into a bacterial plasmid. The recombined plasmid then carries the foreign DNA into a bacterial cell, where it is cloned when the cell reproduces. If the foreign DNA contains a gene for insulin production, each transgenic cell will make insulin.

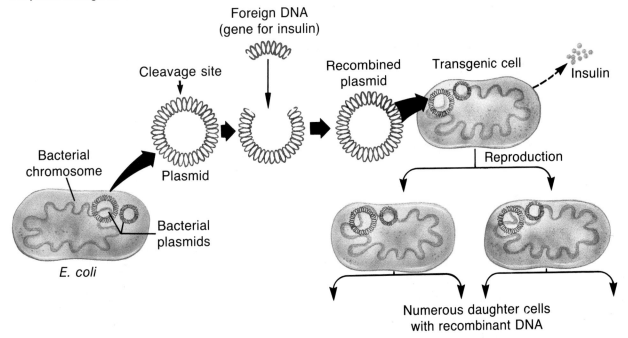

Experimental procedures have been developed that allow recombinant DNA molecules to be engineered in a test tube. From a wide variety of restriction enzymes available, scientists choose one or two that recognize particular sequences of DNA within a longer DNA sequence of a chromosome. The enzymes are added to the DNA, which is cleaved at the recognition sites. Because the cleaved fragments have ends that are available for attachment to complementary strands, the fragments can be added to plasmids or to viral DNA that has been similarly cut. When the fragment has been incorporated into the DNA of the plasmid or virus, it is called recombinant DNA.

Modeling Recombinant DNA

PREPARATION

Problem
How can you model recombinant DNA technology?

Objectives
In this Biolab, you will:
- **Model** the process of preparing recombinant DNA.
- **Analyze** a model for preparation of recombinant DNA.

Materials
paper
colored pencils, red and green
tape
scissors

Safety Precautions
Be careful with sharp objects.

PROCEDURE

1. Cut a lengthwise strip of paper from a sheet of typing paper into a rectangle about 3 cm by 28 cm. This strip of paper represents a long sequence of DNA containing a particular gene that you wish to combine with a plasmid.

2. Cut another lengthwise strip of paper into a rectangle about 3 cm by 10 cm. When taped into a ring, this piece of paper will represent a bacterial plasmid.

3. Use your colored pencils to color the longer strip red and the shorter strip green.

4. Write the following DNA sequence once on the shorter strip of paper and two times about 5 cm apart on the longer strip of paper.

-G-G-A-T-C-C-
| | | | | |
-C-C-T-A-G-G-

5. After coloring the shorter strip of paper and writing the sequence on it, tape the ends together.

6. Assume that a particular restriction enzyme is able to cleave DNA in a staggered way as illustrated.

-G   G-A-T-C-C-
|
-C-C-T-A-G   G-

Cut the longer strand of DNA in both places as shown. You now have a cleaved foreign DNA fragment containing a gene that can be inserted into the plasmid.

7. Once the foreign gene has been cleaved, cut the plasmid in the same way.

8. Splice the foreign gene into the plasmid by taping the paper together where the sticky ends pair properly.

9. The new plasmid represents recombinant DNA.

-G-G-A-T-C-C-   -G-G-A-T-C-C-
| | | | | | | | | | | |
-C-C-T-A-G-G-   -C-C-T-A-G-G-

-G-G-A-T-C-C-
| | | | | |
-C-C-T-A-G-G-

10. Copy the following table. Relate the steps of producing recombinant DNA to the activities of the modeling procedure by explaining how the terms relate to the model.

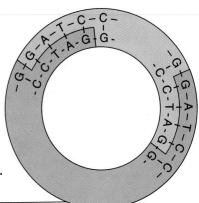

Term	Biolab Model
Gene splicing	
Plasmid	
Restriction enzyme	
Sticky ends	
Recombinant DNA	

ANALYZE AND CONCLUDE

1. **Comparing and Contrasting** How does the paper model of a plasmid resemble a bacterial plasmid?

2. **Comparing and Contrasting** How is cutting with the scissors different from cleaving with a restriction enzyme?

3. **Thinking Critically** Some restriction enzymes cut DNA at particular places but do not leave sticky ends. These enzymes cannot be used to engineer recombinant DNA. Explain why. What function might they serve in a cell?

Going **Further**

Project Design and construct a three-dimensional model that illustrates the process of preparing recombinant DNA. Consider using clay, plaster of paris, or other materials in your model. Label the model and explain it to your classmates.

Jurassic Park
by Michael Crichton

One of the most widely read adventure novels of the 1990s, Michael Crichton's *Jurassic Park* is a story of biology and technology run amok.

Ancient history brought to life

Jurassic Park is an island theme park created by multi-millionaire-philanthropist John Hammond. Hammond has hired an impressive array of biologists, computer scientists, and other technological experts to create the ultimate animal preserve, one populated by previously extinct dinosaurs from the Jurassic Era!

Cloning dinosaurs Hammond's chief geneticist, Dr. Henry Wu, obtained samples of dinosaur DNA from prehistoric insects trapped in amber—insects that bit and drew blood from the dinosaurs that shared the insects' time on Earth. By cloning the dinosaur DNA and using computerized gene sequencers to structure the genetic proteins properly, Dr. Wu succeeded in re-creating long-extinct stegosaurs, triceratops, and other dinosaurs, including the fierce carnivore, *Tyrannosaurus rex.*

The best-laid plans Despite elaborately complex and seemingly fail-safe precautions and security measures, Jurassic Park suffers a terrible catastrophe. Unbeknownst to Dr. Wu, the dinosaurs, which are supposed to be sterile and all female, have begun to breed.

When a powerful tropical storm lashes the island park and electrical power is cut off, the dinosaurs are freed from their individual territories to begin a reign of terrifying death and destruction. In the end, John Hammond is literally consumed by his vision, attacked and killed by scavenger dinosaurs of his own creation.

CONNECTION TO Biology

What scientific insights might be gained from being able to re-create extinct life forms?

The reason recombinant bacteria are used is because a huge population of them can be grown. Each cell makes the desired product. For example, if a bacterial cell were engineered to produce human growth hormone, billions of cloned cells could generate the large quantities of hormone that are needed for patients who require it. It would be impossible to obtain such large amounts of the product from humans because hormones are produced in the body in minute amounts.

Sequencing DNA

Now that large numbers of the desired DNA fragment have been cloned by the procedure just described, the sequence of DNA bases in the fragment can be determined. Sequencing is of importance to biologists because it helps them understand the

Figure 16.4

Sequencing is a technique used to identify the sequence of nucleotides in DNA. The sequence of the bases can be read directly from the bands produced in the gel by electrophoresis. Determining the sequence of a fragment of DNA requires many copies of the fragment; therefore, DNA cloning is needed to produce them.

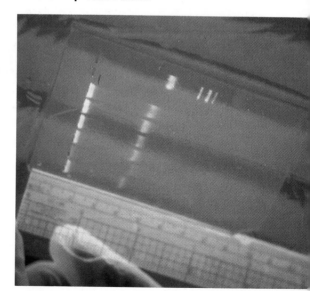

fundamental basis of life—DNA. In DNA sequencing, one of the two strands of DNA is produced by cloning. In a complex procedure, a set of DNA strands of varying lengths is produced. The strands are placed in a gel that has been treated with a dye, which binds to DNA and glows pink in ultraviolet light. Then the gel is subjected to an electric field.

Each DNA strand migrates toward the electrode of opposite charge at a rate determined by its charge and length (larger pieces move more slowly than smaller pieces). The positions of the bands of dye in the gel allow a direct reading of the sequence of bases, as shown in *Figure 16.4.* The process of separating DNA fragments in an electric field is called gel electrophoresis.

Applications of DNA Technology

Once it became possible to transfer genes from one organism to another, large quantities of hormones and other products could be produced. How is this technology of use to humans?

Transgenic bacteria in agriculture

Many species of bacteria have been engineered to produce chemical compounds that are of use to humans. The three main areas proposed for transgenic bacteria are in agriculture, industry, and medicine. One species of bacteria has already been used successfully on agricultural crops. This particular bacterium normally occurs on strawberry plants and promotes frost damage of leaves and fruits because ice crystals form around a protein on the bacterium's surface. After engineering to remove

the gene for this protein, frost damage is prevented.

Farmers hope that another species of bacteria that lives in soil and in the roots of plants such as peas, soybeans, and peanuts can be engineered to increase the rate of conversion of atmospheric nitrogen to nitrates, a natural fertilizer used by plants. If this can be accomplished, farmers will be able to obtain savings by cutting back on fertilizer.

Figure 16.5

The first patented organism, a bacterium capable of breaking down oil, was engineered by Dr. Ananda Chakrabarty. A patent ensures that the originator of the organism or product owns exclusive rights to use the product. The flask in the rear contains oil and natural bacteria. The flask at the front contains the engineered bacteria and is almost free of oil.

Transgenic bacteria in industry

In industry, transgenic bacteria have been engineered to break down pollutants into harmless products. Laboratory experiments with transgenic bacteria first showed that these engineered bacteria could degrade oil more rapidly than the same species of naturally occurring bacteria, as shown in *Figure 16.5.* These transgenic bacteria were used with some success in the Gulf of Mexico to clean up an oil spill off the coast of Texas. Mining companies are interested in bioengineering bacteria to extract valuable minerals from ores.

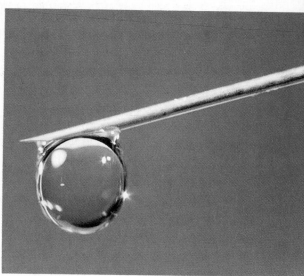

Figure 16.6

The hormone insulin, important in the treatment of diabetes, is produced commercially by bacteria. The human gene for insulin is inserted into a bacterial plasmid by genetic-engineering techniques. Transgenic bacteria produce such large quantities of insulin that they bulge. The insulin is collected as a liquid (right) and purified as crystals (left).

Transgenic bacteria in medicine

In medicine, pharmaceutical companies already are producing molecules made by transgenic bacterial species to treat human disease. Transgenic bacteria are employed in the production of growth hormone to treat dwarfism, interferon to treat cancer, and human insulin used to treat diabetes, *Figure 16.6.* Another strain of transgenic bacteria is being used to produce phenylalanine, an amino acid that is needed to make NutraSweet, an artificial sweetener found in most diet products.

Transgenic plants

Can you think of some ways that plants could be improved by genetic bioengineering? Plants are more difficult to genetically engineer than bacteria because plant cells do not have the plasmids or kinds of viruses needed for taking up foreign pieces of DNA that bacterial cells have.

Because plant cells also are surrounded by thick cell walls, it is difficult to insert DNA into plant cells.

A bacterium that normally causes tumorlike galls in the tissues of certain plants, *Figure 16.7,* has been employed to carry foreign genes into plant cells. Most engineering of plants uses mechanical vectors such as the gene gun. Brief jolts of high-voltage electricity can also be used. The jolts cause temporary pores to form in the plasma membrane, through which the DNA can enter.

Figure 16.7

The bacterium that causes these tumorlike plant galls is the only known biological plant vector. Unfortunately, it does not work on many kinds of plants.

Plants have been genetically engineered to resist herbicides, produce internal pesticides, or increase their protein production, *Figure 16.8.* These plants have been used only in field trials, and they are not yet widely available for crop use. In the future, it is expected that plants will be engineered to be more nutritious, require less fertilizer, produce fruit that ripens later, or have the ability to grow in unfavorable conditions.

Transgenic animals

Like plant cells, most animal cells will not readily accept plasmids into their own DNA. To introduce genes into animal cells, a micropipette is sometimes used to inject DNA into the eggs of animals before they are fertilized. Animal experimentation has not progressed as far as experimentation in bacteria and plants, but some trials have succeeded in introducing genes into livestock in the hope of producing bigger varieties, more milk, and even some human proteins. Transgenic animals include goats that have been engineered to produce milk containing high levels of a human protein that dissolves blood clots.

Although these advances appear to be major breakthroughs, bioengineering is still in its infancy. It is expected that within a decade, several varieties of transgenic animals will be available for farm use.

Figure 16.8

Plants that have been engineered to produce internal pesticides will not suffer damage such as this (left). Transgenic tomato plants have foreign genes that slow the process of spoilage. This trait will make storing, shipping, and stacking tomatoes in supermarkets less of a problem (right).

Section Review

Understanding Concepts
1. How are transgenic organisms different from natural organisms of the same species?
2. How are sticky ends important in making recombinant DNA?
3. Why is it presently more difficult to engineer transgenic plants and animals than bacteria?

Thinking Critically
4. Many scientists consider genetic engineering to be simply an efficient method of selective breeding. What do you think? Explain.

Skill Review
5. **Sequencing** Order the steps in producing recombinant DNA in a bacterial plasmid. For more help, refer to Organizing Information in the *Skill Handbook*.

16.2 The Human Genome

Section Preview

Objectives

Analyze how the effort to completely map and sequence the human genome will advance human knowledge.

Predict future applications of the Human Genome Project.

Key Terms

human genome
linkage map
cell culture
gene therapy

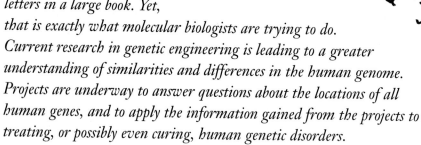

Scientists do not yet know on which chromosomes most human genes occur. It has been estimated that there are 3 billion base pairs in the DNA that makes up your genes. Imagine trying to identify the correct sequence! It's almost like trying to catalog all the letters in a large book. Yet, that is exactly what molecular biologists are trying to do. Current research in genetic engineering is leading to a greater understanding of similarities and differences in the human genome. Projects are underway to answer questions about the locations of all human genes, and to apply the information gained from the projects to treating, or possibly even curing, human genetic disorders.

Mapping and Sequencing the Human Genome

In 1990, scientists in the United States organized the Human Genome Project. It is an international effort to completely map and sequence the **human genome,** the approximately 100 000 genes on the 46 human chromosomes. The project is still underway. Maps currently are being prepared that show the locations of known genes on each chromosome. At the same time, the sequence of the 3 billion base pairs of DNA in the human genome is being analyzed and mapped. Eventually, the two maps will be synchronized.

genome:

genos (GK) offspring

A genome is the total number of genes in an individual.

Linkage maps

Only a few thousand of the total number of known genes have been mapped on particular chromosomes. This means that for most of these genes, scientists don't know the exact or even the approximate locations on chromosomes. The genetic map that shows the location of genes on a chromosome is called a **linkage map.** Such a map for one human chromosome is shown in *Figure 16.9.*

The historical method used to assign genes to a particular human chromosome was to study linkage data from human pedigrees. Recall from your study of meiosis that crossing-over occurs often during

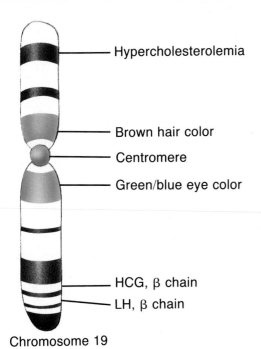

Hypercholesterolemia

Brown hair color

Centromere

Green/blue eye color

HCG, β chain

LH, β chain

Chromosome 19

Figure 16.9

A linkage map for chromosome 19 is shown here. The DNA banding pattern is determined by staining. Identification of some of the known genes on this chromosome is shown, and their locations are indicated. LH, ß chain is a polypeptide chain in a male sex hormone. Hyper-cholesterolemia is a disorder that causes high blood levels of cholesterol. HCG is a female hormone produced during pregnancy.

prophase I. As a result of crossing-over, the offspring have a combination of alleles not found in either parent. The frequency with which these alleles occur together is a measure of the distance between the genes. Genes that cross over frequently must be farther apart than genes that rarely cross over. The percentage of these crossed-over traits appearing in offspring is then used to determine the relative position of genes on the chromosome, and thus to create a linkage map.

Because humans have only a few offspring compared to other species and the generation time is so long, mapping by linkage data is extremely inefficient. Biotechnology has now provided scientists with new methods of mapping genes. By a technique called polymerase chain reaction (PCR), millions of copies of minute DNA fragments can be made in a matter of a few hours. Scientists can copy the DNA from thousands of separate sperm cells produced by one individual and analyze the results of crossing-over that occurred during the meioses that produced the sperm.

Instead of examining actual offspring, they can examine sperm cells—hundreds and even thousands of potential offspring—to create linkage maps. Another method of mapping chromosomes is shown in *Figure 16.10*.

Sequencing the human genome

The difficult job of sequencing the human genome is accomplished by cleaving samples of DNA into fragments using restriction enzymes, as described earlier in this chapter. Each fragment is then individually sequenced. The short fragments are aligned in the proper order by overlapping matching sequences, thus determining the sequence of a longer fragment. Automated machines can perform this work, greatly increasing the speed of map development. It is expected that the entire human genome may be mapped within ten to 15 years.

Magnification: 4000×

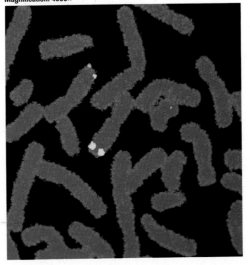

Figure 16.10

Radioactively labeled complementary DNA for the gene to be mapped is made and added to metaphase chromosomes. The labeled DNA binds to the gene on the chromosome and indicates its location as a glowing spot. Why are there four spots in this photo?

Applications of the Human Genome Project

As chromosome maps are made, how can they be used? Improved techniques for prenatal diagnosis of human disorders, use of gene therapy, and development of new methods of crime detection are current areas of research.

Diagnosis of genetic disorders

Once it is clearly understood where a gene is located and the gene's DNA sequence is known, a diagnosis of a genetic disorder may be made before birth. What technique leads to making this diagnosis? The DNA of people with the trait is analyzed to look for common patterns that appear to be associated with the trait. A few cells are obtained from a fetus or from the fluid surrounding the fetus. To obtain a large enough sample of DNA, the cells are grown in a nutrient medium until many cells have formed, a technique known as **cell culture.** Cells in a cell culture all have the same genetic material; that is, they are clones. Thus, when fetal cells are examined and found to have DNA with the pattern associated with the disorder, there is a strong probability that the fetus will develop the trait.

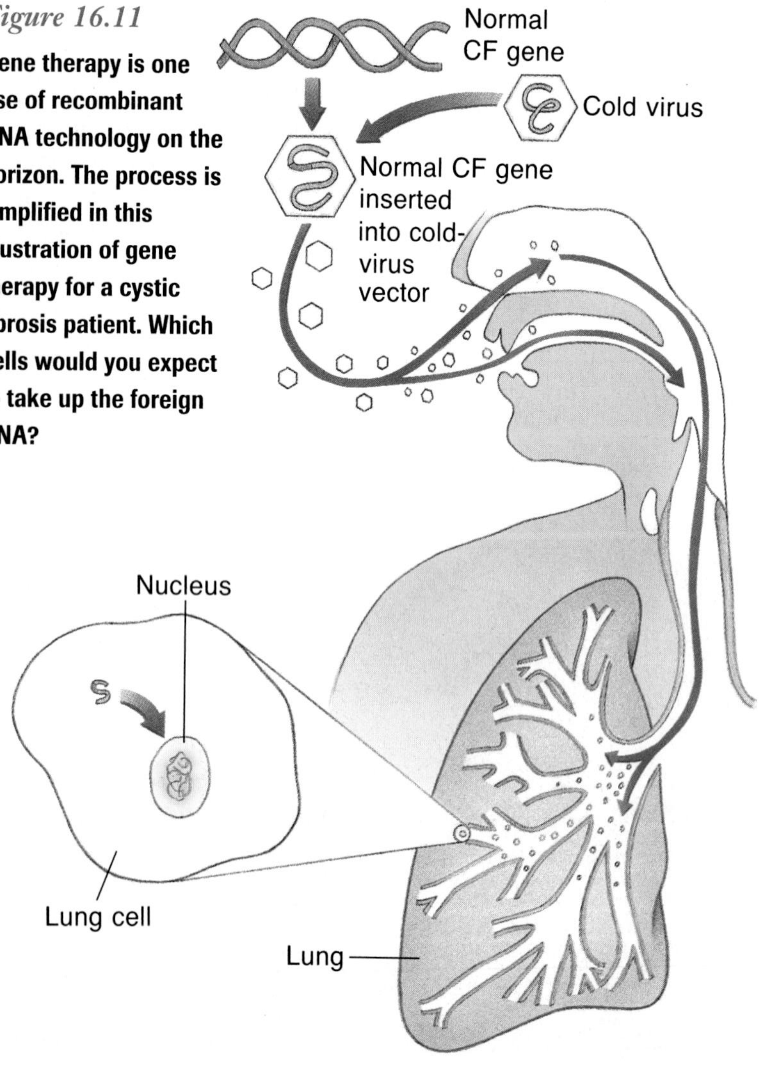

Gene therapy

The next step after diagnosis is gene therapy. **Gene therapy** is the insertion of normal genes into human cells to correct genetic disorders. Gene therapy for humans has already entered trial stages in a number of attempts to treat or cure genetic disorders; the first trials were on patients suffering from cystic fibrosis. The method used in these trials is diagrammed in *Figure 16.11.* It is hoped that copies of the normal gene introduced into the lungs by way of a nasal spray will cause lung cells to produce normal mucus. Results to date remain unclear, but it seems likely that the next decade will see the use of DNA technology to cure genetic disorders.

DNA fingerprinting

Law-enforcement workers use unique fingerprint patterns to determine whether suspects have been at a crime scene. In the past ten years, biotechnologists have developed a method that determines DNA fingerprints. DNA fingerprinting can be used to convict or acquit individuals of criminal offenses because every person is genetically unique.

Figure 16.11

Gene therapy is one use of recombinant DNA technology on the horizon. The process is simplified in this illustration of gene therapy for a cystic fibrosis patient. Which cells would you expect to take up the foreign DNA?

Normal CF gene

Cold virus

Normal CF gene inserted into cold-virus vector

Nucleus

Lung cell

Lung

Figure 16.12

DNA fingerprinting compares the DNA fragments from a known sample to those from suspects. The bands on the fingerprint will match only if the same person donated both specimens. Can you think of a use for DNA fingerprinting other than crime analysis?

Minute DNA samples can be obtained from blood, hair, skin, or semen and copied millions of times using polymerase chain reaction techniques. When an individual's DNA is cleaved with a restriction enzyme, the resulting DNA segments are of different lengths. As shown in *Figure 16.12,* electrophoresis can be used to separate the DNA fragments so they can be compared with those obtained from a crime scene.

The genes themselves follow fairly standard patterns from person to person, but in between the genes are segments of noncoding DNA. This noncoding DNA doesn't code for proteins, but it does follow a distinct pattern in each individual—so dis-

Dino-doubles?

Thanks to a combination of books, movies, and new scientific discoveries, the 1990s may well become known as the Decade of the Dinosaurs. People have always been curious about these giants from the Mesozoic Era that ruled Earth until their disappearance some 65 million years ago.

A chip off the old block No aspect of dinomania so completely captured the imagination of the public as the notion that dinosaurs actually could be brought back to life. The technique by which this revolutionary reincarnation supposedly would be accomplished is called cloning. Cloning was first accomplished with frogs in the 1950s. The nucleus of a frog cell was transplanted into an egg cell, whose nucleus had been removed. The end result was the genetic duplicate of the frog that donated the nucleus. That's what you get when you clone any organism—a genetic duplicate.

A complex process There's a big jump from transplanting a nucleus to transplanting the complete genome of a dinosaur. Suppose scientists were able to find a Mesozoic mosquito preserved in a chunk of amber. Suppose further that this mosquito had almost certainly fed on the blood of an apatosaurus. That should mean there would be DNA from the apatosaurus's blood cells in the mosquito's stomach. But fossil DNA is never found intact, even from fairly simple organisms. An animal the size and complexity of a dinosaur would have had thousands of genes. What's more, even if scientists were able to find a complete genome from a dinosaur cell, they still couldn't clone the animal. They wouldn't have any idea how to turn the genes on and off in the proper sequence to produce a living dinosaur.

Using PCR to clone DNA
Using polymerase chain reaction analysis, millions of copies can be made of even the tiniest piece of DNA. Researchers have used PCR to reproduce DNA  fragments found in 30-million-year-old insects encased in amber. But getting a few genes out of ancient DNA and cloning a living, breathing dinosaur are two different stories.

Thinking Critically
What else might scientists learn from studying fossil DNA?

How do you analyze a DNA fingerprint?

Assume that you have been chosen to sit on a jury in which the following DNA fingerprint evidence is presented. You must analyze the evidence and draw a conclusion about an accused individual's guilt or innocence.

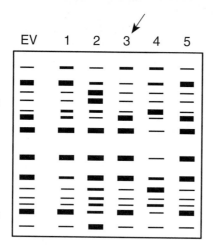

Analysis

The illustration shows the results of a DNA fingerprint analysis. EV represents the electrophoresis results obtained from the evidence found at the scene of the crime. The numbers represent individuals whose DNA fingerprints are shown. Individual 3 is the accused. Analyze the DNA fingerprint patterns to determine which individual's pattern matches the pattern of the DNA obtained from the evidence.

Thinking Critically

Which individual's DNA fingerprint most closely matches the evidence? Would you convict individual 3 of the crime? Why or why not?

tinct, in fact, that DNA patterns can be used like fingerprints to identify the person (or other organism) from whom they came. DNA fingerprinting works because no two individuals (except identical twins) have the same DNA sequences, and because all cells (except gametes) of an individual have the same DNA.

PCR techniques have been used to clone DNA from many sources. Geneticists are cloning DNA from mummies and analyzing it in order to better understand ancient life. Abraham Lincoln's DNA has been taken from the tips of a lock of his hair and studied for evidence of a possible genetic disorder. The DNA from fossils can be analyzed and used to compare extinct species with living related species, or even two extinct species with one another. The uses of DNA technology appear to be unlimited.

Connecting Ideas

We are living in a time of rapidly expanding knowledge about the code of life. The efforts of scientists involved in biotechnology and genetic engineering have resulted in a better understanding of the mechanisms of genetic disorders. As a result of their innovations, scientists may someday develop methods that eliminate inherited disorders and lead to a healthier population. Genetic changes that occur naturally can be important to the evolution of that population.

Section Review

Understanding Concepts

1. How is a linkage map different from a sequencing map?
2. What is gene therapy?
3. Explain why DNA fingerprinting can be used as evidence in law enforcement.

Thinking Critically

4. Describe some possible benefits of the Human Genome Project.

Skill Review

5. **Observing and Inferring** Suppose a cystic fibrosis patient has been treated with gene therapy. Does this person still run the risk of passing the disorder to offspring? Explain. For more help, refer to Thinking Critically in the *Skill Handbook*.

Brave New World

When the Human Genome Project was announced, scientists throughout the world expressed excitement about the potential benefits of the study. Knowing the precise location and function of all the genes would allow scientists to read the underlying instructions that govern all aspects of human biology.

Germ-line therapy   Unlocking the secrets of DNA offers scientists another avenue for experimentation—germ-line therapy. Traditional gene therapy involves replacing defective human genes with copies of normal genes. The implanted genes cannot be passed on to future generations.

Germ-line therapy, however, involves changing the genetic instructions of so-called germ cells—eggs and sperm. Such changes would be passed from one generation to the next.

Different Viewpoints

Few individuals oppose standard gene therapy to treat and prevent disorders. The benefits are obvious and the risks minimal. The quality of life for individuals and society will improve.

Troubling aspects of DNA technology   Many people, however, are troubled by other aspects of gene mapping and therapy.

- Will gene mapping lead to screening of developing fetuses and prospective mates, parents, and employees? Will those found to have genetic "blemishes" be discriminated against?
- Will germ-line therapy allow scientists to engineer human life?
- Will gene therapy create a standard for what is considered physically normal, and will those who differ from this standard suffer prejudice?
- Will health insurance companies use genetic screening to deny coverage to anyone whose genes aren't in perfect working order?

Humankind is on the brink of a biological revolution—one that could either enhance or degrade the quality of human life.

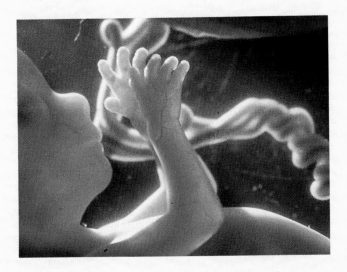

INVESTIGATING   the Issue

1. **Research** Read recent articles to investigate other uses of gene therapy. How are researchers replacing defective genes? With which disorders has the most progress been made?
2. **Debate** Use the four bulleted points above as the basis for a class debate on the future of genetic engineering. After the debate, develop a list of ethical guidelines to control future use of genetic engineering.

16.1 Recombinant DNA Technology

- Scientists have developed methods to move genes from one species into another. The process of genetic engineering uses restriction enzymes to cleave one organism's DNA into fragments. Other enzymes are used to splice the DNA segment into a plasmid or viral DNA. Transgenic organisms are able to manufacture genetic products foreign to themselves using recombinant DNA.
- Genetic engineering has already been applied to bacteria, plants, and animals. The transgenic organisms produced by biotechnology are engineered to be of use to humans in agriculture, industry, and medicine.

16.2 The Human Genome

- International efforts are presently underway to sequence the DNA of the entire human genome and to determine the chromosome location for every gene.
- Applications of the Human Genome Project include goals of detecting, treating, and curing genetic disorders. DNA fingerprinting can be used to identify persons responsible for crimes and to provide evidence that certain persons are not responsible for crimes.

Key Terms

Write a sentence that shows your understanding of each of the following terms.

cell culture	plasmid
clone	recombinant DNA
gene splicing	restriction enzyme
gene therapy	transgenic
genetic engineering	organism
human genome	vector
linkage map	

Understanding Concepts

1. What is recombinant DNA?
2. How do you think genetic engineering differs from most industrial engineering?
3. Why are vectors needed in genetic engineering?
4. List three examples of products that are produced by genetic-engineering techniques.
5. Explain why plants are difficult to bioengineer.
6. List two biological vectors, and explain how they are used in genetic-engineering techniques.
7. What is the goal of the Human Genome Project?
8. How is gel electrophoresis useful in sequencing DNA?

Relating Concepts

9. Make a concept map that relates the following terms and phrases. Supply the appropriate linking words for your map.

   genetic engineering, sticky ends, restriction enzymes, gene cleaving, vector, plasmid, viral DNA, gene splicing, recombinant DNA

Using a Graph

10. The graph on the facing page shows the results of an experiment using natural and bioengineered bacteria of the same species that can break down oil. Each culture had 40 mL of oil added on Day 1. The graph illustrates how much oil remained in each culture at the end of each week. Approximately how much oil had been

converted into harmless products by each kind of bacteria after four weeks? How much more efficient are the bioengineered bacteria than the naturally occurring species?

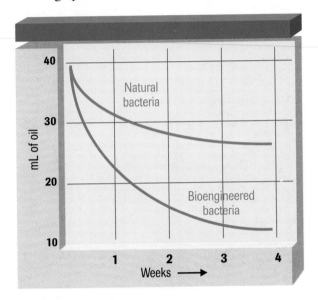

mL of oil

40

Natural bacteria

30

20

Bioengineered bacteria

10

1   2   3   4

Weeks ⟶

Applying Concepts

11. What are some modifications that may be made to plants in the future?
12. What is the potential use of a map showing the sequence of DNA bases in a human chromosome?
13. Why would it be important to be able to have a human gene expressed in a bacterium?
14. Would it be easier to "cure" by gene therapy a person who had a chromosome aneuploidy or a genetic disorder caused by a mutant, recessive allele? Explain.

Biology & Society

15. **Ethics** Which one ethical issue concerning DNA technology do you think is most important? How do you think society should deal with this issue?

Thinking Critically

Extension

16. **Biolab** What would be the next step in your modeling procedure if you were to continue the model of recombinant DNA technology?

Analyzing

17. **Minilab** Identify the palindromes from the following list.
dad; 11234; Madam, I'm Adam; ggcg; ladder; cagcag; atcgcta; 775535577

Extension

18. **Minilab** This graphic illustrates a palindromic sequence of DNA bases. Show how the segment would be cut by restriction enzyme *Eco*RI to produce sticky ends.

-G-A-A-T-T-C-
-C-T-T-A-A-G-

Evaluating

19. How may using biotechnology to engineer many different transgenic organisms alter the course of evolution for a species?

Making Inferences

20. A chimera is a mythical beast that is part lion, part goat, and part serpent. How do you think the word *chimera* is used in DNA technology?

Connecting to Themes

21. **Systems and Interactions** Assume that transgenic organisms can be used to speed the conversion of nitrogen from the air into nitrates that plants can use as fertilizer. How could use of this organism affect an ecosystem?
22. **The Nature of Science** List at least two ways that development of genetic engineering illustrates the nature of science.

Sharks, perhaps the most feared of all predators, are considered to be "living fossils." These fearsome fish are similar to the sharks of 350 million years ago. They've changed little during their long history. In contrast, whales were not always the graceful giants of the sea as we know them today. Long ago, their ancestors lived on land. As ancestors of whales moved into their new aquatic environment, those better adapted for living and moving in water survived. Over many generations, what was once a forelimb adapted for running became modified into an efficient flipper. This process is known as evolution—one of the most important principles of science today.

Scientists have discovered an enormous variety of fossils. What can we learn about evolution from the study of ancient life such as these fossil trilobites?

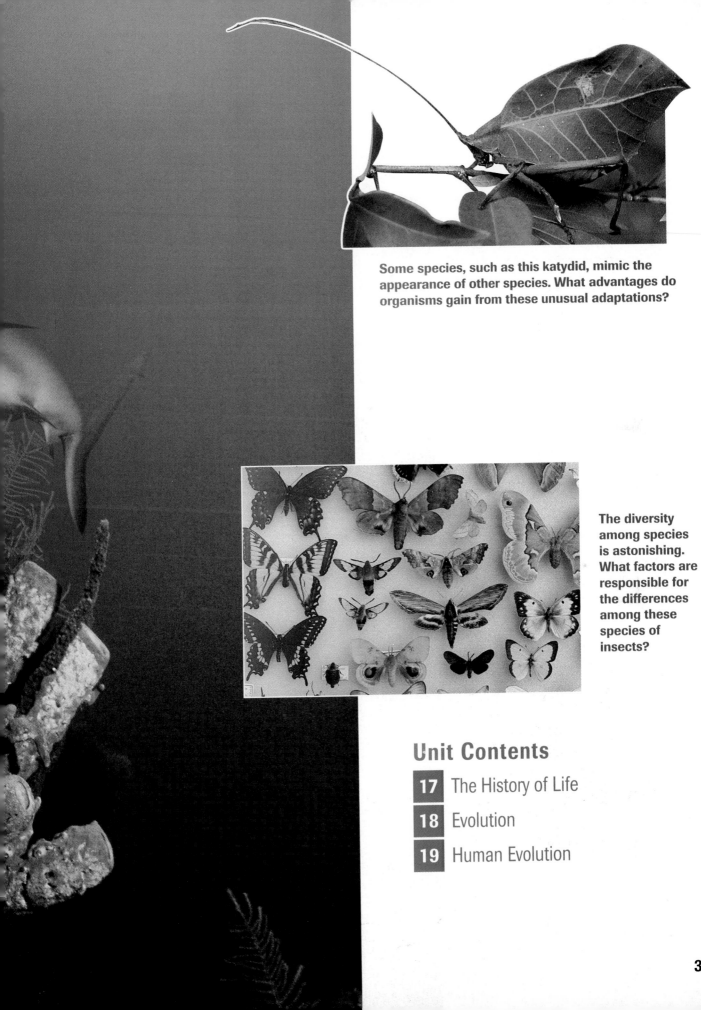

Some species, such as this katydid, mimic the appearance of other species. What advantages do organisms gain from these unusual adaptations?

The diversity among species is astonishing. What factors are responsible for the differences among these species of insects?

Unit Contents

17 The History of Life

In November 1989, NASA scientists launched a satellite designed to help answer one of the most long-standing questions in science: How did the universe begin? From evidence gathered by this satellite, such as the image of the universe at the bottom of this page, scientists believe that the story began about 15 billion years ago. The main characters of the story are the galaxies, stars, and planets of the universe—among them, a small, blue-green planet known as Earth.

On Earth, there is a subplot that poses questions humans have been interested in for thousands of years: What is the history of life on Earth? How did it begin, and how has it changed? In this chapter, you will explore scientific evidence that helps answer these questions.

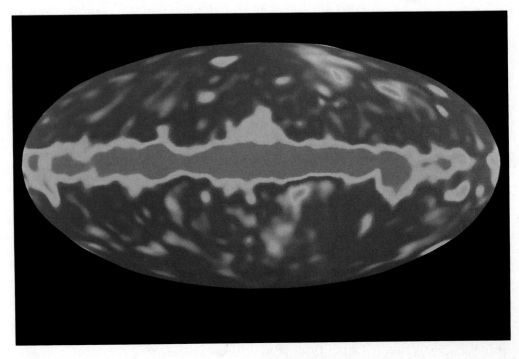

To investigate the origins of life, scientists gather fossil evidence in the field and perform tests and experiments in the laboratory. In the image of the universe (right), the pink areas represent warm temperatures where galaxies are born, and the blue areas are cooler regions.

Many popular movies, television shows, and books have explored the idea of time machines. The main characters in such stories are propelled forward and backward through time, often encountering strange people, odd creatures, and weird environments. There is no reason to expect that organisms and their environments on Earth today would be exactly like those in the past or in the future.

Early History of Earth

Step into your imaginary time machine, punch a few buttons, and get ready to explore a place to which you'll probably never want to return—primitive Earth.

Early Earth was an inhospitable place

Earth is thought to have formed about 4.6 billion years ago. It was very different from today's Earth. *Figure 17.1* illustrates what it may have looked like. Scientists theorize that Earth began as a hot ball of rock. Meteorites bombarded its surface, and volcanoes formed by the high temperatures inside Earth constantly

shook the planet, shooting out gases that formed an atmosphere. Earth was much too hot for life to exist.

About 3.9 billion years ago, Earth had cooled enough for water vapor to condense, and Earth was, for the first time, experiencing violent rainstorms. Eventually, the accumulated rainfall formed Earth's oceans. It is in these oceans, about 3.5 billion years ago, that scientists believe the first living organisms appeared.

Figure 17.1

The conditions on early Earth were not suitable for life. However, geological events, such as volcanic activity, set up conditions that would play a major role in the evolution of life.

A History in the Rocks

Accounts of Earth's formation and early history are based on the best scientific evidence available. However, no rocks have survived to provide direct evidence of Earth's earliest years. The oldest rocks on Earth are only about 3.9 billion years old. Though rocks tell us little about Earth's infancy, they do provide us with a historical record of life.

▼ **Casts** When a mold of an organism is created, it often becomes filled by minerals in the surrounding rock, producing a replica of the original organism.

Fossils—Clues to the past

Anyone who has ever visited a zoo or botanical garden has seen evidence of the tremendous diversity of life. But the millions of species of organisms that live today are only a fraction of all the species that ever lived. Scientists learn about the different types of organisms that have appeared during Earth's history by studying fossils of those organisms. A **fossil** is any evidence of an organism that lived long ago.

Fossils are classified according to the way they are formed. Examples of the main types of fossils are shown in *Figure 17.2*.

Figure 17.2

When you think of fossils, dinosaur bones probably come to mind. However, there are many different types of fossils, each providing clues about ancient organisms.

► **Amber-preserved and frozen fossils**
Sometimes an entire, intact organism can be found frozen in ice or preserved in fossilized tree sap, such as amber. These types of fossils are rare, but valuable to science because even the most delicate parts of the organism are usually preserved.

► **Trace fossils** Trace fossils are the markings or evidence of animal activities. They include footprints, trails, and burrows.

► **Petrified fossils** The hard parts of organisms are sometimes penetrated and replaced by minerals, atom-for-atom. When the minerals harden, an exact stone copy of the original organism is produced.

◄ **Imprints** Sometimes fossils form before sediments harden into rock. Thin objects, such as leaves or feathers, falling into sediments such as mud often leave imprints.

► **Molds** When an organism is buried, it can decay, leaving an empty space in the rock that is the exact shape of the organism.

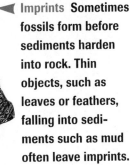

How do scientists interpret fossils?

Scientists learn about organisms that lived in the past by studying fossils. To study the details of mold fossils, casts are produced. In this activity, you will produce and analyze molds of common objects.

Procedure

1. Open completely the top of a paper pint milk container, and grease the inside with petroleum jelly.

2. Mix plaster of paris, and pour the mixture into the milk container until it is half full.

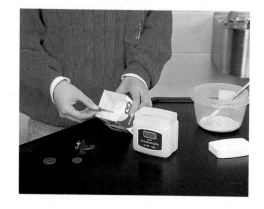

3. When the plaster begins to thicken, grease some small objects with petroleum jelly and press them into the plaster. After the plaster has hardened, remove the objects.

4. Grease the "fossil" mold you have made. Mix and pour another layer of plaster over the hardened mold. When this layer hardens, tear away the milk container and separate the two layers. You now have casts and molds of the objects.

5. Exchange your molds and casts with those of other students. Study the molds and casts, and try to determine which objects were used.

Analysis

1. How do the molds and casts compare?

2. Which was easier to interpret, a mold or a cast? Explain.

3. Compare the way you predicted what the unknown objects were with the way a scientist predicts which organism left a fossil mold or cast.

Paleontologists—Detectives to the past

If you like to solve puzzles, you may enjoy studying fossils. Paleontologists, scientists who study ancient life, may be thought of as detectives who use fossils to understand events that happened long ago. By working with fossils, paleontologists can determine the kinds of organisms that lived in the past and sometimes draw conclusions about their behavior. For example, fossil bones and teeth can indicate how ancient animals moved around and what they ate.

The ancient climate and other environmental conditions also can be determined by studying fossils. If scientists find fossilized pollen that resembles pollen from tropical plants, they may hypothesize that an ancient environment had a tropical climate. *Figure 17.3* shows an example of fossil pollen.

By studying the condition, position, and location of fossils, paleontologists can make deductions about the geography and topography of past environments. For example, if a partial skeleton of an animal is found containing only the heaviest bones, it may mean that the lighter bones were carried away by a stream or river in the area.

Magnification: 800×

Figure 17.3

This fossil pollen came from an extinct seed fern that grew in warm, moist climates about 320 million years ago. *Magnolia grandiflora* (right) probably evolved during the Cretaceous period more than 100 million years ago. It can be found growing in mild climates today.

Fossils occur in sedimentary rocks

In order for fossils to form, organisms usually have to be buried in sediments—small particles of mud, sand, or clay—soon after they die. *Figure 17.4* illustrates this process of fossilization. Over time, sediments build up around and over the organism, preventing it from decaying further. Depending upon future environmental conditions and other factors, a mold, cast, or petrified fossil will then form.

Most fossils will be found in sedimentary rocks because the slow process of depositing sediments prevented damage to the organism. Fossils are unlikely to be found in other types of rock because of the way such rocks are formed. For instance, metamorphic rocks are formed when sedimentary rocks are changed by heat, pressure, and chemical reaction. Any fossils that are in such sedimentary rock are usually destroyed by the change.

Figure 17.4

Fossils are usually found in sedimentary rocks because of the way such rocks form. Follow the steps that illustrate how an organism becomes a fossil in sedimentary rocks.

A If an organism dies, it may fall into the sandy or muddy bottom of a body of water or be carried there by floods.

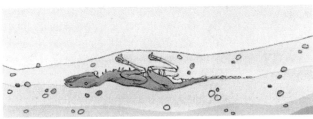

B Over time, sediments carried by the water pile on top of the carcass until it lies underground.

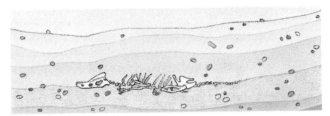

C Eventually, the muds and sands around the carcass are compressed until they form a sequence of sedimentary rocks. Geological processes have an effect on the carcass, and usually only the hardest parts survive.

D After the fossil has become embedded in sedimentary rock, geological events, such as Earth movements and erosion, may bring the fossil to the surface.

E A scientist discovers the fossil, extracts it from the surrounding rock, and brings it to a laboratory for further study.

people in biology

Meet Dr. Robert Bakker, Paleontologist

"We need an army of amateur dinosaurologists combing the hills!" exclaims paleontologist Dr. Robert Bakker. When he isn't leading a group of amateurs on a dinosaur dig, Dr. Bakker might be helping to invent a dinosaur chase video game or advising movie producers about the color and gait of an animated dinosaur.

In the following interview, Dr. Bakker tells about his life as a famous—and sometimes controversial—paleontologist.

On the Job

Q Dr. Bakker, what did you discover on your latest dig in the Laramie Range?

A We uncovered the earliest armor-plated dinosaur skull found anywhere in the world. Right now, we're calling it Boris, the name chosen by the person who unearthed it. She's an amateur paleontologist who took my field course along with her 13-year-old son. Actually, most important finds are made by amateurs.

Q How does it work out to have young people along on your digs?

A Kids are good workers in the field. Their energy really gets focused when they're helping to uncover a bone. Fortunately, there are lots of opportunities for kids as young as ten to go on digs. Kids can call this number to find out about some family digs in the United States: 1-800-DIG-DINO. Or they can ask at libraries or museums.

Q How do you proceed when you find a dinosaur skeleton?

A It's something like investigating the scene of a crime. Here's this dead body, and we ask ourselves, How did it die? We look for broken tooth crowns from meat-eating dinosaurs, which can be direct evidence—"the smoking tooth."

Early Influences

Q How did you first get interested in paleontology?

A I have right here the magazine I first read at my grandfather's house when I was ten years old. It's a *Life* magazine dated September 7, 1953, and contains an article called *The World We Live In: 2 Billion Years of Evolution.* What struck me then and stays with me today is that dinosaurs aren't just a monster circus parade—not just funny animals with bizarre shapes. They're part of the incredible story of evolution, which stretches over 3 billion years. This great story seemed to be something I could commit my life to.

Q Did you have an opportunity to go on any digs when you were a child?

A Back then, there weren't any amateur programs, but I did spend a lot of time visiting what I thought was the greatest dinosaur show in the world at the American Museum. The digs had to wait until I was in college and went to Wyoming to dig dinos.

Q Do you remember the first fossil you found?

A It was some bits of a fossil tortoise in Nebraska. When you pick up your first fossil and hold in your hand what was once living bone, it's an electric experience. Bones are literally pulsating with life. They are quite vivid documents and very aesthetically pleasing, too.

Personal Insights

Q What advice do you give students who are interested in paleontology—dinosaurs in particular?

A The best thing they can do is to contact a local nature center, museum, or zoo and volunteer. I firmly believe in breaking down that false barrier between the zoo and the museum. Life is life, and evolution connects the living with the past. If you want to understand dinosaurs, go to the zoo and look at elephants and ostriches. If you want to understand elephants, go to the museum and look at extinct elephants.

Making your own collection of bones and videotapes is a great way to start. Here's just one idea for a project. Kids can make videos of an animal in zoos and of a brontosaurus skeleton in a museum, and then compare those videos to try to figure out how a brontosaurus would move. Whether or not it *could* move on land is a hot topic today.

CAREER CONNECTION

If you "dig" the distant past, you might want to explore these career paths.

Paleontologist *Master's degree or PhD; courses in geology and biology*

Site Surveyor *Technical institute training*

Expedition Aide *High-school degree; on-the-job training*

Q There's a lot of talk today about bringing dinosaurs back to life with genetic engineering. What do you think about that?

A Right now, I think it's impossible. However, we might see in our lifetime the woolly mammoth brought back because we have entire frozen woolly mammoths. There's a strong possibility of taking woolly mammoth DNA, inserting it into the egg of its close relative, the Indian elephant, and getting a woolly mammoth. It would be fun to comb its hair.

Q If you could go out tomorrow and discover the most exciting find of your career, what would it be?

A It would simply be the *next* thing. There isn't *one* thing I'm looking for. It may be something I've never even thought of before. For example, we found the skulls of turtles that lived with the brontosaurus. These turtles have an important story to tell because they witnessed a big extinction of dinosaurs at the end of the Jurassic period. We ask ourselves, Why would that be? and look for the answer.

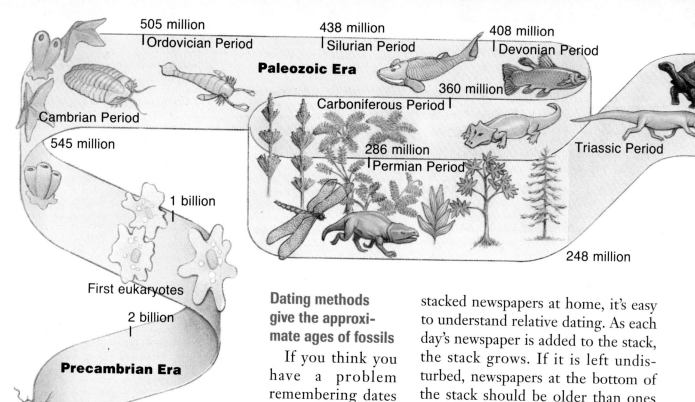

505 million
Ordovician Period

438 million
Silurian Period

408 million
Devonian Period

Paleozoic Era

Cambrian Period

360 million
Carboniferous Period

545 million

286 million
Permian Period

Triassic Period

248 million

1 billion

First eukaryotes

2 billion

Precambrian Era

3 billion

3.5 billion

First life forms

4.6 billion years ago

4 billion

Dating methods give the approximate ages of fossils

If you think you have a problem remembering dates in history, think about a paleontologist who has just unearthed a fossil skull. Scientists use a variety of methods to date fossils in order to find out how old they are.

Perhaps the simplest dating method is a technique called relative dating, as shown in *Figure 17.5*. If you've ever stacked newspapers at home, it's easy to understand relative dating. As each day's newspaper is added to the stack, the stack grows. If it is left undisturbed, newspapers at the bottom of the stack should be older than ones toward the top.

Relative dating works in the same way. If sediments have been left undisturbed, layers closer to the surface should be younger than deeper layers. In this way, scientists can get at least an idea of the order of appearance and disappearance of organisms buried in sediments.

Relative dating techniques can determine only whether one fossil is older than another fossil. This technique cannot determine the actual age of a fossil. To find the absolute ages of fossils, scientists often use

Figure 17.5

Relative dating techniques rely on the geological law of superposition. According to this law, most sedimentary rocks are laid down in horizontal layers with the younger layers closer to the surface and the older layers buried deeper. Using this law, scientists can determine the order of appearance and disappearance of organisms. Which layer is the oldest? Which is the youngest?

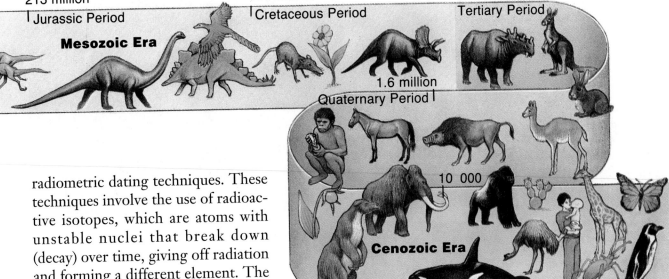

213 million

Jurassic Period

Mesozoic Era

144 million

Cretaceous Period

65 million

Tertiary Period

1.6 million

Quaternary Period

10 000

Cenozoic Era

radiometric dating techniques. These techniques involve the use of radioactive isotopes, which are atoms with unstable nuclei that break down (decay) over time, giving off radiation and forming a different element. The decay rate of each radioactive element is known and continues at a steady rate. Therefore, scientists use them as a type of clock.

The principle of radiometric dating is often compared to watching the passage of time in an hourglass. You can judge how long an hourglass has been running by comparing the amounts of sand in the top and bottom. In the same way, scientists determine the ages of rocks or fossils by comparing the amount of the original radioactive element to the amount of the new element formed from decay. For example, suppose a rock contains a radioactive element that decays to half the original amount in 1 million years. If tests show that the rock contains equal amounts of the original radioactive element and the new element, then the rock is about 1 million years old.

Potassium-40, which decays to form argon-40, is an example of an isotope that is used to date the rocks in which fossils are found. It decays to half its original amount in 1.3 billion years. Fossils and archaeological artifacts that are less than 50 000 years old are dated with carbon-14. Carbon-14 breaks down to half its original amount in 5730 years.

A Trip Through Geologic Time

By dating fossils and examining layers of sediments in Earth's crust, scientists have been able to put together a time scale for the history of life on Earth. This scale, called the Geologic Time Scale, is a type of calendar that allows scientists to communicate about events that have occurred since Earth was formed.

What is the Geologic Time Scale?

The Geologic Time Scale is based on the different types of living organisms that have appeared during Earth's history. *Figure 17.6* shows that it is divided into four eras: Precambrian Era, Paleozoic Era, Mesozoic Era, and Cenozoic Era. An era is the largest division, and each era is subdivided into periods.

The Geologic Time Scale begins at the time of the formation of Earth about 4.6 billion years ago. To appreciate the immensity of this number, it is useful to scale down the history of life on Earth into a calendar year.

Figure 17.6

The Geologic Time Scale is based on the appearance of different types of life during Earth's history. The first life appeared on Earth about 3.5 billion years ago.

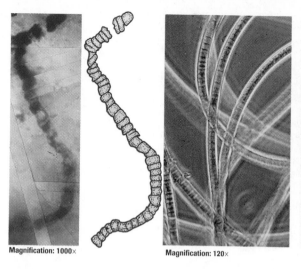

Magnification: 1000× Magnification: 120×

Figure 17.7

Stromatolites (right) provide evidence of photosynthetic bacteria that lived 3.5 billion years ago. Note that these filamentous fossil organisms (left) resemble modern cyanobacteria such as *Oscillatoria* (center). At least 11 different fossil species of these organisms have been identified.

Life began in the Precambrian Era

On this hypothetical calendar, Earth formed on the first of January. The oldest evidence of life on Earth is contained in Precambrian rocks dated to approximately 3.5 billion years ago, about March 20 on the hypothetical calendar. Precambrian rocks found in the hot deserts of western Australia contain fossils of an assortment of spherical and filamentous organisms that resemble present-day species of photosynthetic bacteria called cyanobacteria, as shown in *Figure 17.7*.

Also found here are dome-shaped structures called stromatolites. Stromatolites form today from thick mats of cyanobacteria; thus, their occurrence in Precambrian rocks suggests the presence of cyanobacteria or other photosynthetic bacteria 3.5 billion years ago.

The Precambrian Era accounts for about 87 percent of Earth's history—until about the middle of October in our hypothetical calendar year. Primitive pro-karyotes dominated the early Precambrian, and 2 billion years passed before eukaryotic organisms, organisms with nuclei, appear in the fossil record. By the end of the Precambrian, about 545 million years ago, the oceans were filled with simple, multicellular organisms such as algae, sponges, and jellyfish.

New organisms appear in the Paleozoic Era

The Paleozoic Era, which lasted from about 545 to 248 million years ago, is characterized by the appearance of many more animal and plant phyla. Scientists often refer to a Cambrian explosion of life, which started at the beginning of this era in the Cambrian period. During the period, shallow seas were teeming with many types of invertebrates including worms, echinoderms, and primitive arthropods, as shown in *Figure 17.8*.

Figure 17.8

Primitive arthropods, such as this trilobite, were among the many invertebrate species to appear during the Cambrian explosion.

In the first half of the Paleozoic, fishes, the earliest vertebrates, appeared, and there is evidence of plant life such as ferns and gymnosperms on land. Around the middle of the Paleozoic, amphibians appeared, and during the second half, reptiles also appeared.

Dinosaurs, mammals, and flowering plants appear in the Mesozoic

The Mesozoic Era began about 248 million years ago, which represents about December 10 on our one-year calendar. It is a period of many changes, both in Earth's geology and its inhabitants.

The Mesozoic Era is divided into three periods; the oldest period is the Triassic. During the Triassic period, mammals made their first appearance. Early mammals were small, mouselike animals that scurried around in the shadows of huge fern forests. Likewise, the first dinosaurs appeared.

The middle period of the Mesozoic is called the Jurassic. Beginning about 213 million years ago, mid-December on the hypothetical calendar, the Jurassic period is often referred to as the Age of the Dinosaurs. Scientists believe that toward the end of the Jurassic, the ancestors of modern birds evolved, as shown in *Figure 17.9.*

The last period in the Mesozoic, the Cretaceous, which began about 144 million years ago, is characterized by the radiation of mammals and the evolution of flowering plants such as oak, fig, and elm trees.

Figure 17.9

The evolution of birds is currently under debate. *Archaeopteryx* (left) was once thought to be one of the earliest birds because it had feathers and other birdlike features. In 1992, a 135-million-year-old fossil named *Sinornis* (right) was discovered in China. Some scientists now believe that *Sinornis* represents the earliest bird because it is more birdlike than *Archaeopteryx*.

▶ **The possible appearance of *Sinornis***

▼ **Fossil *Archaeopteryx***

How can you plot the appearance of organisms on a time line?

It may be difficult for you to visualize the 4.6 billion years that have passed during the history of Earth. In this activity, you will construct a time line for the appearance of organisms on the Geologic Time Scale.

Procedure

1. Using a meterstick, draw a continuous line down the middle of exactly 5 m of adding-machine tape.

2. For your time line, use a scale in which 1 m equals 1 billion years. Each millimeter then represents 1 million years.

3. At one end of the tape, draw a straight line and label it *The Present.* Measure to find the spot on the tape that represents 4.6 billion years ago, and label this point *Earth's Beginning.*

4. Using the table shown below, plot each event on your time line. Label the event and the number of years ago it occurred.

Event label	Estimated years ago
Earliest evidence of life	3.5 billion
Paleozoic Era begins	545 million
Mesozoic Era begins	245 million
First land plants	400 million
Triassic period begins	248 million
Jurassic period begins	208 million
First mammals and dinosaurs	225 million
First birds	150 million
Cretaceous period begins	144 million
Dinosaurs become extinct	65 million
Cenozoic Era begins	65 million
Primates appear	60 million
Humans appear	200 000

Analysis

1. Which era is the longest? The shortest?

2. In which era did dinosaurs exist or begin to exist? Mammals?

3. Which lived on Earth the longer time, dinosaurs or mammals?

Figure 17.10

Since the beginning of the Mesozoic, the continents have been drifting apart.

Ⓐ About 245 million years ago, the continents were joined together in a single landmass known as Pangaea. Modern continents are shown in different colors.

Ⓑ During the Mesozoic, Pangaea began to break apart. By about 135 million years ago, the breakup of Pangaea had formed two large landmasses.

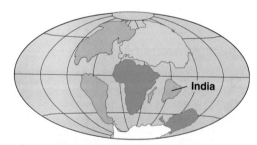

Ⓒ By 65 million years ago, the end of the Mesozoic, the continents had drifted far enough apart that they were in the shapes that we recognize today.

Changes on Earth during the Mesozoic affected species

Geological events during the Mesozoic Era had a large impact on the diversity and distribution of modern species. Among the most important of these was the movement of the continents, as shown in *Figure 17.10.*

The geological explanation for how the continents move is called **plate tectonics.** Earth's crust consists of several rigid plates that float on Earth's thick, liquid interior. These plates are continually moving, spreading apart, and pushing against one another.

The Cenozoic Era—A world with humans

The Cenozoic Era began approximately 65 million years ago, around December 26 on our calendar of Earth history. It is the era in which you now live. Mammals began to flourish during the early part of this era. Among the mammalian groups that evolved was the order of mammals to which you belong, the primates. Primate species have radiated and diversified during this era, with the modern human species evolving perhaps as early as 200 000 years ago—the evening of December 31 on the hypothetical calendar.

Extinctions affect the diversity of species

It may seem that modern Earth has a tremendous diversity of species, but most species that have existed have also become extinct over the history of life. It is estimated that, at times, fossil species became extinct at the rate of one or more per year.

The fossil record also indicates several mass extinctions in the past. You've probably heard about the mass extinction of the dinosaurs at the end of the Cretaceous period,

65 million years ago. Many scientists think that this calamity was caused by the crash of a large meteorite on Earth, sending clouds of dust into the atmosphere and causing climatic changes. Scientists estimate that nearly two-thirds of other species also died during this mass extinction.

tectonics:
tekton (GK) builder
Plate tectonics builds
continents.

ThinkingLab | Interpret the Data

How can a clock represent Earth's history?

By studying fossils and analyzing geological events, scientists have constructed the Geologic Time Scale, a timetable for the appearance of organisms during the history of Earth.

Analysis

The diagram shown here compresses the history of Earth into a 12-hour clock. Note that the formation of Earth occurred at midnight on the clock, and the oceans formed at 2:00 A.M. Use this information to help you answer the questions.

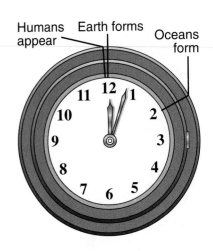

Thinking Critically

Based on fossil evidence, at approximately what time on the clock did prokaryotes evolve? What time did the first eukaryotes appear?

Section Review

Understanding Concepts

1. Describe the environmental conditions of early Earth.
2. Why are most fossils found in sedimentary rocks?
3. Suggest ways that fossils can provide evidence of the diet of animal species?

Thinking Critically

4. Suppose you are a scientist examining layers of sedimentary rock for fossils. In one layer of rock, you discover the remains of an extinct relative of the polar bear. In a layer of rock below this, you discover the fossil of an extinct alligator. What can you determine about changes over time in the climate of this area?

Skill Review

5. **Making and Using Tables** Construct a table listing the four geologic eras, their time ranges, and the major forms of life that appeared during each. For more help, refer to Organizing Information in the *Skill Handbook*.

17.2 The Origin of Life

Will rotting meat give rise to flies? Can mud produce live fish? Will a bag of wheat give birth to mice? In the past, scientists asked such questions in pondering a deep mystery of science: How did life begin? Then as today, scientists such as Louis Pasteur attempted to answer this question by relying on scientific methods of observation, hypothesis, and experimentation.

Origins: The Early Ideas

For early scientists, the ideas that mud produced fish and that rotting meat produced flies were reasonable explanations for what people observed. After all, maggots seemed to simply materialize on meat and then change into flies. These and other observations led scientists to believe in the idea of **spontaneous generation,** a process by which life was thought to be produced from nonliving matter.

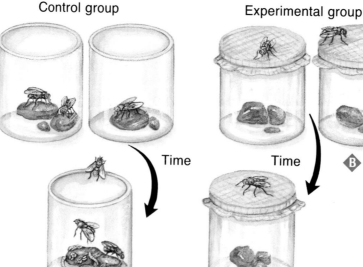

A Rotten meat is placed in two experimental jars and two control jars. Redi used two control jars to ensure accurate results.

Control group

Experimental group

Time

Time

B By placing cloth over the experimental jars of rotting meat, Redi was able to test his hypothesis that only flies produce more flies.

Results

Control

Experimental

Figure 17.11

Francesco Redi's experiment to test the idea of spontaneous generation is a classic example of the scientific method.

C Over time, the control jars were filled with maggots and flies. The experimental jars were free of insects, indicating that only when eggs were laid on the rotting meat were flies produced.

Controlled experiments disprove spontaneous generation

In 1668, an Italian physician, Francesco Redi, designed a controlled experiment to test the idea of spontaneous generation. Redi's experiment is shown in *Figure 17.11.* The results of Redi's experiment convinced many scientists, but the spontaneous generation debate raged on.

At about the same time that Redi carried out his experiment, other scientists began using the latest tool in biology—the microscope. With the microscope, scientists discovered that the world was teeming with microorganisms. Although Redi had disproved spontaneous generation of larger organisms, microorganisms were so numerous and widespread that it was believed that they arose spontaneously from a vital force in the air.

Pasteur shuts the door on spontaneous generation

In the mid-1800s, a French scientist, Louis Pasteur, decided to test this idea of a vital force in air. To disprove spontaneous generation once and for all, Pasteur realized that he would have to set up an experiment in which only air, and not microorganisms, was allowed to come in contact with a nutrient broth. *Figure 17.12* shows how Pasteur carried out his experiment.

Pasteur showed that no microorganisms would arise in the broth, even in the presence of air. With his experiment, Pasteur claimed to have "driven partisans of the doctrine of spontaneous generation into the corner." **Biogenesis,** the idea that living organisms come only from other living organisms, became a cornerstone of biology.

Figure 17.12

Pasteur's experiment to disprove spontaneous generation ended the debate. His flasks are on display today at the Pasteur Institute in Paris. They are still free from growth of microorganisms after almost 150 years.

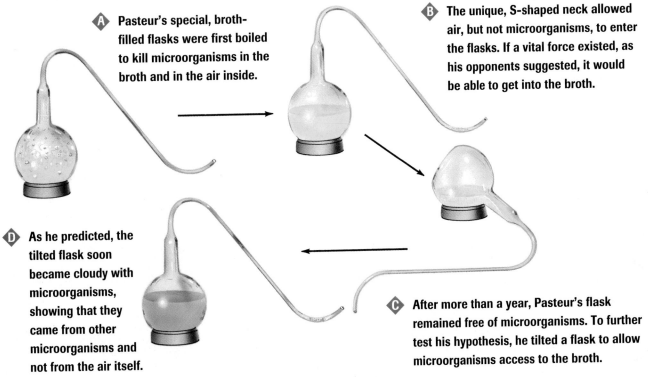

A Pasteur's special, broth-filled flasks were first boiled to kill microorganisms in the broth and in the air inside.

B The unique, S-shaped neck allowed air, but not microorganisms, to enter the flasks. If a vital force existed, as his opponents suggested, it would be able to get into the broth.

C After more than a year, Pasteur's flask remained free of microorganisms. To further test his hypothesis, he tilted a flask to allow microorganisms access to the broth.

D As he predicted, the tilted flask soon became cloudy with microorganisms, showing that they came from other microorganisms and not from the air itself.

Origins: The Modern Ideas

The concept of biogenesis has been accepted by biologists for more than 100 years. However, biogenesis does not answer one of the most basic questions: How did life begin on Earth? Modern hypotheses about the origin of life come from theoretical and experimental analysis, and not from fossils.

Simple organic molecules formed from the primitive atmosphere

For life to come into being, scientists agree that two developments must have occurred: the formation of simple organic molecules important to life, and the organization of these molecules into complex organic molecules such as proteins.

Several billion years ago, Earth's atmosphere had no free oxygen as it does today. Instead, it was composed of water vapor, hydrogen, methane, and ammonia. How these substances could have formed simple organic compounds important to life is a challenging scientific puzzle.

In the 1930s, a Russian scientist, Alexander Oparin, proposed a widely accepted hypothesis that life began in the early oceans. He suggested that energy from the sun and from lightning triggered chemical reactions to produce simple organic compounds from the substances present. Oparin envisioned many chemical reactions occurring in the atmosphere and the products raining down into the oceans to form what is often called a primordial soup.

In 1953, two American scientists, Stanley Miller and Harold Urey, decided to test Oparin's hypothesis by simulating the conditions of early Earth in the laboratory. In an experiment similar to the one shown in *Figure 17.13,* Miller and Urey circulated water in the form of steam with ammonia, methane, and hydrogen, and subjected the mixture to electric

primordial:
primordium (L) origin
The origin of life may have been in the primordial soup.

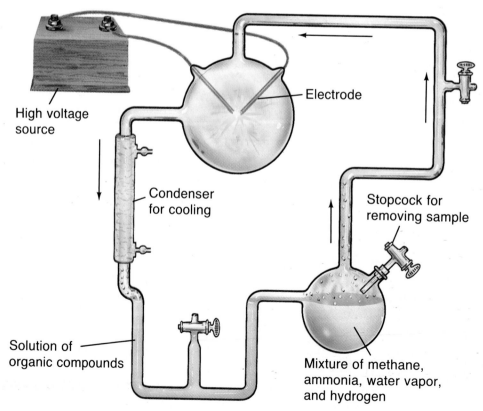

Figure 17.13
The "life-in-a-test-tube" experiment of Miller and Urey remains the cornerstone of the theories of the origin of life. At about the same time that they created amino acids in the lab, it was also learned that DNA carries the code for making proteins from amino acids. The photo (right) shows Stanley Miller with his experimental setup.

High voltage source

Electrode

Condenser for cooling

Stopcock for removing sample

Solution of organic compounds

Mixture of methane, ammonia, water vapor, and hydrogen

sparks to simulate lightning. They also repeatedly heated and cooled the mixture to simulate daily temperature fluctuations. After a week of such treatment, they analyzed the chemicals in the flask and found several kinds of amino acids, sugars, and other organic compounds, just as Oparin had predicted.

The formation of complex organic compounds and pre-cells

The next step in the origin of life, according to most scientists, was the formation of complex organic compounds and their enclosure by some type of bounding membrane. Since the 1950s, experiments have shown that amino acids will link together to form small proteins when heated in the absence of oxygen. Similar processes have also produced ATP and nucleic acids. Such experiments have led scientists to speculate that life may have originated in small pools of water, where amino acids could concentrate and be warmed.

History

The Puzzle of the Origin of Life

The origin of life on Earth is one of the greatest puzzles of science. Aristotle, a Greek philosopher, supported the theory of spontaneous generation.

Seeds of life   In the late 1600s, Jan Swammerdam of Holland proposed that the seeds of all life were formed when Earth was formed. He believed that each generation was contained in the one before, like nesting boxes or dolls. This theory of preexistence was largely ignored, but 100 years later, Charles Bonnet suggested a similar concept called preformation. According to Bonnet, each egg contains within it a tiny, preformed creature, which contains an egg with a tiny, preformed creature in it, and so on.

Did life come from life?   In the 18th century, Linnaeus pointed out relationships among different species. His work opened the door for other scientists to begin thinking of the development of organisms by evolution from a single, primitive life form. At about the same time as Linnaeus, Comte de Buffon, a French naturalist, first proposed that life originated on a warm, primitive Earth. Unfortunately, he believed that Earth had formed only about 60 000 years ago.

Organic soup   By the 20th century, scientists knew that Earth was billions of years old and the origin of life was likewise distant in time. In the 1920s, A. Oparin in Russia and J.B.S. Haldane in England independently hypothesized that organic compounds were created in Earth's primitive atmosphere and became organized to form cells. In 1953, Miller and Urey showed that organic compounds, including amino acids, could be synthesized in a simulation of Earth's primitive atmosphere using sparks instead of lightning. Experiments continue today to determine whether modern living things are the descendants of the amino acids that formed some 4 billion years ago on a primitive Earth.

CONNECTION TO   Biology

Why would the work of Linnaeus cause scientists to begin thinking about a common origin for all life?

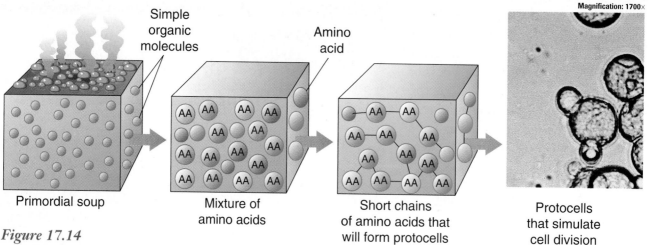

Simple organic molecules

Amino acid

Magnification: 1700×

Primordial soup

Mixture of amino acids

Short chains of amino acids that will form protocells

Protocells that simulate cell division

Figure 17.14

Sidney Fox took the Miller-Urey experiment one step further and showed how short chains of amino acids could cluster to form protocells. Under the microscope, protocells can look so much like living cells that scientists sometimes mistake them for new species of bacteria.

archaebacteria:
archaios (GK) ancient
Archaebacteria are similar to ancient bacteria.

How did these chemicals combine to form the first cells? The work of American biochemist Sidney Fox showed how this may have occurred. As shown in *Figure 17.14,* Fox was able to produce protocells by heating solutions of amino acids. A **protocell** is a large, ordered structure that carries out some activities associated with life such as growth, division, or metabolism.

The Evolution of Cells

Fossils show that by 3.5 billion years ago, life had diversified into several different types of cells. But scientists know that these cells don't represent the earliest cells, which arose sometime between 3.9 and 3.5 billion years ago. What were the earliest cells like, and how did they evolve?

Heterotrophic prokaryotes were the first true cells

Scientists speculate that the first forms of life were prokaryotes, which probably evolved from some type of protocell. Because Earth's atmosphere lacked oxygen, these organisms most likely were anaerobic, getting their energy through glycolysis and fermentation. Bathed in the warm sea, the first prokaryotes probably took in organic molecules for food. Because they obtained food from their surroundings, they were heterotrophs.

Over time, competition for nutrients by heterotrophic prokaryotes led to the evolution of Earth's first autotrophs. Organisms that could make food would have had a distinct advantage over other heterotrophic organisms. Many scientists think that these first autotrophs were similar to present-day archaebacteria as shown in *Figure 17.15.* **Archaebacteria** are a group of prokaryotes that live in harsh conditions with little sunlight and oxygen, such as hot sulfur springs and deep-sea vents. The earliest autotrophs probably made glucose by chemosynthesis as archaebacteria do today rather than by photosynthesis, which requires light-trapping pigments.

Next, photosynthesizing prokaryotes evolved. The cell organization of the 3.5-billion-year-old fossils from Australia indicate that these were among the first photosynthesizing prokaryotes.

Photosynthesizing prokaryotes changed Earth's atmosphere

You will recall that the process of photosynthesis releases oxygen. With the evolution of photosynthesizing prokaryotes, the concentration of oxygen in Earth's atmosphere began to increase. This increase in oxygen made possible the evolution of aerobic respiration, a more efficient method of energy conversion. The change was so drastic that scientists call it the oxygen revolution, as shown in *Figure 17.16* on page 418. An increase in the diversity of prokaryotes in the fossil record indicates that the oxygen revolution was fully underway by 2.8 billion years ago.

Figure 17.15

The earliest true cells were probably heterotrophic prokaryotes. Competition for nutrients probably led to the evolution of chemosynthetic organisms similar to archaebacteria that live in hot springs like the one shown. Ancient microfossils like the ones in the inset photo are similar to archaebacteria.

Magnification: 20 000 ×

The Endosymbiont Hypothesis

Many years ago, scientists noted that cyanobacteria and chloroplasts resemble each other. Likewise, they saw that mitochondria and bacteria look alike. These key observations led to the endosymbiont hypothesis. This hypothesis states that eukaryotes evolved through a symbiotic relationship between primitive prokaryotes.

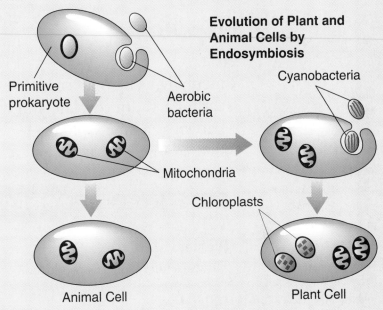

Evolution of Plant and Animal Cells by Endosymbiosis

Primitive prokaryote

Aerobic bacteria

Cyanobacteria

Mitochondria

Chloroplasts

Animal Cell

Plant Cell

Origin of the endosymbiont hypothesis In 1962, scientists discovered DNA in chloroplasts. This finding led American biologist, Lynn Margulis, to develop the endosymbiont hypothesis. In the years since then, she has been compiling evidence to support it. Other observations include the fact that cholorplasts and mitochondria have their own ribosomes and grow and reproduce independently of the cell itself, and that the cell has no means to make either chloroplasts or mitochondria.

Structural evidence How cyanobacteria and bacteria evolved a symbiotic relationship and came to function as cell organelles is largely a matter of scientific speculation. However, compelling structural evidence indicates that the ancestors of mitochondria and chloroplasts were once free-living prokaryotes. For example, mitochondria and bacteria can both reproduce themselves, have similar nucleic acids, are about the same size and shape, and carry out protein synthesis on ribosomes. Chloroplasts and cyanobacteria also share important characteristics, the most important of which is the ability to carry out photosynthesis.

Thinking Critically

Some prokaryotes live in close association with eukaryotes today. Could this fact be used as evidence for the endosymbiont hypothesis? Explain your answer.

17.2 The Origin of Life **415**

Scientists estimate that life formed sometime between 3.9 and 3.5 billion years ago from chemicals present on primitive Earth.

Making Microspheres

One of the most important steps in the origin of life must have been the enclosure of organic compounds by a membrane, creating a protocell. Protocells can be produced experimentally using materials that scientists believe were present on early Earth. In this lab, you will produce a type of protocell known as a microsphere.

PREPARATION

Problem
How are microspheres produced?

Objectives
In this Biolab, you will:
- **Infer** how microspheres may have been produced on primitive Earth.
- **Compare and contrast** microspheres with living cells.

Materials
microscope
microscope slide
coverslip
50-mL graduated cylinder
50-mL Erlenmeyer flasks (2)
400-mL beaker
1% NaCl solution
aspartic acid
glutamic acid
glycine
hot plate
ring stand
ring stand and clamp
tongs
watch or clock with second hand
stirring rod
pipette
apron
oven mitts
goggles

Safety Precautions
Use care when handling hot objects. Be sure to wear oven mitts.

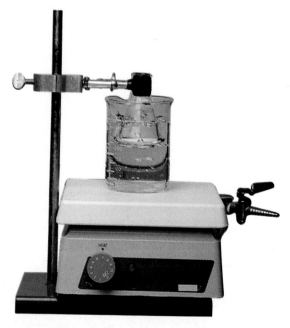

PROCEDURE

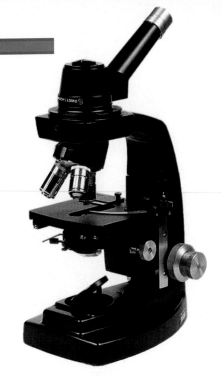

1. Pour 250 mL of water into the beaker, and place the beaker on a hot plate.

2. Clamp one of the flasks to the ring stand.

3. Add 0.5 g of each of the following amino acids to the flask: glycine, aspartic acid, and glutamic acid. Use the stirring rod to mix thoroughly.

4. When the water in the beaker begins to boil, carefully loosen the clamp and lower the flask into the water. Tighten the clamp.

5. Heat the amino acids for 20 minutes, keeping the water in the beaker at a simmer.

6. Measure and pour 5 mL of the 1% NaCl solution into the second flask, and place the flask on the hot plate.

7. When the NaCl solution begins to boil, use tongs to remove it from the hot plate. Slowly add the NaCl solution to the amino acids. Turn off the hot plate.

8. Loosen the clamp on the ring stand, and raise the flask containing the amino acid solution out of the beaker. Allow the contents of the flask to cool for ten minutes.

9. Make a copy of the data table.

10. Prepare a wet mount using one drop of the mixture, and observe under low power. Locate and diagram a microsphere.

11. Switch to high power to examine the microspheres. Draw a diagram in your data table.

12. After you have finished observing the microspheres, clean and return all equipment to its proper place.

Characteristics of Microspheres		
	Drawing	**Description**
Low power		
High power		

ANALYZE AND CONCLUDE

1. **Comparing and Contrasting** How do microspheres resemble cells? How are they different?

2. **Interpreting Observations** What features of living cells do microspheres exhibit?

3. **Thinking Critically** How does the method you used to generate microspheres compare with the conditions on primitive Earth?

Going Further

Thinking Critically
Do you think your microsphere experiment would have worked if you had substituted other amino acids? How could you test this hypothesis?

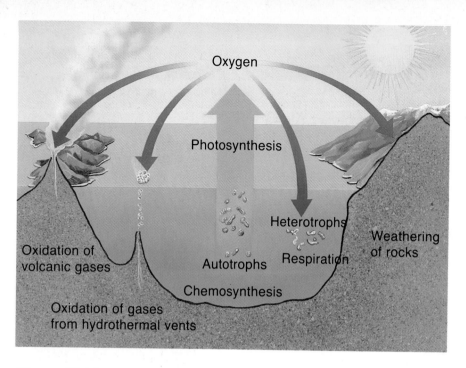

Oxygen

Photosynthesis

Heterotrophs

Oxidation of
volcanic gases

Autotrophs   Respiration

Chemosynthesis

Weathering
of rocks

Oxidation of gases
from hydrothermal vents

The release of oxygen through photosynthesis had another effect. Lightning converted much of the atmospheric oxygen into ozone (O_3) molecules, resulting in an ozone layer 10 to 15 miles above Earth's surface. The ozone layer shielded emerging life forms from the harmful effects of ultraviolet radiation, and thus allowed for the evolution of more complex organisms. The ozone layer still exists today, protecting Earth's surface from ultraviolet radiation.

Figure 17.16

Oxygen buildup in the atmosphere resulted from photosynthesis by early prokaryotes. Oxygen was then available for aerobic respiration. By the end of the Precambrian Era 545 million years ago, oxygen content had reached approximately 10 percent of present-day levels.

Connecting Ideas

Four-and-a-half billion years ago, a story began that is still unfolding—the story of life on Earth. Details of the story are found in the many bits and pieces of the fossil record. The fossil record of life on Earth may not be perfect, but it is unambiguous. Through the careful analysis and dating of fossils, scientists have shown that life evolved from the unicellular, photosynthesizing prokaryotes of several billion years ago, to the soft-bodied animals and plants of several hundred million years ago, to the diversity of species present on Earth today.

How does evolution work, and what are the mechanisms for change? How did the unicellular organisms living in the warm oceans of 3.5 billion years ago evolve into the remarkable diversity of life forms we see today?

Section Review

Understanding Concepts

1. How did Pasteur's experiment finally disprove spontaneous generation?
2. What was Oparin's hypothesis, and how was it tested experimentally?
3. Why do scientists think the first living cells were anaerobic heterotrophs?

Thinking Critically

4. Some scientists speculate that lightning was not present on early Earth. How could you modify the Miller-Urey experiment to reflect this new idea? What energy source would you use to replace lightning?

Skill Review

5. **Sequencing** Make a time line sequencing the evolution of life from protocells to eukaryotes. For more help, refer to Organizing Information in the *Skill Handbook.*

The Impact of Extinctions

New Beetle Now Extinct

THE POST-TRIBUNE   OCTOBER 9, 1994

A team of naturalists following a timber crew in a New Guinea rain forest recently announced the discovery of a new insect species, the horned emerald beetle. Today, the same team reported that this newly discovered species is now extinct because its habitat was destroyed by clear-cutting.

Every living species occupies a unique ecological niche that depends upon other species around it. In the worst cases, the extinction of a single species in the web of life may trigger the extinction of 10 to 30 other species of organisms.

Causes of extinction The most common cause of extinction is destruction of an organism's habitat, either from conversion of woodlands to farmland or from human development of an area.

Ivory-billed woodpeckers have already lost the battle. Some scientists estimate that more than 6000 species disappear yearly from deforestation alone.

Extinction can also be caused by excessive hunting, as in the case of elephants and sperm whales. Insecticides may be toxic to other animals that eat the insects or plants on which the insecticide is sprayed. Bald eagles became endangered in this way.

The need for biodiversity has been used as an argument to protect certain species on the verge of extinction, such as spotted owls and tree frogs. They are organisms that are appealing to humans. But what about other species?

Different Viewpoints

In fact, the extinction of other species less appealing to humans may be much more tragic. Plants, protists, fungi, and microorganisms are essential to food chains and to the recycling of nitrogen, carbon, and phosphorus.

Some conservationists argue that much of the effort spent on preserving popular furry and feathery species should instead go to identifying and studying plants, fungi, and other species. Others say that endangered mammals and birds perform a public-relations function in that they draw the public's attention to all endangered species. What do you think?

INVESTIGATING   the Issue

1. **Writing About Biology** Read "The Diversity of Life," by Edward O. Wilson, *Discover*, September 1992, pp. 45-68. Summarize Wilson's point of view on extinctions.
2. **Analyzing Consequences** Suppose that mice suddenly became extinct for some reason. Make lists of immediate and long-term consequences of this occurrence.

Reviewing Main Ideas

17.1 The Record of Life

- Fossils provide a record of life on Earth. Fossils come in many forms, such as the print of a leaf, the burrow of a worm, or a stone replica of a bone.
- Through the study of fossils, scientists can learn about the diversity of life. Fossils also provide clues about the behavior of organisms and the environments in which they lived.
- Scientists divide Earth's history into a series of eras, periods, and epochs known as the Geologic Time Scale. This scale provides a timetable for the appearance of living organisms during the history of Earth.

17.2 The Origin of Life

- During the 17th, 18th, and 19th centuries, scientists experimented to test the idea of spontaneous generation. As a result of these experiments, scientists accept biogenesis, the idea that life comes from existing life.
- There are several competing theories about the origin of life. Many scientists agree that organic compounds were formed from elements present in Earth's primitive atmosphere and oceans.
- The earliest cells were probably anaerobic, heterotrophic prokaryotes. Over time, the ability to capture energy to make food by photosynthesis evolved, changing the atmosphere and triggering the evolution of aerobic cells and eukaryotes.

Key Terms

Write a sentence that shows your understanding of each of the following terms.

archaebacteria	protocell
biogenesis	spontaneous
fossil	generation
plate tectonics	

Understanding Concepts

1. Why are amber-preserved or frozen fossils of great value to scientists?
2. How can scientists infer past climates from fossilized pollen?
3. Why is it important to know the condition and exact location of a fossil?
4. What types of fossils or objects can be dated with carbon-14?
5. How is the rate of decay of an isotope used in the dating of fossils?
6. Why do scientists think that the 3.5-billion-year-old fossils from Australia were not the first organisms on Earth?
7. What is the relationship between protocells and the first living cells?
8. What is the probable sequence of events in eukaryote evolution?

Using a Graph

9. The following graph illustrates the decay rate of a particular radioactive element. According to the graph, how long does it take for half of the element to decay? How much of the original material is left after 2 billion years? After 4 billion years?

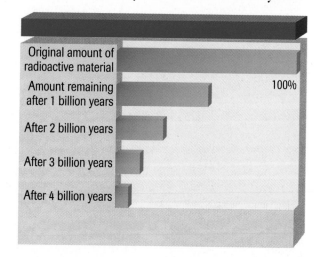

Original amount of radioactive material

Amount remaining after 1 billion years   100%

After 2 billion years

After 3 billion years

After 4 billion years

Relating Concepts

10. Make a concept map that relates the following terms and phrases. Supply the appropriate linking words for your map.

protocell, autotroph, heterotroph, eukaryote, prokaryote, endosymbiont hypothesis, anaerobic, aerobic

Applying Concepts

11. Is the evolution of the first living organism an example of spontaneous generation?
12. Why is a thorough understanding of geology important to paleontologists?
13. How might paleontologists use fossils to draw conclusions about the social behavior of animals? Which social behaviors might they detect?
14. Explain why you think that synthetic life may or may never be formed in the laboratory.
15. How might the movement of continents affect the future evolution of organisms?

A Broader View

16. What advantage would a primitive cell containing mitochondria have over a cell that had no mitochondria?

Thinking Critically

Identifying Variables

17. **Biolab** If you were performing an experiment about the origin of life, what environmental conditions might you subject microspheres to in order to simulate primitive Earth?

Applying Concepts

18. **Minilab** The time line shown here was constructed by a group of students. Why do you think the students made the Precambrian Era the longest era?

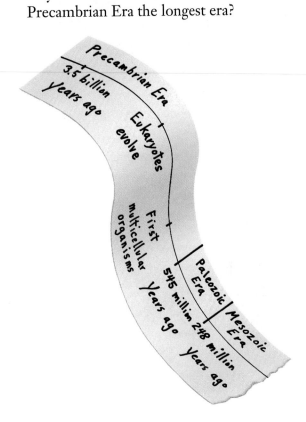

Analyzing Data

19. **Minilab** Why do scientists often make casts from mold fossils? What details can casts show that can't be seen on molds?

Connecting to Themes

20. **Energy** How has the way organisms obtain energy evolved throughout the history of life?
21. **Evolution** How can fossils be used to trace the evolution of particular adaptations?
22. **Patterns of Change** What can fossils indicate about geological changes during Earth's history?

CHAPTER
18 Evolution

On a hot, dry day, just after the rainy season of southern Africa, the common mole-rat, *Cryptomys hottentotus*, drags across the sandy surface in search of food and a new home. Once selected, it begins the lengthy process of digging out its extensive, underground burrow system. Using its large, chisel-like incisors, the common mole-rat starts a small hole as its shortened limbs push the dirt out of the way. When the hole is large enough, it slithers its cylindrical body in—aided by a loose layer of skin that allows it to turn and complete the burrow.

Common mole-rats don't look much like other rodents; the body, skull, and teeth of common mole-rats are adapted to life underground. But there is another variation. Common mole-rats are completely sightless. Limited or reduced vision is not uncommon in the animal kingdom.

How do underground or cave-dwelling species evolve from ancestors with normal vision? In this chapter, you'll learn how the mechanism of evolution can explain the origin of unique species, such as mole-rats.

Members of the cave-dwelling fish genus *Amblyopsis,* the burrowing snake genus *Typhlops,* and the African mole-rat genus *Cryptomys* are examples of sightless animals whose close relatives have normal eyesight. Why are so many cave-dwelling or burrowing animals sightless, and how does evolution work to produce such species?

SECTION
18.1 Natural Selection and the Evidence for Evolution

Section Preview

Objectives

Summarize Darwin's theory of evolution by natural selection.

Relate the idea of natural selection to the origin of structural and physiological adaptations.

Evaluate the various lines of evidence for evolution.

Key Terms

artificial selection
natural selection
mimicry
camouflage
homologous structure
analogous structure
vestigial structure

Observations of living organisms today provoke questions about their evolutionary origins. The theory for how organisms change over time was developed close to 150 years ago, and it is still the one that is favored by most scientists.

Charles Darwin and Natural Selection

The modern theory of evolution is perhaps the most important concept in biology. It's impossible to understand any of the fields of biology without understanding evolution. What is the origin of the concept of evolution?

Fossils interested scientists in evolution

A rich fossil record has been important to biological science since the 18th century and formed the basis of early evolutionary concepts.

Scientists wondered where these species came from, why some no longer existed, and what relationship, if any, they have to modern species.

When it became clear through geological studies that Earth was much more ancient than previously believed, many early 19th-century scientists were convinced that life slowly changed over time, or evolved. But how? Several ideas were proposed during the course of the century, but only one has survived to become universally accepted by scientists today. English scientist Charles Darwin (1809-1882) is considered to be the founder of modern evolutionary theory.

Darwin studied the natural world during the voyage of the *Beagle*

In 1831, at the age of 21, Darwin was recommended for the position of naturalist on *H.M.S. Beagle*, a ship chartered for a five-year collecting and mapping expedition to South America and the South Pacific—a voyage that would forever change the science of biology.

As ship's naturalist, Darwin's job was to collect, study, and store biological specimens discovered along the journey. The voyage of the *Beagle* took him through a number of environments with some of the world's greatest biodiversity.

Darwin's careful study of the animal and plant life gathered along the way helped him develop his theory of how evolution occurs.

Darwin's observations in the Galapagos

For Darwin's hypothesis, observations made in the Galapagos Islands were among his most important. The Galapagos Islands, a group of small islands about 1000 km off the west coast of South America, support a great diversity of animal and plant life. Here Darwin studied and compared the anatomy of many species of reptiles, insects, birds, and flowering plants that are unique to the islands, yet similar to species seen in other parts of the world, as shown in *Figure 18.1.* By the end of his trip, Darwin was convinced that evolution occurs—that species can and do change. However, he still needed to test this idea before he could explain how such changes occur.

Figure 18.1

The voyage of *H.M.S. Beagle,* which lasted for nearly five years, took Darwin from England to South America, around Cape Horn and north to the Galapagos Islands, and west through the South Pacific to Australia. Animal species in the Galapagos possess adaptations that are unique to the islands; however, they are also similar to species found in other parts of the world.

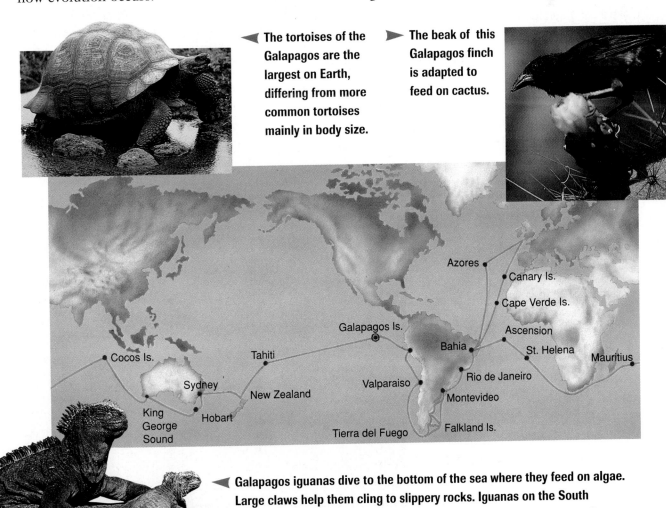

◀ The tortoises of the Galapagos are the largest on Earth, differing from more common tortoises mainly in body size.

▶ The beak of this Galapagos finch is adapted to feed on cactus.

◀ Galapagos iguanas dive to the bottom of the sea where they feed on algae. Large claws help them cling to slippery rocks. Iguanas on the South American mainland have smaller claws for moving efficiently through trees.

18.1   Natural Selection and the Evidence for Evolution   **425**

MiniLab

How variable are traits?

Within a population, individuals vary in many observable traits. For example, individual birds in a population can vary in weight, length, and color. Natural selection acts upon variation, favoring traits that are best suited for a particular environment. In this activity, you will determine the variation in leaves collected from nearby trees.

Procedure

1. Collect a sample of 20 leaves from a nearby tree or bush. Do not pull leaves directly from trees. Use leaves that have fallen to the ground.

2. Using a ruler, measure the length in millimeters of each leaf blade, from the tip of the blade to its union with the petiole, or stalk, as shown here.

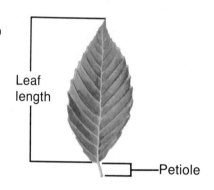

3. Make a bar graph for leaf length. Plot *Leaf Length* on the horizontal axis and *Number of Leaves* on the vertical axis.

4. Count the number of leaves that fall within each 5 mm interval. Plot these data on your graph.

Analysis

1. How would you describe the shape of your graph?

2. Which intervals contain the most leaves? The fewest leaves?

3. What are some possible sources of error in your graph?

Darwin completed his studies in England

Darwin returned from his voyage on the *Beagle* in 1836, and spent the next 22 years studying his collections and conducting experiments.

Darwin was also intrigued by an essay written at the time that suggested that the human population was growing faster than the food supply on Earth. Darwin knew that many species produce large numbers of offspring, and since Earth was not covered with such species, Darwin suspected that there must be a struggle for existence among individuals. Darwin envisioned many kinds of struggles—competition for food and space, escape from predators, and ability to find shelter. Only some of the individuals in any population survive the struggle long enough to produce offspring of their own. But which ones?

Darwin's answer to his question about the struggle for existence came from his own pigeon-breeding experiments. Darwin observed that in populations of plants or animals, individuals have different variations of traits that can be inherited. Guinea pigs, for example, *Figure 18.2,* differ greatly in the patterns and colors of their coats.

Figure 18.2

These guinea pigs are in the same species, yet their coats show drastic differences. The many varieties of guinea pigs seen today are the result of artificially selecting parents with desirable coats.

By breeding pigeons that had desirable variations, Darwin was able to produce offspring with these same features. Breeding experiments are an example of **artificial selection**—a technique in which a breeder selects particular traits. Darwin wondered if there were some force in nature similar to artificial selection.

Darwin's explanation for evolution

Using data he gathered from the natural world, Darwin gradually began to form his well-known idea of evolution by natural selection.

Natural selection is a mechanism for change in populations that occurs when organisms with favorable variations for a particular environment survive, reproduce, and pass these variations on to the next generation. Organisms with less-favorable variations are less likely to survive and pass on traits to the next generation. Therefore, each new generation is made up largely of offspring from parents with the most favorable variations. The idea of natural selection is summarized in *Figure 18.3*

Figure 18.3

Darwin proposed the idea of natural selection to explain how populations of organisms evolve.

Overproduction of offspring

A In nature, there is a tendency toward overproduction of offspring. Fishes, for example, lay thousands, sometimes millions, of eggs.

Variations exist in populations

B In any population of organisms, individuals will exhibit slight variations. Fishes may differ slightly in color, fin and tail size, and speed.

Some variations are favorable; some are not

C Individuals with variations favorable for a particular environment are more likely to survive and pass those variations on to the next generation than individuals with less-favorable variations. For example, a fast fish with a skin color that allows it to blend in better with its surroundings is more likely to survive to reproductive age than a slow fish with more obvious coloring.

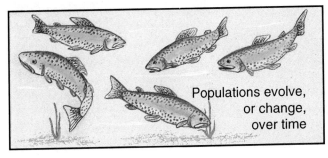

Populations evolve, or change, over time

D Gradually, offspring of survivors will make up a larger proportion of the population. Depending upon environmental factors, after many generations, a population may come to look entirely different.

In the 1850s, Darwin began work on a detailed, multivolume book to present his idea of natural selection to the scientific community. In 1859, Darwin published his book, *On the Origin of Species by Natural Selection.* This book convinced many scientists that evolution does occur through natural selection, and today this theory is the unifying one for all biology.

Natural Selection and Adaptations

Have you ever wondered why some plants have thorns or why some animals have distinctive coloring? How do such adaptations originate? Recall that an adaptation is any trait that aids the chances of survival and reproduction of an organism. Darwin's theory of natural selection can be applied to explain the evolution of adaptations in organisms.

Structural adaptations arise over many generations

How can Darwin's theory be used to explain unique structural adaptations in species, such as those seen in common mole-rats? Common mole-

rats possess a suite of features that adapt them to life underground, but their ancestors lived aboveground and didn't have such characteristics. How did the features of modern-day mole-rats evolve from the ancestral mole-rats? *Figure 18.4* illustrates a likely scenario for the evolution of the common mole-rat.

The adaptations of common mole-rats are examples of structural adaptations—changes in the structure of body parts. Some of the more obvious structural adaptations in organisms are used for defense against predators, such as the thorns of rose-bushes or the spines of sea urchins. However, others are more subtle. For example, **mimicry** is a structural adaptation that provides protection for an organism by copying the appearance of another species, as shown in *Figure 18.5*. Another type of defensive adaptation that involves changes to the color of organisms is camouflage. **Camouflage** is a structural adaptation that enables an organism to blend in with its surroundings. Organisms that are well camouflaged are more likely to escape predators and survive to reproduce.

Figure 18.4

The evolution of the common mole-rat can be explained by Darwin's theory of natural selection.

Ⓐ According to some scientists, the ancestors of common mole-rats resembled the African rock rat.

Ⓑ Over time, variations arose. Large teeth and claws, for example, allowed some individuals to dig deeper holes and avoid predators.

Figure 18.5

Camouflage and mimicry involve changes to the external appearance of an organism.

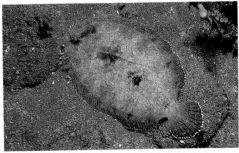

▲ Camouflage coloration enables organisms to blend in with their surroundings. Can you see how the coloration of this flounder allows it to avoid predators?

▲ The bee orchid is a flower that mimics the appearance of females of the bee species that pollinate it. By mimicking females, the flower attracts males, which then spread pollen to other flowers.

C If these individuals were better able to survive and reproduce, useful variations could be passed on. Over time, most individuals in the population would possess these beneficial adaptations.

MiniLab

How is camouflage an adaptive advantage?

Camouflage is a structural adaptation that allows organisms to blend in with their surroundings. In this activity, you'll discover how natural selection operates to produce camouflage adaptations in organisms.

Procedure

1. Working with a partner, punch 100 dots from a sheet of white paper with a paper hole punch. Repeat with a sheet of black paper. These dots will represent black and white insects.
2. Place both white and black dots on a sheet of black paper.
3. Select one student to play the bird. That student must look away from the paper, turn back to it, and then immediately pick up the first dot he or she sees.
4. Repeat the procedure to see how many insects of each color can be collected in two minutes.

Analysis

1. What color were most of the dots that the "bird" collected?
2. How does color affect the survival rate of insects?
3. What might happen over time to a similar insect population in the wild?

D Over many generations, evolution would result in the adaptations seen in common mole-rats of today. Blindness most likely evolved because as vision decreased in importance, sightedness no longer gave the mole-rat a selective advantage. In fact, in this environment, eyes are a handicap due to the potential for infection and injury.

How can natural selection be observed?

The following are data from an actual natural selection experiment by H.B. Kettlewell. Kettlewell studied camouflage adaptations in a population of light- and dark-colored moths that live on the trunks of trees. Kettlewell noticed that pollution in the area caused lichens that usually lived on the bark of the trees to die. The tree bark was darker due to the lack of lichens. Kettlewell was interested in discovering if this played a role in natural selection of the moth population.

Analysis

Kettlewell captured, released, and recaptured moths from the country and from the polluted city. The number of moths recaptured indicates their survival rate in the environment. Shown here are raw data from Kettlewell's experiment.

Location		Numbers of light moths	Numbers of dark moths
Unpolluted country	number released	469	473
	number recaptured	62	30
Polluted city	number released	137	447
	number recaptured	64	154

Thinking Critically

How can you explain the differences in survival rates of moths in the country and city? How is natural selection involved?

Depending on the type of adaptation and the rate of reproduction of an organism, as well as environmental factors, structural adaptations may take millions of years to develop. However, some structural adaptations can evolve relatively quickly in geologic terms. For example, a well-known study has shown that camouflage adaptations can evolve in moth species within 100 years.

Physiological adaptations can develop rapidly

Would you believe that some of the medicines developed during this century may no longer be useful for fighting disease? Only 50 years ago, the antibiotic drug, penicillin, was considered a wonder drug because it could kill many types of disease-causing bacteria. Today, scientists know that penicillin is not as effective as it used to be because many species of bacteria have been selected for physiological adaptations that make them resistant to penicillin, *Figure 18.6.* Physiological adaptations are changes in an organism's metabolic processes.

Scientists have also observed rapid changes in pest organisms such as insects and weeds. After being exposed to particular pesticides, many species of insects and weeds have been selected for physiological resistance to the chemicals.

Figure 18.6

Examples of bacterial resistance to antibiotics, such as penicillin, provide scientists with direct evidence for evolution in progress.

Individual bacteria in a population show variation. Some bacteria possess a gene that makes them resistant to some antibiotics.

When the population is exposed to an antibiotic, some individuals die, but resistant bacteria survive.

Resistant bacteria survive and reproduce. In time, the entire population is resistant to a certain antibiotic.

Evidence for Evolution

The evolution of camouflage in moths and the development of physiological resistance to chemicals in organisms provide examples of the process of evolution. Such examples occur rapidly enough that they can be observed in a laboratory. However, these examples are exceptions; most of the evidence for evolution comes from indirect sources.

Fossils can show evolutionary changes over time

Did you know that the ancestors of whales were once land-dwelling, doglike animals? Scientists have learned about the evolutionary history of whales by studying their fossils. In the last chapter, you learned how fossils provide a record for life. Although the fossil record is incomplete, it does provide evidence that evolution has indeed occurred.

Working with an incomplete fossil record can be compared to completing a jigsaw puzzle with missing pieces. It's great if all the pieces are there, but even if some are missing, you can still understand the overall picture. It's the same with fossils. Although scientists can't trace each and every step in the evolution of some species because intermediate forms of these species cannot be found, they can still understand the general pathway of evolution. When the fossil record is complete enough, scientists can show a step-by-step sequence of evolution. *Figure 18.7* shows the evolution of the camel as indicated by fossil skulls, teeth, and limb bones. As you can see, 40 to 50 million years ago, camels were a species of small, rabbit-sized animals.

Figure 18.7

Fossils indicate that during the Eocene epoch 37 to 54 million years ago, camels had small body size, four-toed feet, and low-crowned teeth. By Miocene times, 30 million years later, species of camels had grown larger and evolved two-toed feet and high-crowned teeth. By studying fossils, scientists have been able to trace the evolution of camels to the forms of present-day species seen today.

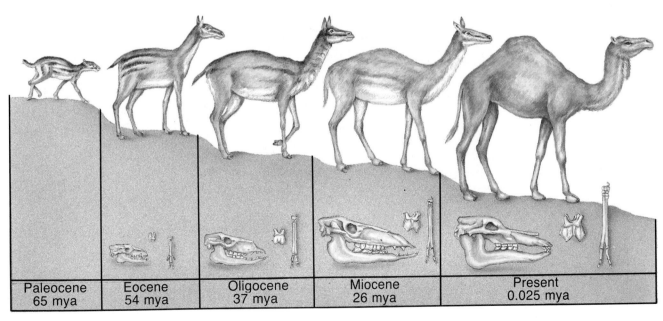

Paleocene 65 mya	Eocene 54 mya	Oligocene 37 mya	Miocene 26 mya	Present 0.025 mya

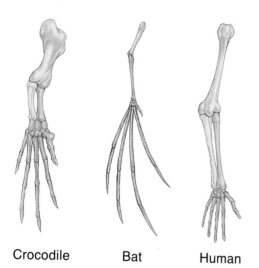

Crocodile   Bat   Human

Figure 18.8

The forelimbs of crocodiles, bats, and humans are examples of homologous structures. Can you see how the bones of the forelimb have become modified in relation to their function?

Anatomical studies indicate evolutionary relationships

Do the forelimbs of the animals shown in *Figure 18.8* look similar to you? If you said yes, then you just made a type of observation that scientists use to establish evolutionary relationships among organisms. As you can see, although the limbs of these animals look strikingly different from the outside and vary in function, details of their skeletons are similar.

Scientists view such similarities as evidence that these organisms evolved from a common ancestor with the same basic limb structure. Over the course of evolution, vertebrates moved into different environments. A modified structure that is seen among different groups of descendants is called a **homologous structure.** Within each environment, animals faced very different needs for survival. Animals that had limbs that were most useful in their environ-ment survived. In this way, species became adapted to different ways of life. Although different species used their forelimbs for different functions in different environments, they are said to be homologous structures.

Homologous structures are often similar in structure, in function, or in both structure and function. How-ever, similarity of function doesn't always mean that two organisms are closely related. Homologous struc-tures are structures with a common evolutionary origin. Compare the structures of the butterfly wing and the bird wing in *Figure 18.9.* Bird and butterfly wings are not similar in structure, but they do have the same function. The wings of birds and insects evolved independently in two distantly related groups of ancestors. Any body structure that is similar in function but different in structure is an **analogous structure.**

Analogous structures can't be used to indicate evolutionary relationships among organisms, but they do pro-vide evidence of evolution. Insect and bird wings must have evolved when their different ancestors indepen-dently adapted to similar ways of life in similar environmental conditions.

Figure 18.9

Although they have the same function, insect and bird wings are not similar in structure. Bird wings are made up of a set of bones, whereas insect wings are mainly composed of a tough substance called chitin.

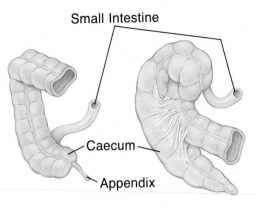

Small Intestine

Caecum

Appendix

Human   Horse

Functionless structures indicate evolutionary pathways

Do you know what your appendix is used for? Your doctor may not be able to tell you what your appendix *is* for, but an evolutionary biologist can tell you what an appendix *was* for. That's because many organisms, such as yourself, contain structures that have reduced functions but were once used by ancestral organisms. Any body structure that is reduced in function in a living organism but may have been used in an ancestor is known as a **vestigial structure.**

Particular structures become vestigial as species change in form and behavior. The structure, although it may serve no function, continues to be inherited as part of the body plan for that species. The eyes of sightless species, such as common mole-rats and cave fish, may be considered vestigial structures because eyes were most likely functional in their ancestors. *Figure 18.10* shows a common vestigial structure in humans.

Embryological development shows evolution from a common ancestor

It's easy to tell the difference between adult birds and adult mammals, but do you think you could pick them out by looking at their embryos? *Figure 18.11* illustrates various stages of embryological

Figure 18.10

Vestigial structures provide evidence of evolution because they show structural change over time. The small, wormlike appendix has reduced function in humans. It is a remnant of the digestive pouch known as the caecum, which is used by plant-eating animals for digestion of cellulose.

development in fish, birds, rabbits, and humans. Notice the similarities in structure among the embryos. In the earliest stage, a tail and gill slits can be seen in all species. As development continues, the embryos become more and more distinct, and in the stages before birth, they attain their distinctive forms. Similarities among vertebrate embryos suggest evolution from a common ancestor.

Figure 18.11

Comparative embryological studies of vertebrates indicate that they share a common ancestry. The presence of gills and tails in early vertebrate embryos supports evidence from the fossil record that aquatic, gill-breathing vertebrates preceded air-breathing, terrestrial species.

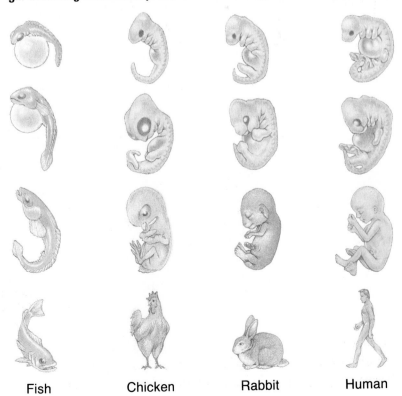

Fish   Chicken   Rabbit   Human

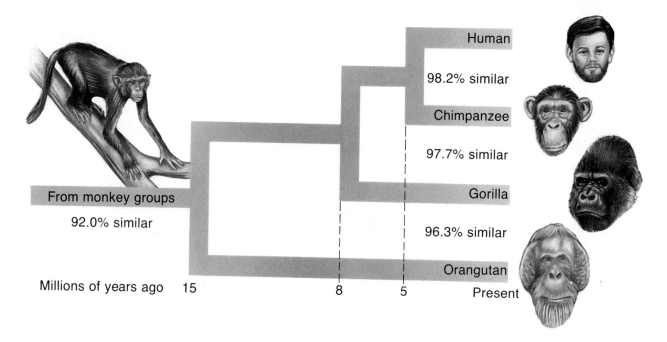

Human

98.2% similar

Chimpanzee

97.7% similar

Gorilla

96.3% similar

Orangutan

From monkey groups

92.0% similar

Millions of years ago 15 8 5 Present

Figure 18.12

Comparisons of DNA have shown that human and chimpanzee DNA is nearly 99 percent identical, and that these organisms are more closely related to each other than they are to other apes.

Genetic comparisons may reveal hidden relationships

Knowledge of genetics can help us understand relationships among organisms. It can also help tell us how different groups are related. The most recent evidence for evolution has come from comparisons of DNA among taxonomic groups. Many scientists consider genetic studies to be more reliable than anatomical studies for sorting out evolutionary relationships among species. It is also believed that DNA studies can reveal when species diverged from their ancestral types.

Since the 1970s, research has shown that by comparing nucleotide sequences in the DNA of species, it's possible to construct evolutionary trees showing relationships among the species.

Figure 18.12 shows how DNA studies have been useful for interpreting the relationships among apes and humans.

Section Review

Understanding Concepts

1. Why is the fact that snakes have skeletal remnants of legs considered evidence for evolution?
2. How does the evolution of camouflage and mimicry in animals illustrate natural selection?
3. How do homologous structures provide evidence for evolution?

Thinking Critically

4. After sequencing a section of DNA from two newly discovered species, you find that the DNA sequences differ by only two percent. Is it reasonable to assume that these two organisms are closely related? Explain.

Skill Review

5. **Sequencing** Whales evolved from ancestors that had legs. Using your knowledge of natural selection, sequence the steps that may have occurred during the evolution of whales from their terrestrial ancestors. For more help, refer to Organizing Information in the *Skill Handbook*.

18.2 Mechanisms of Evolution

I f you visit the Grand Canyon National Park today, you might see an interesting case of speciation. On the north side of the canyon, you will likely see a population of dark-brown squirrels with white abdomens and tails. If you visit the south side, you might see squirrels that are charcoal gray. What has happened here? The formation of the canyon itself resulted in the complete isolation, divergence, and eventual speciation of these two kinds of squirrels.

In the natural world, several mechanisms contribute to the evolution of species. Two factors that are the most consequential are geography and the effects of environmental change.

Section Preview

Objectives

Summarize the effects of the different types of natural selection on gene pools.

Relate mechanisms of speciation to changes in genetic equilibrium.

Explain the role of natural selection in convergent and divergent evolution.

Key Terms

gene pool
allelic frequency
genetic equilibrium
genetic drift
stabilizing selection
directional selection
disruptive selection
speciation
geographic isolation
reproductive isolation
polyploid
gradualism
punctuated equilibrium
adaptive radiation
divergent evolution
convergent evolution

Population Genetics and Evolution

As you recall, adaptations of species, such as the shape of a tooth or the color of a flower, are determined by the genes contained in the DNA code. When Charles Darwin developed his theory of natural selection in the 1800s, he did so without including the idea of the gene. Since that time, a great deal of information about genes has been gathered and Darwin's theory has been modified. The modern understanding of evolution is based on Darwin's theory, but now also includes principles of genetics that were not known at the time.

Populations evolve, individuals don't

Can individual organisms evolve? That is, can they acquire or lose structures or characteristics in response to natural selection? As you know, natural selection can act upon an individual's phenotype, the external expression of genes. If an individual organism possesses a phenotype that isn't adapted to the environment, it may result in the individual's inability to successfully compete. However, within the lifetime of one individual, new features cannot evolve in response to natural selection. Rather, natural selection operates only on populations over many generations.

To understand how the genes of a population can change over time, picture the entire collection of genes among a population as its **gene pool.** Consider, for example, the population of snapdragon plants that is shown in *Figure 18.13.* Snapdragon plants show a pattern of heredity that is known as incomplete dominance, which you learned about in Chapter 14.

If you know the genotypes of all the organisms in a population, you can calculate the **allelic frequency,** the percentage of a particular allele in the gene pool. A population in which frequency of alleles does not change from generation to generation is said to be in **genetic equilibrium,** *Figure 18.13.*

Changes in genetic equilibrium lead to evolution

A population that is in genetic equilibrium is not evolving. Because allelic frequencies remain the same generation after generation, no new traits are gained or lost. Only when the genetic equilibrium of a population is disrupted can evolution occur.

What factors cause changes to genetic equilibrium? Any factor that affects genes can change allelic frequencies, and thus disrupt the genetic equilibrium of populations. You learned in Chapter 13 that one such mechanism for genetic change is mutation. Many mutations are caused by factors in the environment, such as radiation or chemicals, but others happen purely by chance.

Figure 18.13

Incomplete dominance produces three phenotypes, so in these snapdragon plants, red flowers are homozygous (*RR*) for the red trait, white flowers are homozygous (*R'R'*) for the white trait, and pink flowers are heterozygous (*RR'*) for the pink trait. Compare the parent and offspring generations of snapdragon flowers. The generations differ in phenotype frequency, but the allelic frequencies for the R and R' alleles are the same in both generations. This population of snapdragon plants is in genetic equilibrium.

First generation

	Phenotype frequency	Allele frequency
	White = 0	R = 0.75
	Pink = 0.5	R' = 0.25
	Red = 0.5	

RR RR RR' RR' RR RR' RR RR'

Second generation

	Phenotype frequency	Allele frequency
	White = 0.125	R = 0.75
	Pink = 0.25	R' = 0.25
	Red = 0.625	

RR RR' RR RR' RR R'R' RR RR

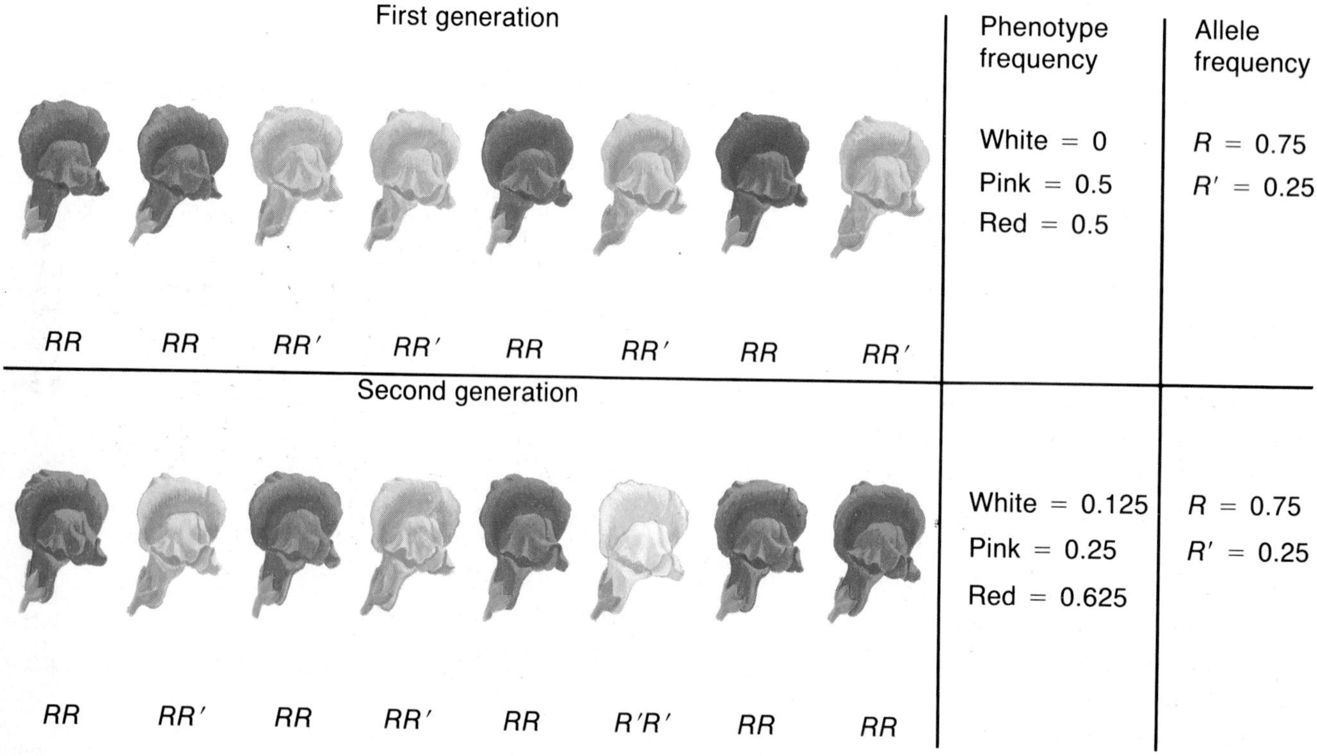

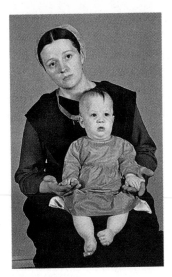

Figure 18.14

Sometimes genetic drift results in an increase, rather than a decrease, of rare alleles in a population. An example of this is seen in the Amish community of about 12 000 in Lancaster County, Pennsylvania. This community was started by about 30 individuals. Some of these people had a recessive allele that causes short arms and legs and extra fingers or toes. In the Amish community, the frequency of this allele is much higher than in other populations—one in 14 rather than one in 1000.

Mutations are important in evolution because they add new genetic material to the gene pool. Many random mutations are harmful, and most have no significance at all. However, once in a great while, they provide a favorable variation to an offspring, and natural selection takes over to distribute this variation to the following generations.

Another mechanism that causes changes to genetic equilibrium is genetic drift. **Genetic drift** is the alteration of allelic frequencies by chance processes. Genetic drift is more likely to occur in small populations than in large ones.

Consider, for example, a small population of Amish, *Figure 18.14.* Because of chance events, such as

Expression of Recessive Traits

When a recessive gene is rare, its phenotype does not show up often in a population. In fact, for a rare recessive to show up, two people who are heterozygous for the trait must mate, and even then there is only a one-in-four chance of a homozygous recessive offspring. Thus, these conditions, such as albinism, are exceedingly rare. However, they are less rare in certain populations—usually ones that are geographically or socially isolated. For example, albinism occurs in European populations in about one in every 20 000 births. But among the San Blas Indians in Panama, a small, isolated group, the ratio is one in 132. This is because the gene pool is much smaller than that in Europe, and the number of potential mates among other tribe members is more limited.

Relatives and royalty Historically, the most common examples of recessive genes being expressed were among royalty. For example, in ancient Egypt, it was the custom for the pharaoh to marry his sister. This severely limits the gene pool, can lead to genetic disorders, and was probably one of the reasons that the dynasty died out. Another famous example is the history of the descendants of Queen Victoria. Many of the women in this family carried a sex-linked gene that causes hemophilia, a disease in which the blood does not clot. The result was that they transmitted the disease to their sons, who were in danger of bleeding to death from minor injuries.

Relatives and religion Most religions prohibit marriage between close relatives. However, some religions are isolated by practice or lifestyle from the larger population. For example, the Amish and Mennonite populations of the United States are farmers with strong beliefs about how they should live. Because this lifestyle requires a willingness to forgo much modern technology, these populations are socially isolated from the larger population. Members tend to marry within their religion, which is a relatively small gene pool. Several genetic studies have shown that these people must be careful about the genetic pedigree of whom they marry.

Thinking Critically
How do breeding-exchange programs at zoos prevent the expression of recessive traits in animal species?

The Mathematics of Evolution

In the early 1900s, a British mathematician, G.H. Hardy, and a German doctor, W. Weinberg, discovered some basic principles about populations and how to describe them mathematically.

The Hardy-Weinberg law The Hardy-Weinberg law says that the frequencies of alleles that make up a gene will remain the same in a stable population. Mathematically, the law is expressed by a binomial equation. The letters p and q represent the frequency of each allele in a gene.

$$p + q = 1 \text{ (or 100\%) for homozygous matings}$$

$$p^2 + 2pq + q^2 = 1 \text{ for heterozygous crosses}$$

Providing that five conditions are met, the frequencies will stay the same in generation after generation of a sexually reproducing population.

Conditions The five conditions are:
1. Mutations cannot occur.
2. Mating must be random.
3. Natural selection cannot occur.
4. No genes may enter or leave the population.
5. The population must be large.

These conditions are probably never met in natural populations. Mutations happen every day, mating is not random, natural selection occurs, individuals and their genes migrate in and out of populations regularly, and many breeding populations are small. So why have a law that describes a situation that almost never happens?

Evolution benchmark The law is critical because it shows that populations tend to remain static without evolution. Thus, the law provides a standard by which evolution (genetic change) can be measured. In the event that gene frequency does change, it delivers solid evidence for evolution at work.

CONNECTION TO Biology

The general population of the United States is getting taller. Assuming that height is genetically controlled, does this observation violate the Hardy-Weinberg Law? Explain.

crossing-over during meiosis and the independent distribution of chromosomes to gametes, it's likely that in a small population, a larger proportion of individuals would inherit the recessive allele that causes short arms and legs and extra fingers or toes. In a large population, the frequency of recessive alleles is much lower. Suppose that these few individuals did not mate. The recessive allele would then soon be lost from the population.

Genetic equilibrium may also be disrupted by the movement of individual organisms in and out of a population. Any time an individual leaves a population, genes are lost. When individuals enter a population, those genes are added to the gene pool.

While mutation, genetic drift, and the migration of individuals can affect allelic frequencies, they usually do not cause significant changes. The factor that causes the greatest change in gene pools is natural selection.

Figure 18.15

These swallow tail butterflies come from different areas in North America. Despite their slight variations, they can interbreed. Therefore, all belong to the same species. Their variations and isolation from others of the same species make them subspecies.

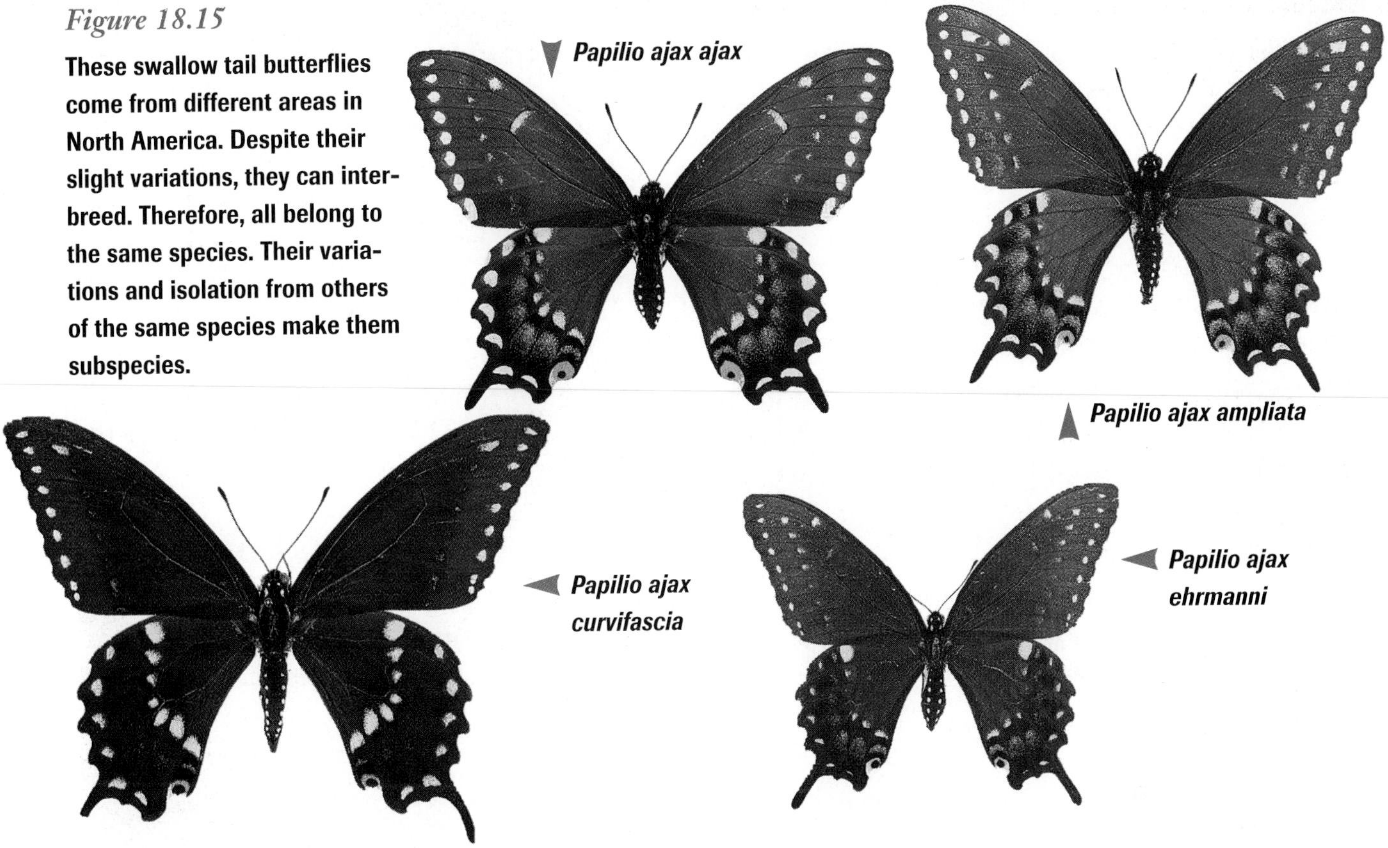

▼ *Papilio ajax ajax*

▲ *Papilio ajax ampliata*

◄ *Papilio ajax curvifascia*

◄ *Papilio ajax ehrmanni*

Natural selection acts upon the variation in populations

As you learned in the last section, the result of natural selection is that some members of a population are more likely to contribute their genes to the next generation than others. Thus, by the action of natural selection, allelic frequencies change from one generation to the next. Each of three types of natural selection—stabilizing, directional, and disruptive—causes changes in gene pools by acting upon the variation in populations.

As you know, all traits show variation. If you were to measure thumb length of students in your biology class, for example, you would find average, long, and short thumbs. Variation such as this can be thought of as the raw material of evolution because natural selection acts upon ranges of variation. *Figure 18.15* shows variation in a species of butterflies.

The type of natural selection that favors average individuals in a population is **stabilizing selection.** Consider a population of spiders in which average size is an advantage in terms of survival and reproduction. Spiders that are larger than average may be at a disadvantage because they can be seen and captured more easily. On the other hand, spiders that are smaller than average might not be able to catch enough prey to survive and reproduce and thus may also be at a disadvantage. In other words, average-sized spiders have a selective advantage in this particular environment.

When one of the extreme forms of a trait is favored by natural selection, it's known as **directional selection.** For example, imagine a population of woodpeckers. Woodpeckers feed by pecking holes in trees in order to get at the insects living under the bark. Suppose that one year, the trees in the woodpeckers' area are invaded by

Figure 18.16

The different types of natural selection act upon the ranges of variation in organisms. The red, bell-shaped curve indicates the normal variation in a population.

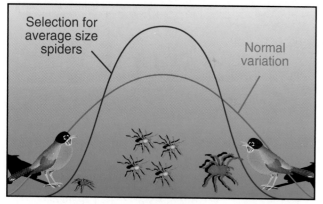

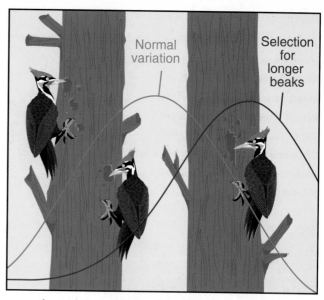

A Stabilizing selection affects genetic equilibrium by favoring average individuals. In this way, variation in a population is reduced.

B By favoring either of the extreme forms of a trait, directional selection can lead to the rapid evolution of a population.

C In disruptive selection, both extreme forms of a trait are favored. In some cases, there may be no intermediate forms, which can lead to the evolution of two new species.

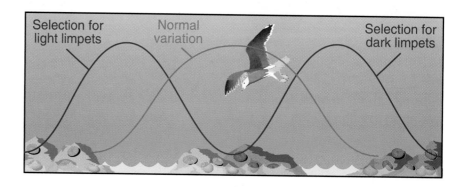

a species of insect that lives deep within trees. Only woodpeckers with long beaks would be able to reach the insects. In this scenario, shown in *Figure 18.16,* woodpeckers with long beaks have the selective advantage over those with short or average-sized beaks.

The third type of natural selection is known as disruptive selection. In **disruptive selection,** individuals with both extreme forms of a trait are at a selective advantage. For example, consider a population of marine organisms known as limpets. The shell color of limpets varies from white to dark brown. Limpets live their adult lives attached to rocks. On light-colored rocks, white-shelled

limpets are at an advantage because birds that prey upon them have a more difficult time locating them. On dark-colored rocks, dark-colored limpets are at an advantage because they are also camouflaged. On the other hand, tan-colored limpets, the intermediate forms, are easily spotted on either the light or dark backgrounds. In other words, disruptive selection, *Figure 18.16,* eliminates the intermediate forms by favoring the extremes.

Natural selection is the most significant factor that alters genetic equilibrium and causes changes in the gene pool of a population. Significant changes in the gene pool can also lead to the formation of new species.

The Evolution of Species

You've seen how natural processes such as mutation, random genetic drift, and natural selection can lead to changes in a population's gene pool, but how does this result in the evolution of new species? Remember that a species is defined as a group of organisms that look alike and have the ability to interbreed and produce fertile offspring in nature. The evolution of new species, **speciation,** can occur only when either interbreeding or the production of fertile offspring is somehow prevented.

Physical barriers can prevent interbreeding

Imagine a population of tree frogs living in the dense rain forests of the Amazon basin. Over time, periodic droughts begin to break up stretches of continuous forest into smaller patches, separating the original tree frog population into smaller groups. Other environmental factors cause shifts in local river courses that further isolate the patches.

In the natural world, physical barriers frequently form and break up large populations into smaller ones. Volcanic eruptions can cause lava flows, which have the potential for splitting a population. Sea-level changes along continental shelves, such as those in New Guinea, can create island environments that impose a water barrier between populations. **Geographic isolation** occurs if a physical barrier separates a population into groups.

Geographic isolation is one of the ways new species form. Think about the population of tree frogs. If small populations of tree frogs were geographically isolated from one another, by rivers or areas of deforestation, they would no longer be able to interbreed, and gene exchange would cease. Over time, each small population would adapt to the local environment through the process of natural selection. Eventually, the gene pools of each group would become so different that they could be considered a new species, as shown in *Figure 18.17.*

Figure 18.17

When populations become geographically isolated, individuals can no longer mate. Gradually, natural selection produces gene pools so distinct that individuals can no longer produce fertile offspring, even if mating did occur.

BioLab

Evolution can be described as the change in allelic frequencies of a gene pool over time. Natural selection can place pressure upon specific phenotypes and cause a change in the frequency of the alleles that produce the phenotypes.

Natural Selection and Allelic Frequency

In this activity, you will simulate the effects of eagle predation on a population of rabbits, where *GG* represents the homozygous condition for gray fur, *Gg* is the heterozygous condition for gray fur, and *gg* represents the homozygous condition for white fur.

PREPARATION

Problem
How does natural selection affect allelic frequency?

Objectives
In this Biolab, you will:
- **Simulate** natural selection by using beans of two different colors.
- **Calculate** allelic frequencies over five generations.
- **Demonstrate** how natural selection can affect allelic frequencies over time.

Materials
colored pencils (2)
paper bag
graph paper
pinto beans
white navy beans

PROCEDURE

1. Make a copy of the data table shown here.
2. Place 50 pinto beans and 50 white navy beans into the paper bag.

3. Shake the bag. Reach into the bag and remove two beans. These represent an individual rabbit's genotype. Set these beans aside, and continue to remove beans until you have 50 "rabbits."

4. Arrange the beans on a flat surface in two columns representing the two possible rabbit phenotypes, gray (*GG* and *Gg*) and albino (*gg*), as shown here.

5. Examine your columns, and remove 25 percent of the gray rabbits and 100 percent of the white rabbits. These numbers represent the selection pressure on your rabbit population. If the number you calculate is a fraction, remove a whole rabbit to make whole numbers.

6. Count the number of pinto and navy beans that remain, and record this number in your data table.

7. Calculate the allelic frequencies by dividing the number of beans of one type by 100. Record these numbers in your data table.

8. Begin the next generation by placing 100 beans into the bag. The proportions of pinto and navy beans should be the same as the percentages you calculated in step 7.

9. Repeat steps 3 through 8 until you have collected data for five generations.

10. Graph the frequencies of each allele over five generations. Plot the frequency of the allele on the vertical axis and the number of the generation on the horizontal axis. Use a different-colored pencil for each allele.

Allele Frequencies

Generation	Allele G			Allele g		
	Number	Percentage	Frequency	Number	Percentage	Frequency
Start	50	50	0.50	50	50	0.50
1						
2						
3						
4						
5						

ANALYZE AND CONCLUDE

1. **Calculating Results** Did allelic frequencies change over time? Why or why not?

2. **Analyzing Data** Did either of the alleles totally disappear? Why or why not?

3. **Thinking Critically** What does your graph show about allelic frequencies and natural selection?

4. **Making Inferences** What would happen to the allelic frequencies if the number of rabbit-eating eagles declined?

Going Further

Project Develop an activity using two traits instead of just one, such as fur length and color. Select a new predator and new selection pressures for the population.

Geographic isolation can lead to differences in mating behavior

When geographic isolation of a population occurs, gene pools are closed, and genetic material is not exchanged between the groups. Over time, as populations become more and more distinct, reproductive isolation can arise. **Reproductive isolation** occurs when formerly interbreeding organisms are prevented from producing fertile offspring.

There are many different types of reproductive isolation. One type occurs when the genetic material of the groups becomes so different that mistakes happen during development of the zygote—if a zygote forms at all. Another type of reproductive isolation is seasonal. For example, if one group of tree frogs evolves the behavior of mating in the fall, while another group mates during the summer months, these two groups are reproductively isolated from one another.

Speciation can occur when chromosome numbers change

Although geographic isolation plays an important role in most cases of speciation, it's not the only factor. Many new species of plants have evolved in the same geographic area as a result of polyploid speciation. Any species with any multiple of the normal set of chromosomes is known as a **polyploid.**

Polyploids arise when mistakes occur during meiosis. Polyploid speciation is perhaps the fastest form of speciation because reproductive isolation is instantaneous. When a polyploid individual mates with a normal diploid individual, because of the change in chromosome number, the resulting zygotes have difficulty developing normally. It's estimated that nearly half of the known flowering plant species originated in this way, as well as many important crops such as wheat, cotton, apples, and bananas. *Figure 18.18* illustrates how polyploid speciation occurs.

Speciation can occur quickly or slowly

You've seen how one form of speciation, polyploidy, can occur after only one generation. However, most speciation does not occur as quickly. What is the tempo of speciation?

When Darwin proposed his theory of evolution, he argued that evolution proceeds at a slow, steady rate, and that small, adaptive changes gradually accumulate over time in populations. **Gradualism** is the idea that species originate through a gradual buildup of new adaptations.

Figure 18.18

Mistakes during meiosis can result in the formation of polyploid species. If chromosomes fail to separate properly during the first meiotic division, the result can be gametes having the diploid (2*n*) condition, rather than the normal haploid (*n*) set of chromosomes. Many cultivated garden flowers, such as this chrysanthemum (below) have been developed by selecting and hybridizing polyploid individuals.

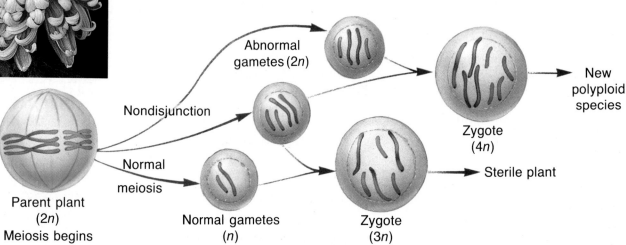

Parent plant (2*n*) — Meiosis begins

Nondisjunction → Abnormal gametes (2*n*)

Normal meiosis → Normal gametes (*n*)

Zygote (3*n*) → Sterile plant

Zygote (4*n*) → New polyploid species

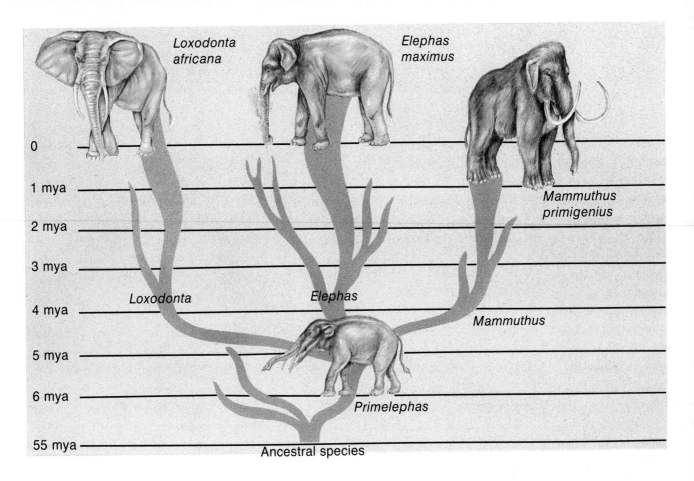

Figure 18.19

Punctuated equilibrium involves rapid evolution of a gene pool. Many supporters of punctuated equilibrium point to the fossil record of elephants to support their view of evolution. As you can see, three different types of ancient elephants evolved from an ancestral population in a short period of time.

Gradualism is supported by evidence from the fossil record. For example, the slow and steady buildup of adaptations seen in the fossils of camels and horses supports this view.

In 1971, another hypothesis about the rate of evolution was proposed. This idea, known as **punctuated equilibrium,** states that speciation occurs quickly in rapid bursts, with long periods of stability in between, *Figure 18.19.* Supporters of this hypothesis argue that populations remain at or close to genetic equilibrium for long periods of time. When environmental conditions change, such as warmer temperatures or the introduction of a new competitive species, rapid genetic change and speciation can occur, interrupting, or punctuating, genetic equilibrium. A new species can form in a span of 10 000 years or less. Like gradualism, punctuated equilibrium is supported by fossil evidence.

Whether the rate of evolution is slow or fast probably won't be resolved by fossils alone. Many scientists conclude that speciation can occur both slowly and rapidly, depending upon the circumstances. It shouldn't be surprising that alternative hypotheses can be offered to explain observations. The nature of science is that hypotheses change when new evidence or technologies become available.

Patterns of Evolution

You've learned how evolution by natural selection ultimately results in the formation of new species. In the natural world, scientists have identified several patterns of evolution occurring throughout the world and in vastly different environments. These patterns add further proof that natural selection is indeed the agent for evolutionary change.

Species diversify when introduced to new environments

The Hawaiian Islands are home to an extraordinary diversity of plants and animals not seen elsewhere. Among them are a family of birds commonly called Hawaiian honeycreepers, and they represent a case study in evolution. What makes this group of birds interesting is that, while similar in body size and shape, they differ sharply in plumage color, and most importantly, beak shape. Each species of honeycreeper occupies its own niche and possesses a beak adapted to the types of foods eaten.

Despite the differences in beak design, however, scientists hypothesize that all honeycreepers evolved from a single ancestral species that invaded Hawaii long ago. The process of evolution of an ancestral species into an array of species that occupy different niches, *Figure 18.20,* is called **adaptive radiation.**

Adaptive radiation is not limited to the Hawaiian Islands; examples can be seen throughout the world. The many species of finches that Darwin discovered on the Galapagos Islands during the voyage of the *Beagle* are a good example of adaptive radiation,

Figure 18.20

Scientists think that the ancestor of all honeycreepers somehow made it across the 2000 miles of Pacific Ocean from the American mainland about 5 million years ago. When this ancestral bird population encountered the diverse niches of Hawaii, speciation occurred, with each species adapting to a different food source.

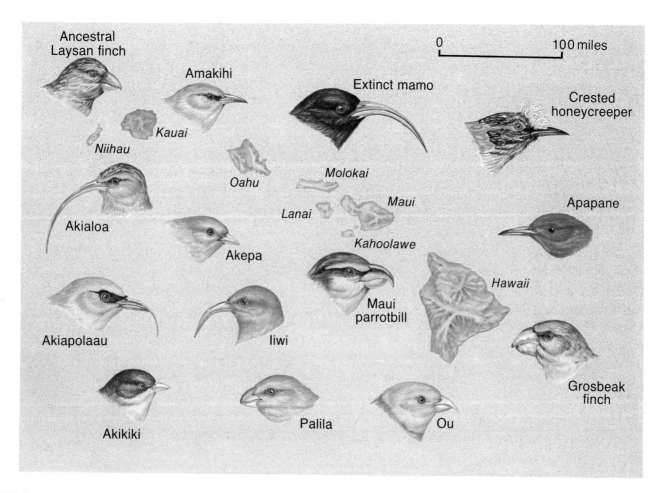

Figure 18.21

Although marsupials and placental mammals have evolved independently of one another since at least the late Cretaceous period 100 million years ago, they are similar in ecology and body form. Notice the similarities between marsupial species on the left side of the diagram, and their placental counterparts on the right.

as are the more than 300 varieties of cichlid fish in Lake Victoria, Africa.

Adaptive radiation is one example of **divergent evolution,** the pattern of evolution in which species once all similar to the ancestral species become more and more distinct. Divergent evolution occurs when species begin to adapt to different environmental conditions and change, becoming less and less alike, according to the pressures of natural selection.

Distantly related species can evolve similar features

Have you ever wondered why dolphins and fishes have similar body shapes? Dolphins and fishes, although both vertebrates, are not closely related animals. The structural design of dolphins and fishes is an example of **convergent evolution,** the pattern of evolution in which distantly related organisms evolve similar traits, *Figure 18.21*.

Figure 18.22

Many species of fishes (left) and dolphins (right) share a similar streamlined body shape. Which of these two species needs to surface to breathe air?

Convergent evolution occurs when unrelated organisms occupy similar environments and face similar selection pressures. The streamlined shape of fishes and dolphins, *Figure 18.22,* is an adaptive response related to the need for moving efficiently through water.

Because convergent evolution has been shown to occur in organisms that evolved from entirely different groups of ancestors, it provides strong proof for natural selection.

Connecting Ideas

Darwin's theory of how evolution occurs explains how organisms have adapted to the land, sea, air, and every imaginable niche in between. Twelve years after *On the Origin of Species by Natural Selection,* Darwin published a book that dealt with the evolution of a group of organisms that have adapted successfully to life in the trees. This book was titled *The Descent of Man,* and it was one of the first books to discuss the evolution of the order of organisms you belong to—the Primates.

Section Review

Understanding Concepts

1. Why is the evolution of resistance to antibiotics in bacteria considered an example of directional selection?
2. How does geographic isolation result in changes to a population's gene pool?
3. Why is rapid evolutionary change more likely to occur in small populations?

Thinking Critically

4. What environmental factors may prevent some plant populations from speciating through geographic isolation mechanisms?

Skill Review

5. **Designing an Experiment** Two squirrels that live on opposite sides of the Grand Canyon are hypothesized to have evolved from a recent, common ancestor. What observations or experiments might you perform to test this hypothesis? For more help, refer to Practicing Scientific Methods in the *Skill Handbook.*

Who owns the fossils?

F.B.I. Nabs Illegal Dinosaur

FOSSIL HUNTER'S JOURNAL   MAY 19, 1992

Agents of the Federal Bureau of Investigation, backed up by members of the South Dakota National Guard, have seized an extraordinary fossil find—the largest and most well-preserved *Tyrannosaurus rex* ever discovered.

In 1990, a geologist made a startling discovery. Protruding from a South Dakota cliff face were several large bones—bones that the geologist realized at once were those of a *Tyrannosaurus rex*, a fierce, carnivorous dinosaur from the late Cretaceous Era.

After purchasing excavation rights from the Native American owner of the land, he and his coworkers began the delicate, painstaking process of easing bone from rock. The geologists are members of a commercial fossil-hunting organization, and hope to display the Tyrannosaur as the centerpiece of its planned museum.

The government acts Two years later, the F.B.I. moved in. Federal agents seized the fossil remains of the Tyrannosaur, along with the group's records, photographs, and related specimens.

The agents were acting to enforce laws that prohibit the removal of fossils from federal land without the approval of the Secretary of the Interior. Although the property on which the fossil was found is part of an Indian reservation, the land is legally held in trust by the government, and nothing can be removed from it.

Different Viewpoints

A federal judge later ruled that the fossil does indeed belong to the government. The judge's ruling has set off a serious debate within the scientific community.

Some paleontologists think that such government action is necessary to prevent commercial fossil hunters from selling scientifically significant specimens to private collectors. Fossils can be big business; a good specimen can be worth several million dollars on the fossil market.

Other paleontologists are wary of too much government intervention and believe that commercial fossil hunters, who have the time and money that most scientists lack to search for fossil evidence of dinosaurs, may actually save some fossils that might otherwise be lost to erosion or human neglect.

INVESTIGATING   the Issue

1. **Research** Contact a local museum of natural history that has fossils in its collection. Ask the curator how the museum stocks its fossil displays. Does the museum deal with commercial fossil hunters? Report your findings to the class.
2. **Debate** Who owns history? Stage a class debate on the subject. Who should own the treasures from centuries-old shipwrecks? The government? The discoverers?

18.1 Natural Selection and the Evidence for Evolution

- Through many years of experimentation and observation of the natural world, Charles Darwin proposed the idea that species evolve through the process of natural selection, an idea that has become universally accepted among scientists.
- Natural selection leads to the evolution of adaptations by favoring variations that are better suited for the environment. Individuals with favorable variations are likely to survive, reproduce, and pass these variations to future generations.
- Evolution has been observed in the lab, but much of the evidence for evolution has been gathered from studies of the fossil record, and through anatomical and genetic comparisons of organisms.

18.2 Mechanisms of Evolution

- Evolution occurs when there are changes in the genetic equilibrium of populations. Mutation, genetic drift, and migration of individuals cause minor disruptions of genetic equilibrium, but natural selection has the greatest effect.

- The separation of populations by physical barriers can prevent interbreeding and lead to speciation. Speciation can also occur when individuals are prevented from producing fertile offspring.
- Many patterns of evolution can be observed in the natural world. Since these patterns are observed in many different environments, they are strong evidence for evolution.

Key Terms

Write a sentence that shows your understanding of each of the following terms.

adaptive radiation
allelic frequency
analogous structure
artificial selection
camouflage
convergent
  evolution
directional selection
disruptive selection
divergent evolution
gene pool
genetic drift
genetic equilibrium
geographic isolation

gradualism
homologous
  structure
mimicry
natural selection
polyploid
punctuated
  equilibrium
reproductive
  isolation
speciation
stabilizing
  selection
vestigial structure

Understanding Concepts

1. How did artificial breeding experiments help Darwin develop his hypothesis about evolution?
2. How does the fossil record indicate evolutionary change over time?
3. What two concepts formed Darwin's basis for the theory of evolution by natural selection?
4. Why is adaptive radiation considered a form of divergent evolution?
5. How is the idea of competition involved in the process of natural selection?
6. How do Hawaiian honeycreepers provide evidence for evolution?
7. How do mutations affect allelic frequency?

Relating Concepts

8. Make a concept map that relates the following terms and phrases. Supply the appropriate linking words for your map.

speciation, reproductive isolation, geographic isolation, natural selection, mutation, genetic drift, migration, gene pool

Using a Graph

9. Variation in tooth size in a population of sharks was plotted on a graph. What might be occurring in this population that explains this pattern?

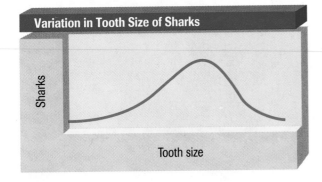

Variation in Tooth Size of Sharks

Sharks

Tooth size

Applying Concepts

10. Males in a population of organisms are often larger than females. Explain how natural selection may be involved in this difference.
11. Many poisonous plants and animals have striking patterns of red and black colors. How and why might this similarity have evolved?
12. Many organisms, such as sharks, have changed very little over the course of evolution. What evolutionary factors may be responsible for keeping sharks relatively unchanged over the generations?
13. Why is it impossible for convergent evolution to result in identical organisms?

Math Connection

14. In a population of clams, shell color is represented by two alleles. The population consists of 10 *TT* clams and 10 *tt* clams. What are the allelic frequencies?

Thinking Critically

Relating Concepts

15. **Biolab** How would a low offspring-survival rate affect allelic frequencies in a population of gray and white rabbits in which white rabbits were also heavily preyed upon by eagles?

Applying Concepts

16. **Minilab** The following graph was plotted after measuring leaf length in a population of maple trees. What does the shape of this graph suggest about natural selection in maple trees?

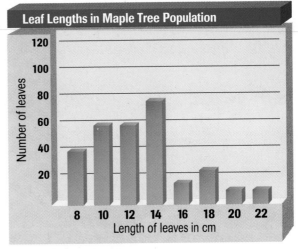

Leaf Lengths in Maple Tree Population

Number of leaves

8 10 12 14 16 18 20 22

Length of leaves in cm

Making Predictions

17. **Minilab** What might happen if white insects that are easily spotted by birds evolved defense mechanisms, such as stingers, that may harm a predator?

Applying Concepts

18. Explain why the wings of bats and butterflies are examples of convergent evolution.

Connecting to Themes

19. **Unity Within Diversity** Why is DNA ideal for determining the relatedness of organisms?
20. **Homeostasis** How is the evolution of resistance to antibiotics in bacteria similar to homeostatic mechanisms in organisms?
21. **Systems and Interactions** How might viruses and the host cells they invade have evolved?

19 Human Evolution

It's mid-morning on the African savannah, and that means it's feeding time for the chimpanzees of Tanzania! In one tree, a female quietly munches on a piece of fruit, occasionally dropping a piece or two for her offspring on the ground below. In a nearby tree, young adult males gather some choice leaves. An older female snaps a dry twig from a small bush and slowly sticks it into a termite mound. Moments later, the twig, now covered with termites, is withdrawn from the nest. The female chimp is using a tool!

Despite this example, tool use is a rarity in the animal kingdom. The only animals to consistently use tools to manipulate the environment are humans. Humans also play sports, make music, and read books. How did these behaviors develop? In this chapter, you'll gain some insights into this question as you consider evidence discovered in the search for our ancestors.

Chimpanzees and sea otters use tools, but humans are the only animals to deliberately manufacture tools for a variety of needs. This behavior developed sometime during the evolution of humans. What is the evidence for human evolution?

Primate Adaptation and Evolution

Section Preview

Objectives

Relate the adaptations of primates to life in the trees.

Compare and contrast the diversity of living primates.

Sequence the evolutionary history of modern primates.

Key Terms

primate
opposable thumb
anthropoid
prehensile tail

Monkeys have always fascinated humans, most likely because of the many structural and behavioral similarities between them and us. But only since 1871 and the publication of Charles Darwin's book, The Descent of Man, have scientists realized the true evolutionary link between monkeys, apes, and humans. Today, scientists examine living and fossil primates for the information they can provide about human evolution.

What Is a Primate?

If you've ever visited the monkey house at a zoo or watched one of the many television programs about apes or monkeys, you have observed primates. A **primate** is a mammal such as a lemur, monkey, ape, or human. As you know, primates come in a variety of shapes and sizes. However, despite the diversity, all primates share some common traits.

Look at *The Inside Story* to see some of the features shared by all primates. Perhaps the most distinctive trait of primates is the shape of the head. Primates have a relatively rounded head and flattened face when compared with other mammals. This is due, in part, to the size of the brain. Relative to body size, primates have the largest brain of any terrestrial mammal. Only marine mammals are comparably brainy in

terms of volume or capacity. Primate brains are also much more complex than those of other animals, and this is reflected in their diverse behaviors and social interactions.

Most primates are arboreal, meaning tree-dwelling, animals, and they have several adaptations that have evolved for survival in trees. For example, all primates have a highly developed sense of vision. Primate eyes face forward rather than to the side as in some other mammals. This eye positioning enables primates to perceive depth and gauge distances—a type of vision known as stereoscopic vision. As you might imagine, this ability is important for any animal that moves around in a complex environment such as a forest. Primates also have color vision. How might the perception of color have an adaptive value to animals that live in trees?

A Primate

Primates are a diverse group of mammals, but they all share some common features. You can see in this diagram that primates have rounded heads and flattened faces, unlike most other mammals.

Chimpanzee

Opposable thumb

Brain

Eyes

Shoulder

Hand

Elbow

Feet

1 Primates have thumbs that enable them to grasp objects. Primates that use their hands for food manipulation have flexible thumbs.

2 Much of primate brain size is related to the reorganization of the cerebrum, the part of the brain involved in thinking, memory, and interpretation.

3 Vision is the dominant sense in primates. In monkeys, apes, and humans, the light-sensitive cells of the retina are packed closely together, allowing good visual perception.

4 The shoulder of arboreal primates is adapted for arm movement in different directions. In some species, such as apes and humans, a ball-and-socket shoulder joint provides great mobility.

5 Primate elbows are flexible, allowing the palm of the hand to be turned in many directions.

6 Primate feet are constructed for grasping. However, there are differences between species, depending on the type of locomotion.

19.1   Primate Adaptation and Evolution   **455**

Figure 19.1

Generally, primate hands and feet are adapted for locomotion in trees. Compare the thumbs of humans to those of other primates. Can you see that the human thumb diverges more than the thumbs of other primates?

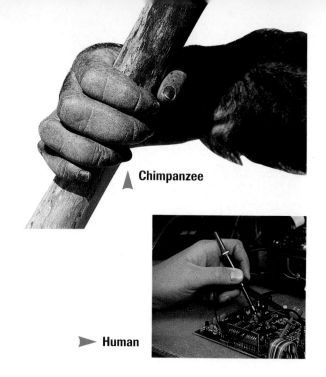

▲ Chimpanzee

► Human

▲ Ring-tailed lemur

The primate skeleton is also adapted for movement among trees. For example, all primates have relatively flexible shoulder and hip joints, features important for the various types of primate locomotion such as climbing, clinging, and branch swinging.

Primate hands and feet are also unique among mammals, as shown in *Figure 19.1.* Both hands and feet are equipped with nails, rather than the claws of other mammals, and the joints are mobile. Also, most primates have an **opposable thumb;** that is, a thumb that can be brought opposite the forefinger, allowing them to grasp and cling tightly to the branches of trees. This opposable thumb allows primates to make finer manipulations with their hands, leading to the ability to use tools.

In humans, the thumb is larger, proportionately, than in other primates, and the tip of the opposable thumb can cross over the palm to touch the other fingers. To illustrate how important this is, tape your thumb to your hand so it points in the same direction as your fingers. Now try to hold a key as if you were going to open a door. Wouldn't it be difficult to carry out your normal activities without an opposable thumb?

Primate Evolution

What are the different types of primates, and how did they evolve? Primates are fairly well represented in the fossil record, and the evolutionary history of primates has been reconstructed through fossils and comparative studies of modern primates.

The earliest primates were small, squirrel-like mammals

Scientists theorize that primates evolved about 65 million years ago, but it has not yet been determined which group of mammals gave rise to primates.

The earliest identified primate from the fossil record is *Purgatorius*, a primate that lived about 60 to 65 million years ago. *Figure 19.2* illustrates how *Purgatorius* and its close relatives may have lived.

Figure 19.2

Primates, like *Purgatorius,* probably evolved in response to selection pressures for life in the trees. According to one hypothesis, primates evolved keen vision and grasping hands as an adaptation for preying on tree-living insects.

MiniLab

How useful are primate adaptations?

Have you ever watched a gymnast on the rings or parallel bars? Two keys to the survival of primates in the trees are opposable thumbs and stereoscopic vision. Humans don't live in trees, but these adaptations are useful to us in a variety of ways. In this activity, you will explore the importance of these two key adaptations.

Procedure

1. Make a dot on a piece of paper. Put the paper on your desk, and move back about 0.5 m.
2. Close one eye, and quickly try to touch the point of a pencil on the dot.
3. Close the other eye and repeat step 2, and then try it with both eyes open.
4. With tape, loosely wrap your dominant hand so that your thumb points in the same direction as your fingers.
5. Try to perform a variety of activities, such as writing with a pen, holding a key, or opening a door.

Analysis

1. How is stereoscopic vision useful to humans?
2. How would the absence of an opposable thumb affect the ability of the hand to perform different tasks?

Today, there are no living species like *Purgatorius;* however, living prosimians may come close. Primates are generally divided into two subgroups: the prosimians and the anthropoids. Prosimians are a group of small-bodied primates that include the lemurs, aye-ayes, and tarsiers, *Figure 19.3.* They can be found in the tropical forests of Africa and Southeast Asia, where they leap and run through the canopy in search of insects, seeds, and small fruits. Most prosimians are nocturnal and have large eyes—important features for their nightly insect-catching activities. According to fossil evidence, prosimians evolved about 50 to 55 million years ago from ancestors similar to the earliest primates.

Figure 19.3

Prosimians are widely distributed, but most are found in areas with tropical environments.

The aye-aye, a prosimian found on Madagascar, uses its long middle finger to dig grubs out of deep holes in trees.

Tarsiers, tiny nocturnal animals found in Borneo, can turn their heads nearly 180°, an advantage to an arboreal primate.

Humanlike primates evolve

Monkeys, apes, and humans form the other subgroup of primates, the **anthropoids,** or humanlike primates. Anthropoids differ from prosimians in many features of the head and skeleton. In particular, anthropoids have more complex brains than prosimians, which gives them increased intelligence and humanlike qualities.

The three major radiations of anthropoids are the New World monkeys of South and Central America; the Old World monkeys of Africa, Asia, and Europe; and the hominoids, which include Asian and African apes as well as humans, as shown in *Figure 19.4.*

New World monkeys are an entirely arboreal group and are diverse in terms of size and ecology. New World monkeys are characterized by a long tail that they use as a fifth limb. This **prehensile tail** is a muscular tail that can grasp or wrap around a branch as the monkey moves from tree to tree. Among the many New World monkeys are the tiny insect-eating marmosets and the larger, fruit-eating spider monkeys.

Old World monkeys are generally larger than New World monkeys and include the arboreal colobus and guenon monkeys and the terrestrial baboons. Old World monkeys use their tails for balance and are adapted to a variety of environments, from the hot, dry savannahs of Africa to the cold mountain forests of Japan.

Apes include gibbons, orangutans, chimpanzees, and gorillas. Unlike prosimians and monkeys, apes lack tails and have unique adaptations for arboreal locomotion. For example, all apes have extremely flexible hips, shoulders, and elbows, as well as long, heavily muscled forelimbs for climbing and branch swinging. The African apes, chimpanzees and gorillas also possess skeletal adaptations for knuckle-walking on the ground.

Brain size and complexity is also increased in apes, as evidenced by their diverse social interactions and increased parental care for offspring.

Similarities among monkeys, apes, and humans suggest that they all share a common ancestor. Fossils indicate that this anthropoid common ancestor evolved from prosimians 37 to 40 million years ago.

Golden lion tamarins are arboreal New World monkeys found in South America. They eat fruit, insects, and other small animals.

Figure 19.4

The three groups of anthropoids include New World monkeys, Old World monkeys, and hominoids.

This mandrill is an Old World monkey found in the forests of West Africa. It is omnivorous, active during the day, and spends most of its time on the ground.

Other mammals	Prosimians	New World monkeys	Old World monkeys	Apes	Humans	
						5
						15
						25
						35
						45
						55
						65
		Purgatorius				Millions of years ago

Anthropoids evolved in Africa

New World monkeys were the first modern anthropoids to evolve. According to the fossil record, *Figure 19.5,* New World monkeys evolved 30 to 35 million years ago when ancestral anthropoids from Africa successfully crossed the Atlantic Ocean to South America. Scientists speculate that they did so either by rafting on floating islands or by crossing small island bridges when sea levels were low.

Old World monkeys evolved a little later in Africa, about 20 to 22 million years ago. Fossils indicate that early Old World monkeys were arboreal. Environmental reconstructions of the Miocene period when Old World monkeys evolved indicate that global temperatures were dropping. These changes led to the shrinking of rain forests, and food sources became scarce, forcing some groups to exploit resources on the ground. This led to the eventual speciation of baboons and other ground-living monkeys.

Figure 19.5

All primates probably evolved from common ancestors. The evolutionary history of primates is based on the fossil record and comparative studies of living primates.

◀ **Gibbons are found throughout Southeast Asia to southern China and south to Indonesia. Gibbons are less closely related to humans than other apes.**

Primates

The family Hominidae contains one species, Homo sapiens. *This species is a large, erect, omnivorous biped that lives on land. In addition, it has a large brain, vocal cords, and good manual dexterity. Biologically, these characteristics, transmitted by our DNA, are what make us human.*

Eleven other families divided into four groups make up the rest of the order Primate.

Prosimians

Small, tree-dwelling primates
These primates look the least like primates. They range in size from that of a mouse to that of a large house cat. Their faces are triangular with large eyes. Because they lack the facial muscles that most primates have, they cannot make the facial changes that other primates use to communicate.

The most famous of the prosimians are the lemurs that live in Madagascar. Although lemurs do spend some time on the ground, most of their lives are spent in trees.

Lemur, *Propithecus verreauxi*

Mouse lemur

Spider monkey

Squirrel monkey

New World monkeys

Monkeys with an extra hand As their name implies, New World monkeys are natives of Central and South America. They all live in trees and are unique in that they have prehensile tails with which they can grasp things. These animals do not usually swing by their tails as they travel. Most often, they travel through the forests by leaping from branch to branch, using their tails as extra hands for grasping.

Old World monkeys

Cold-weather monkeys   Old World monkeys are natives of Africa and Asia. They are a varied group with most species living in trees, some living on the ground, and still others spending time in both environments. They are the only primates, other than humans, that naturally occur outside of the tropics. The Japanese macaque and several other species live in habitats where it snows. However, most species are tropical.

Grey langur

Proboscis monkey

Apes

Primates without tails   Apes are our closest animal relatives. They live in Africa and Southeast Asia, are tailless, and have large brains. There are five genera of apes that include chimpanzees, gibbons, gorillas, orangutans, and siamangs. All apes are herbivores, although chimpanzees have been observed occasionally killing other animals for food. Apes are subject to many of the same diseases as humans, can use simple tools, and have been taught to communicate with humans in a sign language that includes a vocabulary of several hundred words.

Mountain gorilla

EXPANDING YOUR VIEW

1. **Drawing Conclusions** Examine the photos of Old World monkeys. How is having a tail an adaptive advantage for these primates?

2. **Compare and Contrast** Which of the species in this feature probably live in trees? Explain your answer.

Apes evolved in response to dietary pressures

The evolutionary history of anthropoids is particularly important because of the information it can provide about human origins. Can you make a hypothesis to explain the adaptive radiation of modern apes?

According to fossils found in Africa and elsewhere, apes evolved about the same time as Old World monkeys and faced similar environmental changes, shown in *Figure 19.6.* Scientists hypothesize that instead of moving to the ground like Old World monkeys, apes evolved their unique adaptations to more efficiently exploit choice fruits in trees.

Fossils indicate that gibbons were the first ape to evolve, followed by orangutans and then African apes. As you learned in the last chapter, this agrees with the evolutionary tree based on DNA comparisons.

Figure 19.6

Apes and Old World monkeys are adapted to different environments.

▼ **Baboons are ground-dwelling primates that live on the African savanna.**

▼ **Orangutans spend all of their lives in the forests of Borneo and rarely come down to the ground.**

Section Review

Understanding Concepts

1. How do primates show adaptation to life in the trees?
2. What is the evolutionary significance of an opposable thumb?
3. What features distinguish anthropoids and prosimians?

Thinking Critically

4. An unidentified fossil skeleton is brought to your lab. You suspect it's a primate, but you're not sure. What observations might you make in order to identify this fossil as a primate?

Skill Review

5. **Classifying** List the different types of primates, key facts about each group, and how each group is related. For more help, refer to Organizing Information in the *Skill Handbook.*

Ever since fossils of human ancestors were discovered in the 1800s, many myths have been passed on about the behavior of our ancestors. Many people have viewed early humans as dumb brutes incapable of doing anything correctly. But scientists studying the early history of human evolution have changed these views. Today, reconstructions of the anatomy and behavior of our ancestors are based on careful examination of fossils and detailed comparisons of living apes and humans.

TOM the DANCING BUG
©1994 RUBEN BOLLING

JUNE, 3.5 MILLION B.C.
PRICE: TWENTY TERMITES

australopithecine MAGAZINE

THE ONLY MAGAZINE BY AND FOR AUSTRALOPITHECUS AFARENSIS, THE PLIOCENE EPOCH'S HIPPEST APE

5 DELICIOUS WAYS TO EAT GRUBS p.8

SELF TEST p.47?
ANSWER 20 EASY QUESTIONS AND FIND OUT IF YOU'RE CONSCIOUS!

P.18: A FEATURE ON OUR RIVAL, *Australopithecus Robustus*
• Why are they so big?
• Why are they so dumb?
• How many of them does it take to screw in a lightbulb?

OUR PANEL RATES THE LATEST TOOLS p.56
→ BIG STICK passé!
→ CHEWED-UP LEAVES ingenious!
→ HEAVY ROCK essential!

SPECIAL EVOLUTION SECTION

OUR EXPERTS SHOW HOW YOU CAN EVOLVE INTO *Homo Habilis* OR EVEN *Homo Erectus* IN JUST ONE MILLION YEARS! (THE SECRET: NATURAL SELECTION)

FEMALES
• SPICE UP YOUR LIFE! DON'T JUST GATHER-- HUNT! p.66
• BIPEDAL ISN'T ENOUGH! Raise Your Standards In a Mate! p.70

MALES
• DO YOU HAVE THE STUFF TO BECOME THE DOMINANT MALE IN *YOUR* GROUP? p.81
• HOW TO FEIGN INTEREST IN CHILD REARING IN ORDER TO GET YOUR MATE! p.86

COMING TO GRIPS WITH THE SHAME OF LIMITED CRANIAL CAPACITY p.34

Are You Evolving Away From Your Mate?

DIST. BY QUATERNARY FEATURES-P.O. BOX 72-NY-NY-10021. APOLOGIES TO MATT GROENING.

Objectives

Compare and contrast the adaptations of australopithecines with those of apes and humans.

Summarize the major anatomical changes in hominids during human evolution.

Key Terms
bipedal
hominid
australopithecine
Neanderthal
Cro-Magnon

Up from the Apes

The story of human evolution begins between 5 and 8 million years ago in Africa. During this time, scientists propose that a population of ancestral apes diverged into two lineages. One lineage would eventually evolve into the African apes—gorillas and chimpanzees. The other would lead to modern humans. What factors may have led to the divergence of African apes and humans?

There are few fossils that date to the time period when African apes and humans diverged. However, scientists suggest that the human lineage may have begun in response to environmental changes, which forced a population of apes to leave the protection of the trees and to search for new food sources on the ground.

In order to move efficiently in this new environment, scientists speculate that this population of apes eventually became **bipedal;** that is, they evolved the ability to walk on two legs. Walking on two legs leaves the arms and hands free for other activities such as using tools. These apes were the earliest members of the human family of primates known as hominids. A **hominid** is a humanlike, bipedal primate. The fossil record of hominids is fairly complete, and we now have a good picture of the anatomy and behavior of early hominids.

paleoanthropology

paleo (GK) ancient
anthropo (GK) human
logos (GK) study of
Paleoanthropology is the study of fossil humans.

A Family of Anthropologists

History is filled with examples of families that have pursued a common interest with extraordinary results. Science, too, has its famous families. For example, the Leakeys—husband Louis, wife Mary, and son Richard—have discovered a wealth of fossil evidence for the origin and evolution of humans.

The son of missionaries Louis Leakey was born in Kabete, Kenya, in 1903. He began working as an archaeologist in eastern Africa. At Olduvai Gorge in Tanzania during the 1930s, Louis Leakey began discovering animal fossils and simple stone tools. Olduvai Gorge was to become the site of many important discoveries by the Leakey family that would change the world's view of human evolution.

From art to archaeology In 1936, Louis Leakey married Mary Douglas Nicol, an artist. In 1947, Mary found a skull of *Proconsul africanus,* believed to be the apelike ancestor of both apes and early humans. The skull is estimated to be 25 million years old. In 1978, Mary discovered fossil footprints just south of Olduvai Gorge. The footprints were made by primates that lived 3.5 million years ago and walked upright, suggesting that they were humanlike.

In his parents' footsteps Richard Erskine Frere Leakey is one of three sons born to Louis and Mary. At age 19, Richard found a fossil jaw while exploring northeastern Tanzania. Perhaps Richard's most important contribution to anthropology was his discovery in 1967 of a site along the shore of Lake Turkana called Koobi Fora. Here, Richard unearthed the greatest hominid fossil find in history—more than 400 hominid fossils, the remains of perhaps 20 individuals. One nearly complete skull is believed to be a direct ancestor of *Homo sapiens.*

Thinking Critically

Over the last 70 years, China has been closed to scientific research, including archaeology. Recently, China has allowed research teams to begin digging in promising sites. How could fossils found in China change current ideas about human evolution?

Early hominids possessed ape and human characteristics

In 1924, South African scientist Raymond Dart discovered a fossil skull that sent shock waves through the paleoanthropological world. The skull had an apelike braincase and face, but it was unlike any primate Dart had ever seen, *Figure 19.7.* One feature that stood out was the position of the *foramen magnum,* the hole in the skull where the spinal column enters. In the fossil, the hole was located on the bottom of the skull, as in humans. This indicated to Dart that the organism must have walked upright. Dart classified this fossil as a new primate species, *Australopithecus africanus,* meaning "southern ape from Africa." Dating methods have shown the skull to be 2 to 3 million years old. Since Dart's discovery, scientists have recovered many more australopithecine specimens. An **australopithecine** is an early hominid that lived in Africa and had apelike and humanlike qualities.

Figure 19.7

The Taung child was the first australopithecine ever discovered. The braincase, face, and teeth are chimplike, but the position of the *foramen magnum* indicates that it walked upright like humans.

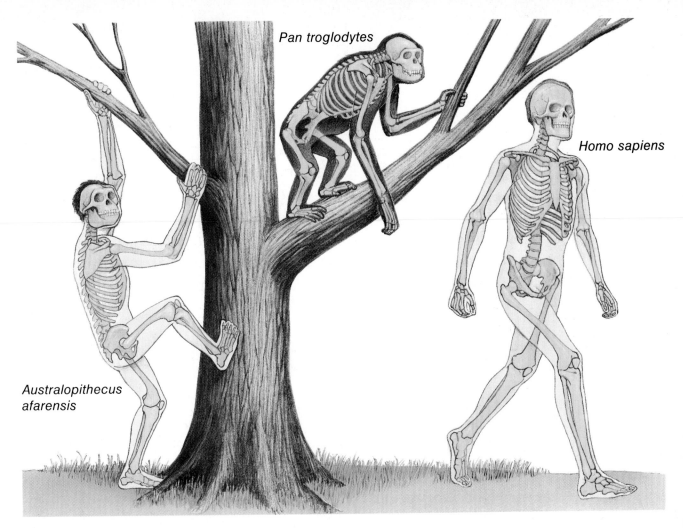

Pan troglodytes

Homo sapiens

Australopithecus
afarensis

Figure 19.8

The skeleton of an australopithecine is in many ways intermediate between those of living apes and humans. Compare the **A. afarensis** skeleton (left) with the chimpanzee (center) and human (right) skeletons.

Much of what scientists know about australopithecines comes from the nearly complete "Lucy" skeleton, discovered in East Africa in 1974 by American paleoanthropologist Donald Johanson. This fossil has been dated to be 3.5 million years old, and was nicknamed "Lucy" after a song by *The Beatles*, a popular rock group around the world, that was often played on a tape deck at the campsite.

The Lucy skeleton was classified as *Australopithecus afarensis*, and other fossil discoveries have shown that this species existed between 3 and 5 million years ago. *A. afarensis* is the ear-liest known hominid species, and as shown in *Figure 19.8,* the structure of the pelvis, legs, and feet indicates that it was bipedal, like humans. On the other hand, the size of the braincase suggests an apelike brain. That is, walking upright occurred before the evolution of a larger brain. The shoulders and forelimbs are also apelike.

Because of this combination of features, scientists propose that *A. afarensis* and other australopith-ecines spent a considerable amount of time climbing in trees. When they came to the ground, they walked bipedally, but clumsily.

BioLab

Australopithecines are the earliest hominids in the fossil record, and in many ways, their anatomy is intermediate between living apes and humans. In this lab, you'll determine the apelike and humanlike characteristics of an australopithecine skull, and compare the skulls of australopithecines, gorillas, and modern humans. The diagrams of skulls shown below are one-fourth natural size. The heavy black lines indicate the angle of the jaw.

Comparing Skulls of Three Primates

PREPARATION

Problem
How do skulls of primates provide evidence for human evolution?

Objectives
In this Biolab, you will:
- **Determine** how paleoanthropologists study early human ancestors.
- **Compare and contrast** the skulls of australopithecines, gorillas, and modern humans.

Materials
metric ruler
protractor
copy of skull diagrams

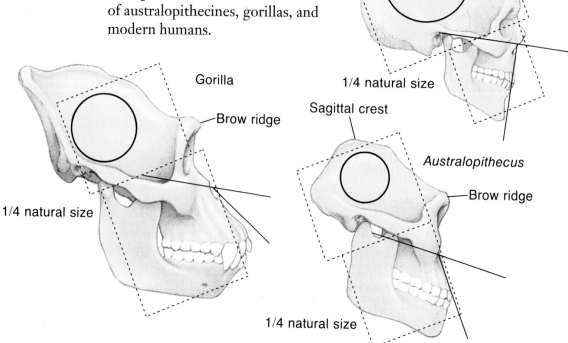

Modern human

1/4 natural size

Gorilla

Brow ridge

1/4 natural size

Sagittal crest

Australopithecus

Brow ridge

1/4 natural size

1. Your teacher will provide you with a copy of skull diagrams of *Australopithecus africanus*, *Gorilla gorilla*, and *Homo sapiens*.

2. The rectangles drawn over the skulls represent the area of the brain (upper rectangle) and face (lower rectangle). On each skull, determine and record the areas of each rectangle (length × width).

3. Measure the diameters of the circles in each skull. Multiply these numbers by 200 cm². The result is the cranial capacity (brain volume) in cubic centimeters.

4. The two heavy lines projected on the skulls are used to measure how far forward the jaw protrudes. Use the protractor to measure the outside angle (toward the right) formed by the two lines.

5. Complete the data table.

Comparison of Gorilla, *Australopithecus*, and Modern Human Skulls

	Gorilla	*Australopithecus*	Modern human
1. Face area in cm²			
2. Brain area in cm²			
3. Is brain area smaller or larger than face area?			
4. Is brain area 3 times larger than face area?			
5. Cranial capacity in cm³			
6. Jaw angle			
7. Does lower jaw stick out in front of nose?			
8. Sagittal crest present?			
9. Browridge present?			

ANALYZE AND CONCLUDE

1. **Comparing and Contrasting** Describe the similarities and differences in face-to-brain area in the three primates.

2. **Interpreting Observations** How do the cranial capacities compare among the three skulls? How do the jaw angles compare?

3. **Drawing Conclusions** Based on your findings, what statements can you make about the place of australopithecines in human evolution?

Going Further

Application
Different parts of the australopithecine skeleton are also intermediate between apes and humans. Obtain diagrams of primate skeletons to determine the similarities and differences.

In addition to *A. afarensis* and *A. africanus*, scientists recognize two, perhaps three, other species of australopithecines that evolved slightly later in time. These species are known from East African and South African fossil sites dated to 1 to 2.5 million years old. Overall, they were similar to the earlier species, but they had larger teeth and jaws that enabled them to eat tough plant materials.

The evolutionary relationships of australopithecines are, thus far, unclear, but it is clear that australopithecines died out by 1 million years ago, and that the early australopithecines played a role in the evolution of modern hominids.

The Emergence of Modern Humans

The discovery of australopithecines ended the long-standing debate on whether large brain size or bipedalism evolved first. When did large brains and other human characteristics, such as stone tool use and culture, develop?

Members of the genus *Homo* made the first stone tools

About 40 years after the discovery of australopithecines, anthropologists Louis and Mary Leakey discovered the skull of another type of hominid, which created excitement in the scientific community.

This skull, shown in *Figure 19.9*, was much more humanlike than australopithecine skulls. In particular, the braincase was much larger in this new skull, and the teeth and jaws were much smaller, like humans. Because of its close similarities to humans, the Leakeys classified this fossil in the genus *Homo*, the same genus in which modern humans are classified. They named the species *Homo habilis*, which means "handy human," because simple stone tools were found near the fossil.

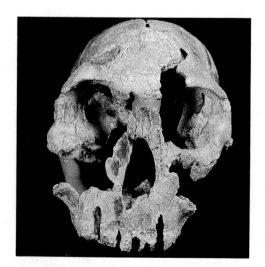

Figure 19.9

The average brain volume of *Homo habilis* has been estimated at 600 to 700 cm³, much smaller than the 1350 cm³ average of modern humans, but larger than the 400 to 500 cm³ of apes and australopithecines.

Homo habilis existed between 1.5 and 2 million years ago, and is considered to be the first hominid to make and use stone tools. According to scientists, *H. habilis* was probably a part-time scavenger and used stone tools to slice off bits of meat from animal carcasses found in the area.

Hunting and fire developed later

Anthropologists suggest that *Homo habilis* gave rise to another hominid species that appeared about 1.6 million years ago. This new species was *Homo erectus*, which means "upright human." This hominid had a larger brain and a more humanlike face than *H. habilis*, *Figure 19.10,* but with large browridges and a lower jaw that sloped back with no chin.

Figure 19.10

The most complete *Homo erectus* fossil ever found was discovered in East Africa in 1985. It was the skeleton of a 12-year-old boy. *H. erectus* had a brain volume of approximately 900 cm³ and long legs like modern humans, indicating efficient bipedal locomotion.

ThinkingLab   Interpret the Data

How can tooth structure identify types of food eaten by primates?

The tooth structure of living primates is often used to infer the dietary behavior of fossil primates. Primates with large molars tend to eat tough foods such as plant stems, seeds, and hard fruits.

Analysis

Examine the following data from a study comparing the molar teeth of apes with those of fossil and living hominids.

Primate	Average molar area in mm²
Gorilla gorilla	1011
Australopithecus africanus	901
A. boisei	1010
Homo erectus	656
Homo sapiens	500

Thinking Critically

What factors may have contributed to the trend of decreasing molar area during human evolution?

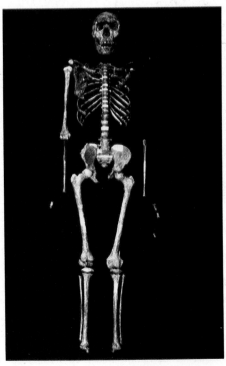

Tracking Early Humans

"Where did I come from?" is a question often asked by young children of their parents. Scientists, too, ask the same question about the human species.

Mother Africa   Fossil evidence presents a convincing case that the earliest hominids evolved on the African continent well over 3 million years ago. About 1 million years ago, hominids, in the form of *Homo erectus,* probably began fanning out from Africa and populating other continents. *Homo erectus* slowly evolved into *Homo sapiens,* modern humans.

Enter Eve   In 1987, a new hypothesis was put forth, called mitochondrial Eve. This hypothesis, based on DNA studies of 147 modern women of different geographic origins, held that all modern humans were descended from a single female who lived in Africa just 200 000 years ago.

Scientists discovered that the mitochondrial DNA (mtDNA) in the 147 women they studied was remarkably similar. The scientists concluded that the first radiation of hominids out of Africa 1 million years ago had ended in failure. The mitochondrial DNA theory suggests that a second radiation occurred 200 000 years ago, after *Homo erectus* in Africa had evolved into *Homo sapiens.* The people from this second radiation displaced those from the first radiation.

New evidence   However, in 1989 and 1990, two skulls were discovered in China. The skulls have features of both *Homo erectus* and *Homo sapiens,* suggesting that modern humans were evolving in Asia at the same time they were evolving in Africa.

In South America, fossil sites have been discovered that may date back nearly 35 000 years. Such sites fly in the face of current theory, which holds that modern humans crossed the now submerged Bering Sea land bridge into North America only about 12 000 years ago.

CONNECTION TO **Biology**

How might the discovery of fossil mtDNA influence the discussion of the evolutionary history of modern humans?

Magnification: 80 000×

Large stone tools, called hand axes, found at some sites indicate that *H. erectus* probably hunted. In addition, charred bones and hearths found at some sites suggest that these were the first hominids to use fire, and perhaps the first to live in caves.

Unlike the earlier hominids, *H. erectus* migrated out of Africa about 1 million years ago. Fossils indicate that this hominid spread through Africa and Asia, and possibly into Europe. Scientists estimate that *Homo erectus* became extinct between 300 000 and 500 000 years ago, but first this species gave rise to hominids that resembled modern humans.

Culture developed in modern humans

Dating the emergence of modern humans is perhaps the most controversial topic in the field of paleoanthropology today, due to an incomplete fossil record and controversial dating of fossils. Nevertheless, ample evidence exists to sketch out a basic picture for the origin of our own species, *Homo sapiens.*

The fossil record indicates that 100 000 to 400 000 years ago, *Homo sapiens* appeared in Europe, Africa, the Middle East, and Asia. The earliest forms of our species are known as archaic *Homo sapiens* because although the skulls resembled those of *Homo erectus,* they had less prominent browridges, more bulging foreheads, and smaller teeth. Also, their brain size of 1000 to 1400 cm^3 was in the modern human range.

Best known among the first *Homo sapiens* were the Neanderthals, illustrated in *Figure 19.11*. The **Neanderthals** were a group of early humans that lived from 35 000 to 100 000 years ago in Europe, Asia, and the Middle East. Neanderthals were a powerfully built group with thick bones and large faces with prominent noses.

Neanderthals lived in caves during the ice ages of the Pleistocene period, but the popular idea that they were brutish cavemen is incorrect. Neanderthal brains were as large as or larger than those of modern humans. Other evidence such as burial grounds suggests that they may have been the first hominids to develop religious views and communication through spoken language.

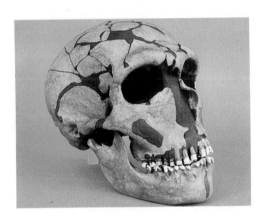

MiniLab

How do human proteins compare with those of other primates?

Scientists use the amino acid sequences in proteins to determine the evolutionary relationships of living species. In this activity, you'll compare amino acid sequences among groups of primates to determine primate evolutionary history.

Amino acid sequences															
Baboon	ASN	THR	THR	GLY	ASP	GLU	VAL	ASP	ASP	SER	PRO	GLY	GLY	ASN	ASN
Chimp	SER	THR	ALA	GLY	ASP	GLU	VAL	GLU	ASP	THR	PRO	GLY	GLY	ALA	ASN
Lemur	ALA	THR	SER	GLY	GLU	LYS	VAL	GLU	ASP	SER	PRO	GLY	SER	HIS	ASN
Gorilla	SER	THR	ALA	GLY	ASP	GLU	VAL	GLU	ASP	THR	PRO	GLY	GLY	ALA	ASN
Human	SER	THR	ALA	GLY	ASP	GLU	VAL	GLU	ASP	THR	PRO	GLY	GLY	ALA	ASN

Procedure
1. Make a copy of the data table.

Primate	Amino acids different from humans	Percent difference
Baboon		
Chimpanzee		
Gorilla		
Lemur		

2. For each primate, count the number of amino acids that are different from the human sequence. Record these numbers in your data table.
3. Calculate the percentage differences by dividing the numbers by 15 and multiplying by 100. Record the numbers in your data table.

Analysis
1. Which primate is most closely related to humans? Least closely related?
2. Construct a phylogenetic diagram of primates that most closely fits your results.

Figure 19.11

Neanderthals were skilled hunters and were the first hominids to develop diversified tool kits that included spears and tools for scraping leather and slicing meat.

What happened to Neanderthals?

The big question in anthropology today is what relationship exists between Neanderthals and present-day humans. Did Neanderthals evolve directly into modern *Homo sapiens*, or were they merely replaced by Cro-Magnons? The **Cro-Magnon** people appeared in Europe 35 000 to 40 000 years ago, and were identical to modern humans in height, skull and teeth structure, and brain size. As shown in *Figure 19.12*, Cro-Magnons were talented toolmakers and artists, and it is almost certain that these people used language.

Did Neanderthals evolve directly into the modern Cro-Magnon people? Recent genetic and archaeological evidence suggests that this is not likely. New fossil dates indicate that anatomically modern *Homo sapiens* were in South Africa and the Middle East approximately 100 000 years ago, before the time of Neanderthals.

Genetic evidence also supports an African origin of modern *Homo sapiens*, perhaps as early as 200 000 years ago. All of this supports the view that Neanderthals were nothing more than a side branch of *Homo sapiens*, unlikely to be ancestral to modern humans, *Figure 19.13*.

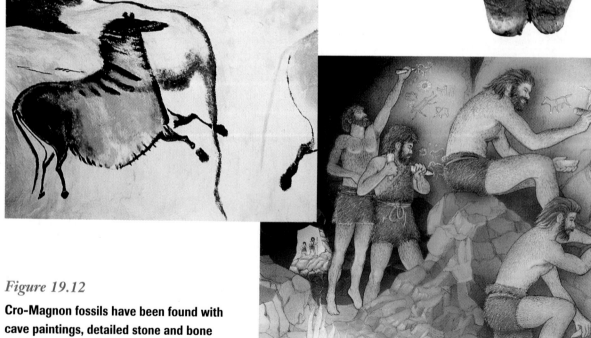

Figure 19.12

Cro-Magnon fossils have been found with cave paintings, detailed stone and bone artifacts, and human figurines. All of this evidence suggests that human culture was quite developed by 30 000 years ago.

Figure 19.13

This diagram represents the most widely accepted view of the evolution of *Homo sapiens*.

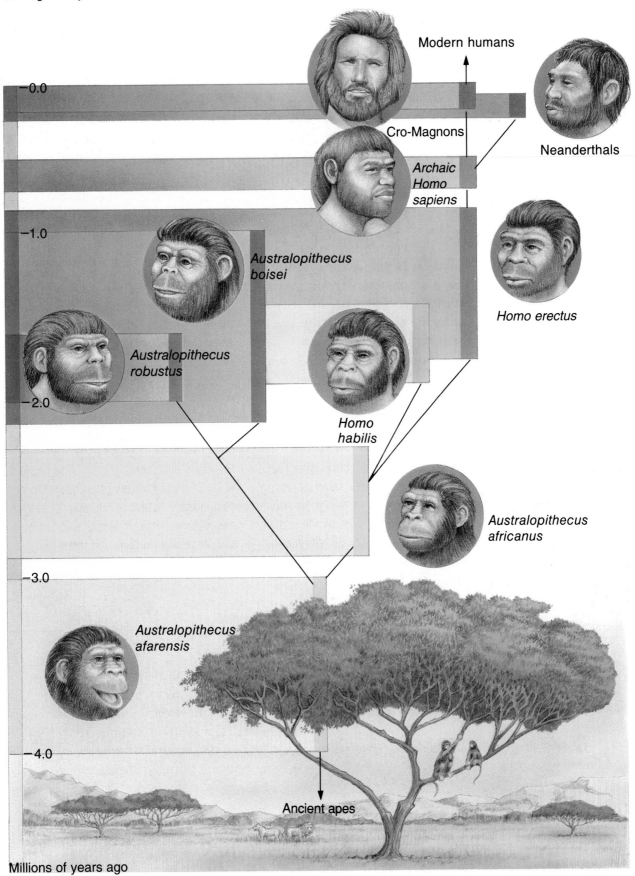

Modern humans

Cro-Magnons

Neanderthals

Archaic Homo sapiens

Australopithecus boisei

Homo erectus

Australopithecus robustus

Homo habilis

Australopithecus africanus

Australopithecus afarensis

Ancient apes

Millions of years ago

-0.0
-1.0
-2.0
-3.0
-4.0

Figure 19.14

Most anthropologists now agree that modern *Homo sapiens* evolved in Africa and spread from Africa throughout the rest of the world. Humans have remained relatively unchanged over the last 200 000 years. Humans reached North America by at least 12 000 years ago, and by 8000 to 10 000 years ago, cultural innovations such as animal domestication, agriculture, and the development of permanent settlements by Native Americans had begun, *Figure 19.14.*

Connecting Ideas

Natural selection led to a diverse group of mammals, the primates, that had characteristics that enabled them to move from life in forests to life in the grasslands. The ability to walk upright on two legs gave anthropoids the freedom to use arms and hands for other needs. By comparing primate fossils with groups of living primates, scientists have been able to understand the 65-million-year history of primates. The evolutionary history of humans is being pieced together from fossils and comparative studies of humans and other living anthropoids.

Section Review

Understanding Concepts

1. How are australopithecines intermediate between apes and humans?
2. How do scientists know that Neanderthals are not ancestors of Cro-Magnon people?
3. Why are Neanderthals considered to be a side branch in human evolution?

Thinking Critically

4. What evolutionary factors may have been responsible for the increase in brain size during human evolution?

Skill Review

5. **Interpreting Scientific Illustrations** By examining figures in this section, draw a time line to show the evolution of hominids. Indicate the species of hominid that evolved, and when and where this evolution occurred. For more help, refer to Thinking Critically in the *Skill Handbook*.

Communication and Computers

You may not realize it, but you are living in an important time in history. It probably will be ranked right alongside the Renaissance and the Industrial Revolution as turning points in human history. You are living in the early days of the Computer Age.

Looms are simple computers Crude computers have been around for nearly 200 years. In the early 1800s, weavers working in modernized textile mills controlled looms that wove fabrics with cards containing punched out holes. Threads of various colors were added or subtracted from the cloth being woven according to the locations of holes in the stacks of cards. Later, computers controlled simple traffic lights. However, it wasn't until the 1960s that electronic computers were widely available, but because of cost, they were purchased only by universities, governments, and big businesses. What really accelerated the pace of the computer revolution was the development of the personal computer.

Power to the people Widespread access to personal computers has empowered ordinary people and given them access to other people and resources all over the world. At first, this was not true because each computer acted alone. However, individuals recently have been given access to networks that allow them to span the globe from their home or school. The largest of these networks is called Internet.

Internet is a computer network Internet is a system that interconnects 11 000 different networks and an estimated 10 million people. Over these connections, scientists can transmit and discuss their latest findings, and businesses can transmit data. Students can even communicate with researchers on the cutting edge of the subject they are studying. Thus, students in Hobson, Montana (population 200), can correspond with their peers in Australia, Japan, and Russia.

From defense to news Internet was originally funded by the Department of Defense. As university researchers discovered its ease, they began to use it for other purposes. Now, anyone can use the network at very low cost, and its potential has expanded tremendously. This has benefited countries where access to communications is severely restricted.

Applications for the Future

One of the problems with Internet is that it is slow and restricted in capacity. Recently, it has been proposed that a fiber-optic "superhighway" for information be constructed. This superhighway would carry messages on a beam of light and vastly expand the network's capacity. Combined with an upgraded telephone system, this system could connect you to every person and source of information on the planet.

INVESTIGATING the Technology

1. **Apply** Design a system to connect your home or school to a computer network.
2. **Discuss** Lead a discussion in your class of how you could use Internet to help you learn more about current research on a particular topic in biology.

Reviewing Main Ideas

19.1 Primate Adaptation and Evolution
- Primates are primarily an arboreal group of mammals. They possess adaptations related to survival in the trees, such as stereoscopic vision, an opposable thumb, and mobile skeletal joints.
- Primates include prosimians, such as lemurs and tarsiers, and anthropoids, which include monkeys, apes, and humans.
- Fossils indicate that primates evolved approximately 65 million years ago. Major trends in primate evolution include increases in brain size, and diversity in diet and methods of locomotion.

19.2 Human Origins
- The human lineage began in Africa approximately 5 to 8 million years ago. Scientists propose that australopithecines, the earliest known hominids, had the ability to walk bipedally and climb trees.
- Fossils show that during human evolution, brain and body size increased, bipedal locomotion became more efficient, and jaws and teeth decreased in size.
- Stone tool use coincides with the appearance of the genus *Homo* in the fossil record about 2 million years ago. Use of fire, tools, culture, and language developed later in larger-brained species of *Homo*.

Key Terms
Write a sentence that shows your understanding of each of the following terms.

anthropoid
australopithecine
bipedal
Cro-Magnon
hominid

Neanderthal
opposable thumb
prehensile tail
primate

Understanding Concepts

1. According to the fossil record, when did primates evolve?
2. Describe the appearance of the earliest primates.
3. Which living primate group is most similar to the earliest primates?
4. How is stereoscopic vision an effective adaptation for primates?
5. What is the significance of increased brain size in primates?
6. Which developed first in primates, bipedalism or increased brain size?
7. Describe the feeding behavior of australopithecines.
8. What traits identified *Homo habilis* as an important species in human evolution?
9. What is the current view of the place of Neanderthals on the human evolutionary tree?

Relating Concepts
10. Make a concept map that relates the following terms and phrases. Supply the appropriate linking words for your map.

    australopithecine, primate, prosimian, *Homo erectus*, *Homo habilis*, *Homo sapiens*, anthropoid, Neanderthal, Cro-Magnon, ape, hominoid

Using a Table

11. The intermembral index is the ratio of forelimb length to hindlimb length. Primates with a high intermembral index are good climbers and branch swingers. Primates with a low index tend to walk on all four legs. Based on the table below, what patterns in intermembral indices can you observe in primates?

Intermembral Index of Some Primates	
Species	Intermembral index $\frac{\text{forelimb length}}{\text{hindlimb length}} \times 100$
Prosimians Indri Slow loris	64 88
New World monkeys Squirrel monkey Black spider monkey	80 105
Old World monkeys Pig-tailed macaque Hanuman langur	92 83
Hominoids Orangutan Common chimpanzee	139 103

Applying Concepts

12. How might human fossils be important for determining the origin of particular diseases?

13. Why is it important for someone interested in human fossils to also understand nonhuman primates?

14. Some scientists suggest that Neanderthals evolved directly into the different human groups of today. How might this hypothesis be tested?

15. Why is ape evolution considered to be an example of adaptive radiation?

16. What factors may have led to the speciation of australopithecines?

A Broader View

17. How could you tell from looking at footprints that an animal walked upright? Explain.

Thinking Critically

Designing an Experiment

18. **Biolab** What tests might you perform to determine the evolutionary significance of large jaws in primates?

Designing an Experiment

19. **Minilab** How would you test the idea that opposable thumbs are excellent adaptations for arboreal mammals?

Drawing Conclusions

20. **Minilab** The following data are from an experiment comparing amino acid sequences in apes. What conclusions can you draw from such data?

Primate	Percentage amino acid difference from humans
Orangutan	3.7%
Chimpanzee	1.8%
Gibbon	5.2%
Gorilla	2.1%

Connecting to Themes

21. **Systems and Interactions** How might primate fossils be used to reconstruct ancient habitats?

22. **The Nature of Science** How is it possible for anthropologists to have two competing views about the origin of modern *Homo sapiens?*

H ow many different types of organisms could you identify in this small patch of forest? Certainly, you could pick out several species of plants. Perhaps you would also find an animal, a colony of mushrooms, or lichen growing on a rotting log. Unseen to you, however, would be the countless species of microscopic bacteria, algae, and fungi that exist in the decaying matter and soil on the forest floor.

Nearly 1.5 million species of organisms have been identified— only a fraction of the nearly 100 million species that some scientists estimate actually exist. A large portion of these uncounted species belong to kingdoms of mostly microscopic organisms that you will study in this unit.

Magnification: 12 000×

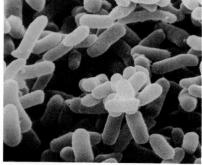

Many species of bacteria cause disease, but others play important roles in their ecosystems. Why are bacteria so important to the existence of life on Earth?

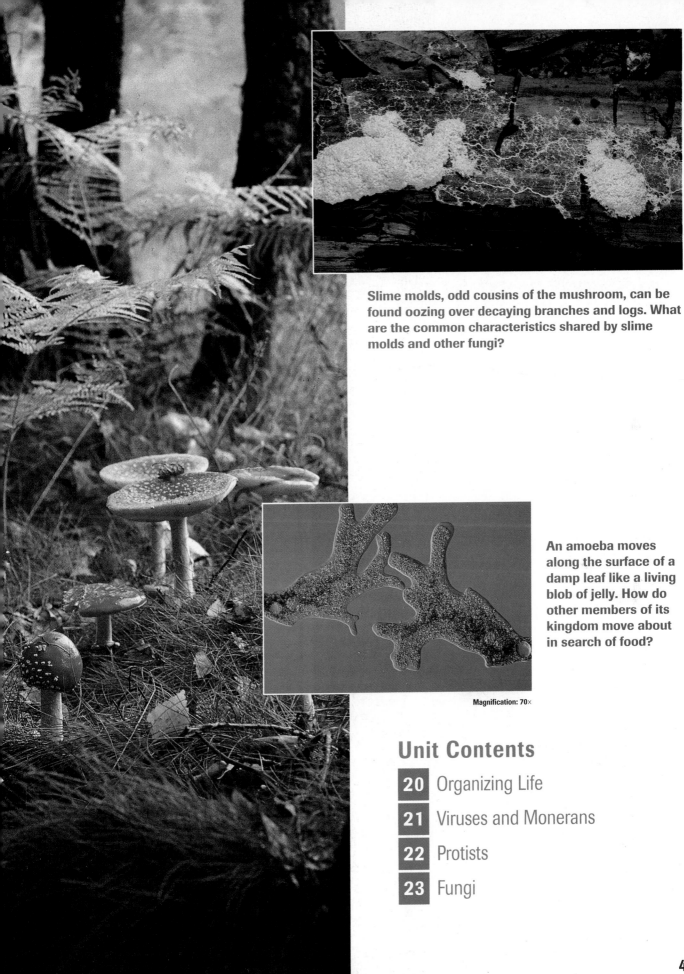

Slime molds, odd cousins of the mushroom, can be found oozing over decaying branches and logs. What are the common characteristics shared by slime molds and other fungi?

An amoeba moves along the surface of a damp leaf like a living blob of jelly. How do other members of its kingdom move about in search of food?

Magnification: 70×

Unit Contents

479

20 Organizing Life

Scientists estimate that between 5 and 30 million species of organisms live on our planet, although fewer than 2 million have been identified and named. Imagine trying to study life on Earth. One major problem is simply the identification of the millions of species. To that end, biologists have organized living things into related groups.

Even though the catalogs and databases of known organisms are large, there are still places on Earth where little is known about the diversity of life. Scientists estimate that only 15 percent of the species in the remaining rain forests have been identified. Perhaps one day you might make an expedition to such a place to discover more about its biodiversity. If you find a new species, you will probably use a system of classification designed by biologists to place it in relationship with other known species.

Think about the last time you visited a zoo. Remember the variety of organisms you saw. In one area, lions yawned in the shade of an acacia plant while vultures searched the grass for food. In a nearby area, flamingos rested by standing on one leg and hiding their heads under their wings. In the reptile exhibit, a cobra slowly lifted its head and expanded its hood as you passed by. The many animals and plants you saw represented only a tiny fraction of the variety of organisms that inhabit Earth.

20.1 Classification

Section Preview

Objectives

Evaluate the history, methods, and purpose of taxonomy.

Demonstrate the use of concepts in classification.

Explain the purpose of a phylogenetic classification.

Key Terms

classification
taxonomy
genus
binomial nomenclature
family
order
class
phylum
kingdom
division
phylogeny

taxonomy:
taxo (GK) to arrange
nomy (GK) ordered knowledge
Taxonomy is the science of classification.

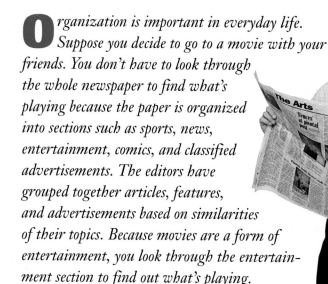

Organization is important in everyday life. Suppose you decide to go to a movie with your friends. You don't have to look through the whole newspaper to find what's playing because the paper is organized into sections such as sports, news, entertainment, comics, and classified advertisements. The editors have grouped together articles, features, and advertisements based on similarities of their topics. Because movies are a form of entertainment, you look through the entertainment section to find out what's playing.

How Classification Began

People have always sought a better understanding of nature; classification is one tool early scientists have used to gain that understanding. **Classification** is the grouping of objects or information based on similarities. Early systems of classification had a purpose. Plants were classified as edible or toxic, based on their effects on people who first ate them. As time passed, the science of taxonomy developed. **Taxonomy** is the branch of biology concerned with the grouping and naming of organisms. Biologists who study taxonomy are called taxonomists.

Aristotle's system

The Greek philosopher, Aristotle (384–322 B.C.), developed the first method of classification. He classified all living things known at that time into two major groups—plants and animals. Plants were classified as herbs, shrubs, and trees depending on their size and structure. Animals were grouped according to where they lived—on land, in the air, or in water. Later observations convinced scientists that Aristotle's system did not work. They observed that some animals, such as frogs, live both on land and in water. Scientists also realized that Aristotle's classification system did not show natural relationships among organisms. According to his system, birds, bats, and flying insects would be grouped together even though they have little in common besides the ability to fly.

Linnaeus designs a classification system

It wasn't until the 18th century that the Swedish botanist, Carolus Linnaeus (1707–1778), developed a method of classification that is still used today. Unlike Aristotle's system,

Linnaeus selected physical characteristics that led to classification based on close relationships of organisms. For example, he based his classification of flowering plants on the numbers and similarities of their reproductive structures, *Figure 20.1.* Linnaeus selected characteristics of organisms that also led eventually to classifications based on evolutionary relationships. Let's look at a bat as an example of an organism classified by this method. Whereas bats fly like birds, their origins are shared by all animals that have hair and feed milk to their young. Therefore, bats are classified with the group known as mammals, rather than with birds.

A species is known by two names

Linnaeus also invented the two-word system used to identify species. In this system, the first word identifies the genus in which the organism is classified. A **genus** consists of a group of closely related species. The genus name is immediately followed by the second word, which is often descriptive of that organism. Thus, the scientific name for each species, called a binomial, is a combination of the genus and descriptive names. The system of naming objects is known as nomenclature. The system devised by Linnaeus that gives each organism two names is called **binomial nomenclature.**

binomial:
bi (L) two
nomen (L) name
A binomial consists of two names.

Figure 20.1

Carolus Linnaeus designed a system to classify organisms based on numbers and similarities of body structures and on physical form such as size, shape, and methods of obtaining food. What features could you use to classify these flowers?

Figure 20.2

The poisonous mushroom on the right, called a destroying angel, and the edible mushroom on the left, a smooth lepiota, look similar. Only a taxonomist or a person who uses taxonomy to identify the two species knows for sure which mushroom can be eaten safely.

Taxonomy—The Study of Classification

Even today, taxonomists continue to search for underlying natural relationships as a basis for classification. They compare the external and internal structures of organisms, as well as their chemical makeup and evolutionary relationships.

Why are living things organized?

Just as the editors of a newspaper organize material into categories based on similarities, taxonomists organize living things into groups by using a set of characteristics or criteria that describe how similar or different organisms are from each other. Classification provides a framework of logic and order so that relationships among living things can be seen easily.

One question that scientists are investigating is whether dinosaurs are more closely related to warm-blooded or to cold-blooded animals. They find that dinosaur bones have large spaces inside, similar to the spaces in the bones of birds, which are warm-blooded. From your knowledge of classification of animals with hollow bone structure, you might conclude that some dinosaurs may have been able to control their body temperature in the same way that warm-blooded animals do. Based on this evidence, would you conclude that dinosaurs were more closely related to cold-blooded lizards or to warm-blooded ostriches?

Taxonomy—A useful tool

Scientists working in agriculture, forestry, and medicine use taxonomy as a basis for their work. Let's assume that a child has eaten one of the mushrooms shown in *Figure 20.2*. Knowing that many poisonous mushrooms resemble harmless species, the child's parents probably would rush the child to the nearest hospital. A taxonomist working at the poison control center could identify the mushroom by examining its structures, and the doctor attending the child would then know whether treatment was necessary.

Knowledge of taxonomy—Its importance to the economy

The discovery of new sources of lumber, medicines, and energy is often the result of explorations by taxonomists. Knowledge of relationships between new and existing species may lead to increased knowledge of these species. For example, we know that pine trees contain resins that can be used as disinfectants. It's possible that other evergreen trees produce useful substances. In fact, the bark of the Pacific yew, *Figure 20.3,* produces taxol, an experimental drug that shows promise in treating some forms of cancer. If we know that one group of plants has a certain kind of chemical substance, we can often assume that related plants contain similar substances.

How Living Things Are Classified

In classification systems, things are categorized for a particular purpose. Just as the classified advertising in the newspaper places all autos for sale into groups with similar characteristics such as domestic or imported, model, brand, and year, biologists classify organisms using similar characteristics.

Taxonomic categories

In taxonomy, organisms are grouped into a series of categories called taxa, each one larger than the previous one. The taxa fit together like nested boxes, one inside another. You already know that organisms which look alike and successfully reproduce among themselves belong to the same species, and that a group of similar species that are alike in general features and are closely related is a genus.

ThinkingLab   Interpret the Data

What fraction of species has been identified?

The numbers of species inhabiting Earth are presently unknown. Estimates vary between 5 and 30 million or more. The number of species identified and cataloged by taxonomists has grown at different rates for different groups of organisms. Some groups of organisms have been studied extensively for a long period of time, whereas other groups have been studied only recently. Consequently, groups that have been studied for a long time are more completely known than groups that have only recently been studied.

Analysis

Examine the graphs below. Compare the percentage of bird species known by 1845 with the percentage of arthropod species (insects, spiders, crayfish) known by 1845. Note the shapes of the graphs.

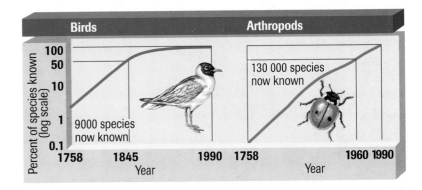

Thinking Critically

What percentage of bird species was known by 1960 compared with the percentage of arthropod species known? If the graph were extended to the year 2000, what probably would happen to the bird graph as compared with the arthropod graph, based on past trends? Explain.

Figure 20.3

The Pacific yew is a rare evergreen tree that grows only in the Pacific Northwest of the United States.

Figure 20.4

A lynx, *Lynx canadensis,* has a short tail with a black tip running all the way around the tail. It also has highly visible tufts of hair on the ears. A bobcat, *Lynx rufus,* has a short tail with black only on top of the tail's tip. It also has inconspicuous ear tufts. In what ways are these two animals alike? You know they are similar in many ways because their classification is the same in all taxa except species.

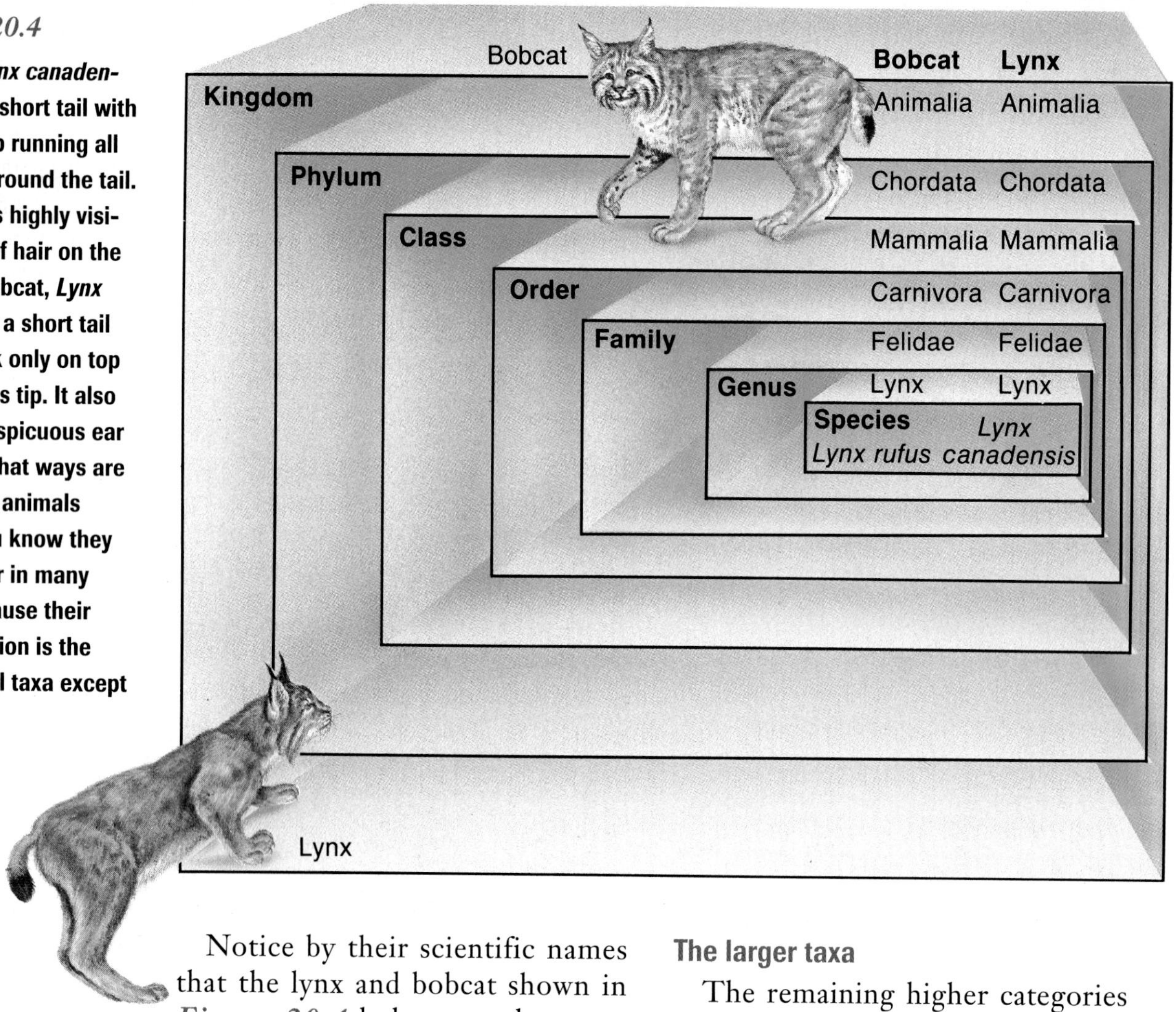

Notice by their scientific names that the lynx and bobcat shown in *Figure 20.4* belong to the same genus. The genus is indicated by the first word of the binomial. In the genus, *Lynx,* the skulls of the animals have 28 teeth. The skulls of lions and mountain lions have a different number of teeth. Therefore, these animals are classified in a different genus, even though they are still obviously cats.

The next larger taxon is the family. A **family** is a group of closely related genera. The bobcat and the lynx both belong to the same family, the cat family, Felidae. Pet house cats also belong to this family. All members of the cat family have short faces, small ears, and most have claws that can be retracted. They also have five toes on each front foot and four toes on each back foot.

The larger taxa

The remaining higher categories of classification are order, class, phylum, and kingdom. An **order** is a group of related families. A **class** is a group of related orders. A **phylum** is a group of related classes, and a **kingdom** is a group of related phyla. In this system of classification, animal groups are called phyla and plant groups are called **divisions.**

As *Figure 20.4* shows, bobcats and lynxes belong to the same order, Carnivora. The carnivores all have similar structures and arrangements of teeth. These two cats also belong to the same class, Mammalia. Mammals have hair or fur and feed their young milk. The phylum Chordata, to which a bobcat and a lynx belong, includes animals with backbones; the kingdom, Animalia, includes all animals.

How Are Relationships Determined?

Species are classified based on their similarities in structure, chemistry, and behavior. Evolutionary relationships are revealed by using all these features for classification.

Relationships can be determined by evolutionary history

You have learned that species share many characteristics, which indicates that they all evolved from a common ancestor. Because members of the cat family resemble each other more than they resemble members of other groups, taxonomists conclude that lynxes and bobcats shared a common ancestor.

When classifying species, taxonomists also compare structures of modern-day life forms with those found in fossils. The relationships revealed by these comparisons demonstrate the evolutionary history of organisms. The evolutionary history of a species is called **phylogeny.** Developing phylogeny can be one purpose for a system of classification. Phylogenetic classifications show the evolutionary history of the classified species.

Biologists find phylogenetic classification useful because organisms belonging to the same group can be expected to share certain characteristics. For example, if you are identifying an animal and find that it has retractile claws, you can infer that it is a member of the cat family and can assume that it has other features common to cats.

Infinite Variety

You might think the seven categories of kingdom, phylum or division, class, order, family, genus, and species would be sufficient to encompass the range of differences among plants and animals, but that is not the case. So seemingly infinite is the variety of characteristics among even closely related species, that further classification is often considered necessary.

Subspecies   Just below the level of species are the subspecies. Generally, subspecies can be defined as morphologically, physiologically, and often geographically distinguishable plant or animal groups that could interbreed if they were ever to come into contact.

The golden whistler birds that inhabit the many islands of the Southwest Pacific are examples of subspecies. Identifiable types of golden whistlers inhabit different islands, a fact that originally prompted zoologists to classify each type of golden whistler as a separate species. When it was discovered that whistlers from different islands would mate and produce offspring when brought together, the 73 species of whistlers were reclassified as subspecies.

Varieties, forms, and cultivars   Below the level of subspecies are the categories of varieties and forms. Varieties and forms are genetically controlled variations. For example, in a population of sunflowers, there may be a variety that has hairier leaves or one with taller stems. A form of one of these varieties may have no hairs at all. These forms may occur only rarely in the population.

Botanists produce cultivars by crossbreeding two varieties. Cultivated varieties are common. Suppose a botanist has two varieties of tomatoes. One variety is exceptionally sweet; another variety has an unusual creamy-orange color. By crossbreeding the two varieties, the botanist may eventually produce a cultivar that displays both the sweetness and the unusual coloration of its parents.

Thinking Critically

Some scientists estimate that Earth may hold millions of as-yet-undiscovered species. Others say billions of species are still to be named. Are these taxonomists all considering the same categories of organism? Explain.

How can you classify seeds?

Seeds are plant structures that contain an embryo plant. They also contain a source of stored food and a protective seed coat. Many seeds have winglike structures or small hooks that aid in their dispersal to new areas, where they may begin to grow.

Procedure

1. Work as a team with four other students. Choose a characteristic that will allow you to classify the ten seeds your teacher gives you into two groups. Size, shape, color, or structure are some possible characteristics.

2. Record the characteristic used for your first groups in a chart provided by your teacher.

3. Within each of the two groups, form smaller categories by choosing characteristics that will allow you to separate the seeds of each group into subgroups. Record these characteristics.

4. Continue following step 3 until only one seed remains in each group.

Analysis

1. What characteristics did you use to group your seeds?

2. Did other teams use the same characteristics? How did their criteria affect their final groups compared with your final groups?

3. Why is it an advantage to scientists to use one standard classification system?

Figure 20.5

In what ways are a guinea pig (left) and a mouse (right) different? One of the most important differences cannot be seen. Of the 51 amino acids in insulin, a protein hormone, guinea pigs and mice have differences in 18. Humans and mice also differ by the same number of amino acids in their insulin molecules. Thus, guinea pigs are no more related to mice than are humans.

Relationships can be determined by development

In addition to similarities in structures between modern and fossil organisms, taxonomists have examined developmental stages of animals for similarities to determine their relationships and phylogeny. Biologists have found that even though adults of certain species may look different, the larval form of one species may show a resemblance to the adult of another species. Taxonomists conclude, based on developmental evidence, that these two animals are related.

Relationships can be determined by biochemistry

Taxonomists also use information from biochemical analyses of organisms to compare and classify species. Look at the guinea pig shown in *Figure 20.5.* Long thought to be rodents, guinea pigs were reclassified in a separate group because the amino acid sequences in some proteins have been found to be significantly different in the two groups. Closely related species have similar DNA, and therefore, similar proteins. In general, the more amino acid or nucleotide base sequences that are shared by two species, the more closely related they are.

Relationships can be determined by behavior

Behavioral patterns are also examined by taxonomists. Sometimes, behavioral patterns provide important clues to relationships. For example, two species of frogs, *Hyla versicolor* and *H. chrysoscelis*, live in the same area and have the same physical appearance. During the breeding season, the males of each species make distinctive calls to attract mates. Scientists used both DNA analysis and analysis of the distinctive mating calls to identify the frogs as being two separate species.

Scientific Names

The common names of organisms do not tell you how the organisms are related or classified. Common names can be misleading. A sea horse is a fish, not a horse. In addition, confusion can occur when an organism has more than one common name. The bird in *Figure 20.6* lives not only in the United States but also in several countries in Europe, and in each country it is identified by a different common name. Suppose an English scientist publishes an article on the bird's behavior and identifies it by its English common name. A Spanish scientist might not recognize the bird as being the same species that lives in Spain.

All newly discovered species are given their scientific names in Latin. Latin is the language chosen by taxonomists because it is no longer spoken and, therefore, does not change as spoken languages do. It is important that scientific names remain the same for years to come. Other naming rules include printing scientific names in italics or underlining them when they are written, and making the first letter uppercase for the genus but not for the second word of the binomial.

Figure 20.6

In the United States, this bird is called the English sparrow, while in England it is called the house sparrow, in Spain the gorrion, in Holland the musch, and in Sweden the hussparf. Because of the potential confusion caused when organisms have more than one common name, scientists from all countries use the same language for names of organisms. The scientific name for this bird is *Passer domesticus.*

Section Review

Understanding Concepts

1. Give two reasons why binomial nomenclature is useful.
2. What kinds of data can be used to classify organisms?
3. What did Linnaeus contribute to the field of taxonomy?

Thinking Critically

4. Why is phylogenetic classification more natural than a system based on characteristics such as usefulness in medicine, or shapes, sizes, and colors of body structure?

Skill Review

5. **Classifying** Make a list of all the furniture in your classroom or your room at home. Classify it into groups based on structure and function. For more help, refer to Organizing Information in the *Skill Handbook.*

Section Preview

Objectives

Compare the five kingdoms of organisms.

Distinguish between the Kingdoms Monera, Protista, Fungi, Plantae, and Animalia.

Key Terms

moneran
protist
fungus

Do you remember looking in the newspaper to find out what movies were in town? Suppose you decide to rent a video instead of going to the movies. When you go to the video store, you find that movies are classified into categories such as comedy, drama, horror, western, and adventure. The numbers of these categories may vary from store to store. Although organisms can be classified in different ways, most taxonomists now group them into five kingdoms.

Phylogenetic Classification: A Model

phylogenetic:

phylon (GK)
  related group
geny (GK) origin
A phylogenetic classification system is based on evolutionary relationships.

Aristotle's classification system and other early systems were developed before scientists had studied geologic time. Therefore, phylogenetic relationships were not originally used by taxonomists; these early systems reflected only differences in structures of organisms. As a result, the number of kingdoms in early classification schemes varied. As biologists began to unravel the trends in evolution of species, these early classification schemes were modified to reflect evolutionary relationships. Aristotle's two categories of plants and animals were eventually replaced by five kingdoms. The five-kingdom system is used by most biologists today because they think it best reflects phylogeny.

Using the fan diagram

Pictorial models like the one you'll see as you turn the page often are helpful tools for understanding scientific concepts. *Figure 20.7* is a model of the five-kingdom classification system. The model looks like a fan with all the modern-day major groups of organisms placed around the outer edges. Thus, the outermost layer of the model represents present time. Within the body of the fan, geologic time is represented by different shaded bands. Using this fan diagram of the five kingdoms, it is possible to identify extinct groups of species. Identify the five kingdoms in this model. Within each kingdom, there are examples of present-day organisms. Recall from Chapter 17 that scientists have divided Earth's history into various eras and periods that make up the Geologic Time Scale. Notice that in this model, the center of the fan's base represents the origin of life. The rays of the fan show the evolution of modern-day species from a common origin. Groups joined close together in the rays of the fan share many characteristics;

groups that are far apart are very different and, therefore, probably not as closely related.

As you study the taxonomic groups in this textbook, refer to this diagram or to the inside of the back cover. You will learn not only about taxonomic groups, but also about their phylogenetic relationships.

How the Five Kingdoms Are Distinguished

As you have seen in the model of phylogenetic classification, the organisms comprising the five kingdoms are monerans, protists, fungi, plants, and animals. In general, the kingdoms can be distinguished by characteristics at the cellular level and by the methods of obtaining food.

Monerans—Microscopic prokaryotes

Kingdom Monera contains all prokaryotes, cells without nuclei bound in membranes. Other kingdoms are eukaryotic. Members of this kingdom, the **monerans,** are microscopic, and almost all are unicellular. They are commonly known as bacteria. Monerans first appeared in the fossil record about 3.5 billion years ago. More than 10 000 species have been identified. Yet because of the extreme environments where these organisms live—such as salt lakes, swamps, deep-ocean hydrothermal vents, and the digestive systems of animals—untold numbers probably remain to be discovered, named, and described. The monerans shown in *Figure 20.8* obtain their food from many diverse metabolic pathways.

The Terrestrial Paradise
by Jan Brueghel the Elder

The Terrestrial Paradise was painted by Jan Brueghel the Elder (1568–1625) in the 16th century. This pictorial inventory of exotic plants and animals was typical of the decorative style of Flemish painters of the time. Northern European artists of this period were known for their realistic style of painting, and some of the greatest paintings in this style were done by artists of Flanders, a region that lies in what is now Belgium and France.

The Terrestrial Paradise was a good representation of the diversity of life, as viewed in that period of time before people began exploring new lands and discovering the existence of vast numbers of species. Brueghel filled his canvas with rich details of the more fantastic, as well as the familiar, mammal and bird species known in Europe. As you can see, he depicts mainly the species that were the most useful or attractive to humans. Most of the animals shown are birds or mammals. The plant life is represented mainly by flowering plants. It would not be until the next century that the great natural scientist Carolus Linnaeus would develop the science of taxonomy, the systematic naming and recording of species.

CONNECTION TO Biology

How do ideas about the diversity of life in the 16th century compare with ideas about the diversity of life today?

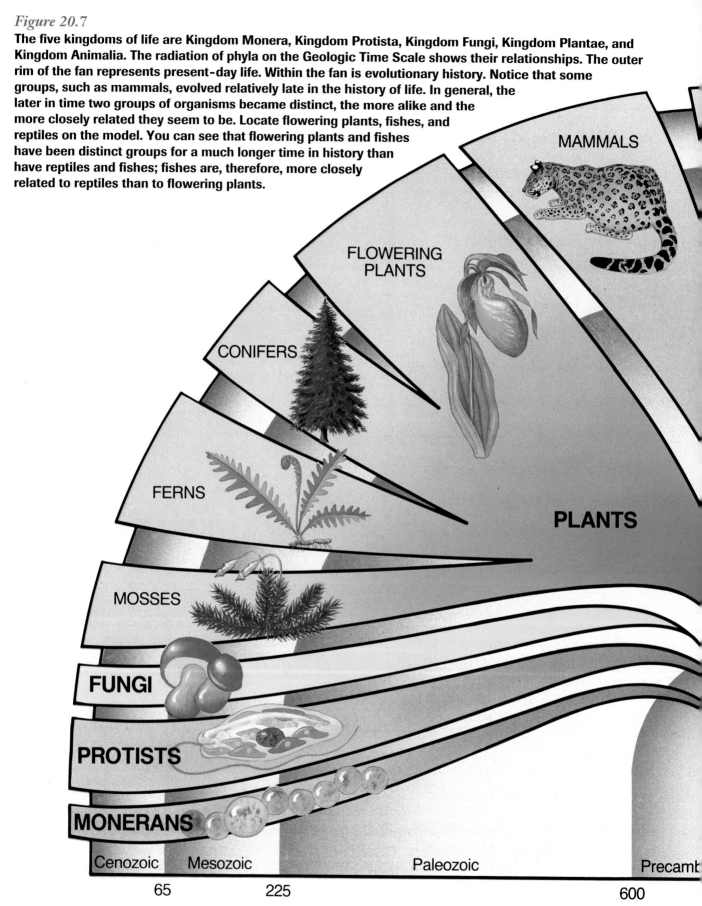

Figure 20.7

The five kingdoms of life are Kingdom Monera, Kingdom Protista, Kingdom Fungi, Kingdom Plantae, and Kingdom Animalia. The radiation of phyla on the Geologic Time Scale shows their relationships. The outer rim of the fan represents present-day life. Within the fan is evolutionary history. Notice that some groups, such as mammals, evolved relatively late in the history of life. In general, the later in time two groups of organisms became distinct, the more alike and the more closely related they seem to be. Locate flowering plants, fishes, and reptiles on the model. You can see that flowering plants and fishes have been distinct groups for a much longer time in history than have reptiles and fishes; fishes are, therefore, more closely related to reptiles than to flowering plants.

MAMMALS

FLOWERING PLANTS

CONIFERS

FERNS

PLANTS

MOSSES

FUNGI

PROTISTS

MONERANS

Cenozoic　Mesozoic　Paleozoic　Precamb

65　　　225　　　600

Eras: shown in millions of years ago

Life's Five Kingdoms

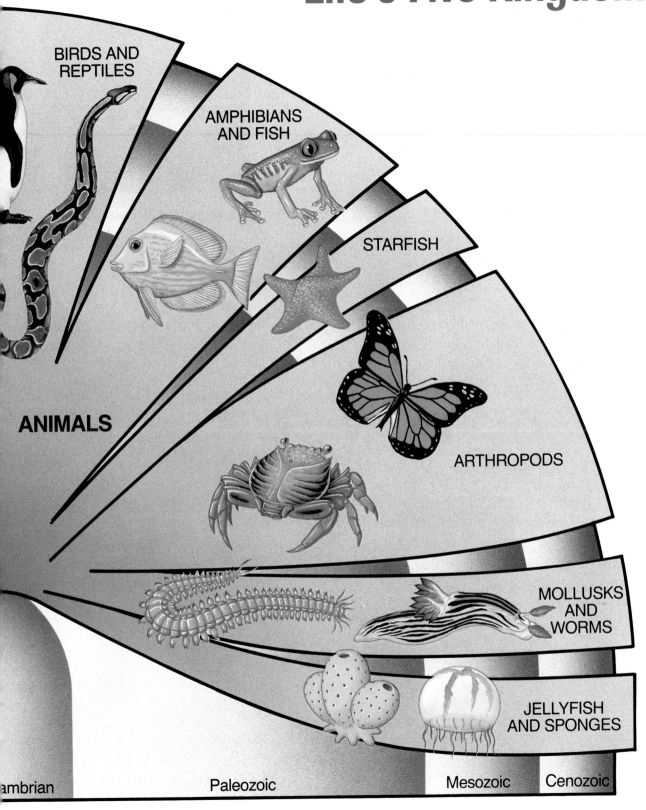

BIRDS AND REPTILES

AMPHIBIANS AND FISH

STARFISH

ANIMALS

ARTHROPODS

MOLLUSKS AND WORMS

JELLYFISH AND SPONGES

ambrian

Paleozoic

Mesozoic

Cenozoic

Figure 20.8

Monerans have a variety of structures and metabolic systems.

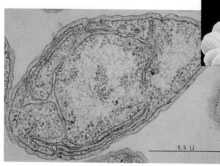

(Magnification: 123 700×)

▲ Some bacteria, such as *Nitrosomonas*, produce food by metabolizing such simple molecules as ammonia and methane.

(Magnification: 40 000×)

▲ The bacterium that causes strep throat must get its food from an outside source.

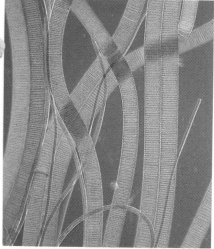

(Magnification: 100×)

▲ Cyanobacteria, such as *Oscillatoria*, live in fresh water and produce food by photosynthesis.

Figure 20.9

These three protists may seem to have little in common. Review the definition of a protist, and speculate about what features they share.

Protists—A diverse group

Kingdom Protista includes eukaryotic unicellular or multicellular organisms with a variety of characteristics—some plantlike, some animal-like, and some funguslike. A **protist** is a eukaryotic organism that lacks complex organ systems and lives in moist environments. Protists first appeared in the fossil record about 1 billion years ago and today number 90 000 to 200 000 species. The protists in *Figure 20.9* show the great diversity in this kingdom.

▼ The paramecium is an animal-like, unicellular protist that can move rapidly as it captures smaller protists for food.

◀ A slime mold can creep along the forest floor like an animal, but later produces reproductive structures, shown here, that are similar to those made by a fungus.

(Magnification: 140×)

▶ You may have seen large kelps washed up on the shore. Although kelp may look like a plant, it does not have tissues organized into organs or organ systems.

Figure 20.10

Morels are edible fungi that are considered delicacies. These gourmet delights grow for only a few days in limited habitats.

Fungi—Earth's decomposers

Organisms from the Kingdom Fungi are consumers that do not move from place to place. A **fungus** is a unicellular or multicellular heterotrophic eukaryote that absorbs nutrients obtained by decomposing dead organisms and wastes in the environment. Fungi first appeared in the fossil record about 400 million years ago. More than 100 000 species have been named. You may be familiar with the fungus in *Figure 20.10.*

Plants—Earth's multicellular oxygen producers

The organisms in the Kingdom Plantae are stationary, multicellular eukaryotes that photosynthesize. Most have cellulose cell walls and tissues organized into organs and organ systems. Because plant materials do not fossilize as easily as the bones of animals, the oldest plant fossils that have been found are only 400 million years old. Half a million species have been described. You are probably familiar with flowering plants like the one shown in *Figure 20.11;* however, the mosses, ferns, and their relatives represent plants with some distinctly different characteristics.

MiniLab

How is a dichotomous key used?

How could you find out the name of a tree you see growing in front of your school? You might consult a local expert, or you could use a manual or field guide that contains descriptive information and keys to aid identification. A key is a set of descriptive sentences divided into steps. A dichotomous key has two descriptions at each step. The steps are followed until the key leads to the name of the tree.

Procedure

1. Using a collection of leaves from local trees and a dichotomous key for trees of your local area, determine the names of the trees from which the leaves came. To use the key, begin with a choice from the first pair of descriptions. Continue following the key until the name of each tree is found.

2. Glue each leaf on a separate sheet of paper. Record the tree name for each leaf.

Analysis

1. What is the function of a dichotomous key?

2. List three different characteristics used in your key.

3. As you used your key, did the characteristics become more general or more specific?

Figure 20.11

You can see the cell wall and chloroplasts typical of plant cells in the enlarged cell of this hibiscus plant. The structures visible in the center of the flower are male and female reproductive organs.

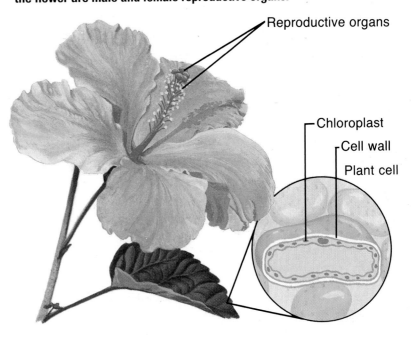

Reproductive organs

Chloroplast

Cell wall

Plant cell

BioLab

Do you remember the first time you saw a beetle? You may have asked someone nearby, "What is it?" You may still have that natural curiosity and want to know the names of insects you find in your yard. To help identify organisms, taxonomists have developed classification keys. A dichotomous key is a set of paired statements that can be used to identify organisms. When you use a dichotomous key, you choose one statement from each pair that best fits the organism. After each choice, you are directed to the next set of statements you should use. When you have made all the choices needed, you will arrive at the name of the organism or the group to which it belongs.

Making a Dichotomous Key

PREPARATION

Problem
How is a dichotomous key made?

Objectives
In this Biolab, you will:
- **Classify** organisms on the basis of structural characteristics.

- **Develop** a dichotomous key.

Materials
sample keys from guidebooks
metric ruler

PROCEDURE

1. Study the drawings of beetles.

2. Choose one characteristic and classify the beetles into two groups based on that characteristic. Take measurements if you wish.

3. Record the chosen characteristic in a diagram like the one shown. Write the numbers of the beetles in each group on your diagram.

4. Continue to form subgroups within your two groups based on different characteristics. Record the characteristics and numbers of the beetles

in your diagram until you end with one beetle in each group.

5. Using the diagram you have made, make a dichotomous key for the beetles. Remember that each numbered step should contain two alternative choices for classification. Begin with 1A and 1B. For help, examine sample keys provided by your teacher.

6. Exchange dichotomous keys with another team. Use their key to classify the beetles.

1 Variegated mud-loving beetle

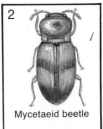

2 Mycetaeid beetle

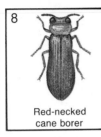

3 Apricot borer

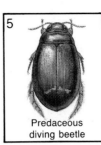

4 Water tiger

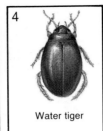

5 Predaceous diving beetle

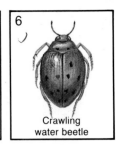

6 Crawling water beetle

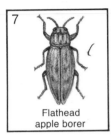

7 Flathead apple borer

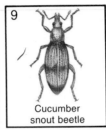

8 Red-necked cane borer

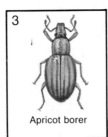

9 Cucumber snout beetle

10 Whirligig beetle

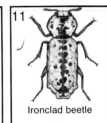

11 Ironclad beetle

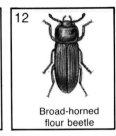

12 Broad-horned flour beetle

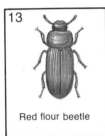

13 Red flour beetle

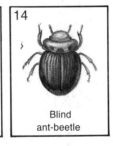

14 Blind ant-beetle

15 False wireworm beetle

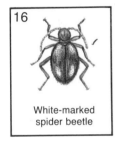

16 White-marked spider beetle

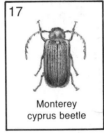

17 Monterey cyprus beetle

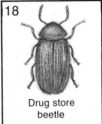

18 Drug store beetle

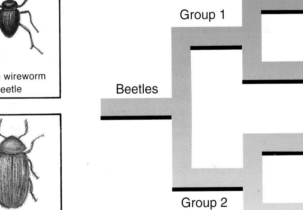

Group 1

Beetles

Group 2

ANALYZE AND CONCLUDE

1. **Comparing and Contrasting** Was the dichotomous key you constructed exactly like those of other students? Explain.

2. **Analyzing Data** What characteristics were most useful for making a classification key for beetles? What characteristics were not useful?

3. **Thinking Critically** Why do keys typically offer two choices rather than a larger number of choices?

Going Further

Application Using the same method, make a dichotomous key to identify the students in your class.

20.2 The Five Kingdoms **497**

The luna moth's antennae are sense organs. With these antennae, the moth can detect minute quantities of chemical odors in the air.

The cheetah's brain, a part of its nervous system, keeps it on course as it speeds through the grasslands using its powerful muscles. What senses might the cheetah use while running?

Figure 20.12

Most animals have well-developed nervous and muscular systems.

Animals–Multicellular consumers

Animals are multicellular consumers that eat and digest other organisms for food. Animal cells have no cell walls. Nearly all animals are able to move and have tissues organized into organs and complex organ systems. As shown in *Figure 20.12,* nervous systems, muscle systems, and sense organs are common to many. Animals first appeared in the fossil record 700 million years ago.

Connecting Ideas

Taxonomists have made logic and sense of much of the information gathered over the past century through the study and observation of living things. As you begin to examine phylogeny and study organisms of the five kingdoms, you will come to appreciate even more how much biology depends on the work of taxonomists. They have given us a useful tool that is continually modified as new information about organisms becomes available. The species that make up the five kingdoms of life will be presented in this text in the order of their phylogenetic relationships. These relationships represent a trend in evolution from unicellular species through the highly complex plants and animals.

Section Review

Understanding Concepts
1. Describe the five kingdoms of life based on methods that organisms use to obtain food.
2. How do monerans differ from the other four kingdoms?
3. How is the fan diagram a useful model for phylogenetic classification?

Thinking Critically
4. If the present system of classification were changed to one with more kingdoms, which of the present kingdoms do you think could be further divided? Explain your answer.

Skill Review
5. **Making and Using Tables** Make a table that compares characteristics of members of each of the five kingdoms. For more help, refer to Organizing Information in the *Skill Handbook.*

Computers and Cladistics

The most meaningful system of taxonomy for biologists is one that reflects phylogeny—the evolutionary history of a species. Traditional taxonomy, however, is based on structural similarities and differences. When a biologist finds a new organism, he or she carefully notes which structures are similar to those of known species and which are unique. The new organism then is grouped with those it most closely resembles. However, this method is subjective and sometimes leads to errors.

Cladistics   One method of establishing relationships among species may be cladistics. Cladistics, from the Greek word *clados*, meaning "branch," is the school of taxonomy that uses only evolutionary relationships to classify organisms.

Cladists hypothesize that every phylum or division originated in a single group of organisms that has branched out into today's species. They attempt to determine the order in which these branches have arisen over time. Then they trace the branches back until they join other branches again and again, ultimately ending up at the original ancestral species. The ladderlike diagram that is formed is called a cladogram. Each branch point, represented by a fork in the cladogram, shows a point in evolution where a unique characteristic developed to separate one group from another.

Cladograms   Look at the cladogram. Each of the five animal groups has many characteristics that were present in the common ancestor, as well as some characteristics unique to each animal group. When retractable claws evolved, cats branched from the common lineage that included lions. Lions and cats branch from the lineage that contains seals because pointed molars evolved. By tracing the branches back, you can see that snakes branched off when hair evolved. Cladistics is concerned not with the degree of these structural differences, but only with the order in which these differences arose.

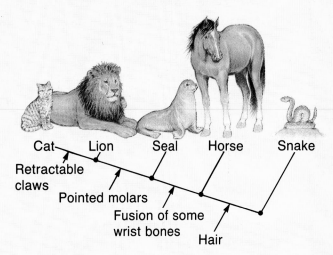

Cat — Lion — Seal — Horse — Snake
Retractable claws
Pointed molars
Fusion of some wrist bones
Hair

Applications for the Future

There are millions of characteristics and millions of species, resulting in billions of possible combinations. How is any human mind able to cope with all these data? The answer is that no human mind has to because a computer can. Enormous quantities of data can be gathered together in a single database. If all of the known quantifiable data are fed into a database of animal characteristics, you could search for and compare separate categories of information one by one. Trying millions of comparisons of characteristics every second, a computer can accomplish in a few hours what might take trained scientists years.

INVESTIGATING the Technology

1. **Model** Make your own database. Choose a subject from around your home or school, and propose criteria that could be used to make a cladogram.

2. **Evaluate** A major problem with organ transplants is getting a close tissue match so that the organ is not rejected by the body. Suggest ways that cladistics might help solve this problem.

Reviewing Main Ideas

20.1 Classification

- Aristotle developed the first classification system based on the similarity of organisms and where they live.
- Modern classification systems began with Linnaeus's organization of living things based on their similarities in structure. This system naturally led to the phylogenetic classification most commonly used today.
- Species are named by a two-word system called binomial nomenclature.
- Classification provides a framework of logic and order so that relationships among living things can be seen easily.
- Organisms are classified by using a hierarchy of taxa. From largest to smallest, the hierarchy is kingdom, phylum, class, order, family, genus, species.
- The criteria used by biologists to classify living things are similarities in phylogeny, structure, development, biochemistry, and behavior.

20.2 The Five Kingdoms

- The five kingdoms are represented by the monerans, prokaryotic cells without nuclei bound in membranes; protists, eukaryotic organisms that lack complex organ systems and live in water or moist environments; fungi, eukaryotes that obtain food by absorption; plants, multicellular eukaryotes that produce their own food; and animals, multicellular consumers whose cells lack cell walls.

Key Terms

Write a sentence that shows your understanding of each of the following terms.

binomial nomenclature	kingdom
	moneran
class	order
classification	phylogeny
division	phylum
family	protist
fungus	taxonomy
genus	

Understanding Concepts

1. If two fungi are in the same class, what other categories must they also share?
2. Why is the job of classifying all living things on Earth not completed?
3. What types of information are useful for identifying relationships among species?
4. What is the relationship between the complexity of organisms from the five kingdoms and their time of appearance on Earth?
5. The second part of the binomial of a moneran is *circinalis*. This moneran belongs to the same genus as *Anabaena flos-aquae*. Write the complete scientific name of the first species.
6. Why is the Linnaean system of classification better than Aristotle's system?
7. How is phylogeny related to taxonomy?

8. Would similarity in protein structure be an indication of a close relationship between two different kinds of organisms? Explain.

Using a Graph

9. A researcher compares DNA nucleotide sequences of a newly discovered animal with sequences from three different animals—A, B, C. Which animal is most closely related to the newly discovered animal? Explain.

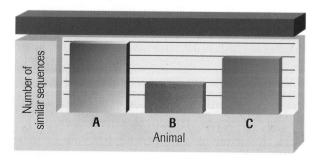

Relating Concepts

10. Make a concept map that relates the following terms and phrases. Supply the appropriate linking words for your map.

    classification, taxonomy, Linnaeus, Aristotle, behavior, biochemistry, structural similarities, development, phylogeny

Applying Concepts

11. Explain why early systems of classification are not used today.
12. An organism is a unicellular producer and has a rigid cell wall and a nucleus. To what kingdom does it belong?
13. You have four organisms. Organism A has wings, six legs, and antennae. Organism B has wings, two legs, and large ears. Organism C has no wings, four legs, and large ears. Organism D has wings, two legs, and no ears. Classify these four organisms to show their relationships.
14. Based on your own knowledge of their characteristics, classify the following organisms by kingdom: human, mushroom, bear, rosebush, clam, seaweed, earthworm, frog, horse, pine tree, pond scum, jellyfish, mosquito, tulip, turtle, mold on fruit, carrot.
15. Compare the classification system used by your school library with the classification system used for living organisms.

Art Connection

16. Why do you think Brueghel showed so few species of plants and animals in his painting?

Thinking Critically

Making Inferences

17. Suppose you find a unicellular organism swimming in a drop of water you have placed under a microscope. It is green and moves fast. You can see the nucleus clearly. In what kingdom does this organism belong? What other features would you predict this organism to have?

Thinking Critically

18. Explain how the work of Linnaeus illustrates the nature of science.

Interpreting Data

19. **Biolab** Tell whether the following two insects are dragonflies or damselflies by using the following key.
    1A. Front and hind wings similar in size and shape, and folded parallel to the body when at rest....damselflies
    1B. Hind wings wider than front wings near base, and extended on either side of the body when at rest.... dragonflies

Making Predictions

20. **Minilab** Could you classify trees in the winter? Make a list of the features you might use to place trees into groups.

Classifying

21. **Minilab** Make sketches of utensils you might find in a kitchen drawer. Classify them into groups based on their functions.

Connecting to Themes

22. **Systems and Interactions** Explain how the behavior of organisms became a part of the classification system used today.
23. **Evolution** In what way is evolution a part of taxonomy?

21 Viruses and Monerans

Did you ever wonder what happens to all the leaves that fall from the trees every autumn? Imagine if they just accumulated on the forest floor. In a few years, the leaves in the woods would be knee-deep, and eventually they would cover the trees themselves! This does not happen because the leaves decompose after they fall from the trees. Many bacteria, along with fungi, are involved in this decomposition. It is estimated that 10 billion bacteria exist in a spoonful of garden soil. Many of them are decomposers. Decomposers are especially important because they return nutrients that are locked up in dead organisms back to the ecosystem. These nutrients, in the form of simple, inorganic molecules, can be taken up by producers and returned to the food chains of the ecosystem.

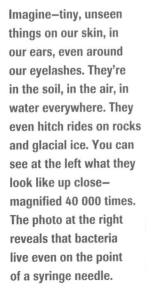

Imagine–tiny, unseen things on our skin, in our ears, even around our eyelashes. They're in the soil, in the air, in water everywhere. They even hitch rides on rocks and glacial ice. You can see at the left what they look like up close— magnified 40 000 times. The photo at the right reveals that bacteria live even on the point of a syringe needle.

Magnification: 560×

21.1 Viruses

Section Preview

Objectives

Categorize the different kinds of viruses.

Compare the different reproductive cycles of viruses.

Key Terms

virus
host cell
bacteriophage
lytic cycle
lysogenic cycle
provirus
retrovirus
reverse transcriptase

How many childhood diseases can you recall having: mumps, measles, German measles? These are called childhood diseases because they used to attack mainly children. A generation or so ago, childhood diseases were common; most children got them. Nowadays, vaccination of most infants and children has made these diseases rare. But the causes of childhood diseases have not been wiped out. They still pose a threat to those who have not been vaccinated.

Structure and Shape of Viruses

The diseases mentioned above are not caused by bacteria. They are not caused by organisms at all, but rather by tiny, nonliving particles called **viruses.** Viruses are about one-half to one-hundredth the size of the smallest bacterium. Most biologists do not consider them to be alive because they don't fulfill all the criteria for life. They do not carry out respiration or grow or move. All viruses can do is reproduce—and they can't even do that by themselves. Viruses reproduce only inside a living cell. The cell in which they reproduce is called the **host cell.**

Because they are nonliving, viruses are not given Latin names, as are cellular organisms. Viruses often are named for the diseases they cause—

for example, rabies virus, polio virus. Others are named for the organ or tissue they infect—for example, adenoviruses were first detected in adenoid tissue at the back of the throat. In some cases, code numbers are used to distinguish several viruses infecting the same host. Seven viruses that infect the common intestinal bacterium, *Escherichia coli,* are named bacteriophage T1 through T7 (*T* stands for *Type*). A **bacteriophage** is a virus that infects bacteria.

Viral structure

A virus consists of an inner core of nucleic acid surrounded by one or two protein coats. Some relatively large viruses, such as the human flu virus, may have another layer, called the viral envelope, surrounding the outer coat. This layer consists primarily of protein.

bacteriophage:

bakterion (GK) small rod
phagein (GK) to devour
A bacteriophage is a virus that infects and destroys bacteria.

The nucleic acid core of a virus contains the virus's genetic material. The nucleic acid in some viruses is DNA and in others it is RNA, but it is never both DNA and RNA. The nucleic acid consists of genes that contain coded instructions only for making copies of the virus. They code for nothing else.

The arrangement of the proteins in the viral coat gives different kinds of viruses different shapes, as shown in *Figure 21.1.* These shapes play a role in the infection process.

Figure 21.1

Viruses have a variety of shapes, determined by the proteins in their coats. A viral coat is called a capsid.

▼ Polyhedral viruses, such as the polio virus, resemble small crystals under the electron microscope. Many polyhedral viruses are polygons with 20 sides.

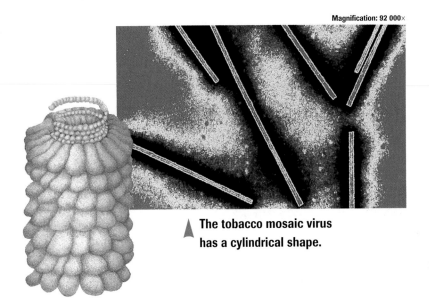

Magnification: 92 000×

▲ The tobacco mosaic virus has a cylindrical shape.

▼ Some viruses, such as influenza and HIV, the virus that causes AIDS, are covered with an envelope that is studded with projections. The diagram shows HIV; the photo shows the influenza virus.

Magnification: 20 000×

Magnification: 95 000×

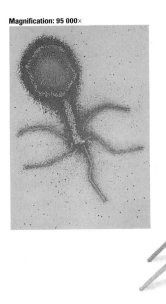

◀ This T4 virus, which infects bacteria, resembles a minuscule lunar-landing module. It has a polyhedral head containing DNA, a protein tail, and protein tail fibers.

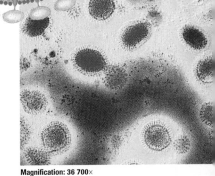

Magnification: 36 700×

What does a bacteriophage look like?

Bacteriophages (*phages,* for short) are viruses that infect bacteria. A typical phage consists of a polyhedral protein head surrounding a DNA core, a tail made of several proteins, and six tail fibers.

Procedure

1. You can build a model of a bacteriophage by attaching two nuts onto the top of a bolt (3.7 cm × 0.7 cm). Screw on the nuts so that they touch the top of the bolt.

2. Take three 14-cm-long pieces of #22 gauge wire, and twist them around the bottom of the bolt. Bend down the ends of the wires so that they resemble the figure below.

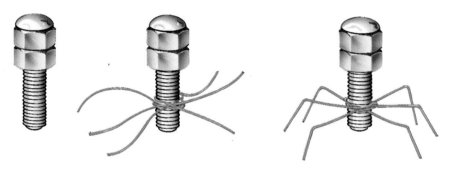

Analysis

1. What bacteriophage structure do the wires represent? What structure do the nuts and the top of the bolt represent? What structure does the body of the bolt represent?

2. Where would the nucleic acid be located in the model?

3. What structure in this model is not generally found in viruses that infect eukaryotic cells?

4. Why would a bacteriophage like the one you have modeled be unable to infect one of your own cells? Refer to a specific structure in the model.

Recognition of a host cell

Viral shape is important in viral reproduction. Before a virus can enter and reproduce in a cell, it must recognize and attach to a specific receptor site on the plasma membrane of the host cell. That's where the viral protein coat plays its role. A protein on the surface of a virus has a three-dimensional shape that matches the shape of a molecule in the plasma membrane of its host cell, like two interlocking pieces of a jigsaw puzzle. In this way, a virus recognizes its host cell. The two interlocking shapes allow the virus to land on the host and lock on, like two spaceships docking. The bacteriophage T4 uses the protein in its tail fibers to attach to the host cell it will infect. In other viruses, the attachment protein is in the envelope that surrounds the virus or in the protein coat.

Attachment is a specific process

Each attachment protein has its own particular shape due to its sequence of amino acids. Therefore, attachment is a specific process. As a result, most viruses can enter and reproduce in only a few kinds of cells. The T4 bacteriophage can infect only certain strains of *E. coli.* T4 cannot reproduce in human, animal, or plant cells, or even in other kinds of bacteria because it cannot attach to these cells. Similarly, tobacco mosaic virus can reproduce only in tobacco plant cells. Not only are viruses species specific, but some also are cell-type specific. For example, polio viruses infect only nerve cells of humans.

The fact that particular viruses can infect only a few kinds of cells has enormous significance. In the 1970s, the World Health Organization eradicated smallpox, a viral disease that has been one of the greatest killers in history. One reason why this eradication was possible was because the smallpox virus could infect only humans. It would have been much more difficult to eradicate a virus that could infect other animals as well.

Viral Reproductive Cycles

Once attached to the plasma membrane of the host cell, the virus must get inside and take over the cell's metabolism. Only then can the virus reproduce. Method of entry into a host cell depends on the shape of the virus. Some viruses are shaped so that they can inject their nucleic acid into the host cell like a syringe, leaving their coat still attached to the plasma membrane. Other viruses have a shape that can barge in by indenting the plasma membrane, which seals off and breaks free, enclosing the virus within a vacuole in the cell's cytoplasm. The virus then bursts out of the vacuole and releases its nucleic acid into the cytoplasm.

Lytic cycle

Once inside the cell, the virus destroys the host's DNA. It reprograms the cell's metabolic activities to copy the viral genes and make viral protein coats using the host's enzymes, raw materials, and energy. The nucleic acid and coats are assembled into new viruses, which burst from the host cell, usually killing it. The newly produced viruses are then free to infect and kill other host cells. This process is called a **lytic cycle.** *Figure 21.2* shows a typical lytic cycle for a bacteriophage.

lytic:
lyein (GK) to break down
The host cell is destroyed during a lytic cycle.

Figure 21.2

Viruses use the host cell's energy and raw materials to make new viruses. A typical lytic cycle takes about 90 minutes and produces about 200 new viruses.

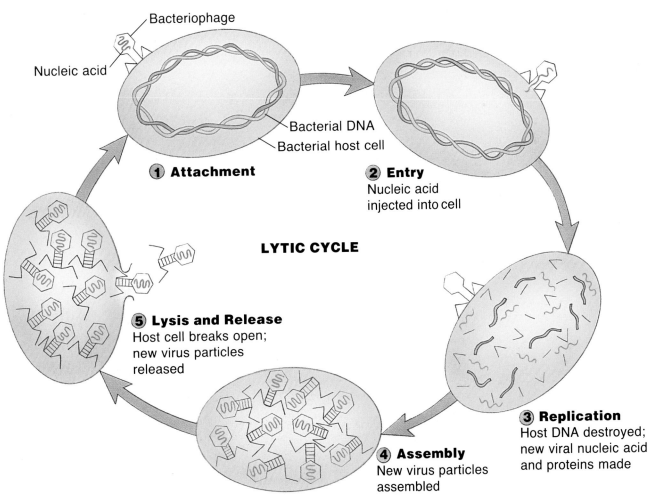

Bacteriophage

Nucleic acid

Bacterial DNA
Bacterial host cell

① Attachment

② Entry
Nucleic acid injected into cell

LYTIC CYCLE

③ Replication
Host DNA destroyed; new viral nucleic acid and proteins made

④ Assembly
New virus particles assembled

⑤ Lysis and Release
Host cell breaks open; new virus particles released

Lysogenic cycle

Viruses vary in the degree of damage they inflict on the host. Not all virus infections are fatal to their host cells. Some viruses can attack but not always kill a cell. These types of viruses go through a **lysogenic cycle,** a viral reproductive cycle in which the viral DNA becomes integrated into the host cell's chromosome.

A lysogenic cycle begins like a lytic cycle. As an example, let's look at a virus that contains DNA. The virus attaches to the plasma membrane of the host cell and injects its DNA into the cell. Events then become different from those in the lytic cycle. Instead of destroying the host DNA and making new viruses, the viral DNA becomes part of the host DNA.

Once the viral DNA is inserted into the host cell chromosome, it is known as a **provirus.** It does not interfere with the normal functioning of the host cell, which is able to carry out its regular metabolic activity. Every time the host cell reproduces, the provirus is replicated right along with the host cell's chromosome. This means that every descendant of the host cell will have a copy of the provirus in its own chromosome. This lysogenic phase can continue undetected for many years. However, the provirus can pop out of the host cell's chromosome at any time and enter a lytic cycle. Then the virus starts reproducing and kills the host cell. A typical lysogenic cycle is shown in *Figure 21.3.*

Figure 21.3

Some viruses do not cause the host cell to burst and be destroyed. In a lysogenic cycle, viral nucleic acid is inserted into the genetic material of the host cell and replicates with it. Eventually, the virus enters a lytic cycle and kills the cell.

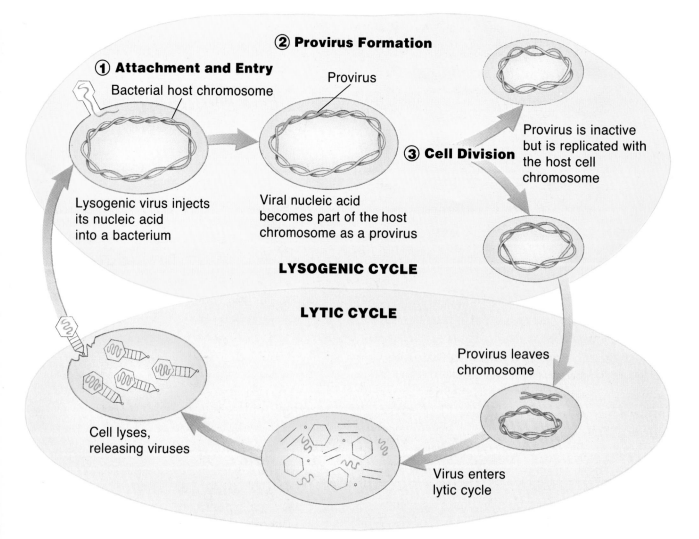

① **Attachment and Entry**
Bacterial host chromosome
Lysogenic virus injects its nucleic acid into a bacterium

② **Provirus Formation**
Provirus
Viral nucleic acid becomes part of the host chromosome as a provirus

③ **Cell Division**
Provirus is inactive but is replicated with the host cell chromosome

LYSOGENIC CYCLE

LYTIC CYCLE

Provirus leaves chromosome

Virus enters lytic cycle

Cell lyses, releasing viruses

Figure 21.4

Influenza is caused by a virus that consists of an RNA core inside a protective protein coat. The virus also has an outer envelope, which it takes from the plasma membrane of the cell it infects. When the virus buds off from the host cell, it wraps itself in a piece of the host's plasma membrane as it leaves the cell.

Proviruses explain symptoms of disease

The lysogenic process explains why cold sores recur. Once you have had a cold sore, the virus causing it— herpes simplex 1—remains as a provirus in one of the chromosomes in your cells even after the cold sore has healed. When the virus pops out and enters a lytic cycle, you get another cold sore. No one knows the exact conditions that cause the provirus to pop out.

New virus particles may be released from the host cell by lysis— bursting of the cell—or they may be released without lysis by a kind of budding process, *Figure 21.4.* This process is a form of exocytosis. The plasma membrane engulfs the virus particles and then opens up to the outside, releasing them.

Sometimes, a provirus nestled in a host cell's chromosome will maintain a low level of activity by repeatedly producing small numbers of new viruses while the host cell continues to function normally. HIV and the influenza virus can act in this manner. Small numbers of HIV viruses bud off from infected white blood cells. These viruses then can infect other white blood cells. However, at this time, the viruses do not kill the host cells.

You can see now why a person can be infected with HIV and appear to be perfectly healthy. As long as most of the infected cells are producing only a few viruses, the person will have no clinical symptoms. It also explains why most people infected with HIV will eventually get AIDS. As more and more cells become infected, and as proviruses begin to pop out, enter a lytic cycle, and kill their host cells, there is a gradual loss of white blood cells, which are important to the body's self-protection processes. This loss eventually results in the breakdown of the body's disease-fighting system that is characteristic of AIDS. It also explains why you can get AIDS from a person who appears to be perfectly healthy. A person infected with HIV who has no symptoms can still transmit the virus.

Lysogenic viruses and disease

Many disease-causing viruses go through lysogenic cycles. These include the herpes viruses that cause cold sores and genital herpes, and the virus that causes hepatitis B. Another such virus is the one that causes chicken pox. An attack of chicken pox, which usually occurs before age ten, gives lifelong protection from another invasion of chicken pox viruses. However, the virus remains dormant within nerve tissues. It may flare up later in life, causing a disease called shingles—a painful infection of the nerves that transmit impulses from certain areas of the skin.

Retroviruses

Many viruses have RNA as their only nucleic acid. The RNA viruses with the most complex reproductive cycles are the **retroviruses.** When retroviruses inject their nucleic acid into the host cell, they also inject a molecule of the enzyme **reverse transcriptase,** which copies viral RNA into DNA. A lysogenic cycle of a retrovirus is shown in *Figure 21.5*.

RNA lytic viruses

Not all RNA viruses are retroviruses. Some viruses contain RNA as their nucleic acid, but they do not have reverse transcriptase. Because a cell has no mechanism and enzymes for copying RNA, these viruses supply an enzyme that can make copies of their RNA molecule by base pairing. Such viruses follow a lytic cycle. They cannot follow a lysogenic cycle because they cannot make a DNA copy of their RNA to insert into the host cell's chromosome.

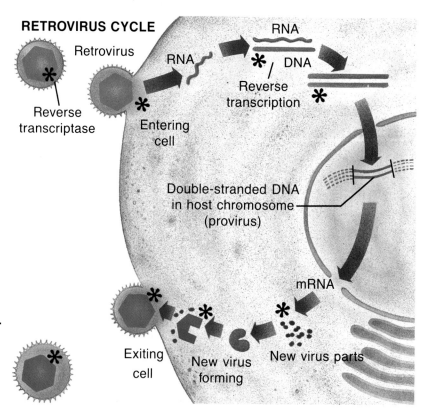

Figure 21.5

Retroviruses have a unique enzyme that transcribes RNA into DNA by base pairing, the reverse of the usual process of transcribing DNA to RNA. This process gives these viruses their name—*retro* means "backward." The viral DNA then can enter the host cell's chromosome, become a provirus, and begin a lysogenic cycle. The viral DNA lodged in the host cell's chromosome can code for small amounts of viral RNA and protein. This results in the production of small numbers of new viruses that bud off.

RETROVIRUS CYCLE

Retrovirus

Reverse transcriptase

RNA

Entering cell

RNA

RNA

DNA

Reverse transcription

Double-stranded DNA in host chromosome (provirus)

mRNA

New virus parts

New virus forming

Exiting cell

Plant Viruses

More than 400 viruses are known to infect plant cells, causing as many as 1000 plant diseases. In fact, the first virus to be identified was a plant virus—tobacco mosaic virus. *Figure 21.6* shows its effects.

Not all plant diseases caused by viruses are fatal or even harmful. For example, some mosaic viruses cause a striking variegation in the flowers of plants such as gladioli and pansies. The diseased flowers exhibit streaks of contrasting colors in their petals, seen in *Figure 21.6.* These viruses are easily spread by cutting infected flower stems and then healthy stems with the same tool.

Origin of Viruses

You might think that because viruses are so simple, they represent an ancient and primitive form of life. This is probably not so. Viruses require host cells for their reproduction. Therefore, scientists believe that viruses originated from their host cells. It has been suggested that viruses are nucleic acids that escaped from their host cells and developed a way to reproduce as parasites of the host cells. If this is correct, then in terms of evolution, viruses are more closely related to their host cells than they are to each other.

Figure 21.6

Tobacco mosaic virus causes yellow spotting on the leaves of tobacco plants (left) and makes them worthless as a cash crop. These Rembrandt tulips (right) get their stripes from a virus that is transmitted from plant to plant.

Section Review

Understanding Concepts

1. Why is a virus not considered to be alive?
2. What is the difference between a lytic and lysogenic cycle?
3. Why can most viruses infect only a few kinds of cells?

Thinking Critically

4. Describe the state of a herpes virus in a person who had cold sores several years ago but who does not have any now.

Skill Review

5. **Making and Using Graphs** A microbiologist added some viruses to a bacterial culture. Every hour from noon to 4 P.M., she determined the number of viruses present in a sample. The numbers of viruses in each sample beginning at noon are 3, 3, 126, 585, 602. Graph these results. How would the graph look if there had been no living bacteria in the culture? For more help, refer to Organizing Information in the *Skill Handbook.*

21.2 Monerans

Section Preview

Objectives

Summarize the history and adaptations of monerans.

Evaluate the economic importance of monerans.

Key Terms

saprobe
obligate aerobe
obligate anaerobe
endospore
toxin
binary fission
conjugation
nitrogen fixation

I magine yourself going back three-and-a-half billion years. You wander around the young Earth and find yourself face-to-face with the first life on this planet. Dinosaurs? Saber-toothed tigers? No. You would be face-to-face with the most ancient and diverse life on Earth—bacteria.

Early microbiologists

Classification of Monerans

Monerans are classified into two very different groups—archaebacteria and eubacteria. Many biochemical differences exist between these groups. Their cell walls and the lipids in their plasma membranes have different structures. The sequence of bases in their tRNA and rRNA is different. Also, they react differently to antibiotics.

Archaebacteria and eubacteria probably diverged from each other several billion years ago, although the exact time is unknown.

Archaebacteria—The earliest monerans

The archaebacteria include three types of bacteria that are found mainly in extreme habitats where little else will live. One group lives in oxygen-free environments and produces methane. A second group can live only in bodies of concentrated salt water, such as the Great Salt Lake in Utah and the Dead Sea in the Middle East. A third group is found in the hot, acidic waters of sulfur springs. *Figure 21.7* introduces you to some of the unique places archaebacteria are found.

Eubacteria—The heterotrophs

Eubacteria, the second main group of monerans, display a wide array of habitats and metabolism. One group of eubacteria, the heterotrophs, are found everywhere. These bacteria need organic molecules as an energy source but are not adapted for trapping the food that contains these molecules. Thus, some live as parasites, absorbing nutrients from living organisms. Others live as **saprobes,** organisms that feed on dead organisms or organic wastes. Saprobes help recycle the nutrients contained in decomposing organisms.

Eubacteria—Photosynthetic autotrophs

A second group of eubacteria are photosynthetic autotrophs. They obtain their energy from light. Cyanobacteria are in this group. They can trap the sun's energy by photosynthesis because they contain a photosynthetic pigment. Most cyanobacteria are blue-green in color, which is why they are commonly called blue-green bacteria. They are not all blue-green, though; some cyanobacteria are red or yellow. Cyanobacteria are common in ponds, streams, and moist areas of land. They are composed of chains of cells—an exception to the rule that all monerans are unicellular.

cyanobacterium:
kyanos (GK) blue
bakterion (GK) small rod
Cyanobacteria are blue-green bacteria.

Figure 21.7

All of the known archaebacteria live without oxygen. They obtain their energy from inorganic molecules or from light.

▶ **Heat- and acid-loving bacteria grow in hot sulfur springs such as this one at Yellowstone National Park. They grow best at temperatures around 60°C and a pH of 1 to 2. This is the pH of concentrated sulfuric acid.**

◀ **Salt-loving bacteria live in saturated salt solutions. They produce an unusual purple pigment that allows them to carry on photosynthesis. They can be seen in this aerial photograph as pink/purple mats. These bacteria are growing in seawater-evaporating ponds near San Francisco Bay. The ponds are used for commercial salt production.**

▶ **Some methane-producing bacteria grow in the stomachs of cows, which lack the enzymes needed to break down cellulose in the grass they eat. These bacteria produce the enzymes that convert cellulose to glucose. Methane producers are also widely distributed in sewage-treatment plants, freshwater swamps, and deep-sea marine habitats. They get their energy from carbon dioxide and hydrogen gas, producing methane as a waste product.**

What are the shapes of bacteria?

Bacteria come in three different shapes: round (coccus), rods (bacillus), and spirals (spirillum). They may appear singly or in pairs, or form chains or clusters. Each species has a typical shape and growth pattern.

Procedure

1. Obtain slides from your teacher showing the three shapes of bacteria.

2. Using low power, locate bacteria of one shape. Then switch to high power. Look for individual cells and observe their shape. Observe also the size of the cells. Then look for groups of bacterial cells to determine their growth pattern.

3. In this manner, observe all three shapes of bacteria. Compare their sizes with those of plant and animal cells on other slides.

4. Draw a diagram of each type of bacteria.

Analysis

1. How did the three bacteria compare in size? How did they compare in size with plant and animal cells?

2. What adaptive advantage might there be for bacteria to form groups of cells?

Figure 21.8

Cyanobacteria, such as *Anabaena* (left), have a blue-green color because the only photosynthetic pigment they contain is chlorophyll *a*, whereas the bacterium *Prochloron* (right) contains both chlorophylls *a* and *b* and is bright green in color. The chlorophyll in cyanobacteria is found free in the cytoplasm, not in chloroplasts.

Magnification: 160×

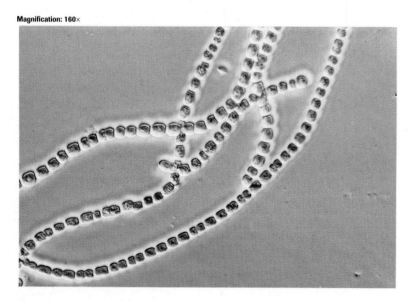

For a long time, the reason why only chlorophyll *a* was present in cyanobacteria was a mystery because plants contain two forms of chlorophyll, *a* and *b*. This mystery was solved with the discovery of *Prochloron*, which contains both chlorophylls *a* and *b*, *Figure 21.8*. The existence of this bacterium indicates that chloroplasts of plants may have evolved from a similar precursor.

Eubacteria—Chemosynthetic autotrophs

A third group of eubacteria are chemosynthetic autotrophs. These bacteria obtain their energy from the chemosynthetic breakdown of inorganic substances such as sulfur and nitrogen compounds. Some of these bacteria are important in converting nitrogen in the atmosphere to forms that can be used readily by plants.

Structure of Monerans

Monerans are the smallest and simplest of living things. In terms of complexity, they fall between the nonliving viruses and the living eukaryotic, cellular organisms.

Magnification: 15 900×

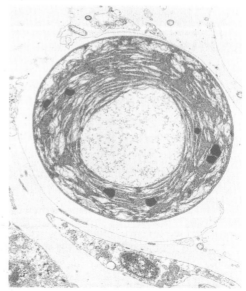

Figure 21.9

Wash, scrub, shampoo, and gargle as you might, these actions remove only a small fraction of the bacteria that live on and in you. These and other bacteria can be classified by their cell shapes.

Magnification: 11 000×

Magnification: 27 000×

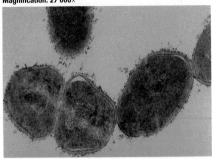

Magnification: 600×

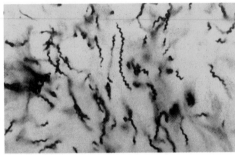

▲ A coccus is a round bacterium. The cocci shown here, *Streptococcus mutans,* are the bacteria that cause tooth decay by converting sugar to an acid that erodes tooth enamel.

▼ A spirillum is a spiral-shaped bacterium. The spirillum shown here, *Treponema pallidum,* causes syphilis, a sexually transmitted disease.

▲ A bacillus is a rod-shaped bacterium that exists as individual cells and short chains. This bacterium, *Clostridium botulinum,* produces a potent poison that results in food poisoning when eaten.

You may recall from Chapter 8 that prokaryotic cells have no membrane-bound organelles such as a nucleus, mitochondria, or chloroplasts. Their ribosomes are smaller than those of eukaryotes. Their inherited information is held in a single circular chromosome, rather than in paired chromosomes.

Bacteria are often classified by the shapes of their cells. The three most common shapes are spheres, rods, and spirals, shown in *Figure 21.9.*

Bacterial cells are also classified by their arrangements. Although some cells live singly, others are typically grouped together, as *Figure 21.10* shows.

Figure 21.10

Bacteria are often arranged in characteristic groups.

Magnification: 14 000×

Magnification: 52 000×

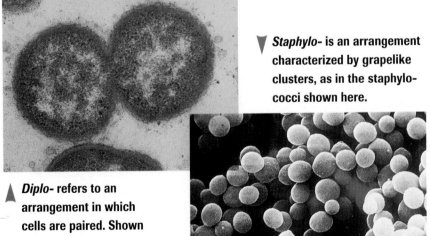

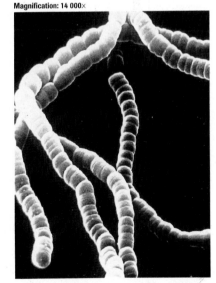

▼ *Staphylo-* is an arrangement characterized by grapelike clusters, as in the staphylococci shown here.

▲ *Diplo-* refers to an arrangement in which cells are paired. Shown here are diplococci.

Magnification: 34 000×

▲ *Strepto-* is an arrangement characterized by long chains, as in the streptococci shown here.

A Typical Bacterial Cell

Bacteria are microscopic, prokaryotic cells. The great majority of bacteria are unicellular. A typical bacterium would have some or all of the structures shown in this diagram of a cell.

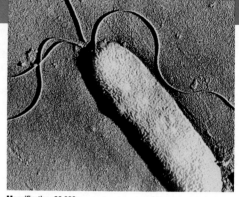

Magnification: 26 000×

This bacterium, *Pseudomonas syringae,* does not have a capsule or pili, but does have flagella.

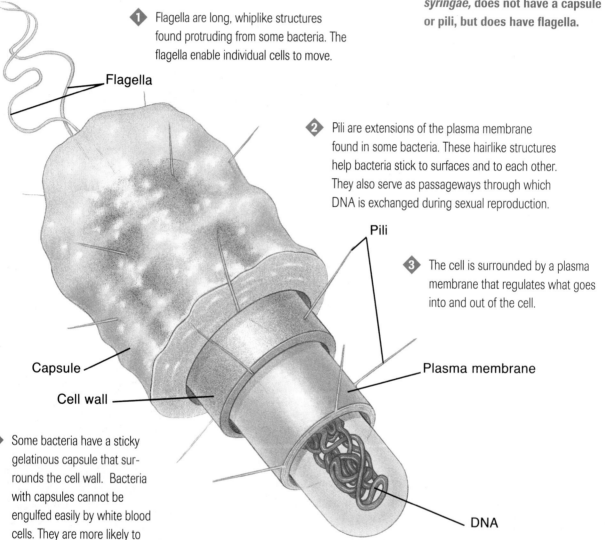

1 Flagella are long, whiplike structures found protruding from some bacteria. The flagella enable individual cells to move.

Flagella

2 Pili are extensions of the plasma membrane found in some bacteria. These hairlike structures help bacteria stick to surfaces and to each other. They also serve as passageways through which DNA is exchanged during sexual reproduction.

Pili

3 The cell is surrounded by a plasma membrane that regulates what goes into and out of the cell.

Capsule

Cell wall

Plasma membrane

6 Some bacteria have a sticky gelatinous capsule that surrounds the cell wall. Bacteria with capsules cannot be engulfed easily by white blood cells. They are more likely to cause disease than are bacteria without capsules.

DNA

5 A cell wall surrounds the plasma membrane. It is rigid, gives the cell its shape, and prevents osmotic rupture.

4 A single DNA molecule arranged as a circular chromosome contains the genetic information of the cell. The DNA is not enclosed in a nucleus. The area of the cell containing the DNA is called the nucleoid.

Protection from osmotic rupture

Bacteria have a cell wall that forms a rigid outer covering of the cell. Unlike the cellulose cell walls of plants, the cell walls of bacteria are made of long chains of sugars linked by short chains of amino acids. One of the functions of the cell wall in bacterial cells is to prevent osmotic rupture. Most bacteria live in a hypotonic environment, in which there is a higher concentration of water molecules outside than inside the cell. Water is always entering a bacterial cell, but because it is surrounded by a cell wall, the cell remains intact. However, should the cell wall be damaged, the water entering the cell would cause it to rupture and die.

Penicillin—Bacterial killer?

Osmosis is also an important factor in the treatment of bacterial diseases. Penicillin kills bacteria by interfering with the enzyme that links the sugar chains in the cell wall. Bacteria growing in penicillin develop holes in their cell walls. As a result, when water enters the cell by osmosis, the bacterium ruptures and dies.

Penicillin is nontoxic to viruses and animals because neither has a cell wall. It is also nontoxic to plants because the structure of the plant cell wall is different from the structure of the bacterial cell wall.

Miracle Cure: The Story of Penicillin

"I need not go into details of the research on which I was engaged at the time but my culture plate was covered with staphylococcal colonies. The particular work involved opening the plate and looking at it under a dissecting microscope and then leaving it for growth to take place. That, of course, was asking for trouble by contamination, as things were always dropping from the air and sure enough, trouble came but it led to penicillin. A mould spore, coming from I don't know where, dropped on the plate. That didn't excite me. I had seen such contamination before, but what I had never seen before was staphylococci undergoing lysis around the contaminating colony. Obviously something extraordinary was happening." —Sir Alexander Fleming, "Antiseptics Old and New," 1946

Penicillium notatum The story of Fleming's accidental discovery of penicillin in 1928 and the events that followed are legendary. The bacteria-killing substance produced by his notorious airborne mold, *Penicillium notatum,* became the world's first antibiotic and prompted a golden age of antibiotics that save millions of lives from infectious disease. However, the development of this antibiotic was 12 years in coming. It was not until 1940, at the beginning of World War II, that a team of English scientists was able to purify the substance produced by Fleming's mold.

World War II production Production of the drug for human use became crucial as the war raged on in Europe. Penicillin moved from small-scale laboratory production in England to the chemical factories of America, where it helped win the war for the Allies.

Penicillin's impact Although penicillin's popularity faded somewhat when resistant bacteria appeared and allergy-induced deaths occurred, its discovery stimulated further research into lifesaving antibiotics. The post-war drug industry produced the second-most-important antibiotic—streptomycin—as well as many others.

CONNECTION TO Biology

Penicillin revolutionized medicine, providing a formidable weapon against killer diseases such as pneumonia and bacterial meningitis. Why is penicillin ineffective against cold, polio, or AIDS viruses?

Doctors need to determine which antibiotic will kill disease-causing bacteria. A test similar to the one in this Biolab can be

How sensitive are bacteria to antibiotics?

used. You will use sterile petri dishes containing nutrient agar and sterile disks that have been impregnated with antibiotics. When a disk is placed on the agar, the antibiotic diffuses into the agar. The clear ring that appears around the disk is called a zone of inhibition and represents an area where sensitive bacteria have been killed.

PREPARATION

Problem

Which antibiotic is most effective in killing a particular kind of bacteria?

Hypotheses

Have your group agree on a hypothesis to be tested. Record your hypothesis.

Objectives

In this Biolab, you will:

- **Compare** the effectiveness of different antibiotics in killing a particular strain of bacteria.
- **Determine** the most effective antibiotic for treating an infection caused by this strain of bacteria.

Possible Materials

cultures of bacteria
sterile nutrient agar petri dishes
antibiotic disks
sterile disks of blank filter paper
marking pen
cotton swabs
forceps
37°C incubator
metric ruler

Safety Precautions 🚫 ☣ 🖐

Although the strains of bacterial cultures you will be working with are not pathogenic, be careful not to spill them. Wash your hands with soap immediately after handling any live bacterial culture. Be sure to clean your work area and dispose of your cultures and petri dishes as directed by your teacher.

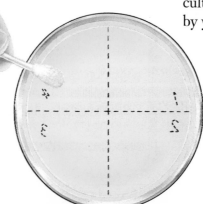

1. Examine the materials provided by your teacher, and study the illustrations on these pages. As a group, make a list of the possible ways you might test your hypothesis.

2. Agree on one way that your group could investigate your hypothesis. Design an experiment that will allow for the collection of quantitative data.

3. Prepare a list of numbered directions. Include a list of materials and the amounts you will need. Try to limit yourself to one petri dish per person.

4. Design and construct a table for recording your data. What do you think will happen around the antibiotic disks as the antibiotic diffuses into the agar? How will you measure this?

Check the Plan

Discuss the following points with other group members to decide the final procedure for your experiment.

1. Determine how you will set up your plates. How many antibiotics can you test on one plate? How will you measure the effectiveness of each antibiotic? What will be your control? Be sure you mark your petri dishes on the *bottom*. Why?

2. Do you think it would be better to add the bacteria or the antibiotic disks to the petri dish first?

3. What precautions will you take to prevent contamination of your petri dish by bacteria from around the room?

4. How often will you observe your plates?

5. ***Make sure your teacher has approved your experimental plan before you proceed further.***

6. Carry out your experiment. Make any needed observations, and complete your data table. Design and complete a graph or visual representation of your results. Do you think a bar graph or a line graph would be more appropriate?

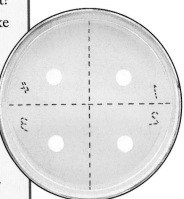

1. **Measuring in SI** How did you measure your zones of inhibition? Why did you do it this way?

2. **Comparing and Contrasting** Which antibiotic caused the largest zone of inhibition? What is the significance of this?

3. **Drawing Conclusions** If you were a physician treating a patient infected with this bacterium, which antibiotic would you use? Why?

4. **Analyzing the Procedure** Can you think of any limitations of this technique? If a real person were involved, what other tests might give you more confidence in your results?

Going Further

Application Do library research to find out how different antibiotics kill bacteria. Penicillin, for example, interferes with the production of bacterial cell walls.

Living on the Edge

In April 1993, some residents of Milwaukee, Wisconsin, became ill after drinking contaminated tap water. Health officials advised residents to boil their water before drinking it. Why? At 100°C, the temperature at which water boils at sea level, the molecules that make up life as we know it are destroyed. Cell walls collapse, DNA unravels, and proteins unfold. Thus, any organisms in the water, including harmful bacteria, are quickly killed. Or are they?

Mighty monerans Over the past three decades, scientists have discovered a variety of simple, unicellular organisms that actually thrive at extreme temperatures. Researchers have found colonies of cyanobacteria growing in a frozen pond just 360 km from the geographic South Pole. And on the ocean floor along the Mid-Atlantic Ridge, large colonies of monerans thrive around volcanic vents where the water temperature soars to almost 400°C.

Did they adapt? Traditional science maintains that heat-loving organisms are conventional life forms that have somehow adapted to life at extreme temperatures. Recently, however, some scientists have challenged this idea. They've begun to speculate that these heat-loving organisms haven't had to adapt to anything because they are, in fact, biological relatives of some of the earliest forms of life on this planet.

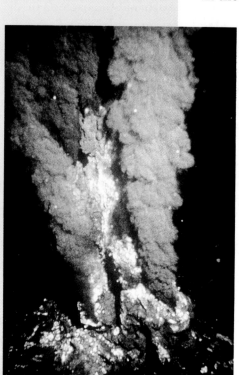

Life in the primordial soup
Scientists think that life began on Earth at a time when the planet was still cooling down from its fiery birth. Extensive volcanic activity was the rule, and any organism that was going to survive would have had to be highly resistant to heat. What's more, they would most likely have been autotrophic.

These are precisely the types of organisms scientists are finding in superhot environments on Earth today. As more is discovered about these unique monerans, additional light will be shed on some of the mysteries surrounding the origin of life on Earth.

Thinking Critically
Why were the earliest forms of life on Earth probably obligate anaerobes as well as being heat tolerant?

Ecology and Adaptations

Some of the oldest known fossils are of monerans. These organisms lived on an Earth with a climate and atmosphere very different from those of today. Some of these fossils contain compounds similar to chlorophyll and so were among the first photosynthetic organisms. As Earth's atmosphere gradually became more oxygen rich, monerans evolved adaptations that allowed them to survive in the changing environment.

Diversity of metabolism

You have already read that eubacteria display a broad variety of metabolic pathways. Various eubacteria obtain nutrition and energy from light, from simple inorganic molecules, and from complex organic molecules.

Most bacteria require oxygen for cellular respiration. These bacteria are called **obligate aerobes.** *E. coli* is an obligate aerobe. Other bacteria cannot use oxygen and are killed by it. These bacteria are called **obligate anaerobes.** The bacterium that causes syphilis is an obligate anaerobe. This explains why syphilis can be transmitted only by intimate contact from one moist membrane to another. You cannot catch syphilis from a contaminated toilet seat because the bacteria would be poisoned by oxygen and die. Still other kinds of bacteria can live either with or without oxygen. They have two types of metabolism and can derive energy aerobically by cellular respiration or anaerobically by fermentation.

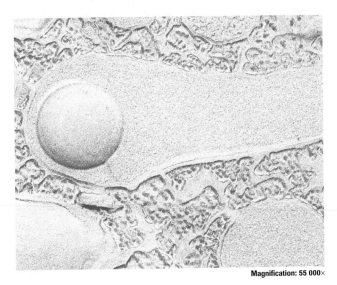

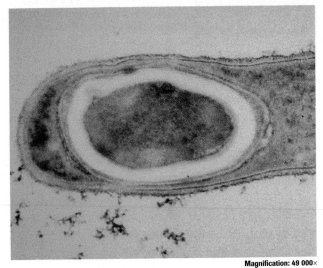

Figure 21.11

The SEM on the left shows an endospore being produced within a bacterial cell. The endospore contains a complete copy of the DNA of the bacterium. After the endospore is formed, the rest of the cell may disintegrate. The TEM on the right shows a cell containing an endospore. Notice the thickness of the wall of the endospore. Of what value to the organism is a thick wall?

Several monerans have evolved unusual biochemical pathways. Green sulfur bacteria, for example, use hydrogen sulfide rather than water for photosynthesis. Instead of producing oxygen, they produce sulfur as a by-product. They can grow only in anaerobic environments such as lake sediments.

Adaptations for survival

Some bacteria, when faced with unfavorable environmental conditions, produce structures called **endospores,** *Figure 21.11.* Endospores have a hard outer covering and are resistant to drying out, boiling, and many chemicals. While in the endospore form, the bacterium is in a state of slow metabolism, and it does not reproduce. When it encounters favorable conditions, the endospore germinates and gives rise to a bacterial cell that resumes growing and reproducing. Some endospores have been found to germinate after thousands of years.

Endospores are of great survival value to bacteria. Because endospores can survive boiling, canned foods and medical instruments must be sterilized under high pressure. This can be done either in a pressure cooker or in an autoclave. When water is boiled under high pressure, it is hotter than the normal boiling point of water. This greater degree of heat can kill endospores.

The clostridia are a group of obligate anaerobes that can form endospores. One member of this group, *Clostridium botulinum,* produces an extremely powerful **toxin,** or poison. Instead of dying when exposed to oxygen, *Clostridium botulinum* forms endospores. These endospores can find their way into canned food. If the canned food has not been properly sterilized, the endospores will germinate and the bacteria will grow and produce their deadly toxin. Although the resulting disease, botulism, is extremely rare, it is often fatal.

How fast can bacteria reproduce?

When they are in a favorable environment, bacteria reproduce extremely rapidly. Under ideal conditions, some bacteria can reproduce every 20 minutes. As conditions become less ideal, the rate of reproduction slows down.

Analysis

Suppose you start with a single bacterium at Hour 0, and it reproduces once an hour. How many bacteria will you have at the end of Hour 24?

To solve this problem, set up a table listing Hours 0 to 24. At Hour 0, you have one bacterium. After one hour, the bacterium will have reproduced once, so you will have two bacteria. At Hour 2, each of these bacteria will have reproduced once, giving a total of four. To find the number of bacteria after 24 hours, continue this doubling 24 times, once for each time the bacteria reproduce. If you can, calculate how many bacteria you will have after two days.

Thinking Critically

How many bacteria were there after one day? Use your knowledge of bacterial growth to explain why boils, which are caused by bacteria, swell quickly to their maximum size but decrease slowly in size.

Another member of the Clostridium group, *Clostridium tetani*, produces a powerful nerve toxin that causes the often-fatal disease, tetanus, shown in *Figure 21.12*. Because endospores of *C. tetani* are found on nearly every surface, they can easily enter a wound. A puncture wound doesn't allow air to enter, so conditions are ideal for growth of anaerobes. The endospores germinate in the wound, and the bacteria reproduce in great numbers. The bacteria produce a toxin, which enters the bloodstream and attacks the nerve cells in the spinal cord. Fortunately, there is an immunization for tetanus. You received this shot as a child. A booster shot is given as a precaution after a puncture wound. Deep wounds are hard to clean and provide ideal conditions for growth of anaerobes.

Reproduction

Bacteria cannot reproduce by mitosis or meiosis because they have no nucleus. Instead, they have evolved different methods of reproduction.

Binary fission

Bacteria usually reproduce asexually by a process known as **binary fission**. The bacterium first copies its single chromosome. The copies attach to the cell's plasma membrane. As the cell grows in size, the two copies of the chromosome separate. The cell then divides in two as a partition forms between the two new cells, as shown in *Figure 21.13*. Each new cell receives one copy of the chromosome. Therefore, the daughter cells are genetically identical to each other and to the parent cell.

Figure 21.12

This painting by Charles Bell shows a British soldier dying of the final stages of tetanus during the Napoleonic wars in the early years of the 19th century. Note the characteristic muscle spasms.

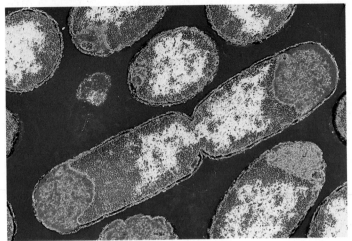

Magnification: 16 500×

Figure 21.13

This *Escherichia coli* cell is starting to divide. The newly forming partition is visible in the center of the cell. Why is cell growth needed for the process of binary fission?

Bacterial reproduction can be extremely rapid. Under ideal conditions, bacteria can reproduce every 20 minutes. Such a rate of reproduction yields enormous numbers of bacteria in a short time. If bacteria reproduced at their maximum rate, they would cover the whole planet within a few weeks. Yet, obviously, this does not happen. The reason is that, like all living things, bacteria don't always find ideal conditions. They run out of nutrients, they are eaten by predators, they dry up when no water is available, and they are poisoned by the wastes they produce.

When you have an infection, billions of bacteria grow in your body. If you are given an antibiotic for the infection, you should take the antibiotic for the full prescribed period—even though you feel better after just one or two days. Shortly after you begin to take the antibiotic, most of the bacteria are killed. However, if you stop taking the antibiotic and even a single bacterium is left, it will start reproducing. A day later, you will have millions of bacteria in your body and you will be sick again. Completing the antibiotic as prescribed ensures that all of the bacteria will be killed so you will not get sick again.

Sexual reproduction

In addition to reproduction by binary fission, some bacteria have a simple form of sexual reproduction called **conjugation**. During conjugation, one bacterium transfers all or part of its chromosome to another cell through a bridgelike structure called a pilus (plural pili) that connects the two cells. *Figure 21.14* shows this genetic transfer.

Figure 21.14

The *E. coli* bacterium on the bottom left is transferring its chromosome to the *E. coli* on the top through a long pilus. What is the advantage to the bacterium of this exchange of genetic material?

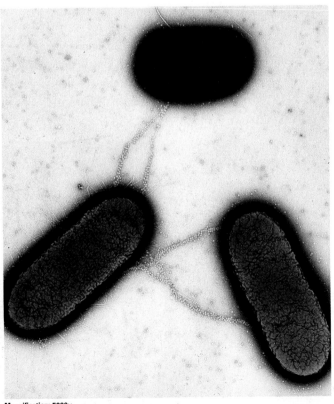

Magnification: 5000×

Economic Importance

Bacteria are important to us in a number of ways. Some of our favorite picnic foods—mellow Swiss cheese, crispy pickles, tangy yogurt—would not be possible without bacteria. And once we have digested and extracted all the nutrients from those foods and eliminated them from our bodies as waste, other bacteria help break down that waste in septic tanks or at sewage-treatment plants. Still other bacteria increase crop yields when used as a biological means to control insect pests. Bacteria can increase crop yields in another way, as well.

Nitrogen fixation

Most of the nitrogen on Earth exists in the form of the nitrogen gas (N_2) that makes up 80 percent of the atmosphere. All organisms need nitrogen to produce the proteins, DNA, RNA, and ATP in their cells. Yet few organisms can use N_2 gas directly. Of all mineral elements, nitrogen is the one that most often limits the growth of plants.

Several species of bacteria have enzymes that convert N_2 gas into ammonia (NH_3) in a process known as **nitrogen fixation.** Other bacteria then convert the ammonia into nitrite (NO_2^-) and nitrate (NO_3^-), which can be used by plants. Bacteria are the only organisms that can fix nitrogen. Some nitrogen-fixing bacteria form symbiotic associations with legumes such as peas, peanuts, and soybeans, *Figure 21.15.* In nitrogen-deficient soils, legumes grow better than plants such as corn that do not have root nodules. Thus, nodules of these bacteria are a distinct advantage to a plant. The relationship is also an advantage to agriculture. When legumes are grown and then harvested, the remaining roots with nodules add considerable usable nitrogen to the soil. The following season, other crops can be grown in the newly nitrogen-rich soil. This is the basis of crop rotation.

Figure 21.15

The nodules on these soybean roots (left) contain *Rhizobium* bacteria (right) that convert nitrogen gas into ammonia. In this symbiotic association, the plant gains a source of usable nitrogen, and the bacteria can use sugars supplied by the plant for their own needs. Cyanobacteria also can fix nitrogen.

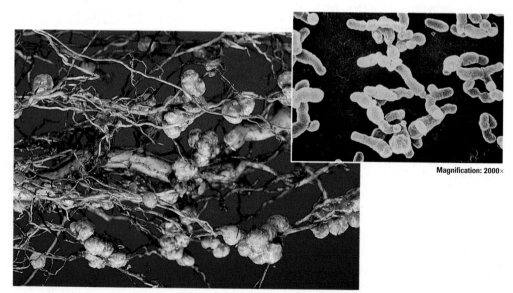

Magnification: 2000×

Figure 21.16

The holes in Swiss cheese are caused by bubbles of carbon dioxide produced by the bacteria that give the cheese its flavor. Cheese is made in huge vats such as the one above. The large industrial fermenter (right) holds 1500 L and is used for growing bacteria in industry.

Recycling of nutrients

You learned in Chapter 3 that life could not exist if bacteria did not break down the organic matter in dead organisms and wastes, returning the nutrients to the environment to be used by producers at the bottom of food chains. Cyanobacteria, along with plants and algae, replenish the supply of oxygen in the atmosphere. Autotrophic bacteria convert carbon dioxide in the air to the organic compounds that are passed to consumers in food chains and webs. All life depends on bacteria.

Food and medicines

Because bacteria are so metabolically diverse, different species produce a wide variety of molecules as the result of fermentation, *Figure 21.16.* Many of these molecules have distinctive flavors and aromas. As a result, the bacteria that produce them are used to make vinegar, yogurt, butter, cheese, pickles, and sauerkraut.

Strains of bacteria have evolved that produce important antibiotics, which are used as medicines to kill other bacteria. Streptomycin, erythromycin, chloromycetin, and kanmycin are some of the antibiotics that are derived from bacteria.

Bacteria cause disease

Although only a few kinds of bacteria actually cause disease, those that do have a great impact on our lives. It is estimated that about half af all human diseases are caused by bacteria.

Meet Dr. Barbara Staggers, Physician

"If you fall between the ages of 11 and 21, you're at risk, no matter who you are or where you live." Those words are frequently spoken by Dr. Barbara Staggers, Director of Adolescent Medicine at Children's Hospital in Oakland, California.

In the following interview, Dr. Staggers talks about the tremendous stresses on young people today and tells how she and her staff try to help teens at her clinic.

On the Job

Q Dr. Staggers, what do you do here at the teen clinic?

A We see students for everything from physicals for sports eligibility to patching them up after a fight. Kids feel comfortable here. We're known as an okay place to be. We also do a lot of outreach in the community. We employ students as peer health-care counselors to do AIDS education in local high schools and also for young people in jail. Kids would rather talk to their peers, no matter how cool I think I am. But kids seem to think I'm okay, too. Their demand for honesty is one thing I really love about adolescents.

Q Are many students interested in becoming peer counselors?

A Yes, and I think that's because it's a way for them to give back something to others. I think we underestimate how much teens need an opportunity to give. Often, because they've been there, peer counselors can be very effective in helping other kids with serious problems.

Q What are some of the major problems that you and the peer counselors see?

A The three leading causes of death in teenagers are motor vehicle injuries (of which 50 percent are alcohol and drug related), homicide, and suicide. The number-one disease in teens is mental-health illness, such as depression. And, of course, there are also the sexually transmitted diseases.

Q How do you work with kids who have such serious problems?

A I refuse to focus solely on their behavior. I try to find out where a teen is and what's happening in his or her life. I hear kids say things like, "Every boyfriend I've dated in the last five years is dead, so if I get pregnant, at least I'll have someone to be with me." Or, "I probably won't live long enough to worry about getting hooked on drugs." Then I need to deal with kids around those issues and have sort of a general life discussion. I think that having a conversation that will prevent someone from shooting himself or herself or someone else is just as valuable as taking a bullet out of a person who's been shot.

Early Influences

Q How did you decide to become a doctor?

A When I was 18 years old, I was torn between being a veterinarian and joining the Dance Theater of Harlem. Then I worked in a summer sports program at a college. I discovered that a lot of young people in the program had never been to a physician. Their access to health care was a lot different from mine because my father was a doctor. That difference made me rethink my career plans. I decided that I wanted to be in the field of medicine and work with kids of color, particularly disadvantaged, urban youth.

Q Has that decision always felt like the right one for you?

A It's clear to me, because it's so comfortable, that this is what I'm supposed to do. Not a day passes that I don't learn from one of the teens I see. They keep me honest. I can't put a price on having a teen flag me down in a mall and thank me for having given him or her some time four years ago.

Personal Insights

Q What advice do you have for teenagers who might want to pursue a medical career?

A The best advice I can give is never to give up your dream. Start by looking away from material wealth and toward human health. To be in medicine, you have to like people. You have to be able to talk to people—even sick, grouchy people—and be willing to invest in them. That part of medicine is most important. Necessary classes such as chemistry may have been hard for me, but I've always loved to work with people.

It's important to learn about yourself. You can't start too young to find out what makes you feel good about yourself and pursue those things.

CAREER CONNECTION

People in these careers work to help keep young people healthy.

Physician *A bachelor's degree, four years of medical school, three years of residency (graduate medical education)*

Social Worker *A bachelor's or master's degree in counseling, rehabilitation, or a related field*

Nurse's Aide or Orderly *High-school diploma; on-the-job training*

Q What makes you, personally, feel good about yourself?

A I've been involved in 50 million zillion things all my life—karate, ballet, jazz, poetry, photography. And I hang out with my family because that's where I get support. I know it's essential for me to have all of those things in place to make me who I am.

In the past, bacterial illnesses took a major toll on human populations. As recently as 1900, life expectancy in the United States was only 47 years. The leading killers of the time were tuberculosis and pneumonia, both bacterial illnesses. In the intervening century, life expectancy has increased to about 75 years. This remarkable 50 percent increase means that, on average, people live nearly 25 years longer today than they did in 1900! This increase in life expectancy is due to many factors. People now have better living conditions. We have less poverty, better public health systems, improved water and sewage treatment, better nutrition, and better medical care. These improvements, combined with the presence of antibiotics, have reduced death rates from bacterial diseases to very low levels. *Table 21.1* lists some bacterial diseases.

Table 21.1		
Diseases Caused by Bacteria		
Tuberculosis	Scarlet fever	Rocky Mountain spotted fever
Bacterial pneumonia	Syphilis	Tetanus
Botulism	Gonorrhea	Ear infections
Strep throat	Chlamydia	Boils
Staph infections	Diphtheria	Lyme disease

Connecting Ideas

Monerans include all organisms with prokaryotic cells. They represent the most ancient forms of life on Earth. They were the only life forms until about 2.1 billion years ago when the first eukaryotic cells evolved. The other four kingdoms—Protista, Fungi, Plantae, and Animalia—all contain organisms with eukaryotic cells. When you compare the metabolic pathways of monerans with those of the other four kingdoms, you may find it surprising that there is so much more metabolic diversity among monerans than among members of the other kingdoms.

Section Review

Understanding Concepts
1. Describe six parts of a typical bacterial cell, and give their functions.
2. How do endospores and pili help bacteria survive?
3. How is a bacterial cell affected by penicillin?

Thinking Critically
4. Why is it reasonable to hypothesize that archaebacteria were the first organisms to evolve on Earth?

Skill Review
5. **Making and Using Tables** Make a table comparing archaebacteria and eubacteria. Include at least three ways they are alike and three ways they are different. For more help, refer to Organizing Information in the *Skill Handbook*.

Helping Nature Help Us

Materials in the natural world are recycled over and over again. It is a fundamental law of nature that matter cannot be created or destroyed by ordinary means.

Bacteria, fungi, and other decomposers live by extracting nutrients from dead bodies and wastes of other organisms. The simpler molecules they release become, in turn, the raw materials for new generations. This system has worked in delicate balance for millions of years. However, modern trends have overwhelmed these natural recycling systems.

Too many people, too many chemicals As humans learned how to synthesize chemicals, they created many materials that natural systems cannot break down. Compounding the problem is the fact that many of these chemicals are toxic to life. In addition, the human population has reached such numbers in some areas that its effluent has overwhelmed the ability of natural processes to decompose it all. Further compounding the problem is the fact that many raw materials are far from population centers and must be transported. This provides many opportunities for accidents such as oil spills that can overwhelm native ecosystems.

Mechanical cleanup fails For many years, humans have tried to clean up the environment by mechanical means. They used chemicals, filters, and other devices in an attempt to return ecosystems to their predamaged condition. This was expensive, time-consuming, and energy-intensive, and in the end, they still had to dispose of the pollutants they recovered. Then it became obvious that nature pointed to a better way. Why not find organisms that would decompose the pollution harmlessly? Thus was born the idea of bioremediation—the destruction of pollution by natural biological processes.

Applications for the Future

Sometimes, just adding a single kind of bacteria won't solve the problem. In these cases, a system may have to be devised. For example, TCE (trichloroethylene) is probably the most widespread pollutant in the United States. It has been used for years as a solvent and degreaser, and then dumped. Certain anaerobic bacteria thrive on TCE, metabolizing it to DCE and vinyl chloride. The problem is that vinyl chloride is even more toxic than TCE. However, if the bacteria are fed an "inducer" such as methane, they produce enzymes that degrade the vinyl chloride into harmless by-products. Using this method, one test reduced the TCE level from 3000 to 100 parts per billion in less than a week. Thus, by injecting an inducer into a contaminated area, a bioremediator can clean the area with minimal side effects and cost.

INVESTIGATING the Technology

1. **Research** Read "Toxic Avengers" by Dawn Stover, *Popular Science*, July 1992, pp. 70-74, 93. Identify one or more of the new technologies in the article that could help solve a problem in your community.
2. **Analyze** How do you think bioremediation will change in the future? What technology will probably be used?

21.1 Viruses

- Viruses are made of nucleic acids and proteins. In order for a virus to reproduce in a cell, it must recognize and attach to a specific receptor molecule on the plasma membrane of the cell.
- During a lytic cycle, a virus attacks and kills the host cell. In a lysogenic cycle, a virus inserts its DNA into a chromosome of the host cell.
- Retroviruses contain RNA, which must be converted to DNA before it can be incorporated into a host cell chromosome.
- Viruses are probably derived from their host cells and do not represent ancient forms of life.

21.2 Monerans

- Monerans can be classified either as archaebacteria or as eubacteria. Archaebacteria are more ancient and inhabit extreme environments.
- All monerans are prokaryotes. They exhibit a tremendous diversity of metabolism including aerobes, anaerobes, autotrophs, and heterotrophs.
- Some bacteria can form endospores that survive under unfavorable conditions.
- Monerans reproduce asexually by binary fission. Some monerans also have primitive forms of sexual reproduction.

Key Terms

Write a sentence that shows your understanding of each of the following terms.

bacteriophage	obligate aerobe
binary fission	obligate anaerobe
conjugation	provirus
endospore	retrovirus
host cell	reverse transcriptase
lysogenic cycle	saprobe
lytic cycle	toxin
nitrogen fixation	virus

Understanding Concepts

1. What is the function of the tail fibers of a bacteriophage?
2. Describe what happens to viral nucleic acid during a lysogenic cycle.
3. Retroviruses infect cells whose chromosomes are made of DNA. How can a retrovirus, which has RNA as its nucleic acid, insert a copy of its genetic material into the chromosome of its host cell?
4. What are the three common shapes of bacteria? What is meant by a staphylococcus?
5. Name three unusual habitats in which archaebacteria are found.
6. What is the function of the cell wall of bacteria?
7. Explain how binary fission results in two new bacteria, each identical to the original bacterium.
8. If you have an ear infection, why is it important for you to continue to take your antibiotic for the full prescribed period?

Relating Concepts

9. Make a concept map that relates the following terms and phrases. Supply the appropriate linking words for your map.

   obligate aerobe, obligate anaerobe, endospore, decomposition, nitrogen fixation, cyanobacteria, bacteria, pili, capsule, photosynthesis, *Clostridium*, toxin

Interpreting Scientific Illustrations

10. Identify each of the bacteria shown here as a coccus, bacillus, or spirillum. Where appropriate, indicate the arrangement of the bacteria, such as *diplo-*, *strepto-*, or *staphylo-*.

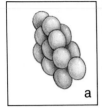 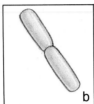 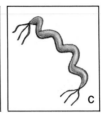

Applying Concepts

11. You have read that monerans are essential to life. Why do you think this statement is true?

12. Discuss two ways that prokaryotic cells are different from eukaryotic cells.

13. Why is *Prochloron* thought to be a precursor of the chloroplasts found in eukaryotic cells? Why are cyanobacteria probably not precursors of chloroplasts?

14. Discuss three factors that limit bacterial growth. Why do they prevent bacteria from taking over the world?

Biology & Society

15. Technology Why might it be better to clean up a polluted site by using bacteria that live there naturally rather than by bringing new bacteria to the site?

Thinking Critically

Making Inferences

16. Thinking Lab If you were offered either a million dollars or a sum equal to a penny that doubles every day for 64 days, which choice would give you more money? Relate this to the growth rate of bacteria.

Interpreting Data

17. Biolab A bacterium isolated from a person with an infection was tested for its sensitivity to three different antibiotics. The results of the test are shown in the petri dish. If you were a physician treating this patient, which antibiotic would you use and why?

Relating Cause and Effect

18. Cases of botulism poisoning are occasionally traced to home-canned food that was not sterilized properly. Explain how improper sterilization might cause botulism.

Analyzing

19. Some bacteria produce antibiotics that kill other bacteria. Explain why antibiotic-producing bacteria would be selected for and would survive in nature.

Connecting to Themes

20. Unity Within Diversity Although monerans display a tremendous variety of metabolic adaptations, they also share many characteristics in common. Name and discuss at least three of these characteristics.

21. The Nature of Science It has been estimated that 99 percent of all monerans have such unusual nutritional requirements that scientists are unable to grow them in the laboratory. How do you think this inability hinders our understanding of monerans?

22. Evolution The oldest moneran fossils are 3.5 billion years old, whereas the oldest fossils of eukaryotic cells are 2.1 billion years old. What kind of relationship does this suggest between monerans and eukaryotes?

22 Protists

Imagine sitting on an old log by the edge of a small pond. Except for the buzzing of a few insects, few sounds can be heard and nothing seems to be moving. Not much going on, right? Not so!

Although you don't see them, you are surrounded by a host of tiny organisms. In the soil and rotting leaves at your feet live countless numbers of protists. In the pond, there are even more—hundreds of different species. Protists are part of nearly every type of moist habitat on Earth, from high mountain peaks and rain forests to rivers, lakes, and oceans.

Of what importance are they? They form the foundation of many ecosystems. Countless other species depend either directly or indirectly on them or on what they do. In this chapter, you'll discover what vital roles protists play, how they interact with other organisms, and how they affect you.

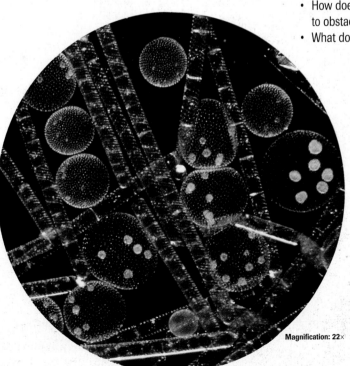

Protists are all around us, but we seldom notice them. Despite their small, often microscopic size, protists are surprisingly complex and are vital components of nearly every type of ecosystem. The algae on the rocks (left) are protists as are *Volvox* and *Spirogyra* (right).

Magnification: 22×

22.1 The World of Protists

In just a few drops of pond water, you can find an amazing collection of protists. Some will be moving actively, searching for food. Others, such as this diatom, will be photosynthesizing, using the sun's energy to transform carbon dioxide and water into food. Still others will be playing an active role in decomposing. The protists are a diverse group.

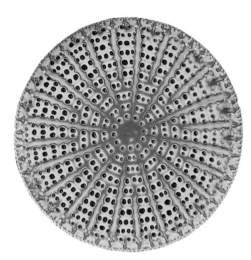

Magnification: 300×

Protist Diversity

The Kingdom Protista is the most diverse of the five kingdoms. In fact, there is no such thing as a typical protist. The more than 200 000 species in the Kingdom Protista come in a broad array of different shapes, sizes, and colors. *Figure 22.1* shows several examples of this kingdom.

Figure 22.1

Members of the Kingdom Protista can be organized into three general groups—animal-like, plantlike, and funguslike protists.

▼ **Animal-like protists are all unicellular heterotrophs. Those that move do so in a variety of ways.**

► **Plantlike protists are photosynthetic autotrophs.**

► **During part of their life cycle, funguslike protists look remarkably like some types of fungi.**

Magnification: 10×

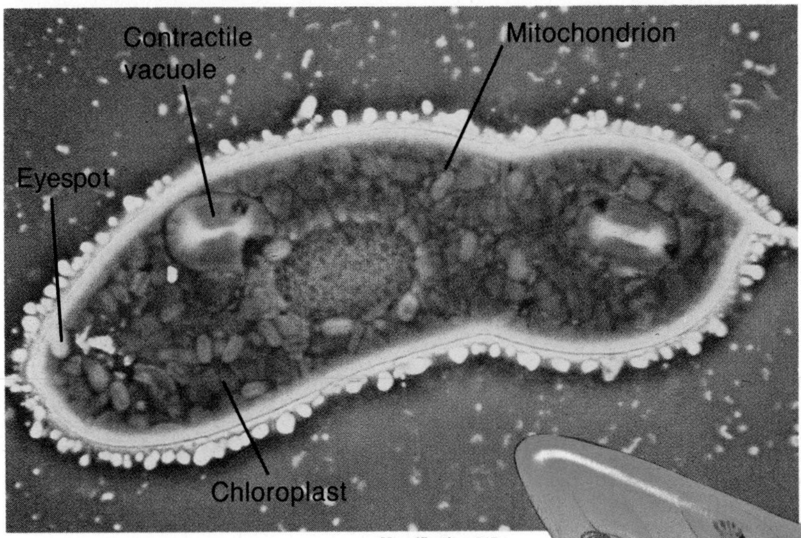

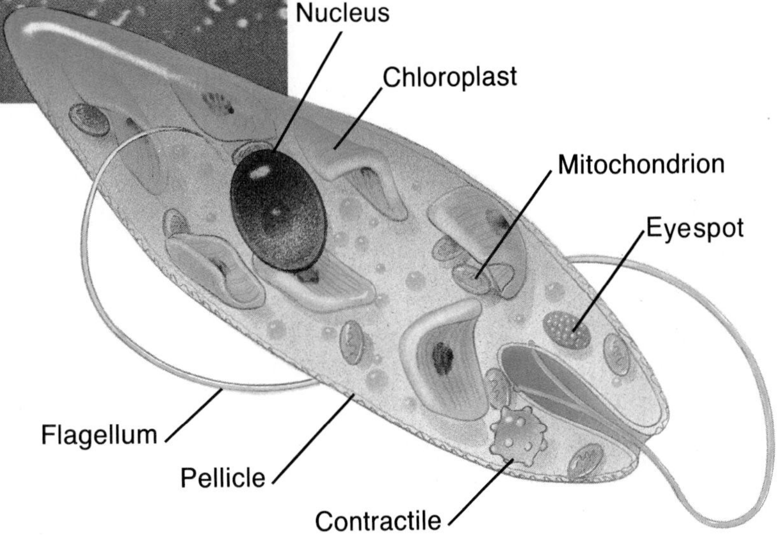

Figure 22.2

Protists are eukaryotic organisms with a wide variety of complex organelles.

Magnification: 315×

Protists range from small, unicellular amoebas that prowl for food among old, wet leaves to giant, multicellular ocean seaweeds that may grow to more than 100 m long. Some protists are slow-moving, jellylike blobs that constantly change shape. Others are rigid, with hard but delicate shells. Still others undergo transformations from living as single cells to becoming part of a coordinated colony of many cells.

At first glance, most protists might look like simple organisms, but appearances can be deceiving. Protists are much more complex than monerans. All protists are eukaryotes, which means that most metabolic processes in these organisms take place in membrane-bound organelles, as shown in *Figure 22.2.*

Protozoans: Animal-like Protists

As you sit by the pond, you notice clumps of dead leaves at the water's edge. Under a microscope, a bit of that wet, decaying leaf litter becomes its own world, very likely inhabited by animal-like protists.

Animal-like protists are known as **protozoans.** All protozoans are unicellular heterotrophs that meet their energy requirements by feeding on other organisms or dead organic matter. Protozoans are grouped into phyla according to the way they move. Some motile protozoans propel themselves with cilia or flagella. Others get around by sending out cytoplasm-containing extensions of their plasma membranes called **pseudopodia.** These protozoans use their pseudopodia to engulf food.

Nonmotile protozoans have no way of pursuing or actively capturing food. They live as parasites. Parasitic protozoans are usually found in a part of a host that has a constant and readily available food supply, such as an animal's bloodstream or intestine.

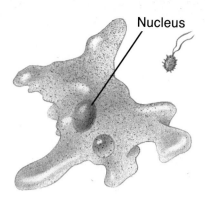

Nucleus

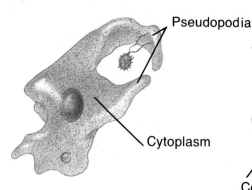

Pseudopodia

Cytoplasm

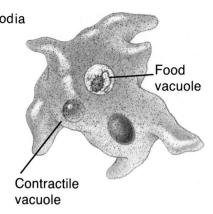

Food vacuole

Contractile vacuole

A An amoeba senses food in its immediate environment. As it approaches the food, pseudopodia begin to extend toward and around it.

B The amoeba engulfs the food, which becomes enclosed in a food vacuole.

C Digestive enzymes secreted into the vacuole break down the food, and the nutrients are released into the rest of the cell.

Figure 22.3

When a sarcodine happens upon a bit of food, pseudopodia stream out to engulf the morsel and take it into the cell to be digested in a food vacuole.

Amoebas: The bloblike protists

The phylum Sarcodina includes hundreds of species of amoebas and amoeba-like organisms. Amoebas are protists that have no wall outside their cell membrane and move by forming pseudopodia. As new pseudopodia form, the shape of the cell constantly changes, so an amoeba looks different from one moment to the next. As they move, amoebas engulf bits of food by flowing around and over them, *Figure 22.3.*

Most members of this phylum are marine, but there are freshwater amoebas that live in the ooze of ponds and slow-moving streams, in wet patches of moss, and even in moist soil. Because amoebas live in water or moist places, nutrients dissolved in the water around them can diffuse directly through their cell membranes. However, freshwater amoebas face the problem of living in a hypotonic environment; they constantly take in water. To solve this problem, they have contractile vacuoles that collect and pump out excess water.

Amoebas with shells

Two forms of sarcodines have shells. Found throughout the oceans, foraminiferans, like the one pictured in *Figure 22.4,* are marine amoebas that have a hard, outer shell made of calcium carbonate. Foraminiferans are extremely abundant. Much of the bottom ooze that covers the seafloor is made up of their shells. Fossil

Figure 22.4

Both foraminiferans (left) and radiolarians (right and far right) are unicellular amoebas that live inside shells. They extend pseudopodia through the tiny holes you can see in the shells. Their pseudopodia act as a sticky net in which bits of food are trapped.

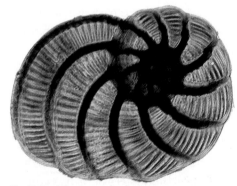

Magnification: 80×

Magnification: 800×

forms of these protists provide geologists with clues to the ages of rocks and sediments.

Most radiolarians have shells made of silica, as shown in *Figure 22.4.* Seen under a microscope, the great complexity of these tiny shells becomes apparent.

Reproductive strategies of sarcodines

Most sarcodines reproduce by **asexual reproduction,** in which a single parent produces one or more identical offspring. When environmental conditions become unfavorable, some types of sarcodines form a cyst that can survive extreme conditions. One cyst-forming species causes dysentery, an unpleasant intestinal illness known for its sharp pains and diarrhea. Amoebic dysentery is one reason people should avoid drinking tap water in less-developed parts of the world where the sanitary conditions may be poor.

Flagellates: Protozoans that move with flagella

The phylum Zoomastigina is made up of protists known as **flagellates,** so named because they have one or more flagella. Flagellated protists move by whipping their flagella from side to side.

Magnification: 160×

Beware the Protists

When Donna Huber left Chicago for a seven-day photographic safari in Kenya, she expected to return home with hundreds of breathtaking shots of African wildlife. Donna got her pictures, but she brought home something unexpected as well.

By the time Donna's plane landed at O'Hare International airport, she was feeling listless and out of sorts. "Jet lag," she told the friend who met her at the plane. "I just need some rest."

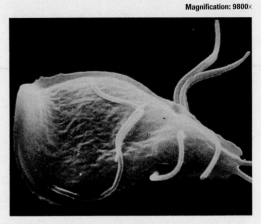

Magnification: 9800×

Giardia

Two days later, however, she awoke in the middle of the night with a dangerously high fever—shivering, sweating, and feeling delirious. Her physician ordered blood tests, and the diagnosis was confirmed. Donna had contracted malaria, a serious and sometimes fatal disease caused by a sporozoan protist, *Plasmodium.*

A plague of protists *Plasmodium* is just one of many protists that can cause deadly disease in humans. Trypanosome, like *Plasmodium,* is a parasite that is transmitted from one host to another by the African tsetse (teht zee) fly. The result is African trypanosomiasis, or sleeping sickness. People with sleeping sickness develop high fevers and swollen lymph nodes. Eventually, the parasites make their way to the brain, where they cause uncontrolled sleepiness.

Giardiasis Muskrats and beavers are responsible for transmitting giardiasis, the symptoms of which include extreme fatigue, cramps, diarrhea, and weight loss. People contract the disease by drinking water contaminated by the protist giardia, which is carried by muskrats and beavers but which harms only humans.

Dysentery Amoebas, too, can cause an unpleasant and, if left untreated, fatal digestive tract malady called amoebic dysentery. Amoebas usually live in pond and stream water; however, the dysentery amoeba prefers life in the large intestine of humans, where it feeds on the intestinal lining, causing bleeding, diarrhea, vomiting, and sometimes death. Amoebic dysentery is spread through infected water and food.

Thinking Critically

What are some ways humans can protect themselves against diseases caused by protists?

How does a swimming paramecium respond to obstacles in its path?

The beating cilia on the surface of a paramecium move in a coordinated way so that the cell normally swims through the water with its front (anterior) end directed forward. But when the front end bumps into an obstacle in the paramecium's path, the organism responds by changing its swimming behavior.

Procedure

1. Observe a *Paramecium* culture that has had boiled, crushed wheat seeds in it for several days.

2. Carefully place a drop of water containing wheat seed particles on a microscope slide, and gently add a coverslip.

3. Using low power (10× or 12.5×), scan across the slide until you locate a paramecium near some wheat seed particles.

4. Watch the paramecium as it swims around among the particles. Record your observations of the organism's responses each time it contacts a particle.

Analysis

1. How does a paramecium respond when it encounters an object?

2. How does it change its swimming direction in response to an obstacle? How long does this response last? What does the cell do?

3. Did you see any changes in the shape of the paramecium as it moved among the particles?

Although some flagellate species are parasites that cause diseases in animals, others are helpful. Termites like those in *Figure 22.5* can survive on a diet of wood. Without the help of a certain species of flagellate, termites could not digest the cellulose in wood. However, within the intestines of termites are flagellates that can. In a mutualistic relationship, these minute protozoans convert cellulose into a carbohydrate that both they and their termite hosts can use.

Ciliates: The most complex protozoans

The roughly 8000 members of the phylum Ciliophora, known as **ciliates,** move by the synchronized beating of cilia that cover their bodies.

Ciliates are found in every kind of aquatic habitat, from ponds and streams to oceans and sulfur springs.

Paramecia usually reproduce asexually, with the cell dividing crosswise and separating into two daughter cells. But when food supplies dwindle or environmental conditions degenerate, paramecia can also undergo a form of conjugation. In this complex process, two paramecia come together and exchange genetic material through their oral grooves. After becoming genetically altered in this way, the two individuals separate, and each goes on to divide asexually.

Magnification: 135×

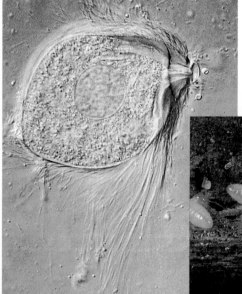

Figure 22.5

The flagellated protozoans (left) that live in the guts of termites (right) possess enzymes that digest wood, making nutrients available to their hosts.

Inside a Paramecium

Paramecia are unicellular organisms, but that does not mean they are simple. Within the cell of a paramecium are a variety of complex organelles and structures that are each adapted to carry out a distinct function.

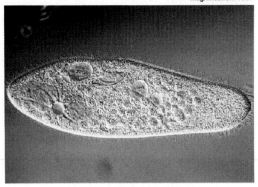

Magnification: 100×

Paramecium caudatum

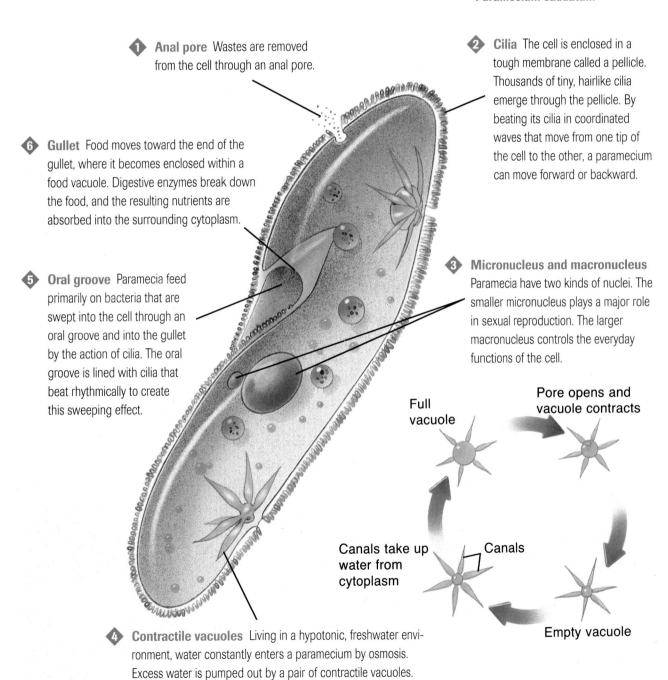

❶ Anal pore Wastes are removed from the cell through an anal pore.

❷ Cilia The cell is enclosed in a tough membrane called a pellicle. Thousands of tiny, hairlike cilia emerge through the pellicle. By beating its cilia in coordinated waves that move from one tip of the cell to the other, a paramecium can move forward or backward.

❻ Gullet Food moves toward the end of the gullet, where it becomes enclosed within a food vacuole. Digestive enzymes break down the food, and the resulting nutrients are absorbed into the surrounding cytoplasm.

❺ Oral groove Paramecia feed primarily on bacteria that are swept into the cell through an oral groove and into the gullet by the action of cilia. The oral groove is lined with cilia that beat rhythmically to create this sweeping effect.

❸ Micronucleus and macronucleus Paramecia have two kinds of nuclei. The smaller micronucleus plays a major role in sexual reproduction. The larger macronucleus controls the everyday functions of the cell.

Full vacuole

Pore opens and vacuole contracts

Canals take up water from cytoplasm

Canals

Empty vacuole

❹ Contractile vacuoles Living in a hypotonic, freshwater environment, water constantly enters a paramecium by osmosis. Excess water is pumped out by a pair of contractile vacuoles.

How do digestive enzymes function in Paramecia?

Paramecia ingest food particles and enclose them within food vacuoles. Each food vacuole circulates in the cell as the food is digested by enzymes that are added to the vacuole. Nutrients made available during digestion are absorbed into the cytoplasm.

Analysis

1. Some digestive enzymes function best at higher pH levels, while others function best at lower (more acid) pH levels.

2. Congo red is a pH indicator dye; it is red when the pH is above 5 and blue when the pH is below 3 (very acid).

3. Yeast cells that contain Congo red can be produced by adding dye to solution in which the cells are growing.

4. When paramecia feed on dyed yeast cells, the yeast is visible inside food vacuoles.

5. Examine the drawing below. The appearance of a yeast-filled food vacuole *over time* is indicated by the colored circles inside the paramecium. Each arrow indicates movement and that time has passed.

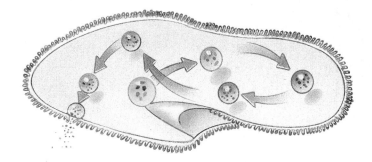

Thinking Critically

Analyze what happens to the pH in the food vacuole over time. Explain your conclusions about the sequence of different digestive enzymes that function in paramecium digestion.

Sporozoans: The parasitic protozoans

Protists that are grouped in the phylum Sporozoa are called **sporozoans.** They are all parasitic, nonmotile protozoans. Sporozoans get their name from the fact that many of them produce spores. A **spore** is a reproductive cell that can produce a new organism without fertilization.

Living as internal parasites in one or more hosts, sporozoans have complex life cycles. Probably the best-known sporozoans are members of the genus *Plasmodium*. Different species of *Plasmodium* cause the disease malaria in people, some other mammals, and birds.

Sporozoans and malaria

Malaria is a disease that is common in tropical climates. Around the world, more than 300 million people are afflicted with this debilitating disease. Malaria is caused by microscopic protozoans that are spread from person to person by mosquitoes. As you can see in *Figure 22.6,* malaria-causing *Plasmodium* species spend part of their life in humans and part in mosquitoes.

For years, medical researchers have been trying to develop a vaccine for malaria, which kills 2 to 4 million people each year. Despite many advances, however, an effective vaccine has not yet been developed.

Figure 22.6

The life cycle of *Plasmodium* begins when a hungry female *Anopheles* mosquito bites a person who already has malaria. When a mosquito bites a person who is already infected with the malaria parasite, it takes in *Plasmodium* reproductive cells with its blood meal.

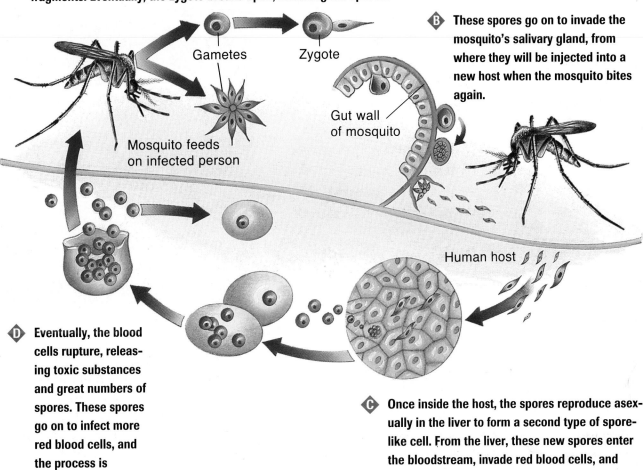

A The reproductive cells move into the mosquito, where they fuse to form a zygote. The zygote divides many times to form numerous sporelike cell fragments. Eventually, the zygote breaks open, releasing the spores.

Gametes

Zygote

Mosquito feeds on infected person

Gut wall of mosquito

B These spores go on to invade the mosquito's salivary gland, from where they will be injected into a new host when the mosquito bites again.

Human host

D Eventually, the blood cells rupture, releasing toxic substances and great numbers of spores. These spores go on to infect more red blood cells, and the process is repeated.

C Once inside the host, the spores reproduce asexually in the liver to form a second type of sporelike cell. From the liver, these new spores enter the bloodstream, invade red blood cells, and multiply rapidly inside them.

Section Review

Understanding Concepts
1. Describe how amoebas move.
2. How do ciliates differ from flagellates?
3. What makes a sporozoan different from other protozoan groups?

Thinking Critically
4. What role do contractile vacuoles play in helping freshwater protozoans maintain homeostasis with their environment?

Skill Review
5. **Sequencing** Retrace the life cycle of *Plasmodium*, identifying the different forms of the sporozoan that are present in the host, and the role each form plays in the disease malaria. For more help, refer to Organizing Information in the *Skill Handbook*.

22.2 Algae: Plantlike Protists

Section Preview

Objectives

Compare the variety of plantlike protists.

Analyze the concept of alternating sporophyte and gametophyte generations in algae.

Key Terms

algae
thallus
colony
fragmentation
alternation of
    generations
gametophyte
sporophyte

As you sit by the pond, a gust of wind blows toward you bringing the scents of flowers and wet vegetation. Each time you inhale, you breathe in oxygen, some of which is being produced right in front of you by the algae in the pond. Algae hold an important position in the varied world of living things. Just about every living thing depends either directly or indirectly on these plantlike protists for oxygen. They also form, along with other eukaryotic autotrophs, the foundation of Earth's food chains.

What Are Algae?

Multicellular, photosynthetic protists are known as **algae.** Even though some kinds of algae look like plants because they are big and green, they have no roots, stems, or leaves. All algae contain one or more of four kinds of chlorophyll, as well as other photosynthetic pigments. These pigments are responsible for the colors you see in different types of algae, from purplish or rusty-red to olive-brown, golden-brown, and yellow. Pigments are also important in the classification of algae. The different species of algae in *Figure 22.7* show the diversity of this group of organisms.

Photosynthesizing protists, known as phytoplankton, are so numerous that they rank as the major producers of nutrients in aquatic ecosystems and as releasers of oxygen in the world. As such, they form the critical first link in aquatic food chains. It's been estimated that algae produce more than half of the oxygen generated by all of Earth's photosynthesizing organisms.

Phyla of Algae

Algae are classified into six phyla. Three of these phyla—the euglenoids, diatoms, and dinoflagellates—are composed of only unicellular species. The other three phyla of algae include some unicellular members, but most are multicellular. These phyla are the green, red, and brown algae.

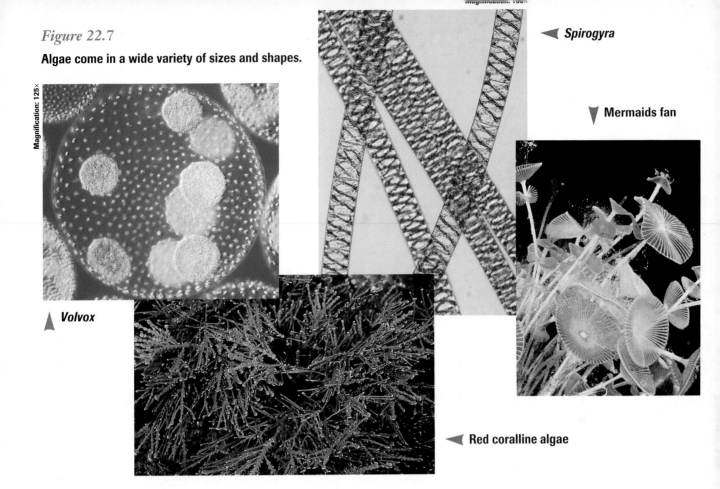

Figure 22.7

Algae come in a wide variety of sizes and shapes.

◀ *Spirogyra*

▼ **Mermaids fan**

Magnification: 125×

▲ *Volvox*

◀ **Red coralline algae**

Euglenoids: Autotrophs as well as heterotrophs

Hundreds of species of euglenoids make up the phylum Euglenophyta. Euglenoids are unicellular, aquatic protists that have traits of both plants and animals. Although they lack a cellulose cell wall as in plant cells, euglenoids have a flexible pellicle made of protein strips inside the cell membrane. Euglenoids are like plants in that most contain chlorophyll and carry out photosynthesis. They are also like animals because they are responsive and move —some very actively—by using one or two flagella for locomotion. When light is not available for photosynthesis, *Euglena* can ingest food from the surrounding water the same way that many protozoans do; in short, they can be heterotrophs. *Figure 22.8* shows an example of a euglenoid.

Diatoms: The golden algae

Diatoms, members of the phylum Bacillariophyta, are unicellular organisms with shells made of silica. Diatoms are photosynthetic autotrophs, and they are abundant in both marine and freshwater ecosystems, where they make up a large component of the phytoplankton.

Figure 22.8

Euglena gracilis **is one of the most famous and well-studied members of this group of protists. Notice the eyespot—a red, light-sensitive structure. This photoreceptor helps** *Euglena* **orient itself toward areas of bright light, where photosynthesis can occur most efficiently.**

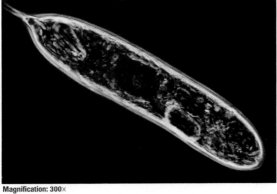

Magnification: 300×

Members of the genus *Paramecium* are ciliated protozoans—unicellular, heterotrophic protists that actively swim around in search of small food particles.

How do *Paramecium* and *Euglena* respond to light?

Euglena are unicellular algae—autotrophic protists that usually contain numerous chloroplasts. How do these two different kinds of protists respond to light in their environment? Are they attracted to or repelled by light? What might be an explanation for their behavior?

PREPARATION

Problem
Do *Paramecium* and *Euglena* respond to light? Do *Paramecium* and *Euglena* respond in different ways?

Hypotheses
Form hypotheses about the potential responses of these two protists to light. Remember that *Paramecium* is heterotrophic. State a testable hypothesis about its possible responses to light and darkness. Remember, too, that *Euglena* is a photosynthetic autotroph. State a testable hypothesis about *Euglena's* possible responses to light and darkness.

Objectives
In this Biolab, you will:
- **Prepare** slides of *Paramecium* and *Euglena* cultures, and observe swimming patterns in the two organisms.
- **Compare** how these two different protists respond to light.

Possible Materials
Euglena culture
Paramecium culture
microscope
microscope slides
dropper
methyl cellulose
coverslips
metric ruler
index cards
scissors
toothpicks

Safety Precautions
Be careful when placing coverslips on slides. They break easily and can produce sharp fragments of glass.

1. Decide on an experimental procedure that you can use in testing your hypothesis. Keep the materials list in mind as you plan your procedure.

2. Record your planned procedure, step by step, and make a list of all the materials you will be using.

3. Design and construct a data table for recording your observations. This table should include enough spaces for both organisms, and for the results of as many repetitions of the experiments as you plan to do.

Check the Plan

Discuss the following points as you work out the details of the procedure for your experiment:

1. What variables will you measure?

2. What will be your control?

3. What will be the shape of the light-controlled area(s) on your slide?

4. Decide who will prepare materials, make observations, and record data.

5. **Make sure your teacher has approved your experimental plan before you proceed further.**

6. In order to carry out your experiments, you will need to mount drops of *Paramecium* culture and *Euglena* culture on microscope slides. This is done by using a toothpick to place a small ring of methyl cellulose on a clean microscope slide. Place a drop of *Paramecium* or *Euglena* culture within the ring of methyl cellulose. Place a coverslip over the ring and drop. Methyl cellulose is a syrupy material that slows movement of organisms for easy observation.

7. Make preliminary observations of swimming *Paramecium* and *Euglena* cells. Then think again about the observation times that you have planned. Maybe you will decide to allow more or less time between your observations.

8. Carry out your experiment.

1. **Checking Your Hypothesis** Did your data support your hypothesis? Why or why not?

2. **Comparing and Contrasting** Compare and contrast the responses of *Paramecium* and *Euglena* to light and darkness.

What explanations can you suggest for their behavior?

3. **Making Inferences** Can you use your results to suggest what sort of responses to light and darkness you might observe using other heterotrophic or autotrophic protists?

Going Further

Project You may want to extend this experiment by varying the shapes or relative sizes of light and dark areas, or by varying the brightness or color of the light. In each case, make hypotheses before you begin. Keep your data in a notebook, and draw up a table of your results at the end of your investigations.

22.2   Algae: Plantlike Protists   **545**

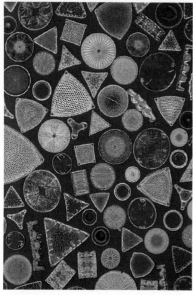

Figure 22.9

Many diatom shells look like delicately carved, microscopic pillboxes. They come in an innumerable array of forms.

The delicate shells of many diatoms, like those in *Figure 22.9,* are like small pillboxes with lids. Each species has its own unique shape, decorated with exquisitely fine grooves and pores. Diatoms are classified into two major groups according to their basic shape. Some are radially symmetrical, while others are elongated and have distinct right and left sides.

Diatom reproduction

When diatoms reproduce asexually, the two halves of the box separate; each half then produces a new half to fit inside itself. With each new generation, half of the offspring are always smaller than the parent cells. When asexually reproducing diatoms reach the point where they are roughly one-quarter of their original size, sexual reproduction takes place. Gametes are produced and released, and fuse with those from another individual to produce a zygote, *Figure 22.10.* The zygote develops into a full-size diatom, which goes on to divide asexually to start the entire process again.

As photosynthesizing autotrophs, diatoms contain chlorophyll as well as other pigments called carotenoids that give the majority of them a golden-yellow color. Carotenoids give carrots and other yellow vegetables their distinctive colors. The food that diatoms manufacture photosynthetically is stored in the form of oils rather than starch. These oils give fish that feed on diatoms an unpleasant, oily taste.

When diatoms die, their glasslike remains sink to the ocean floor to join huge deposits of diatom shells that have been accumulating there for many millions of years. These deposits are dredged up or mined and used as abrasives in tooth and metal polishes, or added to paint to give the sparkle that makes pavement lines more visible at night.

Figure 22.10

Diatoms will reproduce asexually for several generations before sexual reproduction takes place.

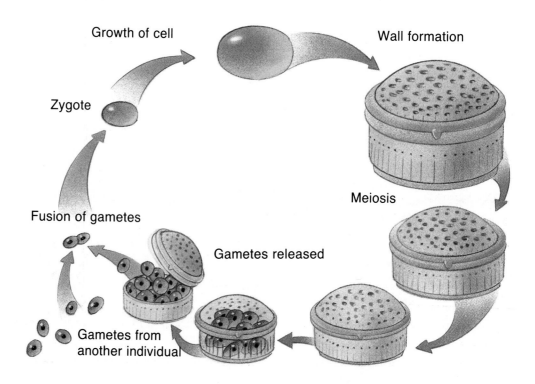

Growth of cell

Wall formation

Zygote

Meiosis

Fusion of gametes

Gametes released

Gametes from another individual

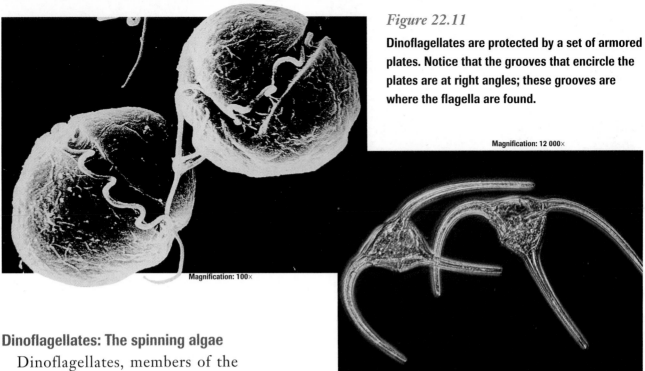

Figure 22.11

Dinoflagellates are protected by a set of armored plates. Notice that the grooves that encircle the plates are at right angles; these grooves are where the flagella are found.

Magnification: 12 000×

Magnification: 100×

Dinoflagellates: The spinning algae

Dinoflagellates, members of the phylum Dinoflagellata, are unicellular algae that have cell walls made up of thick cellulose plates. They come in a great variety of shapes and styles; some resemble helmets, while others look like bizarre suits of armor. *Figure 22.11* shows highly magnified images of two different species of dinoflagellates.

Dinoflagellates have two flagella located in grooves at right angles to each other. When these flagella beat, the cell spins slowly. A few species of dinoflagellates are found in fresh water, but most are marine and are a major component of ocean phytoplankton.

Red tides

Dinoflagellates are autotrophic and contain chlorophyll, carotenoids, and red pigments. Many species live symbiotically with jellyfish, mollusks, and corals. Some free-living species are bioluminescent.

Several species of dinoflagellates produce poisonous toxins. One species, *Gonyaulax catanella*, produces an extremely strong nerve toxin that can be lethal. In the summer, these minute organisms may undergo tremendous population explosions. The dinoflagellates become so numerous that the ocean may actually turn red or orange-red, as you can see in *Figure 22.12*. Toxins released into the water during these red tides kill tons of fish. People who eat shellfish that have fed on these algae may risk being poisoned, too. During red tides, the harvesting of shellfish is usually banned.

Figure 22.12

This toxic red tide off the coast of Baja California, was caused by a bloom of dinoflagellates. In a red tide like this, the concentration of dinoflagellates can reach an incredible 40 to 60 million cells per liter of seawater.

Figure 22.13

There are some 4000 species of red algae. Some species are edible and are popular foods in Japan and other parts of the world.

phaeophyta:

phaios (GK) dusky
phyton (GK) plant

Red algae

Members of the phylum Rhodophyta are the red seaweeds, all of which are multicellular marine organisms. *Figure 22.13* shows examples of red algae. Red algae grow in tropical waters or along rocky coasts in colder water. They attach to rocks by structures appropriately called holdfasts.

Deep-water adaptations

In addition to chlorophyll, red algae also contain red and blue pigments called phycobilins, which are involved in photosynthesis. These pigments absorb green, violet, and blue light waves—the only part of the light spectrum that penetrates water below depths of 100 m. Red algae that live in deep water have a lot of these pigments, and so can carry on photosynthesis where light is limited.

Brown algae

Brown algae make up the phylum Phaeophyta. There are approximately 1500 species of brown algae.

Figure 22.14

Growing close together like trees in a forest, giant kelps stretch dozens of meters from the seafloor to the ocean's surface.

Almost all of these species live in salt water along rocky coasts in cool areas of the world. Brown algae contain chlorophyll as well as a yellowish-brown carotenoid called fucoxanthin, which gives them their brown color. Diatoms also contain the carotenoid fucoxanthin.

The largest and most complex of all brown algae are kelp. Found along cold-water coasts, kelp anchor themselves to rocks or the sea bottom with sturdy holdfasts. The body of a multicellular algae such as a kelp is called a thallus. A **thallus** is a simple plant or algae body without roots, stems, or leaves. Some giant kelps may grow up to 60 m long. In kelp, the thallus is divided into the holdfast, stipe, and blade. Many species of brown algae have air bladders that help keep the thallus afloat near the surface, where it can get the light it needs for photosynthesis.

In some parts of the world, such as off the California coast, giant kelps form dense, underwater forests. These kelp forests are rich ecosystems, *Figure 22.14*, and are home to a wide variety of marine organisms.

Other species of brown algae include Sargasso seaweed, *Sargassum nitans*. This seaweed forms extensive masses that cover the Sargasso Sea in the Atlantic Ocean northeast of the Caribbean.

Green algae

Green algae make up the phylum Chlorophyta. Of all the types of algae, the green algae are the most diverse with more than 7000 species. The major pigment in green algae is chlorophyll, but some species also produce yellow pigments that give them a yellow-green color.

Most species of green algae live in fresh water, but some live in the oceans, moist soil, on tree trunks, in snow, and even in the fur of sloths—large, slow-moving mammals that live in the rain forest canopy.

A large variety of species

Green algae come in a wide variety of forms. *Figure 22.15* shows you two of the different types of green algae. *Chlamydomonas* is a unicellular and flagellated green alga. Other species are multicellular, forming slender filaments (like the hairy pond scum), flat sheets, or colonies. A **colony** is a group of cells that lives together in close association. A *Volvox* colony, *Figure 22.15,* is a hollow ball composed of hundreds, even thousands, of flagellated cells arranged in a single layer. The cells are held together by strands of cytoplasm, and their flagella face outward. When the flagella beat—in a remarkably coordinated fashion—the colony spins through the water.

MiniLab

What do aquarium snails eat?

Many aquatic animals feed on photosynthesizing organisms that live in the water around them. Snails in an aquarium crawl around scraping off and eating the greenish film that forms on the inner surface of the glass walls, the bottom, and objects in the tank.

Procedure

1. Observe the activity of snails in your classroom or school aquarium, and record your observations.

2. Use a toothpick to scrape some green film from the glass and other surfaces. Place each sample on a slide.

3. Mix the scraped material with a drop of water. Carefully place a coverslip on each slide.

4. Look for cells on the slides and record their appearance. Each sample probably will contain several different kinds.

5. Attempt to distinguish photosynthesizing protists from cyanobacteria. The protists will have chloroplasts that will appear as darker-green areas inside the cells. Cyanobacteria are usually uniformly colored because they do not have organized chloroplasts. They are also smaller than protists.

Analysis

1. What evidence did you see that snails were consuming photosynthesizing organisms in the tank?

2. What did the cells that you collected look like? Did they contain chloroplasts?

3. What conclusions can you draw about the photosynthesizing organisms eaten by snails in your aquarium?

Magnification: 15×

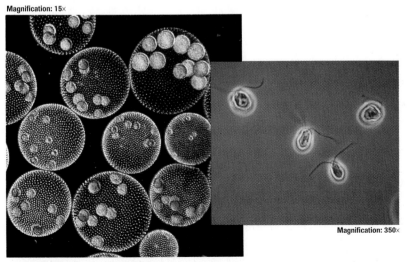

Magnification: 350×

Figure 22.15

The wall of the *Volvox* sphere contains hundreds of individual cells (left). The smaller balls inside the sphere are daughter colonies. When these daughter colonies reach a certain size, the wall of the sphere breaks open. The daughter colonies are released. *Chlamydomonas* is a unicellular species of green algae (right).

Reproductive Strategies of Algae

Green algae reproduce both asexually and sexually. In *Spirogyra*, a multicellular, filamentous green alga, the filaments are haploid. They reproduce asexually by fragmentation. During **fragmentation,** an individual breaks up into pieces and each piece grows into a new individual.

The life cycles of algae such as *Ulva* show **alternation of generations.** These organisms alternate between haploid and diploid generations. The haploid form of an alga is called the **gametophyte** because it produces gametes. The diploid form of the organism—the **sporophyte**—develops from the zygote. Certain cells in the sporophyte undergo meiosis to become haploid spores. Each haploid spore can go on to develop into a gametophyte.

Look at *Figure 22.16* to see the life cycle of *Ulva*, a multicellular alga.

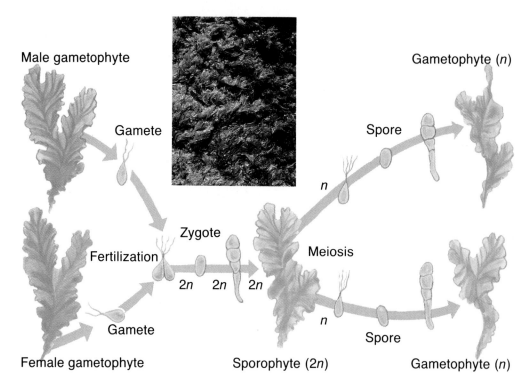

Figure 22.16

In the life cycle of *Ulva*, sea lettuce, notice how the generations alternate from haploid (gametophyte) to diploid (sporophyte). You will also see this pattern of reproduction in fungi and plants.

Male gametophyte

Gamete

Gametophyte (n)

Spore

n

Zygote

Fertilization

Meiosis

$2n$ $2n$ $2n$

n

Gamete

Spore

Female gametophyte

Sporophyte ($2n$)

Gametophyte (n)

Gametophyte (n)

Section Review

Understanding Concepts

1. How are algae important to all living things?
2. Give examples that show why green algae are the most diverse of the algae.
3. How do the sporophyte and gametophyte generations of an alga such as *Ulva* differ from each other?

Thinking Critically

4. Do you think euglenoids should be classified with protozoans or algae? Explain.

Skill Review

5. **Making and Using Tables** Construct a table listing the different phyla of algae. Indicate whether they have one or more cells, their color, and give an example of each. For more help, refer to Organizing Information in the *Skill Handbook*.

22.3 Funguslike Protists

Section Preview

Objectives

Contrast the two types of slime molds with respect to their cellular differences and life cycles.

Discuss the economic importance of the downy mildews and water molds.

Key Term

plasmodium

As you get up from the fallen log you've been sitting on by the pond's edge, a spot of color at its base catches your attention. Turning the log over, you uncover a glistening mass of yellow-orange slime that fans out over the underside of the log. What you've found is a slime mold,

one of a variety of funguslike protists. Slime molds, along with water molds and downy mildews, obtain energy by decomposing organic materials, and play an important role in recycling nutrients in ecosystems.

Different Kinds of Funguslike Protists

Fungi are classified in a distinct kingdom that you will study in Chapter 23. However, certain groups of both the slime molds and water molds show some features of fungi and protists. They form delicate, netlike structures on the surfaces of their food supplies. These organisms obtain energy by decomposing organic materials.

There are three phyla of funguslike protists, and they all obtain energy by decomposing organic materials. Slime molds make up two of these phyla. Slime molds have characteristics of both protozoans and fungi and are classified by the way they reproduce. The third phylum of funguslike protists comprises the water molds and downy mildews. Although we rarely notice in our everyday lives, some disease-causing varieties have the potential to do great damage to crops.

Slime Molds

Many slime molds are beautifully colored, ranging from brilliant yellow or orange to rich blue, violet, and jet black. They live in cool, moist, shady places where they grow on damp, organic matter such as rotting leaves or decaying tree stumps and logs.

There are two major types of slime molds: plasmodial slime molds, which belong to the phylum Myxomycota, and cellular slime molds, which make up the phylum Acrasiomycota. Slime molds are distinctly animal-like during much of their life cycle, moving about and engulfing food in the same way that amoebas do. They reproduce by making spores, a funguslike characteristic.

Figure 22.17

The moving, feeding form of a plasmodial slime mold is a multinucleate blob of cytoplasm. To reproduce, the plasmodium reorganizes itself into stalked, spore-producing structures. Minute spores are released and are carried great distances.

Plasmodial slime molds

Plasmodial slime molds get their name from the fact that they form a **plasmodium**, a mass of cytoplasm that contains many diploid nuclei but no cell walls or membranes. This slimy, multinucleate mass, like the one shown in *Figure 22.17*, is the feeding stage of the organism. It creeps by amoeboid movement, forming a netlike structure on the surfaces of decaying logs or leaves. As

it moves—at the rate of about 2.5 cm per hour—it engulfs microscopic organisms and digests them in food vacuoles.

A plasmodium may reach more than a meter in diameter and contain thousands of nuclei. However, when food becomes scarce or its surroundings dry up, it transforms itself into many separate reproductive structures. Meiosis takes place within these structures. Haploid spores form, which are dispersed by the wind. Spores germinate into either flagellated or amoeboid cells that serve as gametes. The diploid zygote grows into a new plasmodium.

Cellular slime molds

During the feeding stage, cellular slime molds exist as individual, haploid, amoeboid cells that feed, grow, and divide by cell division, *Figure 22.18.* When food becomes scarce, these independent cells come together with hundreds or thousands

Figure 22.18

Cellular slime molds spend part of their life cycle as an independent, amoeba-like cell that feeds, grows, and divides. At times, many hundreds of these amoeboid cells gather to form a multicellular clump. When it moves, this clump of cells compacts itself into a blob that looks like a small garden slug. Eventually, the slug forms a stalked reproductive structure (left). Cellular slime molds are haploid organisms during their entire life cycle.

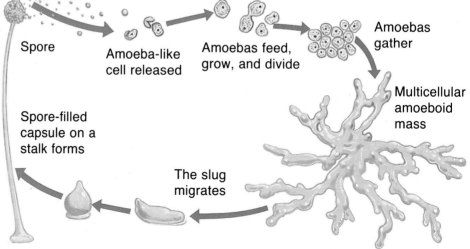

Spore

Amoeba-like cell released

Amoebas feed, grow, and divide

Amoebas gather

Multicellular amoeboid mass

Spore-filled capsule on a stalk forms

The slug migrates

of their own kind to reproduce. Such an aggregation of amoeboid cells superficially resembles a plasmodium. However, this mass is multicellular; it is made up of many individual amoeboid cells, each with a distinct cell membrane.

Phylogeny of Protists

How are the many different kinds of protists related to each other and to fungi, plants, and animals? *Figure 22.19* shows the relationships of protists to each other. Although taxonomists are now comparing the RNA and DNA of these groups, there is little conclusive evidence to indicate whether ancient protists were the evolutionary ancestors of fungi, plants, and animals or whether protists emerged as evolutionary lines that were separate. Many biologists agree that ancient green algae were probably the ancestors of plants.

Figure 22.19

The radiation of the different protist phyla on the Geologic Time Scale shows their relationships.

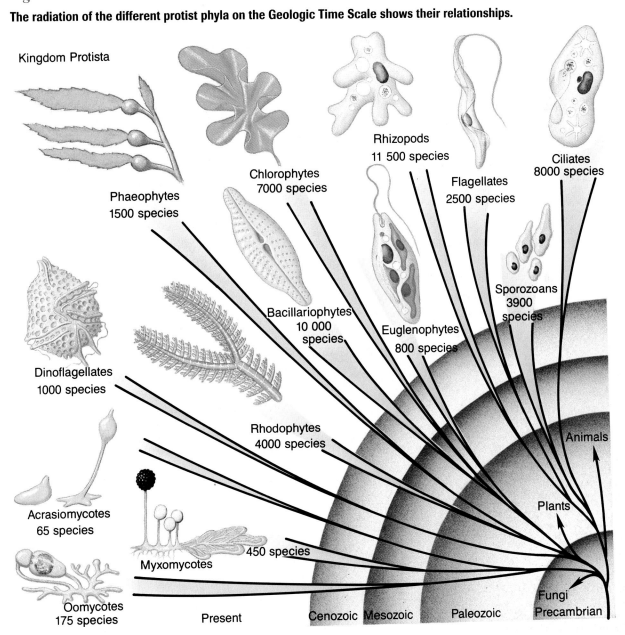

Kingdom Protista

Phaeophytes
1500 species

Chlorophytes
7000 species

Rhizopods
11 500 species

Flagellates
2500 species

Ciliates
8000 species

Bacillariophytes
10 000 species

Euglenophytes
800 species

Sporozoans
3900 species

Dinoflagellates
1000 species

Rhodophytes
4000 species

Animals

Acrasiomycotes
65 species

Plants

Myxomycotes

450 species

Fungi

Oomycotes
175 species

Present

Cenozoic Mesozoic Paleozoic Precambrian

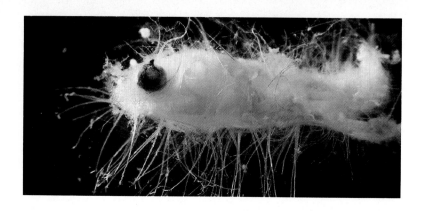

Figure 22.20

Water molds grow over the surface of dead organisms, like this fish, gradually decomposing the tissues and absorbing the nutrients released in the process.

Water Molds and Downy Mildews

Water molds and downy mildews are members of the phylum Oomycota. Most members of this large and diverse group of funguslike protists live in water or moist places. As shown in *Figure 22.20,* some parasitize plants, while others feed on dead organisms, such as the fish.

Most water molds appear as fuzzy, white growths on decaying matter. They resemble a fungus in that they grow as a mass of threads over a substrate, digest it, and then absorb the predigested nutrients. But at some point in their life cycle, water molds produce flagellated reproductive cells—something that true fungi never do. This is why water molds are classified as protists rather than fungi.

Another economically important member of the phylum Oomycota is a downy mildew that causes a serious disease in many plants.

Connecting Ideas

The phylum Protista is made up of a variety of species of unicellular and multicellular eukaryotic organisms. While many protists look like simple organisms, many have complex workings. Protists play some extremely important roles in the environment. Diatoms produce a large percentage of the oxygen we breathe. Unicellular algae form the first link in aquatic food chains.

Large, multicellular species can form habitats for other organisms as well as supply them with food. The funguslike protists—slime molds, water molds, and downy mildews—with their characteristics of both fungi and protists provide a link between the two groups.

Section Review

Understanding Concepts
1. What characteristics of slime molds might cause difficulty in their classification?
2. Explain why plasmodial slime molds are sometimes referred to as acellular slime molds.
3. How would a water mold growing on a fish ultimately kill it?

Thinking Critically
4. What kinds of environments would you consider to be favorable for the growth of slime molds? Explain.

Skill Review
5. **Observing and Inferring** If you know that a plasmodium consists of many nuclei within a single cell, what can you infer about the process that formed the plasmodium? For more help, refer to Thinking Critically in the *Skill Handbook.*

Saving Kelp Beds

California's Undersea Forests Disappearing

OCEAN NEWS   JUNE 10, 1993

Giant kelp beds form a fertile ecosystem that is an essential habitat for a network of inter-dependent plants and animals. Natural and human-made pressures are threatening these undersea forests.

In the last few decades, California's kelp has faced dangers from several sources. California's intense storms in the winter of 1982 were blamed on a phenomenon called El Niño, an equatorial current that brought unusually warm water to the coast. Since kelp is adapted to a cool-water environment, some died. Surviving beds were ripped up by violent weather.

The importance of kelp California's kelp beds provide food for crustaceans, shelter for young fish, and hunting grounds for sea otters and fish. More than 800 species of marine life are supported by these vast beds of tangled seaweed, which cover about 40 000 acres of ocean floor off the state's coastline.

Kelp is an important ingredient in products ranging from ceramics to cosmetics and generates $40 million a year.

Human-produced dangers When untreated sewage pours into the ocean, as it sometimes does after a storm, it can inhibit the growth of kelp. However, sea urchins welcome the odorous outpouring, causing a boom in their populations. As sea urchin numbers increase, they mow down great swaths of kelp, their natural food.

Kelp beds are also damaged by nuclear power plants. Experts claim a 60 percent reduction in the size of one large kelp bed off the southern California coast due to the nearby power plant's cooling system, which sucks up and kills 50 tons of fish each year. When the debris-filled water is released into the ocean, light penetration to the ocean floor is reduced, making photosynthesis by the kelp less efficient.

Different Viewpoints

Although solutions to the human-produced dangers to kelp have been offered, they have been challenged. Abalone fishermen are opposed to reducing the number of sea urchins to save kelp forests. Sea urchins are an important food source for this mollusk, which provides the fishermen with a living. Proponents of nuclear power plants argue that methods to reduce use of cooling water will double utility bills.

INVESTIGATING the Issue

1. **Applying Concepts** Discuss with a small group some ways that you think wildlife and people can both get what they need from the California coastal environment.
2. **Analyzing the Consequences** What do you think could happen to underwater wildlife if California's kelp forests were destroyed completely?

Reviewing Main Ideas

22.1 The World of Protists

- The Kingdom Protista is a diverse group that contains animal-like, plantlike, and funguslike organisms.
- Some protists are heterotrophs, some are autotrophs, and some get their nutrients by decomposing organic matter.
- Sarcodines move by extending pseudopodia. Some marine sarcodines have shells.
- Flagellates have one or more flagella, which they use to propel themselves. Ciliates move by the synchronized beating of cilia. Sporozoans are nonmotile and live as parasites.

22.2 Algae: Plantlike Protists

- Algae are photosynthetic autotrophs. There are both unicellular and multicellular species of algae. Unicellular species include the euglenoids, diatoms, dinoflagellates, and some green algae. Multicellular species include red, brown, and green algae. Euglenoids have characteristics of both plants and animals.

- Green, red, and brown algae, often called seaweeds, have complex life cycles that alternate between haploid and diploid generations.

22.3 Funguslike Protists

- Slime molds, water molds, and downy mildews are funguslike protists that obtain nutrients by decomposing organic material. Both plasmodial and cellular slime molds undergo changes in appearance and behavior to produce reproductive structures.

Key Terms

Write a sentence that shows your understanding of each of the following terms.

algae	gametophyte
alternation of generations	plasmodium
asexual reproduction	protozoan
ciliate	pseudopodia
colony	spore
flagellate	sporophyte
fragmentation	sporozoan
	thallus

Understanding Concepts

1. Different types of protozoans move in different ways. List and describe three types of movement exhibited by these animal-like protists.
2. Why are so many other living things ultimately dependent on algae?
3. During their life cycle, many algae alternate between haploid and diploid forms. Compare gametophyte and sporophyte generations in *Ulva*.
4. Euglenoids have some plantlike and some animal-like characteristics. What is the adaptive advantage for a photosynthetic organism to be able to move from place to place?

5. How are diatoms and dinoflagellates ecologically and economically important?

Relating Concepts

6. Make a concept map that relates the following terms and phrases. Supply the appropriate linking words for your map.

   protists, algae, diatoms, flagellates, protozoans, amoeba, ciliates, slime molds, chlorophyta, rhodophyta

Using a Graph

7. During a summer ecology class, a group of high-school students studied unicellular algae at a site in the middle of a pond. For three days and nights, they measured the number of diatoms in the water at

various depths. They produced the following graph based on their data.

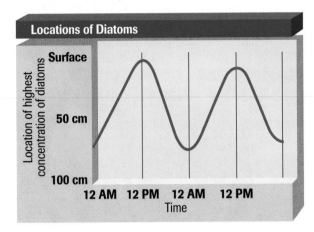

Locations of Diatoms

Location of highest concentration of diatoms

Surface
50 cm
100 cm

12 AM 12 PM 12 AM 12 PM
Time

What pattern in the locations of diatoms in the pond do you observe? What could explain why the diatoms were routinely found concentrated near the surface of the water at midday?

Applying Concepts

8. What do you think a major disadvantage might be for the parasitic behavior of a sporozoan? How might a sporozoan's method of asexual reproduction offset this disadvantage?
9. In what type of ecosystem would you expect plasmodial slime molds to transform themselves into spore-producing structures more frequently—a forest in the Pacific Northwest that has heavy rainfall or a dry, oak forest in the Midwest? Explain your answer.
10. In agricultural regions where farmers apply large amounts of nitrogen fertilizers to their fields, local ponds and lakes often develop a thick, green scum of algae and cyanobacteria in mid to late summer. Speculate why this happens.
11. Up to the late 1800s, malaria was not uncommon in the extreme southeastern part of the United States. In an attempt

to fight the disease, ponds and wetland areas were often filled in or drained. How do you suppose this action helped cut down on the number of malaria cases?

Biology & Society
12. How may sea urchins be important to the entire food chain of a kelp forest?

Thinking Critically

Interpreting Data
13. **Biolab** The following table was presented by a pair of students, showing the data they collected while observing how *Paramecium* and *Euglena* respond to light. How did the reaction of *Euglena* and *Paramecium* differ in response to light?

Response of *Euglena* and *Paramecium* to Light		
Organism	Time in minutes	% of individuals in the light
Paramecium	3	45%
Paramecium	6	30%
Paramecium	9	15%
Euglena	3	55%
Euglena	6	70%
Euglena	9	90%

Making Inferences
14. **Minilab** Why do you suppose many people who own aquariums add snails to them?

Connecting to Themes

15. **Energy** How do autotrophs, heterotrophs, and decomposers obtain their energy?
16. **Homeostasis** Give three examples of organelles that help protists maintain homeostasis in their environments.

23 Fungi

Have you ever had a ring of mushrooms mysteriously appear in your yard overnight? You know they weren't there the day before. How did they form so quickly? And what are they doing?

Members of the Kingdom Fungi are found in nearly every type of habitat on Earth. However, what we see of them is often just the tip of the iceberg. Hidden in the soil beneath those mushrooms in your yard, for example, lies the rest of the fungus that produced them.

Fungi may not always be noticeable, but their actions are. They live by decomposing living and nonliving organic matter.

In one way or another, fungi affect nearly all other forms of life. Some cause diseases and destroy important food crops. But these negative aspects are far outweighed by the importance of fungi as decomposers, and by the crucial role they play in the biosphere as nutrient recyclers.

Mushrooms are a visible sign of fungi at work in the soil. Mushrooms are just one of many different kinds of fungi. As decomposers, fungi form a critical link in the web of life on Earth.

Section Preview

Objectives

Identify the basic characteristics of fungi.

Explain the importance of fungi as decomposers and in the flow of energy and nutrients through food chains.

Key Terms

hypha
mycelium
chitin
haustoria
budding
sporangium

*Y*ou would probably recognize a mushroom growing in your yard as a fungus, but what about the yeast cells in this photograph? Yeasts are fungi too, but they don't look much like mushrooms. More than 100 000 different fungus species have been described by mycologists, scientists who study fungi. Each type of fungus has its own distinctive features. But all fungi are similar in their cellular composition.

Magnification: 70×

General Characteristics

Fungi are everywhere—in the air, in the water, on damp basement walls, in the garden, on foods, even between our toes. Some are large, bright, and colorful, while others are easily overlooked, *Figure 23.1.* Many have descriptive names such as stinkhorn, puffball, rust, or ring-worm. Most species grow best at temperatures between 20° and 30°C, but many thrive at cooler temperatures. How many times have you opened the refrigerator and pulled out a piece of fruit or a chunk of cheese, only to find that it has already become a meal for a thick, furry mass of pink, green, white, or black mold?

Figure 23.1

Fungi come in many different forms, sizes, and colors.

▶ **These coral fungi resemble ocean corals and often sport vivid colors.**

▶ **Bird's nest fungi are appropriately named. The tiny cups look like nests complete with eggs.**

◀ **This insect has been attacked and gradually killed by the parasitic fungus that you can see emerging all over its body.**

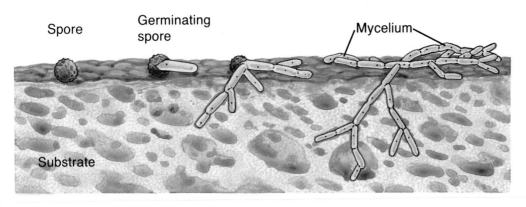

Spore

Germinating spore

Mycelium

Substrate

Figure 23.2

Hyphae are the basic structural units of fungi that grow to form mycelia—complex masses of filaments. Some hyphae in a mycelium are adapted to anchor, invade, or produce reproductive structures.

Fungi used to be classified as members of the plant kingdom. This early classification was based on the fact that, like plants, many fungi grow anchored in soil and have cell walls. However, as biologists came to learn more about fungal structure and how fungi live, they realized that fungi make up a distinct kingdom.

The structure of fungi

A few unicellular types of fungi exist, such as yeasts, but most fungi are multicellular. As you can see in *Figure 23.2*, the basic structural units of a multicellular fungus are threadlike filaments called **hyphae,** which develop from fungal spores. Hyphae elongate at their tips and branch extensively to form a network of filaments called a **mycelium.** With a magnifying glass, you can see individual hyphal filaments in molds that grow on bread. However, the hyphae that make up the mushrooms growing in your yard are much more difficult to see. That's because hyphae in mushrooms are tightly packed to form a dense mass.

Unlike plants, which have cell walls made of cellulose, the cell walls of most fungi contain a complex carbohydrate called **chitin.** As you will see in Chapter 31, chitin is also found in the external skeletons of some animals such as lobsters, crabs, insects, and spiders.

Inside hyphae

In many types of fungi, hyphae are divided by cross walls, called septa, into individual cells that contain one or more nuclei, *Figure 23.3.* Septa usually have holes, or pores, in them. Through these pores, cytoplasm and organelles can flow from one cell to the next. One advantage of this free-flowing cytoplasm is that it helps move nutrients from one part of a fungus to another.

Some fungi have hyphae with no septa. When you look at these hyphae under a microscope, you can see hundreds of nuclei streaming along in an undivided mass of cytoplasm.

Figure 23.3

Many types of fungi have hyphae that are divided into cells by septa (top). Hyphae without such cross walls look like giant, multinucleate cells (bottom).

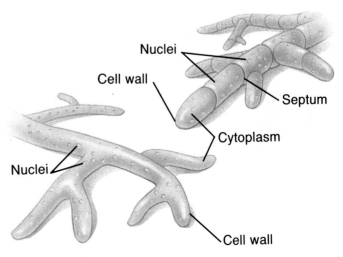

Nuclei

Cell wall

Septum

Cytoplasm

Nuclei

Cell wall

The Fungus that Ate Michigan

It is huge—truly a humongous fungus. This single individual weaves through 38 acres of forest floor and is estimated to weigh at least 100 tons! Located at the western end of Michigan's Upper Peninsula, an enormous example of *Armillaria bulbosa* is estimated to be at least 1500 years old. It is surrounded by neighbors of the same species.

The fungus consists of an underground network of long, connecting bundles of mycelia that give rise to occasional mushrooms. The mushrooms produce and spread spores. Thus, the bulk of the organism is hidden in the soil and, except for the reproductive structures of the mushrooms, it is seldom noticed.

Identity check If the fungus is buried and the mycelia can't be seen, how do we know this is one individual? To answer this question, researchers at the University of Toronto and Michigan Technological University conducted several experiments. In one experiment,

they gathered many specimens from a 75-acre study area, and then analyzed the DNA of 16 genes from each specimen. They found that the DNA in all the specimens from a 38 acre area was identical, but different from surrounding specimens of *A. bulbosa*. This was strong evidence that these specimens were from the same individual.

Even bigger? The Michigan fungus is certainly not the largest of its kind. Two forest pathologists from the state of Washington may have identified an even larger *Armillaria*. About 20 years ago, these scientists found one they believe traverses 1500 acres. They have not performed DNA tests to confirm that it is one individual. Instead, they depended on observations of the organism. One property of this fungus is that the hyphae of an individual will not intermingle with the hyphae of other individuals of the same species. The mycelia of this Washington fungus seemed to be well intermingled, leading the researchers to conclude that all the mycelia were from the same fungus.

Thinking Critically

How can the Michigan fungus and the Washington fungus be said to be over a thousand years old, when we know fungus cells live for only a few years at most? Defend your answers.

In thinking about fungi, it's usually the negative things that come to mind: spoiled food, plant and animal diseases, and poisonous mushrooms. But fungi have an important role in the interactions of organisms on Earth. Fungi decompose a large amount of the world's waste material.

The great decomposers

Fungi perform an indispensable function in their role as decomposers. Imagine a world without fungi. Huge mounds of wastes, dead organisms, and debris would be everywhere. Because of fungi, the leaves that carpet a forest floor each fall are mostly gone in the spring. Slowly but surely, fallen leaves, animal carcasses—whatever dies or becomes waste—is decomposed, *Figure 23.4*. Fungi, along with several species of monerans and protists, are constantly at work transforming complex organic substances into the raw materials that other organisms can use.

Figure 23.4

These patches of lush, green grass are growing where pads of horse dung have been decomposed by fungi. By breaking down animal wastes and other organic matter, fungi make essential nutrients such as nitrogen available for plants.

How Fungi Get Their Food

Fungi cannot produce their own food as do photosynthesizing protists or plants. A fungus depends on other sources for its supply of energy.

Extracellular digestion

Fungi are heterotrophs that obtain nutrients through a process called extracellular digestion; that is, food is digested *outside* a fungus's cells. The products of digestion are then absorbed. Take a fungus-covered orange as an example. As hyphae grow over and into the cells of the orange, they release powerful digestive enzymes. These enzymes break down the large organic molecules in the orange into smaller molecules. These small molecules diffuse into the fungus, where they are used to synthesize materials for growth and repair. The more extensive a mycelium becomes, the more surface area there is through which nutrients can be absorbed.

Different feeding relationships

Different kinds of fungi use different types of food sources. A fungus may be a decomposer, a parasite, or a mutualist that lives symbiotically with another organism, *Figure 23.5.* The most specialized parasitic fungi absorb nutrients from the living cells of their hosts. They do this using specialized hyphae called **haustoria,** which penetrate and grow into host cells without killing them.

Many fungi live in a symbiotic relationship with a plant. These mutualistic fungi absorb nutrients from living hosts, but they also benefit their hosts in some way. For example, a fungus in a mutualistic relationship might help its host retain water, or obtain minerals from the soil in exchange for organic food.

haustoria:
haurire (L) to drink
Haustoria are hyphae that invade the cells of a host to absorb nutrients.

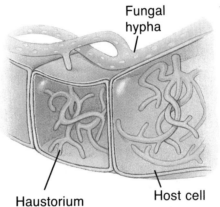

Fungal hypha

Haustorium

Host cell

◀ **Nutrients absorbed across the walls of a haustorium are transported to the rest of the fungus.**

▼ **Organisms attacked by parasitic fungi, such as the American elm tree below, are harmed by the relationship and may eventually die.**

Figure 23.5

Feeding relationships

▲ **What you see along this dead tree branch are the visible parts of a turkey-tail fungus *(Trametes versicolor).* What you don't see deep within the tissue of the branch is the extensive mycelium of this fungus.**

Are there mold spores in your classroom?

Any mold spore that lands in a favorable place can germinate, produce hyphae, and begin developing into a mycelium. Can you demonstrate that there are airborne bread mold spores in your classroom by exposing bread to the air?

Procedure

1. Place two slices of freshly baked bakery bread on a plate. Sprinkle some water on one slice to make its surface moist. Leave both slices uncovered for several hours.

2. Sprinkle a little more water on the moistened slice, and place it in a plastic, self-sealing bag. Place the dry slice of bread in another self-sealing plastic bag. Blow a little air into each bag so that the plastic does not lay against the bread's surface, and then seal the bags.

3. After four or five days, take out the bags and look for mold growing on the bread slices.

4. Remove a small piece of bread mold with a forceps, place it on a slide in a drop of water, and add a coverslip. Observe the mold under a microscope, first on low power and then on high power.

Analysis

1. Did you observe mold growth on the moistened bread? On the dry bread?

2. Describe what you saw on the slide that you made.

3. How does this experiment demonstrate that there are mold spores in your classroom? What conclusions can you draw about conditions required for the growth of bread molds?

Reproduction in Fungi

Depending on the species and on environmental conditions, a fungus may reproduce asexually or sexually. Fungi reproduce asexually by fragmentation, budding, or by producing spores.

sporangium:

sporos (GK) seed
anggeion (GK) vessel
A sporangium
produces spores.

Fragments and buds

In fragmentation, pieces of hyphae that are broken off or torn away from a mycelium are capable of growing into new mycelia. Suppose you are spading up your garden before planting seeds. Every time your shovel cuts down into the dirt, it slices through mycelia. Most of those hyphal fragments can grow and branch to form new mycelia.

Budding is a form of asexual reproduction in which mitosis takes place and a new individual—like the one you can see in *Figure 23.6*—grows out and eventually separates from a parent cell.

Reproducing by spores

As you learned in Chapter 22, a spore is a reproductive cell that germinates and develops into a new organism. Most fungi produce spores. When a fungal spore lands in a place that has all the conditions necessary for growth, a threadlike hypha emerges and begins to grow and branch to form a mycelium.

Eventually, some hyphae may grow upward from the substrate and produce a spore-containing structure called a sporangium. A **sporangium** is a sac or case in which spores are produced. Mushroom caps contain

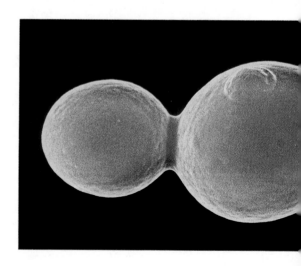

Figure 23.6

Most yeasts reproduce asexually by budding. Wherever a bud pinches off from its parent cell, a tiny bud scar is left behind. Can you spot the bud scar on the cell on the right in this photograph?

one type of sporangium. The tiny black spots you can see on bread mold are another type of sporangium. In fact, sporangia are often the only part of a fungus you can see, and they usually represent only a small portion of the total organism.

Fungi may produce spores by either mitosis or meiosis. The way in which spores are formed during sexual reproduction is the basis on which fungi are classified into their major groups.

Adaptations for Survival

Fungi have evolved a variety of strategies that help them adjust to changing conditions and exploit new environments and food sources. Many of these survival adaptations involve sporangia and spores.

Sporangia protect spores and, in some cases, keep them from drying out until they are ready to be released.

Spores and spore dispersal

Fungal spores are small and extremely lightweight. When released from sporangia, spores are easily swept along by the wind. Look at the puffball releasing its spores in *Figure 23.7.* The slightest breeze may carry such dust-fine spores a great distance. Spores released in one place can be blown many hundreds of kilometers.

Most fungi produce a lot of spores—another adaptation for survival. A single puffball that measures 25 cm in circumference produces roughly 1 trillion spores. Producing such huge numbers of spores improves the chances that at least some will survive.

Fungal spores can also be dispersed by water, as well as by animals such as birds and insects. The fungus that causes Dutch elm disease, a usually fatal disease of elm trees, is carried by bark beetles. Spores of the fungus stick to the bodies of the beetles as they feed on infected trees. When spore-covered beetles move to healthy elms to lay their eggs, they transfer spores in the process.

Figure 23.7

Wind, water, and animals help disperse fungal spores. A passing animal or falling raindrops will cause some fungi, like these puffballs, to discharge a cloud of spores that will be carried away by the wind.

Section Review

Understanding Concepts

1. In what way are pores in septa an advantage for a fungus with a large mycelium?
2. Explain how a fungus obtains nutrients from the environment.
3. What role do fungi play in food chains?

Thinking Critically

4. Explain why you might expect to find several different types of fungi growing in a bird's nest.

Skill Review

5. **Measuring in SI** Outline the steps you would take to calculate the approximate number of spores contained in a puffball fungus that had a circumference of 10 cm. For more help, refer to Practicing Scientific Methods in the *Skill Handbook.*

23.2 The Diversity of Fungi

How are fungi classified? If you took one of the mushrooms that had so abruptly appeared in your yard to a mycologist, he or she would look for certain key features in that mushroom to help identify it. In particular, the mycologist would look at the spores it contained. Fungi are classified into divisions according to the way in which spores are produced. As you'll see, the name of each division is derived from the types of sexual structures that are characteristic of each group.

Zygospore-forming Fungi

Have you ever pulled out the last slice of bread in a bag, only to find that it was covered with rather unappetizing black spots and a bit of fuzz? If so, then you have had a close encounter with *Rhizopus stolonifer*, the common black bread mold. *Rhizopus* is probably the most familiar member of the phylum Zygomycota. Many of the 1500 or so species that belong to this division are decomposers. Zygomycotes reproduce asexually by producing spores, and sexually by forming thick-walled spores. Hyphae of zygomycotes do not have septa.

Growth and asexual reproduction

When a *Rhizopus* spore settles on a piece of bread, it germinates and hyphae begin to grow. Some hyphae called **stolons** grow horizontally along the surface of a food source, in this case the bread, and rapidly produce a mycelium. Other hyphae form **rhizoids** that penetrate down into the food and anchor the mycelium to its substrate. They also are the site of most extracellular digestion and nutrient absorption.

Asexual reproduction begins when certain hyphae grow upward and develop sporangia at their tips, *Figure 23.8*. These sporangia mature to form black, rounded sporangia loaded with asexual spores. When each sporangium splits open, hundreds of spores are released into the air. Those that land on a moist food supply germinate, form new hyphae, and start the asexual cycle again.

Producing zygospores

Suppose that your piece of bread fell behind the kitchen garbage can and began to dry out. This change in environmental conditions could trig-

Figure 23.8

The life cycle of the black bread mold, *Rhizopus stolonifera* Asexual reproduction is most common, but sexual reproduction takes place when different mating types (+ and -) come together and form zygospores.

▶ In this photomicrograph, you can see dark zygospores forming where gametangia have come together and fused during sexual reproduction.

◀ This scanning electron micrograph shows a *Rhizopus* sporangium covered with thousands of haploid spores.

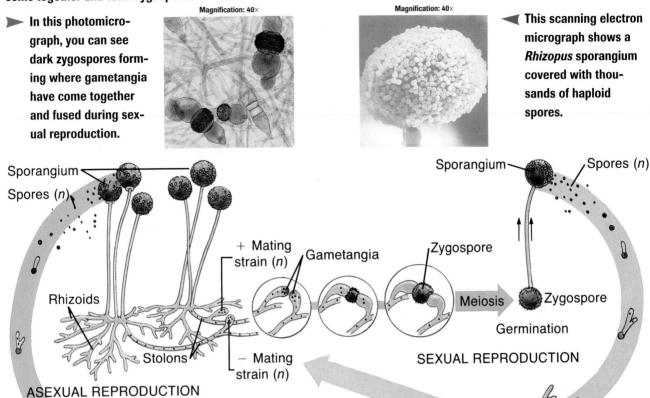

ger the fungus to reproduce sexually and produce **zygospores**—thick-walled spores that are adapted to withstand unfavorable conditions.

Look again at *Figure 23.8.* Sexual reproduction in *Rhizopus* takes place when the tips of hyphae from *two different mycelia* come together and fuse. Where the hyphae fuse, two gametangia form, each with a haploid nucleus inside. A **gametangium** is a structure with a haploid nucleus and in which gametes are produced. When the contents of the two gametangia fuse, a diploid zygote forms. This diploid cell then develops into a thick-walled zygospore.

A zygospore may lie dormant for many months and can survive periods of drought, cold, and heat. When conditions improve, the zygospore absorbs water, undergoes meiosis, and germinates to produce an upright hypha with a sporangium. Each haploid spore formed in this sporangium is capable of growing into a new mycelium.

Sac Fungi

The phylum Ascomycota is the largest group of fungi, with about 30 000 species. The ascomycotes are also known as sac fungi. Both names refer to the little saclike structures, each known as an **ascus,** in which the sexual spores of these fungi develop.

Figure 23.9

Many ascomycotes are cup shaped or have multiple cup-shaped indentations. Asci line the inside of these cup-shaped surfaces.

▶ The scarlet cups of an ascomycote grow on wood.

▶ Morels are often served as a delicacy.

Because they are produced inside an ascus, the spores are called **ascospores**. Ascomycotes can take many different forms, *Figure 23.9*.

During asexual reproduction, ascomycotes produce a different kind of spore. Hyphae rise up from the mycelium and elongate to form **conidiophores**, *Figure 23.10*. Chains or clusters of asexual spores called **conidia** develop from the tips of conidiophores. Once released, these haploid spores are dispersed by wind, water, and animals.

Important ascomycotes

You've probably encountered a few types of sac fungi in your refrigerator or in the grocery store in the form of blue-green, red, and brown molds that decorate decaying foods. Other sac fungi are well-known by farmers and gardeners because they cause plant diseases such as apple scab, Dutch elm disease, and ergot on rye. Quite a few ascomycotes are also the cause of some serious fungal diseases of animals and people.

Not all sac fungi have a bad reputation. Morels, *Figure 23.9,* and truffles are edible members of this phylum. Perhaps the most economically important ascomycotes are the yeasts. Yeasts are unicellular sac fungi that usually do not produce hyphae. Most of the time, yeasts reproduce asexually by budding.

Perhaps the best-known yeasts are those that are used in brewing and baking. Grown anaerobically, these minute cells can ferment sugars to produce carbon dioxide and ethyl alcohol. Some yeasts are used to make wine and beer. Yeasts used in baking produce the carbon dioxide that causes bread dough to rise and take on a light, airy texture.

Figure 23.10

Most ascomycotes reproduce asexually by producing conidia. Conidia and conidiophores differ from species to species, making them important clues in identification.

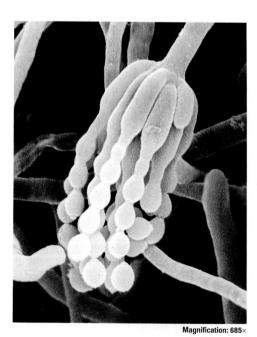

Magnification: 685×

Yeasts are also important tools for research in genetics because they have large chromosomes. A vaccine for the disease hepatitis B is produced by splicing human genes with those of yeast cells. Because yeasts can be produced quickly, they are an important source of the vaccine.

Club Fungi

Of all the different kinds of fungi, some of the 25 000 species in the phylum Basidiomycota are probably most familiar to you. Mushrooms, puffballs, stinkhorns, bird's nest fungi, and bracket fungi are all basidiomycotes. So are the rust and smut fungi, which cause billions of dollars worth of damage to grain crops every year.

Basidia and basidiospores

The spore-producing cells of basidiomycotes are club-shaped hyphae called **basidia.** It's from these

Figure 23.11

Most basidiomycotes produce haploid basidiospores on club-shaped basidia. Typically, four basidiospores develop at the end of each basidium, as you can see here.

Magnification: 7750×

basidia that the members of this division get their more common, general name—the club fungi. Basidia usually develop on a short-lived reproductive structure, which in the case of mushrooms is the mushroom cap and stalk. During sexual reproduction, basidia produce spores called **basidiospores,** *Figure 23.11.*

The reproductive cycle of a typical basidiomycote, such as a mushroom, is complex. You see evidence of this reproductive process only when mushrooms appear on lawns and forest floors. Read *The Inside Story* for an in-depth look at the structure and life of a mushroom.

The Life of a Mushroom

*W*hat we call a mushroom is just the above-ground sexual structure of the fungus. A single mushroom can contain hundreds of thousands of spores, which are produced as a result of sexual reproduction. Most types of mushrooms have no asexual reproductive stages in their life cycle.

1 Above ground, a mushroom typically consists of a stipe that supports a cap.

Cap

Gills

Stipe

2 The undersides of some mushroom caps have hundreds of thin sheets of tissue extending out from the stalk like spokes on a wheel. These are gills. Club-shaped basidia line the gill surfaces.

Basidiospores

Basidium

Basidiospore

Mycelium

7 As basidiospores mature, they break off from basidia and are carried by the wind to new locations.

3 When a basidiospore lands in a suitable environment., it germinates to produce white, threadlike hyphae that grow down into the soil to form a haploid mycelium.

+ Mating type

− Mating type

Two mating types fuse. Nuclei remain separate

New mycelium — each cell with two nuclei

6 A button can develop into a mushroom quickly. Inside each basidium, haploid nuclei come together to form a diploid cell. Meiosis then occurs, producing four haploid nuclei that each become a basidiospore.

4 The mycelia of basidiomycotes have different mating types. When mycelia from two different mating types come in contact, the hyphae fuse.

5 A new mycelium forms. Each cell contains one haploid nucleus from each of the two mating strains. Eventually, compact masses of hyphae, called buttons, are formed just under the soil's surface.

Figure 23.12

Many mushrooms are edible. Many more are parasites or very poisonous.

◀ **These mushrooms belong to the genus *Amanita*. Some types of *Amanita* mushrooms are extremely poisonous; eating just one can be fatal.**

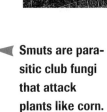

◀ **Smuts are parasitic club fungi that attack plants like corn.**

▲ **Shelf and bracket fungi are basidiomycotes often seen growing on tree branches and fallen logs.**

The visible reproductive structures of members of the phylum basidiomycota come in a variety of different shapes and sizes as *Figure 23.12* shows you.

The Deuteromycotes

Zygomycotes, ascomycotes, and basidiomycotes reproduce both asexually and sexually. About 25 000 species of fungi, the deuteromycotes, have no known sexual stage in their life cycle. These fungi reproduce only asexually or their sexual phase has not yet been observed by mycologists.

Diverse deuteromycotes

If you've ever had strep throat, pneumonia, or some other kind of infection, your doctor probably prescribed penicillin as a treatment. Penicillin is an antibiotic produced from a deuteromycote, one that is commonly seen growing on fruit, *Figure 23.13*. Other deuteromycotes are used in making soy sauce and some kinds of blue-veined cheese. Citric acid, which gives jams, jellies, soft drinks, and fruit-flavored candies their tart taste, is produced in huge quantities using deuteromycote fungi.

Figure 23.13

Many deuteromycotes are useful.

◀ **The antibiotic penicillin is derived from *Penicillium* mold, shown here on an orange.**

◀ **Many people enjoy the strong, distinctive flavors of cheeses such as this Roquefort. The blue veins and splotches are patches of conidia.**

Does temperature affect the rate of carbon dioxide production by yeast? Are there differences in yeast metabolism at cold and warm temperatures?

Does temperature affect the metabolic activity of yeast?

Look at the experimental setup pictured here. As yeast metabolizes in the stoppered container, the carbon dioxide that is produced is forced out through the bent tube into the open tube, which contains a solution of bromothymol blue (BTB). Carbon dioxide causes chemical reactions that result in a color change in BTB. Differences in the time required for this color change provide an indication of relative rates of carbon dioxide production.

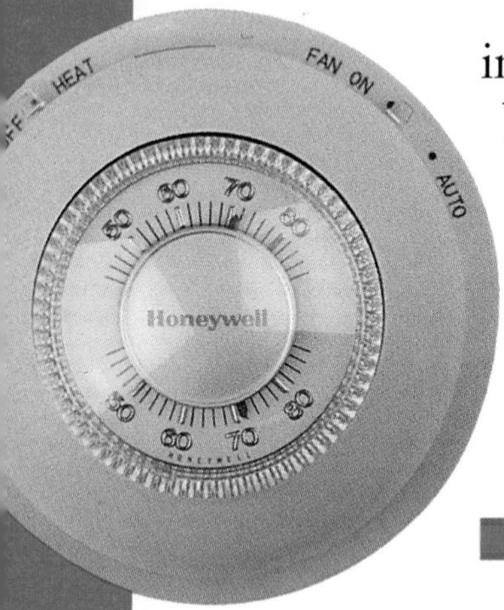

PREPARATION

Problem
Do cold temperatures slow down yeast metabolism? Does warmth speed it up?

Hypotheses
What is your group's hypothesis? One hypothesis might concern the specific effects of cold or heat on yeast metabolism.

Objectives
In this Biolab, you will:
- **Measure** the rate of yeast metabolism using a BTB color change as a rate indicator.
- **Compare** the rates of yeast metabolism at several temperatures.

Possible Materials
bromothymol blue solution (BTB)
straw
small test tubes (4)

large test tubes (3)
one-hole stoppers for large test
  tubes with glass tube inserts (3)
yeast/molasses mixture
water/molasses mixture
water/yeast mixture
test-tube rack
250-mL beakers (3)
ice cubes
Celsius thermometer
hot plate
50-mL graduated cylinder
glass-marking pencil
10 cm rubber tubing (3)
aluminum foil

Safety Precautions 🚫 ♨ ⚠
Be careful in attaching rubber tubing to the glass tube inserts in the stoppers. Avoid touching the top of the hot plate. Wash your hands carefully after cleaning out test tubes at the end of your experiments.

1. Decide on ways to test your group's hypothesis.
2. Record your procedure, and list the materials and amounts of solutions that you plan to use.
3. Design and construct a data table for recording your observations.
4. Pour 5 mL of BTB solution into a test tube. Use a straw to blow gently into the tube until you see a series of color changes. Cover this tube with aluminum foil, and set it aside in a test-tube rack. Record your observations of the color changes caused by carbon dioxide in your breath.

Check the Plan

Discuss the following points with other group members to decide

the final procedure for your experiment.

1. What data on color change and time will you collect? How will you record your data?
2. What variables will have to be controlled?
3. What control will you use?
4. Assign tasks for each member of your group to be carried out during this investigation.
5. **Make sure your teacher has approved your experimental plan before you proceed further.**
6. Carry out your experiment.

30 mL Yeast/molasses solution

5 mL Bromothymol blue solution

ANALYZE AND CONCLUDE

1. **Checking Your Hypotheses** Explain whether your data support your hypothesis. Use specific experimental data to support or reject your hypothesis concerning temperature effects on the rate of yeast metabolism.
2. **Making Inferences** Explain what you infer about the function of molasses in this experiment.
3. **Identifying Variables** Describe some variables that had to be controlled in this experiment. How were they controlled?

4. **Drawing Conclusions** Describe the control used in your experiment and how your experimental results enabled you to make conclusions about the effect of temperature on yeast metabolism. Did your experiment clearly show that differences in rates of yeast metabolism were due to temperature differences?

Going Further

Project To carry this experiment further, you may wish to examine the effects of temperatures as warm as 45°C on the rate of yeast metabolism. Formulate a hypothesis, and suggest an appropriate control for your experiment. Use data from your experiment to make conclusions about the rate of yeast metabolism at a very warm temperature.

How are new antibiotics discovered?

In 1928, Alexander Fleming discovered that a *Penicillium* mold interfered with the growth of bacteria he was studying. This chance observation later led to the discovery of the antibiotic penicillin. Biologists continue to investigate the effects of various fungi on the growth of bacteria in hopes of finding new antibiotics that can be used in human and veterinary medicine.

Analysis

Study the experimental procedures followed in these three sets of experiments.

1. A number of small samples of soil were placed on culture dishes containing soil bacteria and nutrients needed for the bacteria to grow. The cultures were examined after a few days for inhibition zones—areas in which bacteria failed to grow around any of the soil samples.

2. A number of small samples of soil were placed on culture dishes containing bacteria that can infect humans along with nutrients needed for the bacteria to grow. The cultures were examined after a few days for inhibition zones around any of the soil samples.

3. Soil samples were placed in culture dishes that contained nutrients required for the growth of a variety of fungi. Different fungi that grew on these dishes were then isolated, and a sample of hyphae from each fungus was placed on a culture dish in which possible human-infecting bacteria were growing. The bacterial cultures were examined after a few days for inhibition zones around any of the fungus samples.

Thinking Critically

Analyze each of the experiments described above. Explain what information the results of each experiment would give you. Explain how you think each experiment would or would not be helpful in discovering new antibiotics useful in human or veterinary medicine.

Mutualism with Fungi: Mycorrhizae and Lichens

Most trees live in a mutualistic association with fungi. A **mycorrhiza** is a symbiotic relationship in which a fungus lives in close contact with the roots of a plant partner. Most of the fungi that take part in mycorrhizae are basidiomycotes, but some zygomycotes also form these important relationships.

A beneficial partnership

How does a plant benefit from a mycorrhizal relationship? The fungal hyphae increase the amount of nutrients that move into the plant by increasing the absorptive surface of the plant's roots, *Figure 23.14.* Fine, threadlike hyphae from the fungus surround and often grow into the plant's roots without harming them. Phosphorus, copper, and other minerals in the soil are absorbed by the hyphae and then released into the roots. The fungus also may help to maintain water in the soil around the plant. The fungus in a mycorrhiza benefits by receiving organic nutrients, such as sugars and amino acids, from the plant. It has been suggested that evolution of trees has been positively influenced by the presence of mycorrhizae.

mycorrhiza:
mykes (GK) fungus
rhiza (GK) root
Mycorrhizae are fungi that live in close association with the roots of a plant.

Figure 23.14

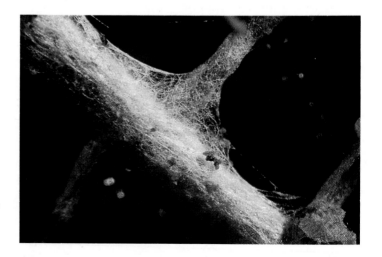

A mycorrhiza is formed with a tree root and a fungus. White strands of fungal hyphae encase the root and help supply it with mineral nutrients.

Figure 23.15

Lichens can be found in a variety of different shapes.

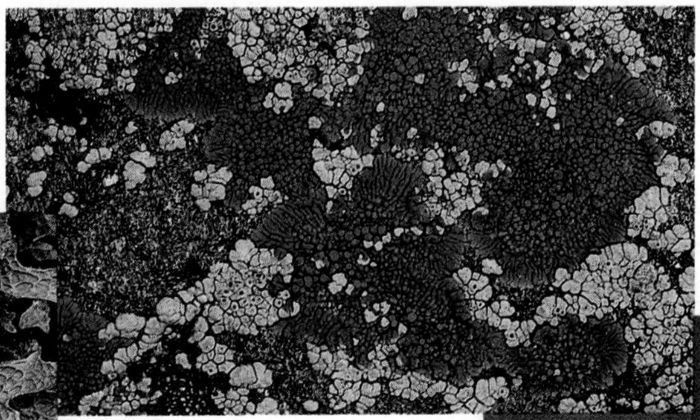

Others form crustlike growths on bare rocks and stone walls.

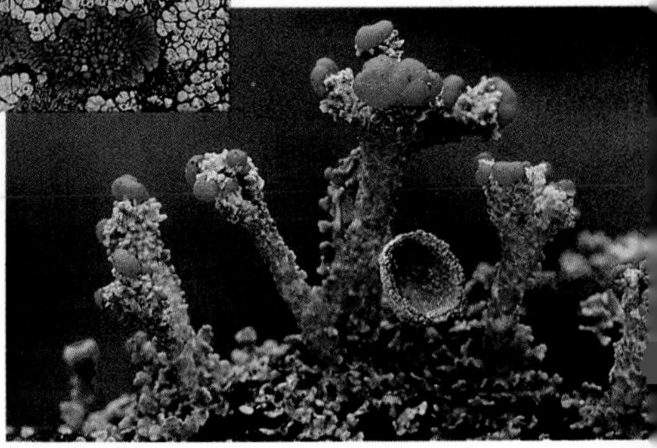

▲ **Some lichens resemble leaves, like this one growing on a dead twig.**

▶ **Each stalk of a British soldier lichen is about 3 cm tall.**

In addition to trees, 80 to 90 percent of all plant species have mycorrhizae associated with their roots. Mycorrhizae are extremely important in agriculture. Having this relationship with fungi makes plants larger and more productive. Without mycorrhizae, most plants do not grow well. In fact, some species cannot survive without them. Orchid seeds, for example, will not germinate without a mycorrhizal fungus to provide them with water and nutrients.

Lichens

It's sometimes hard to believe that the orange, green, and black blotches you see on rocks, trees, and stone walls are alive, *Figure 23.15.* They may look like flakes of old paint or dried moss, but they are lichens. A **lichen** is a symbiotic association between a fungus, generally an ascomycote, and a green alga or cyanobacterium. The fungus portion of a lichen forms a dense web of tangled hyphae in which the algae or cyanobacteria grow. Together, the fungus and its microscopic, photosynthetic partners form a spongy structure that looks like a single organism.

About 20 000 species of lichens exist. They range in size from less than 1 mm to several meters in diameter. Lichens grow slowly, increasing in diameter only 0.1 to 10 mm per year. Very large lichens are thought to be thousands of years old.

Living together

Lichens need only light, air, and a few minerals to grow. The photosynthetic partner in the lichen provides itself and the fungus with food. The fungus, in turn, helps retain moisture, provides the algae with water and minerals absorbed from rainwater and air, and protects them from intense sunlight.

Found worldwide, lichens live in some of the harshest, most barren habitats on Earth. Lichens are true pioneer species. They are among the first to colonize a barren area. You can find lichens in arid deserts, on bare rocks exposed to the blazing sun or bitter-cold winds, and just below the snow line on mountain peaks. On the arctic tundra, where large plants are scarce, lichens are the dominant form of vegetation. Caribou and musk oxen graze on lichens there, much like cattle graze on grass elsewhere.

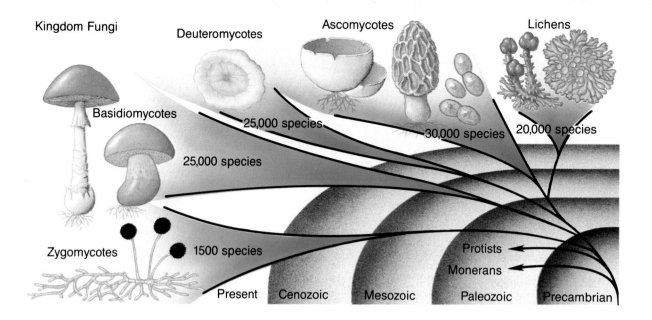

Figure 23.16

The radiation of phyla of fungi on the Geologic Time Scale shows their relationships. Which group showed up latest in the fossil record?

Origins

Many researchers hypothesize that the ascomycotes and the basidiomycotes evolved from a common ancestor and the zygomycotes evolved from an earlier common ancestor, *Figure 23.16*.

Fossils can provide clues as to how organisms evolved, but fossils of fungi are rare due to their composition of soft materials. The oldest fossils that have been identified as definitely being those of fungi are between 450 and 500 million years old.

Connecting Ideas

Fungi have evolved, along with some bacteria and protists, as the major decomposers of the biosphere. Much of what fungi decompose is plant material. As fungi break down organic matter, they release nitrogen compounds and other nutrients into the soil, where these molecules can be picked up and used again by plants. How do plants utilize the nutrients made available by fungi and other decomposers? How have plants been so successful in adapting to their environments?

Section Review

Understanding Concepts

1. What happens between the time when a basidiospore germinates and a button forms?
2. How does a lichen obtain its nutrients?
3. Describe how each partner benefits in a mycorrhizae relationship.

Thinking Critically

4. You are working with a team of archaeologists on Easter Island in the Pacific Ocean. Huge stone statues were carved and erected on the island by a now-vanished civilization. How might you use lichens to help determine when the statues were carved?

Skill Review

5. **Observing and Inferring** Describe how lichens can be used as biological indicators of air quality. For more help, refer to Thinking Critically in the *Skill Handbook*.

Mycological Mystery

Mushroom Disappearance Puzzles Scientists

DAILY HERALD   OCTOBER, 1993

People throughout Europe are asking the same question this autumn: Where have all the mushrooms gone? Mushroom harvesters are reporting serious declines in many species of fungi, a vanishing act that both puzzles and worries scientists.

Chanterelle, truffle, porcini, morel, shiitake—these are just some of the varieties of edible mushrooms prized by gourmets for their heady aromas and earthy flavors. In Italy, fresh porcini mushrooms are often grilled and savored as a main course, not unlike steak. And the French are fiercely proud of their native truffles, rooted out of the earth by specially trained pigs and sold for extravagant prices in the marketplaces of Provence.

Mass extinction?   However, all of this is changing. Mycologists are tracking a serious decline in mushroom populations throughout Europe—so serious that an ecologist at the University of Rochester in New York calls the decline a mass extinction. Although mushroom growth and harvesting is not monitored in the United States as closely as it has been in Europe throughout this century, scientists believe that mushroom species in this country also may be vanishing.

Symbiosis threatened   More than ethnic cuisine will suffer if the mushroom decline isn't halted. Mushrooms and the trees around which they grow have a symbiotic relationship; that is, they enable each other to remain healthy and strong. Trees provide mushrooms with carbohydrates. In turn, mushrooms provide trees with water and minerals.

Different Viewpoints

Several explanations have been offered for the mushroom decline, but the most credible of these differing scenarios have one common theme: pollution. Air pollution is believed to be the culprit. Research shows that as levels of nitrogen, sulfur, and ozone have risen in Europe, mushroom populations have fallen.

Bad fertilizer In Holland, agricultural practices may be contributing to the decline. Dutch farmers use large quantities of nitrogen fertilizer. Some of the nitrogen is blown by the wind into the upper atmosphere, where it mixes with moist air and returns to Earth in rainfall.

Chanterelle

Exactly how the nitrogen from air pollution and fertilizer affects mushroom growth is questioned. The nitrogen may affect the mushrooms directly by poisoning the soil. Or, it may first sicken the tree which, in a weakened state, might well be unable to fulfill its part of the symbiotic bargain—nourishing the mushrooms.

INVESTIGATING   the Issue

1. **Writing About Biology** Prepare a brief report on inedible mushrooms. How many species exist? Where do they grow? What makes them toxic?
2. **Research the Issue** Investigate the most recent studies on the effects of acid rain. Have scientists discovered any new links to polluted rain and the declining mushroom population?

Reviewing Main Ideas

23.1 The Life of Fungi

- The basic structural units of fungi are hyphae that grow to form mycelia. Hyphae have chitin-containing cell walls that may or may not be divided into individual cells by septa.
- Fungi are heterotrophs that carry out extracellular digestion. A fungus may be a saprobe, a parasite, or a mutualist in a symbiotic relationship with another organism.
- Most fungi produce asexual and sexual spores. The way in which sexual spores are produced is the basis for fungal classification.

23.2 The Diversity of Fungi

- Fungi are classified based on reproductive structures. Zygomycotes form asexual spores in a sporangium. They reproduce sexually by producing zygospores.
- Ascomycotes reproduce asexually by conidia and sexually by forming ascospores in saclike asci.
- In basidiomycotes, sexual spores are borne on club-shaped basidia.

- Deuteromycotes reproduce only asexually, usually by producing conidia.
- Fungi play an indispensable role in decomposing organic material and recycling nutrients.
- Lichens, symbiotic associations of fungi and algae or cyanobacteria, can survive in a variety of inhospitable habitats.
- Certain fungi associate with plant roots to form mycorrhizae, a relationship in which both plants and fungi benefit.

Key Terms

Write a sentence that shows your understanding of each of the following terms.

ascospore

ascus	hypha
basidiospore	lichen
basidium	mycelium
budding	mycorrhiza
chitin	rhizoid
conidiophore	sporangium
conidium	stolon
gametangium	zygospore
haustorium	

Understanding Concepts

1. Your neighbor is pulling up mushrooms that are growing in his lawn. He tells you that he heard mushrooms won't come back again if they are quickly removed. What would you tell him?
2. How does a fungus acquire the nutrients it needs from its immediate environment?
3. Describe a mycelium, and explain the relationship between a mycelium and mushrooms that appear on a lawn.
4. Explain how a zygomycote such as *Rhizopus* reproduces sexually.
5. What is the advantage of producing a zygospore?

Interpreting a Photomicrograph

6. The photomicrograph below shows what the fungus looks like under a microscope. To what divisions could this fungus belong? What additional information would you need before being able to place this fungus in the proper division?

Relating Concepts

7. Make a concept map that relates the following terms and phrases. Supply the appropriate linking words for your map.

   reproductive structures, ascus, ascospores, basidium, basidiospores, conidia, conidiophores, phyla, Kingdom Fungi

Applying Concepts

8. While hiking along a trail through a woods near your rapidly growing city, you notice that there are fewer lichens on the rocks and trees than there used to be. How might you interpret this change in the forest ecosystem?
9. In an effort to have healthy lawns, some home owners apply chemicals to their grass to kill fungi. How might these practices result in a nutrient-starved lawn?
10. Why is being able to produce spores that can be dispersed far from their source such an important adaptation for fungi?
11. When you transplant flowers, shrubs, or trees, why do you think it is a good idea to leave the soil intact around a plant's roots?

Biology & Society

12. How might limiting the harvesting of wild mushrooms help protect the decline of various species? Would this have any effect if the cause of their decline is air pollution?

Thinking Critically

Relating Cause and Effect

13. When making bread, yeast is usually activated by combining it with sugar and warm water, then added to the rest of the ingredients. The result is a nicely shaped loaf of bread that has risen well due to the bubbles of carbon dioxide released by the yeast cells. How do you think that mixing yeast with sugar and ice water would affect the way the bread turns out?

Interpreting Data

14. **Biolab** The following experimental results were presented by students after they performed experiments on the metabolic activity of yeasts at various temperatures.

Test Tube #	Temperature of yeast solution	Time elapsed until color change
1	2° C	no color change
2	25° C	44 minutes
3	37° C	22 minutes

   (a) At what temperature was the yeast most active? Explain your choice.
   (b) How was temperature related to the rate of carbon dioxide production?

Making Inferences

15. **Minilab** Suppose you exposed a moistened slice of commercially prepared bread that contained preservatives to the air and then sealed it in a plastic bag. How do you think mold growth in this bread would compare with mold growth on homemade bread?

Making Predictions

16. **Minilab** What do you think would be the result if you used fresh, light-colored mushroom caps in making a spore print?

Connecting to Themes

17. **Systems and Interactions** Both fungi and animals are classified as heterotrophic organisms, but they differ in the way in which they interact with other kinds of organisms in an ecosystem. Contrast the interactions of fungi and plants with the interactions of animals and plants.
18. **Homeostasis** How does the fungus of a lichen contribute to the homeostasis of its photosynthetic partner?

UNIT
7 Plants

O rchids represent the most
advanced level of evolutionary form and function in flowering
plants. Orchids are the largest and
most diverse family of plants,
varying widely in structure, color,
size, and fragrance.

Most orchids are found in the
tropics, but what makes them so
successful is that they have adapted
to a variety of habitats, from the
sandy soils near oceans to the acidity of mountain bogs. For each
locality, you will find an orchid
species with a set of characteristics
best suited for that environment.
A study of orchids would reveal
most of the structural and physiological adaptations characteristic
of the plant kingdom.

**Orchids provide flower lovers
with the widest variety of choices
to adorn a vase, dress, or garden.
How is the structure of a flower
related to the way a plant
species reproduces?**

Many plants live in areas, such as deserts, that would be inhospitable to most organisms. What adaptations enable plants to inhabit such environments?

Plants make food by capturing the energy in sunlight. What structural features allow plants to make the most of this energy?

Unit Contents

581

The damp, misty bank of a rushing stream is the perfect environment for a lush growth of mosses. Mosses may be similar to some of the first plants that lived on land. Like all members of the plant kingdom, mosses probably evolved from green algae that lived in ancient swamps and oceans.

One of the challenges plants faced when they moved to land was the need for water. Plants have evolved a variety of adaptations for obtaining and conserving water. The roots of giant redwood trees obtain water from deep, underground reservoirs. The waxy covering on the leaves of a magnolia tree helps retain moisture. These are just a few of the plant adaptations that enable them to live on land.

Individual moss plants are usually no more than a few centimeters in height, and most parts of the plant are only one cell thick. Giant redwoods, on the other hand, have a circumference of up to 30 m and soar to heights of 90 m or more.

This magnolia tree absorbs water through organized tissues in its roots, which then transport the water through its trunk to its branches and leaves.

24.1 Adapting to Life on Land

When you studied ecology in Chapter 4, you learned that plants are such an important part of life on Earth that they are used to define biomes, ecosystems, and communities. Plants are a major group of Earth's producers. They trap the energy of sunlight and store it as chemical energy in food, which supplies the fuel that makes all life possible. Multitudes of organisms, including humans, rely on plants for both food and shelter. Because plants take in carbon dioxide and release oxygen during photosynthesis, they are responsible for maintaining the supply of oxygen needed by most organisms for respiration.

Origins of Plants

What is a plant? A plant is a multicellular eukaryote, with cells surrounded by cell walls made of cellulose and with a waxy waterproof coating called a cuticle.

A billion years ago, plants had not yet begun to appear on land. No ferns, mosses, trees, grasses, or wild-flowers existed. The land was barren except for some algae at the edges of inland seas and oceans. However, the shallow waters that covered much of Earth's surface at that time were teeming with bacteria; algae and other protists; and simple animals such as corals, sponges, jellyfish, and worms. Among these organisms were those green algae that would slowly become adapted to life on land.

The first plants began to appear around 500 million years ago. These early plants may have looked something like present-day mosses. As you learned in Chapter 3, mosses are pioneer organisms that help turn bare rock into rich soil. Mosslike plants might have helped lead the way for other plants by being among the first soil builders. There is no fossil record of the first land plants, in part because their tissues were probably delicate and decayed too easily to

Figure 24.1

The oldest fossil psilophyte is probably more than 400 million years old. The plant was made up of leafless stems without roots. Underground portions of the stem bore rootlike hairs that absorbed water and nutrients from the soil.

Sporangia

become fossilized. The earliest known plant fossils are those of plants called psilophytes, some of which still exist today, *Figure 24.1.*

All plants probably evolved from filamentous green algae that dwelt in the ancient oceans. Both green algae and plants have cell walls that contain cellulose. Both groups have the same types of chlorophyll used in photosynthesis. Both algae and plants store food in the form of starch. All other major groups of organisms store food in the form of glycogen and other complex sugars.

Adaptations of Plants

Life on land has advantages as well as challenges. A filamentous green alga floating in a pond does not need to conserve water. The alga is completely immersed in a bath of water and dissolved nutrients, which it can absorb directly into its cells. For most land plants, the only available supply of water and minerals is in the soil, and only the portion of the plant that penetrates the soil can absorb these nutrients.

Algae reproduce by releasing unprotected gametes into the water, where fertilization and development take place. The gametes of land plants are protected from drying out by a waterproof covering of thick-walled cells. Land plants must also withstand the forces of wind and weather and be able to grow upright against the force of gravity. Over the past 500 million years or so, plants have developed a huge variety of adaptations that reflect both the challenges and advantages of living on land.

Chemistry

Rayon: A Natural Fiber

Look at the labels of your favorite clothes. Chances are they are made from natural fabrics—cottons, linen, wool, and silk. Cotton comes from the cotton plant. Linen is made from the fibers of the flax plant. Wool is the hair of sheep. Silk is the fiber that silkworm moths make to create their cocoons. Natural fibers "breathe," absorbing moisture so they feel cooler, and they soften with each washing.

What is rayon?   Rayon is a natural fabric, too. In the 1800s, a silk worm epidemic nearly ended the French silk industry; a prize was offered to anyone who could produce an artificial silk. Louis Pasteur's assistant, Chardonnet, created cellulose nitrate from soft wood. This artificial silk, called rayon, was abandoned in the late 1890s due to its flammability, but further research led to the discovery that fabrics could be made from the cellulose fibers that coat all cell walls of plants.

How rayon is made   The most common form of rayon made today is xanthate rayon; it is chemically identical to cotton. It is made by dissolving cotton seed cellulose and forcing the solution through small holes in a nozzle. The extruded cellulose fibers are long, smooth filaments like silk and give rayon a similar texture and shine.

Magnification: 50×

Other uses for rayon   Most lightweight, sheer fabrics in use today, such as satin, are made from rayon. Rayon holds dye well and is found in carpet fibers, coverings for home furnishings, surgical gauze, and tire cords.

CONNECTION TO Biology

Use a microscope to compare textures of natural fibers from plants, such as cotton and linen. How might fiber texture affect the properties of the cloth?

Plant Survival in Arid Conditions

Sahara. Kalahari. Gobi. Mojave. These are the names of some of Earth's great deserts and landscapes where the annual rainfall is less than 25 cm, where the mercury may soar to more than 100°F during the day and plunge to below freezing at night, and where drought conditions may persist for years at a time.

Living with adversity

Plants called xerophytes can survive in hot, dry climates because they have evolved adaptations that ensure their continued existence. Take the ocotillo, for example. This desert shrub seems to be dead during the dry season. When the rains come, the ocotillo springs into action. Leaves quickly sprout and begin photosynthesis. Flowers appear and produce seeds. When the water has evaporated, the ocotillo once again becomes dormant, releasing its seeds into the desert soil. The process will be repeated with the next downpour.

Other survival strategies

Some desert plants obtain what little water there may be by extending their roots deep into the soil. The roots of the mesquite tree, for instance, may extend to depths of more than 75 m. Other plants keep their roots near the surface, but extend them out over great distances to gather as much moisture from as wide an area as possible after a rain.

Succulents, such as the barrel cactus, store water in their fleshy stems or leaves. The barrel cactus swells with water after a rainfall and then slowly shrinks as it replenishes water lost to evaporation. Some succulents can survive for years on the stored water from a single rainstorm.

Most plants lose water through their leaves, so desert plants have leaves that are adapted to desert conditions. Some leaves are coated with a thick, waxy cuticle that slows evaporation. Other plants fold up their leaves or turn them sideways to reduce the leaf area exposed to the sun. Still other plants drop their leaves during dry periods.

Thinking Critically

What is the adaptive value of a green stem and no leaves to a cactus? Explain your answer.

Preventing water loss

If you run your fingers over the surface of an apple, a maple leaf, or the stem of a houseplant, you'll find that it feels smooth and slightly slippery. Most fruits, leaves, and stems are covered with a protective, waxy layer called the **cuticle.** Waxes and oils are lipids, and you read in Chapter 7 that water does not dissolve in lipids. The waxy cuticle helps prevent the water in the plant's tissues from evaporating into the atmosphere.

Figure 24.2

The opening and closing of a stoma is regulated by guard cells that surround the pore.

▼ Guard cells are balloonlike cells with walls that are attached at the ends. During the day, water moves into the guard cells by osmosis and they swell, opening the stoma.

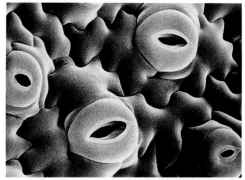

Magnification: 300×

▼ At night, the guard cells lose water into surrounding cells and straighten, partially closing the pore.

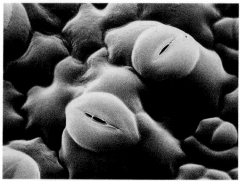

Magnification: 300×

Figure 24.3

Leaves take advantage of the plentiful supply of sunlight and carbon dioxide available to land plants.

Most leaves are thin and allow sunlight to penetrate throughout the organ's tissues.

Most plants have an enormous number of leaves, which provide a large surface area for trapping sunlight and exchanging gases during photosynthesis.

There are openings called **stomata** in the cuticle of the leaf that allow the exchange of gases. The pores of stomata open during the day while photosynthesis is taking place, *Figure 24.2*. Stomata that open during the daytime release water and oxygen and take in the carbon dioxide needed for photosynthesis. During the night, these openings partly close down to prevent too much water loss. Although the plant is coated with a waxy cuticle, plants lose up to 90 percent of the water they contain through these openings in the plant's epidermis, the outer layer of cells.

Leaves carry out photosynthesis

All the cells of a filamentous green alga carry out photosynthesis. But in most land plants, the leaves are the organs usually responsible for photosynthesis. A **leaf** is a broad, flat organ of a plant that traps light energy for photosynthesis, *Figure 24.3*. Leaves also exchange gases through their stomata. Leaves are supported by the stem and grow upward toward sunlight. They have both an upper and a lower surface.

Putting down roots

Most plants depend on the soil as their primary source for water and other nutrients. A **root** is a plant organ that absorbs water and minerals from the soil, transports those nutrients to the stem, and anchors the plant in the ground. Some roots, such as those of radishes or sweet potatoes, also accumulate starch reserves and function as organs of storage.

Mosses and their relatives have rhizoids rather than roots. Rhizoids in mosses are usually only one cell thick and do not extend far into the soil.

How do roots, stems, and leaves compare?

As land plants evolved, they developed structures that carry out different functions. In this lab, you will examine roots, stems, and leaves to identify their similarities and differences.

Procedure

1. Obtain one or more plants from your teacher.

2. Observe your plant carefully. Identify the root, stem, and leaves. Determine where the root ends and the stem begins, and where the stem ends and the leaf begins.

3. Use a hand lens or low-power microscope to observe small parts or features of each structure. Draw or take down a written description of what you see.

Analysis

What differences in shape, size, and color did you observe in the different parts of the plant? What microscopic details did you observe? What similarities and differences did you observe among structures used for photosynthesis, water absorption, and transport of food and water?

Transporting materials

Water moves from the roots of a tree to its leaves—and the sugars produced in the leaves move to the roots—through the stem. A **stem** of a plant provides structural support for upright growth and contains tissues for transporting food, water, and other materials from one part of the plant to another, *Figure 24.4*. Stems may also serve as organs for food storage. Green stems contain chlorophyll and take part in photosynthesis.

The stems of most plants contain vascular tissues made up of tubelike, elongated cells through which water, food, and other materials are transported. Plants that possess vascular tissues are known as **vascular plants.** Most of the plants you are familiar with—including pine and maple trees, ferns, rhododendrons, rye grasses, English ivy, and sunflowers—are vascular plants.

Mosses and several other small, less-familiar plants are nonvascular plants. **Nonvascular plants** are plants that do not have vascular tissues. The tissues of nonvascular plants are usually no more than a few cells thick, and water and nutrients travel from one cell to another by the relatively slow processes of osmosis and diffusion. The evolution of vascular tissues was of major importance in enabling plants to survive in the many habitats they now occupy on land. Vascular plants can live farther

Figure 24.4

Stems such as the trunk of this eucalyptus tree provide support that enables plants to grow to great heights.

away from water than nonvascular plants. Also, because vascular tissues include thickened cells called fibers that help support upright growth, vascular plants can grow much larger than nonvascular plants.

Reproductive strategies

Adaptations in land plants include the evolution of spores and seeds. These structures protect the zygote or embryo and keep them from drying out before they encounter conditions needed for growth and development. A **seed** contains an embryo, along with a food supply, covered by a protective coat. In contrast, as you learned in Chapter 22, a spore consists only of a single haploid cell with a hard, outer wall.

In spore-releasing plants, which include mosses and ferns, the sperm swim through a film of water to reach the egg. In seed-producing plants, which include all conifers and flowering plants, sperm are able to reach the egg without swimming through a film of water. This difference explains why spore-releasing plants require more moist habitats than most seed producers.

Alternation of generations

The lives of all plants consist of two alternating stages, or generations. The gametophyte generation of a plant is responsible for development of gametes. All cells of the gametophyte, including the gametes, are haploid (n). The sporophyte generation is responsible for the production of spores. All cells of the sporophyte are diploid ($2n$). The spores are produced in the sporophyte plant body by meiosis, and therefore are haploid (n).

ThinkingLab   Make a Hypothesis

What is the function of trichomes?

Trichomes are hairlike or scaly outgrowths of the epidermis of stems and leaves of some plants. Leaves with trichomes can feel prickly, sticky, fuzzy, or woolly.

Analysis

Examine the different kinds of trichomes on the leaves shown here and think about what their function might be.

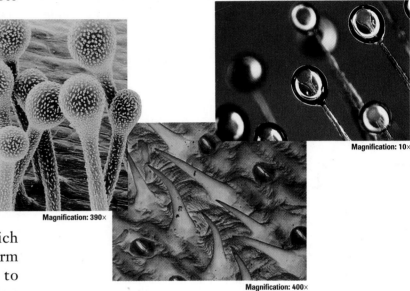

Magnification: 390×

Magnification: 10×

Magnification: 400×

Thinking Critically

Propose a hypothesis about the adaptive value of these structures to the survival of each plant.

All plant life cycles include the production of spores. In spore-producing plants such as ferns, the spores have hard outer coverings and are released directly into the environment, where they grow into gametophytes. In other plants, such as conifers and wildflowers, the spores are retained by the parent plant and develop into gametophytes of only a few cells retained within the sporophyte. These plants, usually called seed-producing plants, release the new sporophytes into the environment in the form of seeds.

Phylogeny of Plants

Many changes have taken place since the first plants became adapted to life on land. Landmasses have moved from place to place over Earth's surface, climates have changed, and bodies of water have formed and disappeared. Hundreds of thousands of plant species have evolved, and countless numbers of these have become extinct as conditions continually changed. Plants have adaptations that enable them to survive in almost every type of environment that is found on Earth. Taxonomists have classified plants into ten divisions. Examples of spore-producing plant divisions are shown in *Figure 24.5* and *Figure 24.6.*

Figure 24.5

The relationships of divisions of spore-producing plants on the Geologic Time Scale show that bryophytes and spore-producing plants are closely related.

Bryophyta

Bryophytes are nonvascular, spore-producing plants that include mosses and liverworts. These groups of organisms are, for the most part, small and limited to moist habitats. Their leaves are only one to two cells thick. Spores of the bryophytes are formed in capsules.

Psilophyta

Psilophytes, also known as whisk ferns, consist of thin, green, leafless stems and are thought to represent the first land-dwelling, spore-producing vascular plants. The stem is covered with small, leaflike scales. Most of the 30 species of psilophytes are tropical or subtropical, although one genus is found in the southern United States.

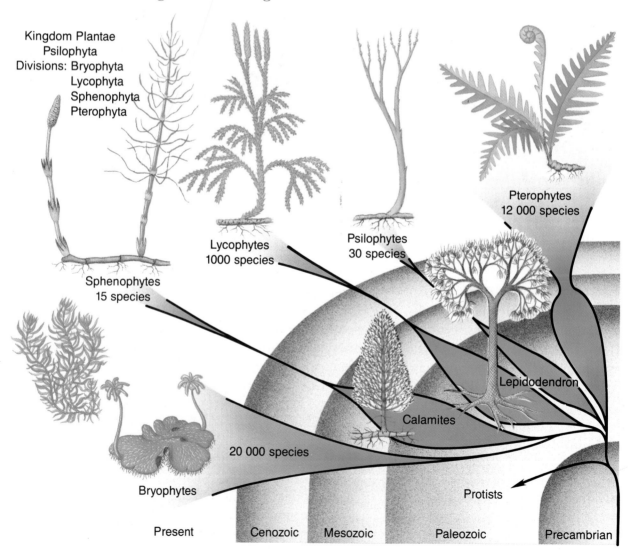

Kingdom Plantae
Psilophyta
Divisions: Bryophyta
  Lycophyta
  Sphenophyta
  Pterophyta

Pterophytes
12 000 species

Lycophytes
1000 species

Psilophytes
30 species

Sphenophytes
15 species

Lepidodendron

Calamites

20 000 species

Bryophytes

Protists

Present   Cenozoic   Mesozoic   Paleozoic   Precambrian

Lycophyta

Lycopods, the club mosses, are simple, spore-producing vascular plants adapted primarily to moist environments. Those species that exist today are only a few centimeters high, but their ancestors grew as tall as 30 m and formed a large part of the vegetation of Paleozoic forests. These ancient forests are now used by people in the form of coal.

Sphenophyta

Sphenophytes, the horsetails, are spore-producing vascular plants. They have hollow, jointed stems surrounded by whorls of scalelike leaves. Although primarily a fossil group, about 15 species of sphenophytes exist today. All present-day horsetails are small, but their fossil relatives were treelike.

Pterophyta

Pterophytes, ferns, are the most well-known and diverse group of spore-producing vascular plants. Ferns were abundant in Paleozoic and Mesozoic forests. Ferns have leaves that vary in length from 1 cm to 500 cm. Most ferns grow in the tropics.

► Tree ferns like this *Cyathea arborea* can still be found growing in damp, tropical forests.

Figure 24.6

The plant kingdom includes five divisions of spore-producing plants. How many of these plants can you classify into the correct division?

◄ *Selaginella,* a spike moss, produces two types of spores that develop into male and female gametophytes.

◄ *Equisetum* has true roots, stems, and leaves, but the stems are hollow and appear jointed.

▼ *Marchantia* is a common liverwort found on damp soil. It has air chambers on the upper surface.

◄ *Psilotum* sporophytes have simple stems but no leaves. They have underground rhizomes that produce rhizoids.

Figure 24.7

Plants classified into the five divisions of seed-producing plants produce seeds covered by tough, protective seed coats. In gymnosperms, the seeds are held on woody scales that form cones. In angiosperms, seeds are protected inside a fruit.

▲ *Welwitschia mirabilis* is found in harsh desert environments in Africa. The leaves of this plant may be 2 m long.

▲ Cycads are often mistaken for ferns or small palm trees.

▶ *Ginkgo biloba,* the maidenhair tree, is no longer found in the wild, although it continues to be cultivated in many countries, including the United States.

Examples of the seed-producing plant divisions are shown in *Figure 24.7.*

Cycadophyta

Cycads were abundant during the Mesozoic Era. Today, there are about 100 species of cycads, all of them short, palmlike trees with scaly trunks. Seeds are produced in cones, with male and female cones on separate trees. Cones of cycads may be as long as 1 m.

Gnetophyta

There are three genera of gnetophytes, each of which is quite distinct. *Gnetum* includes species of trees and climbing vines, *Ephedra* includes shrubby species, and *Welwitschia* is found only in South Africa. *Welwitschia* has a short stem that grows as a large, shallow cap.

The leaves grow from the base of the stem. These plants grow in the desert and can live to be 100 years old.

Ginkgophyta

This division has only one living species, *Ginkgo biloba*, a distinctive tree with fan-shaped leaves. Like cycads, ginkgos have separate male and female trees. The seeds have an unpleasant smell, so ginkgos planted in city parks are usually male trees. Ginkgos are hardy and resistant to insects and to air pollution.

◀ Wildflowers can be found in nearly every environment on Earth. There are more species of anthophytes than any other plant division. This is a wildflower called chickory.

◀ *Pinus banksiana,* the jack pine, keeps its cones closed until a fire passes over them, an example of an adaptation to extreme conditions.

Coniferophyta

These are the conifers—cone-bearing trees such as pine, fir, cypress, and redwood. Conifers are vascular plants that produce naked seeds in cones. Species of conifers can be identified by the characteristics of their needlelike or scaly leaves. The oldest trees in the world are members of this plant division.

Anthophyta

Anthophytes, commonly called the flowering plants, are the largest, most diverse group of plants living on Earth. Fossils of the Anthophyta only date to the Cretaceous period, 130 million years ago. Anthophytes produce seeds enclosed in a fruit. This division has two classes: the mono-cotyledons and dicotyledons.

Section Review

Understanding Concepts

1. What is the primary difference between seeds of conifers and anthophytes?
2. Explain how development of the cuticle, stomata, and the vascular system influenced the evolution of plants on land.
3. List the sequence of events involved in the alternation of generations in land plants.

Thinking Critically

4. Explain why the alternation of generations is of adaptive value for plants living on land.

Skill Review

5. **Comparing and Contrasting** How does the life of a spore-producing, nonvascular plant differ from the life cycle of a flowering plant? For more help, refer to Thinking Critically in the *Skill Handbook.*

Section Preview

Objectives

Identify the structures of a typical bryophyte.

Explain the life cycle of a moss or liverwort.

Key Terms

protonema
antheridium
archegonium
vegetative reproduction
gemmae

An observant hiker in this shady forest is sure to come across patches of soft, feathery mosses covering soil, stones, rotting wood, or tree bark with a velvety layer of green. The hiker might also notice shiny liverworts growing along the stony bank of a stream. Mosses and liverworts are bryophytes. Bryophytes are spore-producing plants that have no vascular tissue. They usually live in moist, cool environments.

Characteristics of Bryophytes

Nonvascular plants are not as common or as widespread as vascular plants because they must carry on photosynthesis, reproduction, and other life functions where there is a steady supply of moisture. Adequate water is not available everywhere, so most bryophytes are limited to moist habitats such as by streams and rivers or in humid tropical forests. Lack of vascular tissue also limits the size of bryophytes. They cannot compete with neighboring vascular plants, which can easily overgrow them and cut them off from sunlight and gases. But even with these limitations, bryophytes are successful in

Figure 24.8

Mosses have a central stem surrounded by small, thin leaves.

▶ **Brown stalks and spore capsules of the sporophyte generation can be seen growing from the green, leafy gametophyte of this moss.**

habitats with adequate water. The division Bryophyta includes mosses and liverworts.

Mosses grow in sheltered places

Mosses are more familiar and more numerous than liverworts. Mosses are small plants with leafy stems and usually grow in dense carpets or tufts. Some have upright stems; others have creeping stems that lie along the ground or hang from steep banks or tree branches. *Figure 24.8* shows some typical mosses.

▼ Mosses with creeping stems form extensive mats that retard erosion on exposed slopes.

MiniLab

How do mosses and green algae compare?

Biologists hypothesize that the first plants to migrate to land had many similarities with filamentous green algae that live in water. In this lab, you will observe similarities and differences between mosses and algae.

Procedure

1. Make a table to compare the following observations of a moss and a filamentous green alga: overall size, structures for obtaining water and nutrients, structures for support, and structures for photosynthesis.

Magnification: 50×

2. Obtain samples of both types of organisms from your teacher.

3. Make wet mounts of the structures used for photosynthesis, and observe them under low and high power. Describe similarities and differences.

4. Make wet mounts of structures used for absorbing water and nutrients, and observe them under the microscope. Describe similarities and differences.

Analysis

How do the structures of algae and mosses compare? What evolutionary advances do mosses show over algae?

▼ This tufted moss is usually found growing on the rocks of mountain slopes. Like most mosses, this plant becomes dormant during dry spells. When rainy weather returns, the plant revives and resumes photosynthesis and reproduction.

▼ Peat moss, *Sphagnum,* is probably the most well-known moss because of its usefulness. Compressed, dead peat moss can be dried and cut into bricks for fuel. Dried peat moss absorbs large amounts of water, so florists and gardeners use it to increase the water-holding ability of soil.

▲ A leafy liverwort like this one is difficult to distinguish from a moss. Leafy liverworts have two rows of larger leaves and another row of smaller ones.

▲ A thallose liverwort has a distinctive appearance. Thallose liverworts called hornworts are of interest to biologists because each cell has just one large chloroplast.

Liverworts have a flattened appearance

Like mosses, liverworts are small plants that usually grow in clumps or masses in moist habitats. However, liverworts occur in many environments, from the Arctic to the Antarctic. Some are found in water and others in deserts. They include two groups: the thallose liverworts and the leafy liverworts, *Figure 24.9.* The body of a thallose liverwort is a thallus. It is a broad, ribbonlike body that resembles a lobed leaf. Leafy liverworts are creeping plants with three rows of flat, thin leaves attached to a stem. Most liverworts have an oily or shiny surface.

Liverworts can respond to small changes in their environments, and as a result, they exhibit a wide variety of forms. The same species may be found as a compact form in its normal habitat, but have a more slender and elongated form in an environment with more moisture or diffuse light.

Reproduction in Bryophytes

Like all plants, the life cycle of bryophytes includes an alternation of generations between the diploid sporophyte and the haploid gametophyte. However, bryophytes are the only plant division in which the gametophyte generation is dominant.

Mosses produce a protonema

In mosses, the haploid spore germinates to form a structure called a protonema. The **protonema** is a small, green filament of cells that develops into either a male or female gametophyte or a gametophyte containing both kinds of reproductive structures. Liverworts have no protonema; the spore germinates and grows directly into the plant body. In both mosses and liverworts, gametophytes produce two kinds of reproductive structures, male and female. The **antheridium** is the reproductive structure in which sperm are produced. The **archegonium** is the reproductive structure in which eggs are produced.

Life Cycle of a Moss

*A*lthough mosses alternate between the haploid gametophyte and the diploid sporophyte, it is fairly easy to find huge carpets of mosses made up only of gametophytes.

Haircap moss

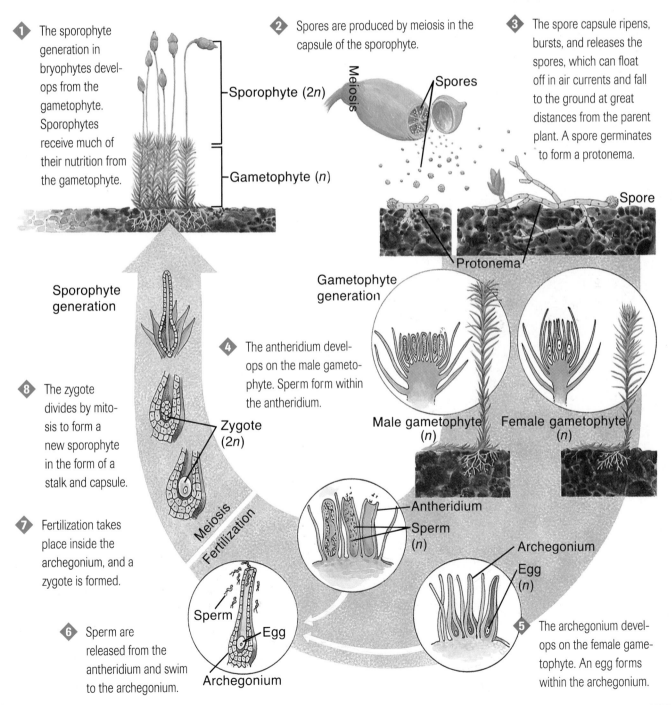

1 The sporophyte generation in bryophytes develops from the gametophyte. Sporophytes receive much of their nutrition from the gametophyte.

Sporophyte (2*n*)

Gametophyte (*n*)

2 Spores are produced by meiosis in the capsule of the sporophyte.

Meiosis

Spores

3 The spore capsule ripens, bursts, and releases the spores, which can float off in air currents and fall to the ground at great distances from the parent plant. A spore germinates to form a protonema.

Spore

Protonema

Sporophyte generation

Gametophyte generation

8 The zygote divides by mitosis to form a new sporophyte in the form of a stalk and capsule.

Zygote (2*n*)

4 The antheridium develops on the male gametophyte. Sperm form within the antheridium.

Male gametophyte (*n*)

Female gametophyte (*n*)

Meiosis

Fertilization

Antheridium

Sperm (*n*)

Archegonium

Egg (*n*)

7 Fertilization takes place inside the archegonium, and a zygote is formed.

Sperm

Egg

Archegonium

6 Sperm are released from the antheridium and swim to the archegonium.

5 The archegonium develops on the female gametophyte. An egg forms within the archegonium.

24.2 Bryophytes **597**

BioLab

Most plants have life cycles involving *alternation of generations.* This term means that a gametophyte, or gamete-bearing genera-

Alternation of Generations in Mosses

tion, alternates with a sporophyte, or spore-bearing generation. In more complex plants, the gametophyte is greatly reduced in size and the sporophyte predominates. In mosses, both generations are visible to the naked eye, although the gametophyte is larger. A moss gametophyte is the conspicuous, leafy, green plant you are most familiar with. The sporophyte is the stalklike structure with a capsule at the tip.

PREPARATION

Problem
What are the similarities and differences of the gametophyte and sporophyte reproductive structures in mosses?

Objectives
In this Biolab, you will:
- **Distinguish** between gametophyte and sporophyte generations in a moss plant.
- **Diagram** the alternation of generations in mosses.

Materials
microscope
microscope slides (4)
single-edged razor blade
water
forceps
dropper
moss plants with male and female gametophytes and sporophytes
paper towels

Safety Precautions
Use care in working with razor blades.

PROCEDURE

Part A: Reproductive Structures of the Gametophyte
1. Obtain moss gametophyte plants with both male and female reproductive structures.

2. With forceps, remove all the leaves from the upper 1-cm portion of the stem of each plant. Be careful not to remove the reproductive structures at the tip of the stem.

3. Place each plant onto opposite ends of a microscope slide. With a razor blade, cut each plant stem 0.5 cm from the tip end. **CAUTION:** *The razor blade is sharp. Cut away from your body.*

4. Add several drops of water to each tip end. Place a second glass slide over the first slide. Press down firmly on the top of the slide with your thumb so that each tip end is slightly squashed. To prevent thumbprints on the slide, place a piece of paper towel over the slide before pressing on it.

5. Place the stacked slides on the stage of the microscope, keeping both slides together. Use the stage clips to keep the slides from slipping.

6. Observe the moss tip ends *only under low-power magnification.*

7. Identify each of your moss tips as antheridium or archegonium.

8. Draw and label the following structures: male gametophyte, female gametophyte, antheridium, archegonium.

Part B: Reproductive Structure of the Sporophyte

1. Obtain another specimen of moss plant with sporophyte stage. This plant has a thin stalk and a capsule sticking out from the tip end of the gametophyte.

2. Remove the small capsule from the tip of the stalk, and mount it in several drops of water on a microscope slide. Add a second slide as you did for the gametophyte, and squash the capsule.

3. Observe the squashed capsule *under low-power magnification.*

4. Draw and label the following structures: sporophyte, capsule, spore.

ANALYZE AND CONCLUDE

1. **Observing and Inferring** What type of reproductive cell is formed by the archegonium? By the antheridium? By the sporophyte? Which of these cells are haploid and which are diploid?

2. **Formulating Models** Using your observations of moss reproductive structures and information from the text, diagram the life cycle of a moss. Make sure your diagram includes the following: gametophyte, archegonium, antheridium, egg, sperm, zygote, sporophyte, spore capsule, spore. Label each stage as haploid or diploid, and mark where meiosis and fertilization take place.

3. **Observing and Inferring** Which kind of gamete is produced in greater numbers, eggs or sperm? What is the adaptive advantage of this?

Going Further

Application If petri dishes and a growth medium (agar) are available, germinate moss spores and examine the developing moss gametophyte under the microscope.

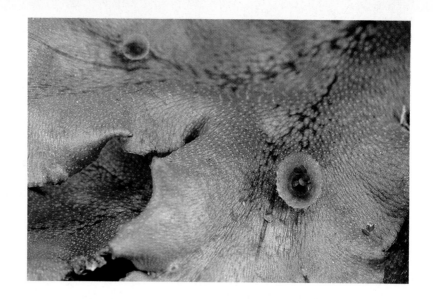

Figure 24.10

Small cups filled with tiny gemmae have formed on the thallus of this liverwort. Rainfall washes the gemmae onto the soil, where they develop rhizoids and grow into new thalluses.

Vegetative reproduction in bryophytes

Like all plants, mosses and liverworts can also reproduce asexually, as shown in *Figure 24.10.* In asexual, or **vegetative reproduction,** a plant gives rise to new individuals without going through the alternation of generations. Some mosses can break up into pieces when the plant is dry and brittle. With the arrival of wetter conditions, these pieces each become a whole plant. Vegetative reproduction in liverworts and some mosses involves the formation of **gemmae,** which are tiny, haploid bodies that grow in cups on the surface of the gametophyte.

Connecting Ideas

Bryophytes represent the only division of plants that lack vascular tissue. This lack of a transport mechanism for water and nutrients limits the size and distribution of mosses and liverworts. Vascular plants, on the other hand, have tissues adapted for such transport. The ability to transport needed materials from the soil to plant parts above ground level enabled vascular plants to grow taller and led to further adaptations to life on land.

Section Review

Understanding Concepts
1. How can you tell a moss and a thallose liverwort apart?
2. In what way is the sporophyte generation of a moss dependent on the gametophyte generation?
3. How is the formation of spores an adaptation to life on land?

Thinking Critically
4. Explain why reproduction in a moss depends upon water.

Skill Review
5. **Sequencing** Sequence the events in the life cycle of a moss, beginning with the protonema. For more help, refer to Organizing Information in the *Skill Handbook.*

Bioengineered Food: Do We Need to Know?

Future Crops: No Farmers Needed?

THE DAILY TRIBUNE   MAY 3, 1993

By this time next year, researchers say, farmers will be able to purchase at the hardware store a genetically engineered corn seed that plants itself, produces its own fertilizer and pesticide, requires no water, and grows tortilla chips, high-fructose sweetener, and diesel fuel in place of the usual kernels.

Although most of the seeds mentioned above do not yet exist, you can find some extraordinary tomatoes in the produce section of your local grocery store today. They look no different from regular tomatoes, but inside their DNA—the genetic code that determines the way they look, feel, and taste—is an extra gene. Scientists have added a gene that is a copy of the gene for a ripening enzyme, but it is in reverse order. This antisense gene makes backward RNA, which binds to the normal RNA for fruit softening, thereby rendering the gene inactive. The enzyme does not form, and this slows down the rotting process. Able to stay on the vine longer, these high-tech tomatoes taste better and have a longer shelf life than regular tomatoes.

Different Viewpoints

This agricultural revolution is led by bioengineers who use gene-splicing technology as a means to solve world hunger and poverty. Nutrition-enhanced, longer-lasting produce and drought-, frost-, and pest-resistant crops can relieve famine and improve economies in many developing countries.

Plant breeders who choose genetic engineering over traditional crossbreeding are able to get desired traits faster and with more consistency.

How safe are engineered foods?   The safety of genetically engineered foods has been the critical issue in bringing these products to market. Consumers who have food allergies are concerned that genetically altered foods will cause allergic reactions. For example, say a gene from a wheat plant is used to provide resistance to disease in corn plants. Will a person who is allergic to wheat products have a reaction after eating the genetically altered corn? Vegetarians have concerns about consuming genetically engineered plants that may include genes derived from animals.

What are the ecological consequences?
Environmentalists are concerned about the ecological consequences of altering plant genetic material. What would happen if altered plant or animal species without natural predators were introduced into the environment?

INVESTIGATING   the Issue

1. **Writing About Biology** Would you want to know if the food you were eating had been genetically altered? Why or why not?
2. **Applying Concepts** What do traditional artificial selection and bioengineering techniques have in common?
3. **Recognizing Cause and Effect** What are the benefits of creating insect-resistant corn?

24.1 Adapting to Life on Land

- Plants are multicellular eukaryotes with a cuticle and cells that are surrounded by cell walls. Plants have chlorophyll for photosynthesis and store food in the form of starch.
- All plants on Earth probably evolved from filamentous green algae that dwelt in ancient oceans. The first plants to make the move from water to land may have been leafless like mosses.
- Adaptations for life on land include a waxy cuticle that helps prevent water loss; stomata that open to allow gas exchange; development of leaves, roots, and stems; development of spores and seeds; and alternation of the gametophyte and sporophyte generations in the life cycle.
- The plant kingdom includes five divisions of spore-producing plants and five divisions of seed-producing plants.

24.2 Bryophytes

- Bryophytes are spore-producing plants that have no vascular tissue and reproduce by forming spores. They usually live in moist, cool environments, and the gametophyte generation is dominant.
- Mosses are small plants with leafy stems. Spores germinate to form a protonema, from which the gametophyte grows.
- Thallose liverworts resemble a leathery, ribbonlike or lobed leaf. Leafy liverworts have flat, narrow leaves attached to a stem.
- Liverwort spores germinate to form the plant body; there is no protonema as in mosses.

Key Terms

Write a sentence that shows your understanding of each of the following terms.

antheridium	root
archegonium	seed
cuticle	stem
gemmae	stomata
leaf	vascular plant
nonvascular plant	vegetative
protonema	reproduction

Understanding Concepts

1. List at least four major differences between vascular and nonvascular plants.
2. Compare and contrast roots and rhizoids.
3. Compare and contrast seeds and spores.
4. Explain the difference between a sporophyte and a gametophyte.
5. Name at least two important characteristics that all bryophytes have in common.
6. Explain how the photosynthesizing cells of a bryophyte obtain water, and compare this method with the way in which the leaves of a vascular plant obtain water.
7. Explain why algae are thought to be the ancestors of plants.
8. Would you expect to find mosses in a desert? Explain your answer.
9. How is a protonema formed?
10. Name the distinguishing characteristics of the organism below that identify it as a plant.

Using a Graph

11. Moss is often used to improve the water-retention properties of soil. Students evaluated three different soil mixes, one of which contained moss. They poured 100 mL of water into a sample of each mix, measured the water that was not absorbed by each mixture, and used their data to produce the graph below. Which of the soil mixtures contained moss?

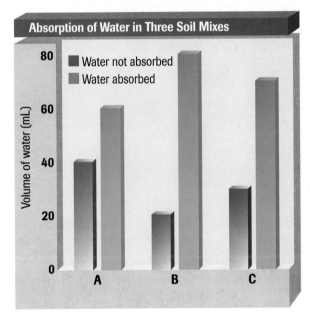

Absorption of Water in Three Soil Mixes

- Water not absorbed
- Water absorbed

Volume of water (mL)

A   B   C

Relating Concepts

12. Make a concept map that relates the following terms. Supply the appropriate linking words for your map.

    sporophyte, gametophyte, spore, diploid, haploid, antheridium, archegonium, gamete, meiosis, mitosis, fertilization

Applying Concepts

13. Compare what might happen to a moss and a liverwort after exposure to a long period of drought.

14. Centuries ago, dried mosses were used to fill in the cracks between timbers in log buildings and between the planks of boats. Explain why you think moss would work well for these purposes.

15. How might the distribution of mosses and liverworts on Earth be affected by a significant increase in worldwide temperatures?

16. What is the advantage to mosses of living in dense populations?

Biology & Society

17. Explain why the altered DNA of bioengineered foods should have no ill effects on people who eat them.

Thinking Critically

Making Predictions

18. **Biolab** Predict whether the reproductive cells found in the gemma cup of a liverwort are produced by meiosis or mitosis.

Observing and Inferring

19. **Minilab** In parts of Australia, moss species with a normal height of 20 cm can sometimes be found growing as tall as 70 cm. In what type of environment would you search for these especially tall specimens?

20. **Minilab** You have collected a twig from an unknown tree. It has huge thorns sticking out from all directions around the stem. What function do these thorns have?

Connecting to Themes

21. **Evolution** Explain why biologists think the first plants to adapt to life on land may have been similar to mosses and liverworts.

22. **Systems and Interactions** In what ways are bryophytes important to the ecosystem of a forest?

25 Ferns and Gymnosperms

ow is this stately sugar pine different from a tiny moss? The largest of all the pines, this sugar pine can grow to a height of 75 m, with a trunk diameter of more than 3 m. The sugar pine gets its name from resin that oozes from cracks in the bark and dries into white patches resembling sugar crystals. Huge cones hanging at the ends of branches are almost 0.5 m long and may weigh several kilograms. Each cone contains meaty seeds that will grow into new trees or perhaps provide a meal for a squirrel. Hikers beware—the sugar pine's cones make a for-midable crash when they fall to the ground, and you don't want to be in the way!

The sugar pine is one example of the success with which the seed plants have adapted to life on land. In this chapter, your investigation of plant evolution continues with the study of vascular plants, including the spore-producing ferns and the seed-producing gymnosperms.

Ferns are vascular plants, but they share two important similarities with nonvascular bryophytes. Ferns produce spores, and they have free-swimming sperm that must travel through a surface film of water to reach the egg. Possession of vascular tis-sue has enabled ferns to adapt to a larger variety of habitats than their nonvascular cousins. Pines are gymnosperms. Unlike ferns, they produce seeds in cones. Pines can grow to great heights due to vascular tissue. How are ferns and gymnosperms alike?

25.1 Seedless Vascular Plants

*I*magine traveling back in time to look at the vascular plants that formed Earth's forests more than 300 million years ago. The land is damp and swampy. There are huge flying insects and giant amphibians, but no birds or mammals. Even dinosaurs won't appear

for another 50 million years or so. Everywhere you look, there are leafy plants and many incredibly tall, unusual-looking trees. These plants will eventually be transformed into the coal that will provide humans with fuel. How do we know what these plants looked like? Many of these ancient plants were preserved as fossils. There are plants living on Earth today that resemble these fossils of ancient vascular plants. Among them are the club mosses, horsetails, and ferns.

Lycophyta

The Lycophyta, club mosses and spike mosses, first appeared on Earth about 390 million years ago. Ancient species grew as tall as 30 m and were extremely abundant in the warm, moist forests that dominated Earth during the Carboniferous period. Most species of lycophytes died out about 280 million years ago, when Earth's climate became drier and cooler.

Lycophytes are small vascular plants

Modern lycophytes are much smaller than their ancestors. They grow close to the ground and are found mostly in damp forests, though some live in desert or mountain climates. The lycophytes are commonly called club mosses and spike mosses because their leafy stems resemble moss gametophytes, and their reproductive structures are club shaped, as shown in *Figure 25.1.* However, these plants are not mosses, nor are they closely related to mosses. As with other vascular plants, but unlike mosses, the sporophyte generation of the lycophytes is dominant and has roots, stems, and leaves.

Leaves are adapted for reproduction

A major advance in this group of vascular plants was the adaptation of leaves into structures that protect the reproductive cells. Spore-bearing

Figure 25.1

Lycophytes are simple vascular plants with upright or creeping stems. Roots usually grow from the base of the stem. A single vein of vascular tissue runs through the center of each narrow leaf.

◀ **This club moss, *Lycopodium*, is commonly called ground pine because it is evergreen and has a treelike growth habit. The plant shown here is the sporophyte. The gametophyte is smaller and remains buried in the soil.**

▲ **This spike moss, *Selaginella*, grows on prairie soils or dry, rocky outcrops. During dry periods, the plant appears dead, but after a rainfall the leaves turn green and the plant continues its life cycle.**

leaves form a compact cluster called a **strobilus,** which is located at the end of the stem. Leaves of lycophytes occur in spirals, whorls, or pairs. For a more detailed look at the structure of a strobilus, see *Figure 25.2.* In a lycophyte life cycle, the spore germinates to form a gametophyte, which is called a **prothallus.** The prothallus is relatively small, lives in or on the soil, and produces either antheridia or archegonia. Sperm swim from an antheridium, through a film of water on the surface of the prothallus, to the egg in an archegonium. The sporophyte plant grows from the fertilized zygote and then becomes larger and dominant.

Figure 25.2

The lycophyte life cycle is similar to that of all other spore-producing plants.

▼ **Spores are produced in sporangia at the bases of small, modified leaves that form the conelike strobilus.**

▼ **Some species of lycophytes have two kinds of sporangia, each of which produces a different type of spore. Small spores germinate to form a male prothallus. Large spores germinate to form a female prothallus.**

▼ **A young sporophyte grows from the archegonium in a female prothallus. The prothallus with its rhizoids resembles a root with root hairs.**

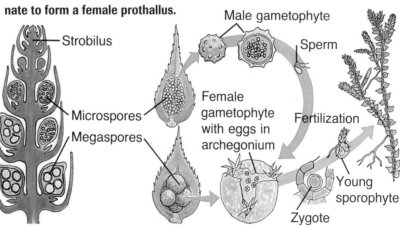

Strobilus

Microspores

Megaspores

Male gametophyte

Sperm

Female gametophyte with eggs in archegonium

Fertilization

Young sporophyte

Zygote

Figure 25.3

This cross section of the stem of a club moss shows tissues of xylem and phloem. Both xylem and phloem are located in the center of the stem. The central mass of phloem tissue is interrupted by irregular shaped strands of xylem tissue.

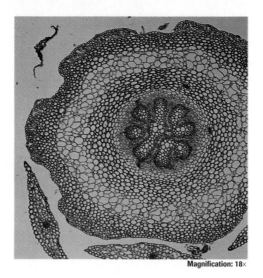

Magnification: 18×

Vascular tissues transport materials

Unlike bryophytes, lycophytes contain vascular tissues. **Vascular tissues** are tissues that transport materials from one part of a plant to another. As shown in *Figure 25.3,* vascular tissues consist of xylem and phloem. **Xylem** tissue is made up of a series of dead tubular cells, joined end to end, that transport water and dissolved minerals upward from roots to leaves. **Phloem,** made up of a series of tubular cells that are still living, transports sugars from the leaves to all parts of the plant. Both xylem and phloem extend from close to the ends of the root tips, through the stem, and into the leaves of a vascular plant.

Sphenophyta

Sphenophytes, horsetails, represent a second group of ancient vascular plants. Like the lycophytes, early horsetails were tree-sized members of the forest community. There are only about 15 species in existence today, all of the genus *Equisetum.* The name *horsetail* refers to the bushy appearance of some species. These plants are also called scouring rushes because they contain silica, an abrasive substance, and were used to scour cooking utensils. If you run your finger along a horsetail stem, you can feel how rough it is.

Sphenophytes have jointed stems

Today's sphenophytes are much smaller than their ancestors, usually growing to about 1 m tall. Most horsetails, like the one shown in *Figure 25.4,* are found in marshes, shallow ponds, stream banks, and other areas with damp soil. Some species are common in the drier soil of fields and roadsides. The stem structure of horsetails is unlike most other vascular plants. The stem is ribbed and hollow, and appears jointed. At each joint, there is a whorl of tiny, scalelike leaves.

Figure 25.4

This is the sporophyte generation of a horsetail, *Equisetum*. It has thin, narrow leaves that circle each joint of the slender, hollow stem. Sporangia-bearing leaves form a strobilus at the tips of some stems.

Reproduction similar to Lycophyta

Sphenophytes, like other vascular plants, exhibit an alternation of generations. Their reproduction is similar to that of lycophytes. Sphenophytes produce spores that grow into the gametophytes. A gametophyte of a sphenophyte is also called a prothallus. Eggs are produced in the archegonia, and sperm are produced in the antheridia. Sperm swim along the surface of the prothallus to the archegonia, where fertilization takes place. If more than one egg is fertilized, several sporophytes may grow from one prothallus.

Pterophyta

Ferns first appeared in the fossil record nearly 400 million years ago, at about the same time that club mosses and horsetails were the prominent members of Earth's plant population. Ferns, division Pterophyta, grew tall and treelike, forming vast fern forests. You are probably more familiar with ferns, *Figure 25.5,* than with club mosses and horsetails, primarily because ferns evolved into many more species and are more abundant. Ferns can be found in many types of environments.

Figure 25.5

There are about 12 000 species of living ferns. Many require damp, shady environments, but others live in drier, sunnier habitats.

▶ **Most modern ferns are fairly small and leafy, but many species of tall tree ferns still exist, primarily in the tropics.**

▲ **This deer fern is found only in wet, shady, coastal evergreen forests.**

◀ **The bracken fern thrives in the partial sun of open forests. One of the most common ferns in the world, it often takes over large areas of abandoned pasture or agricultural land. Bracken ferns usually grow about 1.5 m tall, though they can reach a height of 5 m.**

What does a fern prothallus look like?

The gametophyte generation in spore-producing vascular plants is much smaller than the gametophytes of nonvascular bryophytes. In this lab, you will investigate the structure of the fern gametophyte.

Procedure

1. Obtain a mature fern prothallus. Make a wet mount, with coverslip, and observe the prothallus at low power under the microscope.
2. Near the notch in the heart-shaped prothallus, locate the reproductive structures that produce eggs.
3. Near the pointed base of the prothallus, locate the reproductive structures that produce sperm. Use the eraser end of a pencil to press gently on the coverslip. Watch closely to see if sperm are released. If they are, observe their movement.

Analysis

1. What is the name of the structure that produces eggs?
2. What is the name of the structure that produces sperm?
3. Is the prothallus haploid or diploid?
4. What type of cell division takes place within the prothallus?
5. If you were able to observe sperm, describe their movement.

Figure 25.6

Most ferns in warm climates are perennial plants that live from year to year, gradually enlarging in size. The fronds of ferns that live in temperate climates die back during the winter.

▼ The creeping, underground rhizome of a fern is a modified, condensed stem, with roots growing downward from each joint of stem and frond. The rhizome contains many starch-filled cells and is a storage organ for overwintering.

▼ A fern frond has a stemlike stipe and green, often finely divided leaflets called pinnae.

▼ Young fern fronds unfurl as they grow and are called fiddleheads because their shape is similar to the neck of a violin.

Sporophyte is dominant in ferns

As with all vascular plants, it is the sporophyte generation of the fern that has roots, stems, and leaves, and this is the plant we commonly recognize. The gametophyte in most ferns is a thin, flat structure that is independent of the sporophyte. In most ferns, the main stem is underground, *Figure 25.6.* This thick, underground stem is called a **rhizome.** The leaves of a fern are called **fronds** and grow upward from the rhizome. The fronds of ferns are often divided into leaflets called pinnae. Ferns are the first of the vascular plants to have evolved leaves with branching veins of vascular tissue. The branched veins in ferns transport water and food to and from all the cells.

Life Cycle of a Fern

*I*n the life cycle of a fern, the sporophyte generation is independent of the gametophyte. As in mosses, meiosis in a fern takes place in sporangia, and sperm must swim through a surface film of water to reach the egg.

Bracken fern

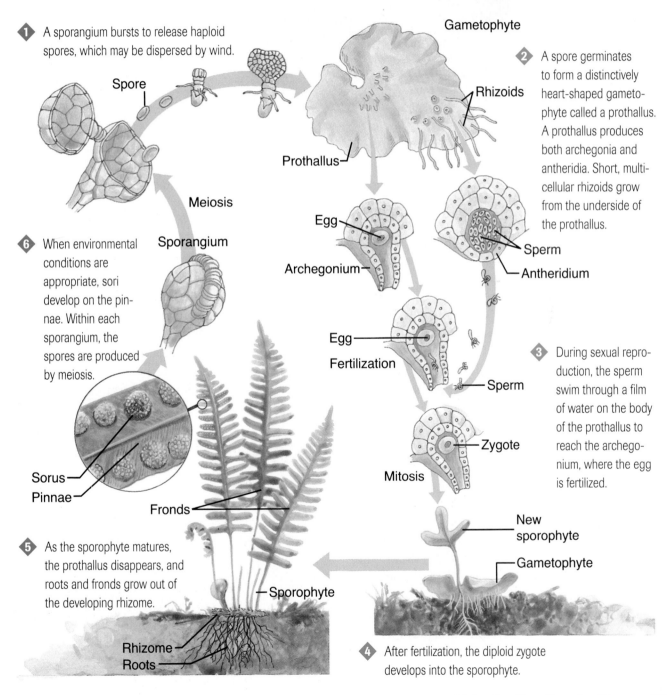

❶ A sporangium bursts to release haploid spores, which may be dispersed by wind.

Spore

Meiosis

Sporangium

❻ When environmental conditions are appropriate, sori develop on the pinnae. Within each sporangium, the spores are produced by meiosis.

Sorus
Pinnae
Fronds

❺ As the sporophyte matures, the prothallus disappears, and roots and fronds grow out of the developing rhizome.

Sporophyte

Rhizome
Roots

Gametophyte

❷ A spore germinates to form a distinctively heart-shaped gametophyte called a prothallus. A prothallus produces both archegonia and antheridia. Short, multicellular rhizoids grow from the underside of the prothallus.

Rhizoids

Prothallus

Egg

Archegonium

Sperm
Antheridium

Egg
Fertilization

Sperm

❸ During sexual reproduction, the sperm swim through a film of water on the body of the prothallus to reach the archegonium, where the egg is fertilized.

Zygote

Mitosis

New sporophyte

Gametophyte

❹ After fertilization, the diploid zygote develops into the sporophyte.

Figure 25.7

Sori found on the underside of fern fronds look like brown- or rust-colored dust.

▶ Most sori are found as round clusters on the pinnae, although the shape of the clusters and arrangement on the pinnae vary with the species.

◀ Some species of ferns have sori on the edges of the pinnae, such as this *Osmunda* fern.

Most ferns require moist soil

Most modern ferns are much smaller than their ancestors. You may have seen shrub-sized ferns, such as those pictured in *Figure 25.5,* on the damp forest floor or along stream banks. Some species of ferns float in water or are rooted in mud, whereas others cling to the sides of rocky cliffs or live in cold regions above the Arctic Circle. Some ferns inhabit dry areas, becoming dormant when moisture is scarce and resuming growth and reproduction when water again becomes available.

Fern spores are held in sori

A fern life cycle is similar to that of other spore-producing vascular plants. Spores are produced in structures called sporangia. Clusters of sporangia form a structure called a **sorus** (plural *sori*). The sori are usually found on the pinnae, *Figure 25.7,* but in some ferns, spores are borne on whole, modified fronds.

Section Review

Understanding Concepts

1. Why do most spore-producing vascular plants live in moist habitats?
2. What are the major differences between spore-producing vascular plants that exist today and those that lived in the Carboniferous forests?
3. Compare and contrast the structure of the sporophyte in the lycophytes and pterophytes.

Thinking Critically

4. Why do you think there are fewer spore-producing vascular plants on Earth today than there were 300 million years ago?

Skill Review

5. **Sequencing** Sequence the stages in the life cycle of a fern. For more help, refer to Organizing Information in the *Skill Handbook.*

25.2 Gymnosperms

Although Earth's ancient forests were dominated by spore-producing vascular plants, early seed-bearing plants had also evolved. About 280 million years ago, when club mosses, ferns, and other spore producers had reached their greatest numbers and diversity, Earth's climate changed. Long periods of drought and freezing weather caused many spore-producing plants to become extinct, but a few of the seed-bearing plants had adaptations that enabled them to survive.

What Are Gymnosperms?

Among the early seed-bearing plants were the gymnosperms. **Gymnosperms** are vascular plants that produce seeds on the scales of woody strobili called **cones.** The seeds of gymnosperms are not protected by a fruit, nor do gymnosperms produce flowers. The term *gymnosperm* is used to describe four divisions of plants that bear naked seeds: Cycadophyta, Ginkgophyta, Gnetophyta, and Coniferophyta.

Gymnosperms produce spores

Gymnosperms are similar to all other plants in that spores are produced by the sporophyte generation. **Microspores** are produced in the male cone and give rise to the male gametophyte, eventually developing into pollen grains. **Megaspores** are produced in the female cone and give rise to the female gametophyte. Megaspores are contained within the ovule and eventually develop into archegonia with egg cells. In conifers, pollen is carried by wind to the ovule and the male gametophyte contained in the pollen grain produces a pollen tube, which grows into the archegonium and provides a way for the sperm cell to reach the egg. This occurs without water.

After fertilization, the zygote develops into an embryo. An **embryo** is an organism at an early stage of growth and development. In plants, an embryo is the young, diploid sporophyte of a plant. A gymnosperm embryo has many cotyledons. **Cotyledons** are food-storage organs of a plant embryo that become the plant's first leaves. As the embryo develops, the tissues of the ovule form the food supply and seed coat of the pine seed.

Figure 25.8

A pine seed shows the structures and some of the adaptive advantages of seeds.

▶ Both the embryo and its food supply are protected by a tough, outer seed coat. The seed coat breaks down, and growth begins only when certain conditions of temperature, moisture, and light are met.

◀ The plant embryo is surrounded by seven to nine cotyledons, which will serve as an initial food supply when growth begins.

▲ The winged shapes of these pine seeds aid in their dispersal. The wind may carry them several feet away from the parent tree.

Advantages of seeds

A seed consists of an embryo with a food supply enclosed in a tough, protective coat, *Figure 25.8.* Plants that bear seeds have several important advantages over spore producers. The seed contains a supply of food to nourish the young plant during the early stages of growth. This food is used by the plant until its leaves are developed sufficiently to carry out photosynthesis. In gymnosperms, the food supply is stored in cotyledons. The embryo is protected during harsh conditions by a tough seed coat. The seeds of many species are also adapted for easy dispersal to new areas, so the young plant does not have to compete with its parent for sunlight, water, soil nutrients, and living space.

Fertilization without water

Seed plants also have another important advantage over spore producers: fertilization does not require water. Spore-producing vascular plants are not fully adapted to life on dry land because the sperm needs a surface film of water in which to swim to the egg during sexual reproduction. In gymnosperms, the male gametophyte develops inside a structure called a **pollen grain,** which includes sperm cells, nutrients, and a protective outer covering. The female gametophyte develops inside a structure called an ovule. An **ovule** contains a megaspore cell, one or two layers of tissue, and a protective covering. Sperm are carried in the pollen grain to the ovule by wind rather than water.

Cycadophyta

Cycads are seed plants that shared Earth's forests with the dinosaurs during the Triassic and Cretaceous periods. About 100 species exist today, exclusively in the tropics and subtropics. The only present-day species that grows wild in the United States is found in Florida, although you may see cycads cultivated in greenhouses or botanical gardens. The trunk and leaves of many species resemble palm trees, as you can see in *Figure 25.9,* but cycads are not closely related to palms.

Figure 25.9

Cycads have a terminal rosette of leaves and bear seeds in cones. All cycads have separate male and female plants.

▲ **The male plant bears cones that produce pollen grains, which are released in great masses into the air.**

▲ **The female plant bears cones that contain ovules with eggs. The moisture in the ovule is sufficient for the sperm, which are released from the pollen grain, to swim to the eggs.**

Earth Science

How Coal Was Formed

More than 300 million years ago, the huge leaves of ferns swayed in the moist breezes of vast bogs and swamps. As the lush vegetation died and was replaced by succeeding generations of plants, rich layers of plant material collected on the bottom of the swamp. In some places, there was not enough oxygen or bacteria to produce decay. Over a long period of time, these swamps drained and filled again, producing many layers of plant material and sediment pressing down on one another. This combination of plant material and pressure produced a fossil fuel called coal.

Plants turned into fuel More than half of all the known reserves of coal are found in the U.S. The plant material that accumulated 300 million years ago during the Carboniferous period ultimately became coal. Pterophytes, gymnosperms, and lycopods were the dominant plants during this period.

From peat to lignite to coal As the plants died and fell into boggy waters lacking oxygen and bacteria, the layers of packed vegetation became peat. As more and more dead plant material built up on top of the peat, it became compressed and began to harden, turning into a brown material called lignite coal. Under continuous pressure, lignite coal becomes first bituminous and, finally, anthracite coal—a shiny, hard, black material that is burned in coal-fired furnaces to produce heat and electricity today.

Sand and water in coal Mud or silt often was washed or blown into the swamp during plant growth. This produced a higher ash content in today's coal beds. Vertical fractures filled with mineral deposits from groundwater also can be found in coal beds. Impurities such as these result in lower-grade coal, that is, coal that burns less efficiently.

CONNECTION TO Biology

Where does the energy contained in coal come from?

What are the similarities and differences of spores and pollen grains?

Reproduction in ferns involves the release of spores, and the gymnosperm life cycle includes the release of pollen. Are the roles of these reproductive cells the same or different? In this lab, you will compare the structure and function of spores and pollen.

Procedure

1. Obtain a male pine cone and a fern frond with mature sporangia. Prepare separate wet mounts of pine pollen and fern spores.

2. Observe each cell type under a microscope, and make diagrams of each.

3. Prepare a data table that compares the structure and function of pollen and spores. Use the text and your observations to include answers to the following questions in your table.

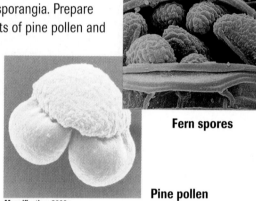

Magnification: 114×

Fern spores

Magnification: 3000×

Pine pollen

Analysis

1. By what cell division process is each cell formed?

2. Is each cell haploid or diploid?

3. Is each cell part of the sporophyte or the gametophyte generation?

Ginkgophyta

Members of the Ginkgophyta were most numerous during the Jurassic period, about 200 million years ago. They were an important part of the vegetation that lived alongside the dinosaurs but, like the dinosaurs, most members of the Ginkgophyta died out by about 65 million years ago. Today, the division is represented by only one living species, *Ginkgo biloba*, *Figure 25.10*. The ginkgo tree is considered sacred in China and Japan, and has been cultivated in temple gardens for thousands of years. These Asian temple gardens seem to have prevented the tree from becoming extinct.

▼ **In addition to being attractive, ginkgo trees are resistant to air pollution, so they are often planted in urban parks and gardens.**

Figure 25.10

The ginkgo is sometimes called the maidenhair tree because its lobed leaves resemble the fronds of a maidenhair fern.

► **Like cycads, ginkgos bear male and female cones on separate plants. The male ginkgo produces pollen in strobilus-like cones that grow from the bases of leaf clusters.**

Figure 25.11

Like the cycads and ginkgos, most gnetophytes have separate male and female plants.

▲ One species of *Ephedra,* also known as Mormon tea, is a shrubby, cone-bearing plant with scalelike leaves similar to those of horsetails. It is found in dry regions such as the white sands of New Mexico. Members of this genus are a source of ephedrine, a medicine used to treat asthma, emphysema, and hay fever.

▲ The single species of *Welwitschia* is a bizarre-looking desert dweller that grows close to the ground. It has a large tuberous root, and though it may live 100 years, this plant has only two leaves, which continue to lengthen and become tattered by weathering as the plant grows older.

Gnetophyta

Fossil gnetophytes are unknown, except for pollen from Permian, upper Cretaceous, and tertiary rock formations. Most living gnetophytes can be found in the deserts or mountains of Asia, Africa, and Central or South America. The division Gnetophyta contains only three genera, which are all different in structure and adaptations. The genus *Gnetum* is composed of tropical climbing plants. The genus *Ephedra* contains shrublike plants and is the only gnetophyte genus that can be found in the United States. The third genus, *Welwitschia*, is found only in South Africa. *Ephedra* and *Welwitschia* are pictured in *Figure 25.11.*

◀ The female *Ginkgo* bears the seeds, which develop a fleshy outer covering and resemble orange-yellow cherries as they ripen. The male trees are usually preferred by gardeners because the fleshy seed coat of ginkgos has an unpleasant rancid smell.

Figure 25.12

Conifers are named for the woody cones in which the seeds of most species develop.

► The Douglas fir, a member of the pine family, is one of the most important lumber trees in North America. It grows straight and tall, to a height of 100 m.

▲ Yews are popular ornamental trees because they are graceful in shape and have attractive, glossy, green needles. A *Taxus* species from the Pacific Northwest is used to produce a cancer-fighting drug called taxol. Yews do not have cones. They produce single seeds with a fleshy covering.

Coniferophyta

The sugar pine is one of many familiar-looking forest trees that belong to the division Coniferophyta. The conifers—the largest, most diverse division of the gymnosperms—are trees and shrubs with needle- or scalelike leaves. Most conifers produce seeds in woody cones. They are abundant in forests throughout the world, and include pine, fir, spruce, juniper, cedar, redwood, yew, and larch. A few representative conifers are shown in *Figure 25.12*.

The first conifers probably emerged around 280 million years ago. During the Jurassic period, 150 million years ago, conifers became prominent forest inhabitants and remain so today. Although the fossil record indicates that many species have become extinct, *Figure 25.13*, the conifers continue to evolve and flourish.

Figure 25.13

Petrified Forest National Park in Arizona contains fossilized remains of an extinct species of conifer from the Triassic period. When the trees died, they were buried in mud and volcanic ash. Instead of decaying, the organic matter was replaced with minerals. The wood has become petrified, or turned to stone.

Conifers are adapted to cold climates

Conifers form large forests in many regions of the world and have many adaptations that enable them to live in cold or dry habitats. Under dry or freezing conditions, plant roots cannot absorb water from the soil. The needlelike leaves of conifers are covered with a thick cuticle that helps retain water in the tissues of the tree, as shown in *Figure 25.14.* The bark of conifers and other seed-bearing trees also helps reduce water loss by forming a protective covering over the stem.

Conifers are evergreens

Most conifers are evergreen plants. **Evergreen plants** retain their leaves all year. Although individual leaves drop off as they age or are damaged, the tree never loses all its leaves at once. Pine needles, for example, may remain on the tree for anywhere from two to 40 years, depending on the species. Trees that retain their leaves can begin photosynthesis in the early spring as soon as the temperature warms. They usually have a heavy coating of cutin. Evergreens are often found where the warm growing season is short, so keeping leaves year-round gives these plants a head start on growth. They are abundant where nutrients are scarce; evergreens thus eliminate the need to grow a whole new set of leaves each year. The branches and needles of conifers are extremely flexible. They allow snow and ice to slide off the tree, rather than build up on branches causing them to break.

ThinkingLab | Analyze the Procedure

How much acid in rain is too much?

Scientists hypothesize that acid rain may have contributed to a decline in the number of red spruce trees in forests in the northeastern United States. When exposed to acid rain, needles of the tree turn brown and fall off, and new ones grow in slowly. Eventually, the tree may die. Scientists wanted to know exactly what concentrations of acid caused damage.

Analysis
Groups of spruce seedlings were misted with acidic water at pH levels of 2.5, 3.0, 3.5, 4.0, 4.5, 5.0, and 5.5, and effects were noted in all groups. If fewer effects were noted in plants misted with water at a higher pH, can scientists say conclusively that acidity causes problems?

Thinking Critically
What would need to be done in order to make the above experimental procedure scientifically sound?

Figure 25.14

In cross section, the needle of a conifer resembles a cylindrical leaf.

▼ **The narrow shape reduces the surface area from which water can evaporate.**

Stomata

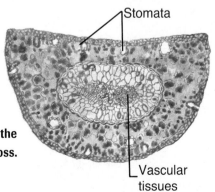

► **The stomata are sunken within folds of the epidermis, which helps prevent water loss.**

Vascular tissues

The Lodgepole Pine and Fire

During the summer of 1988, Yellowstone National Park, the oldest national park in the United States, was swept by fire. When the conflagration finally died down, nearly 20 percent of the park had been touched by the flames, although just one percent of the area suffered severe burning.

Although many people were sorry to see the ashes of this once-beautiful park, fire plays an important role in maintaining forest and prairie ecosystems.

New life from ashes The lodgepole pines of Yellowstone provide a vivid example of how fire renews life. Like all pine trees, lodgepole pines produce both male and female cones, but lodgepole pines have two types of female cones. One type develops on the tree for two years and then opens and deposits its seeds on the ground.

The second type of female cone on the lodgepole pine is called a serotinous cone. It is covered with a resin that seals the cone shut and traps the seeds inside. The cone remains sealed until fire sweeps through the forest. Heat from the blaze dissolves the resin and dries out the cone. Then the dry cone opens up and releases its seeds onto the forest floor. The fire burns away the dense forest canopy as well as the vegetation on the forest floor, giving the seeds an excellent chance to germinate and grow.

Wire grass to the rescue Another type of pine that benefits from fire is the longleaf pine that grows in the southeastern United States. Longleaf pines grow in sandy soil surrounded by thick wire grass.

When lightning strikes the wire grass, it burns the hardwood saplings that grow in between the pines also. The fire never becomes intense enough to harm the pines, but does clear the ground so that pine seedlings can flourish. Wire grass seeds also germinate after a fire, returning the ecosystem to its original state—until, of course, the next fire, when the process of rebirth and renewal will begin again.

Thinking Critically Explain the adaptive value to the lodgepole pine of having two types of female cones.

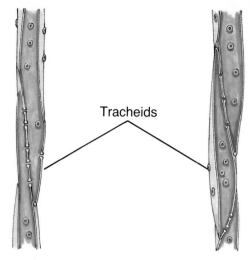

Figure 25.15

This view of tracheids shows that these cells have an elongated shape and closed, tapered ends. The average length of a tracheid cell is about 4 mm. Water moves from one tracheid to another by passing through the pits in connecting tracheid walls.

Deciduous trees lose their leaves

A few conifers, including larches and bald cypress trees, are deciduous plants. **Deciduous plants** lose all their leaves at the same time. Dropping all leaves is another adaptation for reducing water loss when water is unavailable during the winter. Plants lose most of their water through the leaves; very little is lost through bark or roots. However, a tree with no leaves cannot photosynthesize and must remain dormant during the winter months.

How do conifers grow so tall? The tissue that makes up much of the trunk of a conifer—the tissue we usually refer to as wood—is composed of thick-walled, nonliving cells called **tracheids**, *Figure 25.15*. They form the xylem of the vascular systems of ferns and all gymnosperms. In addition to providing support, tracheids transport water and dissolved minerals from the roots to all other parts of the plant.

THE INSIDE STORY

Life Cycle of a Pine

A pine, which is a common conifer found all over the northern hemisphere, can be used as a model for the seed-production process in all gymnosperms. The pine shown here is a white pine.

White pine

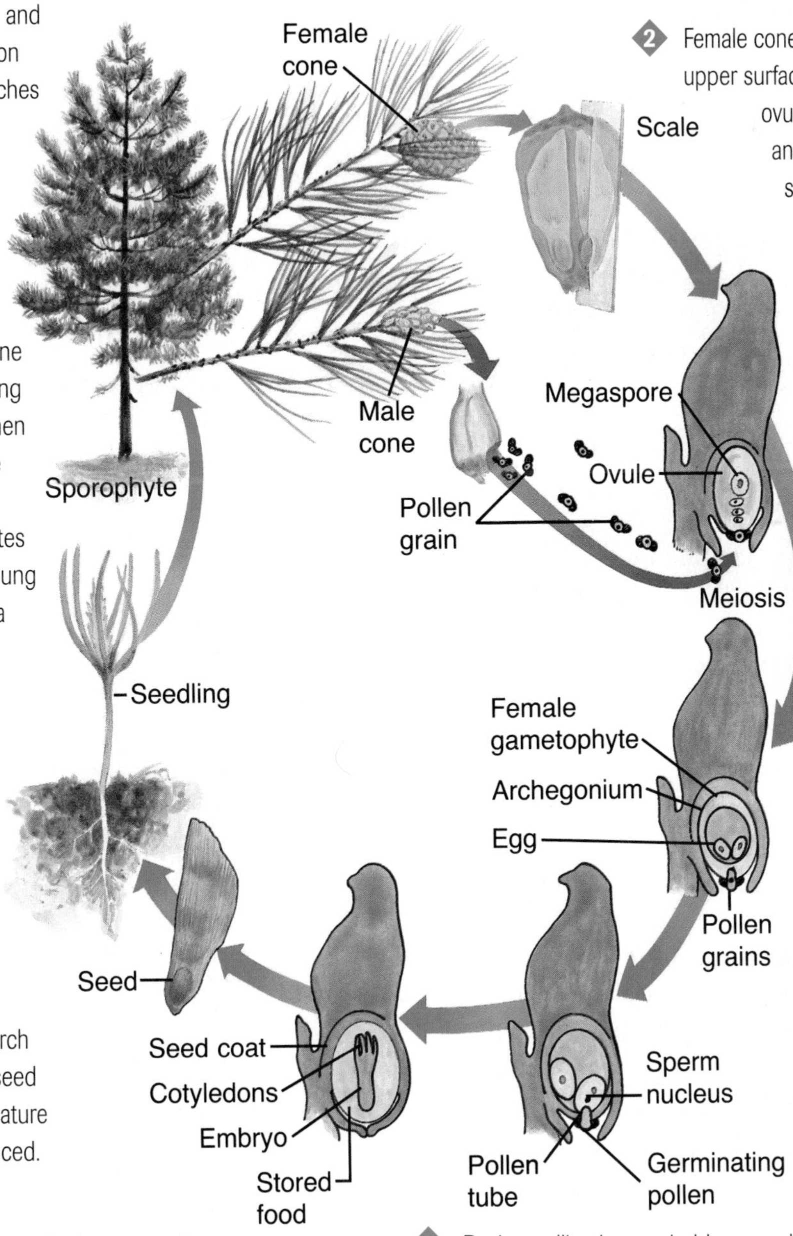

1 The adult sporophyte develops male and female cones on separate branches of the tree.

2 Female cones develop two ovules on the upper surface of each cone scale. Each ovule contains haploid megaspores and remains attached to the sporophyte.

3 Male cones produce microspores by meiosis; the microspores develop into pollen grains. Each pollen grain, with its hard, water-resistant outer covering, is a four-celled male gametophyte.

4 The female gametophyte grows, producing two or more archegonia, each of which contains an egg. Scales of the female cone are closed during this time.

5 During pollination, a wind-borne pollen grain falls near the opening in one of the ovules of the female cone. Each male gametophyte forms a pollen tube that penetrates the tissue of the female gametophyte.

6 When the pollen tube has grown into the archegonium, a sperm cell from the male gametophyte fertilizes the egg.

7 The zygote develops into an embryo with many cotyledons, the ovule generates starch tissue and a seed coat, and a mature seed is produced.

8 The female cone opens, releasing the seeds. When conditions are favorable, the seed germinates into a new, young sporophyte—a pine tree seedling.

Female cone

Scale

Megaspore

Ovule

Meiosis

Pollen grain

Male cone

Sporophyte

Seedling

Female gametophyte

Archegonium

Egg

Pollen grains

Sperm nucleus

Germinating pollen

Pollen tube

Seed

Seed coat

Cotyledons

Embryo

Stored food

Most conifers have cones and needlelike or scalelike leaves. Different species have cones of different sizes, shapes, and thicknesses. The leaves of different species also have different characteristics. How would you go about identifying a conifer you are unfamiliar with? You would probably use a biological identification key. Biological keys list features of related organisms in a way that allows you to determine each organism's scientific name. Below is an example of the selections that could be found in a key that might be used to identify trees.

How can you make a key for identifying conifers?

Needles grouped in bundles
Needlelike leaves
Needles not grouped in bundles
Leaves composed of three or more leaflets
Flat, thin leaves
Leaves not made up of leaflets

PREPARATION

Problem
What kinds of characteristics can be used to create a key for identifying different kinds of conifers?

Hypotheses
State your hypothesis in terms of the kinds of characteristics you think will best serve to distinguish among several conifer groups. Explain your reasoning.

Objectives
In this Biolab, you will:
- **Compare** structures of several different conifer specimens.
- **Identify** which characteristics can be used to distinguish one conifer from another.
- **Communicate** to others the distinguishing features of different conifers.

Possible Materials
twigs, branches, and cones from several different conifers that have been identified for you

1. Make a list of characteristics that could be included in your key. You might consider using shape, color, size, habitat, or other factors.

2. Determine which of those characteristics would be most helpful in classifying your conifers.

3. Determine in what order the characteristics should appear in your key.

4. Decide how to describe each characteristic.

Check the Plan

1. The traits described at each fork in a key are often pairs of contrasting characteristics. For example, the first fork in a key to conifers might compare "needles grouped in bundles" with "needles attached singly."

2. Someone who is not familiar with conifer identification should be able to use your key to correctly identify any conifer it includes.

3. **Make sure your teacher has approved your experimental plan before you proceed further.**

4. Carry out your plan by creating your key.

1. **Checking Your Hypothesis** Have someone outside your lab group try using your key to identify your conifer specimens. If they are unable to make it work, try to determine where the problem is and make improvements.

2. **Making Inferences** Is there only one correct way to design a key for your specimens? Explain why or why not.

3. **Relating Concepts** Give one or more examples of situations in which a key would be a useful tool.

Going Further

Project Design a different key that would also work to identify your specimens. You may expand your key to include additional conifers.

Conifers of the World

Conifers were probably among the first seed-bearing plants to evolve. They have been part of Earth's landscape for 300 million years, which shows how well adapted they must be. Some of the most remarkable trees in the world are conifers.

Many people tend to think of conifers in connection with cold, snowy climates. Although it's true that many species of conifers grow in the northern areas, several are found in Africa, India, and other warm regions.

The tallest

The coast redwoods are the tallest trees in the world, measuring over 117 m in height. By contrast, the cones of these tall trees measure only about 2.5 cm or less in length. The coast redwoods thrive in an area from San Francisco to southern Oregon. They require high humidity, which drifts in from the ocean in the form of fog. Redwoods take thousands of years to grow, so efforts are being made to protect some of the tallest groves of trees.

A living fossil

The dawn redwood was first described in 1941 from fossil remains. A few years later, botanists found living trees growing in the Szechwan province of China. Unlike most conifers, the dawn redwood is deciduous.

The biggest

For a time, the California big tree, *Sequoiadendron giganteum,* was classified as a coast redwood. However, the differences between the two trees were enough to place the big tree in a separate genus. These trees don't grow as tall as the coast redwoods; they grow to a height of only about 99 m. However, their trunks may measure as much as 9 m in diameter. They are the most massive living things in the world and are also among the oldest. Some of the trees living today are more than 3000 years old. They grow in a narrow area in central California. Few of these trees are left, and many areas where they grow are protected.

A southern beauty

Norfolk Island pine, although not a member of the pine family, is known to most of the world as a striking houseplant, but on Norfolk Island, this exotic conifer grows into a magnificent tree.

The survivors

The bitter cold of northern Europe supports a vast forest that stretches from Norway to Siberia. Beautiful larch trees grow in these forests. Unlike other conifers, they turn golden in autumn and shed their needles.

The oldest

The bristlecone pine grows in the mountains of the southwestern United States. A slow-growing tree, bristlecones may take as long as 3000 years to reach their full height of 14 m. The oldest living bristlecones have been growing for more than 4000 years, making them the oldest living things on Earth.

EXPANDING YOUR VIEW

1. **Hypothesize** The dawn redwoods survived for millions of years in a small area in China. Why did they become extinct elsewhere in the world?

2. **Discuss** What conditions may contribute to the incredible growth of coast redwoods?

▲ Two seeds develop at the base of each of the woody scales that make up a female cone. Cones differ in size depending on species.

▲ Male cones are made up of thin, papery scales that open to shed clouds of sulfur-colored pollen grains into the wind.

Figure 25.16

Each scale of a cone is a modified leaf or branch.

Conifers are named for their cones

The reproductive structures of most conifers are produced in cones. Conifers have two types of cones: male cones that produce pollen and female cones that ultimately produce seeds, as shown in *Figure 25.16*. Male cones are small and easy to overlook. They usually drop off the tree soon after the pollen they contain has been released. Female cones are much larger. They stay on the tree until the seeds have matured, which may take from several months up to two years. In the genus *Pinus*, the pollen tube grows part of the way to the egg in one year, then stops until the next spring, when it begins to grow again. Most conifers bear male and female cones on different branches of the same tree.

Connecting Ideas

With the evolution of vascular seedless plants, the prominence of the gametophyte generation was lost. The sporophyte became dominant, and roots, stems, and leaves evolved. The gametophyte generation became modified in the first seed-bearing plants to just a few cells within the pollen grain and the ovule. Fertilization without a surface film of water became possible. In the gymnosperms, plants finally became fully adapted to life on dry land. There was one further development in the evolution of land plants: development of the flowering plants—the most diverse, widespread, and numerous plants living on Earth today.

Section Review

Understanding Concepts

1. Compare the structure of the seed in ginkgos and conifers.
2. Name two ways that seeds are an important adaptation to life on land.
3. How are needlelike leaves an adaptation to life in cold climates?

Thinking Critically

4. How do you think the development of the seed might have affected the lives of herbivorous animals living in Earth's ancient forests?

Skill Review

5. **Comparing and Contrasting** Compare the formation of a spore in ferns and a seed in conifers. For more help, refer to Thinking Critically in the *Skill Handbook*.

Should we let fires burn?

Letting Fires Burn in National Parks

THE DAILY SENTINEL   MAY 1993

Devastating fires that swept through Yellowstone National Park in the summer of 1988 continue to fuel controversy over the National Park Service's "let-it-burn" policy. A Park Service spokesperson denies that such a policy exists and maintains that every fire is managed. Yet not all fires are managed; some still develop into uncontrolled wildfires.

In 1988, more than one-third of Yellowstone National Park's 2.2 million acres burned. Other national forests across the U.S. also experienced massive fires that year. Homes burned and tourism dropped significantly as a result. The firefighters, planes, and bulldozers needed to stop the blazes cost millions of dollars. By the end of the summer, many people were convinced that forest-management policies had failed.

Policy and nature   Until 1972, forest fires were put out upon detection. Forestry officials and biologists have since realized that fire plays a critical role in forest regeneration. Low-intensity ground fires, which do not burn tree branches or crowns, clean the forest floor of debris that can encourage disease, insects, and pests and fuel larger and more damaging fires.

These discoveries led to a "let-it-burn" policy, allowing some naturally ignited fires (usually lightning fires) to burn unless they threaten developed areas or get out of control. Complicated formulas are used to determine if a fire is out of control, including classifications of temperature, humidity, wind speed, and the type of fuel. When a fire exceeds any of these parameters, a crew is sent to contain it.

Different Viewpoints

After the 1988 fires, the people most directly affected—home owners, small businesses, and towns dependent on tourists—protested that the "let-it-burn" policy was inadequate. They argued that if the fires had been put out immediately, they would not have become so costly or damaging. These fires have also left the land unprotected from erosion and stream sedimentation.

Deliberately set fires reduce risks   Forest Service managers support management-set fires as one way to reduce major fire risks. The burning of small areas over many years divides the forest into growth areas of different ages. The younger growth areas help to contain fires that begin in older, drier areas.

INVESTIGATING   **the Issue**

1. **Debate** Find an instance of prescribed burning where you live. Debate the benefits and hazards to the community of this practice.
2. **Research** Go to the library and research the 1988 Yellowstone fires. If all fires had been suppressed immediately, would the damage have been less severe?

25.1 Seedless Vascular Plants

- The spore-producing vascular plants were prominent members of Earth's ancient forests. All are represented by modern species.
- Vascular tissues include xylem and phloem and provide the structural support that enables vascular plants to grow taller than nonvascular plants.
- Spore-producing vascular plants include the Lycophyta, or club mosses; the Sphenophyta, or horsetails; and the Pterophyta, or ferns.
- Fertilization in seedless vascular plants requires a surface film of water through which sperm swim to the egg.

25.2 Gymnosperms

- The term *gymnosperm* is used to describe the four divisions of plants that bear unprotected seeds: Cycadophyta, the cycads; Gnetophyta, the gnetophytes; Ginkgophyta, the ginkgos; and Coniferophyta, the conifers.

- Seeds contain a supply of food to nourish the young plant, protect the embryo during harsh conditions, and provide methods of dispersal to new areas.
- Fertilization in seed-bearing plants does not require water.
- The xylem in conifers and other gymnosperms is composed of thick-walled, nonliving cells called tracheids, which transport water and dissolved nutrients and provide support.

Key Terms

Write a sentence that shows your understanding of each of the following terms.

cone	phloem
cotyledon	pollen grain
deciduous plant	prothallus
embryo	rhizome
evergreen plant	sorus
frond	strobilus
gymnosperm	tracheid
megaspore	vascular tissue
microspore	xylem
ovule	

Understanding Concepts

1. List two new structures that evolved among the spore-releasing vascular plants, and explain the importance of each structure.
2. Contrast the structure of the gametophyte in bryophytes with that of gymnosperms.
3. What characteristics do cycads share with club mosses?
4. What external structures would you use to distinguish between a fern and a gymnosperm?
5. List the similarities and differences in the adaptations for support in ferns and conifers.
6. Describe at least three ways in which male and female pine cones differ.
7. Explain why seeds are an important adaptation for life on land.
8. Explain the functions of xylem and phloem.
9. How does the presence of xylem tissue enable conifers to grow extremely tall?

Relating Concepts

10. Make a concept map that relates the following terms and phrases. Supply the appropriate linking words for your map.

    vascular plants, seeds, spores, lycophytes, sphenophytes, pterophytes, cycads, ginkgos, gnetophytes, conifers

Using a Graph

11. The germination rate of a batch of seeds is the percentage of planted seeds that eventually sprout to produce a new plant. The seeds of the bristlecone pine must be exposed to cold temperatures before they will sprout. Based on the graph below, how long would you keep bristlecone pine seeds under refrigeration before planting?

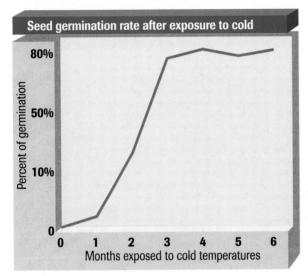

Seed germination rate after exposure to cold

Applying Concepts

12. What is the adaptive advantage of fertilization that no longer requires water for the sperm to reach the eggs?

13. Why do you think conifers are abundant in Canada, Alaska, and Siberia?

14. Explain how evergreen and deciduous plants differ, and describe the adaptive value of each.

15. What might be some evolutionary advantages of having a gametophyte that is dependent on the sporophyte?

16. You are looking for land because you want to build a house. You are shown a beautiful plot of land that is lush with ferns. It also has some club mosses and horsetails. Do you think it would be wise to build a house on this land? Why or why not?

Focus On Conifers

17. Explain why conifers are often found in areas with extreme environmental conditions such as the taiga, remote windy mountaintops, or foggy coastal areas.

Thinking Critically

Interpreting Data

18. Biolab Use the information in the table to develop a key to identify the five different species of pine.

Pine Species	Number of Needles per Bundle	Length (cm)	Type of Spines on Cones
Red	2	10-15	none
Longleaf	3	30-45	small
Pitch	2	8-12	sharp
Ponderosa	2 or 3	15-25	sharp
White	5	8-14	none

Drawing Conclusions

19. Minilab Fern gametophytes must live in moist places. List at least two characteristics of the gametophyte that require the presence of water.

Comparing and Contrasting

20. Compare the functions of strobili on a club moss with the functions of the sori on a fern frond.

Connecting to Themes

21. Systems and Interactions What are the niches of pines and ferns in the forests in which they are found?

22. Evolution Describe how the presence of vascular tissues and seeds enables plants to occupy more land niches.

26 Flowering Plants

eople in temperate regions of the world associate springtime with flowers. As days get longer and warmer, bright crocuses push up through the melting snow. Dry, brown grass turns green and soft again, and new leaves and young flowers unfurl on bare tree branches. The flowering plants are the most widespread, diverse, and colorful members of the plant kingdom. Flowering plants have become adapted to all corners of the world, and their species far outnumber the nonflowering plants.

Most of the plants you are familiar with are flowering plants. Fruit trees, vegetables, cereal grains, and wildflowers are all flowering plants. Why are flowering plants so successful? In this chapter, you will investigate the enormous variety of adaptations that have enabled flowering plants to fill virtually any kind of habitat, from dry salt deserts to snowy mountains and steamy tropics.

How are the wildflowers, the early leaf buds, and the crocus alike? The spring crocus emerges long before the soil has become warm enough for most seeds to sprout. Wildflowers grow year after year in the same field. New leaves and flowers on a tree begin the process that eventually leads to production of seeds in a fruit. All of these organisms are flowering plants.

26.1 What Is an Angiosperm?

Section Preview

Objectives

Differentiate between gymnosperms and angiosperms.

Compare and contrast structures of monocots and dicots.

Key Terms

angiosperm
monocotyledon
dicotyledon
annual
biennial
perennial

Did you have cereal or fruit for breakfast this morning? You've probably already eaten several flowering plants today. Check the labels on your clothing; if your clothes contain cotton or linen fibers, you are wearing flowering plants as well. A common name for all flowering plants is angiosperms.

Diversity of Angiosperms

Angiosperms are the most well-known plants on Earth today. Like gymnosperms, angiosperms have seeds, roots, stems, and leaves. But unlike the other seed plants, an **angiosperm** produces flowers and develops seeds enclosed in a fruit, as shown in *Figure 26.1.* One of the advantages of producing fruit-enclosed seeds is the added protection the fruit provides for the newly developing plants. The flowering plants are classified in the Division Anthophyta.

► Carnations belong to the family Caryophyllaceae. This family includes cultivated flowers such as pinks and baby's breath.

▼ Peanuts are the seeds of this flowering plant. Classified in the Family Fabaceae, peanuts are related to peas, soybeans, alfalfa, and beans. They are often called legumes.

Figure 26.1

There are more than 230 000 species of Anthophyta. Most of the plants you are familiar with probably belong to this division.

Figure 26.2

Monocots have leaves with parallel veins and flower parts in multiples of three.

The grass family Poaceae includes important cereal grains, such as rice and wheat, as well as bamboo and the sugarcane shown here.

▲ The lily family Liliaceae includes asparagus and onions as well as the ornamental lilies grown by many home gardeners.

Anthophytes are made up of two classes, the monocotyledons, or monocots, and the dicotyledons, or dicots. The two classes are named for the number of seed leaves, or cotyledons, contained within the seed. **Monocotyledons** have one seed leaf; **dicotyledons** have two seed leaves. Monocots are the smaller group, with about 60 000 species that include families such as grasses, orchids, lilies, and palms, *Figure 26.2*. Dicots make up the majority of flowering plants with about 170 000 species. They include nearly all the familiar shrubs, trees (except conifers), wildflowers, garden flowers, and herbs. Familiar dicots are shown in *Figure 26.3*.

Figure 26.3

Dicots have leaves with netted veins and flower parts in multiples of four or five.

▼ The rose family Rosaceae includes blackberries, raspberries, apples, plums, peaches, pears, and hundreds of cultivars of garden roses.

▶ The daisy family Asteraceae includes sunflowers, lettuce, dandelions, chrysanthemums, and goldenrod.

◀ The mustard family Brassicaceae includes many important food plants such as cabbage, mustard, broccoli, radish, turnip, collards, and kale.

26.1 What Is an Angiosperm? **633**

Beans of the World

Beans are vegetables that are the seeds of plants with pods or legumes. The Fabaceae family includes more than 14 000 species, but only 22 are grown in quantity for human food. Even poor soils can produce a bean crop and become more fertile at the same time. This is because the root systems of beans absorb nitrogen from the air and leave it in the soil.

Beans for protein   Beans are a good source of the protein that a human body needs. Protein from legumes is just as healthful as protein from animal sources. However, no one type of bean has all of the amino acids needed by humans. That is why vegetarians eat two or three types of beans in a single meal.

Beans around the world   The following are just a few of the legumes enjoyed by people. Black-eyed peas are used in many African and Indian dishes. Cannellini beans are an important part of Italian cuisine. Channa dal are grown and eaten in India and parts of Southeast Asia. Chickpeas are India's most important legume. Lentils provide protein for the people in North Africa and Asia. Lima beans spread from Central America to North America, where Native Americans combined them with corn to make succotash. Peas originated in the Middle East around 6000 B.C. and are now eaten almost everywhere. Pinto beans are used in South American and Mexican dishes. Soybeans have been cultivated in northern China for more than 5000 years and are often made into tofu.

CONNECTION TO Biology

Why might it be more efficient to obtain our protein from beans instead of from animal products?

Angiosperm Adaptations

Flowering plants may be trees, shrubs, herbs, vines, floating plants, epiphytes, and even parasitic plants that may or may not have their own chlorophyll. They can be found in deserts, tropical forests, temperate forests, tundra, snowy mountains, lakes, ponds, and freshwater and salt-water marshes. Angiosperms are the most successful plants on Earth because they have evolved sophisticated adaptations that enable them to survive and reproduce in almost any kind of habitat.

Roots and stems have become adapted to storing food during periods of drought, cold, or limited sunlight. Roots and stems that function

Figure 26.4

Roots and stems are sometimes modified into storage organs that enable a plant to survive through a winter season underground.

▼ **Tulips, daffodils, and crocuses grow from bulbs that have stored food in the bases of last year's leaves.**

Figure 26.5

Angiosperms are more complex, and more adaptable, than any other plant group.

▲ Some grasses form dense mats of roots that hold water and enable the plants to survive long periods of drought.

▼ Mistletoe is an evergreen plant with roots that penetrate the tissues of the branches of deciduous trees. This parasite absorbs water and nutrients from the host tree.

▲ Many species of orchids are epiphytes that live on the branches of trees in a tropical forest.

as food-storage organs are called bulbs, corms, and tubers. A bulb is a short stem enclosed in fleshy leaf bases. A corm is a short, thickened, underground stem that is not enclosed in leaf scales. A tuber is a swollen root or stem with buds that will sprout new plants. *Figure 26.4* illustrates some of the plants that form food-storage organs.

Leaves of different shapes and sizes have evolved in response to the amount of sunlight and moisture available. *Figure 26.5* describes a few of the adaptations that have contributed to the survival and success of flowering plants.

▼ Growers plant cut sections of potato tubers, another kind of underground stem, that have one or more eyes, or buds, each of which will grow into a new shoot and eventually a full-grown plant.

◀ The corm of a gladiolus is a thickened, underground stem from which leaf and flower buds arise.

▼ The rhizome of an iris or a ginger plant is an underground stem.

How can you tell monocots from dicots?

The two classes of flowering plants are monocots and dicots. An easy way to tell them apart is by looking at the pattern of veins in their leaves. The veins of monocot leaves are parallel; dicot leaves have a network of branching veins.

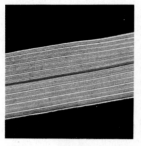

Procedure

1. Examine the pattern of veins in the leaves of each plant your teacher has provided.

2. Determine whether each plant is a monocot or dicot. Make a table to record your answers.

3. Examine the prepared microscope slide of monocot and dicot stems. Describe what you see.

Analysis

1. What differences did you observe between the external appearance of monocot and dicot leaves?

2. What differences did you observe between the internal organization of monocot and dicot stems?

Figure 26.6

Anthophytes may be annuals, biennials, or perennials.

▲ **Vegetable gardeners grow biennial Swiss chard for its leaves, which taste a little like spinach. During cold winter months, the leaves and stems die back, but long, branching roots survive to support the growth of flowers the following spring.**

Life spans of anthophytes

Why do some plants live longer than people, while others live only a few weeks? The life span of a plant reflects its strategies for surviving periods of cold, drought, or other harsh conditions.

Annual plants live for only a year or less. They sprout from seed, grow, reproduce, and die in a single growing season. Most annuals are herbaceous, which means their stems are green and do not contain woody tissue. Many food plants such as corn, wheat, peas, beans, and squash are annuals, as are many weeds of the temperate garden. Annuals survive winter by forming drought-resistant seeds.

Biennials have a life span that lasts two years. Many biennials are plants that develop large storage roots, such as carrots, beets, and turnips. During the first year, biennials grow many leaves and develop a strong root system. Over the winter, the aboveground portion of the plant dies back, but the roots remain alive. Underground roots are able to survive conditions that leaves and stems cannot endure. During the second spring, food stored in the root is used to produce new shoots that bear flowers and seeds.

Perennials live for several years, producing flowers and seeds periodically—usually once each year. They survive harsh conditions by dropping

Woody perennials, like this maple, drop their leaves and become dormant during the winter. New leaves grow when the weather warms and food stored in the tree's roots and trunk moves up into the branches.

Herbaceous perennials often have underground storage organs used for overwintering.

These blue lupines and orange poppies are annual plants that live out their lives during spring and summer, leaving behind seeds for next year's growth.

their leaves or dying back to soil level, while their woody stems or underground storage organs remain intact and dormant. Perennials include the deciduous trees of the world's temperate forests. Although many perennials are large, woody trees and shrubs, there are also herbaceous perennials, including numerous species of grasses and spring wildflowers. What kinds of plants make up a grassy lawn? *Figure 26.6* gives examples of annual, biennial, and perennial anthophytes.

Section Review

Understanding Concepts
1. Why do most early spring flowers grow from bulbs, corms, or tubers?
2. Explain the adaptive value for a maple tree to lose its leaves in the fall.
3. Explain the adaptive value of annual, biennial, and perennial life spans.

Thinking Critically
4. What characteristics would you look for if you were examining an unknown plant to determine whether it is an angiosperm or a gymnosperm?

Skill Review
5. **Observing and Inferring** You are examining a plant with long, thin leaves and flowers that have six petals. Is this plant a monocot or a dicot? For more help, refer to Thinking Critically in the *Skill Handbook*.

Angiosperm Structures and Functions

You've learned that vascular plants have roots, stems, and leaves, and that these structures have become adapted to distinct functions. Recall from Chapter 24 that roots absorb water and minerals from the soil, the stem supports the plant and transports materials, and photosynthesis takes place in the leaves.

Plants are not as active or as complex as animals, but like all multicellular organisms, plants are made up of a variety of tissues. These tissues have functions that support the growth and development of flowering plants.

Roots

Roots are the underground parts of a plant. They anchor the plant in the ground, absorb water and minerals from the soil, and transport these materials to the base of the stem. Some plants such as carrots also accumulate and store food in their roots. The total surface area of a plant's roots may be as much as 50 times greater than the surface area of its leaves. As *Figure 26.7* illustrates, roots may be short or long, thick or thin, massive or threadlike. Some roots even extend above ground.

Figure 26.7

Root systems vary according to the needs of the plant and the texture and moisture content of the soil. The two main types of root systems are taproots and fibrous roots.

▶ **Corn is a monocot with shallow, fibrous roots. As the plants grow to maturity, roots called adventitious roots grow from the stem to help keep the tall plants upright.**

▼ **The fleshy taproot of the biennial beet plant serves as a food-storage organ.**

Figure 26.8

The root structure of dicots and monocots differs in the arrangement of xylem and phloem.

▶ In dicots, xylem forms a star-shaped mass at the center of the root, with phloem nestled between the rays of the star.

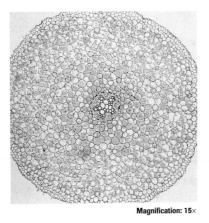

Magnification: 15×

▶ In monocots, strands of xylem alternate with strands of phloem. Monocots usually have a central core of cells called pith. Pith is also composed of parenchyma.

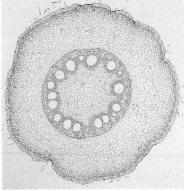

Magnification: 14×

The structure of roots

If you look at the cross section of a typical dicot root in *Figure 26.8,* you can see that the **epidermis** forms the outermost cell layer. A **root hair** is a tiny extension of a single epidermal cell that increases the surface area of the root and its contact with the soil, and absorbs water and dissolved minerals. The next layer is the **cortex,** which is involved in the transport of water and ions into the vascular core at the center of the root. The cortex is made up of packed cell layers of **parenchyma,** a tissue that sometimes acts as a storage area for food and water. Parenchyma occurs throughout most plants.

At the inner limit of the cortex lies the **endodermis,** a single layer of cells that forms a waterproof seal that surrounds the root's vascular tissue. The endodermis controls the flow of water and dissolved ions into the root. *Figure 26.9* traces the two pathways by which water and mineral ions move into the root. Just below the endodermis is the pericycle. The **pericycle** is a tissue that gives rise to lateral roots.

Figure 26.9

Water and mineral ions enter the root either by absorption into root hairs and across the cells or by flowing between the cells of the epidermis.

▼ Mineral ions and water molecules enter root hair cells and travel through the cells of the cortex by osmosis (A). Water may also flow between the cells of the cortex.

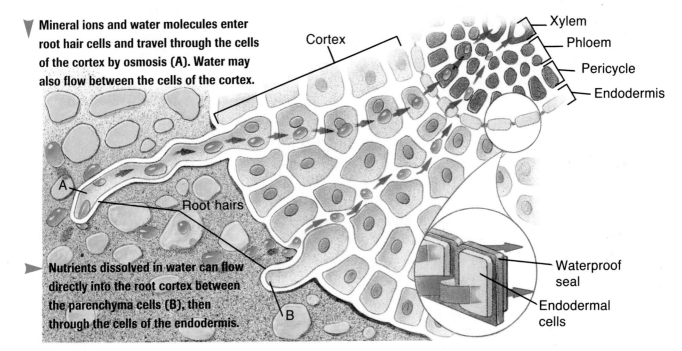

▶ Nutrients dissolved in water can flow directly into the root cortex between the parenchyma cells (B), then through the cells of the endodermis.

Cortex

Root hairs

A

B

Xylem
Phloem
Pericycle
Endodermis

Waterproof seal
Endodermal cells

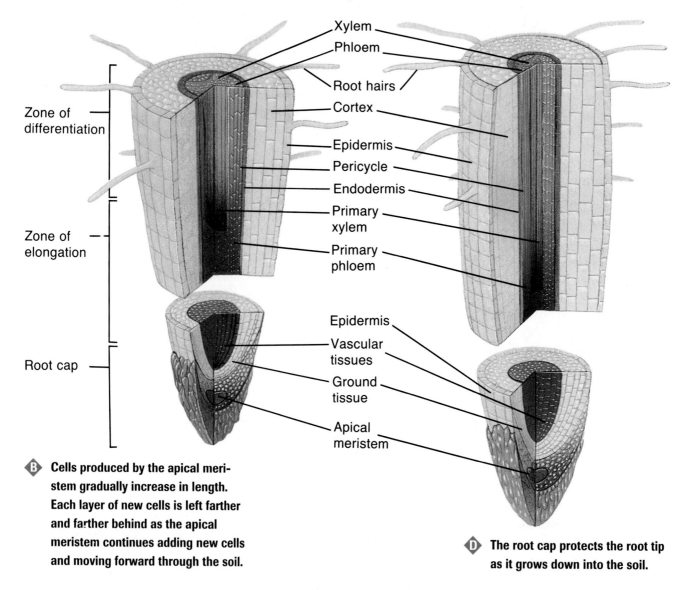

A As root cells mature, they begin to differentiate. Epidermal cells develop root hairs, and internal cells develop into cortex, endodermis, pericycle, cambium, and vascular tissues. Cells that form the cambium soon begin contributing to the root's growth by adding cells that increase its size.

C As cells of the root cap wear away, new cells are added by the tip end of the apical meristem. Cells produced further behind the apical meristem will differentiate into other types of root tissue.

Zone of differentiation

Zone of elongation

Root cap

Xylem
Phloem
Root hairs
Cortex
Epidermis
Pericycle
Endodermis
Primary xylem
Primary phloem

Epidermis
Vascular tissues
Ground tissue
Apical meristem

B Cells produced by the apical meristem gradually increase in length. Each layer of new cells is left farther and farther behind as the apical meristem continues adding new cells and moving forward through the soil.

D The root cap protects the root tip as it grows down into the soil.

Figure 26.10

Roots develop by both cell division and growth. As cells build up, the root increases in size.

Xylem and phloem are located in the center of a root. Between the xylem and phloem tissues, there is a single layer of cells called the vascular cambium. **Cambium** is growth tissue that produces additional xylem and phloem cells.

Root growth

Remember from Chapter 11 that the cells responsible for root growth

in an onion root lie just behind the tip. Cells that are capable of mitosis form a growth tissue that remains just behind the root tip and is called the **apical meristem.** As cells produced by the apical meristem begin to mature, they develop root hairs, as *Figure 26.10* indicates. The tip of each root in a plant is covered by a tough, protective layer of living parenchyma cells called the **root cap.**

Stems

Stems are the aboveground parts of plants that support leaves and flowers. Their form ranges from the thin, herbaceous stems of daisies, which die back every year, to the massive woody trunks of trees that may live for centuries. Green, herbaceous stems are soft and flexible and usually carry out some photosynthesis. Petunias, marigolds, impatiens, and carnations are examples of plants with herbaceous stems. Trees, shrubs, some woody perennials such as oaks, maples, lilacs, and roses have woody stems. Woody stems are hard and rigid and contain a large number of strands composed of xylem.

Stems have several important functions. They provide support for all the aboveground parts of the plant. The vascular tissue that runs through the stem includes xylem and phloem and transports water, mineral ions, and sugars to and from roots and leaves. Leaves and flower buds are produced in the apical meristem of the shoot. The buds at each node located at intervals along the length of the stem produce lateral stems or flowers.

Internal structure of stems

As you learned in Chapter 24, xylem is the vascular tissue that transports water upward from the roots, and phloem is the vascular tissue that transports sugars made during photosynthesis away from the leaves. Although the stem contains many of the same tissues as a root, the tissues are arranged differently. As you can see in *Figure 26.11,* monocots and dicots differ in the arrangement of vascular tissues in their stems.

Woody stems

Most herbaceous stems live only for a single growing season. They remain soft and green because they never develop woody tissue. The stems of gymnosperms and some herbaceous and perennial dicots survive for several years. As the stems of these plants grow in height, they also grow in thickness. This added thickness, or secondary growth, results from cell division in the vascular cambium of the stem. Cell division in the vascular cambium produces new vascular tissue that increases the thickness, or girth, of the stem and is

Figure 26.11

One of the primary differences between roots and stems is the bundled arrangement of vascular tissues within a surrounding mass of parenchyma tissue.

▼ **The vascular bundles in a monocot are scattered throughout the stem.**

► **In young herbaceous dicot stems and those that do not increase in thickness, xylem and phloem are arranged in vascular bundles in the cortex. In older stems, including all woody stems, the vascular tissues form a continuous cylinder between the cortex and the pith.**

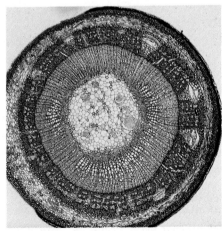

Magnification: 20×

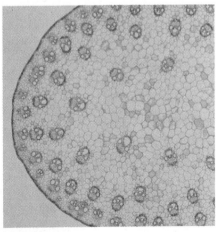

Magnification: 20×

Figure 26.12

Xylem carries water up from roots to leaves. Phloem transports sugars from the source in the leaves to sinks located throughout the plant.

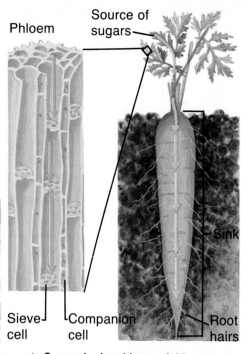

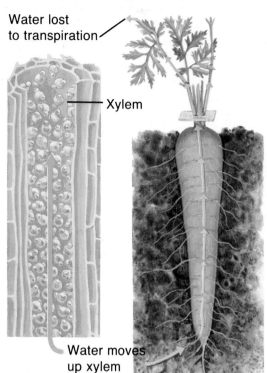

Water lost to transpiration

Xylem

▶ The open ends of xylem vessel cells form complete pipelike tubes. The water molecules form an unbroken column within the xylem. As transpiration from the leaves pulls water molecules from the top of the water column, the rest of the water in the xylem is pulled up through the plant.

Water moves up xylem

Source of sugars

Phloem

Sink

Sieve cell

Companion cell

Root hairs

▲ Sugars in the phloem of this carrot plant are moving to sinks. Companion cells help manage the one-way traffic that flows through the sieve cells of the phloem.

called secondary growth. You can read about secondary growth in woody stems in *The Inside Story.* Some annual stems such as sunflowers start out as herbaceous, but over the growing period also develop woody tissues.

As secondary growth continues, the outer portion of a woody stem develops bark. **Bark** is made up of phloem tissue and a tough, corky layer that protects the phloem from damage by burrowing insects and herbivores.

Vascular tissue

Whenever a plant produces more sugars during photosynthesis than are needed for plant processes, some may be stored for future use. Any portion of the plant that uses or stores these sugars is called a **sink.** Vascular tissues contain fibers for support and parenchyma for storage.

Both xylem and phloem are composed of tubular cells joined end to end. The xylem of angiosperms is made up of vessel cells and fibers. A **vessel cell** is an elongated cell with open ends. Fibers are elongated, elastic cells that contain lignin, a polymer that makes cell walls more rigid. In angiosperms, vessels conduct water and fibers provide support. Water diffuses into and out of the xylem.

Phloem is made up of **sieve cells,** which are thin-walled cells with sievelike plates on their end walls. Sieve cells have cytoplasm but no nuclei. Each sieve cell has a **companion cell** that does contain a nucleus and helps control movement through the sieve cell. Sugars move into and out of the phloem by active transport and osmosis. The structure of vascular tissue cells and the movement of water and sugars through them is shown in *Figure 26.12.*

642 Flowering Plants

Growth Rings of a Tree

The inner portion of the trunk of a tree is composed primarily of dead xylem cells from the growth of previous years. A woody stem also includes bark, an outer layer of cells that protects phloem cells.

Growth rings

1 After its first year of growth, a tree trunk is stiff and green but without bark. The xylem is on the inside adjacent to the pith, whereas the phloem is on the outside adjacent to the cortex.

2 During the second year of the stem's growth, the vascular cambium produces a second layer of xylem tissue. As the cambium is pushed outward by the newly formed layer of xylem tissue, the phloem is also pushed outward.

3 Another layer of cambium, called the cork cambium, develops outside the ring of phloem. The cork cambium produces cork cells, which replace the epidermis and form the tough, waterproof, outer covering of the bark.

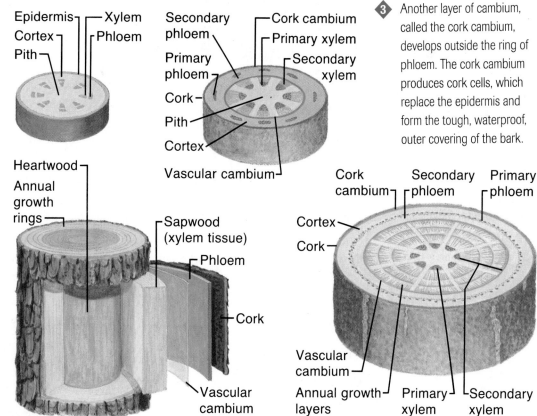

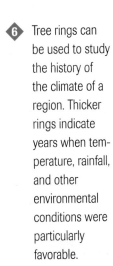

6 Tree rings can be used to study the history of the climate of a region. Thicker rings indicate years when temperature, rainfall, and other environmental conditions were particularly favorable.

5 Over time, the xylem cells toward the center of the trunk stop functioning as vascular tissue. This tissue forms the darker heartwood of the tree. The outer layers of cells that remain actively involved in transport form the lighter-colored sapwood of the tree.

4 The vascular cambium does not divide during long periods of extreme cold or drought. Xylem cells produced in the spring tend to be larger than those produced later in the growing season because there is more water available in the spring. This alternation in the sizes of xylem cells produces a pattern of annual growth rings that in temperate regions reveals the age of the tree.

What do stomata look like?

The lens-shaped openings in the epidermis of a leaf allow gas exchange and help control water loss.

Procedure

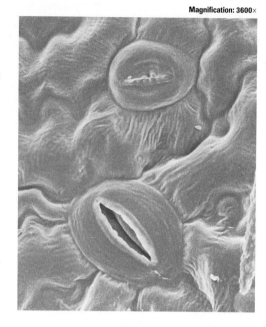

Magnification: 3600×

1. Make a wet mount by tearing a leaf at an angle to expose a thin section of epidermis. Use tap water to make wet mounts of both the upper and lower epidermis.

2. Examine each of your slide preparations under the microscope. Draw or take down a written description of what you see.

3. Make another wet mount using a five percent salt solution instead of tap water. Examine the slide under the microscope and record your observations.

Analysis

1. What do the cells of the leaf epidermis look like? What is their function?

2. How do the epidermal cells differ from guard cells? Which cells, if any, contain chloroplasts?

3. What differences in the stomata did you notice when you used a salt solution to prepare your wet mount? Can you explain what happened in terms of osmosis?

Growth of the stem

Primary growth in a stem is similar to primary growth in a root. The tissue responsible for this growth is the apical meristem, which lies at the tip of a stem. Lateral meristem, located at nodes along the stem and at the axils of leaves, gives rise to new branches, which have their own apical meristem.

Leaves

The primary function of the leaves is to trap light energy for photosynthesis. Most leaves have a relatively large surface area so they can receive plenty of sunlight. They are also often flattened, so sunlight can penetrate to the photosynthetic tissues just beneath the surfaces of the leaf.

Types of leaves

When you think of a leaf, you probably think only of the flat, broad, green structure known as the leaf blade. Some leaves are joined directly to the stem, such as grass blades. In other leaves, there is a stalk that joins the leaf blade to the stem. Also part of the leaf, this stalk is called the **petiole.** The petiole contains vascular tissues that extend into the leaf to form veins. If you look closely, you will notice these veins as lines or ridges running along the leaf blade. *Figure 26.13* gives one example of the variety of shapes in leaves.

Figure 26.13

A simple leaf has one entire leaf blade.
A compound leaf has a divided leaf blade.
The leaves of maple trees are simple leaves.
The leaves of walnut trees, below, are compound leaves.

Internal structure of leaves

You read in Chapter 24 that leaves have a waxy cuticle and stomata that help prevent water loss. The evaporation of water from these stomata in the leaves is called **transpiration.** Plants lose up to 90 percent of all the water they transport from the roots through transpiration. To help reduce water loss from transpiration, the size of the pore is reduced by guard cells. **Guard cells** are cells that surround and control the size of the opening in a stomata. The operation of guard cells is described in *Figure 26.14.*

Figure 26.14

Guard cells prevent a plant from drying out. They regulate the size of the openings of the stomata according to the amount of water in the plant.

ThinkingLab   Analyze the Procedure

What determines the number of stomata on leaves?

Stomata are openings on the surface of a leaf that enable a plant to exchange gases. Assume that you have examined a large number of leaves from individual trees of the same species in your area under the microscope. You observe that they all have about the same number of stomata in a certain area. You are curious to know whether that number is determined genetically or if it varies for trees of the same species that grow in different environments. You hypothesize that in drier climates, there would be fewer stomata on the leaves of individual trees in this species. You think that fewer stomata in a drier climate would prevent the leaves from drying out.

Analysis

You ask a relative who lives in a drier state to send you some leaves from the same species of tree. After examination, you find that there are fewer stomata on the leaves of these trees.

Thinking Critically

Can you conclude, based on your data, that dry climates cause fewer stomata to form on leaves? Why or why not? If not, what else would you need to do to determine whether the variation is related to the climate?

A Guard cells are modified cells of the leaf epidermis that contain chloroplasts. These cells also contain microfibrils made of cellulose that are arranged in belts around the circumference of the cells.

B When there is plenty of water in surrounding cells, guard cells take in water by osmosis. When water enters the guard cells, microfibrils prevent them from expanding in width, so they expand in length. Because the two guard cells are attached end to end, this expansion in length forces them to bow out and the pore opens.

C When the plant becomes dry, there is less water in tissues surrounding the guard cells. Water leaves the guard cells, thus lowering turgor pressure. The cells return to their previous shape, reducing the size of the pore.

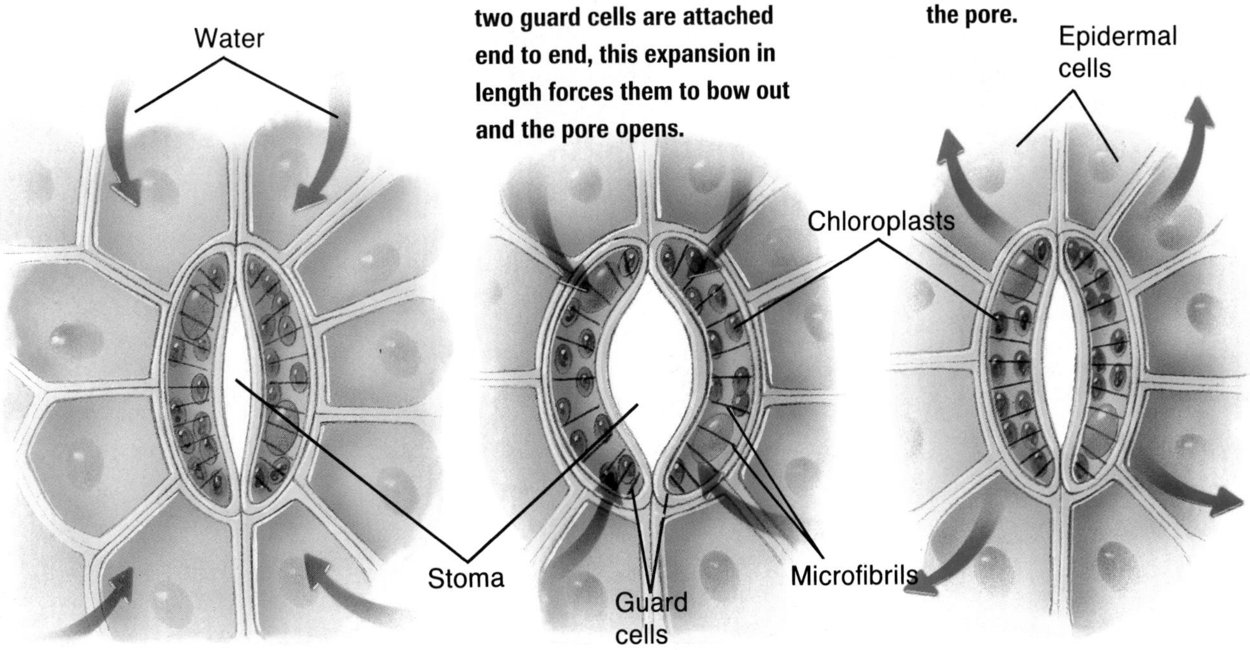

Water

Epidermal cells

Chloroplasts

Stoma

Guard cells

Microfibrils

BioLab

Plant stems contain living tissues that grow and change as the plant matures. Major stem functions include support for the plant as it grows and transport of food, water, and minerals throughout the plant. Vascular tissues of stems continue to grow as the tree matures. The woody stem of a tree provides protection of plant tissues and contains a record of the tree's age.

Growth of Stems

PREPARATION

Problem
How do the stems of one-, two-, and three-year-old trees compare?

Objectives
In this Biolab, you will:
- **Identify** stem tissues in one-, two-, and three-year-old tree stems.
- **Relate** the functions of these stem tissues.

Materials
art of one-, two-, and three-year-old stems

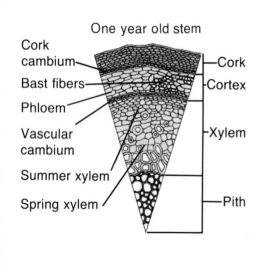

One year old stem

- Cork cambium
- Bast fibers
- Phloem
- Vascular cambium
- Summer xylem
- Spring xylem

- Cork
- Cortex
- Xylem
- Pith

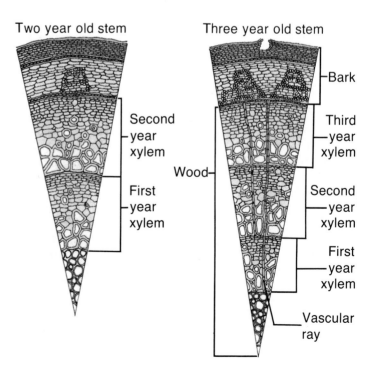

Two year old stem

- Second year xylem
- First year xylem

Three year old stem

- Bark
- Third year xylem
- Wood
- Second year xylem
- First year xylem
- Vascular ray

1. Examine the diagram of a cross section through a one-year-old tree stem.

2. Study the descriptions of stem tissue given in the table, and identify these tissues on the diagram.

3. Compare the first diagram with the diagram of a cross section through a two-year-old stem.

4. Note that a difference in the size of spring and summer xylem cells creates a line separating one year's xylem from the next year's xylem.

5. Compare the diagram of the three-year-old tree stem with the other two diagrams.

6. Identify the stem tissues shown on the three-year-old tree stem, including bark and wood.

Tissue	Description
Cork	Outermost layer, about eight cells thick
Cork cambium	Single layer of cells inside cork layer
Cortex	First layer inside cork cambium, about ten cells thick
Pith	Tissue at center of stem with large, thin-walled cells
Xylem	Thick layer of cells next to pith, widest layer of cells in stem
Vascular cambium	Single layer of cells at top edge of xylem
Phloem	Groups of thin-walled cells inside cortex
Bast fibers	Groups of thick-walled cells that surround phloem

ANALYZE AND CONCLUDE

1. **Drawing Conclusions** How can you tell the age of a tree that has grown in a temperate climate by looking at a cross section of its stem?

2. **Identifying Variables** Annual rings in a tree stem vary in thickness in response to environmental factors that influence growth. What kinds of factors during a year might influence a tree's growth?

3. **Drawing Conclusions** Give a possible reason why the cell diameter of spring xylem cells is greater than the cell diameter of summer xylem cells.

Going Further

Application
Examine a cross section of twigs from local trees with a hand lens. Determine differences in anatomy of different species.

The Fall of the Food Factories

When evenings become crisp, you can feel that autumn is near. In a few weeks, trees will be surrounded by piles of dry, papery leaves in shades of scarlet to brown. This yearly event in northern temperate regions of the world is an adaptation that

keeps deciduous trees from dehydrating during winter when roots cannot absorb water from the frozen ground.

Leaf colors don't change The process of leaf drop, called abscission, begins with nitrogen and minerals moving from leaves for storage in stems and roots. Photosynthesis slows as the amount of sunlight decreases, plants stop producing chlorophyll, and water and nutrients stop moving through the plant. Fall colors result when the dominant pigment, chlorophyll, which is green, stops being produced. Other pigments that are always present in the leaf are then revealed. Carotenoid pigments present in a leaf reflect the colors red, orange, and yellow.

Hormones and enzymes After leaves change color, what causes them to fall off the tree? The answer has to do with the plant's production of a gas called ethylene, which increases in the fall. When you put unripe fruit in a paper bag, the fruit continues to produce ethylene and continues to ripen. In a tree, ethylene causes swelling on the stem side of a leaf, pushing it away from contact with the petiole. At the same time, in the abscission layer, an enzyme attacks cellulose in cell walls, making the layer loose and weak; the leaf usually falls of its own weight, and a protective layer is left to guard against disease.

Time to go Leaf abscission in deciduous trees is environmentally controlled by the shorter day length and cooler temperatures that are characteristic of autumn. City trees near streetlights may actually retain their leaves longer than usual because the lights alter the plant's response to changes in day length.

Thinking Critically

How could you determine whether it is temperature, day length, or availability of water that causes leaves to fall?

Just beneath the cuticle-covered epidermis are two layers of mesophyll. **Mesophyll** is the photosynthetic tissue of a leaf; it is made up of two types of parenchyma. The veins of vascular tissue run through the mesophyll of leaves, *Figure 26.15.* You can read more about the internal structures of the leaf and how they function in photosynthesis in *The Inside Story.*

Figure 26.15

The mesophyll of leaves contains veins of vascular bundles. The pattern veins make in a leaf can help to identify the species.

▼ **Leaves of corn plants have parallel veins, a characteristic of monocots.**

▼ **Leaves of lettuce plants have veins that form a branching network, a characteristic of dicots.**

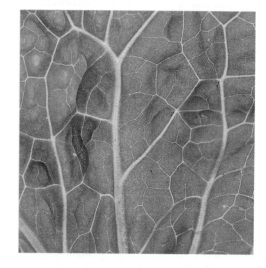

Internal Anatomy of a Leaf

*T*he tissues of the leaf are adapted for photosynthesis, gas exchange, limiting water loss, and transporting sugars. Most leaves have an upper and lower surface and are attached to small branches.

Pin oak

2 Sunlight penetrates the cuticle-covered epidermis to reach the first type of mesophyll tissue, the palisade mesophyll, which is made up of column-shaped cells with many chloroplasts. Most photosynthesis takes place in the palisade mesophyll.

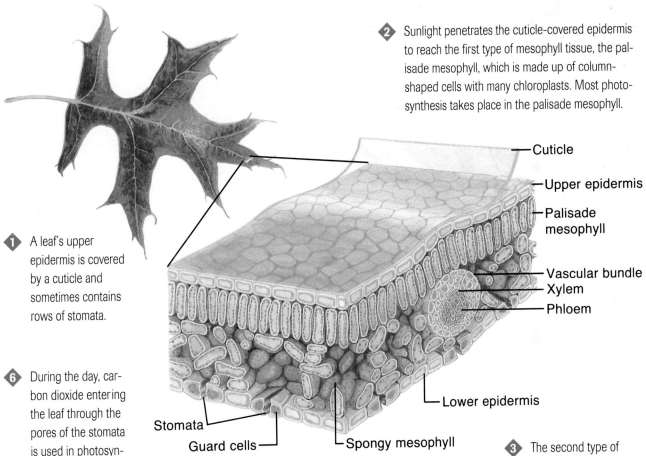

Cuticle

Upper epidermis

Palisade mesophyll

Vascular bundle
Xylem
Phloem

Lower epidermis

Stomata
Guard cells
Spongy mesophyll

1 A leaf's upper epidermis is covered by a cuticle and sometimes contains rows of stomata.

6 During the day, carbon dioxide entering the leaf through the pores of the stomata is used in photosynthesis. Oxygen also enters the pores for the process of respiration. The products of photosynthesis, water vapor and oxygen, exit through the stomata.

4 Below the spongy mesophyll is the lower epidermis, with many stomata. Spongy mesophyll allows gases to move between palisade cells and out of the leaf.

5 Within the spongy mesophyll can be seen cross sections of bundles of vascular tissue that compose the veins of the leaf. The veins are usually surrounded by a sheath of parenchyma that helps regulate the flow of materials into and out of the vascular tissue.

3 The second type of mesophyll tissue is the spongy mesophyll, which is made up of loosely packed, irregularly shaped cells surrounded by air spaces. There are fewer chloroplasts in the spongy layer because this tissue is usually on the underside of the leaf where less light is absorbed.

Figure 26.16

Modified leaves serve many functions in addition to photosynthesis.

▼ The tendrils of pea plants are leaflets that are modified for climbing.

▲ The leaves of the pitcher plant are modified for trapping insects.

▶ The thick leaves of this *Aloe vera* plant are adapted to store water in a dry desert environment.

Modified leaves

Many plants have leaves that are modified for other functions besides photosynthesis. For example, cactus spines are leaves that protect the plant from herbivores and help reduce water loss. The fleshy leaf bases of onion and daffodil bulbs are modified for food storage. Some additional examples of modified leaves are shown in *Figure 26.16*.

Connecting Ideas

Angiosperms are the most diverse group of plants, including more than 230 000 species. These plants have moved entirely away from a dependence on water for reproduction and are adapted to life in nearly every environment found on Earth.

Angiosperms have adaptations that enable them to prevent water loss from their tissues and store food under adverse environmental conditions. They also have evolved ways to attract other organisms to help in both reproduction and seed dispersal.

Section Review

Understanding Concepts

1. Compare and contrast the arrangement of xylem and phloem in dicot roots and stems.
2. In a plant with leaves that float on water, such as a water lily, where would you expect to find stomata? Explain.
3. You are given the leaf of an unknown plant and asked to determine whether it is a monocot or a dicot. Explain how you would determine this.

Thinking Critically

4. Some animals such as squirrels damage trees by stripping off portions of the bark. Of what use is bark to a squirrel?

Skill Review

5. **Observing and Inferring** You are looking at a slide of the cross section of a leaf. The label has fallen off. How can you tell which way to orient the slide on the microscope stage? Hint: Look at the numbers of chloroplasts in cells. For more help, refer to Thinking Critically in the *Skill Handbook*.

Who is responsible for the rain forests?

Bolivia's Rain Forest Shrinking Fast

SCIENCE QUARTERLY   JANUARY, 1994

Some are calling it the ecological disaster of the century. Others say it's much worse than that. The rain forests along Earth's equator are systematically being destroyed at an alarming rate. In Bolivia, 227 000 hectares of forest are being cleared each year.

During the 19th century, a Bolivian prophet, Noko Guahi, told his people of a holy hill— Loma Santa—where outsiders would never find them and where the animals and the fruits of the rain forest would provide for all their needs.

Logging and farming destroy rain forests Today, many indigenous Bolivians are again searching for Loma Santa as they attempt to escape the relentless destruction of their forest homelands by logging and agricultural interests.

Many countries are affected Brazil, Peru, Colombia, Panama, Haiti, and Venezuela are among the Latin American countries where rain forests are being destroyed. Rain forests in Malaysia, the Philippines, Thailand, India, Nigeria, and Ghana are also changing rapidly as logging continues in these fragile ecosystems.

A vital resource Although the rain forests cover just two percent of Earth's surface, they are critically important ecosystems. They act as Earth's lungs, taking in huge amounts of carbon dioxide and releasing oxygen in return. They also filter water. Each year, one-fifth of Earth's fresh water is recycled through the forests along the Amazon and back to the Atlantic Ocean.

More than one-half of all plant and animal species can be found in the rain forests. At last count, the Amazon alone contains 3000 species of fishes and 50 000 species of plants.

Different Viewpoints

Because the rain forests are such a vital link in Earth's water cycle, further deforestation could interrupt weather patterns and contribute to the rise of atmospheric carbon dioxide.

Economics rules the other side of the issue. Many countries with significant rain forests are either extremely poor or seriously in debt to other nations. The rain forest land can be used to generate needed income.

INVESTIGATING the Issue

1. **Research the Issue** Find out what recent scientific discoveries have been made in the rain forest. Report to the class on your research results.
2. **Debating the Issue** Hold a class discussion about the rain forest dilemma. Use new facts to support each side of the debate.

Reviewing Main Ideas

26.1 What Is an Angiosperm?

- Angiosperms make up the Division Anthophyta, which contains two classes, the Monocotyledons and Dicotyledons.
- Angiosperms are an extremely diverse group of plants that includes species adapted for virtually all kinds of habitats.

26.2 Angiosperm Structures and Functions

- Roots grow downward into the soil as cells elongate. Cell division in the apical meristem produces new cells that add to the length of the root. Cell division in the vascular cambium adds xylem and phloem tissue.
- Stems also contain apical meristem, vascular cambium, and vascular tissue. The primary function of the stem is to support upright growth and transport food and water from one part of the plant to another.
- Leaves contain layers of mesophyll tissue that have chloroplasts and perform photosynthesis. Guard cells control the opening and closing of stomatal pores.

Key Terms

Write a sentence that shows your understanding of each of the following terms.

angiosperm	mesophyll
annual	monocotyledon
apical meristem	parenchyma
bark	perennial
biennial	pericycle
cambium	petiole
companion cell	root cap
cortex	root hair
dicotyledon	sieve cell
endodermis	sink
epidermis	transpiration
guard cell	vessel cell

Understanding Concepts

1. What are the characteristics that distinguish the Anthophyta from other divisions of vascular plants?
2. List at least three plant structures in which food is stored. Identify whether each is a modification of the root, stem, or leaf.
3. List two ways in which leaves may be modified for functions other than photosynthesis.
4. What is the adaptive advantage of a root cap?
5. Explain the difference between primary and secondary growth.
6. What is the function of vascular cambium?
7. Explain the function of the apical meristem of a root.
8. What is bark?

Interpreting Scientific Illustrations

9. This diagram was drawn based on the observation of a cross section of a plant structure through the microscope. Identify the plant structure, and identify the lettered parts.

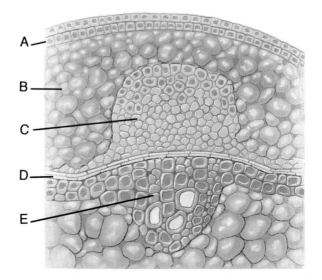

Relating Concepts

10. Make a concept map that relates the following terms and phrases. Supply the appropriate linking words for your map.

    cortex, root hair, xylem, active transport, endodermis, phloem, mineral ions, simple diffusion, epidermis, water

Applying Concepts

11. Compare the expected rates of water movement in the xylem during the day and at night in a plant.
12. What is the difference between palisade and spongy mesophyll?
13. A tree in the park grew around a piece of barbed-wire fence until the fence was covered by tree tissues. Explain how this happened.
14. Compare the movement of water through the xylem to the movement of juice through a soda straw.
15. When you transplant a flowering tree such as a dogwood, you are told to be careful not to disturb the roots. Explain how damage to the roots could harm the tree.
16. The sundew is a carnivorous plant that captures and digests insects. The plant has chlorophyll and produces sugars by photosynthesis. Why do you think this plant captures insects?

Social Studies Connection

17. Many farmers grow wheat in their fields for a few years followed by a crop of beans, even though they don't plan to harvest the beans. Explain why they plant beans.

Thinking Critically

Applying Concepts

18. One spring, after the tulips in your garden bloomed, you cut off the leaves that were still green to tidy up the garden. The next spring, only a few of the tulips bloomed; the rest just grew a few weak leaves. Explain why this happened.

Relating Concepts

19. **Biolab** Foresters work on tree farms called plantations in which they manage trees by cutting down some to allow others to grow straighter. One technique foresters use to remove trees involves removing the bark in a circle around the trunk of the tree. This process is called girdling a tree. How does girdling eventually kill a tree?

Interpreting Data

20. **Thinking Lab** A scientist examined leaves from the same species of daisies from two different states. Examine the data in the graph below, and identify which plant grew in the state with higher amounts of rainfall.

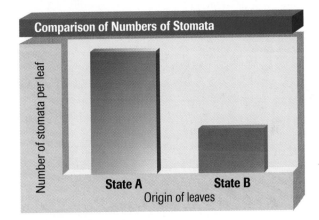

Connecting to Themes

21. **Systems and Interactions** Every spring, sap is collected from sugar maple trees to make maple syrup. Why is sap so full of sugars in the spring?
22. **Evolution** What are the adaptive advantages to the plant of having the palisade mesophyll near the top surface of the leaf, and a larger number of stomata in the lower epidermis than in the upper epidermis of the leaf?

CHAPTER 27
Reproduction in Flowering Plants

Flowering plants are the most colorful as well as the most widespread and diverse members of the plant kingdom. This enormous variety of colors, sizes, shapes, and fragrances may be due to the close evolutionary partnership between flowers and insects. The pollen and sweet nectar produced by flowers provide food for butterflies, bees, and other insects. Brightly colored or fragrant petals help guide the insects to the nectar. While feeding, they carry pollen from flower to flower, causing fertilization and contributing to the success of seed production.

As flowers and insects evolved, they adapted to each other, as well as to other factors in their environment. In this chapter, you will study the structure of a flower, learn how flowers are pollinated, and explore the adaptations of flowers that encourage animals to help with pollination. You will also investigate the formation, dispersal, and germination of seeds.

Mammals and flowering plants appeared on Earth at about the same time. Like insects, mammals also have a close evolutionary partnership with anthophytes. Fruit provides food for mammals and, by eating the fruit, mammals help disperse seeds. Why is it an adaptive advantage for a plant to produce fruit that looks and tastes good to mammals?

27.1 What Is a Flower?

How would you choose flowers for a bouquet or a garden? Perhaps you would start with fragrant roses, jasmine, or gardenias. You might add color with tall spikes of gladioli, cushions of marigolds, bright tulips, or daisies. Grasses would contribute a graceful shape, though their flowers may be so small we completely overlook them. All of these flowers are beautiful to look at and some have delicate scents as well. In what other ways are all of these flowers alike?

The Structure of a Flower

The process of sexual reproduction in flowering plants takes place in the flower, which is a complex structure made up of several parts. Some parts of the flower are directly involved in fertilization and seed production, whereas other parts have functions in pollination. There are probably as many different shapes, sizes, colors, and configurations of flower parts as there are species of flowering plants. In fact, features of the flower are often used in plant identification. *Figure 27.1* shows some examples of the variety in flower forms.

Even though there is an almost limitless variation in flower shapes and colors, all flowers share a simple, basic structure. A flower is made up of four kinds of organs: sepals, petals, stamens, and pistils.

The flower parts you are probably most familiar with are the petals. **Petals** are leaflike, usually colorful structures arranged in a circle called a corolla around the top of a flower stem. **Sepals** are also leaflike, usually green, and encircle the flower stem beneath the petals. Inside the circle of petals are the stamens. A **stamen** is the male reproductive structure of a flower. At the tip of the stamen is the **anther,** which produces pollen containing sperm.

At the center of the flower, attached to the tip of the flower stem, lie one or more pistils. The **pistil** is the female structure of the flower. The bottom portion of the pistil enlarges to form the **ovary,** the part of the flower in which the ovules containing eggs are formed. As you read in Chapter 25, the female gametophyte develops inside the ovule.

The *Inside Story* on page 658 shows how these structures are arranged in a typical flower.

Modifications in Flower Structure

A flower that has all four organs—sepals, petals, stamens, and pistils—is called a **complete flower.** The morning glory and tiger lily shown in *Figure 27.1* are examples of complete flowers, as is the phlox shown in *The Inside Story.* A flower that lacks one or more organs is called an **incomplete flower.** For example, squash plants have separate male and female flowers. The male flowers have stamens but no pistils; the female flowers bear pistils but no stamens. Plants like sweet corn that are adapted for pollination by wind and not animal pollinators have no corolla.

Adaptations for Pollination

You learned in Chapter 12 that pollination is the process of transferring pollen grains from the anther to the stigma. Plant reproduction is most successful when the rate of pollination is high, which means that the pistil of a flower receives enough pollen of its own species to fertilize the egg in each ovule.

Figure 27.1

The diversity of flower forms is evidence of the success of flowering plants.

◀ **The male flowers of the walnut tree form long catkins.**

▶ **The spotted petals of the tiger lily curl away from the reproductive structures at the center of the flower.**

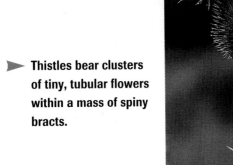

▶ **Thistles bear clusters of tiny, tubular flowers within a mass of spiny bracts.**

▼ **The petals of the morning glory are fused together to form a bell shape.**

Parts of a Flower

O f the four major organs of a flower, only two— the stamens and pistils—are fertile structures directly involved in seed development. Sepals and petals support and protect the fertile structures and help attract pollinators. The structure of a typical flower is illustrated here by a phlox flower.

Blue phlox

1 Petals are usually brightly colored and often have perfume or nectar at their bases to attract pollinators. The corolla is made up of the flower's petals. In many flowers, the corolla also provides a surface for insect pollinators to rest on while feeding. Petals may be fused to form a tube, or shaped in ways that make the flower more attractive to pollinators.

2 The stigma, at the top of the pistil, is a sticky or feathery surface on which pollen grains land and grow. The style is the slender stalk of the pistil that connects the stigma to the ovary. The pollen tube grows down the length of the style to reach the ovary. The ovary, which will eventually become the fruit, contains the ovules. Each ovule, if fertilized, will become a seed.

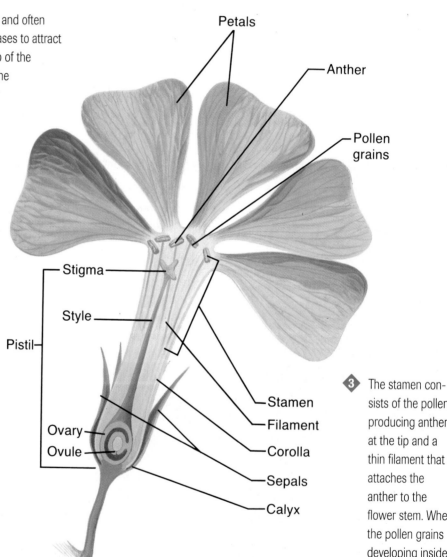

Petals

Anther

Pollen grains

Stigma

Style

Pistil

Stamen

Filament

Ovary

Corolla

Ovule

Sepals

Calyx

3 The stamen consists of the pollen-producing anther at the tip and a thin filament that attaches the anther to the flower stem. When the pollen grains developing inside the anther reach maturity, the anther splits open to release them.

4 The ring of sepals makes up the outermost portion of the flower. All of the sepals together form the calyx. The calyx serves as a protective covering for the flower bud, helping to protect it from insect damage and prevent it from drying out. Sepals sometimes are colored and resemble petals.

Pollination by wind is random

As you know, gymnosperms depend only on the wind for pollination. So do many flowering plants. But wind scatters pollen randomly, and pollen grains can land many places besides the stigma of the proper flower. Many anthophytes have elaborate mechanisms that help ensure that pollen grains are deposited in the right place. These pollination mechanisms involve relationships with animals, including beetles, butterflies, moths, bees, flies, hummingbirds, and even bats. These relationships have helped make flowering plants the most successful plant group on Earth.

Nectar attracts animal pollinators

Most flowers produce nectar, which serves as a valuable, highly concentrated food for animal pollinators. Nectar is a liquid, made up of proteins and sugars, that may be produced by any of the flower organs. It usually collects in the cuplike area at the base of the petals. For example, butterflies and moths unwind their long, curled proboscises to suck nectar from flowers that are often bell or tube shaped. The stamens of these flowers may extend past the petals. As the animal positions itself for a meal, it brushes against the anthers. During a visit to another flower, some of the pollen grains sticking to the pollinator's body are brushed off onto the stigma of a different flower, resulting in pollination.

Pollination in Orchids

Orchids make up the largest family of flowering plants. Orchids are found in every habitat except the frozen polar regions. Some orchid flowers are many centimeters across, while some are extremely small. There are orchids that grow in dense clusters, while others are borne singly on a thick stalk. Many are exquisitely beautiful, some are green, and others can only be described as bizarre.

A unique flower Orchid flowers vary tremendously in shape, size, and color. But they all have three sepals and three petals. However, one of the petals is different from the other two. This special petal, called the lip, is usually large. It may also sport distinctive markings, be shaped into a tube or pouch, or be decorated with hairs or warts.

Orchid flowers are unique in that the male and female reproductive structures—the stamen and the pistil—are fused together into a single structure called the column. Pollen grains in orchids are not loose and powdery like they are in most other flowering plants. They are clumped together in waxy masses called pollinia.

Master of deception A group of tropical American orchids that belongs to the genus *Coryanthes* is known as bucket orchids. The lip of the flower forms a sort of bucket that is filled with liquid, and bees

are attracted to the flower by a sweet-smelling oil that coats the rim of this structure. Invariably, as a bee tries to collect the tempting oil, it slips and falls into the liquid. The drowning bee eventually discovers the only escape route—a narrow tunnel that contains the pollinia. As the bee struggles to safety, the pollinia adhere to its back.

Several different kinds of orchids, including *Trichoceros antennifera* and some members of the genus *Ophrys,* have flowers that mimic the look and smell of female flies, bees, or wasps. When the appropriate male insect comes along, he vigorously tries to mate with the flower, picking up pollen in the process. When he forces his attentions on the next flower, pollination takes place.

Thinking Critically
Many orchid flowers are adapted for pollination by only a particular insect species, such as a single type of fly or wasp. What potential disadvantage can you see in having only one type of pollinator?

Red Poppy

by Georgia O'Keefe (1887–1986)

"When you take a flower in your hand and really look at it," she said, cupping her hand and holding it close to her face, "it's your world for the moment. I want to give that world to someone else. Most people in the city rush around so, they have no time to look at a flower. I want them to see it whether they want to or not."

American artist Georgia O'Keefe attracted much attention when the first of her many floral scenes was exhibited in New York in 1924. Everything about these paintings—their color, size, point of view, and style—overwhelmed the viewer's senses, just as their creator had intended.

The viewer's eye is drawn into the flower's heart In this early representation of one of her familiar poppies, O'Keefe directed the viewer's eye down into the poppy's center, much as the flower naturally attracts an insect for reproduction purposes. By contrasting the light tints of the outer ring of petals with the darkness of the poppy's center, the viewer's eye is pulled beelike into the heart of the flower. The overwhelming size and detailed interiors of O'Keefe's flowers give an effect similar to the photographer's close-up camera angle.

By the time of her death in New Mexico in 1986, O'Keefe's gargantuan blossoms—including irises, lilies, and poppies—had become her signature work.

CONNECTION TO Biology

Color plays a prominent role in the artistic effect of O'Keefe's flowers. What role does color play in the life of a real flower?

Flower colors signal appropriate pollinators

Nectar-feeding pollinators are attracted to a flower by its color or scent, or both. Flowers that attract butterflies usually have petals with bright, vivid colors, such as daisies, phlox, rhododendrons, and zinnias. Butterflies alight while feeding, so they visit flowers with a platform or cluster of petals. As moths are active at night, flowers that attract moths stay open all night. They include tobacco, many orchids, night-blooming cereus, and honeysuckle. Moth-pollinated flowers are usually pale in color but have a strong, sweet scent that attracts the insects from some distance away. Moths hover while feeding, so they can visit blossoms that do not have a landing platform. Bees collect pollen as well as nectar. They are attracted to yellow or blue flowers with a sweet scent, such as peas, mints, primroses, irises, and lupine.

Figure 27.2

The shape, color, and size of a flower reflect its relationship with a pollinator.

▼ **The butterfly uses its long proboscis to sip nectar that the shorter tongues of bees and flies cannot reach.**

Scent attracts other pollinators

Flowers pollinated by beetles and flies have a strong scent but are often dull in color and may not produce nectar. Examples include magnolias, calla lilies, and wild parsley. Beetles and some fly species chew on flower parts, picking up and depositing pollen as they crawl about. The flowers aren't seriously harmed by this activity because they have such a large number of stamens and pistils that many are left undamaged. The scent of some fly-pollinated flowers such as skunk cabbage or trillium resembles rotting meat and attracts female flies searching for places to lay their eggs. *Figure 27.2* shows how flowers are adapted to attract their pollinators.

Pollination and Variation

You learned in Chapter 14 that sexual reproduction, which involves the processes of meiosis and fertilization, provides a mechanism for mixing the genetic material in a population. In anthophytes, the process of pollination contributes to the possibility of variations in genetic information. Pollination includes self-pollination and cross-pollination.

The wind-pollinated flowers of this ragweed plant are small and green with no petals to block wind currents. Wind-pollinated plants produce large amounts of pollen, as anyone with a pollen allergy can tell you.

Flowers pollinated by hummingbirds are often colored bright red or yellow but may have little scent, as birds do not have a well-developed sense of smell.

Bats sip nectar from night-blooming flowers with a strong, musty odor, such as bananas and some cacti. Notice how the anthers are positioned to rub against the bat's head as the animal reaches for nectar at the base of the flower.

Examining the Structure of a Flower

Flowers are the reproductive structures of anthophytes. Seeds that develop within the flower are carried inside a fruit. Seeds enable the plant to reproduce. Flowers come in many colors and shapes. Often their colors or shapes are related to the manner in which pollination takes place. The major organs of a flower include the petals, sepals, stamens, and pistils. Some flowers are incomplete, which means they do not have all four kinds of organs. You will study a complete flower.

PREPARATION

Problem
What do the parts of a flower look like, and how are they arranged?

Objectives
In this Biolab, you will:
- **Observe** the structures of a flower.
- **Identify** the functions of flower parts.

Materials
flower—any complete flower that is available locally, such as phlox, lily, or tobacco flower
2 microscope slides
water
dropper
single-edged razor blade
2 coverslips
microscope
hand lens (or stereomicroscope)
colored pencils (red, blue, green)

Safety Precautions
Razor blades are sharp. Always use caution with razor blades.

PROCEDURE

1. Examine your flower. Locate the sepals and petals. Note their numbers, size, color, and arrangement on the flower stem.

2. Remove the sepals and petals from your flower by gently pulling them off the stem. Locate the stamens, each of which consists of a thin filament with a pollen-filled anther on top. Note the number of stamens.

3. Locate the pistil. The stigma at the top of the pistil is sticky. The style is a long, narrow structure that leads from the stigma to the ovary.

4. Place an anther from one of the stamens onto a microscope slide and add a drop of water. Cut the anther into several pieces with the razor blade. **CAUTION:** *Always take care when using a razor blade.*

5. Examine the anther under low and high power of your microscope. The small, dotlike structures are pollen grains.

6. Slice the ovary in half lengthwise with the razor blade. Mount one half, cut side facing up, on a microscope slide.

7. Examine the ovary section with a hand lens or stereomicroscope. The many small, dotlike structures that fill the two ovary halves are ovules. Each ovule contains an egg cell that is not visible under low power. A tiny stalk, called a funiculus, connects each ovule to the ovary wall.

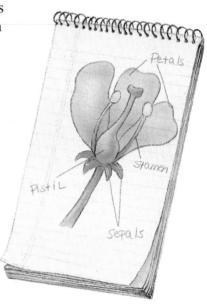

8. Identify the ovary, ovules, and a funiculus.

9. Make a diagram of the flower, labeling all its parts. Color the female reproductive parts red. Color the male reproductive parts green. Color the remaining parts blue.

ANALYZE AND CONCLUDE

1. **Observing** How many stamens are present in your flower? How many pistils, ovaries, sepals, and petals?

2. **Comparing and Contrasting** Make a reasonable estimate of the number of pollen grains in an anther and the number of ovules in an ovary of your flower.

3. **Interpreting Data** Are there more pollen grains produced by one anther than ovules produced by one ovary? Give a possible explanation for your answer.

Going Further

Project Use a field guide to identify common wildflowers in your area. Most field identifications are made on the basis of color, shape, numbers, and arrangement of flower parts. If collecting is permitted, pick a few common flowers to press and make into a display of local flora.

▲ The lower petal of the flower serves as a landing platform for a bee.

▲ The weight of the insect crawling into the flower causes the anther to drop and deposit pollen on the bee's back.

Figure 27.3

The anthers of this *Salvia* flower mature before the stigma. The immature stigma is the long, thin structure on the top lip of the flower in the photo.

Self-pollination is less adaptive than cross-pollination

If a stigma receives pollen from the same flower, or from another flower on the same plant, it is self-pollinated. Self-pollination results in progeny with the same genetic makeup as the parent, although the progeny will not be identical to the parent genetically due to independent assortment of chromosomes and crossing-over. If a stigma receives pollen from another flower of the same species, the flower is cross-pollinated. Cross-pollination results in an exchange of genetic material. Most flowers have adaptations that favor cross-pollination over self-pollination. For example, the male flowers of wind-pollinated sweet corn are borne at the top of the plant, while the female flowers are farther down the plant stem. This arrangement gives pollen grains a chance to be blown away from the parent plant before drifting down onto a female flower. In plants with both male and female flowers, the males on one plant often develop before the female flowers do. Self-pollination is prevented because the pollen has already been dispersed to other plants by the time the female flowers appear.

Structural adaptations favor cross-pollination

Plants with complete flowers may have several structural adaptations for cross-pollination. In some species, chemical factors may prevent successful fertilization with pollen from the same plant. Anthers may release pollen long before the stigma matures, ensuring that the stigma will receive pollen from a different plant's flower. The stigma may extend farther from the petals than the anthers, making it difficult for pollen to reach the stigma of the same flower.

Insect pollination is particularly effective in favoring cross-pollination. The stamens, pistils, and nectaries of insect-pollinated flowers are usually arranged in such a way that insects pick up pollen when visiting one flower and deposit it when visiting another. *Figure 27.3* illustrates one such relationship between a flower and its insect pollinator.

Photoperiodism

The relative length of day and night has a significant effect on the rate of growth and the timing of flower production in many species of flowering plants. For example, chrysanthemums produce flowers only during the fall, when the days are getting shorter and the nights longer. A grower who wants to produce chrysanthemum flowers in the middle of summer drapes black cloth over the plants to artificially increase the length of night. The response of flowering plants to the difference in the duration of light and dark periods in a day is called **photoperiodism.**

Photoperiodism depends on length of night

Plants can be placed into three categories, depending on the day length they require for flower production, as shown in *Figure 27.4.* **Short-day plants,** such as strawberries, some

Figure 27.4

Photoperiodism refers to a plant's sensitivity to the changing length of night.

▶ **Most plants are day-neutral. Flowering in cucumbers, tomatoes, and corn, for example, is influenced more by temperature than by day length.**

▲ **Spinach and lettuce are long-day plants that flower in midsummer, when days are longer than nights.**

◀ **Short-day plants such as pansies and goldenrod flower in late summer and fall or early spring, when days are shorter than nights.**

Meet Flora Ninomiya, Horticulturist

The weather in Richmond, California, is ideal for growing all kinds of flowers. Flora Ninomiya has inherited her father's—and before him, her grandfather's—rose-growing business, but before she and her brother and sister could run it successfully, they first were sent to colleges to train in horticulture and business studies. Although Ms. Ninomiya spends much of her time today managing the greenhouses, she's not sure of the future of the small flower-growing business in the United States.

In the following interview, Ms. Ninomiya describes some of the problems and pleasures of growing flowers.

On the Job

Q Ms. Ninomiya, when did Ninomiya Nursery get started?

A My grandfather, Jiro Ninomiya, started growing roses here in 1913, and my father continued the business. Then, during World War II, President Roosevelt signed an executive order that forced all Japanese-Americans to leave the West Coast and live in internment camps. My family couldn't return to our business for three years. While we were gone, we were lucky to have good friends and neighbors, the Aebis, who kept the nursery going for us. Frederick Aebi had come to the United States from Switzerland about the same time my grandfather came from Japan.

Q Could you tell us about your duties in the nursery?

A My brother, my sister, and I share the running of the business. I'm in charge of the horticulture part. We have 600 000 rosebushes—mostly red—and almost a million square feet of greenhouses. All our flowers are grown in greenhouses. Our roses are shipped all over the United States and Canada. I schedule production so that we'll always have the right amount and mix of cut flowers available for market. I also have to watch out for plant diseases and oversee watering, fertilizing, and spraying. In addition, we have recently begun a hybridization program to produce new rose varieties.

Q What problems are there in your business?

A I have to check constantly for insects or diseases. Whatever sprays we use must be recorded and reported monthly to the Department of Agriculture.

A big problem is the nematodes, or roundworms, in the soil. Many of the chemicals we used to use have been taken off the market because they're hazardous. Right now, we're working on a project with our farm adviser to test different materials to see what is most effective against nematodes.

One problem that is growing each year is competition from cut-flower growers in places like Mexico, Ecuador, and Colombia. They can grow flowers cheaply and bring them into the United States at a lower price than we can produce them. Ten years ago, only about two percent of cut roses were imported. Now, the numbers have reached 50 percent. Only cut flowers, not rooted plants, can legally be brought into the United States because plants could bring soil-borne diseases and parasites.

Early Influences

Q Did you get an early start in the flower business?

A My brother and sisters and I all worked in the family business when we were young. It was just expected that we would pitch in. My parents named me Flora in honor of the business. It wasn't enough to learn the trade from my father and grandfather. When I grew up, I went to college and studied horticulture.

Q What do you especially remember from your childhood in a flower-growing family?

A I remember watching my grandfather painstakingly making his own rose plants by grafting. Grafting stopped when my father ran the business, but now that flower growing has become so competitive, we have come back full circle, and we're developing our own plants to cut costs. Of course, modern grafting techniques are much faster, but I still think of my grandfather while we're doing it. I also remember as a little girl going with my father to the flower market in San Francisco to sell our roses to florists. The market opened very early in the morning— about 2 A.M.—and it was a colorful and bustling scene.

Personal Insights

Q Can you tell us about one of the highlights in your life of growing roses?

A Every once in a while, a rosebush will for some reason send out a flower of an entirely different color, called a sport. One time, a raspberry-colored bush produced a beautiful white flower. It was a pleasing little genetic accident. We named the flower after my mother, Hayame, and marketed it for a while.

Q Do you have any advice for students who are interested in horticultural careers?

A One thing I would say is to try to expose yourself to all kinds of plants and learn as much as you can about growing things. I wish that I had as a horticultural student taken more business courses, too, because a flower grower has lots of business decisions to make.

Figure 27.5

The responses of flowering plants to day length influence both their adaptations for pollination and where they are able to grow and reproduce.

▲ How would the seed production of these daisies be affected if they bloomed in early spring when butterflies were still just caterpillars?

▲ Ragweed is never found growing in northern Maine. Can you explain why in terms of photoperiodism?

chrysanthemums, cockleburs, primroses, ragweed, and poinsettias bloom when days are shorter than the nights. **Long-day plants,** such as carnations, peppermint, petunias, clover, potatoes, and garden peas, bloom as days get longer than nights. Most plant species are **day-neutral,** which means their blooming times are controlled by temperature, moisture, or other environmental factors rather than by day length.

The effect of photoperiodism on pollination

Fields of wildflowers bloom during late spring and summer, when bees and butterflies are most active and numerous, *Figure 27.5.* The photoperiodism of wildflowers may ensure that a plant produces its flowers at a time when there is an abundant population of pollinators. Photoperiodism also affects the distribution of flowering plants.

Section Review

Understanding Concepts
1. Name the four organs of a flower and the functions of each.
2. How is photoperiodism of flowers related to pollination?
3. What is the adaptive value of cross-pollination for plants?

Thinking Critically
4. What benefit does an insect pollinator receive from a flower? What benefit does

the flower receive from the pollinator? What type of ecological relationship is this?

Skill Review
5. **Comparing and Contrasting** Compare and contrast important features of wind-pollinated, bee-pollinated, and moth-pollinated flowers. For more help, refer to Thinking Critically in the *Skill Handbook.*

27.2 Flowers and Reproduction

Section Preview

Objectives

Outline the processes of seed and fruit formation and seed germination.

Explain strategies for seed dispersal.

Describe the effects of hormones on plant growth.

Key Terms

micropyle
double fertilization
endosperm
fruit
dormancy
germination
tropism
nastic movement
hormone

Transferring pollen from anther to stigma is just one step in the life cycle of a flowering plant. How does pollination lead to the development of seeds enclosed in a fruit? How do sperm cells in the pollen grain reach the egg cells in the ovary? These steps in the reproductive cycle of anthophytes take place without water—an evolutionary step that enabled flowering plants to occupy nearly every environment on Earth.

Pollen Growth and Fertilization

As you learned in Chapter 24, the life cycle of all plants, from bryophytes to anthophytes, involves the alternation of a diploid sporophyte generation with a haploid gametophyte generation. The life cycle of flowering plants resembles that of gymnosperms in that both produce seeds. In both groups, small gametophytes are retained within the body of the sporophyte.

Pollen grains grow pollen tubes

Once a pollen grain has reached the stigma, several events take place before fertilization occurs. Inside each pollen grain are two haploid sperm cells and one haploid tube cell. The tube cell elongates, forming a pollen tube that grows down the length of the style and into the ovary.

The two sperm cells move through the pollen tube into the ovule through a tiny opening in the ovule called the **micropyle.**

Double fertilization occurs inside the ovule

Inside the ovule is the female gametophyte, the embryo sac, which contains several cells surrounded by cytoplasm. One of the cells is a haploid egg cell. Another is the central cell, which contains two haploid nuclei. When the tube cell reaches the female gametophyte, the nucleus of one of the sperm cells unites with the nucleus of the egg cell to form a diploid zygote. The second sperm unites with the central cell to form a cell with a triploid ($3n$) nucleus. This process, in which two sperm cell nuclei unite with two cell nuclei of the female gametophyte, is called **double fertilization.** The triploid

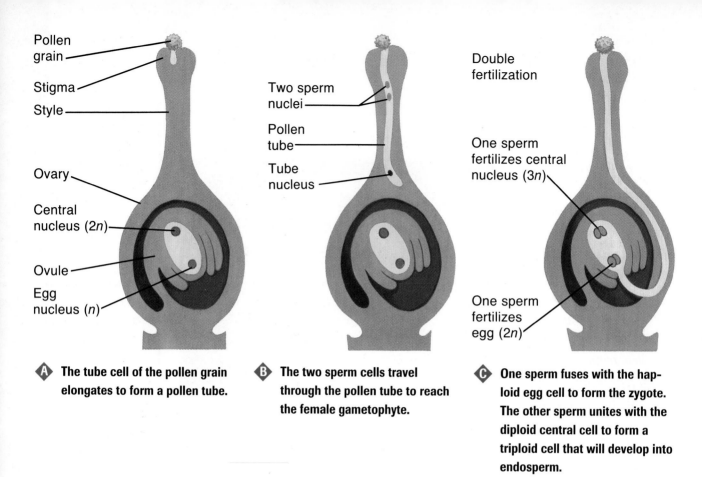

Pollen grain

Stigma

Style

Ovary

Central nucleus (2n)

Ovule

Egg nucleus (n)

Two sperm nuclei

Pollen tube

Tube nucleus

Double fertilization

One sperm fertilizes central nucleus (3n)

One sperm fertilizes egg (2n)

A The tube cell of the pollen grain elongates to form a pollen tube.

B The two sperm cells travel through the pollen tube to reach the female gametophyte.

C One sperm fuses with the haploid egg cell to form the zygote. The other sperm unites with the diploid central cell to form a triploid cell that will develop into endosperm.

Figure 27.6

In flowering plants, the male gametophyte grows through the pistil to reach the female gametophyte. Double fertilization involves the fusion of two sperm cell nuclei with two cell nuclei of the female gametophyte.

nucleus will divide many times, eventually forming the endosperm of the seed. The **endosperm** is food-storage tissue that supports development of the embryo.

Many flowers contain more than one ovule. Pollination of these flowers requires that at least one pollen grain land on the stigma for each ovule contained in the ovary. In a watermelon plant, for example, hundreds of pollen grains are required to pollinate a single flower. In most plants, pollen tube growth and fertilization take place fairly quickly. In barley, less than an hour elapses between pollination and fertilization. In corn, the process takes about 24 hours; in tomatoes, about 50 hours;

and in cabbages, about five days. *Figure 27.6* shows the processes of pollen tube formation and double fertilization.

Seed Formation

After fertilization takes place, most of the flower parts die and the seed begins to develop. The wall of the ovule becomes the hard seed coat, which helps protect the embryo until it begins growing into a new plant. Inside the ovule, the zygote divides and grows into the plant embryo. The triploid central cell develops into the endosperm. The ovary then develops into a fruit.

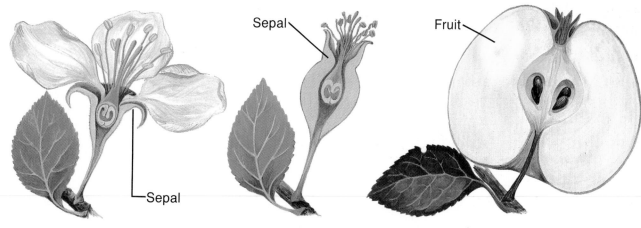

A When the eggs in the ovules of an apple flower have been fertilized, the petals, stamens, and upper portion of the pistils wither and fall away.

B The walls of the ovary in an apple harden and the supporting stem becomes fleshy and grows up and around the ovary as the seeds develop inside each ovule.

C The remains of the sepals can be seen at the end of the fruit. The fruit is attached to the plant at the opposite, or stem, end of the fruit.

Fruit Formation

As the seed develops, the ovary that surrounds the seeds enlarges and becomes the fruit. A **fruit** is the structure that contains the seeds of an anthophyte. *Figure 27.7* shows how the fruit of an apple tree develops from the ovary inside the flower.

Figure 27.8

A fruit is the ripened ovary of a flower that contains the seeds of the plant. The most familiar fruits are those we consume as food.

Fruits can be fleshy or dry

A fruit is as unique to a plant as its flower, and many plants can be identified by examining the structure of their fruit. You are familiar with plants that develop fleshy fruits, such as apples, grapes, melons, tomatoes, and cucumbers. Other plants develop dry fruits such as peanuts and walnuts. Some plant foods that we call vegetables or grains are actually fruits, as shown in *Figure 27.8.*

Figure 27.7

A fruit consists of the seeds enclosed in the mature ovary of a flowering plant.

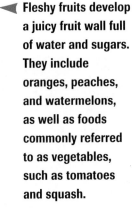

◄ Fleshy fruits develop a juicy fruit wall full of water and sugars. They include oranges, peaches, and watermelons, as well as foods commonly referred to as vegetables, such as tomatoes and squash.

◄ Dry fruits have dry fruit walls. The ovary wall may start out with a fleshy appearance, as in hickory nuts or bean pods, but when the fruit is fully matured, the ovary wall is dry. Dry fruits include pecans, walnuts, and other nuts, as well as grains such as wheat, barley, and rice.

How do fruits and flowers compare?

Some parts of the flower develop into fruits and seeds, while others wither and fall away. The form of a particular flower is reflected in the structure of the fruit it produces.

Procedure

1. Carefully examine the fruits and flowers provided by your teacher.
2. Locate and identify the major structures of the flower.
3. Examine the fruit to see if any flower structures are still attached.
4. Carefully cut open both flower and fruit, and compare their internal structures.

Analysis

1. Which flower structures can be seen on the exterior of the fruit?
2. Diagram the interior of a flower and its fruit. Use a color code to label similar structures on each diagram.
3. What kind of correlation can you make between the number of seeds in the fruit and the internal structure of its flower?

Seed Dispersal

A fruit not only protects the seeds inside it, but also may aid in dispersing those seeds away from the parent plant and into new habitats. Dispersal of seeds, *Figure 27.9,* is important because it reduces competition for sunlight, soil, and water between the parent plant and its offspring. Animals such as raccoons, deer, bears, and birds help distribute many seeds by eating fleshy fruits. They may carry the fruit some distance away from the parent plant before consuming it and spitting out the seeds. Or they may eat the fruit, seeds and all. Seeds that are eaten pass through the digestive system unharmed and are deposited in the animal's wastes. Squirrels, birds, and

Figure 27.9

A wide variety of seed-dispersal mechanisms have evolved among flowering plants.

▼ **Natives of Russia, tumbleweeds are common plants of the American Southwest. When seeds mature, the dried leaves and stems break off and seeds are scattered as the plant tumbles across the prairie landscape.**

◀ **Wind-dispersed seeds have adaptations that enable them to be held aloft while they drift away from their parent plant. Milkweed seeds have hairy plumes that act as parachutes. Elm and maple trees produce winged fruit.**

▶ **Clinging fruits, like those of the cocklebur, produce seeds in achenes, single-sided dry fruits with hooks that stick to the fur of passing animals or the clothes of passing humans.**

Figure 27.10

Seeds can remain dormant for long periods of time. Lupine seeds like these germinated after remaining dormant for 10 000 years.

they can remain in the soil until conditions are favorable for growth and development of the new plant. This period of inactivity in a mature seed is called **dormancy**.

Dormant seeds survive unfavorable conditions

Metabolic activity slows during dormancy, which may last for a few months or many years. Some seeds, such as willow, poplar, magnolia, and maple, remain dormant for only a few weeks after they mature. These seeds cannot survive harsh conditions for long periods of time. If conditions remain unfavorable for growth, the embryo within the seed may die. Most flower and vegetable seeds can survive for two or three years if kept cool and dry. Seeds of plants that live in dry or cold environments, such as desert wildflowers or conifers including spruce, fir, and pine, can survive dormant periods of 15 to 20 years. The seeds of evening primrose and curly dock, two common weeds, can survive for 100 years. The seed coat acts as a mechanical barrier in some plants. In other plants, dormancy is related to chemical inhibitors in the seed coat.

The longest-lived seeds are reported, by some scientists, to be lupine seeds, *Figure 27.10.* Found buried in the dry, frozen burrows of small arctic tundra mammals, these seeds were estimated to be 10 000 years old. They were, however, not radiocarbon dated.

other nut gatherers may drop and lose some of the seeds they collect, or even bury them only to forget where.

Plants that live in or near water, such as water lilies and coconut palms, produce fruits or seeds with air pockets in the walls, which enable them to float and drift away from the parent plant. The ripened fruits of many plants split open to release seeds designed for dispersal by the wind or by clinging to animal fur. Orchid seeds are so tiny that they resemble dust grains and are easily blown away in the wind. The fruit of the poppy flower forms a seed-filled capsule that releases sprinkles of tiny seeds like a salt shaker as it bobs about in the wind. Tumbleweed seeds are scattered by the wind as the whole plant rolls along the ground.

Seed Germination

At maturity, seeds are fully formed. The seed coat dries and hardens, enabling the seed to survive conditions that are unfavorable to the parent plant. Once seeds are dispersed,

What does a germinating seed look like?

Seeds are made up of a plant embryo and food-storage tissue inside a seed coat. Monocot and dicot seeds differ in their internal structure.

Procedure

1. Obtain from your teacher a soaked, ungerminated corn kernel (monocot), a bean seed (dicot), and corn and bean seeds that have begun to germinate.

2. Remove the seed coats from each of the ungerminated seeds, and examine the structures inside. Use low-power magnification. Locate and identify each structure of the embryo and any other structures you observe.

3. Examine the germinating seeds. Locate and identify the structures you observed in the dormant seeds.

Analysis

1. Diagram the dormant embryos in the soaked seeds, and label their structures.

2. Diagram the germinating seeds, and label their structures.

3. List at least three major differences you observed in the internal structures of the corn and bean seeds.

Requirements for germination

Dormancy ends when the seed is ready to germinate. **Germination** is the beginning of the development of the seed into a new plant. The absorption of enough water and the presence of oxygen and favorable temperatures usually end dormancy, but there may be other requirements. Water is important because it activates the embryo's metabolic system. Once metabolism has begun, the seed must continue to receive water or it will die. Just before the seed

coat breaks open, the plant embryo begins to respire rapidly. Many seeds germinate best at temperatures between 25°C and 30°C. Arctic species germinate at lower temperatures than tropical species. At temperatures below 0°C or above 45°C, most seeds won't germinate at all.

Some seeds have special requirements for germination, *Figure 27.11.* For example, some germinate more readily after they have passed through the acid environment of an animal's digestive system. Others require a period of freezing temperatures such as apple seeds, extensive soaking in water such as coconut seeds, or certain day lengths. You may remember reading in Chapter 25 that the seeds of some conifers will not germinate unless they have been

Figure 27.11

The seeds of the desert tree *Cercidium floridum* have hard seed coats that must be cracked open. This occurs when the seeds tumble down arroyos in sudden rainstorms.

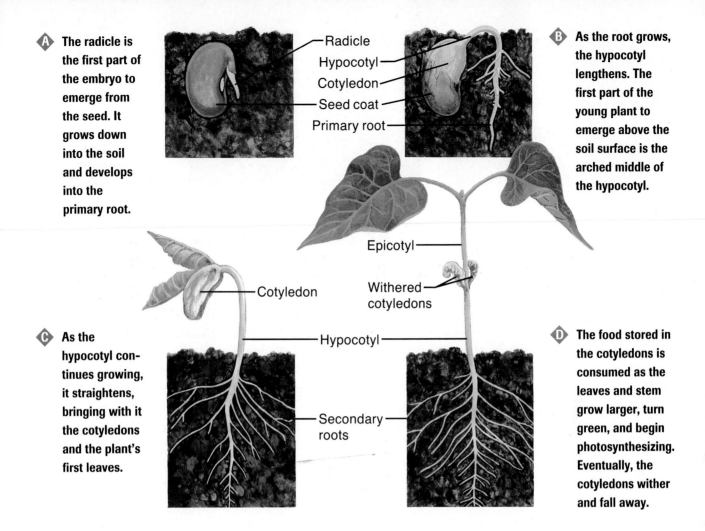

A The radicle is the first part of the embryo to emerge from the seed. It grows down into the soil and develops into the primary root.

Radicle
Hypocotyl
Cotyledon
Seed coat
Primary root

B As the root grows, the hypocotyl lengthens. The first part of the young plant to emerge above the soil surface is the arched middle of the hypocotyl.

Epicotyl

Cotyledon

Withered cotyledons

C As the hypocotyl continues growing, it straightens, bringing with it the cotyledons and the plant's first leaves.

Hypocotyl

Secondary roots

D The food stored in the cotyledons is consumed as the leaves and stem grow larger, turn green, and begin photosynthesizing. Eventually, the cotyledons wither and fall away.

exposed to fire. The same is true of certain wildflower species, including lupines and gentians. *Figure 27.12* illustrates the germination of a typical plant embryo.

Growth and Development

Why do roots grow down into the soil and stems grow up into the air? Although plants lack a nervous system and cannot make quick responses to stimuli, they do have mechanisms that enable them to respond to their environment. Plants grow; produce flowers and seeds; and shift the position of roots, stems, and leaves in response to environmental conditions such as gravity, sunlight, temperature, and day length.

Responsive movement in plants

A **tropism** is a plant's response to an external stimulus that comes from a particular direction. If the tropism is positive, the plant grows toward the stimulus. If the tropism is negative, the plant grows away from the stimulus. The tendency of stems to bend toward light is called phototropism. The positive phototropism exhibited by stems is an adaptation that enables the plant to obtain a maximum amount of sunlight.

There is another tropism associated with the upward growth of stems and the downward growth of roots. The stimulus is gravity and the tropism is called gravitropism. Stems show negative gravitropism, while roots show positive gravitropism. Do leaves exhibit gravitropism?

Figure 27.12

Germination of a bean seed is stimulated by water, which softens the seed coat, and warm temperatures.

is touched, as shown in *Figure 27.13*. Because nastic movements do not involve growth, they are reversible.

Plant hormones include auxins

Plants, like animals, have hormones that regulate growth. A **hormone** is a chemical that is produced in one part of an organism and transported to another part, where it causes a physiological change. Only a small amount of the hormone is needed to make this change.

Growth hormones called auxins affect many aspects of plant development. For example, auxins are responsible for regulating phototropism in plants. *Figure 27.14* describes how auxins regulate this response.

Figure 27.13

Mimosa pudica is also known as the sensitive plant. When touched, it folds its leaves in less than one-tenth of a second. This nastic response might provide some protection from insect damage or help reduce water loss in drying winds.

Because tropisms involve growth, they are not reversible. For example, the position of a stem that has grown several inches in a particular direction cannot be changed. But if the direction of the stimulus is changed, the stem will begin growing in another direction.

A responsive movement of a plant that is not dependent on the direction of the stimulus is called a **nastic movement.** An example of a nastic movement is the sudden drooping of the leaves of a *Mimosa* when the plant

Ⓐ Stems grow toward light, or show positive phototropism, because auxin collects in the cells on the shaded side of the stem.

Ⓑ Auxin is a hormone that stimulates the elongation of cells, so the cells on the auxin-rich, shaded side of the stem grow longer than the cells on the other side, causing the stem to bend toward the light.

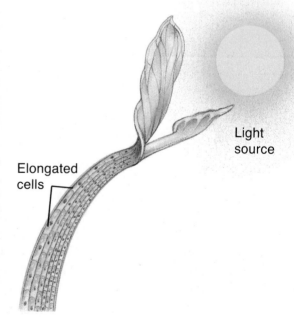

Elongated cells

Light source

Figure 27.14

Tropisms are the result of unequal stimulation, as when light comes to a plant from only one side.

Auxins have a number of other effects on plant growth and development. Have you ever seen a gardener pinch the tip off a plant to encourage the growth of side branches? Auxin produced in the apical meristem inhibits the growth of side branches, *Figure 27.15.* Pinching the tip reduces the amount of auxin present in the stem and allows side branches to form. High concentrations of auxin promote development of fruit and inhibit the dropping of fruit from the plant. When auxin concentrations decrease in the autumn, ripened fruit falls to the ground and trees begin to drop their leaves.

Gibberellins promote growth

Gibberellins are growth hormones that cause plants to grow taller because, like auxins, they stimulate cell elongation. Gibberellins also increase the rate of seed germination and bud development, and may stimulate the formation of flowers and fruits in some plants.

Figure 27.15

Side branches developed on this bush when the apical meristem of the main stem was removed.

ThinkingLab · Make a Hypothesis

How does fruit ripen?

Ethylene is a gas made by plants that causes ripening of fruit. Ethylene is also given off in small amounts by the fruits themselves, causing ripening to occur even after they are picked.

Analysis

Assume that you buy peaches from the supermarket and find that they are not completely ripe. Someone tells you that if you put them in a paper bag with a ripe banana, they will ripen more quickly than if you just leave them out on your counter.

Thinking Critically

Make a hypothesis about what you think might happen to unripe peaches if you put them in a paper bag with a banana compared to what would happen if you left them on the counter. Explain your reasoning.

Some hormones inhibit growth

While auxins and gibberellins stimulate growth, other plant hormones, called inhibitors, have the opposite effect. One inhibitor is abscisic acid. Inhibitors are responsible for producing dormancy in seeds and buds during periods of unfavorable environmental conditions. Abscisic acid also plays a role in closing stomata during periods of water shortages. This hormone is often thought of as the hormone produced to protect the plant in times of stress.

Figure 27.16

As flowers wither, rose plants produce their seeds in ovaries known as rose hips. What causes flowers to wither?

▲ Rose hips are collected and used in tea. They are a source of vitamin C.

◄ Roses such as this *Rugosa* belong to the family Roseaceae.

Ethylene gas promotes ripening

Ethylene gas, a simple compound of carbon and hydrogen, is a hormone that speeds the ripening of fruits. This hormone is produced on the plasma membrane. Ethylene is also responsible for the withering of flower parts after fertilization takes place, *Figure 27.16.* It also promotes the dropping of leaves in the fall.

Connecting Ideas

You've had an opportunity to take a broad look at the plant kingdom. The journey started with the simplest nonvascular plants, continued through the ferns and gymnosperms, and ended with the anthophytes—the most complex, colorful, and numerous plants on Earth today. Now it's time to take a similar phylogenetic journey through the animal kingdom, starting with the invertebrates.

Section Review

Understanding Concepts

1. How does pollination of a flower lead to development of the seed?
2. What parts of the flower become the fruit?
3. Name two plant hormones, and describe how each one influences growth and development.

Thinking Critically

4. What does a pollen grain have in common with a spore?

Skill Review

5. **Making and Using Tables** Make a table that indicates whether each structure of a flower is involved in pollination, fruit formation, seed production, seed dispersal, or growth and development. For more help, refer to Organizing Information in the *Skill Handbook.*

Growing Plants Without Soil

Since the beginning of agriculture, the planting process has been much the same—plant seeds in the ground and nurture them until harvest. But what about areas where the soil is poor or the climate harsh? How can an expanding population in these areas be sustained?

What is hydroponics? One solution may be the science of hydroponics. Hydroponics is the growing of plants without soil. In nature, some soils are rich in nutrients but do not support the plants well. Other soils, such as sand and small rocks, support plants well but are almost devoid of available nutrients. Hydroponics solves these problems by supplying the nutrients needed by plants in carefully compounded solutions, and by supporting plants on substrates other than soil.

Out of the lab Hydroponics was developed by botanists who studied the nutritional needs of plants. They planted their seeds in an inert substance such as gravel or sand, and then flooded the young roots periodically with a carefully controlled nutrient solution. By varying the amount of each nutrient in the solution and observing the plants' reactions, they were able to determine the best mix for optimum plant growth.

Extending the growing season But hydroponics did not remain in the laboratory for long.

The practice soon spread to Israel and India, where poor soil and harsh conditions make this system ideal for producing fresh food. In addition, when carried on in greenhouses, hydroponics allows the growing season to be extended year-round. Thus, the fresh tomatoes you eat in January or the beautiful roses you give your sweetheart on Valentine's Day may have been grown hydroponically.

Applications for the Future

The science of hydroponics represents much more than just growing plants with optimum nutrients. Today, the science encompasses controlling all of the conditions that affect plant growth. For example, many plants can benefit from extended availability of light. By connecting a photocell to a microprocessor and a timer, artificial light can be used to supplement sunlight at a minimal energy cost.

Controlling acidity The acidity of the growing solution is important to plants. Modern pH meters can be used to control the release of buffering compounds to keep the pH of the growing solution within the optimum range. More sophisticated instruments in the future will adjust the concentration of individual nutrients as needed, possibly resulting in systems that are totally automated.

INVESTIGATING the Technology

1. **View** Watch a videotape of the movie *Silent Running*. This is the story of a giant hydroponics operation in the future. List two technological developments shown in the film.
2. **Investigate** Find out the nutrient requirements of a plant of your choice and grow it hydroponically.

Reviewing Main Ideas

27.1 What Is a Flower?

- Flowers are made up of four organs: sepals, petals, stamens, and pistils.
- Flowering plants have a wide variety of adaptations for pollination.
- Most flowers have adaptations that favor cross-pollination.
- The relationship between the number of hours of daylight and darkness in a day, called photoperiodism, affects the timing of flower production in many anthophytes.

27.2 Flowers and Reproduction

- A pollen tube grows downward from the stigma through the style to the ovary. Sperm move into the ovule, where fertilization takes place.
- Flowering plants undergo double fertilization. One sperm cell nucleus joins with the egg cell nucleus to form the zygote. The second sperm cell nucleus joins with the central cell nucleus in the ovule to form a triploid cell.
- The zygote develops into the plant embryo. The triploid cell develops into the endosperm. The ovary wall becomes the fruit, which may be fleshy or dry.

- When the seed matures, it separates from the ovary and may be dispersed by animals or by the wind.
- Seeds can remain dormant for weeks, months, or years. Dormancy ends and germination begins when water availability, temperature, and other factors become favorable for growth.
- Auxins and gibberellins are plant hormones that stimulate cell elongation. These and other hormones regulate flower and seed development, the ripening of fruit, and the timing of fruit and leaf drop in the autumn.

Key Terms

Write a sentence that shows your understanding of each of the following terms.

anther	micropyle
complete flower	nastic movement
day-neutral plant	ovary
dormancy	petal
double fertilization	photoperiodism
endosperm	pistil
fruit	sepal
germination	short-day plant
hormone	stamen
incomplete flower	tropism
long-day plant	

Understanding Concepts

1. What parts might be missing from an incomplete flower?
2. One plant produces heavy, spiky pollen grains. A second plant produces very light, smooth pollen. What conclusions can you draw about the reproductive strategies of each plant?
3. What parts of a flower develop into a fruit?
4. Explain the adaptive advantage of a sticky or feathery stigma.
5. Distinguish between pollination and fertilization in flowering plants.
6. Explain the adaptive advantage to a plant of producing a fruit.
7. The leaves of a Venus's-flytrap close when an insect lands on them. Is this a tropism or a nastic response? Explain.
8. What does a pollen grain have in common with a spore?
9. Where is food stored in the seeds of dicots?
10. What structures are formed from the microspore and megaspore of an anthophyte?

Relating Concepts

11. Make a concept map that relates the following terms and phrases. Supply the appropriate linking words for your map.

    pollen, ovule, anther, stigma, style, insect, ovary, sperm, pollen tube, double fertilization, embryo, triploid cell, zygote, endosperm, seed, fruit

Applying Concepts

12. You eat peas, beans, corn, peanuts, and cereals. Why are seeds a good source of food?
13. How does dormancy contribute to the survival of a plant species in a desert ecosystem?
14. Gardening books recommend that gardeners with small yards plant corn in blocks of at least 12 plants, spaced a foot or so apart. If fewer plants are grown, cobs do not develop a full set of kernels. What do you think is the basis for this advice?

A Broader View

15. Many orchids are epiphytes that live high above the ground on trees of the tropical rain forest. These orchids produce millions of tiny seeds. What is the adaptive value to an epiphytic orchid of producing millions of seeds?

Thinking Critically

Making Predictions

16. **Biolab** A team of plant ecologists determines the average number of pollen grains produced by each anther of several different species of flowers. The team also determines the average number of ovules within each ovary of the same species. When the scientists compile their data, what will they discover about the relationship between the number of pollen grains and the number of ovules produced by any given species?

Designing an Experiment

17. Explain how you would conduct an experiment to test whether or not auxin inhibits leaf abscission.

Developing a Hypothesis

18. **Minilab** Form a hypothesis that explains why the primary root is the first part of the plant to emerge from a germinating seed.

Interpreting Graphs

19. The graph below provides data from an experiment that tested the effects of ionizing radiation on the germination of seeds. Explain the results of the experiment.

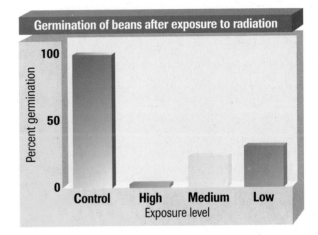

Germination of beans after exposure to radiation

Connecting to Themes

20. **Systems and Interactions** In what ways are flowering plants important to the lives of insects, birds, and mammals?
21. **Evolution** Explain why a scientist might hypothesize that the eating habits of herbivorous mammals affected the evolution of fruits in flowering plants.
22. **Homeostasis** How do hormones help plants maintain homeostasis?

Focus On

Plants for People

Agriculture was probably the single most significant development in human history. It is not an accident that the beginnings of civilization occurred in productive farming areas. Today, many farmers concentrate on growing just one species of food crop, often a grain, in their fields. Monocultures enable farmers to concentrate on one species of plant and allow them to use heavy equipment for planting, cultivating, and harvesting. However, monocultures also leave farming areas vulnerable to disease. A pest infestation can attack a monoculture and spread quickly from field to field, wiping out an entire season's crop. Farmers have begun to use more native species in their fields. Native species are less susceptible to pests and are more hardy than other crops used in monocultures. The preservation of native species for use as crops is gaining attention from farmers and biologists alike.

Agriculture

Corn More land in the United States is used to grow corn, family Poaceae, than any other crop. Most of the corn is used for livestock feed, but a significant portion is used to manufacture starch, oil, sugar, meal, breakfast cereals, and alcohol.

Rice For the majority of humanity, rice means survival. In Asia, rice, family Poaceae, is the basis of almost all diets. More than 95 percent of the world's rice crop is used to feed humans. Rice is the only grain that can grow submerged in water.

Taro The roots of this plant provide an important staple to Asia and Pacific Ocean island areas. Taro, a member of the arum family Araceae, can be made into cakes that can be baked or toasted.

Oats Oats, family Poaceae, are an important food for both man and beast. Containing from ten to 16 percent protein, oats are low in fat but high in carbohydrates, proteins, B vitamins, fiber, and minerals. They make an excellent food for growing humans and young livestock. A native of northern Europe, oats grow well in poor soils and cool, wet climates.

Wheat Nearly one-third of all the land in the world used for crop production is planted in wheat. Wheat, family Poaceae, probably originated in the Middle East and was an important food of the ancient Mesopotamian, Egyptian, and Indus civilizations.

Barley Barley, family Poaceae, is probably the oldest grain crop used by humans. The Egyptians grew barley in 6000 B.C., and it is the fourth-largest cereal crop today. Of all the grains, it grows the fastest and can stand the toughest conditions such as in Lapland and high in the Himalayas.

Potato This South American native arrived in the United States by a circuitous route—via Europe. The potato, family Solanaceae, is very nutritious, containing more of the essential amino acids than whole wheat. In addition, potatoes carry appreciable amounts of vitamins B and C, as well as minerals such as calcium and iron.

Sorghum Since prehistoric times, sorghum has been a major food crop in Africa. Exceptionally resistant to drought because of its extensive root system, sorghum, family Poaceae, provides not only food for people but also hay for cattle.

Silviculture

Silviculture, the growing of trees, is sometimes confused with forestry. Silviculture includes the growing of trees for lumber and paper, but it also refers to the growing of trees for food crops—for example, apple and pecan orchards. Oranges, nuts, olives, and cloves are some of the foods that silviculture provides. Trees also are sources of medicine, such as quinine that was used to combat malaria, or aspirin, originally derived from the willow tree. Trees even help replenish the oxygen we breathe.

Bark The bark of various trees provides us with cork, spices, and even drugs to fight disease. Spices such as cinnamon and drugs such as aspirin are derived from the bark of trees. Recently, an extract from the bark of a Pacific Coast yew tree (taxol) has been used to fight cancer.

Tree fluids Fluids from trees are the source material for maple syrup, latex rubber, and turpentine used as solvent in paint. Tree fluids can be extracted year after year, providing a continuous crop with minimal injury to the tree.

Fruits An obvious, and delicious, product of trees is fruit—both dry fruit and fleshy fruit. Dry fruits include acorns, walnuts, pecans, and other nuts used as food by wildlife and humans. Fleshy fruits include oranges, apples, pears, apricots, and cherries. Prized for centuries, fruit trees have been the object of much research, which has resulted in plants that bear staggering yields. Some of these new varieties can also provide better nutrition as their vitamin and mineral contents have been enhanced.

Chemicals Wood is the source of many chemicals. Wood alcohol, the rosin used on the bows of violins, cellulose used to make paper, charcoal, and rayon are just a few of the chemicals derived from trees.

Lumber When most people think of forest products, they think of lumber. It is one of our most common building materials. Easy to work, durable, relatively abundant, and lightweight, it is close to ideal for construction of homes and other small buildings.

Fuel Over much of the world, wood is the major source of fuel. It is high in heat content and easy to transport and store. However, in some overpopulated areas, the need is so great that whole countries have been denuded of trees.

Fodder When you think of woody plants, you probably don't think of them as food. But to many wild and domestic animals, they are a major source of nutrition. Young trees are a source of nutrition to wildlife as diverse as deer and giraffes and to domesticated species such as cattle and goats, as well.

Horticulture

Most of the plants we are familiar with were originally brought back from distant lands by naturalists and great explorers of the world. Using these exotic plants, horticulturists have left us a legacy of beautiful and useful varieties of flowers and food plants by their dedication to selective-breeding programs, grafting methods, and more recently, genetic engineering.

Vegetables Vegetables were some of the first plants cultivated by humans. Many of those belonging to the mustard family—such as green cabbage, watercress, and radishes—were known to the Egyptians and Romans in the Bronze Age. Herbs such as parsley, chives, and mint were also cultivated by early horticulturists and used to season meats and fish.

Root crops such as beets, carrots, and turnips have always been especially valuable because they could be stored easily during the winter months. They also provide excellent nutrition and lots of fiber.

Fruit plants Small fruits such as raspberries, strawberries, and blueberries are important commercial crops that can also be grown by the home gardener.

Shrubs These are short, woody plants that are used in landscaping. Some are left to grow naturally, some are trimmed to shape, and some are even used to form living sculptures or outdoor structures such as this maze.

Legumes Some plants improve the soil. Legumes such as peas, beans, alfalfa, and clover have nitrogen-fixing bacteria in small nodules on their roots. They "fix" nitrogen in the soil. Having clover growing in your lawn benefits your grass.

Legumes provide some of the most nutritious foods that can be grown. Peas, beans, peanuts, and lentils are all major sources of proteins. When combined with grains such as rice and corn, they offer all the proteins humans need for good health. In areas of the world where meat is expensive or unavailable, this can be an important consideration.

Flowers Flowering bulbs, roses, scented herbs, and garden flowers of all kinds have been a source of joy and delight to people for generations.

Medicinal plants Plants, in eras past, supplied medicines for people who could not afford doctors. For example, people used peppermint to soothe upset stomachs, willow bark for headaches, and licorice for coughs.

EXPANDING YOUR VIEW

1. **Understanding** Take a piece of paper and make two columns. In the left column, list all of the products essential to your life that come from plants and/or agriculture. In the right column, list all of the essential products that *do not* come from plants and/or agriculture. Now write a paragraph summarizing the role of plants in your life.

2. **Applying** Look around your community and identify an area that could be improved by applying your knowledge of agriculture, silviculture, or horticulture. Devise a plan to improve the area.

3. **Writing** Pick out a plant from this Focus On Plants for People and write a poem about its color, form, and fragrance.

Invertebrates

*S*and-colored starfish inch along on tube-shaped feet over the barnacle- and mussel-encrusted rocks. Tiny sand fleas hop among scattered patches of salt-encrusted seaweed.

A tidal pool along the edge of a beach is an excellent place for an exploration of different kinds of life. That's because tide pools and the oceans that sustain them are home to many of the world's invertebrates, the most common and diverse forms of life on Earth. They live in almost every habitat imaginable—from the bottom of the deepest oceans to the tops of the highest mountains. Thus, it's not surprising that invertebrates comprise about 95 percent of all animal species.

Mussels and many other invertebrate species remain in one place during their adult lives. How do these animals obtain food?

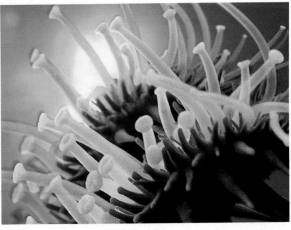

Tube feet and spiny skins are common characteristics of starfishes and many of their relatives. What other features make spiny-skinned animals unique?

The hard, outer skeleton of this fiddler crab protects it from predators. What advantages and problems do exoskeletons present for arthropods, the most diverse of the invertebrate groups?

Unit Contents

28 What Is an Animal?

Dusk falls and soon shadows flit quickly by street-lights. Insects attracted to light become food for bats as they steer a zigzag course across the nighttime sky. Flying silently on leathery wings as they scoop up their insect dinners, bats avoid obstacles by using sounds that humans cannot hear.

On the ground, night spells sleep for a chipmunk. These striped rodents work hard all day, collecting seeds and berries and storing them in hidden burrows. By evening, the chipmunks settle down underground for the night, safe from predators like the owl that waits to snatch up their nocturnal mouse cousins.

Supported by warm currents, a jellyfish floats along, tentacles trailing behind. A small fish brushes past one tentacle; suddenly it is caught, immobilized, and pulled up toward the jellyfish's mouth. Even jellyfish must find and capture food in order to eat.

You know that the bat, chipmunk, owl, mouse, and jellyfish all are animals, but can you name the characteristics they share? There are many different species of animals, yet each species has several features in common with all other animals. In this chapter, you will see what characteristics animals share, and how some of these characteristics provide clues to the evolutionary relationships among them.

Section Preview

Objectives

Compare the characteristics of animals.

Sequence the development of a typical animal.

Key Terms

sessile
blastula
gastrula
ectoderm
endoderm
protostome
deuterostome
mesoderm

When you hear the word animal, *do you picture an organism with hair or fur and a bony skeleton? More than 1.5 million species of animals have been described, yet 95 percent of these have neither bones nor hair.*

If you saw the organism in the photograph for the first time, would you classify it as an animal? This organism is a sponge, *an animal that remains attached to rocks or coral reefs in the ocean for all of its adult life. It doesn't move around in search of food. It doesn't have a bony skeleton or hair. It doesn't even have a mouth, a stomach, or intestines, yet it is still an animal.*

Characteristics of Animals

All animals have several characteristics in common. Animals are multicellular organisms that feed on other organisms and have ways of moving that enable them to obtain food. Once they have consumed their food, they break it down for use as energy or as raw material for building tissue. Unlike plants, animals are composed of cells that do not have cell walls.

Methods for obtaining food vary

Examine the animals shown in *Figure 28.1.* One characteristic common to all animals is that they are heterotrophic, meaning they obtain energy and nutrients from outside sources. Animals such as frogs and birds can move from place to place in an active search for food. Other animals such as barnacles remain stationary and are adapted to draw food toward them.

Figure 28.1

The lizard, hummingbird, and barnacle each get their food in different ways.

▼ **A lizard captures flies with its long, sticky tongue.**

◄ **A hummingbird flies to flowers to drink nectar.**

▼ **A barnacle extends bristles from its shell to catch small organisms as they drift by in the water.**

An animal's ability to move about is directly related to its method of obtaining food. Although the heterotrophic way of life and the ability to move are characteristics shared by animals, there are many different methods of locomotion as shown in *Figure 28.2*. The fish, osprey, and sidewinder snake are examples of vertebrates, animals with backbones, whereas the starfish is an invertebrate, an animal without a backbone. Yet all these animals have to find food to survive. As you can see, methods of locomotion have been refined and elaborated over the course of evolution.

Figure 28.2

Animals move in a remarkable variety of ways.

▼ **The fish uses its fins and powerful tail muscles to swim after smaller fish or other prey.**

◀ **A sidewinder rattlesnake barely touches the ground as it follows the trail of a mouse.**

▶ **A starfish moves slowly along the ocean floor using a unique system of tube feet that act like tiny suction cups.**

▲ **The soaring osprey dives to snatch fish from the waters of a lake or stream.**

Figure 28.3

Because adult sponges and corals are sessile, they rely on water currents to bring food to them.

▲ Corals, with tentacles extended, snag food drifting by in the water.

◄ Sponges filter food out of the water by setting up a stream of water through their bodies.

Some water-dwelling animals, such as the sponges and corals shown in *Figure 28.3,* are able to move about only during the early stages of their lives. They hatch from eggs into free-swimming larval forms, but when they become adults, they attach themselves to rocks or other objects. Organisms that don't move from place to place are known as **sessile** organisms.

Animals must digest food

Animals break down, or digest, their food once they have consumed it. In some animals—including earthworms, frogs, and monkeys—digestion takes place in an internal cavity; in other animals, such as sponges and flatworms, digestion is carried out within individual cells. Much of the food an animal consumes and digests is stored as fat or glycogen and used as an energy supply when food is not available.

Examine the structure of the digestive tracts of a flatworm and an earthworm in *Figure 28.4.* Notice that there is only one opening to the flatworm's digestive tract. An earthworm has a more complex digestive tract with two openings, one at either end.

Figure 28.4

Flatworms and earthworms are animals that digest food in separate digestive tracts.

► Have you ever turned over a rock in the shallow water of a lake or stream? If you looked closely, you might have seen tiny flatworms attached to the underside of that rock. Flatworms feed on small, live organisms or on the dead bodies of larger animals. The flatworm's digestive tract has only one opening through which food enters and wastes exit.

Digestive tract

Mouth

Pharynx

Animal cell adaptations

Animals are multicellular. Just as animal bodies are adapted to a variety of environments, most animal cells are adapted for different functions. For example, in the human body nerve cells conduct information, red blood cells transport oxygen, muscle cells make movement possible, and cells lining the stomach secrete digestive juices.

▼ Earthworms ingest soil and digest the organic matter contained in it. The earthworm's digestive system has two openings and contains several structures for storing, grinding, or dissolving food. Food travels along the digestive tract in only one direction, and indigestible waste is eliminated at the second opening. Do you think a digestive tract with two openings is more efficient than one with a single opening?

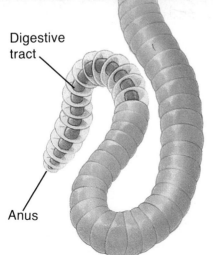

Digestive tract

Anus

History

Gardens of Intelligence

Have you ever been amused by the antics of the chimpanzees at a zoo? Human beings have always enjoyed a close kinship with and appreciation for other species in the animal kingdom.

Ancient zookeepers As early as 4500 B.C., the people who inhabited the area now known as Iraq kept pigeons in captivity. Egyptian Queen Hatshepsut established a royal zoo in 1500 B.C. In 1000 B.C., the Chinese emperor Wen Wang created a 1500-acre park for wild animals. He named the park *Ling-Yu,* or Garden of Intelligence.

Modern zoos The advent of modern zookeeping is usually traced back to 1752 and the establishment of the Schönbrunn Zoo in Vienna, Austria. In the United States, the first zoo to open was New York City's Central Park Zoo, in 1864.

Today, the object of most zoos is to recreate each animal's natural habitat. Monkeys and apes scramble up and down real trees in outdoor settings, and large cats sun themselves on natural rock outcroppings. Wildlife parks are being created so that animals can roam freely.

Maintaining diversity Most modern zoos cooperate with other zoos in captive-breeding programs. Zoos that are successful in breeding cheetahs might trade a few for animals like the golden-lion tamarin. Keeping track of the heritage of captive animals ensures that zoos maintain genetic diversity in breeding programs. Zookeepers are especially interested in breeding endangered species, with the hope of releasing these animals back into the wild in the future.

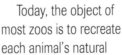

CONNECTION TO Biology

Why would zookeepers want to know the pedigree of animals to be used in captive-breeding programs?

BioLab | Design Your Own Experiment

What do you think of when you hear the word *animal?* A lion, tiger, dinosaur, or perhaps a bird? An animal is defined as a multicellular, eukaryotic heterotroph whose cells lack cell walls. Animals can be distinguished by the way they feed, digest food, respire, respond to stimuli, move, excrete, transport materials in their bodies, and reproduce. In this Biolab, you will examine characteristics of animals that are used to place them in the animal kingdom.

What is an animal?

PREPARATION

Problem
What characteristics enable you to identify organisms as animals?

Hypotheses
Make a hypothesis, based on the general features of animals, that best describes why a particular animal can be placed in the animal kingdom and how that animal can be specifically identified. Consider behavior, structure, and physiology.

Objectives
In this Biolab, you will:
- **Observe** behavior of animals.
- **Compare** features of animals.

Possible Materials
microscope and depression slides
mm ruler
variety of live animals such as mouse, earthworm, frog, hermit crab, sea urchin, cricket, goldfish, lizard, planaria, hydra, *Daphnia*, snail
suitable containers for each animal such as cages, pans, watch glasses, clear deli containers with lids, culture dishes

Safety Precautions
Treat animals in a humane manner at all times. Wash your hands after handling animals.

1. Choose four animals you wish to investigate.

2. In your group, brainstorm a list of possible characteristics that could be used to categorize and identify your animal.

3. Construct a data table for the features that identify your organisms as animals and that identify them specifically.

Check the Plan

1. Recall the demonstration done by your teacher at the beginning of this investigation. Have you incorporated these ideas into your table?

2. Give reasons why you made your table the way you did.

3. What quantitative data do you plan to collect?

4. *Make sure your teacher has seen your data table before you proceed further.*

5. Carry out your experiment.

6. Record measurements and observations in your data table.

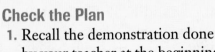

1. **Identifying Variables** On what basis did you identify your animals generally and specifically?

2. **Checking Your Hypothesis** Did your results confirm your hypothesis? Why or why not?

3. **Comparing and Contrasting** Compare your table with that of another group. How is it the same? How is it different?

4. **Drawing Conclusions** What did you learn about animals in this Biolab that you did not know before?

5. **Thinking Critically** How are animals identified?

Going Further

Project Based on this lab experience, design an experiment that would help you to answer a question about an unfamiliar animal.

Development of Animals

Most animals develop from a single, fertilized egg cell called a zygote. But how does a zygote develop into the many different kinds of cells that make up a snail, a fish, or a human?

During development, the zygotes of different species of animals all have similar stages of development. The stages of development from zygote to fully formed embryo are shown in *Figure 28.5*. In this figure, the developing sea urchin embryo is used as an example. However, not all animals follow exactly the same patterns as the sea urchin.

Division of the egg

Initially, the unicellular zygote divides, forming two new cells in a process called cleavage. These two cells divide to form four cells and so on, until a hollow ball of cells called a blastula is formed. The **blastula** is a single layer of cells surrounding a fluid-filled space and is formed early in the development of an animal embryo. Formation of a blastula is complete about ten hours after fertilization in sea urchin development.

Forming a gastrula

The cells on one side of the blastula then fold inward to form a two-layered gastrula. The **gastrula** is a structure that is made up of two cell layers. Gastrula formation can be compared to the way a potter creates a cup or bowl from a lump of clay, *Figure 28.6.* First, the clay is formed into a solid ball. Then, the potter presses in on the top of the ball to form a cavity that will become the interior of the bowl. In the same way, the cells at one end of the blastula fold inward, forming a cavity lined with a second layer of cells. The layer of cells on the outer surface of the gastrula is called the **ectoderm.** The layer of cells lining the inner surface is called the **endoderm.**

Figure 28.5

Development of the sea urchin from a zygote is shown as an example of the sequence in which animal embryos develop.

A Blastula formation is the earliest developmental stage of an embryo. Continuous cell divisions in sea urchin development result in a 32-cell blastula. The total number of cells present in blastula stages of different classes of animals varies. Notice that during these early developmental stages, the total amount of cytoplasm has not increased. Each cell division results in smaller cells.

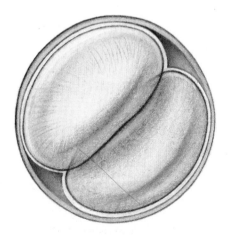
Two-celled stage

Eight-celled stage

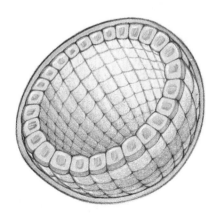
Blastula

Figure 28.6

You can think of a blastula as a hollow ball of cells. By pushing in on one side, a gastrula is formed.

The ectoderm cells of the gastrula continue to grow and divide, and eventually they develop into the skin and nervous tissue of the animal. The endoderm cells develop into the lining of the animal's digestive tract.

Protostomes and deuterostomes

In some classes of animals, the opening of the indented space in the gastrula becomes the mouth. These animals, which include earthworms and insects, are called protostomes. A protostome is an animal with a mouth that develops from the opening in the gastrula.

In animals with more complex tissues and organ systems, such as fish, birds, and humans, the opening of the gastrula develops into an anus, an opening through which digestive wastes are eliminated from the animal's body. A **deuterostome** is an animal in which the anus develops from the opening in the gastrula. The mouth of a deuterostome develops from cells elsewhere on the blastula.

Determining whether an animal is a protostome or deuterostome can help biologists determine the phylogeny of the animal's group. For example, both sea urchins and fishes are deuterostomes and are, therefore, more closely related than you might conclude from comparing their adult body structures.

Mesoderm develops later

The development of the gastrula progresses until a layer of cells called the mesoderm is formed. The **mesoderm**, illustrated in *Figure 28.5,* is the third cell layer formed in the developing embryo. The term *meso* means "middle," so mesoderm is found in the middle of the developing embryo; it lies between the ectoderm and endoderm.

protostome
proto (GK) before
stoma (GK) mouth

deuterostome
deutero (GK) secondary
stoma (GK) mouth
The protostomes and deuterostomes differ in the location of the cells that become the mouth.

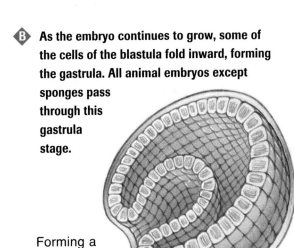

B As the embryo continues to grow, some of the cells of the blastula fold inward, forming the gastrula. All animal embryos except sponges pass through this gastrula stage.

Forming a gastrula

Mesoderm

Endoderm

Mesoderm breaks off

Ectoderm

C The mesoderm in the gastrula develops between the endoderm and ectoderm. In deuterostomes, the mesoderm is formed from clumps of cells that break off from the endoderm. The mesoderm gives rise to muscles, reproductive organs, and circulatory vessels.

28.1   Typical Animal Characteristics   **699**

How does mesoderm influence development of nervous tissue?

Biologists wanted to determine what causes the ectoderm on the dorsal side of the embryo to develop into nervous tissue. They hypothesized that perhaps the mesoderm lying directly beneath the ectoderm is the key. To test this idea, they removed a piece of mesoderm from beneath the dorsal surface of the ectoderm of one embryo and transplanted it to another embryo. The transplant was placed in a spot on the ventral surface of the second embryo where the mesoderm had been removed. The ectoderm was sealed back over the top of the mesoderm transplant.

Analysis

Put in order the following illustrations that show the sequence of the steps the biologists took to perform the experiment.

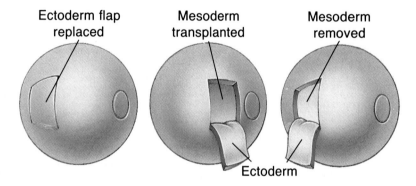

Ectoderm flap replaced Mesoderm transplanted Mesoderm removed

Ectoderm

Thinking Critically

Hypothesize what might happen as the embryo develops if, as the scientists hypothesized, mesoderm influences development in the ectoderm.

Forming a larva

As cells in the developing embryo change shape and become specialized to perform different functions, the sea urchin grows into its larval form, *Figure 28.7.* Some cells begin secreting calcium, an element that is important in the formation of spines in the adult sea urchin.

Figure 28.7

A free-swimming larva (top) develops from a fertilized sea urchin egg in just 48 hours. These larvae can be found as part of the plankton floating near the surface of oceans. The larvae will develop into adult sea urchins over the next few months. An adult sea urchin (bottom) has spines that it uses for protection from predators as it moves slowly along the ocean floor.

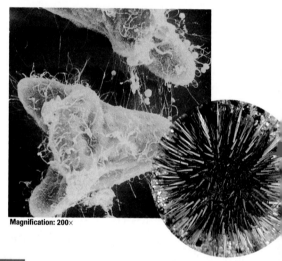

Magnification: 200×

Section Review

Understanding Concepts

1. Describe the characteristics that make a mouse an animal.
2. How does an animal with only a mouth get rid of undigested waste materials?
3. Explain the difference between a protostome and a deuterostome.

Thinking Critically

4. Identical twins occur in humans when a single zygote separates into two equal halves that go on to develop into individual organisms. Explain at what point in development this separation must occur in order to result in two complete embryos.

Skill Review

5. **Sequencing** Place the following words in sequence, beginning with the earliest stage: gastrula, larva, adult, fertilized egg, blastula. For more help, refer to Organizing Information in the *Skill Handbook.*

28.2 Body Plans and Adaptations

Objects made by a potter represent many different shapes and sizes. There may be a basic plan for making each type of pottery. One plan is for a bowl, another for a vase, another for a flat plate. Each piece of pottery is suited for a particular function. Animal bodies also have basic plans or body structures that are suited to a particular way of life. In this section, you will study a variety of animal body plans and see how a specific body structure is adapted for life in a particular environment.

Objectives

Distinguish among the body plans of different classes of animals.

Trace the phylogeny of animal body plans.

Compare body plans of acoelomate, pseudocoelomate, and coelomate animals.

Key Terms

symmetry
asymmetry
radial symmetry
bilateral symmetry
anterior
posterior
dorsal
ventral
acoelomate
pseudocoelom
coelom
exoskeleton
invertebrate
endoskeleton
vertebrate

What Is Symmetry?

Look at the animals shown in *Figure 28.8.* You know that all animals share certain characteristics, but these animals certainly don't look like they have much in common. The sponge seems to have no particular shape, whereas the fish has a head, body, fins, and a tail. The jellyfish doesn't have a head or tail, but is circular in form. Each animal has a different shape or form because each exhibits a different kind of symmetry. **Symmetry** refers to a balance in proportions of an object or organism. All animals have some kind of symmetry. Different kinds of symmetry enable animals to move about and find food in different ways.

Figure 28.8

The sponge (left), the fish (center), and the jellyfish (right) all exhibit different kinds of symmetry, yet each animal is able to find and digest food.

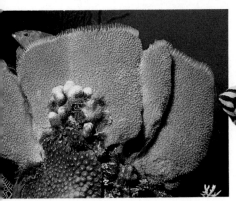

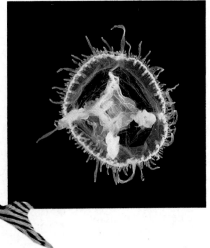

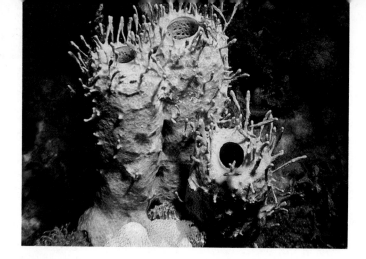

Figure 28.9

This irregularly shaped sponge is an example of an animal with an asymmetrical body plan. A sponge's structure is adapted for filtering food from the water. The two cell layers have distinct functions. The outer layer of cells protects the sponge from predators. The cells of the inner layer keep water moving with the whiplike motion of their flagella. These inner cells also trap and digest food.

Asymmetry in a sponge

Most sponges are adapted to living on the bottom of the ocean, as seen in *Figure 28.9.* Many sponges have an irregularly shaped body. An animal that is irregular in shape has a body plan that exhibits **asymmetry.** Asymmetrical animals often are sessile organisms that do not move from place to place.

The bodies of most sponges consist of two layers of cells. Unlike most other classes of animals, the sponges' embryonic development does not include formation of endoderm and mesoderm, or a gastrula stage. Sponges first evolved about 600 million years ago and represent one of the oldest groups of animals on Earth—proof that their two-layer body plan makes them well adapted for life on the ocean floor.

Radial symmetry in a hydra

The hydra, a tiny predator pictured in *Figure 28.10,* feeds on small animals it snares with its tentacles. A *Hydra* has radial symmetry. Its tentacles radiate out from around its mouth. As you can see, animals with **radial symmetry** can be divided along any plane, through a central axis, into roughly equal halves. Radial symmetry is an adaptation in *Hydra* that enables the animal to detect and capture prey coming toward it from any direction.

Have you ever had your groceries double bagged at the store? The body plan of a *Hydra* can be compared to a sac within a sac. These sacs are cell layers organized into tissues with distinct functions.

Figure 28.10

An example of an animal with radial symmetry, a hydra (left) feeds on tiny animals it immobilized with venomous stinging cells found along its tentacles. How might stinging cells be a helpful adaptation for a predator with limited ability to move?

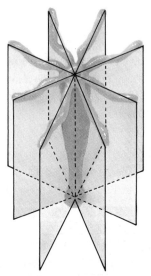

Hydra shows radial symmetry

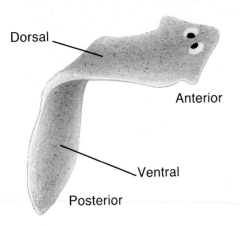

Figure 28.11

In bilaterally symmetrical animals, such as a flatworm, sensory tissue is commonly concentrated in the head, or front, end. The head detects light, temperature, and chemical makeup of the new environment.

Bilateral symmetry

The flatworm in *Figure 28.11* has bilateral symmetry. An organism with **bilateral symmetry** can be divided down its length into similar right and left halves that form mirror images of one another. In such animals, the **anterior,** or head end, has a different appearance than the **posterior,** or tail end. The **dorsal,** or back surface, also looks different from the **ventral,** or belly surface. Animals with bilateral symmetry can find food and mates and avoid predators more effi-

ciently than animals with radial symmetry because they have more muscular control.

Bilateral Symmetry and Body Plans

Animals that are bilaterally symmetrical also share other characteristics. What other characteristic do the animals in *Figure 28.12* have in common? All of these animals have body cavities in which internal organs are found. The development of body cavities made it possible for animals to grow larger and to move and feed more efficiently. Without a body cavity, animals depend on diffusion to take food into cells and eliminate wastes.

Acoelomate flatworms have no body cavities

The earliest group of animals in which a mesoderm evolved may have been similar to flatworms. Recall that the mesoderm of flatworms develops from cells near the opening in the gastrula. Flatworms may also demonstrate the first class of animals in which muscles and other organs evolved from the mesoderm.

Figure 28.12

Grasshoppers (left), muskrats (center), and manatees (right) are bilaterally symmetrical animals that have body cavities in which internal organs are found.

How do body plans affect locomotion?

Roundworms have a pseudocoelom with one set of muscles arranged along the length of the body. Earthworms have a coelom with two sets of muscles—one arranged along the length of the body and the other arranged in a circular fashion around the body.

Procedure

1. Observe nematode movement in their culture medium through a hand lens or binocular microscope.

2. Place an earthworm on your tabletop. Observe its movement.

Analysis

1. Describe the movements of a nematode.

2. Describe the movements of an earthworm.

3. Compare the movements of a nematode and an earthworm in terms of their differences in body plans.

Flatworms have solid, compact bodies as shown in *Figure 28.13*. Animals that have three cell layers with a digestive tract but no body cavities are called **acoelomate** animals. Flatworms are the oldest animal group that shows different tissues organized into organs. The organs of flatworms are embedded in the solid tissues of their bodies.

Water and particles of digested food travel through a solid acoelo-

mate body by the slow process of diffusion. A flattened body allows for diffusion of nutrients, water, and oxygen quickly enough to supply all body cells. Notice, too, that the digestive tract extends throughout the body to help in the process of diffusion.

Pseudocoelomates have a body cavity

A roundworm also has bilateral symmetry. However, the body plan of a roundworm is more complex than that of a flatworm. A roundworm has a body cavity, called a pseudocoelom, that develops between the endoderm and mesoderm. A **pseudocoelom** is a fluid-filled body cavity partly lined with mesoderm. Although the prefix *pseudo* means "false," a pseudocoelom is a real body cavity.

The pseudocoelom enables animals to move more efficiently. How does this happen? Think about the way your muscles work. The muscles in your arm lift your hand by pulling against your arm bones. If there were no rigid bones in your arms, your muscles would not be able to do any work. Although the roundworm has no bones, it does have a rigid, fluid-filled space, the pseudocoelom, which its muscles can attach to and

Figure 28.13

Animals with acoelomate bodies (left) usually have a thin, somewhat flattened shape. The evolution of the pseudocoelom (center) made it possible for animals to become larger and thicker in body shape than their acoelomate ancestors. The coelom (right) provides a space for complex internal organs.

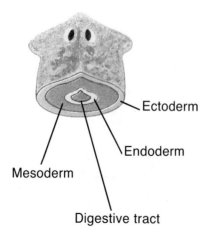

An Acoelomate Flatworm

Ectoderm

Endoderm

Mesoderm

Digestive tract

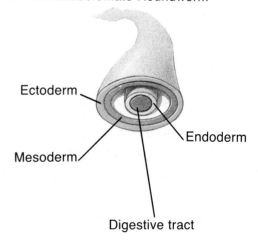

A Pseudocoelomate Roundworm

Ectoderm

Endoderm

Mesoderm

Digestive tract

brace against. The pseudocoelom thus makes it possible for pseudocoelomate animals to move more quickly than acoelomate animals.

A one-way digestive tract with two openings first evolved in roundworms. This innovation enabled various parts of the digestive tract to take on specific functions: the mouth opening takes in food, the middle region digests, and the anus expels waste.

The coelom provides space for internal organs

Earthworms have a more complex body plan than roundworms. The body cavity of an earthworm forms as a fluid-filled space inside the mesoderm as shown in *Figure 28.13* (below). This space is called the coelom. A **coelom** is a body cavity completely surrounded by mesoderm. Humans, insects, fish, and many other animals have a coelomate body plan.

In coelomates, the digestive tract and other internal organs are attached by double layers of mesoderm and are suspended within the fluid-filled coelom. Like the pseudocoelom, the coelom acts as a kind of watery skeleton against which muscles can work. The coelom provides space for more complex organs.

A Coelomate Segmented Worm

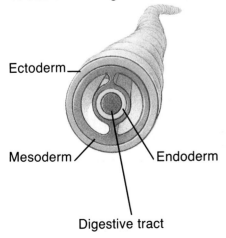

Ectoderm

Mesoderm   Endoderm

Digestive tract

The First Landlubbers

If you could travel backward in time, you might be surprised to discover that Earth's oceans were the birthplace of the first organisms on this planet. By 500 million years ago, the seas were alive with red, brown, and green algae and a variety of invertebrates including sponges and crabs.

New habitats appear Life, however, did not remain in the oceans. Earth's active geology was transforming the planet, creating new continents and islands. By 300 million years ago, plants, amphibians, insects, and reptiles began populating these new landforms.

How were organisms that had evolved in water over millions of years able to survive in a completely different environment on land?

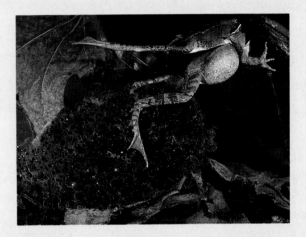

In and out of water Amphibians are animals that spend most of their time on land but must return to water to reproduce. Like fish, amphibians have external fertilization in which eggs and sperm require water for transport. Amphibian eggs also have no protective covering; if laid on land, they would simply dry up.

A different type of egg evolved in the reptiles—an amniotic egg that surrounds the developing embryo with a protective membrane and a shell. Amniotic eggs are fertilized by internal fertilization; water is not necessary to carry sperm to the eggs.

Further adaptations Other adaptations allowed animals to survive on land. The smooth, moist skin of amphibians gave way to the horny scales or plates that form the dry skin of reptiles, to the feathers of birds, and to the hair of mammals. Reptile skin prevented water loss, whereas feathers enabled flight, and hair provided warmth and protection.

Thinking Critically

Invertebrates also moved from the sea to the land, but their adaptations varied. Mosquitoes and many other insects still lay eggs in water. What adaptations do these organisms have that allow them to survive on dry land?

28.2   Body Plans and Adaptations   **705**

Focus On

Whales

Killer whale

They grow to lengths of up to 30 m and may weigh more than 136 000 kg. They can travel at 56 km/hr, and may, during their annual migrations, navigate distances as great as 9000 km. For nearly 50 million years, they have ruled the seas from Antarctica to the polar north. They are the whales.

Two kinds of whales

Baleen and toothed whales   All whales belong to the order Cetacea and are usually classified into two groups, the toothed whales and the baleen whales. Toothed whales include the familiar dolphins, porpoises, and killer whales, whereas baleen whales include the blue whale, humpback whale, and bowhead whale. As you can tell from their name, toothed whales have teeth, but baleen whales filter or strain tiny organisms out of ocean water with baleen—fringed, bony, comblike plates.

How Baleen Works

Baleen

Water with plankton enters mouth

Tongue

Water enters as whale opens its mouth.

Plankton caught on baleen

Water

Tongue

Tongue moves up, water exits, and plankton are caught and swallowed.

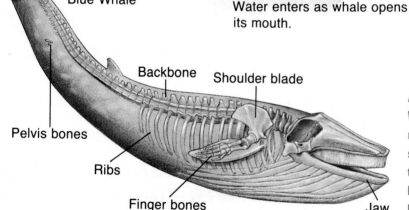

Blue Whale

Pelvis bones

Ribs

Finger bones

Backbone

Shoulder blade

Jaw

A return to the sea

Whales once lived on land   Whales were not always sea dwellers. Modern whales show skeletal evidence of a previous existence on dry land. Buried deep in a whale's hip muscles are two small bones—all that is left of the whale's pelvis and hind legs.

Warm-blooded giants

Whales give birth Whales are mammals—warm-blooded creatures that, after a gestation period of ten to 16 months, will give birth to one calf. When the calf emerges, the mother nudges it to the surface to take its first breath. A whale inhales and exhales through one or more blowholes at the top of the head, taking as much as 2000 L of air into its lungs. Most baleen whales breathe every five to 15 minutes. In contrast, sperm whales can go nearly two hours between breaths.

Humpback whale with calf

Beluga whale

Dolphins

Social animals

Living and working together Toothed whales are highly social animals. They swim in groups called pods, herds, or schools, sometimes including as many as 1000 animals. Whales seem to sense when other members of the group need help. In addition to females assisting each other during birth, whales are known to gather around and support a sick or injured companion. Mothers also actively shield their calves from danger.

EXPANDING YOUR VIEW

1. **Understanding Concepts** How might the social structure of toothed whales be an advantage for these animals in obtaining food?
2. **Journal Writing** Write a short essay that could be used to persuade nations that still hunt whales to cease this practice.

Figure 28.14

The exoskeleton of this crab has been shed in order for the crab to grow. A new exoskeleton forms in a few hours to again provide protection.

Animal Protection and Support

During the course of evolution, as development of body cavities resulted in more complex animal body plans, many animals became adapted to life on land. Some of these classes of animals developed exoskeletons. An **exoskeleton** is a hard, waxy covering on the outside of the body that provides a framework for sup-

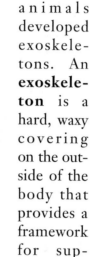

port, *Figure 28.14.* Exoskeletons prevent water loss and provide protection. Exoskeletons extend into the body, provide a place for muscle attachment, and are common in invertebrates. An **invertebrate** is an animal that does not have a backbone. Some invertebrates such as crabs, spiders, grasshoppers, dragonflies, and beetles have exoskeletons, while others do not.

An endoskeleton, or internal skeleton, is a characteristic of vertebrates. An **endoskeleton** is a support framework housed within the body. A **vertebrate** is an animal with a backbone. Examples of vertebrates include fish, birds, reptiles, amphibians, and mammals including humans. The endoskeleton protects internal organs and provides an internal brace for muscles to pull against.

Connecting Ideas

Animal body plans increased in complexity during the course of evolution from asymmetrical sponges to the bilaterally symmetrical flatworms, roundworms, and all other classes of animals. Animals can be differentiated depending upon the presence or absence of body cavities. Some animals have exoskeletons that provide protection and support, whereas others have endoskeletons that serve the same functions. How similar, and how different, animals are will be the focus of the following chapters.

Section Review

Understanding Concepts

1. Explain the difference between radial and bilateral symmetry in animals, and give an example of each type.
2. Compare the body plans of acoelomate and coelomate animals. Give an example of an animal with each type of body plan.
3. Explain how an adaptation such as an exoskeleton enabled animals to survive on land.

Thinking Critically

4. Make a list of five household items that are asymmetrical, five that have radial symmetry, and five that have bilateral symmetry.

Skill Review

5. **Making and Using Tables** Construct a table that compares the body plans of the sponge, hydra, flatworm, roundworm, and earthworm. For more help, refer to Organizing Information in the *Skill Handbook.*

Marine Biologists and the Battle Against Birth Defects

From 1959 to 1962, thousands of babies throughout Europe were born with missing arms or legs or with just stubby, flipperlike appendages where their arms and legs should have been. Medical investigators determined the cause of this disaster to be a sedative called thalidomide, which had been prescribed to expectant mothers as a sleep aid. For the first time, medical science was faced with compelling evidence that severe birth defects could be caused by an artificial substance.

Teratology is the study of birth defects In the decades following the thalidomide tragedy, a new branch of medical research developed—teratology, the study of birth defects. Teratologists search the environment for teratogens, natural and artificial compounds that can wreak havoc on a developing fetus. Thalidomide is a teratogen; so is alcohol, which some researchers think may be responsible for up to 20 percent of all mental retardation in the United States.

Using crabs to test teratogens Most teratological research is done on mammal embryos. However, Duke University marine biologists Anthony Clare and John Costlow think they have found a simpler, better way. Their unlikely accomplice in this biological quest is the star of seafood restaurants around the world—the crab.

Regenerating claws are the key A crab has the ability to shed a leg, leaving a hungry predator who thinks he has captured a full meal with little more than an appetizer. Later, the crab simply grows a new limb. So, using a larval mud crab, Clare and Costlow pinch off a claw, bathe the crab in a solution containing a suspected genetic toxin, and wait for the claw to grow back—a process that takes about two weeks. If the regenerated claw is smaller than expected, or

malformed in some way, the biologists can classify the toxin as a potential human teratogen.

Applications for the Future

The crab screening test has important implications for the future of medical research. Obviously, identifying human teratogens will eventually reduce the number of babies born with serious birth defects. Also, regeneration experiments with crabs reduce the need for more complex procedures requiring rabbits, guinea pigs, and other mammals. Finally, the study of limb regeneration in crabs may eventually reveal the biological secrets of this remarkable process.

INVESTIGATING the Technology

1. **Research** Go to the library and research the process of limb regeneration in animals. Which species regenerate lost limbs and other tissues? What have researchers discovered about the process?
2. **Debate** Have a debate about the use of crabs versus other species in biomedical research. Are some species more acceptable as research candidates? Explain your reasoning.

Reviewing Main Ideas

28.1 Typical Animal Characteristics

- Animals are heterotrophs, digest their food inside the body, typically have a type of locomotion, and are multicellular. Animal cells have no cell walls.
- Embryonic development from a fertilized egg is similar in many animal phyla. The sequence after division of the fertilized egg is: blastula, with one layer of cells; a gastrula with two cell layers, ectoderm and endoderm; and finally the formation of mesoderm, a layer of cells in between the ectoderm and endoderm.

28.2 Body Plans and Adaptations

- Animals have a variety of body plans and types of symmetry that are adaptations.
- Animal symmetry includes radial symmetry, bilateral symmetry, and asymmetry.
- Acoelomate animals such as flatworms have flattened, solid bodies with no body cavities.
- Animals with pseudocoeloms such as roundworms have a body cavity that develops between the endoderm and mesoderm.

- Coelomate animals such as humans and insects have internal organs suspended in a body cavity that is completely surrounded by mesoderm.
- Exoskeletons provide a framework of support on the outside of the body, whereas endoskeletons provide internal support.

Key Terms

Write a sentence that shows your understanding of each of the following terms.

acoelomate	gastrula
anterior	invertebrate
asymmetry	mesoderm
bilateral symmetry	posterior
blastula	protostome
coelom	pseudocoelom
deuterostome	radial symmetry
dorsal	sessile
ectoderm	symmetry
endoderm	ventral
endoskeleton	vertebrate
exoskeleton	

Understanding Concepts

1. List the main characteristics of animals.
2. In spite of the fact that animals have features in common, they are also diverse. Give two examples of this diversity in the ways animals obtain food.
3. How is food getting related to an animal's method of locomotion? Give two examples.
4. Compare the digestive systems of a flatworm and an earthworm. Explain how each system is adapted to the animal's way of life.
5. Explain the development of endoderm. What structures develop from these cells?
6. Explain the difference between development of protostomes and deuterostomes.

7. In what ways is an acoelomate body plan adapted to a specific way of life?
8. Why isn't the digestive tract of a flatworm considered to be a body cavity?

Interpreting Illustrations

9. Assume that you are a biologist observing the development of a sea urchin egg. The following sequence is what you observe. In which step does the process begin to show abnormal development?

Relating Concepts

10. Make a concept map that relates the following terms and phrases. Supply the appropriate linking words for your map.

    blastula, ectoderm, gastrula, endoderm, protostome, deuterostome, mesoderm, fertilized egg, development, larva, adult

Applying Concepts

11. As you watch development of an egg, you notice that a blastula forms, followed by formation of a gastrula. The mouth of the embryo develops from the opening in the gastrula. Could this egg have been from a bird or fish? Explain.

12. You are looking at a preserved specimen of an unidentified animal. It is easy to see that it has radial symmetry. What would you predict about its speed of movement when alive? Explain.

13. You observe under a microscope an organism from a tide pool that is moving rapidly. You offer live invertebrate food and see that your unknown organism seems to pursue and consume this prey. How could you tell if the organism is a protist or an animal?

14. Why are asymmetrical organisms often sessile? Explain.

15. How is obtaining food different in animals and fungi?

Focus On Whales

16. In what way are sponges and baleen whales similar? Explain.

Thinking Critically

Making Predictions

17. **Biolab** Examine the body plans to the right and infer which would be capable of more complex and powerful movement. Explain.

18. **Minilab** Assume that you are a zookeeper working with the reptile exhibit of a large zoo. You are given a group of rare lizards about which little is known. In order to provide an adequate habitat, you must find out what temperature the lizards prefer. Design an experiment that would determine their preferred environmental temperature.

19. **Minilab** If you were to observe two worms with your naked eye and found that one moved in a thrashing motion and the other moved in a pushing-forward motion, which one would be a roundworm? Which would be an earthworm relative? Explain.

20. **Thinking Lab** The development of certain structures is influenced by other tissues located nearby. Formation of the frog eardrum is influenced by the tough cartilage that lies beneath it. If a mutation occurred in the cartilage so that it did not develop properly, predict what might happen to the eardrum. Explain your prediction.

Connecting to Themes

21. **Systems and Interactions** In what way are body plans of animals related to the environment in which they live? Give two examples.

22. **Evolution** Explain in evolutionary terms how the development of more complex body plans accompanied the move from water to a land environment. Refer to diagrams 1, 2, and 3 below.

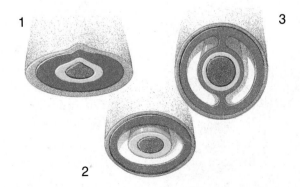

1

2

3

CHAPTER 29 Sponges, Cnidarians, Flatworms, and Roundworms

What kinds of animals would you expect to see if you could visit a 600-million-year-old zoo? Would the animals resemble those found in today's zoos? A 600-million-year-old zoo would probably be called an aquarium today. That's because the earliest animals—sponges, jellyfish, and worms—were found only in the warm, shallow seas that covered most of Earth's surface.

The ancestors of all modern invertebrates and vertebrates had simple body plans. Sponges, jellyfish, and worms that lived in water obtained most of their food, oxygen, and other materials directly from their surroundings. But to scientists, the living descendants of these early animals are indicators of how successful these simple body plans were for food gathering, reproduction, and digestion because they still exist in animals today. By studying these surviving species, it is possible to trace the evolutionary history of body systems.

The moon jellyfish and the sponge share common traits. Both are invertebrates with simple body plans that live in water. Do all invertebrates live in water?

29.1 Sponges

Section Preview

Objectives

Relate the sessile life of sponges to their food-gathering adaptations.

Relate the structure of sponges to their reproductive adaptations.

Key Terms

filter feeding
hermaphrodite
external fertilization
internal fertilization

I s this red organism a plant or an animal? At first glance, it may look like a plant because it is colorful and doesn't move from place to place, but it is an animal. How do we know this organism is an animal? Like snakes, tigers, and you, this organism is multicellular and a heterotroph—two characteristics that place it in the Animal Kingdom. This sessile animal is a sponge.

What Is a Sponge?

Sponges are asymmetrical animals that are a variety of colors, shapes, and sizes. Many are bright shades of red, orange, yellow, or green. Some sponges are ball-shaped; others have many branches. Sponges can be as small as a quarter or as large as a door. Although sponges do not resemble more familiar animals, they carry on the same life processes as all animals. *Figure 29.1* shows a natural sponge harvested from the ocean.

Figure 29.1

This heavy bath sponge is dark brown or black in its natural habitat. After harvest, it is washed and dried in the sun. When the process is complete, only a pale, lightweight skeleton made of spongin remains.

Where do sponges live?

Sponges are classified in the invertebrate phylum Porifera, which means "pore-bearer." Of the 5000 species of sponges, most live in the ocean, but a few species can be found in freshwater environments. Most occupy shallow waters, but some species have been discovered at depths up to 8500 m.

No matter where sponges live, they are sessile organisms. Remember, a sessile organism is one that remains permanently attached to a surface for all of its adult life. Because adult sponges are sessile, they can't travel in search of food. Sponges get their food by **filter feeding,** a method in which organisms feed by filtering small particles of food from water as it passes by or through some part of the organism. Filter feeding is a common feeding strategy in marine organisms.

Inside a Sponge

Sponges have no tissues, organs, or organ systems. The body plan of a sponge is simple, being made up of only two cell layers with no body cavity. Between these two cell layers is a jellylike substance that contains cells, as well as the components of the sponge's internal support system. Sponges have four types of cells that perform all the functions necessary to keep them alive.

Orange tube sponges

1 Water and wastes are expelled through the osculum, the large opening at the top of the sponge. A sponge no bigger than a pen can move more than 20 L of water through its body per day.

Osculum

Pore cells

2 Surrounding each pore is a single pore cell. Pore cells bring water carrying food and oxygen into the sponge's body.

Collar cells

4 Lining the interior of sponges are collar cells. Each collar cell has a flagellum that whips back and forth, drawing water through the pores of the sponge.

Pore cell

Direction of water flow

5 Between the two cell layers of a sponge's body are amoeba-shaped cells called amoebocytes. Amoebocytes carry nutrients to other cells, aid in reproduction, and also produce chemicals that help make up the spicules of sponges.

Amoebocytes

Epithelial cells

Spicules

3 Epithelial cells are thin and flat. They contract in response to touch or to irritating chemicals, and in so doing, close up pores in the sponge.

6 Spicules are structures produced by other cells that form the hard support systems of sponges. Spicules are small, needlelike structures located between the cell layers of a sponge.

From one cell to cell organization

If you took a living sponge and put it through a sieve, a type of filter, you would witness a rather remarkable event. Not only would you see the sponge's many cells alive and separated out, but within an hour you would be able to see these same cells coming together to form a whole sponge once again. Many biologists hypothesize that sponges evolved directly from colonial, flagellated protists, such as *Volvox*. More importantly, however, sponges demonstrate what appears to have been a major step in the evolution of animals—the change from a unicellular life to a division of labor among groups of organized cells.

Reproduction in Sponges

Sponges reproduce both sexually and asexually. Sponges reproduce asexually by fragments that break off from the parent animal and form new sponges, or by forming external buds. Buds may break off and float away, and eventually settle and become separate animals. Sometimes buds remain attached to the parents, forming a colony of sponges. You can see a colony in *Figure 29.2.*

Most sponges reproduce sexually. Sponges are **hermaphrodites;** that is, an individual sponge can produce both eggs and sperm, though at different times. Eggs and sperm are formed from amoebocytes. During reproduction, sperm released from one sponge are carried by water currents to other sponges where fertilization occurs.

Fertilization in sponges may be either external or internal. In **external fertilization,** the eggs and sperm are both released into the water, and fertilization occurs outside the animal's body. In **internal fertilization,** eggs remain inside the animal's body, and sperm are carried to the eggs. In sponges, the collar cells collect the sperm, transfer them to amoebocytes, and the amoebocytes transport the sperm to ripe eggs. Most sponges reproduce sexually through internal fertilization. The result is the development of a free-swimming, flagellated larva (plural, larvae), *Figure 29.3.* Sponges also have the ability to regenerate lost body parts. In regeneration, lost body parts are replaced by mitosis.

Figure 29.2

Sponge colonies are the result of asexual reproduction. How would these sponges compare genetically?

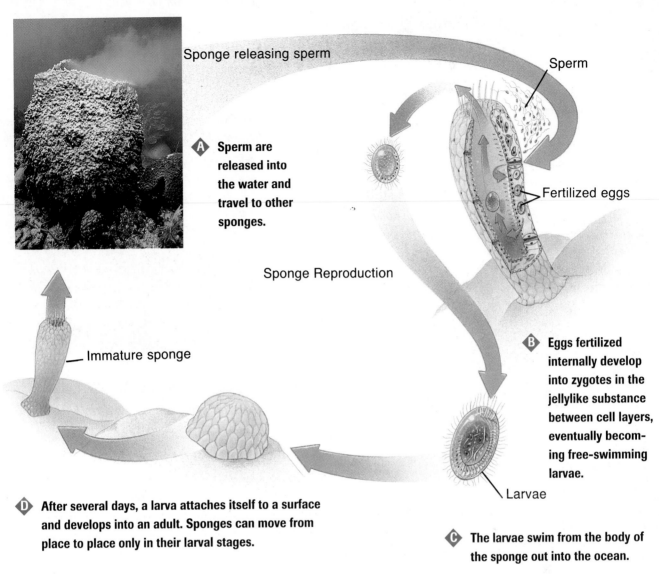

Sponge Reproduction

A Sperm are released into the water and travel to other sponges.

Sponge releasing sperm

Sperm

Fertilized eggs

B Eggs fertilized internally develop into zygotes in the jellylike substance between cell layers, eventually becoming free-swimming larvae.

Larvae

Immature sponge

D After several days, a larva attaches itself to a surface and develops into an adult. Sponges can move from place to place only in their larval stages.

C The larvae swim from the body of the sponge out into the ocean.

Figure 29.3

Sponges reproduce sexually when the sperm from one sponge fertilize the eggs of another sponge.

Section Review

Understanding Concepts
1. How does a sponge obtain food?
2. Relate the functions of the four cell types in sponges.
3. Describe the steps involved in the sexual reproduction of sponges.

Thinking Critically
4. What advantages do multicellular organisms such as sponges have over unicellular organisms? Explain.

Skill Review
5. **Making and Using Tables** Make a table listing the cell types and other structures of sponges along with their functions. For more help, refer to Organizing Information in the *Skill Handbook*.

29.2 Cnidarians

Section Preview

Objectives

Distinguish the different classes of cnidarians.

Relate the polyp and medusa forms to the reproductive cycle of jellyfish.

Analyze the adaptations of cnidarians for obtaining food.

Key Terms

polyp
medusa
nematocyst
gastrovascular cavity
nerve net

What's the largest structure ever built by living organisms? Is it the Sears Tower in Chicago? How about the Great Pyramid in Egypt? Actually, the largest structure ever built is the Great Barrier Reef, which extends for more than 2000 km off the northeastern coast of Australia, and it wasn't built by humans! This structure was built over many centuries by a colony of small marine invertebrate animals called corals. Corals and their relatives, jellyfishes and sea anemones, all belong to the phylum Cnidaria.

cnidarian:
knid (GK) nettle, a plant with stinging hairs. *Cnidarian* is pronounced with a silent *c*.

What Is a Cnidarian?

Cnidarians are a group of marine invertebrates made up of about 9000 species of jellyfishes, corals, sea anemones, and hydras. Most cnidarians are found worldwide, but some species of coral prefer the warmer oceans of the South Pacific and the Caribbean.

Though cnidarians are a diverse group, all possess the same basic body structure, which supports the theory that they had a single origin.

All cnidarians have radial symmetry. Like sponges, cnidarian bodies have only one body opening, and they are made up of two cell layers. Unlike sponges, however, the cell layers of cnidarians are organized into separate tissues with specific functions. Cnidarians have simple nervous systems, and both cell layers have cells that can contract like muscles. The two cell layers are derived from the ectoderm and endoderm of the embryo. The outer layer of cells serves to protect the cnidarian, whereas the inner layer is involved mostly in digestion.

Cnidarians display only two basic body forms at different stages of their life cycles, polyp and medusa (plural, medusae), *Figure 29.4.* A **polyp** is the stage that has a tube-shaped body form with a mouth surrounded by tentacles. A **medusa** is the stage that has a body form shaped like an umbrella with the tentacles hanging down.

Figure 29.4

If you turn the polyp form of a cnidarian (left) upside down, you have the basic structure of a medusa (right).

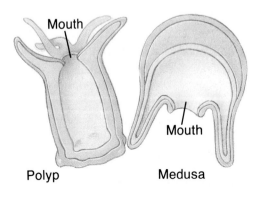

Mouth

Mouth

Polyp

Medusa

Inside a Cnidarian

Hydra

Cnidarians display a remarkable variety of shapes and sizes. Some, such as corals and hydras, are as small as the tip of a pencil, while others, such as some jellyfishes, can be as long as half of a football field! All cnidarians, except sea anemones and corals, go through both polyp and medusa stages at some point in their life cycles. The flowerlike forms of sea anemones are often brilliant shades of red, purple, and blue, and the bodies of some jellyfishes contain glowing pigments.

2 A medusa is a free-swimming form of a cnidarian. It possesses a bell-shaped, floating body with the mouth pointing down. Jellyfishes have the medusa body form.

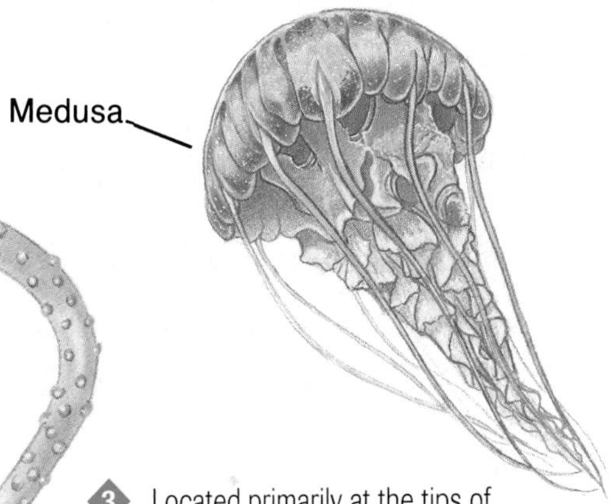

Tentacles

Medusa

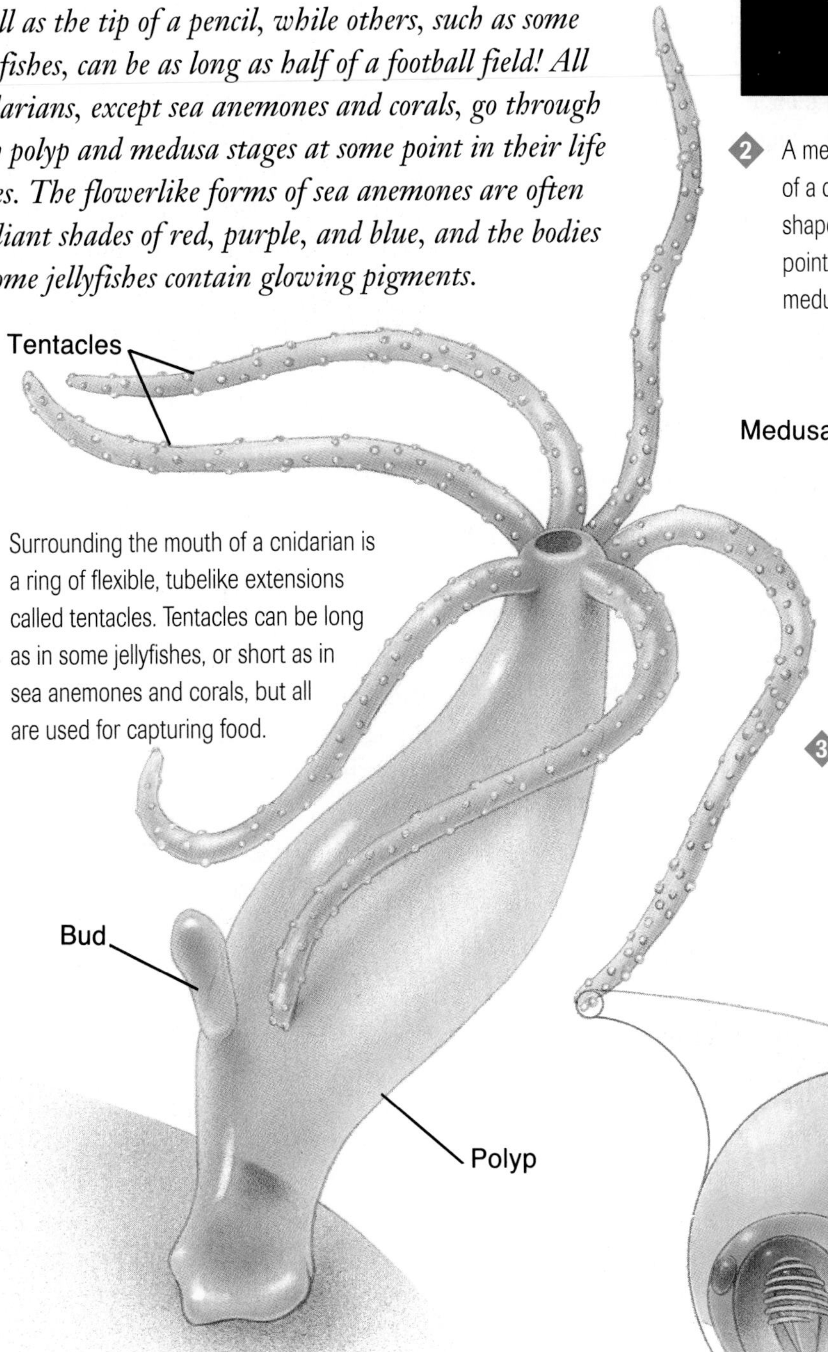

1 Surrounding the mouth of a cnidarian is a ring of flexible, tubelike extensions called tentacles. Tentacles can be long as in some jellyfishes, or short as in sea anemones and corals, but all are used for capturing food.

3 Located primarily at the tips of the tentacles are stinging cells that contain nematocysts. When prey touches the tentacles, the nematocyst releases a coiled tube that ejects a toxin that paralyzes the prey.

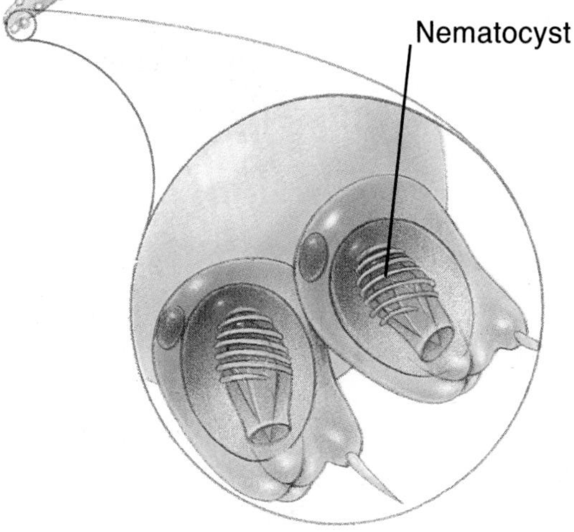

Bud

Nematocyst

Polyp

4 Polyps are the sessile form of cnidarians. Polyps have mouths that point upward. Examples of polyps include sea anemones, corals, and hydras.

How do nematocysts function?

Some of the cells in the body surface of cnidarians contain nematocysts – tiny, harpoonlike structures with sharp barbs. These barbs are released by touch and when stimulated by certain chemicals. In this lab, you will observe nematocyst discharge in a hydra.

Procedure

1. Place a hydra on a clean, flat slide, and cover gently with a coverslip.
2. Place a drop of methylene blue at the edge of the coverslip. Draw the stain under the coverslip by placing a piece of paper towel on the opposite edge of the coverslip.
3. With a compound microscope, focus on the tentacles and nematocysts. Next, place a drop of vinegar at the edge of the coverslip, and draw it under the coverslip as you did in step 2. Vinegar will stimulate the release of nematocysts.

Analysis

1. Of what survival value are nematocysts? How do they function?
2. Why are toxins important adaptations in animals not capable of pursuing prey?
3. Draw a diagram of nematocysts before and after discharge.

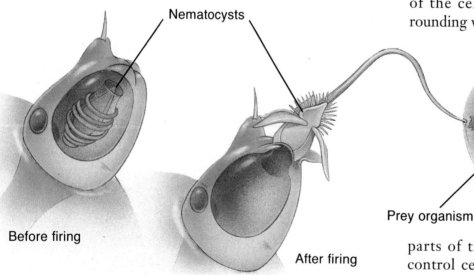

Before firing

After firing

Prey organism

Nematocysts

Figure 29.5

Nematocysts, located in special stinging cells in the outer cell layer of cnidarians, are discharged like toy popguns in response to touch or chemicals. Captured by nematocysts on the ends of tentacles, prey is brought to the mouth by contraction of the tentacles.

Digestion in cnidarians

Cnidarians are predators that capture or poison their prey with nematocysts. A **nematocyst** is a capsule that contains a coiled, threadlike tube. The tube may be sticky or barbed, or it may contain toxic substances, *Figure 29.5*. In cnidarians, you begin to see the origins of a digestive process similar to that of more complex animals. Cells adapted for digestion inside the bodies of cnidarians release enzymes over the newly captured prey. The inner cell layer of cnidarians surrounds a space called a **gastrovascular cavity** in which digestion takes place. Any undigested materials are ejected back out through the mouth.

Oxygen enters cells directly

Because of a cnidarian's simple, two-cell-layer body plan, no cell in its body is ever far from water. Oxygen diffuses directly into the body cells from water, and carbon dioxide and other wastes diffuse out of the cells directly into the surrounding water.

Nervous regulation in cnidarians

Cnidarians do not have a nervous system as in complex animals; rather, cnidarians possess a nerve net. A **nerve net** conducts nerve impulses from all parts of the body, but there is no control center such as the brain of more complex animals. The impulses from the nerve net bring about contractions of musclelike cells in the tentacles and bodies of cnidarians. For example, when touched, a hydra will react rapidly by contracting the musclelike cells of its body.

Reproduction in cnidarians

Cnidarians have the ability to reproduce both sexually and asexually. Polyps, such as hydras, reproduce asexually by a process known as budding, as shown in *Figure 29.6*.

The medusa form of cnidarians, such as seen in a jellyfish, is the sexual stage, which alternates back and forth between generations with the asexual polyp stage of the life cycle. As shown in *Figure 29.7*, medusae reproduce sexually to produce polyps, which, in turn, reproduce asexually to form new medusae.

Figure 29.6

The main form of reproduction in polyps is budding. During this process, small buds grow as extensions of the body wall. As the buds grow, a mouth with tentacles develops and the new polyp breaks away from the parent.

Figure 29.7

A free-swimming larva develops into a polyp. The structure of this larva gives scientists clues about the origin of cnidarians.

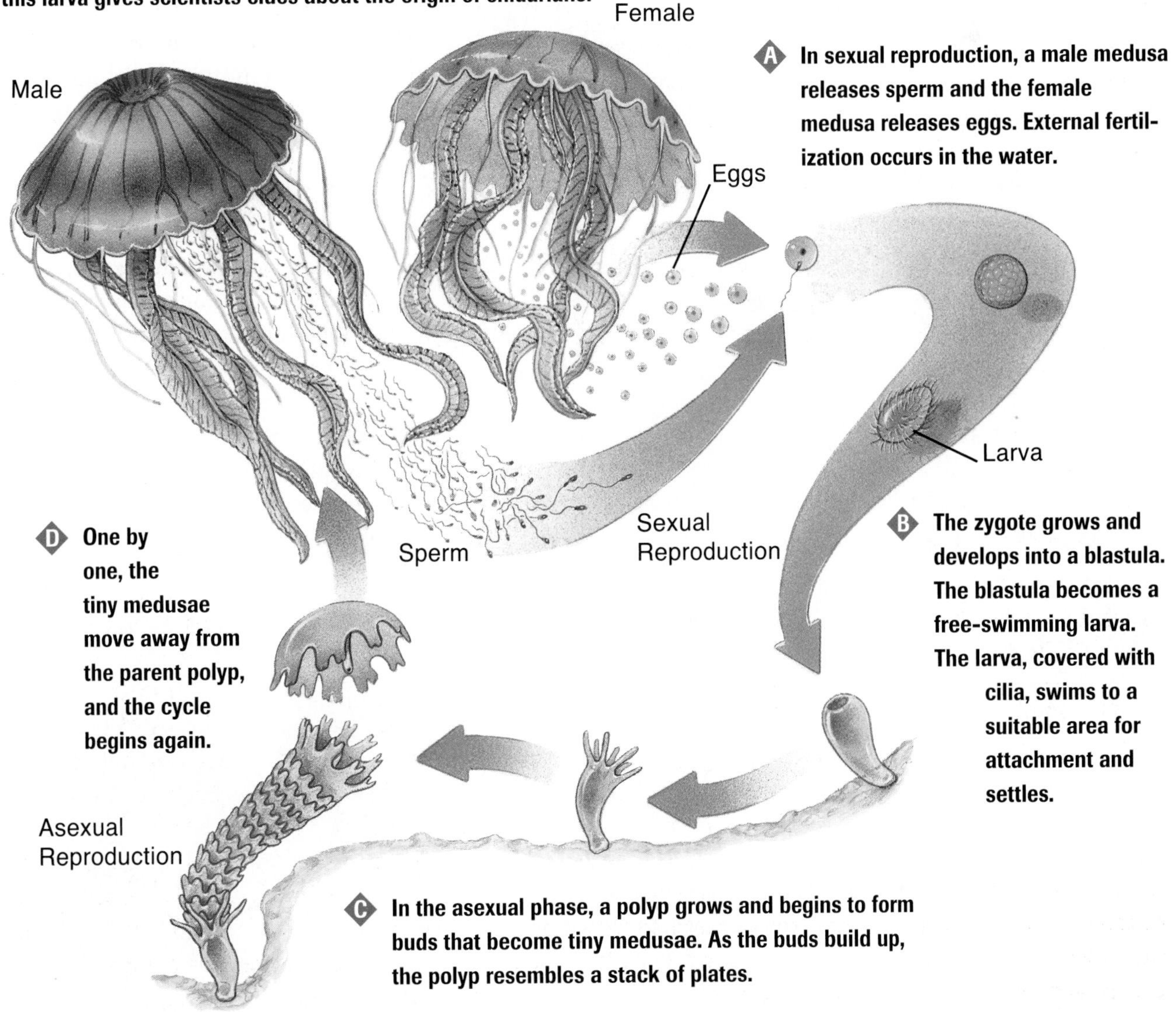

Female

Male

Eggs

A **In sexual reproduction, a male medusa releases sperm and the female medusa releases eggs. External fertilization occurs in the water.**

Larva

Sperm

Sexual Reproduction

D **One by one, the tiny medusae move away from the parent polyp, and the cycle begins again.**

B **The zygote grows and develops into a blastula. The blastula becomes a free-swimming larva. The larva, covered with cilia, swims to a suitable area for attachment and settles.**

Asexual Reproduction

C **In the asexual phase, a polyp grows and begins to form buds that become tiny medusae. As the buds build up, the polyp resembles a stack of plates.**

Mass Spawning of Corals

In the spring of 1981, a group of post-graduate students was working on reproduction in corals on the Great Barrier Reef. One night, they became the first people to witness one of the most spectacular reproductive events in nature—the synchronized mass spawning of corals on the Great Barrier Reef!

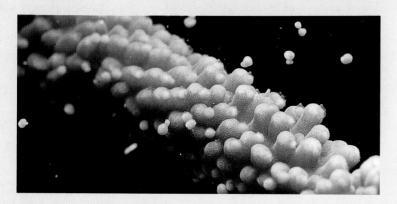

Developing eggs and sperm In the months before spawning, eggs and sperm cells begin to grow inside the tiny bodies of coral polyps. In hermaphroditic corals, the eggs and sperm usually clump together to form a colorful, compact ball called an egg bundle. At this point, the corals are ready to spawn. Now they have only to wait for their cue to release their bundles into the water.

The cue to spawn Although water temperature and tides play a role, the major cue for coral spawning comes from the moon. Every year on the Great Barrier Reef, a few nights after the full moon in late spring or early summer, billions and billions of coral polyps all release their eggs and sperm together. Brightly colored bundles of eggs and sperm pop out of the mouths of polyps and float up to the water's surface. If you were diving on the reef on this special night, you would be in the middle of a colorful snowstorm underwater—except the "snow" would be going up instead of coming down!

A new generation Fertilized eggs develop into tiny, free-swimming larvae that drift through the ocean for several days. Eventually, some of the survivors settle down on a reef as young coral polyps. Each of these new polyps can form the basis of a new coral colony.

Thinking Critically

Many coral reef inhabitants, including the thousands of fish that roam the reef by day, find coral eggs and larvae a tasty treat. By spawning at the same time that so many other corals do, how does an individual coral reduce the risk that all of its eggs will be devoured by reef predators? What do you think might be the particular advantage of spawning at night?

Adaptations and Ecology

Each of the 9000 known species of cnidarians belongs to one of three classes: Hydrozoa, Scyphozoa, or the class of coral reefs, Anthozoa.

Hydrozoans form colonies

It's difficult to believe that the organism shown in *Figure 29.8* is actually a closely associated group of individual animals. The Portuguese man-of-war is an example of a hydrozoan colony.

Each individual in the colony has a different function that helps the entire organism to survive. For example, just one individual forms a large, blue, gas-filled float. Regulation of the gas in the float allows the colony to dive to lower depths or rise to the surface. Other polyps hanging from the float have different functions, such as reproduction and feeding.

Figure 29.8

The Portuguese man-of-war, *Physalia,* is really a collection of related individuals working together to form one organism. The Portuguese man-of-war captures fish by injecting venom from nematocysts. Feeding polyps then secrete digestive enzymes over the captured prey. The venom in a large colony such as this is powerful enough to kill a human.

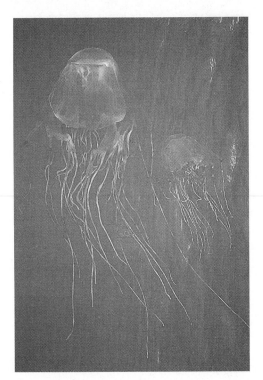

Figure 29.9

The Australian box jelly is invisible to swimmers and can kill within minutes of the sting.

Scyphozoans include jellyfish

The fragile and sometimes luminescent bodies of jellyfish can be beautiful, but most people know about jellyfish through their painful stings. Shown in *Figure 29.9* is the Australian box jelly. Many people consider this the most venomous sea animal known.

Anthozoans build coral reefs

Corals are anthozoans that live in colonies of polyps in warm ocean waters around the world. They build the beautiful coral reefs that serve as food sources and shelter for many other species of animals, including the sea anemones, *Figure 29.10*, another group of anthozoans.

Figure 29.10

Sea anemones may look like flowers, but they are animals that capture prey. Sea anemones exhibit only the polyp form.

How does coral reef bleaching change over time?

Reef-building corals normally contain microscopic algae. The corals provide protection for the algae, and the corals benefit by obtaining oxygen and food produced by the algae. Sometimes, though, environmental conditions cause corals to expel their algae. The corals become pale and appear to be bleached. Corals lacking algae do not build reefs and cannot survive solely on organisms captured by nematocysts.

Analysis

Examine the map provided, and explain the trend in coral reef bleaching by answering the following question.

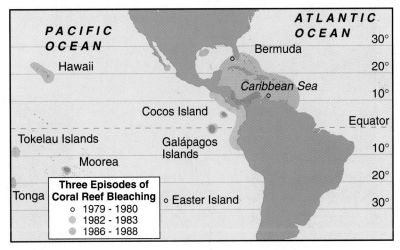

Thinking Critically

How did coral reef bleaching change over the three time periods represented on the map?

Coral Reefs

Magic kingdoms in the sea

Coral reefs come in all shapes, sizes, and colors In addition to producing the structural base of a reef, corals—along with algae and marine protists—provide food for a bewildering array of animals, from snails, starfish, and sea urchins to hundreds of species of brilliantly colored reef fish. These animals, in turn, are preyed upon by other reef inhabitants such as rays and sharks. Coral reefs teem with life. They are magic kingdoms of color and pattern, of remarkable adaptations and unexpected relationships.

Many biologists consider tropical rain forests to be the richest terrestrial ecosystems in terms of biological diversity. However, a similar level of biodiversity can be found underwater on and around coral reefs.

Coral reefs are found only in warm, tropical seas between the latitudes of 30 degrees north and 30 degrees south. The world's largest coral reef complex is the Great Barrier Reef, which consists of more than 2600 individual reefs and stretches roughly 2000 km along the northeastern coast of Australia. Coral reefs are also found off the coast of Belize in Central America, around many islands in the Caribbean, in the Pacific and Indian Oceans, and in the Red Sea.

Coral polyps are predators

Polyps also harbor unicellular organisms At night (left), most coral polyps extend their tentacles and use them to snare minute planktonic organisms that drift past. Coral polyps also harbor unicellular organisms called zooxanthellae inside their bodies. Zooxanthellae use sunlight to manufacture food for themselves and their coral hosts. During the day (right), most coral polyps withdraw into the safety of their limestone cups and let their zooxanthellae make food for them.

What are nudibranchs?

Bright colors warn predators they are poisonous
Nudibranchs are related to snails, but they lack a shell.
Most are brilliantly colored with distinctive stripes or spots, and many species are poisonous. When do you suppose nudibranchs are most active: during the day or at night?

From tiny polyp to massive reef

Corals are base of the reef   Each polyp in a reef-building colony secretes a cuplike, limestone skeleton around itself, which is joined to that of its neighbors in the colony. When polyps die, their fragile bodies decay but their hard skeletons are left behind, becoming the foundation upon which other corals build. A thin layer of living coral grows on top of the stony remains of previous generations.

Stars of the sea

Sea stars are vibrant colors   This cobalt blue starfish (sea star) is common on Pacific coral reefs. They are usually found in fairly shallow water in sheltered areas. Unlike most species of starfish, they are active and exposed during the day, and fish never seem to bother them. Do you think these starfish are good to eat?

Fish are abundant near a reef

Fish find food easily   More species of fish can be found on a coral reef than in any other type of habitat in the ocean. In just one hectare of reef, you might find 200 different fish species.

EXPANDING YOUR VIEW

1. **Understanding Concepts** In tropical areas, local shops often sell jewelry made from coral. Why is collecting "dead" coral from a reef destructive?

2. **Journal Writing** In a short paragraph, explain the benefits to a fish of living among the tentacles of a sea anemone.

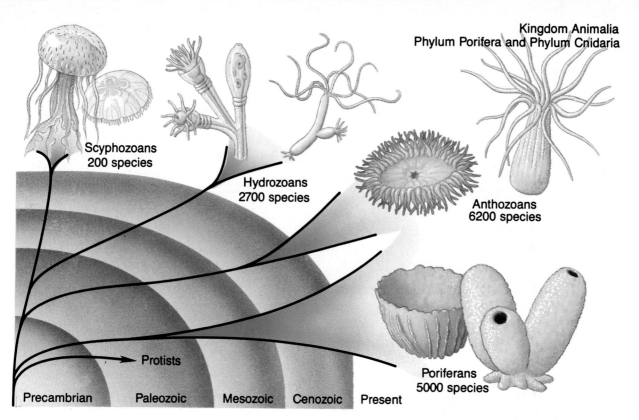

Kingdom Animalia
Phylum Porifera and Phylum Cnidaria

Scyphozoans
200 species

Hydrozoans
2700 species

Anthozoans
6200 species

Protists

Poriferans
5000 species

Precambrian Paleozoic Mesozoic Cenozoic Present

Figure 29.11

Sponges and cnidarians evolved early in geologic time. Sponges probably were the first to appear, followed by the classes of cnidarians.

Origins of Sponges and Cnidarians

As shown in *Figure 29.11*, sponges represent an old animal phylum. The earliest fossil evidence for sponges dates this group to the Paleozoic Era, about 600 million years ago. From the similarity of a group of flagellated protists that resemble the collar cells of sponges,

scientists infer that sponges may have evolved directly from these protists.

The earliest known cnidarians date to the Precambrian Era, about 630 million years ago. Because cnidarians are soft-bodied animals, they do not preserve well as fossils, and their origins are not well understood. The larval form of cnidarians resembles protists, and most scientists agree that cnidarians evolved from protists.

Section Review

Understanding Concepts

1. Compare the structures of medusa and polyp forms of cnidarians.
2. Diagram the reproductive cycle of a jellyfish.
3. What are the advantages of a two-layered body in cnidarians?

Thinking Critically

4. Coral reefs are being destroyed at a rapid rate as coral is harvested. What effect

would you expect the destruction of a large coral reef to have on other ocean life?

Skill Review

5. **Making and Using Tables** In a table, distinguish the three main groups of cnidarians, list their characteristics, and give examples of a member from each group. For more help, refer to Organizing Information in the *Skill Handbook*.

Magnification: 80 ×

Imagine the ultimate couch potato among living organisms—an organism that never has to move with its own muscle power, is always carried by another organism, is surrounded by food that is already digested, and never has to expend much energy. This describes a parasite called a tapeworm. The parasitic way of life has advantages as well as disadvantages.

Section Preview

Objectives

Distinguish the adaptive structures of parasitic flatworms and planarians.

Explain how parasitic flatworms are adapted to their way of life.

Key Terms

pharynx
scolex
proglottid

What Is a Flatworm?

To most people, the word *worm* brings to mind a long, slimy creature without eyes and limbs. Many animals have this general structure, but now it is understood that worms can be classified into many different phyla.

Worms have more complex structures than sponges and cnidarians, but the least complex worms belong to the phylum Platyhelminthes, *Figure 29.12*. These flatworms are acoelomates with thin, solid bodies. The most well-known members of this phylum are the parasitic tapeworms (Class Cestoda) and flukes (Class Trematoda), which cause diseases in humans. The most commonly studied flatworms in biology classes are the free-living planarians (Class Turbellaria). Flatworms range in size from 1 mm up to several meters.

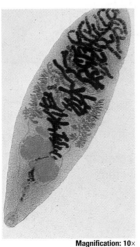

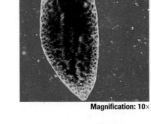

Magnification: 10×

Magnification: 10×

Figure 29.12

Tapeworms (left) are parasites that invade and live in host organisms. Flukes (center) usually require two hosts in a complex life cycle. Planarians (right) are not parasitic, nor do they cause diseases.

Inside a Planarian

If you've ever waded in a shallow stream and turned over some rocks, you may have found tiny, black organisms stuck to the bottom of the rocks.

These organisms were most likely planarians. Planarians have many characteristics common to all species of flatworms. The bodies of planarians are flat, with both dorsal and ventral surfaces. Unlike sponges and jellyfish, all flatworms have bilateral symmetry.

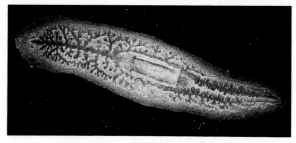

Common planarian, *Dugesia*

4 Eyespots are sensitive to light and enable the animal to respond to the amount of light present. Eyespots can't form images as do more complex eyes.

1 In contrast to sponges and cnidarians, flatworms have a clearly defined head. The head is responsible for sensing and responding to changes in the environment.

2 The pharynx is a muscular tube that can be extended outside the body. It is used to suck food into the planarian's gastrovascular cavity through the mouth.

5 Located on the sides of the head, sensory pits are used to detect food, chemicals, and movements in the environment.

3 Hairlike cilia are located on the ventral surface of planarians. Cilia help the worm to pull itself along.

6 Excess water is removed from the planarian's body by a system of flame cells. The water from flame cells collects in tubules and leaves the body through pores on the body surface. Flame cells are so named because the constant movement of the cilia inside the flame cell resembles the flickering of a candle's flame.

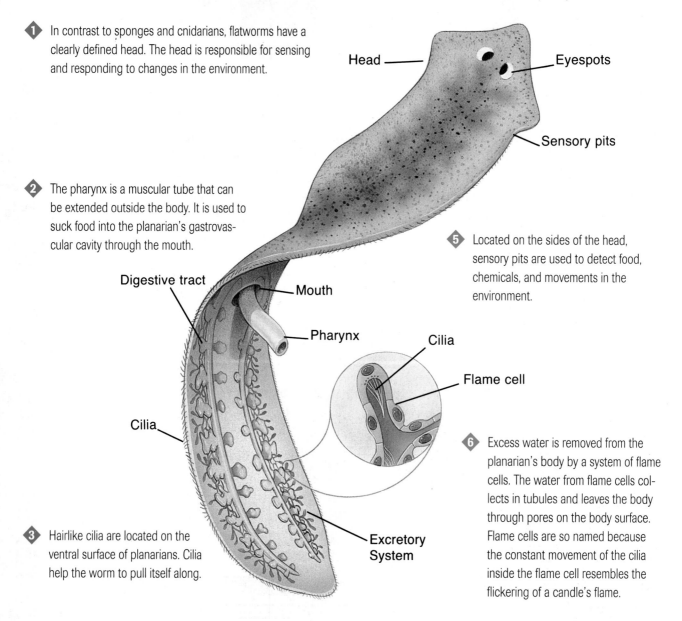

Head — Eyespots

Sensory pits

Digestive tract — Mouth

Pharynx

Cilia

Flame cell

Cilia

Excretory System

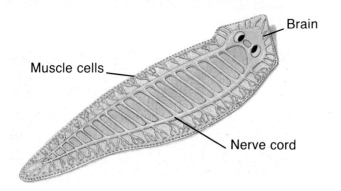

Brain

Muscle cells

Nerve cord

Figure 29.13

Messages from the nerve cords trigger responses in the planarian's muscle cells. In this way, planarians are able to make adjustments to stimuli in the environment.

cut into two pieces may grow into two new organisms—a form of asexual reproduction.

Feeding and digestion

Planarians feed on dead or slow-moving organisms, including small planarians. To feed, a planarian extends a tubelike, muscular organ, called the **pharynx,** out of its mouth. Enzymes released by the pharynx begin digestion of the food before it is sucked into the digestive system.

Nervous control in planarians

In addition to eyespots and other cells that are sensitive to the environment, planarians possess a brainlike structure called a ganglion. Located in the head, the ganglion sends messages, received from the eyespots and sensory pits, along two nerve cords that run the length of the body, *Figure 29.13.*

Reproduction in planarians

Like many of the organisms studied in this chapter, most flatworms are hermaphrodites. During sexual reproduction, planarians exchange sperm, and fertilization occurs internally. The zygotes are released in capsules into the water, where they hatch into tiny planarians.

Planarians can also reproduce asexually by fission. When a planarian is damaged, it has the ability to regenerate new body parts. If a planarian is cut horizontally, the head piece will grow a new tail, and the tail piece will grow a new head. Thus, a planarian that is accidentally

Parasitic Flatworms

Planarians are free-living organisms, but most flatworm species, such as flukes and tapeworms, are parasitic.

Although the basic structure of parasitic flatworms is similar to that of planarians, parasitic worms are adapted to obtaining their nutrients from inside the bodies of one or two kinds of hosts. Parasitic worms have mouthparts with hooks that serve to hold the worm's position inside its host. Because they are surrounded by nutrients, there is less demand for complex nervous or muscular tissue.

Planaria sports

Planarians live in temperate climates under submerged stones on the bottom of streams and ponds. In order to survive, they must detect changes in their environment such as presence of food, amount of light, and water temperature. In this lab, you will determine how planarians respond to these conditions.

How Planarians Respond to Stimuli

PREPARATION

Problem
How do planarians respond to stimuli?

Objectives
In this Biolab, you will:
- **Determine** stimuli that cause responses in planarians.
- **Interpret** data in terms of adaptations to changes in the environment.

Materials
planaria culture
small pebbles
eyedropper
test tube with pond water
test tube with ice
test tube with heated water
thermometer
petri dishes
pieces of cooked egg white
black construction paper
toothpick
hot plate

Safety Precautions
Be careful handling the heated test tube, and be sure to wash your hands before and after handling planarians.

1. Make a data table with the stimuli in one column and the responses in the other column.

Stimulus	Response		Time Spent
	Positive	Negative	
pebbles			
black paper			
white paper			
ice			
heat			
egg white			

2. Hypothesize whether the planarian will have a positive or negative response to each stimulus.

3. Draw a line on the bottom of the petri dish to divide it into two halves, and fill halfway with pond water.

4. With a toothpick, transfer a planarian to the dish.

5. Place your petri dish on a white surface and add three or four pebbles to one side. Time, for two minutes, how long the worm spends on each side of the line, and record your data in the data table.

6. Remove the pebbles. Place the dish on a piece of black paper so that the dish is half on black paper and half on white. Time, for two minutes, how long the worm spends on each side of the dish.

7. Remove the black paper.

8. Place a test tube, with ice, on the inside edge of one side of the dish and a heated test tube on the opposite side of the dish. Time how long the worm spends on each side of the dish, and record in the data table.

9. Remove the two test tubes.

10. Gently stir the water in the dish to equalize temperature. Place a small piece of egg white on one side of the dish. Record how long the worm spends on each side of the dish during a two-minute period.

11. Record your data in your table.

1. **Identifying Variables** Tell whether the planarian responded positively or negatively to each stimulus in your table. Explain.

2. **Checking Your Hypothesis** Was your hypothesis supported by your data? Why or why not?

3. **Drawing Conclusions** What might be the survival value of each of the observed responses?

Going Further

Project Design and conduct an experiment to test food preferences of planarians.

Pigs and Parasites

Throughout history, people who adhere to certain religious beliefs have avoided pigs and pork products. The pig's unsavory reputation among these groups probably originated from the experiences of generations of people who learned to relate eating pork to getting sick. It is now well-known that pigs are unwilling hosts to *Taenia solium,* a parasitic flatworm (tapeworm) that can cause serious illness and sometimes even death in humans who eat contaminated and improperly prepared pork.

From host to host   The life cycle of *Taenia* is complex. Tapeworms hatch from eggs in the intestine of an intermediate host such as a pig. Worms in this larval stage are called oncospheres. The oncospheres burrow through the intestines and settle in muscle tissue. Heat or extreme cold can kill the oncospheres, so pork meat, which is muscle tissue, should be cooked before it is eaten. When the intermediate host is eaten by another animal, the oncosphere either develops into a second larval stage or into a mature tapeworm.

Larvae enter bloodstream   In the second larval stage, *Taenia* enters the bloodstream from the intestines of its new host and is carried to the brain. There, the larvae cause inflammation and nerve damage, often accompanied by seizures. Sometimes, the larvae even trigger a form of epilepsy.

Oncospheres also can develop into adult tapeworms that reside in the intestines. These worms can range in size from 1 mm to 15 m. They attach themselves to their host's intestines by means of suckers and hooks in their heads.

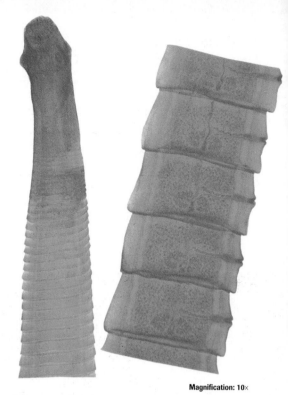

Magnification: 10×

Figure 29.14

The scolex and proglottids are adaptations to the parasitic life of a tapeworm. The scolex is covered with hooks and suckers that attach to the intestinal lining of the host.

Tapeworm adaptations

Tapeworms are parasitic flatworms of the Class Cestoda. These parasites live in the intestines of many vertebrates including dogs, cats, cattle, monkeys, and people who live in countries where sanitation is poor. Some adult tapeworms can grow to more than 10 m in length. The body of a tapeworm is made up of a head and individual repeating sections called proglottids, *Figure 29.14.* The knob-shaped head of a tapeworm is called a **scolex. A proglottid** is a detachable section of a tapeworm that contains muscles, nerves, flame cells, and male and female reproductive organs. Each proglottid may contain up to 100 000 eggs, and some tapeworms consist of 2000 proglottids.

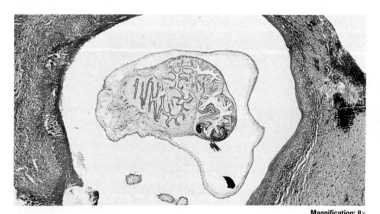

Magnification: 8×

CONNECTION TO Biology

In what ways can you avoid becoming the unwilling host to a tapeworm?

The life cycle of flukes

Flukes are parasitic flatworms that invade the digestive systems of vertebrates such as humans and sheep. They gain their nutrition by embedding themselves in tissues that line the intestine, where they feed on cells, blood, and other fluids of a host organism.

Blood flukes of the genus *Schistosoma*, shown in *Figure 29.15*, cause a disease in humans known as schistosomiasis. In this disease, the body is robbed of nutrition, and symptoms such as intestinal and urinary problems may occur. Blood flukes are common in parts of Africa, the Middle East, South America, and Asia—places where certain species of snails (one of the hosts) are found.

Figure 29.15

Blood flukes have a complex life cycle that often includes two or more hosts. The *Schistosoma* fluke requires two hosts to complete its life cycle.

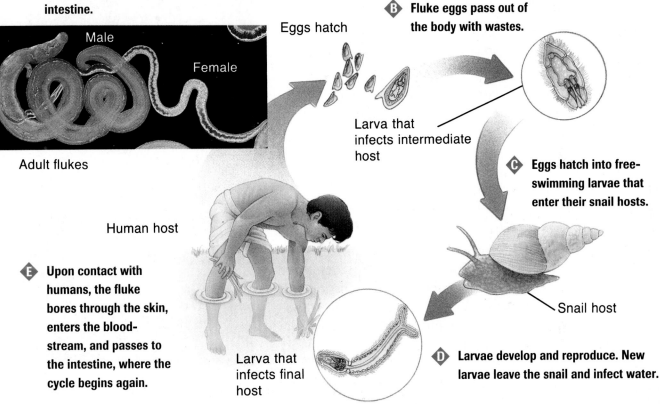

A **Adult flukes are about 1 cm long and live in the blood vessels of the human intestine.**

Male

Female

Adult flukes

Human host

E **Upon contact with humans, the fluke bores through the skin, enters the bloodstream, and passes to the intestine, where the cycle begins again.**

Eggs hatch

B **Fluke eggs pass out of the body with wastes.**

Larva that infects intermediate host

C **Eggs hatch into free-swimming larvae that enter their snail hosts.**

Snail host

Larva that infects final host

D **Larvae develop and reproduce. New larvae leave the snail and infect water.**

Section Review

Understanding Concepts

1. Diagram and label the structures of a planarian.
2. Describe the adaptations of a parasitic flatworm at different stages of its life cycle.
3. What is the adaptive advantage of a nervous system to a free-living flatworm?

Thinking Critically

4. Examine the life cycle of a parasitic fluke, and suggest ways to prevent infection on a rice farm where workers, during harvest and planting, must walk into water in the rice paddies.

Skill Review

5. **Observing and Inferring** What can you infer about the way of life of an organism that has no mouth or digestive system, but is equipped with a sucker? For more help, refer to Thinking Critically in the *Skill Handbook*.

29.4 Roundworms

Section Preview

Objectives

Compare the structural adaptations of round-worms and flatworms.

Identify the character-istics of four human roundworm parasites.

Key Terms

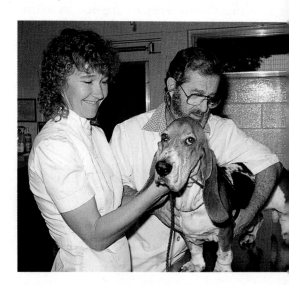

Have you ever been to the veterinarian to have your dog tested for heartworms? Perhaps you were once warned not to eat uncooked pork products. Flatworms are not the only type of worms that can cause harm to humans and other vertebrates. It has been estimated that about one-third of the human population suffers from problems caused by roundworms.

What Is a Roundworm?

Roundworms belong to the phylum Nematoda. Roundworms are widely distributed, living in soil, animals, and both freshwater and saltwater environments. Most roundworm species are free-living, but many are parasitic, *Figure 29.16.* In fact, virtually all plant and animal species are affected by parasitic roundworms.

Roundworms tend to be smaller than flatworms and are tapered at both ends. They have a thick outer

Figure 29.16

Roundworm parasites invade humans through a variety of methods. *Ascaris,* for example, is contracted by eating food grown in soil contaminated by *Ascaris* eggs. *Trichinella* is contracted by eating foods such as uncooked pork. Other parasitic roundworms, such as hookworms, can be contracted just by walking barefoot on soil containing eggs.

▲ ***Ascaris*** mainly infects children. People swallow eggs when they put dirty hands into their mouths or eat vegetables that have not been washed. The eggs hatch in the intestines, move to the bloodstream, then to the lungs where they are coughed up and swallowed to begin the cycle again.

Magnification: 100×

covering that protects them from being digested by their host organisms. On a flat surface, roundworms look like tiny, wriggling bits of sewing thread. Lacking circular muscles, they have pairs of lengthwise muscles. As one muscle of a pair contracts, the other muscle relaxes. This alternating contraction and relaxation of muscles causes roundworms to move in a thrashing fashion.

Roundworms have a pseudocoelom and are the simplest animals with a tubelike digestive system. Unlike flatworms, roundworms have two body openings—a mouth and an anus. The free-living species have well-developed sense organs, such as eyespots, although these are reduced in parasitic forms.

MiniLab

How can you recognize a roundworm?

Free-living roundworms feed on bacteria in all moist soils. They recycle soil nutrients and destroy some agricultural pests. One square meter of garden soil will contain millions of roundworms. In this lab, you will examine garden soil with a hand lens to find roundworms.

Procedure
1. Place a small amount of garden soil in a petri dish.
2. While observing with the hand lens, tease the soil apart gently with a toothpick.
3. Roundworms will appear as tiny white, threadlike animals.

Analysis
1. Describe the movement of the roundworms you find.
2. How many roundworms did you see?
3. Explain how roundworms are adapted to their habitat.

Hookworm commonly infects humans in warm climates and in areas of poor sanitation. Hookworms cause people to feel weak and tired due to blood loss. Hookworms are contracted by walking in bare feet on contaminated soil.

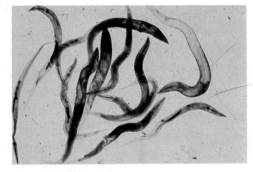

Magnification: 3×

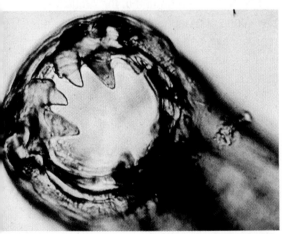

Magnification: 110×

◀ *Trichinella* is not as common in the United States as it once was because of stricter meat inspection standards. *Trichinella* worms do still exist in areas of the world where infected, uncooked meat scraps are fed to hogs, which in turn are eaten by humans.

▶ Pinworms are the most common parasites in children. Pinworms invade the intestinal tract when children eat something that has come in contact with contaminated soil. Female pinworms lay eggs near the anus, and reinfection is common because the worms cause itching.

Magnification: 100×

▶ The common soil nematode, *Rhabditis,* invades roots of plants grown for food, such as tomatoes, and causes a slow decline of the plant.

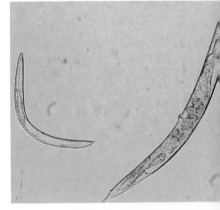

Figure 29.17

Roundworms that are plant parasites usually enter the roots and affect their ability to absorb water.

Magnification: 50×

Parasitic Roundworms

Roundworms are found as parasites in most organisms on Earth. Nematodes can infect and kill pine trees, cereal crops, and food plants such as potatoes. Nematodes are particularly attracted to plant roots. About 1200 species of nematodes cause diseases in plants, *Figure 29.17.* They also infect fungi and form symbiotic associations with bacteria.

Approximately half of the known roundworms are parasites, and about 50 species of these are human parasites. The most common human parasites are *Ascaris*, hookworm, *Trichinella*, and pinworm.

Connecting Ideas

Sponges and cnidarians are acoelomate animals that have simple body plans with two cell layers and only one body opening. Flatworms are also acoelomate animals with a mouth, but they have a simple nervous system and are able to respond to stimuli in the environment. Roundworms, however, have two body openings and a pseudocoelom; they are so successful that they are found everywhere on Earth.

Increasing complexity of body plans and structures of animals enables these organisms to survive in widely diverse environments. Development of a pseudocoelom with a digestive tract enabled roundworms to invade even inhospitable environments such as the frozen ground of the Arctic. Development of more complex digestive, reproductive, and nervous systems, in turn, enables other animals to survive and thrive in the oceans, on land, and in the air.

Section Review

Understanding Concepts

1. Compare the body structures of roundworms and flatworms.
2. Why do parents teach children to wash their hands before eating?
3. Describe the method of infection of one human roundworm parasite.

Thinking Critically

4. An infection of pinworms is spreading to children who attend the same preschool. Make a list of precautions that could be taken to help prevent its continued spread.

Skill Review

5. **Making and Using Tables** Make a table of the characteristics of four roundworm parasites, indicating the name of the worm, how it is contracted, the action of the parasite in the body, and means of prevention. For more help, refer to Organizing Information in the *Skill Handbook*.

Sunscreens from Coral

Sun exposure, skin cancer, and ultraviolet radiation are topics that are in the news a lot these days. Ultraviolet (UV) radiation is an invisible component of sunlight and a powerful form of energy that is harmful to most living things. When your skin is exposed to sunlight, for example, it isn't long before UV rays begin to cause a sunburn. Prolonged and repeated exposure to ultraviolet radiation also can damage your skin cells in such a way that you may eventually develop skin cancer.

Do corals get sunburned? The corals that grow on the upper part of coral reefs live in fairly shallow water. They are exposed to intense tropical sunshine, and therefore high levels of ultraviolet radiation, nearly every day. Yet they show no signs of damage from UV rays. Several years ago, researchers in Australia began to ask why.

Coral samples were tested A team of marine biologists collected samples of many different kinds of corals from the Great Barrier Reef. Each sample was then ground up and chemically analyzed. The researchers discovered that most of the samples contained one or more of several closely related chemical compounds that absorb ultraviolet radiation. These compounds act as natural sunscreens for corals by preventing harmful UV rays from damaging cells in the delicate bodies of coral polyps.

Applications for the Future

If these compounds work as sunscreens for corals, why not for people, too? After the researchers had isolated and identified the natural sunscreen compounds in corals, they synthesized them in the laboratory. Now an Australian pharmaceutical company is working on developing a sunscreen lotion that incorporates these compounds as its active ingredients. In research trials, this experimental sunscreen lotion is effective in absorbing UV rays—it has an SPF (sun protection factor) of at least 50.

Compounds are now being tested It probably will be a few years before this new sunscreen lotion is available in stores. In the meantime, researchers are exploring other possible uses for these natural UV-absorbing chemicals. The compounds are being tested in paints, varnishes, and plastics to determine if they can protect these products from the damage that UV rays can cause.

INVESTIGATING the Technology

1. **Compare** How does the coral sunscreen story support arguments for preserving habitats like rain forests that contain an enormous variety of as-yet-unstudied species of plants and animals?
2. **Apply** Ultraviolet radiation coming from the sun can penetrate seawater down to a maximum depth of 20 m. Would you expect corals that live below this depth to contain natural sunscreen compounds?

Reviewing Main Ideas

29.1 Sponges
- A sponge is an aquatic, sessile, asymmetrical, filter-feeding invertebrate.
- Sponges are made of four types of cells. Each cell type contributes to the survival of the organism.
- Sponges are hermaphrodites with free-swimming larvae.

29.2 Cnidarians
- All cnidarians are radially symmetrical, aquatic invertebrates that display two basic forms—medusa and polyp.
- Cnidarians feed by stinging or entangling their prey with cells called nematocysts, usually located at the ends of their tentacles.
- The three classes of cnidarians include the hydrozoans, hydras; scyphozoans, jellyfish; and anthozoans, corals and anemones.

29.3 Flatworms
- Flatworms are acoelomates with thin, solid bodies belonging to the phylum Platyhelminthes. They are grouped into three classes: free-living planarians, parasitic flukes, and tapeworms.

- Planarians have well-developed nervous and muscular systems. These systems are reduced in parasitic flatworms, but flukes and tapeworms have structures adapted to their parasitic existence.

29.4 Roundworms
- Roundworms are pseudocoelomate, cylindrical worms with lengthwise muscles and relatively complex digestive systems with two body openings.
- Roundworm parasites include parasites of plants, fungi, and animals, including humans. *Ascaris*, hookworms, *Trichinella*, and pinworms are roundworm parasites of humans.

Key Terms
Write a sentence that shows your understanding of each of the following terms.

external fertilization	nematocyst
filter feeding	nerve net
gastrovascular cavity	pharynx
hermaphrodite	polyp
internal fertilization	proglottid
medusa	scolex

Understanding Concepts

1. How do sponge cells illustrate the concept of a division of labor?
2. Why are sponges filter feeders?
3. Compare the adaptations of the three classes of cnidarians.
4. How are cnidarians adapted for a carnivorous life?
5. Why do planarians look different from other flatworms?
6. Describe how a parasite's body is adapted to its way of life.
7. Describe how one kind of roundworm parasite is contracted by humans.
8. Some species of sponges, as adults, are attached to the shells of crabs. Explain how this attachment might be beneficial to the sponge. Could the crab benefit as well?
9. What features seen in roundworms are not present in flatworms?

Relating Concepts
10. Make a concept map that relates the following terms and phrases. Supply the appropriate linking words for your map.

    flatworm, roundworm, parasite, free-living, worm, planarian, fluke, tapeworm, hookworm, *Ascaris*, pinworm

Interpret the Diagram

11. Which part of the life cycle for a beef tapeworm is missing?

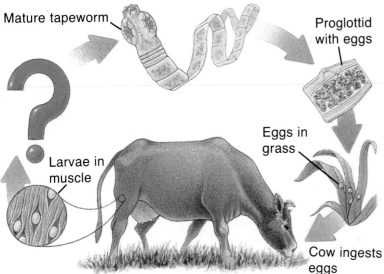

Mature tapeworm

Proglottid with eggs

Eggs in grass

Larvae in muscle

Cow ingests eggs

Applying Concepts

12. In what ways are cnidarians more complex than sponges?

13. Compare the methods for obtaining food in cnidarians and sponges.

14. You are examining an organism found in the intestines of a sheep. It has a head with tiny hooks. What kind of worm is it?

15. At what points could the life cycle of a blood fluke be interrupted so that disease would be prevented?

16. Of what advantage is hermaphroditism to an animal such as a sponge?

Health Connection

17. What could you do to ensure that people eating your pork roast dinner would *not* become hosts to pork tapeworm?

Thinking Critically

Interpreting Data

18. Thinking Lab Explain the trend in coral bleaching as shown in this map.

Relating Cause and Effect

19. Minilab While examining soil from the bottom of a pond, you notice tiny red worms wriggling aimlessly on a petri dish. What kind of worms are they? Explain.

Making Predictions

20. Biolab If a summer is much hotter than normal and stream temperatures increase where planarians are usually found, what might happen to the planarian population? Explain your answer.

Making Predictions

21. Minilab Which of the following substances would be most likely to induce nematocyst discharge from cnidarians: milk, sugar water, lemon juice, soapy water? Explain.

Connecting to Themes

22. Systems and Interactions Describe the features that would be important for a predator of jellyfish.

23. Homeostasis How do flatworms maintain a water balance?

24. Evolution Why is the phylogeny of cnidarians so little understood?

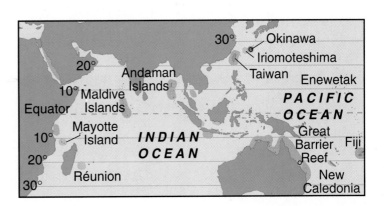

30° Okinawa
Iriomoteshima
20° Andaman Islands
Taiwan
Enewetak
10° Maldive Islands
PACIFIC OCEAN
Equator
Mayotte Island
INDIAN OCEAN
10°
20°
Great Barrier Reef
Fiji
30° Réunion
New Caledonia

CHAPTER 30 Mollusks and Segmented Worms

Animals often leave evidence showing they have been in a certain place. Broken twigs and fur on a branch indicate that a bear rested at the base of a tree. The distinctive tracks in the sand show that a sidewinder snake passed by. The fine, silvery path across a leaf is proof that a slug lives in your garden.

Garden slugs move slowly along a path of mucus. Both the mucous path and the thick mucus that covers the slug are produced by the slug. Mucus is an adaptation that allows the animal to adhere to and move across a surface.

The secretion of mucus is a characteristic not only of slugs, but of other animals as well. You may remember the first time you picked up an earthworm. It was soft and slippery from the mucus, without which the earthworm would not be able to burrow efficiently past rocks and through soil.

Slugs and earthworms are examples of two very different groups of animals. In this chapter, you will learn why these animals are classified as two different phyla. Snails belong to the same phylum as slugs. Snails also secrete a mucous trail over which to move. How do you suppose the snail is most like the slug, and how is it different?

Mucus is an important adaptation for earthworms, as well as for slugs and snails. In addition to allowing earthworms to move through soil, mucus holds two earthworms together as they mate.

If you are a shell collector, a walk on the beach as high tide begins to recede reveals bountiful treasures. The shell sizes, shapes, and colors are clues to the many different kinds of animals that once inhabited these structures. How could the marine animal that lived in the fan-shaped shell be related to the common garden slug?

What Is a Mollusk?

A slug, snails, and animals that once lived in the shells on the beach are all mollusks. These organisms belong to the phylum Mollusca. Members of this phylum range from the slow-moving slug to the jet-propelled octopus. While most species live in the ocean, others live in fresh water and on land. Some have shells, while others, like slugs and squid, are adapted to life without a hard covering. All mollusks have bilateral symmetry, a coelom, two body openings, a muscular foot for movement, and a mantle. The **mantle** is a thin membrane that surrounds the internal organs of the mollusk. In shelled mollusks, the mantle secretes the shell.

Although the phylum is diverse, mollusks all share similar developmental patterns. The larval stages of all mollusks are similar, *Figure 30.1,* but they have different appearances as adults. Three classes of mollusks are shown in *Figure 30.2.*

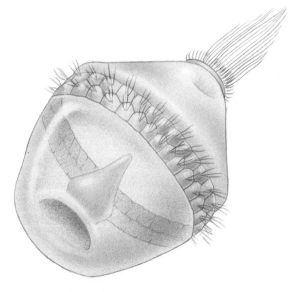

Figure 30.1

Larvae of most mollusks resemble a top with tufts of cilia. Most of these larvae are free-swimming before settling to the ocean floor for adult life. These larvae serve as food for many other organisms.

Figure 30.2

With 100 000 species, phylum Mollusca is second in size only to insects and their relatives.

▶ **Bivalves:** Oysters, clams , and this scallop have two, hinged shells and are known as bivalves. These animals strain their food from the ocean water.

▼ **Gastropods:** One-shelled mollusks, called gastropods, make up the largest class of mollusks and exhibit a huge array of shell shapes and sizes. Members of this group include lung-breathing land snails and slugs, and the marine limpet shown here.

Where Do Mollusks Live?

Mollusks live in a wide variety of habitats. Most live in marine or freshwater habitats; some live on land. A few species of mollusks can be found in cold, polar regions, and many are common in warm, tropical areas. Some aquatic mollusks, such as oysters and mussels, live firmly attached to the ocean floor or to the base of docks and wooden boats. Others, such as the octopus, swim freely in the ocean. Land-dwelling slugs and snails can be found most frequently in moist tropical and temperate climates.

▲ **Cephalopods:** Predatory squids and the octopus, shown here, are cephalopods, have sharp eyesight, muscular tentacles, jet-propelled swimming, a complex brain, and the ability to learn. They often apply their intelligence to capturing prey or avoiding harmful situations.

Classes of Mollusks

Within the large phylum of mollusks, there are seven classes. The three classes that include the most common and well-known species are the classes gastropoda, bivalvia, and cephalopoda.

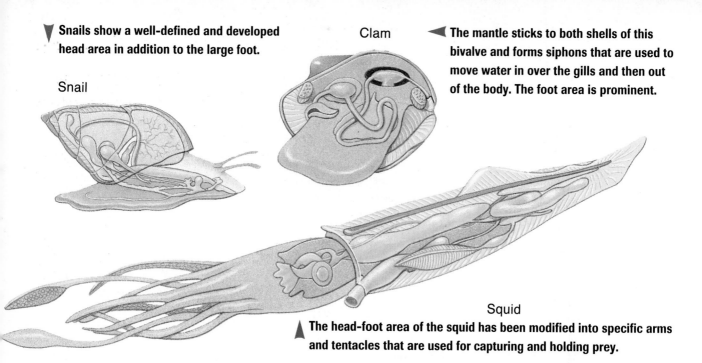

▼ Snails show a well-defined and developed head area in addition to the large foot.

Snail

Clam

◄ The mantle sticks to both shells of this bivalve and forms siphons that are used to move water in over the gills and then out of the body. The foot area is prominent.

Squid

▲ The head-foot area of the squid has been modified into specific arms and tentacles that are used for capturing and holding prey.

Key

Visceral mass

Mantle

Shell

Foot

Figure 30.3

Mollusks have soft bodies composed of a foot, a mantle, a shell, and a visceral mass that contains internal organs. Comparing these three body sections shows how the bodies of the three classes of mollusks are adapted for survival.

While members of these three classes of mollusks look different from each other on the outside, they share many internal similarities. You can see the similarities and the differences in these body areas in the drawings in *Figure 30.3* as you compare the clam, snail, and squid.

Gastropods: One-shelled mollusks

The largest class of mollusks is Gastropoda, or stomach-footed mollusks. The name comes from the way the animal's large foot is positioned under the rest of its body. Most species of gastropod have a single shell and are sometimes called univalves. Other gastropod species have no shell. Snails, slugs, and sea slugs belong to this class.

Gastropod adaptations

You may have watched a snail clean algae from the sides of an aquarium with a rasping structure called a radula. A **radula** is a tongue-like organ with rows of teeth located within the mouth of the gastropod. The radula is used to scrape, grate, or cut food.

The nervous system of gastropods is simple. Gastropods have a small brain. Associated nerves coordinate the animal's movements and behavior.

A heart is part of the gastropod's well-developed circulatory system. Blood is pumped by the heart in an open circulatory system. In an **open circulatory system,** the blood moves through vessels and into open spaces around the body organs. This adaptation exposes body organs directly to blood that contains nutrients and oxygen and removes metabolic wastes.

Mollusks are the first animals to have evolved respiratory structures. These structures are called gills. Gill tissue increases the surface area through which gases can diffuse, and it contains a rich supply of blood for transport of gases. While aquatic gastropods have gills, the mantle cavity of land snails and slugs has evolved into a primitive lung.

Mollusks are also the first animals to have evolved excretory structures called nephridia. **Nephridia** are organs that remove metabolic wastes from an animal's body. Gastropods usually have one nephridium.

Many gastropods that live on land are hermaphrodites. The ability to produce both eggs and sperm is an adaptation commonly found in slow-moving animals. Most aquatic gastropods have either male or female reproductive organs. No mating occurs in these species. Eggs and sperm are released at the same time into the surrounding water, where external fertilization takes place.

Shelled gastropods

Snails, abalones, and conches are examples of shelled gastropods. They may live in fresh water, salt water, or on land. Shelled gastropods may be creeping plant eaters, dangerous predators, or parasites. *Figure 30.4* shows two examples of shelled gastropods.

Gastropods without shells

Slugs are one type of gastropod without shells. Instead of being protected by a shell, the body of the slug is protected by a thick layer of mucus. The colorful nudibranchs, also called sea slugs, are protected in another way. When certain species of sea slugs feed on jellyfish, they incorporate the poisonous nematocysts of the jellyfish into their own tissues without causing the cells to discharge. Any fish trying to eat the sea slugs are repelled when the nematocysts discharge into the unlucky predator. Other sea slugs secrete a strong, unpleasant-smelling mucus. Still others secrete a poisonous or strongly acidic mucus. The bright colors of these gastropods function to warn predators of the potential danger.

ThinkingLab Design an Experiment

What determines a snail's size before reproduction?

Biologists studying snails found that the size a particular species of snail reaches before beginning reproduction depends on the environment. If the water in which the snails live comes from an area in which snails were preyed on by crayfish, the snails grow to be about 10 mm in length before reproducing. If the water comes from an area containing only snails or only crayfish or only crushed snails, the snails grow to a length of about 4 mm before beginning reproduction.

Analysis
Biologists know that larger snails are not eaten by crayfish. Growing large quickly before reproducing has survival value for the snail, as it will not be eaten by the crayfish. How might the young snail detect the presence of crayfish in the water?

Thinking Critically
Design an experiment that will provide information about how snails detect the presence of crayfish in their water.

Figure 30.4

Shelled gastropods vary from petite, thin-shelled species to large, thick-shelled ones.

▶ **A small, delicate gastropod species is the smooth dove shell. These organisms can be found in the Florida Keys and West Indies.**

◀ **The pink conch is a large gastropod with a thick shell. Members of this species can measure up to 30 cm high. The meat of these gastropods is an important food source in the West Indian Islands. Overfishing for food and souvenirs has depleted this species in several areas.**

What path does food and water take in a clam?

Clams are filter feeders, taking in water, small invertebrates, and other organic materials, and filtering out the food before it lets the water out. In this lab, you will observe the path water takes as it enters and exits a clam. These structures are called the incurrent and excurrent siphons.

Procedure

1. Find the incurrent and excurrent siphons of a live clam. See the photograph below.

2. Place a live clam in a beaker of water so that about 6 cm of water covers the clam. Let the clam rest undisturbed for about five minutes.

3. Place two drops of carmine powder suspension on the side of the clam with the siphons.

4. Observe what happens to the carmine suspension.

Analysis

1. Explain what happens to the carmine suspension.

2. Explain what happens during filter feeding.

3. With what other animal could you use carmine suspension in water to observe water entering and exiting during the process of filter feeding?

Figure 30.5

Members of the class Bivalvia have some interesting names such as cockles, arks, angel wings, jewel boxes, and jingle shells. Whether thick or thin shelled, smooth or spiny, all these organisms have two shells held together by a hinge.

Bivalves: Two-shelled mollusks

Two-shelled mollusks such as clams, oysters, and scallops belong to the class Bivalvia. Most bivalves are marine, but a few species live in fresh water. As seen in *Figure 30.5,* bivalves show a range of sizes. Some are less than 1 mm in length and others, such as the tropical giant clam, may be 1.5 m long. Bivalves have no distinct head, or radula. Most use their large, muscular foot for burrowing in the mud or sand at the bottom of the ocean or lake. A strong ligament, like a hinge, holds their shells together; muscles allow the shell to open and close over the soft body.

One of the main differences between gastropods and bivalves is that bivalves are filter feeders. Bivalve mollusks have several adaptations for filter feeding. They have cilia that beat to draw water in through an incurrent siphon. The water moves over gills and exits through the excurrent siphon. As water moves over the gills, food and sediments become trapped in mucus. Cilia that line the gills push food particles to the stomach. Cilia also act as a sorting device. Large particles, sediment, and anything else that is rejected is transported to the mantle or to the foot where it is eliminated.

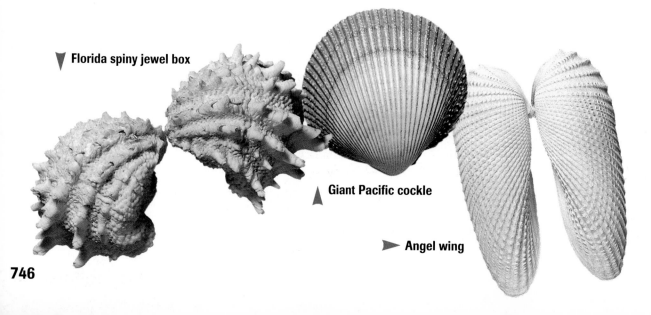

▼ **Florida spiny jewel box**

▲ **Giant Pacific cockle**

➤ **Angel wing**

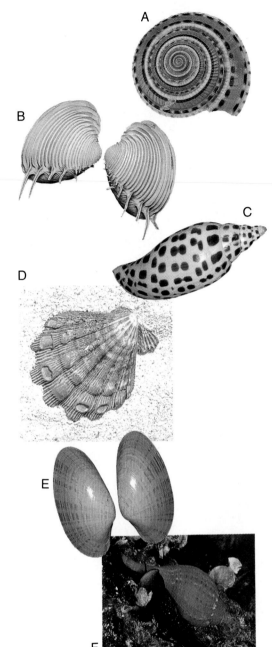

How are mollusks identified?

Have you ever taken a walk on the beach and filled your pockets with shells, and then as you examined them later, wondered what they all were? The shells pictured are ones that you might have picked up if you walked along the Florida coast. Use the following dichotomous key to determine the names of the shells.

Procedure

1. To use a dichotomous key, begin with a choice from the first pair of descriptions.

2. Follow the instructions for the next choice. Notice that either a scientific name can be found at the end of each description, or directions will tell you to go on to another numbered set of choices.

1A	One shelled......Gastropods	see 2
1B	Two shelled.......Bivalves	see 3
2A	Flat coil..Sundial shell:	*Architectonica nobilis*
2B	Thick coil...See 4	
3A	Spines on posterior..Elegant venus shell:	*Pitar dione*
3B	No spines...See 5	
4A	Spotted surface....Junonia shell:	*Scaphella junonia*
4B	Lined surface...Banded Tulip Shell:	*Fasciolaria hunteria*
5A	Polished surface...Sunray shell:	*Macrocallista mimbosa*
5B	Rough surface...Lion's paw shell:	*Lyropecten nodosus*

Analysis

1. Why is a dichotomous key used for a variety of organisms?

2. What features were easy to pick out using the key for shells? What features were more difficult? Why?

3. What was the most general feature you used to identify shells?

Cephalopods: Head-footed mollusks

The head-footed mollusks are in the class Cephalopoda. This class includes the octopus, squid, and chambered nautilus, as shown on the following page in *Figure 30.6.* Scientists consider the cephalopods to be the most complex and most recently evolved of all mollusks. The only cephalopod with a shell is the chambered nautilus. All cephalopods are marine organisms. Instead of a single, muscular foot, cephalopods have tentacles with suckers for movement and capturing food. The radula and sharp, beaklike jaws tear apart fish, bivalves, and crabs.

These marine predators move nutrients and oxygen through a closed circulatory system. In a **closed circulatory system,** blood moves through the body enclosed entirely in a series of blood vessels. A closed system provides a more efficient means of gas exchange within the body.

Figure 30.6

Literature

Natural Acts
by David Quammen

"I'm not a scientist," says naturalist writer David Quammen. "What I am is . . . a haunter of libraries and snoop." In spite of his disclaimer, Quammen, a keen observer of everything around him, has written dozens of essays on a wide variety of scientific topics. Abandoning his earlier writing style, the political thriller, Quammen seems to have settled into his favorite mode: asking an intriguing science question and then setting out to answer it in an entertaining way. In his collection of essays, called *Natural Acts: A Sidelong View of Science and Nature,* Quammen poses these kinds of questions: What are the redeeming merits, if any, of the mosquito? Are crows too intelligent for their station in life?

The eye of the octopus Another topic Quammen wondered about was the enormous eyes of the octopus, which seem to stare back at human observers in a somewhat unnerving fashion. "These animals don't just gape at you glassily, like a walleye. They make eye contact, as though they are someone you should know," he writes. The reason, he continues, is that an octopus is among the most intelligent of animals that live in the sea. In laboratory tests, an octopus combines that intellect with acute eyesight to do well in the mazes that scientists devise. Besides, adds Quammen, they have eyelids, "so they can wink at us fraternally."

More questions It seems unlikely that Quammen will run out of questions—big questions, such as "Is sex necessary?" and smaller ones, such as "Why are there so many different species of beetles?" A fellow writer calls reading about Quammen's pursuit of the answers "a wild ride on a slightly unsettled roller coaster."

CONNECTION TO Biology

What are some other questions about the natural world that you would like an entertaining writer like David Quammen to try answering?

The class cephalopoda contains octopods, squid, and the only shelled member, the chambered nautilus.

▲ The genus, *Nautilus,* is the only remaining living example of cephalopods with a shell. All other members of this class are extinct. The shell of a *Nautilus* is divided into a series of chambers.

◄ An octopus, as well as all other cephalopods, has separate sexes. One tentacle of a male octopus is adapted to transferring sperm into the body of the female. Fertilization is internal, but eggs are laid outside the body.

▼ A squid has large eyes, a well-developed nervous system, and moves by jet propulsion. With this system of movement, squids can attain speeds of 20 m per second!

Origins of Mollusks

Fossil records show that mollusks lived in great numbers as long as 500 million years ago. Some, like the chambered nautilus, appear to have a similar structure to the related species that lived 500 million years ago. Other early species have shown radiation into a variety of forms that include land slugs and exotic marine slugs. *Figure 30.7* shows the radiation of mollusks and the closely related phylum of segmented worms.

Figure 30.7

Both mollusks and segmented worms share the same pattern of early cell division. Based on this and other evidence, biologists think that mollusks and segmented worms are closely related.

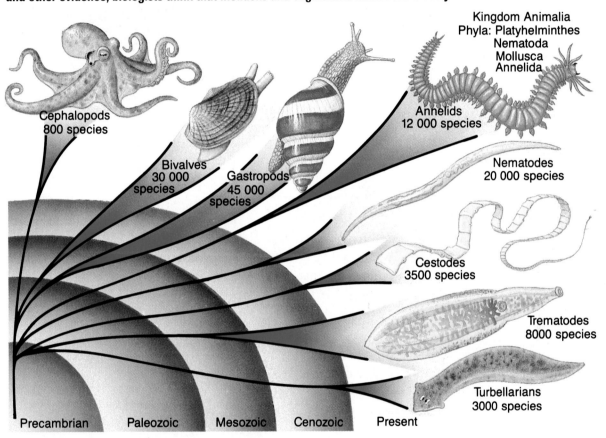

Kingdom Animalia
Phyla: Platyhelminthes
Nematoda
Mollusca
Annelida

Cephalopods
800 species

Bivalves
30 000 species

Gastropods
45 000 species

Annelids
12 000 species

Nematodes
20 000 species

Cestodes
3500 species

Trematodes
8000 species

Turbellarians
3000 species

Precambrian Paleozoic Mesozoic Cenozoic Present

Section Review

Understanding Concepts

1. Describe how mucus is important to some mollusks.
2. What adaptations make the cephalopods effective predators?
3. Compare filter feeding with getting food by using a radula.

Thinking Critically

4. How are the methods of movement for the snail, clam, and squid related to the structure of their foot?

Skill Review

5. **Classifying** Construct a key to identify the three classes of mollusks discussed. For more help, refer to Organizing Information in the *Skill Handbook*.

Section Preview

Objectives

Describe the characteristics of segmented worms, and describe their importance to the survival of these organisms.

Compare and contrast the classes of segmented worms.

Key Terms
gizzard
nocturnal

***D**o earthworms have a front and a back end? Yes, they do. In fact, if you have watched one move, you know that it crawls by first stretching the front of its body forward, and then pulling the back of its body up to the front. A worm in motion looks a little like an accordion playing.*

What Is a Segmented Worm?

Members of the phylum Annelida are segmented worms. They include the bristleworms, earthworms, and leeches as shown in *Figure 30.8*. The term *annelid* means "tiny rings." Segmented worms, like mollusks, are bilateral and have a coelom and two body openings. Some have a larval stage similar to the larval stages of certain mollusks.

Figure 30.8

The phylum Annelida contains more than 8700 species, which are placed in three classes. While having a definite anterior end, the earthworm does not have a distinct head as the bristleworms do. Earthworms, members of the class Oligochaeta, have only a few setae on each segment.

► The class Hirudinae is made up of leeches that may live in marine, freshwater, or land habitats. They lack parapodia and setae. All leeches have 32 segments.

◄ The bristleworm belongs to the class Polychaeta, and most members of its class are marine organisms that have a distinct head with eyes and tentacles. Each body segment has a pair of appendages called parapodia, which have many bristlelike structures called setae.

Figure 30.9

Segmentation is easily seen in the class Oligochaeta. While most species in this class are 5 to 10 cm long, the giant earthworm of Australia can be more than 3 m long.

All annelids are made up of segments

The most distinguishing characteristic of annelids is their cylindrical bodies that appear to be divided into a series of ringed segments, as seen in the worms in *Figure 30.9.* This segmentation continues internally as each segment is separated from the others by a body partition. Segmentation is an important advantage because each segment has its own muscles, allowing shortening and lengthening of the body for movement.

The segments also allow for specialization. Groups of segments may be specialized for a particular function. If you examine each segment of most annelids, you find that most of the body is made of identical segments. Although most segments contain excretory organs and nerve centers, a few also contain organs for digestion and reproduction.

The **gizzard** is a sac with muscular walls and hard particles that grind soil before it passes into the intestine.

Some organs of the digestive system, nervous system, and circulatory system do run the length of the annelid's body. The internal structure of an earthworm can be seen in *The Inside Story.*

Where do segmented worms live?

Segmented worms live everywhere except in the frozen soil of the polar regions and the dry sand and soil of the deserts. You may be familiar with the earthworms in your garden, but these are just one of about 9000 species of segmented worms that live in soil, fresh water, and the sea. The delicate and beautiful fan worms in *Figure 30.10* live by the shores of tropical and temperate seas. Their bodies are protected by the hard tubes they build of sand grains. The sea mouse in *Figure 30.11* is another segmented worm that lives in the ocean.

Figure 30.10

The fan worm traps food in the mucus on its "fans." Disturbances in the water, such as a change in the direction of the current, the passage of a shadow overhead, or the passing by of an organism, cause these worms to quickly withdraw into their tubes.

The Earthworm

As an earthworm burrows through soil, it loosens, aerates, and fertilizes the soil. Burrows provide passageways for plant roots and improve drainage.

A common earthworm, _Lumbricus terrestris_

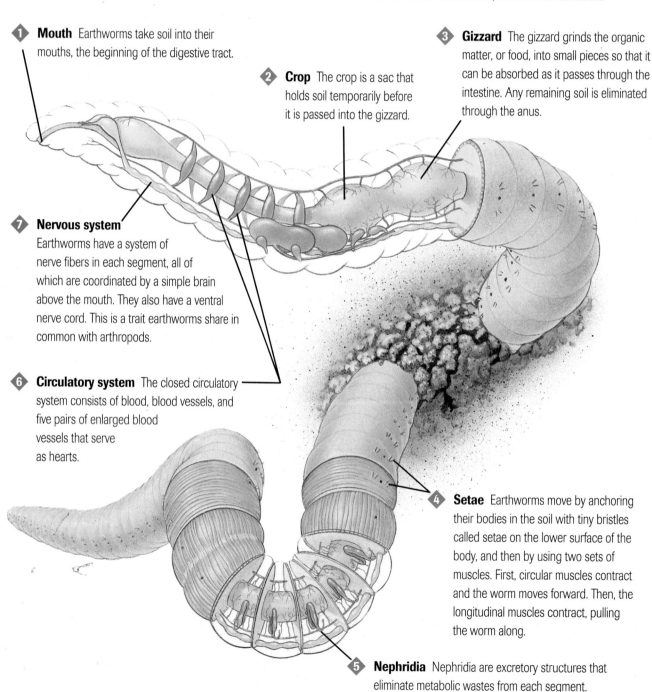

1 Mouth Earthworms take soil into their mouths, the beginning of the digestive tract.

2 Crop The crop is a sac that holds soil temporarily before it is passed into the gizzard.

3 Gizzard The gizzard grinds the organic matter, or food, into small pieces so that it can be absorbed as it passes through the intestine. Any remaining soil is eliminated through the anus.

7 Nervous system Earthworms have a system of nerve fibers in each segment, all of which are coordinated by a simple brain above the mouth. They also have a ventral nerve cord. This is a trait earthworms share in common with arthropods.

6 Circulatory system The closed circulatory system consists of blood, blood vessels, and five pairs of enlarged blood vessels that serve as hearts.

4 Setae Earthworms move by anchoring their bodies in the soil with tiny bristles called setae on the lower surface of the body, and then by using two sets of muscles. First, circular muscles contract and the worm moves forward. Then, the longitudinal muscles contract, pulling the worm along.

5 Nephridia Nephridia are excretory structures that eliminate metabolic wastes from each segment.

Earthworms are nocturnal animals that live in burrows. A **nocturnal** animal is one that moves about primarily at night. At night, earthworms come to the surface but stay close to their burrows. The cool, moist soil provides protection for the worms during the day. Water in the soil is a source of oxygen that diffuses into an earthworm's body through the skin.

Figure 30.11

The sea mouse is a species of marine polychaete. The sea mouse is well-known due to its distinctive furry appearance, although it is not commonly seen.

How do annelids reproduce?

Earthworms are hermaphrodites. During mating, two worms exchange sperm. Each worm forms a capsule for the eggs and sperm. The eggs are fertilized in the capsule, and the capsule is slipped off the worm into the soil. In two to three weeks, young earthworms emerge from the fertilized eggs.

Most polychaetes reproduce sexually, although mating occurs in only a few species. Eggs and sperm are released into the seawater where fertilization occurs. All members of the class Hirudinea are hermaphrodites. Eggs and sperm are transferred from one animal to another in a way similar to the oligochaetes.

Polychaetes: Feather Dusters and Christmas Trees

Polychaetes, segmented worms that live in the sea, are like all other annelids. Each segment has a pair of swimming or crawling appendages and external gills. It's hard to imagine a worm being beautiful, but many polychaetes are spectacular as they show iridescent colors and often unusual shapes.

Worms of the shallows Many polychaetes live in shallow areas of the ocean. The name, tube worm, describes the shell-like tube that the sedentary species secrete and in which they live. Many of these worms have highly developed fans of feathery tentacles, banded in beautiful colors. This has led to descriptive names such as Magnificent Feather Duster, Star Feather, and Christmas Tree. Unlike most other annelid worms, polychaetes are not hermaphroditic and have separate sexes. Eggs and sperm are shed into the water and join to form free-swimming trochophore larvae. Many polychaetes can regenerate lost or damaged parts.

Varied feeding relationships Polychaetes are abundant in some areas and form an important link in certain marine food chains. Others have a commensal relationship with sponges, mollusks, or crustaceans—feeding on debris that the other animals leave. A few polychaetes are parasitic, and some can bite or even deliver a painful sting.

Thinking Critically
How do you think the adaptations that these species of worms have developed help them survive in their environments?

Featherduster worm

Starfeather worm

Blue Plume worm

BioLab | Design Your Own Experiment

An earthworm spends its time eating its way through soil, digesting organic matter, and passing inorganic matter through the

How do earthworms respond to their environment?

system and out of the body. Because earthworms are dependent on soil for food and shelter, they respond to stimuli in a way that will ensure a continuous supply of food and a safe place in which to live. These responses are genetically controlled. In this lab, you will design an experiment to determine the responses of earthworms to various stimuli.

PREPARATION

Problem
How do earthworms respond to light, different surfaces, moist and dry environments, and warm and cold environments?

Hypotheses
Place your worm in a tray with some moist soil. Watch your worm for about five minutes, and record what you observe. Make a hypothesis based on your observations about what the worm might do under conditions of light and dark, rough and smooth surface, moist and dry surface, and warm and cold conditions. Limit your investigation as time permits.

Objectives
In this Biolab, you will:
- **Measure** the sensitivity of earthworms to different stimuli including light, water, and temperature.
- **Interpret** earthworm responses in terms of adaptive value to their lifestyle.

Possible Materials
live earthworm	water
paper towels	dropper
glass pan	penlight
sandpaper	ice
culture dishes	ruler
warm tap water	black paper
thermometer	
hand lens or stereomicroscope	

Safety Precautions
Be sure to treat the earthworm in a humane manner at all times. Wash your hands when your experiment is complete.

1. As a group, make a list of possible ways you might test your hypothesis. Keep the available materials in mind as you plan your procedure.

2. Be sure to design an experiment that will test one variable at a time. Plan to collect quantitative data.

3. Record your procedure and list materials and amounts you will need. Design and construct a data table for recording your findings.

Check the Plan

Discuss the following points with other group members to decide the final procedure for your experiment.

1. What data will you collect, and how will it be recorded?

2. Does each test have one variable and a control? What are they?

3. Each test should include measurements of some kind. What are you measuring in each test?

4. How many trials will you run for each test?

5. Assign roles for this investigation.

6. ***Make sure your teacher has approved your experimental plan before you proceed further.***

7. Carry out your experiment.

1. **Checking Your Hypothesis** Which surface did the worm prefer? Explain.

2. **Interpreting Observations** In which temperature was the worm most active? Explain.

3. **Observing and Inferring** How did the earthworm respond to light? Of what survival value is this behavior?

4. **Observing and Inferring** How did the earthworm respond to dry and moist environments? Of what survival value is this behavior?

5. **Drawing Conclusions** Were your hypotheses supported by your data? Why or why not?

Going Further

Project Based on your experiment, design another experiment that would help to answer a question that arose from your work. You might want to try other variables similar to the ones you used, or you might choose to investigate a completely different variable.

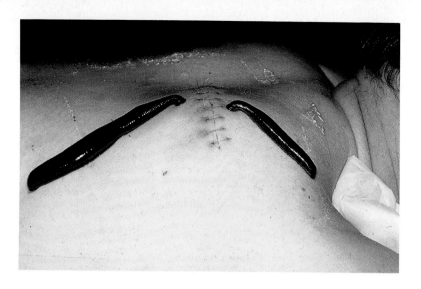

Figure 30.12

Besides anesthetic and anticlotting agents, the saliva of leeches also contains chemicals to dilate blood vessels to increase blood flow. Because of these characteristics, they have many medical uses.

bodies and no bristles. Although these animals can be found in many different habitats, most leeches live in fresh water. Most species are parasites that suck blood or other body fluids from the bodies of their hosts such as ducks, turtles, fishes, and people. Front and rear suckers enable leeches to attach themselves to their hosts.

You may cringe at the thought of being bitten by a leech, but the bite is not painful. This is because the saliva of the leech contains chemicals that act as an anesthetic. Other chemicals prevent the blood from clotting. A leech can ingest two to five times its own weight in one meal. Once fed, it may not eat again for a year.

Leeches

Leeches, as seen in *Figure 30.12,* are segmented worms with flattened

Connecting   Ideas

In this chapter, you studied the features of two groups of invertebrate organisms, the mollusks and annelid worms. Both of these groups of protostomes show increasing specialization. All mollusks have an identifiable body plan that includes a foot, head, mantle, and gills. The major classes of mollusks include the gastropods, bivalves, and cephalopods.

The phylum Annelida includes common earthworms, marine polychaetes, and

leeches. All members of this phylum show external segmentation. The annelid worms also have a closed circulatory system and a specialized digestive system. The segmentation of annelid worms was a significant step in evolution. The arthropods, descendants of annelids, became an extremely successful group due to the specialization of feeding structures and appendages that segmentation allowed.

Section Review

Understanding Concepts

1. What is the most distinguishing characteristic of the bristleworm, earthworm, and leech? Why is it important?
2. Describe how earthworms reproduce.
3. How do earthworms improve soil fertility?

Thinking Critically

4. Suppose a patient was not responding well to anticoagulant drugs administered after

microsurgery. Explain why the doctor might prescribe the use of leeches.

Skill Review

5. **Interpreting Scientific Illustrations** Using *The Inside Story*, interpret how the two types of muscles in the earthworm are used to move the animal through the soil. For more help, refer to Thinking Critically in the *Skill Handbook*.

Superglue from the Sea

One goal of biotechnology is to duplicate naturally occurring substances. When potentially useful products are found in marine organisms, they are extracted and tested for other uses. Mussels, which are species of bivalves, excrete a gooey protein from a groove in their foot that allows them to attach themselves to rocks and ship hulls. The glue is extremely strong and adheres even underwater.

Mussel glue for profit A biochemist studying mussels discovered a commercial value for this adhesive protein. He realized that the adhesive would be perfect for medical uses where an adhesive is needed in areas impossible to keep dry. This material is used as a bonding adhesive and sealant in dentistry, to repair torn corneas in the eye, and as an adhesive for fractured bones.

Maximum glue production It takes 3000 mussels to yield 1 g of mussel protein adhesive. Rather than harvest these sea creatures in such numbers, researchers utilized genetic engineering techniques. The gene sequence that codes for the glue protein is inserted into the genome of a strain of harmless bacteria or yeast. These organisms, each carrying the encoded sequence, multiply quite rapidly. Although carrier organisms each produce only a minute amount of the adhesive, they reproduce so rapidly and their numbers become so great that they produce enough of the protein for use in medical technology. This procedure is more cost-effective and easier to con-trol than sea harvesting. Sea harvesting in such large quantities would have detrimental effects on the mussel population.

Applications for the Future

Through genetic engineering, modifications of the amino acid sequence of the mussel-produced glue have been made. Changes in the code have produced a more efficient glue that hardens faster, bonds more securely, and can be broken down rapidly if necessary. Another superglue with a marine origin is being studied as well. This glue is produced by bacteria that form a symbiotic relationship with oysters. This glue will be genetically engineered also for commercial use. This material shows promise for similar applications as the mussel glue. In producing these glues through genetic engineering, biotechnology is helping to protect the environment.

INVESTIGATING the Technology

1. **Research** Read "Drugs from the Sea" by R. Lewis, *Discover*, May 1988, pp. 62-69. What other materials are being drawn from the sea for medical use?
2. **Apply** What feature of the genetic code makes it possible to insert the genes for the glue protein into bacteria?

30.1 Mollusks

- Mollusks have bilateral symmetry and a coelom. Many also have shells and similar larvae.
- Mollusks live in marine, freshwater, and land ecosystems.
- Most one-shelled mollusks have a shell, mantle, radula, open circulatory system, gills, and nephridia. Slugs are protected by a slime covering.
- Bivalve mollusks have two shells and are filter feeders.
- Cephalopods include the octopus, squid, and chambered nautilus. They have tentacles with suckers, a beaklike mouth with a radula, and a closed circulatory system.
- Fossil remains of mollusks show that they first lived 500 million years ago.

30.2 Segmented Worms

- The phylum Annelida includes the bristleworms, earthworms, and leeches. They are bilateral and have a coelom and two body openings; some have larvae that look like the larvae of mollusks. Their bodies are cylindrical and segmented.
- Earthworms are nocturnal and have complex digestive, excretory, muscular, and circulatory systems.
- Leeches are flattened, segmented worms. Most are aquatic parasites.

Key Terms

Write a sentence that shows your understanding of each of the following terms.

closed circulatory system
gizzard
mantle
nephridia

nocturnal
open circulatory system
radula

Understanding Concepts

1. How is an earthworm similar to a land snail?
2. Why is the fossil record of mollusks more complete than that of worms?
3. What characteristics distinguish the species within the three classes of mollusks?
4. How do leeches differ from other annelids?
5. What are the functions of gills in mollusks?
6. Explain the differences between bivalves and gastropods by their methods of feeding.
7. Explain how mucus is used by mollusks and segmented worms for locomotion.
8. How has the circulatory system become more complex in mollusks and worms? Explain how the system has evolved.

Interpreting Scientific Illustrations

9. Explain what is happening in the photograph below. What animal is involved? How is the body of this animal adapted to this way of life?

Relating Concepts

10. Make a concept map that relates the following terms and phrases. Supply the appropriate linking words for your map.

    tentacle, filter feeder, gastropod, radula, cephalopod, mantle, slug, clam, octopus, bivalve, mollusk, snail

Applying Concepts

11. Compare how a cnidarian and an octopus use their tentacles to get food.
12. Explain how sponges and bivalves have a similar way of obtaining food.
13. Why is a snail a good animal to keep in an aquarium?
14. Compare the adaptations gastropods, bivalves, and cephalopods have developed to protect themselves.

Literature Connection

15. Why do you think David Quammen is so successful in his writing about science and nature?

Thinking Critically

Interpreting Data

16. **Biolab** The following graph shows how a number of different animals respond to light. Tell which ones might be nocturnal. Explain.

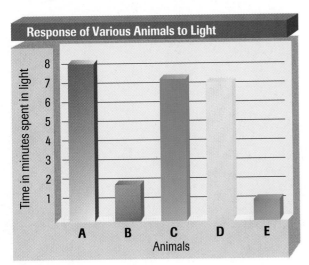

Response of Various Animals to Light

Making Inferences

17. **Thinking Lab** The following graph illustrates data obtained on the growth rate of a particular tropical snail when grown in water from a tank that also contains certain shrimps, and the growth rate in water in which there were only other snails. What do you think the relationship is between the shrimp and the snails? Explain.

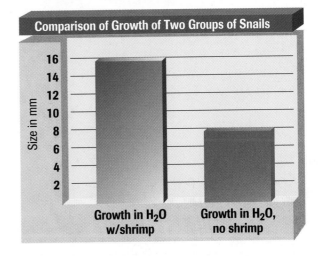

Comparison of Growth of Two Groups of Snails

Interpreting Scientific Illustrations

18. **Minilab** Use the key on page 747 to determine to which shells in the Minilab this shell is most closely related.

Connecting to Themes

19. **Homeostasis** Explain how nephridia enable mollusks and segmented worms to maintain homeostasis.
20. **Evolution** Explain why origins of worms are not as well understood as origins of mollusks.
21. **Systems and Interactions** Explain how certain bivalves in salt marshes are important for all the organisms that live there.

31 Arthropods

Y ou spread a tablecloth on the picnic table as the aroma of barbecued hamburgers attracts yellow jackets, and flies buzz around the macaroni salad. A procession of ants carries discarded potato chip pieces several times their size, while a spider descends and hovers on a glossy thread from a tree branch, waiting for its next meal. After you eat, you wade in a nearby stream and find crayfishes flipping their tails in backward lurches. Meanwhile, you hear the tiny whine of a mosquito near your ear. You knew you were going to a picnic with your friends. Did you know you would be sharing it with perhaps 1500 or more species of arthropods—the average number in a suburban garden?

Arthropods include insects, centipedes, millipedes, spiders, ticks, scorpions, mites, lobsters, shrimps, crabs, and crayfishes. They range in size from the 0.2-mm-long hairy beetle to the giant Japanese spider crab, which measures 4 m across. Phylum Arthropoda includes a diverse group of animals.

There are about 1 million known species of arthropods, and many more remain unidentified. In fact, one scientist estimates that there may be up to 30 million species of insects in the world's tropical rain forests. How can we explain the enormous diversity of arthropods such as spiders, yellow jackets, and crayfish? What are the adaptations that have allowed arthropods to dominate the world's habitats?

Characteristics of Arthropods

Section Preview

Objectives

Relate the structural and behavioral adaptations of arthropods to their ability to live in different habitats.

Analyze the adaptations that make arthropods an evolutionarily successful phylum.

Key Terms

appendage
molting
cephalothorax
tracheal tube
spiracle
book lung
pheromone
simple eye
compound eye
mandible
Malpighian tubule
parthenogenesis

Two out of every three animals living on Earth today are arthropods. You can find arthropods deep in the ocean and on high mountaintops. They live in polar regions and in the tropics. Arthropods are adapted to living in air, on land, and in freshwater and saltwater environments. This water flea, Daphnia, *lives in freshwater lakes and filters microscopic food from the water with its bristly legs.*

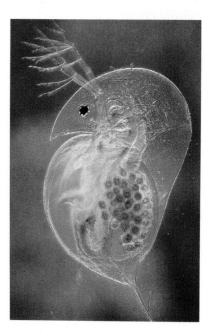

What Is an Arthropod?

Despite the enormous diversity of arthropods, they all share some common characteristics. How can you recognize an arthropod?

A typical arthropod is an invertebrate animal with bilateral symmetry, a coelom, an outer covering called an exoskeleton, and jointed structures called appendages. An **appendage** is any structure, such as a leg or an antenna, that grows out of the body of an animal. In arthropods, appendages are adapted for a variety of purposes including sensing, walking, feeding, and mating. *Figure 31.1* shows some of these adaptations.

The advantage of jointed appendages

Arthropods were the first invertebrates to evolve jointed appendages. Joints are advantageous because they allow for more powerful movements during locomotion, and they enable an appendage to be used in many different ways. For example, the second pair of appendages in spiders is used for sensing and for mating. In scorpions, this pair of appendages is used for seizing prey.

Arthropod exoskeletons give protection

The success of arthropods as a group can be attributed in part to the presence of an exoskeleton. The exoskeleton is a hard, thick, outer covering made of protein and chitin. Chitin is the same substance found in fungal cell walls. In some species, the exoskeleton is a continuous covering over most of the body. In other species, the exoskeleton is made of separate plates held together by hinges. The exoskeleton protects and supports internal tissues and provides places for attachment of muscles. In many species that live on land, the exoskeleton is covered by a waxy layer that provides additional protection against water loss.

arthropod:

arthron (GK) joint
pous (GK) foot
Arthropods have jointed appendages.

Why arthropods must molt

Exoskeletons are an important adaptation for arthropods, but they also have their disadvantages. First, they are relatively heavy structures. Many terrestrial and flying arthropods are adapted to their habitats by having a thinner, lighter-weight exoskeleton, which offers less protection but allows the animal more freedom to fly and jump.

More importantly though, exoskeletons cannot grow, so arthropods must shed them periodically. Shedding of the old exoskeleton is called **molting.** Before an arthropod molts, a new exoskeleton develops beneath the old one. When molting occurs, the animal contracts muscles in the rear part of its body, forcing blood forward. The forward part of the body swells, causing the old exoskeleton to

Figure 31.1

The development of jointed appendages was a major evolutionary step that led to the success of the arthropods.

ThinkingLab Draw a Conclusion

What determines the type of waxy coating produced by grasshoppers?

Grasshopper exoskeletons are covered with a thin, waxy coating. This coating helps to protect against water loss in hot, dry climates in which certain grasshoppers live. Biologists raised grasshoppers at temperatures of 29°, 32°, and 34°C to simulate three different climates in which grasshoppers can live. After the grasshoppers molted, the scientists tested the discarded exoskeletons to find what temperature would melt the waxy layer.

Analysis
The data from the experiment were plotted in this graph.

Thinking Critically
Draw a conclusion about what may determine the type of waxy layer a grasshopper exoskeleton might produce.

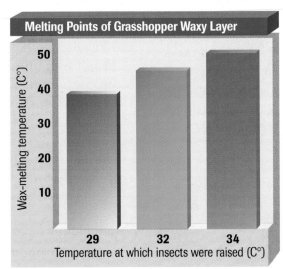

Melting Points of Grasshopper Waxy Layer

y-axis: Wax-melting temperature (C°)
x-axis: Temperature at which insects were raised (C°)

◀ **Spiders hold their prey with jointed mouthparts while feeding.**

▼ **The short, flattened appendages of the lobster, called swimmerets, are used like flippers for swimming. Large claws are modified for defense.**

▲ **The powerful jointed legs of this crab are adapted for walking.**

◀ **The antennae of a moth are adapted for the senses of touch and smell.**

Figure 31.2

Figure 31.2
Arthropods molt several times during their development. The old exoskeleton is discarded after a new one is formed from chitin-secreting cells beneath the old exoskeleton. Before the new exoskeleton hardens, the animal swallows air or water to puff itself up in size. Thus, the new exoskeleton hardens in a larger size, allowing some room for the animal to grow.

split open, as *Figure 31.2* shows. The animal then wiggles out of its old exoskeleton.

Most arthropods molt four to seven times in their lives, and during these periods, they are particularly vulnerable to predators. Arthropods are soft and have little muscle strength after molting, so many species hide for a few hours or days while the new exoskeleton hardens.

Segmentation in arthropods

Most arthropods do not have as many segments as are found in segmented worms, but their bodies show the markings of segments. In arthro-pod bodies, segments have become fused into one to three body sections—head, thorax, and abdomen. In most groups of arthropods, several segments are fused, forming the head as shown in *Figure 31.3*. Other segments are fused, forming a thorax and an abdomen. In other groups of arthropods, there is an abdomen with a fused head and thorax. The fused head and thorax of an arthropod is called a **cephalothorax.**

Arthropods have efficient gas exchange

Arthropods are generally quick, active animals. They crawl, run, climb, dig, swim, and fly. In fact, some flies beat their wings 1000 times per second. As you would expect, arthropods have efficient respiratory structures that ensure rapid oxygen delivery to cells. This large oxygen demand is needed to sustain the high levels of metabolism required for rapid movements.

Figure 31.3

Fusion of the body segments is related to movement and protection. Many species such as shrimps and lobsters have a fused head and thorax, which protects the internal structures in this sensitive region but which limits movement. Other species have separate head and thorax regions that allow for more flexibility.

▶ In the camel cleaner shrimp, the head and thorax are fused into a cephalothorax. The animal also has an abdomen.

◀ A stag beetle shows fusion of body segments into a distinct head, thorax, and abdomen.

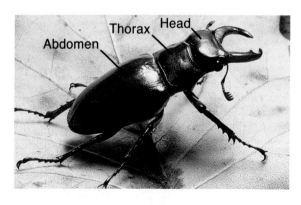

Arthropods have evolved three types of respiratory structures for taking oxygen into their bodies: gills, tracheal tubes, and book lungs. In some arthropods, air diffuses right through the body wall. Aquatic arthropods exchange gases through gills, which extract oxygen from water and release carbon dioxide into the water, *Figure 31.4.* Land arthropods have either a system of tracheal tubes or book lungs. Most insects have **tracheal tubes,** branching networks of hollow air passages that carry air throughout the body. Muscle activity helps pump the air through the tracheal tubes. Air enters and leaves the tracheal tubes through openings on the thorax and abdomen called **spiracles.**

Most spiders and their relatives have **book lungs,** air-filled chambers that contain leaflike plates. The stacked plates of a book lung are arranged like pages of a book and serve for gas exchange.

MiniLab

How strong is a spiderweb?

Imagine a fiber strong enough to withstand the impact of a bullet. Spiders produce a flexible protein thread that is five times stronger than steel. Research is underway to use spider silk in helmets, parachute cords, and other light but strong equipment, even bulletproof vests.

Procedure

1. Collect spiderwebs from the corners of your ceilings, windowsills, basement, or garden by using a cotton swab or small twig. Place the web in a jar of water for transport. Make sure you do not disturb the webs of black widows or brown recluses.
2. Wet a few chemically untreated human hairs, at least 6 cm long, in a beaker of water.
3. Measure the lengths of the hair and a strand of the web in their relaxed positions before stretching.
4. Stretch the hair on your ruler until it breaks. Do the same with the strand of web. Note the length each stretches before breaking.

Analysis

1. Which stretched more before breaking, the spider silk or the hair?
2. Speculate about the advantage of strong, stretchable silk to a spider.
3. Why did you wet the human hair?

Figure 31.4

As you examine these arthropod respiratory structures, evaluate which are best suited for land life and which are best suited for water life.

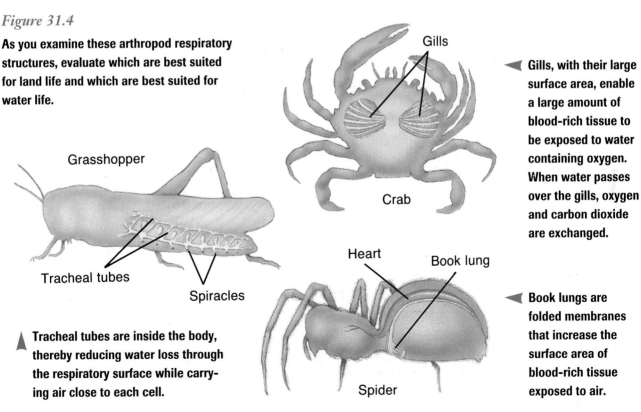

Grasshopper

Tracheal tubes

Spiracles

Gills

Crab

Heart

Book lung

Spider

▲ **Tracheal tubes are inside the body, thereby reducing water loss through the respiratory surface while carrying air close to each cell.**

◀ **Gills, with their large surface area, enable a large amount of blood-rich tissue to be exposed to water containing oxygen. When water passes over the gills, oxygen and carbon dioxide are exchanged.**

◀ **Book lungs are folded membranes that increase the surface area of blood-rich tissue exposed to air.**

Figure 31.5

The compound eyes of this dragonfly cover most of its head and consist of about 30 000 lenses. Compound eyes form as many image parts as there are lenses. As a result, the dragonfly can see simultaneously in all directions. Multiple lenses enable a flying arthropod to analyze a fast-changing landscape during flight. Dragonflies must capture other insects in midair, a fact that may explain why their eyes have more lenses than eyes of flying insects that do not capture prey in flight. Would you expect a mosquito to have more or fewer lenses than a dragonfly?

Arthropods have acute senses

Quick movements that are the result of strong, muscular contractions enable arthropods to respond to a variety of stimuli. Movement, sound, or chemicals can be detected with great sensitivity by antennae, stalklike structures that detect changes in the environment.

Antennae are also used for communication between animals. Have you ever watched as a group of ants efficiently took apart and carried home a crumb or small piece of food? The ants were able to work together as a group because they were communicating with each other by **pheromones,** chemical odor signals given off by animals. Antennae sense the odors of pheromones, which signal animals to engage in a variety of behaviors. Some pheromones are used as scent trails, such as in the group-feeding behavior of ants, and many are important in the mating behavior of arthropods.

Accurate vision is also important for the active lives of arthropods. Most arthropods have one pair of large compound eyes, shown in *Figure 31.5,* and from three to eight simple eyes. A **simple eye** is a visual structure with only one lens. Simple eyes are used for detecting light. A **compound eye** is a visual structure with many lenses. Each lens registers light from a tiny portion of the field of view. The total image that is formed is made up of thousands of parts, somewhat like the image of dots produced on a television screen. Although the image formed by a compound eye is not as detailed as one formed by a human eye, it is better for detecting motion. Compound eyes can detect even the slightest movements of prey, mates, or predators, and can also detect colors. However, the images will be fuzzy.

Arthropod nervous systems are well developed

Arthropods have a well-developed nervous system that processes information coming in from the sense organs. The nervous system consists of a double ventral nerve cord, an anterior brain, and several ganglia. Although earthworms have a ganglion for each segment, some segmental ganglia have become fused in arthropods. These ganglia act as control centers for the body section in which they are located.

Arthropods have evolved other complex body systems

Arthropod blood is pumped by one or more hearts in an open circulatory system with vessels that carry blood away from the heart. The blood

Figure 31.6

Mouthparts of arthropods exhibit a tremendous variation among species. How might this variation have contributed to the success of arthropods?

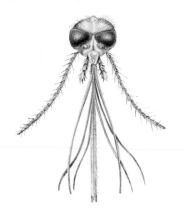

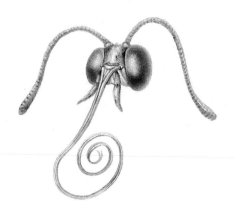

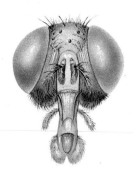

▲ Sand flies and other insects that feed by drawing blood have piercing blades or needlelike mouthparts.

▲ The rolled-up sucking tube of moths and butterflies can reach nectar at the bases of long, tubular flowers.

▲ The sponging tongue of the housefly has an opening between its two lobes through which food is lapped.

flows out of the vessels, bathes the tissues of the body, and returns to the heart through open body spaces.

In addition to a complex circulatory system, arthropods have a complete digestive system with a mouth, stomach, intestine, and anus, together with various glands that produce digestive enzymes. The mouthparts of most arthropod groups, shown in *Figure 31.6*, include a variety of jaws—called **mandibles.** Mouthparts are adapted for holding, chewing, sucking, or biting the various foods eaten by arthropods.

Most terrestrial arthropods excrete wastes through **Malpighian tubules.** In arthropods, the tubules are all located in the abdomen rather than in each segment, as they are in segmented worms. Malpighian tubules are attached to and empty into the intestine.

Another well-developed system in arthropods is the muscular system. *Figure 31.7* shows the differences in muscle attachment in vertebrate and in arthropod systems. An arthropod muscle is attached to the exoskeleton on both sides of the joint.

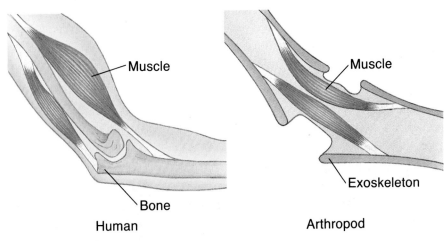

Human

Arthropod

Figure 31.7

In a human limb (left), muscles are attached to the outer surfaces of internal bones. In an arthropod limb (right), the muscles are attached to the inner surface of the exoskeleton. A flea can jump a distance 100 times its own length, equivalent to a six-foot human doing a 600-foot standing high jump. Relative to body weight, a flea has stronger muscles than those of humans.

In many tropical countries, a mosquito bite can result in far more serious symptoms than an itching welt. Malaria, a potentially fatal disease, is caused by a protist that is carried by mosquitoes. Insecticides are used in the United States to control mosquito populations, but in many countries these control measures are too expensive. A less-expensive method might be to find plants or plant parts that produce insect-repelling substances. Keep in mind that some insecticides are made from chemicals extracted from plant parts. From plants such as these, skin creams could be produced that would repel the disease-carrying mosquitoes.

Do flowers produce insect repellents?

PREPARATION

Problem
Which flowers have natural insect repellents?

Hypotheses
In your group, brainstorm a list of possible hypotheses and discuss the evidence on which each hypothesis is based. Select the best hypothesis from your list.

Objectives
In this Biolab, you will:
- **Analyze** insect behavior to determine which flowers may have insect repellents.
- **Infer** why a flower might produce an insect-repelling substance.

Possible Materials
local insects such as crickets or
  grasshoppers
petri dishes
flower extract-treated filter paper, cut
  in half
plain filter paper, cut in half
cotton swabs
forceps

Safety Precautions
Use forceps to handle filter paper. Be sure to keep your hands away from your face after handling filter paper. Wash your hands at the end of the lab.

Cave cricket

1. Make a list of the possible ways you might test your hypothesis using the materials your teacher has provided.

2. Plan to test one variable, collect quantitative data, and have a control.

3. How might you measure a particular insect behavior when the insects are exposed to the flower extract?

4. Make a table to record your data.

Check the Plan

1. Which insect behaviors will you measure and how will you measure them?

2. What controls will you use?

3. Consider the sample size you will use and how many trials you will record in each experiment.

4. **Make sure your teacher has approved your experimental plan before you proceed further.**

5. Carry out your experiment. Make a graph of your data.

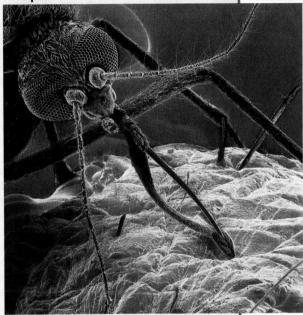

Mosquito

1. **Checking Your Hypothesis** Was your hypothesis supported by your data? Why or why not?

2. **Observing and Inferring** Why might a flower attract a specific insect and repel others?

3. **Drawing Conclusions** Could your experimental findings be used to help develop an insecticide? Explain.

4. **Relating Concepts** Why is repelling insects important to people?

5. **Thinking Critically** What advantage might natural repellents have over synthetic repellents?

Going Further

Application Design another experiment to determine if other parts of a plant produce insect-repelling chemicals.

Arthropods reproduce sexually

Most arthropod species have separate males and females and reproduce sexually. Fertilization is usually internal in land species but is often external in aquatic species. A few species are hermaphrodites, animals with both male and female reproductive organs. Some species exhibit **parthenogenesis,** a form of asexual reproduction in which an organism develops from an unfertilized egg.

Ecology of Arthropods

Arthropods are found in so many habitats because of the enormous variety of adaptations that have evolved for obtaining and digesting different foods. Name any food source you can think of and there will probably be an arthropod that uses it.

Most arthropods obtain their own food, but many are parasites. As shown in *Figure 31.8,* some lay their eggs on other insects.

Arthropods and humans

Arthropods are beneficial to humans in a variety of ways. They pollinate a great many of the flowering plants and crop plants on Earth and provide food, honey, shellac, wax, and silk. Some arthropods, such as the ladybird beetle and certain spiders, provide alternatives to chemical control of insects. Research on arthropods has led to advances in the fields of genetics, evolution, and biochemistry. Research is underway to develop a chemical based on scorpion venom that will kill harmful insects but won't bother other animals. From crab shells, scientists have made artificial skin, surgical sutures, and antifungal medication.

Insects cause problems for humans by eating important crops. They also spread plant and animal diseases, including deadly human diseases such as malaria and yellow fever.

The Origins of Arthropods

Arthropods most likely evolved from the annelids. As arthropods evolved, body segments became reduced in number when they fused and became adapted for certain functions such as locomotion, feeding, and sensing the environment. Segments show more complexity of organization in arthropods than in annelids, and the head shows greater development of nerve tissue and sensory organs such as eyes.

The exoskeleton of arthropods is harder and provides more protection than the cuticle of annelids. Because arthropods have many hard parts, much is known about their evolutionary history. The trilobites shown in *Figure 31.9* were once an important group of ancient arthropods, but they have been extinct for 250 million years.

Figure 31.8

Insect parasites often lay their eggs on the larvae of other insects. You can see that as the insects hatch from the eggs, they will have an instant "fast-food" meal.

Figure 31.9

The radiation of classes of arthropods on the Geologic Time Scale shows their relationships. Which group is largest? Which are the oldest living arthropods?

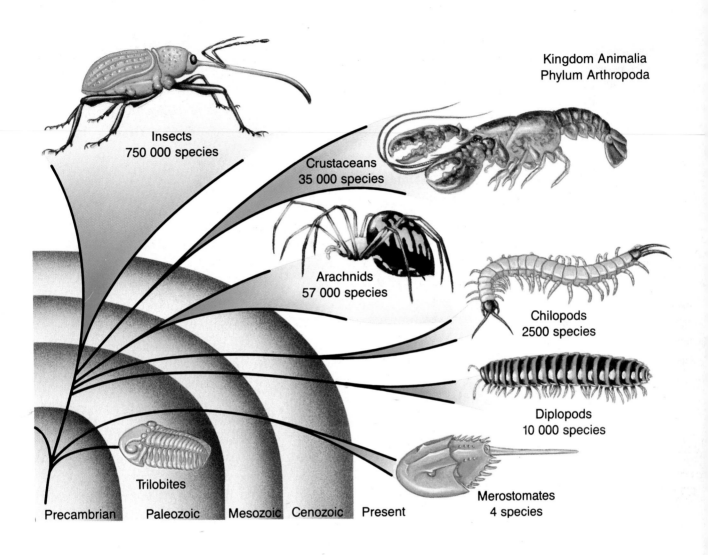

Kingdom Animalia
Phylum Arthropoda

Insects
750 000 species

Crustaceans
35 000 species

Arachnids
57 000 species

Chilopods
2500 species

Diplopods
10 000 species

Trilobites

Merostomates
4 species

Precambrian   Paleozoic   Mesozoic   Cenozoic   Present

Section Review

Understanding Concepts

1. Describe the pathway taken by the blood as it circulates through an arthropod's body.
2. Describe two features that are unique to arthropods.
3. What are the advantages and disadvantages of an exoskeleton?

Thinking Critically

4. What characteristics of arthropods might explain why they are the most successful animals in terms of population sizes and numbers of species?

Skill Review

5. **Comparing and Contrasting** Compare the adaptations for gas exchange in aquatic and land arthropods. For more help, refer to Thinking Critically in the *Skill Handbook*.

31.2 The Diversity of Arthropods

Section Preview

Objectives

Compare and contrast the similarities and differences between the major groups of arthropods.

Explain the adaptations of insects that contribute to their success.

Key Terms

chelicerae
pedipalp
spinneret
metamorphosis
larva
pupa
nymph

Female mosquitoes drink an average of 2.5 times their body weight in blood every day. Imagine yourself, weighing 120 pounds, sitting down to a steak dinner and getting up weighing 300 pounds. Other arthropods feed on nectar, dead organic matter, oil, and just about any other substance you can imagine. The varied eating habits of arthropods reflect their huge diversity.

Boll weevil

Arachnids

Do you remember the last time you saw a spider? Did you draw back with a quick, fearful breath, or did you move a little closer, curious to see what it would do next? Of the 30 000 species of spiders, only about a dozen are dangerous to humans. In North America, you need to watch out for only the two illustrated in *Figure 31.10*—the black widow and the brown recluse.

What is an arachnid?

Spiders, scorpions, mites, and ticks belong to the class Arachnida. Spiders are the largest group of arachnids. Whereas most arthropods have three body regions, the arachnids have no distinct thorax and have only two body regions. Arachnids have six pairs of jointed appendages. The first pair of appendages, called **chelicerae,** is located near the mouth. In arachnids, chelicerae are often modified as pincers to hold

Figure 31.10

The black widow spider (left) is shiny black with a red, hourglass-shaped spot on the underside of the abdomen. The brown recluse (right) is brown to yellow and has a violin-shaped mark on its body. A bite from either spider can make a person sick, but if the person gets medical treatment, the bites are rarely fatal.

Inside a Spider

The garden spider weaves an intricate and beautiful web, dribbles sticky glue on the spiraling silk threads, and waits for insects to crash into them. Spiders are predatory animals, feeding almost exclusively on other arthropods. Each spider species builds a unique web, which is effective in trapping flying insects. Many of the structural adaptations of spiders are related to this hunting process.

Garden spider

1 The four pairs of walking legs are located on the cephalothorax of the spider.

2 Spiders have six or eight simple eyes that, in most species, detect light but do not form images. Spiders have no compound eyes.

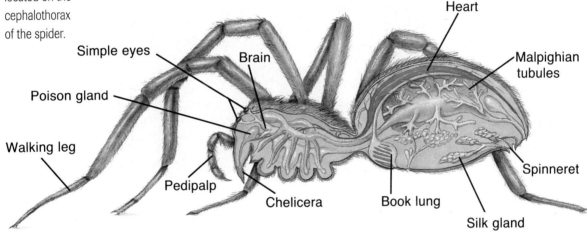

Simple eyes

Poison gland

Walking leg

Pedipalp

Chelicera

Brain

Heart

Malpighian tubules

Spinneret

Book lung

Silk gland

6 A pair of pedipalps is used to hold and move food and also to function as sense organs. In males, pedipalps are bulbous and are used to carry sperm.

5 Chelicerae are the two biting appendages of arachnids. In spiders, they are modified into fangs. Located near the tips of the fangs are poison glands.

3 Gas exchange in spiders takes place in book lungs.

4 Spiders have between two and six silk glands. Silk is first released from silk glands as a liquid. It then passes through as many as 100 small tubes before being spun into thread by the spinnerets.

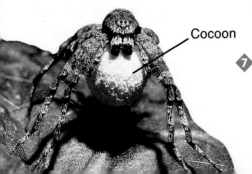

Cocoon

7 Female spiders wrap their eggs in a silken sac or cocoon, where the eggs remain until they hatch. Some spiders lay their eggs and never see their young. Others carry the sac around with them until the eggs hatch.

What Little Miss Muffet Didn't Know

In a popular children's nursery rhyme, Little Miss Muffet's snack of curds and whey is rudely interrupted by a spider. Muffet, unwilling to share either her snack or her seat with the eight-legged, fanged visitor, flees in terror. Had she known what you're about to discover about the hunting habits of spiders, she surely would have fled even faster. Spiders are carnivores; that is, they eat other animals. Occasionally, they even eat each other. Spiders can be classified according to the way they trap their prey.

The web spinners Web-spinning spiders use their silk to fashion geometric and sometimes elaborate webs in which they trap their victims. Insects walk or fly into a web and become enmeshed in its sticky strands. Unable to escape, the unfortunate insects soon become dinner for their hungry host.

Web-spinning spiders create webs of many sizes and shapes to capture prey. For example, orb weavers build round webs that are as beautiful as they are complex. The weaver may wait in the center of the web for prey to become trapped, or it may extend a trapline from the web to its nest nearby. When prey becomes tangled in the web, the trapline vibrates, and the spider rushes from the nest to snatch its meal.

A stick spider carries its web with it. The spider first weaves a rectangular web. Placing its four front legs at the corners of the web, the spider stretches it to several times its original size. Then, waving its legs in tandem, the stick spider sweeps the web over its prey like a fisher casting a net wide across the water. Just as fish get caught in the net, insects get caught in the web.

The hunt is on Hunting spiders don't weave webs to trap prey. Instead, they take a more direct approach. They lie in wait for an unsuspecting meal to wander by and then either pounce on it or run it down. The bolas spider produces a single strand of silk with a drop of glue at the end. It resembles the rope-and-ball bolas used as a weapon by Argentinian cowboys. The spider swings its bolas at prey and reels in its meal, which is stuck to the glob of glue.

Thinking Critically

Given what you know about spiders' hunting practices, which of the two types of spiders, web-spinners or hunters, seems to work harder for its meals?

food, or as fangs that can inject prey with poison. Spiders have no mandibles for chewing food. Using a process of extracellular digestion, digestive enzymes from the spider's tiny mouth liquefy the internal organs of the captured prey. The spider then sucks up the liquefied food.

The second pair of appendages, called the **pedipalps,** are adapted for handling food and for sensing. In male spiders, pedipalps are further modified to carry sperm during reproduction. The four remaining appendages in arachnids are modified as legs for locomotion. Arachnids have no antennae.

Most people know spiders for their ability to make elaborate webs. Although all spiders spin silk, not all make webs. Spider silk is secreted by silk glands in the abdomen. As silk is secreted, it is spun into thread by structures called **spinnerets,** located at the rear of the spider.

Close relatives of spiders

Ticks and mites differ from spiders in that they have only a single body section, *Figure 31.11.* The head, thorax, and abdomen are completely fused. Mites are so small that they often are not visible to the naked eye. However, you can certainly feel them bite if mites called chiggers get under your clothing while you are camping.

A walk in the woods may require a check for ticks. Ticks feed on blood from reptiles, birds, and mammals. They are small but capable of expanding up to 1 cm or more after a blood meal.

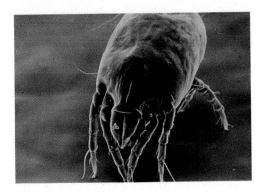

Figure 31.11

Mites, close relatives of spiders, are distributed throughout the world and in just about every habitat. House-dust mites feed on discarded skin cells that collect in dust on floors, in bedding, or on clothing. Some people are allergic to mite waste products.

Scorpions are easily recognized by their many abdominal body segments and enlarged pincers. Related to scorpions are horseshoe crabs, members of the class Merostomata. Horseshoe crabs are considered to be living fossils because they have remained relatively unchanged since the Cambrian period, about 500 million years ago. They are similar to trilobites in that they are heavily protected by an extensive exoskeleton and live a bottom-dwelling existence.

Crustaceans

Most crustaceans are aquatic and exchange gases as water flows over feathery gills. All crustaceans have mandibles for crushing food, two pairs of antennae for sensing, and two compound eyes, which are usually on movable stalks. Unlike the up-and-down movement of your jaws, crustacean mandibles open and close from side to side. Five pairs of walking legs are used for walking, for seizing prey, and for cleaning other appendages.

Health

Terrible Ticks

Every American city, it seems, has its claim to fame. Chicago is recognized for its outstanding architecture. New York has long been thought of as the cultural center of the United States. Los Angeles is home to television production and to the nation's legendary movie industry. One American city, however, would probably just as soon forget *its* claim to fame. Lyme, Connecticut, will forever be associated with Lyme disease, a crippling bacterial malaise that was first identified in this town in 1975.

A progressive disease   Lyme disease manifests itself in humans in three distinct stages. First, a circular, bull's-eye rash appears. Sometimes the rash is accompanied by chills, fever, and aching joints. This is the mildest form of the disease. If left untreated, Lyme disease progresses to a second stage. The joint pains become more severe and may be joined by neurological symptoms, such as memory disturbances and vision impairment. Stage three is the most severe form of the disease. Crippling arthritis, facial paralysis, heart abnormalities, and memory loss may result.

Tick transmission   The cause of this debilitating disease is *Borrelia burgdorferi,* a corkscrew-shaped bacterium that is transmitted to humans through the bite of ticks. The bacterium infects mostly deer and white-footed mice. Ticks pick up the bacteria by sucking the blood from these animals. When the same ticks bite humans, the bacteria are passed on, and the result is Lyme disease.

Antibiotics to the rescue
The only good news about Lyme disease is its response to treatment. Like most bacteria, *Borrelia burgdorferi* responds to antibiotics. Early treatment with tetracycline will usually prevent the disease from progressing to its second or third stages.

CONNECTION TO **Biology**

Since the turn of the century, the deer population in the United States has been increasing steadily. How might this increase affect the incidence of Lyme disease? Why?

Figure 31.12

Barnacles are distinct from other arthropods in terms of structure. Most are sessile and are covered with thick plates. A gluelike substance anchors barnacles to surfaces. Barnacles are filter feeders that trap food by extending their feathery legs out of the shell.

Crabs, lobsters, shrimps, crayfishes, barnacles, water fleas, and pill bugs are members of the class Crustacea, *Figure 31.12.* Some crustaceans have three body sections, and others have only two.

Pill bugs, the only land crustaceans, must live where there is moisture, which aids in gas exchange. They are frequently found in damp areas around building foundations.

Centipedes and Millipedes

Centipedes, which belong to the class Chilopoda, and millipedes, members of the class Diplopoda, are shown in *Figure 31.13.* If you have ever turned over a rock on a damp forest floor or kicked a pile of damp, dead leaves, you may have seen the flattened bodies of centipedes wriggling along on their many tiny, jointed legs. Centipedes are carnivorous and eat soil arthropods, snails, slugs, and worms. The bite of a centipede is painful to humans. Like spiders, millipedes and centipedes have Malpighian tubules for excreting wastes. In contrast to spiders, centipedes and millipedes have tracheal tubes rather than book lungs for gas exchange.

A millipede, the so-called thousand-legger, eats mostly plants and dead material on damp forest floors. Millipedes do not bite, but they can spray obnoxious-smelling fluids from their defensive stink glands. You may have seen their cylindrical bodies walking with a slow, graceful motion rather than the wriggling, scurrying motion of a centipede.

Figure 31.13

A centipede (left) may have from 15 to 181 body segments—always an odd number. Each segment has only one pair of jointed legs. The first body segment has a pair of poison claws that secretes a toxic substance from a pair of poison glands. A millipede (right) may have more than 100 segments in its long abdomen, each with two spiracles and two pairs of legs.

Insects

Have you ever launched an ambush on a fly with your rolled-up newspaper? You swat with great accuracy and speed, yet your prey is now firmly attached upside down on the kitchen ceiling. How does a fly do this?

The fly approaches the ceiling right-side up at a steep angle. Just before impact, it reaches up with its front legs. The forelegs grip the ceiling with tiny claws and sticky hairs, while the other legs swing up into position. The flight mechanism shuts off, and the fly is safely out of swatting distance. Adaptations that enable flies to land on ceilings are among the many that make insects the most successful arthropod group.

Flies, grasshoppers, lice, butterflies, bees, and beetles are just a few members of the class Insecta, by far the largest group of arthropods. *Figure 31.14* shows several insects.

MiniLab

Which insects are attracted to light?

Have you ever heard a June bug bump against your screen at night, or seen moths fluttering around your porch light? You know that some insects are nocturnal and others are active in the daytime.

Procedure

1. Construct and set up a light trap for insects by putting a paper sack with a 5-cm opening over your porch light. You may also use a flashlight in a box with a 5-cm hole.

2. Leave your light trap set up for several hours after dark.

3. When you have trapped some insects, cover the opening of the trap with tape and place the trap into a freezer for an hour. This will temporarily immobilize the insects.

4. Empty your trap, and examine the insects with a magnifying glass or dissecting microscope. Count the number of each kind of insect in your trap. Use insect identification books to try to identify them.

Analysis

1. How is nocturnal life an adaptation for some insects?

2. Why might insects be attracted to light?

3. Examine the bodies of the insects you captured. Predict the diet and habitat of each, based on its structural adaptations.

Figure 31.14

Insects generally have three pairs of legs, one pair of antennae, and three body regions.

▶ **Bees, ants, and wasps are social animals and live in large colonies.**

▼ **Fireflies are grouped with more than 300 000 species of beetles in the largest insect order.**

▼ **Butterflies and moths are grouped in an insect order with more than 100 000 known species.**

▶ **Water striders belong to the order of true bugs.**

Meet Dr. May R. Berenbaum, Entomologist

In "The Deadly Mantis," a huge praying mantis gobbles up most of the population of the East Coast. The movie was a popular attraction at the Insect Fear Film Festival organized by Dr. Berenbaum at the University of Illinois. The annual attraction aims at entertaining the audience while pointing out entomological errors in popular films and interspersing some information about real-life insects.

In the following interview, Dr. Berenbaum tells about her study of insects, which, together with their close relatives, make up nearly 75 percent of the animal species alive today.

On the Job

Q Dr. Berenbaum, what is your favorite class of those you teach at the university?

A I teach a course called "Insects and People," which shows insects' influence on art, literature, and history. For example, both batik and the classic method of casting bronze would be impossible without beeswax. The product of silkworms is used in many fabrics and tapestries. Insects show up in all kinds of literature, including *Make No Bones* by Aaron Elkins, a contemporary murder-mystery writer. And disease-carrying insects have caused more deaths during war than bombs and bullets ever have.

Q What about your specialty, the interrelationships between insects and plants?

A I like to remind my students that, if it weren't for insects, the world's favorite hamburger would have no special sauce, no lettuce, no cheese, no pickles, and no onions. There would just be the bun, because wheat is wind-pollinated. The contribution of a single kind of insect, the honeybee, is worth about $19 billion to American agriculture.

My work with plants and insects began because as an undergraduate, I was strongly interested in plant chemistry. Plants contain an extraordinary diversity of chemicals including many that are toxic. I decided to combine my interests and work on insects that eat plants. Part of my work is studying how some insects switch foods. For instance, a shift in an insect's preferred food from carrots to citrus is of crucial importance to farmers.

Q Are you also involved in laboratory research?

A Yes. One project that I'm involved in uses molecular biology to study how insects cope with insecticides. We want to learn how insects have been able to develop resistance so quickly. For instance, DDT's insecticidal properties were discovered in 1939, and only six years later, there was documented resistance in insect populations. We're trying to understand the molecular basis for the evolution of resistance so we can keep insects susceptible to various toxins.

Early Influences

Q How did you get involved in the study of insects?

A I actually used to be afraid of insects. When I was a freshman in college, I decided that my fear stemmed from ignorance. I thought an entomology class would at least teach me what insects I *should* be afraid of. I had a professor who showed me that insects really are absolutely amazing. Up to that point, I was truly phobic. The class altered my life.

Q Was there a particular book that helped foster your interest?

A I found Howard E. Evans's book, *Life on a Little-Known Planet,* to be a fascinating look at the diversity of the insect world. It was a great pleasure to me that Evans later wrote a blurb for the cover of my own book, *Ninety-Nine Gnats, Nits, and Nibblers,* which profiles insects that people encounter in everyday life, such as cockroaches, houseflies, and tiny insects nicknamed no-see-ums.

Personal Insights

Q What advice do you have for students about choosing a field of study?

A Having chosen a career that I never thought I would even consider, my best advice would be never to rule *anything* out! To me, insects are endlessly entertaining, even better than cable TV. Insects are everywhere, too. For example, three-fourths of the human population harbor tiny mites that live inside hair follicles. No amount of scrubbing is going to get rid of them. It's a good thing that they're harmless.

Q Are there any careers related to insects that some people might find surprising?

A A forensic entomologist can collect insects on a corpse and calculate how long it would take the species to reach that developmental stage. Then law enforcement officers have a very good idea about when a death occurred. This university is interested in developing a program in forensic entomology.

Q Could you give us just one more amazing insect fact?

A There are almost 1 million species of known insects, and at least twice that number are as yet undiscovered. There's still plenty of work left to be done by future entomologists.

Inside a Grasshopper

Grasshoppers make rasping sounds either by rubbing their wings together or by rubbing small projections on their legs across a scraper on their wings. Most calls are made by males. Some aggressive calls are made when other males are close. Other calls attract females, and still others serve as an alarm to warn nearby grasshoppers of a predator in the area.

Grasshopper structure is typical of many insects. Grasshoppers have three main body sections, a pair of antennae, two pairs of wings, and six legs.

Grasshopper

1 Insects have one pair of antennae, which is used to sense vibrations, food, and pheromones in the environment. Sensory hairs, which are sensitive to touch, cover the exoskeleton and antennae.

2 Insects are the only invertebrates that can fly. With the ability to fly, insects can move large distances, find new places to live, discover new food sources, escape quickly from predators, and find mates.

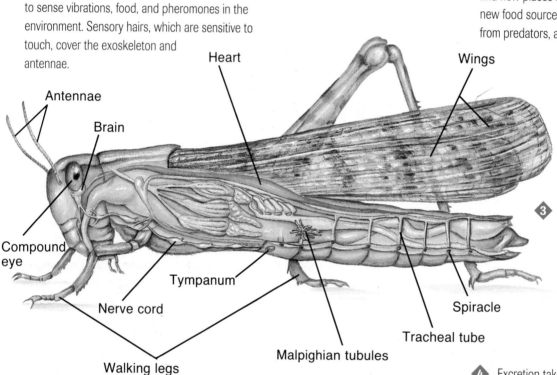

Antennae
Brain
Heart
Wings
Compound eye
Tympanum
Nerve cord
Walking legs
Malpighian tubules
Spiracle
Tracheal tube

3 The spiracles in the abdomen open into a series of tracheal tubes used in gas exchange.

7 Grasshoppers have two compound eyes and three simple eyes. Compound eyes do not produce clear images, but they are effective for spotting movement of prey.

6 Most insects have six legs. By looking at an insect's legs, you can sometimes tell how it moves about and what it eats. The grasshopper has long, thick legs for jumping large distances.

5 The structure used for hearing by an insect is a flat membrane called a tympanum.

4 Excretion takes place by Malpighian tubules. In the grasshopper and other insects, nitrogenous wastes are in the form of dry crystals of uric acid. Producing dry waste helps insects to conserve water.

Insect reproduction

Insects mate once, or at most only a few times, during their lifetime. The eggs are fertilized internally, and shells form around them. Many female insects are equipped with an appendage that is modified for digging a hole below the surface of the ground or in wood. The female lays large numbers of fertilized eggs in the hole. Laying large numbers of eggs increases the chances that some offspring will survive long enough to reproduce.

Metamorphosis—Change in body shape and form

Growing insects undergo a series of changes in body structure as they develop. This series of changes, controlled by chemical substances in the animal, is called **metamorphosis.**

Figure 31.15

During complete metamorphosis, an insect undergoes a series of developmental changes from egg to adult.

Most insects go through four stages on their way to adulthood—egg, larva, pupa, and adult. The **larva** is the free-living, wormlike stage of an insect, often called a caterpillar. As the larva eats and grows, it molts several times.

The **pupa** stage of insects is a period of reorganization in which the tissues and organs of the larva are broken down and replaced by adult tissues. After a period of time, a fully formed adult emerges from the pupa.

The series of changes that occur as an insect goes through the egg, larva, pupa, and adult stages is known as complete metamorphosis. The complete metamorphosis of a butterfly is illustrated in *Figure 31.15.* Other insects that undergo complete metamorphosis include ants, beetles, flies, and wasps.

Complete metamorphosis is an advantage for arthropods because larvae do not compete with adults for the same food. For example, caterpillars feed on leaves, but adult butterflies feed on nectar from flowers.

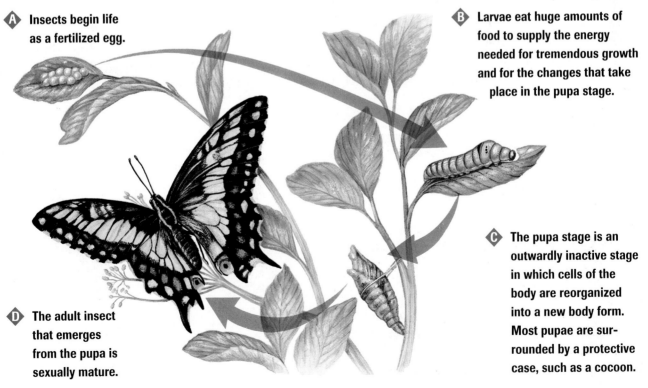

A Insects begin life as a fertilized egg.

B Larvae eat huge amounts of food to supply the energy needed for tremendous growth and for the changes that take place in the pupa stage.

C The pupa stage is an outwardly inactive stage in which cells of the body are reorganized into a new body form. Most pupae are surrounded by a protective case, such as a cocoon.

D The adult insect that emerges from the pupa is sexually mature.

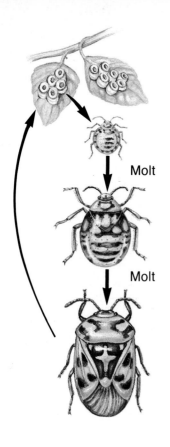

A The fertilized egg is surrounded by a protective shell that contains food.

B Because movement is limited, many nymphs have colorings that resemble their surroundings for camouflage.

C Nymphs continue to grow and molt, gradually increasing in size.

D After several molts, nymphs become adults with wings and reproductive capabilities.

Molt

Molt

Figure 31.16

Shown here is the incomplete metamorphosis of the harlequin bug.

Incomplete metamorphosis has three stages

Many insect species undergo a gradual or incomplete metamorphosis, in which the insect goes through three stages of development—egg, nymph, and adult, as shown in *Figure 31.16*. A **nymph,** which hatches from an egg, has the same general appearance as the adult but is smaller. Nymphs lack certain appendages, such as wings, and they cannot reproduce. As the nymph eats and grows, it molts several times. With each molt, it comes to resemble the adult more and more. Wings begin to form, and an internal reproductive system develops. Gradually, the nymph becomes an adult. Grasshoppers and cockroaches are insects that undergo incomplete metamorphosis.

Connecting Ideas

Arthropods have been enormously successful in establishing themselves over the entire surface of Earth. Their ability to exploit just about every habitat is unequaled in the animal kingdom. The success of arthropods can be attributed in part to their varied life cycles, high reproductive output, and structural adaptations such as small size, a hard exoskeleton, and jointed appendages. Another group of animals, the echinoderms, followed a different evolutionary path toward success. Echinoderms are marine animals that have evolved unusual structural adaptations, which contribute to this group's success in the seas.

Section Review

Understanding Concepts
1. How are centipedes different from millipedes?
2. How are insects different from spiders?
3. Describe three sensory adaptations of insects.

Thinking Critically
4. Why might complete metamorphosis have greater adaptive value for an insect than incomplete metamorphosis?

Skill Review
5. **Recognizing Cause and Effect** Some plants produce substances that prevent insect larvae from forming pupae. How might this chemical production be a disadvantage to the plant? For more help, refer to Thinking Critically in the *Skill Handbook*.

Milking Spiders

In an Arizona laboratory, a spider is milked for its venom. Stimulated with an electric needle, the spider vomits and secretes a tiny bit of venom. A clean venom sample is quickly collected with a pipette, as the vomit is suctioned away. Some species require as many as 1000 milkings to obtain 100 µL (2 drops).

Spider venom is a complex mixture of chemicals. Poisonous spiders produce venom laden with toxic chemicals that affect nerve pathways in specific ways.

Blocked channels   Spider venom blocks the action of the amino acid glutamate, a neurotransmitter that increases nerve cell activity. Neurotransmitters are released by nerve cells and cause other nerve cells to respond.

Abundant in the brain, glutamate attaches to receptor sites on the plasma membrane of the receiving cell. As it does so, channels in the plasma membrane open to allow ions to flow into the cell. This causes the receiving cell to send glutamate to the next nerve cell. Thus, the nerve impulse is relayed from one nerve cell to the next. Some spider venoms contain specific toxins that plug these channels, making it possible to study glutamate transmission more thoroughly.

Applications for the Future

Glutamate is known to play a major role in causing death after strokes. In a stroke, blood and oxygen are cut off when a blood clot blocks a blood vessel in the brain. Glutamate opens the channels in the damaged nerve cells to a flood of ions that kill the cells by overworking them. Scientists hope to limit stroke damage with glutamate-blocking toxin, which would give the damaged cells a chance to recover.

An excess of glutamate may also play a major role in severe forms of epilepsy. Epileptic seizures are a result of hyperactive nerve cells. Experimental drugs that block glutamate receptors have been developed, but they cause side effects because they affect the nervous system too generally. The specificity of glutamate-blockers in spider venom has great potential for managing these seizures without side effects.

The venom of another spider has been found to block calcium channels in the brain. Calcium is important in the nervous system because it triggers the action of neurotransmitters. Without calcium, neurotransmitters can't work.

Current research shows that the calcium-blocking toxin in spider venom may prevent the brain cell death that leads to paralysis after head injuries. It also may prevent the brain cell death that is typical of chronic degenerative diseases such as Alzheimer's disease. Both paralysis and Alzheimer's disease result, at least partially, from an excess of calcium in damaged brain cells.

INVESTIGATING   the Technology

1. **Apply** Read the article "Spider Man," by Kathryn Phillips, *Discover,* June 1991, pp. 48-53. How do toxins in spider venom block ion channels?
2. **Research** Research and describe one venomous animal other than spiders that is found in your location. Describe the effects of this animal's toxin and any antidotes for humans.

Reviewing Main Ideas

31.1 Characteristics of Arthropods

- Arthropods have jointed appendages; exoskeletons; varied life cycles; and body systems adapted to life on land, water, or air.
- Arthropods are members of the most successful animal phylum in terms of diversity. This can be attributed in part to their structural and behavioral adaptations.

31.2 The Diversity of Arthropods

- Spiders have two body regions with four pairs of walking legs. They spin silk. Ticks and mites have only one body section. Scorpions have many abdominal segments, enlarged pincers, and a stinger at the end of the tail.
- Most crustaceans are aquatic and exchange gases in gills. They include crabs, lobsters, shrimps, crayfishes, barnacles, and water fleas.

- Centipedes are carnivores with flattened, wormlike bodies. Millipedes are herbivores with cylindrical, wormlike bodies.
- Insects are the most successful arthropod class in terms of diversity. They have many structural and behavioral adaptations that allow them to exploit all habitats.

Key Terms

Write a sentence that shows your understanding of each of the following terms.

appendage	nymph
book lung	parthenogenesis
cephalothorax	pedipalp
chelicerae	pheromone
compound eye	pupa
larva	simple eye
Malpighian tubule	spinneret
mandible	spiracle
metamorphosis	tracheal tube
molting	

Understanding Concepts

1. How are jointed appendages important to arthropod success?
2. How can arachnids, crustaceans, and insects each be distinguished?
3. Identify a major difference between incomplete metamorphosis and complete metamorphosis.
4. Why are horseshoe crabs considered to be living fossils?
5. Describe the defense mechanisms of arthropods.
6. Explain molting and discuss its adaptive value.
7. What structural evidence can be used to trace the evolution of arthropods from annelids?
8. Describe some advantages of compound eyes to arthropods.

Interpreting Scientific Illustrations

9. Identify each of the arthropods below as an arachnid, crustacean, or insect. Explain how you can tell.

a   b

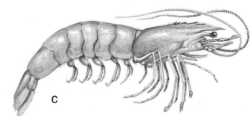

c

Relating Concepts

10. Make a concept map that relates the following terms and phrases. Supply the appropriate linking words for your map.

    insect, appendage, leg, antennae, head, thorax, abdomen, spiracles, tracheal tubes, Malpighian tubules, compound eye, sensory hairs

Applying Concepts

11. Many insects are pests to humans when they are larvae but are beneficial when they are adults. Explain.
12. Why do you think most barnacles are hermaphrodites?
13. Relate differences in exoskeleton structure to the various modes of arthropod locomotion.
14. In what ways are wings important to the success of insects?
15. Discuss some factors that limit the size of land- and air-dwelling arthropods.

Health Connection

16. Evidence shows that household pets, as well as deer and mice, may harbor the bacterium that causes Lyme disease. Suggest a way pets might become infected.

Thinking Critically

Analyzing

17. **Biolab** What is the advantage to a plant of producing an insect repellent?

Making a Hypothesis

18. **Minilab** During an experiment to test spiderweb strength, you compare webs built by wild spiders with those of captive-bred members of the same species. You find that the spiders kept in captivity build webs that are not as strong and flexible. Make a list of possible hypotheses that you could test to determine why this is so.

Interpreting Data

19. **Thinking Lab** The melting points of the waxy layers on certain insect exoskeletons are shown in the graph below. What can you tell about the temperatures of the environments in which insects A, B, and C live?

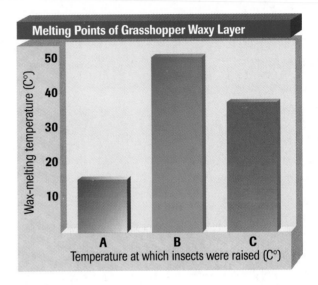

Melting Points of Grasshopper Waxy Layer

Wax-melting temperature (C°) / Temperature at which insects were raised (C°)

Making Inferences

20. **Minilab** Suppose you catch insects from a field at different times of the day. The insects you catch at midday are different from the types you catch at night. What might you infer about the habits of the insects you catch in the daytime?

Analyzing

21. Of what advantage might movable, stalked eyes be to a crustacean that has a cephalothorax?

Connecting to Themes

22. **Systems and Interactions** What might be an effect on other animal life if all insects were to die suddenly?
23. **Homeostasis** Give three examples of organs in arthropods that enable the animals to maintain homeostasis with their environment. Explain your examples.

32 Echinoderms and Invertebrate Chordates

Beneath the cold ocean waters off the California coast, the rocky bottom is covered with brightly colored sponges, anemones, snails, and other organisms. Atop one rock, a spiny sea urchin browses on a piece of algae. A huge, purple starfish emerges from a nearby crevice and glides smoothly toward the urchin. This starfish has 24 arms—far more than the usual five—and measures almost two feet in diameter. It moves along on hundreds of tiny tube feet that line the underside of its body, coming to a stop only after it has completely covered the spiny urchin.

Now positioned for a meal, the starfish extends its stomach from its mouth and slowly engulfs the urchin. Digestion begins. Hours later, the starfish draws its stomach back in and moves away. All that's left of the urchin is the bumpy globe you see here. Even its spines are gone.

Feather stars

Sea urchin skeleton

Like the sea urchin and its starfish predator, the graceful feather star also is a member of the Phylum Echinodermata. Whereas sea urchins and starfishes eat algae or small animals, feather stars live on tiny particles of organic matter that drift to the ocean bottom.

*T*hink about what the best defense might be for a small animal that moves slowly on the bottom of tide pools on the seashore. Did you think of armor, spines, or perhaps poison as methods of protection? Sea urchins are masters of defense—some use all

three methods. The sea urchin looks very different from the starfish and feather star pictured on the previous pages, yet all three belong to the same phylum. What characteristics do they have in common? What features determine whether or not an animal is an echinoderm?

What Is an Echinoderm?

Echinoderms have a number of unusual characteristics that easily distinguish them from members of any other phylum. Nowhere else in the animal kingdom will you find creatures that move by means of hundreds of hydraulic, suction cup-tipped appendages or that have skin covered with tiny, jawlike pincers. Echinoderms live only in salt water and are found in all the oceans of the world.

Echinoderms have an internal skeleton

If you were to examine the skin of several different echinoderms, you would find that they all have a hard, spiny, or bumpy endoskeleton covered by a thin epidermis. The long, pointed spines on a sea urchin are obvious. Some starfishes may not appear spiny at first glance, but a close look reveals that their long, tapering arms, called **rays,** are covered with short, rounded spines. The spiny skin of a sea cucumber consists of soft tissue embedded with small, platelike structures that barely resemble spines at all. The endoskeleton of all echinoderms is made primarily of calcium carbonate, the compound that makes up limestone.

Some of the spines found on starfishes and sea urchins have become modified into pincerlike appendages called **pedicellarias.** These jawlike pedicellarias are used for protection.

Echinoderms have radial symmetry

You may remember that radial symmetry is an advantage to animals that are stationary or move slowly. Radial symmetry enables these animals to sense potential food, predators, and other aspects of their environment from all directions. Observe the radial symmetry, as well as the various sizes and shapes of spines, of each echinoderm pictured in *Figure 32.1.*

Echinoderms have a water vascular system

Another characteristic unique to echinoderms is the system that enables them to move, exchange gases, capture food, and excrete wastes. Look at the close-up of the ventral side of a starfish on the next page. From the area of the starfish's mouth run grooves filled with tube feet. **Tube feet** are hollow, thin-walled tubes that

each have a suction cup on the end. Tube feet look somewhat like miniature eyedroppers. The round, muscular structure called the **ampulla,** which is located on the opposite end from the suction cup, corresponds to the bulb of the eyedropper. The ampulla has muscles that contract and relax, similar to the squeezing movement of the eyedropper. Each tube foot works independently of the others, and the animal moves along slowly by alternately pushing out and pulling in its tube feet.

Tube feet function in gas exchange and excretion, as well as in locomotion. Gases are exchanged and wastes are eliminated by diffusion through the thin walls of the tube feet.

pedicellaria:
pediculus (L) little foot
Pedicellarias resemble little feet.

Figure 32.1

All echinoderms have radial symmetry and an endoskeleton composed primarily of calcium carbonate.

▶ A sea lily's stalk and feathery rays are composed of calcified skeletal plates covered with a thin epithelium. The plates give the stalk a jointed appearance.

▶ A brittle star's long, snakelike rays are composed of overlapping, calcified plates covered with a thin layer of skin cells.

◀ A living sand dollar has a solid, immovable skeleton composed of flattened plates that are fused together.

▲ A sea cucumber may not appear to have an endoskeleton at all. Its spines have been reduced to tiny, calcified plates embedded in its fleshy skin.

madreporite:

mater (L) mother
poros (GK) channel
the main channel
through which water
flows into and out of a
starfish

Water for the operation of an echinoderm's tube feet comes from the animal's water vascular system, *Figure 32.2*. The **water vascular system** is a hydraulic, or water pressure, system that regulates locomotion, gas exchange, food capture, and excretion for an echinoderm. Water enters and leaves the water vascular system of a starfish through the **madreporite,** a sievelike, disc-shaped opening in the echinoderm's body. You might think of this disc as the little strainer that fits into the drain in a sink and keeps large particles out of the pipes.

Figure 32.2

Tube feet enable starfishes and other echinoderms to creep along the ocean bottom or to pry open the shells of bivalves. The tube feet attach to the two halves of the shell by suction, and the starfish pulls open the shell just enough to insert its stomach.

▶ The muscular ampulla contracts and relaxes with an action similar to the squeezing of an eyedropper bulb. When the ampulla contracts, it pushes water into the suction-cup portion of the tube foot, causing it to lengthen and stick tightly to the surface it is touching. When the ampulla relaxes, water flows back out of the suction cup, causing it to shorten and release its grip.

▶ The starfish's water vascular system provides the water pressure that operates the animal's tube feet. From the madreporite, water moves into the ring canal, then into the rays through radial canals, and finally to the tube feet. The canals are like a network of water pipes attached to the tube feet. Water also exits the body through the madreporite.

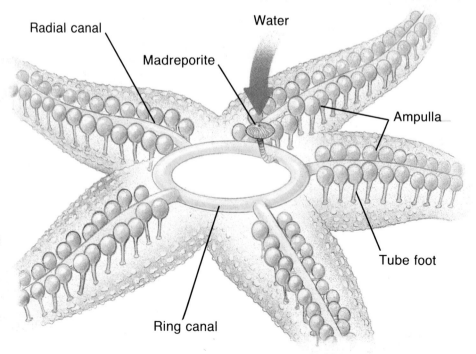

Radial canal

Water

Madreporite

Ampulla

Tube foot

Ring canal

Inside a Starfish

I f you ever tried to pull a starfish from a rock where it is attached, you would be impressed with how unyielding and rigid the animal seems. Yet at other times, the animal shows great flexibility, such as when it rights itself after being turned upside down.

Starfish

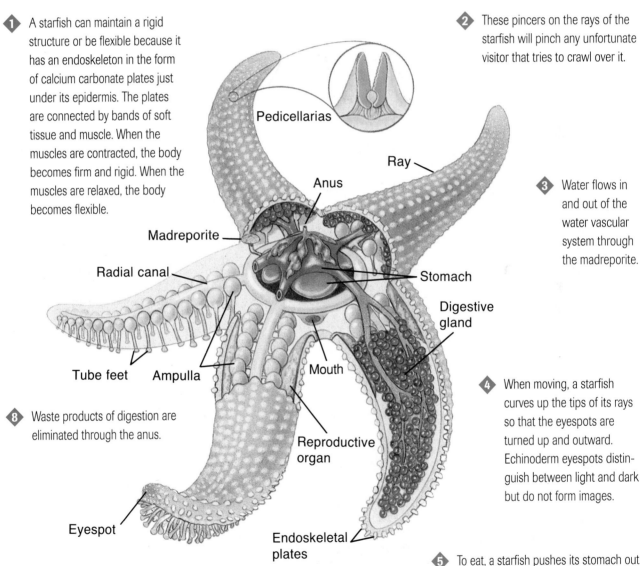

1 A starfish can maintain a rigid structure or be flexible because it has an endoskeleton in the form of calcium carbonate plates just under its epidermis. The plates are connected by bands of soft tissue and muscle. When the muscles are contracted, the body becomes firm and rigid. When the muscles are relaxed, the body becomes flexible.

2 These pincers on the rays of the starfish will pinch any unfortunate visitor that tries to crawl over it.

3 Water flows in and out of the water vascular system through the madreporite.

4 When moving, a starfish curves up the tips of its rays so that the eyespots are turned up and outward. Echinoderm eyespots distinguish between light and dark but do not form images.

5 To eat, a starfish pushes its stomach out of its mouth and spreads the stomach over the food. Powerful enzymes secreted by the stomach turn solid food into a soupy liquid that the animal can easily absorb. Then the stomach is pulled back into the starfish's body.

6 The digestive gland gives off chemicals for digestion.

7 The suction action of tube feet, caused by the contraction and relaxation of the ampulla, is so strong that the starfish's muscles can open a clam or oyster shell.

8 Waste products of digestion are eliminated through the anus.

Pedicellarias

Ray

Anus

Madreporite

Radial canal

Stomach

Digestive gland

Tube feet   Ampulla

Mouth

Reproductive organ

Eyespot

Endoskeletal plates

MiniLab

How do starfishes open mollusk shells?

If you have ever pried open the shell of an oyster or clam, you know it's not an easy job. Starfishes feed on clams by wrapping their rays around the mollusk and using thousands of tube feet to apply suction to the shell. Then, starfish muscles exert force to pry open the tightly closed shell just far enough so that the starfish stomach can be inserted into the clam. The clamshell is held together by powerful muscles that must be overcome by the starfish. In this lab, you will see how starfish muscles overcome the force exerted by mollusk muscles.

Procedure

1. Hold your arm straight out, palm up. Your arm muscle represents the clam's muscles.

2. Place a heavy book on your hand. The book represents the force applied by the starfish.

3. Have a partner time how long you can hold your arm up with the book on it.

Analysis

1. Explain how this method of getting food works for the starfish.

2. Why do you think the mollusk shell opens wider as digestion progresses?

3. Which might take longer for a starfish to open, a small clam or a large clam? Why?

Echinoderm larvae have bilateral symmetry

Tube feet, rays, and pedicellarias may make echinoderms seem completely unrelated to chordates, the animal phylum you will begin to study later in this chapter. However, if you examine the larval stages of echinoderms, you will find that they, like chordates, have bilateral symmetry. The ciliated larva that develops from the fertilized egg of an echinoderm is shown in *Figure 32.3*.

Echinoderm classes have varied nutrition

All echinoderms have a mouth, stomach, and intestines, but their methods of food getting vary. Starfishes are carnivorous and prey on worms or on mollusks such as clams. Most sea urchins are herbivores and graze on algae. Brittle stars, sea lilies, and sea cucumbers feed on dead and decaying matter called **detritus** that drifts down to the ocean floor. Sea lilies capture detritus with their tentacle-like tube feet and move it to the mouth.

Echinoderms have a simple nervous system

Echinoderms have no head or brain, but they do have a nerve net and nerve ring. Most echinoderms have cells that detect light and touch, but do not have sensory organs. Starfishes are an exception. A starfish's body consists of long, tapering rays that extend from the animal's central disc. At the tip of each ray, on the ventral surface, is an eyespot, a sensory organ consisting of a cluster of light-detecting cells. When walking, starfishes curve up the tips of their rays so that the eyespots are turned up and outward. This enables a starfish to detect the intensity of light coming from every direction.

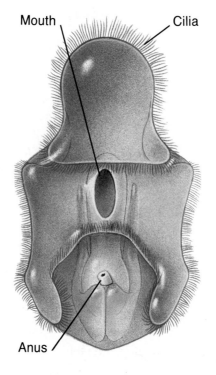

Mouth Cilia

Anus

Figure 32.3

This sea-urchin larva is only 1 mm in size. The larval stage of echinoderms is bilateral, even though the adult is radial. Through metamorphosis, the free-swimming larvae make dramatic changes both in body parts and in symmetry. The bilateral symmetry of echinoderm larvae indicates that echinoderm ancestors also may have had bilateral symmetry, suggesting a close relationship to the chordates.

Phylogeny of Echinoderms

The origin of echinoderms is subject to much speculation. Many biologists believe that the earliest echinoderms were members of an extinct class of organisms that had bilateral symmetry in their adult forms. Like modern sea lilies, members of this extinct class were sessile and lived attached to the ocean floor by stalks. Other biologists believe that modern echinoderms evolved from extinct species that were bilateral and free-swimming. Today, all adult echinoderms have radial symmetry, but nearly all their larvae have bilateral symmetry. The development of bilateral larvae is one piece of strong evidence biologists have for placing echinoderms in the evolutionary record as the closest invertebrate relatives of the chordates.

Perhaps the strongest evidence for placing the echinoderms close to the chordates is the fact that echinoderms have deuterostome development. You read in Chapter 28 that most invertebrates show protostome development, whereas deuterostome development is seen mainly in chordates. The echinoderms represent the only major group of deuterostome invertebrates. This type of development, as well as bilateral larvae and an internal skeleton, is the most important evidence for the phylogenetic relationship between echinoderms and chordates.

Echinoderms, as a group, date from the Paleozoic Era, *Figure 32.4.* Because the endoskeletons of echinoderms were easily fossilized, there is a good record of this phylum. More than 13 000 fossil species have been identified.

Figure 32.4

Sea cucumbers, sea urchins, and starfishes have all been found as fossils from the early Paleozoic Era. Fossils of brittle stars, on the other hand, are found beginning at a later period. From which echinoderm group are the brittle stars most likely to have evolved?

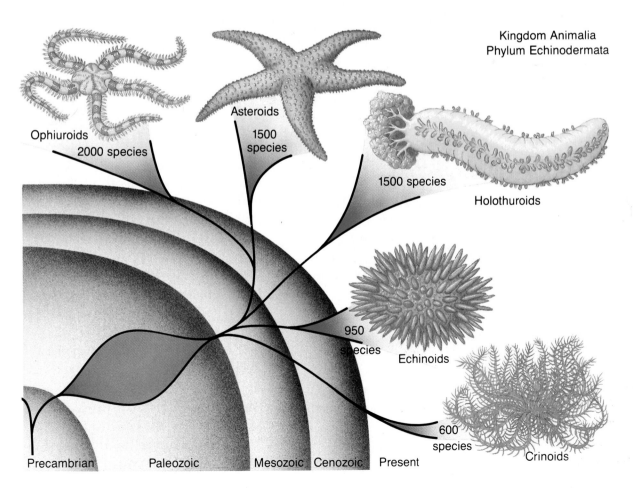

Kingdom Animalia
Phylum Echinodermata

Ophiuroids
2000 species

Asteroids
1500 species

1500 species

Holothuroids

950 species

Echinoids

600 species

Crinoids

Precambrian Paleozoic Mesozoic Cenozoic Present

Regeneration in Echinoderms

Many members of the phylum Echinodermata are well-known for their ability to regenerate body parts. In fact, the regenerative powers of some echinoderms are so dramatic they are hard to believe.

Bodies regenerating parts   Despite their delicate appearance, feather stars can regenerate rays to replace missing ones. Brittle stars can spontaneously cast off rays that are seized by predators—or handled by careless scuba divers! In fact, it is this characteristic brittleness that gives brittle stars their name. As long as a brittle star's central disc is not damaged, any rays it loses will eventually be regrown.

A gut reaction   Many kinds of sea cucumbers eviscerate when harassed. They explosively expel some of their internal organs into the water around them. In some species, the tentacles, pharynx, and parts of the intestine come shooting out of the animal's anterior end through a rupture in the body wall. In other species, parts of the respiratory, digestive, and reproductive tracts erupt from the posterior end of the body. In either case, eviscerated sea cucumbers are able to survive the loss of these vital organs for as long as it takes to regenerate new ones. Like a lizard letting go of its tail, evisceration may be a way for some sea cucumbers to use replaceable body parts to distract a predator long enough to make an escape.

A part regenerating a body   Of all the echinoderms, starfishes may have the most remarkable powers of regeneration. One or even several missing rays are readily regenerated as long as part of the central disc is present. The award for "most dramatic regeneration" must go to starfishes that belong to the genus *Linckia*. A single, cast-off ray from one of these asteroids can regenerate a complete starfish. From the end where the ray was detached, a new central disc and four new rays will grow.

Thinking Critically

Being able to regenerate lost body parts has great survival value. Feather stars, brittle stars, sea cucumbers, and starfishes live in constant danger from predators. What connection might there be between these echinoderms' inability to move fast and their ability to regenerate lost body parts?

The Diversity of Echinoderms

Approximately 6000 species of echinoderms exist today. More than one-third of these species are in the class Asteroidea, to which the starfishes belong. The four other classes of living echinoderms are Ophiuroidea, the brittle stars; Echinoidea, the sea urchins and sand dollars; Holothuroidea, the sea cucumbers; and Crinoidea, the sea lilies and feather stars.

Starfishes

Most starfishes have five rays, but some species may have as many as 40 or more. The rays are tapered and come out gradually from the central disc. You have been reading about the characteristics of starfishes as a typical example of an echinoderm.

Brittle stars

As their name implies, brittle stars are extremely fragile. If you try to pick up a brittle star, parts of its rays will break off in your hand. This is an adaptation that helps the brittle star survive an attack by a predator. While the predator is busy with the broken-off ray, the brittle star can escape. A new ray will regenerate within weeks. **Regeneration** is the replacement or regrowth of missing parts; it is a common feature of echinoderms.

Brittle stars do not use their tube feet for locomotion. Instead, they use the snakelike, slithering motion of their flexible rays to propel them. The tube feet are used to pass particles of food along the rays and into the mouth in the central disc.

Sea urchins and sand dollars

Sea urchins and sand dollars are globe- or disc-shaped animals covered with spines, as *Figure 32.5* shows. They do not have rays. A living sand dollar looks very different from the dead, dried-out specimens you may have seen in seashell collections or washed up on the beach. Their circular, flat skeletons look like clay sculptures with a five-petaled flower pattern on the surface. When alive, the sand dollar is covered with minute, hairlike spines that are lost when the animal dies. The living sand dollar has tube feet that protrude from the petal-like markings on its dorsal surface. These tube feet are modified to act as gills. Tube feet on the animal's ventral surface aid in bringing food particles to the mouth.

Although sand dollars live on the sandy ocean bottom, sea urchins inhabit primarily rocky areas. They look like living pincushions, bristling with long, usually pointed spines. They have long, slender tube feet that, along with the spines, aid the animal in locomotion.

Figure 32.5

Echinoderms have adapted to life in a variety of habitats.

▼ **Sand dollars burrow into the sandy ocean bottom. They feed on tiny organic particles found in the sand.**

ThinkingLab — Design an Experiment

What makes sea cucumbers release gametes?

The orange sea cucumber lives in groups of 100 or more per square meter. In the spring, these sea cucumbers produce large numbers of gametes, which they shed in the water all at the same time. The adaptive value of such behavior is that fertilization of many eggs is assured. When one male releases sperm, the other sea cucumbers in the population, both male and female, also release their gametes. Biologists do not know whether the sea cucumbers release their gametes in response to a seasonal cue, such as increasing day length or increasing water temperature, or whether they do this in response to the release of sperm by one sea cucumber.

Analysis

Design an experiment that will help to determine if sea cucumbers release eggs and sperm in response to the release of sperm from one individual or in response to a seasonal cue.

Thinking Critically

If you find that female sea cucumbers release 200 eggs in the presence of male sperm and ten eggs in the presence of warmer water, what would you do in your next experiment?

◀ **Sea urchins often burrow into rock to protect themselves from predators and rough water. Most urchins browse on algae.**

▼ **Basket stars, a kind of brittle star, live on the soft substrate found below deep ocean waters. They feed by using their tube feet to pass particles of detritus to the mouth.**

Figure 32.6

Sea cucumbers (left) trap detritus particles by sweeping their mucous-covered tentacles over the ocean bottom. Sea lilies and feather stars (right) use their feathery rays to capture downward-drifting organic particles.

The sea urchin's spines protect it from predators. In some species, sacs located near the tips of the spines contain a poisonous fluid that is injected into an attacker, further protecting the urchin. The spines also aid in locomotion and in burrowing. Burrowing species move their spines in a circular motion that grinds away the rock beneath them. This action, which is aided by a chewing action of the mouth, forms a depression in the rock that helps protect the urchin from predators and from wave action that could wash it out to sea.

Sea cucumbers

Sea cucumbers are so called because of their vegetable-like appearance, *Figure 32.6*. Their leathery covering allows them to be more flexible than other echinoderms; they pull themselves along the ocean floor using tentacles and tube feet. When sea cucumbers are threatened, they show a curious behavior. They may expel a tangled, sticky mass of tubes through the anus, or they may rupture, releasing some internal organs that are regenerated in a few weeks.

Sea lilies and feather stars

Sea lilies and feather stars resemble plants in some ways. Sea lilies are the only sessile echinoderms. Feather stars are sessile only in larval form. The adult feather star uses its feathery arms to swim from place to place.

Section Review

Understanding Concepts

1. How does a starfish move?
2. Describe the differences in symmetry between larval echinoderms and adult echinoderms.
3. How are sea cucumbers different from other echinoderms?

Thinking Critically

4. How does the variety of defense mechanisms among the echinoderm classes help them to deter predators?

Skill Review

5. **Classifying** Prepare a key that distinguishes between classes of echinoderms. Include information on presence and shape of rays, presence of spines, body shape, and other features you may find significant. For more help, refer to Organizing Information in the *Skill Handbook*.

The brightly colored object pictured here is a sea squirt. As one of your closest invertebrate relatives, it is placed, along with humans, in the phylum Chordata. At first glance, this sea squirt may seem to resemble a sponge more than its fellow chordates. It is sessile, and it filters food particles from water it takes in through the large opening in the top of its body. What characteristics could a human—or a fish or a lizard, for that matter—possibly share with this squishy, colorful, ocean-dwelling creature?

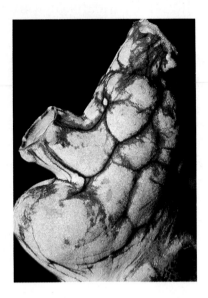

Section Preview

Objectives
Summarize the characteristics of chordates, and show how invertebrate chordates are related to vertebrates.
Distinguish between sea squirts and lancelets.

Key Terms
notochord
dorsal nerve cord
gill slit

What Are Invertebrate Chordates?

The chordates most familiar to humans are the vertebrate chordates —chordates that have backbones. However, a few invertebrate chordates have no backbones. What characteristics do these invertebrate animals share with vertebrates? Several features diagrammed in *Figure 32.7* are shared by all chordates at some time during their development.

All chordates have notochords

The term *chordata* refers to the long, semirigid, rodlike structure called a **notochord,** which is common to all members of the phylum Chordata. During the embryologic development of vertebrate chordates, this structure becomes the backbone. Invertebrate chordates have a notochord but do not develop a backbone.

The physical support of a notochord enables invertebrate chordates to make powerful side-to-side movements of the body. These movements propel the animal through the water at a much faster speed than would be possible if the body had no support.

Chordates have a dorsal nerve cord

The **dorsal nerve cord** is a bundle of nerves housed in a fluid-filled canal that lies above the notochord. In most adult chordates, the posterior portion of the dorsal nerve cord develops into the spinal cord. The anterior portion develops into the brain. A pair of nerves goes from the cord to each block of muscles.

Figure 32.7

Chordate innovations—the notochord, dorsal nerve cord, gill slits, and muscle blocks—had a dramatic effect on the evolution of animals. In addition, all chordates have bilateral symmetry, a well-developed coelom, and segmentation.

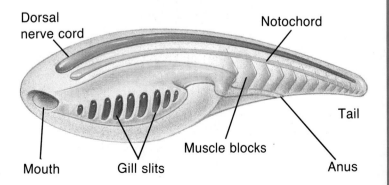

Dorsal nerve cord
Notochord
Tail
Mouth
Gill slits
Muscle blocks
Anus

BioLab

Sea urchins are echinoderms that move slowly on the bottom of the sea by means of spines and tube feet. They have separate sexes, which shed their eggs and sperm into the water, where fertilization takes place. The fertilized eggs become free-swimming larvae with bands of cilia extending onto their long, graceful arms. The bilaterally symmetrical, free-swimming larvae of echinoderms are one indication of their evolutionary relationship to chordates.

Comparing Sea Urchins and Lancelets

Lancelets belong to the phylum Chordata. They are scaleless, fishlike animals that spend most of their time partly buried in the sand with only their mouths protruding.

PREPARATION

Problem
How do sea urchin larvae and lancelet adults compare?

Objectives
In this Biolab, you will:
- **Analyze** the stages of sea urchin development.
- **Identify** the features of chordates.
- **Compare** a sea urchin larva with a lancelet.

Sea urchin embryo

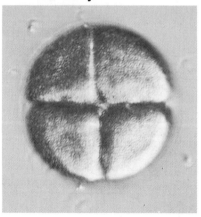

Materials
compound microscope
prepared slides of sea urchin development
prepared slide of *Amphioxus*

Safety Precautions
Use care when handling the microscope.

Sea urchin larvae

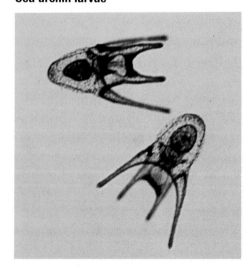

PROCEDURE

Part A: Sea Urchin Development

1. Obtain prepared slides showing sea urchin eggs at different stages of development.

2. Prepare a data sheet consisting of eight circles, each about 5 cm in diameter.

3. Observe slides of the unfertilized egg, zygote, 2-cell stage, 4-cell stage, 8-cell stage, blastula, gastrula, and sea urchin larva.

4. Examine the slides under low power. When you have one stage in focus, switch to high power.

5. Draw a diagram of each stage. Label each diagram with the stage name. Examine the stages in the order listed above, and draw your diagrams in the same order.

6. Note how the size of the blastula compares with the size of the fertilized egg. Note also the size of the cells at each stage.

Part B: Lancelet Structure

1. Examine a prepared slide of a lancelet, *Amphioxus.*

2. Find the following structures: dorsal nerve cord, notochord, gill slits, muscle blocks, and tail.

3. Draw a diagram of the lancelet on a second data sheet. Label your diagram with the name of the animal and the parts listed in step 2.

4. Note the fishlike form of the lancelet.

Lancelet

ANALYZE AND CONCLUDE

1. **Comparing and Contrasting** How does the size of the sea urchin blastula compare with that of the fertilized egg? How does the size of the cells change as development proceeds?

2. **Comparing and Contrasting** At what stage do the cells look different from one another?

3. **Comparing and Contrasting** Look at a photo of an adult sea urchin in this chapter, and compare the symmetry of the larva with the symmetry of the adult. How do they compare?

4. **Making Inferences** What does the symmetry of the larva imply about the evolutionary relationship between echinoderms and chordates?

5. **Interpreting Observations** What is the function of each part you have identified in the lancelet? What is the importance of these features in placing the lancelet in the phylum Chordata?

Going Further

Application
Examine prepared slides of protostome development in *Ascaris,* and compare it with deuterostome development in sea urchins. What differences do you notice?

What does a notochord look like?

The notochord is a flexible rod that forms on the dorsal side of the early embryo of all chordates. The notochord provides support for the animal. In lancelets, the notochord is present throughout the life of the animal. In adult vertebrates, the notochord is replaced by a backbone.

Procedure

1. Examine a prepared slide of a cross section of *Amphioxus,* and observe the position of the notochord.
2. Locate the dorsal nerve cord above the notochord and compare it with the notochord.

Analysis

1. How can you tell the difference between the notochord and the dorsal nerve cord?
2. What is the function of the notochord?
3. What is the significance of the notochord in terms of evolution of chordates?

All chordates have gill slits

The **gill slits,** or gill pouches, of a chordate are paired openings located in the pharynx, behind the mouth. Many chordates have several pairs of gill slits only during embryonic development. Invertebrate chordates that have gill slits as adults use these structures to strain food from the water. In some vertebrates, especially the fishes, the gill slits develop into internal gills and become modified for gas exchange.

All chordates have muscle blocks

Muscle blocks are modified body segments that consist of stacked muscle layers. You can see these muscle blocks in the easily separated flakes of meat of a cooked fish. Muscle blocks are anchored by the notochord, which gives the muscles a firm structure to pull against. As a result, chordates tend to be more muscular than members of other phyla. The muscle blocks aid in movement of the tail. At some time during their lives, all chordates have a muscular tail. In humans, the tail appears only in the developing embryo.

Sea Squirts and Lancelets

Sea squirts, also called tunicates, are members of the subphylum Urochordata. The larval stage, as shown in *Figure 32.8,* has a tail that makes it look similar to a tadpole.

Adult sea squirts retain only gill slits as indicators of their chordate relationship. These small, tubular animals range in size from microscopic to several centimeters long. Most adult sea squirts live attached to objects on the seafloor. If you remove one from its sea home, it might squirt out a jet of water—hence the name *sea squirt.*

Figure 32.8

As larvae, sea squirts possess all the features common to chordates but have only gill slits and muscle blocks as adults. Sea squirt larvae (left) are about 1 cm long and swim freely through the water. As adults (right), sea squirts become stationary filter feeders enclosed in a tough, baglike layer of tissue called a tunic.

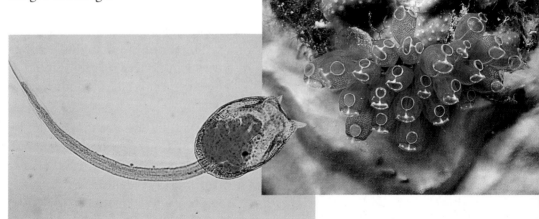

Inside a Sea Squirt

Sea squirts, or tunicates, are a group of about 1250 species that live in the ocean. They may live near the shore or at great depths. They may live individually, or several animals may share a tunic to form a colony.

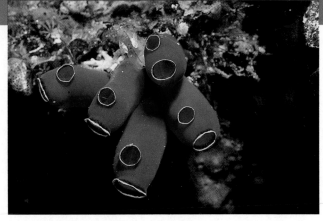

Sea squirt colony

2 Water comes into the animal through the mouth, or incurrent siphon.

1 Water leaves the body of the animal through the excurrent siphon. When a sea squirt is disturbed, it may forcefully spout water from its mouth and excurrent siphon simultaneously.

3 During filter feeding, food is trapped by mucus secreted in a ciliated groove. The food and mucus are digested in the intestine. Some free-living tunicates build a house of mucus around their bodies. When large amounts of food are available, these animals build up in large numbers. Scuba divers say that diving through these areas is like swimming through a snowstorm with walnut-sized snow.

6 The pharynx is lined with gill slits and cilia. The beating of the cilia causes a current of water to move through the animal. Food is filtered out, and dissolved oxygen is removed from the water. You may remember this form of filter feeding in a clam.

4 The heart of the tunicate is unusual because it pumps blood in one direction for several minutes and then reverses direction. This reversal of blood flow occurs only in tunicates.

Water
Mouth
Ciliated groove
Excurrent siphon
Gill slits
Anus
Pharynx
Heart
Tunic
Stomach
Intestine
Reproductive organs

5 Sea squirts are covered with a layer of tissue called a tunic. Some tunics are thick and tough, and others are thin and translucent. All serve to protect the animal from predators.

Figure 32.9

Lancelets, such as this *Amphioxus,* are shaped like fishes and are capable of swimming freely. But they usually spend most of their time buried in the sand with only their heads sticking out so they can filter tiny morsels of food from the water. Why is filter feeding such a successful adaptation for aquatic animals?

Lancelets belong to the subphylum Cephalochordata. They are small, streamlined, marine animals, usually about 5 cm long, as *Figure 32.9* shows. They spend most of their time buried in the sand with only their heads sticking out. Like tunicates, lancelets are filter feeders.

Unlike tunicates, however, lancelets retain all their chordate features throughout life.

Because sea squirts and lancelets have no bones, shells, or other hard parts, their fossil record is incomplete. Biologists are not sure where sea squirts and lancelets fit in the phylogeny of chordates. According to one hypothesis, echinoderms, invertebrate chordates, and vertebrates all arose from an ancestral sessile animal that fed by capturing food in its tentacles. This hypothesis proposes that the ancestral organism may have evolved into a sessile filter feeder that eventually gave rise to the lancelets and tunicates. Perhaps modern vertebrates arose from the free-swimming larval stages of ancestral invertebrate chordates.

Connecting Ideas

Studying the animal phyla in order, from simplest to most complex, gives a feeling for the progress of events during evolution. It almost seems as though evolution follows the pattern of a dramatic story line as it moves from sponges and flatworms, through mollusks and segmented worms, and on to arthropods. But evolution doesn't always follow a straight path. The echinoderms add a side plot that goes off in a different direction. You've learned that we must study the larvae of both echinoderms and invertebrate chordates to continue following evolution's story. The structural adaptations of the invertebrate chordates led to the evolution of all vertebrates.

Section Review

Understanding Concepts
1. Describe the four features common to all chordates.
2. How are invertebrate chordates different from vertebrates?
3. Compare the physical features of sea squirts and lancelets.

Thinking Critically
4. What features of invertebrate chordates suggest that you are more closely related to invertebrate chordates than to echinoderms?

Skill Review
5. **Designing an Experiment** Assume that you have found some tadpolelike animals in the water near the seashore and that you can raise them in a laboratory. Design an experiment in which you will determine whether the animals are larvae or adults. For more help, refer to Practicing Scientific Methods in the *Skill Handbook.*

Too Many Starfishes?

Scientists Monitor Starfish Populations

QUEENSLAND COURIER   JANUARY 12, 1994

Teams of scientists are spending weeks at sea this summer counting crown-of-thorns starfishes on reefs along the Great Barrier Reef. This starfish survey is an attempt to determine whether another outbreak of this coral-eating echinoderm is likely to occur in the near future.

In the early 1960s and again in the late 1970s, huge numbers of crown-of-thorns starfishes appeared on coral reefs along part of Australia's Great Barrier Reef. Millions of these starfishes advanced across the reefs, consuming every coral polyp in their path. Eventually, the armies of starfishes dwindled and disappeared.

An elusive explanation Coral reefs do eventually recover from crown-of-thorns outbreaks, but the causes of outbreaks remain a mystery. Are these periodic population explosions a natural event? Or are they the result of human activities on or near coral reefs?

Different Viewpoints

Many coral-reef biologists think that crown-of-thorns starfish outbreaks are natural events. Each female crown-of-thorns starfish produces 40 to 60 million eggs during every spawning season. The vast majority of larvae that develop from fertilized eggs are eaten by predators. But because there are so many larvae initially, a slight increase in their survival rate could cause a huge increase in the number of adult starfishes.

When crown-of-thorns starfishes die, tiny pieces of their skeletons become buried in reef sediments. Sediment cores that are obtained by drilling deep into reefs and then analyzed for the presence of these skeletal remains suggest that crown-of-thorns starfishes may have undergone natural periodic population explosions on the Great Barrier Reef for hundreds of years.

Unnatural events Some biologists think crown-of-thorns outbreaks may be caused by human activities. Outbreaks may occur because shell collectors have nearly wiped out one of the starfish's few natural enemies—the giant triton. Pollution may kill off predators that eat crown-of-thorns larvae, or dredging and heavy fishing may disrupt the reef ecosystem. Perhaps sewage discharged into the sea or fertilizer washed off from agricultural lands adds nutrients to the water. Artificially high nutrient levels cause an increase in phytoplankton, which, in turn, provides food to support the survival of higher-than-normal numbers of crown-of-thorns larvae.

INVESTIGATING the Issue

1. **Research** Find out why crown-of-thorns starfishes are so much more efficient at eating corals than are other types of starfishes.
2. **Analyzing Consequences** If crown-of-thorns outbreaks are eventually shown to be natural events, what, if any, actions should be taken?

Reviewing Main Ideas

32.1 Echinoderms

- Echinoderms have spines or bumps on their endoskeletons, radial symmetry, and water vascular systems. Most move by means of the suction action of tube feet.
- Deuterostome development, an internal skeleton, and bilaterally symmetrical larvae are indicators of the close phylogenetic relationship between echinoderms and chordates.
- Echinoderms include starfishes, sea urchins, sand dollars, sea cucumbers, sea lilies, and feather stars.

32.2 Invertebrate Chordates

- Chordates have a dorsal nerve cord, notochord, gill slits, and a tail composed of muscle blocks at some stage during development.
- Sea squirts and lancelets are invertebrate chordates.

Key Terms

Write a sentence that shows your understanding of each of the following terms.

ampulla	pedicellaria
detritus	ray
dorsal nerve cord	regeneration
gill slit	tube feet
madreporite	water vascular
notochord	system

Understanding Concepts

1. How do starfishes feed?
2. How is the larva different from the adult starfish, and how is this difference important in terms of evolution?
3. What is the advantage of a spiny outer covering in echinoderms?
4. Why are fossils of echinoderms more common than those of worms?
5. How are lancelets and sea squirts protected from predators?
6. What is the relationship of sea squirts and lancelets to vertebrates?
7. Describe how gas exchange and excretion take place in starfishes.

Using Scientific Illustrations

8. Examine *Figure 32.4* on page 793, and suggest which group of echinoderms is the oldest.

Relating Concepts

9. Make a concept map that relates the following terms and phrases. Supply the appropriate linking words for your map.

   echinoderm, radial symmetry, tube feet, water vascular system, spines, starfish, madreporite, larva, sea urchin, sand dollar

Applying Concepts

10. Why can't a starfish move as rapidly and smoothly as an earthworm?
11. If you were an oyster farmer, why would you be advised not to break apart and throw back any starfishes that were destroying the oyster beds?
12. How are the invertebrate chordates adapted to their way of life?
13. Relate the various functions of the water vascular system to the environment in which echinoderms live.

A Broader View

14. How might the ability of echinoderms to regenerate be of use to medical scientists?

Thinking Critically

Interpreting Scientific Diagrams

15. **Biolab** Examine the following diagrams, and tell which is a cross section of a lancelet. The intestines are shown in red and the nerve cords are blue.

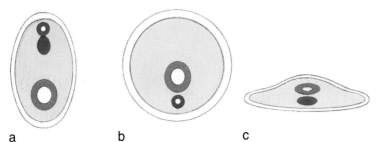

a   b   c

Comparing and Contrasting

16. Compare the pedicillarias of echinoderms with the nematocysts in cnidarians.

Analyzing

17. Suggest how the unusual defense mechanism of the sea cucumber may have evolved.

Making Inferences

18. **Thinking Lab** Tiny crustaceans eat the eggs of sea cucumbers that are poisonous to most other invertebrates. What mechanism may have evolved that prevents the sea cucumber from becoming extinct due to predation?

Making Inferences

19. **Minilab** If over the course of thousands of years, starfishes evolved tube feet and muscles that could exert even more pressure on clams, what would you expect might happen to clams over the same time period?

Connecting to Themes

20. **Systems and Interactions** In what ways are echinoderms important to marine ecosystems?
21. **Homeostasis** Explain how a sea squirt maintains homeostasis.

Every summer, the few visitors allowed to view the annual congregation of grizzly bears at McNeil River Falls, Alaska, are treated to a thrilling sight. At this unique sanctuary, a single sweep of the eyes may reveal up to sixty Alaskan grizzly bears playing and feeding on the river's migrating salmon. Anxious gulls wait for scraps of discarded fish.

Bears, salmon, gulls—and humans—belong to the group of animals known as vertebrates. Their ability to survive the rigors of changing environments such as the Alaskan tundra, and engage in complex behaviors such as migration, are just a few of the characteristics that distinguish vertebrates from other members of the animal kingdom.

After spending years at sea, salmon are able to find their way back to the streams in which they were hatched in order to spawn. What are some of the amazing sensory abilities of vertebrates that allow them to achieve such feats?

Mammals such as bears inhabit a wide range of environments. What characteristics of mammals allow them to survive in such diverse climates?

Physical obstacles such as waterfalls, rocks, and hungry bears present few problems for seagulls. What adaptations make bird flight possible?

Unit Contents

807

33 Fishes and Amphibians

On a bright spring day in Idaho, a female chinook salmon brushes the clean river gravel into a depression nest and deposits about 6000 eggs. Soon a male salmon swims past, depositing his sperm. Eighteen months later, the young salmon begin their 4800-km migration downstream to the Pacific Ocean. As they migrate, the salmon store memories of the chemical odors in the stream water so that they may return two or three years later to the place where they hatched on that bright spring day. There they will begin the cycle again.

Like salmon, amphibians also undergo a series of structural changes during development that will enable them to survive in a different environment as adults. When leopard frog eggs hatch, they are wiggling tadpoles with bulging heads and squirming tails. A few months later, they become the four-legged, jumping, croaking, swamp-roving animals that have inspired cartoonists and writers of folktales throughout history.

Many species of fishes lay great numbers of eggs. This ocean sunfish is said to lay more than 300 million eggs. Why do fishes lay so many eggs?

Tadpoles go through many changes on the way to becoming frogs. What happens to the tadpole's tail as it becomes an air-breathing animal? The development of frogs from tadpoles provides clues about the origin of land vertebrates.

*H*ave you ever visited an aquarium to see the amazing diversity of fishes? As you pass tank after tank, you can see fishes of all shapes, sizes, and colors. What's interesting is that even though fishes share a common environment, they have evolved a variety of different adaptations. But although fishes may show considerable variety in structure and behavior, they all share common characteristics.

What Is a Fish?

Fishes, like all vertebrates, are classified in the phylum Chordata. This phylum includes three subphyla: Urochordata, the sea squirts; Cephalochordata, the lancelets; and Vertebrata, the vertebrates. In addition to fishes, the subphylum Vertebrata includes amphibians, reptiles, birds, and mammals. You remember from Chapter 32 that all chordates have three traits in common—a notochord, gill slits, and a dorsal nerve cord. In vertebrates, the notochord of the embryo is replaced by a backbone. All vertebrates are bilaterally symmetrical, coelomate animals that have endoskeletons, closed circulatory systems, nervous systems with complex brains and sense organs, and efficient respiratory systems.

Three classes of fishes

Fishes comprise three classes of the subphylum Vertebrata: Class Agnatha, the lampreys and hagfishes; Class Chondrichthyes, the sharks and rays; and Class Osteichthyes, the bony fishes. Far more variety can be found among the classes of fishes than in any other vertebrate group.

The diversity of fishes

Fishes inhabit nearly every type of aquatic environment on Earth. They are adapted to living in shallow, warm water and deeper cold and sunless water. They are found in fresh and salt water, and some fishes can survive heavily polluted water.

Fishes range in size from the tiny dwarf goby that is less than 1 cm long, to the huge whale shark that can reach a length of 15 m—the length of three school buses.

More than 30 000 species of fishes exist. In fact, there are more fish species than all other kinds of vertebrates added together!

Agnathans Are Jawless Fishes

Lampreys and hagfishes, shown in *Figure 33.1,* belong to the class Agnatha. Though they do not have jaws, they are voracious feeders. Hagfishes, for example, have a slit-like, toothed mouth and feed on dead or dying fish by drilling a hole and sucking the blood and insides from the animal. Parasitic lampreys attack other fish and attach themselves by their suckerlike mouths. They use their sharp teeth to scrape away the flesh and then suck out the prey's blood. The skeletons of agnathans, as well as sharks and their relatives, are made of a tough, flexible material called **cartilage.**

Agnathans breathe using gills

Agnathans, like all fishes, have gills made up of feathery gill filaments that contain tiny blood vessels. Gills are an important adaptation for fishes and other vertebrates that live in water. As a fish takes water in through its mouth, water passes over the gills and then out through slits at the side of the fish. Oxygen and carbon dioxide are exchanged through the capillaries in the gill filaments.

Agnathans reproduce sexually

Like most fishes, lampreys and hagfishes have separate sexes and reproduce sexually. Fertilization is external in most fishes, with eggs and sperm deposited in protected areas such as on floating aquatic plants or shallow nests of gravel on stream bottoms. Although most fishes produce large numbers of eggs at one time, hagfish produce small numbers of large eggs.

Two-chambered hearts

Agnathans have two-chambered hearts, *Figure 33.2,* like all fishes. One chamber receives deoxygenated blood from the body tissues, whereas the second chamber pumps blood directly to the capillaries of the gills, where oxygen is picked up and carbon dioxide released. Oxygenated

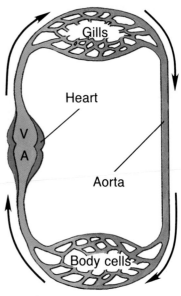

Circulation in a Fish

Figure 33.2

In a fish's heart, deoxygenated blood flows from the first chamber to the second chamber, then on to the gills where it picks up oxygen. Blood in a fish flows in a one-way circuit throughout the body.

Figure 33.1

Hagfishes (left) and their relatives, the lampreys (right), have long, tubular bodies without paired fins, and no scales. When touched, a hagfish's skin gives off a tremendous amount of mucus, thus allowing the fish to slither away without becoming a meal.

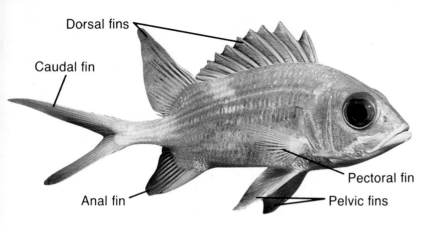

Dorsal fins

Caudal fin

Anal fin

Pectoral fin

Pelvic fins

Figure 33.3

The paired fins of a fish include the pectoral fins and the pelvic fins. Fins found on the dorsal and ventral surfaces include the dorsal fin and anal fin. A fish may have all of these fins or just a few. Most fishes use their body fins like the rudders and stabilizers of boats.

Figure 33.4

You can see how jaws evolved from the cartilaginous gill arches of early jawless fishes in this series of illustrations. Teeth evolved from skin.

blood is carried from the gills to body tissues. Blood flow through the body of a fish is relatively slow because most of the heart's pumping action is used to push blood through the gills.

Sharks and Rays Are Cartilaginous Fishes

Sharks, skates, and rays, like agnathans, possess skeletons composed entirely of cartilage. Sharks, skates, and rays belong to the class Chondrichthyes. Because living sharks, skates, and rays are classified in the same genera as species that swam the seas more than 100 000 years ago, they are considered living fossils.

Sharks and rays have paired fins

Fishes in the class Chondrichthyes have paired fins. **Fins** are fan-shaped membranes, supported by stiff spines

called rays, that are used for balance, swimming, and steering. Fins are attached to and supported by the endoskeleton and are important in locomotion. The paired fins of fishes, *Figure 33.3*, foreshadowed the development of limbs for movement on land and wings for flying.

Jaws evolved in fish

Perhaps one of the most important events in vertebrate evolution was the evolution of jaws in primitive fishes. The advantage of a jaw is that it enables an animal to grasp and crush its prey with great force. Sharks are able to eat large chunks of food. This, among other factors, explains why some early fishes were able to reach such great size. *Figure 33.4* shows the evolution of jaws in fishes.

When you think of a shark, do you imagine gaping jaws and rows of razor-sharp teeth? Sharks have six to 20 rows of teeth that are continually replaced. The teeth point backwards, preventing prey from escaping once caught. Sharks are among the most streamlined of all fishes and are well adapted for life as predators.

Sharks and rays have developed sensory systems

Success as a predator is also the result of the shark's highly developed sensory system. Like salmon, a shark has an extremely sensitive sense of smell and can detect small amounts

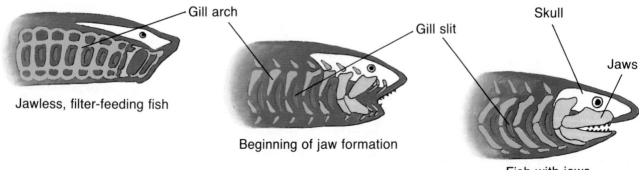

Gill arch

Jawless, filter-feeding fish

Gill slit

Beginning of jaw formation

Skull

Jaws

Fish with jaws

Figure 33.5

Fishes can be classified by the type of scales present in the skin.

▼ Diamond-shaped scales are common to primitive bony fishes, such as a gar, that evolved early in geologic history about 200 million years ago.

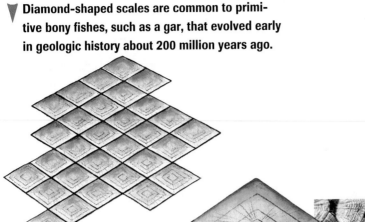

▲ Tooth-shaped scales are characteristic of the shark group.

► Bony fishes that evolved later, such as chinook salmon, have either cone-shaped or round scales.

of chemicals in the water. Sharks can follow a trail of blood through the water for several kilometers. This ability helps them locate their prey.

Another adaptation that allows sharks and other fishes to detect their prey is the lateral line system. The **lateral line system** is a line of fluid-filled canals running along the sides of a fish that detects movement and vibrations in the water.

Like sharks, most rays are predators and feed on or near the ocean floor. Rays have flat bodies and broad pectoral fins on their sides. By slowly flapping their fins up and down, rays can glide, searching for mollusks and crustaceans, along the ocean floor. Some species of rays have sharp spines with poison glands on their long tails for defense, as anyone who has stepped on one can attest. Other species have organs that generate electricity to kill both prey and predators.

Cartilaginous fishes have scales

Most fishes have skin covered by overlapping rows of scales. **Scales** are made of bone formed from the skin.

Scales, *Figure 33.5,* can be tooth-like, diamond-shaped, cone-shaped, or round. Shark scales are similar to teeth found in other vertebrates. The age of some species of fishes can be estimated by counting annual growth rings in their scales.

Sharks and rays have internal fertilization

Some fishes have internal fertilization. Sharks and rays, for example, may produce as few as 20 eggs and keep them inside their bodies until they have hatched and developed to about 40 cm in length. These young, when released, behave like miniature adults, and a large portion survive.

Bony Fishes Have Skeletons Made of Bone

The majority of the world's fishes belong to the class Osteichthyes, or bony fishes. Bony fishes are a successful and widely distributed class, differing greatly in habitat, size, feeding behavior, and shape, as *Figure 33.6* shows.

All bony fishes have a bony skeleton, gills, paired fins, and highly developed sense organs. As the name indicates, these fishes have skeletons made of bone, rather than the cartilage skeletons of the other classes of fishes. Bone is the hard, mineralized, living tissue that makes up the endoskeleton of most vertebrates. The appearance of bone was important for the evolution of fishes and vertebrates in general because it allowed fishes to adapt to a variety of aquatic environments, and finally even to land.

Vertebrae provide flexibility

The evolution of a backbone composed of separate, hard segments called vertebrae was important; the backbone is the major support structure of the vertebrate skeleton. Separate vertebrae provide great flexibility. This is especially important for fish locomotion, which involves continuous flexing of the backbone.

You can see how modern bony fishes propel themselves in water in *Figure 33.7.* Some fishes are effective predators, in part because of the fast speeds they can attain due to a flexible skeleton.

Figure 33.6

Bony fishes vary in appearance, behavior, and way of life.

▶ Predatory bony fishes, such as a pike, have sleek bodies with powerful muscles and tail fins for fast swimming.

▼ Stargazers shovel themselves into the gravel of the sea bottom, then wait for passing fishes. A wormlike filament inside the jaw encourages fishes that are prey to come closer.

▼ Eels are long and snakelike and can wriggle through mud and crevices in search of food.

◀ Sea horses move slowly through the underwater plants and algae where they live. They are unusual in that the males brood their young in stomach pouches.

Development of a swim bladder

Another key to the evolutionary success of bony fishes was the evolution of the swim bladder. A **swim bladder** is a thin-walled, internal sac found just below the backbone in bony fishes, which can be filled with gases that diffuse out of a fish's blood. Fish with a swim bladder control their depth by regulating the amount of gas in the bladder. The gas, which is mostly oxygen, works like the gas in a blimp that adjusts the height of the blimp above the ground.

Fishes that live in oxygen-poor water or in ponds or rivers that dry up in the hot season often have other ways to get oxygen. The African lungfish, for example, has a special structure that allows it to obtain oxygen by gulping air. These structures are modified swim bladders. The modified swim bladder is connected to the fish's mouth by a tube.

Reproduction in bony fishes

In most bony fishes, sexes are separate and fertilization is external. Breeding in fishes and some other animals is called **spawning.** During spawning, some female bony fishes, such as cod, produce as many as 9 million eggs, of which only a small percentage will survive.

Some bony fish species are live-bearers; that is, offspring are born fully developed. In these species, such as guppies, mollies, and swordtails, fertilization is internal and young fishes develop within the mother's body. After hatching, other fishes, such as the mouth-brooding cichlids, stay with their young and scoop them into their mouths if they are threatened.

Figure 33.7

Most bony fishes swim in one of three possible ways.

MiniLab

How do fishes swim?

As fishes have been swimming for about 500 million years, it is not surprising that they have mastered this method of movement in a watery world. Most fishes move in one of three ways. Some swim in an S-shaped pattern, alternately tightening muscles on either side of the body. These fishes do not have well-developed fins. Another swimming pattern is to keep the front portion of the body rigid and move the tail and back portion of the body back and forth. The third swimming pattern is to keep the entire body rigid and move only the tail back and forth.

Procedure

1. Observe several different fishes from the top of an aquarium for several minutes each to determine which swimming patterns they follow.

2. Add the normal food to the aquarium and observe any changes in swimming patterns.

Analysis

1. Compare the swimming patterns of the fishes you observed.

2. What might be the survival value for each type of swimming pattern?

3. How might the swimming patterns of fishes reflect their feeding habits?

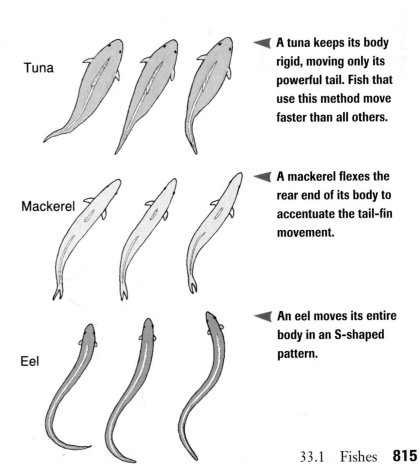

Tuna

A tuna keeps its body rigid, moving only its powerful tail. Fish that use this method move faster than all others.

Mackerel

A mackerel flexes the rear end of its body to accentuate the tail-fin movement.

Eel

An eel moves its entire body in an S-shaped pattern.

A Bony Fish

The bony fishes, class Osteichthyes, include some of the world's most familiar fishes, such as the bluegill, trout, minnow, bass, swordfish, and tuna. Though diverse in general appearance and behavior, bony fishes share some common adaptations with other fish classes.

Rainbow trout

1 When fishes swim past obstacles, pressure changes occur in the water. Fishes can detect these minute changes of pressure with their lateral lines. The lateral line system enables a fish to swim in the dark and to navigate in a complex coral reef.

2 The mass of a fish's tissue is greater than that of water; therefore, without a swim bladder, a fish would not be able to float. Gas pressure in the swim bladder alters the specific gravity of the whole body, enabling the fish to float at any depth. Only bony fishes have swim bladders.

3 Gills are thin, blood vessel-rich tissues where gases are exchanged.

4 Scales are covered with slippery mucus, allowing a fish to move through water with minimal friction. Fishes in the class Agnatha do not have scales.

5 Fishes display a variety of fin shapes. The structure and arrangement of fins are related to the type of locomotion. Tropical fishes that live among coral reefs tend to have small fins capable of maneuvering in this complex, three-dimensional environment, whereas a tuna has large, broad fins for moving quickly through open water.

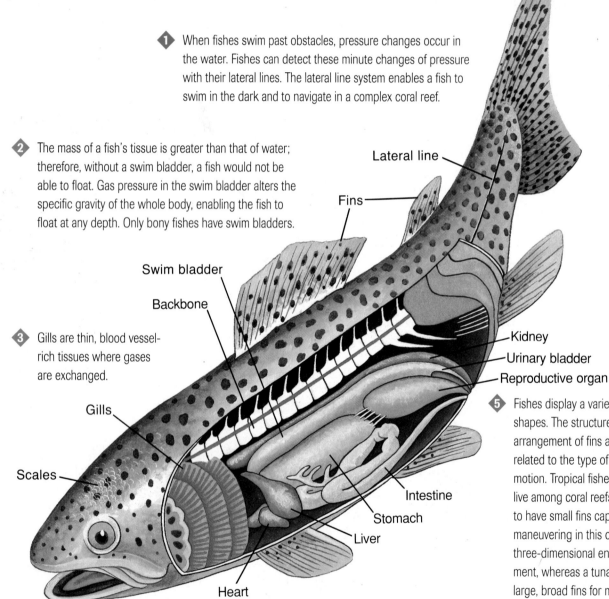

Lateral line

Fins

Swim bladder

Backbone

Kidney

Urinary bladder

Reproductive organ

Gills

Scales

Intestine

Stomach

Liver

Heart

Three subclasses of bony fishes

Scientists recognize three subclasses of bony fishes. *Figure 33.8* shows one subclass, the lungfishes. Another subclass is the lobe-finned fishes, represented by only one living species. The third subclass, the ray-finned fishes such as catfish, perch, salmon, and cod, are probably the most familiar to you because many are fishes eaten by humans. Ray-finned fishes have fins supported by rays.

Lobe-finned fishes, or coelacanths, are an ancient group, appearing in the fossil record about 395 million years ago. They are characterized by lobelike, fleshy fins, and live at great depths where they are difficult to find. The skeletal structure of the fleshy fins of coelacanths shows the ability for locomotion on land. Scientists believe that either the lungfishes or lobe-finned fishes were directly ancestral to amphibians, the first land vertebrates.

Figure 33.8

Lungfishes represent ancient subclasses, having arisen close to 400 million years ago. Lungfishes have both gills and lungs. One type of lungfish burrows in the mud and breathes with lungs when streams dry out. Another lungfish survives in oxygen-poor water by gulping air at the surface.

Variety in the Sea

Four hundred million years ago, a type of jawless fish lived in the bottom of the ocean. It moved slowly, rooting out food from the mud. One of the descendants of that ancient fish is the lamprey. It is also jawless, but it doesn't grub about in the mud for food. Lampreys are parasites of other fishes. Parasitism is probably the reason they survived.

The alien that invaded America The sea lamprey is an unpleasant creature in both appearance and behavior. Lampreys have no jaws, but they have up to 125 sharp teeth set in a circle. Using these teeth, they rasp through the scales of their host fish. Then, with the aid of a powerful sucker, they attach themselves permanently. The lamprey sucks the blood and, eventually, the life out of the fish.

Aquatic dinosaur Another link to the past is a fish that most people thought was long extinct. No one believed the story of the strange fish brought in to East London, South Africa, in December 1938. The description of the paired, lobed fins and large, round scales sounded like

an organism scientists had previously believed to be extinct. The coelacanth had died out during the Cretaceous period. Or had it? Ichthyologists who examined the fish found that this was indeed a coelacanth. No other fishes they had seen, except as fossils, had the lobed fins and the strange tuft at the end of its tail. Lobe-finned fishes were believed to represent a group of fishes that eventually evolved into amphibians.

A fish that fishes But what about modern fishes? How have they survived and adapted to life in the abyss? The angler fish has perfected a unique way to catch a meal. They live deep in the sea where there is total darkness, and use a "rod and bait" to attract and catch other fish. The rod is actually a bony growth that hangs over the fish's mouth. The fleshy bait looks like a morsel of food dangling from the rod. Making the lure even more effective is its luminescence, which can be switched on and off.

Thinking Critically

Scientists were astonished when they discovered living coelacanths. Hypothesize how a fish could survive when many of the animals that evolved from it on land are long gone.

Origins of Fishes

Scientists have identified fossils of fishes that existed during the early Devonian period 400 million years ago. Although the fossil record for fishes is incomplete, most scientists agree that the relationships shown in *Figure 33.9* represent the best fit for the available evidence. To which group of fishes might amphibians be closely related?

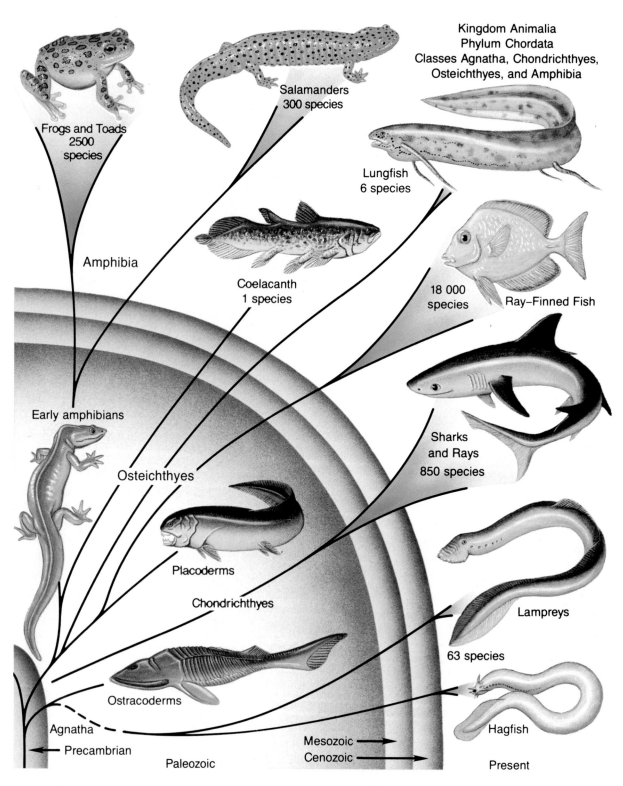

Kingdom Animalia
Phylum Chordata
Classes Agnatha, Chondrichthyes,
Osteichthyes, and Amphibia

Salamanders
300 species

Lungfish
6 species

Frogs and Toads
2500 species

18 000 species

Ray-Finned Fish

Amphibia

Coelacanth
1 species

Early amphibians

Sharks and Rays
850 species

Osteichthyes

Placoderms

Chondrichthyes

Lampreys

63 species

Ostracoderms

Hagfish

Agnatha

Precambrian

Mesozoic
Cenozoic

Paleozoic

Present

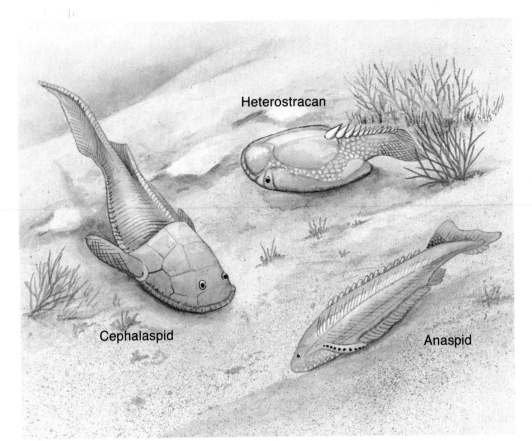

Heterostracan

Cephalaspid

Anaspid

Figure 33.10

Ostracoderms, the earliest vertebrate fossils found, were characterized by bony, external plates covering the body and a jawless mouth. Lacking jaws, ostracoderms obtained food by sucking up bottom sediments and sorting out the nutrients.

Weighed down by heavy, bony external armor, ancestral fishes, shown in *Figure 33.10,* were fearsome-looking animals that swam sluggishly over the murky seafloor. The development of bone in these animals was an important evolutionary step because bone provides a place for muscle attachment, which improves locomotion. In ancestral fishes, bone that formed into plates provided protection as well.

Scientists hypothesize that the jawless ostracoderms were the common ancestors of all fishes, including the present-day jawless fishes. Modern cartilaginous and bony fishes evolved later from an ancient class of cartilaginous, jawed fishes known as placoderms.

Section Review

Understanding Concepts

1. List three characteristics of fishes.
2. Compare how jawless fishes and cartilaginous fishes feed.
3. Why was the evolution of a swim bladder important to fishes?

Thinking Critically

4. Why was the development of jaws an important step in the evolution of fishes?

Skill Review

5. **Making and Using Tables** Construct a table to compare the characteristics of the jawless, cartilaginous, and bony fishes. For more help, refer to Organizing Information in the *Skill Handbook.*

33.2 Amphibians

Section Preview

Objectives

Relate the demands of a terrestrial environment to the adaptations of amphibians.

Relate the evolution of the three-chambered heart to the amphibian lifestyle.

Key Terms

ectotherm
vocal cords

If an alien visitor to our planet were to watch our television programs and read our children's literature, it might return home with wondrous stories of how frogs on Earth can talk and change by the touch of a kiss into princes. Frogs and toads don't talk, but they do change—from fishlike tadpoles to four-legged animals with bulging eyes, long tongues, loud songs, and remarkable jumping ability.

The Move to Land

Imagine a time 350 million years ago when the inland, freshwater seas were filled with carnivorous fishes. Any fish that could move itself or its eggs to the land, even for a brief time, may have had more chance of survival. At the same time, the climate was dominated by alternating floods and drought. Any fish that could move from the mud of a drying stream to another water source might survive. Most likely, these fishes breathed air as they crawled clumsily on stumpy fins to another pond.

Challenges of life on land

Life on land held many advantages for early amphibians. There was a large food supply, shelter, and no predators. In addition, there was much more oxygen in air than in water. However, land life also held many dangers. Unlike the temperature of water, which remains fairly constant, air temperatures vary a great deal. In addition, without the support of water, the body was clumsy and heavy. Some of the efforts at moving on land by early amphibians were like movements of modern-day salamanders. The legs extended out from the body, then down, allowing for better support than lobed fins. You can see in *Figure 33.11* why the bellies of these animals dragged on the ground.

Adaptations improved success

The success of inhabiting the land depended on adaptations that would provide support and efficient locomotion, protection for membranes involved in respiration, and efficient circulation.

What Is an Amphibian?

The striking transition from a completely aquatic larva to an air-breathing, semiterrestrial adult gives the class Amphibia its name, which means "double life." The class Amphibia includes three orders: Urodela, the salamanders and newts; Anura, the frogs and toads; and Apoda, the legless caecilians, as shown in *Figure 33.12*. Although most adult amphibians are capable of a terrestrial existence, nearly all of them rely on water for breeding. Amphibian eggs lack protective

Figure 33.12

Caecilians, order Apoda, are long, limbless amphibians that often have no tails. They look like worms but have eyes that are covered by skin.

Figure 33.11

Adaptation to life on land involves positioning of limbs. The lobed fins of early fishes evolved into the limbs of early amphibians. The evolution of primitive limbs in amphibians led to the diversification of land vertebrates.

membranes and shells and must be laid in water to keep them moist. Fertilization in most amphibians is external. Therefore, water is needed as a medium for transporting sperm.

▼ In mammals, the fastest-moving land animals, the body is raised above the ground, with legs positioned underneath.

▲ The salamander has legs that extend straight out from the body.

◄ Reptiles have legs on the sides of their bodies, but the limbs are flexed.

Is egg size related to survival of salamander larvae?

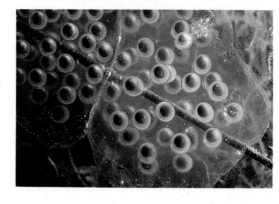

Salamanders and other amphibians are decreasing in numbers worldwide. Studies are underway to determine the cause. Biologists wanted to find out if egg size in salamanders had any effect on survival of the larvae under adverse conditions. They grew large and small salamander eggs from the same species under constant conditions, and under conditions that simulated a drying pond in the summer.

Analysis

Biologists found that the survival rates for larvae from larger eggs were higher than for larvae from small eggs under similar conditions. No difference in survival rates was observed in the drying treatment.

Thinking Critically

From the results of this experiment, explain whether larger eggs enable larvae of salamanders to survive adverse conditions in a pond.

Amphibians undergo metamorphosis

Unlike fish, amphibians go through the process of metamorphosis. Fertilized eggs hatch into tadpoles, the aquatic stage of most amphibians. Tadpoles possess fins, gills, and a two-chambered heart as in fishes. As tadpoles grow into adult frogs and toads, they develop legs, lungs, and a three-chambered heart. *Figure 33.13* shows this life cycle. Young salamanders resemble adults, but they have gills and usually have a tail fin. Amphibians have thin, moist skin and no claws. Most adult salamanders lack gills and fins. Instead, they breathe through their moist skin or with lungs. Up to one-fourth of all salamanders have no lungs and breathe only through their skin. Salamanders also have four legs for moving about.

Figure 33.13

The amphibian life cycle includes an aquatic tadpole stage and a terrestrial adult stage. How could having two different forms in a life cycle have survival value for a species?

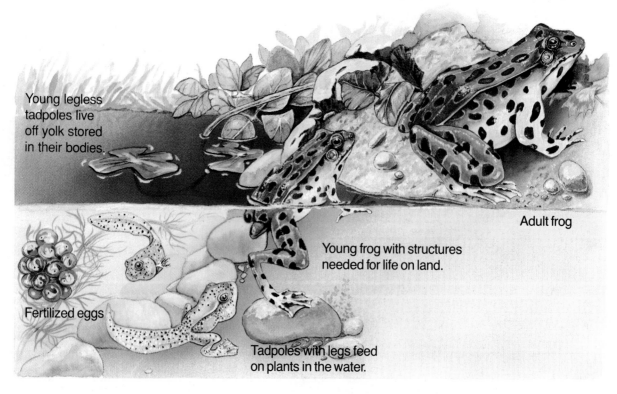

Young legless tadpoles live off yolk stored in their bodies.

Fertilized eggs

Tadpoles with legs feed on plants in the water.

Young frog with structures needed for life on land.

Adult frog

Amphibians are ectotherms

Amphibians are more common in regions that have warm temperatures all year because they are ectotherms. An **ectotherm** is an animal in which the body temperature changes with the temperature of its surroundings. Because many biological processes require particular temperature ranges in order to function, amphibians become dormant in regions that are too hot or cold for part of the year. During such times, many amphibians burrow into the mud and stay there until suitable conditions return.

Walking requires more energy

The laborious walking of early amphibians required a great deal of energy from food and large amounts of oxygen for aerobic respiration. The evolution of the three-chambered heart in amphibians ensured that cells receive the proper amount of oxygen. This was an important evolutionary transition from the simple circulatory system of fishes.

In the three-chambered heart of amphibians, one chamber receives oxygen-rich blood from the lungs and skin, and another chamber receives oxygen-poor blood from the body tissues. Blood from both chambers then moves to the third chamber, which pumps oxygen-rich blood to body tissues and oxygen-poor blood back to the lungs and skin so it can pick up more oxygen. In amphibians, the skin is much more important than the lungs as an organ for gas exchange.

MiniLab

How does leg length affect how far a frog can jump?

Frogs have long legs relative to their body weight—a key adaptation that allows them to jump long distances and move away from predators rapidly. In this activity, you will determine how leg length is related to the jumping ability of a frog.

Procedure

1. Measure the length of one hind leg of a frog by gently extending the leg. The measurement should start on the ventral surface where the frog leg joins the body and extend to the tip of the longest toe.

2. Place a meterstick on the floor.

3. Place your frog so the anterior part of its head is at the beginning of the meterstick as it sits on the floor.

4. Make a hypothesis about how leg length affects the length of the jump.

5. Lightly tap the rear portion of the frog with the eraser end of your pencil until the frog jumps a measurable distance. The distance recorded should be marked at the position of the frog's nose. Conduct four trials with each frog.

6. Place your data in a table on the chalkboard. Make a graph of the class data. Draw conclusions.

Analysis

1. Use class data or your data from several frogs to explain how leg length is correlated to the distance a frog can jump.

2. What other factors might be important in determining how far a frog can jump?

3. Describe stimuli in the natural habitat of a frog that might cause it to jump.

4. Was your hypothesis supported by your data?

Because the skin of an amphibian must stay moist to exchange gases, most amphibians are limited to life on the water's edge. However, some newts and salamanders remain totally aquatic. Amphibians such as toads have thicker skin. Although toads live primarily on land, they still must return to water to breed.

BioLab

Development of Frog Eggs

Most frogs breed in water. The male releases sperm over the female's eggs as she lays them. Some frogs lay up to a thousand eggs. A jellylike casing protects the eggs as they grow into embryos. The embryos then hatch and develop into aquatic larvae commonly called tadpoles. Tadpoles feed by scooping algae from the water or by scraping algae from water plants with small, toothlike projections in their mouths.

During metamorphosis, tadpoles lose their tails and develop legs. Many internal changes take place as well. Gills are reabsorbed by the body, and lungs form. Development and metamorphosis to the adult frog take from three weeks up to several years, depending upon the species.

PREPARATION

Problem
How does temperature affect the development of frog eggs?

Objectives
In this Biolab, you will:
- **Compare** development of frog eggs at varying temperatures.
- **Distinguish** among various stages of development.

Materials
Ringer's solution
4 culture dishes
light bulbs with source of electricity
thermometer
binocular microscope
frog eggs, *Rana pipiens* or *Xenopus laevis*
flashlight

Safety Precautions
Wash hands after each observation.

1. Obtain four culture dishes of Ringer's solution and fertilized frog eggs.

2. Make a chart similar to the one shown for drawing sketches of stages of development of the eggs.

3. Observe your eggs and determine what stage of development you have. At room temperature, you should see the two-cell stage about 3.5 hours after fertilization, the eight-cell stage at 5.7 hours, the 32-cell stage at 7.5 hours, the late-gastrula stage at 42 hours, and a visible head area between 72 and 84 hours. Hatching will occur at about 140 hours, or five to six days, depending on the species used. Development time varies according to the species.

4. Set up the appropriate numbers of light bulbs of different wattages over the water to keep the temperatures at 20, 25, and 30°C.

5. Place a dish of eggs in the refrigerator. Measure the temperature of your refrigerator. Keep a flashlight on in your refrigerator at all times so that you have only one variable, the temperature.

6. Make a hypothesis about how temperature will affect development of the eggs.

7. Set up a time schedule for observations and making sketches based on what stage you are observing. Make observations until the eggs hatch. Record your observations in a journal.

8. Observe your eggs under the microscope according to the schedule you have made. Draw sketches of your eggs in the table.

Temperature	Day 1	Day 2	Day 3
30 degrees			
25 degrees			
20 degrees			
refrigerator:			
–degrees			

1. **Interpreting Observations** Which eggs develop the fastest? The slowest? Explain.

2. **Interpreting Observations** Did your data support your hypothesis?

3. **Drawing Conclusions** What advantage is it for frogs to have eggs that develop at different rates that correspond to different temperatures?

4. **Thinking Critically** What would happen if frog eggs developed when the weather was still cold in the spring?

Going Further

Developing a Hypothesis Make a hypothesis about what other environmental factors would affect the development of frog eggs. Explain your reasoning.

Figure 33.14

Most male frogs have throat pouches that increase the volume of their calls.

Chemistry

Killer Frogs

The most colorful frogs in the world are found in South and Central America. Poisonous frogs, including 130 species of the Dendrobatidae family, range in size from 1 to 5 cm. Although all frogs have glands that produce secretions, these frogs secrete toxic chemicals through their skin. A predator will usually drop the foul-tasting frog when it feels the numbing or burning effects of the poison in its mouth. The frogs advertise their poisonous personalities by bright coloration; they may be red or blue, solid colored, marked with stripes or spots, or have a mottled appearance. The poison secreted by these frogs is used by native peoples to coat the tips of the darts they use in their blow guns for hunting.

Biochemistry of the toxin The secretions of these frogs are alkaloid toxins. An alkaloid toxin is a compound that includes a ring consisting of five carbon atoms and one nitrogen atom. The toxins secreted by poisonous frogs act on an ion channel between nerve and muscle cells. Normally, the channel is open to allow movement of sodium, potassium, and calcium ions. The toxins can block the flow of potassium and stop or prolong nerve impulse transmission and muscle contraction. One group of alkaloids affects the transport of calcium ions, which are responsible for muscle contraction. Current research indicates that these alkaloids may have clinical applications for muscle diseases.

CONNECTION TO Biology

Research on newly discovered organisms such as poisonous frogs may result in drugs to treat specific disorders in human patients. Find out what human diseases are caused by problems in the transmission of nerve impulses and write an essay identifying the disorders that might be treated by toxins from poisonous frogs.

The Diversity of Amphibians

The movement of early amphibians from water to a life on land has not yet been completed. Because amphibians still complete part of their life cycle in water, they are limited to the edges of ponds, lakes, streams, and rivers or to areas that remain damp during part of the year.

Frogs and toads belong to the order Anura

Frogs and toads have vocal cords that are capable of producing a wide range of sounds. **Vocal cords** are sound-producing bands of tissue in the throat. As air moves over the vocal cords, they vibrate and cause molecules in the air to vibrate. In many male frogs, air passes over the vocal cords, then passes into a pair of vocal sacs lying underneath the throat, *Figure 33.14.*

Salamanders belong to the order Urodela

Unlike a frog or toad, a salamander has a long, slender body with a neck and tail. Salamanders resemble lizards. However, salamanders have smooth, moist skin and lack claws. Some salamanders are totally aquatic, while others live in damp places on land.

A Frog

Many species of frogs look similar. As adults, they have short, bulbous bodies with no tails. This adaptation allows them to jump more easily.

Green frog, *Rana clamitans*

1 Vibrations from water or air are picked up by the tympanic membrane and transmitted to the inner ear and then to the brain. The ear is connected to the mouth by a channel called the eustachian tube. Eustachian tubes help to maintain equal air pressure on both sides of the tympanic membrane.

2 Some frogs' eyes protrude from the tops of their heads—an adaptation that enables them to stay submerged in the water with only their eyes above water. Frog eyes possess one upper eyelid that does not move and a lower movable eyelid that is folded to form the nictitating membrane, which helps to keep the eye moist.

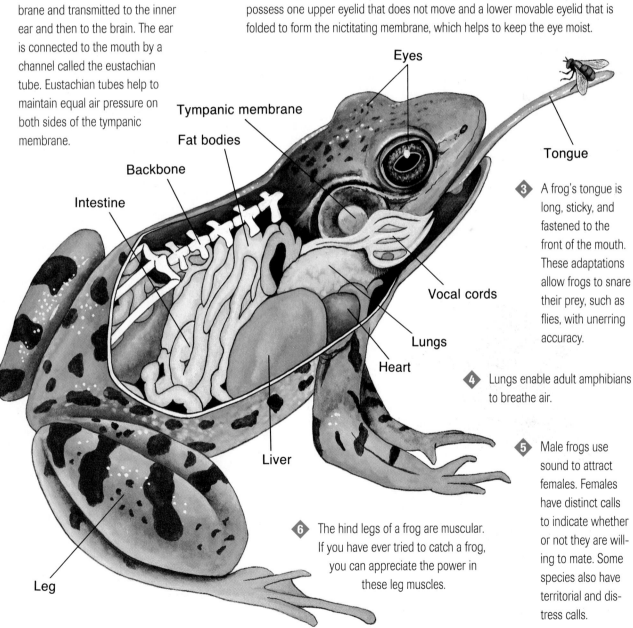

Eyes

Tympanic membrane

Fat bodies

Backbone

Intestine

Tongue

Vocal cords

Lungs

Heart

Liver

Leg

3 A frog's tongue is long, sticky, and fastened to the front of the mouth. These adaptations allow frogs to snare their prey, such as flies, with unerring accuracy.

4 Lungs enable adult amphibians to breathe air.

5 Male frogs use sound to attract females. Females have distinct calls to indicate whether or not they are willing to mate. Some species also have territorial and distress calls.

6 The hind legs of a frog are muscular. If you have ever tried to catch a frog, you can appreciate the power in these leg muscles.

Frogs and Toads

F rogs and toads, members of the order Anura, exhibit a wide variety of sizes and colors. They can be found in many different habitats, from the tops of tropical rain forest trees to mountain peaks. All anurans lay eggs in water, but that water may be located inside a bromeliad plant, the hollow of a tree, or even on the back of a parent frog.

Red-eyed tree frog

Tadpoles that drop in A stranger to the forest floor, the red-eyed tree frog spends its life in the trees, climbing with the aid of large toe pads. At breeding time, the female lays her eggs on leaves overhanging water. When the tadpoles hatch, they drop into the water where they undergo metamorphosis.

Fire-bellied toad

A toad that flashes The coloration of this fire-bellied toad serves a dual purpose. The grey-green top aids concealment. When threatened, the toad arches its back and raises its palms and feet to reveal the bright coloration on its belly, frightening off predators.

Horned frog

Flashy frogs that bite Horned frogs like these marbled-green bullies have distinctive fleshy "horns" above their eyes. Known for their nasty, aggressive behavior, horned frogs will eat almost anything—including one another! These amphibians are also capable of inflicting serious bites on humans.

African bullfrog

A frog that eats mice It will take nearly 28 years for this African bullfrog to reach full adult size. Once it does, it will likely supplement its diet of earthworms, locusts, beetles, and cockroaches with more substantial fare—perhaps an occasional mouse or rat. Caninelike projections on the frog's lower jaw enable it to nab and swallow such prey.

Surinam toad

Tadpoles that hitch a ride on mother This toad has a flattened body and pointed snout. Its front feet end in starlike toes, and its back feet are huge and webbed. During breeding, the Surinam toad female releases a few small eggs that are fertilized by the male, then pushed into the skin of her back.

Madagascan tomato frog

These fat tomato-red frogs are found in northwestern Madagascar. They range in length from 5-7 cm. Madagascan tomato frogs are an endangered species. Fortunately, they have been bred successfully in captivity.

EXPANDING YOUR VIEW

1. **Writing About Biology** In a short essay, explain how natural selection could have led to the brighter colors of poisonous frogs.

2. **Understanding Concepts** Tropical tree frogs may complete their life cycles without ever leaving the tree. How can this occur if amphibians must lay their eggs in water?

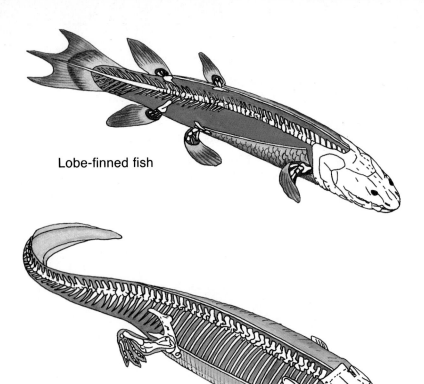

Lobe-finned fish

Early amphibian

Origins of Amphibians

Amphibians first appeared about 350 million years ago. It is widely believed that amphibians evolved from lobe-finned fishes, *Figure 33.15*, around the middle of the Paleozoic Era. At that time, the climate on Earth became warm and wet, a climate ideally suited for an adaptive radiation of amphibians. Amphibians, able to breathe through their lungs, gills, or skin, became a transitional group of vertebrates. Many new species evolved, and for a time, amphibians were the dominant vertebrates on the land.

Figure 33.15

Based on fossils from the Devonian period, this artist's rendition of an early amphibian shows that they had better-developed bones in their limbs than lobed-finned fishes.

Connecting Ideas

Early fishes evolved from mud-sucking swimmers to air-gulping animals that dragged themselves on their bellies from one pond to another, and finally to amphibians that could push themselves with legs over land. Amphibians faced dangers of dehydration and temperature extremes.

You've read about the adaptations that allow amphibians to survive on land. Yet amphibians have not completely made the transition from water to land. The need to protect young and keep eggs moist prevents amphibians from striking out to inhabit dry places on land.

Section Review

Understanding Concepts

1. Describe the events that may have led early animals to move to land.
2. List three characteristics of amphibians.
3. Name two ways that amphibians depend on water.

Thinking Critically

4. How does a three-chambered heart enable amphibians to maintain homeostasis?

Skill Review

5. **Sequencing** Trace the evolutionary development of amphibians from lobe-finned fishes. For more help, refer to Organizing Information in the *Skill Handbook*.

Living Organisms as Pollution Detectors

Amphibian Species Disappearing

THE DAILY NEWS   OCTOBER 20, 1993

Hundreds of species of frogs, toads, and salamanders are experiencing steep declines in their populations according to experts from the Smithsonian Institution. From the New York suburbs to the rain forests of Central and South America, amphibian species are being wiped out.

One of the first people to warn of the ongoing problem with pollution and the unrestricted use of pesticides was Rachel Carson. In her history-changing book, *Silent Spring*, Dr. Carson warned that careless use of pesticides was having a disastrous effect on bird populations. Recent studies have shown her concerns to apply to other species as well. Hundreds of species of amphibians all over the world are in a steep population decline. Why are amphibian species declining?

Amphibians have moist, thin skin, which is used in gas exchange; thus, they are susceptible to changes in water quality and chemistry. Because they spend part of their lives on land and part in water, amphibians are exposed to pollutants in soil and air, as well. Amphibians also are noticeable because they are part of the wild chorus people hear at night.

Loss of habitat is another problem Amphibians are also at risk due to habitat destruction. Most amphibians live in wetlands—areas that remain wet for all or part of a year. Wetlands provide a crucial service by filtering pollutants out of water, and providing food and shelter for young fish and amphibians. But the land has been drained to use for other purposes, destroying amphibian habitats and, as a result, reducing their populations.

Different Viewpoints

Once a species is gone, it is gone forever. When this happens, we all are poorer—the world is less diverse, a gene pool that might be useful is gone, and we have failed in our role as the stewards of Earth. Every species is part of a complex system of relationships. In destroying one species, we could set off a chain of events that threatens all life.

Let them adapt Species such as salamanders and frogs have outlived the dinosaurs. Species adapt, and what we are seeing are these species adapting to changing conditions. When the weak individuals die, the stronger ones survive, and a stronger species evolves.

INVESTIGATING the Issue

1. **Debating the Issue** Discuss the issue of disappearing wetlands. Have each side take a stand on developing wetland areas, and prepare research data to back up each side's arguments.
2. **Recognizing Cause and Effect** Amphibians eat insects, especially those that are found in wetland areas, such as mosquitoes. Write a paragraph describing the effect on a wetland ecosystem if all the amphibians died.

Reviewing Main Ideas

33.1 Fishes

- Fishes are vertebrates with backbones and nerve cords that have expanded into brains.
- Fishes belong to three groups: the jawless lampreys and hagfishes, the cartilaginous sharks and rays, and the bony fishes. Bony fishes are divided into three groups: the lobe-finned fishes, the lungfishes, and the ray-finned fishes.
- Early jawless fishes may be the ancestors of present-day jawless fishes. Cartilaginous and bony fishes may have evolved from ancient placoderms.

33.2 Amphibians

- Land animals face problems of dehydration, gas exchange in the air, and support for heavy bodies. Amphibians possess adaptations well suited for life on land.
- Adult amphibians have three-chambered hearts that provide oxygen to body tissues, but most gas exchange is done through the skin.
- Amphibians may have evolved from lobe-finned fishes.

Key Terms

Write a sentence that shows your understanding of each of the following terms.

cartilage
ectotherm
fin
lateral line system

scale
spawning
swim bladder
vocal cords

Understanding Concepts

Using a Graph

1. The following data indicate estimates of the number of leopard frogs in a wetland on a farm. One year, there was a prolonged drought in the farmer's area. What year did the drought occur? Explain.

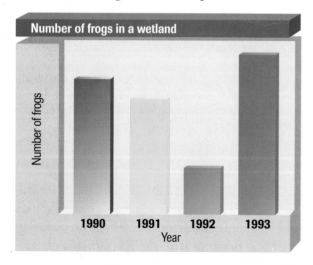

Number of frogs in a wetland

(y-axis: Number of frogs; x-axis: Year — 1990, 1991, 1992, 1993)

Relating Concepts

2. Make a concept map that relates the following terms and phrases. Supply the appropriate linking words for your map.

   fishes, lateral line, fins, cartilage, scales, bone, rays, gills, swim bladder

3. Sequence the evolutionary development of the different groups of fishes.
4. Compare the feeding adaptations of jawless and bony fishes.
5. In what ways are fishes important to humans?
6. Describe the evolutionary link between fishes and amphibians.
7. Compare the locomotion of frogs and salamanders.
8. How do frogs, toads, and salamanders differ?
9. What were the advantages and disadvantages to ancestral fishes as they first moved to land environments?

Applying Concepts

10. What accounts for the different body shapes of fishes?
11. In what ways are fishes and young amphibians alike?
12. What are the advantages and disadvantages of internal fertilization?
13. Describe the importance of the evolution of bone in fishes to the evolution of vertebrates.
14. The male sea horse incubates its eggs in a brood pouch. The codfish lays its eggs in the open sea. Which of these two fishes do you think lays more eggs? Why?

Chemistry Connection

15. While visiting South America, your friend notices a beautiful blue frog with orange spots sitting underneath a rotting log. Should he pick up the frog? Explain.

Thinking Critically

Interpreting Data

16. **Biolab** You have decided to begin a frog farm. Based on the following data, what temperature would you select for the solution in which to grow the frog eggs? Why?

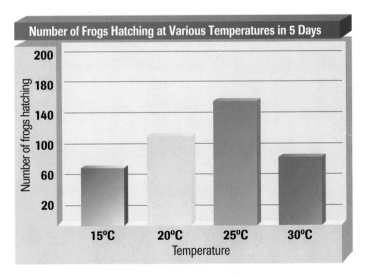

Number of Frogs Hatching at Various Temperatures in 5 Days

Drawing Conclusions

17. **Minilab** You decide to determine if there is a correlation between how far a frog can jump and its body mass. You find that frog A has a mass of 5 g and jumps an average of 50 cm. Frog B has a mass of 3 g and jumps an average of 40 cm. Frog C has a mass of 2 g and jumps an average of 20 cm. What conclusion could you draw?

Designing an Experiment

18. **Thinking Lab** Suppose that you have unlimited resources and funds to find out why salamander larvae from small and large eggs differ in survival rates. Design an experiment that would test a hypothesis. Make sure you design an experiment in which you have a control and gather quantitative data.

Connecting to Themes

19. **Evolution** Explain why biologists think that early forms of fishes evolved into amphibians.
20. **Systems and Interactions** In what ways are fishes important to aquatic habitats?
21. **Homeostasis** Explain the advantages to a species that has a life cycle with alternate forms. Use the frog, sea urchin, arthropod, and butterfly as examples.

34 Reptiles and Birds

The setting sun turns daylight to dusk. You are walking along a shrubby knoll when you hear a sinister, rattling buzz coming from a large rock nearby. Rattlesnakes inhabit the area, and you know that they move onto rock surfaces in the evening because the rocks retain the warmth of the day as the evening air cools. When you look around carefully, you see the nested coils with raised rattle, and the menacing head with a flicking tongue. Knowing that the venom of this snake can kill a full-grown human, you slowly back away. Calming yourself, you recall that more people in the United States are hit by lightning than are bitten by poisonous snakes.

What does a legless, cold-blooded viper with poisonous fangs have in common with a bird? Scientists think that reptiles were the ancestors of birds. How do they know? The next time you see a bird up close, look at its scaly legs and the claws at the end of its toes. Think about the shelled eggs it lays. These are all characteristics of the early lizardlike animals from which birds evolved.

Concept Check

You may wish to review the following concepts before studying this chapter.
- Chapter 1: adaptation, evolution
- Chapter 28: vertebrate
- Chapter 30: gizzard
- Chapter 33: ectotherm

Chapter Preview

34.1 Reptiles
What Is a Reptile?
Reptile Adaptations
The Diversity of Reptiles
Origins of Reptiles

34.2 Birds
What Is a Bird?
Bird Adaptations
Origins of Birds

Laboratory Activities

Biolab: Design Your Own Experiment
- What is the ideal length and width for a bird's tail?

Minilabs
- How do down feathers and contour feathers compare?
- What types of foods do local birds prefer?

Reptiles and birds share other characteristics besides scales and claws. You can see that this is a nest. Can you tell that it was made by a crocodile? Both reptiles and birds lay shelled eggs.

34.1 Reptiles

You may remember seeing an adventure movie in which a ferocious crocodile devours a villain who is trying to swim across a jungle river. Moviemakers often use crocodiles, alligators, lizards, and snakes in their films to convey a sense of fear to the audience. Fear of these "cold-blooded" reptiles may have been inspired more by legends, stories, and movies than by actual experiences. However, a few snakes are capable of killing humans. Of the approximately 120 species of snakes found in the United States, the poisonous ones include the rattlesnakes, moccasins, copperheads, and coral snakes. Even people who collect and study reptiles have a healthy respect for snakes.

What Is a Reptile?

At first glance, it may be difficult to determine how a legless snake is related to a tortoise. Snakes, turtles, alligators, and lizards seem to be an extremely diverse group of animals, yet all share certain traits that place them in the class Reptilia.

Early reptiles, such as the cotylosaur shown in *Figure 34.1,* evolved important innovations that freed them from dependence on water. Reptiles have adaptations that enable them to complete their life cycles entirely on land. These adaptations released the cotylosaurs and other reptiles from the need to return to swamps, lakes, rivers, ponds, or oceans and allowed them to evolve into successful land animals.

Reptiles have scaly skin

Unlike the moist, thin skin of amphibians, reptiles have a dry, thick skin covered with scales. Scaly skin, *Figure 34.2,* prevents the loss of body moisture and provides additional protection from predators. Because gas exchange cannot occur through scaly skin, reptiles are entirely dependent on lungs as their primary organ of gas exchange.

Figure 34.1

Cotylosaurs, the earliest known reptiles, were probably the ancestors of the long-extinct dinosaurs as well as today's living reptiles, birds, and mammals.

Figure 34.2

Scales on a reptile's skin overlap like tiles on a roof.

▲ The scales of reptiles, unlike the separate glossy scales of fish, are part of the skin itself. The scales are all connected to one another by hinges of skin.

◄ To grow, young reptiles molt. They lose their scaly skins when they molt.

Some reptiles have four-chambered hearts

Most reptiles, like amphibians, have three-chambered hearts. Some reptiles, notably the crocodilians, have a four-chambered heart that completely separates the supply of blood with oxygen from blood without oxygen. The separation enables more oxygen to reach body tissues. This separation is an adaptation to life on land that supports a higher level of energy use in reptiles than that found in amphibians.

Skeletal changes in reptiles

Look again at the illustration of the cotylosaur. This reptile had legs that were placed more under the body rather than out to the sides as in early amphibians. This positioning of the legs provides greater body support and makes walking and running on land easier for the reptiles. They have a better chance of catching prey or avoiding predators. Reptiles also have claws that facilitate food getting and protection. Additional changes in the structure of the jaws and teeth of reptiles allowed them to exploit other resources and niches on land.

Reptiles reproduce on land

Reptiles reproduce by laying eggs on land. Unlike amphibians, reptiles have no aquatic larval stage, and thus are not as vulnerable to water-dwelling predators as young amphibians are. Reptile hatchlings look just like adults, only smaller.

Although all of the adaptations discussed so far enabled reptiles to live successfully on land, the evolution of the amniotic egg was the adaptation that liberated reptiles from a dependence on water for reproduction. An **amniotic egg** provides nourishment to the embryo and contains membranes that protect it while it develops in a terrestrial environment. The egg functions as the embryo's total life-support system.

All reptiles have internal fertilization. The eggs are laid after fertilization, and embryos develop after eggs are laid. Most reptiles lay their eggs under rocks, bark, grasses, or other surface materials, but a few dig holes or collect materials for a nest. Most reptiles provide no care for hatchlings, but female crocodiles have been observed guarding their nests from predators.

The Amniotic Egg

*T*he evolution of the amniotic egg was a major step in reptilian adaptations to land environments. Amniotic eggs enclose the embryo in amniotic fluid; provide a source of food in the yolk; and surround both embryo and food with membranes and a tough, leathery shell. These innovations in the egg help prevent injury and dehydration of the embryo as it develops on land.

Leatherback turtle

1 The amnion is a membrane filled with fluid that surrounds the developing embryo. The fluid-filled amnion cushions the embryo and prevents dehydration.

2 The reptile egg is encased in a leathery shell. Most reptiles lay their eggs in protected places beneath sand, earth, gravel, or bark.

6 The chorion is a membrane that forms around the yolk, allantois, amnion, and embryo. It allows gas exchange.

3 The embryo's nitrogenous wastes are excreted into the allantois, a membranous sac that is associated with the embryo's gut. When a reptile hatches, it leaves behind the allantois with its collected wastes.

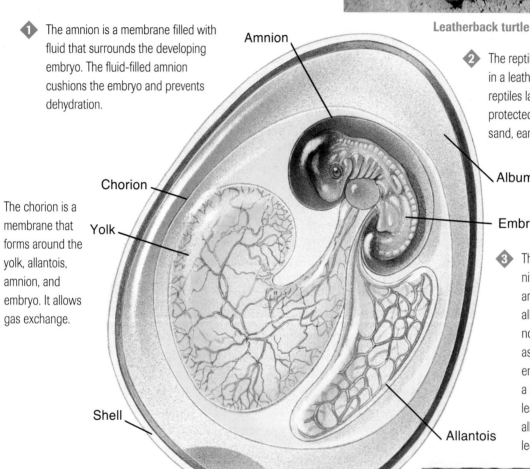

Amnion

Chorion

Yolk

Shell

Albumen

Embryo

Allantois

5 The main food supply for the embryo is the yolk, which you see each time you crack open an egg. The yolk is enclosed in a sac that is also attached to the embryo. The clear part of the egg is albumen, a source of additional food and water for the developing embryo.

4 A reptile hatches by breaking its shell with the horny tooth on its snout.

► A chuckwalla keeps its temperature constant by lying in the shade of a large rock.

▼ A bearded lizard suns itself to get warm in the morning.

Figure 34.3

Different reptiles regulate their body temperatures by a variety of behaviors.

Reptiles are ectotherms

Even though reptiles are different from amphibians in many ways, they are similar in one way. Both amphibians and reptiles are ectotherms. Their body temperatures depend on the temperatures of their environments. In the cool morning, a turtle might pull itself out of the pond or swamp and bask on a log in the sunlight until noon. Then, when the temperature gets a little too warm, the turtle may slip back into the cool water. This example shows that even though reptiles cannot control their body temperatures themselves, they can use behavioral adaptations to compensate for changes in environmental temperature. **Figure 34.3** shows other examples of behavioral adjustment of body temperature.

Because reptiles are dependent on the temperatures of their surroundings, they do not inhabit extremely cold regions. Reptiles are numerous in temperate and tropical regions, where climates are warm, and in hot desert climates. Many reptiles become dormant in moderately cold environments such as in the northern United States.

Reptile Adaptations

Like other animals, reptiles have adaptations that enable them to find food and to sense the world around them.

How reptiles obtain food

Most turtles and tortoises are too slow to be effective predators, but that doesn't mean they go hungry. Most are herbivores, and those that are predators prey on worms and mollusks. Snapping turtles, however, are extremely aggressive. They attack fish, amphibians, and also pull ducklings underwater.

Lizards eat insects primarily. The marine iguana of the Galapagos Islands is one of the few herbivorous lizards, feeding on marine algae. The Komodo dragon, the largest lizard, is an efficient predator, sometimes even of humans. Although lizards like the Komodo dragon look slow, they are capable of bursts of speed to catch their prey.

Figure 34.4

Many reptiles are skillful predators that obtain prey in a variety of ways.

◄ The snapping turtle is a common turtle in North America. Look at its head. Can you figure out why the snapping turtle is an effective predator?

► Some snakes can swallow eggs or whole animals that are larger than their heads because their lower jaws are loosely attached to their skulls and can be unhinged when opening over the food.

▲ The Komodo dragon is a predator that captures its prey in large jaws. It can kill animals as large as a deer or even a water buffalo.

Snakes are also effective predators. Some, like the rattlesnake, have poison fangs that they use to subdue or kill their prey. A constrictor wraps its body around its prey, tightening its grip each time the prey animal exhales. Several of these reptiles are shown in *Figure 34.4.*

Figure 34.5

Snakes have sense organs that enable them to detect prey or identify chemicals in the environment.

How reptiles use their sense organs

Reptiles have a variety of sense organs that can detect danger or potential prey. How does a rattlesnake know you are nearby? The heads of some snakes, as shown in *Figure 34.5,* have heat-sensitive organs or pits that enable them to detect tiny variations in air temperature brought about by the presence of warm-blooded animals.

Snakes and lizards are equipped with a keen sense of smell. Remember the rattlesnake's flicking tongue? The tongue is picking up chemicals in the air. The snake then draws its tongue back into its mouth and inserts it into a structure called **Jacobson's organ,** *Figure 34.5.*

▼ Rattlesnakes have heat-sensitive pits below their eyes that enable them to detect prey in total darkness. The snake can tell the exact distance and direction in which to strike because the pits are paired.

▼ Jacobson's organ is a pitlike sense organ in the roof of a snake's mouth that picks up airborne chemicals. The long, flexible tongue of a snake picks up molecules in the air and transfers them to the Jacobson's organ for chemical analysis.

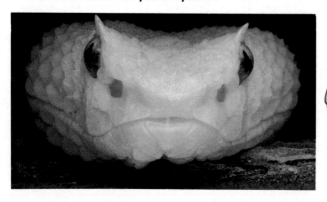

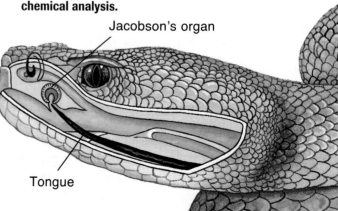

Jacobson's organ

Tongue

The Diversity of Reptiles

Gracefully gliding snakes and quickly darting lizards are grouped together in the order Squamata. Turtles, slowly plodding and carrying heavy shells, belong to the order Chelonia. Basking crocodiles and alligators, classified in the order Crocodilia, may look clumsy but are surprisingly quick hunting machines that snap up fishes and lunge out at antelopes and other large animals that come to the river to drink.

Turtles have shells

Some turtles are aquatic, and some live on land. Turtles that live on land are called tortoises. Most turtles can draw their limbs, tail, and head into their shells for protection against predators. Although turtles have no teeth, they do have powerful jaws with a beaklike structure that is used to crush food.

Figure 34.6

In the past, sailors killed Galapagos tortoises for food. As a result, their numbers declined rapidly. When just a few tortoises were left, they rarely met one another, and reproduction nearly ceased. Scientists collected the remaining tortoises and put them in an enclosure where they could easily find a mate. Soon, adult and hatchling tortoises were released back to their home areas. The tortoises are now protected by law.

ThinkingLab | Analyze the Procedure

Why are tortoises becoming frail?

The number of desert tortoises of the southwestern United States has recently begun to decline. These tortoises are not growing as large as tortoises in the past, and their shells and bones are more frail and easily fractured. Scientists hypothesize that the reason for this frailty is a deficiency of calcium in the tortoises' diets. Calcium is important for the growth and development of shells and bones. Over the last century, many of the native weeds and succulent plants in the tortoises' habitat have been replaced with cultivated plants.

Analysis

In order to find out if tortoises' diets are lacking in calcium, decide which of the following procedures should be followed .

1. Measure the calcium in native and cultivated grass.

2. Measure the calcium in the bones of tortoises.

3. Offer a group of tortoises a calcium-rich diet in one dish and a calcium-deficient diet in another dish, and measure the amount eaten from each dish.

Thinking Critically

Analyze the above possibilities for ways to determine if turtles have calcium-deficient diets. Choose the procedure that will produce data that will support the hypothesis. Explain why you made your choice.

Tortoises live on land, foraging for fruit, berries, and insects. The largest tortoises in the world, shown in *Figure 34.6,* are found on the Galapagos Islands off the coast of Ecuador.

Some adult marine turtles swim enormously long distances to lay their eggs. Like salmon, these turtles return from their feeding grounds to the place where they hatched. For example, green turtles travel from the coast of Brazil to Ascension Island in the Atlantic, a distance of more than 4000 km.

Sea Turtle Migration

When you hear the word *migration,* do you think of flocks of geese? Surprising though it may be, sea turtles, like many birds, undertake long-distance migrations.

Long-distance travelers When baby sea turtles hatch, they struggle up through the sand that covers their nest and scramble down the beach toward the surf. The moment the hatchlings hit the water, they swim away from land and toward the open sea.

Once sea turtles reach adulthood, they leave their feeding grounds and migrate back across vast expanses of seemingly featureless ocean to breed, usually returning to the same beach where they were hatched.

Navigational cues How do sea turtles find their way on these remarkable, round-trip migrations? Some researchers think that sea turtles, like pigeons and some other animals, can detect Earth's magnetic field and use it as a sort of magnetic compass to orient themselves on their long-distance migrations. Sea turtles also appear to use the direction of waves and deep ocean swells as navigational cues as they migrate between feeding grounds and nesting sites.

Researchers have also shown that tiny turtle hatchlings can detect the chemical makeup of their home beach. As mature adults, they may be able to find their nesting beaches by retracing a trail of these chemicals carried on ocean currents.

Thinking Critically

In some countries, sea turtle breeding grounds and nesting beaches have become protected sanctuaries. These conservation efforts have helped improve nesting success. How might the lessons we are learning about protecting sea turtles be applied to other migratory animals?

Crocodiles include the largest living reptiles

In contrast to marine turtles, crocodiles don't migrate. They may spend their days alternately basking in the sun on a riverbank and floating like motionless logs. Only their eyes and nostrils are above water. Crocodiles can be identified by their long, slender snouts, whereas alligators have short, broad snouts, as shown in *Figure 34.7.* Both animals have powerful jaws with sharp teeth that can drag prey underwater and hold it there until it drowns. One species of alligator is found throughout many of the water habitats of the southeastern United States. The one species of crocodile in the United States can be found only in southern Florida. Alligators and crocodiles lay eggs in nests on the ground. Unlike other reptiles, these animals stay close to their nests and guard them from predators.

Figure 34.7

Alligators and crocodiles use their powerful tails to swim, moving so rapidly that few swimming animals can escape. On land, the tail is a formidable weapon, able to knock most opponents to the ground. Even though these animals have stubby legs, they can move fast (up to 17 km per hour) over short distances.

Figure 34.8

Lizards have many adaptations enabling them to live in a variety of different habitats.

▶ Geckos are small, nocturnal lizards that live in warm climates, such as that of the southern United States, West Africa, and Asia. The toe pads of some geckos enable them to walk across walls and ceilings.

▲ Chameleons are tree-dwelling lizards that can change color.

◀ Only the Gila monster of the southwestern United States and Mexico and the beaded lizard of Mexico are poisonous lizards.

Snakes and lizards are found in many environments

Lizards, shown in *Figure 34.8,* are found in many types of habitats in all but the polar regions of the world. Some live on the ground, some burrow, some live in trees, and some are aquatic. Many are adapted to hot, dry climates.

Snakes, in contrast to most vertebrates, have no limbs and lack the bones to support limbs, except for pythons and boas which retain bones of the pelvis. The many vertebrae of snakes permit fast undulations through grass and over rough terrain. Some snakes even swim and climb trees!

Snakes usually kill their prey in one of three ways. Remember that constrictors wrap themselves around their prey. If you ever watch someone handle a constrictor, you will notice that the handler never lets the snake start to wrap around his or her body. The snake is always held carefully so that its tail does not cross over its own head to begin a coil.

Common constrictors include boas, pythons, and the anacondas. Venomous snakes—known as vipers—such as rattlesnakes, cobras, and sea snakes inject poison from venom glands as shown in *Figure 34.9.* Most snakes are neither constrictors nor poisonous. They get food by grabbing it with their mouths and swallowing it whole. Snakes eat rodents, amphibians, insects, fish, and eggs.

Figure 34.9

Many poisonous snakes have venom glands and hollow fangs for injecting venom. Venom may either paralyze the prey so it cannot run away or kill the prey immediately.

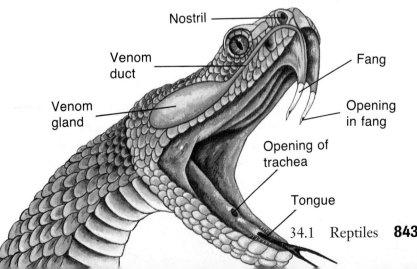

Nostril

Venom duct

Fang

Venom gland

Opening in fang

Opening of trachea

Tongue

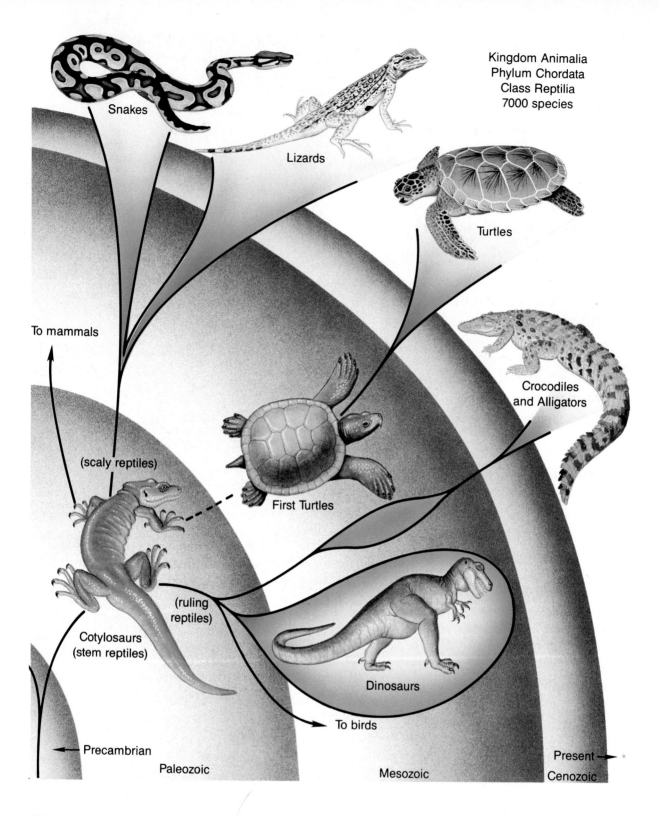

Kingdom Animalia
Phylum Chordata
Class Reptilia
7000 species

Snakes

Lizards

Turtles

Crocodiles
and Alligators

To mammals

(scaly reptiles)

First Turtles

(ruling
reptiles)

Cotylosaurs
(stem reptiles)

Dinosaurs

To birds

Precambrian

Paleozoic

Mesozoic

Present

Cenozoic

Figure 34.10

**Reptiles evolved from amphibians 300 million years ago during the late
Paleozoic Era. Most orders of reptiles have remained similar to their
fossil ancestors. The radiation of orders of reptiles on the Geologic
Time Scale shows their relationships. Which group of modern reptiles
is most closely related to dinosaurs?**

Origins of Reptiles

You may have marveled about dinosaurs ever since you were very young. These animals were the most numerous land animals during the Mesozoic Era. Some were the size of chickens, and others were the largest land dwellers that ever lived. What do today's reptiles have in common with these giants of the past?

The ancestors of snakes and lizards are traced to a group of early reptiles, called scaly reptiles, that branched off from the ancient cotylosaurs. The name "scaly reptiles" may be misleading because it implies that other reptiles lacked scales—which is not true. Although the evolutionary history of turtles is incomplete, scientists have suggested that they may also be descendants of cotylosaurs. Dinosaurs and crocodiles are the third group to descend from cotylosaurs, *Figure 34.10*. Although scientists used to think that birds arose as a separate group from this third branch, many now theorize that birds actually are the living descendants of the dinosaurs. The fourth order of reptiles, *Rhynchocephalia*, is represented by one living species, the tuatara, *Figure 34.11*. The tuatara is the only survivor of a primitive group of reptiles that became extinct 100 million years ago.

Figure 34.11

The tuatara, *Sphenodon punctatus,* is found only in New Zealand. It has ancestral features, including teeth fused to the edge of the jaws, and a skull structure similar to early Permian reptiles.

Section Review

Understanding Concepts

1. Choose one adaptation of early reptiles and explain how it enabled these animals to live on land.
2. Describe two ways in which turtles protect themselves.
3. How do snakes use the Jacobson's organ for finding food?

Thinking Critically

4. Explain how the development of a dry, thick skin was an adaptive advantage for reptiles.

Skill Review

5. **Classifying** Set up a classification key that allows you to identify a reptile as a snake, lizard, turtle, or crocodile. For more help, refer to Organizing Information in the *Skill Handbook*.

34.2 Birds

Have you ever seen a robin tug on a worm that is struggling to stay in the soil? Maybe you have had a chance to see the amusing antics of a chickadee hanging upside down on a snow-covered branch in the forest. Almost everyone admires birds. The brilliant flash of a bluebird's wings, the uplifting sound of a bird's song that fills the woods on a spring morning, and the effortless soaring of a redtail hawk have always fascinated and delighted people.

What Is a Bird?

After conquering the sea and land, vertebrates took to the air where there was a huge source of insect food and a refuge from land-dwelling predators. The existence of 9000 species of modern birds, classified in the class Aves, shows that flight was a successful strategy for survival.

Except for domestic animals and humans, the most common vertebrates you see in your daily life are birds. Biologists sometimes refer to birds as feathered dinosaurs. Fossil evidence seems to indicate that birds have evolved from small, two-legged, lizardlike animals, called thecodonts, *Figure 34.12*. While present-day reptiles and birds have very different physical appearances, some slight resemblances can still be seen. Like reptiles, birds have clawed toes and scales on their feet. Fertilization is internal and shelled amniotic eggs are produced in both groups.

Figure 34.12

Most scientists agree that birds evolved from a group of reptiles called thecodonts. The skeletons of birds and thecodonts are similar.

Birds have feathers

Birds can be defined simply as the only organism with feathers. A **feather** is a lightweight, modified scale that provides insulation and enables flight, *Figure 34.13*. You may have seen a bird running its bill through its feathers while sitting on a tree branch or on the shore of a pond. This process, called preening, keeps the feathers in good flying condition. The bird also uses its beak to rub oil from a gland near the tail onto the feathers. This process is especially important for water birds as a way to waterproof the feathers.

Even with good care, feathers wear out and must be replaced. The shedding of old feathers with the growth of new ones is called molting. Most birds molt in late summer. However, most do not lose their feathers all at once and are able to fly while they are molting. Flight and tail feathers are usually lost in pairs so that the bird can maintain its balance in flight.

Figure 34.13

Feathers streamline a bird's body, making it possible to fly.

MiniLab

How do down feathers and contour feathers compare?

Birds have two kinds of feathers. Contour feathers used for flight are found on a bird's body, wings, and tail. Down feathers lie under the contour feathers and insulate the body.

Procedure

1. Examine a contour feather with a hand lens, and make a sketch of how the feather filaments are hooked together.

2. Examine a down feather with a hand lens. Draw a diagram of the filaments of the down feather.

3. Fan your face with each feather separately. Note how much air is moved past your face by each type of feather.

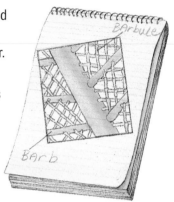

Analysis

1. How does the structure of a contour feather help a bird fly?

2. How does the structure of a down feather help keep a bird warm?

3. How can you explain the differences you felt when fanning with each feather?

◄ **Fluffy down feathers have no hooks to hold the filaments together. Down feathers act as insulators to keep a bird warm. Have you ever worn a down coat or used a down bed cover? If you have, you know how warm down feathers can be.**

◄ **A large bird can have 25 000 or more flight feathers that have a million tiny hooks that interlock and make the feathers hold together. To see how feathers are held together, turn one palm up and make a hook of your fingers on that hand. Turn your other palm down and make a hook of the fingers; then hook the fingers of both hands together. All the filaments in the feather are held together by hooks, making a "fabric" suited for flight.**

Hot Birds

Birds are models of efficiency in motion. They are able to fly thousands of miles, soar high in the atmosphere, and hover for long periods in front of a flower. Of course, no one bird can do all this, but it does demonstrate the versatility exhibited by this fascinating class of organisms. In order to do all these high-energy tasks, birds must respire with maximum efficiency, and this means that they must control their temperatures within narrow limits. How do birds keep their temperatures within the necessary range?

Two types of feathers help insulate   Most birds live in environments where the temperature is lower than their body temperature. Heat energy tends to move from hot areas to cold. Any heat lost by a bird must be replaced by burning more food. Birds need to retain as much heat as possible. One of the most efficient insulators on Earth is the layer of down feathers close to a bird's skin. These are covered by a tight coat of contour feathers. Working together, these two trap a layer of air and prevent heat from moving from the skin outward.

Circulation helps regulate body temperatures   When it is cold, the blood vessels in a bird's skin contract and lessen the flow of blood. When the vessels contract, less blood goes to the surface. This cuts heat loss. Conversely, if the bird is too hot, the blood vessels dilate, releasing heat to the body surface. Birds can also hold their contour feathers apart to release heat.

Air sacs increase oxygen supply   Birds have an efficient respiratory system. It includes not only lungs but a number of air sacs. Water in the air sacs evaporates to cool birds. When the air is exhaled, the heat is carried away with water vapor. This is similar to the way you perspire. The difference is that you perspire on the outside, but the bird does it internally.

CONNECTION TO   Biology

People often use the expression, "You eat like a bird!" to mean that a person eats very little. From a biological viewpoint, explain why this expression doesn't make sense.

Birds' bodies are adapted for flight

A second adaptation for flight in birds is the modification of the front limbs into wings. Powerful flight muscles are attached to a large breastbone called the **sternum** and to the upper bone of each wing. The sternum looks like a blade-shaped anchor and is important because it supports the enormous thrust and power produced by the muscles as they move to generate the lift needed for flight.

Flight requires high levels of energy. Several factors are involved in maintaining these high energy levels. First, a bird's four-chambered, rapidly beating heart moves oxygenated blood quickly throughout the body. This efficient circulation supplies the cells with the oxygen needed to produce energy. Second, birds are able to maintain high energy levels because they are endotherms. An **endotherm** is an animal that maintains a constant body temperature that is not dependent on the environmental temperature.

Birds have a variety of ways to save or give off their body heat in order to maintain a constant body temperature. Feathers reduce heat loss in cold temperatures. The feathers fluff up to trap a layer of air that limits the amount of heat lost. Responses to high temperatures include flattening the feathers and holding the wings away from the body. Birds will also pant to increase respiratory heat loss.

A major advantage of being endothermic is that birds can live in all environments, from the hot tropics to the frigid Antarctic. However, birds and other endotherms must eat large amounts of food to sustain these higher levels of energy.

Flight

You may have envied birds and their ability to fly. Humans have always dreamed of being able to rise free of earthbound problems. The popularity of hang gliding and parachute jumping may reflect these dreams. In order for birds to fly, they were subjected to complex selective pressures that resulted in the evolution of many adaptations.

Blue jay

2 Birds' bones are thin and hollow, thereby maintaining low weight and making flight easier. The hollow bones of birds are strengthened by bony crosspieces. The sternum is the large breastbone to which powerful flight muscles are attached.

Hollow bone

3 Birds have horny beaks. The lack of a bony jaw reduces the birds' weight even further. Birds do not have teeth.

Wing

Beak

1 Birds have a variety of wing shapes and sizes. Some birds have wings adapted for soaring on updrafts, whereas others have wings adapted for quick, short flights among forest trees.

Lung

Air sacs

Sternum

Gizzard

4 The digestive system of a bird is adapted for dealing with large quantities of food that must be eaten to maintain the needed level of energy for flight. As birds have no teeth, many swallow small stones that help to grind up food in the gizzard.

Bone

Intestine

Leg

6 About 75 percent of the air inhaled by a bird passes directly into the air sacs rather than into the lungs. When a bird exhales, oxygenated air in the air sacs passes into the lungs. Birds receive oxygenated air when they breathe in and when they breathe out.

5 The legs of birds are made up of skin, bone, and tendons. The feet are adapted to perching, swimming, walking, or catching prey.

34.2 Birds **849**

Have you ever watched a bird balance itself on a telephone pole wire? Some birds have feet that are adapted for gripping tree branches. This adaptation prevents them from falling off a branch even when they are asleep.

What is the ideal length and width for a bird's tail?

The tail of a bird is an important structure for balance. When a bird first touches down on a branch, it flicks its tail up and down a few times. The bird uses its tail to find its balance point. Ducks have short tails because a duck lands in the water. Magpies have tails as long as their bodies. When a magpie lands on the ground, it must land with precision to avoid falling over on its bill. The long tail gives precise balance. In this lab, you'll design the best tail for a perching bird.

PREPARATION

Problem
For a perching bird of a particular size, what type of tail will be best for balance? Will a broad tail be better than a narrow tail? Will a long tail be better than a short tail?

Hypotheses
Make a hypothesis about the kind of tail that would best balance a medium-sized bird on a thin branch. Explain the evidence you have used to develop your hypothesis.

Objectives
In this Biolab, you will:
- **Determine** some of the factors in a bird's tail that are important for balance.
- **Compare** the importance of width, length, and angle of attachment for a bird's tail.

Possible Materials
tagboard
tracings of birds' bodies and tails
pencil
paper fastener
protractor
mm ruler
scissors

PLAN THE EXPERIMENT

1. Trace the outline of the bird's body and tail onto tagboard.

2. Punch out the holes indicated by the dotted lines, and attach the tail to the body with a paper fastener.

3. Insert a pencil through the larger hole. Make sure the bird can move easily on the pencil.

4. Bend the tail back and forth until the bird balances. Measure the angle at which the balance point is achieved.

5. Design an experiment in which you test the effect of different tail lengths or widths on the balance point.

Check the Plan

1. Make sure that you are using only one variable.

2. How many trials will you conduct?

3. How will you make your measurements consistent from one trial to the next?

4. What is your control?

5. How will you show your data graphically?

6. ***Make sure your teacher has approved your experimental plan before you proceed further.***

7. Carry out your experiment.

8. Record your data in a table you have prepared.

ANALYZE AND CONCLUDE

1. **Drawing Conclusions** In your experiment, how did length or width of the tail affect the balance point?

2. **Checking Your Hypothesis** Was your hypothesis supported by your data? Why or why not?

3. **Comparing and Contrasting** Compare your results with those of another group. How do they compare?

4. **Making Inferences** Live birds are three dimensional. Would this fact alter your results? Explain.

5. **Thinking Critically** Name functions, other than balancing, of some birds' tails.

Going Further

Changing Variables
Conduct another experiment with bird tails, but this time choose another variable you may have thought about during your experiment. These might include body size, weight of the tail itself, or weight on various parts of the tail.

Owl

Ptarmigan

Adele penguins

Figure 34.14

Examine these birds and infer where they live and how they are adapted to their environments.

Bird Adaptations

Unlike reptiles that take on a wide variety of forms, from legless snakes to shelled turtles, birds are all very much alike in their basic structure. You have no difficulty recognizing a bird.

In spite of the basic uniformity of birds, they do exhibit specific adaptations, depending on the environment in which they live and the food they eat. As shown in *Figure 34.14,* ptarmigans have feathered legs that serve as snowshoes in the winter, making it easier for the birds to walk in the snow. Penguins are flightless birds with wings and feet modified for swimming and a body surrounded with a thick layer of insulating fat. Large eyes, an acute sense of hearing, and sharp claws make owls well-adapted, nocturnal predators, able to swoop with absolute precision onto their prey. Can you guess why owls' eyes aren't on the sides of their heads?

Bird beaks are a reflection of the food they eat. For example, a swallow is adapted to feeding on dead insects but not fish or nectar from flowers. Other birds spear fish with long, pointed beaks. Hummingbirds, *Figure 34.15,* have long beaks that are used for dipping into flowers to obtain nectar. Hawks have large, curved beaks that are adapted for tearing apart prey. Pelicans have huge beaks with pouches that they use as nets for capturing fish. The short, stout beak of the goldfinch is adapted to cracking and eating seeds.

Brown pelican

Calliope hummingbird

Figure 34.15

The beaks of hummingbirds, hawks, pelicans, and goldfinches are adapted to different kinds of food. Based on what they eat, how do you think their feet might be modified to assist in getting food?

Galapagos hawk

American goldfinch

Big Birds

O f the 130 species of birds known to have become extinct over the past 300 years, many were strange island birds that developed into flightless giants in isolation from predators. Carnivores, which quickly came to dominate the animal kingdom on the continents, were not native to islands such as New Zealand and Madagascar, home of the largest birds known to man. These fearless birds were able to gain great size by foraging on the ground for food. Over time, they lost the ability to fly, and their wings became vestigial.

The biggest eggs

Elephant bird eggs
Compared to the common hen's egg, the egg laid by the extinct elephant bird was huge (about 8 kg and able to hold up to 8 L). Fragments of eggs and entire eggs are still being found washed up on the beaches and lakeshores in Madagascar.

Which is the largest bird?

The African ostrich and the elephant bird The largest currently living bird, the African ostrich, is no match for the extinct elephant bird of Madagascar, the largest bird ever known. Three meters tall and weighing up to 456 kg, the elephant bird weighed at least three times as much as the African ostrich. Standing more than 2.5 m tall and weighing more than 150 kg, the African ostrich is the world's fastest bird, with speeds up to 37 kph.

What is a dodo?

The trusting bird When 16th-century sailors landed on the beach at Mauritius, an island in the Indian Ocean, they found a curious, flightless bird they called dodo, or silly. Weighing some 22 kg, it seemed to have no fear of people. These early colonists caught dodos for food and brought predators such as dogs, which quickly decimated dodo populations. Within 200 years, all of the dodos were killed.

Largest nests built

Eagles add on yearly The endangered bald eagle is the builder of the largest tree nest in the world. Measuring up to 3 m wide and 6 m deep, an eagle's nest can be active for more than 20 years. Unfortunately, pesticide use has severely lowered the success rate of the bald eagle's reproductive efforts. About 1200 breeding pairs of bald eagles exist in the United States today.

The greatest wingspan

The wandering albatross The wandering albatross has the greatest wingspan of any living bird, measuring nearly 3.5 m.

Too fat to fly

The great auks Like the dodo, the great auk was another clumsy, flightless bird, but one with remarkable swimming skills. The great auks had plentiful supplies of fat, for which they were hunted to extinction. Standing about 0.5 m high, with great webbed feet and black-and-white markings, the great auk looked much like a penguin. The auks were unrelated to penguins and were isolated on islands in the northern hemisphere. On Eldey Island off the coast of Iceland, the last of the great auks was killed by sailors on June 4, 1844.

EXPANDING YOUR VIEW

1. **Understanding Concepts** How is the evolution of flightless birds related to their environment?
2. **Journal Writing** In a short paragraph, explain the major differences between the modern flightless birds such as penguins and an extinct bird such as the dodo.

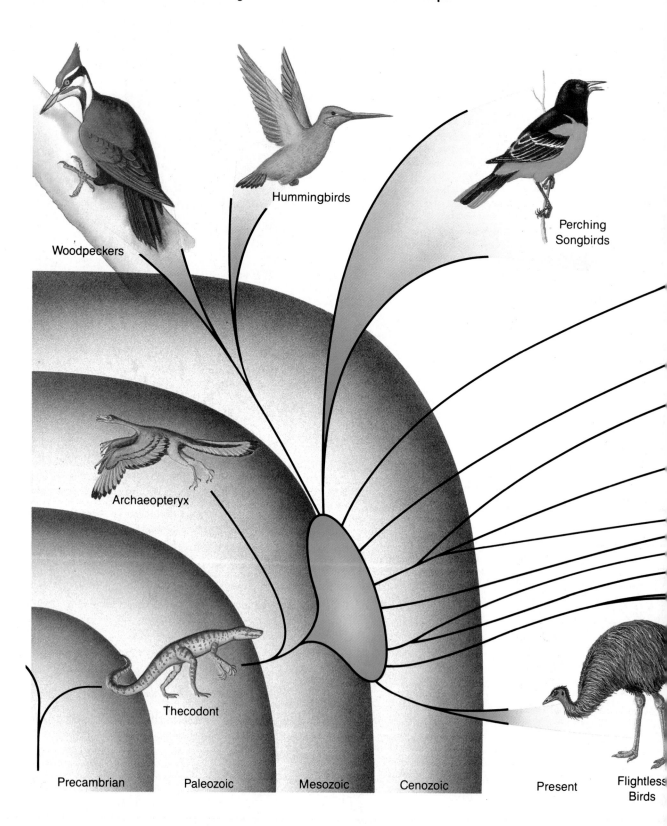

Figure 34.16

The complete evolutionary history of birds is not clear. The fossil record of birds is incomplete because bird skeletons are light and delicate and, therefore, are easily destroyed by carnivores and other natural causes before they can be preserved. The radiation of orders of birds on the Geologic Time Scale shows their relationships.

Woodpeckers

Hummingbirds

Perching Songbirds

Archaeopteryx

Thecodont

Precambrian Paleozoic Mesozoic Cenozoic Present Flightless Birds

The Orders of Birds

Kingdom Animalia
Phylum Chordata
Class Aves
8600 species

Owls

Quails, Turkeys, and Chickens

Parrots

Pigeons

Gulls, Sandpipers, and Puffins

Hawks, Eagles, and Falcons

Penguins

Swans, Geese, and Ducks

Pelicans

Archaeopteryx   Protoavis

Origins of Birds

Current thoughts about bird evolution are illustrated in *Figure 34.16.* You know that some scientists think today's birds are related to an evolutionary line of dinosaurs that did not become extinct. *Figure 34.17* shows the earliest known bird in the fossil record, *Archaeopteryx.* At first, scientists thought that *Archaeopteryx* was a direct ancestor of modern birds; however, some paleontologists now think that it most likely did not give rise to any other bird groups. *Archaeopteryx* was about the size of a crow and had feathers and wings like a modern bird. But it also had teeth, a long tail, and clawed front toes, much like a reptile.

Another fossil discovered in 1984 has hollow bones and a well-developed breastbone with a keel, which are characteristics of birds. Named *Protoavis*, this fossil is thought to be similar to birds, yet it is 75 million years older than fossils of *Archaeopteryx.* However, *Protoavis* has hind legs, a pelvis, and a bony tail which resemble those of ground-dwelling dinosaurs, *Figure 34.17.*

Connecting Ideas

You have seen how reptiles dominated the Mesozoic Era. Modern reptiles also have forms as varied as legless snakes, turtles with rocklike shells, and crocodiles with huge snapping jaws.

Birds, however, are characterized by their uniformity. Flying necessitates uniformity; feathers, lightweight bones, air sacs, beaks, and other features are modifications for flight. Flight restricts the diversity commonly seen in other vertebrates. For example, mammals include such diverse groups as whales, mice, porcupines, bats, and buffalo.

Section Review

Understanding Concepts
1. What body structure found in reptiles but not in birds does the gizzard replace?
2. Explain why a bird's respiratory system is extremely efficient.
3. Why does being an endotherm have adaptive value?

Thinking Critically
4. Large, flightless birds are most common in areas that do not have large, carnivorous animals. What hypothesis can you suggest for the evolution of large, flightless birds?

Skill Review
5. **Making and Using Tables** Make a table that summarizes the adaptations birds have that enable them to fly. For more help, refer to Organizing Information in the *Skill Handbook.*

Zoos: Preserves or Breeding Grounds?

Your Local Zoo: An Endangered Animal

SCIENCE REPORTS   AUGUST 24, 1992

As biologists often comment, when species are faced with a change in the environment, they can either adapt or die. And so it will be with zoos.

Your local zoo is fast becoming endangered. Faced with rising costs on top of negative press from critics who claim they are no better than prisons and should be shut down, modern zoos are struggling to survive. Many metropolitan zoos are questioning their existence and redefining their mission. The most dedicated zoos—and the richest—have started natural habitat programs and are experimenting with captive breeding.

Zoos help educate Originally created for public entertainment, zoos began educational programs and species conservation in the 1940s. Zoo reform began in the 1980s in response to public outrage over poor conditions. Today's most modern zoos—such as the San Diego and Cincinnati Zoos—are using the latest reproductive techniques (artificial insemination and test-tube fertilization) to save endangered species and improve the gene pools of others. Through such innovative efforts, in-bred female Siberian tigers have successfully been used as surrogate mothers for Bengal tigers.

Less need to capture wild animals Through captive-breeding programs, zoos are able to lessen their reliance on wild populations for their exhibits. The more sophisticated breeding laboratories are able to freeze and save genetic materials from endangered species for future use.

Different Viewpoints

Many biologists argue that the mission modern zoos set for themselves is unattainable. Of the 10 million species on the planet, one-third are likely to become extinct over the next century. They argue that biodiversity is more likely to be preserved through public education than through captive-breeding programs, whose success rate they feel is exaggerated.

Animal rights activists accuse zoo managers of species favoritism. Unable to save all endangered species, they reason that only the most attractive or popular are selected. The most strident zoo critics feel extinction is better than confinement in an artificial environment.

INVESTIGATING the Issue

1. **Debating the Issue** What should be the role of the modern zoo?
2. **Analyzing Cause and Effect** Most zoos are financially unable to reform their programs to meet their new mission as wildlife conservationists. What kinds of fund-raising efforts do you think are appropriate for zoos?
3. **Writing About Biology** Write a paper researching the history and development of a famous metropolitan zoo such as the Bronx Zoo in New York City.

34.1 Reptiles

- Reptiles are ectotherms that have dry, scaly skin; legs under the body; internal fertilization; and amniotic eggs. Most reptiles have three-chambered hearts. Some reptiles have four-chambered hearts.
- Present-day reptiles belong to one of four groups. Turtles have shells and no teeth. Crocodiles and alligators have streamlined bodies and powerful, toothed jaws. Lizards have a variety of adaptations including long bodies, tails, and short limbs. Snakes have no limbs.
- The ancestors of present-day reptiles arose from ancient cotylosaurs.

34.2 Birds

- Birds have adaptations for flight including feathers; sternum; four-chambered heart; endothermy; thin, hollow bones; a beak; and air sacs.
- Birds may be related to a line of dinosaurs that did not become extinct. The fossils *Protoavis* and *Archaeopteryx* reflect features of modern birds.

Key Terms

Write a sentence that shows your understanding of each of the following terms.

amniotic egg	Jacobson's organ
endotherm	sternum
feather	

Understanding Concepts

Using a Scientific Illustration

1. Examine the diagram of the amniotic egg in *The Inside Story*, and explain the relationships among the amnion, chorion, yolk sac, and allantois.

Relating Concepts

2. Make a concept map that relates the following terms and phrases. Supply the appropriate linking words for your map.

   amnion, chorion, allantois, yolk, shell, internal fertilization, reptile, terrestrial

3. How are some snakes adapted to eating prey larger than themselves?

4. What characteristic of reptiles enables them to be more active than amphibians?

5. Describe two ways in which birds are similar to reptiles.

6. Describe two adaptations in birds that enable them to maintain high energy levels.

7. Explain the difference between flight and down feathers.

Using a Graph

8. A biologist is comparing yearly censuses of owls and mice found in one area. The data obtained are represented in the graph below. Explain what might be causing the fluctuating population of owls.

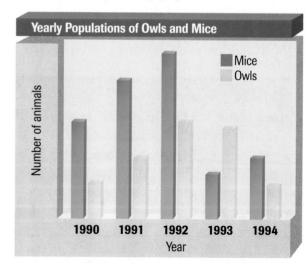

Yearly Populations of Owls and Mice

Number of animals

Mice
Owls

1990 1991 1992 1993 1994
Year

Applying Concepts

9. Part of your parakeet's diet contains gravel. Explain why.

10. Most reptiles lay between one and 200 eggs at a time. Amphibians lay thousands of eggs at a time. Is there an adaptive value to laying fewer eggs on land? Explain.

11. Why are birds able to inhabit more diverse environments than reptiles?

12. Why don't birds' feet freeze when they perch on tree branches in the winter?

13. No snakes live on the Hawaiian Islands. How would introducing snakes to these islands affect the environment there?

A Broader View

14. Newly developed industries that locate on shorelines often release chemicals into the ocean. How would this affect sea turtles that migrate to these areas to breed? Explain.

Thinking Critically

Developing a Hypothesis

15. **Biolab** Assume that you are walking through an aviary in a zoo. You notice that the cranes have long, thin legs. Make a hypothesis about the type of habitat in which these birds might live. Justify your hypothesis.

Relating Cause and Effect

16. **Minilab** Every year since you were a child, your family has taken a vacation near the Florida everglades. Over the years, modern hotels, homes, and roads have been built in the area. Last year, you noticed that you rarely see the herons, ibises, and other birds that are native to the area. What might cause the reduced number of sightings?

Making Inferences

17. **Minilab** A biologist counts the feathers on the bodies of two different species of birds. The data are represented in the graph below. What might be inferred about the type of environments in which the birds live?

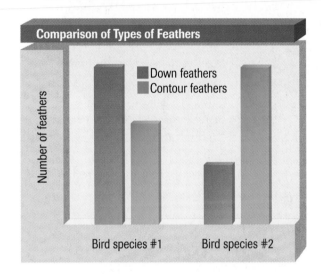

Connecting to Themes

18. **Homeostasis** Explain how the amniotic egg maintains homeostasis.

19. **Evolution** Discuss why the fossils of *Archaeopteryx* and *Protoavis* are significant in explaining the evolutionary history of birds.

20. **Energy** Compare and contrast the methods that reptiles and birds use to stay warm.

21. **Systems and Interactions** Describe reptile adaptations to life on land.

Dinosaurs

The worlds of the dinosaurs

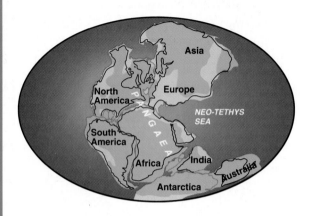

One landmass During the late Triassic period, about 210 million years ago, the continents as we know them today did not exist. The surface of Earth included vast seas and one large landmass, Pangaea. Almost imperceptibly, however, Pangaea was splitting apart, and separate continents began drifting across the globe. New species of animals began to emerge. These animals, descended from ancient reptiles and destined to rule Earth for the next 130 million years, were the dinosaurs. The large dinosaur at the far right is a *Plateosaurus*. The two smaller ones are *Ceolophysis*. An early crocodile is in the water.

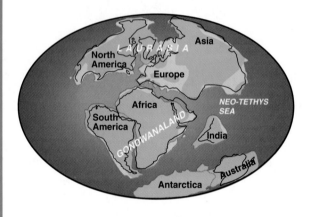

New species emerge By the late Jurassic period, the modern continents were beginning to take shape, although Africa still clung stubbornly to South America and Australia remained attached to Antarctica. The temperature was warming up, enabling life to flourish in new areas around the world. The dinosaurs, too, were flourishing. The dominant plant life during this time included tree ferns and conifers. Examples of dinosaurs living during the Jurassic period were *Stegosaurus* (middle), *Diplodocus* (background), and *Allosaurus* (right).

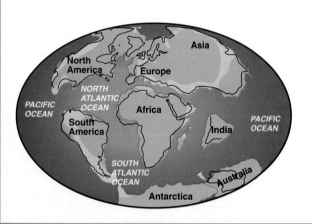

The beginning of the end By the late Cretaceous period, 100 million years ago, the continents had almost assumed the shape and position they have today. New vegetation, including flowering plants, and new species of dinosaurs, such as *Tyrannosaurus rex* (right) appeared. Other dinosaurs that lived during this period included *Pachycephalosaurus* and the carnivorous *Deinonychus* (left). *Sinornis,* a seven inch long link to modern birds, is also shown. About 65 million years ago, the dinosaurs died out. Exactly why is still a mystery. Today, only fossil evidence remains to tell their fascinating story.

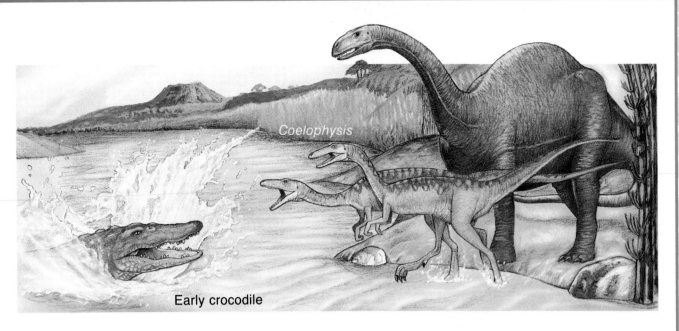

Coelophysis

Early crocodile

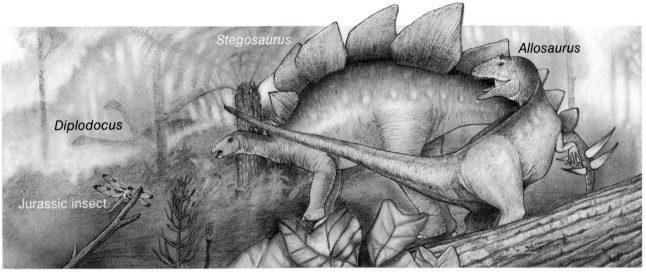

Stegosaurus

Allosaurus

Diplodocus

Jurassic insect

Deinonychus

Sinornis

Tyrannosaurus rex

Pachycephalosaurus

Classifying dinosaurs

Paleontologists have identified several hundred species of dinosaurs, some as small as the seven-inch *Sinornis* and others as huge as the *Siesmosaurus,* more than 100 feet long. Some were meat eaters, and others browsed on vegetation; some had relatively smooth skin, while others were covered with horns and bony plates.

Despite such diversity, scientists have devised a straightforward means of classifying dinosaurs. The system first classifies dinosaurs into two main groups according to the skeletal structure of their hips: the ornithischians and the saurischians.

Ornithischians: The bird-hipped dinosaurs Although descended from ancient reptiles, ornithischians developed a hip structure similar to that of modern birds, with a process of the pubic bone, shown here in red, angled backwards.

Saurischians: The reptilian-hipped dinosaurs The saurischians retained their reptilian hip structure, with the pubic bone (shown in red) projecting forward. Ironically, modern birds are thought to be more closely related to reptile-hipped saurischians, not the bird-hipped ornithis-chians!

Typical of the ornithischians was *Iguanodon,* an early Cretaceous species.

Two hundred million years ago, the saurischian *Dilophosaurus* roamed North America.

Finding, excavating, and reconstructing dinosaurs

At the site Dinosaur fossils are found in sedimentary rock, layers of rock that have built up over millions of years. When paleontologists stumble across a major fossil find, they set up a dig site that can include everything from dynamite to sophisticated biotechnical instruments. Workers may camp at the site for many months as they painstakingly extract fossils from rock.

Tools of the trade Freeing fossils from sedimentary rock requires patience and the right equipment. Tools include hammers, saws, chisels, brushes, and, for extremely delicate work, dental picks. Workers wear goggles to shield their eyes, gloves to protect their hands, and sometimes masks to avoid breathing in too much dust or fumes from toxic chemicals used to preserve fossils.

On display This *Diplodocus* stands in the Museum of Natural History, Houston, Texas. Paleontologists often disagree about how to pose dinosaur skeletons, but as new information and new discoveries come to light, such reconstructions offer people an increasingly accurate picture of these prehistoric creatures.

Back to the lab Today, paleontologists have access to sophisticated biomedical equipment with which to study their finds. This *Nanotyrannus* skull was x-rayed using a CT scanner that produces three-dimensional images. The pictures revealed the size and location of the carnivore's braincase.

Focus On

Dinosaur behavior

Feeding time Some dinosaurs were carnivores, others were herbivores, and some, like humans, were omnivores. Carnivores, like *Allosaurus* (top), dined on fish, insects, birds, reptiles, and, of course, other dinosaurs. Their jaws were equipped with rows of curved, sawlike teeth.

Herbivores, like *Diplodocus* (bottom), often browsed treetop vegetation such as evergreens. Lacking many teeth and unable to chew their food, most herbivores swallowed vegetation whole. In the stomach, the food would be ground up by gizzard stones and shed teeth the dinosaur had swallowed, and fermented by bacteria.

Nesting Dinosaurs laid their hard-shelled eggs in hollowed-out nests. Sometimes, several dinosaurs would nest close together, indicating that some dinosaurs lived in herds.

Duck-billed dinosaurs probably covered their nests with vegetation which would decay and thereby provide warmth to hatch the eggs.

Social or solitary? Fossil bones and footprints strongly suggest that many dinosaur species traveled in great herds. Paleontologists have discovered many bone beds of ornithiscian dinosaurs. Some contain the remains of more than 10 000 individuals. Trackways that record the passage of hundreds of dinosaurs have been discovered.

Offense and defense Dinosaurs were well equipped for attacking prey and for defending themselves against attacks from predators. *Velociraptor* used its razor-sharp, sickle-shaped claws to slash at prey. *Stegosaurus* had sharp tail spikes that could be whipped back and forth to protect it from predators.

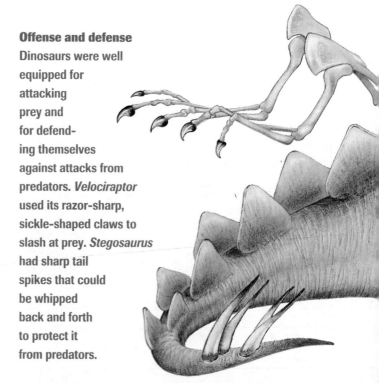

Theories about dinosaurs

Allosaurus

Warm or cold? Scientists long believed that dinosaurs, like lizards and other reptiles, were cold-blooded creatures dependent on the sun and shade for warming and cooling. However, noted paleontologist Robert Bakker is convinced that all dinosaurs were warm-blooded, energetic animals. Dinosaur bones, like those of birds, are laced with small canals. Reptile bones have no such features. Furthermore, dinosaur fossils have been found on land that would have once been within the Arctic circle, the wrong environment for cold-blooded land dwellers but perfectly hospitable to warm-blooded creatures.

Extinction Near the end of the Cretaceous period, about 65 million years ago, dinosaurs suddenly disappeared from Earth's landscape. Why? Some scientists are convinced that a large meteorite struck Earth, filling the atmosphere with huge clouds of dust and chemical pollutants that may have circled the planet and blocked sunlight from reaching the ground for several years. As vegetation disappeared, the herbivores would have died out. Lacking herbivores on which to feed, the carnivores would have disappeared also. Compelling evidence exists for this theory, but no conclusive proof has been found.

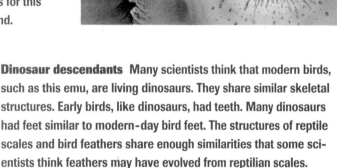

Dinosaur descendants Many scientists think that modern birds, such as this emu, are living dinosaurs. They share similar skeletal structures. Early birds, like dinosaurs, had teeth. Many dinosaurs had feet similar to modern-day bird feet. The structures of reptile scales and bird feathers share enough similarities that some scientists think feathers may have evolved from reptilian scales.

EXPANDING YOUR VIEW

1. **Understanding Concepts** Explain some ways in which dinosaurs were adapted to their environment. What was the principal result of these advantageous adaptations?

2. **Research** Investigate the theory that the dinosaurs disappeared as the result of a mass extinction caused by a giant meteor colliding with Earth. What evidence exists to support this theory? Why do some scientists doubt its validity? Share your findings with the class in an oral report. Provide graphic aids to enhance the impact of your report.

The platypus, a mostly aquatic animal, looks as if it were designed by a committee. It has a broad, flat tail, much like that of a beaver. Its rubbery snout resembles the bill of a duck. The platypus has webbed front feet for swimming through water, but it also has sharp claws on its front and hind feet for digging and burrowing into the soil. Much of its body is covered with thick, brown fur.

The spiny anteater is only slightly less strange in appearance. It has coarse, brown hair, and its back and sides are covered with sharp spines that it can erect for defensive purposes when threatened by enemies. From its mouth, located at the end of a long snout, the anteater extends its long, sticky tongue to catch insects.

Figure 35.12

The duck-billed platypus seems like a jumble of physical features that belong to a variety of other animals. Only one species survives today.

Monotremes

When is a mammal not exactly like a mammal? When it's a monotreme, of course.

A living link to the past Monotremes are evolutionary oddities within the class *Mammalia.* Possessing characteristics of both mammals and reptiles, monotremes offer biologists a rare opportunity to study an animal that seems almost frozen in time—the time when mammals were evolving from reptiles into a distinct class of their own.

Duckbills and anteaters The order *Monotremata* includes just three living species: platypuses, also known as duckbills, and two species of echidnas, or spiny anteaters. Platypuses are found in eastern Australia and Tasmania; echidnas are native to both Australia and New Guinea.

Mammal or reptile? Monotremes are grouped with mammals because they share important biological similarities. Like other mammals, monotremes have hair, a large brain, and a diaphragm. They also have mammary glands from which they secrete milk to feed their young. Unlike other mammals, however, monotremes have no teats. The milk oozes directly from pores in the monotreme's skin.

Monotremes lay eggs Unlike all other mammals, monotremes do not give birth to fully developed young. Like reptiles, female platypuses and spiny anteaters lay tough, leathery eggs. Platypuses lay their eggs, usually just two, on the ground and then curl their bodies around them for warmth. This incubation period lasts about 12 days before the young hatch. Spiny anteaters develop an abdominal pouch during their breeding season. A single egg is laid into this pouch, where it incubates for about ten days. After hatching, the newborn anteater lives in its mother's pouch for a brief period of time.

Thinking Critically

Marsupial young, such as kangaroos, also live in a maternal pouch for a time after birth. How does this unusual evolutionary adaptation confer an advantage to marsupials?

Placental Mammals

Most of the more than 4300 species of mammals are placental mammals. Although all placental mammals develop within the mother's uterus, the condition of a newborn placental mammal varies. Newborn gazelles can run fast enough to keep up with the herd shortly after birth. Young kittens are blind and helpless. A human baby spends many years with its parents before it can take care of itself. There are 17 orders of placental mammals, of which 13 are shown here. The remaining four orders are pictured in Figure 35.13 following this feature.

Order Insectivora

Although they all eat insects, insectivores are a mixture of mammals that have few other common characteristics. Like other insectivores, the shrew has an extremely high metabolic rate.

Order Chiroptera

All bats are classified in this order. The bat is the only flying mammal. It has skin that stretches from the body, legs, and tail to the arms and fingers to form thin, membranous wings.

Bats continuously emit high-pitched sounds while in flight in a process called echolocation. Upon hitting an object, the sound waves bounce back to the bat. The waves are picked up by the bat's ears, allowing it to accurately locate the object. The vampire bats of South America suck blood from domestic animals such as cattle and horses. They carry and transmit diseases such as rabies to these animals.

Order Primate

Pygmy marmosets live in the upper canopy of the Amazon rain forest. Except for nursing, the father takes care of the offspring. The outstanding characteristic of these mammals is their keen intelligence. Most primates have complex social lives. Members of this order can be divided into three groups: lemurlike, monkeylike, and humanlike.

Order Edentata

Giant anteaters have no teeth and feed on insects. They have strong forelegs and claws suited for digging in termite mounds. Anteaters, armadillos, and tree sloths are the only living members of this order. They are confined to Central and South America and to the southern regions of North America.

Order Lagomorpha

Pikas live in large communities in mountains and grasslands. They collect stacks of dried vegetation that they place next to their rocky homes. Rabbits, pikas, and hares are among the most numerous mammals. Lagomorphs are characterized by a fusion of bones in their hind legs where they move against the ankle bones. This fusion allows the bounding and leaping movement seen in these animals.

Order Rodentia

The golden hamster is a popular pet. Larger species have been known to destroy entire fields of wheat in Europe and Russia. The largest order of mammals, rodents live in all environments. They are identified by their continuously growing, razor-sharp teeth. They must gnaw on hard seeds, trees, bark, twigs, and roots. Rodents include beavers, porcupines, and chipmunks, as well as the more common mice and rats.

Focus On

Order Cetacea

The killer whale is the fastest swimmer and most maneuverable of all whales. They hunt for seals and porpoises in groups known as pods. Dolphins, porpoises, and whales are placed in this order. They have little or no hair, and breathe through blowholes on the tops of their heads.

Order Carnivora

Like all carnivores, a weasel has long, pointed canines and incisors, strong jaws, and long claws. A weasel can kill an animal several times its own size. Some of the best-known mammals, including dogs and cats, are placed in this group. While most of these mammals are flesh eaters, some, such as bears, do consume plant material as well.

Order Pinnipedia

The walrus is a huge mammal that lives in the Arctic. Males use their tusks to find mollusks and sea urchins, which they swallow whole. They also swallow pebbles to grind up the food in their stomachs. This order includes seals, sea lions, and walruses. They mostly live in cold-water habitats.

Order Proboscidea

Proboscideans have flexible trunks that are used to gather vegetation for eating and suck up water for drinking. One pair of incisors is modified into large tusks used for digging up roots and stripping bark from trees. The largest land animals now living, African elephants can be distinguished from Asian elephants by their larger ears. They spend most of their time eating.

Order Perissodactyla

The wild Przewalski's horse from Mongolia looks similar to the ancestor of all modern horses. These hairy horses have thick legs and sturdy bodies. Zoos throughout the world have breeding programs to save this species. Several species of mammals have toenails that have become modified into hooves. Mammals with an odd number of toes belong to the Order Perissodactyla. Most hoofed mammals are herbivores with molars used for grinding.

Order Sirenia

Dugongs live in the oceans. They have tails and front flippers like whales, distinct heads with a snout that points downward, and short necks. Nicknamed sea cows, dugongs and manatees are aquatic mammals that are distantly related to elephants. Only four species survive today.

Order Artiodactyla

Found along African rivers, hippos eat enormous amounts of vegetation at night. They spend the warm days laying in the water so that their skin doesn't dry out. About 200 species of these hoofed, even-toed mammals are alive today. These mammals have multiple stomachs.

EXPANDING YOUR VIEW

1. **Understanding Concepts** Explain, by using examples, how mammals have become so successful.

2. **Writing** Go to a toy store and inventory the kinds and numbers of mammals in the stuffed animal section. Write a report detailing what the data might reflect about the view of society concerning certain mammals.

Figure 35.13

Four smaller orders of placental mammals are represented by only one or a very few species.

◄ Hyraxes, order Hyracoidea, look like rabbits but have teeth like rhinoceroses and hooves on their toes.

▼ Aardvarks, order Tubulidentata, dig up termites and eat them.

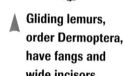

▲ Gliding lemurs, order Dermoptera, have fangs and wide incisors.

▲ Pangolins, order Pholidota, are anteaters that are covered with horny scales. They are powerful animals.

Connecting Ideas

Mammals are an integral part of your life. The gorilla intrigues and delights you with its humanlike antics. You laugh at the kitten playing with a ball of string. You may have ridden a horse or taken part in a whale watch. It's astonishing to think that all modern mammals evolved from the ancestors of insectivores in the time when dinosaurs moved across the land. During the evolution of mammalian structures, how do you suppose the evolution of behavior contributed to the success of mammals?

Section Review

Understanding Concepts

1. Describe the characteristics of placental mammals.
2. Compare monotremes and marsupials.
3. Name three characteristics that are used to classify mammals into one of the orders illustrated in the Focus On Mammals feature.

Thinking Critically

4. You are a zoologist observing animals with a wildlife photographer in Australia. A local rancher brings you a rabbitlike animal and asks you to identify it. By looking at the skin covering the animal's abdomen, you know it is not a rabbit. How did you know? Explain.

Skill Review

5. **Observing and Inferring** You find a mammal fossil and observe the following traits: hooves, flattened teeth, skeleton the size of a large dog. What can you infer about its way of life? For more help, refer to Thinking Critically in the *Skill Handbook*.

What Price Beauty?

FDA Backs Animal Testing

THE DAILY NEWS   OCTOBER 3, 1992

In a move sure to fuel the debate over animal rights, the Food & Drug Administration (FDA) today issued a formal Position Paper reaffirming its support for the use of animals to test the safety of health-and-beauty aids.

Although the Food, Drug, & Cosmetic Act of 1938 does not require cosmetic manufacturers to test their products before marketing them, the FDA strongly urges such testing. Before this law was passed, cosmetics commonly contained such toxic chemicals as arsenic, mercury, and quinine.

Protecting consumers Manufacturers of cosmetics and other toiletries have complied with the FDA's request for testing in order to ensure public safety and avoid potentially costly lawsuits. Consumers, manufacturers, and the government all agree that such testing is essential. What they don't agree on is the type of tests.

Guinea pigs For decades, companies specializing in health-and-beauty aids routinely tested

their products on animals. During the 1980s, many people began to question this. Some concerned groups asked the Cosmetic, Toiletry & Fragrance Association (CTFA) to urge its member companies to suspend all tests on animals. A few manufacturers agreed to halt all animal testing; others decreed a moratorium, or temporary delay, in such testing; and others continued using animals to test for toxicity.

Different Viewpoints

People who claim that using animals to test the safety of cosmetics is unnecessary and cruel point to new procedures, including computer modeling and the use of cultured artificial skin, as possible alternatives. Some companies are exploring the use of protists and other unicellular organisms as test subjects. But it is still not known if results from such tests can be applied directly to human use of the product. Some scientists have never believed that testing products on animals gives reliable data for human use.

Where do consumers stand on the issue? A 1990 consumer survey of 1000 adults found that 60 percent opposed using animals to test the safety of cosmetics and toiletries. Furthermore, 89 percent said that they would buy products that had not been tested on animals.

INVESTIGATING the Issue

1. **Debating the Issue** Along with your classmates, research the issue of testing cosmetics and toiletries on animals, and stage a debate that focuses on facts rather than emotions.
2. **Writing About Biology** Develop a public opinion survey that will uncover views on the use of cosmetics. Tabulate your data according to the ages, gender, education, and occupations of your respondents, and prepare a written analysis of the results.

Reviewing Main Ideas

35.1 Mammal Characteristics
- The first mammal-like animals were therapsids.
- Mammals are endotherms with hair, diaphragms, modified limbs and teeth, highly developed nervous systems and senses, and mammary glands.

35.2 The Diversity of Mammals
- Mammals are classified into three subclasses—monotremes, marsupials, and placentals—based on the way they reproduce.
- Monotremes are egg-laying mammals found only in Australia, Tasmania, and New Guinea.
- Marsupials carry partially developed young in pouches on the outside of the mother's body. Most marsupials are found in Australia and surrounding islands.
- Placental mammals carry young inside the uterus until birth. The young are nourished through an organ called the placenta.

- Most mammal species alive at present are placental mammals.
- There are 17 orders of placental mammals that include animals found on land, in the air, and in the sea.
- The orders of placental mammals are: Insectivora, Chiroptera, Primate, Edentata, Lagomorpha, Rodentia, Sirenia, Cetacea, Proboscidea, Carnivora, Pinnipedia, Artiodactyla, Perissodactyla, Hyracoidea, Pholidota, Tubulidentata, and Dermoptera.

Key Terms
Write a sentence that shows your understanding of each of the following terms.

cud chewing	monotreme
diaphragm	placenta
gestation	placental mammal
gland	therapsid
mammary gland	uterus
marsupial	

Understanding Concepts

1. Compare the adaptations of an ocean-dwelling mammal with those of a fish.
2. How are monotremes unlike other mammals?
3. Why do rodents damage the woodwork in your house?
4. What is the adaptive value of a young zebra being able to run within a few hours of its birth?
5. Explain why most bats have large ears.
6. What makes you a primate?
7. What influences the number of young a certain species of mammal might have?

Interpreting a Scientific Illustration
8. Examine the teeth in the diagram of a skull and infer what the mammal might eat.

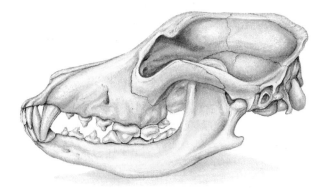

Relating Concepts

9. Make a concept map that relates the following terms and phrases. Supply the appropriate linking words for your map.

   mammal, rodent, monotreme, marsupial, placental, kangaroo, duck-billed platypus, gorilla, beaver, diaphragm, endothermy

Applying Concepts

10. What is the advantage of bearing live young over laying eggs?
11. What evidence shows that monotremes and marsupials had different ancestors?
12. African elephants live in hot grasslands of Africa. They can often be seen using their trunks to slap watery globs of mud onto their heads and backs. Make a hypothesis that explains this behavior.
13. What is the adaptive value of an arm and hand modified into a flipper for a seal?
14. You find the skeleton of an animal. What features would indicate that it is a mammal rather than a reptile?

A Broader View

15. Like reptiles, platypuses lay eggs on land and have claws. Why aren't platypuses considered to be reptiles? Explain.

Thinking Critically

Developing a Hypothesis

16. **Biolab** A marine biologist discovers a whale that has washed up on the shore. The blubber layer is thin throughout the body of the whale. Make a hypothesis about why the whale might have died.

Making Inferences

17. **Minilab** Assume that you find that the stride length of a mammal's tracks in the mud is 15 cm. Would this mammal be about the size of a house cat, a bear, or a mouse? Explain.

Interpreting Graphs

18. **Thinking Lab** A biologist obtains the following data on three animals. Which one would most likely be a hoofed herbivore? Explain.

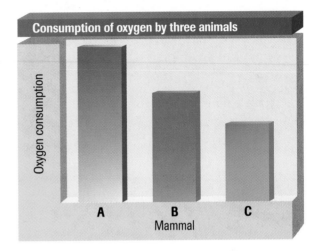

Consumption of oxygen by three animals

Connecting to Themes

19. **Homeostasis** Explain the relationship between homeostasis and endothermy.
20. **Unity Within Diversity** Mammals and insects are both considered to be extraordinarily successful animals. Explain the criteria used in both cases that give them this distinction.
21. **Evolution** Trace the main paths of evolution from fishes to mammals that explain how mammals became adapted to life on land.

36 Animal Behavior

The sun comes up over the prairie and makes the dew look like shining crystals. A herd of antelope wanders slowly over the landscape, heads down, as they begin their daily foraging. Nearby, a furry head pops up out of a hole in the ground. The prairie dog is soon joined by a host of its neighbors, all alert for possible danger before coming out of their underground burrows. Up in the brightening sky, a vulture slowly circles, searching the ground below for a dead or dying animal that could become its morning meal.

A female coyote suddenly appears from behind a pile of rocks. She stretches, yawns, and, with ears pricked, canters toward the prairie dog town. Several prairie dogs bark, and the rest dive for cover. The noise alerts an antelope; head raised, he turns and signals the herd with tail up and white rump showing. The herd springs into action, leaping quickly away. Soon, the coyote seems to be alone on the prairie. Overhead, the vulture wheels away to look for breakfast somewhere else.

Concept Check

You may wish to review the following concepts before studying this chapter.
- Chapter 1: adaptation, evolution, response, stimulus
- Chapter 2: experiment, hypothesis

Prairie dogs live in a community. Vultures can often be seen feeding together on dead or dying animals. Coyotes pursue prey. Each animal has set patterns of behavior that allow it to survive and pass on its genes to offspring. How much of this behavior is directly inherited from parents? How much of it is learned in a lifetime of trial and error?

36.1 Innate Behavior

Have you ever watched a bird feed its young? Baby birds greet a parent returning to the nest with cries and open beaks. Parent birds practically stuff the food down their offsprings' throats, then fly off to find ever more food. Why do baby birds open their beaks wide? Why do parent birds respond to open beaks by feeding their offspring? These actions are examples of behavior that appears in birds without being taught or learned. Animals exhibit many kinds of behavior in nature.

What Is Behavior?

A peacock displaying his colorful tail, a whale spending the winter months in the ocean off the coast of southern California, and a lizard seeking shade from the hot, desert sun are all examples of animal behavior. **Behavior** is anything an animal does in response to a stimulus in its environment. The presence of a peahen stimulates a peacock to open its tail feathers and strut. Environmental cues, such as change of temperature and length of daylight, might be the stimuli that cause the whale to leave its summertime arctic habitat. Heat stimulates the lizard to seek shade. Look at the stimuli that cause the animals in *Figure 36.1* to behave the way they do.

Figure 36.1

Animals exhibit a variety of behavioral responses.

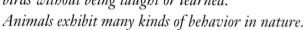

▼ **This butterfly exposes eyespots on its wings, and a predatory bird stops its pursuit of the insect. The eyespots look like the eyes of an owl.**

▶ **The onset of short days and cold weather stimulates squirrels to collect acorns and walnuts and store them. What is the adaptive value of the squirrel's behavior?**

Animals carry on many activities—such as getting food, avoiding predators, caring for young, finding shelter, and attracting mates—that enable them to survive. These behavior patterns, therefore, have adaptive value. For example, a parent sea gull that is not incubating eggs or chicks joins a noisy flock of gulls to dive for fish. If the parent did not catch fish, not only would it die, but its chicks would not survive either. Therefore, this feeding behavior has adaptive value for the sea gull.

Behavior that Is Inherited

Inheritance plays an important role in the ways animals behave. You don't expect a duck to tunnel underground or a mouse to fly. Yet, why does a mouse run away when a cat appears? Why does a mallard duck fly south for the winter? These behavior patterns are genetically programmed. An animal's genetic makeup determines how that animal reacts to certain stimuli.

Natural selection favors certain behaviors

Often, a behavior exhibited by an animal species is the result of natural selection. The variability of behavior among individuals affects their ability to survive and reproduce. Individuals with behavior that makes them more successful at surviving and reproducing will produce more offspring. These offspring will inherit the genetic basis for the successful behavior. Individuals without the behavior will die or fail to reproduce.

Inherited behavior of animals is called **innate behavior.** Consider the sea gulls in *Figure 36.2.* If the female is not able to recognize an appropriate mate of her own species, and is

Figure 36.2

Each species has its own courtship ritual. Male and female black-headed gulls dance in unison side by side and turn their heads away from each other as if doing a tango. The female taps the male's bill, who then regurgitates a fish into her mouth. Soon the dancing and dining are over, and the pair mate.

confused by several other species of sea gulls that also live in the same location, she will not be successful in reproducing. Courtship in the form of a ritualized dance helps the female recognize an appropriate mate.

Genes form the basis of behavior

Through experiments, scientists have found that an animal's hormonal balance and its nervous system—especially the sense organs responsible for sight, touch, sound, or odor identification—affect how sensitive the individual is to certain stimuli. Because genes control the production of an animal's hormones and development of its nervous system, it's logical to conclude that genes indirectly control behavior. Innate behavior includes both automatic responses and instinctive behaviors.

Meet Lisa Stevens, Assistant Zoo Curator

The giant panda cannot speak for itself to TV and newspaper reporters, so Lisa Stevens becomes its voice. Acting as press contact for the black-and-white animal is just one of the duties of this assistant curator for mammals at the National Zoo, which is part of the Smithsonian Institution in Washington, D.C.

In the following interview, Ms. Stevens describes her profession and the important role of today's zoos.

On the Job

Q Ms. Stevens, could you tell us about a typical day on the job for you?

A The first thing to understand about this profession is that there *isn't* a typical day. As an assistant curator in the mammal department, I supervise the day-to-day care of three groups of animals—the primates, the camels, and the giant pandas. Here's an idea of my other duties. The first thing tomorrow morning, I'll oversee the release of a new group of monkeys to an island exhibit. Then I'm chairing a committee of our education council, which evaluates the zoo's educational programs. Later, I have a meeting about setting up a new exhibit in the primate house.

Q What will be in that new exhibit?

A It's called the "Think Tank." In this exhibit, we'll address the question of whether or not animals have a capacity to think. It will be an interactive exhibit designed to increase people's appreciation and respect for animals' cognitive abilities.

Q Does the zoo have a captive-breeding program?

A Like other zoos in North America, the National Zoo participates in over thirty Species Survival Plans. One of our captive-breeding programs involves the white-cheeked gibbon, which has a natural habitat in southern China and Vietnam. We've also had gorillas born in 1991 and 1992. They were our first gorilla births in 20 years. However, we won't, in most cases, be able to release animals into the wild because their habitats are gone. The goal of these programs is to maintain a diverse genetic population as a hedge against extinction in the wild.

Early Influences

Q How did you get interested in animals?

A As a child, I wasn't afraid of anything. I was the kind of kid who let bugs crawl up and down my arms and caught lizards and snakes. I had the good fortune to spend part of my childhood in the tropics, in Thailand and Okinawa. My exposure to all kinds of animals was probably quadrupled in that environment. My parents took me places where I could fulfill my interest in animals. For example, once we visited a Buddhist snake temple in Penang, Malaysia, that had all kinds of snakes draped in artificial trees with gold leaves. As a child, I also got to go on a safari in Nairobi National Park, in Kenya.

Q How did you first come to the National Zoo?

A I got a college degree in zoology and wanted to have some firsthand experience with animals before entering veterinarian school. I had worked for a vet and in stables and pet stores. I thought, Hmmm, I've never worked in a zoo, so maybe that would be kind of interesting. Actually, I was going to volunteer, but before I could do that, I got hired as a keeper of lions, tigers, and bears. I first thought zookeepers were sort of like janitors. I had no idea that there was such a level of professionalism involved. Ultimately, I realized that the zoo was a whole world in itself, and I've never left that world.

Personal Insights

Q Do you have a favorite animal?

A I'm asked that a lot, but it's really important to be flexible. I'm lucky that I'm able to work with all different kinds of animal species. I truly appreciate and find delight in the diversity of species.

Q What advice do you have for students who are interested in careers related to animals?

A Often, school programs steer people with animal interests into veterinary medicine, but there are many other animal-related careers to explore. It's important to think of the complex issues that are facing wildlife, especially the issue of disappearing animal habitats. For example, there's a need for people to work in the field of human

population biology and the economics of agriculture. Biologists now have to think in terms of helping the people who live in countries where many species of wildlife also live and of the interrelations of species, both animal and human.

For any kind of a career, volunteering may be an opportunity to look at a profession. Here at our zoo, we use teenage volunteers to answer visitors' questions. The young volunteers enjoy their work and learn a lot about both animals and people.

Automatic Responses to Stimuli

What happens if something quickly passes in front of your eyes or if something is thrown at your face? Your first reaction is to blink. Even if a protective clear shield is placed in front of you, you can't keep your eyes open when the object is thrown. This eye blink is the simplest form of innate behavior called a reflex. A reflex is a simple, automatic response that involves no conscious control. *Figure 36.3* shows two examples of reflexes.

The adaptive value of another automatic response is obvious. Think about a time when you were suddenly scared. Immediately, your heart began to beat faster. Your skin got cold and clammy, your respiration increased, and maybe you trembled. You were having a fight-or-flight response. A **fight-or-flight response** mobilizes the body for greater activity. Your body is being prepared to either fight or run from the danger. A fight-or-flight response is automatic and controlled by internal chemical mechanisms.

Figure 36.3

Reflexes have survival value for animals.

▼ When you accidentally touch a hot stove, you jerk your hand away from the hot surface. The movement saves your body from serious injury.

▼ When a mollusk is touched, it withdraws into its shell. The touch stimulus may be that of a mollusk predator.

Instinctive Behavior Patterns

Compare the mating dance of the sea gull with your fight-or-flight response. The dance does not happen as quickly as a reflex does. The mating dance is an example of an instinct. An **instinct** is a complex pattern of innate behavior. A reflex can happen in less than a second. Instinctive behavior patterns may have several parts and may take weeks to complete. Instinctive behavior begins when the animal recognizes a stimulus and continues until all parts of the behavior have been performed.

As shown in *Figure 36.4,* greylag geese instinctively retrieve eggs that have rolled from the nest and will go through the motions of egg retrieval even when the eggs are taken away. You can see that survival of the young may be dependent on this behavior.

Courtship behavior ensures reproduction

Much of an animal's courtship behavior is instinctive. **Courtship behavior** is a behavior that males and females of a species carry out before mating. Like other instinctive behaviors, courtship has evolved through natural selection. Imagine what would happen to the survival of a species if members were unable to recognize other members of that same species. Individuals often can recognize one another by the behavior patterns each performs. In courtship, behavior ensures that members of the same species find each other and mate. Obviously, such behavior has an adaptive value for the species. Different species of fireflies, for example, can be

Figure 36.4

The female greylag goose instinctively retrieves an egg that has rolled out of the nest by arching her neck around the stray egg and moving it like a hockey player advancing a puck. The female goose will retrieve many objects outside the nest, including baseballs and tin cans.

ThinkingLab — Design an Experiment

Why do starlings use wild carrot leaves in their nests?

Starlings build their nests in the cavities of hollow trees. They construct a nest of dried grasses and twigs, but they also include freshly gathered wild carrot leaves. Biologists were interested in why starlings seemed to decorate their nests with this greenery. They hypothesized that the birds may have been camouflaging their nests from predators, insulating the eggs with material that would keep in warmth, or using material that would poison or repel parasites that might attack nestlings.

Analysis

Design an experiment that would test one of the hypotheses. Make sure your experiment has a control and a method to gather quantitative data.

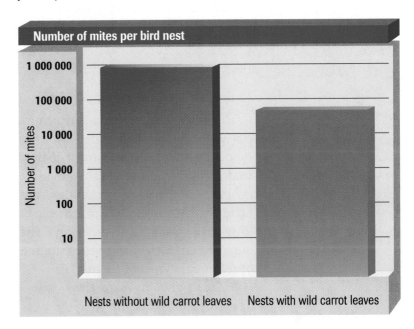

Thinking Critically

Biologists tested the hypothesis that the plant material may act as an insect repellent. They obtained the data in the graph by counting the mites in nests with wild carrot leaves and nests without wild carrot leaves. Explain the graph in terms of how wild carrot leaves affect mites in nests.

seen at dusk flashing distinct light patterns. However, female fireflies of one species respond only to those males exhibiting the species-correct flashing pattern.

Some courtship behaviors help prevent females from killing males before they have had the opportunity to mate. For example, in some spiders, the male is smaller than the female and risks the chance of being eaten if he approaches her. Before mating, the male presents the female with a nuptial gift, an insect wrapped in a silk web. While the female is unwrapping and eating the insect, the male is able to mate with her without being attacked. After mating, however, the male may be eaten by the female anyway.

Figure 36.5

Female hanging flies instinctively favor the male that supplies the largest nuptial gift—in this case, a moth. The amount of sperm the female will accept from the male is determined by the size of the gift. What adaptive value does this behavior have?

In some species, nuptial gifts play an important role in allowing the female to exercise a choice as to which male to choose for a mating partner. The hanging fly, shown in *Figure 36.5,* is such a species.

Territoriality reduces competition

You may have seen a chipmunk chase another chipmunk away from seeds on the ground under a bird feeder. The chipmunk was defending its territory from an invader, the other chipmunk. A **territory** is a physical space that contains the breeding area, feeding area, shelter, or potential mates of an animal.

Animals that have territories will defend their space by driving away other individuals of the same species. For example, a male sea lion will patrol the area of beach where his harem of female sea lions rests. He will not bother a neighboring male that has a harem of his own because both have marked their territories, and each respects the common boundaries. But if an unattached, young male tries to enter the sea lion's territory, the owner of the territory will attack and drive the intruder away from his harem.

Although it may not appear so, setting up territories actually reduces conflicts, controls population growth, and provides for efficient use of environmental resources by spacing out animals so they don't compete for the same resources within a limited space. This behavior improves the chances of survival of the young produced, and, therefore, survival of the species. If the male has selected an appropriate site and the young survive, they may inherit his ability to select an appropriate territory. Therefore, territorial behavior has survival value, not only for individuals, but also for the species. The male stickleback shown in *Figure 36.6* is another animal that exhibits territoriality, especially during breeding season.

Figure 36.6

The male three-spined stickleback displays a red belly to other breeding males near his territory. The red belly is a signal or stimulus indicative of the breeding condition of the fish. The male will instinctively respond to other red-bellied males by attacking and driving them away.

Figure 36.7

In many species, such as bighorn sheep, individuals have innate inhibitions that make them fight in relatively harmless ways among themselves.

Remember from Chapter 31 that pheromones are chemicals that communicate information between individuals of the same species. Many animals produce pheromones to mark territorial boundaries. For example, wolf urine contains pheromones that warn other wolves to stay away. The male pronghorn antelope uses a pheromone secreted from facial glands. One advantage of using pheromones is that they work during both the day and night.

Aggression threatens other animals

Animals occasionally engage in aggression. **Aggression** is behavior that is used to intimidate another animal of the same species. Animals fight or threaten one another in order to defend their young, their territory, or a resource such as food. Aggressive behaviors, such as bird singing, teeth baring, or growling, deliver the message to others of the same species to keep away.

When a male gorilla is threatened by another male moving into his territory, for example, he does not kill the invader. Animals of the same species rarely fight to the death. The fights are usually symbolic, as shown in *Figure 36.7*. Male gorillas do not

usually even injure one another. Why does aggression rarely result in serious injury? One answer is that the defeated individual shows signs of submission to the victor. These signs inhibit further aggressive actions by the victor. Continued fighting might result in serious injury for the victor; thus, its best interests are served by stopping the fight.

Submission leads to dominance hierarchies

Do you have an older or younger sibling? Who wins when you argue? In animals, it is usually the oldest or strongest that wins the argument. But what happens when several individuals are involved in the argument? Sometimes, aggression among several individuals results in a grouping in which there are different levels of dominant and submissive animals. A **dominance hierarchy** is a form of social ranking within a group in which some individuals are more subordinate than others. Usually, one animal is the top-ranking, dominant individual. This animal often leads others to food, water, and shelter. There might be several levels in the hierarchy, with individuals in each level subordinate to the one above.

Figure 36.8

A dominance hierarchy often prevents energy from being wasted on continuous fighting because submissive birds give way peacefully in confrontations. The hierarchy also may be adaptive in providing a way for females to choose the best males.

The term *pecking order* comes from a dominance hierarchy that is formed by chickens, *Figure 36.8.* The top-ranking chicken can peck any other chicken. The chicken lowest in the hierarchy is pecked at by all other chickens in the group.

Behavior resulting from internal and external cues

Some instinctive behavior is exhibited in animals in response to internal, biological rhythms. Behavior based on a 24-hour day/night cycle is one example. Many animals, humans included, sleep at night and are awake during the day. Other animals, such as owls, reverse this pattern and are awake at night. A 24-hour cycle of behavior is called a **circadian rhythm.** Most animals come close to this 24-hour cycle of sleeping and wakefulness. Experiments have shown that in laboratory settings with no windows to show night and day, animals continue to behave in a 24-hour cycle.

Rhythms also can occur yearly. Migration is a yearly rhythm. **Migration** is the instinctive, seasonal movement of animals, *Figure 36.9.* In the United States, about two-thirds of bird species fly south in the fall to areas such as South America where food is available during the winter. The birds fly north in the

Figure 36.9

A variety of animals respond to the urge to migrate.

▶ Adult monarch butterflies fly southward. In the spring, their young fly back north.

▲ Both the freshwater eel and all species of salmon migrate to their spawning grounds.

spring to areas where they breed during the summer. Whales migrate seasonally, as well. Change in day length is thought to stimulate the onset of migration in the same way that it controls the flowering of plants.

Migration calls for remarkable strength and endurance. The arctic tern migrates between the Arctic Circle and the Antarctic, a one-way flight of almost 18 000 km.

Animals navigate in a variety of ways. Some use the positions of the sun and stars to navigate. They may use geographic clues, such as mountain ranges. Some bird species seem to be guided by Earth's magnetic field. You might think of this as being guided by an internal compass.

Biological rhythms are clearly governed by a combination of internal and external cues. Animals that migrate might be responding to colder temperatures and shorter days, as well as to hormones. You can easily see why animals migrate from a cold place to a warmer place, yet most animals do not migrate. How animals cope with winter is another example of instinctive behavior.

▼ **Canadian and Alaskan caribou migrate from their winter homes in the boreal forests to the tundra for the summer.**

Earth Science

Finding Their Way Home

The arctic tern, a seabird, breeds on the coasts and islands of northern North America. Yet every August, the terns and their offspring begin a monumental journey to their winter home in Antarctica, a flying distance of almost 18 000 km! Come spring, the terns begin the return flight back to the Arctic. Among birds, the arctic tern is the gold medalist of long-distance flight. Many other birds also migrate annually. How do birds navigate such great distances without losing their way?

Finding home Homing pigeons are birds trained to return to their home lofts from wherever they are released—even at distances up to 80 km. They are bred for speed; their owners often hold races to see whose pigeons can return home most quickly.

Backup systems Scientists have found that pigeons use a variety of environmental and sensory cues to guide them on their journeys. Pigeons use the sun as a compass; they allow their internal clocks to compensate for the sun's movement during the day. But how does a pigeon locate its own home so precisely? Apparently, pigeons also use Earth's magnetic field to find home on cloudy days. But pigeons raised near areas where the magnetic field is disturbed seem to rely on many other factors to find their way home. Among these are smells, winds, pressure changes, infrasonic sounds, ultraviolet light, and polarized light patterns. Perhaps all migrating birds rely on an array of navigational aids to find their way home.

CONNECTION TO Biology

Some migrating birds follow river valleys, coastlines, or mountain ranges, whereas others seem to navigate by the stars. How could scientists test which navigational cues a particular bird species uses for migration?

Figure 36.10

The golden-mantled ground squirrel has a body temperature of around 37°C during normal activity. When the surrounding temperature drops to 0°C, the ground squirrel's heart rate drops. Its temperature drops to 2°C, and it goes into hibernation. Hibernation conserves the animal's energy.

You know that many animals store food in burrows and nests. But other animals survive the winter by undergoing physiological changes that reduce their need for energy. Many mammals, some birds, and a few other types of animals go into hibernation during the cold winter months. **Hibernation** is a state in which the body temperature drops substantially, oxygen consumption decreases, and breathing rates decline to a few breaths per minute. Animals that hibernate typically eat vast amounts of food before entering hibernation. This extra food fuels the animal's body while it is in this state. The golden-mantled ground squirrel shown in *Figure 36.10* is an example of an animal that hibernates.

What happens to animals that live year-round in hot environments? Some of these animals respond in a way that is similar to hibernation. **Estivation** is a state of reduced metabolism that occurs in animals living under intense heat. Desert animals appear to estivate sometimes in response to lack of food or periods of drought. However, Australian long-necked turtles, *Figure 36.11,* will estivate even if kept in a laboratory with constant food and water. Clearly, estivation is an innate behavior that depends on both internal and external cues.

Figure 36.11

Australian long-necked turtles are among the reptiles and amphibians that respond to hot and dry summer conditions by estivating.

Section Review

Understanding Concepts
1. What is behavior?
2. How is a reflex different from instinct?
3. Explain by example two types of innate behavior.

Thinking Critically
4. How is innate behavior an advantage to a species in which the young normally hatch after the mother has left?

Skill Review
5. **Designing an Experiment** Design an experiment to test what stimulus causes an earthworm to return to its burrow. For more help, refer to Practicing Scientific Methods in the *Skill Handbook*.

36.2 Learned Behavior

*Y*ou were born knowing how to cry, but were you born knowing how to tie your shoes or read? Behavior controlled by instinct, as you now know, occurs automatically. However, some behavior is the direct result of the previous experiences of an animal. A dog that has been hit by a car and survives may avoid streets. You may know how to build a model ship after being shown only once or it may take a hundred trials. These behaviors are a result of learning.

Section Preview

Objectives

Distinguish among types of learned behavior.

Demonstrate by example types of learned behavior.

Key Terms
habituation
imprinting
trial-and-error learning
motivation
conditioning
insight
communication
language

What Is Learned Behavior?

Learning, or learned behavior, takes place when behavior changes through practice or experience. The more complex an animal's brain, the more elaborate the patterns of its learned behavior. As you can see in *Figure 36.12*, innate types of behavior are more common in invertebrates, and learned types of behavior are more common in vertebrates. In humans, many behaviors are learned. Speaking, reading, writing, and playing a sport are all learned.

Learning has survival value for all animals in a changing environment because it permits behavior to change in response to varied conditions. Learning allows an animal to adapt to change, an ability that is especially important for animals with long life spans. The longer an animal lives, the greater the chance that its environment will change and that it will encounter unfamiliar situations.

Figure 36.12

Examine the graph and tell which groups of animals demonstrate the most learned behavior. Which groups of animals demonstrate the most innate behavior?

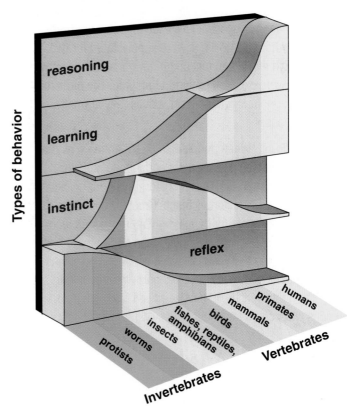

36.2   Learned Behavior   **907**

Imprinting

Have you ever seen a mother duck waddling down to a pond with a line of fluffy ducklings following close behind her? It's an endearing sight, but also an example of a learned behavior called imprinting. Many kinds of birds and mammals do not innately know how to recognize members of their own species. Instead, they learn to make this distinction early in life.

In birds such as ducks, imprinting takes place during the first day or two after hatching. A duckling rapidly learns to recognize and follow the first conspicuous moving object it sees. Normally, that object is the duckling's mother. Learning to recognize mother and follow her ensures that food and protection will always be nearby.

Imprinting experiments In the 1930s, behavioral biologist Konrad Lorenz studied imprinting in geese. In his experiments, Lorenz showed that if newly hatched goslings were kept isolated for two or three days, they would imprint on just about any conspicuous, moving object that they were exposed to during that time. The birds followed and became attached to this object, just as they normally would to a mother goose.

Practical applications When people hand raise animals, as they do in many captive-breeding programs for endangered species, imprinting becomes an important issue. California condor chicks, for example, are isolated as soon as they hatch. Great care is taken that the young birds do not imprint on people or inappropriate objects. With their bodies and faces hidden, workers feed and care for each chick using hand puppets that look like adult condors. The chicks imprint on the puppets, and so learn to recognize their own kind. When released into the wild, these captive-raised birds are independent of humans and can interact successfully with other condors.

Thinking Critically

Why is it important that captive-raised birds destined for release into the wild do not form attachments to humans? Explain your answer.

Kinds of Learned Behavior

Just as there are several types of innate behavior, there are several types of learned behavior. Some learned behavior is simple and some is complex. Which group of animals do you think carries out the most complex type of learned behavior?

Habituation is one of the simplest forms of learning

Horses normally shy away from an object that suddenly appears from the trees or bushes, yet after awhile they disregard cars that speed by the pasture and honk their horns. This lack of response is called habituation. **Habituation,** *Figure 36.13,* occurs when an animal is repeatedly given a stimulus that is not associated with any punishment or reward. An animal has become habituated to a stimulus when it finally ceases to respond to the stimulus.

Figure 36.13

Habituation is a loss of sensitivity to certain stimuli. Young horses are often afraid of cars and noisy streets. Gradually horses become habituated to the city and ignore normal sights and sounds.

Imprinting involves a permanent attachment

Have you ever seen young ducklings following their mother? This behavior is the result of imprinting. **Imprinting** is a form of learning in which an animal, at a specific critical time of its life, forms a social attachment to another object. Imprinting takes place only during a specific period of time in the animal's life and is usually irreversible. For example, birds that leave the nest immediately after hatching, such as geese, imprint on their mother. They learn to recognize and follow her within a day of hatching.

Learning by trial and error

Have you ever watched a young child learning to ride a bicycle? The child tried many times before being able to successfully complete the task. Nest building, like riding a bicycle, may be a learning experience. The first time a jackdaw builds a nest, it uses grass, bits of glass, stones, empty cans, old light bulbs, and anything else it can find. With experience, the bird finds that grasses and twigs make a better nest than do light bulbs. The jackdaw has used **trial-and-error learning,** a type of learning in which an animal receives a reward for making a particular response. When an animal tries one solution and then another in the course of obtaining a reward, in this case a suitable nest, it is learning by trial and error.

Learning happens more quickly if there is a reason to learn or be successful. **Motivation,** an internal need that causes an animal to act, is necessary for learning to take place. In most animals, motivation often involves satisfying a physical need, such as hunger or thirst. If an animal isn't motivated, it won't learn.

Animals that aren't hungry won't respond to a food reward. Mice living in a barn, *Figure 36.14,* discover that they can eat all the grain they like if they first chew through the container the grain is stored in.

Figure 36.14

Mice soon learn where grain is stored in a barn and are motivated to chew through the storage containers.

In an innovative experiment on the Seney National Wildlife Refuge in Michigan, sandhill cranes were hatched from eggs and

What makes a good feeding puppet for sandhill crane chicks?

raised by puppet mothers, scientists with their arms in puppets that resembled the mother sandhill crane. This experiment was undertaken to develop a strategy for future use in hand raising

endangered whooping cranes and releasing them into the wild. It would be important for hand-raised cranes not to become imprinted on humans who raised them. Otherwise, they would have no fear of humans. Therefore, the cranes were fed by humans covered with sheets the color of cranes who used hand puppets. Biologists thought that they could perfect the technique with sandhill cranes, which are not endangered but are similar to whooping cranes.

PREPARATION

Problem
What characteristics must be present for you to make a successful feeding puppet for sandhill crane chicks?

Objectives
In this Biolab, you will:
- **Analyze** the features of sandhill crane heads and distinguish important characteristics.
- **Design** a hand puppet suitable for feeding insects to young sandhill cranes.

Possible Materials
photos of sandhill cranes
black fabric and black plastic trash
  bags
red fabric
white fabric
needle and thread
long, gray socks
spring-type clothes pins
black markers
stapler
tape
tagboard
various colored beads
various colored map tacks
glue
scissors

Safety Precautions
Use care with the scissors and needles.

1. Examine carefully the photos or diagrams of sandhill cranes.

2. Make a list of the important features of the head of the crane.

3. Brainstorm a list of ideas about how a hand puppet for feeding young cranes could be constructed.

4. Make a detailed diagram of exactly how your puppet will look. Label it with the material you will use.

5. Describe in the form of a list exactly how you will put your puppet together.

Check the Plan

1. Have you considered all the important features of the crane's head?

2. With your hand in the puppet, will you be able to feed insects to young cranes?

3. *Make sure your teacher has approved your plan before you proceed further.*

4. Put your puppet together.

1. **Interpreting Scientific Diagrams** What features of the crane head were important for identification of a sandhill crane?

2. **Making Predictions** How did you ensure your puppet's ability to feed insects to a young sandhill crane?

3. **Applying Concepts** Why would it be important for endangered whooping cranes being hand raised and released into the wild not to see humans?

Going Further

Project Contact an environmental group that works with endangered birds. Ask a representative of the group to visit the class or send materials that explain his or her work and the results seen.

Figure 36.15

In 1900, Ivan Pavlov, a Russian biologist, first demonstrated conditioning in dogs.

A He noted that dogs salivate when they smell food. Responding to the smell of food is a reflex, an example of innate behavior.

B By ringing a bell each time he presented food to a dog, Pavlov established an association between the food and the ringing bell. The dog smelled food while a bell was rung. The dog salivated.

C Eventually, the dog salivated at the sound of the bell alone. The dog had been conditioned to respond to a stimulus that it did not normally associate with food.

Conditioning is learning by association

When you first got a new kitten, it would meow and rub against your ankles as soon as it smelled the aroma of cat food in the can you were opening. After a few weeks, the sound of the can opener alone attracted your kitten, causing it to meow and rub against your ankles. Your kitten had become conditioned to respond to a stimulus other than the smell of food. **Conditioning** is learning by association. A well-known example of an early experiment in conditioning is illustrated in *Figure 36.15.*

Insight is the most complex type of learning

In a classic study of animal behavior, a chimpanzee was given two bamboo poles, neither of which was long enough to reach some fruit placed outside its cage. By connecting the two short pieces to make one longer pole, the chimpanzee learned to solve the problem of how to reach the fruit. This type of learning is called insight. **Insight** is learning in which an animal uses previous experience to respond to a new situation.

Much of human learning is based on insight. When you were a baby, you learned a great deal by trial and error. As you grew older, you relied more on insight. Solving math problems is a daily instance when you use insight. Probably your first experience with mathematics was when you learned to count. Based on your concept of numbers, you then learned to add, subtract, multiply, and divide. Years later, you continue to solve problems in mathematics based on your past experiences. When you encounter a problem you have never experienced before, you solve the problem through insight.

The Role of Communication

When you think about the interactions that happen between animals as a result of their behavior, you realize that some sort of communication has taken place. **Communication** is an exchange of information that results in a change of behavior. The black-headed sea gulls visually communicate their availability for mating with instinctive courtship behavior. The pat on the head after the dog retrieves the stick signals a job well done.

Most animals communicate

Animals have several channels of communication open to them. You remember from Chapter 2 that elephants use sounds to communicate over long distances. Other animals signal each other by sounds, sights, touches, or smells. Sounds radiate out in all directions and can be heard a long way off. The sounds of the humpback whale can be heard 1200 km away. Sounds such as songs, roars, and calls are best for communicating a lot of information quickly. For example, the song of a male cricket tells his sex, his location, his social status, and, because communication by sound is usually species specific, his species.

Signals that use odors may be broadcast widely and carry a general message. Ants, *Figure 36.16,* leave odor trails that are followed by other members of their nest. Some odors may also be species specific. As you know, pheromones, such as in moths, may be used to attract mates. Because only small amounts are needed, other animals, especially predators, don't detect the odor.

MiniLab

What is a habit?

How many things do you do automatically without even thinking about how to do them? Do you have to think about how to brush your teeth or button a shirt? Once you learn how to carry out a procedure that is repeated often, it becomes a habit. Doing something by habit has value because you don't waste time relearning the procedure each time you perform it. The same is true for an animal in the wild.

Procedure

1. Without actually tying your shoes or even looking at the laces, write a set of directions that explain step by step how to tie a shoe.

2. Have your partner read your directions and then try to follow exactly your procedure to tie one shoe.

Analysis

1. What is the advantage of forming a habit?

2. Why was it difficult to describe how to tie a shoe?

3. How is forming a habit like habituation?

Figure 36.16

Did you ever watch as a group of ants moved in single file to a piece of potato chip left on a picnic table? Ants follow chemical trails left by other ants to find food sources. In what way does this behavior have adaptive value?

Social Insects

F amily life for most insects does not exist, but for ants, termites, some wasps, and most bees—the social insects—family life is everything. Social insects live in organized colonies where all members depend on one another. In these colonies, there is a division of labor where all work for the benefit of the entire colony.

Bees

Honeybees—A family affair
The queen bee deposits one fertilized egg in each wax cell of the hive. An adult nurse worker continuously feeds a larva for six days until it fills the cell. The cell is then sealed. About 12 days later, the adult bee emerges. Two-thirds of a bee's life is spent doing in-hive duties. The last third is spent foraging for nectar and pollen. Bees have a life span of about six weeks.

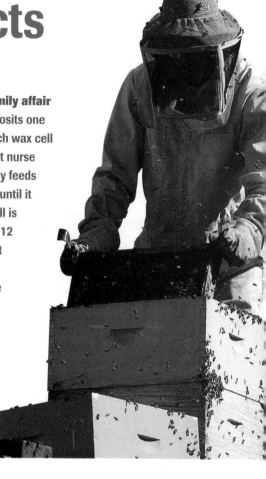

Honey production Bees carry thin, watery nectar from flowers back to the hive in a pouch on their bodies. In the hive, the nectar is broken down into simple sugars and allowed to evaporate and thicken into honey. Enzymes are added that give honey its flavor. Honey is food for the bees in the hive.

Bees pollinate crops Bees usually visit one species of flower at a time. As the bee gathers nectar, pollen grains cling to the hair on its body and fall off at the next flower. Honeybees pollinate crops such as apples, plums, cherries, cucumbers, carrots, celery, and onions.

Dancing bees

Where's the nectar? One of the most complex types of animal communication occurs when honeybees return to the hive and communicate the location of nectar to other members of the hive. If the location is close, the bee does a round dance. A waggle dance is performed if the location is more than 50 m away.

1 Worker bees leave the hive and forage for pollen in nearby flowers. Sometimes, the bees must travel quite a distance from the hive.

2 The worker bee returns to the hive with pollen-laden baskets on her legs.

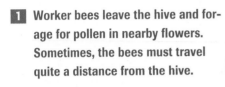

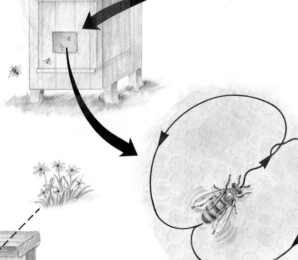

5 If the straight run is done on a vertical wall inside the hive, the angle between the straight run and a vertical plane through the wall (40° in the diagram) is the same as the angle between the sun and the flowers.

3 While other bees crowd around, the returning bee makes a series of figure eight patterns.

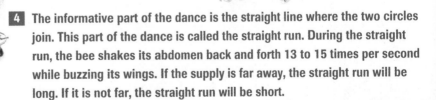

4 The informative part of the dance is the straight line where the two circles join. This part of the dance is called the straight run. During the straight run, the bee shakes its abdomen back and forth 13 to 15 times per second while buzzing its wings. If the supply is far away, the straight run will be long. If it is not far, the straight run will be short.

Ants

Jungle nomads—The army ants Army ant colonies include up to 20 million individual ants. There are about 240 different species of army ants, including species found in jungles in Central and South America, Africa, and Southeast Asia. Members of these colonies inflict strong stings and bites, but as the whole colony moves only about one foot per minute, many animals are able to escape them.

Army ant family structure Army ants, unlike other species of ants, have no permanent home. The behavior of the colony alternates between a quiet, stationary phase when the queen is preparing to lay eggs, and an exploratory, nomadic phase when food is being obtained. A single queen does all the egg laying. During food-getting raids, individual workers explore an area and lay down odor trails. These trails guide other workers. Prey that is caught is carried back along the column of ants and collected in "booty caches."

Leaf-cutter ants are gardeners Another species of ant found in Central and South America, the leaf-cutter ants actually grow their own food supply. These colonies are composed of the queen and large and small workers. Unlike army ants, leaf-cutter ants live in huge, underground nests. The large workers cut pieces of leaves from plants and transport them to the colony where they are chewed to pulp. This pulp is used as the substrate for the fungi that are grown for food.

Termites

Termites have complex societies Of all the social insects, termites have the largest and most complex societies. One species of West African termites builds nests with up to 5 million members. The nests are extremely large, complex structures built with systems of tunnels to provide ventilation and cooling. The nests have just one queen and one king. The queen lays an egg about every two to three seconds; she must be fed continuously by the workers of the nest. Many trails branch from the nest. These trails are guarded by large soldier termites. Most termites feed on rotting wood.

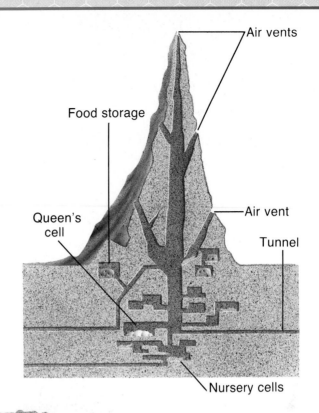

Air vents

Food storage

Queen's cell

Air vent

Tunnel

Nursery cells

Termite nests vary in size
Termites build nests of varying shapes and sizes. West African termites, *Macrotermes bellicosus,* build huge, chimney-shaped structures. Below the tower is a cavity about ten feet in diameter where the termites live. The termites also dig channels deep into the ground to obtain water. These structures are so well designed that with slight adjustments to air passages, the temperature inside the nest can be regulated to within 1°. Another species of termite, *Cubitermes speciosus,* builds mushroom-shaped nests that resist heavy jungle rainfall.

EXPANDING YOUR VIEW

1. **Understanding Concepts** How is the organization of a society such as a termite nest related to its environment?
2. **Journal Writing** In a short paragraph, explain the benefits for insects of living together in a group.

Figure 36.17

English and other languages are made of words that have specific meanings. People who use the language put the words together to convey messages and ideas. An unlimited number of meanings can be communicated using the same words.

behavior. Yet members of the same species that live in different regions learn different variations of the song. They learn to sing with a regional dialect.

Some animals use language

Language, the use of symbols to represent ideas, is present primarily in animals with complex nervous systems, memory, and insight. Humans, with the help of spoken and written language, can benefit by what other people and cultures have learned and don't have to experience everything for themselves. They can use accumulated knowledge as a basis on which to build new knowledge, *Figure 36.17.*

Using both innate and learned behavior

Some communication is a combination of both innate and learned behavior. Male songbirds automatically sing when they reach sexual maturity. Their songs are specific to their species, and singing is innate

---

Connecting Ideas

The environment selects for genes that direct the development of animals' nervous systems in ways that promote certain behavior patterns. In all animals, some behaviors are innate, and others are learned. Whether the behavior is innate or

learned, it increases the animal's ability to adapt and pass on its genes to its own offspring. Humans, like other animals, have complex nervous systems and other body systems that work together to increase the individual's chances of survival.

Section Review

Understanding Concepts
1. What is the difference between imprinting and other types of learned behavior?
2. How does learning have survival value in a changing environment?
3. Explain by example the difference between trial-and-error learning and conditioning.

Thinking Critically
4. How would a cat respond if the mice in a barn no longer entered at the usual places?

Skill Review
5. **Observing and Inferring** Two dog trainers teach dogs to do tricks. One trainer gives her dog a bone-shaped treat whenever the dog correctly performs the trick. The other trainer does not. Which trainer will be more successful at dog training? Why? For more help, refer to Thinking Critically in the *Skill Handbook.*

Guide Dogs Help Disabled

Dogs have provided a richer and more fulfilling life for many people with disabilities. Dog companions can retrieve objects, guide people who are visually and hearing impaired, assist with money transactions in shops, and choose safe travel routes through congested areas.

Both dogs and owners require training Most dogs require more than four months of training to learn a series of 60 commands, followed by a month of training with their new owners. A dog and a person with disabilities must learn to live as a team and trust one another. The dogs are taught to walk in straight lines, stop at curbs and obstructions, manage bus rides, and ignore distractions. Some dogs can be trained to lock and unlock wheelchairs. They must also be able to identify right and left, up and down, and in and out. Because the life of the disabled person may depend on the animal, a special dog is required.

Dogs that are ears For people who are hearing impaired, dogs are ears. They alert their companions to telephones, doorbells, smoke detectors, timers, and crying babies. Welsh corgis and Border collies excel at this kind of signaling.

A shopper's helper In a store, a dog can act as hands for a shopper who is wheelchair-bound. The dog scans the contents of the store when commanded to "look," and takes the desired item when the person commands "get it." The dog can then place the item in the lap of its companion or carry it to the counter where it drops the item on the counter at the commands "up" and "give." The dog then gives the clerk a credit card or wallet to complete the sale.

Dogs are keys to freedom For many people with disabilities, canine companions provide a mobility they have never had. During an exercise in which a woman who is visually impaired was practicing with her first dog on a trip around the block, the instructor found her in tears. She told him, "This is the first time in ten years that I've gone anywhere on my own." For her, the dog represented freedom.

Applications for the Future

Although some dog trainers and breeders did not think dogs could be taught to help people who are disabled live independently, the success of several different training programs has proven that dogs are intelligent and capable enough to perform many helpful tasks. Dogs may be trained to help with physical therapy and even for work with individuals who are emotionally disturbed.

INVESTIGATING the Technology

1. **Apply** Imagine that you are disabled and recently received your first dog companion. Write a letter to the editor of the local paper describing your newfound freedom.
2. **Research** How do dogs communicate with their blind masters? Do some research to find out how dogs learn to navigate curbs, traffic, and congested areas.

36.1 Innate Behavior

- Behavior is anything an animal does in response to a stimulus. Behaviors have adaptive value and are shaped by natural selection.
- Innate behavior is inherited. Innate behaviors include automatic responses and instincts. Automatic responses include reflexes and fight-or-flight responses.
- An instinct is a complex pattern of innate behavior.
- Behaviors such as courtship rituals, displays of aggression, territoriality, dominance hierarchies, hibernation, and migration are all forms of instinctive behavior.

36.2 Learned Behavior

- Learning takes place when behavior changes through practice or experience. Learned behavior has adaptive value.
- Learning includes habituation, imprinting, trial and error, and conditioning. The most complex type of learning is learning by insight.
- Some animals use language, whereas most communicate by either visual, auditory, or chemical signals.

Key Terms

Write a sentence that shows your understanding of each of the following terms.

aggression	hibernation
behavior	imprinting
circadian rhythm	innate behavior
communication	insight
conditioning	instinct
courtship behavior	language
dominance hierarchy	migration
estivation	motivation
fight-or-flight	territory
response	trial-and-error
habituation	learning

Understanding Concepts

1. What is the adaptive value of variable behavior?
2. Which type of learning is more complex: trial and error or insight? Why?
3. Compare one advantage and one disadvantage of innate behavior and learned behavior.
4. Name a behavior you learned by trial and error, by conditioning, and by insight.
5. By accident, a gull drops a snail on the road. The snail's shell breaks, and the gull eats the snail. The gull continues to drop mollusks on the road. What type of behavior is this?
6. Chameleons can catch insects on the first try. Is this behavior innate or learned?
7. Explain how courtship rituals could aid in species recognition.
8. Bighorn rams are rarely injured or killed when they fight. Why?
9. In a pack of dogs, how can you tell which dog is the dominant one? Explain.

Relating Concepts

10. Make a concept map that relates the following terms and phrases. Supply the appropriate linking words for your map.

    innate behavior, learned behavior, imprinting, reflex, habituation, courtship behavior, insight, migration, conditioning, trial-and-error learning

Applying Concepts

11. When Charles Darwin visited the Galapagos Islands in 1835, he was amazed that the animals would allow him to touch them. Why were they not afraid?

12. What would be the advantage of a dominance hierarchy in a species that does not defend its territory?

13. You want your dog to wake you up when the alarm clock rings in the morning. How would you train your dog to do this?

14. You notice that birds sitting on a telephone wire always maintain an equal distance from one another. If a new bird approaches and perches on the wire, all the birds move slightly so that this distance is again established. What kind of instinctive behavior are they exhibiting? Explain.

15. If you found a nest of five-day-old goslings, would they imprint on you and follow you home if you stayed close to the nest for a while? Explain.

Earth Science

16. On a cloudy day, pigeons raised in an environment that includes distinct smells have no trouble returning home. Would it be easier or more difficult for these pigeons to return if their nostrils were numbed temporarily? Explain.

Thinking Critically

Developing a Hypothesis

17. **Biolab** You have prepared a hand puppet for hand feeding young sandhill cranes. You have made it look similar to the female crane's head. You try to feed beetles to the young chicks, and they refuse the food. Make two hypotheses about why the chicks refuse the food.

Interpreting Data

18. **Thinking Lab** The following bar graph represents mite infestations of different nests made with or without an aromatic green plant in the nesting materials. Explain what the data might show.

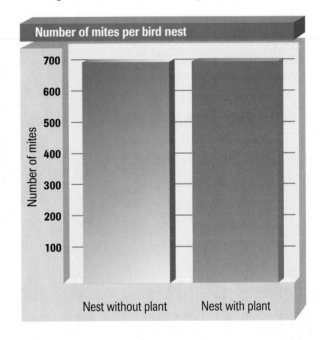

Number of mites per bird nest

Making Comparisons

19. **Minilab** Ducklings display an alarm reaction when a model of a hawk is flown over their heads, and no alarm reaction when a model of a goose is flown over their heads. If the models are flown over the ducklings for several days, eventually neither model causes any reaction. Compare the effects of the two models during the first two days with the effects of the same models two weeks later.

Connecting to Themes

20. **Evolution** Explain by example how natural selection is important in determining behavior of animals.

21. **Systems and Interactions** Explain by example how behavior has adaptive value.

22. **The Nature of Science** Explain how Ivan Pavlov used the methods of science to study conditioning behavior.

*T*he crowd falls silent as the starter raises the pistol. With a bang, the runners are off—their leg muscles flexing powerfully as they race down the track. The crowd cheers. As the leader crosses the finish line, her heart is pounding hard. For a few moments, she finds it difficult to catch her breath, but when she recovers, she's exalted and ready to run another race.

Winning a race may be the result of individual effort, but any successful dash to the finish line requires the interaction of all of the body's systems—systems that don't work independently, but are guided by hundreds of complex interactions.

Magnification: 5000×

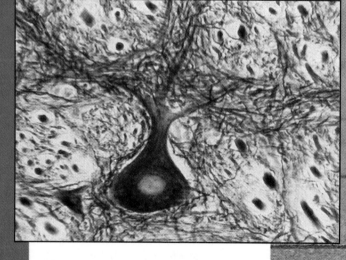

Nerve cells are part of the body's control center. How might nerve cells function during a complex activity such as running?

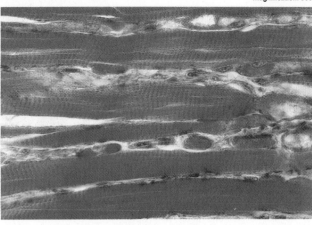

Muscles provide the force needed for running. How do muscle cells contract so that movement occurs?

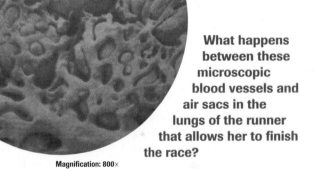

What happens between these microscopic blood vessels and air sacs in the lungs of the runner that allows her to finish the race?

Magnification: 800×

Unit Contents

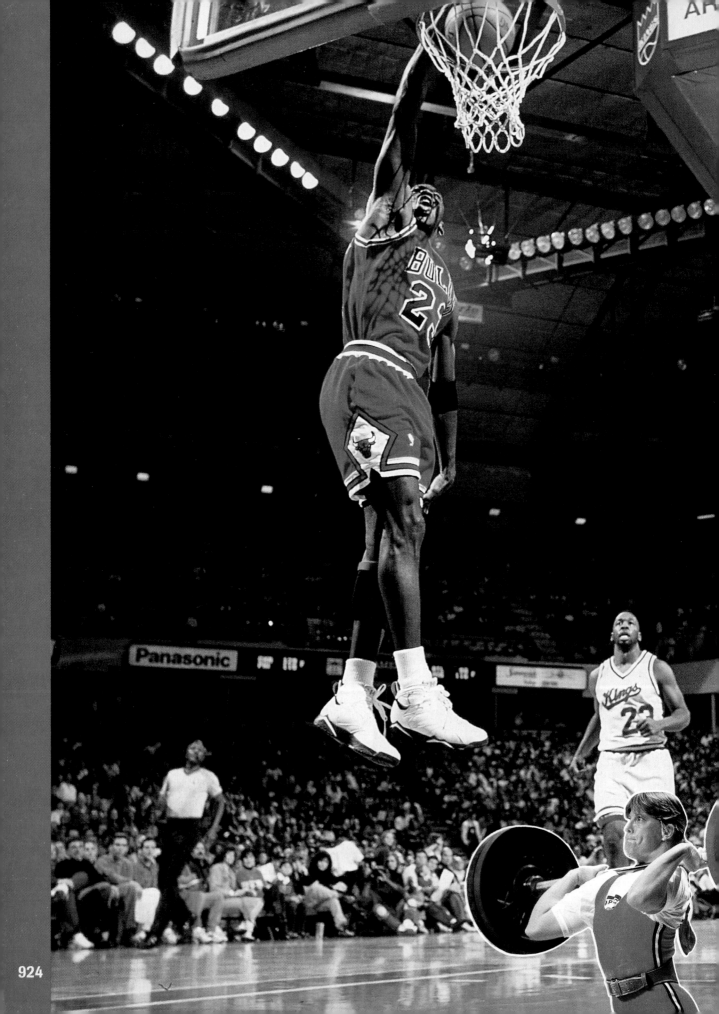

37 Protection, Support, and Locomotion

Some people believe Michael Jordan can defy the laws of gravity. With his body drenched in sweat, the team feeds him the ball; he dribbles, lunges, and hangs in the air, seemingly forever. Down comes his hand. SLAMMMM! Another basket and the fans are on their feet.

How do you suppose Jordan is able to perform such feats? His conditioning probably involves running to increase muscle tone and strengthen parts of his skeletal system, as his body's natural cooling system, skin, works overtime. Perhaps he lifts weights to increase the mass of the muscles that move his skeleton into the incredible positions that you see. Jordan may be an athlete who has reached peak physical condition, but even if you are not an athlete, your bones, muscles, and skin work in the same way. In this chapter, you'll investigate the structure of these important parts and learn how they function to protect, support, and move your body.

Animals differ from plants and fungi in that they are able to move from place to place. In vertebrates, the muscular and skeletal systems are mainly responsible for this function of locomotion, and skin protects muscles and bones and helps maintain homeostasis. How are bones, muscles, and skin adapted for these important processes?

37.1 Skin: The Body's Protection

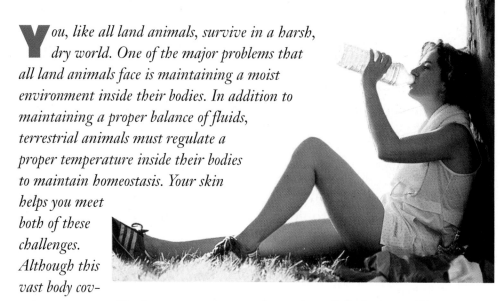

You, like all land animals, survive in a harsh, dry world. One of the major problems that all land animals face is maintaining a moist environment inside their bodies. In addition to maintaining a proper balance of fluids, terrestrial animals must regulate a proper temperature inside their bodies to maintain homeostasis. Your skin helps you meet both of these challenges. Although this vast body covering may seem like just a wrapping on the surface of the body, you'll see that it is a complex organ that performs a variety of functions.

Structure and Function of the Skin

Skin is composed of many layers of cells of each of the four types of body tissues: epithelium, connective, muscle, and nerve. You may recall from Chapter 28 the origins of different tissue types during embryo development. Epithelial cells are derived from the ectoderm layer of the embryo and function to cover surfaces of the body. Connective tissue is a fibrous tissue usually made of collagen and elastin protein fibers, and connects cell layers to each other. Nerve tissue helps to detect stimuli from the external environment. Muscles in skin are associated with hairs and respond to some stimuli like cold or fright. As a result of these four types of tissue, skin is an elastic, flexible, and responsive organ.

Skin is composed of two principal layers—the epidermis and dermis. Each layer performs a different function in the body.

Epidermis—The outer layer of skin

The layer of skin that you see covering your body is called the epidermis. The epidermis is the outer, thinner portion of the skin. It's composed of layers of both dead and living cells. The top layer consists of 25 to 30 layers of flattened, dead cells that are continually shed. Although dead, these cells still serve an important function. Dead epidermal cells contain a protein called **keratin** that helps waterproof and protect the living cell layers underneath.

epidermis:
epi (GK) on
derma (GK) skin
The epidermis covers other layers of skin.

The inner layer of the epidermis contains living cells that continually divide by mitosis to replace the dead cells. These cells contain **melanin,** a cell pigment that colors the skin and protects the cells from damage by solar radiation. As the newly formed cells are pushed up toward the surface, the nuclei degenerate and the cells die. Eventually, as they reach the top, they are shed.

Look at your fingertips. The epidermis on the fingers and palms of your hands and on the toes and soles of your feet contains ridges and grooves formed before birth, *Figure 37.1.* The epidermal ridges are important for gripping because they increase friction. Prints of these patterns are used to identify individuals.

Dermis—The inner layer of skin

The second major component of the skin is the dermis. The **dermis** is the inner, thicker portion of the skin. The thickness of the dermis varies in different parts of the body, depending on the function of that part. For example, dermis is 3 to 4 mm thick on the palms of the hands and soles of the feet, providing padding and protection. The 0.5-mm thickness of skin on the surface of the eye allows it to be transparent.

The dermis is adapted to a broader range of functions than the epidermis. These adaptations include structures such as blood vessels, nerves, nerve endings, sweat glands, and oil glands. Have you ever wondered why some areas of your skin are looser and more flexible than others? It's because fat deposits lie underneath the dermis in the subcutaneous layer. The amount of fat deposited in this layer varies in different areas of the body and among individuals. These deposits function to cushion the body, to insulate and help the body retain heat, and to store food for long periods of time.

Although skin is a well-adapted organ of protection, you know that skin is not a continuous, unbroken layer. Hair, for example, grows out of narrow cavities in the dermis called **hair follicles.** As hair follicles develop, they are supplied with blood vessels and nerves and become attached to muscle tissue.

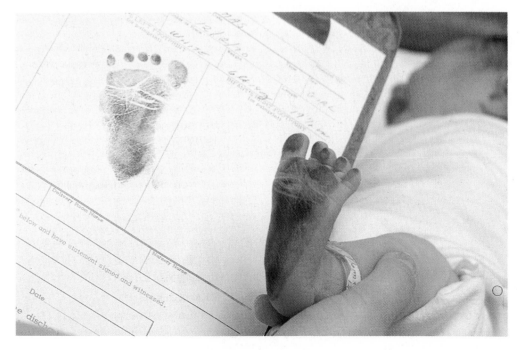

Figure 37.1

Babies have their footprints recorded at birth to establish an identification record. The patterns of fingerprints, footprints, and even ear prints are genetically determined and are often used for identification.

Skin Deep

The skin is an organ because it consists of tissues joined together to perform specific activities. It is the largest organ of the body; the average adult's skin occupies approximately seven square meters.

Surface view of skin

1 All people have the same number of melanin-filled cells. Differences in skin color are due to the amount of pigment produced by the cells. Exposure to sunlight causes an increase in melanin production, and the skin becomes darker.

2 Most oil glands are connected to hair follicles. Skin oil is a mixture of fats, cholesterol, proteins, and inorganic salts. Oil keeps hair from drying out and keeps the skin soft and pliable. Oil also inhibits the growth of certain bacteria.

4 Hair's primary function is protection of the skin from injury and sun. When hairs stand up, as occurs when you get goose bumps, they provide an insulating layer of air on the surface of the skin, thus reducing heat loss.

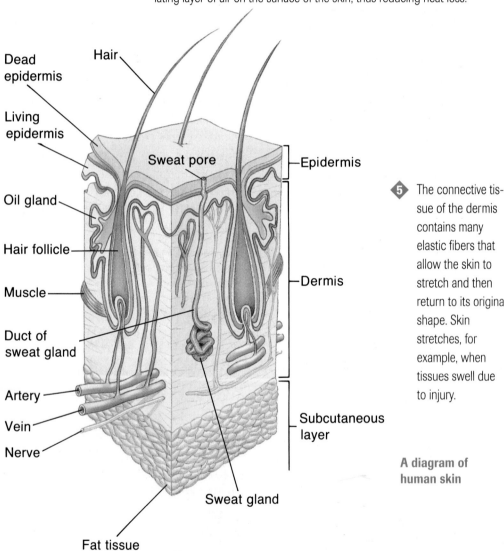

5 The connective tissue of the dermis contains many elastic fibers that allow the skin to stretch and then return to its original shape. Skin stretches, for example, when tissues swell due to injury.

A diagram of human skin

3 Sweat glands are located in the dermis and open up through pores onto the surface of the skin. Sweat is a mixture of water, salt (mostly NaCl), sugar, lactic acid, ascorbic acid (vitamin C), and small amounts of organic waste products. A person usually loses about 900 mL of sweat each day, depending on the type of activity and the temperature and humidity of the environment.

The skin's vital functions

Skin, shown in *Figure 37.2,* has several vital roles in maintaining homeostasis. Think about how your body warms up as you exercise. A major function of skin is to regulate your body temperature. When your body heats up, the many capillaries in the dermis dilate, blood flow increases, and body heat is lost by radiation. This mechanism also works in reverse. When you are cold, the blood vessels in the skin constrict and heat is conserved.

Another noticeable thing that happens to your skin as your body heats up is that it becomes wet. Glands in the dermis produce sweat in response to increases in body temperature. As sweat evaporates, the body cools because when water changes state from liquid to vapor, heat in the form of calories is lost.

Another function of skin is to serve as a protective layer to underlying tissues. Skin protects the body from physical and chemical damage and from the invasion of microbes.

Of course, anyone who has ever stepped on a sharp object or been burned by the sun knows that skin also functions as a sense organ. Nerve cells in the dermis receive stimuli from the environment and relay information about pressure, pain, and temperature.

Magnification: 100×

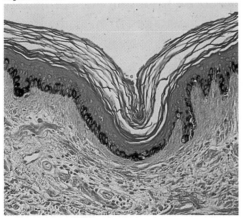

Skin also functions to maintain the balance of chemicals in the body. When exposed to ultraviolet light, dermis cells produce vitamin D, a nutrient that aids the absorption of calcium into the bloodstream. Because an individual's exposure to sunlight varies, a daily intake of vitamin D may be needed.

Figure 37.2

Photomicrograph of human skin

Acne

Have your ever had a pimple or blackhead? Acne is a disorder of the hair follicles and oil glands that affects almost everyone at some time or another. The exact cause of acne is unknown, but it is believed that four factors are involved: hormones, oil, bacteria, and abnormal development within hair follicles.

How acne starts An acne lesion begins when the epidermis produces too many cells surrounding a hair follicle. These cells then stick to each other, forming a mass that mixes with oil and blocks the hair follicle. A blackhead results if the opening of the hair follicle is blocked by the accumulating cells and oil. A pimple develops if the wall of the hair follicle ruptures, stimulating an immune response that results in a red pimple filled with pus. Acne is more of a problem during the teen years because the increase in sex hormones increases the secretions of oil into the hair follicles.

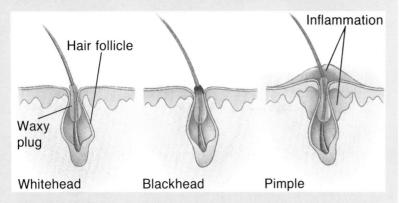

Whitehead Blackhead Pimple

Keeping acne at bay Acne can be treated by regular cleansing of the skin. By keeping skin clean, acne-causing bacteria living on the skin can be destroyed, and dead cells and oil can be reduced.

Sometimes physicians prescribe drugs to control acne. Antibiotics to manage bacteria and synthetic hormones to decrease the production of oil are often used. Over-the-counter medicines that produce peeling of the skin are also available. These include salicylic acid, benzoyl peroxide, and synthetic vitamin A.

Thinking Critically
Consider each form of treatment. How does each control acne?

Figure 37.3

When skin is injured into the dermis, the first reaction of the body is to restore the continuity of the skin. This prevents the invasion of harmful bacteria that live on the skin.

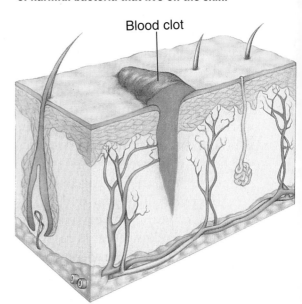

Blood clot

A Blood flows out of the wound until a clot forms over it.

Skin Injury and Restoration of Homeostasis

If you've ever had a mild scrape on your arms or legs, you know that it doesn't take long for the scrape to heal. When small injuries occur to the epidermis, such as a scrape, epidermal cells divide by mitosis and fill in the gap left by the abrasion. However, when the skin is injured into the dermis, bleeding usually occurs and the skin goes through a series of stages to heal itself and maintain homeostasis. *Figure 37.3* shows the stages involved in skin repair.

Have your ever suffered a painful burn? Burns can result from exposure to the sun or contact with chemicals or hot objects. Burns are rated according to their severity. Most people have received a first-degree burn at one time or another.

B The wound is closed by the formation of a scab. The scab unites the wound edges to prevent bacteria from entering. Blood vessels dilate, and infection-fighting cells speed to the wound site.

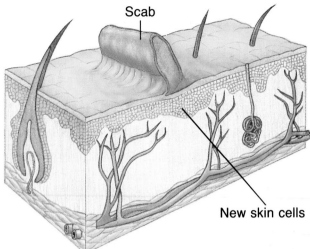

Scab

New skin cells

C Skin cells beneath the scab begin to multiply and fill in the gap. Eventually, the scab falls off to expose new skin. If a wound is large, a scar may result due to the formation of large amounts of dense connective tissue fibers.

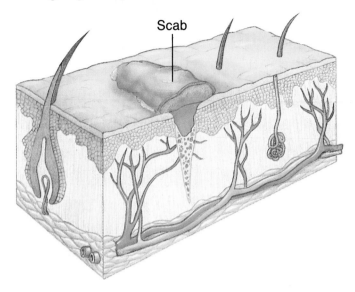

Scab

First-degree burns are characterized by redness and mild pain and involve the death of epidermal cells. Second-degree burns involve damage to skin cells of the dermis and can result in blistering and scarring. The most severe burns are third-degree burns, which destroy the epidermis and dermis. With this type of burn, skin function is lost, and regrowth of the skin is slow with much scarring. Skin grafts may be required to replace lost skin. In most cases, skin can be removed from another area of the patient's body and transplanted to a burned area.

As people get older, the appearance and function of the skin changes. Aging of the skin is evidenced by an increase in wrinkles and sagging. Wrinkles appear because the skin becomes less elastic with age. As aging progresses, the oil glands produce less oil and the skin becomes drier. These changes are natural, but prolonged exposure to ultraviolet rays from the sun can damage skin cells and accelerate the aging process.

Section Review

Understanding Concepts
1. Compare the structures of the epidermis and dermis.
2. How does skin control body temperature?
3. How does skin protect the underlying tissue layers from environmental factors?

Thinking Critically
4. Why is it uncomfortable to wear wool and nylon in hot climates and cotton in cold climates?

Skill Review
5. **Sequencing** Outline the steps that occur when a cut in the skin heals. For more help, refer to Organizing Information in the *Skill Handbook*.

37.2 Bones: The Body's Support

Section Preview

Objectives

Summarize the structure and functions of the skeleton.

Compare the types of movable joints.

Explain how the skeleton forms.

Key Terms

axial skeleton
appendicular skeleton
joint
ligament
bursa
tendon
osteoblast
compact bone
spongy bone
marrow

Because bones are hard, many people believe that they are not living tissue. Would you believe that you had more bones when you were born than you have now? That's because bones are living tissue and have grown together since you were born. Your head, for example, had soft spots when you were an infant. Your head feels solid now because the bones of the skull have fused together. Remodeling of the skeleton occurs throughout life. In fact, your skeleton hasn't completely fused yet. You will not have a solid, fused skeleton until about age 25.

Structure and Function of the Skeletal System

The adult human skeleton has 206 named bones and is composed of two main parts, as shown in *Figure 37.4*. The **axial skeleton** includes the skull and the bones that support it, such as the vertebral column, ribs, and sternum. The **appendicular skeleton** includes the bones of the arms and legs and structures associated with them, such as the shoulders and pelvic girdles.

Joints—Where bones meet

Every time you open a door, you're using a joint. A door is connected to a door frame by a hinge joint. In vertebrates, a **joint** is where two or more bones meet. Most joints allow movement, but some, such as the joints of the skull, are fixed. Fixed joints are joints that do not move.

Figure 37.4 shows the types of movable joints in the skeleton and the different actions that they allow.

Joints are supported and surrounded by structures that aid in their movement. The joints are held together and often enclosed by ligaments. A **ligament** is a tough band of connective tissue that connects bones to bones. Joints with large ranges of motion, such as the knee, typically have more ligaments surrounding them. In addition, the ends of bones are covered with a layer of cartilage, which allows for smooth movement between them. In movable joints such as the shoulder and knee joints, there is also a fluid-filled sac called a **bursa** located between the bones. The bursa acts as a cushion to absorb shock and keep bones from rubbing against each other. **Tendons,** which are thick bands of connective tissue, attach muscles to bones.

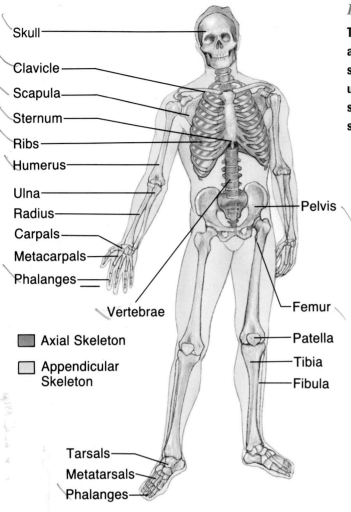

Skull
Clavicle
Scapula
Sternum
Ribs
Humerus
Ulna
Radius
Carpals
Metacarpals
Phalanges
Vertebrae

Pelvis
Femur
Patella
Tibia
Fibula

■ Axial Skeleton
□ Appendicular Skeleton

Tarsals
Metatarsals
Phalanges

Figure 37.4

The axial skeleton includes the bones of the head, back, and chest. The bones of the skull and face contain the sinuses, which are cavities lined with cells that are continuous with the nasal cavity. Bones in the appendicular skeleton are related to movement of the limbs. To which skeletal group do the phalanges belong?

◆ Body movements are made possible by joints that move in several different ways.

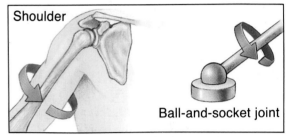

Shoulder

Ball-and-socket joint

▲ Ball-and-socket joints allow rotational movements. The joints of the hips and shoulders are ball-and-socket joints.

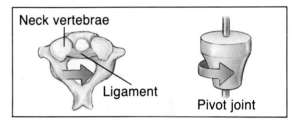

Neck vertebrae

Ligament

Pivot joint

▲ Pivot joints allow bones to twist against each other. The joint between the first two vertebrae of the neck is a pivot joint.

Elbow   Hinge joint

▲ Hinge joints are present in the elbows, knees, fingers, and toes and allow back-and-forth movement like a door hinge.

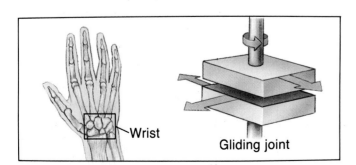

Wrist
Gliding joint

▲ Gliding joints occur in the wrists and ankles and allow bones to slide against each other.

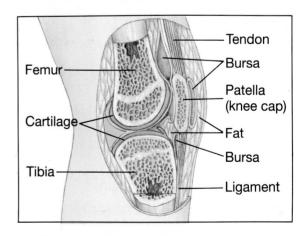

Tendon
Bursa
Patella (knee cap)
Fat
Bursa
Ligament
Femur
Cartilage
Tibia

▲ Structures of the knee aid in its movement and reduce wear and tear.

X rays—The Painless Probe

X rays are a form of light emitted by X-ray tubes and by some astronomical objects such as stars. The short wavelength and high energy of X rays make it possible to discern a broad range of objects from the tiniest atoms to the largest galaxies. X-ray machines are so common that you have probably had contact with one recently. Dentists use them to examine teeth, doctors to inspect bones and organs, and airports to look inside your carry-on baggage.

Non-invasive diagnosis In medicine, X rays are passed through soft tissues to photographic film placed behind the area; bones and other dense objects show up as white areas on the film. The position or nature of a break is clearly visible. The contours of organs such as the stomach are seen when a patient ingests a high-density liquid; other organs can be marked with special dyes. CAT scans are 3-D images made by X-raying an area many times at different angles. They are useful for viewing the brain and observing tissues shadowed by overlying organs.

Radiation treatments As X rays bombard atoms of tissues, electrons are knocked from their orbits, resulting in damage to the exposed tissue cells. To protect healthy tissues, absorptive metals are used as shields. You've probably had a dental X ray where the dental assistant spread a heavy lead apron across your chest. The destructive nature of X rays has proven useful in the treatment of cancers, where cancerous cells are targeted and destroyed.

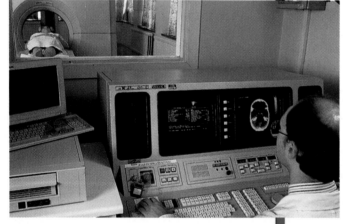

CONNECTION TO Biology

Why would an X ray be useful to diagnose a fractured bone but not a sprained ligament?

Forcible twisting of a joint, called a sprain, can result in injury to the bursa, ligaments, and tendons. A sprain most often occurs at joints with large ranges of motion such as the wrist, ankle, and knee.

Besides injury, joints can also be subjected to disease. One common joint disease is arthritis, an inflammation of the joints. It can be caused by infections, aging, or injury. One kind of arthritis results in large bone spurs or bumps of bone inside the joints. Such arthritis is painful, and joint movement may become limited. The causes of all types of arthritis are not completely known.

The formation of bone from cartilage

The skeleton of a vertebrate embryo is made of cartilage. By the ninth week of human development, bone begins to replace cartilage. The cartilage framework formed during embryonic life is covered by a membrane. Blood vessels penetrate the membrane and stimulate cells in the cartilage to enlarge and become potential bone cells called **osteoblasts.** These bone cells secrete a substance in which minerals are deposited. The deposition of calcium salts and other ions hardens in the cartilage and transforms it to bone.

The adult skeleton is almost all bone, with cartilage found in regions such as the nose tip, external ears, discs between vertebrae, and lining the movable joints.

Bones grow in both length and diameter. Growth in length occurs at the ends of bones in cartilage plates. Growth in diameter occurs on the outer surface of the bone. The increased sex hormones produced during the teen years cause the

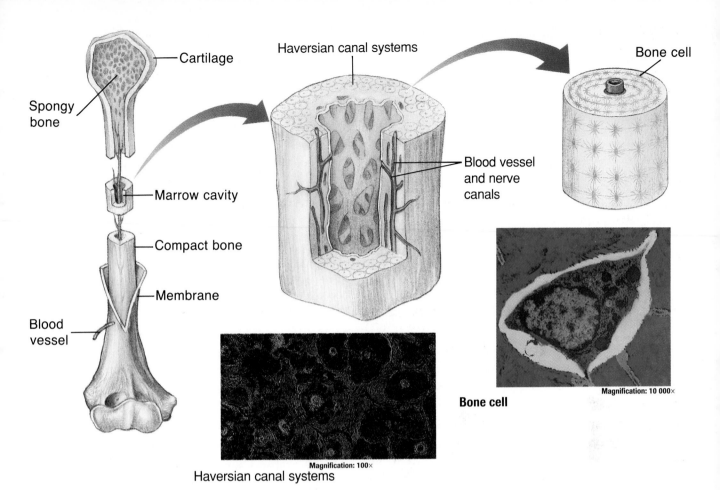

Spongy bone

Cartilage

Haversian canal systems

Bone cell

Marrow cavity

Blood vessel and nerve canals

Compact bone

Membrane

Blood vessel

Magnification: 10 000×

Bone cell

Magnification: 100×

Haversian canal systems

Figure 37.5

A bone has several components, including compact bone, spongy bone, and a surrounding membrane. Notice in the cross section (center) that the compact bone looks like a series of rings. The rings, along with the canals they surround, make up what are called Haversian canal systems. Bone cells (right) receive oxygen and nutrients from small blood vessels running within the Haversian canals. Nerves in the canals conduct impulses to and from each bone cell.

osteoblasts to divide more rapidly, resulting in a growth spurt. By age 20, 98 percent of the growth of the skeleton is completed. However, these same hormones also cause the growth centers at the ends of bones to degenerate. As these cells die, growth slows. After growth in length stops, bone-forming cells are involved mainly in repair and remodeling of bone.

The compact and spongy structure of bone

Although bones may appear uniform, they are actually composed of two different types of bone tissue, as shown in *Figure 37.5*. Surrounding every bone is a layer of hard bone, or **compact bone.** Compact bone is covered by a nerve and blood vessel-filled membrane that supplies nutrients and oxygen to bone cells. Compact bone surrounds less-dense bone known as **spongy bone,** so called because it is filled with many holes and spaces like a sponge. The center cavity of a bone is filled with a soft tissue called **marrow.** Marrow fills the cavities of the ribs, sternum, vertebrae, skull, and the long bones of the arms and legs.

What structures can be seen in compact bone?

The skeletal system of vertebrates is composed of bones like those in your skeleton. Bone-forming cells originate in cartilage and secrete bone tissue. Pathways for blood vessels and nerves form in the bone.

Procedure

1. Obtain a slide of compact bone, and use the low-power objective to focus on the tissue.

2. Compact bone is made up of many Haversian canal systems. Look at a Haversian canal system shown in the illustration, and locate a similar structure on your slide.

3. A Haversian canal is located in the center of each system. Locate these structures on your slide. Blood vessels and nerves are found in each Haversian canal.

4. Bone cells are embedded in mineral salts in layers around Haversian canals. Find these layers on your slide. Within each layer are small canals called canaliculi that carry fluids between the blood vessels and bone cells.

5. Draw the Haversian canal system as observed on your slide. Label a Haversian canal, a bone cell, and canaliculi.

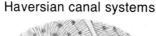

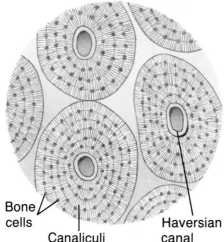

Haversian canal systems

Bone cells

Canaliculi

Haversian canal

Analysis

1. What is the function of the Haversian canal? The canaliculi?

2. How does the Haversian canal system allow for efficient delivery of oxygen and nutrients to bone cells?

3. Where do you find the majority of calcium salts in the Haversian canal system?

Figure 37.6

Dairy products and leafy vegetables are good sources of calcium.

Skeletal system functions

The primary function of your skeleton is to provide a framework for the tissues of your body, much like a building's inner, steel framework. The skeleton is also designed to protect the internal organs including the heart, lungs, and brain.

Besides support and protection, the skeleton is designed for efficient movement. Joints allow for this movement, and muscles that move the skeleton need firm points of attachment to pull against so they can work efficiently.

Bones are also responsible for producing blood cells. Red marrow—found in the humerus, femur, sternum, ribs, vertebrae, and pelvis—produces red blood cells, some white blood cells, and cell fragments that are involved in blood clotting. Yellow marrow found in many other bones stores fats and aids in producing red blood cells when there is a massive blood loss due to severe injury.

Finally, the bones of the skeleton serve as storehouses for minerals, including calcium and phosphorus. Calcium is a critical part of the diet for healthy, strong bones. Sources of calcium are shown in *Figure 37.6*. Calcium is also important for transmission of nerve impulses and muscle contraction. However, most of these minerals are used during growth and formation of the skeleton.

Homeostasis, Aging, and the Skeletal System

Bones will remodel themselves throughout life. While osteoblasts maintain their action of depositing calcium, another group of bone cells acts to remove calcium. The calcium removed from bone is needed by many tissues of the body, such as

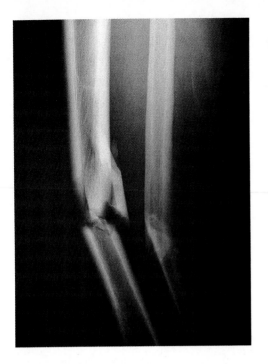
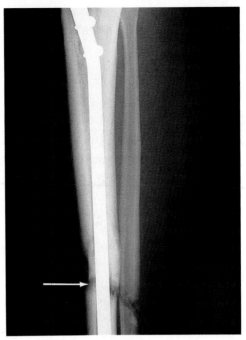

nerves and muscles, to perform their functions. The removal of calcium also prevents bone from becoming thick and heavy. This remodeling of the skeleton occurs as you age, gain or lose weight, or change activity.

Bone cells are also involved in the process of repair when a bone is broken. Perhaps you have injured a bone and had the injured area set. When bones are broken, *Figure 37.7,* a doctor moves them back into place, and they are kept immobile until the bone tissue regrows and replaces the damaged area.

The composition of bone changes as a person ages. Minerals are contin- ually deposited in the bones. These minerals, such as calcium and phos- phorus, make the bones hard. A child's bones have more collagen pro- tein and fewer minerals than the bones of adults and, as a result, are less brittle. Bones tend to become more brittle as their composition changes with age. For example, a dis- ease called osteoporosis is a condition in which there is loss of bone mass. The reduced bone mass results in the bones becoming more porous and brittle. Osteoporosis is most com- mon in older women because they produce lesser amounts of a hormone that aids in bone formation.

Section Review

Understanding Concepts

1. Distinguish between the appendicular skeleton and the axial skeleton.
2. List the five main kinds of joints and pro- vide an example of each.
3. In what way do bones help regulate min- eral levels in the body?

Thinking Critically

4. Why would it be impossible for bones to grow from within?

Skill Review

5. **Sequencing** Outline the steps involved in bone formation from cartilage to bone. For more help, refer to Organizing Information in the *Skill Handbook.*

37.3 Muscles for Locomotion

Section Preview

Objectives

Distinguish among the three types of muscles.

Explain the structure of a myofibril and summarize the sliding filament theory.

Key Terms

smooth muscle
involuntary muscle
cardiac muscle
skeletal muscle
voluntary muscle
myofibril
myosin
actin
sarcomere
sliding filament theory

Perhaps you have seen the Olympic games on television. Think about the different athletes involved in the games. You may be able to tell what sports some of them participate in just by looking at their body shapes. For example, swimmers and ice skaters have different shapes because ice skaters develop strong leg muscles over many months of training, whereas swimmers develop larger shoulder muscles.

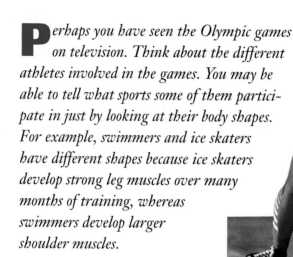

Three Types of Muscles

A muscle consists of a mass of fibers grouped together. Almost all of the muscle fibers you will ever have were present at birth. Nearly half of your body mass is muscles. The muscles in the body are of three main kinds, *Figure 37.8.* One type of muscle, **smooth muscle,** is found in internal organs and blood vessels. Smooth muscle is made up of sheets of cells that are ideally shaped to form a lining for or-

gans such as the trachea, digestive tract, and reproductive tract. The most common function of smooth muscle is to squeeze, exerting pressure on the space inside the tube or organ it surrounds. Contractions of smooth muscle are slow and prolonged compared with the contractions of the other two kinds of muscle. These contractions are not under conscious control, so smooth muscle is an example of an **involuntary muscle.**

Figure 37.8

Muscles are either under voluntary or involuntary control and differ in their structure and appearance.

▶ **Smooth muscle cells that make up involuntary muscle are spindle shaped and have a single nucleus.**

▼ **Cardiac muscle cells that also form involuntary muscle, appear striated or striped when magnified.**

Another type of involuntary muscle is the muscle that makes up the heart, or **cardiac muscle.** Cardiac muscle cells are interconnected and form a network that helps the heart muscle contract more efficiently. Cardiac muscle is found only in the heart, and is adapted to conduct the electrical impulses necessary for rhythmic contraction. The third type of muscle makes up the larger mass of muscle in the body, skeletal muscle. **Skeletal muscle** is the type of muscle that is attached to bones and moves the skeleton. Skeletal muscle cell contractions are short and strong, providing the force needed for movement. The majority of the muscles in your body are skeletal muscles and, as you know, you can control their contraction. A muscle that contracts under conscious control is called **voluntary muscle.**

Skeletal Muscle Contraction

Whether you are playing tennis, pushing a lawn mower, or writing, your muscles are contracting as they do work. When the muscle contracts, the bones are pulled by tendons. *Figure 37.9* shows the movement of the lower arm.

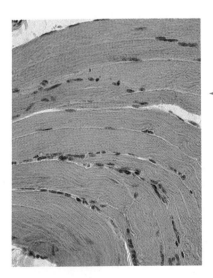

◄ Skeletal muscle is also striated but makes up voluntary muscle. Each skeletal muscle cell has more than one nucleus and may be up to 30 cm long in tall humans.

How is rigor mortis used to estimate time of death?

When a person dies, the respiratory and circulatory systems that obtain and deliver oxygen to the muscles stop functioning. Rigor mortis is the stiffening of both voluntary and involuntary muscles due to lack of ATP after death. This condition generally develops about an hour or two after death, and full rigor mortis takes 10 to 12 hours. Stiffening usually proceeds from the upper body to the lower body, and it is believed to be related to fiber length. Rigor mortis affects smaller fibers in the face area first; medium-length fibers in the neck, chest, and arms next; and longer fibers in the legs last. As time proceeds, rigor mortis gradually subsides in the same order it develops. It disappears completely after 24 to 36 hours.

Analysis
Coroners determine the time of death, based on the degree of rigor mortis, by examining the body and its tissue under a microscope.

Thinking Critically
A murder victim is found with relaxed face muscles, but a stiff upper and lower body. How long might it have been since the crime?

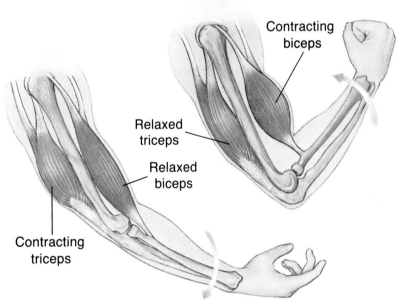

Contracting biceps

Relaxed triceps

Relaxed biceps

Contracting triceps

Figure 37.9

When the biceps muscle contracts (top), the lower arm is moved upward. When the triceps muscle on the back of the upper arm contracts (bottom), the lower arm moves downward. Muscles work in opposing pairs, like the biceps-triceps pair, to move bones.

The movement of body parts results from the contraction and relaxation of muscles. In this process, muscles use energy from aerobic respiration and lactic acid fermentation. When exercise is continued for a long period of time, the waste products of fermentation accumulate and muscle fibers are stressed, causing fatigue. Fatigue affects the various muscles differently. It also affects individuals differently, even when they are performing the same tasks. Muscular strength, muscular endurance, and the amount of effort required to perform a task are variables to consider.

Does fatigue affect the ability to do work?

PREPARATION

Problem
How does fatigue affect the number of repetitions of an exercise you can accomplish? How do different amounts of resistance affect rate of fatigue?

Hypotheses
Hypothesize whether muscle fatigue affects the amount of work muscles can accomplish. Consider whether fatigue can occur within minutes or hours.

Objectives
In this Biolab, you will:
- **Hypothesize** whether muscle fatigue affects the amount of work muscles can accomplish.
- **Measure** the amount of work done by a group of muscles.
- **Prepare** a graph to show the amount of work done by a group of muscles.

Possible Materials
stopwatch or clock with second hand
graph paper
small weights

PLAN THE EXPERIMENT

1. Design an exercise for a group of muscles that can be counted over time.

2. Work in pairs with one member of the team being a timekeeper and the other member performing the exercise.

3. Consider setting up your experiment so that the amount of resistance is the independent variable. Compare your results with other groups.

Check the Plan

1. Be sure that the exercises are ones that can be done rapidly and cause a minimum of disruption to other groups in the classroom.

2. Consider how long you will do the activity and how often you will record measurements.

3. **Make sure your teacher has approved your experimental plan before you proceed further.**

4. Make a table that records the number of exercise repetitions versus time intervals.

5. On the graph paper, plot the number of repetitions on the vertical axis and the time on the horizontal axis.

6. Carry out the experiment.

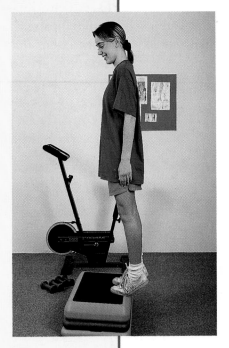

ANALYZE AND CONCLUDE

1. **Making Inferences** What effect did repeating the exercise over time have on the muscle group?

2. **Comparing and Contrasting** As you repeated the exercise over time, how did your muscles feel?

3. **Recognizing Cause and Effect** What physiological factors are responsible for fatigue?

4. **Thinking Critically** How well do you think your fatigued muscles would work after 30 minutes of rest? Explain your answer.

Going Further

Project Design an experiment that will enable you to measure the strength of muscle contractions.

How muscles contract is related to their structure. Skeletal muscle tissue is made up of muscle cells or fibers. Each cylindrical muscle fiber is made up of smaller fibers called **myofibrils.** Myofibrils are composed of even smaller protein filaments. Filaments can be either thick or thin. The thick filaments are made of the protein **myosin,** and the thin filaments are made of the protein **actin.** The arrangement of myosin and actin gives skeletal muscle its striated appearance. Notice in *Figure 37.10* that each myofibril appears to be divided into sections. Each section of a myofibril is called a **sarcomere** and is the functional unit of muscle.

The sliding filament theory is currently the best explanation for how muscle contraction occurs. The **sliding filament theory** states that the actin filaments within the sarcomere slide toward one another during contraction. The myosin filaments do not move.

myofibril:

mys (GK) muscle
fibrilla (L) small fiber
A myofibril is a small muscle fiber.

Figure 37.10

Locomotion is an essential body function that results from the contraction and the relaxation of muscles. The structure of muscle helps explain the sliding filament theory of how muscles contract.

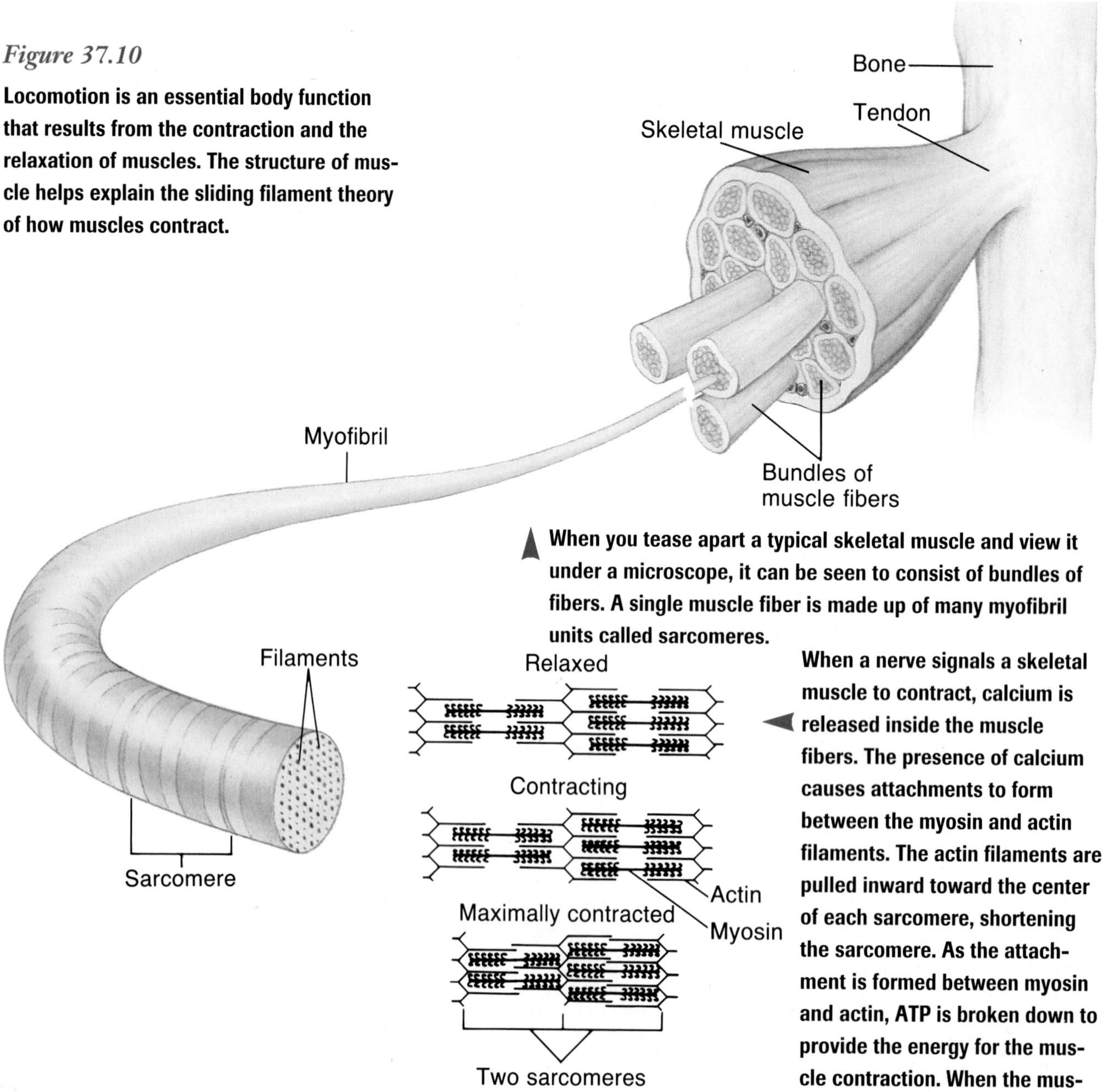

Bone

Tendon

Skeletal muscle

Bundles of muscle fibers

Myofibril

Filaments

Sarcomere

When you tease apart a typical skeletal muscle and view it under a microscope, it can be seen to consist of bundles of fibers. A single muscle fiber is made up of many myofibril units called sarcomeres.

Relaxed

Contracting

Actin

Maximally contracted

Myosin

Two sarcomeres

When a nerve signals a skeletal muscle to contract, calcium is released inside the muscle fibers. The presence of calcium causes attachments to form between the myosin and actin filaments. The actin filaments are pulled inward toward the center of each sarcomere, shortening the sarcomere. As the attachment is formed between myosin and actin, ATP is broken down to provide the energy for the muscle contraction. When the muscle relaxes, the filaments slide back into their original positions.

Figure 37.11

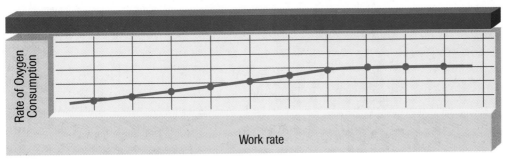

As individuals increase the intensity of their exercise activity, their need for oxygen goes up in almost exact, predictable increments.

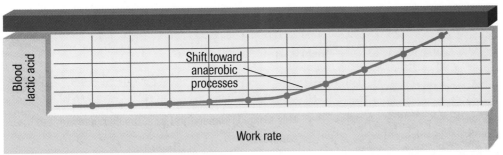

Shift toward anaerobic processes

At a certain intensity, which differs among individuals, the body shifts from aerobic respiration to the anaerobic process of lactic acid fermentation for the energy needed to accomplish the activity. Because lactic acid is produced, an upswing in its presence in the bloodstream can be used to indicate the point at which lactic acid fermentation dominates.

Muscle Strength and Exercise

How can you increase the strength of your muscles? Muscle strength does not depend on the number of fibers in a muscle. It has been shown that this number is basically fixed before you are born. Rather, muscle strength depends on the thickness of fibers and how many of them contract at one time. Thicker fibers are stronger and contribute to muscle mass. Regular exercise stresses muscle fibers slightly, and to compensate for this added workload, the fibers increase in size.

Recall from Chapter 10 that ATP is produced during cellular respiration. Muscle cells are continually supplied with ATP from both aerobic and anaerobic energy systems. However, the aerobic energy system for producing ATP dominates when adequate oxygen is delivered into a muscle cell to meet its energy production needs, as when a muscle is at rest or during slow or moderate activity. When an inadequate supply of oxygen is available to meet the muscle cell's energy needs, the anaerobic process of lactic acid fermentation, *Figure 37.11,* becomes the primary source of ATP during vigorous activity.

Think about what happens when you are running in gym class or around the track at school. When your muscles are working hard, they are not able to get oxygen fast enough to sustain aerobic respiration to supply adequate ATP. Thus, the amount of ATP available becomes limited. For your muscle cells to get the energy they need, they must rely on lactic acid fermentation as well.

Bionics
BIOlogy + ElectroNICS = BIONICS

Accidents, wars, disease, and genetic disorders—no matter how hard we try to prevent them, they still happen. When they do, people are frequently left without limbs or the ability to move their muscles. However, teamwork between biologists, physicists, and engineers is beginning to bring hope and mobility to people with these types of disabilities.

When the use of a limb is lost, it is generally for one of two reasons: the limb itself has been lost or the nerves that control the limb have been severed. For example, car accidents and sports injuries sometimes damage the spinal cord in the neck (cervical) area. As a result, the victim may be paralyzed from the point of damage down.

Mechanical intervention

Bionics at work A functional electrical stimulation (FES) system is currently being developed. The process used by this system is illustrated below.

1 The user generates a signal by moving unparalyzed muscles in the shoulder.

2 This signal is transmitted from the command unit to the computer.

3 The computer interprets the message and sends a radio signal to an antenna.

4 The radio signal energizes the stimulator.

5 The stimulator sends electrical impulses through wires to electrodes.

6 The embedded electrodes in the muscles are required for action.

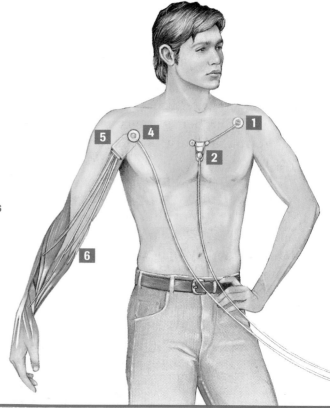

Lost messages

Nerves serve as lines of communication. Nerves are the information highways of the body. They coordinate and direct everything that goes on inside complex animals with electrical impulses. When nerves are severed, the muscles they serve lose contact with the brain and are helpless. It's similar to when wires in an electrical circuit within a house are cut. However, while wires can be reconnected and the function of all appliances within the circuit restored, nerves do not regenerate or heal. Thus, nerve damage is usually permanent.

Researchers are now working on alternative methods of activating and coordinating muscles served by severed nerves. Developments in bionics, the science of designing instruments modeled after living organisms, may hold the key to these researchers' success.

Replacing the irreplaceable

Prostheses—Artificial limbs When limbs are lost, the remaining nerves may still function, but the muscles they command are gone. Today's solution to this problem is to replace the lost limb with an artificial one (prosthesis). The most sophisticated of these are connected to existing muscles and use wire and springs to approximate natural movement. The dream of the future is to use servo motors connected directly to undamaged nerves to provide natural movement and delicate control, as in fingers.

Bionic fact and fiction

Fact Bill Denby is a Vietnam veteran who lost both his legs. When he returned home, he wanted to participate fully in one of his favorite activities—playing basketball. Modern prostheses, made of a high-tech plastic, allow him to run and shoot again.

Fiction Captain Picard of the *Enterprise* was injured in a fight with a group of Norsecans when he was a young officer just out of Star Fleet Academy. His life was saved when his heart was replaced with a bionic unit. Although Picard's unit has required maintenance and replacement, it could be the difference between life and death if it were available today. Perhaps this is the route of the future.

EXPANDING YOUR VIEW

1. **Applying Concepts** How could people paralyzed from the neck down use a system similar to this?

2. **Journal Writing** While Bill's new legs did not improve his game, speculate on improvements that could come in the future. Write your response in your journal.

3. **Brainstorming** What could be some alternatives to a mechanical heart?

During exercise, lactic acid builds up in muscle cells. As the excess lactic acid is passed into the bloodstream, the blood becomes more acidic and rapid breathing is stimulated. As you catch your breath following exercise, oxygen is supplied adequately to your muscles and lactic acid is broken down. Regular exercise can result in improved functioning of muscles. *Figure 37.12* shows how aerobic training, for example, can affect a muscle's ability to share and use energy.

Figure 37.12

Muscles that are exercised regularly become stronger and able to store more fuel for sustained activity.

Connecting Ideas

The skin of animals plays an important role in maintaining homeostasis of the body. Because skin is in direct contact with the external environment, it often responds first to any changes in the environment. However, the majority of responses are made by the muscular and skeletal systems to produce movement from place to place. The size and arrangement of muscles, and the structure of bones and joints, work to produce an efficient mode of locomotion for animals. Animals move for a variety of reasons. Perhaps most important of these is to support their heterotrophic way of life. The energy you require for all the muscle activity in your daily life is delivered from the food you eat.

Section Review

Understanding Concepts

1. Compare the structure and functions of three main types of muscles.
2. Summarize the sliding filament theory of muscle contraction.
3. How does exercise change muscle strength?

Thinking Critically

4. Why would a disease that causes paralysis of smooth muscles be life threatening?

Skill Review

5. **Interpreting Scientific Illustrations** Outline the composition of muscle fibers as shown in *Figure 37.10*. For more help, refer to Thinking Critically in the *Skill Handbook*.

Hip Replacement Surgery

Hip replacement surgery to relieve advanced osteoarthritis is the most common adult orthopedic procedure performed in America today, with 200 000 implants installed each year. Concerned health-care experts are pushing for injury-prevention therapies and improved treatment for arthritis to help stem the hip implant tide. Of particular interest are technological improvements to increase the life of an implant and limit second surgeries.

Structure of implants

An artificial hip implant consists of both elements of the ball-and-socket hip joint. During surgery, the defective socket is replaced with a plastic cup. Then, the metal ball end of the implant is fitted into the new socket. Finally, the implant's stem is fitted into the thighbone and affixed either with or without cement.

Recipients of implants

Whether or not to use cement usually depends on the age and activity level of the recipient. Cemented implants do best when the wearer is older than age 75 and relatively inactive. This is because the cement breaks down under continued stress after about five years.

Research and technology

With the increasing number of patients resulting from an aging U.S. population and the average lifetime of the recipient being extended by improved health care and healthier lifestyles, stronger and more durable installations are needed. Therefore, technological research has centered on improving cementless implants, which rely on a precise fit and bone regrowth to hold them in place.

Applications for the Future

To improve the performance of cementless implants, manufacturers now make custom-fitted implants. Chemical engineers have developed types of high-tech porous coatings that promote ingrowth or direct bonding of the bone to an implant's stem. Use of a robotic arm to drill a cavity in a human leg for cement-less implants may improve the accuracy of surgeons. Other technological advances focus on the implant's moving parts and seek to reduce the wear of the metal ball on the plastic socket.

INVESTIGATING the Technology

1. **Research** Read "The Thighbone's Connected to the. . . Artificial Hip Bone," *The New York Times*, November 29, 1992. What factors may contribute to the long-term success or failure of a hip implant?

2. **Apply** Following hip replacement surgery, doctors usually recommend exercise therapy to maintain muscle tone around the joint. What activities or exercises do you think would benefit a hip implant recipient and why?

37.1 Skin: The Body's Protection

- Skin plays a major role in maintaining homeostasis in the body. It regulates body temperature, protects the body, functions as a sense organ, and produces vitamin D.
- Skin responds to injury by producing new cells by mitosis and signaling a response to fight infection.

37.2 Bones: The Body's Support

- The skeleton is made up of the axial and appendicular skeletons. The skeleton supports the body, provides a place for muscle attachment, protects vital organs, manufactures blood cells, and serves as a storehouse for calcium and phosphorus.
- Bones remodel themselves throughout life.

37.3 Muscles for Locomotion

- There are three types of muscles: smooth, cardiac, and skeletal.
- Skeletal muscles contract due to sliding myofibrils made of actin and myosin.

- Muscle strength is due to muscle fiber thickness and the number of fibers contracting.

Key Terms

Write a sentence that shows your understanding of each of the following terms.

actin
appendicular skeleton
axial skeleton
bursa
cardiac muscle
compact bone
dermis
hair follicle
involuntary muscle
joint
keratin
ligament
marrow
melanin
myofibril
myosin
osteoblast
sarcomere
skeletal muscle
sliding filament theory
smooth muscle
spongy bone
tendon
voluntary muscle

Understanding Concepts

1. Refute the idea that the skin is just an outer wrapping for an animal.
2. Why do blood vessels dilate when the body heats up?
3. What are the functions of bone marrow?
4. How does bone composition change with age?
5. Explain how the muscles and bones function together in locomotion.
6. How do muscles work in pairs?
7. What happens to muscle filaments when the muscle relaxes?
8. What causes lactic acid to build up in muscles?

Using a Graph

9. A muscle physiologist studied the effect of load or stress on the shortening of a muscle. The more a muscle shortens, the more work a muscle can do. Based on the graph, describe the relationship between muscle shortening and load.

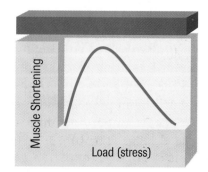

Muscle Shortening

Load (stress)

Relating Concepts

10. Make a concept map that relates the following terms and phrases. Supply the appropriate linking words for your map.

    actin, cardiac muscle, involuntary muscle, myosin, myofibrils, sarcomere, skeletal muscle, smooth muscle, voluntary muscle

Applying Concepts

11. Why is recovery from third-degree burns slower than from first-degree burns?
12. A bruise is due to a broken blood vessel. What part of a bone can be bruised?
13. How would the destruction of red marrow affect other systems of the body?
14. How could you use the skeleton to determine the age of a person?

A Broader View

15. How might regular cleansing of skin help prevent acne?

Thinking Critically

Comparing and Contrasting

16. **Biolab** If you were to compare your results with those of other groups that used identical procedures, how could you explain differences?

Observing and Inferring

17. **Minilab** What, other than fingerprints, could be used to identify individuals?

Recognizing Cause and Effect

18. **Thinking Lab** A coroner arrives at a crime death scene. The body is found to be warm, and stiff from only the neck up. What can the coroner conclude about the time of death?

Designing an Experiment

19. A new study with rat bone cells has shown that bone-building cells will mature and grow masses of tendrils when placed in small ceramic cubes that are permeated with tunnels. The filled ceramic cubes can then be inserted into a bone injury where the cells will continue to build new bone. Suppose you discover that human bone cells grow slower than rat bone cells. Design an experiment to test why the results in the two species differ.

Interpreting Graphs

20. The graph below shows the time relationship between muscle force and calcium level inside the muscle cell. Relate the cause-and-effect relationship between development of muscle force and calcium level with increasing load.

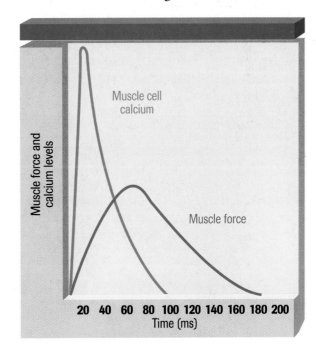

Connecting to Themes

21. **Systems and Interactions** Describe how the muscle and skeletal systems interact to produce efficient locomotion.
22. **Homeostasis** The skin is an important body organ in maintaining homeostasis. Outline how the skin helps the body maintain body temperature.
23. **Evolution** How is the ability to produce ATP by lactic acid fermentation an adaptive advantage for animals?

sleep
parkinsons
high blood
paranoid
delusion
arythmia
Heart attack
Suicide attempts

38 Digestion and Nutrition

Have you ever been sitting in a quiet classroom full of students, all silently working, when your stomach started to rumble and gurgle? As the noise got louder and louder, you may have wished you could crawl under your desk. As common and natural as this event is, it still can be quite embarrassing. What causes an empty stomach to be so noisy?

Maybe you've also experienced butterflies in your stomach—a tense feeling accompanied by mild nausea. Perhaps the butterflies occurred before you had to take a big exam, deliver a speech, or play in an important game. What makes your stomach feel this way?

Your stomach is just one part of your digestive system—the group of organs that is responsible for bringing food into your body and breaking that food into microscopic, usable molecules. How they accomplish that change is the story of digestion.

You probably don't pay much attention to your stomach—unless it begins to rumble like an earthquake or to feel like a nest of butterflies. When this embarrassed student has an opportunity to eat a good, nutritious meal, his stomach (inset) will stop gurgling and start digesting.

Section Preview

Objectives

Summarize the digestive functions of the organs of the digestive system.

Outline the pathway food follows through the digestive tract.

Key Terms

amylase
esophagus
peristalsis
epiglottis
stomach
pepsin
small intestine
pancreas
liver
bile
gallbladder
villus
large intestine
rectum

Sometimes it may seem like your stomach has a mind of its own. But it's really one part of a system that is, ultimately, controlled by your brain and by hormones that are secreted along the digestive pathway. As in many animals you have studied, the human digestive system is a tube that develops in the embryo. Different parts of the tube have evolved into different digestive organs, each with a special function to perform as food passes along.

Functions of the Digestive System

The digestive system provides several functions in the preparation of food for cellular utilization. First, the system ingests food. The next two functions take place at the same time; the system digests food as it is forced along the digestive tract. Fourth, the system absorbs the digested food into the cells of your body, and fifth, it eliminates undigested materials from your body.

Refer to *Figure 38.1* as you read about the digestive system and follow a meal through the organs of digestion.

Figure 38.1

All the digestive organs work together to break down food into simpler compounds that can be absorbed. As you read about each digestive organ, use this diagram to locate its position within the system.

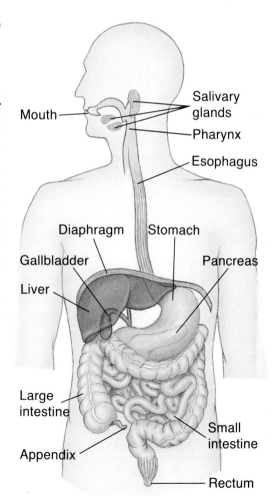

Remember from Chapter 8 how a cell is like a place of business in which all essential parts are assembled to form an efficient working body? The digestive system works in the opposite way. Food is disassembled into its component molecules so that the body can have a supply of resources for building new cells.

The Mouth

The first step in the digestive disassembly line is in the mouth. Suppose it's lunchtime, and you just prepared a bacon, lettuce, and tomato sandwich. The first thing you do is bite off a piece and chew it for ten to 20 seconds.

What happens as you chew?

In your mouth, your tongue moves the food around and helps position it between your teeth as you chew. Chewing is a part of mechanical digestion, the physical process of breaking food into smaller pieces.

Mechanical digestion increases the surface area of food particles and prepares them for chemical digestion. Chemical digestion is the process of structurally changing food molecules through the action of enzymes. Digestive enzymes, listed in *Table 38.1,* act to break the molecules apart.

Chemical digestion begins in the mouth

Some of the nutrients in your sandwich are starches, which are large polysaccharides. As you chew your bite of sandwich, salivary glands around your mouth, shown in *Figure 38.1,* secrete saliva. Saliva is a slightly alkaline solution that adds water and a digestive enzyme to the food. This enzyme, called **amylase,** breaks down starch into smaller sugar molecules called disaccharides. Even though food is swallowed too quickly for all the starches to be reduced in your mouth, amylase in the swallowed food continues to digest the starches in the stomach for about 30 minutes.

Table 38.1 Digestive Enzymes			
Organ	**Enzyme**	**Acts on**	**Product**
Salivary glands	Salivary amylase	Starch	Disaccharide
Stomach	Pepsin	Proteins	Peptides
Pancreas	Pancreatic amylase	Starch	Disaccharide
	Trypsin	Protein	Peptides
	Pancreatic lipase	Fats	Fatty acids and glycerol
	Nucleases	Nucleic acids	Sugar and nitrogen bases
Small intestine	Maltase	Disaccharide	Monosaccharide
	Sucrase	Disaccharide	Monosaccharide
	Lactase	Disaccharide	Monosaccharide
	Peptidase	Peptides	Amino acids
	Nuclease	Nucleic acids	Sugar and nitrogen bases

Inside Your Mouth

Your mouth houses many structures that are involved in other functions besides digestion. For example, these structures protect against foreign materials invading your body and help you taste the food you eat.

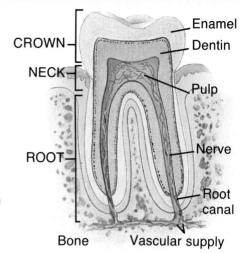

Structure of a tooth

Teeth are made mainly of dentin, a bonelike substance that gives a tooth its shape and strength. The dentin encloses a cavity filled with pulp, a tissue that contains blood vessels and nerves. The dentin of the crown is covered with an enamel that consists mostly of calcium salts. Tooth enamel is the hardest substance in the body.

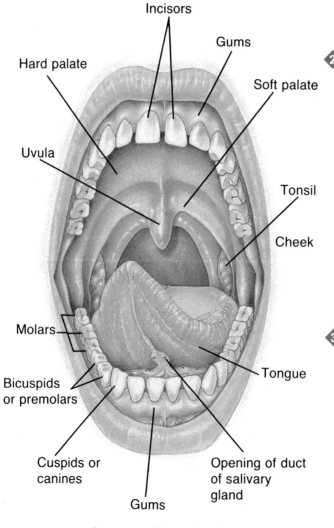

1 The incisors are adapted for cutting food. The cuspids, or canines, tear or shred food. The three sets of molars can crush and grind food. Often, there is not enough room for the third set of molars, called wisdom teeth, which then have to be removed.

2 A pair of tonsils are located at the back of the mouth. They help remove bacteria that have entered the mouth and nose.

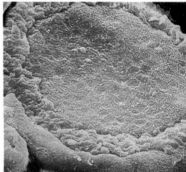

Magnification: 520×

3 Attached to the floor of the mouth is the tongue. It is made of skeletal muscle covered with a mucous membrane.

The upper surface of the tongue has many projections that form the taste buds, shown in the micrograph above, which contain numerous taste receptor cells.

Structures of the oral cavity

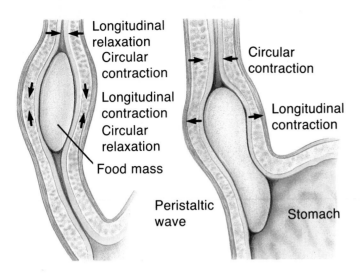

Longitudinal
relaxation
Circular
contraction

Longitudinal
contraction
Circular
relaxation

Food mass

Circular
contraction

Longitudinal
contraction

Peristaltic
wave

Stomach

Figure 38.2

Smooth muscle contractions are responsible for moving food through the esophagus. The contractions occur in waves; first, circular muscles relax and longitudinal muscles contract, then circular muscles contract and longitudinal muscles relax.

Swallowing your food

Once you've thoroughly chewed your bite of sandwich, your tongue shapes it into a ball and moves it to the back of your mouth to be swallowed. Swallowing forces food from your mouth to the esophagus. The **esophagus** is the muscular tube that connects the mouth to the stomach. Like a busy expressway, the esophagus moves the food into the stomach in five to eight seconds. *Figure 38.2* shows how the esophagus moves food along by **peristalsis,** a series of involuntary muscle contractions along the walls of the digestive tract.

Have you ever had food go down the wrong way? When you swallow, the food is diverted by the larynx and passes over the windpipe, the tube that leads to the lungs. Usually when you swallow, a flap of skin called the **epiglottis** covers the opening to the windpipe, and breathing is temporarily interrupted. After the food passes into your esophagus, the epiglottis opens again. But if you talk or laugh as you swallow, the epiglottis opens and the food may enter the windpipe. Your response, a reflex, is to choke and cough, forcing the food out of the windpipe.

The Stomach

When the chewed food reaches the bottom of your esophagus, a valve lets the food enter the stomach. The **stomach** is a muscular, pouch-like enlargement of the digestive tract.

Muscular churning

Three layers of muscles, lying across one another, make up the wall of the stomach. When these muscles contract, as in *Figure 38.3,* they mix the food in the stomach — another step in mechanical digestion.

peristalsis:
peri (GK) around
stellein (GK) to
  draw in
Peristalsis propels
food in one direction.

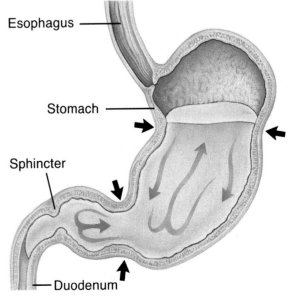

Esophagus

Stomach

Sphincter

Duodenum

Figure 38.3

When food enters the stomach, peristaltic movements pass through the stomach walls every 15 to 25 seconds. These waves soften the food, mix it with digestive juices, and reduce it to a thin liquid. The waves slowly become more vigorous and force a small amount of liquid out of the muscular opening at the lower end of the stomach. What type of muscle makes up the walls of the stomach?

Chemical digestion in the stomach

The lining of the stomach contains millions of glands that secrete a mixture of chemicals called gastric juice. Gastric juice contains hydrochloric acid and pepsin. **Pepsin** is a digestive enzyme that begins the chemical digestion of proteins in food. The enzyme pepsin works only in an acidic environment. This necessary environment is provided by the hydrochloric acid. The hydrochloric acid reduces the pH of the stomach contents to 2.

Knowing that the stomach secretes acids and enzymes, you may be wondering why the stomach doesn't digest itself. It does to an extent; however, the stomach lining secretes a mucus that forms a protective layer that limits damage, and cells lining the stomach that are damaged are constantly replaced.

Food remains in your stomach for two to four hours. Serving as a holding chamber, the stomach slowly releases food into the small intestine. By the time the food is ready to leave the stomach, it is in the form of a thin liquid about the consistency of tomato soup.

The Small Intestine

From your stomach, the partially liquid food moves into your **small intestine,** a muscular tube about 6 m long. This section of the intestine is called *small*, not because of its length, but because it has a narrow diameter of only 2.5 cm. Digestion of your meal is completed within the small intestine. Muscle contractions contribute to further mechanical breakdown, and at the same time, carbohydrates and proteins undergo further chemical digestion.

Chemical action

The first 25 cm of the small intestine is called the duodenum. Although the inner walls of the duodenum secrete enzymes, most of the enzymes and chemicals that function in the duodenum enter it through a duct from the pancreas and the liver. These organs, shown in *Figure 38.4,* have important roles in digestion, even though food does not pass through them.

Secretions of the pancreas

The **pancreas** is a soft, flattened gland that secretes both digestive enzymes and hormones. The mixture of enzymes it secretes breaks down carbohydrates, proteins, and fats. The pancreas also secretes sodium hydrogen carbonate, which makes the pancreatic juice alkaline. Recall that the liquid leaving the stomach—a mixture of partially digested food and gastric juice—is acidic. The alkalinity of pancreatic juice neutralizes this acidity and stops any further action of pepsin.

Figure 38.4

Both the pancreas and the liver produce chemicals that are needed for digestion in the small intestine. The pancreatic juices secreted from the pancreas empty into the bile duct that leads from the gallbladder. This duct leads to the duodenum.

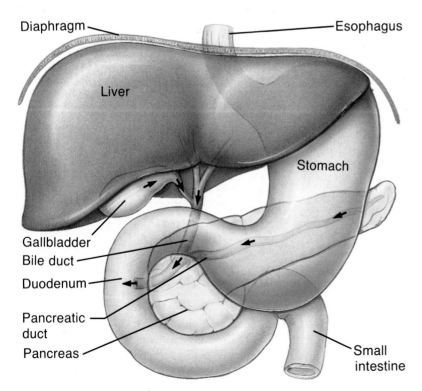

Diaphragm — Esophagus
Liver
Stomach
Gallbladder
Bile duct
Duodenum
Pancreatic duct
Pancreas
Small intestine

Secretions of the liver

The **liver** is a complex gland that, among its many important functions, produces bile. **Bile** is a chemical that breaks fats into small droplets and is also involved in neutralizing stomach acids. Once made in the liver, bile is stored in a small organ called the **gallbladder,** from which bile passes into the duodenum. Although bile is a chemical, this step is part of mechanical digestion. Large drops of fat are broken apart into smaller droplets, but the fat molecules are not changed chemically by the bile.

Absorption of food

Liquid food stays in your small intestine for three to five hours and is slowly moved along by peristalsis. As food moves through the small intestine, it passes over thousands of fingerlike structures called villi. A **villus** is a projection on the lining of the small intestine that functions in the absorption of digested food. The structure and number of the villi greatly increase the surface area of the small intestine.

Notice in *Figure 38.5,* on the next page, that a network of blood vessels and a lymph duct go into and out of each villus. The villi, then, are the link between the digestive system and the circulatory system, and between the digestive system and the lymphatic system, described in Chapters 39 and 42, respectively.

As your lunch comes to the end of its passage through the small intestine, only the materials from your sandwich that could not be digested remain in the digestive tract.

Lactose Intolerance

Enzymes found in the gastrointestinal tract play an important role in the digestive process. One such enzyme is lactase, which breaks down lactose, the natural sugar found in dairy products. A deficiency of lactase causes milk-induced indigestion, the severity of which depends on your degree of enzyme deficiency. If you find yourself suffering from diarrhea, gas, cramps, and/or bloating after drinking milk or eating dairy products, you may be one of many people who are lactose intolerant.

Digestion   Let's say you go for ice cream after a movie. The natural milk sugar in your ice cream, the lactose, is really two simpler sugars, glucose and galactose, linked together. During digestion, the linkage is broken by the enzyme lactase, and the sugars become absorbed into the blood as ready energy. If you don't make enough lactase to split the sugars, the undigested lactose becomes food for bacteria living in your intestine. These bacteria produce hydrogen gas as they digest the sugar, which can quickly make you uncomfortable. The quantity of undigested lactose in your intestine also causes water and salt retention, leading to diarrhea.

Incidence of intolerance
Although lactose intolerance is usually inherited (affecting most ethnic groups including Asians, Hispanics, and African Americans), it can occur as the result of gastrointestinal disease or surgery affecting the area where the enzyme is made. Some degree of lactose intolerance is thought to be part of the natural aging process; as we grow older, we produce less of the active enzyme. Children and infants rarely suffer from lactase deficiency.

Treatment   The most common treatment for lactose intolerance is to limit your consumption of dairy products. However, depending on your degree of deficiency, you can have your cheesecake and eat it too by taking an over-the-counter preparation of the enzyme before indulging.

Thinking Critically

Doctors recommend that you avoid milk and dairy products if you suffer from lactose intolerance. What vitamin deficiency could result from such a diet? Can you think of ways to supplement this vitamin?

Figure 38.5

Once food has been fully digested in the small intestine (cross section shown on left), it is composed of molecules small enough to enter the body's bloodstream. Most monosaccharides and amino acids are actively transported into the epithelial cells of the villi. The molecules diffuse into the blood vessels that supply each villus (right). Within the cells of each villus, the fat molecules diffuse into the lymphatic system, which provides the body with tissue fluids.

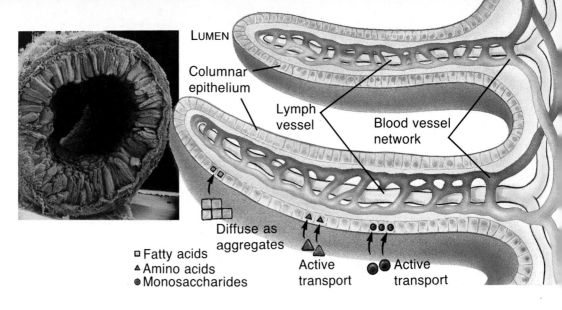

LUMEN

Columnar epithelium

Lymph vessel

Blood vessel network

Diffuse as aggregates

□ Fatty acids
▲ Amino acids
● Monosaccharides

Active transport

Active transport

appendix:

ad (L) to
pendere (L) to hang
The appendix hangs
from the intestine.

The Large Intestine

The indigestible material from your meal passes into your **large intestine,** a muscular tube that is also called the colon. Even though the large intestine is about 1.5 m long, and therefore much shorter than the small intestine, it is much wider—about 6.5 cm. The appendix, an extension at the junction of the small and large intestines, is an evolutionary remnant from herbivorous ancestors. It does not function in digestion.

Water absorption and vitamin synthesis

Water is absorbed from the indigestible mixture through the walls of the large intestine, leaving behind a more solid material. Anaerobic bacteria residing in the large intestine digest more of this material and also synthesize some B vitamins and vitamin K, which are absorbed as needed by the body.

Elimination of wastes

After 18 to 24 hours in the large intestine, the remaining indigestible material, now called feces, reaches the rectum. The **rectum** is the last section of the digestive system. Feces are eliminated from the rectum through the anus. The journey from the beginning of the digestive system to the end has taken your meal from 24 to 33 hours.

Section Review

Understanding Concepts

1. Sequence the organs of your body through which food passes.
2. In which two parts of the digestive system are starches digested?
3. How do villi of the small intestine enhance the rate of nutrient absorption?

Thinking Critically

4. What could happen to a person if the number of bacteria in the intestine were lowered because of chronic diarrhea?

Skill Review

5. **Making and Using Graphs** Prepare a pie graph representing the time food remains in each part of the digestive tract. For more help, refer to Organizing Information in the *Skill Handbook*.

The Control of Digestion and Homeostasis

Section Preview

Objectives

Explain how the nervous system and hormones control the digestive process.

Identify the functions of the glands that affect digestion.

Key Terms

exocrine gland
endocrine gland
target tissue
thyroid gland
parathyroid gland

*R*emember the predicament of the student embarrassed in class by his rumbling stomach? The rumbling can be explained when we examine how digestion is controlled by the interaction of the nervous system, the endocrine system, and the digestive system.

Control of Digestion

When you smell food, look at a picture of food when you're hungry, or just think about your favorite food, you may notice that your mouth waters. The smell, sight, or thought of food increases the secretions of saliva in preparation for eating. This response involves a part of the brain and the memories that are associated with food.

The secretion of gastric juice in the stomach is under the control of both the nervous and endocrine systems. The sight or smell of food stimulates the secretion of gastric juices. This response is controlled by the nervous system. Once food is in the stomach, nerve receptors respond to the stretching of the stomach and signal the medulla of the brain. The medulla then stimulates the stomach glands to continue the secretion of gastric juices. Recall that these gastric juices are responsible for beginning the digestion of proteins.

Protein in the stomach stimulates the stomach lining to secrete gastrin. Gastrin is a hormone that is absorbed into the blood and further stimulates the glands in the stomach to secrete gastric juices. As food passes into the intestine, hormones are secreted by the intestine. These hormones inhibit the secretion of gastric juices in the stomach, thus slowing its action, and stimulate pancreatic secretions. They also stimulate the production and release of bile.

Endocrine Control of Homeostasis and Metabolism

Just as hormones of the digestive tract are directly involved in digestion, hormones also play key roles in maintaining the necessary levels of nutrients in the blood and controlling the energy exchange between cells in the body.

What are the effects of glucagon and insulin during exercise?

Exercise represents a special example of rapid fuel mobilization in the body. The body must gear up to supply great amounts of glucose and oxygen for muscle metabolism, while also providing a steady delivery of glucose to the central nervous system. The glucose use in a resting muscle is generally low but changes dramatically with exercise. Within ten minutes after exercise has started, glucose uptake from the blood may increase up to fifteenfold, and by 60 minutes up to thirtyfold.

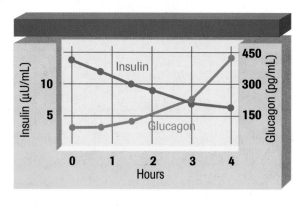

Analysis

The graph here shows the effects of prolonged exercise on plasma insulin and glucagon in humans.

Thinking Critically

Explain why glucagon concentration goes up and insulin concentration goes down during exercise, and how these actions help get glucose to the body cells.

Glucose regulation: The pancreas

The maintenance of blood glucose is accomplished by two means—by eating meals at regular intervals during the day and by hormones making finely tuned adjustments in the blood. Digested sugars from a meal just eaten may pass directly to the liver where they are absorbed and converted into glycogen. Glycogen is a polysaccharide and is the form in which sugars are stored in the liver as well as in skeletal muscle. When your body needs energy, glycogen is broken down into glucose, which is transported by blood to fuel cell processes throughout the body. The release of sugar into the bloodstream is controlled by a portion of the pancreas that produces the hormones glucagon and insulin.

Recall that the pancreas is involved in the digestive process by secreting enzymes into the small intestine. In this role, it functions as an **exocrine gland,** secreting its enzymes through ducts. The portion of the pancreas that secretes hormones functions as an endocrine gland. An **endocrine gland** is a ductless organ that releases hormones directly into the bloodstream. The hormones released by endocrine glands convey information to specific cells in the body called **target tissue.** *Figure 38.6* illustrates how the hormones of the pancreas regulate the body's blood sugar level before and after a meal. Physical activity also will cause secretions of these hormones to vary.

Figure 38.6

The endocrine portion of the pancreas is made of about 1 million clusters of cells called islets. The islets contain alpha and beta cells. The alpha cells secrete glucagon, which raises blood sugar level. The beta cells secrete insulin, which lowers blood sugar level.

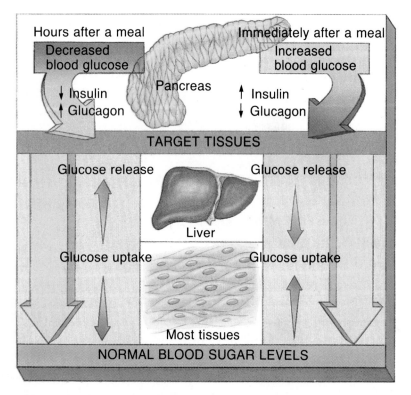

Glucagon increases blood glucose levels by accelerating the conversion of glycogen in the liver into glucose. The liver then releases the glucose into the blood, and the blood sugar level rises. Secretion of glucagon is directly controlled by the level of blood sugar. When the blood sugar level falls below normal, as occurs between meals, chemical sensors in the alpha cells stimulate the cells to secrete glucagon.

Insulin acts to lower the level of glucose in the blood by the opposite action of glucagon. Insulin accelerates the transport of glucose from the blood into cells, especially muscle fibers. Insulin also accelerates the conversion of glucose into glycogen. The regulation of insulin, like that of glucagon, is directly determined by the level of sugar in the blood.

Metabolic control: The thyroid gland

The **thyroid gland** regulates metabolism and energy balance, growth and development, and the general activity of the nervous system. The main metabolic and growth hormone of the thyroid is thyroxine. This hormone affects the rate at which the body uses energy and determines your food intake requirements.

The thyroid also secretes the hormone calcitonin. Calcitonin is one of two hormones that regulate calcium and phosphate levels in the blood. The other hormone is secreted by the parathyroid glands.

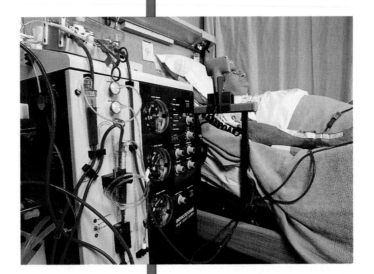

Diabetes—A Sugar Problem

To understand diabetes, you need to understand how sugar and other carbohydrates are metabolized.

Insulin is the key During digestion, carbohydrates and sugar are changed to glucose and travel through the bloodstream nourishing the cells of the body. But glucose can't do its job without insulin. It's an important hormone because it enables glucose to enter the cells. Sometimes the pancreas doesn't produce enough insulin, or the cells of the body become resistant to insulin. When either happens, the glucose level in the blood rises, and large amounts of sugar are excreted in the urine.

The hypothalamus is also affected by the lack of insulin. The hypothalamus is a gland located at the base of the brain that regulates the appetite. A lack of insulin prevents glucose in the blood from entering the cells of the hypothalamus, so the person with diabetes gets hungry and eats constantly, but he or she is actually starving because glucose can't enter the cells and nourish the body. Rapid weight loss, along with great thirst and hunger, are the first signs of this disorder.

Causes of diabetes Diabetes can occur in young children. When it does, it is probably due to genetics, a virus infection, or an autoimmune reaction. An antoimmune reaction is one in which the body's immune system attacks its own tissue, the pancreas in this case, as if it were a foreign substance.

Diabetes can also affect adults in their middle years. Unlike the children with diabetes, they have higher-than-normal amounts of insulin as well as sugar in their blood. How does this happen? The pancreas is producing insulin normally, but the body cells are resistant to insulin.

CONNECTION TO Biology

Hypothesize how genetic engineering could be used to attack childhood diabetes.

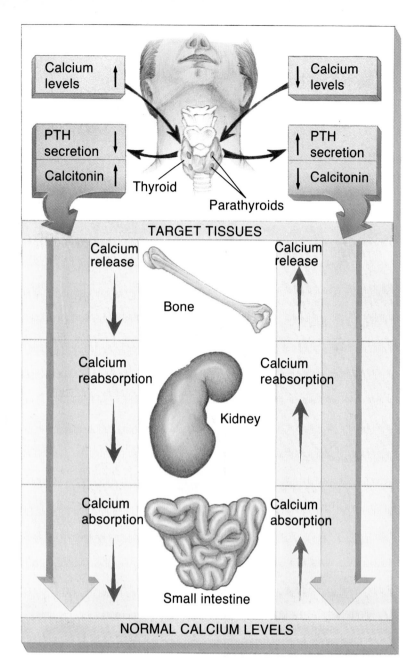

Calcium regulation: The parathyroid glands

Recall that calcium is necessary for blood clotting, formation of bones and teeth, and normal muscle function. It is also needed for normal nerve activity. Parathyroid hormone (PTH) from the **parathyroid glands** increases the rate of calcium, phosphate, and magnesium absorption in the intestines and causes the release of calcium and phosphate from bone tissue. It also increases the rate at which the kidneys remove calcium and magnesium from urine and returns them to the blood. The overall results of secretion of parathyroid hormone are the decrease of blood phosphate level and the increase in calcium and magnesium levels. *Figure 38.*7 shows the effects of the interaction of parathyroid hormone and calcitonin from the thyroid on the calcium level of blood.

Parathyroid hormone also increases the amount of vitamin D that the body makes. Vitamin D is required for the intestines to absorb calcium ions.

*Figure 38.*7

Parathyroid hormone and calcitonin have opposite actions in regulating calcium levels in the bloodstream.

Section Review

Understanding Concepts

1. Outline the process by which digestion is controlled.
2. Explain how the pancreas is both an exocrine and an endocrine gland.
3. What is the effect of parathyroid hormone on bone tissue?

Thinking Critically

4. Hormones continually make adjustments in blood glucose. Why must the blood glucose level be kept fairly constant?

Skill Review

5. **Comparing and Contrasting** What effects do calcitonin and PTH have on blood calcium levels? For more information, refer to Thinking Critically in the *Skill Handbook.*

ZiGGY®

PROCESSED CHEESE FOOD?... GIMME A BREAK!

© 1993, Ziggy and Friends, Inc. Dist. by Universal Press Syndicate

7/28

What's your *favorite food? Does it taste salty, sour, sweet, bitter, or a combination of these? Does it have a wonderful aroma? Of what nutritional value is it? To be considered a food, a substance must provide energy or building materials, or it must assist in some body process. In other words, it must contain at least one of six essential nutrients.*

Section Preview

Objectives

Summarize the role of the six classes of nutrients in body nutrition.

Relate Calories and metabolism.

Key Terms

mineral
vitamin
Calorie

The Vital Nutrients

Six kinds of nutrients can be found in foods: carbohydrates, fats, proteins, minerals, vitamins, and water. These substances are essential to chemical reactions in the body. You supply your body with these nutrients when you eat foods from the five main food groups, shown in *Figure 38.8.*

Carbohydrates: The body's preferred energy source

Perhaps your favorite food is pasta, fresh-baked bread, or corn on the cob. If so, your favorite food contains carbohydrates. Recall from Chapter 7 that carbohydrates are starches and sugars. Starches are complex carbohydrates found in bread, cereal, potatoes, rice, corn, beans, and pasta. Sugars are simple carbohydrates found mainly in fruits such as plums, strawberries, and oranges, as well as syrups and jellies. Table sugar

Figure 38.8

Select foods from the five food groups and you'll have a healthful diet that supplies the six essential nutrients your body needs.

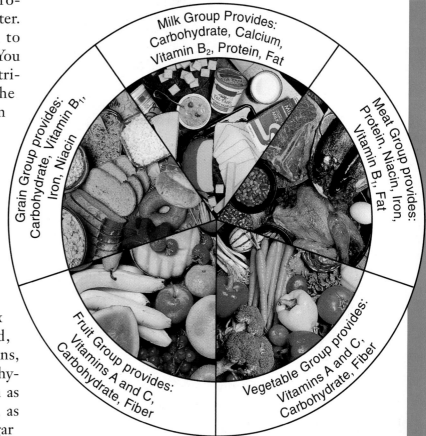

Milk Group Provides: Carbohydrate, Calcium, Vitamin B$_2$, Protein, Fat

Meat Group provides: Protein, Niacin, Iron, Vitamin B$_1$, Fat

Grain Group provides: Carbohydrate, Vitamin B$_1$, Iron, Niacin

Vegetable Group provides: Vitamins A and C, Carbohydrate, Fiber

Fruit Group provides: Vitamins A and C, Carbohydrate, Fiber

Where does the food in a spaghetti dinner fit in the food pyramid?

A food pyramid describes pictorially the number of servings a person should have from each food group daily. The small top of the pyramid contains fat, oils, and sweets because these foods should be eaten most sparingly. Most Americans' diets are too high in fat. Note that some fat and sugar symbols are shown in all the food groups. This is because even foods that are naturally low in fat are often prepared in ways that add fat or sugar.

Procedure

1. Construct a food pyramid showing the foods that make up the following dinner:
   spaghetti and meatballs, garlic bread, glass of milk
2. Compare your pyramid with the pyramid shown.

Analysis

1. What food groups are missing in the dinner?
2. What could be added to this dinner to make it fit the food pyramid better?
3. Overall, do you think the spaghetti was a nutritious dinner? Explain.

Fat (naturally occurring and added during cooking)

Sugars (added to foods)

USE SPARINGLY

2-3 SERVINGS

2-3 SERVINGS

3-5 SERVINGS

2-4 SERVINGS

6-11 SERVINGS

Cellulose, another complex carbohydrate, is found in all plant cell walls and is not digestible by humans. However, cellulose is still an important item to include in your diet. It provides roughage, often called fiber. By eating such cellulose-containing foods as oranges, celery, and spinach, you'll provide bulk in your diet that stimulates your digestive tract and helps in the elimination of wastes.

Fats: Energy sources and building materials

To many people, eating fat means getting fat. Yet fats are an essential nutrient. Fats provide energy for your body and are also used as building materials. Recall from Chapter 7 that fats are essential building blocks of the cell membrane. Fats also are important in the synthesis of hormones, for protecting body organs against injury, and for insulating the body from cold. Sources of fat in the diet include meats, nuts, and dairy products, as well as cooking oils. In the digestive system, fats are broken down into fatty acids and glycerol and absorbed by the villi of the small intestine. Eventually, some of these fatty acids end up in the liver. The liver converts them to glycogen or stores them as fat. Unused fats are stored as fat deposits throughout your body.

Proteins: Building materials

Your body has many uses for proteins. Enzymes, antibodies, many hormones, and chemicals that help the blood to clot are proteins. Proteins are part of muscles and many cell structures, including the cell membrane.

is derived from sugar cane or sugar beet, and is generally not considered an essential nutrient. Carbohydrates are important sources of energy for your body cells.

During digestion, complex carbohydrates are broken down into simple sugars such as glucose, fructose, and galactose. Absorbed into the bloodstream through the villi of the small intestine, these sugar molecules circulate to fuel body functions. Remember that some sugar is carried to the liver where it is converted to glycogen.

During digestion, proteins are broken down into amino acids. After the amino acids have been absorbed by the small intestine, they enter the bloodstream and are carried to the liver. The liver can convert amino acids to fats or glucose, both of which can be used for energy. However, your body uses amino acids for energy only if other energy sources are depleted. Most amino acids are absorbed by cells and used for protein synthesis. The human body needs 20 different amino acids to carry out protein synthesis, but it can make only 12 of them. The rest must be consumed in the diet and are called essential amino acids. Sources of essential amino acids include meats, dried beans, whole grains, eggs, and dairy products.

Minerals and vitamins

When you think of minerals, you may picture substances that people mine, or extract from the earth. The same minerals can also be extracted from foods and put to use by your body.

A **mineral** is an inorganic substance that serves as a building material or takes part in chemical reactions in the body. Minerals make up about four percent of your total body weight. Most of the mineral content of your body is in your skeleton. Calcium and phosphorus form much of the structure of bone. *Table 38.2* lists the functions of some of the minerals that are needed by humans and the foods that provide them. Notice that the body does not use minerals as energy sources.

Table 38.2 Vital Minerals		
Mineral	**Function**	**Source**
Calcium	formation of bones and teeth, blood clotting, normal muscle and nerve activity	milk, cheese, nuts, whole grains
Phosphorus	formation of bones and teeth, regulation of blood pH, muscle contraction and nerve activity, component of enzymes, DNA, RNA, and ATP	milk, whole-grain cereals, meats, vegetables
Iron	component of hemoglobin (carries oxygen to body cells) and cytochromes (ATP formation)	liver, egg yolk, peas, enriched cereals, whole grains, meat, raisins, leafy vegetables
Iodine	part of thyroid hormone, required by thyroid gland	seafood, eggs, milk, iodized table salt
Sodium	regulation of body fluid pH, transmission of nerve impulses	bacon, butter, table salt, vegetables
Potassium	transmission of nerve impulses, muscle contraction	vegetables, bananas, ketchup
Magnesium	muscle and nerve function, bone formation, enzyme function	potatoes, fruits, whole-grain cereals, vegetables
Fluorine	teeth structure	fluoridated water
Manganese	enzyme activator for carbohydrate, protein, and fat metabolism; important in growth of cartilage and bone tissue	wheat germ, nuts, bran, leafy green vegetables
Copper	ingredient in several respiratory enzymes, needed for development of red blood cells	kidney, liver, beans, whole-meal flour, lentils
Sulfur	component of insulin; builds hair, nails, skin	nuts, dried fruits, barley, oatmeal, eggs, beans, cheese

Table 38.3 Vitamins

Vitamin	Function	Source
Fat-soluble		
A	maintain health of epithelial cells; formation of light-absorbing pigment; growth of bones and teeth	liver, broccoli, green and yellow vegetables, tomatoes, butter, egg yolk
D	absorption of calcium and phosphorus in digestive tract	egg yolk, shrimp, yeast, liver, fortified milk; produced in the skin upon exposure to ultraviolet rays in sunlight
E	formation of DNA, RNA, and red blood cells	leafy vegetables, milk, butter
K	blood clotting	green vegetables, tomatoes; produced by intestinal bacteria
Water-soluble		
B_1	sugar metabolism; synthesis of neurotransmitters	ham, eggs, green vegetables, chicken, raisins, seafood, soybeans, milk
B_2 (riboflavin)	sugar and protein metabolism in cells of eye, skin, intestine, blood	green vegetables, meats, yeast, eggs
Niacin	energy-releasing reactions; fat metabolism	yeast, meats, liver, fish, whole-grain cereals, nuts
B_6	fat metabolism	salmon, yeast, tomatoes, corn, spinach, liver, yogurt, wheat bran, whole-grain cereals and breads
B_{12}	red blood cell formation; metabolism of amino acids	liver, milk, cheese, eggs, meats
Pantothenic acid	aerobic respiration; synthesis of hormones	milk, liver, yeast, green vegetables, whole-grain cereals and breads
Folic acid	synthesis of DNA and RNA; production of red and white blood cells	liver, leafy green vegetables, nuts, orange juice
Biotin	aerobic respiration; fat metabolism	yeast, liver, egg yolk
C	protein metabolism; wound healing	citrus fruits, tomatoes, leafy green vegetables, broccoli, potatoes, peppers

Unlike minerals, a **vitamin** is an organic nutrient that is required in small amounts to maintain growth and metabolism. Vitamins do not provide energy or building materials for the body; rather, they regulate processes in the body.

The two main groups of vitamins are fat-soluble vitamins and water-soluble vitamins, *Table 38.3.* Fat-soluble vitamins must be dissolved in fat before they are absorbed into the small intestine. Fat-soluble vitamins can be stored in the liver, where excess amounts can become toxic. On the other hand, water-soluble vitamins dissolve readily in water and cannot be stored in the body.

Vitamin D, a fat-soluble vitamin, is synthesized in your skin. Vitamin K and some B vitamins are made by bacteria in your intestine. The rest of the vitamins must be consumed in your diet. *Table 38.3* summarizes the roles of vitamins and some of their food sources.

Water: Abundant and essential

Water is the most abundant substance in your body. It makes up 60 percent of red blood cells and 75 percent of muscle cells. Water plays a role in many chemical reactions in the body and is necessary for the breakdown of foods in digestion. Water is an excellent solvent, and oxygen and nutrients from food could not enter your cells without water.

Water absorbs and releases heat slowly. It is this characteristic that helps water regulate your body's temperature. A large amount of heat is needed to raise the temperature of water. Because the body contains so much water, this slow-to-heat property of water helps keep body temperature nearly constant. Your body loses about 2.5 L of water per day through exhalation during breathing, and through sweat and urine. As a result, water must be replaced constantly.

Metabolism and Calories

Think of all the chemical reactions you've learned about that are involved in your body's metabolism. Some of these chemical reactions break down materials such as the bacon in your sandwich into fatty acids, glycerol, and amino acids. Other chemical reactions involve synthesizing. For example, once in your cells, the amino acids from your digested sandwich are put together to form the proteins that make up your body.

The energy content of food is measured in units of heat called **Calories**—representing a kilocalorie, or 1000 calories (written with a small *c*).

MiniLab

Does a bowl of soup provide a complete meal?

As a consumer, you will be bombarded by advertising that promotes the nutritional benefits of specific food products. Choosing a food to eat on the basis of such ads may not make nutritional sense. By examining the ingredients of processed foods, you can learn important things about their nutritional content.

Percentage of Daily Value (DV)

Carbohydrates	60%
Fat	30%
Saturated Fats	10%
Cholesterol	1.5%
Protein	10%
Total Calories	2000

NUTRITION FACTS

Serving Size: 2 cups (452g)
Serving Per Container: 1

Amount Per Serving

Calories 140   Calories from Fat 54

	% Daily Value*
Total Fat 8g	12%
Saturated Fat 6g	30%
Cholesterol 20mg	7%
Sodium 1640 mg	68%
Total Carbohydrate 22g	7%
Dietary Fiber 5g	20%
Sugars 5g	
Protein 6g	

Vitamin A	50%	Vitamin C	4%
Calcium	2%	Iron	2%

*Percent Daily Values are based on a 2,000 calorie diet. Your daily values may be higher or lower depending on your calorie needs:

		Calories	2,000	2,500
Total Fat	Less than		65g	80g
Sat Fat	Less than		20g	25g
Cholesterol	Less than		300mg	300mg
Sodium	Less than		2,400mg	2,400mg
Total Carbohydrate			300g	375g
Fiber			25g	30g

Calories per gram:
Fat 9   *   Carbohydrates 4   *   Protein 4

Procedure

1. Examine the information in the above table listing the daily value (DV) of various nutrients. DV expresses what percent of Calories should come from certain nutrients. For instance, in the proposed diet of 2000 Calories, 60 percent of the Calories should come from carbohydrates.

2. Examine the nutritional information on the soup can label and compare it with the DV table.

Analysis

1. Does your bowl of soup provide more than 30 percent of any of the daily nutrients? Which ones?

2. Is soup a nutritious meal? Explain your answer.

3. What could be eaten along with the soup to provide more carbohydrates? Protein?

BioLab

As heterotrophs, humans need to take in vital nutrients. Carbohydrates, such as sugars and starches, are used by the body for energy. Lipids, or fats, may be used directly as an energy source, or they may be stored as a future source of energy. Proteins are used in metabolic reactions and as building blocks to make new cells or repair old ones, or they are converted to fat or glucose to be used as a source of energy.

Testing for Nutrients in Foods

Simple chemical or physical tests can be performed on foods to see which of these nutrients they contain.

PREPARATION

Problem
What nutrients are found in common foods?

Objectives
In this Biolab, you will:
- **Hypothesize** which nutrients are present in each food tested.
- **Demonstrate** by means of chemical tests the presence of carbohydrates and proteins in foods.
- **Determine** by a physical test whether foods contain lipids.

Materials
food samples (3)
mortar and pestle
test tubes (8)
droppers (4)
water
soapy water
test-tube rack
test-tube holder
brown paper (3 pieces)
iodine solution in dropper bottle
biuret solution in dropper bottle
laboratory apron
safety goggles

Safety Precautions 🧤 👓 ⚗️
If solutions are spilled, rinse with water and call your teacher immediately.

PROCEDURE

Part A: Preparation of Samples

1. Prepare tables for Parts B and C using the heads illustrated below.
2. Choose three food samples. Grind each solid with about 5 mL of water. Return the ground food to its container.
3. Label your food samples 1, 2, and 3.
4. Make a hypothesis to predict which nutrients are present in each food tested.

Part B: Test for Starch

1. Label four test tubes 1, 2, 3, and 4. Using a different dropper for each food sample, add ten drops of food sample 1 to test tube 1, ten drops of food sample 2 to test tube 2, and ten drops of food sample 3 to test tube 3. Add ten drops of water to test tube 4.
2. Record the color of each test tube's contents in your Data Table.
3. Add three drops of iodine solution to each test tube.
4. Record the color of each tube's contents. If the color of the solution is deep blue, starch is present.

Part C: Biuret Test for Proteins

1. Label four test tubes, and add the food samples and water as in step 1 of Part B.
2. Record the color of each test tube's contents.
3. Add ten drops of biuret reagent to each test tube. **CAUTION:** *Biuret reagent is extremely caustic to the skin and clothing.*
4. Record the color of each tube's contents in your data table. If the color of the solution is a shade of purple, protein is present.

Part D: Brown Paper Test for Lipids

1. Label four pieces of brown paper 1, 2, 3, and 4.
2. Place one drop of each food on each piece of paper.
3. Wait for the papers to dry.
4. Hold the pieces of paper up to a light, and observe each to see if light passes through it. If light passes through it, the spot is translucent and lipids are present. Record this for each food.

Test tube number	Food	Color before adding solution	Color after adding solution	Nutrient present? (+ or −)

ANALYZE AND CONCLUDE

1. **Analyzing Data** Which of the food samples gave a positive test for starches? For proteins? For lipids?
2. **Analyzing Data** Which foods contain more than one nutrient?
3. **Checking Your Hypothesis** Do your data support your hypothesis? Why or why not?

Going Further

Making Predictions Predict which nutrients are present in one of your school lunches, and then test the food.

Figure 38.9

Shifts in the energy equation determine whether a person gains weight, loses weight, or maintains weight.

Calories consumed   Calories burned

A calorie is the amount of heat required to raise the temperature of 1 mL of water by 1°C. Some foods, especially those with fats, contain many more Calories than others. In general, 1 g of fat contains nine Calories, while 1 g of carbohydrate or protein contains four Calories.

The actual number of Calories needed each day varies from person to person depending on a person's body mass, age, sex, and level of physical activity. In general, males need more Calories per day than females, teenagers need more Calories than adults, and active people need more Calories than inactive people.

What happens if you eat more Calories than your body can metabolize? As *Figure 38.9* shows, you'll store the extra energy as body fat and gain weight. If you eat fewer Calories than your body can metabolize, you will use some of the energy stored in body fat and lose weight.

Connecting Ideas

Your body needs a constant supply of energy, yet you don't need to eat constantly. As a meal is digested and the nutrients are absorbed, your blood has more than enough nutrients in it. When blood passes from the intestine to the liver, the liver absorbs extra nutrients from the blood. Hormones released directly into blood regulate the blood's levels of nutrients and keep it fairly constant. What other roles does blood play in the body? How does the body get the oxygen it needs along with food to produce energy? And how else does it rid itself of wastes it cannot use? Answers to these questions lie in an understanding of the circulatory, respiratory, and excretory systems of the human body.

Section Review

Understanding Concepts

1. In what ways are proteins used in the body?
2. Why can an excess of fat-soluble vitamins be toxic?
3. A person can live several weeks without food, but can live only days without water. Why is the constant intake of water necessary for the body?

Thinking Critically

4. How does the liver help maintain homeostasis?

Skill Review

5. **Classifying** Prepare a chart of food groups high in each of the six nutrients. For more help, refer to Organizing Information in the *Skill Handbook*.

Megavitamins

Vitamin C Believed to Ward Off Disease

The Daily News   September 6, 1994

A recent review from Finland of all studies done so far backs Linus Pauling and his theory on vitamin C and colds; and evidence now seems overwhelming that vitamins C and E do have value in preventing cancer.

Almost half of the American population takes vitamins. Multivitamins are the most popular, followed by vitamins C and E, and B complexes. This extraordinary and widespread use of vitamins is not for fear of traditional diet-deficiency diseases like rickets; vitamin D was added to milk and margarine to prevent rickets. Most supplement users are trying to enhance their already-adequate diets for optimum health—greater energy levels, better complexions, stronger hair and nails, and the hope for longer lives.

RDAs Traditionally, doctors recommended supplements only if a person's diet was deficient in specific vitamins. The U.S. Recommended Daily Allowances (USRDAs), the nutritional numbers used to prevent deficiencies, were used as guidelines. Today, the Federal Drug Administration (FDA) believes that almost one out of every ten persons who use supplements is taking more than the RDA. Currently, the benefits and risks of taking large doses of vitamins, or megavitamins, are at the center of a heated debate among scientific experts.

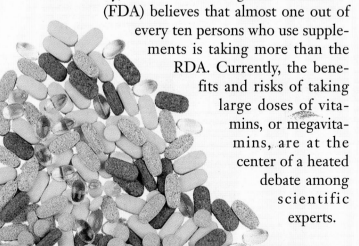

Different Viewpoints

Pauling and vitamin C Two-time Nobel Prize-winning chemist Linus Pauling was the first well-known scientist to favor the megavitamin-as-preventive-medicine theory. Two decades ago, he claimed that large doses of vitamin C could ward off the common cold and help prevent heart disease and cancer. His theories, which challenge our current understanding of vitamins, have been supported by recent studies where large daily doses of vitamin C cut heart disease deaths by nearly half in men and one-fourth in women, and added more than five years to life expectancy.

Antioxidants and free radicals Other research substantiates the fact that megavitamins may offer protection from chronic disease and the effects of aging, as well as boost the immune system. High doses of antioxidants—vitamins such as C and E and beta-carotene—have been found to trap and destroy substances called free radicals, which can damage cells and lead to cancer.

How much is too much? Some doctors caution that megavitamins should be treated much as all medicines are. Some vitamins, when taken in large doses and/or over a long period of time, can be toxic. For example, in some cholesterol patients who were not carefully monitored by their doctors, extended use of niacin has caused serious liver damage.

INVESTIGATING the Issue

1. **Apply** Make a vitamin table that lists at least three vitamins, their adult RDAs, food sources, benefits, and risks.
2. **Journal Writing** Research the use and abuse of amino acids as dietary supplements. Investigate the laws regulating their sale and distribution. Do you think amino acids should be labeled as drugs? Respond in your journal.

Reviewing Main Ideas

38.1 Following Digestion of a Meal

- Digestion begins in the mouth with both mechanical and chemical action on food. The esophagus transports food from the mouth to the stomach.
- Chemical and mechanical digestion continue in the acidic environment of the stomach.
- In the small intestine, digestion is completed and food is absorbed. The liver and pancreas play key roles in digestion.
- The large intestine absorbs water before indigestible materials are eliminated.

38.2 The Control of Digestion and Homeostasis

- Digestion is controlled by both the nervous system and the endocrine system.
- Hormones regulate the blood levels of products of digestion.

38.3 Nutrition

- Carbohydrates are the main source of energy for the body. Fats are used as a source of energy and as building blocks. Proteins are used as building materials, enzymes, hormones, antibodies, and a source of energy. Minerals serve as structural materials or take part in chemical reactions. Vitamins are needed for growth and metabolism. Water serves many vital functions in the body.
- Metabolic rate determines how quickly energy is burned. Together with Calorie intake, it helps determine a person's weight.

Key Terms

Write a sentence that shows your understanding of each of the following terms.

amylase	pancreas
bile	parathyroid gland
Calorie	pepsin
endocrine gland	peristalsis
epiglottis	rectum
esophagus	small intestine
exocrine gland	stomach
gallbladder	target tissue
large intestine	thyroid gland
liver	villus
mineral	vitamin

Understanding Concepts

1. Why is it important when eating to chew food thoroughly?
2. What keeps the stomach from digesting itself?
3. Why is the pancreas considered part of both the digestive system and the endocrine system?
4. How does the brain initiate the process of digestion?
5. Why does the body need a diet high in carbohydrates?
6. Why is fat an important part of the diet?
7. How do Caloric intake and metabolism relate?

Using a Graph

8. The relationship between parathyroid hormone (PTH) secretion and plasma calcium is shown in the graph below. What level does the calcium plasma of the blood have to fall to in order to get maximum parathyroid hormone secretion?

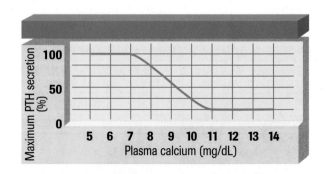

Relating Concepts

9. Make a concept map that relates the following terms and phrases. Supply the appropriate linking words for your map.

   liver, gallbladder, bile, duodenum, small intestine, villus, peristalsis, pancreas

Applying Concepts

10. Achlorhydria is a condition in which the stomach fails to secrete hydrochloric acid. How would this condition affect digestion?
11. Compare the digestion of regular soda pop with sugar-free soda pop. Is sugar-free soda pop a food?
12. Various medical conditions can make it necessary to remove portions of the large intestine. What consequences would this have on the digestive process?
13. How could removal of the parathyroid glands affect muscle contraction?

Health Connection

14. Provide a sample menu for one day (breakfast, lunch, dinner) that would be healthful for an individual with diabetes.

Thinking Critically

Analyze the Procedure

15. **Biolab** What was the purpose of adding the solution or reagent to water in test tube 4 when testing for starch and protein?

Comparing and Inferring

16. **Minilab** Think about your favorite meal. Determine the nutritional content of the meal and compare it to the DV table in the Minilab. Evaluate the nutritional value of your favorite meal.

Identifying Patterns

17. **Minilab** Write down the contents of your meals and snacks for a day. Compare them with the food pyramid in the Minilab. Is there a pattern to your meals? Does the pattern fit the recommended food pyramid?

Drawing Conclusions

18. **Minilab** Evaluate the content of the following bowl of cereal with two percent milk, considering the DV table and the food pyramid.

1 cup of bran with raisins and
1/2 cup two percent milk
Calories   125
Protein   10%
Carbohydrate   9%
Fat   2%

Recognizing Cause and Effect

19. **Thinking Lab** What will happen to glucagon and insulin levels when exercise stops?

Connecting to Themes

20. **Evolution** Why was the development of a tubular digestive system with a mouth and anus important to the success of animals?
21. **Homeostasis** How is the liver's role in glucose regulation important for homeostasis?
22. **Systems and Interactions** How do the nervous, endocrine, and digestive systems interact?

CHAPTER
39 Respiration, Circulation, and Excretion

These mountain climbers have to be in good physical condition. But hiking up the mountain has begun to have its effect on their bodies. Their breathing has become deeper as their lungs strive to take in more vital oxygen. Their hearts now beat a little faster to help transport oxygen and nutrients to cells of their bodies. This enables them to continue to produce the energy needed to complete their adventure. As the breathing rates and heart rates of the climbers increase, so do wastes within their body cells, the end products of energy production. A slow, methodical pace, with rest stops to catch one's breath and enjoy the view, is the safest way to hike the mountains.

As the hikers climb higher, the demands on their bodies will increase. How does the human body adapt to these changing conditions? The answers lie in the workings of the human respiratory, circulatory, and urinary systems.

Concept Check

You may wish to review the following concepts before studying this chapter.
- Chapter 9: diffusion
- Chapter 10: respiration
- Chapter 14: multiple alleles, blood groups

Chapter Preview

39.1 The Respiratory System
Passageways and Lungs
The Mechanics of Breathing
Control of Respiration

39.2 The Circulatory System
Your Blood: Fluid Transport
ABO Blood Types
Your Blood Vessels: Pathways of Circulation
Your Heart: The Vital Pump

39.3 The Urinary System
Kidneys: The Body's Janitors
The Urinary System and Homeostasis

Laboratory Activities

Biolab
- Measuring Respiration

Minilabs
- What effect does swallowing have on pulse rate?
- What structures are visible in a sheep's kidney?

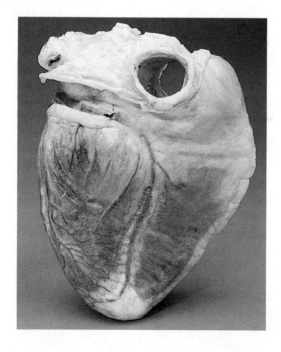

The view at the top will be the hikers' reward for all the hard work their bodies have done. Of course, their feet and legs have brought them this far. But without the work of their lungs, heart, and other organs of their respiratory and circulatory systems, their feet and legs wouldn't have the energy to take them any further.

39.1 The Respiratory System

Breathing underwater is easy—if you're a fish. With their gills, fishes are adapted to extract oxygen from water for their life processes. Humans, on the other hand, are not. But we are equipped to extract oxygen from air using our own adapted respiratory system.

Passageways and Lungs

Your respiratory system is made up of a pair of lungs, a series of passageways, and a thin sheet of smooth muscle called the diaphragm, shown in *Figure 39.1*. When you hear the word *respiration*, you probably think of breathing. But breathing is just part of the process of respiration that an oxygen-dependent organism carries out. Respiration includes all the mechanisms involved in getting oxygen to the cells of your body and getting rid of carbon dioxide. You have also learned in Chapter 10 that respiration involves the formation of ATP within cells.

The path air takes

The first step in the process of respiration involves taking air into your body through either your nose or your mouth. It passes to the pharynx, moves past the epiglottis, and passes through the larynx. Air then travels down the **trachea,** the passageway that leads to the lungs. Recall that at the time of swallowing food, the epiglottis covers the trachea. It prevents food and other large materials from getting into the air passages.

Cleaning dirty air

The air you breathe is far from clean. It is estimated that an individual living in an urban area breathes in 20 million particles of foreign matter each day. To prevent most of this material from reaching your lungs, the trachea and bronchi are lined with cilia and cells that secrete mucus. The cilia constantly beat upward in the direction of your throat where foreign material can be expelled or swallowed.

Alveoli: The place of gas exchange

The trachea divides into two narrower tubes called bronchi. Each bronchus branches into bronchioles, which in turn branch into numerous microscopic tubules that eventually expand into thousands of thin-walled sacs called alveoli. **Alveoli** are the sacs of the lungs where oxygen and carbon dioxide are exchanged by diffusion between the air and blood. Diffusion takes place easily because the wall of each alveolus is only one cell thick. External respiration, as shown in *Figure 39.1*, is the exchange of oxygen or carbon dioxide between the air that is in the alveoli and the blood that supplies these air sacs.

Figure 39.1

As air passes through the respiratory system, it travels through narrower and narrower passageways until it reaches the alveoli. The clusters of alveoli are surrounded by networks of tiny blood vessels. Blood in these vessels has come from the cells of the body and contains wastes from cellular respiration. What are the products of cellular respiration?

Blood transport of gases

Once oxygen from the air diffuses into the blood vessels surrounding the alveoli, it is transported to the cells of your body, where it is used for cellular respiration. In Chapter 10, you learned that cellular respiration uses oxygen during the process by which glucose is broken down and results in energy in the form of ATP. Carbon dioxide and water are waste products of this process. The water can stay in the cell or diffuse into the blood. The carbon dioxide diffuses into the blood, which carries it back to the lungs.

As a result, the blood that comes to the alveoli from the body's cells is high in carbon dioxide and low in oxygen. As carbon dioxide from the body diffuses from the blood into the air in the alveoli to be carried out of your body, oxygen diffuses from the air in the alveoli into the blood, making the blood rich in oxygen. This oxygen-rich blood then leaves the lungs and is pumped by the heart to the body cells again.

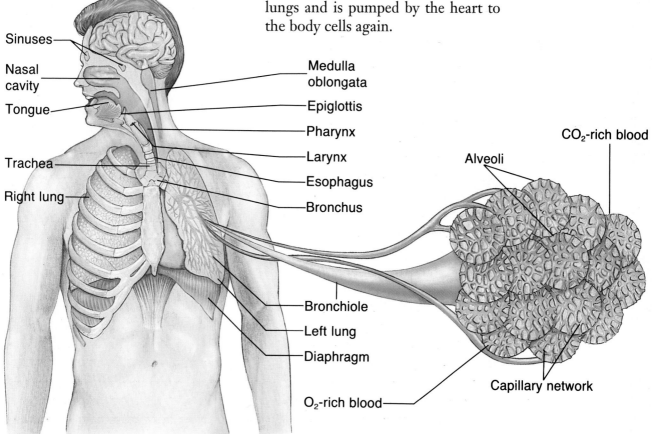

Sinuses
Nasal cavity
Tongue
Trachea
Right lung
Medulla oblongata
Epiglottis
Pharynx
Larynx
Esophagus
Bronchus
Bronchiole
Left lung
Diaphragm
O_2-rich blood
Alveoli
CO_2-rich blood
Capillary network

BioLab

Measuring Respiration

The exchange of oxygen and carbon dioxide among the atmosphere, the blood, and the body cells is commonly called respiration. This external respiration should not be confused with cellular respiration, the chemical reactions that take place within cells to provide energy. In external respiration, the lungs serve as the area of gas exchange between the atmosphere and the blood. The amount of air exchanged during breathing can be measured using a clinical machine called a spirometer. It also can be measured, although less accurately, using a balloon.

PREPARATION

Problem
How can you measure volume of respiration?

Objectives
In this Biolab, you will:
- **Measure** the resting breathing rate.
- **Measure** tidal volume by exhaling into a balloon.
- **Calculate** the amount of air inhaled per minute.

Materials
round balloon
string (1 m)
metric ruler
clock or
   watch with
   second hand

PROCEDURE

Part A: Breathing Rate at Rest
1. Copy Table 1.
2. Have your partner count the number of times you inhale in 30 s.
3. Repeat step 2 two more times.
4. Calculate the average number of breaths.
5. Multiply the average number of breaths by two to get the average resting breathing rate per minute.

Table 1 Resting Breathing Rate	
Trial	Inhalations in 30 s
1	
2	
3	
Average	
Inhalations per minute	

Part B: Tidal Volume

1. Copy Table 2.
2. Take a regular breath and exhale normally into the balloon. Pinch the balloon closed.
3. Have a partner fit the string around the balloon at the widest part.
4. Measure the length of string, in centimeters, used to go around the balloon and record.
5. Repeat steps 2–4 four more times.
6. Calculate the average circumference of the five measurements.
7. Calculate the average radius of the balloon by dividing the average circumference by 6.28 ($\approx 2\pi$).
8. Tidal volume is the amount of air expelled during a normal breath. Tidal volume can be determined using the balloon radius and the formula for determining the volume of a sphere

$$\text{Volume} = \frac{4\pi r^3}{3}$$

   where r = radius and π = 3.14. Calculate the average tidal volume using the average balloon radius.
9. Your calculated volume will be in cubic centimeters; 1 cm^3 = 1 mL.

Table 2 Tidal Volume	
Trial	String measurement
1	
2	
3	
4	
5	
Average circumference	
Average radius	
Average tidal volume	

Part C: Amount of Air Inhaled

1. Copy Table 3.
2. Multiply the average tidal volume by the average number of breaths per minute to calculate the amount of air you inhale per minute.
3. Divide the number of milliliters of air by 1000 to get the number of liters of air you inhale per minute.

Table 3 Amount of Air Inhaled	
mL/min	
L/min	

ANALYZE AND CONCLUDE

1. **Making Comparisons** Compare average breaths and tidal volume per minute with other students.
2. **Thinking Critically** An average amount of air inhaled per minute for an adult is 6000 mL. Compare your average volume of air with this figure. What factors could be responsible for any differences?
3. **Making Predictions** What would you predict would happen to your resting breathing rate right after you exercise?

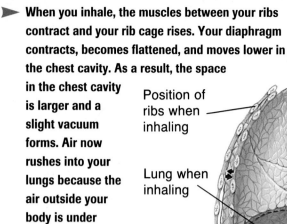

When you inhale, the muscles between your ribs contract and your rib cage rises. Your diaphragm contracts, becomes flattened, and moves lower in the chest cavity. As a result, the space in the chest cavity is larger and a slight vacuum forms. Air now rushes into your lungs because the air outside your body is under more pressure than the air inside your lungs.

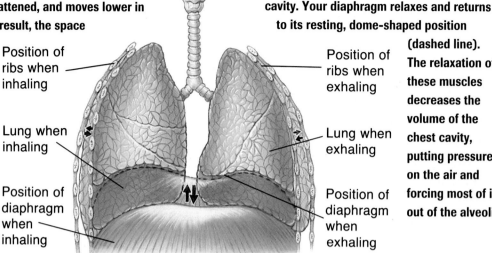

Position of ribs when inhaling

Lung when inhaling

Position of diaphragm when inhaling

When you exhale, the muscles over the ribs relax, and your ribs drop down in the chest cavity. Your diaphragm relaxes and returns to its resting, dome-shaped position (dashed line). The relaxation of these muscles decreases the volume of the chest cavity, putting pressure on the air and forcing most of it out of the alveoli.

Position of ribs when exhaling

Lung when exhaling

Position of diaphragm when exhaling

Figure 39.2

The pressure in the lungs is varied by changes in the volume of the chest cavity. When relaxed, your diaphragm lies in a dome shape under your lungs.

The Mechanics of Breathing

The action of your diaphragm and the muscles between your ribs enable you to breathe in and breathe out. *Figure 39.2* shows how air is drawn in or is forced out of the lungs as a result of the diaphragm's position.

The alveoli in healthy lungs are elastic, like balloons. They stretch as you inhale and return to their original size as you exhale. Like a balloon that has had the air let out of it and does not go totally flat, the alveoli still contain a small amount of air when you exhale.

Control of Respiration

Breathing is usually an involuntary process. It is controlled by the chemistry of your blood as it interacts with a part of your brain called the medulla oblongata. The medulla oblongata helps maintain homeostasis. It responds to higher levels of carbon dioxide in your blood by sending nerve signals to the rib muscles and diaphragm. As a result, these muscles contract and you inhale. As breathing becomes more rapid, as during exercise, a more rapid exchange between air and blood occurs.

Section Review

Understanding Concepts
1. Outline the path of an oxygen molecule from your nose to a cell.
2. Compare and contrast external respiration and cellular respiration.
3. Explain the process by which gases are exchanged in the lungs.

Thinking Critically
4. During a temper tantrum, four-year-old Josh tries to hold his breath. His parents are afraid that he will be harmed by this behavior. How will Josh be affected by holding his breath?

Skill Review
5. **Sequencing** What is the sequence of muscle actions during inhaling and exhaling? For more help, refer to Organizing Information in the *Skill Handbook*.

39.2 The Circulatory System

Blood flowed freely from this injury until direct pressure was applied. This stopgap measure is needed only until the blood's adaptive ability to clot takes over. Your blood has other life-supporting qualities. As it travels around your body, it carries oxygen from your lungs and nutrients from your digestive system to the cells of your body, then hauls away cell wastes. Together, your blood, your heart, and a mazelike network of passageways make up the components of your circulatory system.

Section Preview

Objectives

Distinguish between the various components of blood and between blood types.

Trace the route blood takes through the body and heart.

Explain how heart rate is controlled.

Key Terms

plasma
red blood cell
hemoglobin
white blood cell
platelet
antigen
antibody
artery
capillary
vein
atrium
ventricle
vena cava
aorta
pulse
blood pressure

Your Blood: Fluid Transport

Your blood is a tissue composed of fluid, cells, and fragments of cells. *Table 39.1* summarizes information about these human blood components. The fluid portion of blood in which blood cells move is called **plasma.** Plasma is straw colored and makes up about 55 percent of the total volume of blood. Blood cells—both red and white—and cell fragments are suspended in plasma.

Red blood cells: Oxygen carriers

The round, disk-shaped cells in blood are **red blood cells.** Red blood cells carry oxygen to body cells. They make up 44 percent of your blood.

Red blood cells are produced in the red bone marrow of your ribs, humerus, femur, sternum, and other long bones. At the same time, old red blood cells are being destroyed in your liver and spleen, an organ of the lymphatic system.

Table 39.1 Blood Components	
Components	**Characteristics**
Red blood cells	Transport oxygen and some carbon dioxide; lack a nucleus; contain hemoglobin
White blood cells	Large; several different types; all contain nuclei; defend the body against disease
Platelets	Cell fragments needed for blood clotting
Plasma	Liquid; contains proteins; carries red and white blood cells, platelets, nutrients, enzymes, hormones, gases, and inorganic salts

Figure 39.3

Red blood cells in humans have nuclei only at an early stage in development. The nucleus is lost as the cell enters the bloodstream. As a result, red blood cells have a limited life span. They remain active in the bloodstream for only about 120 days before they break down and are removed as wastes.

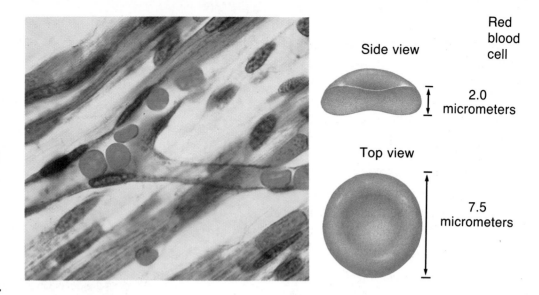

Side view

Red blood cell

2.0 micrometers

Top view

7.5 micrometers

Oxygen in the blood

How is oxygen carried by the blood? Red blood cells, *Figure 39.3,* are equipped with an iron-containing protein molecule called **hemoglobin**. Hemoglobin picks up oxygen after it enters the lungs. There, oxygen becomes loosely attached to hemoglobin and is carried to the body's cells. As blood passes tissues where oxygen concentration is low, the oxygen attached to the hemoglobin in the red blood cells diffuses into the tissues.

Carbon dioxide in the blood

Hemoglobin carries some carbon dioxide as well as oxygen. You know that once biological work has been done in a cell, wastes in the form of carbon dioxide diffuse into the blood and are carried in the bloodstream to the lungs. About 70 percent of this carbon dioxide combines with water and sodium in the blood plasma to form sodium hydrogen carbonate. The remaining 30 percent travels back to the lungs dissolved in the plasma or attached to hemoglobin.

White blood cells: Infection fighters

Shown in *Figure 39.4,* **white blood cells** make up only one percent of your blood, yet they play a major role in protecting your body from foreign substances and from organisms that cause disease. The role of white blood cells and their function in immunity will be discussed in detail in Chapter 42.

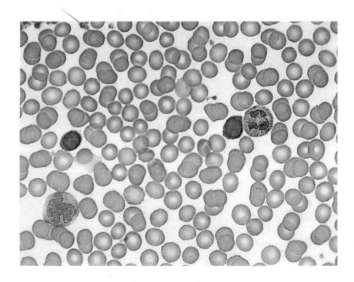

Figure 39.4

Compared with red blood cells, white blood cells have a nucleus, are larger, and are far fewer in number.

Blood clotting

Think about what happens when you cut yourself. If the cut is slight, you usually bleed for a short while and then stop. That's because in addition to red and white blood cells, your blood also contains some cell fragments called **platelets,** which help blood clot after such an injury, *Figure 39.5.* Platelets are produced from cells in red bone marrow. They have a short life span, remaining in the blood for only about one week.

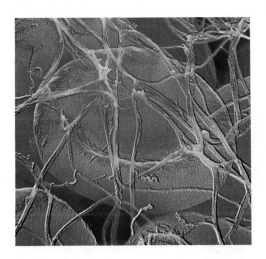

Figure 39.5

When you cut yourself, platelets in your blood help clot the blood, preventing you from bleeding to death. A sticky network of protein fibers called fibrin forms a web over the wound, trapping escaping blood components. Eventually, a dry, leathery scab forms.

ABO Blood Types

If a person is injured so severely that a massive amount of blood is lost, a transfusion of blood from a second person may be necessary. Whenever blood is transfused from one person to another, knowing blood types becomes important. In Chapter 15, you learned of the four human blood types—A, B, AB, and O. You've inherited one of these blood types from your parents.

Literature

Julius Caesar
by William Shakespeare

Amidst all the talk of treason and intrigue in his play *Julius Caesar,* William Shakespeare interjected a tender scene between Brutus and his wife, Portia. Brutus says to Portia, "You are my true and honorable wife; As dear to me as are the ruddy drops that visit my sad heart." These words must have puzzled the people in the audience watching this play in 1599. Perhaps they thought, "Ruddy drops of blood going to the heart? Nonsense! Doesn't Shakespeare know about Galen?"

Galen's theory Galen, a Greek physician who died more than 1000 years before Shakespeare began writing, had originated the accepted theory for the circulation of blood. In his view, the liver changed food into blood. Then the blood flowed through the veins to other parts of the body, where it was used up.

Galen's theory stood until 1628, when English physician William Harvey published a study about blood circulation. Unlike Galen, Harvey made firsthand observations of the circulatory system and dissected dozens of animals and humans. Harvey's experiments demonstrated that the heart acts as a pump and that blood circulates endlessly through a system of arteries and veins.

A meeting of the minds? How did Shakespeare, writing 29 years before Harvey, come to make an accurate comment about the flow of blood? Perhaps the quote was just poetic, or maybe it was a guess about the workings of the body. There's another possibility, too. Shakespeare and Harvey *may* have been in London at the same time, probably between 1592 and 1593, when Harvey was a student at nearby Cambridge and Shakespeare was a London actor and playwright. London's taverns have always been places where ideas are exchanged and theories are argued. Could Shakespeare have overheard Harvey's theory of blood circulation being discussed? Or might the two great men even have discussed these ideas? Of course, we'll probably never know, but it's intriguing to speculate.

CONNECTION TO Biology

Read further about the works of William Harvey. Why do you think Harvey's theory on blood circulation withstood attacks by followers of the ancient Greek physician Galen?

Figure 39.6

Blood contains both antigens and antibodies. The plasma in your blood contains specific antibodies that will *not* react to antigens on your own red blood cells.

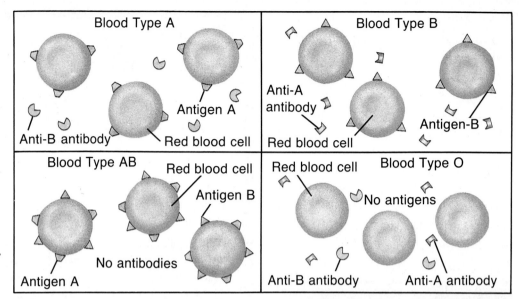

Blood Type A — Antigen A, Anti-B antibody, Red blood cell

Blood Type B — Anti-A antibody, Red blood cell, Antigen-B

Blood Type AB — Red blood cell, Antigen B, Antigen A, No antibodies

Blood Type O — Red blood cell, No antigens, Anti-B antibody, Anti-A antibody

▶ **For example, if you have type A blood, you have the A antigen on your red blood cells and the anti-B antibody in your plasma. You do *not* have anti-A antibodies because they would react with your red blood cells and your blood would clot. Antigen-antibody reactions are used to type blood.**

▶ **If you have type A blood and anti-A is added to it by way of transfusion, the blood will form clumps as shown on the left. If you add anti-B to type A blood, no clumps will form (right).**

Antigens determine blood type

Differences in blood type are due to the presence or absence of proteins, called antigens, on the membranes of red blood cells. **Antigens** are normally foreign substances that stimulate an immune response in the blood. The letters *A* and *B* stand for the type of antigen found in a person's blood.

Blood plasma may contain other proteins with shapes that match those of the different antigens, like matching pieces of a jigsaw puzzle. These proteins are called **antibodies.** A type of antibody will react with each antigen on a red blood cell. However, each type of blood contains antibodies for the antigens found only on the red blood cells of the other blood types—*not* of its own red blood cells. *Figure 39.6* explains why.

Rh factor in the blood

Another blood group can cause complications in some pregnancies. Illustrated in *Figure 39.*7, this problem involves the antigen called Rh, or Rhesus factor, also an inherited characteristic. People are Rh positive (Rh^+) if they have the Rh antigen factor on their red blood cells. They are Rh negative (Rh^-) if they don't.

Treatment for this problem is now available. At 28 weeks and shortly after each Rh^+ baby is born, the Rh^- mother is given a substance that binds and removes the antibodies in her bloodstream. As a result, the next fetus will not be in danger.

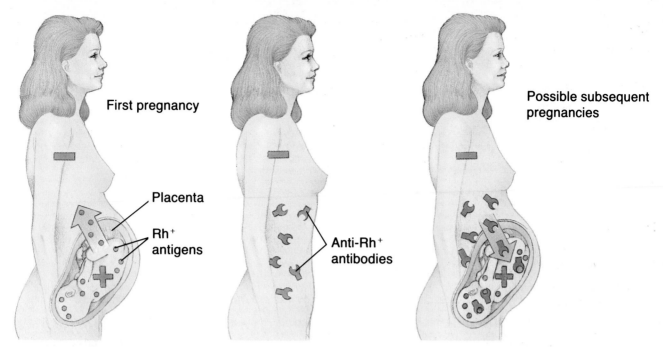

First pregnancy

Placenta

Rh⁺ antigens

Anti-Rh⁺ antibodies

Possible subsequent pregnancies

▲ Rh⁺ baby's blood mixes with Rh⁻ blood of mother.

▲ Upon exposure to the baby's Rh⁺ antigen factor, the mother will make anti-Rh⁺ antibodies.

▲ Should the mother become pregnant again, these anti-bodies will cross the pla-centa. If the new fetus is Rh⁺, the anti-Rh⁺ antibodies from the mother will destroy red blood cells in the fetus.

Figure 39.7

If a baby is Rh⁺ as a result of an Rh⁺ father and the mother is Rh⁻, problems may develop if the bloodstreams of the mother and baby mix during birth.

Your Blood Vessels: Pathways of Circulation

Because blood is fluid, it must be channeled through blood vessels, *Figure 39.8.* The three types of blood vessels are arteries, capillaries, and veins. Each is different in structure and function.

Arteries are large, thick-walled, muscu-lar, elastic vessels that carry blood away from the heart. The blood that they carry is under great pres-sure. As blood travels through an artery, the artery's elastic walls expand slightly. Then, the vessel shrinks a bit, pushing on the blood. This action causes a steady pulsating of the blood. Without the alternating elastic stretch and recoil of arterial walls, the movement of blood would be stop and go, like bumper-to-bumper traffic on the expressway.

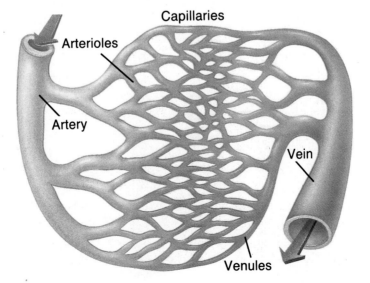

Capillaries

Arterioles

Artery

Vein

Venules

Figure 39.8

Arteries carry blood away from the heart, whereas veins carry blood toward the heart. Capillaries form an extensive web in the tissues. How are arter-ies different from veins?

Maintaining Homeostasis of Body Temperature

Birds and mammals are warm-blooded, also known as endothermic. Their body temperatures are maintained internally by a group of reflex responses integrated by the hypothalamus of the brain rather than being adjusted by the external environment.

Heat generation   The breakdown of food in the body generates heat. The higher the metabolic rate, the faster foods are broken down, and the more heat is produced. In the resting state, the liver, heart, brain, and most of the hormone-secreting glands produce large amounts of heat. Each skeletal muscle does not produce much heat itself, but because half of the body mass is skeletal muscle, the muscle contributes about 20 to 30 percent of all the body's heat. Exercise affects metabolic rate and heat production, increasing them by as much as 40 times their resting levels.

Heat regulation   The hypothalamus monitors body temperature and thereby maintains homeostasis. When you are too warm, the hypothalamus responds by allowing you to perspire and lose heat by evaporation. Blood vessels in your skin also dilate, resulting in heat loss by radiation. Heat can also be lost directly through the skin by means of convection. When you lose too much heat, your hypothalamus stimulates an increase in metabolism. Blood vessels will also constrict away from the skin surface to conserve heat.

Too much heat   Heatstroke can occur when the temperature and humidity are high. During this condition, body temperature can reach 110°F, as the body loses its ability to rid itself of the heat that has built up. Notably, the skin becomes hot and dry, and blood will rush to the head and face. Convulsions, brain damage, and death may follow.

Too little heat   Hypothermia results when the body temperature falls below 95°F. It can be due to cold, stress, drugs, burns, or malnutrition. If the body's core temperature falls below 90°F, the heart can't pump blood and will begin to contract erratically.

Thinking Critically
When you are cold, you may shiver. How does shivering help maintain body temperature?

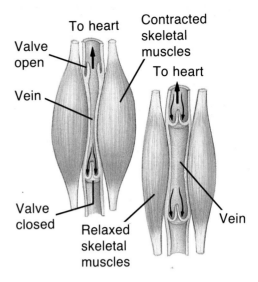

Figure 39.9

Veins contain one-way valves that work in conjunction with skeletal muscles. When the skeletal muscles contract, the valves open, and blood is forced toward the heart. When the skeletal muscles relax, the valves close to prevent the backflow of blood away from the heart.

After the arteries branch off from the heart, they divide into smaller arteries that, in turn, divide into even smaller vessels called arterioles. Arterioles enter tissues and branch into the smallest blood vessels, capillaries. **Capillaries** are microscopic blood vessels with walls that are only one cell thick. This feature enables nutrients and gases to diffuse easily between the blood and tissues.

As blood leaves the tissues, the capillaries join to form slightly larger vessels called venules. The venules merge to form **veins,** the large blood vessels that carry blood from the tissues back toward the heart. Blood in veins is not under pressure as great as that in the arteries. In some veins, especially those in your arms and legs, blood travels uphill against gravity. These veins are equipped with valves that prevent blood from flowing backward. *Figure 39.9* shows how these valves function.

Your Heart: The Vital Pump

The thousands of blood vessels in your body would be of little use if there were not a way to move blood through them. The main function of the heart is to keep blood moving constantly throughout the body. Being adapted to this job, the heart is a large organ made of cardiac muscle cells that are rich in mitochondria.

All mammalian hearts, including yours, have four chambers. The two upper chambers of the heart are the **atria.** The two lower chambers are the **ventricles.** The walls of each atrium are thinner and less muscular than those of each ventricle. As you will see, the ventricles perform more work than the atria, a factor that contributes to the thickness of their muscles. In addition, the muscle walls of the left ventricle are thicker than the muscles of the right ventricle, so your heart is somewhat lopsided.

Blood's path through the heart

Blood enters the heart through the atria and leaves it through the ventricles. Both atria fill up with blood at the same time. The right atrium receives oxygen-poor blood from the head and body through two large veins called the **venae cavae.** The left atrium receives oxygen-rich blood from the lungs through four pulmonary veins. These veins are the only veins that carry blood rich in oxygen. After they have filled with

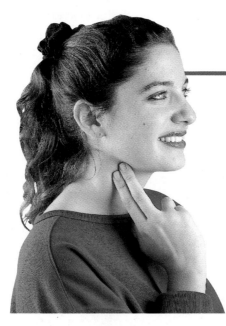

MiniLab

What effect does swallowing have on pulse rate?

The heart pumps blood to all the cells of the body. An important nerve that affects the heart rate is the vagal nerve, which sends inhibitory impulses to the heart. The vagal nerve is also affected by swallowing.

Procedure

1. Locate your pulse by placing your index and middle fingers on the carotid artery. This artery is located in your neck.
2. Take your partner's reference pulse rate for one minute.
3. Repeat step 2 three times and find the average.
4. Have your partner slowly sip a glass of water while you take his or her pulse rate.

Analysis

1. What was your average resting pulse rate?
2. What effect did swallowing water have on the average resting pulse rate?
3. What factors might cause heart rate to increase?

blood, the two atria then contract, pushing the blood down into the two ventricles.

Then both the ventricles contract. When the right ventricle contracts, it pushes the oxygen-poor blood from the right ventricle against gravity, out of the heart, toward the lungs through the pulmonary arteries. These arteries are the only arteries that carry blood poor in oxygen. At the same time, the left ventricle forcefully pushes oxygen-rich blood from the left ventricle out of the heart through the **aorta** to the arteries. The aorta is the largest blood vessel in the body.

vena cava

vena (L) vein
cava (L) empty
Each vena cava empties blood into the heart.

Your Heart

Your heart is about 12 cm by 8 cm—roughly the size of your fist. It lies in your chest cavity, just behind the breastbone and between the lungs, and is essentially a large muscle completely under involuntary control.

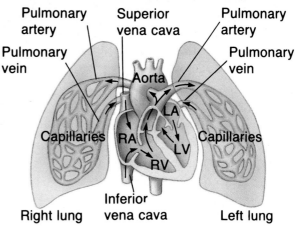

Pulmonary artery · Superior vena cava · Pulmonary artery
Pulmonary vein · Aorta · Pulmonary vein
Capillaries · LA · Capillaries
RA · LV
RV
Right lung · Inferior vena cava · Left lung

① The heart is enclosed by a membrane called the pericardium. This membrane confines the heart in its fluid-filled cavity, yet it provides enough freedom of movement so that the heart can beat vigorously and rapidly.

② If you were to trace the path of a drop of blood through the heart, you could begin with blood coming back from the body through a vena cava. The drop travels first into the right atrium, then into the right ventricle, and through a pulmonary artery to one of the lungs. In the lungs, the blood drops off its carbon dioxide and picks up oxygen. Then it moves through a pulmonary vein to the left atrium, into the left ventricle, and finally out to the body through the aorta, to eventually return once more to your heart.

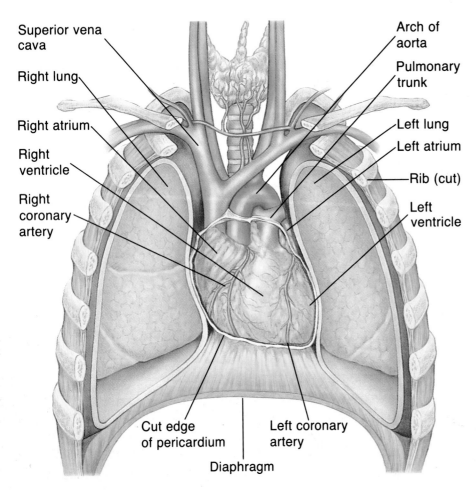

Superior vena cava · Right lung · Right atrium · Right ventricle · Right coronary artery

Arch of aorta · Pulmonary trunk · Left lung · Left atrium · Rib (cut) · Left ventricle

Cut edge of pericardium · Left coronary artery · Diaphragm

③ Between the atria and ventricles are one-way valves that keep blood from flowing back into the atria. Sets of valves also lie between the ventricles and the arteries leaving the heart.

Heartbeat regulation

Each time the heart beats, a surge of blood flows from the left ventricle into the aorta and then the arteries. Because the radial and carotid arteries are fairly close to the surface of the body, the surge of blood can be felt as it moves through them. This surge of blood through an artery is called a **pulse.**

The heart rate is set by the pacemaker, a group of cells at the top of the right atrium. This pacemaker generates an electrical impulse that spreads over both atria. The impulse signals the two atria to contract at almost the same time. The pacemaker also triggers a second set of cells at the base of the right atrium to send the same electrical impulse over the ventricles, causing them to contract. The pattern of a heartbeat can be drawn or traced by a machine. *Figure 39.10* shows the record that is produced.

Control of the heart

Whereas the pacemaker controls the heartbeat, the medulla oblongata regulates the rate of the pacemaker, speeding or slowing its activity. If the heart beats too fast, sensory cells in arteries near the heart become stretched. A signal is sent to the medulla via nerves from these cells. The medulla then slows the pacemaker. If the heart slows down too much, pressure drops in the arteries, and the medulla signals the pacemaker to speed up.

ThinkingLab   Draw a Conclusion

How does exercise affect heart rate?

The volume of blood ejected from the left ventricle into the aorta per minute is called the cardiac output. Cardiac output is determined by measuring the amount of blood pumped out by the left ventricle during each heartbeat and the number of beats per minute. Actual cardiac output depends on how much blood enters the ventricle during the diastolic phase, which can be as much as 70 percent. As the heart rate increases, the length of time the ventricle spends in the diastolic phase becomes shorter and shorter.

Analysis

When you exercise, your cardiac output may increase to as much as four to six times its normal rate. You might think that cardiac output would just continue to increase, but in actual measurements, the heart rate rises and then levels off.

Thinking Critically

Why do you think your heart rate levels off and does not continue to rise as you exercise?

Figure 39.10

An electrocardiogram, or EKG, is a record of the electrical changes in the heart. The EKG is an important tool in diagnosing abnormal heart rhythms or patterns. Each peak or valley in the diagram represents a particular electrical activity during a heartbeat.

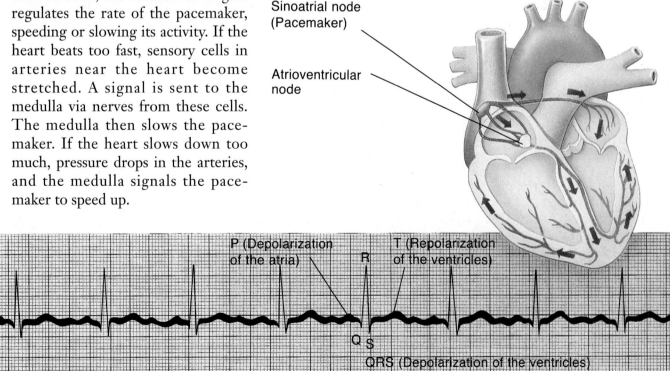

Sinoatrial node (Pacemaker)

Atrioventricular node

P (Depolarization of the atria)

R

T (Repolarization of the ventricles)

Q S

QRS (Depolarization of the ventricles)

Focus On

Care of Your Heart

Heart disease is the leading cause of death for humans. If you were to apply the principle of cause and effect to the incidence of this disease, you could trace the causes in many instances to an individual's decisions and behaviors. To reduce your chances of acquiring heart disease, consider the following areas of risk.

Physical activity

Stay active   Many victims of heart disease are sedentary or inactive for long periods of time. Inactivity, particularly when coupled with overeating, can cause one to become over-weight and result in a strained and weakened cardiovascular system. People who exercise regularly, however, reduce their risk of heart disease. You can develop and maintain good cardiac health by doing as little as walking briskly each day for about 30 minutes. Of course, if you are active in vigorous sports such as basketball, running, or tennis, your heart could be even more fit.

Diet

Eat healthy meals   Consuming a diet high in fat is a major cause of atherosclerosis. In this disease, the muscle cells that line the arteries pick up fat from the blood and swell. This narrows, and sometimes closes, the opening in the artery. If the blocked artery is one that feeds the heart muscle, that part of the muscle will die. This is called a heart attack, and it kills more Americans each year than any other disease. You can eliminate this cause of heart disease by reducing the amount of fat in your diet and by eating plenty of vegetables, fruits, and grain products.

Stress

Recognize stress Stress increases the strain on every part of your body, but especially your heart. What causes stress? Two factors seem to be constant pressure to produce and lack of control over how much work has to be done. These two factors can raise blood pressure, squeeze the arteries that carry blood to the heart muscle, and increase the level of chemicals that cause blood clots. If this happens often enough, it can weaken the arteries and heart muscle and lead to a heart attack.

Manage stress Reducing and managing stress is vital to good cardiac health. Learning techniques such as relaxation can help you cope with stress, and good organization and planning can help you avoid panic situations. Learning conflict resolution methods is also a good exercise that can reduce stress. Occasional stress is part of life and is unavoidable. However, constant stress can be a life-threatening danger.

Smoking

Avoid harmful habits Tobacco products are not conducive to good cardiac health. Nicotine causes the blood vessels to constrict and the heart to work harder. At the same time, smoking cuts down on the amount of oxygen in the blood and damages the lungs so that even less can enter. This risk of disease increases with the amount smoked. An ex-smoker's risk of heart disease can eventually be reduced to that of nonsmokers; however, protection against disease is best provided by not smoking at all.

EXPANDING YOUR VIEW

1. **Apply** Plan a personal fitness program that will help you have a healthy heart.
2. **Analyze** Write down all the things you have eaten over the last two days. Which foods were bad for your heart, and what could you substitute for them?
3. **Identify** One research study has identified loneliness as a major cause of stress. What do you do that increases your loneliness, and how can you correct it?

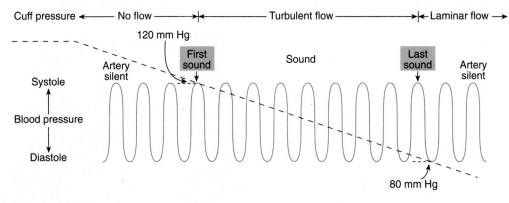

Figure 39.11

The solid line on the graph represents the pressure variation between systolic and diastolic blood pressure. The dashed line represents the cuff pressure, which is slowly being released. When the cuff pressure equals the systolic pressure the first sound of blood flow is heard. The sound stops when the cuff pressure equals the diastolic pressure.

Blood pressure

A pulse beat represents the pressure that blood exerts as it pushes against the walls of an artery. **Blood pressure** is the force that the blood exerts on the vessels of the body. As *Figure 39.11* shows, this pressure rises and falls as the heart contracts and then relaxes.

Blood pressure rises sharply when the ventricles contract. The high pressure is called systolic pressure. Blood pressure then drops dramatically as the ventricles relax. The lowest pressure occurs just before the ventricles contract again and is called diastolic pressure.

Pressure is exerted on all vessels throughout the body, but the term *blood pressure* is usually used to refer to pressure in the arteries where it is the greatest, as in the brachial artery of the upper arm.

As you have seen, blood plays a critical role in supplying nutrients to and removing wastes from the body's cells. It can function as an efficient supply and sanitation medium only because cellular wastes are constantly removed by the urinary system.

Section Review

Understanding Concepts
1. Summarize the distinguishing features and role of each of the four components of blood.
2. Distinguish between an artery and a vein.
3. Outline the path taken by a red blood cell as it passes from the left atrium to the right ventricle of the heart.

Thinking Critically
4. The level of carbon dioxide in the blood affects breathing rate. It also affects the heart rate. How would you expect high levels of carbon dioxide to affect the heart rate?

Skill Review
5. **Making and Using Graphs** Make a pie graph showing the relative proportions of the components of blood. For more help, refer to Organizing Information in the *Skill Handbook.*

*W*ater consumption helps speed the filtering process of the kidneys and maintain their efficiency. Because the function of the kidneys is so essential in maintaining the balance of fluids in the body, any disruption of this function is potentially serious. The kidneys are the most important organs of the human urinary system. They perform a major cleanup job for the body.

Section Preview

Objectives

Describe the structures and functions of the urinary system.

Explain the kidneys' role in maintaining homeostasis.

Key Terms

kidney
ureter
urinary bladder
nephron
urine
urethra
hypothalamus

Kidneys: The Body's Janitors

The urinary system is made up of two kidneys, a pair of ureters, the urinary bladder, and the urethra, which you can see in *Figure 39.12*. The **kidneys** filter collected wastes from the blood and maintain the homeostasis of body fluids. Each kidney is connected to a tube called a **ureter,** which leads to the urinary bladder. A smooth muscle bag, the **urinary bladder** stores a solution of wastes called urine.

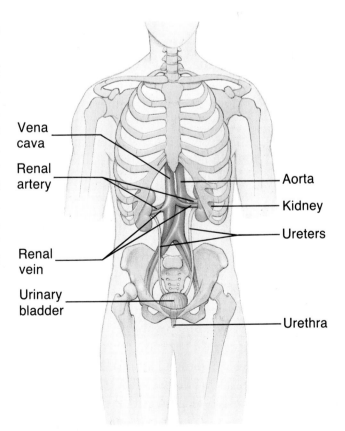

Vena cava
Renal artery
Renal vein
Urinary bladder
Aorta
Kidney
Ureters
Urethra

Figure 39.12

The paired kidneys are reddish organs that resemble kidney beans in shape. They are found just above the waist, behind the stomach—one on each side of the spine. Ribs partially protect the kidneys.

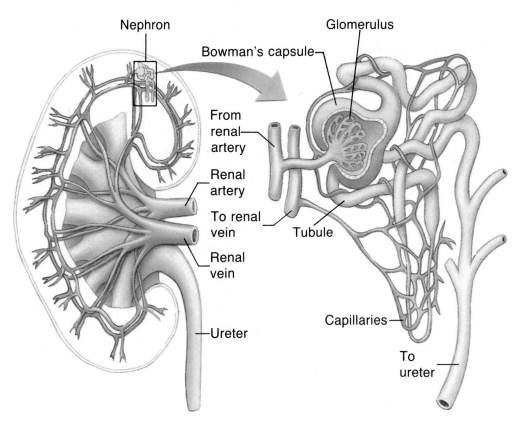

Nephron

Bowman's capsule

Glomerulus

From renal artery

Renal artery

To renal vein

Renal vein

Tubule

Ureter

Capillaries

To ureter

Figure 39.13

The kidney receives a large blood supply through the renal artery for filtering. The blood leaves the kidney by the renal vein. Blood enters the nephron from a branch of the renal artery and immediately enters a capillary bed called the glomerulus, which is enclosed in the Bowman's capsule. Here, the first step in the filtering process begins.

Nephron: The unit of the kidney

Have you ever seen an air filter on a car or a water filter in an aquarium? A filter is a device that removes impurities from a substance. Within your body, each kidney is made up of about 1 million tiny filters. Each filtering unit in the kidney is called a **nephron.** *Figure 39.13* shows the parts of a typical nephron.

Blood entering a nephron in a kidney carries cell wastes. As blood enters the nephron, it is under high pressure, and it immediately flows into a bed of blood capillaries. Because of the pressure, water, glucose, vitamins, amino acids, protein waste products, salts, and ions from the blood pass out of the capillaries into a part of the nephron called a

Bowman's capsule. Blood cells and most proteins are too large to pass through the walls of a capillary, so these components stay within the blood vessels.

The liquid forced into the Bowman's capsule passes through a narrow, U-shaped tubule. As the liquid passes along the tubule, most of the ions and water, and all of the glucose and amino acids, are reabsorbed into the bloodstream. This second movement of substances is the process by which the body's water is conserved and homeostasis is maintained. Small molecules, such as water, diffuse back into the capillaries. Other molecules, such as ions, are moved back into the capillaries by active transport.

nephron:

nephros (GK) kidney
A nephron is a unit of a kidney.

The formation of urine

The liquid that remains in the tubules—composed of excess water, waste molecules, and excess ions—is **urine**, *Figure 39.14.* You produce about 2 L of urine a day. This waste fluid flows out of the kidneys through the ureter to the urinary bladder, where it may be stored. Urine then passes from the urinary bladder out of the body through a tube called the **urethra.**

The Urinary System and Homeostasis

The major waste products of cells are nitrogenous wastes, which come from the breakdown of proteins. These wastes include ammonia and urea. Both compounds are toxic to your body and, therefore, have to be removed from the blood regularly.

In addition to removing these wastes, the kidneys control the level of sodium in blood by removing and reabsorbing ions from it. This helps control the osmotic pressure of the blood. The kidneys also regulate the pH of blood by excreting hydrogen ions and reabsorbing sodium bicarbonate.

Recall from Chapter 38 that parathyroid hormone increases the rate of reabsorption of calcium and magnesium from the urine. Aldosterone is another hormone that stimulates reabsorption of sodium and chloride ions. If a person takes in large amounts of sodium and chloride ions, aldosterone production decreases and more sodium is eliminated.

Figure 39.14

In the nephron, filtration and reabsorption take place.

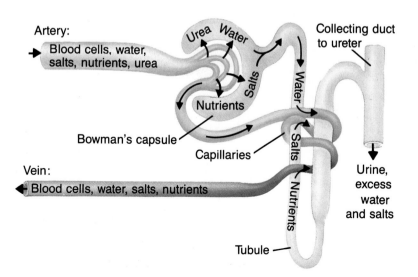

Urine regulation

Involved in the regulation of wastes in the kidneys is the **hypothalamus,** a small area of the brain that controls homeostatic activities. Antidiuretic hormone (ADH) is a hormone produced by the hypothalamus. It stimulates the reabsorption of water in a nephron. This keeps both the fluid level of the body and blood pressure from decreasing. Illustrated in *Figure 39.15,* the amount of urine produced by the kidneys is a result of an interaction between body fluid levels and ADH.

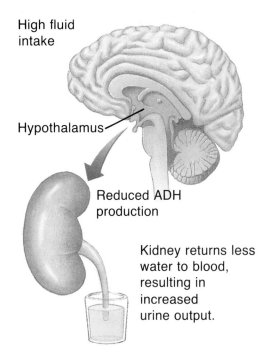

High fluid intake

Hypothalamus

Reduced ADH production

Kidney returns less water to blood, resulting in increased urine output.

Figure 39.15

If you drink a large amount of water, the fluid level of the body increases. This increase in water triggers the hypothalamus to slow up production of ADH. As a result, more water is eliminated. When the water level in the blood drops too low, the hypothalamus will again produce more ADH.

Connecting Ideas

The respiratory, circulatory, and urinary systems play a central role in your ability to cope with your environment and maintain homeostasis. They are responsible for the intake and distribution of vital oxygen and water, and the removal and excretion of carbon dioxide and other wastes. Yet these systems are largely beyond the voluntary control of a person.

How *are* the intricate processes of each of these three systems regulated? What controls the cycle of intake, distribution, and excretion? The answers lie in the system that controls all other body systems—the nervous system.

Section Review

Understanding Concepts
1. What is the function of a nephron in the kidney?
2. Identify the major components of urine, and explain why it is considered a waste fluid.
3. What is the kidney's role in maintaining homeostasis in the body?

Thinking Critically
4. It is a hot day. You are sweating, and you feel thirsty. When you drink, you replace water in your body. What other substance might you need to replace? Explain.

Skill Review
5. **Sequencing** Trace the sequence of urinary waste from a cell to the outside of the body. For more help, refer to Organizing Information in the *Skill Handbook.*

Secondhand Smoke and Individual Rights: The Ban on Public Smoking

No-Smoking Restrictions at Ballpark

BEST IN SPORTS   JANUARY 18, 1993

The Baltimore Orioles announced last week that they would limit smoking to certain designated areas of their open-air stadium and prohibit it in the seats.

No-smoking restrictions are nothing new. They have been adopted in 44 states and 500 cities, limiting public smoking in government buildings, movie theaters, hospitals, and even prisons. The airline industry has enforced no-smoking restrictions on most of their flights for years. What is new is the current movement of American businesses to totally restrict their employees from smoking at the workplace.

Secondhand smoke: An environmental hazard
Early in 1993, the Environmental Protection Agency (EPA) formally classified secondhand tobacco smoke as a human carcinogen. The report concluded that environmental smoke is a serious health risk for nonsmokers, responsible for 3000 adult deaths through lung cancer annually, and increased respiratory infections in

young children. EPA officials also suggested that employers could be legally liable if they knowingly expose workers to a health hazard.

Different Viewpoints

Favoring public smoking bans Anti-smoking groups that support bans on public smoking feel that research over the past decade from national health authorities—such as the Surgeon General, the National Academy of Sciences, the Centers for Disease Control, and most recently the EPA—provides overwhelming evidence that secondhand smoke is a health risk for nonsmokers. They feel that smokers willing to quit should be provided with social and medical support, whereas those unwilling to do so should be prohibited from endangering the lives of others.

Protesting smoking bans Traditionally, the argument against restricting public smoking has focused on the rights of individuals to choose to indulge in behavior that harms no one but themselves. Those against a public smoking ban, particularly the tobacco industry itself, insist that efforts to restrict smoking are a violation of a smoker's civil liberties. They also claim that the EPA report is inconclusive because the evidence is based on chronic exposure studies—the kind found in long-term marriages of nonsmokers with heavy smokers.

INVESTIGATING the Issue

1. **Recognizing Cause and Effect** Why would large employers in particular adopt no-smoking bans?
2. **Applying Concepts** Given research indicating the hazards of secondhand smoke, what steps can you take to minimize your personal risk?

39.1 The Respiratory System

- External respiration involves taking in air through the passageways of the respiratory system and exchanging gases in the alveoli of the lungs.
- Breathing involves contraction of the diaphragm, air rushing into the lungs, relaxation of the diaphragm, and air being pushed out of the lungs.
- Breathing is controlled by the chemistry of the blood.

39.2 The Circulatory System

- Blood is composed of red and white blood cells, platelets, and plasma. These components carry oxygen, carbon dioxide, and other substances through the body.
- Blood cell antigens determine blood type and are important in blood transfusions.
- Blood is carried by arteries, veins, and capillaries.
- Blood is pushed through the vessels by the heart.

39.3 The Urinary System

- The nephrons of the kidneys filter wastes from the blood.
- The urinary system helps maintain the homeostasis of body fluids. The hypothalamus helps regulate this process.

Key Terms

Write a sentence that shows your understanding of each of the following terms.

alveoli	platelet
antibody	pulse
antigen	red blood cell
aorta	trachea
artery	ureter
atrium	urethra
blood pressure	urinary bladder
capillary	urine
hemoglobin	vein
hypothalamus	vena cava
kidney	ventricle
nephron	white blood cell
plasma	

Understanding Concepts

1. Compare external respiration and cellular respiration.
2. Describe the steps in inhalation and exhalation.
3. Explain how breathing is controlled.
4. How are red blood cells different from other body cells?
5. Why must blood types be known before a transfusion?
6. What is the role of the pacemaker?
7. What is the function of heart valves?
8. What stimulates an increase in ADH secretion?

Using a Graph

9. The following graph shows blood pressure fluctuation in blood vessels. Explain why the blood pressure changes periodically in the aorta through the small arteries, but not in the capillaries and veins.

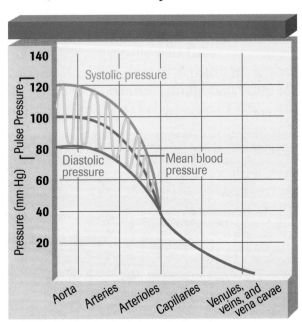

Relating Concepts

10. Make a concept map that relates the following terms and phrases. Supply the appropriate linking words for your map.

    aorta, arteries, atria, blood pressure, capillaries, pulse, ventricles, venae cavae

Applying Concepts

11. Explain why all blood in all arteries is not oxygen rich. Where in the circulatory system do arteries carry oxygen-poor blood?

12. Carbon monoxide combines with hemoglobin more strongly than does oxygen. Because carbon monoxide is found in car exhaust, explain why sitting in a traffic jam with the windows open could be dangerous.

13. Think about the structure of a nephron, and suggest a reason that high blood pressure can damage the kidneys.

14. Alcohol inhibits the hypothalamus from secreting antidiuretic hormone. What is the effect on the body when this occurs?

Focus On Heart Disease

15. A healthy diet for the heart is usually low in saturated fats and cholesterol. Why?

Thinking Critically

Recognizing Cause and Effect

16. **Biolab** What activities besides exercise could affect breathing rate?

Designing an Experiment

17. **Minilab** Design an experiment that would test the effect of various types of music on heart rate.

Drawing a Conclusion

18. Explain the relationship between the total cross-sectional area of the various blood vessels and the velocity of blood flow, using the graph.

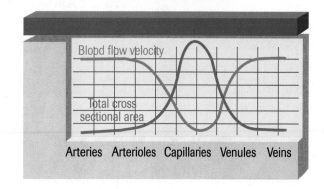

Interpreting Tables

19. Explain the differences between the quantities shown for Bowman's capsule filtrate and the urine output in the following table.

	Bowman's Capsule Filtrate	Per-Day Urine Output
Water	180 000 mL	1500 mL
Protein	10-20 g	0 g
Chloride	630 g	5 g
Sodium	540 g	3 g
Glucose	180 g	0 g
Urea	53 g	25 g

Connecting to Themes

20. **Evolution** The adaptation of the mammalian kidney allowed mammals to live efficiently on land and lose little water. Explain how the mammalian kidney conserves water.

21. **Systems and Interactions** Briefly describe the interactions between the respiratory and the circulatory systems.

22. **Energy** The cells of heart muscle are loaded with mitochondria. Explain the adaptive advantage of this.

CHAPTER 40

The Nervous System and the Effects of Drugs

Zap! Pow! Slam! With lightning-quick speed, or so you hope, you push the buttons in response to the moving lights and electronic sounds of your video game. After all, if you don't respond quickly enough, you might lose this one fight or even the entire game.

Your success at a video-arcade game depends on your response time to the stimuli in the game. In the time you take to respond, electrical impulses have bolted from your eyes and ears, to your brain, and down through your arms to your fingertips. All the while, similar electrical impulses have been coursing through your body, controlling critical body functions. Unlike your moves in the game, body actions such as your heartbeat and breathing are beyond your control.

How does your body manage all this activity? Any complex machine requires a system of control. In your body, that system is the nervous system.

Playing a video-arcade game requires not only fast-moving fingers, but also the use of millions of microscopic nerve cells and sensory organs such as your eyes and ears. How do all these parts of your body work together to help you defeat your video foe?

Magnification: 64×

40.1 The Nervous System

Section Preview

Objectives

Explain how nerve impulses travel in the nervous system.

Summarize the functions of the major parts of the nervous system.

Key Terms

neuron
dendrite
axon
synapse
neurotransmitter
central nervous system
peripheral nervous
  system
cerebrum
cerebellum
medulla oblongata
somatic nervous system
reflex
autonomic nervous
  system
sympathetic nervous
  system
parasympathetic
  nervous system

What do you use the telephone for? To talk, or more precisely, to communicate. You probably know that your voice is transmitted as an electrical charge across telephone wires. Would it surprise you to know that a similar electrical charge travels through your body, helping different parts of your nervous system communicate and control other body parts?

Neurons: Basic Units of the Nervous System

The basic unit of structure and function in the nervous system is the neuron. As shown in *Figure 40.1*, **neurons** are cells that conduct impulses throughout the nervous system. You can see that a neuron is made up of dendrites, a cell body, and an axon.

Dendrites are branchlike extensions of the neuron that receive impulses and carry them toward the cell body. The **axon** is a single extension of the neuron that carries impulses away from the cell body. The axon of one neuron may branch extensively, sending its signal to many other neurons, muscles, or glands. In turn, the multiple dendrites of one neuron can receive signals from the axons of many different neurons. While receptor sites on the

Figure 40.1

The cell body of a neuron contains a nucleus, cytoplasm, and organelles such as mitochondria and Golgi apparatus. Dendrites and axons are extensions of the cytoplasm that branch out from the cell body.

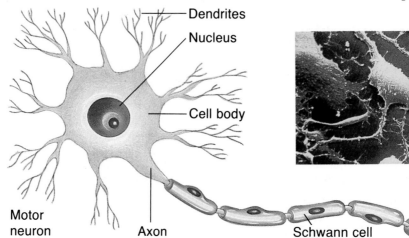

Motor neuron

Dendrites
Nucleus
Cell body
Axon
Schwann cell
Axon endings

Magnification: 2500×

ends of dendrites usually are identified as the point of origin for the signals, the signals may originate at various points on a neuron.

Neurons are classified as sensory neurons, motor neurons, or interneurons. Sensory neurons carry impulses from outside and inside the body to the brain and spinal cord. Interneurons are found within the brain and spinal cord. They process these incoming impulses and may pass response impulses to motor neurons. Motor neurons carry the response impulses away from the brain and spinal cord. How do neurons carry these messages?

How neurons work

Suppose you're in a crowded, noisy store, and you feel a tap on the shoulder. Turning your head, you see the smiling face of a good friend. How did the tap on the shoulder work to get your attention? The touch stimulated sensory receptors in your shoulder, starting an impulse in these neurons. The sensory impulse was carried to the spinal cord and then up to your brain. From your brain, an impulse was sent out causing motor neurons to transmit an impulse to muscles in the neck, and you turned to look. *Figure 40.2* shows how stimuli, such as the tap on your shoulder, are transmitted through your nervous system.

How does impulse transmission take place? Start with a resting neuron—one that is not transmitting an impulse. You have learned that the plasma membrane controls the concentration of ions inside a cell. Although sodium ions (Na^+) and potassium ions (K^+) are on both sides of the membrane, sodium ions exist in a larger concentration outside the cell, and potassium ions are in a higher concentration inside the cell.

Figure 40.2

The nervous system receives information, transmits the information, sorts and interprets the incoming information, determines a response, transmits the response, and activates the response.

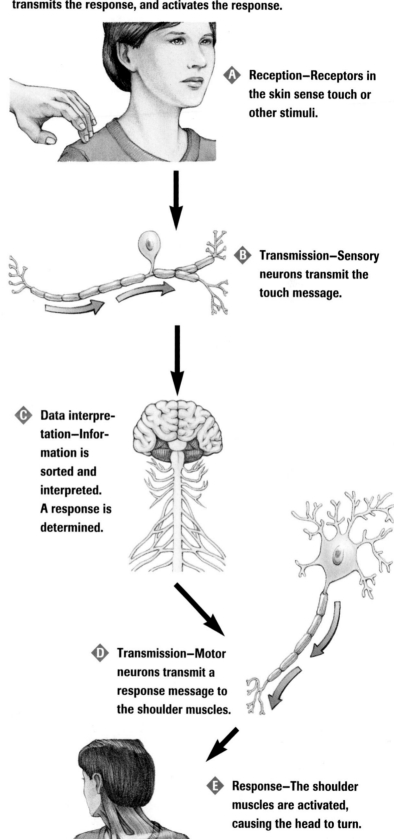

A Reception—Receptors in the skin sense touch or other stimuli.

B Transmission—Sensory neurons transmit the touch message.

C Data interpretation—Information is sorted and interpreted. A response is determined.

D Transmission—Motor neurons transmit a response message to the shoulder muscles.

E Response—The shoulder muscles are activated, causing the head to turn.

These differences occur because the membrane is more permeable to potassium than it is to sodium.

The membrane also contains an active transport set of proteins called the Na^+/K^+ pump. As you can see in *Figure 40.3,* the action of the pump increases the concentration of positive charges on the outside of the membrane. In addition, the presence of many negatively charged proteins and organic phosphates means that the inside of the membrane is more negative than the outside.

Under these conditions, the membrane is said to be polarized, or at rest. A polarized membrane has the potential to do work. In this case, work is the transmission of an impulse.

How an impulse is transmitted

When a stimulus excites a neuron, sodium channels in the membrane open up, and sodium ions rush into the cell. As the positive sodium ions build up inside the membrane, the inside of the cell becomes more positively charged than the outside. This change in charge, called depolarization, sets up a wave similar to an ocean wave. A wave of changing charges moves down the length of the axon, *Figure 40.4.* Because the impulse is generated anew at each stage along the axon, it does not diminish in strength during transmission. As the wave passes, the membrane immediately behind it returns to its resting state, with the inside of the cell negatively charged and the outside positively charged.

This wave of depolarization is actually the transmission of an impulse along the complete length of the axon. An impulse can move down the axon only when the stimulation of the neuron is strong enough to reach a certain threshold level. If the threshold level is not reached, the impulse quickly dies out. This is referred to as the *all-or-none principle.*

For a short time immediately after depolarization occurs, the neuron cannot be restimulated. This happens because the gated sodium channels are closed and the gated potassium channels are open.

Figure 40.3

The membrane of an axon contains open channels that allow movement of sodium (Na⁺) and potassium (K⁺) ions into and out of the cell, as well as gated sodium and potassium channels. In addition, a Na⁺/K⁺ pump uses ATP to pump three sodium ions out for every two potassium ions it pumps in. What effect does this action have on the charge of the normal resting neuron?

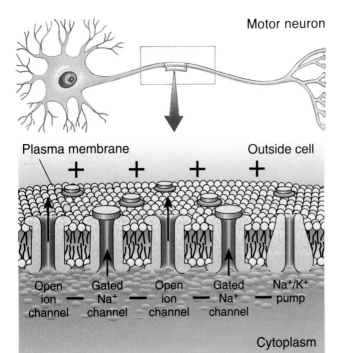

Motor neuron

Plasma membrane · Outside cell

Open ion channel — Gated Na⁺ channel — Open ion channel — Gated Na⁺ channel — Na⁺/K⁺ pump

Cytoplasm

Figure 40.4

A wave of depolarization moves down the axon of a neuron.

A Gated sodium channels open, allowing sodium ions to enter and make the inside positively charged and the outside negatively charged.

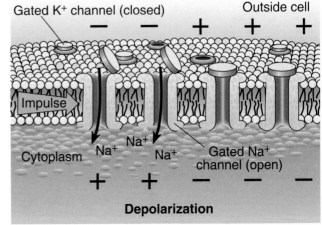

Gated K⁺ channel (closed) · Outside cell

Impulse

Cytoplasm · Na⁺ · Na⁺ · Na⁺ · Gated Na⁺ channel (open)

Depolarization

White matter and gray matter

Most axons, especially those outside the brain and spinal cord, are surrounded by a white covering called the myelin sheath. The sheath forms when Schwann cells grow and wrap around the axon like a jellyroll. The Schwann cells insulate the axon, hindering the movement of ions through the axon's plasma membrane. The ions are forced to move quickly down the axon until reaching a gap in the sheath that occurs between each of the individual Schwann cells. At this point, the ions can pass through the membrane and depolarization occurs. As a result, the impulse does not travel continuously down the length of the axon but jumps from gap to gap, greatly increasing the speed at which the impulse travels. The myelin sheath gives neurons a white appearance. In the brain and spinal cord, masses of myelinated neurons make up what is called white matter. The absence of myelin in masses of neurons accounts for the gray matter of the brain.

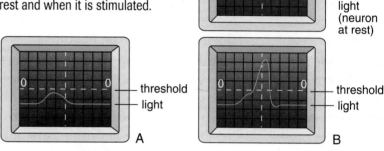
B As soon as the wave passes, that part of the axon returns to a resting state. This occurs because the gated sodium channels close, stopping the influx of sodium ions, and gated potassium channels open, allowing potassium to flow out of the neuron.

C Then, the Na⁺/K⁺ pump takes over again, moving sodium ions back out through the membrane and potassium ions in. The charges return to the same condition as in a resting neuron.

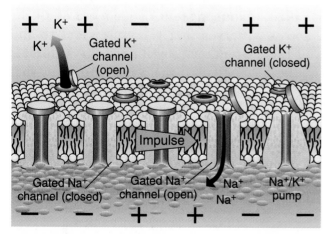

K^+
Gated K^+ channel (open)
Gated K^+ channel (closed)
Impulse
Gated Na⁺ channel (closed)
Gated Na⁺ channel (open)
Na⁺
Na⁺
Na⁺/K⁺ pump

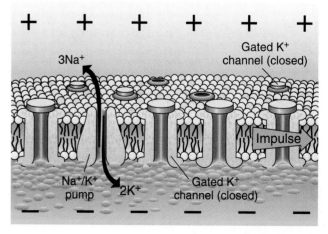

$3Na^+$
Gated K^+ channel (closed)
Na⁺/K⁺ pump
$2K^+$
Gated K^+ channel (closed)
Impulse

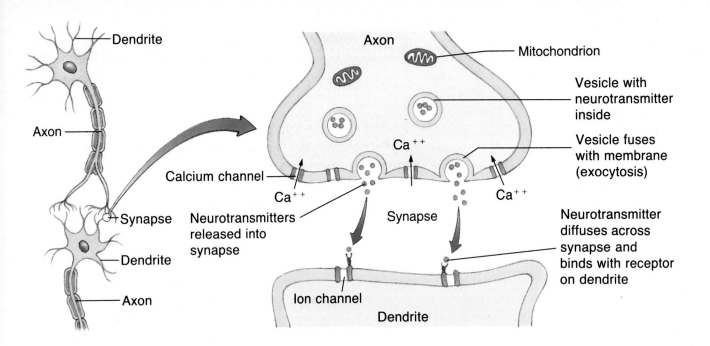

Figure 40.5

Neurotransmitters are released into a synapse by the process of exocytosis. The dendrite receiving the impulse contains receptor sites that are linked to ion channels. When the neurotransmitters diffuse across the synapse and bind to the receptors, the ion channels open, changing the polarity of the receiving dendrite. In this way, nerve impulses move from neuron to neuron.

Connections between neurons

Although the neurons of the nervous system lie end to end in bundles, axons to dendrites, their ends don't touch. A tiny space called a **synapse** lies between one neuron's axon and another neuron's dendrites. Impulses moving to and from the brain must move across these synapses. How does an impulse make this jump?

As an impulse reaches the end of an axon, the wave of changing charges opens calcium channels, allowing calcium to enter the end of the axon. As shown in *Figure 40.5,* the calcium causes small packages of chemicals first to fuse with the membrane and then to burst open into the synapse, releasing their chemicals. These chemicals, called **neurotransmitters,** diffuse across the synapse to the dendrite of the next neuron. There they stimulate polarity changes in the neuron.

A synapse is a one-way passage, allowing a nerve impulse to pass from axons to dendrites. The continuous relay of an impulse is prevented because neurotransmitters typically are broken down quickly by enzymes in the synapse.

The Central Nervous System

When you make a call to a friend, your call travels over wires to a control center, where it is switched over to wires that connect with your friend's telephone. In the same manner, an impulse traveling through neurons in your body usually reaches the control center of the nervous system—your brain. The brain and the spinal cord together make up the **central nervous system,** which acts as your body's control center and which coordinates your body's activities.

Two systems work together

Another division of your nervous system, called the **peripheral nervous system,** is made up of all the nerves that carry messages to and from the central nervous system. It is similar to the telephone wires that run between a phone system's control center and the phones in the homes of the community. Together, the central nervous system (CNS) and the peripheral nervous system (PNS), shown in *Figure 40.6,* make rapid changes in your body in response to stimuli in your environment.

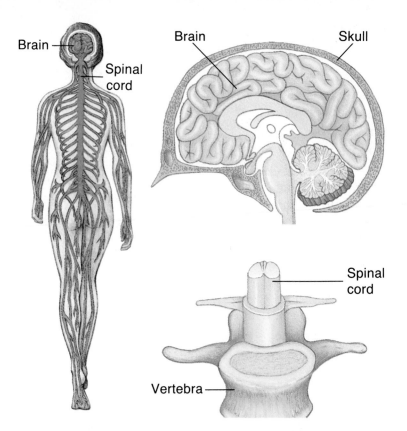

Brain

Spinal cord

Brain

Skull

Spinal cord

Vertebra

Figure 40.6

The human nervous system (left) is divided into the CNS, in red, and the PNS, in blue. The brain receives and transmits impulses by way of the spinal cord, which connects the brain to the peripheral nervous system. The brain and spinal cord are protected by your skull and the vertebrae (right).

Anatomy of the brain

The brain is the control center of the entire nervous system. For descriptive purposes, it is useful to divide the brain into three main sections: the cerebrum, the cerebellum, and the brain stem, *Figure 40.*7.

The **cerebrum** is divided into two halves, called hemispheres, that are connected by nerve tracts. Your conscious activities, intelligence, memory, language, skeletal muscle movements, and senses are all controlled by the cerebrum. The cerebrum is wrinkled with countless folds and grooves and is covered with an outer layer of gray matter called the cerebral cortex. Has anyone ever told you to "use your gray matter"? While the saying means that you should use your powers of reasoning, it refers to neurons that are not covered by myelin sheaths.

The **cerebellum,** located at the back of your brain, controls your balance, posture, and coordination. If the cerebellum is injured, your movements become jerky.

The brain stem is made up of the medulla oblongata, the pons, and the midbrain. The **medulla oblongata** is the part of the brain that controls involuntary activities such as breathing and heart rate. The pons and midbrain act as pathways connecting various parts of the brain with each other. You will find out more about the brain's structure and its fascinating functions in Focus On the Brain following this chapter.

*Figure 40.*7

The large surface area created by the many folds in the cerebrum results in more neurons being located in the cerebral cortex. This increase was important in the evolution of human intelligence. The portion of the brain that receives and sends sensory signals to the cerebrum is the thalamus. It receives all sensory information except smell.

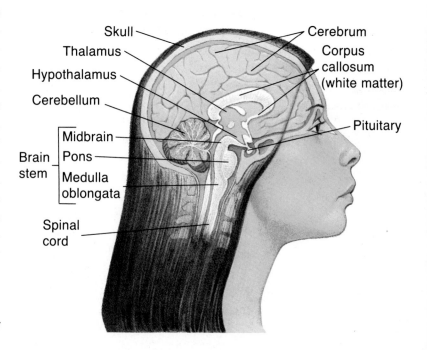

Skull

Thalamus

Hypothalamus

Cerebellum

Midbrain

Brain stem — Pons

Medulla oblongata

Spinal cord

Cerebrum

Corpus callosum (white matter)

Pituitary

The Peripheral Nervous System

Remember that the peripheral nervous system carries impulses between the body and the central nervous system. For example, when a stimulus, such as cold air, is picked up by receptors in your skin, it initiates an impulse in the sensory neurons. The impulse is carried to the CNS. There, the impulse transfers to motor neurons that carry the impulse away from the CNS to a muscle and causes the muscle to contract. The peripheral nervous system consists of two separate divisions—the somatic nervous system and the autonomic nervous system.

The somatic nervous system

The **somatic nervous system** is made up of 12 pairs of cranial nerves from the brain, 31 pairs of spinal nerves from the spinal cord, and all of their branches. These nerves are actually bundles of neurons bound together by connective tissue, similar to a telephone cable. Some nerves contain only sensory neurons, and some contain only motor neurons, but most nerves contain both types.

The nerves of the somatic system relay information mainly between your skin, the CNS, and skeletal muscles. This pathway is voluntary and under conscious control, meaning that you can decide to move or not to move body parts under the control of this system.

Reflex arcs in the somatic system

Sometimes a stimulus will result in an automatic, unconscious response within the somatic system. When you touch something hot, you automatically jerk your hand away. Such an action is a **reflex,** an automatic response to a stimulus. Reflexes are the result of the shortest nerve pathways in the body called reflex arcs, as shown in *Figure 40.8.*

Figure 40.8

Reflex responses are carried out at the level of the spinal cord without assistance from the brain. The impulse travels directly to the spinal cord from the affected body part, crosses over to a small interneuron, and then moves to a motor neuron that transmits it to a muscle. The muscle contracts. The brain becomes aware of the reflex only after it occurs.

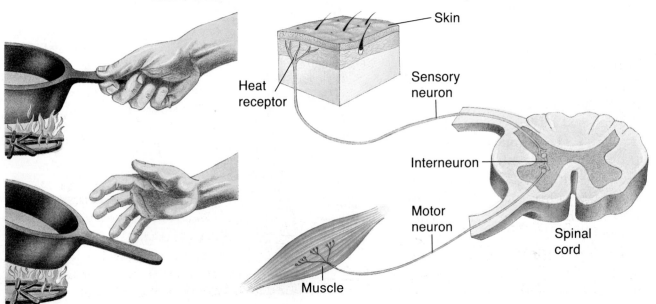

Skin

Heat receptor

Sensory neuron

Interneuron

Motor neuron

Spinal cord

Muscle

The autonomic nervous system

Imagine that you are spending the night alone in a creepy, old, deserted house. Suddenly, a creak comes from the attic. You think you hear footsteps. Your heart begins to pound. Your breathing becomes rapid. Your thoughts race wildly as you try to figure out what to do—stay and confront the unknown, or run fast to get out! The reactions that make up your response in this scary situation are being controlled by your autonomic nervous system.

The **autonomic nervous system** carries impulses from the CNS to internal organs. These impulses produce responses that are involuntary and not under conscious control. For example, glands in your stomach pour out hydrochloric acid without any voluntary decision on your part.

There are two divisions of the autonomic nervous system—the sympathetic nervous system and the parasympathetic nervous system. The **sympathetic nervous system** controls many internal functions during times of stress. It is responsible for the fight-or-flight response shown in *Figure 40.9*. Without a conscious decision on your part, the body sends glucose to muscles and

Figure 40.9

During a fight-or-flight response, the sympathetic nervous system causes the hormone epinephrine, commonly called adrenalin, to be released. This hormone causes an increase in the heart and breathing rates of an individual.

Figure 40.10

Understanding the organization of the human nervous system can be made easier by illustrating the relationships of the different divisions in a chart.

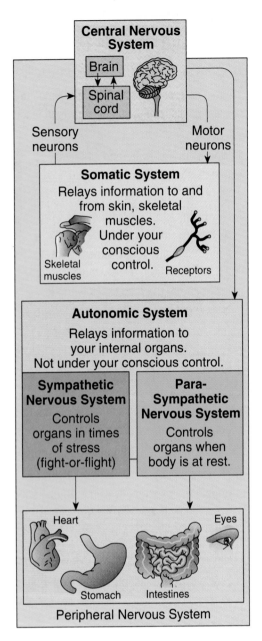

Central Nervous System

Brain

Spinal cord

Sensory neurons

Motor neurons

Somatic System
Relays information to and from skin, skeletal muscles. Under your conscious control.

Skeletal muscles

Receptors

Autonomic System
Relays information to your internal organs. Not under your conscious control.

Sympathetic Nervous System
Controls organs in times of stress (fight-or-flight).

Para-Sympathetic Nervous System
Controls organs when body is at rest.

Heart

Eyes

Stomach

Intestines

Peripheral Nervous System

nerve tissue where extra energy is required, your heart rate increases, and blood flow is directed to your arms and legs, away from the digestive system. Can you see that your sympathetic nervous system would be at work in the scary house or when you're suddenly faced with a life-threatening emergency?

The **parasympathetic nervous system** controls many internal functions of the body at rest. The neurons in this system produce effects opposite to those of the sympathetic nervous system. Your parasympathetic nervous system is in control when you are relaxing on a warm summer day after a picnic or listening to the soothing melody of a lullaby.

Both the sympathetic and parasympathetic systems send signals to the same internal organs. The resulting activity of the organ depends on the intensities of the opposing signals. Thus, the sympathetic system increases your heart rate while you're in the scary house, and the parasympathetic slows it down when you exit.

The different divisions or subsystems of your nervous system are presented in *Figure 40.10.* Each system plays a key role in communication and control within your body.

Section Review

Understanding Concepts

1. Summarize the charge distribution that exists inside and outside a resting neuron.
2. Outline the functions of the three major parts of the brain.
3. Contrast the functions of the two divisions of the autonomic nervous system.

Thinking Critically

4. Why is it nearly impossible to stop a reflex from taking place?

Skill Review

5. **Sequencing** Sequence the events as a nerve impulse moves from one neuron to another. For more help, refer to Organizing Information in the *Skill Handbook.*

Picture yourself in a park on a beautiful summer day. Stretching out on the grass, you look up and see white clouds float by against a background of blue. You feel grass tickling your toes and hear a breeze rustling through the trees. Sipping on a cup of lemonade, you enjoy its lemony smell and sweet, tangy taste. All these sensations are made possible by your senses: sight, touch, hearing, smell, and taste. Senses enable you to detect and respond to your environment.

Section Preview

Objectives

Explain how senses detect chemicals.

Explain how the eye detects light.

Key Terms

taste bud
retina
rods
cones
cochlea
semicircular canal

Senses that Detect Chemicals

How are you able to smell and taste the lemonade? The senses of smell and taste depend on receptors in your body that respond to chemical molecules. The receptors for smell are hairlike nerve endings located in the upper portion of your nose that respond to molecules in the air, as shown in *Figure 40.11*.

The senses of taste and smell are closely linked. Think about what your sense of taste is like when you have a cold and your nose is stuffed up. You can smell little, if anything. Because much of what you taste depends on your sense of smell, it too may be dulled. The sensation of taste occurs when chemicals dissolved in saliva make contact with sensory receptors on your tongue called **taste buds**.

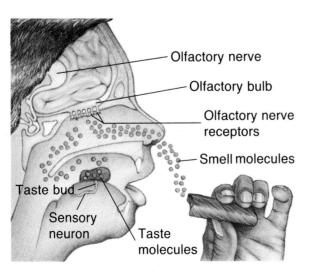

Olfactory nerve

Olfactory bulb

Olfactory nerve receptors

Smell molecules

Taste bud

Sensory neuron

Taste molecules

Figure 40.11

Chemicals acting on the hairlike nerve endings of your nose initiate impulses in the olfactory nerve. In the cerebrum, the signal is interpreted, and you notice a particular odor. Signals from your taste buds travel along a nerve to the medulla, and finally to the cerebrum. There the signal is interpreted, and you experience a particular taste.

Tastes that you experience can be divided into four basic sensations: sour, salty, bitter, and sweet. Certain regions of your tongue react more strongly to a particular taste. Bitter is most likely to be sensed at the back of the tongue, sour on the sides, and sweet and salty on the tip.

A Sense that Detects Light

How are you able to see? The sense of sight depends on receptors in your eyes that respond to light energy. The **retina,** found at the back of the eye, is a layer of nerve tissue made up of these sensory neurons. Follow the pathway of light to the retina in *Figure 40.12*.

The retina contains two types of cells—rods and cones. **Rods** are light receptors adapted for vision in dim light. They help you detect shape

and movement. **Cones** are light receptors adapted for sharp vision in bright light. They also help you detect color. At the back of the eye, retinal tissue comes together to form the optic nerve, which leads to the thalamus and then to the cerebrum.

Close one eye. Everything that you can see with one eye open is the visual field of that eye. The visual field of each eye can be divided into two parts: a lateral and a medial part. As shown in *Figure 40.12,* the lateral half of the visual field projects onto the medial part of the retina, and the medial half of the visual field projects onto the lateral portion of the retina.

The projections and nerve pathways are arranged so that images entering the eye from the right half of each visual field project to the left half of the brain, and vice versa. Each eye sees about two-thirds of the total visual field, so the visual fields of the eyes partially overlap. This overlapping allows your brain to judge the depth of your visual field.

Figure 40.12

A cross section through the human eye shows where light goes as it enters the eye.

▼ Light enters the cornea, passes through the aqueous humor, and through the pupil. The colored iris muscle regulates the amount of light entering the eye. Light then passes through the lens and the jellylike vitreous humor. The lens focuses the light on the back of the eye, where it strikes the retina.

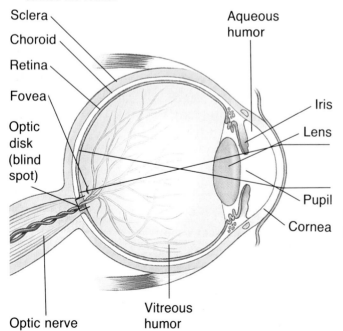

Sclera
Choroid
Retina
Fovea
Optic disk (blind spot)
Optic nerve
Aqueous humor
Iris
Lens
Pupil
Cornea
Vitreous humor

▼ The left halves of the retinas in each eye are connected to the left side of the visual cortex in the cerebrum. The right halves of both eyes send signals to the right side of the brain. Images are interpreted in the cerebrum.

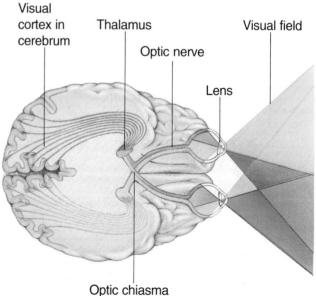

Visual cortex in cerebrum
Thalamus
Visual field
Optic nerve
Lens
Optic chiasma

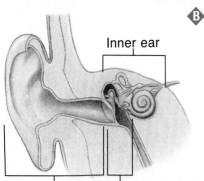

Inner ear

Outer ear   **Middle ear**

Ⓐ Sound waves enter the outer ear and travel down to the end of the ear canal, where they strike a membrane called the eardrum and cause it to vibrate.

Figure 40.13

The internal structure of the human ear is divided into three areas: the outer ear, middle ear, and inner ear. Follow the pathway sound waves take as they move through your ears.

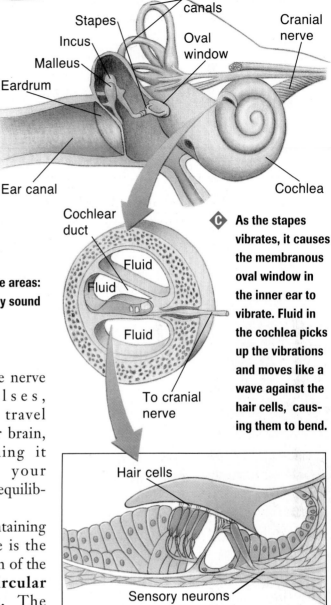

Ⓑ The vibrations then pass to three bones in the middle ear— the malleus, the incus, and the stapes.

Semicircular canals

Stapes

Incus

Malleus

Eardrum

Ear canal

Oval window

Cranial nerve

Cochlea

Cochlear duct

Fluid

Fluid

Fluid

To cranial nerve

Ⓒ As the stapes vibrates, it causes the membranous oval window in the inner ear to vibrate. Fluid in the cochlea picks up the vibrations and moves like a wave against the hair cells, causing them to bend.

Hair cells

Sensory neurons

Ⓓ The movement of the hairs initiates the movement of impulses along the auditory nerve. Impulses travel along this nerve to the sides of the cerebrum, and you hear the sound.

Senses that Detect Mechanical Stimulation

How are you able to hear the leaves rustle and feel the grass as you relax in the park? These senses depend on receptors that respond to mechanical stimulation.

Your sense of hearing

Every sound causes the air around the source of the sound to vibrate. These vibrations are known as sound waves. When the sound waves enter your ear, they set up a response that stimulates a nerve impulse. This impulse then moves along the auditory nerve to the brain. It is in the **cochlea,** a fluid-filled, snail-shaped structure in the inner ear, that the mechanical stimulation of sound is converted into a nerve impulse. Trace the pathway of sound waves to the cochlea in *Figure 40.13.*

Your sense of balance

The ear also converts the physical signal of the position of your head into the nerve impulses, which travel to your brain, informing it about your body's equilibrium.

Maintaining balance is the function of the **semicircular canals.** The semicircular canals also are filled with a thick fluid and lined with hair cells. When you tilt your head, the fluid and hairs move. The mechanical movements of the hairs stimulate the neurons to carry an impulse to the cerebellum and the cerebrum. There, impulses from motor neurons stimulate muscles in the head and neck to readjust the position of your head.

Meet Dr. Benjamin Carson, Neurosurgeon

"Sit back and say, What have I always been good at. Ask other people what they see that you're good at. This is thinking big," says Dr. Benjamin Carson, chief of pediatric neurosurgery at Johns Hopkins Hospital in Baltimore, Maryland. He looked at his list of things he was good at and decided he could be a neurosurgeon.

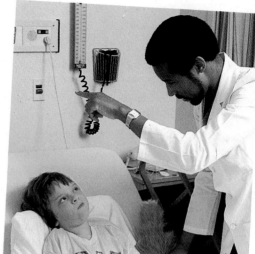

In the interview that follows, Dr. Carson talks about both his work as a brain surgeon and his efforts to provide a much-needed role model for young people.

On the Job

Q Dr. Carson, what do you consider the most difficult brain surgery you've done?

A Believe it or not, even though news magazines in 1987 called it the most complex operation in the history of the world, it wasn't the separating of the Siamese twins who were joined at the head. I find that operating on a patient's brain stem is much more taxing—especially when you do it for 16 hours at a stretch. It's like trying to defuse a time bomb.

Q Those news magazines you mentioned also report that in spite of your extremely full schedule, you make time to talk to young people. Why is that so important to you?

A I believe kids need appropriate role models. Young people often look up to sports stars, but the chances of succeeding in that field are incredibly tiny. I talk not so much about kids becoming brain surgeons as much about utilizing their intellectual ability and the things in their lives that they can control. For instance, I tell kids that if they put in the appropriate amount of work to be an engineer, that's what they'll be. The correlations are completely different with sports stars because only one basketball player in a million can be an NBA star.

Q What is one of the most frequent questions kids ask you about your profession?

A They often want to know if I'm frightened when operating on the brain. I tell them it's just as if you were a tightrope walker. If you're experienced, walking the tightrope isn't all that frightening, but if you haven't done it before, it will scare you to death. Brain surgeons must have a very healthy respect for the brain because they are dealing with the substance that makes people who they are. Brains don't tolerate mistakes of the knife very well! I need to be extremely aware of what I'm doing and of the consequences.

Early Influences

Q How did you get interested in the medical field?

A From the time I was eight years old, I wanted to be a missionary doctor. The stories I heard in church made missionary doctors seem like the noblest people in the world. At great personal sacrifice, they brought not only physical but spiritual healing to people. But then, having grown up in poverty, I decided at 13 that I'd rather be rich and changed my goal to becoming a psychiatrist. I didn't know any, but the ones I saw on TV had mansions and sports cars. It looked as though all they had to do was sit around and talk to people. I majored in psychology at college. When I got to medical school, I had to ask myself, Well, what do I want to do now? I made a list of the things I had going for me: good eye-hand coordination, aptitude for thinking in three dimensions, ability to be very careful. I put all those together and said, What would that make me? A neurosurgeon, of course!

Q Were there certain people who had a great influence on you?

A My mother had only an elementary education, but she always told me "Mr. I-Can't is dead." She would never accept an excuse for not being able to do something. She worked several jobs at a time—scrubbing floors, washing windows, whatever she could find. When I complained about not having expensive clothes like my friends, she invited me to take the money she had earned, pay the bills, and use what was left to buy all the fancy clothes I wanted. I tried and soon discovered that my mother was a financial genius in keeping us housed, fed, and clothed.

Personal Insights

Q Most people have problems in high school. Did you?

A Of course. My grades plummeted for awhile because I gave in to peer pressure. Then I began to think that the here and now was not as important as the future. I decided that I could make my future better and also make other people's futures better. As a young person, I knew that time was on my side.

Q What advice do you have for students about careers in the medical field?

A I tell them that medicine is teamwork and that people in careers like nursing, medical technology, and physical therapy are crucial members of surgical teams. I also tell them that science and math, which you use all the time in medicine, are logical, not difficult. The major key to science and math is taking things one step at a time.

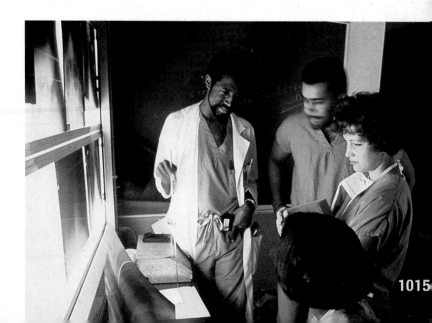

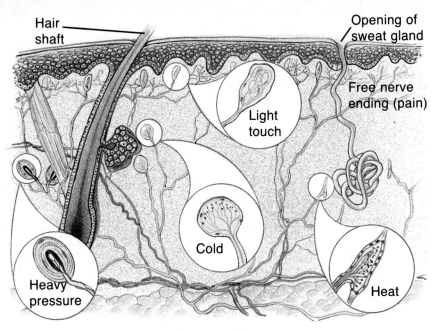

Hair shaft

Opening of sweat gland

Light touch

Free nerve ending (pain)

Cold

Heavy pressure

Heat

Figure 40.14

Many kinds of receptors are located throughout the dermis of your skin. When stimulated, some receptors detect gentle touches; others respond to heavy pressure. The sensations of pain, heat, and cold are sensed by other kinds of receptors.

Have you ever felt dizzy after a carnival ride that spun you around and around? When you stopped, your brain was receiving conflicting signals. The fluid in your semicircular canals was still moving, sending impulses to your brain indicating that you were still spinning.

Your sense of touch

Like the ear, your skin also responds to mechanical stimulation with receptors that convert the stimulation into a nerve impulse. Receptors in the dermis of the skin respond to changes in temperature, touch, pressure, and pain. It is with the help

of these receptors, shown in *Figure 40.14*, that your body meets and responds to its environment.

Although some receptors are found all over your body, those responsible for responding to particular stimuli are usually concentrated within certain areas of your body. For example, touch receptors are numerous on your fingertips, eyelids, lips, the tip of your tongue, and the palms of your hands. When these receptors are stimulated, you perceive sensations of fine or light touch.

Pressure receptors are found inside your joints, in muscle tissue, and in certain organs. They are also abundant on the skin of your palms and fingers and on the soles of your feet. When these receptors are stimulated, you perceive heavy pressure.

Free nerve endings extend into the lower layers of the epidermis. These are most important as pain receptors. Two receptors respond to temperature. Heat receptors are found deep in the dermis, while cold receptors are found closer to the surface of your skin. Can you infer why it would be important for your body to perceive cold skin temperature?

The Effects of Drugs on the Body

*N erves frazzled? Feeling
tense? For a person
trying to quit a smoking
habit, the desire for a cigarette
is almost always overpowering.
Without realizing that it
happened, the lighting of an
occasional cigarette has grown
into a pack-a-day habit. The body
now cries out for a nicotine fix.*

Section Preview

Objectives

Summarize the medic-
inal uses of drugs.

Explain how addictive
drugs affect the ner-
vous system.

Key Terms

drug
narcotic
stimulant
depressant
addiction
tolerance
withdrawal
hallucinogen

*Nicotine affects the body in dozens of ways—most of them harmful. In fact, nico-
tine is such a dangerous substance that a mere thimbleful could kill an adult if
taken all at once. Why, then, do people knowingly take nicotine into their bodies?*

Drugs Act on the Body

You probably hear the word *drug*
used often, maybe even every day. A
drug is a chemical that is capable of
reacting with an animal's body functions. Most drugs combine with some
molecular structure on the surface of
or within a cell. This molecular

structure is a receptor site. In fact,
the receptors with which drugs combine are assumed to be the same as
the receptors for neurotransmitters
of the nervous system or hormones
of the endocrine system. How some
drugs act to modify the transmission
of neurotransmitters at a synapse is
illustrated in *Figure 40.15*.

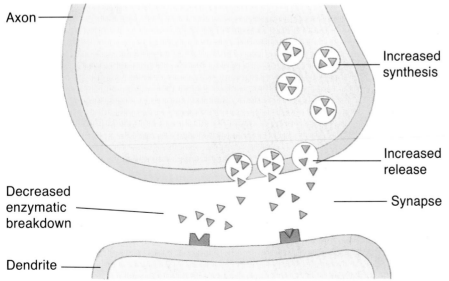

Figure 40.15

**Some drugs can increase
the amount of a neurotransmitter at the synapse by
increasing the rate at which
the transmitter is synthesized and released, or by
slowing the breakdown of
the transmitter by enzymes.**

The Medicinal Uses of Drugs

A medicine is a drug that, when taken into the body, helps prevent, cure, or relieve a medical problem. Medicines that are used to prevent disease are called vaccines, and those that are used to cure disease may be antibiotics or antiviral drugs. You'll study these medicines in Chapter 42. Some of the many kinds of medicines that are used to relieve medical problems are discussed below.

Relieving pain

Headache, muscle ache, cramps— these are some pain sensations that a person sometimes feels. You just studied how pain receptors in your body send signals of pain to your brain. Medicines that relieve pain affect these receptors. Pain relievers that do not cause a loss of consciousness are called analgesics. If you've ever taken aspirin for a headache, you've used such an analgesic. Medicines that relieve pain but also cause sleep because of their effect on the central nervous system are called **narcotics.** Many narcotics are made from the opium poppy flower, which is shown in *Figure 40.16.*

Treating circulatory problems

Many drugs have been developed to treat heart and circulatory problems such as high blood pressure. These medicines are called cardiovascular drugs. In addition to treating high blood pressure, cardiovascular drugs may be used to normalize an irregular heartbeat, increase the heart's pumping capacity, or enlarge small blood vessels.

Treating nervous disorders

Several kinds of medicines are used to help relieve symptoms of problems with a person's nervous system. Among these medicines are stimulants and depressants.

Drugs that increase the activity of the central and sympathetic nervous systems are **stimulants.** Amphetamines are synthetic stimulants that increase the output of CNS neurotransmitters. Amphetamines are seldom prescribed because they can lead to dependence, which you'll read about in detail later. However, because they increase wakefulness and alertness, amphetamines are sometimes used to treat patients with sleep disorders.

Drugs that lower, or depress, the activity of the nervous system are **depressants.** The primary medicinal uses of depressants are to encourage calmness and produce sleep. These drugs are commonly called sedatives. For some people, the symptoms of anxiety become so extreme that they interfere with the person's ability to function effectively. By slowing down the activities of the CNS, a depressant can temporarily relieve some of this anxiety.

cardiovascular:
kardia (GK) the heart
vasculum (L) small vessel
Cardiovascular drugs treat problems associated with blood vessels of the heart.

Figure 40.16

The sticky sap of the fruit of an opium poppy is used to make drugs called opiates. Opiates can be useful in medical therapy because only these narcotics are able to relieve severe pain.

The Misuse and Abuse of Drugs

With all drugs, there are responsibilities as well as risks. Drug misuse is using a medicine in a way that was not intended. For example, giving your prescription to someone else, not following the prescribed dosage by taking too much or too little, and mixing medicines are all instances of drug misuse.

On the other hand, drug abuse is the inappropriate self-administration of a drug for nonmedical purposes. Drug abuse may involve use of an illegal drug, such as cocaine; use of an illegally obtained medicine, such as a depressant; or excessive use of a legal drug, such as alcohol or nicotine. Abused drugs have powerful effects on the nervous system as well as other systems of the body, *Figure 40.17.* In addition, abused drugs may lead to tolerance and addiction.

Addiction to drugs

When a person believes he or she needs a drug in order to feel good or function normally, that person is psychologically dependent on the drug. When a person's body actually develops a chemical need for the drug in order to function normally, the person is physiologically dependent. Psychological and physiological dependence are the same as **addiction.**

MiniLab

What information can be found on the label of an over-the-counter drug?

One common misuse of drugs is not following the instructions that accompany a medicine. Over-the-counter medicines have the potential of being harmful, even fatal, if not used as directed. The information on the label is required by the Food and Drug Administration to help the consumer use the medicine properly and safely.

Procedure

1. The photograph below shows a label from an over-the-counter drug. Read it carefully.

2. Make a data table like the one shown, and complete the table using the information on the label.

People with these conditions should avoid this drug	Possible side effects	This drug should not be taken with these medicines	Symptoms this drug will relieve	Correct dosage

Information from a Drug Label

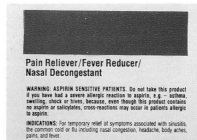

Pain Reliever/Fever Reducer/Nasal Decongestant

WARNING: ASPIRIN SENSITIVE PATIENTS. Do not take this product if you have had a severe allergic reaction to aspirin, e.g. – asthma, swelling, shock or hives, because, even though this product contains no aspirin or salicylates, cross-reactions may occur in patients allergic to aspirin.

INDICATIONS: For temporary relief of symptoms associated with sinusitis, the common cold or flu including nasal congestion, headache, body aches, pains, and fever.

DIRECTIONS: *Adults and children 12 years of age and older:* Take 1 caplet every 4 to 6 hours while symptoms persist. If symptoms do not respond to 1 caplet, 2 caplets may be used but do not exceed 6 caplets in 24 hours, unless directed by a doctor. The smallest effective dose should be used. Take with food or milk if occasional and mild heartburn, upset stomach, or stomach pain occurs with use. Consult a doctor if these symptoms are more than mild or if they persist. *Children:* Do not give this product to children under 12 years of age except under the advice and supervision of a doctor.

WARNINGS: Do not take for colds for more than 7 days or for fever for more than 3 days unless directed by a doctor. If the cold or fever persists or gets worse or if new symptoms occur, consult a doctor. These could be signs

of serious illness. As with aspirin and acetaminophen, if you have any condition which requires you to take prescription drugs or if you have had any problems or serious side effects from taking any non-prescription pain reliever, do not take this product without first discussing it with your doctor. IF YOU EXPERIENCE ANY SYMPTOMS WHICH ARE UNUSUAL OR SEEM UNRELATED TO THE CONDITION FOR WHICH YOU TOOK THIS PRODUCT, CONSULT A DOCTOR BEFORE TAKING ANY MORE OF IT. If you are under a doctor's care for any serious condition, consult a doctor before taking this product. Do not exceed recommended dosage because at higher doses nervousness, dizziness or sleeplessness may occur. Do not take this product if you have high blood pressure, heart disease, diabetes, thyroid disease or difficulty in urination due to enlargement of the prostate gland, except under the advice and supervision of a doctor.

DRUG INTERACTION PRECAUTION: Do not take this product if you are presently taking a prescription drug for high blood pressure or depression without first consulting your doctor. Do not combine this product with other non-prescription pain relievers. Do not combine this product with any other ibuprofen-containing product.

As with any drug, if you are pregnant or nursing a baby, seek the advice of a health professional before using this product. IT IS ESPECIALLY IMPORTANT NOT TO USE THIS PRODUCT DURING THE LAST 3 MONTHS OF PREGNANCY UNLESS SPECIFICALLY DIRECTED TO DO SO BY A DOCTOR BECAUSE IT MAY CAUSE PROBLEMS IN THE UNBORN CHILD OR COMPLICATIONS DURING DELIVERY. Keep this and all drugs out of the reach of children. In case of accidental overdose, seek professional assistance or contact a poison control center immediately.

Store at room temperature. Avoid excessive heat above 40°C (104°F).

Active Ingredients: Each caplet contains ibuprofen 200 mg and pseudo-ephedrine HCL 30 mg.

Inactive Ingredients: Cellulose, Corn Starch, Glyceryl Triacetate, Hydroxypropyl Methylcellulose, Silicon Dioxide, Sodium Lauryl Sulfate, Sodium Starch Glycolate, Stearic Acid, Titanium Dioxide, Red #40 Aluminum Lake.

Analysis

1. What is a side effect? What side effects are caused by this drug?

2. Why should a person never take more than the recommended dosage of an over-the-counter drug?

3. How are over-the-counter drugs different from prescription drugs?

Figure 40.17

The use of anabolic steroids without the careful monitoring by and a prescription from a physician is illegal. In spite of the very dangerous side effects, including cardiovascular disease, kidney damage, and cancer, some male adolescents and professional athletes use the drugs to increase muscle mass and body strength.

Depending on their chemical composition, drugs affect different parts of your body. Stimulants and depressants are drugs that

What drugs affect the heart rate of *Daphnia?*

affect the central nervous system. These drugs also affect the autonomic nervous system. Stimulants tend to increase the activity of the "fight-or-flight" sympathetic nervous system. They cause an increase in your breathing rate and in your heart rate. Depressants, on the other hand, increase the activity of the parasympathetic nervous system and decrease your breathing and heart rates.

PREPARATION

Problem
What legally available drugs are stimulants to the heart? What legal drugs are depressants? Because these drugs are legally available, are they less dangerous?

Hypotheses
From your knowledge and reading of the chapter, which drugs are stimulants? Which drugs are depressants? How will they affect the heart rate in *Daphnia?* Make a hypothesis concerning how each of the drugs listed will affect heart rate.

Objectives
In this Biolab, you will:
- **Measure** the resting heart rate in *Daphnia*.
- **Compare** the resting heart rate with the heart rate when a drug is applied.

Possible Materials
dilute solutions of coffee, tea, cola, ethyl alcohol, tobacco, and cough medicine (destromethorphen hydrobromide)
Daphnia culture
microscope
dropper
microscope slide
aged tap water

Safety Precautions
Do not drink any of the beverages used during this lab.

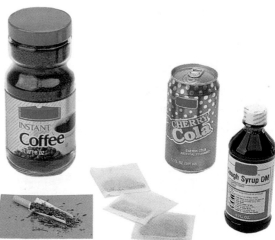

1. Using a dropper, place a single *Daphnia* crustacean on a slide.

2. Observe on low power and find the heart.

3. Design an experiment to measure the effect on heart rate of four of the drugs listed in the Possible Materials list.

4. Design and construct a data table for recording your data.

Check the Plan

1. Be sure to consider what you will use as a control measurement.

2. Plan to add two drops of a drug directly to the slide.

3. When finished testing one drug, you will need to flush the used *Daphnia* with the solution into a beaker of aged tap water provided by your teacher. Plan to use a new *Daphnia* for each drug.

4. ***Make sure your teacher has approved your experimental plan before you proceed further.***

5. Carry out your experiment.

Magnification: 40×

Daphnia

1. **Making Inferences** Which drugs are stimulants? Depressants?

2. **Checking Your Hypothesis** Compare your predicted results with the experimental data.

Explain whether your data support your hypotheses.

3. **Drawing Conclusions** How do the drugs affect the rate of the heartbeat of an animal?

Going **Further**

Changing Variables Many other over-the-counter drugs are available. You may wish to test their effect on heart rate in *Daphnia*.

40.3   The Effects of Drugs on the Body   **1021**

A Plant that Changed the World

Agriculture is the basis of our civilization. Until humans learned to farm, we could not progress past the point of hunting and gathering. Thus, it is ironic that one agricultural product that changed our world is neither an effective source of nutrition nor particularly good for us.

The noxious weede   This is how King James I described the tobacco brought back from his American colony of Virginia in 1604. A militant nonsmoker, he tried to pass laws prohibiting its use. However, sailors and merchants saw great profit in this native American import. Descendants of King James soon accepted its use because tobacco taxes became a major source of government revenue. First promoted as a medicine, tobacco rapidly gained wide use because of the addictive properties of nicotine. As more people around the world learned of the profit to be made from tobacco, it became widely cultivated.

The tobacco trade   The economy of colonial Virginia was built on the tobacco trade. Thomas Jefferson and George Washington made their fortunes as tobacco planters. After harvesting, the tobacco was dried and then packed in barrels for transport to England. There the tobacco was processed and packaged for sale throughout the world. Thus, this native American plant became the basis of some of the world's great fortunes, triggered the building of great merchant fleets, and opened trade with the far corners of the world.

CONNECTION TO Biology

All over the world, land is used to grow crops of high commercial, but limited nutritional, value. With the malnutrition and starvation in the world, discuss how this imbalance could be corrected.

Tolerance and withdrawal

Addiction becomes evident if a drug user experiences tolerance or withdrawal. **Tolerance** occurs when a person needs larger and/or more frequent doses of a drug to achieve the same effect. The drug increases are needed because the body becomes more efficient at eliminating the drug from its systems. Nerve cells may also become less responsive to the dose. **Withdrawal** occurs when the person stops taking the drug and actually becomes ill, both psychologically and physiologically.

Classes of Drugs

Each class of drug produces a special effect on the body by working on body systems in its own way. Each class of drug produces its own symptoms of withdrawal, as well.

Stimulants: Cocaine, amphetamines, caffeine, and nicotine

You already know that stimulants increase the activity of the central nervous system and the sympathetic nervous system. The CNS stimulation can be seen in behaviors that range from the mild elevation of alertness to increased nervousness, anxiety, and even convulsions.

Cocaine stimulates the CNS by working on the part of the inner brain that governs emotions and basic drives such as hunger and thirst. It affects this reward center by interfering with the neurotransmitters, such as dopamine and norepinephrine. Cocaine causes levels of the neurotransmitters in the brain to increase, which results in a false

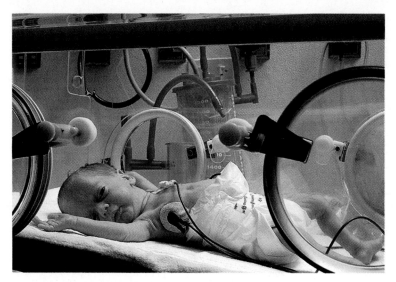

message being sent to the reward center indicating that a basic drive has been satisfied. The user quickly feels a euphoric high called a rush. This sense of intense pleasure and satisfaction cannot be maintained, and the effects of the drug change. Physical hyperactivity follows, and the user is unable to sit still. Anxiety and depression set in.

Cocaine also causes the sympathetic system to disrupt the body's circulatory system. It initially causes a slowing of the heart rate followed by a great increase in heart rate and a narrowing of blood vessels, called vasoconstriction. The result is high blood pressure. Heavy use of this drug causes weight loss, a compromise of the immune system, and eventually heart abnormalities. Eventually, death may occur. Cocaine affects more than just the people who use it. As *Figure 40.18* shows, babies are sometimes born already addicted to this drug.

As you've already learned, amphetamines are stimulants that increase CNS neurotransmitters. Like cocaine, amphetamines cause vasoconstriction, a racing heart, and increased blood pressure. Other adverse side effects of amphetamine abuse include irregular heartbeat, chest pain, paranoia, hallucinations, convulsions, and coma.

Not all stimulants are illegal or even difficult to get. One stimulant in particular is as close as the nearest coffee maker or soft-drink machine. Caffeine—a substance found in coffee, cola-flavored drinks, cocoa, and tea—is a CNS stimulant. Its effects include increased alertness and some mood elevation. Caffeine also causes an increase in heart rate and urine production, *Figure 40.19*.

Figure 40.19

The condition called tachycardia, when the heart beats more than 100 times per minute, can be triggered by caffeine.

Figure 40.20

Fruits such as grapes, berries, and apples may be covered with a coating of wild yeast. The yeast is an important component of the alcoholic fermentation process that is used to produce wine.

While the moderate use of caffeine does not appear to cause problems in most people, there is little question that such use is habit forming. Chronic coffee drinkers who suddenly abstain often experience the withdrawal symptoms of headaches and a general feeling of fatigue.

Nicotine, a substance found in tobacco, is a stimulant. By increasing the release of the hormone epinephrine, nicotine increases heart rate, blood pressure, breathing rate, and stomach acid secretion. While nicotine is the substance in tobacco that leads to addiction, it is only one of about 3000 known chemicals found in cigarettes, many of which are also harmful. Smoking cigarettes is legal for adults. But our ever-increasing knowledge of its effects on the bodies of nonsmokers as well as smokers has made tobacco use a much less acceptable drug habit than in the past.

Depressants: Alcohol and barbiturates

As you already know, depressants slow down the activities of the CNS. All CNS depressants relieve anxiety, but most also produce noticeable sedation.

One of the most widely abused drugs in the world today is alcohol.

Easily produced from various grains and fruits, one of which is shown in *Figure 40.20,* this depressant is distributed throughout the body by the blood. In organs with high blood flow, such as the brain, liver, lungs, and kidneys, it is rapidly distributed.

Unlike other drugs, alcohol probably acts on the brain by dissolving through the membranes of neurons rather than acting on specific receptors. Once inside a neuron, alcohol disrupts important cellular functions. It appears to block the movement of sodium and calcium ions that are responsible for transmitting impulses and releasing neurotransmitters.

The degree to which alcohol disrupts the normal body functions is directly proportional to the quantity consumed. Small amounts can result in disorientation, loss of muscle coordination, and diminished judgment.

Tolerance to the effects of alcohol develops as a result of heavy consumption. A number of organs are adversely affected by chronic alcohol use. For example, cirrhosis, a hardening of the tissues of the liver, is common. Addiction to alcohol—alcoholism—can also occur. Addiction can result in the destruction of nerve cells and brain damage. Research is ongoing to understand the

genetics of alcoholism. Most researchers agree that addiction has some hereditary basis, but it has not yet been linked to a specific gene.

Barbiturates are sedatives and antianxiety drugs. A barbiturate abuser generally is sluggish and has difficulty thinking, as well as paying attention. Chronic use results in both tolerance and addiction. When barbiturates are used in excess, the user's respiratory and circulatory systems are also depressed.

Narcotics: Opiates

As mentioned earlier, narcotics are opiates used to relieve severe pain. They act directly on the brain. In fact, narcotic receptors have been identified in the brain. The most abused narcotic in the United States is heroin. It depresses the CNS, slows breathing, and lowers the heart rate. Tolerance develops quickly, and withdrawal from heroin is painful.

Hallucinogens: Natural and synthetic

Natural hallucinogens have been known and used since ancient times, but the abuse of hallucinogenic drugs did not become widespread in the United States until the 1960s when new synthetic types became widely available.

Hallucinogens stimulate the CNS, altering moods, thoughts, and sensory perceptions. Quite simply, the user sees, hears, feels, tastes, or smells something that is not there.

Drugs and Crime

Drugs and crime have long been a part of the 20th-century American landscape. In 1900, some 1 million Americans were addicted to cocaine, opium, and other narcotics. Beginning with the 1909 ban on opium importation and prohibition in the 1920s, the government has tried to control the flow of addictive drugs into the country and restrict their use. By the time organized crime took over the international heroin industry in the 1960s, drug trafficking had become a multibillion-dollar-a-year business.

The drug-crime link Most addicts resort to criminal activity to pay for their drug habits. One recent study showed that more than 75 percent of crimes committed in major cities are done under the influence of drugs. Such drug-related crimes cost the nation about $7 billion each year.

Crack and the rise of urban gangs The most compelling example of the link between drugs and crime is the rise of violent drug gangs in the cities. A new form of organized crime, these gangs do open battle on the city streets for control of the trade in heroin, cocaine, and crack. Gang members are responsible for thousands of crimes, including the murder of innocent bystanders and police officers.

Swamped courts and repeat offenders The strategies to end domestic drug-related crime have not proved effective. Beefed-up police squads and crime-watch groups have succeeded in more drug arrests and convictions, but overcrowded prisons result in reduced sentences for most criminals. State court systems are swamped with drug cases, most of which involve repeat offenders. Most experts insist that only a drop in the demand for illegal drugs can make a long-lasting change.

Thinking Critically

The legalization of illicit drugs has been debated as one answer to the drug-crime link. Do you think making drugs legal would be a solution?

Mushrooms of the genus *Psilocybe* contain the CNS hallucinogen psilocybin. These mushrooms are known as sacred mushrooms to certain Native Americans, who use them in religious rites.

Ergot, a mold disease of cereal grains that contains a hallucinogen, killed thousands of people in the Middle Ages when they ate flour made from ergotized rye.

Figure 40.21

Some hallucinogens are found in nature.

Hallucinogens also increase heart rate, blood pressure, respiratory rate, and body temperature, and sometimes cause sweating, salivation, nausea, and vomiting. After large enough doses, convulsions may occur.

Unlike the hallucinogens shown in *Figure 40.21*, LSD is a synthetic drug. The mechanism by which LSD produces hallucinations is being debated but may be due to the repression of a CNS neurotransmitter.

Connecting Ideas

The nervous system is an electrical communication and control system that senses the environment, integrates the information, and issues a command for action. Motor neurons help carry out the commands of the central nervous system by carrying an impulse to a muscle or gland. Drugs may interfere with the natural sensation-integration-command process.

How does such a complex system develop? What effect does it have on the development of a new human? Answers can be found in an understanding of human reproduction and growth.

Section Review

Understanding Concepts
1. What are the ways in which a drug can be used to treat a cardiovascular problem?
2. What is the difference between an analgesic and a narcotic?
3. What effects does nicotine have on the CNS?

Thinking Critically
4. Suggest why a physician and not a pharmacist is able to write a prescription.

Skill Review
5. **Comparing and Contrasting** Distinguish between a stimulant and a depressant, and compare their effects on the body. For more help, refer to Thinking Critically in the *Skill Handbook*.

Living with Disabilities: Access to Public Places and Jobs

Business Complies with the Americans with Disabilities Act

NEWS AND VIEWS   OCTOBER 12, 1992

Citibank is equipping its automated-teller machines with high-resolution graphics and easy-to-read symbols for people with disabilities, thus conforming to the Americans with Disabilities Act.

Perhaps you've experienced an injury that disabled you in some way—a temporary loss of hearing, even a broken arm. Simple activities like holding a pencil or answering the phone suddenly become formidable challenges. You'd probably be frustrated with how inaccessible the world had suddenly become.

Gaining access!   All persons with a disability share issues of accessibility to public buildings and transportation, to jobs and schools, to adequate health care, and particularly to lives of dignity and respect.

Different Viewpoints

Discrimination and the ADA   Discrimination of persons with disabilities is nothing new. To eliminate job discrimination, Congress passed the Americans with Disabilities Act (ADA) in 1990. The ADA requires businesses to remove barriers to employment. Compliance includes providing wheelchair access into office buildings, sign-language interpreters, and braille keyboards for computers. The law provides tax incentives to businesses that comply. However, it also provides

a loophole that disqualifies the requirement if undue hardship would result. Many small businesses, for example, claim that the costs of conversion would put them out of business.

Stephen Hawking   One example of the power of access is British physicist Stephen Hawking. A lecturer at Cambridge University and a best-selling author, Dr. Hawking is also severely physically impaired. In spite of his disabilities, he is considered one of the finest scientific minds in the world today.

INVESTIGATING   the Issue

1. **Applying Concepts** Ask your teacher to plan a field trip to a public place and allow at least one class member to attend in a wheelchair. What problems did your classmate encounter?
2. **Writing About Biology** Write a paper on the accomplishments of a well-known person who lived/lives with a disability.
3. **Interview** Talk to a student, friend, or relative with a disability. Is access an issue in the daily life of that person? Has he or she felt discriminated against because of the disability?

40.1 The Nervous System
- The neuron is the basic unit of structure in the nervous system. Impulses move along a neuron in a wave of changing charges.
- The central nervous system consists of the brain and spinal cord.
- The peripheral nervous system brings messages to and from the central nervous system. It consists of the somatic and autonomic systems.

40.2 The Senses
- Taste and smell are senses that respond to chemical stimulation.
- Sight is a sense that responds to light stimulation.
- Hearing and touch are senses that respond to mechanical stimulation. Receptors in the semicircular canals of the ears produce a sense of balance.

40.3 The Effects of Drugs on the Body
- Drugs act on receptor sites on neurons of the body.
- Some medicinal uses of drugs include relieving pain and treating cardiovascular problems and nervous disorders.

- The misuse of drugs involves using a medicine in a way that was not intended. Drug abuse involves using a drug for a nonmedical purpose.
- Harmful, commonly abused drugs include stimulants, depressants, narcotics, and hallucinogens.

Key Terms
Write a sentence that shows your understanding of each of the following terms.

addiction	parasympathetic
autonomic	nervous system
nervous system	peripheral nervous
axon	system
central nervous	reflex
system	retina
cerebellum	rods
cerebrum	semicircular canal
cochlea	somatic nervous
cones	system
dendrite	stimulant
depressant	sympathetic nervous
drug	system
hallucinogen	synapse
medulla oblongata	taste bud
narcotic	tolerance
neuron	withdrawal
neurotransmitter	

Understanding Concepts

1. How is the surface of the cerebrum related to intellect?
2. What is the state of charges on the membrane of a resting neuron?
3. Contrast the functions of the two divisions of the autonomic nervous system.
4. What are the functions of the fluid and hair cells in the cochlea?
5. How are the senses of smell and taste related?

6. Distinguish between psychological and physical addiction.
7. How does cocaine affect the brain?

Relating Concepts

8. Make a concept map that relates the following terms and phrases. Supply the appropriate linking words for your map.

   neuron, neurotransmitter, axon, dendrite, synapse, depolarization, central nervous system, peripheral nervous system

Using a Graph

9. An electrode that measures charges was placed inside a neuron. The neuron was stimulated, and the change in charges was measured as an impulse passed by. Explain the shape of the curve that was obtained in terms of changes in the concentrations of ions inside and outside the cell.

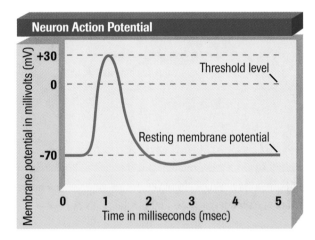

Applying Concepts

10. What might cause seasickness?
11. A chemical stops the breakdown of enzymes that remove neurotransmitters at a synapse. How would this affect a person's body?
12. Describe the route an impulse travels from a stimulus to a skeletal muscle contraction.
13. Tetrodotoxin, a chemical produced by the puffer fish, blocks sodium channels. How does the toxin help fish capture prey?

Biology & Society

14. What are some provisions a business might make for a hearing-impaired customer?

Thinking Critically

Drawing a Conclusion

15. Some pain impulses travel along myelinated neurons. The speed of these impulses is higher than other pain impulses, as shown in the table below. Explain why it is important for a person's survival that these pain sensations have a faster impulse speed.

Sensation	Speed of Impulse	Myelinated
Aching pain	1 m/s	no
Pricking pain	18 m/s	yes
Deep pressure	30 m/s	yes

Making Predictions

16. **Biolab** Suppose you applied both ethyl alcohol and coffee to the *Daphnia* heart. What would you predict the overall effect on heart rate to be compared with that of ethyl alcohol alone?

Making Inferences

17. **Minilab** A medicine has this precaution: "Avoid driving a motor vehicle while taking this medicine. This medication may cause drowsiness." What type of drug does this medicine contain?

Connecting to Themes

18. **Energy** Explain why the nervous system is a high energy user compared with other systems in the body.
19. **Systems and Interactions** Why could you consider the nervous system a part of each of the other body systems?
20. **Homeostasis** Explain how alcohol disturbs the homeostasis inside a neuron.

The Brain

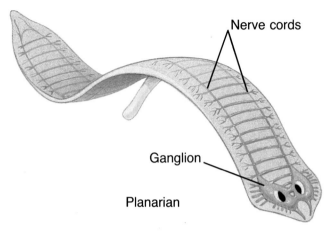

Nerve cords

Ganglion

Planarian

As animals have changed and evolved over hundreds of millions of years, what has really set humans apart from other animals is the evolution of the brain. From a bundle of nerve cells to a staggeringly complex cerebral cortex, this evolution has progressed from organisms that react instantly to changes in the environment or immediate danger to humans capable of thinking and reasoning.

Evolution of the brain

The simplest brain Flatworms are the simplest animals that have an identifiable brain. A planarian, for example, has a mass of nerve tissue called a ganglion that lies beneath the eye-spots.

Millions of years later Jumping ahead by millions of years to when the vertebrates emerged, the five brains shown here illustrate how evolution has transformed a simple ganglion to a complex brain. As the brain evolved, areas that control senses, instinct, and coordination became predominant.

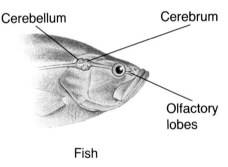

Cerebellum

Cerebrum

Olfactory lobes

Fish

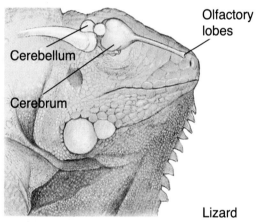

Olfactory lobes

Cerebellum

Cerebrum

Lizard

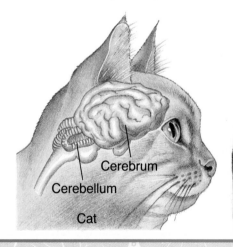

Cerebrum

Cerebellum

Cat

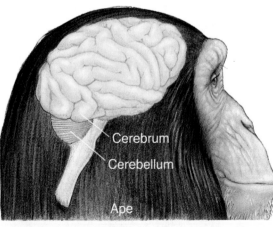

Cerebrum

Cerebellum

Ape

Notice that in humans (right), the brain is much larger, and the area dedicated to thinking (cerebrum) covers up and dominates everything else.

Human brain development

21 days

When a human embryo is just 21 days old and nothing but a hollow ball of cells, a groove begins to form along the outer surface. This groove is the start of the nervous system. Folds of cells come together across the groove, forming a neural tube. The top portion of the tube swells, becoming the brain, and the lower portion becomes the spinal cord. During this stage of development, the brain generates approximately 360 million new cells a day.

40 days

The cerebellum, medulla, and the frontal lobes of the brain begin to develop.

60 days

The head is almost half the size of the entire embryo.

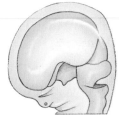

4 months

The cerebral cortex of the fetus begins to develop the ridges and grooves of the mature brain pattern.

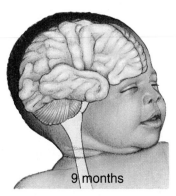

9 months

The brain can now carry on one of its most important functions—maintaining homeostasis.

The adult brain

The largest part of the human brain is the cerebrum. The cerebrum is divided into two hemispheres, left and right, connected by a band of millions of nerve cells called the corpus callosum. But the feature that makes the cerebrum unique is an outer, folded layer less than 5 mm thick—the cerebral cortex. Because of the cortex, you can remember, reason, organize, communicate, understand, and create.

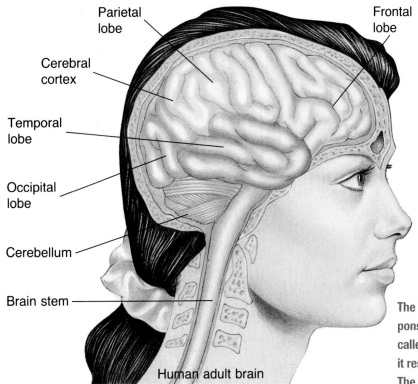

Human adult brain

- Parietal lobe
- Frontal lobe
- Cerebral cortex
- Temporal lobe
- Occipital lobe
- Cerebellum
- Brain stem

When you see a gold-medal performance by an Olympic gymnast or figure skater, you are watching the work of a well-trained cerebellum. It is here that muscles are coordinated and the memories of physical skills are stored. The cerebellum has tripled in size over the last million years.

The brain stem consists of the medulla, pons, and midbrain, and is sometimes called the reptilian brain. This is because it resembles the entire brain of a reptile. The brain stem controls homeostasis.

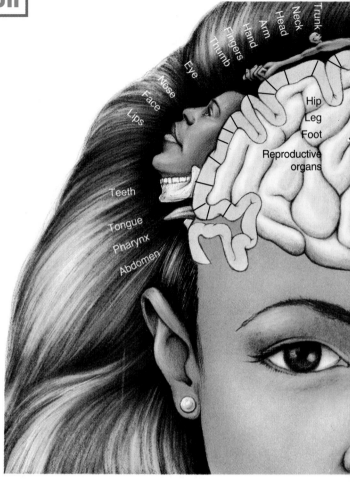

Focus On

The cerebral cortex

Left and right brain Certain higher functions are centered in either one or the other hemisphere. For example, in most right-handed people, language and speech are controlled by specific areas in the left cerebral hemisphere, while the ability to recognize and duplicate certain types of spatial relationships is located in the right hemisphere.

Speech and spatial perception are not solely determined by one or the other hemisphere. The corpus callosum allows you to perceive things such as shapes and then describe them in words.

Different thinking skills are also located in separate hemispheres. For example, logic, mathematical skills, and reasoning are the domain of the left brain. Insight (that flash of understanding), imagination, and awareness of the beauty of art and music are centered in the right brain. However, this does not make one hemisphere more important than the other. It is the blending of these abilities that makes each of us unique.

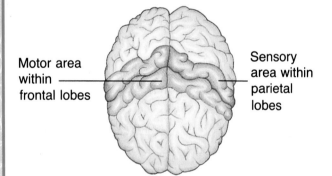

Motor area within frontal lobes

Sensory area within parietal lobes

Somatic sensory cortex

Almost every part of the body sends messages to its own part of the cerebral cortex. The amount of cortex dedicated to the face and limbs is large because control of these areas requires great precision. The areas of the sensory cortex labeled here on the right hemisphere are also present in the left hemisphere.

Using the brain to study the brain—A chronology

circa 275 B.C.
Erasistratus Greek physician and anatomist—made detailed studies of the human brain comparing the convolutions to those of animals. Erasistratus related brain complexity to intelligence.

circa 1600
Descartes French philosopher and mathematician—defined thinking as a range of intellectual thoughts, feelings, sensations, and will. He believed the mental process went on even during sleep.

1871
Cajal Spanish physician and histologist—refined a method of staining nerve cells so they could be studied. He was first to establish that neurons are the basic structure of the nervous system.

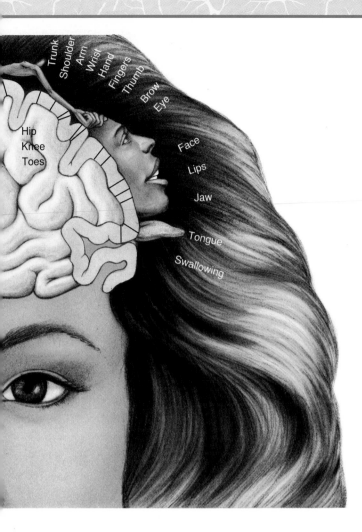

Trunk
Shoulder
Arm
Wrist
Hand
Fingers
Thumb
Brow
Eye

Hip
Knee
Toes

Face

Lips

Jaw

Tongue

Swallowing

1992 ————————

Ronnett, *et al.*
Cultured human
brain cells. Team of
researchers at Johns
Hopkins University
succeeded in cultur-
ing human brain
cells for more than
19 months.

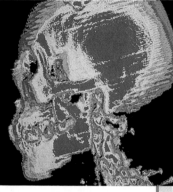

1961

Hounsfield and Cormack
Developed the computerized
axial tomograph (CAT scanner) that revolutionized the
diagnosis of brain lesions,
blood clots, tumors, and cerebral damage.

Motor cortex

When you walk, write, type, or talk, you must
control millions of muscle fibers very precisely. This is where the motor control areas
of your cerebral cortex take over. The areas
labeled here on the left hemisphere are also
present in the right hemisphere.

1950s

Gazzaniga and Sperry
Demonstrated the role of the corpus
callosum in the exchange of information between brain hemispheres.
Sperry won the Nobel prize for split-
brain research.

1930

Luria Soviet psychologist—studied the relationship between injured areas of the brain and apha-
sia (loss of the power to use or understand words).

1921
Loewi German
pharmacologist—
isolated acetyl-
choline, a chemi-
cal from the brain
that transmits
nerve impulses.

1929
Berger German neu-
rologist—first used
technology to record
a paper record of
human brain waves
on an electroen-
cephalograph.

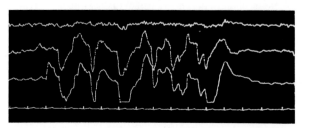

Focus On

Discovering the mind inside the brain

Have a nice day It's Saturday—a beautiful day to relax. You feel wonderful! Your cerebral cortex sends an impulse that activates a part of the limbic system that produces joyful feelings. However, if it rains, an impulse may be sent to another part of the system that activates somber feelings. Other areas in the limbic system can bring on feelings of anger and fear.

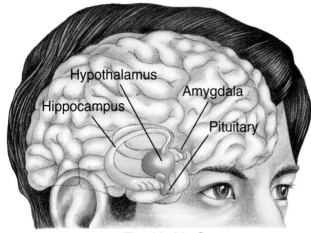

The Limbic System

Many emotions are reactions to events that happen outside the body. A sunny day, a friend's visit, doing chores—almost everything in our daily lives affects our emotions. These stimulate a central structure in the brain called the limbic system.

War and peace It's the night before the game. You are under stress and finding it difficult to sleep. A chemical produced by the passage of nerve currents from the brain results in an over production of adrenalin, which causes rapid heart rate and trembling. Doctors call this a fight-or-flight response. How do you find peace in the midst of this turmoil? By closing your eyes, breathing slowly and deeply, and visualizing a pleasant scene, you can often change the chemicals produced by the brain. A chemical called dopamine produced in the limbic system helps regulate emotions.

To sleep, perchance to dream To an onlooker, sleep seems to be a time of peaceful rest when the brain and the body are totally relaxed. Sleep is actually an active process. Some people roll, toss, and talk in their sleep. Researchers have learned from monitoring brain waves of wired-up sleepers that after an initial period of deep sleep, alternating periods of light and deep sleep occur. During deep sleep, EEG patterns show slow wave activity. In light sleep, the EEG patterns are similar to those recorded during wakefulness. Rapid eye movements occur during light sleep, so it is sometimes called REM (rapid eye movement) sleep. The most vivid dreams often occur during REM sleep.

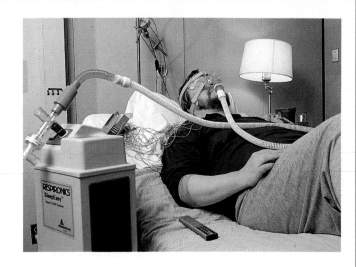

What causes stress? Many stresses exist in our lives, but the most stressful experiences are threatening events over which we have no control. Certain types of people seem especially prone to stress. They are constantly in a hurry, aggressive, competitive—unable to just sit still and relax. Such people are often rushing to a heart attack. This type of stressful personality has been called the Type A personality.

Drugs and the brain The brain is dependent on chemical messengers called neurotransmitters. Mental health may be severely affected if there is an excess or deficiency of any of these chemicals. For example, an excess of dopamine may be the cause of a severe mental illness called schizophrenia. One of the ways to treat some mental illnesses is with carefully monitored prescription drugs that restore normal chemical activity in the brain.

EXPANDING YOUR VIEW

1. **Journal Writing** Write a short story about what happens within your nervous system when you perceive something that is a danger to you.
2. **Research** Find and read materials on current theories about the function of sleep.

Think for a moment what it would be like to ride in the space shuttle. You float slowly, turning and moving effortlessly through space. The shuttle monitors your oxygen and temperature, providing a comfortable environment. Your food has been prepared and packaged for you. What a life!

Like an astronaut in the shuttle, at one time we all dwelt in a capsule, floating and turning in an inner, protected space. Inside your mother's uterus, you did not worry about food, water, or warmth. Your needs were provided for by your mother's body as your own body developed and grew. Of course, once outside your mother's body, you continued and are continuing to grow and develop.

How does the microscopic union of two sex cells eventually develop into a fully mature human being—one who is himself or herself capable of reproducing?

It's not difficult to see what this astronaut and a human fetus have in common. Not only are they both fully protected in a controlled environment, but they also both developed from a single, fertilized egg. The story of how each developed is the story of human reproduction, growth, and development.

Magnification: 1500×

Human Reproductive Systems

G*arfield appears to be up against stiff competition as he tries to woo the female cat away from his macho-looking, muscle-bound rival. The cartoonist was able to illustrate the essence of maleness in the rival cat by exaggerating one of his secondary sex characteristics, specifically, muscle development. In real life, how are secondary sex characteristics related to the reproductive process, and how are they controlled?*

Human Male Anatomy

The ultimate goal of the reproductive process is the formation and union of egg and sperm, development of the fetus, and birth of the infant. The organs, glands, and hormones of the male reproductive system are instrumental in meeting this goal. Their main functions are the production of sperm—the male sex cells—and their delivery to the female. *Figure 41.1* shows the organs and glands that make up the male reproductive system.

Where sperm form

Sperm production takes place in the testes. You will see in *Figure 41.1* that the testes are in the scrotum. The **scrotum** is a sac located outside of the male's body. Since sperm can develop only in an environment with a temperature that is about 3°C lower than normal body temperature, the exterior position of the scrotum provides an ideal location.

Located within each testis, a fine network of highly coiled tubes are the production facilities for sperm. Sperm are produced through the meiosis of cells lining these tubes. Recall from Chapter 12 that meiosis produces haploid cells. One cell divides by meiosis and produces four cells that mature into sperm. In human males, the production of mature sperm takes about 74 days. A sexually mature male can produce about 300 million sperm per day, each day of his life.

As you can see in *Figure 41.2,* a sperm is highly adapted for reaching and penetrating the female egg. It can live for about 48 hours inside the female reproductive tract.

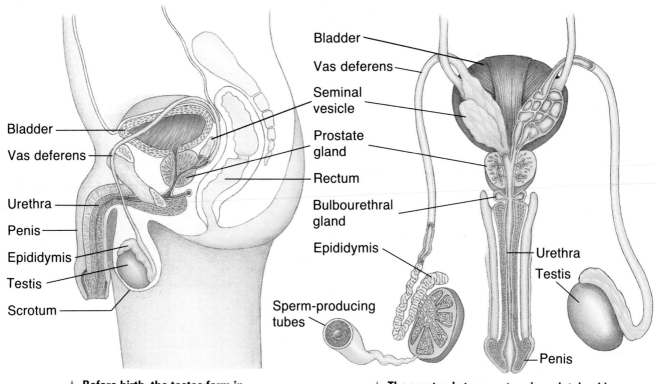

Bladder

Vas deferens

Seminal vesicle

Prostate gland

Rectum

Bulbourethral gland

Epididymis

Sperm-producing tubes

Bladder

Vas deferens

Urethra

Penis

Epididymis

Testis

Scrotum

Urethra

Testis

Penis

▲ Before birth, the testes form in the embryo's abdominal wall and then descend into the scrotum.

Figure 41.1

The organs and glands of the male reproductive system are shown in a front and side view.

▲ The scrotum's temperature is maintained by the movement of muscles. They contract in cold weather, pulling the scrotum closer to the body to maintain warmth. In warm weather, the muscles relax to lower the scrotum, thus allowing air to circulate and cool the sperm.

How sperm leave the testes

Before the sperm mature, they move out of the testes through a series of coiled ducts that empty into a single tube called the epididymis. The **epididymis** is a coiled tube within the scrotum in which the sperm complete their maturation. Mature sperm remain in the epididymis until they are released from the body.

When sperm are released from the epididymis, they enter the vas deferens. The **vas deferens** is a duct that transports sperm from the epididymis toward the ejaculatory ducts and the urethra. Peristaltic contractions of the vas deferens force the sperm along. The urethra is the tube that transports sperm out of the male's body. Notice in *Figure 41.1* that the

urethra also transports urine from the urinary bladder. A muscle located at the base of the bladder prevents urine and sperm from mixing.

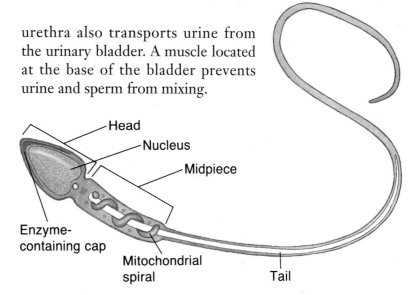

Head

Nucleus

Midpiece

Enzyme-containing cap

Mitochondrial spiral

Tail

Figure 41.2

A sperm is composed of a head, a midpiece, and a tail. The head contains the nuclear material and is covered by a cap containing enzymes that facilitate penetration of the egg. A number of mitochondria are found in the midpiece; they provide energy for locomotion. The tail is a typical flagellum that propels the sperm along its way.

Figure 41.3

Puberty results in many physical and emotional changes. Generally, males undergo puberty sometime between the ages of 13 to 16.

Fluids that help transport sperm

As sperm travel from the testes, they mix with the fluids secreted by several different glands. The **seminal vesicles** are a pair of glands located at the base of the urinary bladder. They secrete a mucouslike fluid into the vas deferens. The fluid is rich in the sugar fructose, which provides energy for the sperm cells.

Another structure, the **prostate gland,** is a single, doughnut-shaped gland that surrounds the top portion of the urethra. The prostate secretes a thinner, alkaline fluid that helps sperm move and survive. Two tiny **bulbourethral glands** are located beneath the prostate. These glands secrete a clear, sticky, alkaline fluid that protects sperm by neutralizing the acidic environment of the vagina. The combination of sperm and all of these fluids is called **semen.**

Hormonal Control

Recall that in Chapter 38 you learned how hormones, which play a key role in the regulation of digestion, metabolism, and homeostasis, are released by glands of the endocrine system. Hormones also are the control mechanism for the male's reproductive system.

Hormones and male puberty

As you watch young children, it's obvious from their physical appearance that they are not sexually mature. However, in the early teen years, changes begin to occur to their bodies. Puberty, *Figure 41.3,* begins. **Puberty** refers to the time when secondary sex characteristics begin to develop so that sexual maturity—the potential for sexual reproduction—is reached. The changes associated with puberty are controlled by the sex hormones that are secreted by the endocrine system.

Hormones and the male reproductive system

In males, the onset of puberty causes the hypothalamus to produce several kinds of hormones that interact with the pituitary gland. The **pituitary** is a gland, located at the base of the hypothalamus, that secretes hormones used to influence many different physiological processes of the body. As shown in *Figure 41.4,* the hypothalamus secretes a hormone that causes the anterior lobe of the pituitary to release two other hormones: follicle-stimulating hormone (FSH) and luteinizing hormone (LH). When released into the bloodstream, FSH and LH are transported to the testes. In the testes, FSH causes the production of sperm cells. LH causes endocrine cells in the testes to produce the male hormone—testosterone.

Testosterone is the steroid hormone responsible for the growth and development of secondary sex characteristics in a male. These characteristics include the growth and maintenance of male sex organs; the production of sperm; an increase in body hair, especially on the face, under the arms, and in the pubic area; an increase in muscle mass and in the growth rates of the long bones of the arms and legs; and the deepening of the voice. Evidence suggests that testosterone may also be responsible for an increase in aggressive behavior.

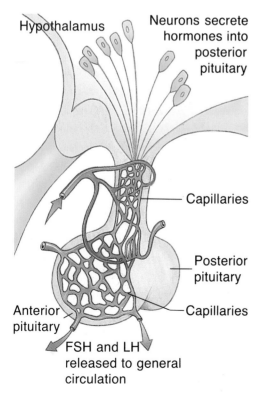

Hypothalamus

Neurons secrete hormones into posterior pituitary

Capillaries

Posterior pituitary

Anterior pituitary

Capillaries

FSH and LH released to general circulation

ThinkingLab — Draw a Conclusion

Why is the level of testosterone an advantage for the dominant male baboon?

In baboon tribes, a social structure of dominant and subordinate males exists. The dominant males have better access to food, to the best resting spots, and to the female baboons. In contrast, the subordinate male baboons must laboriously search for food, often having it stolen by the dominant male.

Analysis

In males, the hormone testosterone regulates sexual behavior and aggression as well as increases the rate at which glucose reaches the muscles. The graph shows testosterone levels of dominant and subordinate male baboons. When the male baboons are at rest, the testosterone levels are essentially equal. After being exposed to the same stress, the reactions of the dominant and subordinate males differ sharply for the first few hours.

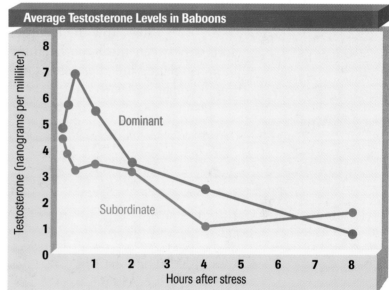

Average Testosterone Levels in Baboons

Dominant

Subordinate

Testosterone (nanograms per milliliter)

Hours after stress

Thinking Critically

Explain the adaptive advantage of higher levels of testosterone in the dominant male during times of stress.

Figure 41.4

Because the body is controlled by two systems, the nervous system and the endocrine system, there must be coordination between the two. This coordination is centered in the hypothalamus and pituitary, which are located close together in the brain. In the hypothalamus, neurons secrete releasing and inhibiting hormones. These hormones travel in the bloodstream to the pituitary gland, where they speed up or slow down the secretion of FSH and LH.

Maintaining the correct level of hormones

In your home, the thermostat regulates the room temperature. When the temperature falls below a certain level, the lack of heat energy is sensed by the thermostat. The thermostat's control unit signals the heating unit to begin the production of heat. When enough heat is produced and the temperature increases to a certain point, the thermostat signals the heating unit to shut off.

In the body, the endocrine system is self regulated by its own **negative-feedback system**, which is similar to the thermostat. In the body's system, the level of the hormone or its effect is signaled back to the hypothalamus. *Figure 41.5* shows how the feedback circuits between the hypothalamus, pituitary gland, and the testes function in human males.

Human Female Anatomy

The main functions of the female reproductive system are to produce eggs, the female sex cells, and to provide an environment for a fertilized egg to develop. Egg production takes place in the two ovaries. Each ovary is about the size and shape of an almond. One ovary is located on each side of the lower part of the abdomen.

As you can see in *Figure 41.6,* the open end of an oviduct is located close to each ovary. The **oviduct** is a tube that transports eggs from the ovary to the uterus. Peristaltic contractions of the muscles in the wall of the oviduct combine with beating cilia to move the egg in the tube.

You learned earlier that female mammals have a uterus in which the fetus develops during pregnancy. The uterus is situated between the urinary

Figure 41.5

In human males, signaling molecules from the hypothalamus cause the pituitary gland to release FSH and LH, which in turn stimulate the production of sperm and testosterone. The solid blue line shows that, within the testes, cells that help in the formation of sperm then send signals back to the hypothalamus to shut off or slow down the production of FSH. The dashed red line shows that an increased level of testosterone sends a signal back to shut off or slow the production of LH. By constantly monitoring the signals, the body maintains the correct levels of hormones.

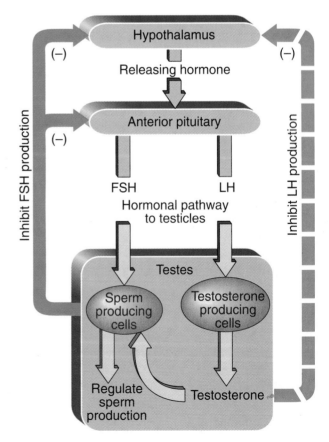

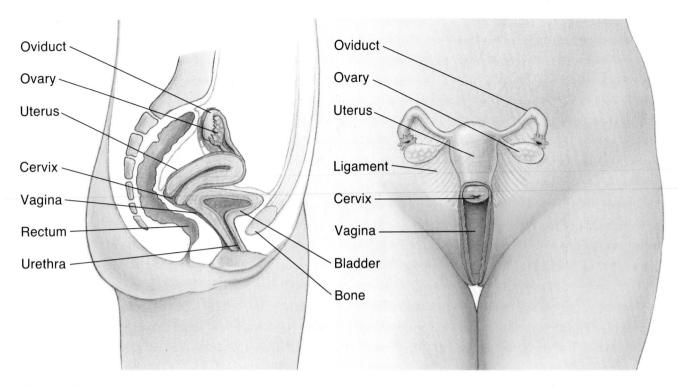

Oviduct
Ovary
Uterus
Cervix
Vagina
Rectum
Urethra

Oviduct
Ovary
Uterus
Ligament
Cervix
Vagina
Bladder
Bone

Figure 41.6

The female reproductive system includes two ovaries, two oviducts—sometimes called fallopian tubes—the uterus, and the vagina. The ovaries produce eggs and female sex hormones. The uterus is composed of two layers: a thick, muscular layer and a thin lining called the endometrium.

bladder and the rectum and is shaped like an inverted pear. Its lower end, called the **cervix,** tapers to a narrow opening into the vagina. The **vagina** is a passageway between the uterus and the outside of the female's body.

Puberty in Females

Just as its onset occurs in males, puberty in females begins when the hypothalamus signals the anterior lobe of the pituitary to produce and release the hormones FSH and LH. These are the same hormones that are produced in males; however, in females, FSH stimulates the development of a follicle in the ovary. A **follicle** is a group of epithelial cells that surround an undeveloped egg cell. FSH also causes the release of

the hormone estrogen from the ovary. Estrogen is the steroid hormone responsible for the secondary sex characteristics of females. These characteristics include the growth and maintenance of female sex organs; an increase in body hair, especially under the arms and in the pubic area; an increase in the growth rates of the long bones of the arms and legs; a broadening of the hips; an increase in fat deposits in the breasts, buttocks, and thighs; and the onset of the menstrual cycle.

Production of eggs

Recall that sperm production does not begin in males until they reach puberty, after which time it continues for the rest of their lives. Egg production is different. Even before a female is born, her body begins to

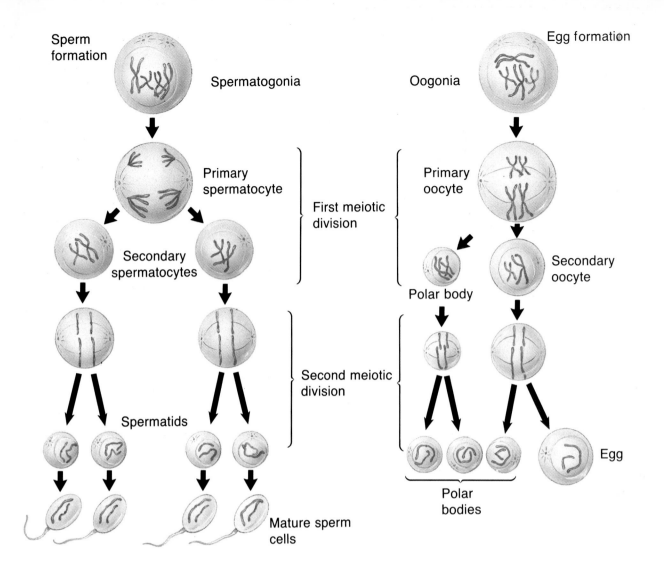

Sperm formation

Spermatogonia

Primary spermatocyte

Secondary spermatocytes

Spermatids

Mature sperm cells

First meiotic division

Second meiotic division

Egg formation

Oogonia

Primary oocyte

Secondary oocyte

Polar body

Polar bodies

Egg

Figure 41.7

During sperm formation, the potential sperm cell divides by meiosis, resulting in the formation of four mature sperm. During egg formation, the potential egg divides by meiosis, resulting in the formation of one egg and three smaller bodies that eventually degenerate.

develop eggs. During this prenatal period, cells in her ovaries divide until the first stage of meiosis, prophase I, is reached. At this point, the cells go into a resting stage. *Figure 41.7* compares the process that produces sperm with the process that produces eggs. At birth, a female has all the potential eggs she will ever have.

How eggs are released

About once a month in a sexually mature female, the process of meiosis starts up again in one of the prophase I cells. The result is the production of an egg that ruptures from the ovary and passes into the oviduct. This process of the egg rupturing through the ovary wall and moving into the

oviduct is called **ovulation.** Fertilization occurs in the oviduct if the egg and sperm unite. *Figure 41.8* shows the process leading to ovulation.

The Menstrual Cycle

All the activities of the human female reproductive system are part of a cycle. You learned that in a sexually mature female, ovulation occurs about once a month. Once the egg has been released, the remaining part of the follicle develops into a structure called the **corpus luteum.** The corpus luteum secretes the steroid hormone, progesterone. Progesterone causes changes to occur in the lining of the uterus,

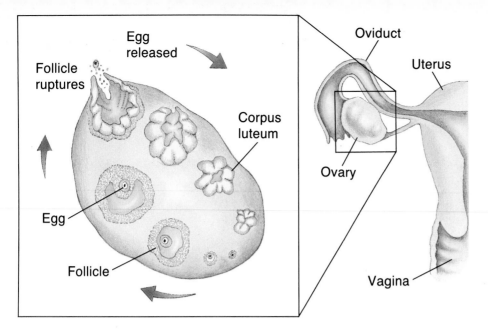

Figure 41.8

Once a female reaches puberty, follicles within her ovaries begin to mature. Usually, one follicle matures each month. As it matures, the follicle secretes the female hormone estrogen. During ovulation, the follicle ruptures, releasing the egg from th ovary. The remains of the fol le become the corpus luteum, which secretes some estrogen and the hormone progesterone.

which prepare it for receiving a fertilized egg. The series of changes in the female reproductive system that include producing an egg and preparing the uterus for receiving it is known as the **menstrual cycle.** The menstrual cycle begins during puberty and continues for 30 to 40 years until menopause. At this time, the female stops releasing eggs and hormone secretion slows.

The length of the menstrual cycle varies from female to female, with the average length being 28 days. If the egg is not fertilized, the lining of the uterus will be shed, and the cycle will begin again. The entire menstrual cycle can be divided into three phases: the flow phase, the follicular phase, and the luteal phase. The phases of the menstrual cycle are illustrated in *Figure 41.9.*

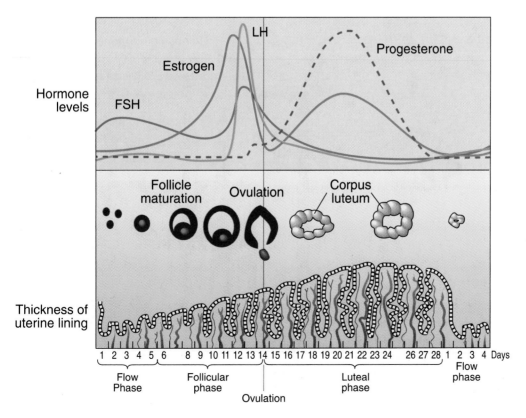

Figure 41.9

The three phases of the menstrual cycle correlate with hormone output from the anterior pituitary. Notice that estrogen levels increase during the follicular phase and progesterone levels increase during the luteal phase. LH levels increase sharply at ovulation. Because the average cycle is 28 days, the graph of phases is based on 28 days, with the first day beginning the flow phase.

Flow phase

Day 1 of the menstrual cycle is the day menstrual flow begins. Menstrual flow is the shedding of the blood, tissue fluid, mucus, and epithelial cells that made up the lining of the uterus. This flow passes from the uterus through the cervix and the vagina to the exterior. Generally, menstrual flow ends by day 5 of the cycle. During the flow phase, FSH is beginning to rise, and a follicle in one of the ovaries is continuing its development as meiosis of the prophase I cell proceeds.

Follicular phase

The second phase of the menstrual cycle varies in length more than the other phases. It lasts from day 6 to about day 14 in a 28-day cycle. FSH and LH released from the pituitary stimulate the ovary to produce more estrogen, and this increase in estrogen stimulates the repair of the lining of the uterus. The lining undergoes mitosis and begins to thicken. The increase in estrogen feeds back to the hypothalamus and decreases its stimulation of the pituitary. The follicle continues to develop and mature. The presence of estrogen also stimulates a sharp increase in the release of LH. About day 14, this increase causes the follicle to rupture, releasing the egg into the oviduct. At this time, a detectable change in the female's body temperature occurs—about +0.5°F. In addition, the cervical cells produce large amounts of mucus in response to the increasing levels of estrogen. Some females also experience discomfort in the area of one or both of the ovaries around the time of ovulation.

Luteal phase

The last stage of the menstrual cycle, from days 15 to 28, is named for the corpus luteum. LH stimulates

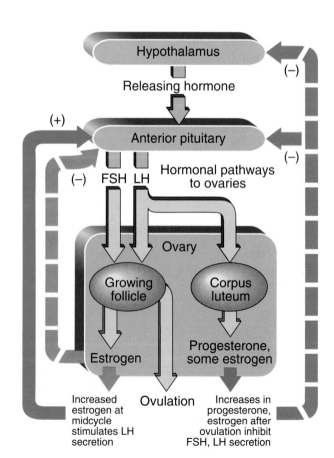

Figure 41.10

Negative feedback controls the levels of hormones in the female during her menstrual cycle. If fertilization occurs, this cycle is broken and the levels of hormones change.

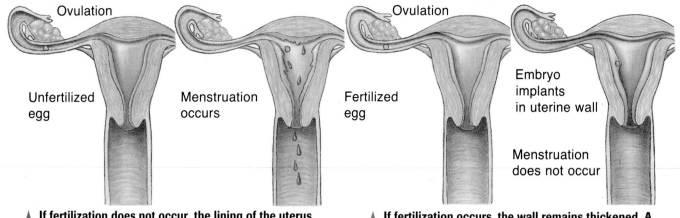

Ovulation

Unfertilized egg

Menstruation occurs

Ovulation

Fertilized egg

Embryo implants in uterine wall

Menstruation does not occur

▲ If fertilization does not occur, the lining of the uterus deteriorates because blood vessels in the uterine wall constrict, reducing the blood supply to the cells.

▲ If fertilization occurs, the wall remains thickened. A fluid rich in nutrients for the embryo is secreted.

Figure 41.11

Events in the uterine wall after ovulation depend on whether fertilization has occurred.

the corpus luteum to develop from the ruptured follicle. The corpus luteum produces both progesterone and estrogen, but progesterone is the dominant hormone during the luteal phase. This hormone causes the uterine lining to thicken, increases its blood supply, and accumulates fat and tissue fluid. These changes correspond to the anticipated arrival of a fertilized egg. Through negative feedback, progesterone prevents the production of LH.

As shown in *Figure 41.10,* if the egg is not fertilized, the rising levels of progesterone and estrogen from the corpus luteum inhibit the release of FSH and LH by inhibiting the hypothalamus. This negative feedback causes the corpus luteum to degenerate, and the progesterone levels drop. The thick lining of the uterus then begins to be shed. *Figure 41.11* shows what happens to the lining of the uterus if fertilization occurs.

Section Review

Understanding Concepts

1. Describe the pathway of an unfertilized egg as it travels through the female reproductive system.
2. Summarize the negative-feedback system in a male, including the roles of the hypothalamus and pituitary gland.
3. What is the function of the menstrual cycle?

Thinking Critically

4. What might happen to sperm production if a male has a high fever?

Skill Review

5. Interpreting Scientific Illustrations Study *Figure 41.1.* Using the terms *dorsal, ventral, anterior, posterior, superior,* and *inferior,* describe where the epididymis is in relation to the vas deferens. Describe where the prostate is in relation to the testes. For more help, refer to Thinking Critically in the *Skill Handbook.*

Section Preview

Objective

Summarize the events during each trimester of pregnancy.

Key Terms

implantation
umbilical cord
genetic counseling
amniocentesis

What do you have in common with a period at the end of a sentence? You were once about the same size. You started out life as a single, microscopic fertilized egg. That one cell went through numerous mitotic divisions to produce the trillions of cells that make up your body today. It all began when an egg from your mother was fertilized by a sperm from your father.

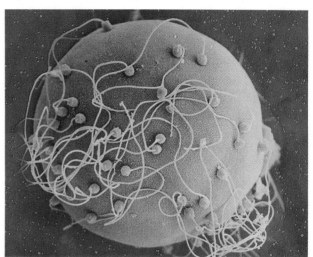

Magnification: 480×

Fertilization and Implantation

After an egg ruptures from a follicle, it is able to stay alive for about 24 hours. For fertilization to occur, sperm must be present in the oviduct at some point during those first hours after ovulation. Sperm enter the female's reproductive system when strong, muscular contractions ejaculate semen from the male's body. As many as 350 million sperm are forced out of the male's penis and into the female's vagina during intercourse. Since sperm can live for 48 hours after ejaculation, fertilization can occur if intercourse occurs anywhere from a few days before to a day after ovulation.

One sperm plus one egg

How is it possible that of the millions of sperm released into the vagina during ejaculation, only one fertilizes the mature egg? One reason is that the fluids secreted by the vagina are acidic and destroy most of the delicate sperm. Yet, some sperm survive because of the buffering effect of semen. The surviving sperm swim up the vagina into the uterus. Of the sperm that reach the uterus, only a few hundred pass into the two oviducts. The egg is present in one of them.

Recall that the head of the sperm contains enzymes that help the sperm penetrate the egg. As the sperm penetrates the egg, it loses its midpart and tail. Once one sperm penetrates the egg, the egg's membrane changes its electrical charge, thus preventing other sperm from entering. After the one sperm penetrates the egg, its nucleus combines with the egg's nucleus to form a zygote.

Hormonal maintenance of pregnancy

Remember that estrogen, and especially progesterone, cause the uterine lining to thicken in preparation for implantation. Once the blastocyst implants, the chorion membrane of the embryo starts to secrete the hormone chorionic gonadotropin. This hormone keeps the corpus luteum alive so that it continues to secrete progesterone. By the third month, the placenta has taken over for the corpus luteum, secreting enough estrogen and progesterone to maintain the pregnancy.

Fetal Development

When you think of an embryo developing within the mother's body, you probably don't realize that the development involves three different processes: growth, development, and cellular differentiation. Growth refers to the actual increase in the number of cells. But the cells must also move within the embryo's body and arrange themselves into specific organs. In addition, the cells become specialized to perform specific tasks and functions. All three processes begin with fertilization.

Pregnancy in humans usually lasts about 280 days, calculated from the first day of the mother's last menstrual period. The baby actually develops for about 266 days, calculated from the time of fertilization to birth. This time span is divided into three trimesters, each equal to three months. Each trimester brings significant advancement in the development of the embryo.

First trimester—Body systems form

During the first trimester, all the body systems of the embryo begin to form. A five-week embryo is shown in *Figure 41.14.* During this time of development, the woman may not even realize she is pregnant. Yet, the first seven weeks after fertilization are critical because the embryo is more sensitive to outside influences, such as alcohol, smoking, and other drugs that cause malformations, than at any other time.

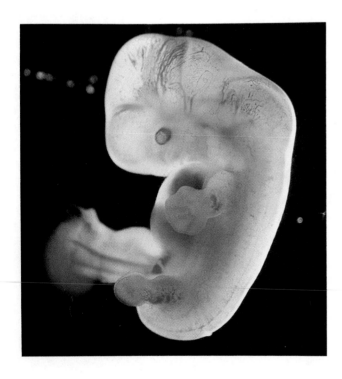

7 mm

Figure 41.14

A five-week-old embryo is about 7 mm long. The heart—the large, red, circular structure protruding out of the embryo—begins as two muscular tubes. It starts to beat around the 25th day of development. You can see the arms and legs beginning to bud, as well as the head. Note that the tissue that will form the eyes is beginning to darken.

How long is an embryo?

You started out as a single cell. That cell divided by the process of mitosis to produce body systems that were able to maintain an independent existence outside your mother's uterus. During the time you were in your mother's uterus, great changes occurred. One of these changes involved your height or length.

Procedure

1. Prepare a graph that plots time on the horizontal axis and length in centimeters on the vertical axis. Equally divide the horizontal axis first into nine months. Then equally divide each of the first three months into four weeks.

2. Plot the data in the table on your graph.

Growth in Length of a Fetus	
Time after fertilization	Size
3 weeks	3 mm
4 weeks	6 mm
6 weeks	12 mm
7 weeks	2 cm
8 weeks	4 cm
9 weeks	5 cm
3 months	7.5 cm
4 months	15 cm
5 months	25 cm
6 months	30 cm
7 months	35 cm
8 months	40 cm
9 months	51 cm

Analysis

1. When is the fastest period of growth?
2. What structures are developing during this period of growth?
3. At what point does growth begin to slow down?

Figure 41.15

In a two-month-old fetus, the two muscular tubes have fused together to form a four-chambered heart. The limbs are beginning to elongate, and the fingers and toes appear. Notice how the eyelids have formed, and the face looks distinctly human. Bones are beginning to ossify, and nearly all muscles have appeared. As a result, the fetus can move spontaneously.

By the eighth week, all the body systems are present, and the embryo is now referred to as a fetus. You can see this stage of fetal development in *Figure 41.15.* At the end of the first trimester, the fetus weighs about 28 g and is about 7.5 cm long from the top of its head to its buttocks. The sex of the fetus can be determined by the appearance of the external sex organs.

Second trimester—A time of growth

For the most part, fetal development during the next three months is limited to body growth. Growth is rapid at the beginning of the trimester, but then slows by the beginning of the fifth month. At this point, the fetus can survive outside the uterus with a great amount of medical assistance, but the mortality rate of such young fetuses is high. The fetus's body metabolism cannot yet maintain a constant temperature, and its lungs have not matured enough to provide a regular respiratory rate. *Figure 41.16* shows a fetus during the second trimester.

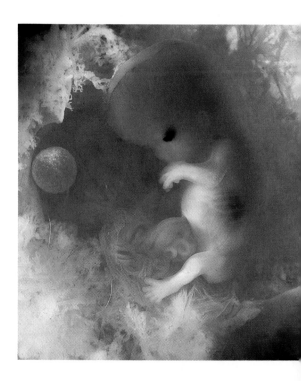

Figure 41.16
The facial features of a second-trimester fetus are well formed. Its skin is covered by a white, fatty substance that protects it against the amniotic fluid. Kicks and movements are commonly felt by the mother during this time as the fetus exercises its muscles. It can also suck its thumb, preparing the muscles that will allow the baby to feed later. By the end of the second trimester, the fetus weighs about 650 g and is about 34 cm long.

Third trimester—Continued growth

During the last trimester, the mass of the fetus more than triples. By the beginning of the seventh month, the fetus kicks, stretches, and moves freely within the amniotic cavity, somewhat like the astronaut in the space shuttle at the beginning of the chapter. During the eighth month, its eyes open.

During the last weeks of pregnancy, the fetus has grown large enough to fill the space within the embryonic membranes. Sometime in the ninth month, the fetus rotates its position so that its head is down, partly as a result of the shape of the uterus, but also because the head is the heaviest part of the body. By the end of the third trimester, the fetus weighs about 3300 g and is about 51 cm long. All of its body systems have developed, and it can now survive independently outside the uterus.

Genetic Counseling

Most expectant parents desperately want just one thing—a healthy, normal baby. With our increasing knowledge of human heredity and advancing technology, determining that a newborn will be healthy and normal is much more possible today than in the past.

Genetic disorders can be predicted

Generally, people in the industrialized nations are more aware of the possible genetic disorders that can affect a child. For many, this awareness has made them eager to know if a potential child will be healthy. As advances have been made in the detection and treatment of genetic disorders, including those you examined in Chapter 15, the demand for genetic services, especially prenatal testing, has increased.

Prenatal Care

Medical experts agree that early prenatal care can save lives. By identifying and monitoring possible complications—like poor nutrition, multiple fetuses, and maternal high blood pressure—healthier babies are born and survive their first year of life. In addition, expectant mothers can be treated for medical problems that may result from pregnancy.

The U.S. infant mortality rate   Low birth weight (less than 5.5 pounds) is the primary contributor to the nation's infant mortality rate. At 9.1 deaths per 1000 births, America currently ranks 19 among all other nations in infant deaths. The infant mortality rate is also influenced largely by the number of premature births. Babies born before the 37th week of gestation are considered to be premature. These children, if they survive, also have an increased risk for health problems later in life.

Prenatal checkups
Experts believe low birth weight or prematurity can be prevented through early and regular prenatal care during pregnancy. Expectant mothers must also make lifestyle changes, like improving nutrition and giving up smoking, drugs, and alcohol—all of which contribute to low birth weight, premature birth, and birth defects. Blood and urine tests taken during regular prenatal checkups monitor both baby and mother for common problems such as iron-deficiency, known as anemia, and Rh incompatibility. Both conditions are treatable.

Thinking Critically

Unfortunately, those who need prenatal care the most—poor women, teenagers, substance abusers—are the ones least likely to get it. What factors, do you think, contribute to the problems these women have in getting proper prenatal care, and what can be done to improve the care they get?

Most people do not even think about genetics until they are considering having children or are already expecting one. If there is no history of genetic disorders in either family of the prospective parents, there may be no need for genetic services. However, if one or both prospective parents have a family history of some genetic disorder, both will likely want to get additional information before proceeding with a pregnancy.

The job of a genetic counselor

Couples who seek information from trained professionals about the probabilities of hereditary disorders and what can be done if they occur are receiving **genetic counseling.** Genetic counselors have a medical background with additional training in genetics. Sometimes, a team of professionals works with prospective parents. The team may include geneticists, clinical psychologists, social workers, and other consultants.

How do genetic counselors go about their work? First, they develop medical histories of both families. These histories may include pedigrees, biochemical analyses of blood, and karyotypes. Once the counselor has collected and analyzed all the available information, he or she explains the risk factors of having offspring with genetic disorders. If the probabilities of having a severely affected child are high, a couple must decide whether or not to have children of their own.

Prenatal testing

How can prospective parents learn whether a child they are expecting is developing normally? A number of prenatal tests now exist that give prospective parents and their doctors valuable information about the fetus while in the mother's uterus. In fact, prenatal testing can detect more than 100 genetic disorders.

Amniocentesis provides cells

A common prenatal procedure that allows genetic analysis of fetal cells is **amniocentesis.** In amniocentesis, a long needle is passed through the abdominal wall of the pregnant woman to withdraw a small amount of the fluid that surrounds the 14- to 16-week-old fetus. The fluid contains cells that have sloughed off from the fetus. Since only a few fetal cells are in the fluid, the results of the tests are delayed for several weeks while the cells grow and multiply in a cell culture. Once enough cells for testing purposes are grown, a karyotype, biochemical tests, or both procedures are carried out. In *Figure 41.17,* you can see how amniocentesis is done. This procedure is often used to determine if a child will have a genetic disorder or chromosomal abnormality such as Down syndrome.

Figure 41.17

The process of amniocentesis can be used to diagnose chromosomal and genetic problems; however, the test is not risk free (left). Infection or injury to the fetus may occur. Ultrasonic techniques help locate the fetus during testing so that injury can be avoided (right).

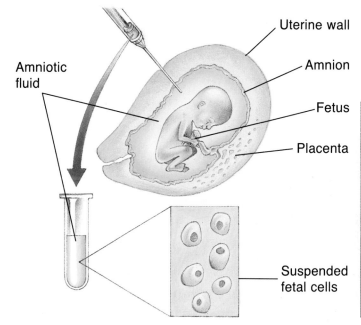

Uterine wall

Amnion

Fetus

Placenta

Amniotic fluid

Suspended fetal cells

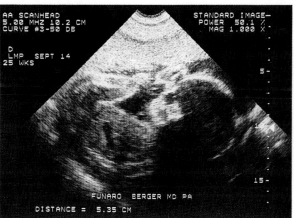

```
AA SCANHEAD
5.00 MHZ 10.2 CM
CURVE #3-50 DB
D
LMP  SEPT 14
25 WKS

STANDARD IMAGE—
POWER  50.1 %
MAG 1.000 X

5-

15-

FUNARO  BERGER MD PA
DISTANCE = 5.35 CM
```

Section Review

Understanding Concepts

1. What changes occur in the zygote as it passes along the oviduct and into the uterus?
2. What is the function of the placenta?
3. Why is an embryo between two and seven weeks old most vulnerable to drugs and other harmful substances taken by its mother?

Thinking Critically

4. Compare the functions of human embryonic membranes with those inside a bird's egg.

Skill Review

5. **Sequencing** Prepare a table listing the events in the three trimesters of pregnancy. For more help, refer to Organizing Information in the *Skill Handbook*.

Section Preview

Objectives

Describe the three stages of birth.

Summarize the developmental stages of humans after they are born.

Key Term

labor

How often have you heard the comment, "My you've grown since I last saw you"? It may seem that you have grown a lot in the last few years. Yet the most rapid stage of growth in the life cycle of a human takes place within the uterus.

From fertilization to birth, mass increases about 1300 times. Even so, although growth slows after birth, changes certainly do not stop. The human body changes throughout life.

Birth

Birth is the process by which a fetus is pushed out of the uterus and the mother's body and into the outside world. What triggers the onset of birth is not fully understood. However, it occurs in three recognizable stages: dilation, expulsion, and the placental stage.

Dilation of the cervix

The physiological and physical changes a female goes through to give birth are called **labor.** Labor begins with a series of mild contractions of the uterine muscles. These contractions are stimulated by oxytocin, a peptide hormone released by the posterior lobe of the pituitary. The contractions open, or dilate, the cervix to allow for passage of the

baby, as shown in *Figure 41.18.* As labor progresses, the contractions begin to occur at regular intervals and intensify as the time between them shortens. This first stage of labor is usually the longest, sometimes lasting up to 24 hours.

Expulsion of the baby

Expulsion occurs when the involuntary uterine contractions become so forceful that they push the baby through the cervix into the birth canal. The mother assists with expelling the baby by contracting her abdominal muscles in time with the uterine contractions. As shown in *Figure 41.18,* the baby moves from the uterus, through the birth canal, and out of the mother's body. The expulsion stage usually lasts from 20 minutes to an hour.

A When the opening of the cervix is about 10 cm, it is fully dilated. Usually, the amniotic sac ruptures and releases the amniotic fluid through the vagina, which is referred to as the birth canal.

Figure 41.18

The stages of birth are dilation, expulsion, and the placental stage.

Uterus

Umbilical cord

Birth canal

Cervix

Dilation

B The birth canal is the passage through which the baby will be expelled from the mother's body, or born. The baby's head rotates as it moves through the birth canal, making it easier for the body to be expelled.

Expulsion

Placenta detaching

C During the placental stage, the placenta and umbilical cord are expelled from the mother's body.

Umbilical cord

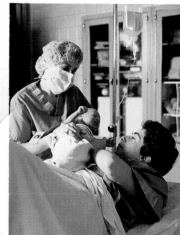

Placental

Placental stage

As shown in *Figure 41.18,* within ten to 15 minutes after the birth of the baby, the placenta separates from the uterine wall and is expelled with the remains of the embryonic membranes. Collectively, these materials are known as the afterbirth. The uterine muscles continue to contract forcefully, constricting uterine blood vessels to prevent the mother from hemorrhaging. After the baby is born, the umbilical cord is clamped and cut near the baby's abdomen. The bit of cord that is left dries up and falls off, leaving an abdominal scar called the navel.

Growth and Aging

Once a baby is born, growth continues and learning begins. Human growth does not continue in a steady, linear process. It varies with age and with the individual and is somewhat sex dependent.

A hormone controls growth

Human growth is regulated in large part by human growth hormone (hGH), a protein secreted by the anterior pituitary. While hGH causes all body cells to grow, it acts principally on the skeleton and skeletal muscles. The hormone works by

Human growth is the result of more than one hormone. Human growth hormone, thyroid hormones, and the reproductive hormones that are produced during puberty are all important in human growth at various ages. Together these hormones stimulate the growth of bone and cartilage, protein synthesis, and the addition of muscle mass. Because the reproductive hormones are involved in human growth, perhaps there is a difference in the growth rate between males and females.

Average Growth Rate in Humans

PREPARATION

Problem
Is average growth rate the same in males and females?

Objectives
In this Biolab, you will:
- **Graph** the average growth rates in males and females.
- **Determine** if there are differences in the average growth rates of males and females.

Materials
graph paper
red and blue pencils

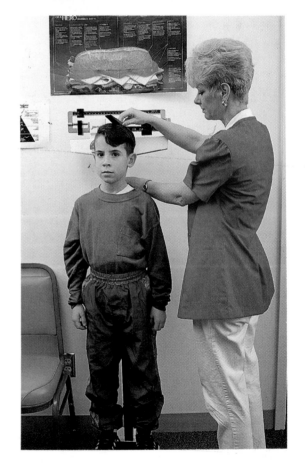

1. Construct a graph that plots mass on the vertical axis and age on the horizontal axis.

2. Plot the data in the table for the average female growth in mass from ages eight to 18. Connect the points with a red line.

3. On the same graph, plot the data for the average male growth in mass from ages eight to 18. Connect the points with a blue line.

4. Construct a separate graph that plots height on the vertical axis and age on the horizontal axis.

5. Plot the data for the average female growth in height from ages eight to 18. Connect the points with a red line. Plot the data for the average male growth in height from ages eight to 18. Connect the points with a blue line.

Averages for Growth in Humans				
	Mass (kg)		Height (cm)	
Age	Female	Male	Female	Male
8	25	25	123	124
9	28	28	129	130
10	31	31	135	135
11	35	37	140	140
12	40	38	147	145
13	47	43	155	152
14	50	50	159	161
15	54	57	160	167
16	57	62	163	172
17	58	65	163	174
18	58	68	163	178

ANALYZE AND CONCLUDE

1. **Analyzing Data** During what ages do females and males increase the most in mass? In height?

2. **Thinking Critically** How can you explain the differences in growth between males and females?

3. **Analyzing Data** Interpret the data to find if the average growth rate is the same in males and females.

Going Further

Applying Concepts
Correlate the range of heights in your biology class to the statistical average.

increasing the rate of protein synthesis inside the cells and by increasing the metabolism of fat molecules.

Like the sex hormones, testosterone and estrogen, the levels of human growth hormone within your body are controlled by a negative-feedback system. *Figure 41.19* shows what occurs if levels of hGH are abnormally high or low.

The first stage of growth—Infancy

During the first two years of life, or infancy, a child shows tremendous growth as well as increased physical coordination and mental development. Generally, an infant will double its birth weight by five months and triple its weight in a year. By two years of age, most infants weigh approximately four times their birth weight. During this time, the infant learns to control its limbs, roll over, sit, crawl, and walk. By the end of infancy, the child also utters his or her first words.

From child to adult

Childhood is the period of growth and development that extends from infancy to adolescence, when puberty begins. Physically, the childhood years are a period of relatively steady growth. Mentally, a child develops the ability to reason and to solve problems.

Figure 41.19

The amount of hGH secreted during childhood determines the height of an individual.

▼ **If a person's pituitary fails to produce enough hGH during childhood, a condition called pituitary dwarfism results.**

◄ **Large amounts of hGH secreted during childhood result in very tall individuals.**

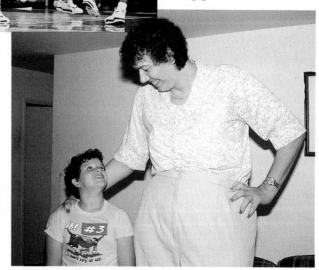

▼ **Abnormally large amounts of hGH result in a condition called gigantism.**

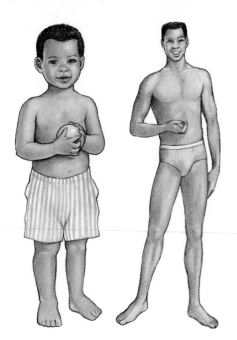

Figure 41.20

Notice the relative sizes of the head, arms, and legs as a person matures. In infancy, the head accounts for almost 25 percent of the body's length.

Adolescence follows childhood. At puberty, the onset of adolescence, there is often a growth spurt, sometimes quite a dramatic one! Increases of 5 to 8 cm of height in one year are not uncommon in teenage boys. During the teen years, adolescents reach their maximum physical stature, which is determined by heredity, nutrition, and their environment. By the time a young person reaches adulthood, his or her organs have reached their maximum mass, and physical growth is complete. You can see in *Figure 41.20* how the physical appearance of a person changes from birth to adulthood.

Art

Emmie and Her Child
by Mary Stevenson Cassatt (1855–1926)

Mary Cassatt's sensitive portrayals of the intimate relationship between mothers and their children brought her world fame. As a young woman, she chose to leave her close-knit family in Philadelphia and settle permanently in Paris. There she met Edgar Degas, and at his invitation, she became the only American to exhibit with the French Impressionists. Her achievement is all the more impressive because she was an unmarried, professional woman living in the male-dominated late 19th century.

Her own individualized art   Unable to attend the better French art schools, which accepted only men, Cassatt's training actually took place as she imitated the Great Masters that hung in the Louvre, the famous art museum in Paris. After learning and perfecting their techniques, Cassatt began experimenting with different materials, such as pastels. She was able to combine her training with modern materials and attitudes to create her own individualized art—exemplified in her mother-child series. Never before had this age-old motif been so realistically rendered by an artist.

A simple embrace   Mary Cassatt's favorite subjects were ordinary people in everyday settings doing everyday activities. She especially enjoyed painting mothers and children performing their daily rituals. In the beautiful painting entitled "Emmie and Her Child," Cassatt sensitively portrays a simple embrace between a mother and her child. The head and back positions of the loving pair communicate not only the tenderness of their embrace, but feelings of complete trust and support, deep warmth, unquestioned security, and unconditional love.

CONNECTION TO Biology

Cassatt communicates the psychological benefits of the maternal instinct very well in her paintings. How might the maternal instinct serve a biological function in human reproduction?

Figure 41.21

Age has not deterred this 90-year-old woman from trying a new experience—skydiving.

An adult ages

As an adult ages, his or her body undergoes many distinct changes. Metabolism and digestion become slower. The skin loses some of its elasticity, while less pigment is produced in the hair follicles; that is, the hair turns white. Bones become thin-ner and more brittle, resulting in an increased risk of fracture. Stature may shorten because the discs between the vertebrae become compressed. Vision and hearing might diminish, but, as *Figure 41.21* shows, many people continue to be both intellectually and physically active.

Connecting | **Ideas**

The human reproductive system is unique among body systems in that there are two—male and female. Together, the reproductive system and the endocrine system work to produce sex cells and the fully grown humans that develop from them. In fact, in order for your body to maintain homeostasis, all systems must function together. In order for your body to be healthy, whether you are an adolescent or an aged adult, your body systems must be free of infections and other disorders.

Section Review

Understanding Concepts
1. What events occur during dilation?
2. How does the human growth hormone produce growth?
3. How does the human body change during childhood?

Thinking Critically
4. Compare the birth of a human baby with that of a marsupial mammal.

Skill Review
5. **Recognizing Cause and Effect** Someone tells you that as people age, their personalities normally change. Do you think this statement is valid? Why or why not? For more help, refer to Thinking Critically in the *Skill Handbook*.

Bioengineered Human Growth Hormone

Heightened Concern Over Growth Hormone

SCIENTIFIC NEWS   DECEMBER 8, 1990

In 1985, the United States and England approved the use of recombinant human growth hormone—a genetically engineered version of a compound normally produced in the body—for children suffering from dwarfism. Researchers are now investigating the drug's potential for short children, even though their bodies are not lacking the hormone.

While most ten-year-old boys dream of being brave and strong, Joshua just wants to be taller. Not taller than his friends, but simply as tall. At 3 feet 11 inches, Joshua is at the bottom of the growth curve. Joshua believes the teasing he undergoes will stop if he grows to 5 feet.

Nightly injections   Joshua's parents are concerned also. Not only do they worry about the effect being small has on his self-image, they also are concerned with its influence on Joshua's ability to compete—now and in the future. So concerned are his parents that they inject Joshua nightly with a new and powerful drug that stimulates bone growth. The synthetic form of human

growth hormone (hGH) has been found to significantly add inches to children who are deficient in the natural hormone by replacing what the body is not producing. The effect on children like Joshua with no such deficiency is as yet unknown. In fact, after many years of costly treatment, Joshua may not be any taller than he would have been if left to grow naturally.

Different Viewpoints

Cosmetic treatment rather than medical
Should short but otherwise healthy children be given a drug to help them grow taller? At issue is the question of exploiting the advances made in bioengineering to treat children for cosmetic rather than medical purposes. Experts also worry that treating nonhormone-deficient children with additional growth hormone may produce serious side effects.

Other uses   Today, the market for hGH has expanded as researchers have uncovered its properties of promoting new muscle tissue growth and inhibiting fat accumulation. These properties seem to temporarily reverse the aging process, a fact that interests many in the growing population of senior citizens.

The advances made in bioengineering technology are certain to continue at a dynamic pace. With those advances, society will be forced to make moral decisions concerning where those advances may lead us.

INVESTIGATING the Issue

1. **Write** Prepare a survey that will determine people's opinions on short stature. Give the survey to your family and classmates. Analyze the data you collect.
2. **Research** Use reference materials to find out about other bioengineered drugs being used in medicine.

41.1 Human Reproductive Systems

- The male reproductive system produces sperm and the female reproductive system produces eggs.
- Through the control of the hypothalamus and pituitary, hormones act on the reproductive system as well as other body systems. Their levels are regulated by negative feedback.
- Changes in males and females at puberty are the result of the production of FSH, LH, and other sex hormones.
- Under the control of hormones, the menstrual cycle produces a mature egg and prepares the uterus for receiving a fertilized egg.

41.2 Development Before Birth

- Fertilization occurs in the oviduct. The ball of cells that develops from the fertilized egg implants in the uterine wall.
- The embryo changes from a small ball of cells to a well-developed fetus over the course of nine months.
- Genetic counseling offers people information about their chances of having a child with a genetic disorder.

41.3 Birth, Growth, and Aging

- Birth involves dilation of the cervix, expulsion of the baby, and release of the afterbirth.
- Infancy, childhood, adolescence, and adulthood are the stages of human development. Human growth hormone (hGH) produces growth in all body cells, especially in cells of the skeleton and muscles.

Key Terms

Write a sentence that shows your understanding of each of the following terms.

amniocentesis	oviduct
bulbourethral gland	ovulation
cervix	pituitary
corpus luteum	prostate gland
epididymis	puberty
follicle	scrotum
genetic counseling	semen
implantation	seminal vesicle
labor	umbilical cord
menstrual cycle	vagina
negative-feedback system	vas deferens

Understanding Concepts

1. What is the adaptive advantage of external testes?
2. Compare egg and sperm production.
3. Explain the general effects of FSH and LH on females during puberty.
4. Compare the events of the first trimester of pregnancy with those of the third trimester.
5. How do the hypothalamus and pituitary gland work together?
6. What are secondary sex characteristics, and which hormones produce them?
7. When is human development most rapid?

Using a Graph

8. This graph indicates changes in cardiac output and heart rate in a woman over the course of pregnancy. Explain the changes.

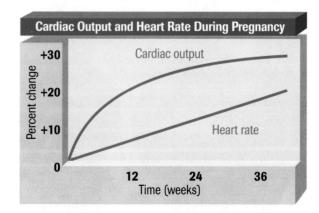

Cardiac Output and Heart Rate During Pregnancy

Relating Concepts

9. Make a concept map that relates the following terms and phrases. Supply the appropriate linking words for your map.

   negative-feedback system, FSH, LH, testosterone, testes, semen, vas deferens, epididymis, puberty

Applying Concepts

10. If the testes were removed from an adult male, what would happen to the signal from the hypothalamus to the pituitary? To the levels of FSH and LH?

11. A woman tells her physician that she is 50 days past the first day of her last menstrual period. How many days has the embryo been developing?

12. Why do pregnancy tests check for chorionic gonadotropin?

Art Connection

13. How might an artist's own experiences with family influence his/her paintings of children or childhood?

Thinking Critically

Interpreting Tables

14. The table shows the critical times during fetal development when certain abnormalities can result from exposure of the mother to various problems. As a doctor, what would you advise a patient who smokes about how her habit might affect her child?

Relating Cause and Effect

15. **Biolab** In gigantism, the pituitary gland produces too much hGH during childhood, resulting in excessively long bones. Where does the negative-feedback system appear to fail in the case of gigantism?

Relating Cause and Effect

16. **Thinking Lab** If a dominant baboon's hypothalamus were damaged so that it secreted lowered levels of FSH and LH, what effects would be seen in the behavior of the dominant baboon?

Connecting to Themes

17. **Systems and Interactions** Explain what role the circulatory system has regarding the sex hormones produced by the endocrine system.

18. **Evolution** What is the adaptive advantage for an embryo to be formed from an egg and a sperm?

19. **Homeostasis** How is the placenta important in maintaining homeostasis in the developing fetus?

Environmental Effects on Fetal Development		
Type of exposure	Time of fetal development (weeks)	Damage to fetus
Nutrition: Poor nutrition	weeks: 20-38	brain damage
Disease: German measles	weeks: 0-6	malformed organs
Drugs: Thalidomide (tranquilizer)	weeks: 0-12	missing/deformed limbs
Environment: Smoking	weeks: 19-38	low birth weight 30 percent greater risk of infant death 50 percent greater risk of heart abnormalities

CHAPTER
42 Immunity from Diseases

Schooled in the art of self-defense, a tae kwon do expert can fend off almost any attacker—even more than one attacker at a time. Through skillful kicks and strikes with her hands, feet, elbows, and knees, she can protect herself from harm.

Your body, too, has a system of self-defense. The attackers it must defend against are disease-producing organisms, which can strike anywhere and at any time. Your body's chief experts in self-defense are its white blood cells, which can do everything from engulfing these foreign invaders to producing antibodies that destroy them.

How do these attacking organisms invade your body? How does your body work to defend itself against them? Your immune system holds the answers to these questions.

Magnification: 6000×

This martial arts expert actually has two systems of self-defense: one is the art of tae kwon do and the other is her immune system. However, if an attacking disease organism should happen to temporarily elude her white blood cells and other lines of defense, she may occasionally succumb to a cold or other more serious infection.

42.1 The Nature of Disease

Occasionally everyone gets a cold. Cold viruses enter your body by way of your nose and are swept to the back of your throat by hairlike cilia. Some are washed down your esophagus and destroyed by your digestive system. Others lodge against the lining of your nasal passage, binding tightly to cell receptors. These viruses enter your nasal cells and unleash their genes, taking over your cells' reproductive machinery. Soon you have a sore throat, a stuffed and runny nose, a headache, and a mild fever. How did the cold virus get in your nose in the first place? How did the infection produce these symptoms?

What Is an Infectious Disease?

The cold virus is an example of a microbe that causes a disease—a change that disrupts the homeostasis in the body. Disease-producing agents such as bacteria, protozoa, fungi, viruses, and other parasites are called **pathogens.** The main sources of pathogens are soil, contaminated water, and infected people or animals. Any disease caused by the presence of pathogens in the body is called an **infectious disease.** Some of the infectious diseases that occur in humans are shown in *Table 42.1.*

Not all microorganisms are pathogenic. In fact, the presence of some microorganisms in your body is beneficial. You are microbe-free before birth. At birth, microbes establish themselves on your skin and in your upper respiratory system, lower urinary tract, reproductive tract, and intestinal tract. *Figure 42.1* shows some common microorganisms that live on your skin.

Figure 42.1

These microorganisms establish a more-or-less permanent residence in or on your skin, but do not cause disease under normal conditions. They have a symbiotic relationship with your body. However, if you become weakened or injured, these same organisms are potential pathogens.

Magnification: 17 000×

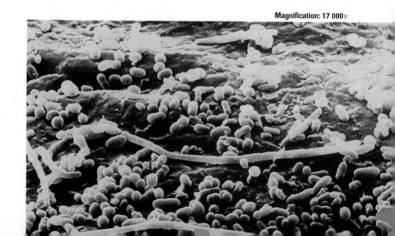

Table 42.1 Human Infectious Diseases

Disease	Cause	Affected Body System	Transmission
Smallpox	Virus	Skin	Droplet
Chicken Pox	Virus	Skin	Droplet
Cold Sores	Virus	Skin	Direct Contact
Rabies	Virus	Nervous System	Animal Bite
Poliomyelitis	Virus	Nervous System	Contaminated Water
Infectious Mononucleosis	Virus	Salivary Glands	Direct Contact
Colds	Viruses	Respiratory System	Direct Contact
Influenza	Viruses	Respiratory System	Droplet
HIV/AIDS	Virus	Immune System	Exchange of body fluids
Hepatitis B	Virus	Liver	Direct Contact
Measles	Virus	Skin	Droplet
Mumps	Virus	Salivary Glands	Droplet
Tetanus	Bacteria	Nervous System	Deep wound
Food Poisoning	Bacteria	Digestive System	Contaminated Food
Tuberculosis	Bacteria	Respiratory System	Droplet
Whooping Cough	Bacteria	Respiratory System	Droplet
Spinal meningitis	Bacteria	Nervous System	Droplet
Impetigo	Bacteria	Skin	Direct Contact

Determining What Causes a Disease

One of the first problems scientists face when studying a disease is finding out what causes the disease. Not all diseases are caused by pathogens. Disorders such as hemophilia, caused by a recessive allele on the X chromosome and discussed in Chapter 15, or rheumatoid arthritis are inherited. Others, such as osteoarthritis, are caused by wear and tear on the body as it ages. Only infectious diseases are caused by pathogens. In fact, about half of all human diseases are infectious. In order to determine which pathogen causes a specific disease, scientists follow a standard set of procedures.

First pathogen identified

The first proof that a microbe actually caused a disease came from the work of Robert Koch in 1876. Koch, a German physician, was working on the cause of anthrax, a deadly disease that mainly affects cattle and sheep but can also occur in humans. Koch discovered a rod-shaped bacterium in the blood of cattle that had died of anthrax. He cultured the bacteria on nutrients and then injected samples of the culture into healthy animals. When these animals became sick and died, Koch isolated the bacteria in their blood and compared them with the bacteria originally isolated. He found that the two sets of blood cultures contained the same bacteria.

A procedure to establish the cause of a disease

Koch established experimental steps, shown in *Figure 42.2*, for directly relating a specific microbe to a specific disease. These steps today are known as **Koch's postulates:**

1. The pathogen must be found in the host in every case of the disease.
2. The pathogen must be isolated from the host and grown in a pure culture.
3. When the pathogen from the pure culture is placed in a healthy host, it must cause the disease.
4. The pathogen must be isolated from the new host and be shown to be the original pathogen.

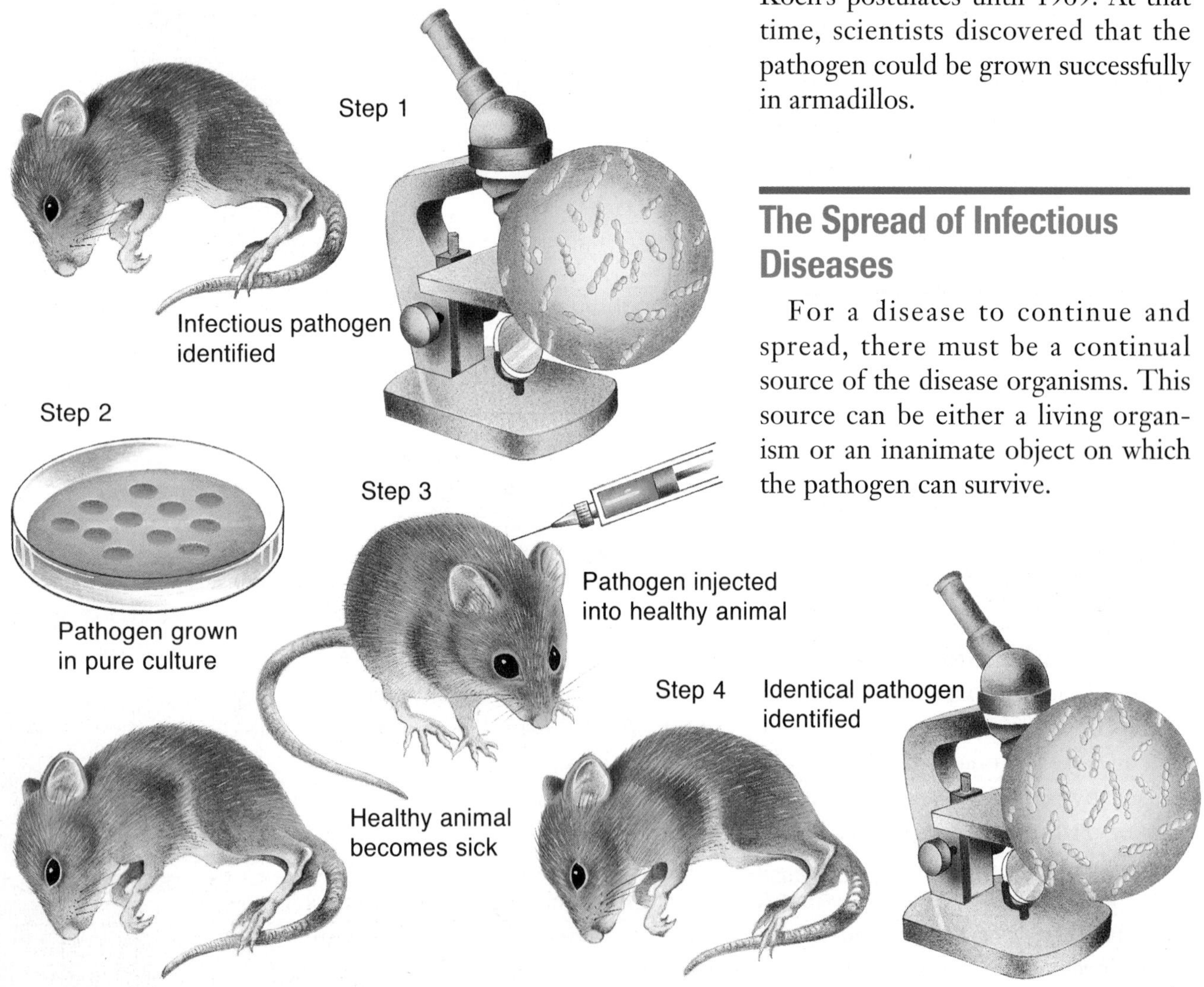

Figure 42.2

Koch's postulates are steps used to identify an infectious pathogen. Step 3, injecting the pathogen into a healthy host, is not morally acceptable using human subjects, so other susceptible hosts are used.

Exceptions to Koch's postulates

Although Koch's postulates are useful in determining the cause of most diseases, some exceptions exist. Some organisms, such as the pathogenic bacterium that causes the sexually transmitted disease syphilis, have never been grown on an artificial medium. Viral pathogens also cannot be cultured this way because they multiply only within cells. Finding living tissue for a culture medium other than human tissue can be difficult. For example, the bacterial pathogen that causes leprosy, a potentially fatal disease that afflicts a person's skin and nerves, was first isolated in 1870. Because the pathogen will not grow on an artificial medium, there was a problem carrying out Koch's postulates until 1969. At that time, scientists discovered that the pathogen could be grown successfully in armadillos.

The Spread of Infectious Diseases

For a disease to continue and spread, there must be a continual source of the disease organisms. This source can be either a living organism or an inanimate object on which the pathogen can survive.

Step 1
Infectious pathogen identified

Step 2
Pathogen grown in pure culture

Step 3
Pathogen injected into healthy animal
Healthy animal becomes sick

Step 4
Identical pathogen identified

Reservoirs of pathogens

The main source of human disease pathogens is the human body itself. In fact, the body can be a reservoir of disease-causing organisms. Many people harbor pathogens and transmit them directly or indirectly to other people. Sometimes, people can harbor pathogens without exhibiting any signs of the illness and unknowingly transmit the pathogens to others. These people are called *carriers* and are a significant reservoir of infectious diseases.

Other people may unknowingly pass a disease during its first stage, before they begin to experience symptoms. This symptom-free period, while the microbes are multiplying within the body, is called an incubation period. Humans can unknowingly spread colds, streptococcal throat infections, and sexually transmitted diseases (STDs) such as gonorrhea and HIV/AIDS during the incubation periods of these diseases.

Animals are other living reservoirs of microorganisms that cause disease in humans. For example, some types of influenza, or flu, and rabies are often transmitted to humans from animals.

The major nonliving reservoirs of infectious diseases are soil and water. Soil harbors pathogens such as fungi and the bacterium that causes botulism, a type of food poisoning. Water contaminated by feces of humans and other animals is a reservoir for several pathogens, especially those responsible for intestinal diseases.

Transmission of disease

How are pathogens transmitted from a reservoir to a human host? Pathogens can be transmitted from reservoirs in four main ways: by direct contact, by an object, through the air, or by an intermediate organism called a vector. *Figure 42.3* illustrates each way.

Figure 42.3

Diseases can be transmitted to humans from reservoirs in various ways.

 **Insects and arthropods are the most common vectors.**

 **The most common ways for a disease to be transmitted by direct contact involve touching, kissing, and/or sexual contact.**

▲ **Common inanimate objects such as food, water, drugs, toys, dishes, and intravenous needles can harbor and transmit pathogens.**

 **Airborne transmission by droplets of water or dust spreads pathogens from the reservoir to the host.**

1071

How are diseases spread?

Among microorganisms, the ability to move from place to place is generally limited. Microorganisms cannot travel over long distances by themselves or fly or climb on their own. Therefore, unless they are transferred from one animal or plant to another, infections will not spread. One method of transference is by direct contact with an infected animal or plant.

Procedure

1. Label plastic bags 1 to 4.

2. Put a fresh apple in bag 1 and seal the bag.

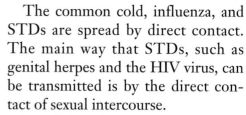

3. Rub a rotting apple over the entire surface of the remaining three apples. The rotting apple is your source of microorganisms.

4. Put one of the apples in bag 2.

5. Drop one apple to the floor from a height of about 2 m. Put this apple in bag 3.

6. Use a cotton ball to spread alcohol over the last apple. Let the apple air-dry for a short time, and then place it in bag 4.

7. Store all of the bags in a dark place for one week. **CAUTION:** *Make sure to wash your hands after handling microorganisms.*

8. At the end of the week, compare all the apples. Record your observations. **CAUTION:** *Give all apples to your teacher for proper disposal.*

Analysis

1. What was the purpose of the fresh apple in bag 1?

2. Explain what happened to the rest of the apples.

3. Why is it important to clean a wound with a chemical like alcohol?

The common cold, influenza, and STDs are spread by direct contact. The main way that STDs, such as genital herpes and the HIV virus, can be transmitted is by the direct contact of sexual intercourse.

Food poisoning is a common example of a disease transmitted by an object. This disease is often transmitted by contamination of the food by the food handler. In order to help prevent transmission of these types of diseases, restaurants have equipment that cleans and disinfects their dishes and utensils. Today, laws require food handlers to wash their hands thoroughly before preparing food, and frequent inspections of restaurants help prevent unsanitary conditions.

Diseases transmitted by vectors are most commonly spread by insects and arthropods. In Chapter 22, you learned about the protozoan disease malaria, which is transmitted by mosquitoes, and in Chapter 31, you read about Lyme disease transmitted by ticks. The bubonic plague—a horrible disease that swept through Europe in the 1600s, killing one-fourth to one-third of the population—was transmitted from infected rats to humans by fleas. Flies also are significant vectors of disease, as shown in *Figure 42.4.*

Figure 42.4

Houseflies transmit disease when they land on infected materials, such as animal wastes, and then land on fresh food that is eaten by humans.

What Causes the Symptoms of a Disease?

When a pathogen invades your body, it initially encounters your immune system. If the pathogen overcomes your defense system, it can cause damage directly in the tissues it has invaded. As the pathogens metabolize and multiply, they can kill host cells.

Damage to the host by viruses and bacteria

As you learned in Chapter 21, viruses invade cells and take over a host cell's genetic and metabolic machinery. Many viruses also cause the eventual death of the cells they invade.

Most of the damage done to host cells by bacteria is done by toxins. Toxins are poisonous substances that are sometimes produced by microorganisms. These poisons are transported by the blood and can cause serious and sometimes fatal effects. Some toxins produce fever and cardiovascular disturbances. Toxins can also inhibit protein synthesis, destroy blood cells and blood vessels, and disrupt the nervous system, causing spasms.

For example, the toxin produced by the tetanus bacteria affects nerve cells and produces uncontrollable muscle contractions, *Figure 42.5.* If the condition is left untreated, paralysis and death occur. Tetanus bacteria are normally present in soil. If dirt with the bacteria is transferred into a deep wound on your body, the bacteria begin to produce the toxin in the wounded area. A small amount of this toxin, about the same amount as the ink to make a period on this page, would kill 30 people. That is why you should be vaccinated for tetanus.

Figure 42.5

Conditions of a battlefield are ideal for the tetanus bacteria. Before the days of modern medicine, wounded soldiers faced the additional, deadly danger of becoming infected with tetanus bacteria.

How does the herpes simplex virus spread?

Herpes simplex virus, which causes cold sores, infects a person for life, occasionally reproducing and then spreading to other cells in the body of its host. Scientists have been interested for a long time in how the herpes virus actually enters a cell.

Analysis

Scientists have found that the herpes virus infects a cell in one of two possible ways. It may latch onto a cell receptor with its own glycoprotein spike, or it may use this spike to grab a growth factor molecule that latches onto the receptor, as shown in the diagram.

Thinking Critically

Design an experiment to determine which method the herpes virus uses to get into a cell.

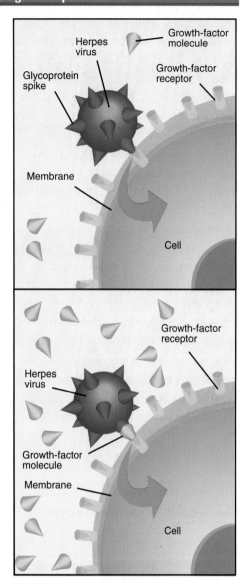

Patterns of Diseases

In today's highly mobile world, diseases can spread rapidly. Contaminated water, for example, can affect many thousands of people quickly. Therefore, identifying a pathogen, its method of transmission, and the geographic distribution of a disease is a major concern of government health departments. The Centers for Disease Control (CDC), the central source of disease information in the United States, publishes a weekly report about the incidence of specific diseases.

Some diseases, such as typhoid fever, occur only occasionally in the United States. An outbreak occurs periodically and is often due to someone traveling in a foreign country and bringing the disease back to the states. On the other hand, many diseases are constantly present in the population. Such a disease is called an **endemic disease.** The common cold is an endemic disease.

Sometimes, an epidemic breaks out. An **epidemic** occurs when many people in a given area are afflicted with the same disease in a relatively short period of time, *Figure 42.6.* Influenza is a disease that often achieves epidemic status, sometimes spreading to many parts of the world.

Figure 42.6

A polio epidemic spread across the United States in the 1950s. Victims of the disease were paralyzed or died when the polio virus attacked the nerves of the brain and spinal cord. Many survived only after being placed in an iron lung— a machine that allowed the patient to continue to breathe.

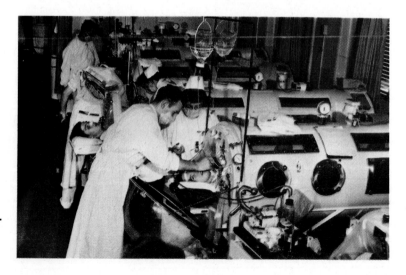

Treating Diseases

When a person becomes sick, the disease often can be treated with medicinal drugs, such as antibiotics. Any substance produced by a microorganism that, in small amounts, will kill or inhibit the growth and reproduction of other microorganisms is an **antibiotic**. Antibiotics are produced naturally by various species of bacteria and fungi. Although they work to cure some bacterial infections, antibiotics do not destroy viruses.

A problem that sometimes occurs with the continued use of antibiotics is that the bacteria become resistant to the drugs. That means the drugs become ineffective. Penicillin, an antibiotic produced by a fungus, was used for the first time in the 1940s and is still one of the most effective antibiotics used today. However, over the past 50 years that penicillin has been used, more and more types of bacteria have evolved that are resistant to it. One example of resistance is shown in *Figure 42.7*.

The use of antibiotics is only one method that helps your body fight off infections. In addition to medicinal drugs, your body has its own built-in defense system that continually works to keep you healthy.

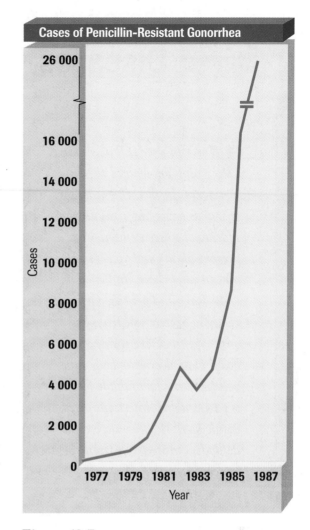

Figure 42.7

Bacteria that are resistant to penicillin produce an enzyme that breaks down this antibiotic. In certain infections, such as gonorrhea, an STD, this resistance has caused a problem because penicillin has been the best drug to treat the infection. Notice the increase in the number of reported cases of this infection in the United States.

Section Review

Understanding Concepts
1. What are the major reservoirs of pathogens?
2. In what way can a family member who is in the incubation period for strep throat be a threat to your good health?
3. When does a disease become an epidemic?

Thinking Critically
4. Many patients enter the hospital with one medical problem but contract an infection while in the hospital. What are possible ways in which a disease might be transmitted to a hospital patient?

Skill Review
5. **Designing an Experiment** Design an experiment to determine whether a recently identified bacterium causes a type of pneumonia. For more help, refer to Practicing Scientific Methods in the *Skill Handbook*.

Y*ou can't see it, but a war is going on around these teenagers. In fact, the same sort of war is occurring around you. Hordes of unseen enemies are present all around you—in the air, on your chair, and even on your books and pencils. Defenders ready to protect you from the onset of attack are inside your body. How does your body save you from the formidable foes that try to establish infectious diseases inside you? Which microscopic defenders protect you from these unseen enemies?*

The Lymphatic System

Your lymphatic system is one line of defense your body has against outside pathogens. In fact, this system performs three basic functions. Within your body, the lymphatic system maintains homeostasis by keeping body fluids at a constant level. This system also absorbs fats from the digestive tract, as described in Chapter 38, and it helps the body defend itself against disease. *Figure 42.8* shows the major organs and vessels that make up the lymphatic system.

Pathways through the lymphatic system

The cells of your body are constantly bathed with fluid. This **tissue fluid** is the fluid that forms when water and dissolved substances diffuse from the bloodstream into intercellular spaces and the surrounding tissues. The tissue fluid then collects in open-ended lymph capillaries. Once the tissue fluid enters these lymphatic vessels, it is called **lymph.**

Lymphatic capillaries meet to form larger vessels called lymph veins. The flow of lymph is only toward the heart, so there are no lymph arteries. The lymph veins converge to form two lymph ducts. These ducts return the lymph to the bloodstream in the shoulder area. However, before it is returned, the lymph has been filtered through various lymph organs.

Organs of the lymphatic system

At locations along the lymphatic pathways, the lymph vessels pass

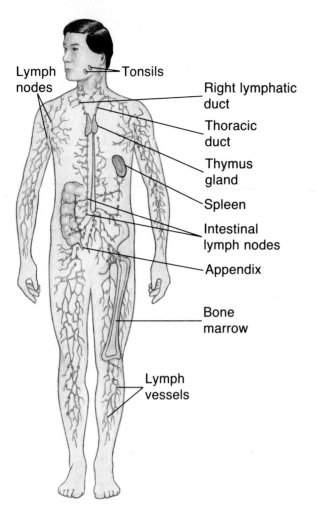

Lymph nodes
Tonsils
Right lymphatic duct
Thoracic duct
Thymus gland
Spleen
Intestinal lymph nodes
Appendix
Bone marrow
Lymph vessels

Figure 42.8

The lymphatic system consists of organs and vessels throughout the body. Each plays a role in filtering out harmful or waste particles. As you read about the various vessels, tissues, and organs of the lymphatic system, locate them on this diagram.

through lymph nodes. A **lymph node** is a small mass of tissue that filters lymph, as shown in *Figure 42.9.* Lymph nodes are made of an interlaced fiber network that holds lymphocytes. A **lymphocyte** is a type of white blood cell that defends the body against foreign substances.

Have you ever had a sore throat caused by infected tonsils? The tonsils are unusually large groups of lymph nodes located at the back of the mouth cavity and at the back of the throat. They form a protective ring around the openings between the nasal and oral cavities. Your tonsils provide protection against bacteria and other harmful material that enters your nose and mouth.

The spleen detects and responds to foreign substances in the blood. It also filters out and destroys bacteria and worn out red blood cells and acts as a blood reservoir. Unlike the lymph nodes, the spleen does not filter lymph.

The thymus gland, located above the heart, is another lymphatic organ. The size of your thymus gland differs markedly depending on age. In newborns and young children, it is quite prominent, and it continues to grow until puberty, although not as rapidly as other body structures. After puberty, it gradually decreases in size. The thymus gland processes some of the lymphocytes that are involved in the body's defense system. It is here that the lymphocytes mature and differentiate into cells that fight specific pathogens.

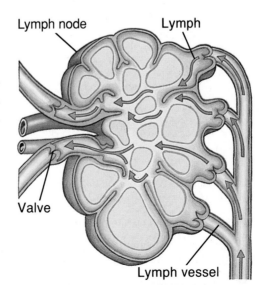

Lymph node
Lymph
Valve
Lymph vessel

Figure 42.9

When lymph filters through a lymph node, the fiber network of the node traps microorganisms. The presence of the microorganisms stimulates the lymphocytes to divide and increase in numbers.

42.2   Defense Against Infectious Diseases   **1077**

What are the percentages of different kinds of white blood cells?

There are five types of white blood cells that defend your body: neutrophils, lymphocytes, monocytes, eosinophils, and basophils.

Procedure

1. Make a data table to record numbers of cells.

2. Mount a slide of blood cells on the microscope, and focus on low power. Turn to high power and examine the white blood cells.

3. Find a neutrophil. Its nucleus has several lobes, usually three. Neutrophils are phagocytic cells that arrive first at a wound site.

4. Find a lymphocyte. Lymphocytes have nuclei that nearly fill the cell.

5. Find a monocyte. This phagocyte is two to three times larger than the other cells. Its round nucleus fills about half of the cell.

6. Find a basophil. These histamine-releasing cells are covered with granules that are stained purple. Eosinophils are phagocytic cells covered with granules that are stained pink.

7. Count a total of 50 white blood cells, and record how many of each type you see.

8. Calculate the percentage by multiplying the number of each cell type by two. Record the percentages. Diagram each cell type.

Type of white blood cell	Number counted	Percent	Diagram

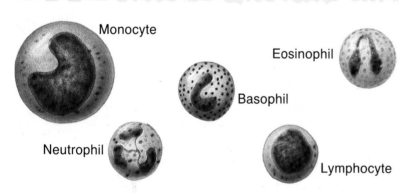

Analysis

1. Which type of white blood cell was most common? Second most common?

2. What is the major difference between red and white blood cells?

3. Why would you expect the white blood cell count to go up during an infection?

Nonspecific Defense Mechanisms

You know that pathogens are constantly bombarding your body. Yet your body is also constantly working to fight them off. Some of the defense mechanisms your body uses are effective against a wide variety of pathogens. In other words, these defense mechanisms are nonspecific. Nonspecific defense mechanisms include several kinds of both physical and chemical barriers.

Skin—The first line of defense

When a potential pathogen contacts your body, the first organ that it confronts is often your skin. Like the walls of a castle, intact skin is a formidable physical barrier to the entrance of microorganisms. Recall that the outer skin consists of many layers of keratinized cells that are closely packed together. Skin is also normally populated by millions of microorganisms that inhibit the multiplication of pathogens that land on the skin.

Secretions that destroy microbes

Besides the skin, pathogens encounter your body's secretions of mucus, sweat, tears, and saliva. The main function of mucus is to prevent various areas of the body from drying out. Because mucus is slightly viscous (thick), it also traps many microbes and other foreign substances that enter the respiratory and digestive tracts. Mucus is continually swallowed and passed to the stomach, where acidic gastric juice (made of hydrochloric acid and other fluids) destroys most bacteria and their toxins. Sweat, tears, and saliva all contain the enzyme lysozyme, which is capable of breaking down the cell walls of some bacteria.

Phagocytosis of microbes

If a pathogen manages to get past the skin and body secretions, your body has several nonspecific defense mechanisms remaining that can destroy the pathogen and restore homeostasis. Some foreign cells encounter body cells that carry on phagocytosis. Recall from Chapter 9 that phagocytosis occurs when a cell engulfs a particle. A **phagocyte** is a type of white blood cell that ingests and destroys pathogens by surrounding and engulfing them. When a pathogen is present, phagocytic cells migrate out of your blood capillaries to the infected areas. One type of phagocyte, called a **macrophage,** that combats invading pathogens is shown in *Figure 42.10.* Known as the giant scavengers or big eaters, macrophages develop from maturing monocytes.

After macrophages engulf large numbers of microbes and damaged tissue, they eventually die. After a few days, the infected area harbors a collection of dead white blood cells and various body fluids called **pus.** Pus formation usually continues until the infection subsides.

Inflammation of body tissues

When microbes damage body tissues, inflammation may result. Inflammation is a reaction to any type of injury, not just infection. Physical force, chemical substances, extreme temperatures, and radiation can also inflame body tissues. Inflammation is characterized by four

Figure 42.10

Macrophages migrate out of capillaries by squeezing between the cells of the capillary wall. Macrophages will attack anything they recognize as foreign, including microbes and dust particles that are breathed into the lungs. Once the macrophage has ingested the foreign material, lysosomal enzymes that digest the foreign matter are released.

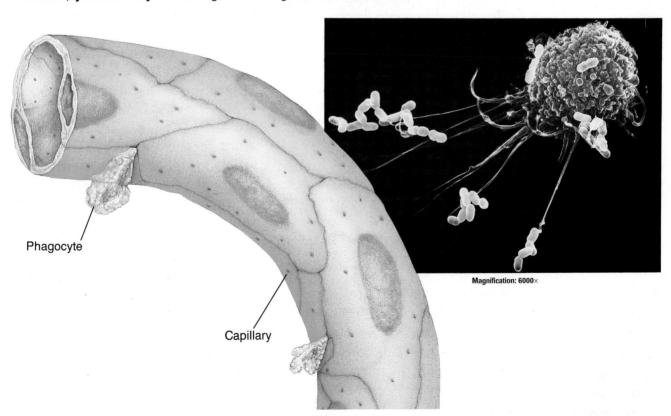

Phagocyte

Capillary

Magnification: 6000×

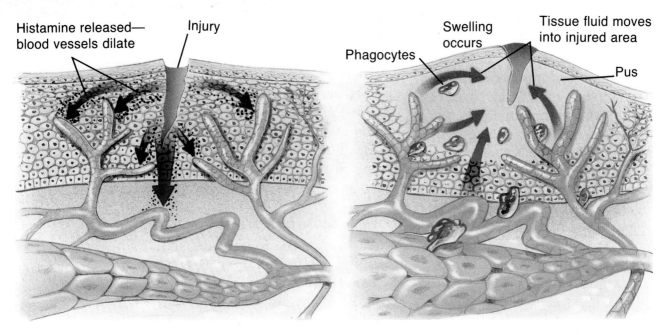

Histamine released— blood vessels dilate

Injury

Swelling occurs

Tissue fluid moves into injured area

Phagocytes

Pus

Figure 42.11

Histamine is produced when tissue damage occurs. Histamine causes blood vessels in the injured area to dilate (left). Therefore, they become more permeable to tissue fluid. This increase in tissue fluid in the injured area helps the body destroy toxic agents and restore homeostasis. The dilated blood vessels cause the redness of an inflamed area (right). The increase in tissue fluid causes the swelling.

symptoms—redness, swelling, pain, and heat. It begins when damaged cells release histamine, *Figure 42.11.*

As inflammation proceeds, phagocytes migrate into the injured area and begin to ingest pathogens. They also release a chemical that causes the hypothalamus to reset the body's temperature, causing a fever. Up to a certain point, fever is actually a defense against disease. A higher body temperature interferes with the metabolism of some microorganisms and speeds up the body's reactions, possibly helping body tissues repair themselves more quickly.

Antimicrobial substances

Another nonspecific defense is called complement. **Complement** is a group of proteins found in the blood that attach themselves to the surfaces of pathogens. When these protein molecules attach to path-

ogens, they help your body destroy the pathogens by damaging their plasma membranes and attracting an increased number of phagocytes.

When an infection is caused by a virus, your body faces a problem. Phagocytes cannot destroy the virus. Recall that a virus multiplies within a host cell. If phagocytes engulf the virus, they are destroyed after the virus multiplies within their cells. One way your body does counter viral infections is with interferons. Interferons are proteins that are host-cell specific; that is, human interferons will protect human cells but will do little to protect cells of other species.

Interferon is produced by a body cell that has been infected by the virus. The interferon diffuses into uninfected neighboring cells, which then produce antiviral proteins that often disrupt viral multiplication.

Specific Defense Mechanisms

When your body is invaded by a pathogen, its nonspecific defenses begin. The nonspecific defenses are an immediate general defense while specific forces are being mobilized—a process that takes several days. Defense against a specific pathogen by building up a resistance to it is called **immunity.** When the body recognizes a specific foreign substance, it develops an immune response to inactivate or destroy it. Specific defense is the job of the lymphatic system, and it includes the production of two kinds of immunity: antibody immunity and cellular immunity.

Initiating antibody immunity

Normally, the immune system recognizes components of the body as self and foreign matter as nonself. Immunity occurs when the system recognizes a foreign substance and responds to it by producing specialized lymphocytes, which then produce specific antibodies. Recall from Chapter 39 that organisms or substances that provoke such a response are called antigens. Antigens are usually proteins and are present on the surfaces of whole organisms, such as bacteria, or on parts of organisms, such as the pollen grains of plants. When a foreign antigen is introduced into your body, it causes the production of antibodies.

Antibody immunity is a type of chemical warfare within your body. Several types of cells are involved. You can see them illustrated and follow the steps of antibody immunity in *Figure 42.12.*

Allergies

Are you finding the great outdoors more and more a headache? Do you sneeze and sniffle your way through every spring and on into fall? If so, you may be one of 50 million Americans who suffer from allergies. You can be allergic to airborne substances like pollen or to pets, foods, medicines, and insect bites. Some people have a genetic tendency toward allergies, or they may develop after repeated exposures to an offending allergen. Whatever the cause, allergies are a major health concern.

What is an allergy? An allergy is the immune system's misguided attack on a foreign, and otherwise harmless, substance called an allergen. When our immune system encounters allergens, our blood produces an antibody that binds to histamine-producing mast cells and causes an allergic reaction. When this occurs in the nose, a runny nose, itchy eyes, and scratchy throat result. When it occurs in the lungs, asthma results. When it occurs throughout the body, anaphylactic shock results, a life-threatening reaction that requires immediate medical attention.

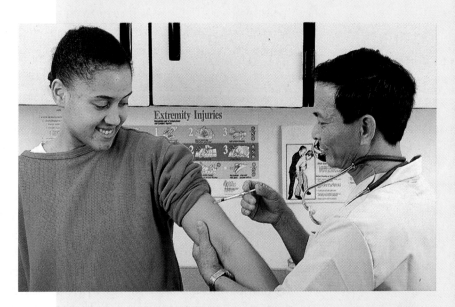

Allergy treatment: Avoidance, medications, and immunotherapy Treatments for allergies include avoiding irritants and doing things such as finding a new home for the family cat or exchanging your goose feather pillow for a synthetic one. Medications like antihistamines and decongestants are also commonly prescribed. Chronic sufferers may need immunotherapy or allergy shots. Injections of specific allergens block the immune system's response and reduce symptoms.

Thinking Critically
Some allergy sufferers find that their food allergies are more severe during the hay fever season. Can you explain why?

42.2   Defense Against Infectious Diseases   **1081**

Figure 42.12
Antibody immunity

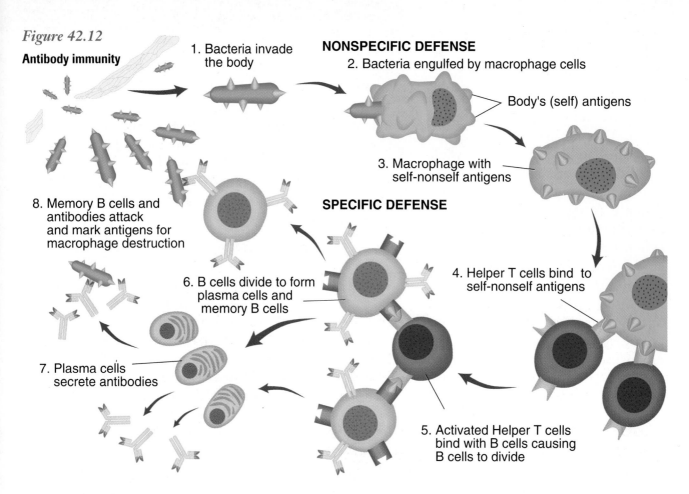

1. Bacteria invade the body

NONSPECIFIC DEFENSE
2. Bacteria engulfed by macrophage cells

Body's (self) antigens

3. Macrophage with self-nonself antigens

SPECIFIC DEFENSE

8. Memory B cells and antibodies attack and mark antigens for macrophage destruction

6. B cells divide to form plasma cells and memory B cells

4. Helper T cells bind to self-nonself antigens

7. Plasma cells secrete antibodies

5. Activated Helper T cells bind with B cells causing B cells to divide

▲ Antibody immunity utilizes B cells in defending your body against invading pathogens.

Bacterial antigen

Viral antigen

▲ An antibody molecule has four chains organized into a Y shape. Sections at the top of the Y are variable, allowing the molecule to recognize and bind with a specific antigen.

When a pathogen invades your body, it is engulfed by a macrophage. Portions of the antigen are taken up by the macrophage and positioned on its plasma membrane. At this time, a type of lymphocyte called a T cell becomes involved. A **T cell** is a lymphocyte that is produced in bone marrow and processed in the thymus gland. Different kinds of T cells play different roles in immunity.

One kind of T cell, called a helper T cell, interacts with B cells. A **B cell** is a lymphocyte that produces antibodies when activated by one of the T cells. B cells are produced from stem cells in bone marrow and released to lymphatic organs.

When an antibody is produced by a B cell, it is released into the bloodstream. Antibodies, *Figure 42.12,* recognize those antigens to which they can fit and bind. This binding results in an antigen-antibody complex. The antigen-antibody complex protects your body in several ways. For example, a complex can neutralize bacterial toxins by blocking their active sites. It can deactivate viruses by attaching to them and preventing them from attaching to host cells. Or it can fix complement, which leads to the lysis of invading cells. Ultimately, the antigen-antibody complex is removed from your body when phagocytes destroy it.

Initiating cellular immunity

Like antibody immunity, cellular immunity also involves T cells and macrophages with antigens on their surfaces, but cellular immunity involves direct contact between the two. T cells, after being processed by the thymus gland, enter the lymphatic system and are stored in the lymph nodes and tonsils. Some of the T cells also circulate in the blood. However, unlike B cells, T cells do not secrete antibodies. Instead, they have antibody-like molecules called antigen receptors attached to their surfaces. These antigen receptors allow T cells to recognize and react to many different kinds of antigens.

Your body produces three main types of T cells: the helper T cells that are involved in antibody immunity, cytotoxic—or killer—T cells, and suppressor T cells. Like B cells, individual T cells seem to be specific for a single antigen, but they do not respond to antigens in the same way that B cells do. Follow *Figure 42.13* to see the steps of T cell response.

Figure 42.13

Cellular immunity

▼ In cellular immunity, a macrophage engulfs an antigen, breaks it down, and places part of the foreign substance on its own cell surface. The macrophage then binds to the antigen receptor on helper T cells, activating the cytotoxic T cells to differentiate and produce identical clones. Some T cells remain behind in the lymph organs as memory T cells. Like memory B cells, these T cells are able to respond rapidly to a second attack. Other T cells travel out to the infected tissue to destroy the pathogen. At the infection site, cytotoxic T cells may destroy the pathogen directly or release chemicals that attract other macrophages to the site.

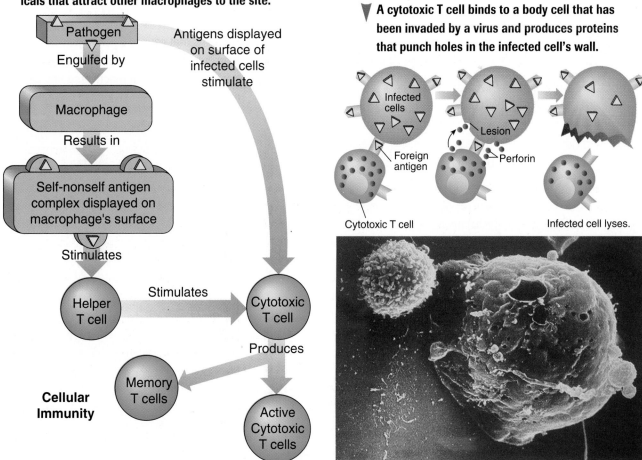

▼ A cytotoxic T cell binds to a body cell that has been invaded by a virus and produces proteins that punch holes in the infected cell's wall.

Magnification: 5000×

Protecting the Future

In some countries, as many as half of the babies die before they can grow up. Immunity for several childhood diseases is now possible when a child is properly vaccinated. The table below shows the diseases that can be prevented by childhood immunization. Some of the information may be new to you because the number of recommended immunizations has been increased. In some cases, immunizations have different names because they have been improved or combined to reduce the number of shots.

Recommended Childhood Immunizations		
Immunization	**Agent**	**Protection Against**
Acellular DPT or Tetrammune	bacteria	diptheria, pertussis (whooping cough), tetanus (lockjaw)
MMR	virus	measles, mumps, rubella
OPV	virus	poliomyelitis (polio)
HBV	virus	hepatitis B
HIB or Tetrammune	bacteria	*Haemophilis influenzae B* (spinal meningitis)

The old, the new, and the improved   Old favorites that are still used are oral polio vaccine (OPV), which prevents the crippling disease poliomyelitis, and MMR, which protects against measles, mumps, and German measles (rubella). The newest recommended vaccine (HBV) is for hepatitis B, a serious disease that affects the liver.

Improved vaccines   Acellular DPT is an enhanced version of the classic DPT shot that protects against diphtheria, whooping cough, and tetanus. Produced by improved methods, this version reduces the side effects, such as pain, fever, and fussiness, caused by the immunization. Another new twist on the classic DPT shot is called Tetrammune. It combines DPT with immunization for *Haemophilis influenzae B* and eliminates the need for a separate injection.

CONNECTION TO   Biology

Vaccines are often made from killed or weakened pathogens. Why might a genetically engineered virus be safer to use for immunization than a weakened one?

Immunity from infectious diseases

Perhaps you had chicken pox as a child, *Figure 42.14.* Most children have had chicken pox by the time they enter school. Why don't you have the disease over again, as you do a cold? The answer is that you have become immune to this virus. Immunity from a disease may be either passive or active. Passive immunity develops as a result of passively acquiring antibodies. Active immunity develops as a result of exposure to antigens, which results in the production of antibodies and memory immune cells.

A glance at the immunization list shows that a chicken pox vaccine is missing. While one exists and may be approved by the FDA soon, a controversy exists about its use. The vaccine doesn't provide permanent immunity, and some scientists are concerned that the disease then will strike adults.

Figure 42.14

The virus that causes chicken pox can remain latent in the body for many years. Years later, it may be reactivated and cause the painful skin disease known as shingles.

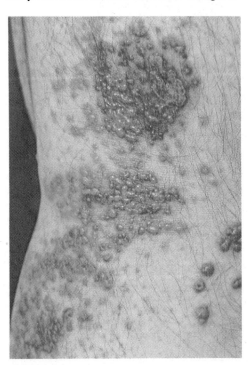

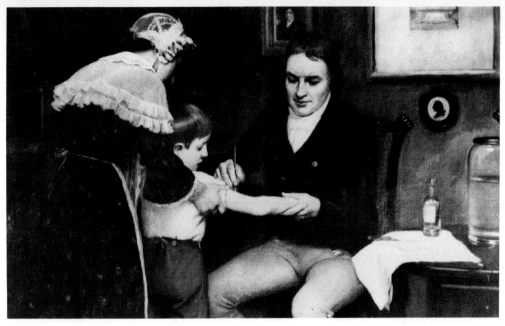

Figure 42.15

Smallpox killed thousands of people up until the middle of the 20th century. A worldwide attack on the disease through vaccinations brought an end to it. Because of the effort by the World Health Organization, smallpox has been eliminated.

Passive immunity may develop in two ways. Natural passive immunity develops when antibodies are transferred from a mother to her unborn baby through the placenta or through mother's milk while nursing the infant. Artificial passive immunity involves injecting antibodies into the body. These antibodies come from an animal or a human who is already immune to the disease. For example, a person who is bitten by a snake might be injected with antibodies from a horse that is immune to the snake venom.

Active immunity is obtained when a person is exposed to antigens. As you've just learned, when a person acquires a disease, the body produces antibodies as well as memory B cells and memory T cells. Once the person recovers, he or she will be immune to the disease if exposed to the pathogen again.

In order to produce active immunity, vaccines have been developed. A **vaccine** is a substance consisting of weakened, dead, or parts of pathogens or antigens that, when injected into the body, cause immunity. This occurs because the body reacts as if it were naturally infected.

In 1798, Edward Jenner, an English country doctor, demonstrated the first safe vaccination procedure. Jenner knew that dairy workers who acquired cowpox from infected cows were resistant to catching smallpox during epidemics. Cowpox is a disease similar to but milder than smallpox. To test whether immunity to cowpox also caused immunity to smallpox, Jenner infected a young boy with cowpox. The boy developed a mild cowpox infection. Six weeks later, Jenner scratched the skin of the boy with viruses from a smallpox victim, shown in *Figure 42.15.* The viruses for cowpox and smallpox are so similar that the immune system cannot tell them apart. The boy, therefore, did not get sick because he had developed antibodies and memory cells.

The HIV that leads to AIDS does not itself cause the life-threatening symptoms associated with the disease. Instead, the virus

AIDS and Its Effect on the Immune Response

weakens a person's immune response to other pathogens that invade the body. This happens because the virus destroys the T cells that help in the production of antibodies. Thus, the immune system's ability to fight disease organisms is severely impaired. The person usually dies from the continually reoccurring secondary infections and cancers.

PREPARATION

Problem
How does AIDS affect the normal immune response?

Objectives
In this Biolab, you will:
- **Plot** graphs that demonstrate a healthy body's normal immune response and an AIDS-infected body's immune response.
- **Compare and interpret** these two different immune responses.

Materials
colored pencils (red and blue)
graph paper (2 pieces)

Magnification: 16 000×

Part A: The Normal Immune Response

1. Make a graph of the data in Table 1. Number the vertical axis from 0 to 10 000 in multiples of 500. Label the horizontal axis with the number of days from 0 to 10. In red, plot the number of microbes on the vertical axis against the number of days. In blue, plot the immune response on the vertical axis against the number of days.

2. Label your graph *Normal Immune Response*.

Part B: The Immune Response in a Person with AIDS

1. On a second piece of graph paper, construct a graph of the data in Table 2. Number the axes as in your first graph, but label the horizontal axis with the number of years instead of days. In red, plot the number of viruses against the number of years. In blue, plot the immune response against the number of years.

2. Label your graph *AIDS Immune Response*.

Table 1 Normal Immune Response

Days	Number of viruses	Immune response (units)
0	1	0
1	1000	0
2	10 000	1
3	1000	100
4	10	1000
5	1	1000
6	0	1000
7	0	100
8	0	10
9	0	1
10	0	1

Table 2 Immune Response in Person with AIDS

Years	Number of viruses	Immune response (units)
0	1	0
1	100	1
2	10 000	50
3	10	1000
4	1	10 000
5	1	10 000
6	1	1000
7	10	500
8	100	100
9	1000	1
10	10 000	1

1. **Analyzing Data** Summarize what happens during the normal immune response and during the immune response of a person with AIDS.

2. **Comparing and Contrasting** How is the AIDS graph similar to that of the normal immune response? How is it different?

3. **Thinking Critically** Why does the number of AIDS viruses increase during years seven through ten?

Going Further

Applying Concepts
How could the HIV antibody be used for blood screening?

Figure 42.16

Kaposi's sarcoma is characterized by blue-violet or brown spots on the patient's body.

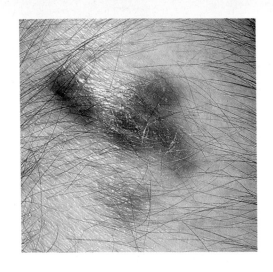

AIDS and the Immune System

In 1981, an unusual cluster of cases of a rare pneumonia caused by a protozoan appeared in the San Francisco area. Medical investigators soon related the appearance of this disease with the incidence of a rare form of skin cancer called Kaposi's sarcoma, *Figure 42.16.* Both diseases seemed associated with a general disease of the body's immune system. By 1983, the pathogen causing this immune system disease had been identified as a retrovirus, now known as Human Immunodeficiency Virus, or HIV. HIV kills helper T cells and leads to the disorder known as Acquired Immune Deficiency Syndrome, or AIDS.

HIV is transmitted when body fluid of an infected person is passed to an uninfected person through direct contact and by contaminated objects.

Intimate sexual contact, contaminated intravenous needles, and blood-to-blood contact such as in transfusions of contaminated blood are methods of transmission. Since 1985, careful screening measures have been instituted by blood banks in the United States to help keep HIV-infected blood from being given to those people who need transfusions. A pregnant woman infected with the virus can also transmit it to her fetus. HIV is not transmitted by social contact.

The incubation period of the AIDS virus may last a long time. The first symptoms of AIDS may not appear until as long as 11 years after the initial infection of the HIV. After the initial stages, many infected persons exhibit symptoms of AIDS-Related Complex (ARC), *Figure 42.17.* Among the symptoms of ARC are swollen lymph nodes, a loss of appetite and weight, fever, rashes, night sweats, and fatigue.

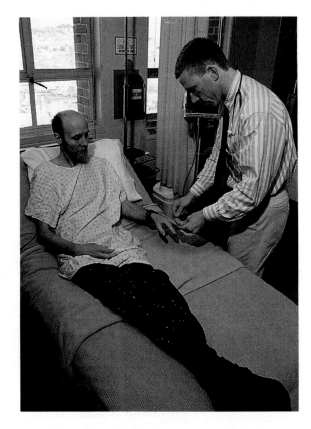

Figure 42.17

ARC (AIDS-Related Complex) may develop after the number of helper T cells in the body is depleted.

The AIDS Epidemic

Research is ongoing into the structure of HIV and the mechanism by which it infects the human body. This knowledge will perhaps help scientists develop a drug treatment or even a vaccine and put an end to the AIDS epidemic.

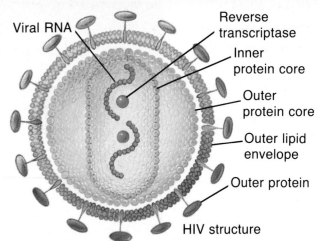

Viral RNA
Reverse transcriptase
Inner protein core
Outer protein core
Outer lipid envelope
Outer protein

HIV structure

2 The HIV attaches one of its outer proteins, clearly visible here (1) to the plasma membrane of a helper T cell (2). The virus then penetrates the cell (3) and leads to the fusion of the two membranes (4). The virus can remain for months. The HIV is a retrovirus that contains RNA and the enzyme reverse transcriptase. Recall from Chapter 21 that the HIV uses its reverse transcriptase to synthesize viral DNA in the host cell.

1 The HIV is basically a set of genes wrapped in proteins. Many researchers are concentrating on the knoblike proteins that lie on the outer envelope that surrounds the virus. They think that this area is important in finding a way to stop the virus.

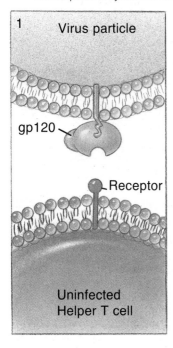

1 Virus particle

gp120

Receptor

Uninfected Helper T cell

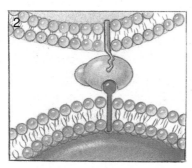

2

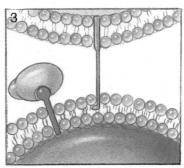

3

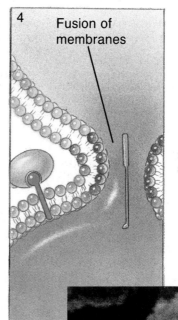

4 Fusion of membranes

3 The virus spreads from cell to cell in a unique manner. After viral DNA has been made, the host cell then begins to make viral proteins. One viral protein, called gp120, is incorporated into the host's plasma membrane. The gp120 protein causes the host cell to attach to other host-cell receptors. This allows cell-to-cell transmission of the virus without the virus leaving the interior of a cell.

Magnification: 1.2 million×

Figure 42.18

The HIV virus does not respect a person's social, economic, racial, or celebrity standing. Power, money, and intellectual, artistic, or athletic talent cannot defeat the disease.

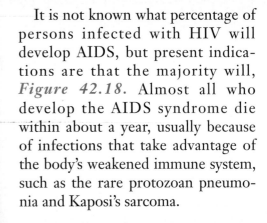

It is not known what percentage of persons infected with HIV will develop AIDS, but present indications are that the majority will, *Figure 42.18.* Almost all who develop the AIDS syndrome die within about a year, usually because of infections that take advantage of the body's weakened immune system, such as the rare protozoan pneumonia and Kaposi's sarcoma.

Connecting Ideas

The immune system is not a system that can keep the body healthy by itself. It relies on the circulation of blood and lymph as well as communication by the nervous and endocrine systems. Like the other body systems you have studied, the immune system works to help maintain homeostasis in your body.

Section Review

Understanding Concepts

1. What role do phagocytes play in defending the body against disease?
2. What role does a lymph node play in defending your body against microorganisms?
3. What is the difference between artificial passive immunity and artificial active immunity?

Thinking Critically

4. Why is it adaptive for memory cells to remain in the immune system after an invasion by pathogens?

Skill Review

5. **Sequencing** Sequence the events that occur in the formation of antibody immunity. For more help, refer to Organizing Information in the *Skill Handbook*.

Human Guinea Pigs

AIDS Researcher Faulted in Vaccine Deaths

MEDICAL DISPATCH   MAY 1, 1993

The National Institutes of Health has imposed serious penalties on a French AIDS researcher after the death of three participants in his AIDS vaccine trials in Paris. The NIH criticized the researcher for failing to report the deaths and for using children as experimental subjects in earlier AIDS vaccine trials.

From the time it was first identified in 1981, AIDS quickly became the medical scourge of the late 20th century. Shortly after HIV was identified as the cause of AIDS, scientists began searching for vaccines that might prevent HIV infection or render the virus harmless in those already infected. The French scientist hoped to be the first to demonstrate the effectiveness of such a vaccine.

Unsuccessful trials   The researcher first tested a crude version of his vaccine in Africa. The vaccine had been prepared for use in animals; however, he chose to test it on humans. His subjects were children, most of them too young to give their informed consent to the procedure. Although the vaccine proved unsuccessful, additional trials were conducted in Paris, which ultimately led to three fatalities.

Different Viewpoints

Pressure for a medical breakthrough   Largely because of intense lobbying by citizens concerned about the slow methods of science to find a cure for AIDS, society has been forced to confront the issue of when to begin human trials on newly developed vaccines, drugs, and other medical procedures.

Some researchers contend that any experimental AIDS drugs or vaccine should first be tested on nonhuman primates. Only if the drug or vaccine helps these primates should human trials proceed. However, some people with AIDS favor early human testing. With so many people dying, these advocates say that time should not be wasted on animal studies that might not apply to humans. In answer to the question of whether trying untested AIDS drugs on humans is ethical, they suggested that they have little to lose.

INVESTIGATING   the Issue

1. **Research** Read articles in newspapers and magazines. Prepare an oral report on how AIDS activists are trying to force drug companies and the Food and Drug Administration to speed up the time it takes to approve experimental drugs for human use.
2. **Write** In a reflective essay, state your views on using humans to test unproven drugs and other therapies. If you suffered from a fatal disease, would you agree to participate in such tests? Explain your reasoning.

Reviewing Main Ideas

42.1 The Nature of Disease
- Infectious diseases are caused by the presence of pathogens in the body.
- The cause of an infection can be established by following Koch's postulates.
- Animals, including humans, and nonliving objects can serve as reservoirs of pathogens. Pathogens can be transmitted by direct contact, through an object, through the air, or by a vector.
- Symptoms of a disease are caused by direct damage or toxins from the pathogen.
- Some diseases are periodic, while others are endemic. Occasionally, a disease reaches epidemic proportions.
- Some infectious diseases can be treated with antibiotics, but pathogens may become resistant to drugs.

42.2 Defense Against Infectious Diseases
- The lymphatic system consists of the lymphatic pathways and the lymph nodes, tonsils, spleen, and thymus.
- Nonspecific defense mechanisms provide general protection against various pathogens.
- Specific defense mechanisms provide a way of fighting particular pathogens by recognizing invaders as nonself. Specific immunity includes the production of antibodies and cellular immunity.
- Caused by the HIV, which damages the immune system, AIDS is a disorder in which other infections invade the body, leading to death.

Key Terms
Write a sentence that shows your understanding of each of the following terms.

antibiotic	lymph node
B cell	lymphocyte
complement	macrophage
endemic disease	pathogen
epidemic	phagocyte
immunity	pus
infectious disease	T cell
Koch's postulates	tissue fluid
lymph	vaccine

Understanding Concepts

1. Why haven't Koch's postulates been applied to all infectious diseases?
2. How can an organism that normally lives on your skin become a pathogen?
3. Tears and saliva contain lysozyme. What does this chemical do?
4. What role does complement play in antibody immunity?
5. In what ways are diseases transmitted?
6. How is immunity different from nonspecific defense mechanisms?
7. How does a person acquire AIDS?
8. Once you have had a disease, why do you normally not get it again?

Relating Concepts

9. Make a concept map that relates the following terms and phrases. Supply the appropriate linking words for your map.

   infectious disease, pathogen, incubation period, AIDS, antibody, antigen, vaccine

Using a Graph

10. Typhoid fever is a bacterial disease of the blood. It is spread by poor sanitation, especially the improper disposal of sewage. The graph below shows the number of cases of typhoid fever in the United States from 1955 to 1987. During what years was the disease in epidemic proportions? What is the approximate endemic level of the disease?

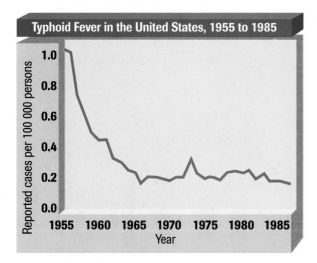

Typhoid Fever in the United States, 1955 to 1985

Applying Concepts

11. If the bacteria that cause tetanus are easily killed by penicillin, why doesn't penicillin cure the disease tetanus?
12. A man went into a pet store and handled a parakeet that had a respiratory illness. One month later, he experienced severe pain in his legs, followed by chills, a fever, diarrhea, and a headache. After two weeks of taking antibiotics, he recovered. Upon returning to the pet store, investigators found many ill parakeets. How could they find out if the man had the same disease as the parakeets?
13. Why must severe burn victims be kept in pathogen-free isolation?
14. Cholera, a waterborne disease, often reaches epidemic proportions after a flood. Explain why this is so.

Health Connection

15. Why might parents decide not to have their child immunized?

Thinking Critically

Interpreting Data

16. When you are vaccinated against a disease, an immune response occurs, creating memory cells. Using the data presented, explain why the second exposure to the pathogen does not cause a disease.

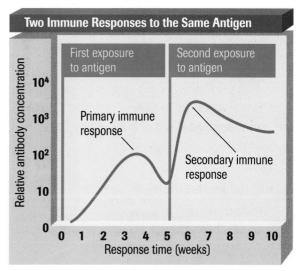

Two Immune Responses to the Same Antigen

Making Inferences

17. Why don't you develop an immunity to colds as you do to mumps?

Applying Concepts

18. **Biolab** Why is an HIV infection itself not the cause of death in an AIDS victim?

Making Predictions

19. **Minilab** What eventually happens to a fresh apple put in a plastic bag? Explain.

Connecting to Themes

20. **Homeostasis** In what way does AIDS upset the homeostasis in the body?
21. **Systems and Interactions** What type of interaction does skin bacteria have with the body? What happens when this interaction is upset?

43 Biology and the Future

As you've discovered through-out this course, biology is very much a part of your life. In fact, you are living at a time when biology is beginning to affect human lives in ways not imagined just a decade or two ago. Advances in the fields of medicine, genetics, and agriculture promise a future of exciting new developments.

Solutions to the future's challenges will come about only as biologists continue to learn about the natural world and develop ways to apply this knowledge. Whether or not you pursue a career in science, you will most certainly play a part in shaping the future.

Doctors will soon have the technology to cure diseases by replacing defective genes with normal ones. What biological studies have led up to this technology?

Researchers at many institutions are rushing to perfect their versions of artificial skin, organs, and other body parts. How will such technology benefit society?

People use oil for energy and a variety of important products, but oil spills can wreak havoc on ecosystems. Can we strike a balance between protecting people's livelihoods and protecting the environment?

Chapter Preview

Section Preview

Objectives

Relate some of the unanswered questions in biology to their importance for the future.

Demonstrate by example the importance of basic research in solving complex biological problems.

Key Terms
none

In October 1993, after analyzing data gathered by NASA's interplanetary probe Galileo, *scientists announced that they had found evidence of an interesting combination of chemicals present on a planet within our solar system. The scientists determined that the chemical signature of this planet indicated signs of life. The name of the planet was Earth.*

Figure 43.1

The Galileo probe experiments showed that we have the technology to discover life on other planets. Life may exist elsewhere in the universe, but it's unlikely that it would resemble anything close to the diversity of life that has evolved on Earth.

The Continuing Search for Answers

As you can imagine, the finding that there is life on Earth came as no surprise to anyone, *Figure 43.1.* Biology is all around us. It's in the fruits and vegetables you buy at the market; it's in the behavior of the dogs, cats, and other animals in our homes, parks, and zoos; and, of course, biology is a part of you, one of the many millions of organisms on this planet.

Figure 43.2

Biology, like all sciences, is an ongoing, dynamic process. In the future, biologists will continue to search for answers about the living world, just as they always have. These scientists are studying sagebrush populations in southeast Oregon.

What we know and don't know

During this biology course, you've learned many facts about the natural world—the unifying concepts of genetics, evolution, and cell structure that underlie the amazing diversity of living things, the complex web of ecological relationships that tie organisms to one another, and the means by which this great diversity of life has evolved and is still evolving.

As you reflect back on this biology course, consider some of the questions you may have had along the way. Were you surprised at some of the unique ways organisms have become adapted to their environments? Did some of the more unexpected ecological interactions of organisms challenge your thinking? Were you amazed to find that the differences between humans and other organisms could be reduced to just a few chemical changes in DNA?

If you were left with many questions about the natural world, then you've just taken the first step toward learning more about biology. In fact, it shouldn't surprise you that some of the questions you may have had about biology are also the questions to which professional biologists continue to seek answers. While biologists have accumulated a vast amount of knowledge about Earth and its organisms, they have seen only the tip of the iceberg when it comes to knowing about the living world, *Figure 43.2.*

Biology in the 21st century

As the science of biology moves into the 21st century, many of the unanswered questions will become critically important. Many of the challenges humans face in the future, such as overcrowding, depletion of critical resources, destruction of the environment, and finding cures to new or old diseases, will require biological solutions. But only by continuing the search for answers in biology can we expect to find such solutions.

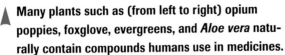

Many plants such as (from left to right) opium poppies, foxglove, evergreens, and *Aloe vera* naturally contain compounds humans use in medicines.

Products produced from plant compounds are used to treat such things as simple burns and chapped skin (*Aloe vera*), common colds, and aches and pains (camphor and aspirin).

Figure 43.3

According to some estimates, about 25 percent of all prescription drugs contain chemicals extracted from plants, 13 percent come from microorganisms, and three percent are derived from animals. The preservation of Earth's organisms will be one of the most challenging problems of the 21st century.

Diversity of Life

You've learned that scientists have identified and studied about 1.5 million different species of organisms, but were you surprised to find out that even more may exist? One of the great unanswered questions in biology is how many different types of organisms inhabit Earth. Biologists have not yet discovered the full extent of Earth's biodiversity.

The question about biodiversity is not just one of basic human curiosity. Biodiversity is perhaps our most valuable, but least appreciated, resource. Knowledge of biodiversity increases our understanding of genetics and evolution by revealing the ranges of variability in species. Biodiversity also directly affects human society. Many of the medicines, *Figure 43.3,* foods, and raw materials humans use are derived from organisms.

As ecosystems shrink due to the growing demands of the human population, the quest for new species will likely grow in importance. By studying Earth's biodiversity, scientists will undoubtedly develop other applications for organisms, and at the same time, gain a better understanding of the complexities of ecosystems.

Complex Interactions

Perhaps you were surprised by the multipurpose roles of bacteria and other microorganisms in ecosystems, or intrigued by the mutually beneficial relationships of insects and plants. The full extent of interactions within ecosystems is something that also remains somewhat of a puzzle to biologists, and ecology is another area that will gain importance as scientists struggle to protect Earth's ecosystems.

Interdependence of organisms

To understand why exploring the full breadth of relationships within ecosystems is important, consider the relationships between insects and flowering plants, *Figure 43.4*. As you've learned, the phylum Arthro-poda contains close to 750 000 different types of insects, and flowering plants constitute close to 18 percent of all known organisms.

This great diversity of insects and flowering plants shows some rather complex and important symbioses. Insects use virtually every anatomical part of flowering plants for food and shelter, while many plants depend on insects for pollination and reproduction. Ultimately, plants owe insects their very lives because many species would not be pollinated without the close relationship developed with an insect. So important are these relationships that many scientists are convinced that if insects were to disappear, so too would humanity. Identifying the important relationships within ecosystems is the first step toward saving them.

Figure 43.4

The important and complex relationships between insects and flowering plants have been evolving for millions of years. Such relationships are indicative of why scientists need to learn more about ecosystems.

▶ **Pipevine swallowtail on thistle**

▼ **Eastern lubber grasshopper on daylily**

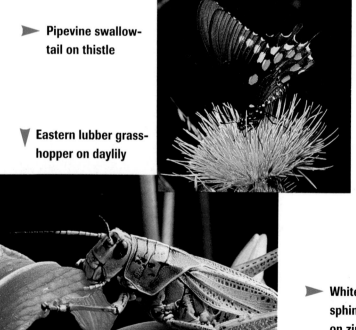

▶ **Locust borer beetle on goldenrod**

▶ **White-lined sphinx moth on zinnia**

Figure 43.5

Through major conservation efforts, scientists were able to restore the sea otter populations and with it the original habitat and biodiversity. By identifying the keystone species in the world's habitats, biologists will be able to concentrate on developing the best methods for preserving ecosystems.

Protecting keystone species

An important area of research that will continue in the future will be the identification of what ecologists call keystone species. Keystone species are those that play central, key roles in ecosystems. Consider the sea otter shown in *Figure 43.5.* Sea otters once thrived in great numbers among the large kelp beds that stretched along the California coast until, by the end of the 19th century, they were hunted to near extinction. In areas where sea otters disappeared completely, an unexpected sequence of events occurred. Sea urchins, a major prey of sea otters, exploded in numbers and proceeded to consume the kelp beds, an important source of shelter and food for many other organisms. Eventually, the ocean floor along the coast was reduced to a barren, desertlike terrain, practically devoid of its original biodiversity.

The existence of keystone species is also indicative of why we need to continue exploring Earth's biodiversity. It's possible that many species heading for extinction at an alarming rate are also unidentified keystone species. The loss of keystone species may have as much influence on habitat destruction as humans themselves. Only by exploring the full extent of interactions within ecosystems will biologists be able to understand the evolutionary factors that influenced the great diversity of life on Earth.

Genetics and Evolution

Did it amaze you to find out that human and chimpanzee DNA is about 99 percent identical, even though humans and chimps have such obvious external differences? If it did, you're in good company. Determining how a few small changes in the chemical composition of DNA can explain the differences among organisms is a challenging research area that will be of central interest in the future.

The differences among species

Knowing how genes contribute to the appearance and functioning of organisms is a critical part of understanding how species evolve. By analyzing genetic differences among species, biologists can differentiate between changes that were influenced by natural selection and those that developed through random mutation. In the case of chimpanzees and humans, scientists have determined that the major differences between these two species evolved through selection for larger brain size in humans.

Genetic engineering and medicine

The applications of genetic studies are important to the development of biotechnology and new medical techniques that improve human lives. You've seen how scientists today are already using genetic-engineering methods to produce medicines for human and veterinary use, hormones such as insulin and growth hormone, and bumper crops of agricultural products, *Figure 43.6.*

Even more exciting for the future of biology will be the medical applications of the Human Genome Project. The goal of this project is to sequence all the DNA in the human genome. While this map will not determine how each gene sequence is expressed, it will give scientists information with which they may be able to manipulate DNA to cure diseases.

Figure 43.6

Imagine a cow that produces skim milk, pigs that grow faster and leaner, sterile pest insects, or naturally decaffeinated coffee beans. Such organisms may sound like science fiction, but they are real possibilities in the growing marriage of biotechnology and agriculture. Biotechnology is still in its infancy, but it's the fastest growing area in science. This trend is expected to continue well into the 21st century.

Coffee beans

Controlled pollination

Genetically-engineered livestock

The Complex Nature of Biology

Genetic engineering and the Human Genome Project offer great possibilities for future human society. Why were they not developed sooner?

Answers to complex problems take time

As you know, biology, like other sciences, is a continuous process that builds upon itself over time. The key word here is *time*. Think how many years it took Gregor Mendel and Charles Darwin to complete their studies. Solutions to major biological questions often involve observations from many different fields of biology. For instance, you now know that the modern understanding of evolution, which has taken over a century and a half to develop, is actually a synthesis of ideas from such diverse biological areas as genetics, ecology, paleontology, and taxonomy, as well as other sciences.

The integration of observations from diverse fields to formulate hypotheses is common to all sciences. But biology is also complicated by a factor that's not as much of a problem in other sciences—the element of change! As you know, organisms, unlike nonliving things, grow, develop, and maintain homeostasis. All of these characteristics involve some form of change, and change takes time, *Figure 43.7*. Because scientists must take these changes into account, even the most simple biological studies sometimes take years to complete.

◄ Jane Goodall did extensive behavioral studies on chimpanzees in Tanzania. She is shown here with Gremlin and her baby Gallahad.

▼ In order not to interfere with natural behaviors, most behavioral studies on animals are done from a distance.

*Figure 43.*7

Behavioral studies of wild animal populations, such as chimpanzees, take many years because behavior changes with age and a host of other factors. In order to make more accurate statements about animal behavior, biologists must observe their subjects over extended periods of time.

Value of basic research

In the future, the production of genetically engineered tomatoes, bacteria, and other organisms will seem rather commonplace as scientists develop even more useful and exciting applications for biotechnology. But one can't forget that genetic engineering itself is the culmination of close to 150 years of basic genetic research beginning with the simple pea plant experiments of Gregor Mendel in the mid-1800s.

All of this points to the value of basic research in science, *Figure 43.8.* As you've seen, scientific studies of even the smallest scale are often integral components in the solutions of major problems. All of the big questions in the future of biology will require contributions from different biological fields, just as they do today.

Figure 43.8

Basic scientific research is necessary before problems can be solved or applications made.

▲ **By making observations and collecting data, biologists accumulate great quantities of information. Much of this information can be applied to solve problems.**

◀ **By sampling and monitoring water quality, biologists can determine problems and protect water quality for aquatic organisms and for drinking.**

Section Review

1. Write an essay about the area of biological research that interested you the most. Why is it interesting to you? In what ways has this research benefited human society? What are some of the unanswered questions in this area of research, and how do you think this field of biology will change in the future?

2. Consult newspapers and science magazines to prepare a report on an application of biotechnology that really impresses you. What basic research was involved in the development of this technology? What are the benefits of this technology, and what, if any, are the risks?

Biologists' Views of the Future

Most biologists would probably agree with this scientist's comment about the future: "I love my life, and I'm so glad I live when I do. My only regret is that I won't see all the exciting changes in science that will happen in the next 100 years." On these pages, a dozen biologists look ahead to the 21st century and speculate on the changes that they hope will develop in their specific fields.

Dr. May Berenbaum, Entomologist Worldwide, the leading cause of death of children under five years of age is from insect-borne diseases. I think a definite possibility in the near future is using genetic engineering to make mosquitoes unable to carry malaria. Other radical, new forms of environmentally compatible control of insects will also improve the quality of life.

Dr. Benjamin Carson, Neurosurgeon I think the marriage of biomedical technology and neurosurgery will be wonderful. As computers become smaller and more biocompatible, it will probably be possible to transplant microcameras instead of artificial eyes to give some blind persons sight. I think microchips will also be inserted in the spinal cord so that paralytic nerves can work again.

Dr. Robert Bakker, Paleontologist Many kids ask me, "Will there be anything left to discover when I'm grown up?" I always tell them that the more we dig, the more we realize we have just scratched the surface of Earth, quite literally. I think that 99 percent of the dinosaurs are still waiting to be discovered.

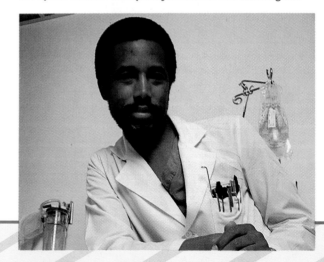

Dr. Marian Diamond, Professor of Anatomy and Museum Director I think the whole field of biology will change tremendously as we learn to understand the human brain, as well as how we interact with other forms of life on Earth. After all, there's a reason it's called the universe; we're all in this together.

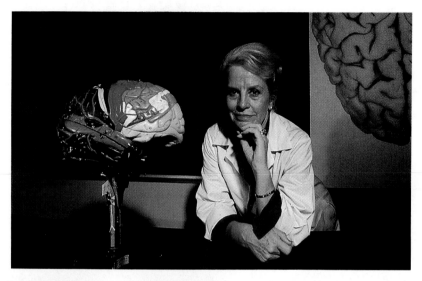

Dr. Robert Murray, Geneticist I see great possibilities in gene therapy, which could mean replacing patients' missing or defective genes with normal genetic components. As I tell my colleagues, genetics holds the secret of life. Some people think that someday we'll be able to synthesize genes and use them to produce new life forms. It's an area of enormous uncertainty because we still don't know a lot about life forms that already exist in nature.

Ms. Flora Ninomiya, Horticulturist Plant growers will continue experiments on less-toxic ways to control disease and pests, such as using wasps to control spiders. We'll also rely more on computers to control temperature, humidity, and watering in greenhouses.

Dr. Baldomero Olivera, Biochemist I think there will be lots of "flavors" of biochemists in the future. A lot of the action in biochemistry will be at the boundaries of different disciplines, such as that between biochemistry and genetics and between biochemistry and ecology.

Ms. Katharine Payne, Naturalist I see increased attempts to design conservation plans that are sensitive to the rural people who live near wild and semi-wild areas. There is a tremendous urgency now to identify all forms of life in the wild places of the world before they vanish due to development. In the past, "conservation" and "development" have been considered separately, but now it's understood that the needs of people and the needs of other life forms and their habitats all must be considered.

Dr. Barbara Staggers, Physician What I would *like* to see is a change in medical care from reactive to proactive. I hope that health-care providers will be encouraged to have important conversations with teens—conversations that could save lives by influencing behavior.

Ms. Lisa Stevens, Assistant Zoo Curator More and more, zoos will be a kind of repository, a genetic bank. Some scientists refer to zoos as places of phylum preservation where gametes are stored for future use. If people can correct some of the environmental errors of the past, perhaps zoo specimens could be used to replenish habitats.

Mr. John Swallow, Forest Technician I see an increasing emphasis on multiple uses for forests. The old mentality that saw forests as just a source of timber is changing as we try to strike a balance between user demands and protecting the environment.

Dr. Sharyn Richardson, Environmental Studies Professor The Human Genome Project, which is attempting to map all the genes, is uncovering some stunning facts about humans as a species and also as individuals. The mapping of the human genome offers great challenges and ethical problems but also our greatest opportunities to make a difference. We're at the forefront of an exciting expansion in the way we view this world and all the living things in it.

EXPANDING YOUR VIEW

1. **Applying Concepts** Choose one field of biology, such as biochemistry, genetics, zoology, or ecology, and suggest how you as a biologist might have an influence on its future.

2. **Journal Writing** What new developments in biology would YOU like to see during your lifetime? Write a paragraph or two to describe them.

SECTION
43.2 Technology and Society: Keeping the Balance

Section Preview

Objectives

Demonstrate by example how different sciences work together to solve critical problems for society.

Identify the benefits, risks, and social concerns of new technology.

Key Terms

Just about everyone has seen or used Velcro, but do you know the origin of this useful product? Actually, Velcro has an innocent, almost amusing origin. One day, while walking through the woods, a Swiss engineer noticed several cockleburs stuck to his clothing. Being interested in this unusual property of cockleburs, he examined them under a microscope and found hundreds of tiny hooks. As a result of this biological observation, Velcro was born!

Velcro

Magnification: 6×

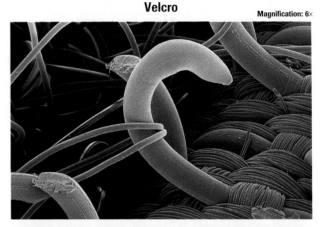

Integration of the Sciences

The development of Velcro can't be attributed only to biological science; it also required contributions from chemistry and physics. Velcro represents a perfect example of how different sciences can merge to solve problems or produce useful products.

Discovering different or better ways to fasten clothing isn't a challenge that humans will face in the future. However, as you've learned, society must address a number of critical problems in the future. These larger problems such as supplying food to the growing human population, protecting the environment, finding cures for human diseases, conserving critical resources, and improving the quality of people's lives aren't simple biological problems. The sciences of biology, chemistry, physics, and mathematics all must come into play to solve various problems that arise in our world. Just as in the development of Velcro, finding solutions to complex biological problems requires the expertise of more than one science.

Figure 43.9

The *Exxon Valdez* spill was the biggest in U.S. history, but oil spills are not uncommon. In the United States alone, about 10 000 oil spills of various types and sizes occur each year. Oil spills happen—the problem is how to clean them up.

Scientists work together to protect the environment

When the oil tanker *Exxon Valdez* spilled more than 11 million gallons of crude oil into the pristine waters of Alaska's Prince William Sound on the frigid morning of March 24, 1989, it resulted in the biggest oil spill in U.S. history, *Figure 43.9.*

As you may well imagine, oil spills have devastating effects on the environment and the organisms that live there. The immediate effects are on the animals, plants, and microorganisms that live near the spill. As shown in *Figure 43.10,* birds, mammals, and fish swimming in the water become covered with thick layers of oil, affecting their ability to move and maintain homeostasis. For the many species of plants, bacteria, fungi, and algae that live near or surround the area of the spill, the poisoning effects of oil are even more devastating because they absorb toxins more readily than animals. The toxins can also persist for years in the soil, affecting future generations.

Figure 43.10

As you've learned, the feathers of birds and the fur of mammals are adaptations that enable them to regulate their internal environments. Oil disrupts the proper functioning of fur and feathers, and these animals slowly freeze to death.

Perhaps worse than the immediate effects are the long-term effects of oil spills. First, the oil itself is a problem. As you know, oil doesn't mix with water, it doesn't disappear, and as long as it remains in the environment, it continues its destructive effects on organisms.

Oil spills can have disastrous consequences in the living world. Although oil spills are essentially biological problems, cleaning up oil and restoring the ecosystems requires technologies from many different scientific disciplines, as *Figure 43.11* indicates.

◀ **Chemistry** Chemists develop methods for cleaning up oil. One important technology that is being developed by chemical engineers is the use of tiny glass beads coated with titanium dioxide (bottom row). These coated beads induce oxygen to attach to the oil, speeding its disintegration. Another type of bead actually takes up oil and forms a floating mass that can be suctioned away (top row).

▶ **Physics** Physicists are involved chiefly in the containment of oil spills. Booms—a type of floating, flexible fence—are placed around a spill to prevent it from spreading.

Figure 43.11

To clean up after large-scale oil spills like the *Exxon Valdez* disaster, a number of techniques have been developed within a variety of scientific disciplines.

Oil spills are not the only ecological disasters scientists must deal with. Similar ecological problems occur in areas decimated by fires, floods, storms, toxic waste and acid rain contamination, and a host of other human-influenced or natural disasters. These problems require diverse solutions too, and as scientists perfect their techniques for protecting the environment, it's possible that the ravaging effects of environmental disasters like oil spills will become a thing of the past.

Magnification: 6800×

► **Biology** Perhaps the biggest contribution of biologists is to study the ecological effects of small-scale spills and then apply this knowledge to larger ones. Biology has also been important in developing methods to use large populations of naturally occurring bacteria that consume oil such as *Pseudomonas putida,* shown here, to help clean up oil spills.

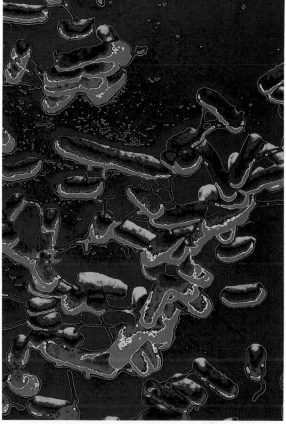

▼ **Earth science** Geologists deal with the effects of oil on the physical environment, such as how oil affects the water supply. Geological studies are also important in the discovery of safer routes for transport.

Integrating disciplines in medicine

Environmental protection is critical for maintaining the long-term health of the human species, but for help in dealing with current health issues, people turn to medicine.

In the field of medicine, biology has always had a long-standing relationship with other sciences. The development of drugs, surgical techniques, X rays, and a variety of other medical instruments and products are just some of the examples of technology that has been created through interdisciplinary research. However, the face of medicine is changing rapidly. Today, scientists from a variety of new fields are developing products and tools that promise a high-tech future for medicine.

One of the newest and perhaps the most exciting medical technologies to come out of this scientific teamwork is in the field of artificial organ development, or bionics. Using technology from such diverse fields as computer science, electronics, chemistry, physics, and biology, scientists have already developed a large assortment of artificial body parts, including artificial tendons, synthetic skin, replacement elbow and hip joints, and implants for the ears and eyes.

But are scientists close to producing the bionic men and women of movie and television fame? Actually, this may be a few years away, but scientists predict that within the not-too-distant future, as you can see in *Figure 43.12,* an artificial component will be able to take the place of almost any portion of human anatomy, with the large exception of the brain.

Figure 43.12

The high-tech body parts of the present and future will combine the technologies of many scientific fields.

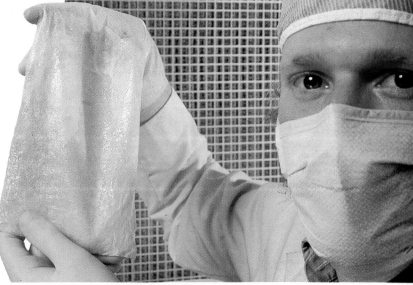

▲ **Skin** Biologists hope to produce sheets of real skin cultured from human skin cells, rather than relying on this artificial skin made of silicone.

◄ **Nerve chips** Computer microchips similar to this one will be able to pick up impulses from nerve fibers in the stump of a limb and transmit messages to computer-driven artificial limbs.

Figure 43.13

Associated with the benefits of new technologies are risks. Since the application of genetic-engineering techniques to agriculture is so new, can scientists be sure that food products are not harmful to humans, or that the release of genetically engineered pests poses no threat to the delicate balances of ecosystems?

Technology and Society

Technological advances in the fields of medicine, genetics, and ecology offer great hope for the future of humans and all life on Earth; however, new technologies sometimes raise serious social and ethical concerns for humans, as well as risks.

New technologies raise questions about benefits and risks

Genetically engineered plants and pests will transform the agriculture industry, but does this technology have limits? By genetically engineering sterile pests, spraying insecticides on crops will become a thing of the past. This will decrease pollution, but what might happen if the newly engineered traits were accidentally transferred to the wild relatives of donor species, *Figure 43.13?* How can scientists be sure that such technology will not be harmful?

▲ **Heart** The first artificial heart, the Jarvik-7 shown here, has been replaced by more sophisticated models.

◀ **Replacement parts** Today some body parts, such as blood vessels, can be replaced using artificial ones made of Dacron or Teflon.

Ethical concerns of technology

Besides risks, new technologies sometimes also raise important social questions. For instance, biologists have been using genetic-engineering techniques to clone organisms such as plants, mice, frogs, and cattle for years, *Figure 43.14,* but now scientists report that it may be possible to clone human embryos as well. Has this technology gone too far? Science has also developed many different techniques that make it possible for infertile couples to conceive. Some of these involve freezing and storing human embryos. Many controversies have already arisen due to this technology.

Protecting the rain forests and other delicate ecosystems is something that is critical for preserving Earth's biodiversity, but even this raises controversy. Rain forests, for example, are diminishing at astonishing rates mainly because the cleared land is being used for agriculture. Some would say it is not fair for highly industrialized nations such as the United States to ask developing nations, many of which are in areas with the greatest biodiversity, to suppress their own poor economies in order to save animals and plants.

Finding the proper balance between preserving the environment and protecting people's livelihoods is a societal problem that has existed for centuries. We encounter this problem in our homes by deciding which products to buy and which cars to drive. And if you choose to pursue a career in science, you'll be making critical decisions about this matter on a day-to-day basis like Mr. Swallow in the People in Biology feature.

Figure 43.14

In vitro fertilization now makes it possible for women unable to conceive to raise families, but many people simply feel uneasy with the idea of tampering with life (right). Cloning plants and some animals is now possible (left). Will this technology ever be applied to humans? What may some of the ramifications be?

Figure 43.15

Only by educating yourself and staying informed on issues can you develop your own viewpoint. You can have an impact on the future by staying interested and involved.

Stay informed . . . Be involved

The ethical and social questions of biotechnology are outside the realm of science; however, you can ensure that science plays a positive role in your life by staying informed and involved.

School is a great place to start becoming an informed citizen. Each of the scientific disciplines offers a wealth of knowledge about you and your world. Reading newspapers and magazines and watching science programs on television are other ways you can stay informed. The more informed you are, the less complex and mysterious the world will seem.

Finally, as a citizen of the future, it's critical that you remain involved, *Figure 43.15.* You will soon be able to vote, and your vote can truly make a difference. Search out candidates who reflect your views on important scientific issues. Above all, remember that the benefits of science can be directed by the continuing efforts of concerned citizens like yourself if you stay informed and involved.

Section Review

1. Write an essay about your feelings on genetic engineering. How will genetic engineering improve people's lives? What concerns, if any, do you have about this technology? What are scientists doing to lessen societal apprehension about genetic engineering?

2. Consult newspapers, science magazines, or television programs to prepare a report about an area of technology that involved the expertise of more than one scientific discipline. What are the benefits of this technology? What sciences were involved in this technology, and what were their contributions?

Meet John Swallow, Forest Technician

From few office windows can you see bighorn sheep and white-tailed deer wander among stands of ponderosa pine. John Swallow, a forest technician in South Dakota's Black Hills National Forest, is glad to have this view from his desk.

In the following interview, Mr. Swallow tells about his work in the United States Forest Service.

On the Job

Q Mr. Swallow, could you explain your responsibilities as a forest technician?

A A forest technician is basically a hands-on field person. One part of my job is setting up timber sales. I flag an area that will be logged, marking the trees to be saved. Much of the cutting is done for thinning, to reduce competition between the trees for light and nutrients and to promote quicker growth. The majority of our lumber is used in building homes in the Michigan area.

Q What factors do you have to consider in choosing trees for cutting?

A I need to identify sensitive or endangered plants or wildlife in the immediate area. I also have to consider the drainage and the slope and determine if the trees to be cut are in a place where lumberjacks can work safely. In addition, I have to know about any archaeological sites in the area. There could be anything from a 50-year-old cabin, which might be an important cultural resource, to remains from prehistoric times.

Q Do national forest visitors from cities ask you questions?

A They often want to know about shooting deer in the forest. I tell them that it is a necessary part of population control, since natural predators like the mountain lion and grizzly bear are mostly gone. I've hunted all my life, but I only take what I need and use every part of the deer. I do my own butchering and use or give away the meat. The hide is often used for moccasins and buckskin ceremonial dance costumes.

Early Influences

Q Could you tell us how you got into forestry?

A I'm an Oglala Sioux. I grew up on the Pine Ridge Indian Reservation, which is way out in an isolated area about 40 miles from the nearest town, Rapid City, South Dakota. After high school, I heard on the radio that workers were needed for a trail crew to develop new hiking trails. Later, I was a recreation campground cleaner and then a forest firefighter. On that job, I was on the line of a California forest fire for almost three weeks, my first time away from home. I've had my present job since 1983.

Q Has there been a particular person who influenced you in choosing a career?

A Besides being a rancher, my dad was a tribal ranger, a wildlife conservation officer chosen by the tribe to establish hunting seasons and licensing. Big game—deer, pronghorn antelope, and elk—and smaller game, such as prairie dogs and rabbits, live on our reservation. The tribe wanted to end the year-round, hunt-all-you-want system, particularly among hunters coming in from the outside. Someday, I'd like to go back to work as a conservation officer and deal with wildlife management on the Pine Ridge Reservation. My dad taught me that once you get committed to something, hang in there!

Personal Insights

Q Of all the animals you see in the forest, do you have a favorite?

A I admire the wolf. I guess that's because wolves' behavior is in some ways similar to the old ways of Indians. In earlier times, everything was geared toward survival of the group, not just the individual. The loss of that feeling is a problem today among Native American tribes.

Q Is maintaining the traditions of your tribe important to you?

A Yes. I attend the summer Oglala Nation powwow. It's a chance to get together and have fun. Lately, I've been visiting the elders of our tribe to listen to their sto-

ries. They're the last contact between the old-timers and us, so it's important to remember what they have to say.

Q What advice do you have for students who might be interested in working in a national forest?

A First of all, stay in school! Everything is important, including what you learn in grade school. The discipline of hanging in there and learning will be useful in later life. Right now, I'm involved in an outreach program that visits schools on reservations to inform students about the Forest Service. The Forest Service offers a large variety of opportunities for many young people. All they need do is find out what they're interested in, and we'll provide them with the information to match them with that kind of work.

43.1 Biology: The Dynamics of Life

- Identifying and studying Earth's biodiversity, understanding the complexities of ecosystems, and learning new techniques to manipulate DNA are among the most important issues biologists will deal with in the future.
- Basic research, whether on a large or small scale, is key for the solutions of major biological problems, just as it is in all sciences.

43.2 Technology and Society: Keeping the Balance

- The integration of biology and other sciences has been important for past technologies, such as those in medicine and agriculture, and will be critical for solving the challenging problems of the future.

- New technologies often introduce complex questions to society. By staying informed about the changing face of biology, you will be able to make the critical decisions that will affect your life in the future.

Understanding Concepts

1. Describe the sciences involved in the field of bionics.
2. Why is it important for humans to preserve biodiversity?
3. What are keystone species, and how are they important in ecosystems?
4. Describe some ecological effects of oil spills.
5. Why is studying living things sometimes more difficult than studying nonliving things?
6. What are some potential benefits of genetic engineering? What are some ethical questions raised by genetic engineering?
7. Why is staying informed of new technologies important for any citizen?

Thinking Critically

8. What critical decisions about technology and society must Mr. Swallow make in his job as a forest technician?

9. How will understanding the phenotypic expressions of genes be important in the medicine of the future?

10. How might studying the environmental effects of small-scale oil spills be useful for cleaning up after large spills?

11. Provide an example from your own experience of how chemists work with biologists. Physicists with biologists.

12. What unanswered questions in biology would you like to see answered in your lifetime? Explain your answer.

13. Describe your feelings about cloning. Should this technology be banned? Why or why not?

14. Describe some possible consequences of marine oil spills on inland organisms.

15. List some biological studies that you believe were important for the development of bionics.

16. How will studying the complex interactions between organisms in ecosystems be important for the future?

17. What does Dr. Bakker's view of the future tell about the diversity of life in the past and present?

Connecting to Themes

18. **Energy** How do oil spills disrupt the flow of energy in ecosystems?

19. **Systems and Interactions** Why does the existence of keystone species increase our need to preserve biodiversity?

20. **Evolution** How do you think the close relationship between flowering plants and insects has affected their evolution?

21. **Homeostasis** How do oil spills affect the abilities of animals, plants, and microorganisms to maintain homeostasis?

Appendices   *Contents*

The classification used in this text is one that combines information gathered from the systems of many different fields of biology. For example, phycologists, biologists who study algae, have developed their own system of classification, as have mycologists, biologists who study fungi. The naming of animals and plants is controlled by two completely different sets of rules. The five-kingdom system, although not yet ideal in reflecting the phylogeny of all life, is the most widely accepted. Taxonomy is an area of biology that evolves just like the species it studies. In this Appendix, for each phylum listed, at least one genus is named as an example. For more information about each phylum, refer to the chapter in the text in which the group is described.

Kingdom Monera

Archaebacteria (Ancient Bacteria)

Phylum Aphragmabacteria (Thermoacidophiles)
  Example: *Mycoplasma*

Phylum Halobacteria (Halophiles)
  Example: *Halobacterium*

Phylum Methanocreatrices (Methanogens)
  Example: *Methanobacillus*

Eubacteria (True Bacteria)

Phylum Actinobacteria
  Example: *Mycobacterium*

Phylum Omnibacteria
  Example: *Salmonella*

Phylum Spirochaetae (Spirochaetes)
  Example: *Treponema*

Phylum Chloroxybacteria (Grass-green Bacteria)
  Example: *Prochloron*

Phylum Cyanobacteria (Blue-green Algae)
  Example: *Nostoc*

Kingdom Protista

Animal-like Protists

Phylum Rhizopoda (Amoebas)
  Example: *Amoeba*

Phylum Ciliophora (Ciliates)
  Example: *Paramecium*

Phylum Sporozoa (Sporozoans)
  Example: *Plasmodium*

Phylum Zoomastigina (Flagellates)
  Example: *Trypanosoma*

Plantlike Protists

Phylum Euglenophyta (Euglenoids)
  Example: *Euglena*

Phylum Bacillariophyta (Diatoms)
  Example: *Navicula*

Phylum Dinoflagellata (Dinoflagellates)
  Example: *Gonyaulax*

Phylum Rhodophyta (Red Algae)
  Example: *Chondrus*

Phylum Phaeophyta (Brown Algae)
  Example: *Laminaria*

Phylum Chlorophyta (Green Algae)
  Example: *Ulva*

Funguslike Protists

Phylum Acrasiomycota (Cellular Slime Molds)
  Example: *Dictyostelium*

Phylum Myxomycota (Plasmodial Slime Molds)
  Example: *Physarum*

Phylum Oomycota (Water Molds, Mildews, Rusts)
  Example: *Phytophthora*

Kingdom Fungi

Phylum Zygomycota (Sporangium Fungi)
  Example: *Rhizopus*

Phylum Ascomycota (Cup Fungi and Yeasts)
  Example: *Saccharomyces*

Phylum Basidiomycota (Club Fungi)
  Example: *Amanita*

Phylum Deuteromycota (Imperfect Fungi)
  Example: *Penicillium*

Phylum Mycophycota (Lichens)
  Example: *Cladonia*

Kingdom Plantae

Spore Plants

Division Bryophyta (Mosses and Liverworts)

Class Mucopsida (Mosses)
  Example: *Polytrichum*

Class Hepaticopsida (Liverworts)
Example: *Marchantia*

Division Psilophyta (Whisk Ferns)
Example: *Psilotum*

Division Lycophyta (Club Mosses)
Example: *Lycopodium*

Division Sphenophyta (Horsetails)
Example: *Equisetum*

Division Pterophyta (Ferns)
Example: *Polypodium*

Seed Plants

Division Ginkgophyta (Ginkgos)
Example: *Ginkgo*

Division Cycadophyta (Cycads)
Example: *Cycas*

Division Coniferophyta (Conifers)
Example: *Pinus*

Division Gnetophyta
Example: *Welwitschia*

Division Anthophyta (Flowering Plants)

Class Dicotyledones (Dicots)

Family Magnoliaceae (Magnolias)
Example: *Magnolia*

Family Fagaceae (Beeches)
Example: *Quercus*

Family Cactaceae (Cacti)
Example: *Opuntia*

Family Malvaceae (Mallows)
Example: *Gossypium*

Family Brassicaceae (Mustard)
Example: *Brassica*

Family Rosaceae (Rose)
Example: *Rosa*

Family Fabaceae (Pea)
Example: *Arachis*

Family Aceracea (Maple)
Example: *Acer*

Family Lamiaceae (Mint)
Example: *Thymus*

Family Asteraceae (Daisies)
Example: *Helianthus*

Class Monocotyledones (Monocots)

Family Poaceae (Grasses)
Example: *Triticum*

Family Palmae (Palms)
Example: *Phoenix*

Family Liliaceae (Lilies)
Example: *Asparagus*

Family Orchidaceae (Orchids)
Example: *Cypripedium*

Kingdom Animalia

Invertebrates

Phylum Porifera (Sponges)
Example: *Spongilla*

Phylum Cnidaria (Corals, Jellyfishes, Hydras)

Class Hydrozoa (Hydroids)
Example: *Hydra*

Class Scyphozoa (Jellyfishes)
Example: *Aurelia*

Class Anthozoa (Sea Anemones, Corals)
Example: *Corallium*

Phylum Platyhelminthes (Flatworms)

Class Turbellaria (Free-living Flatworms)
Example: *Dugesia*

Class Trematoda (Flukes)
Example: *Fasciola*

Class Cestoda (Tapeworms)
Example: *Taenia*

Phylum Nematoda (Roundworms)
Example: *Trichinella*

Phylum Mollusca (Mollusks)

Class Gastropoda (Snails and Slugs)
Example: *Helix*

Class Bivalvia (Bivalves)
Example: *Arca*

Class Cephalopoda (Octopuses, Squid)
Example: *Nautilus*

Phylum Annelida (Annelids)

Class Polychaeta (Polychaetes)
Example: *Nereis*

Class Oligochaete (Earthworms)
Example: *Lumbricus*

Class Hirudinea (Leeches)
Example: *Hirudo*

Phylum Arthropoda (Arthropods)

**Class Arachnida
(Spiders, Mites, Scorpions)**
Example: *Latrodectus*

Class Merostomata (Horseshoe Crabs)
Example: *Limulus*

**Class Crustacea
(Lobsters, Crayfish, Crabs)**
Example: *Homarus*

Class Chilopoda (Centipedes)
Example: *Scutigerella*

Class Diplopoda (Millipedes)
Example: *Julus*

Class Insecta (Insects)
Example: *Bombus*

Phylum Echinodermata (Echinoderms)

**Class Crinoidea
(Sea Lilies, Feather Stars)**
Example: *Ptilocrinus*

Class Asteroidea (Starfishes)
Example: *Asterias*

Class Ophiuroidea (Brittle Stars)
Example: *Ophiura*

**Class Echinoidea
(Sea Urchins and Sand Dollars)**
Example: *Arbacia*

Class Holothuroidea (Sea Cucumbers)
Example: *Cucumaria*

Vertebrates

Phylum Chordata (Chordates)

Subphylum Urochordata (Tunicates)
Example: *Polycarpa*

Subphylum Cephalochordata (Lancelets)
Example: *Branchiostoma*

Subphylum Vertebrata (Vertebrates)

Class Agnatha (Lampreys and Hagfishes)
Example: *Petromyzon*

Class Chondrichthyes (Sharks, Rays)
Example: *Squalus*

Class Osteichthyes (Bony Fishes)
Example: *Hippocampas*

Subclass Crossopterygii (Lobe-finned Fishes)
Example: *Latimeria*

Subclass Dipneusti (Lungfishes)
Example: *Neoceratodus*

Subclass Actinopterygii (Ray-finned Fishes)
Example: *Acipenser*

Class Amphibia (Newts, Frogs, Toads)
Example: *Rana*

Class Reptilia (Turtles, Snakes, Lizards, Crocodiles, Alligators)
Example: *Anolis*

Class Aves (Birds)

Order Anseriformes (Ducks, Geese, Swans)
Example: *Olor*

Order Falconiformes (Hawks, Eagles)
Example: *Falco*

Order Galliformes (Ground Birds)
Example: *Perdix*

Order Passeriformes (Perching Birds)
Example: *Spizella*

Class Mammalia (Mammals)

Order Monotremata (Monotremes)
Example: *Ornithorhynchus*

Order Marsupialia (Marsupials)
Example: *Didelphis*

Order Insectivora (Insect Eaters)
Example: *Scapanus*

Order Chiroptera (Bats)
Example: *Desmodus*

Order Carnivora (Carnivores)
Example: *Ursus*

Order Rodentia (Rodents)
Example: *Cavia*

Order Cetacea (Whales, Dolphins)
Example: *Delphinus*

Order Primates (Primates)
Example: *Gorilla*

Appendix B
Origins of Scientific Terms

This list of Greek and Latin roots will help you interpret the meaning of biological terms. The column headed *Root* gives many of the actual Greek (*GK*) or Latin (*L*) root words used in science. If more than one word is given, the first is the full word in Greek or Latin. The letter groups that follow are forms in which the root word is most often found combined in science words. In the second column is the meaning of the root as it is used in science. The third column shows a typical science word containing the root from the first column. Most of these words can be found in your textbook.

Root	Meaning	Example
A		
a, an (GK)	not, without	anaerobic
abilis (L)	able to	biodegradable
ad (L)	to, attached to	appendix
aequus (L)	equal	equilibrium
aeros (GK)	air	anaerobic
agon (GK)	assembly	glucagon
aktis (GK)	ray	actin
allas (GK)	sausage	allantois
allelon (GK)	of each other	allele
allucinari (L)	to dream	hallucinate
alveolus (L)	small pit	alveolus
amnos (GK)	lamb	amnion
amoibe (GK)	change	amoebocyte
amphi (GK)	both, about, around	amphibian
amylum (L)	starch	amylase
ana (L)	away, onward	anaphase
andro (GK)	male	androgens
anggeion, angio (GK)	vessel, container	angiosperm
anthos (GK)	flower	anthophyte
anti (GK)	against, away, opposite	antibody
aqua (L)	water	aquatic
archaios, archeo (GK)	ancient, primitive	archaebacteria
arthron (GK)	joint, jointed	arthropod
artios (GK)	even	artiodactyl
askos (GK)	bag	ascospore
aster (GK)	star	Asteroidea
autos (GK)	self	autoimmune
B		
bakterion (GK)	small rod	bacterium
bi, bis (L)	two, twice	bipedal
binarius (L)	pair	binary fission
bios (GK)	life	biology
blastos (GK)	bud	blastula
bryon (GK)	moss	bryophyte
bursa (L)	purse, bag	bursa

Root	Meaning	Example
C		
caedere, cide (L)	kill	insecticide
capillus (L)	hair	capillary
carn (L)	flesh	carnivore
carno (L)	flesh	carnivore
cella, cellula (L)	small room	protocells
cervix (L)	neck	cervix
cetus (L)	whale	cetacean
chaite, chaet (GK)	bristle	oligochaeta
cheir (GK)	hand	chiropteran
chele (GK)	claw	chelicerae
chloros (GK)	pale green	chlorophyll
chondros (GK)	cartilage	Chondrichthyes
chondros (GK)	grain	mitochondrion
chorda (L)	cord	urochordata
chorion (GK)	skin	chorion
chroma, chrom (GK)	colored	chromosome
chronos (GK)	time	chronometer
circa (L)	about	circadian
cirrus (L)	curl	cirri
codex (L)	tablet for writing	codon
corpus (L)	body	corpus luteum
cum, col, com, con (L)	with, together	convergent
cuticula (L)	thin skin	cuticle
D		
daktylos (GK)	finger	perissodactyl
de (L)	away, from	decompose
decidere (L)	to fall down	deciduous
degradare (L)	to reduce in rank	biodegradable
dendron (GK)	tree	dendrite
dens (L)	tooth	edentate
derma (GK)	skin	epidermis
deterere (L)	loose material	detritus
dia, di (GK)	through, apart	diastolic
dies (L)	day	circadian
diploos (GK)	twofold, double	diploid
dis, di (GK)	twice, two	disaccharide
dis, di (L)	apart, away	disruptive

Root	Meaning	Example
dormire (L)	to sleep	dormancy
drom (GK)	running, racing	dromedary
ducere (L)	to lead	oviduct
E		
echinos (GK)	spine	echinoderm
eidos, old (GK)	form, appearance	rhizoid
ella (GK)	small	organelle
endon, en, endo (GK)	within	endosperm
engchyma (GK)	infusion	parenchyma
enteron (GK)	intestine, gut	enterocolitis
entomon (GK)	insect	entomology
epi (GK)	upon, above	epidermis
equus (L)	horse	Equisetum
erythros (GK)	red	erythrocyte
eu (GK)	well, true, good	eukaryote
evolutus (L)	rolled out	evolution
ex, e (L)	out	extinction
exo (GK)	out, outside	exoskeleton
extra (L)	outside, beyond	extracellular
F		
ferre (L)	to bear	porifera
fibrilla (L)	small fiber	myofibril
fissus (L)	a split	binary fission
flagellum (L)	whip	flagellum
follis (L)	bag	follicle
fossilis (L)	dug up	microfossils
fungus (L)	mushroom	fungus
G		
gamo, gam (GK)	marriage	gamete
gaster (GK)	stomach	gastropoda
ge, geo (GK)	the Earth	geology
gemmula (L)	little bud	gemmule
genesis (L)	origin, birth	parthenogenesis
genos, gen, geny (GK)	race	genotype
gestare (L)	to bear	progesterone
glene (GK)	eyeball	euglenoid
globus (L)	sphere	hemoglobin
glotta (GK)	tongue	epiglottis
glykys, glu (GK)	sweet	glycolysis
gnathos (GK)	jaw	Agnatha
gonos, gon (GK)	reproductive, sexual	gonorrhea
gradus (L)	a step	gradualism
graphos (GK)	written	chromatograph
gravis (L)	heavy	gravitropism

Root	Meaning	Example
gymnos (GK)	naked, bare	gymnosperm
gyne (GK)	female, woman	gynoecium
H		
haima, emia (GK)	blood	hemoglobin
halo (GK)	salt	halophile
haploos (GK)	simple	haploid
haurire (L)	to drink	haustorium
helix (L)	spiral	helix
hemi (GK)	half	hemisphere
herba (L)	grass	herbivore
hermaphroditos (GK)	combining both sexes	hermaphrodite
heteros (GK)	other	heterotrophic
hierarches (GK)	rank	hierarchy
hippos (GK)	horse	hippopotamus
histos (GK)	tissue	histology
holos (GK)	whole	Holothuroidea
homo (L)	man	hominid
homos (GK)	same, alike	homologous
hormaein (GK)	to excite	hormone
hydor, hydro (GK)	water	hydrolysis
hyper (GK)	over, above	hyperventilation
hyphe (GK)	web	hypha
hypo (GK)	under, below	hypotonic
I		
ichthys (GK)	fish	Osteichthyes
instinctus (L)	impulse	instinct
insula (L)	island	insulin
inter (L)	between	internode
intra (L)	within, inside	intracellular
isos (GK)	equal	isotonic
itis (GK)	inflammation, disease	arthritis
J		
jugare (L)	join together	conjugate
K		
kardia, cardia (GK)	heart	cardiac
karyon (GK)	nut	prokaryote
kata, cata (GK)	break down	catabolism
kephale, ceph (GK)	head	cephalopoda
keras (GK)	horn	chelicerae
kinein (GK)	to move	kinetic
koilos, coel (GK)	hollow, cavity, belly	coelom
kokkus (GK)	berry	streptococcus
kolla (GK)	glue	colloid

Appendix B Origins of Scientific Terms

Root	Meaning	Example
kotyl, cotyl (GK)	cup	cotylosaur
kreas (GK)	flesh	pancreas
krinoeides (GK)	lilylike	Crinoidea
kyanos, cyano (GK)	blue	cyanobacterium
kystis, cyst (GK)	bladder, sac	cystitis
kytos, cyt (GK)	hollow, cell	lymphocyte

L

Root	Meaning	Example
lagos (GK)	hare	lagomorph
leukos (GK)	white	leukocyte
libra (L)	balance	equilibrium
logos, logy (GK)	study, word	biology
luminescere (L)	to grow light	bioluminescence
luteus (L)	orange-yellow	corpus luteum
lyein, lysis (GK)	to split, loosen	lysosome
lympha (L)	water	lymphocyte

M

Root	Meaning	Example
makros (GK)	large	macrophage
marsupium (L)	pouch	marsupial
meare (L)	to glide	permeable
megas (GK)	large	megaspore
melas (GK)	black, dark	melanin
meristos (GK)	divided	meristem
meros (GK)	part	polymer
mesos (GK)	middle	mesophyll
meta (GK)	after, following	metaphase
metabole (GK)	change	metabolism
meter (GK)	a measurement	diameter
mikros, micro (GK)	small	microscope
mimos (GK)	a mime	mimicry
mitos (GK)	thread	mitochondrion
molluscus (L)	soft	mollusk
monos (GK)	single	monotreme
morphe (GK)	form	lagomorph
mors, mort (L)	death	mortality
mucus (L)	mucus, slime	mucosa
multus (L)	many	multicellular
mutare (L)	to change	mutation
mykes, myc (GK)	fungus	mycorrhiza
mys (GK)	muscle	myosin

N

Root	Meaning	Example
nema (GK)	thread	nematology
nemato (GK)	thread, threadlike	nematode
neos (GK)	new	Neolithic
nephros (GK)	kidney	nephron

Root	Meaning	Example
neuro (GK)	nerve	neurology
nodus (L)	knot, knob	internode
nomos, nomy (GK)	ordered knowledge	taxonomy
noton (GK)	back	notochord

O

Root	Meaning	Example
oikos, eco (GK)	household	ecosystem
oisein, eso (GK)	to carry	esophagus
oligos (GK)	few, little	oligochaeta
omnis (L)	all	omnivore
ophis (GK)	serpent	Ophiuroidea
ophthalmos (GK)	referring to the eye	ophthalmologist
organon (GK)	tool, implement	organelle
ornis (GK)	bird	ornithology
orthos (GK)	straight	orthodontist
osculum (L)	small mouth	osculum
osteon (GK)	bone	osteocyte
ostrakon (GK)	shell	ostracoderm
oura, ura (GK)	tail	anura
ous, oto (GK)	ear	otology
ovum (L)	egg	oviduct

P

Root	Meaning	Example
palaios, paleo (GK)	ancient	paleontology
pan (GK)	all	pancreas
para (GK)	beside	parenchyma
parthenos (GK)	virgin	parthenogenesis
pathos (GK)	disease, suffering	pathogenic
pausere (L)	to rest	decompose
pendere (L)	to hang	appendix
per (L)	through	permeable
peri (GK)	around	peristalsis
periodos (GK)	a cycle	photoperiodism
pes, pedis (L)	foot	bipedal
phagein (GK)	to eat	phagocyte
phainein (GK)	to show	phenotype
phaios (GK)	dusky	phaeophyta
phase (GK)	stage, appearance	metaphase
pherein, phor (GK)	to carry	pheromone
phloios (GK)	inner bark	phloem
phos, photos (GK)	light	phototropism
phyllon (GK)	leaf	chlorophyll
phylon (GK)	related group	phylogeny
phyton (GK)	plant	epiphyte
pinax (GK)	tablet	pinacocytes
pinein (GK)	to drink	pinocytosis

Root	Meaning	Example
pinna (L)	feather	pinniped
plasma (GK)	mold, form	plasmodium
plastos (GK)	formed object	chloroplast
platys (GK)	flat	platyhelminthes
plax (GK)	plate	placoderm
pleuron (GK)	side	dipleurula
plicare (L)	to fold	replication
polys, poly (GK)	many	polymer
poros (GK)	channel	porifera
post (L)	after	posterior
pous, pod (GK)	foot	gastropoda
prae, pre (L)	before	Precambrian
primus (L)	first	primary
pro (GK and L)	before, for	prokaryote
proboskis (GK)	trunk	proboscidean
producere (L)	to bring forth	reproduction
protos (GK)	first	protocells
pseudes (GK)	false	pseudopod
pteron (GK)	wing	chiropteran
punctus (L)	a point	punctuated
pupa (L)	doll	pupa

R

Root	Meaning	Example
radius (L)	ray	radial
re (L)	again	reproduction
reflectere (L)	to turn back	reflex
rhiza (GK)	root	mycorrhiza
rhodon (GK)	rose	rhodophyte
rota (L)	wheel	rotifer
rumpere (L)	to break	disruptive

S

Root	Meaning	Example
saeta (L)	bristle	Equisetum
sapros (GK)	rotten	saprobe
sarx (GK)	flesh	sarcomere
sauros (GK)	lizard	cotylosaur
scire (L)	to know	science
scribere, script (L)	to write	transcription
sedere, ses (L)	to sit	sessile
semi (L)	half	semicircle
skopein, scop (GK)	to look	microscope
soma (GK)	body	lysosome
sperma (GK)	seed	angiosperm
spirare (L)	to breathe	spiracle
sporos (GK)	seed	microspore
staphylo (GK)	bunch of grapes	staphylococcus
stasis (GK)	standing, staying	homeostasis

Root	Meaning	Example
stellein, stol (GK)	to draw in	peristalsis
sternon (GK)	chest	sternum
stinguere (L)	to quench	extinction
stolo (L)	shoot	stolon
stoma (GK)	mouth	stoma
streptos (GK)	twisted chain	streptococcus
syn (GK)	together	systolic
synapsis (GK)	union	synapse
systema (GK)	composite whole	ecosystem

T

Root	Meaning	Example
taxis, taxo (GK)	to arrange	taxonomy
telos (GK)	end	telophase
terra (L)	land, Earth	terrestrial
thele (GK)	cover a surface	epithelium
therme (GK)	heat	endotherm
thrix, trich (GK)	hair	trichocyst
tome (GK)	cutting	anatomy
trachia (GK)	windpipe	tracheid
trans (L)	across	transpiration
trematodes (GK)	having holes	monotreme
trope (GK)	turn	gravitropism
trophe (GK)	nourishment	heterotrophic
turbo (L)	whirl	turbellaria
tympanon (GK)	drum	tympanum
typos (GK)	model	genotype

U

Root	Meaning	Example
uni (L)	one	unicellular
uterus (L)	womb	uterus

V

Root	Meaning	Example
vacca (L)	cow	vaccine
vagina (L)	sheath	vagina
valvae (L)	folding doors	bivalvia
vasculum (L)	small vessel	vascular
venter (L)	belly	ventricle
ventus (L)	a wind	hyperventilation
vergere (L)	to slant, incline	convergent
villus (L)	shaggy hair	villus
virus (L)	poisonous liquid	virus
vorare (L)	to devour	carnivore

X

Root	Meaning	Example
xeros (GK)	dry	xerophyte
xylon (GK)	wood	xylem

Z

Root	Meaning	Example
zoon, zo (GK)	animal	zoology
zygotos (GK)	joined together	zygote

Appendix C | Safety in the Laboratory

The biology laboratory is a safe place to work if you are aware of important safety rules and if you are careful. You must be responsible for your own safety and for the safety of others. The safety rules given here will protect you and others from harm in the lab. While carrying out procedures in any of the **Biolabs**, notice the safety symbols and caution statements. The safety symbols are explained in the chart on the next page.

1. Always obtain your teacher's permission to begin a lab.
2. Study the procedure. If you have questions, ask your teacher. Be sure you understand all safety symbols shown.
3. Use the safety equipment provided for you. Goggles and a safety apron should be worn when any lab calls for using chemicals.
4. When you are heating a test tube, always slant it so the mouth points away from you and others.
5. Never eat or drink in the lab. Never inhale chemicals. Do not taste any substance or draw any material into your mouth.
6. If you spill any chemical, wash it off immediately with water. Report the spill immediately to your teacher.

7. Know the location and proper use of the fire extinguisher, safety shower, fire blanket, first aid kit, and fire alarm.
8. Keep all materials away from open flames. Tie back long hair.
9. If a fire should break out in the classroom, or if your clothing should catch fire, smother it with the fire blanket or a coat, or get under a safety shower. **NEVER RUN.**
10. Report any accident or injury, no matter how small, to your teacher.

Follow these procedures as you clean up your work area.
1. Turn off the water and gas. Disconnect electrical devices.
2. Return materials to their places.
3. Dispose of chemicals and other materials as directed by your teacher. Place broken glass and solid substances in the proper containers. Never discard materials in the sink.
4. Clean your work area.
5. Wash your hands thoroughly after working in the laboratory.

First Aid in the Laboratory

Injury	Safe Response
Burns	**Apply cold water. Call your teacher immediately.**
Cuts and bruises	**Stop any bleeding by applying direct pressure. Cover cuts with a clean dressing. Apply cold compresses to bruises. Call your teacher immediately.**
Fainting	**Leave the person lying down. Loosen any tight clothing and keep crowds away. Call your teacher immediately.**
Foreign matter in eye	**Flush with plenty of water. Use eyewash bottle or fountain.**
Poisoning	**Note the suspected poisoning agent and call your teacher immediately.**
Any spills on skin	**Flush with large amounts of water or use safety shower. Call your teacher immediately.**

Safety Symbols

DISPOSAL ALERT
This symbol appears when care must be taken to dispose of materials properly.

ANIMAL SAFETY
This symbol appears whenever live animals are studied and the safety of the animals and the students must be ensured.

BIOLOGICAL HAZARD
This symbol appears when there is danger involving bacteria, fungi, or protists.

RADIOACTIVE SAFETY
This symbol appears when radioactive materials are used.

OPEN FLAME ALERT
This symbol appears when use of an open flame could cause a fire or an explosion.

CLOTHING PROTECTION SAFETY
This symbol appears when substances used could stain or burn clothing.

THERMAL SAFETY
This symbol appears as a reminder to use caution when handling hot objects.

FIRE SAFETY
This symbol appears when care should be taken around open flames.

SHARP OBJECT SAFETY
This symbol appears when a danger of cuts or punctures caused by the use of sharp objects exists.

EXPLOSION SAFETY
This symbol appears when the misuse of chemicals could cause an explosion.

FUME SAFETY
This symbol appears when chemicals or chemical reactions could cause dangerous fumes

EYE SAFETY
This symbol appears when a danger to the eyes exists. Safety goggles should be worn when this symbol appears.

ELECTRICAL SAFETY
This symbol appears when care should be taken when using electrical equipment.

POISON SAFETY
This symbol appears when poisonous substances are used.

PLANT SAFETY
This symbol appears when poisonous plants or plants with thorns are handled.

CHEMICAL SAFETY
This symbol appears when chemicals used can cause burns or are poisonous if absorbed through the skin.

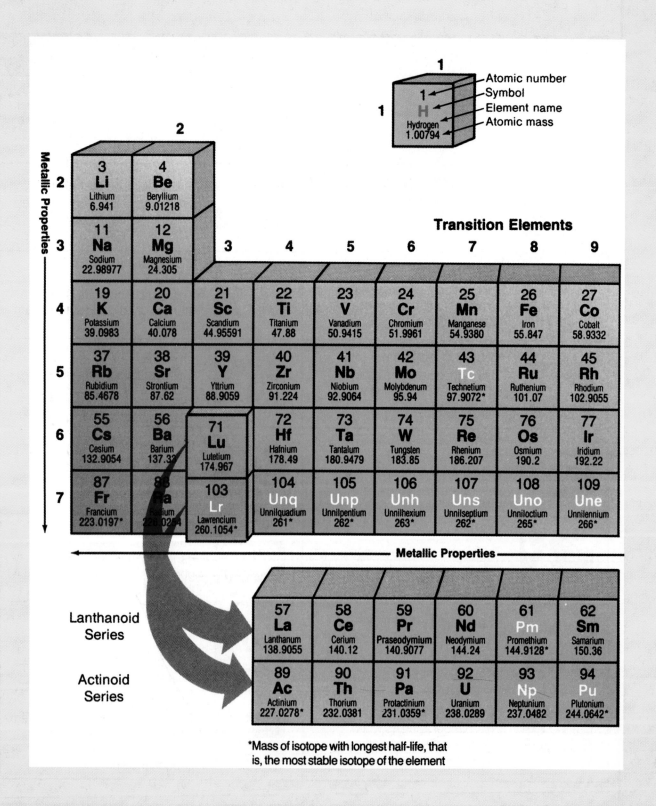

*Mass of isotope with longest half-life, that is, the most stable isotope of the element

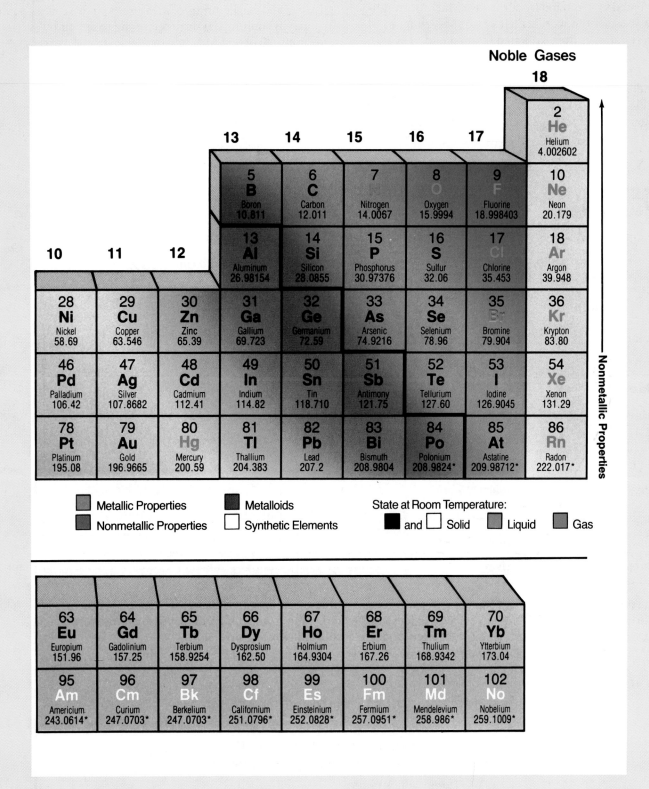

Noble Gases

					18

13	14	15	16	17	2 **He** Helium 4.002602
5 **B** Boron 10.811	6 **C** Carbon 12.011	7 N Nitrogen 14.0067	8 O Oxygen 15.9994	9 F Fluorine 18.998403	10 **Ne** Neon 20.179
13 **Al** Aluminum 26.98154	14 **Si** Silicon 28.0855	15 **P** Phosphorus 30.97376	16 **S** Sulfur 32.06	17 Cl Chlorine 35.453	18 **Ar** Argon 39.948

10	11	12						
28 **Ni** Nickel 58.69	29 **Cu** Copper 63.546	30 **Zn** Zinc 65.39	31 **Ga** Gallium 69.723	32 **Ge** Germanium 72.59	33 **As** Arsenic 74.9216	34 **Se** Selenium 78.96	35 Br Bromine 79.904	36 **Kr** Krypton 83.80
46 **Pd** Palladium 106.42	47 **Ag** Silver 107.8682	48 **Cd** Cadmium 112.41	49 **In** Indium 114.82	50 **Sn** Tin 118.710	51 **Sb** Antimony 121.75	52 **Te** Tellurium 127.60	53 **I** Iodine 126.9045	54 **Xe** Xenon 131.29
78 **Pt** Platinum 195.08	79 **Au** Gold 196.9665	80 Hg Mercury 200.59	81 **Tl** Thallium 204.383	82 **Pb** Lead 207.2	83 **Bi** Bismuth 208.9804	84 **Po** Polonium 208.9824*	85 **At** Astatine 209.98712*	86 Rn Radon 222.017*

Nonmetallic Properties

■ Metallic Properties   ■ Metalloids   State at Room Temperature:
■ Nonmetallic Properties   □ Synthetic Elements   ■ and □ Solid   ■ Liquid   ■ Gas

63 **Eu** Europium 151.96	64 **Gd** Gadolinium 157.25	65 **Tb** Terbium 158.9254	66 **Dy** Dysprosium 162.50	67 **Ho** Holmium 164.9304	68 **Er** Erbium 167.26	69 **Tm** Thulium 168.9342	70 **Yb** Ytterbium 173.04
95 Am Americium 243.0614*	96 Cm Curium 247.0703*	97 Bk Berkelium 247.0703*	98 Cf Californium 251.0796*	99 Es Einsteinium 252.0828*	100 Fm Fermium 257.0951*	101 Md Mendelevium 258.986*	102 No Nobelium 259.1009*

Appendix E — SI Measurement

The International System (SI) of Measurement is accepted as the standard for measurement throughout most of the world. Four of the base units in SI are the meter, liter, kilogram, and second. The size of a unit can be determined from the prefix used with the base unit name. For example: *kilo* means one thousand; *milli* means one-thousandth; *micro* means one-millionth; and *centi* means one-hundredth. The tables below give the standard symbols for these SI units and some of their equivalents.

Larger and smaller units of measurement in SI are obtained by multiplying or dividing the base unit by some multiple of ten. Multiply to change from larger units to smaller units. Divide to change from smaller units to larger units. For example, to change 1 km to meters, you would multiply 1 km by 1000 to obtain 1000 m. To change 10 g to kilograms, you would divide 10 g by 1000 to obtain 0.01 kg.

Common SI Units

Measurement	Unit	Symbol	Equivalents
Length	1 millimeter	mm	1000 micrometers (μm)
	1 centimeter	cm	10 millimeters (mm)
	1 meter	m	100 centimeters (cm)
	1 kilometer	km	1000 meters (m)
Volume	1 milliliter	mL	1 cubic centimeter (cm^3 or cc)
	1 liter	L	1000 milliliters (mL)
Mass	1 gram	g	1000 milligrams (mg)
	1 kilogram	kg	1000 grams (g)
	1 tonne	t	1000 kilograms (kg) = 1 metric ton
Time	1 second	s	
Area	1 square meter	m^2	10 000 square centimeters (cm^2)
	1 square kilometer	km^2	1 000 000 square meters (m^2)
	1 hectare	ha	10 000 square meters (m^2)
Temperature	1 Kelvin	K	1 degree Celsius (°C)

The top of the thermometer is marked off in degrees Fahrenheit (°F). To read the corresponding temperature in degrees Celsius (°C), look at the bottom side of the thermometer. For example, 50°F is the same temperature as 10°C. You may also use the formulas shown here for conversions.

Conversion of Fahrenheit to Celsius
$$°C = \frac{5}{9}(°F - 32)$$

Conversion of Celsius to Fahrenheit
$$°F = \left(\frac{9}{5}°C\right) + 32$$

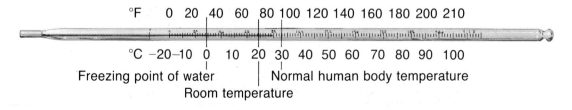

°F 0 20 40 60 80 100 120 140 160 180 200 210

°C −20 −10 0 10 20 30 40 50 60 70 80 90 100

Freezing point of water   Normal human body temperature
Room temperature

Observing and Inferring

Scientists try to make careful and accurate observations. When possible, they use instruments, such as microscopes and tape recorders, to extend their senses. Other instruments, such as a thermometer or a pan balance, are used to measure observations. Measurements provide numerical data, a concrete means of comparing collected data that can be checked and repeated.

When you make observations in science, you may find it helpful first to examine the entire object or situation. Then, using your senses of sight, touch, and hearing, examine the object in detail. Write down everything you observe.

Scientists often use their observations to make inferences. An inference is an attempt to explain or interpret observations or to determine what caused what you observed. For example, if you observed a CLOSED sign in a store window around noon, you might infer that the owner is taking a lunch break. But, perhaps the owner has a doctor's appointment or has a business meeting in another town. The only way to be sure your inference is correct is to investigate further.

When making an inference, be certain to make accurate observations and to record them carefully. Collect all the information you can. Then, based on everything you know, try to explain or interpret what you observed. If possible, investigate further to determine whether your inference is correct. What can you infer from observing the behavior of a cat that is crouched and ready to pounce?

Comparing and Contrasting

Observations can be analyzed and then organized by noting the similarities and differences between two or more objects or situations. When you examine objects or situations to determine similarities, you are comparing. Contrasting is looking at similar objects or situations for differences.

Suppose you were asked to compare and contrast a grasshopper and a dragonfly. You would start by making your observations. You then divide a piece of paper into two columns. List ways the insects are similar in one column and ways they are different in the other. After completing your lists, you report your findings in a table or in a graph.

Similarities you might point out are that both have three body parts, two pairs of wings, and chewing mouthparts. Differences might include large hind legs on the grasshopper, small legs on the dragonfly; wings held close to the body in the grasshopper, wings held outspread in the dragonfly.

Recognizing Cause and Effect

Have you ever observed something happen and then tried to figure out why or how it came about? If so, you have observed an event and inferred a reason for the event. The event or result of action is an effect, and the reason for the event is the cause.

Suppose that every time your teacher fed fish in a classroom aquarium, she tapped the food container on the edge. Then, one day she tapped the edge of the aquarium to make a point about an ecology lesson. You observe the fish swim to the surface of the aquarium to feed.

What is the effect and what would you infer would be the cause? The effect is the fish swimming to the surface of the aquarium. You might infer the cause to be the teacher tapping on the edge of the aquarium. In determining cause and effect, you have made a logical inference based on careful observations.

Perhaps the fish swam to the surface because they reacted to the teacher's waving hand or for some other reason. When scientists are unsure of the cause for a certain event, they often design controlled experiments to determine what caused their observations. Although you have made a sound judgment, you would have to perform an experiment to be certain that it was the tapping that caused the effect you observed.

Interpreting Scientific Illustrations

Illustrations are included in your textbook to help you understand, interpret, and remember what you read. Whenever you encounter an illustration, examine it carefully and read the caption. The caption explains or identifies the illustration.

Some illustrations are designed to show you how the internal parts of a structure are arranged. Look at the illustrations of a squash. The squash has been cut lengthwise so that it shows a section that runs along the length of the squash. This type of illustration is called a longitudinal section. Cutting the squash crosswise at right angles to the length produces a cross section.

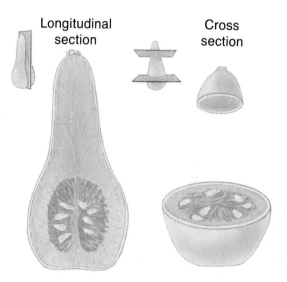

Longitudinal section

Cross section

In your reading and examination of the illustrations, you will sometimes see terms that refer to the orientation of an organism. The word *dorsal* refers to the upper side or back of an animal. *Ventral* refers to the lower side or belly of the animal. The illustration of the shark shows that it has both dorsal and ventral sides.

Symmetry refers to a similarity or likeness of parts. Many organisms and objects have symmetry. When something can be divided into two similar parts lengthwise, it has bilateral symmetry. Look at the illustration of the butterfly. The right side of the butterfly looks very similar to the left side. It has bilateral symmetry.

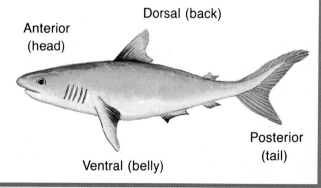

Dorsal (back)

Anterior (head)

Posterior (tail)

Ventral (belly)

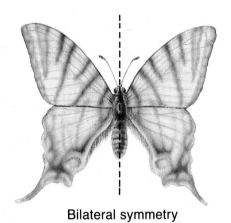

Bilateral symmetry

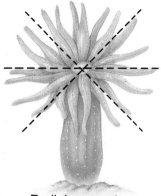

Radial symmetry

Asymmetry

Other organisms and objects have radial symmetry. Radial symmetry is the arrangement of similar parts around a central point. The anemone in the figure has radial symmetry. It can be divided anywhere through the center into similar parts.

Some organisms and objects cannot be divided into two similar parts. If an organism or object cannot be divided, it is asymmetrical. Study the sponge. Regardless of how you try to divide a sponge, you cannot divide it into two parts that look alike.

Calculating Magnification

Objects viewed under the microscope appear larger than normal because they are magnified. Total magnification describes how much larger an object appears when viewed through the microscope.

Look for a number marked with an × on the eyepiece, the low-power objective, and the high-power objective. The × stands for how many times the lens of each microscope part magnifies an object.

To calculate *total* magnification, multiply the number on the eyepiece by the number on the objective. For example, if the eyepiece magnification is 4×, the low-power objective magnification is 10×, and the high-power objective magnification is 40×:

(a) then total magnification under low power is 4× for the eyepiece times 10× for the low-power objective = 40× (4 × 10 = 40).

(b) then total magnification under high power is 4 × 40 = 160.

To measure the field of view of a microscope, you must use a unit called a micrometer. A micrometer equals 0.001 mm; in other words, there are 1000 micrometers in a millimeter. Place a millimeter section of a plastic ruler over the central opening of your microscope stage. Using low power, locate the measured lines of the ruler in the center of the field of view. Move the ruler so that one of the lines representing a millimeter is visible at one edge of the field of view.

Remember that the distance between two lines is one millimeter, and estimate the diameter in millimeters of the field of view on low power. Calculate the diameter in micrometers. For example, if the distance is 1.5 mm, then the diameter of the field of view at low power is 1500 µm. [1.5 × 1000]

To calculate the diameter of the high-power field, divide the magnification of your high power (40×) by the magnification of the low power (10×); 40 ÷ 10 = 4. Then, divide the diameter of the low-power field in micrometers (1500 µm) by this quotient (4). The answer is the diameter of the high-power field in micrometers. In this example, the diameter of the high-power field is 1500 ÷ 4 = 375 µm.

You can calculate the diameters of microscopic specimens, such as pollen grains or amoebas, viewed under low and high power by estimating how many of them could fit end to end across the field of view. Divide the diameter of the field of view by the number of specimens.

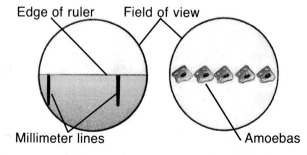

Edge of ruler   Field of view

Millimeter lines   Amoebas

You might say that the work of a scientist is to solve problems. But when you decide how to dress on a particular day, you are doing problem solving, too. You may observe what the weather looks like through a window. You may go outside and see if what you are wearing is warm or cool enough.

Scientists generally use experiments to solve problems and answer questions. An experiment is a method of solving a problem in which scientists use an organized process to attempt to answer a question.

Experimentation involves defining a problem and formulating and testing hypotheses, or proposed solutions, to the problem. Each proposed solution is tested during an experiment, which includes making careful observations and collecting data. After analysis of the collected data, a conclusion is formed and compared to the hypothesis.

Care of the Microscope

1. Always carry the microscope holding the arm with one hand and supporting the base with the other hand.
2. Don't touch the lenses with your fingers.
3. Never lower the coarse adjustment knob when looking through the eyepiece lens.
4. Always focus first with the low-power objective.
5. Don't use the coarse adjustment knob when the high-power objective is in place.
6. Store the microscope covered.

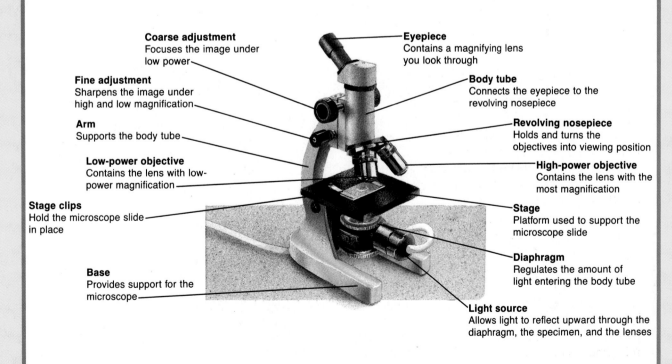

Coarse adjustment
Focuses the image under low power

Fine adjustment
Sharpens the image under high and low magnification

Arm
Supports the body tube

Low-power objective
Contains the lens with low-power magnification

Stage clips
Hold the microscope slide in place

Base
Provides support for the microscope

Eyepiece
Contains a magnifying lens you look through

Body tube
Connects the eyepiece to the revolving nosepiece

Revolving nosepiece
Holds and turns the objectives into viewing position

High-power objective
Contains the lens with the most magnification

Stage
Platform used to support the microscope slide

Diaphragm
Regulates the amount of light entering the body tube

Light source
Allows light to reflect upward through the diaphragm, the specimen, and the lenses

Using a Microscope

1. Place the microscope on a flat surface that is clear of objects. The arm should be toward you.
2. Look through the eyepiece. Adjust the diaphragm so that light comes through the opening in the stage.
3. Place a slide on the stage so that the specimen is in the field of view. Hold it firmly in place by using the stage clips.
4. Always focus first with the coarse adjustment and the low-power objective lens. Once the object is in focus on low power, turn the nosepiece until the high-power objective is in place. Use ONLY the fine adjustment to focus with this lens.

Making a Wet Mount Slide

1. Carefully place the item you want to look at in the center of a clean glass slide. Make sure the sample is thin enough for light to pass through.
2. Use a dropper to place one or two drops of water on the sample.
3. Hold a clean coverslip by its edges and place it at one side of the drop of water. Slowly lower the coverslip onto the drop of water until it lies flat.
4. If you have too much water or a lot of air bubbles, touch the edge of a paper towel to the edge of the coverslip to draw off extra water and force

air out. Do not lift the coverslip to add or remove water. As the water evaporates from the slide, add another drop of water by placing the tip of the medicine dropper next to the edge of the coverslip and squeezing the bulb slowly.

Measuring in SI

Some observations describe something using only words. These observations are called qualitative observations. If you were making qualitative observations of a dog, you might use words such as *furry, brown, short-haired,* or *short-eared.*

Other observations describe how much of something there is. These are quantitative observations, and numbers as well as words are used in the description. Tools or equipment are used to make measurements. Quantitative observations of a dog might include a mass of 45 kg, a height of 76 cm, ear length of 11 cm, and an age of 412 days.

You are probably familiar with the metric system of measurement. The metric system is a uniform system of measurement developed in 1795 by a group of scientists. The development of the metric system helped scientists avoid problems of using different units of measurement by providing an international standard of comparison for measurements. A modern form of the metric system called the International System, or SI, was adopted for worldwide use in 1960.

You will find that your text uses metric units in all its measurements. In the Biolabs and Minilabs, you will use the metric system of measurement.

Base Metric Units

The metric system is easy to use because it has a systematic naming of units and a decimal base. For example, meter is the base unit for measuring length, gram for measuring mass, and liter for measuring volume. Unit sizes vary by multiples of ten. When changing from smaller units to larger, you divide by a multiple of ten. When changing from larger units to smaller, you multiply by a multiple of ten. Prefixes are used to name larger and smaller units. Look at the following table for some common metric prefixes and their meanings.

Metric Prefixes		
Prefix	**Symbol**	**Meaning**
kilo-	k	1000 (thousand)
hecto-	h	100 (hundred)
deka-	da	10 (ten)
deci-	d	0.1 (tenth)
centi-	c	0.01 (hundredth)
milli-	m	0.001 (thousandth)

Do you see how the prefix *kilo-* attached to the unit *gram* is *kilogram* or 1000 grams, or the prefix *deci-* attached to the unit *meter* is *decimeter* or one-tenth (0.1) of a meter?

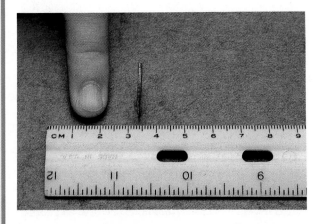

Units of Length

The meter is the SI unit used to measure distance. To visualize the length of a meter, think of a baseball bat, which is about one meter long. When measuring smaller distances, the meter is divided into smaller units called centimeters and millimeters. A centimeter is one-hundredth (0.01) of a meter, which is about the width of the fingernail on your index finger. A millimeter is one-thousandth of a meter (0.001), about the thickness of a dime. The photograph shows this comparison.

Most metric rulers have lines indicating centimeters and millimeters. The centimeter lines are the longer numbered lines, and the shorter lines between the centimeter lines are millimeter lines.

When using a metric ruler, you must first decide on a unit of measurement. You then line up the 0 centimeter mark with the end of the object being measured, and read the number of the unit where the object ends.

Units of length are also used to measure surface area. The standard unit of area is the square meter (m^2), or a square one meter long on each side. Similarly, a square centimeter (cm^2) is a square one centimeter long on each side. Surface area is determined by multiplying the number of units in length times the number of units in width.

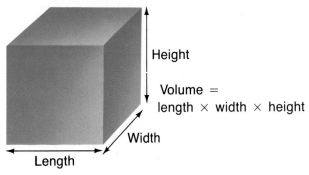

Height

Volume = length × width × height

Width

Length

Units of Volume

The cubic meter (m³) is the standard SI unit of volume. A cubic meter is a cube one meter long on each side. You can determine the volume of rectangular solids by multiplying length times width times height.

Liquid volume is measured using a unit called a liter. A liter has the volume of 1000 cubic centimeters. Because the prefix *milli-* means thousandth (0.001), a milliliter equals one cubic centimeter. One milliliter of liquid would completely fill a cube measuring one centimeter on each side.

You will measure liquids using beakers and graduated cylinders marked in milliliters. A graduated cylinder is marked with lines from bottom to top. Each graduation represents one milliliter.

Units of Mass

You will likely use a beam balance when you want to find the masses of objects. Often, you will measure mass in grams. On one side of the beam balance is a pan and on the other side is a set of beams. Each beam has an object of a known mass called a rider that slides on the beam.

Before you find the mass of an object, you must set the balance to zero by sliding all the riders back to the zero point. Check the pointer to make sure it swings an equal distance above and below the zero point on the scale. If the swing is unequal, find and turn the adjusting screw until you have an equal swing.

You are now ready to use the balance to find the mass of the object. Place the object on the pan. Slide the rider with the largest mass along the beams until the pointer drops below the zero point. Then move it back one notch. Repeat the process on each beam until the pointer swings an equal distance above and below the zero point. Read the masses indicated on the beams. The sum of the masses is the mass of the object.

Never place a hot object or pour chemicals directly on the pan. Instead, find the mass of a clean container, such as a beaker or a glass jar. Place the dry or liquid chemicals you want to measure into the container. Next, find the combined mass of the container and the chemicals. Calculate the mass of the chemicals by subtracting the mass of the empty container from the combined mass.

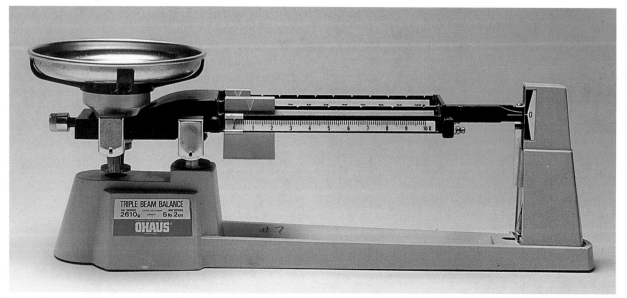

Forming a Hypothesis

Suppose you wanted to earn a perfect score on a spelling test. You think of several ways to accomplish a perfect score. You base these possibilities on past experiences and observations of your friends' results. All of the following are hypotheses you might consider that could explain how it would be possible to score 100 percent on your test:

If the test is easy, then I will get a good grade.

If I am intelligent, then I will get a good grade.

If I study hard, then I will get a good grade.

Scientists use hypotheses that they can test to explain the observations they have made. Perhaps a scientist has observed that fish activity increases in the summer and decreases in the winter. A scientist may form a hypothesis that says: If fish are exposed to warmer water, their activity will increase.

Designing an Experiment to Test a Hypothesis

Once you have stated a hypothesis, you probably want to find out whether or not it explains an event or an observation. This requires a test. To be valid, a hypothesis must be testable by experimentation. Let's figure out how you would conduct an experiment to test the hypothesis about the effects of water temperature on fish.

First, obtain several identical, clear glass containers, and fill them with the same amount of tap water. Leave the containers for a day to allow the water to come to room temperature. On the day of your experiment, you fill another container with an amount of aquarium water equal to that in the test containers. After measuring and recording the aquarium water temperature, you heat and cool the other containers, adjusting the water temperatures in the test containers so that two have higher temperatures and two have lower temperatures than the aquarium water temperature.

You place a guppy in each container. You count the number of horizontal and vertical movements each guppy makes during five minutes and record your data in a table. Your data table might look like this:

Number of Guppy Movements		
Container	Temperature (°C)	Number of movements
Aquarium water	38	56
A	40	61
B	42	70
C	36	46
D	34	42

From the data you recorded, you will draw a conclusion and make a statement about your results. If your conclusion supports your hypothesis, then you can say that your hypothesis is reliable. If it did not support your hypothesis, then you would have to make new observations and state a new hypothesis, one that you could also test. Do the data above support the hypothesis that warmer water increases fish activity?

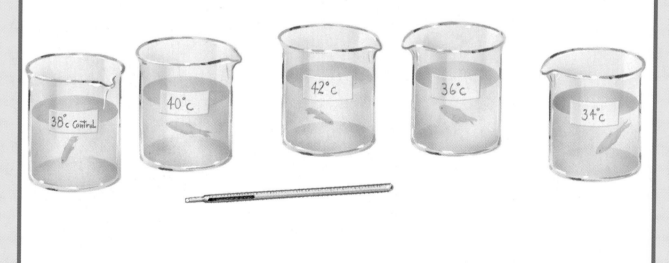

Separating and Controlling Variables

When scientists perform experiments, they must be careful to manipulate or change only one condition and keep all other conditions in the experiment the same. The condition that is manipulated is called the independent variable. The conditions that are kept the same during an experiment are called constants. The dependent variable is any change that results from manipulating the independent variable.

Scientists can only know that the independent variable caused the change in the dependent variable if they keep all other factors the same in an experiment. Scientists also use controls to be certain that the observed changes were a result of manipulation of the independent variable. A control is a sample that is treated exactly like the experimental group except that the independent variable is not applied to the control. After the experiment, the change in the dependent variable of the control sample is compared with any change in the experimental group. This allows scientists to see the effect of the independent variable.

What are the independent and dependent variables in the guppy experiment? Because you are changing the temperature of the water, the independent variable is the water temperature. Because the dependent variable is any change that results from the independent variable, the dependent variable is the number of movements the guppy makes during five minutes.

What factors are constants in the experiment? The constants are using the same size and shape containers, filling them with equal amounts of water, and counting the number of movements during the same amount of time. What was the purpose of counting the number of movements of a guppy in an identical container filled with aquarium water? The container of aquarium water is the control. The number of movements of the guppy in the aquarium water will be used to compare the movements of the guppies in water of different temperatures.

Why is it important to know the best temperature for a fish to survive? If you have an aquarium at home, you have probably learned how different fishes need different conditions to survive. They probably came from many different parts of the world and are adapted to living in very different environments.

Classifying

You may not realize it, but you impose order on the world around you. If your shirts hang in the closet together, your socks take up a corner of a dresser drawer, or your favorite CDs are stacked in groups according to recording artist, you have used the skill of classifying.

Classifying is grouping objects or events based on common features. When classifying, you first make careful observations of the group of items to be classified. Select one feature that is shared by some items in the group but not others. Place the items that share this feature in a subgroup. Place the remaining items in a second subgroup. Ideally, the items in the second subgroup will have some feature in common with one another. After you decide on the first feature that separates the items into subgroups, examine the items for other features and form further subgroups until the items can no longer be distinguished enough to identify them as distinct.

How would you classify a collection of CDs? Classify the CDs based on observable features. You might classify CDs you like to dance to in one subgroup and CDs you like to listen to in another. The CDs you like to dance to could be subdivided into a rap subgroup and a rock subgroup. Note that for each feature selected, each CD fits only one subgroup. For example, you wouldn't place a CD into both rap and rock categories. Keep selecting features until all the CDs are classified. The chart shows one method of classification.

Remember, when you classify, you are grouping objects or events for a purpose. The purpose could be general such as for ease of finding an item. The classification of books in a library is a general-purpose classification. The classification may have a special purpose. For example, plants may be classified as poisonous or harmless to humans.

Sequencing

A sequence is an arrangement of things or events in a particular order. A common sequence with which you may be familiar is the order of steps you must follow to make an omelette. Certain steps of preparation have to be followed in order for the omelette to taste good.

When you are asked to sequence things or events, you must identify what comes first. You then decide what should come second. Continue to choose things or events until they are all in order. Then, go back over the sequence to make sure each thing or event logically leads to the next.

Suppose you wanted to watch a movie that just came out on videotape. What sequence of events would you have to follow to watch the movie? You would first turn the television set to Channel 3 or 4. You would then turn the videotape player on and insert the tape. Once the tape has started playing, you would adjust the sound and picture. Then, when the movie is over, you would rewind the tape and return it to the store. What would happen if you did things out of sequence, such as adjusting the sound before putting in the tape?

Concept Mapping

If you were taking an automobile trip, you would probably take along a road map. The road map shows your location, your destination, and other places along the way. By examining the map, you can understand where you are in relation to other locations on the map.

A concept map is similar to a road map. But, a concept map shows the relationship among ideas (or concepts) rather than places. A concept map is a diagram that visually shows how concepts are related. Because the concept map shows the relationships among ideas, it can clarify the meaning of ideas and terms and help you to understand better what you are studying.

A Network Tree

Notice how some words in the concept map below called a **network tree** are circled. The circled words are science concepts. The lines in the map show related concepts, and the words written on the lines describe relationships between the concepts.

When you are asked to construct a network tree, state the topic and select the major concepts. Find related concepts and put them in order from general to specific. Branch the related concepts from the major concept, and describe the relationship on the lines. Continue to write the more specific concepts. Write the relationships between the concepts on the lines until all concepts are mapped. Examine the concept map for relationships that cross branches, and add them to the concept map.

An Events Chain

An **events chain map** is used to describe ideas in order. In science, an events chain map can be used to describe a sequence of events, the steps in a procedure, or the stages of a process.

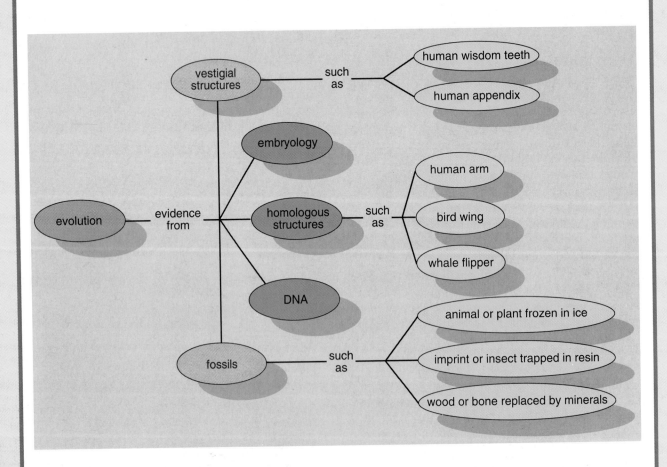

Initiating event:

| Mother asks you to wash dishes. |

↓

Event 2:

| You clear the table. |

↓

Event 3:

| You wash the dishes in soapy water. |

↓

Event 4:

| You rinse the dishes in hot water. |

↓

Event 5:

| You dry the dishes. |

↓

Final outcome:

| You put the dishes away. |

When making an events chain map, you first must find the one event that starts the chain. This event is called the initiating event. You then find the next event in the chain and continue until you reach an outcome. Suppose your mother asked you to wash the dinner dishes. An events chain map might look like the one shown here. Notice that connecting words may not be necessary.

Cycle Concept Map

A **cycle concept map** is a special type of events chain map. In a cycle concept map, the series of events do not produce a final outcome. The last event in the chain relates back to the initiating event. Since there is no outcome and the last event relates back to the initiating event, the cycle repeats itself. Follow the stages shown in the cycle map of insect metamorphosis.

There is usually not one correct way to create a concept map. As you are constructing a map, you may discover other ways to construct the map that show the relationships among concepts better. If you do discover what you think is a better way to create a concept map, do not hesitate to change it.

Concept maps are useful in understanding the ideas you have read about. As you construct a map, you are organizing knowledge. Once concept maps are constructed, you can use them again to review and study and to test your knowledge. The construction of concept maps is a learning tool.

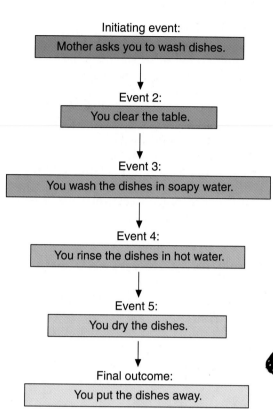

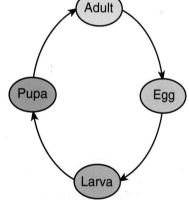

Adult → Egg → Larva → Pupa → Adult

Making and Using Tables

Browse through your textbook, and you will notice many tables both in the text and in the labs. The tables in the text arrange information in such a way that it is easier for you to understand. Also, many labs in your text have tables to complete as you do the lab. Lab tables will help you organize the data you collect during the lab so that it can be interpreted more easily.

Most tables have a title telling you what is being presented. The table itself is divided into columns and rows. The column titles list items to be compared. The row headings list the specific characteristics being compared among those items. Within the grid of the table, the collected data are recorded. Look at the following table, and then study the questions that follow it.

Effect of Excercise on Heart Rate

Pulse Taken	Heart Rate	
	Individual	Class average
At rest	73	72
After exercise	110	112
1 minute after exercise	94	90
5 minutes after exercise	76	75

What is the title of this table? The title is "Effect of Exercise on Heart Rate." What items are being compared? The heart rate for an individual and the class average are being compared at rest and for several durations after exercise.

What is the average heart rate of the class one minute after exercise? To find the answer, you must locate the column labeled "class average" and the row "1 minute after exercise." The data contained in the box where the column and row intersect is the answer. Whose heart rate was 110 after exercise? If you answered "the individual's," you have an understanding of how to use a table.

Making and Using Graphs

After scientists organize data in tables, they often manipulate and organize and then display the data in graphs. A graph is a diagram that shows a comparison between variables. Since graphs show a picture of collected data, they make interpretation and analysis of the data easier. The three basic types of graphs used in science are the line graph, bar graph, and pie graph.

A **line graph** is used to show the relationship between two variables. The variables being compared go on two axes of the graph. The independent variable always goes on the horizontal axis, called the x-axis. The independent variable such as temperature is the condition that is manipulated. The dependent variable always goes on the vertical axis, the y-axis. The dependent variable such as growth is any change that results from manipulating the independent variable.

Suppose a school started a peer-study program with a class of students to see how it affected their science grades.

Average Grades of Students in Study Program	
Grading Period	Average Science Grade
First	81
Second	85
Third	86
Fourth	89

You could make a graph of the grades of students in the program over a period of time. The grading period is the independent variable and should be placed on the *x*-axis of your graph. Instead of four grading periods, we could look at average grades for the week or month or year. In this way, we would be manipulating the independent variable. The average grade of the students in the program is the dependent variable and would go on the *y*-axis.

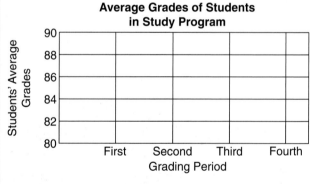

Average Grades of Students in Study Program

Plain or graph paper can be used to construct graphs. After drawing your axes, you would label each axis with a scale. The *x*-axis simply lists the grading periods. To make a scale of grades on the *y*-axis, you must look at the data values provided in the data table above. Since the lowest grade was 81 and the highest was 89, you know that you will have to start numbering at least at 81 and go through 89. You decide to start numbering at 80 and number by twos spaced at equal distances through 90.

You next must plot the data points. The first pair of data you want to plot is the first grading period and 81. Locate "First" on the *x*-axis and 81 on the *y*-axis. Where an imaginary vertical line from the *x*-axis and an imaginary horizontal line from the *y*-axis would meet, place the first data point. Place the other data points the same way. After all the points are plotted, connect them with a smooth line.

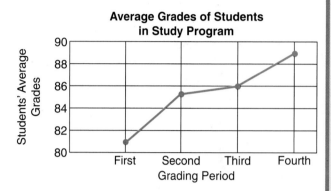

Average Grades of Students in Study Program

What if you wanted to compare the average grades of the class in the study group with the grades of another class? The data of the other class can be plotted on the same graph to make the comparison. You must include a key with two different lines, each indicating a different set of data.

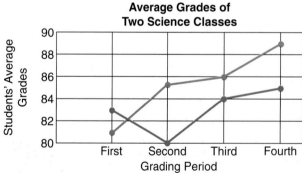

Average Grades of Two Science Classes

KEY: Class of study students ⎯⎯ Regular class ⎯⎯

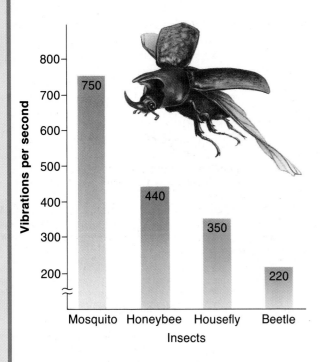

Bar graphs are similar to line graphs, except they are used to show comparisons between data or to display data that does not continuously change. In a bar graph, thick bars show the relationships between data rather than data points.

To make a **bar graph**, set up the x-axis and y-axis as you did for the line graph. The data is plotted by drawing thick bars from the x-axis up to an imaginary point where the y-axis would intersect the bar if it were extended.

Look at the bar graph comparing the wing vibration rates for different insects. The independent variable is the type of insect, and the dependent variable is the number of wing vibrations per second. The number of wing vibrations for different insects is being compared.

A **pie graph** uses a circle divided into sections to display data. Each section represents a part of the whole. When all the sections are placed together, they equal 100 percent of the whole.

Suppose you wanted to make a pie graph to show the number of seeds that germinated in a package. You would have to determine the total number of seeds and the number of seeds that germinated out of the total. You count the seeds and find that the package contains 143 seeds. Therefore, the whole pie will represent this amount.

You plant the seeds and determine that 129 seeds germinate. The group of seeds that germinated will make up one section of the pie graph, and the group of seeds that did not germinate will make up another section.

To find out how much of the pie each section should take, you must divide the number of seeds in each section by the total number of seeds. You then multiply your answer by 360, the number of degrees in a circle. Round your answer to the nearest whole number. The number of seeds that germinated would be determined as follows:

$$\frac{143}{129} \times 360 = 324.75 \text{ or } 325°$$

To plot this data on the pie graph, you need a compass and a protractor. Use the compass to draw a circle. Then, draw a straight line from the center to the edge of the circle. Place your protractor on this line, and use it to mark a point on the edge of the circle at 325°. Connect this point with a straight line to the center of the circle. This is the section for the group of seeds that germinated. The other section represents the group of seeds that did not germinate. Complete the graph by labeling the sections of your graph and giving the graph a title.

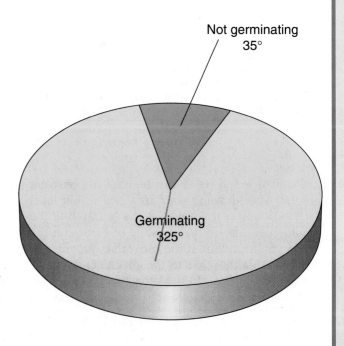

Glossary

This glossary defines each key term that appears in **bold type** in the text. It also shows the page number where you can find the word used.

abiotic factors: nonliving parts of the environment such as air currents, temperature, soil, light, and moisture. (Chap. 3, p. 57)

acid: any substance that forms hydrogen ions in water; acidic solutions have a pH below 7. (Chap. 7, p. 172)

acid precipitation: rain or snow more acid than unpolluted rainwater; leaches valuable nutrients from soil, causing tree death and plant tissue injury; damages stone buildings and statues. (Chap. 6, p. 142)

acoelomate: animal without a body cavity; water and digested food pass into its solid tissues by diffusion; flatworms are acoelomates. (Chap. 28, p. 704)

actin: structural protein in muscle cells; together with the protein myosin, functions in muscle contraction. (Chap. 37, p. 942)

active transport: process requiring energy by which cells move materials against a concentration gradient. (Chap. 9, p. 230)

adaptation: evolution of structural, internal, or behavioral features that help an organism better survive in an environment; the large eyes of nocturnal animals are an adaptation. (Chap. 1, p. 16)

adaptative radiation: type of divergent evolution in which an ancestral species can develop into an array of species, each specialized to fit into different niches. (Chap. 18, p. 446)

addiction: psychological or physiological drug dependence. (Chap. 40, p. 1019)

ADP: adenosine diphosphate; molecule occurring from the breaking off of a phosphate group from ATP, resulting in the release of energy to do biological work. (Chap. 10, p. 239)

aerobic process: oxygen-requiring biological process. (Chap. 10, p. 251)

age structure: proportions of any given population that are either in their pre-reproductive, reproductive, or post-reproductive years. (Chap. 5, p. 126)

aggression: threatening but rarely lethal behavior used to intimidate animals of the same species. (Chap. 36, p. 903)

alcoholic fermentation: anaerobic process in which pyruvic acid is changed to ethyl alcohol and CO_2; carried out by many bacteria and fungi such as yeasts. (Chap. 10, p. 254)

algae: photosynthetic plantlike protists important in oxygen production; along with other eukaryotic autotrophs, they form the foundation of Earth's food chain. (Chap. 22, p. 542)

allele: gene form for each variation of a trait of an organism; for example, Mendel's pea plants had two alleles for height— an allele for tallness and an allele for shortness. (Chap. 12, p. 288)

allelic frequency: percentage of a specific allele in the gene pool. (Chap. 18, p. 436)

alternation of generations: in plants and some protists, the reproductive life cycle in which a gametophyte (haploid, or *n*) generation alternates with a sporophyte (diploid, or *2n*) generation. (Chap. 22, p. 550)

alveoli: in the lungs, tiny, thin-walled sacs surrounded by capillaries; CO_2 diffuses from the blood into the alveoli, and oxygen diffuses from the alveoli into the blood. (Chap. 39, p. 977)

amino acids: basic building blocks of protein molecules. (Chap. 7, p. 180)

amniocentesis: process in which a small amount of amniotic fluid is withdrawn from around the fetus to diagnose chemical, chromosomal, or genetic abnormalities. (Chap. 41, p. 1055)

amniotic egg: in reptiles, birds, and mammals, a major adaptation to land environments; encloses the embryo in amniotic fluid and membranes for protection. (Chap. 34, p. 837)

ampulla: in echinoderms, the bulblike muscular structure that contracts to push water into the tube foot and relaxes to let water flow out. (Chap. 32, p. 789)

amylase: digestive enzyme secreted by the salivary glands and pancreas; breaks starches into smaller sugars. (Chap. 38, p. 953)

anaerobic process: simple biological process that does not require oxygen. (Chap. 10, p. 251)

analogous structures: structures without a common evolutionary origin that are similar in function but not in structure; wings of birds and insects are examples. (Chap. 18, p. 432)

anaphase: stage of mitosis during which the centromeres split and sister chromatids are pulled apart to opposite poles of the cell. (Chap. 11, p. 272)

aneuploidy: abnormal number of chromosomes, usually as a result of accidents during meiosis; can result in Down syndrome and Turner syndrome. (Chap. 15, p. 369)

angiosperms: Anthophyta, flowering plants, extremely diverse plants with seeds enclosed in a fruit; most complex and highly adapted plants on Earth. (Chap. 26, p. 632)

annual: anthophyte that lives for a year or less; examples are corn and blue lupines. (Chap. 26, p. 636)

anterior: the head end in bilaterally symmetrical animals, where sensory tissue is commonly located. (Chap. 28, p. 703)

Glossary

anther: in flowers, the male reproductive structure in which sperm-containing pollen grains develop and from which they burst when mature. (Chap. 27, p. 656)

antheridium: in bryophytes and pteridophytes, the male reproductive structure where sperm develops in the male gametophyte. (Chap. 24, p. 596)

anthropoids: humanlike primates; three major groups include New World monkeys, Old World monkeys, and hominids. (Chap. 19, p. 458)

antibiotic: microbial or fungal product that kills or inhibits the growth of other microorganisms. (Chap. 42, p. 1075)

antibodies: proteins in the plasma produced by B cells and plasma cells in reaction to foreign substances, or antigens. (Chap. 39, p. 984)

antigens: foreign substances that stimulate the production of antibodies in the blood. (Chap. 39, p. 984)

aorta: largest blood vessel in the human body; receives oxygen-rich blood from the left ventricle and sends it out to the body. (Chap. 39, p. 987)

aphotic zone: marine biome that never receives sunlight because of water depth; has intense water pressure; fish populating an aphotic zone have adapted to scarce food and life in darkness. (Chap. 4, p. 91)

apical meristem: terminal growth tissue of roots and stems that adds new cells for growth in length. (Chap. 26, p. 640)

appendage: structure growing out of an animal's body, such as a leg; arthropods were the first invertebrates to evolve jointed appendages. (Chap. 31, p. 762)

appendicular skeleton: one of the two main parts of the human skeleton; includes bones related to limb movement. (Chap. 37, p. 932)

archaebacteria: group of prokaryotes, similar to ancient fossil bacteria that produce glucose by chemosynthesis rather than by photosynthesis. (Chap. 17, p. 414)

archegonium: in bryophytes and pteridophytes, the female reproductive structure that develops eggs in the female gametophyte. (Chap. 24, p. 596)

artery: elastic, thick-walled vessel that transports blood away from the heart with a steady, pulsating rhythm. (Chap. 39, p. 985)

artificial selection: breeding of organisms selected for specific traits in order to produce offspring with those desired characteristics. (Chap. 18, p. 427)

ascospore: sexual spore of ascomycotes. (Chap. 23, p. 568)

ascus: saclike structure of ascomycotes in which the sexual spores develop. (Chap. 23, p. 567)

asexual reproduction: any reproductive process, such as budding, that does not involve the fusion of gametes. (Chap. 22, p. 537)

asymmetry: body plan exhibited by animals, such as sponges, that are irregular in shape. (Chap. 28, p. 702)

atoms: basic building blocks of all matter; smallest particles of an element with all the characteristics of that element. (Chap. 7, p. 161)

ATP: adenosine triphosphate; energy-storing molecule that serves as the cell's "energy currency"; stored energy of glucose is used to attach phosphate groups to ADP to form ATP. (Chap. 10, p. 239)

atria: dual, thin-walled upper chambers of mammalian heart through which blood enters. (Chap. 39, p. 987)

australopithecine: early African hominid; genus *Australopithecus;* the first discovered example is the Taung child, which has a chimplike braincase, face, and teeth but walked upright like humans. (Chap. 19, p. 464)

autonomic nervous system: (ANS) portion of the peripheral nervous system that transmits messages to the internal organs from the central nervous system. (Chap. 40, p. 1009)

autosomes: in humans, the 22 pairs of matching homologous chromosomes. (Chap. 14, p. 336)

autotrophs: organisms that are able to synthesize food using sun energy or energy stored in chemical compounds; plants are the most common autotrophs; also called producers. (Chap. 3, p. 64)

axial skeleton: one of the two main parts of the human skeleton; includes bones of the head, chest, and spine. (Chap. 37, p. 932)

axon: cytoplasmic extension of a neuron; transmits impulses away from the cell body. (Chap. 40, p. 1002)

B

B cells: specialized lymphocytes produced from stem cells in the bone marrow; secrete antibodies that bind with antigens to form a pathogen-fighting antigen-antibody complex. (Chap. 42, p. 1082)

bacteriophages: also called phages; viruses that infect bacteria. (Chap. 21, p. 504)

bark: outer layer of a woody stem that includes phloem tissue. (Chap. 26, p. 642)

base: substance that forms hydroxide ions in water; solutions with a pH above 7 are basic. (Chap. 7, p. 173)

basidia: club-shaped hyphae of basidiomycotes; produce spores during sexual reproduction. (Chap. 23, p. 569)

basidiospore: sexual spore of basidiomycotes. (Chap. 23, p. 569)

behavior: response of an animal to an environmental stimulus; innate, or inherited, behavior is more common among invertebrates, and learned behavior is more common among vertebrates. (Chap. 36, p. 896)

biennial: anthophyte that lives for two years; examples include carrots and beets. (Chap. 26, p. 636)

bilateral symmetry: body plan in which the right and left halves form mirror images when divided down the animal's length; flatworms are bilaterally symmetrical. (Chap. 28, p. 703)

bile: chemical produced by the liver and stored in the gallbladder; breaks fat into small droplets by mechanical digestion. (Chap. 38, p. 957)

Glossary

binary fission: asexual reproduction in which the cell divides into two equal parts. (Chap. 21, p. 522)

binomial nomenclature: two-word classification system originated by Linnaeus for naming species; the first word is the genus name and the second word is the descriptive name or specific epithet. (Chap. 20, p. 483)

biodegradable: solid wastes that can be broken down by natural processes; examples include wood products and food. (Chap. 6, p. 148)

biogenesis: idea that living organisms arise only from other living organisms. (Chap. 17, p. 411)

biology: science of life that seeks to provide an understanding of the natural world. (Chap. 1, p. 6)

biome: large areas with the same type of climax community; examples include deserts and tropical rain forests. (Chap. 4, p. 91)

biosphere: life-supporting portions of Earth composed of air, land, fresh water, and salt water. (Chap. 3, p. 56)

biotic factors: all the living organisms inhabiting any of Earth's many different environments. (Chap. 3, p. 56)

bipedal: having the ability to walk on two legs; in humans, bipedalism may have evolved in response to environmental changes that forced a population of apes to leave the trees and obtain food on the ground. (Chap. 19, p. 463)

blastula: hollow ball of cells in a single layer enclosing a fluid-filled space; an animal embryo after cleavage, before the gastrula is formed. (Chap. 28, p. 698)

blood pressure: the force blood exerts on the vessels; rises when ventricles contract and falls when ventricles relax. (Chap. 39, p. 992)

book lungs: in spiders, air-filled spaces with leaflike plates that function in gas exchange and that resemble book pages. (Chap. 31, p. 765)

budding: asexual reproduction involving mitosis in which a group of cells pinches off from a parent cell and matures into a new individual; yeasts and hydras reproduce asexually by budding. (Chap. 23, p. 564)

bulbourethral glands: in human males, the tiny glands beneath the prostate gland; secrete an alkaline, sperm-protecting fluid into the urethra. (Chap. 41, p. 1040)

bursa: cushioning, fluid-filled sac found between the bones of movable joints. (Chap. 37, p. 932)

Calorie: unit of heat used in measuring the energy content of food. (Chap. 38, p. 967)

Calvin cycle: set of reactions during photosynthesis in which simple sugars are formed from CO_2 using ATP and hydrogen from the light reactions. (Chap. 10, p. 243)

cambium: mytotic growth layer in vascular plants; vascular cambium produces xylem and phloem cells, and can result in secondary growth; cork cambium produces cork cells, which form the waterproof covering of bark. (Chap. 26, p. 640)

camouflage: structural adaptation that involves an individual's external appearance, allowing it to blend in with its surroundings and avoid predators. (Chap. 18, p. 428)

cancer: uncontrolled cell division and death caused by the interaction between environmental factors and changes in the production of enzymes involved in the cell cycle. (Chap. 11, p. 276)

capillaries: smallest blood vessels with walls of only one cell in thickness, through which nutrients and gases diffuse between blood and tissues. (Chap. 39, p. 986)

carbohydrate: organic compound used by cells to store and release energy; composed of carbon, hydrogen, and oxygen. (Chap. 7, p. 178)

cardiac muscle: involuntary muscle found only in the heart; contains interconnected cells that appear striated as a result of proteins in the cells; conducts electrical impulses that produce rhythmic contractions. (Chap. 37, p. 939)

carrier: heterozygous individual that appears phenotypically the same as a homozygous dominant individual, but which has a recessive allele for an undesirable trait. (Chap. 14, p. 344)

carrying capacity: maximum, stable population size an environment can support over time. (Chap. 5, p. 118)

cartilage: in animals such as agnathans, sharks, and skates, the tough flexible material forming the entire skeleton. (Chap. 33, p. 811); tissue from which bone develops in vertebrates. (Chap. 37, p. 932)

cell: building block of both unicellular and multicellular organisms; all living things are made of cells. (Chap. 8, p. 190)

cell culture: growth of cells in a nutrient medium; can be used to diagnose genetic disorders prenatally. (Chap. 16, p. 388)

cell cycle: continuous sequence of growth (interphase) and division (mitosis) of a cell, which is controlled by key enzymes. (Chap. 11, p. 267)

cell theory: the theory that (1) all living things are composed of one or more cells, (2) the cell is the basic unit of organization of organisms, and (3) all cells come from preexisting cells. (Chap. 8, p. 191)

cell wall: firm, fairly inflexible structure outside the plasma membrane of plants, fungi, most bacteria, and some protists; provides support and protection. (Chap. 8, p. 196)

cellulose: polysaccharide made of glucose units hooked together; forms plant cell walls and provides structural support for plants. (Chap. 7, p. 178)

central nervous system: (CNS) in humans, the body's control center; composed of the brain and spinal cord. (Chap. 40, p. 1006)

centrioles: pair of cylindrical structures composed of microtubules that duplicate during interphase and move to opposite ends of the cell during prophase; found in animal cells but seldom in plant cells. (Chap. 11, p. 269)

centromere: cell structure that joins two sister chromatids of a chromosome. (Chap. 11, p. 268)

cephalothorax: structure formed from the fused head and thorax; found in some arthropods, such as shrimps. (Chap. 31, p. 764)

Glossary

cerebellum: portion of the human brain that maintains balance and muscle coordination. (Chap. 40, p. 1007)

cerebrum: largest portion of the human brain; divided into two hemispheres; controls such functions as conscious activities and memory; covered with the highly folded cerebral cortex or gray matter. (Chap. 40, p. 1007)

cervix: in females, the lower end of the uterus that tapers to an opening into the vagina. (Chap. 41, p. 1043)

chelicerae: biting appendages of arachnids. (Chap. 31, p. 772)

chemosynthesis: process to obtain energy and produce food from inorganic compounds used by some organisms, such as methane-producing bacteria. (Chap. 10, p. 249)

chitin: complex carbohydrate found in the cell walls of most fungi and in the exoskeleton of some animals such as insects, lobsters, and crabs. (Chap. 23, p. 561)

chlorophyll: embedded in the inner membrane of chloroplasts, this green pigment traps light energy from the sun and gives some protists and the leaves and stems of plants their green color. (Chap. 8, p. 202)

chloroplasts: chlorophyll-containing organelle found in green plants and some protists; site where light energy is converted into chemical energy, which is stored in food molecules. (Chap. 8, p. 202)

chromatin: long tangled strands of DNA found in the interphase nucleus of eukaryotic cells. (Chap. 8, p. 198)

chromosomal mutation: mutation affecting gene distribution to gametes during meiosis, most commonly by deletions, insertions, inversions, or translocations. (Chap. 13, p. 325)

chromosome: cell structure that carries the genetic material. (Chap. 11, p. 266)

cilia: short, numerous, hairlike projections on a cell's surface that are composed of microtubules; their "beating" activity propels unicellular organisms and moves fluids over the cell surface in multicellular organisms. (Chap. 8, p. 204)

ciliates: protozoans that move through their aquatic habitats by the beating of cilia in coordinated waves; paramecia are ciliates. (Chap. 22, p. 538)

circadian rhythm: daily, 24-hour cycle of behavior in response to internal biological cues. (Chap. 36, p. 904)

citric acid cycle: cyclic series of chemical reactions that take place during aerobic respiration; produces ATP and releases electrons. (Chap. 10, p. 252)

class: taxonomic grouping of related orders. (Chap. 20, p. 486)

classification: grouping together of objects or information based on similarities. (Chap. 20, p. 482)

climax community: stable, mature community that undergoes little ecological succession. (Chap. 4, p. 87)

clones: genetically identical DNA, cells, or organisms. (Chap. 16, p. 378)

closed circulatory system: system within an organism in which blood moves through the body enclosed entirely within blood vessels; found, for example, in cephalopods. (Chap. 30, p. 747)

cochlea: fluid-filled structure in the ear in which sound vibrations are converted into nerve impulses. (Chap. 40, p. 1013)

codominant alleles: equal expression of both alleles; cause the phenotypes of both homozygote parent organisms to be produced in the heterozygote offspring. (Chap. 14, p. 335)

codon: in the genetic code, the set of three nitrogen bases representing a specific amino acid. (Chap. 13, p. 317)

coelom: body cavity totally surrounded by mesoderm in which the digestive tract and other internal organs are suspended; humans, fish, and insects are coelomates. (Chap. 28, p. 705)

colony: group of unicellular or multicellular organisms living together in close association. (Chap. 22, p. 549)

commensalism: symbiotic relationship in which one species benefits and the other species is neither harmed nor helped; an orchid growing on the branch of a larger plant is an example. (Chap. 3, p. 68)

communication: information exchange among animals by signals such as sounds, sights, touches, or smells, resulting in a change in behavior. (Chap. 36, p. 913)

community: several interacting populations that inhabit a common environment and are able to function because each organism within the ecosystem depends on other organisms. (Chap. 3, p. 60)

compact bone: hard, dense bone tissue made up of Haversian canal systems containing blood vessels and nerves; protects spongy bone. (Chap. 37, p. 935)

companion cell: in vascular plants, the nucleated cell that helps manage the transport of food through the sieve cells of the phloem. (Chap. 26, p. 642)

complement: group of proteins that attaches to the surfaces of pathogens, damages their plasma membranes, and attracts phagocytes. (Chap. 42, p. 1080)

complete flower: flower with sepals, petals, stamens, and pistils; examples include phlox and delphinium. (Chap. 27, p. 657)

compound: composed of atoms of different substances that are chemically combined; for example, one atom of sodium (Na) combines with one atom of chlorine (Cl) to form table salt (NaCl), or sodium chloride. (Chap. 7, p. 164)

compound eye: in arthropods, a complex eye composed of multiple lenses that effectively spot moving prey but produce fuzzy images. (Chap. 31, p. 766)

compound light microscope: an instrument that magnifies living cells, small organisms, and preserved cells by passing visible light through the object, then through two or more glass lenses; can magnify up to about $1500\times$. (Chap. 8, p. 190)

Glossary

conditioning: learning a type of behavior by association, for example, feeding associated with the ringing of a bell. (Chap. 36, p. 912)

cone: type of light receptor in the retina; responsible for color detection and vision under bright illumination. (Chap. 40, p. 1012); woody strobili of gymnosperms; scales support male or female reproductive structures and are the site of seed production. (Chap. 25, p. 613)

conidiophores: in ascomycotes, specialized hyphal branches that develop asexual spores in their tips. (Chap. 23, p. 568)

conidium: asexual spore of ascomycotes. (Chap. 23, p. 568)

conjugation: in some bacteria and protists, a simple type of sexual reproduction in which genetic material is transferred from one organism to another through a connecting tube or strand of cytoplasm. (Chap. 21, p. 523)

conservation: planned management of wildlife habitats and other natural areas to prevent exploitation or destruction. (Chap. 6, p. 149)

contractile vacuole: organelle of some protists such as *Paramecium* that collects excess water from the cell, then contracts and expels the water through a plasma membrane pore. (Chap. 9, p. 228)

control: the part of the experiment in which all conditions are kept constant. (Chap. 2, p. 29)

convergent evolution: evolution of similar traits in distantly related organisms resulting from adaptations to similar environments. (Chap. 18, p. 447)

corpus luteum: in human females, the structure stimulated by luteinizing hormone (LH) to develop from the ruptured follicle after the egg has been released; produces progesterone and estrogen. (Chap. 41, p. 1044)

cortex: in plant stems and roots, the tissue between epidermis and vascular core involved in transport and storage of water, sugar, and mineral ions. (Chap. 26, p. 639)

cotyledon: structure of a seed plant embryo that functions as a food-storage organ; sometimes becomes the first leaves of the plant. (Chap. 25, p. 613)

courtship behavior: instinctive behavior patterns of male and female members of a species that occur before mating. (Chap. 36, p. 900)

covalent bond: chemical bond formed when two atoms combine by sharing electrons. (Chap. 7, p. 165)

Cro-Magnons: early *Homo sapiens;* people identical to modern humans in skull and teeth structure and brain size; appeared in Europe about 35 000 years ago and left behind cave paintings and detailed stone artifacts. (Chap. 19, p. 472)

crossing over: exchange of genetic material by non-sister chromatids during late prophase I of meiosis, resulting in new combinations of alleles. (Chap. 12, p. 302)

cud chewing: adaptation of many hoofed mammals enabling swallowed food matter to be brought up and rechewed. (Chap. 35, p. 874)

cuticle: waxy, protective coating on the outer surface of the epidermis of most fruits, stems, and leaves; adaptation that helps prevent water loss. (Chap. 24, p. 586)

cytoplasm: clear fluid in eukaryotic cells surrounding the nucleus and organelles; site of many important chemical reactions. (Chap. 8, p. 199)

cytoskeleton: in eukaryotic cells, a network in the cytoplasm; usually composed of microtubules and microfilaments; provides support for organelles and is important in cell locomotion. (Chap. 8, p. 203)

data: information obtained by observation, particularly during experiments; data are considered valid only if repeating the experiment several times yields similar results. (Chap. 2, p. 34)

day-neutral plant: plant in which flowering is influenced more by environmental conditions than by day length; most plants are day-neutral. (Chap. 27, p. 668)

deciduous plant: plant that loses all of its leaves at the same time and can no longer perform photosynthesis and must remain dormant during winter. (Chap. 25, p. 620)

decomposers: organisms, such as many bacteria and most fungi, that play beneficial roles in all ecosystems by breaking down and absorbing nutrients from dead and decaying organic matter. (Chap. 3, p. 65)

deductive reasoning: "if...then" reasoning used after forming a hypothesis; suggests that something may be true about a specific case based on a known rule. (Chap. 2, p. 28)

demography: study of populations by mathematicians called demographers, who tally characteristics such as birthrates and death rates, fertility rates, age structure, and geographic distribution. (Chap. 5, p. 124)

dendrite: cytoplasmic extension of a neuron; transmits impulses to the cell body. (Chap. 40, p. 1002)

density-dependent factors: factors that limit population density; examples include predation, disease, and competition for food, water, and territory. (Chap. 5, p. 120)

density-independent factors: most often weather-related occurrences such as storms and floods that affect populations, regardless of their density. (Chap. 5, p. 121)

dependent variable: in a controlled experiment, the measurable condition that results from changing the independent variable. (Chap. 2, p. 30)

depressant: drug that produces sedation by slowing down the activities of the central nervous system and increasing the activities of the parasympathetic nervous system; alcohol is a depressant. (Chap. 40, p. 1018)

dermis: inner portion of skin; contains structures such as sweat glands and capillaries and produces vitamin D when exposed to ultraviolet light. (Chap. 37, p. 927)

desert: driest biome south of the taiga; generally receives less than 25 cm of annual rainfall, producing a range of vegetation from shrub communities to drifting dunes; animals include lizards, owls, and coyotes. (Chap. 4, p. 102)

Glossary

detritus: dead and decaying organic matter that provides food for small organisms. (Chap. 32, p. 792)

deuterostome: animal whose anus develops from the opening in the gastrula; examples include humans, fish, and birds. (Chap. 28, p. 699)

development: rapidly or slowly occurring changes due to growth in an organism. (Chap. 1, p. 14)

diaphragm: in mammals, the sheet of muscle in the chest cavity that contracts during inhalation and relaxes during exhalation. (Chap. 35, p. 872)

dicotyledon: plant with flower parts in multiples of four or five, seeds with two cotyledons, and a branching network of veins in the leaves; garden flowers and herbs are examples. (Chap. 26, p. 633)

diffusion: net, random movement of particles from an area of higher concentration to an area of lower concentration, eventually resulting in even distribution. (Chap. 9, p. 224)

dihybrid cross: fertilization between two organisms to study the inheritance of two different traits. (Chap. 12, p. 291)

diploid: a cell with two copies of each type of chromosome is considered to be a diploid ($2n$) cell. (Chap. 12, p. 298)

directional selection: selection favoring individuals with extreme forms of a trait; can lead to rapid evolution of a population. (Chap. 18, p. 439)

disaccharide: two-sugar carbohydrate formed by linking two monosaccharide molecules; sucrose is a disaccharide. (Chap. 7, p. 178)

disruptive selection: selection favoring individuals at both ends of extreme forms of a trait; can lead to the evolution of two new species. (Chap. 18, p. 440)

divergent evolution: evolution in which highly distinct species were once both similar to an ancestral species. (Chap. 18, p. 447)

division: taxonomic grouping of related classes for plants; the equivalent category of phylum in animal taxonomy. (Chap. 20, p. 486)

DNA: deoxyribonucleic acid; complex biological polymer; master copy of an organism's information code, which is passed on each time a cell divides and also from one generation to the next. (Chap. 7, p. 184)

dominance hierarchy: "pecking order" among individuals of a group in which there are several levels of dominance and submission; often prevents continuous fighting and may provide females a way to select the best mate. (Chap. 36, p. 903)

dominant: visible, observable trait of an organism that masks a recessive form of the trait. (Chap. 12, p. 289)

dormancy: in seeds, the period of inactivity allowing them to survive conditions unfavorable for growth. (Chap. 27, p. 673)

dorsal: back surface of bilaterally symmetrical animals. (Chap. 28, p. 703)

dorsal nerve cord: nerve bundle that in most chordates develops into the spinal cord and brain. (Chap. 32, p. 797)

double fertilization: in anthophytes, sexual reproduction resulting in the formation of a diploid ($2n$) zygote and a cell with a triploid ($3n$) nucleus that divides to eventually form the endosperm, or food-storage tissue for the developing embryo. (Chap. 27, p. 669)

double helix: in DNA, the two twisted, ladder-shaped nucleotide strands held together by hydrogen bonds between the bases. (Chap. 13, p. 312)

drug: chemical substance that can react with and alter body functions; examples include aspirin and caffeine. (Chap. 40, p. 1017)

dynamic equilibrium: condition of continuous movement but no overall change in concentration; movement of materials into and out of the cell at equal rates maintains its dynamic equilibrium with its environment. (Chap. 9, p. 224)

ecology: scientific study of interactions between organisms and their environments; for example, ecologists study how day length influences migrating bird behavior. (Chap. 3, p. 55)

ecosystem: populations in a community and abiotic factors with which they interact; examples are terrestrial and marine ecosystems. (Chap. 3, p. 61)

ectoderm: embryonic outer layer of cells in a gastrula; develops into skin and nervous tissue. (Chap. 28, p. 698)

ectotherm: animal with a body temperature regulated by its environment; amphibians and reptiles are ectotherms. (Chap. 33, p. 823)

egg: female sex cell or gamete. (Chap. 12, p. 300)

electron microscope: an instrument that allows scientists to magnify, view, and photograph dead cells or organisms by using a beam of electrons instead of light; its resolving power may be 1000 times greater than that of a light microscope. (Chap. 8, p. 194)

electron transport chain: series of molecules along which electrons are transferred; as the electrons travel, they release energy that is stored in the bonds of ATP. (Chap. 10, p. 244)

embryo: the early stage of growth and development of a plant or animal that followed fertilization of the egg by the sperm cell and the formation of a zygote. (Chap. 25, p. 613)

emigration: movement of individuals out of a population. (Chap. 5, p. 128)

endangered species: species with numbers of individuals so low that it is in danger of extinction; examples include manatees, the Florida panther, and loggerhead turtles. (Chap. 6, p. 138)

endemic disease: disease continually present in a population; the common cold is an example. (Chap. 42, p. 1074)

endocrine gland: ductless gland that releases hormones directly into the bloodstream; the thyroid functions as an endocrine gland in metabolism. (Chap. 38, p. 960)

endocytosis: active transport process by which large particles enter a cell. (Chap. 9, p. 231)

endoderm: embryonic inner layer of cells in a gastrula that gives rise to the lining of the digestive tract. (Chap. 28, p. 698)

endodermis: in plant roots, the waterproof, single layer of cells that controls the flow of solutes into the root. (Chap. 26, p. 639)

endoplasmic reticulum: folded, complex system of membranes forming a type of transport system in the cytoplasm of eukaryotic cells; can be either rough (with ribosomes) or smooth (without ribosomes). (Chap. 8, p. 199)

endoskeleton: internal skeleton of vertebrates; provides a support framework and protects internal organs. (Chap. 28, p. 708)

endosperm: in anthophytes, the triploid ($3n$) food-storage tissue used by the developing embryo. (Chap. 27, p. 670)

endospore: in bacteria, the structure with a hard, protective outer covering formed when conditions are unfavorable; under favorable conditions, the endospore germinates and the bacterium resumes growth and reproduction. (Chap. 21, p. 521)

endotherm: animal able to maintain a constant body temperature based on metabolic processes rather than being regulated by environmental conditions; for example, birds are endotherms. (Chap. 34, p. 848)

energy: ability to do work or move things; powers life processes. (Chap. 1, p. 17)

environment: external conditions, such as weather, or internal conditions, such as the presence of infection, to which an organism must continually adjust. (Chap. 1, p. 15)

enzymes: proteins that accelerate chemical reactions but do not change themselves in the reaction; enzymes enable molecules to undergo chemical change, forming new substances called products. (Chap. 7, p. 181)

epidemic: disease pattern in which large numbers of people in a given area are sickened by the same disease over a short period of time; influenza often produces epidemics. (Chap. 42, p. 1074)

epidermis: in plants, the outermost layer of cells of the plant body; also forms hairs on the plant's surfaces. (Chap. 26, p. 639); in humans and other animals, the protective outer portion of skin composed of two cell layers; the top layer continually sheds dead cells and the inner layer forms new living cells. (Chap. 37, p. 926)

epididymis: in human males, the single, coiled tube in which sperm complete maturation; also provides storage for sperm. (Chap. 41, p. 1039)

epiglottis: flap of skin that covers the opening of the windpipe during swallowing. (Chap. 38, p. 955)

esophagus: muscular tube that moves food from the mouth to the stomach by smooth muscle contractions. (Chap. 38, p. 955)

estivation: state similar to hibernation; affects many desert animals in response to heat or drought; an innate behavior governed by both internal and external cues. (Chap. 36, p. 906)

estuary: coastal body of water in which both fresh water and salt water mix; provides excellent food supply and shelter for young fish. (Chap. 4, p. 92)

ethics: study of right and wrong; in the scientific arena, ethical issues must be decided by all of society. (Chap. 2, p. 45)

eukaryote: cell having a true nucleus and membrane-bound internal organelles; the majority of cells are eukaryotic. (Chap. 8, p. 195)

evergreen plant: plant that retains its leaves year-round; most conifers are evergreens; often found in areas that have a warm, short growing season and scarce nutrients. (Chap. 25, p. 619)

evolution: change in the gene pool of a population in response to various stimuli exhibited by a species over time. (Chap. 1, p. 20)

exocrine gland: gland that releases its secretions through ducts; the pancreas functions as an exocrine gland during digestion. (Chap. 38, p. 960)

exocytosis: active transport process by which materials are expelled or secreted from a cell. (Chap. 9, p. 232)

exoskeleton: hard outer covering common in some invertebrates such as crabs and spiders; provides support, protection, and a place for muscle attachment. (Chap. 28, p. 708)

experiment: procedure by which scientists determine the validity of a hypothesis by collecting information under controlled conditions. (Chap. 2, p. 29)

exponential growth: explosive population growth in which the number of reproducing individuals increases by an ever-increasing rate. (Chap. 5, p. 115)

external fertilization: fertilization that occurs in water into which both eggs and sperm have been released, as in sponges, frogs, and fishes. (Chap. 29, p. 716)

extinction: occurs when the last members of a species die; may be a natural process or the result of human activity such as hunting, urbanization, and the destruction of natural habitats. (Chap. 6, p. 136)

facilitated diffusion: passive transport of materials, such as sugars and amino acids, across the plasma membrane by way of transport proteins. (Chap. 9, p. 230)

family: taxonomic grouping of closely related genera. (Chap. 20, p. 486)

feather: lightweight, external covering of birds; down feathers insulate and contour feathers are used for flight. (Chap. 34, p. 847)

Glossary

fertilization: fusion of male and female gametes; in Mendel's pea plants, occurred when the male gamete in the pollen grain fused with the female gamete in the ovule. (Chap. 12, p. 286)

fetus: the developing mammal from nine weeks until birth. (Chap. 15, p. 362)

fight-or-flight response: automatic, chemically controlled response to a stimulus that mobilizes the individual for greater activity, such as running or fighting—for example, by increasing the heart and breathing rates. (Chap. 36, p. 900)

filter feeding: process in which food particles are filtered from water as it travels through or around the organism; for example, sponges and bivalve mollusks are filter feeders. (Chap. 29, p. 714)

fins: in fishes, the external fan-shaped membranes attached to the endoskeleton and supported by stiff rays; fins function in locomotion. (Chap. 33, p. 812)

flagella: long, threadlike structures composed of microtubules; project from within the plasma membrane and propel cells and organisms by a whiplike motion. (Chap. 8, p. 204)

flagellates: protozoans that move by means of one or more flagella, which they whip from side to side. (Chap. 22, p. 537)

fluid mosaic model: property of the plasma membrane wherein many similar molecules are free to move sideways within their lipid bilayer. (Chap. 9, p. 220)

follicle: in females, the epithelial cells that surround an undeveloped egg cell in the ovary; when the egg is released, the remaining part of the follicle develops into the corpus luteum. (Chap. 41, p. 1043)

food chain: a possible route for the transfer of matter and energy through an ecosystem from autotrophs through heterotrophs and decomposers. (Chap. 3, p. 70)

food web: shows all the possible feeding relationships in a community at each trophic level; represents a network of interconnected food chains. (Chap. 3, p. 72)

fossil: any evidence of organisms that lived in the past; usually found in sedimentary rock; scientists study fossils to look for evolutionary clues. (Chap. 17, p. 399)

fossil fuels: coal, oil, and natural gas formed from the buried remains of organisms. (Chap. 6, p. 136)

fragmentation: asexual reproduction in which an organism breaks up into two or more parts, each of which forms a new organism; seen, for example, in *Spirogyra*. (Chap. 22, p. 550)

frameshift mutation: error in the DNA sequence that adds or deletes a single base, causing nearly all amino acids following the mutation to be changed. (Chap. 13, p. 325)

frond: the leaflike organ of a fern that grows upward from the rhizome; fern fronds have a stemlike stipe and green pinnae. (Chap. 25, p. 610)

fruit: seed-containing, ripened ovary of an anthophyte; examples are dry fruits such as pecans or fleshy fruits such as peaches. (Chap. 27, p. 671)

fungus: heterotrophic, eukaryotic consumer that absorbs nutrients from decomposing wastes and dead organisms. (Chap. 20, p. 495)

gallbladder: small bile-storing organ of the digestive system. (Chap. 38, p. 956)

gametangium: in zygospore-producing fungi, the structure with a haploid nucleus in which gametes are produced. (Chap. 23, p. 567)

gametes: male and female sex cells; sperm and eggs. (Chap. 12, p. 286)

gametophyte: haploid (*n*) form of a plant that produces gametes. (Chap. 22, p. 550)

gastrovascular cavity: cavity in which digestion occurs; in cnidarians, which have a simple, two-cell-layer body plan, this space is surrounded by the inner cell layer. (Chap. 29, p. 720)

gastrula: developmental stage of animal embryos in which the blastula folds inward forming a cavity with two cell layers, ectoderm and endoderm; a middle layer of cells, the mesoderm, is formed later. (Chap. 28, p. 698)

gemmae: tiny haploid (*n*) structures of liverworts and some mosses; grow in cups on the surface of gametophytes during vegetative reproduction. (Chap. 24, p. 600)

gene: a segment of DNA located on the chromosome; directs the protein production that controls the cell cycle. (Chap. 11, p. 276)

gene pool: sum of all genes among a population. (Chap. 18, p. 436)

gene splicing: in recombinant DNA technology, the rejoining of cut DNA fragments. (Chap. 16, p. 378)

gene therapy: a goal of recombinant DNA technology, presently in trial stages; inserts normal genes into human cells to correct genetic disorders. (Chap. 16, p. 388)

genetic counseling: provided by trained professionals in various disciplines to couples seeking information about the probability of having hereditary disorders occur in their offspring. (Chap. 41, p. 1054)

genetic drift: changes in allelic frequency by chance events; results in alteration of genetic equilibrium. (Chap. 18, p. 437)

genetic engineering: process using restriction enzymes to cleave an organism's DNA into smaller fragments in order to move genes from one organism to another of the same or different species; has produced transgenic plants, bacteria, and animals. (Chap. 16, p. 376)

genetic equilibrium: condition in which allelic frequency remains constant over generations; populations in genetic equilibrium are not evolving. (Chap. 18, p. 436)

genetic recombination: major source of genetic variation resulting from crossing over or random assortment. (Chap. 12, p. 304)

genetics: branch of biology that studies heredity. (Chap. 12, p. 286)

Glossary

genotype: an organism's gene combination. (Chap. 12, p. 291)

genus: first word of the two-part scientific name used to classify a group of closely related species. (Chap. 20, p. 483)

geographic isolation: occurs when populations become isolated by a factor such as deforestation, which results in individuals no longer being able to mate; can lead to the formation of new species. (Chap. 18, p. 441)

germination: process by which a seed begins to develop into a new organism. (Chap. 27, p. 674)

gestation: full term of pregnancy during which the young of placental mammals develop in the uterus; length varies from species to species. (Chap. 35, p. 883)

gill slits: in chordates, the paired openings found in the pharynx, behind the mouth; in some vertebrate chordates, the gill slits develop into internal gills used for gas exchange. (Chap. 32, p. 800)

gizzard: muscular sac containing hard particles for grinding food-containing material into small pieces so it can be absorbed by the intestine; found, for example, in annelids. (Chap. 30, p. 751)

gland: fluid-secreting cell or group of cells; in mammals, various glands produce fluids such as enzymes, hormones, and milk. (Chap. 35, p. 871)

glycogen: polysaccharide with very highly branched chains of glucose units; animals store food as glycogen. (Chap. 7, p. 178)

glycolysis: anaerobic process that splits glucose, forming two molecules of pyruvic acid; also produces hydrogen ions and electrons. (Chap. 10, p. 251)

Golgi apparatus: membrane sacs that receive, chemically modify, and repackage proteins into forms the cell can use, expel, or keep stored. (Chap. 8, p. 200)

gradualism: idea that species originate gradually over time through the accumulation of small adaptive changes. (Chap. 18, p. 444)

grasslands: biome composed of large communities of grasses and other small plants; characterized by hot summers, cold winters, and uncertain rainfall; high humus content of soil is ideal for growing oats, rye, and wheat; home to buffalo, prairie dogs, and many bird and insect species. (Chap. 4, p. 103)

greenhouse effect: a natural phenomenon by which carbon dioxide and other atmospheric gases prevent heat from escaping into space; without the greenhouse effect, Earth would be too cold for life to exist. (Chap. 6, p. 144)

groundwater: fresh water from rain and surface streams that accumulates in underground reservoirs. (Chap. 6, p. 146)

growth: process that results in structural changes and increased living material in an organism. (Chap. 1, p. 14)

guard cells: pairs of modified cells of the leaf epidermis that control the opening and closing of stomatal pores. (Chap. 26, p. 645)

gymnosperms: nonflowering vascular plants with seeds produced on the scales of cones; the four divisions are Cycadophyta, Ginkgophyta, Gnetophyta, and Coniferophyta. (Chap. 25, p. 613)

H

habitat: collection of niches in which an organism lives its life. (Chap. 3, p. 62)

habituation: simple form of learned behavior. (Chap. 36, p. 908)

hair follicles: groups of cells in the dermal skin layer from which hair grows. (Chap. 37, p. 927)

hallucinogen: mood-altering substance; examples are LSD and psilocybin. (Chap. 40, p. 1025)

haploid: a cell of an organism that has half the number (n) of chromosomes; one of each type of chromosome that makes up the genotype. (Chap. 12, p. 298)

haustoria: hyphae of parasitic fungi adapted to penetrate the cells of hosts and absorb nutrients. (Chap. 23, p. 563)

hemoglobin: iron-containing molecule of red blood cells that transports oxygen and some CO_2. (Chap. 39, p. 982)

heredity: passing on of characteristics from parents to offspring; first major studies of heredity were done by Gregor Mendel in the 1800s. (Chap. 12, p. 286)

hermaphrodite: organism that can produce both eggs and sperm but is not necessarily self-fertilizing. (Chap. 29, p. 716)

heterotrophs: organisms unable to make their own food, they rely on autotrophs as their nutrient and energy source; examples are rabbits and cows; also called consumers. (Chap. 3, p. 64)

heterozygous: having nonidentical alleles for a particular trait. (Chap. 12, p. 291)

hibernation: state in which metabolic needs are greatly reduced; exhibited by many mammals, some birds, and other types of animals during cold winter months. (Chap. 36, p. 906)

homeostasis: equilibrium of an organism's internal environment that maintains conditions suitable for life; an example is human sweating, which helps the body maintain its proper temperature. (Chap. 1, p. 16); the balance of nature in ecosystems. (Chap. 3, p. 23)

hominid: bipedal, humanlike primate; *Australopithecus afarensis* is the earliest known example of a hominid and was bipedal with an apelike braincase. (Chap. 19, p. 463)

homologous chromosomes: paired chromosomes with genes for the same traits arranged in the same order. (Chap. 12, p. 299)

homologous structures: structures having a common evolutionary origin; examples are the forelimbs of bats, crocodiles, and humans. (Chap. 18, p. 432)

homozygous: having identical alleles for a particular trait. (Chap. 12, p. 291)

Glossary

hormone: chemical secreted by one part of a plant or animal that affects another part of the organism; in plants, hormones such as auxins and gibberellins promote growth. (Chap. 27, p. 676); in humans, hormones play a key role in regulating metabolism of digestion, growth, and reproduction. (Chap. 41, p. 1040)

host cell: cell in which a virus reproduces. (Chap. 21, p. 504)

human genome: approximately 100 000 genes, on 46 human chromosomes, made of about 3 billion DNA base pairs, which when mapped and sequenced may enable the treatment or cure of genetic disorders. (Chap. 16, p. 386)

hybrid: offspring produced when two varieties of plants or animals, or closely related species, are mated; hybrids often exhibit greater vigor and size than their parents. (Chap. 14, p. 347)

hydrogen bond: weak bond formed by attraction of opposite charges between hydrogen and other atoms; helps to hold large molecules, such as proteins, together. (Chap. 7, p. 168)

hypertonic solution: in cells, solution in which the concentration of dissolved materials is higher in the solution surrounding the cell than the concentration inside the cell, which will shrink as water leaves the cell by osmosis. (Chap. 9, p. 229)

hyphae: threadlike filaments that form the basic structural units of multicellular fungi. (Chap. 23, p. 561)

hypothalamus: area of the brain directly above the pituitary that produces antidiuretic hormone (ADH) and releasing factors that cause the pituitary to secrete hormones, which control reproduction. (Chap. 39, p. 996)(Chap. 41, p. 1041)

hypothesis: testable explanation of a question or problem; a hypothesis may be formed by extensive reading, observation, reasoning, and knowledge of earlier experiments. (Chap. 2, p. 27)

hypotonic solution: in cells, solution in which the concentration of dissolved materials is lower in the solution surrounding the cell than the concentration inside the cell, which will swell and possibly burst as water enters the cell by osmosis. (Chap. 9, p. 227)

immigration: movement of individuals into a population; in the last 165 years, more people immigrated to the United States than to any other country. (Chap. 5, p. 127)

immunity: overall ability of an organism to resist a specific pathogen. (Chap. 42, p. 1081)

implantation: in females, the attachment of the fertilized egg, or blastocyst, to the uterine lining. (Chap. 41, p. 1049)

imprinting: species-specific learned behavior that occurs only at a certain, critical time in an animal's life and forms a permanent social attachment. (Chap. 36, p. 909)

inbreeding: mating between closely related individuals to produce pure lines; however, undesirable recessive traits can appear more often than in random matings. (Chap. 14, p. 347)

incomplete dominance: inheritance pattern in which the phenotype of the heterozygote is intermediate between those of the two homozygotes; neither allele of the pair is completely dominant but they combine to give a new trait. (Chap. 14, p. 334)

incomplete flower: flower lacking one or more organs; examples include sweet corn and squash. (Chap. 27, p. 657)

independent variable: in a controlled experiment, the one condition that is changed. (Chap. 2, p. 30)

inductive reasoning: most common type of scientific reasoning used in developing a hypothesis; produces a general rule based on a set of observations. (Chap. 2, p. 27)

infectious disease: disease caused by pathogens in the body; chicken pox and tuberculosis are examples. (Chap. 42, p. 1068)

innate behavior: genetically programmed behavior pattern of an animal; includes both automatic responses and instinctive behaviors. (Chap. 36, p. 897)

insight: most complex type of learning resulting in the formation of a concept or idea that can be applied to new situations. (Chap. 36, p. 912)

instinct: pattern of innate, or inherited, animal behavior, such as courtship rituals. (Chap. 36, p. 900)

internal fertilization: fertilization of eggs by sperm inside an animal's body; in snakes, birds, and mammals. (Chap. 29, p. 716)

interphase: growth period of a cell during which chromosomes are duplicated. (Chap. 11, p. 267)

intertidal zone: part of the shoreline between the high and low tide lines; productivity of intertidal ecosystems is limited because of wave action even though levels of nutrients, sunlight, and oxygen are high. (Chap. 4, p. 92)

invertebrate: animal lacking a backbone; examples include crabs and grasshoppers. (Chap. 28, p. 708)

involuntary muscle: muscle whose contractions are not under conscious control; smooth muscle is involuntary muscle. (Chap. 37, p. 938)

ion: an atom or molecule that gains electrons and carries a positive electrical charge or loses electrons and has a negative electrical charge. (Chap. 7, p. 165)

ionic bond: bond formed by the mutual attraction of two ions of opposite charge. (Chap. 7, p. 166)

isomers: compounds with the same formula but different three-dimensional arrangement of the atoms, resulting in the molecules having different chemical properties. (Chap. 7, p. 176)

isotonic solution: solution in cells in which dissolved materials and water occur in the same concentration as inside the cell. (Chap. 9, p. 226)

isotopes: atoms of the same element differing in the numbers of neutrons in the nucleus; examples are carbon-12 and carbon-14; some radioactive isotopes are used in medicine. (Chap. 7, p. 163)

Jacobson's organ: pitlike olfactory organ of most reptiles that is used to detect airborne chemicals. (Chap. 34, p. 840)

joint: point where two or more bones meet; can be fixed or allow movement; examples include ball-and-socket joints and hinge joints. (Chap. 37, p. 932)

karyotype: charted arrangement of chromosomes possessed by an individual; helpful in locating aneuploidies in humans such as Down syndrome. (Chap. 15, p. 369)

keratin: protein formed by the outer epidermal skin layer; helps protect underlying skin cells. (Chap. 37, p. 926)

kidneys: organs of the vertebrate urinary system; remove nitrogenous wastes, control the sodium level of blood, and regulate the pH of blood. (Chap. 39, p. 993)

kingdom: taxonomic grouping of related phyla. (Chap. 20, p. 486)

Koch's postulates: a set of steps for proving that a specific pathogen causes a specific disease. (Chap. 42, p. 1070)

labor: sum of physical and physiological changes that occur to the mother during the birth process. (Chap. 41, p. 1056)

lactic acid fermentation: anaerobic process in which pyruvic acid changes to lactic acid; occurs in some bacteria, plants, and most animals; can create an oxygen debt in exercised muscles because of a buildup of lactic acid, resulting in fatigue. (Chap. 10, p. 254)

language: form of communication using symbols to represent ideas. (Chap. 36, p. 918)

large intestine: muscular tube into which indigestible material is passed to the rectum for elimination; site of water absorption and synthesis of B and K vitamins. (Chap. 38, p. 958)

larva: in insects, the stage of metamorphosis in which the organism is wormlike, free-living, and molts several times. (Chap. 31, p. 781)

lateral line system: fluid-filled canals along the sides of a fish that serve as sensory receptors, enabling detection of prey and navigation in the dark. (Chap. 33, p. 813)

law of independent assortment: Mendelian principle explaining that different traits are inherited independently if on different chromosomes. (Chap. 12, p. 294)

law of segregation: Mendelian principle explaining the disappearance of a specific trait in the F_1 generation and its reappearance in the F_2 generation. (Chap. 12, p. 290)

leaf: plant organ that provides a surface area for trapping sunlight and exchanging gases through stomata during photosynthesis. (Chap. 24, p. 587)

lichen: organism comprising a symbiotic association between a fungus and a photosynthetic partner. (Chap. 23, p. 575)

ligament: strong band of connective tissue that holds joints together and connects bones to other bones. (Chap. 37, p. 932)

light reactions: highly complex reactions by which light energy is converted to chemical energy during photosynthesis; results in the splitting of water and release of oxygen. (Chap. 10, p. 243)

limiting factor: any factor limiting the survival and productivity of organisms; for example, the lack of water could limit grass growth in a grassland. (Chap. 4, p. 84)

linkage map: genetic map showing gene location on chromosomes. (Chap. 16, p. 386)

lipids: organic compounds commonly called fats and oils; insoluble in water, lipids are the major components of membranes surrounding all living cells. (Chap. 7, p. 179)

liver: gland that produces many chemicals, including several needed for digestion, that are delivered to the small intestine. (Chap. 38, p. 957)

long-day plant: plant that flowers in midsummer, when days are longer than nights; examples are carnations and spinach. (Chap. 27, p. 668)

lymph: tissue fluid collected in lymphatic vessels to be returned to the blood. (Chap. 42, p. 1076)

lymph node: small tissue mass with a fiber network that holds lymphocytes. (Chap. 42, p. 1077)

lymphocytes: white blood cells that mature and differentiate into cells that destroy specific pathogens. (Chap. 42, p. 1077)

lysogenic cycle: viral reproductive cycle; the host cell's chromosome becomes integrated with viral DNA; the provirus formed is replicated each time the host reproduces; the host cell is not killed until the lytic cycle is entered. (Chap. 21, p. 508)

lysosomes: membrane-bound organelles containing enzymes that digest food particles, viruses and bacteria, worn-out cell parts, and sometimes the cell itself. (Chap. 8, p. 201)

lytic cycle: viral reproductive cycle; host cell DNA is destroyed by the virus, which then forms new viruses that burst from the host cell, killing it. (Chap. 21, p. 507)

macrophage: type of phagocyte that attacks anything recognized as foreign, including microbes and dust in the lungs. (Chap. 42, p. 1079)

madreporite: in echinoderms, the disc-shaped, sievelike opening through which water flows in and out of the water vascular system. (Chap. 32, p. 790)

Glossary

Malpighian tubules: waste-excreting, abdominal structures of most terrestrial arthropods. (Chap. 31, p. 767)

mammary glands: milk-secreting glands of female mammals enabling them to feed their young until they are mature enough to forage for food. (Chap. 35, p. 875)

mandible: in most arthropods, mouthpart adapted for piercing, sucking, lapping, and chewing of food. (Chap. 31, p. 767)

mantle: thin, outer membrane of mollusks; in mollusks with shells, the mantle secretes the shell. (Chap. 30, p. 742)

marrow: soft tissue filling the cavities of most bones; functions include red blood cell production and fat storage. (Chap. 37, p. 934)

marsupial: mammal in which the young develop inside the body, followed by longer development outside the body in a pouch of skin and hair. (Chap. 35, p. 883)

medulla oblongata: portion of the human brain that controls involuntary activities such as breathing. (Chap. 40, p. 1007)

medusa: in the life cycle of cnidarians, the free-swimming, bell-shaped stage. (Chap. 29, p. 718)

megaspores: female spores in plants that develop eventually into archegonia with egg cells. (Chap. 25, p. 613)

meiosis: cell division in which one diploid (2n) cell produces four haploid (n) cells called sex cells or gametes, which have half the number of chromosomes as a body cell of the parent. (Chap. 12, p. 300)

melanin: cell pigment formed by the inner epidermal skin layer; colors the skin and helps protect against ultraviolet radiation. (Chap. 37, p. 927)

menstrual cycle: in human females, the monthly cycle in which the lining of the uterus is prepared to receive an egg and is shed if the egg is not fertilized. (Chap. 41, p. 1045)

mesoderm: middle layer of embryonic cells in a gastrula, between the ectoderm and endoderm; develops into reproductive organs, muscles, and circulatory vessels. (Chap. 28, p. 699)

mesophyll: photosynthetic tissue of a leaf, either palisade or spongy. (Chap. 26, p. 648)

messenger RNA (mRNA): carries protein synthesis information from DNA to the ribosomes. (Chap. 13, p. 319)

metabolism: total of all chemical reactions that occur within a living organism. (Chap. 7, p. 169)

metamorphosis: chemically controlled series of changes in body structure development in insects; can be complete or incomplete. (Chap. 31, p. 781)

metaphase: second stage of mitosis in which chromosomes move to the equator of the spindle and chromatids are each attached by centromeres to a separate spindle fiber. (Chap. 11, p. 272)

microfilaments: thin, solid protein fibers present in the cytoskeleton of eukaryotic cells; play a role in cell structure and motion. (Chap. 8, p. 203)

micropyle: in an anthophyte, the tiny opening in the ovule through which two sperm cells enter. (Chap. 27, p. 669)

microspores: male spores in plants that develop eventually into pollen grains. (Chap. 25, p. 613)

microtubules: hollow, thin, protein cylinders found in the cytoskeleton of eukaryotic cells; important in cell structure and locomotion. (Chap. 8, p. 203)

migration: yearly rhythm cycle of behavior resulting in the seasonal movement of animals; governed by internal and external biological cues. (Chap. 36, p. 904)

mimicry: structural adaptation evolved in some organisms resulting in the mimicing appearance to other organisms for protection or other advantages. (Chap. 18, p. 428)

mineral: inorganic substance essential to chemical reactions or as building materials in the body; examples include calcium and potassium. (Chap. 38, p. 965)

mitochondrion: eukaryotic membrane-bound organelle in which food molecules are broken down to produce ATPs; contains highly folded inner membrane that produces energy-storing molecules. (Chap. 8, p. 201)

mitosis: cell division during which chromosomes are equally distributed to the two identical daughter cells that are formed; results in growth. (Chap. 11, p. 267)

mixture: combination of substances that do not combine chemically but retain their individual properties. (Chap. 7, p. 166)

molecule: group of atoms held together by covalent bonds. (Chap. 7, p. 165)

molting: in arthropods, periodic shedding of the protective exoskeleton, allowing for a size increase. (Chap. 31, p. 763)

monerans: microscopic prokaryotes with a wide variety of structures and metabolic demands; examples include *Nitrosomonas* and *Oscillatoria*. (Chap. 20, p. 491)

monocotyledon: plant with flower parts in multiples of three seeds with one cotyledon, and parallel leaf veins; grasses and orchids are examples. (Chap. 26, p. 633)

monosaccharide: a simple sugar such as glucose or fructose. (Chap. 7, p. 178)

monosomy: absence of a chromosome; most monosomic organisms do not survive. (Chap. 13, p. 328)

monotreme: egg-laying mammal; only three species survive today. (Chap. 35, p. 884)

motivation: need such as hunger or thirst that causes an animal to act. (Chap. 36, p. 909)

multicellular: organisms made up of many cells that are highly specialized to perform specific functions of metabolism. (Chap. 8, p. 205)

multiple alleles: presence of more than two alleles for a given genetic trait in a species. (Chap. 14, p. 336)

mutation: random error or change in the DNA sequence that may affect whole chromosomes or just one gene. (Chap. 13, p. 324)

mutualism: symbiotic relationship beneficial to both species; acacia trees and ants have a mutualistic relationship. (Chap. 3, p. 69)

mycelium: in fungi, complex mass of branching hyphae, some of which anchor, invade, or produce reproductive structures. (Chap. 23, p. 561)

mycorrhiza: symbiotic association between fungi and roots of vascular plants; hyphae supply roots with water and nutrients and fungi receive organic nutrients from the plant. (Chap. 23, p. 574)

myofibril: small contractile muscle part composed of thick myosin protein filaments and thin actin protein filaments. (Chap. 37, p. 942)

myosin: a structural protein in muscle cells; together with the protein actin, functions in muscle contraction. (Chap. 37, p. 942)

narcotic: pain reliever that also causes sleep; narcotics are often derived from the opium poppy. (Chap. 40, p. 1018)

nastic movement: reversible movement of a plant in response to a stimulus; the folding of a *Mimosa's* leaves in response to being touched is an example. (Chap. 27, p. 676)

natural resources: renewable or nonrenewable parts of the natural environment such as soil, crops, and water that are used by humans. (Chap. 6, p. 134)

natural selection: a mechanism that explains how populations evolve; changes in populations occur when organisms with favorable variations for a particular environment survive, reproduce, and pass these variations on to the next generation; can be stabilizing, directional, or disruptive. (Chap. 18, p. 427)

Neanderthals: early *Homo sapiens;* powerfully built hominids; were skilled hunters who may have been the first to develop religious views and to have spoken language. (Chap. 19, p. 471)

negative-feedback system: in the endocrine system, the means of self-regulation by which the body maintains correct hormone levels. (Chap. 41, p. 1042)

nematocysts: in cnidarians, the tiny, harpoonlike structures, primarily found at the tips of tentacles, that are used to poison, entangle, or stab prey. (Chap. 29, p. 720)

nephridia: excretory structures, first evolved in mollusks, that remove metabolic wastes from an animal's body. (Chap. 30, p. 744)

nephron: a filtering unit in the kidney; a human kidney has about 1 million nephrons. (Chap. 39, p. 994)

nerve net: conducts nerve impulses, resulting in contraction of musclelike cells; lacks a control center or brain; found, for example, in cnidarians. (Chap. 29, p. 720)

neuron: basic structural and functional unit in the nervous system; composed of dendrites, a cell body, and an axon. (Chap. 40, p. 1002)

neurotransmitters: chemicals that diffuse across the synapses between neurons, changing the polarity of the receiving dendrite, resulting in movement of nerve impulses from neuron to neuron. (Chap. 40, p. 1006)

niche: role of a particular species in a community regarding food, space, reproduction, and how it interacts with abiotic factors. (Chap. 3, p. 62)

nitrogen base: component of DNA or RNA along with a sugar and a phosphate group; can be adenine, guanine, cytosine, thymine, or uracil. (Chap. 13, p. 310)

nitrogen fixation: metabolic process in which bacteria use enzymes to convert atmospheric nitrogen gas into ammonia. (Chap. 21, p. 524)

nocturnal: describes animals that are active primarily at night; for example, earthworms and owls. (Chap. 30, p. 753)

nonbiodegradable: types of wastes that are not easily broken down and can exist in the environment for many years; examples include radioactive residues and plastics. (Chap. 6, p. 148)

nondisjunction: failure of homologous chromosomes to separate during meiosis, resulting in gametes with too few or too many chromosomes. (Chap. 13, p. 327)

nonrenewable resources: resources available in limited amounts that cannot be replaced and cannot be recycled quickly by natural means. (Chap. 6, p. 135)

nonvascular plants: plants that lack vascular tissue to efficiently transport water and nutrients, limiting their size and distribution; bryophytes are nonvascular plants. (Chap. 24, p. 588)

notochord: rodlike structure from which the backbone in vertebrate chordates develops; in invertebrate chordates, the notochord enables side-to-side propulsion through water. (Chap. 32, p. 797)

nucleic acid: complex macromolecule, such as DNA and RNA, that stores information in cells in coded form. (Chap. 7, p. 181)

nucleolus: region within the nucleus of eukaryotic cells that produces ribosomes. (Chap. 8, p. 198)

nucleotides: subunits of nucleic acid formed from a simple sugar, a nitrogen base, and a phosphate group. (Chap. 7, p. 184)

nucleus: positively charged center of an atom; contains neutrons, positively charged protons, and is surrounded by a negatively charged electron cloud. (Chap. 7, p. 161); in eukaryotes, the largest membrane-bound organelle; contains the cell's DNA and manages cell functions. (Chap. 8, p. 195)

nymph: immature stage of many insect species; the hatchling looks like the adult form but smaller and goes through successive molts, eventually becoming a sexually mature adult with wings. (Chap. 31, p. 782)

obligate aerobes: bacteria that cannot survive without oxygen for cellular respiration. (Chap. 21, p. 520)

obligate anaerobes: bacteria that cannot use oxygen and are killed by it. (Chap. 21, p. 520)

Glossary

open circulatory system: system within an organism in which blood is not confined solely to vessels but also circulates in spaces around organs; found, for example, in gastropods. (Chap. 30, p. 744)

opposable thumb: in most primates, a thumb that can be brought opposite the forefinger, allowing the hand to be used for grasping, climbing, and using tools. (Chap. 19, p. 456)

order: taxonomic grouping of related families. (Chap. 20, p. 486)

organ: group of two or more tissues that perform an activity together; examples include plant leaf and mammalian heart. (Chap. 8, p. 208)

organ system: group of organs that work together to perform a major life function; examples include vascular system in plants and circulatory system in vertebrates. (Chap. 8, p. 208)

organelles: internal membrane-bound structures in a cell. (Chap. 8, p. 195)

organism: any unicellular or multicellular form exhibiting all the characteristics of life. (Chap. 1, p. 12)

organization: orderly structure shown by living things. (Chap. 1, p. 13)

osmosis: diffusion of water molecules through a selectively permeable membrane depending on the concentration of solutes on either side of the membrane. (Chap. 9, p. 226)

osteoblast: potential bone-forming cell; originates in the cartilage. (Chap. 37, p. 934)

ovary: the female reproductive organ; in flowers, the structure formed at the lower end of the pistil; eggs are formed in the ovary. (Chap. 27, p. 656)

oviduct: in females, the tube that transports eggs from ovary to uterus by means of peristaltic contractions and beating cilia; sometimes called a fallopian tube. (Chap. 41, p. 1042)

ovulation: in females, the process in which the follicle ruptures, releasing the egg from the ovary. (Chap. 41, p. 1044)

ovule: in seed plants, the structure in which the female gametophyte develops and contains a megaspore cell; forms a seed after fertilization. (Chap. 25, p. 614)

ozone layer: protective layer of ozone at the top of the stratosphere; absorbs most of the sun's harmful radiation; chlorofluorocarbons (CFCs) have caused the ozone layer to thin. (Chap. 6, p. 143)

pancreas: gland that produces both hormones and digestive enzymes; pancreatic juice is alkaline, which stops further digestive action of pepsin. (Chap. 38, p. 956)

parasitism: symbiotic relationship in which one species benefits at the expense of the other species; examples are ticks and tapeworms. (Chap. 3, p. 69)

parasympathetic nervous system: (PNS) portion of the autonomic nervous system that acts mainly during times of relaxation. (Chap. 40, p. 1010)

parathyroid glands: produce parathyroid hormone, which acts to decrease blood phosphate levels and increase calcium and magnesium levels. (Chap. 38, p. 962)

parenchyma: plant tissue found throughout most plants; can store food and water. (Chap. 26, p. 639)

parthenogenesis: in some arthropods, a type of asexual reproduction in which a new individual is produced from an unfertilized egg; in plants, results in seedless fruit. (Chap. 31, p. 770)

particulates: solid particles of soot contained in smoke released by burning fossil fuels. (Chap. 6, p. 141)

passive transport: in cells, movement of particles across cell membranes requiring no expenditure of energy by the cell; examples are diffusion and osmosis. (Chap. 9, p. 230)

pathogens: disease-causing agents such as bacteria, fungi, or viruses; can be transmitted by direct contact, by an object, through the air, or by a vector. (Chap. 42, p. 1068)

pedicellarias: modified spines of echinoderms that are adapted into jawlike appendages used for protection. (Chap. 32, p. 788)

pedigree: graphic representation showing patterns of inheritance in a family or breeding group; yields genetic information about a related group. (Chap. 14, p. 345)

pedipalps: in arachnids, appendages used to hold and transport food and also to serve as sense organs; in male spiders, pedipalps are adapted to carry sperm. (Chap. 31, p. 774)

pepsin: digestive enzyme secreted by the stomach that acts on proteins to produce peptides. (Chap. 38, p. 955)

peptide bond: covalent bond linking amino acids. (Chap. 7, p. 180)

perennial: anthophyte that lives for several years; examples include oaks, chrysanthemums, and numerous grass species. (Chap. 26, p. 637)

pericycle: in roots, the plant tissue from which lateral roots arise. (Chap. 26, p. 639)

peripheral nervous system: (PNS) transmits impulses to and from the body and the central nervous system; composed of the somatic and autonomic nervous systems. (Chap. 40, p. 1006)

peristalsis: waves of smooth muscle contraction in the digestive system that move food along the digestive tract. (Chap. 38, p. 955)

permafrost: layer of permanently frozen ground found under the topmost layer of soil in the tundra. (Chap. 4, p. 100)

petals: leaflike flower organs; usually brightly colored; can attract pollinators with perfume, nectar, or color patterns. (Chap. 27, p. 656)

petiole: leaf part that joins the leaf blade to the stem; contains vascular tissue. (Chap. 26, p. 644)

pH: symbol to describe how acidic or basic a solution is; the pH scale ranges from 0 to 14; a solution with a pH of 7 is neutral, above 7 is basic, and below 7 is acidic. (Chap. 7, p. 172)

phagocyte: white blood cell that migrates from the capillaries to the pathogen, then engulfs and destroys it. (Chap. 42, p. 1079)

pharynx: in planarians, the tubelike, enzyme-releasing organ that can extend out of the mouth and suck food into the gastrovascular cavity; in humans, the area at the back of the mouth. (Chap. 29, p. 729)

phenotype: outward appearance of an organism, regardless of its genes. (Chap. 12, p. 291)

pheromone: chemical odor given off by animals that cues a specific behavior. (Chap. 31, p. 766)

phloem: in vascular plants, the tissue composed of living tubular cells joined end to end in a series; conducts sugars from the leaves to all plant parts. (Chap. 25, p. 608)

phospholipid: membrane lipid having an organic section attached to a phosphate group; plasma membranes are formed of a bilayer of phospholipids with embedded proteins. (Chap. 9, p. 220)

photic zone: portion of the marine biome shallow enough for the sun to penetrate; has abundant life and high productivity; includes coastal areas formed by rocky shores, sandy beaches, and mud flats. (Chap. 4, p. 91)

photolysis: splitting of water molecules during photosynthesis. (Chap. 10, p. 245)

photoperiodism: a plant's response to the difference in day and night length. (Chap. 27, p. 665)

photosynthesis: process by which autotrophs produce simple sugars from water and carbon dioxide using energy absorbed from sunlight by chlorophyll; oxygen is one of the end products. (Chap. 10, p. 242)

phylogeny: evolutionary history of a species based on comparative relationships of structures, and comparisons of modern life forms with fossils. (Chap. 20, p. 487)

phylum: taxonomic grouping of related classes. (Chap. 20, p. 486)

pistil: in flowers, the female reproductive structure attached to the top of the flower stem; its lower portion forms the ovary. (Chap. 27, p. 656)

pituitary: gland situated beneath and stimulated by the hypothalamus; releases FSH and LH, which serve such functions as stimulating production of sperm and testosterone in males, and estrogen release and regular development of follicles in the ovary in females. (Chap. 41, p. 1040)

placenta: organ developed during pregnancy through which the young of placental mammals are supplied with nourishment; also involved in passing oxygen and removing wastes. (Chap. 35, p. 883)

placental mammal: mammal that carries its young inside the uterus until ready for birth. (Chap. 35, p. 883)

plankton: microscopic organisms found floating in the photic zone; form the base of all marine food chains. (Chap. 4, p. 93)

plasma: straw-colored, fluid portion of blood containing, for example, blood cells, platelets, hormones, and nutrients. (Chap. 39, p. 981)

plasma membrane: serves as the boundary between the cell and its external environment and allows materials such as oxygen and nutrients to enter and waste products to leave. (Chap. 8, p. 196)

plasmid: in bacterial cells, the small ring of DNA that can serve as a vector; usually carries nonessential, "accessory" genes. (Chap. 16, p. 378)

plasmodium: multinucleate cytoplasmic mass lacking cell walls or membranes that is the moving and feeding stage of a plasmodial slime mold. (Chap. 22, p. 552)

plasmolysis: process resulting from a drop in turgor pressure as a result of water loss in a cell, causing the plasma membrane to shrink away from the cell wall. (Chap. 9, p. 229)

plastids: plant organelles, some of which contain pigment molecules giving fruits and flowers their color; other plastids store starch and lipids. (Chap. 8, p. 202)

plate tectonics: geological explanation for the movement of continents over Earth's thick, liquid interior. (Chap. 17, p. 408)

platelet: short-lived cell fragment contained in plasma; needed for blood clotting. (Chap. 39, p. 983)

point mutation: error in the DNA sequence that affects only a single base pair; can interfere with protein function. (Chap. 13, p. 324)

polar molecule: a molecule with a positive end and a negative end, resulting in an unequal distribution of charge. (Chap. 7, p. 168)

pollen grain: in seed plants, the structure in which the male gametophyte develops and contains nutrients for the sperm cells. (Chap. 25, p. 614)

pollination: in a flower, the process of transfer of pollen grains from the anther to the stigma. (Chap. 12, p. 286)

pollution: contamination of air, water, or land by wastes produced in such excess that they cannot be recycled by natural processes. (Chap. 6, p. 140)

polygenic inheritance: determination of a given trait, such as skin color or height, produced by the interaction of many genes. (Chap. 14, p. 338)

polymer: large molecule formed by bonding many smaller molecules together, most often in long chains; spider silk is a biological polymer. (Chap. 7, p. 177)

polyp: in the life cycle of cnidarians, the sessile, tube-shaped stage. (Chap. 29, p. 718)

polyploid: form of a species that can be formed from mistakes during meiosis, resulting in diploid ($2n$) gametes rather than haploid (n) gametes; many flowering plants and important crops originated by polyploidy. (Chap. 18, p. 444)

polysaccharides: largest carbohydrate molecules, these polymers are composed of numbers of monosaccharide subunits; examples are cellulose and glycogen. (Chap. 7, p. 178)

Glossary

population: interbreeding individuals of one species that compete with one another for food, water, and mates and live in the same place at the same time. (Chap. 3, p. 58)

posterior: the tail end of bilaterally symmetrical animals. (Chap. 28, p. 703)

prehensile tail: muscular tail that functions as a fifth limb; can be used to grasp or wrap around an object; New World monkeys have a prehensile tail. (Chap. 19, p. 458)

preservation: keeping an organism or an area from harm or destruction by the establishment of parks, wildlife habitats, and other refuges. (Chap. 6, p. 149)

primary succession: development of living communities from bare rock, where pioneer organisms such as lichens are often followed by small plants, shrubs, and then trees. (Chap. 4, p. 87)

primate: group of mammals including monkeys and humans; evolved from a common ancestor; share such characteristics as fingernails, flexible shoulder joints, flattened face, an opposable thumb or big toe, and large, complex brain. (Chap. 19, p. 454)

proglottid: parasitically adapted, detachable section of a tapeworm; contains male and female reproductive organs, flame cells, muscles, and nerves. (Chap. 29, p. 732)

prokaryote: cell lacking a true nucleus or membrane-bound internal organelles. (Chap. 8, p. 195)

prophase: first phase of mitosis during which chromatin coils to form visible chromosomes. (Chap. 11, p. 268)

prostate gland: doughnut-shaped gland of males that surrounds the top of the urethra and secretes an alkaline fluid that transports sperm. (Chap. 41, p. 1040)

protein: large, complex polymer essential to all life composed of amino acids made of carbon, hydrogen, oxygen, nitrogen, and sometimes sulfur; important in muscle contraction, transporting oxygen in the bloodstream, and providing immunity. (Chap. 7, p. 180)

prothallus: the gametophyte generation in the life cycle of spore-producing, vascular plants; produces either antheridia or archegonia. (Chap. 25, p. 607)

protist: eukaryotic plantlike, animal-like, or funguslike organism lacking complex organ systems and living in a moist environment; examples include protozoans, slime molds, and seaweeds. (Chap. 20, p. 494)

protocells: ordered structures formed by heating amino acid solutions, producing clusters resembling living cells that carry out some, but not all, of the functions of life. (Chap. 17, p. 414)

protonema: in mosses, filamentous structure formed by germination of the haploid spore; develops into either a male or female gametophyte or into a gametophyte containing both male and female structures. (Chap. 24, p. 596)

protostome: animal in which the mouth develops from the opening in the gastrula; animals such as insects and earthworms show protostome development. (Chap. 28, p. 699)

protozoans: animal-like protists with a variety of complex organelles; grouped into phyla according to their method of locomotion. (Chap. 22, p. 535)

provirus: virus integrated into a host cell's chromosome; can remain dormant or become activated at any time and enter a lytic cycle. (Chap. 21, p. 508)

pseudocoelom: body cavity partly lined with mesoderm and filled with fluid; enables animals such as roundworms to move more efficiently. (Chap. 28, p. 704)

pseudopodia: temporary cytoplasmic extensions used by amoebas and amoebalike organisms for movement and for engulfment of food. (Chap. 22, p. 535)

puberty: in both males and females, the period characterized by development of secondary sex characteristics as a result of production of FSH, LH, and other sex hormones. (Chap. 41, p. 1040)

pulse: rhythmic surge of blood through an artery. (Chap. 39, p. 989)

punctuated equilibrium: idea that periods of speciation occur rapidly with long periods of no speciation in between. (Chap. 18, p. 445)

pupa: in insects, the stage of metamorphosis from which the larva emerges as a fully mature adult. (Chap. 31, p. 781)

pus: thick, fluid substance formed from dead white blood cells, dead microorganisms, and body fluids in an infected area. (Chap. 42, p. 1079)

R

radial symmetry: body plan exhibited by animals, such as adult starfish, that can be divided along any plane, through a central axis, into roughly equal halves; enables slow-moving or stationary animals to sense potential food and predators from all directions. (Chap. 28, p. 702)

radula: in animals such as gastropods, the rasping, tongue-like organ used for cutting and grating food. (Chap. 30, p. 744)

rays: long, tapering arms of some echinoderms such as starfishes and the sea lily; rays are covered with spines or plates composed primarily of calcium carbonate protected with a thin epidermal layer. (Chap. 32, p. 788)

recessive: hidden trait of an organism that is masked by a dominant trait. (Chap. 12, p. 289)

recombinant DNA: produced when a cleaved DNA fragment is incorporated into the DNA of a plasmid or virus. (Chap. 16, p. 376)

rectum: final segment of the digestive system; passes feces out of the body through the anus. (Chap. 38, p. 958)

red blood cell: hemoglobin-containing cell in humans that transports oxygen and some CO_2; loses its nuclei as it enters the bloodstream, resulting in a limited life span. (Chap. 39, p. 981)

reflex: rapid, automatic, unconscious response to a stimulus. (Chap. 40, p. 1008)

Glossary

regeneration: ability to replace or regrow a missing body part; for example, sea cucumbers can eviscerate and later replace the lost vital organs. (Chap. 32, p. 794)

renewable resources: resources replaced or recycled by natural processes; for example, oxygen is replenished during photosynthesis. (Chap. 6, p. 134)

replication: process in which the two strands of the double helix separate and bases pair with free nucleotides to form two molecules of DNA, each identical to the original molecule. (Chap. 13, p. 313)

reproduction: production of offspring. (Chap. 1, p. 13)

reproductive isolation: occurs when organisms that formerly interbred are prevented from producing offspring—for example, by developing different mating times. (Chap. 18, p. 444)

respiration: process in which cells break down molecules of food to release energy; cellular respiration can be either aerobic or anaerobic. (Chap. 10, p. 251)

response: reaction to an internal or external stimulus. (Chap. 1, p. 15)

restriction enzymes: bacterial proteins that cleave DNA at specific points in the nucleotide sequence. (Chap. 16, p. 377)

retina: layer of the eye containing rods and cones; light entering the cornea is focused by the lens on the back of the eye, where it hits the retina. (Chap. 40, p. 1012)

retroviruses: viruses containing a unique enzyme, reverse transcriptase, which transcribes viral RNA into DNA, enabling the viral DNA to enter the host cell's chromosome. (Chap. 21, p. 510)

reverse transcriptase: enzyme that transcribes viral RNA into viral DNA. (Chap. 21, p. 510)

rhizoids: fungal structures formed by hyphae that anchor the mycelium to its food source or substrate; site of most extracellular digestion and absorption of nutrients. (Chap. 23, p. 566); hairlike extensions of a moss or liverwort gametophyte that anchor the plant to its substrate. (Chap. 24, p. 587)

rhizome: underground, horizontal stem of vascular plants such as ferns; may function as a food-storage organ. (Chap. 25, p. 610)

ribosomal RNA (rRNA): the RNA that composes ribosomes. (Chap. 13, p. 319)

ribosomes: eukaryotic organelles involved in protein synthesis. (Chap. 8, p. 198)

RNA: ribonucleic acid; forms a copy of DNA for use in protein synthesis. (Chap. 7, p. 184)

rods: type of light receptor in the retina; responsible for vision in low illumination. (Chap. 40, p. 1012)

root: plant organ normally found below ground that anchors the plant, absorbs water and minerals, and transports water and nutrients to the stem. (Chap. 24, p. 587)

root cap: layer of tough parenchyma cells that protects the root tip as its grows down into soil. (Chap. 26, p. 640)

root hair: in plant roots, a single-celled extension of the epidermis that absorbs water and dissolved minerals. (Chap. 26, p. 639)

S

safety symbol: warns against specific hazards such as radiation and high-voltage electricity (see Appendix D for safety symbols used in this textbook). (Chap. 2, p. 34)

saprobes: heterotrophic eubacteria that feed on nonliving organic matter or wastes and help recycle nutrients. (Chap. 21, p. 512)

sarcomere: each section of a myofibril in a striated muscle. (Chap. 37, p. 942)

scales: skin covering of reptiles and fishes; can be diamond-shaped, cone-shaped, tooth-shaped, or round in fishes; agnathans lack scales. (Chap. 33, p. 813)

scavenger: animal such as a vulture that plays a positive role in the ecosystem by consuming dead organisms and their refuse. (Chap. 3, p. 65)

scientific methods: common procedures used by biologists and other scientists to gather information used in problem solving and experimentation. (Chap. 2, p. 26)

scolex: parasitically adapted, knob-shaped head of a tapeworm; covered with suckers and hooks that embed in the host's intestinal lining. (Chap. 29, p. 732)

scrotum: testes-containing sac of males; located externally; maintains sperm at a lower temperature than body temperature by means of muscle contraction and relaxation. (Chap. 41, p. 1038)

seed: adaptive reproductive structure of land plants; a protective coat prevents drying out of the embryo; also contains a food supply. (Chap. 24, p. 589)

selective permeability: property of a plasma membrane that maintains the cell's homeostasis by the taking in of needed substances, the elimination of wastes, and the prevention of harmful substances from entering. (Chap. 9, p. 217)

semen: combination of sperm and sperm-carrying fluids of the male reproductive system. (Chap. 41, p. 1040)

semicircular canals: fluid-filled structures in the ear involved in maintaining the body's balance. (Chap. 40, p. 1013)

seminal vesicles: in males, the paired glands at the base of the urinary bladder; produce a fructose-containing fluid that nourishes sperm. (Chap. 41, p. 1040)

sepals: leaflike structures at the base of a flower; protect the flower while in bud. (Chap. 27, p. 656)

sessile: organism that stays attached permanently to a surface during its adult life; sponges and corals are examples. (Chap. 28, p. 694)

sex chromosomes: in humans, the 23rd pair of chromosomes, which controls the inheritance of sex characteristics and differs in males and females. (Chap. 14, p. 336)

sex-linked trait: inherited trait, such as color blindness, controlled by genes located on the sex chromosomes. (Chap. 14, p. 336)

Glossary

sexual reproduction: reproductive pattern in which haploid gametes fuse to produce a diploid zygote, which then develops by mitosis into a new organism. (Chap. 12, p. 300)

short-day plant: flowers in late summer, when days are becoming shorter; examples are primroses and chrysanthemums. (Chap. 27, p. 665)

sieve cell: in vascular plants, the cytoplasmic, nonnucleated cell of the phloem involved in transporting sugars throughout the plant. (Chap. 26, p. 642)

simple eyes: in arthropods, the single-lens structures used to focus images. (Chap. 31, p. 766)

sink: in vascular plants, storage area for excess sugars produced by a plant during photosynthesis. (Chap. 26, p. 642)

sister chromatids: identical halves of the duplicated parent chromosome formed before the onset of cell division; these exact copies are joined at a centromere. (Chap. 11, p. 268)

skeletal muscle: muscle attached to bones; functions under voluntary control to move the skeleton; composed of striated, multinucleated cells. (Chap. 37, p. 939)

sliding filament theory: theory that actin filaments slide toward each other during muscle contraction, whereas the myosin filaments remain still. (Chap. 37, p. 942)

small intestine: narrow, muscular tube in which digestion is completed; connects the stomach to the large intestine. (Chap. 38, p. 956)

smog: type of urban air pollution resulting from a combination of chemical pollutants, particulate matter, and sulfur dioxide. (Chap. 6, p. 141)

smooth muscle: muscle composed of sheets of spindle-shaped cells; lines internal organs and blood vessels; produces involuntarily controlled contractions. (Chap. 37, p. 938)

solution: mixture in which a substance (solute) is dissolved easily in another substance (solvent). (Chap. 7, p. 167)

somatic nervous system: portion of the peripheral nervous system that transmits messages between the skin, the central nervous system, and skeletal muscles. (Chap. 40, p. 1008)

sorus: in ferns, the structure formed on the surface of fronds by clusters of sporangia. (Chap. 25, p. 612)

spawning: act of breeding in fishes and some other animals; results in release of large numbers of eggs to the environment. (Chap. 33, p. 815)

speciation: process by which a new species is formed; occurs when individuals of a population are unable to interbreed or produce fertile offspring. (Chap. 18, p. 441)

species: population of interbreeding organisms capable of producing fertile offspring. (Chap. 1, p. 14)

sperm: male sex cell or gamete. (Chap. 12, p. 300)

spindle: thin fiber structure that forms between the two poles or centrioles during prophase and shortens during anaphase, pulling the sister chromatids apart. (Chap. 11, p. 269)

spinnerets: in spiders, structures that spin silk into thread. (Chap. 31, p. 774)

spiracles: openings on the abdomen and thorax of most insects through which air enters and exits the tracheal tubes. (Chap. 31, p. 765)

spongy bone: bone tissue containing numerous holes and spaces; is less dense than the compact bone surrounding it. (Chap. 37, p. 935)

spontaneous generation: idea that living organisms can arise from nonliving matter. (Chap. 17, p. 410)

sporangium: in zygomycotes, the saclike structure in which asexual spores are formed at the tips of some hyphae. (Chap. 23, p. 564)

spore: in sporozoans, the haploid (*n*) reproductive cell with a hard outer wall that develops into a new organism without the fusion of gametes. (Chap. 22, p. 540)

sporophyte: in algae and plants, the diploid stage in an alternation of generations; develops from the zygote. (Chap. 22, p. 550)

sporozoans: parasitic, nonmotile protozoans, many of which reproduce by the production of spores. (Chap. 22, p. 540)

stabilizing selection: selection favoring average individuals, resulting in the decline of variations in a population. (Chap. 18, p. 439)

stamen: in flowers, the male reproductive structure consisting of an anther and attaching filament. (Chap. 27, p. 656)

starch: the polysaccharide consisting of highly branched chains of glucose units used as food storage in plants. (Chap. 7, p. 178)

stem: plant structure that provides support for leaves and reproductive structures; vascular plant stems contain tissues for transport of nutrients and water. (Chap. 24, p. 588)

sternum: in birds, the large breastbone to which flight muscles are attached; supports the muscular thrust of wings as they produce the power for flight. (Chap. 34, p. 848)

stimulant: drug that increases the activity of the central nervous system and sympathetic nervous system; nicotine is a stimulant. (Chap. 40, p. 1018)

stimulus: any adjustment-requiring condition of an organism's environment. (Chap. 1, p. 15)

stolons: hyphae that grow horizontally across the surface of a food source, such as bread, and produce rhizoids that grow down and reproductive hyphae that grow up. (Chap. 23, p. 566)

stomach: pouchlike, muscular digestive organ that secretes acids and enzymes; food leaves the stomach in a thin, acidic liquid. (Chap. 38, p. 955)

stomata: openings in a leaf's epidermis that release water and oxygen to the air and take in oxygen and carbon dioxide for respiration and photosynthesis. (Chap. 24, p. 587)

strobilus: in lycophytes, the conelike spore-producing leaf cluster at the tip of the stem. (Chap. 25, p. 607)

succession: natural, orderly process in the community of an ecosystem characterized by population growth or reduction. (Chap. 4, p. 86)

swim bladder: gas-filled, internal sac of bony fishes that regulates buoyancy. (Chap. 33, p. 815)

symbiosis: permanent, close association between two or more organisms of different species. (Chap. 3, p. 68)

symmetry: balance in body proportions of animals. (Chap. 28, p. 701)

sympathetic nervous system: portion of the autonomic nervous system that acts mainly during times of stress. (Chap. 40, p. 1009)

synapse: space between neurons across which impulses are chemically transmitted from axons to dendrites. (Chap. 40, p. 1006)

T

T cells: specialized lymphocytes produced in the bone marrow and processed in the thymus gland; have antibody-like antigen receptors on their surfaces that allow them to react to many types of antigens. (Chap. 42, p. 1082)

taiga: biome located just south of the tundra; characterized by acidic, mineral-poor topsoil, long, harsh winters, and short, mild summers; abundant fir and spruce trees provide food and shelter for such animals as moose and lynx. (Chap. 4, p. 101)

target tissue: cells specifically affected by endocrine hormones, which act to convey information. (Chap. 38, p. 960)

taste bud: sensory receptor of the tongue that sends messages to the cerebrum, resulting in experiencing a specific taste. (Chap. 40, p. 1011)

taxonomy: branch of biology dealing with grouping and naming organisms based on their similarities, chemical makeup, and evolutionary relationships. (Chap. 20, p. 482)

technology: scientific research to solve society's needs and problems; can have both beneficial effects and problematic side effects. (Chap. 2, p. 45)

telophase: final stage of mitosis during which the two daughter cells become separated. (Chap. 11, p. 272)

temperate forests: biome in which an even amount of precipitation falls in all four seasons, averaging from 70 to 150 cm annually; temperate forest type is determined by the dominant tree species, which typically includes oak, beech, maple, birch, and hickory; animal life includes squirrels, rabbits, and bears. (Chap. 4, p. 104)

tendon: cord of connective tissue that attaches muscles to bones. (Chap. 37, p. 932)

territory: physical area containing an animal's breeding and feeding areas that is actively defended against members of the same species; territoriality has survival value for the individual and the species. (Chap. 36, p. 902)

testcross: breeding technique used to determine whether an individual is homozygous dominant or heterozygous for a particular trait; often employed by plant and animal breeders to test for such traits as blindness and disease vulnerability. (Chap. 14, p. 344)

thallus: undifferentiated plant or algal body that has no roots, stems, or leaves. (Chap. 22, p. 548)

theory: results when a hypothesis is repeatedly verified over time and through many separate experiments; valid theories enable scientists to predict new facts and relationships of natural phenomena. (Chap. 2, p. 40)

therapsids: mammal-like reptilian ancestors of all mammals. (Chap. 35, p. 881)

threatened species: species that have rapidly decreasing numbers of individuals; examples include African elephants and grizzly bears. (Chap. 6, p. 137)

thyroid gland: produces thyroxine, a hormone that regulates metabolism and growth, and calcitonin, a hormone that helps regulate calcium and phosphate levels. (Chap. 38, p. 961)

tissue: group of cells that function together to carry out an activity; examples include leaf and nerve tissue. (Chap. 8, p. 205)

tissue fluid: fluid that bathes the cells of the body. (Chap. 42, p. 1076)

tolerance: state resulting when a drug increasingly loses its effect, necessitating larger doses for the same effect. (Chap. 40, p. 1022)

toxin: a poison; for example, the obligate anaerobe *Clostridium botulinum* forms endospores when exposed to oxygen; when the endospores germinate, the resultant bacteria produce the powerful toxin that is the causative agent of botulism. (Chap. 21, p. 521)

trachea: the windpipe; lined with constantly beating cilia that prevent foreign particles from reaching the lungs. (Chap. 39, p. 976)

tracheal tubes: in most insects, the internal, air-carrying passages. (Chap. 31, p. 765)

tracheid: water transport cell in plants, forms xylem tissue in ferns and gymnosperms. (Chap. 25, p. 620)

trait: inherited characteristic; in simple Mendelian inheritance, can be either dominant or recessive. (Chap. 12, p. 286)

transcription: the process by which enzymes make an RNA copy of a DNA strand. The process is similar to DNA replication, but a single-stranded molecule of transfer RNA is produced. (Chap. 13, p. 319)

transfer RNA (tRNA): delivers amino acids to the ribosome for protein synthesis. (Chap. 13, p. 322)

transgenic organism: genetically engineered organism containing recombinant DNA that gives the organism the ability to create products foreign to itself; for example, some transgenic plants can produce internal insecticides. (Chap. 16, p. 376)

translation: process in which the order of bases in mRNA codes for the order of amino acids in a protein. (Chap. 13, p. 322)

Glossary

transpiration: in plants, the evaporation of water from the stomata of leaves. (Chap. 26, p. 645)

transport proteins: channel proteins and carrier proteins embedded in the lipid bilayer of the plasma membrane; move ions and molecules across the membrane. (Chap. 9, p. 230)

trial-and-error learning: occurs when an animal makes various attempts before being able to successfully complete a task. (Chap. 36, p. 909)

trisomy: presence of an extra chromosome; trisomic organisms often survive into maturity. (Chap. 13, p. 327)

trophic level: link represented by each organism in a food chain; represents a feeding step in the transfer of energy and matter in an ecosystem. (Chap. 3, p. 71)

tropical rain forest: most biologically diverse terrestrial biome; receives from 200 to 400 cm of annual rainfall and has year-round warm temperatures; plants include ferns, orchids, and trees; animal life includes parrots, monkeys, snakes, and tarantulas. (Chap. 4, p. 105)

tropism: irreversible, responsive movement in plants toward or away from such external stimuli as light (phototropism) or gravity (gravitropism). (Chap. 27, p. 675)

tube feet: in echinoderms, hydraulic, suction cup-tipped appendages that function in locomotion, gas exchange, excretion, and capture of food. (Chap. 32, p. 789)

tundra: treeless biome south of the ice cap of the north pole; has a short growing season; temperatures never rise above freezing, and its permafrost soil layer never thaws; supports grasses, small annuals, and reindeer moss; animal life includes lemmings, arctic foxes, mosquitoes, and reindeer. (Chap. 4, p. 100)

turgor pressure: internal pressure of a cell due to water held there by osmotic pressure. (Chap. 9, p. 227)

umbilical cord: in placental mammals, the cordlike structure that attaches the developing embryo to the uterine wall; the embryo receives nutrients and oxygen and eliminates wastes through the blood of the umbilical cord. (Chap. 41, p. 1050)

unicellular: organism that carries out all its life processes within its single cell. (Chap. 8, p. 205)

ureter: tube that transports urine from each kidney to the urinary bladder. (Chap. 39, p. 993)

urethra: tube through which urine is eliminated from the body. (Chap. 39, p. 995)

urinary bladder: smooth muscle bag in which urine is stored until it leaves the body through the urethra. (Chap. 39, p. 993)

urine: liquid composed of excess H_2O, ions, and waste molecules that is filtered from the blood by the kidneys, stored in the urinary bladder, and eliminated through the urethra. (Chap. 39, p. 995)

uterus: in female placental mammals, the muscular, hollow organ in which young are developed and protected. (Chap. 35, p. 882)

vaccine: immunity-producing substance formed from weakened, dead, or parts of pathogens or antigens. (Chap. 42, p. 1085)

vacuole: membrane-bound, fluid-filled space within the cytoplasm; temporarily stores food, enzymes, and wastes. (Chap. 8, p. 200)

vagina: in females of placental mammals, the passageway that leads from the uterus to the outside of the body. (Chap. 41, p. 1043)

vas deferens: in males, the duct through which sperm move, by peristaltic contractions, from the epididymis toward the urethra, which releases the sperm from the body. (Chap. 41, p. 1039)

vascular plants: plants that contain vascular tissues specially adapted to transport water and dissolved materials, enabling taller growth and survival in land habitats. (Chap. 24, p. 588)

vascular tissues: in vascular plants, the tissues made up of tubular cells that transport water and dissolved nutrients from one part of a plant to another; the two types are xylem and phloem. (Chap. 25, p. 608)

vector: can be biological or mechanical; viruses and plasmids are biological vectors with which DNA fragments are recombined before introduction into the host cell. (Chap. 16, p. 377)

vegetative reproduction: asexual reproduction in plants; examples are bulbs, tubers, gemmae, and rhizomes. (Chap. 24, p. 600)

veins: large blood vessels that return blood from the tissues back to the heart; contain one-way valves to prevent backflow of blood. (Chap. 39, p. 986)

venae cavae: two large veins that empty oxygen-poor blood from the body and head into the right atrium of the mammalian heart. (Chap. 39, p. 987)

ventral: belly surface of bilaterally symmetrical animals. (Chap. 28, p. 703)

ventricles: thick-walled, muscular lower chambers of the mammalian heart that receive blood from the atria and send it toward the lungs and arteries. (Chap. 39, p. 987)

vertebrate: animal with a backbone; examples include mammals, birds, and reptiles. (Chap. 28, p. 708)

vessel cell: open-ended, tubular cell that makes up vessels; in angiosperms, vessels conduct water, which diffuses into and out of the xylem. (Chap. 26, p. 642)

vestigial structure: body structure with reduced function that may have been useful to an earlier stage of the species; provides evidence of evolution. (Chap. 18, p. 433)

Glossary

villus: fingerlike projection on the lining of the small intestine that increases the surface area for absorption of digested food. (Chap. 38, p. 956)

virus: disease-causing, nonliving particle composed of an inner core of nucleic acid enclosed by one or two protein coats; reproduces only in living cells. (Chap. 21, p. 504)

vitamin: organic substance that regulates processes in the body; examples include vitamin E and pantothenic acid. (Chap. 38, p. 966)

vocal cords: in frogs and mammals, the bands of tissue in the throat that vibrate to produce a wide variety of sounds such as mating calls. (Chap. 33, p. 823)

voluntary muscle: muscle whose contractions are under conscious control; skeletal muscle is voluntary muscle. (Chap. 37, p. 939)

water vascular system: in echinoderms, provides water pressure to operate the tube feet; regulates locomotion, excretion, gas exchange, and capture of food. (Chap. 32, p. 790)

white blood cell: large, nucleated, infection-fighting cell in mammalian blood. (Chap. 39, p. 982)

withdrawal: psychological or physiological illness resulting from cessation of drug use. (Chap. 40, p. 1022)

xylem: in vascular plants, the tissue composed of dead tubular vessels or tracheids laid end to end; conducts water and dissolved minerals upward from the plant roots to the leaves. (Chap. 25, p. 608)

zygospore: in zygomycotes, the thick-walled, sexually produced, resting spore adapted to withstand poor environmental conditions. (Chap. 23, p. 567)

zygote: fertilized egg; has a diploid ($2n$) number of chromosomes; develops into a multicellular organism by mitosis. (Chap. 12, p. 300)

Index

A

Abiotic factor, 57
  succession, 88-89
    puddle community, 88-89
ABO blood type, 983-984
Acid, chemical reaction, 172-173
Acid precipitation, 142, *illus.* 143, 619
  effects, 142-143
Acne, 930
Acoelomate flatworm, body cavity,
    703-704, *illus.* 704
Actin, 942
Active transport, 230-232
  large particles, 231-232
Adaptation
  amphibian, 820
  angiosperm, 634-637
  bird, *illus.* 852, 852-853, *illus.* 853
  cnidarian, 722-723
  defined, 16, *illus.* 16
  fungi, 565
  to life on land, 584-593
  moneran, 520-522
  natural selection, 428-430
  primate, 454-462
  reptile, 839-840, *illus.* 840
  tapeworm, 732, *illus.* 732
Adaptive radiation, 446, *illus.* 446
Addiction, 1019
Adenine, *illus.* 310
Adenosine diphosphate. *See* ADP
Adenosine triphosphate. *See* ATP
Adolescence, 1061
ADP, 239-241
Aerobic process, 251
Aerobic respiration, photosynthesis,
    compared, 255-256
Age structure, 126-127
Aggression, 903
Aging, 1062
  skeletal system, 936-937
  skin, 931
Agnathan, *illus.* 811, 811-812
Agriculture, *illus.* 682-683
  transgenic bacteria, 383
AIDS
  epidemic, *illus.* 1089
  immune response, 1086-1087
  immune system, 1088-1090
  virus, *illus.* 505, 509-510
AIDS-related complex, 1088, *illus.*
    1088
Alcohol, 1024-1025
Alcoholic fermentation, 254, *illus.* 255

Algae, 5, 542-554. *See also* Specific type
  alternation of generations, 550
  defined, 542
  fragmentation, 550
  phyla, 542-549
  reproduction, 550
Allele, 288
Allelic frequency, 436
  natural selection, 442-443
Allergy, 1081
Alligator, *illus.* 152, 842
Alternation of generations
  algae, 550
  plants, 589
Alveoli, gas exchange, 976-977, *illus.*
    977
Amino acid, 180, *illus.* 180
Amniocentesis, 1055, *illus.* 1055
Amniotic egg, 837, *illus.* 838
Amniotic fluid, 838, 1050, 1055,
    illus. 1057
Amoeba, illus. 479, *illus.* 536, 536-537
Amphetamine, 1022-1024
Amphibian, 705, 820-830
  adaptation, 820
  defined, *illus.* 821, 821-825
  diversity, 826
  ectotherm, 823
  metamorphosis, 822, *illus.* 822
  move to land, 820
  origins, 830, *illus.* 830
  phylogeny, *illus.* 818
Ampulla, 789
Amylase, 953
Anabolic steroid, *illus.* 46, *illus.* 1019
Anaerobic process, 251
Analogous structure, 432
Anaphase, *illus.* 271, 272, *illus.* 301, 302
Aneuploidy, 369, 370
Angiosperm
  adaptation, 634-637
  defined, 632
  diversity, 632-633
  functions, 638-650
  structures, 638-650
Animal, *illus.* 492-493, *illus.* 498, 498,
    691-708
  body plan, 701
  cell adaptations, 695
  characteristics, 692-700
  defined, 696-697
  development, 698-700
  methods of obtaining food, *illus.* 692,
    692-694, *illus.* 693, *illus.* 694
  symmetry, 701, *illus.* 701
Animal behavior, 895-918
Animal pollinator, 659
Animal testing, 891
Annelid, *illus.* 750, 750-751, *illus.* 751
  reproduction, 753
Annual plant, 636

Ant, *illus.* 916
Antelope, running speeds, 872
Anther, 656
Antheridium, 596
Anthophyta, 593
Anthozoan, 723, *illus.* 723
Anthropoid, 458, 459
Antibiotic, 1075
  bacteria, sensitivity, 518-519
  discovery, 517, 574
Antibody, 984, *illus.* 984
Antibody immunity, infectious disease,
    1081-1082, *illus.* 1082
Antigen, blood type, 984, *illus.* 984
Antimicrobial substance, 1080
Aorta, 987
Ape, *illus.* 461
  evolution, 462
Aphotic zone, 91, 93
Apical meristem, 640
Appendage, 762, *illus.* 763
Appendicular skeleton, 932
Aquatic biome, 91-96
Arachnid, *illus.* 772, 772-775
  defined, 772
Archaebacteria, 414, 512, *illus.* 513
Archegonium, 596
Aristotle, 413, 482
Artery, 985, *illus.* 985
Arthropod, *illus.* 760-761, 761-782
  beneficial to humans, 770
  body systems, 766-767
  characteristics, 762-771
  defined, 762
  diversity, 772-782
  ecology, 770
  exoskeleton, 762
  gas exchange, 764-765, *illus.* 765
  metamorphosis, 781-782
  molting, 763-764, *illus.* 764
  nervous system, 766
  origins, 770, *illus.* 770
  parthenogenesis, 770
  phylogeny, *illus.* 771
  segmentation, 764, *illus.* 764
Artificial selection, 427
Ascomycote, 567-568, *illus.* 568
Ascospore, 568
Ascus, 567-568
Asexual reproduction, bacteria, 522-
    523, sarcodine, 537
Asymmetry, sponge, 702, *illus.* 702
Atom, 161-162
  structure, 161, *illus.* 162
ATP, 238-241, *illus.* 240
  energy to make, 250-256
Atrium, 987
Attachment protein, 506
Australian frog, *illus.* 7
Australopithecine, 464-468, *illus.* 465
Automatic response, behavior, 900

Index

Index

F

Index

Flowering plant, 631-651, 656-679
  parts, *illus.* 658
  pollination, 657-665
  reproduction, 655-679
Fluid mosaic model, 220-221
Fluke, life cycle, 733, *illus.* 733
Food, nutrient, 968-969
Food chain, *illus.* 70, 70-71
  trophic level, *illus.* 70-71, 71
Food groups, *illus.* 963
Food pyramid, 964, *illus.* 964
Food web, 72, *illus.* 72
Forest, *illus.* 60. *See also* Specific type
  community, *illus.* 60
Fossil, *illus.* 394, *illus.* 396-397, 399,
    illus. 399, *illus.* 584
  dating methods, *illus.* 404, 404-405
  evolution, 424, 431, *illus.* 431
  ownership, 449
  sedimentary rocks, 401, *illus.* 401
Fossil fuel, 135-136
Fox, Sidney, 414
Fragmentation
  algae, 550
  fungi, 564
Frameshift mutation, 325
Franklin, Rosalind, 312
Freshwater biome, 96, *illus.* 96
Frog, *illus.* 60, *illus.* 827, *illus.* 828-829
Frog egg, development, 824-825
Frostbite, 233
Fruit, 671, *illus.* 671
  flower, compared, 672
  formation, 671, *illus.* 671
  ripening of, 677
Fungi, 559-577. *See also* Specific type
  adaptation, 565
  budding, 564, *illus.* 564
  characteristics, *illus.* 560, 560-562
  decomposition, 562, *illus.* 562
  diversity, 566-576
  extracellular digestion, 563
  fragmentation, 564
  function, 562, *illus.* 562
  mutualism, 574-576
  origins, 576
  phylogeny, *illus.* 576
  reproducing by spores, 564-565
  structure, 561
Funguslike protist, 551-554

G

Galápagos Islands, 425
Gallbladder, 957
Gametangium, 567
Gamete, 286
Gametophyte, 550, 589

Gas chromatograph, *illus.* 31
Gas exchange
  alveoli, 976-977, *illus.* 977
  arthropod, 764-765, *illus.* 765
Gastrin, 959
Gastropod, *illus.* 743, 744-745
  adaptation, 744-745
  shelled, *illus.* 744, 745
  without shells, 745
Gastrula, animal, 698-699, *illus.* 699
Gel electrophoresis, *illus.* 31
Gene, 298-300
  behavior, 897
  cancer, 276
  function, 316
  manipulation, 376-383
Gene cloning, 378-379, *illus.* 379, 382
Gene-control operon, 327
Gene pool, 436
Gene splicing, 378, 379
Gene therapy, human genome project,
    388, *illus.* 388
Genetic code, 317, *table* 317
Genetic counseling, 1053-1055
Genetic disorder, 356-362, 365,
    368-370, 1053-1054
Genetic drift, *illus.* 437, 437-438
Genetic engineering, 376
  medicine, 1101
Genetic equilibrium, evolution,
    436-438
Genetic recombination, 303-304
Genetic screening, ethics, 371
Genetic variation, meiosis, *illus.* 303,
    303-304, *illus.* 304
Genetics, 286
  applied, 344-350
Genotype, 291
  determination, 292-293, 344-346
  testcross, 344, *illus.* 345
Genus, 483
Geographic isolation, evolution,
    441-442
Giardiasis, 537
Gibberellin, 677
Gill, 765
Gill slits, chordate, 800
Ginkgophyta, 592, 616, *illus.* 616
Gizzard, 751
Gland, 871
Glossary, 1149
Glucagon, exercise, 960
Glucose, *illus.* 960, 960-961
Glycerol, 179, *illus.* 179
Glycogen, 178
Glycolysis, steps, *illus.* 250, 251
Gnetophyta, 592, 617, *illus.* 617
Goldenrod, *illus.* 54
Golgi apparatus, 200, *illus.* 200
Goose, 8, *illus.* 8

Gorilla, *illus.* 43
Gradualism, 444
Graph, 38
  making and using, 1146
Grasshopper, *illus.* 780
  waxy coating, 763
Grassland, *illus.* 103, 103-104, *illus.* 104
Gray matter, 1005
Green algae, 549, *illus.* 549
  moss, compared, 595
Greenhouse effect, 144, *illus.* 144
Grizzly bear, 806, *illus.* 806, *illus.* 807
Groundwater pollution, 146, *illus.* 146
Growth, 1057-1061
  hormone, 1057-1060, *illus.* 1060
Growth rate, 1058-1059
Growth ring, tree, *illus.* 643
Guanine, *illus.* 310
Guard cell, *illus.* 586, 645, *illus.* 645
Guide dog, 919
Gymnosperm, 613-626
  cone, 613
  defined, 613
  spore, 613

H

Habitat, *illus.* 62
Habituation, 908, *illus.* 908
Hair, mammal, 870, *illus.* 870, 871
Hair follicle, 927
Haldane, J. B. S., 413
Hallucinogen, 1025-1026
Haploid, 298
Hardy, G. H., 438
Hardy-Weinberg law, 438
Haustoria, 563
Hawking, Stephen, 1027
Hearing, *illus.* 29, 1013, *illus.* 1013,
    1016
Heart, human, 987-992, *illus.* 988
  blood, 987
  care, *illus.* 990-991
  control, 989
  exercise, *illus.* 990
  nutrition, *illus.* 990
  smoking, *illus.* 991
  stress, *illus.* 991
Heart rate
  drug, 1020-1021
  exercise, 989
Heartbeat regulation, 989
Heat- and acid-loving bacteria,
    512-513
Heat of vaporization, water, 171
Hemophilia, 368
Herbivore, defined, 64

Index

Index

Meiosis, 298-304
  crossing over, 302, 386-387
  genetic variation, *illus.* 303, 303-304,
    illus. 304
  Mendel's results, 304
  mistakes, 369-370
  phases, 300-303, *illus.* 301
Melanin, 927
Membrane selectivity, 225
Mendel, Gregor, 286-297
Mendel's laws of heredity, 286-297
Menstrual cycle, 1044-1048, *illus.* 1045
  flow phase, 1046
  follicular phase, 1046
  luteal phase, 1046-1047, *illus.* 1047
Mesoderm, 699, *illus.* 699
  nervous tissue, 700
Mesophyll, 648
Mesozoic Era, *illus.* 407, 407-408,
    illus. 408
Messenger RNA, 319
Metabolism, 169
  bacteria, 512-514, 520-521
  Calories, 967-970
  cell, 216
  endocrine control, 959-962
  thyroid gland, 961
Metamorphosis
  amphibian, 822, *illus.* 822
  insect, *illus.* 781, 781-782, *illus.* 782
Metaphase, *illus.* 271, 272, *illus.* 301,
    302
Methane-producing bacteria, 249,
    512-513
Microfilament, 203
Micropyle, 669
Microscope, *illus.* 31
  care and use of, 1137
  history, 190-193
  resolution, 194
Microsphere, 416-417
Microspore, 613
Microtubule, 203
Migration, *illus.* 904, 904-905,
    illus. 905
  sea turtle, 842, *illus.* 842
Miller, Stanley, 412
Millipede, 776, *illus.* 776
Mind, *illus.* 1034
Mineral, 965, *illus.* 965
Mite, 774, *illus.* 775
Mitochondria, energy, 201, *illus.* 201,
    252-253
Mitosis, *illus.* 260, 267
  phases, 268-273
Mixture, 166-167
Mobility, population size, 127-128
Mold spore, 564
Molecular chain, carbon, 177

Mollusk, 742-749, *illus.* 743
  classes, 743-748
  defined, 742
  habitat, *illus.* 743
  identification, 747
  larvae, 742, *illus.* 742
  origins, 749
  phylogeny, *illus.* 749
Molting, arthropod, 763-764, *illus.* 764
Moneran, 491, *illus.* 492-493, *illus.* 494,
    512-528
  adaptation, 520-522
  classification, 512-514
  ecology, 520-522
  metabolism diversity, 520-521
  structure, 514-517
Monkey, *illus.* 308-309
Monocotyledon, 633, *illus.* 633
  dicotyledon, compared, 636
Monohybrid cross, 287-290
  Punnett square, 295-296, *illus.* 296
Monosaccharide, 178
Monosomy, 328
Monotreme, 884-885, *illus.* 885
Mosquito, *illus.* 52, 53, *illus.* 769
Moss, *illus.* 582-583, *illus.* 594-595,
    594-600
  alternation of generations, 598-599
  green algae, compared, 595
  life cycle, *illus.* 597
  protonema, 596
  reproduction, 596
Motivation, 909
Motor cortex, *illus.* 1033
Mouth, 953-955, *illus.* 954
  chemical digestion, 953
Mudskipper, *illus.* 6
Multiple alleles, 335-336, 363-364
Murray, Robert, 360-361, 1105
Muscle
  for locomotion, 938-947
  types, *illus.* 938, 938-939
Muscle block, chordate, 800
Muscle strength, exercise, *illus.* 943,
    943-946, *illus.* 946
Mushroom, *illus.* 558-559
  disappearance of, 577
  gills, 569
  life cycle, *illus.* 570
Mussel, *illus.* 120, *illus.* 688, 757
Mutation
  causes, 328
  DNA, 324-325
Mutualism, 69, *illus.* 69
  fungi, 574-576
Mycorrhiza, *illus.* 574, 574-575
Myofibril, 942
Myosin, 942

Narcotic, 1018, 1025
Nastic movement, 676, *illus.* 676
Natural history, 55
Natural resource, 133-153. *See also*
    Specific type
  demand vs. supply, 139
  effects of human activities, 134-139
Natural selection, 424-428
  adaptation, 428-430
  allelic frequency, 442-443
  variation in populations, *illus.* 439,
    439-440, *illus.* 440
Neanderthal, *illus.* 471, 471-472
Nectar, pollination, 659
Nematocyst, 720, *illus.* 720
Nephridia, 744
Nephron, 994, *illus.* 994
Nerve cell, *illus.* 241, *illus.* 922.
    See also Neuron
Nervous system, 1002-1010, *illus.* 1010
  arthropod, 766
  echinoderm, 792
  functions, 1003, *illus.* 1003
Nervous tissue, mesoderm, 700
Neuron, *illus.* 1002, 1002-1006.
    See also Nerve cell
  impulse transmission, *illus.* 1003,
    1003-1004
  neurotransmitter, 1006, *illus.* 1006
Neurotransmitter, neuron, 1006,
    illus. 1006
Neutron, *illus.* 163
New World monkey, 458, *illus.* 458,
    illus. 460
Nicotine, 1022-1024
Ninomiya, Flora, 666-667, 1105
Nitrogen base, 310
Nitrogen cycle, homeostasis, 77,
    illus. 77
Nitrogen fixation, bacteria, 524,
    illus. 524
Nocturnal animal, 753
Nonbiodegradable material, 148
Nondisjunction, 326-327, 369-370
Nonrenewable resource, 135
Nonvascular plant, 588
Northern coniferous forest, 101,
    illus. 101
Notochord
  chordate, 797
  identification, 800
Nucleic acid, structure, 181-184
Nucleolus, 198, *illus.* 198
Nucleotide, 184, *illus.* 184, *illus.* 310,
    310-311, *illus.* 311
  chains, 311-312

Index

Plasma membrane, 196, *illus.* 197, 216-222
  lipid bilayer, 220-221
  maintaining balance, 216-217
  selective permeability, 216-217, *illus.* 217
  structure, 217-222, *illus.* 220
Plasmid, 378, *illus.* 378
Plasmodial slime mold, 552, *illus.* 552
Plasmodium, 552
Plasmolysis, 229
Plastic, recycling, 147
Plastid, 202
Plate tectonics, 408
Platelet, 983
Point mutation, 324, *illus.* 325
Polar molecule, water, 168
Pollen grain, 669, *illus.* 670
  spore, compared, 616
Pollen growth, fertilization, 669-670
Pollen tube, 669, *illus.* 670
Pollination, 286
  animal, 659
  flower colors, 660
  flowering plant, 657-665
  nectar, 659
  scent, 661
  wind, 659
Pollution
  air, 141-142
  biosphere, 140-148
  defined, 140
  at home, 149
  land, *illus.* 147, 147-148
  living organisms as detectors, 831
  nitrogen, *illus.* 140
  pesticide, 145
  water, 144-146, *illus.* 145, 153
Polychaete, 753
Polygenic inheritance, 338-339, 364
Polymer, 177, 180
Polyp, 718
Polyploid species, 444
Polysaccharide, 178, *illus.* 179
Population, *illus.* 59
  defined, 58
  interactions, 58, 66
Population biology, 113-129
Population genetics, evolution, 435-441
Population growth, 114-121. *See also* Human population growth
  age, 126-127, *illus.* 127
  carrying capacity, 118
  density-dependent factor, 120-121
  density-independent factor, 121
  environmental limits, 118, 120-121
  exponential growth, 115, *illus.* 115
  life-history pattern, 118-119
  S-shaped curve, *illus.* 118

Population size
  competition, 123
  emigration, 128
  immigration, 127, *illus.* 128
  interactions among organisms limiting, 122-123
Pothole effect, 172
Prairie, 103
Praying mantis, *illus.* 6
Precambrian Era, 406, *illus.* 406
Precipitation, biome, 97-99
Predation, population size, 122
Pregnancy, hormonal maintenance, 1051
Prehensile tail, 458
Prenatal care, 1054
Prenatal testing, 1055
Preservation, natural areas, 149
  species, 152
Priestley, Joseph, 251
Primate, *illus.* 460-461. *See also* Specific type
  adaptation, 454-462
  characteristics, 454-456, *illus.* 455
  defined, 454
  evolution, 456-462
  phylogeny, *illus.* 459
  skull, 466-467
Principle, 40
Probability, genetics, 295
Problem-solving method, 26-40
Progesterone, 1044
Prokaryote, *illus.* 194, 194-195, 491, 512-528
Prophase, 268-269, *illus.* 269, *illus.* 270, 300-302, *illus.* 301, *illus.* 302
Prosimian, 456-457, *illus.* 457, *illus.* 460
Prostate gland, 1040
Prosthesis, 945
Protein, 964-965
  gene, 316
  gene mutation, 326
  passive transport, 230
  structure, 180-181
  translation, 322
Protein storage, structures, 200
Prothallus, 607
Protist, *illus.* 492-493, 494, *illus.* 532-535
  diversity, *illus.* 534, 534-535, *illus.* 535
  phylogeny, 553
Protonema, moss, 596
Protostome, 699, 793
Protozoan, 535-541
Provirus, 508, *illus.* 509, 509-510
Pseudocoelomate, body cavity, 704-705
Pseudopodia, 535
Psilophyta, *illus.* 585, 586, 590

Pterophyta, 591-592, *illus.* 592, *illus.* 609, 609-612, *illus.* 610
Puberty, 1040
  female, 1043-1044
  male, 1040
Pulse rate, swallowing, 987
Punctuated equilibrium, 445, *illus.* 445
Punnett, Reginald, 295-297
Punnett square, 295-297
  dihybrid cross, 296, *illus.* 297
  monohybrid cross, 295-296, *illus.* 296
Pupa, 781
Pure science, 44
Purine, *illus.* 310
Pyrimidine, *illus.* 310
Pyruvic acid, breakdown, 252, *illus.* 252

Quantitative research, biology, 41

Rabbit, *illus.* 13, *illus.* 14, *illus.* 15
Radial symmetry
  echinoderm, 788-789, *illus.* 789
  hydra, 702, *illus.* 702
Radula, 744
Raimondi, Pete, 12-16
Rain forest, 50, *illus.* 50, *illus.* 105, 105-108, *illus.* 106, 651
  biodiversity, 107
  destruction, *illus.* 51
  ecosystem, 50
Ray, *illus.* 812, 812-813, *illus.* 813
  internal fertilization, 813
Rayon, 586, *illus.* 586
Reaction time, distracting stimulus, 1009
Recessive autosomal heredity, human heredity, 357-362
Recessive trait, 289, 437
Recombinant DNA, 376-385
  modeling, 380-381
Rectum, 958
Recycling, 148-149, *illus.* 149, 529
  plastic, 147
Red algae, 548, *illus.* 548
Red blood cell, oxygen, *illus.* 982
Red tide, 547, *illus.* 547
Redi, Francesco, 411
Reflex, 900, *illus.* 900
  somatic nervous system, 1008, *illus.* 1008
Regeneration, 729, 794
Renewable resource, 134

Index

Index